P9-DHB-710

FOREST ECOSYSTEMS

FOREST ECOSYSTEMS

Second Edition

DAVID A. PERRY
Department of Forest Science and
Cascade Center for Ecosystems Management
Oregon State University

RAM OREN
Division of Environmental Science and Policy
Nicholas School of the Environment and Earth Sciences
Duke University

STEPHEN C. HART
School of Forestry
Merriam-Powell Center for Environmental Research
Northern Arizona University

THE JOHNS HOPKINS UNIVERSITY PRESS | *Baltimore*

Printed in the United States of America on acid-free paper
9 8 7 6 5 4 3 2 1

The Johns Hopkins University Press
2715 North Charles Street
Baltimore, Maryland 21218-4363
www.press.jhu.edu

Library of Congress Cataloging-in-Publication Data

Perry, David A.
Forest ecosystems / by David A. Perry, Ram Oren, and Stephen C. Hart—2nd ed.
p. cm.
Includes bibliographical references and index.
ISBN-13: 978-0-8018-8840-3 (hardcover : alk. paper)
ISBN-10: 0-8018-8840-9 (hardcover : alk. paper)
1. Forest ecology. I. Oren, Ram, 1952– II. Hart, Stephen C., 1961– III. Title.
QK938.F6P46 2008
577.3—dc22 2007042295

A catalog record for this book is available from the British Library.

Special discounts are available for bulk purchases of this book. For more information, please contact Special Sales at 410-516-6936 or specialsales@press.jhu.edu.

The Johns Hopkins University Press uses environmentally friendly book materials, including recycled text paper that is composed of at least 30 percent post-consumer waste, whenever possible. All of our book papers are acid-free, and our jackets and covers are printed on paper with recycled content.

What is needed on our part is the capacity for listening to what the Earth is telling us.

Thomas Berry

Let us bring you songs from the wood

Jethro Tull

CONTENTS

PREFACE TO THE SECOND EDITION

Much has happened in ecology and forestry since this book first appeared in 1995. In ecology, new concepts of meta-communities and meta-ecosystems have extended and affirmed the importance of spatial interactions in ecological dynamics; a new theory of the forces that create biodiversity has joined an already crowded field; progress has been made on the perennially contentious issue of the relation between biodiversity and stability; the importance of positive interactions in ecosystem dynamics has entered the mainstream; a science of community and ecosystem evolution has emerged; the view of the nitrogen cycle has been revised; long-term data sets have begun to yield information about ecosystem dynamics that play out over decades; new tools have produced insights into aspects of nature ranging from the molecular diversity within ecosystems through the structure of canopies, to the teleconnections that tie the global ecosystem together.

The past decade has been characterized by increasing synthesis of previously separate disciplines, deepening and enriching our understanding of how nature works. By asking questions about what species do in ecosystems, ecologists have initiated the long-overdue coupling of community and ecosystem ecology; the discipline of *ecohydrology* has emerged to study the interface between hydrology and ecology, and a mega-integrative new discipline, earth system science, has been born from the marriage of the biologic and earth sciences.

Overlaying and permeating all of these developments are three factors in particular: quantification of the crucial role played by forests in providing ecosystem services, allowing values that until recently were not traded in the marketplace to be compared on a more even footing with wood and other readily marketable values; a growing awareness of the earth as a single, highly interactive system (Gaia is real); and climate change, which can no longer be denied and is already manifesting in many ways, including insect outbreaks, fires, and droughts at scales unprecedented in recorded history.

In forestry, sustainability no longer takes a backseat to wood production but has become a central issue, made more so by the unfolding threats of major extinctions, disruptions due to climate change, and extensive forest loss to other land uses. With regard to actually achieving sustainability, we are on the steep part of the learning curve, but on the curve nonetheless. The past few years have seen the formation of the United Nations Forum on Forests, the National Commission on Sustainable Forestry in the United States, a growing network of model forests (born in Canada but spread widely) dedicated to exploring approaches to sustainability, and the publication of key books such as *Conserving Forest Biodiversity* by Lindenmayer and Franklin, *Towards Sustainable Management of the Boreal Forest* by a consortium of Canadian scientists, and *Designing Sustainable Forest Landscapes* by Bell and Apostol. Green certification of forestry operations has grown explosively worldwide. Various criteria and indicators have been developed to provide guidelines. Community forestry, though facing challenges, is including forest-dependent communities as partners in developing sustainable approaches.

There are some things that are given; they have not changed and will not change. One of these is a bedrock principle of ecology that basically says "what goes around comes around"; in other words, outcomes are determined by interactions, and interactions by outcomes. In forestry, the central unchanging reality is that forests have multiple values for humans, and focusing on one while excluding others leads to both social and biologic problems.

Achieving sustainability is far more than an ecologic and management problem. No workable solution can ignore social, economic, and political factors. It is through our understanding of the workings of nature that we derive workable options and help society understand the consequences of choosing one course over another. With luck, such understanding will lead to wise decisions.

It is impossible to acknowledge all the people who have contributed in one way or another to this book. Everyone we have interacted with over the years—teachers, students, colleagues, forest managers, members of the environmental community, and private citizens concerned with the stewardship of forests—has helped to shape, test, and refine our knowledge of forest ecosystems. We thank all who have walked in the woods with us, both literally and figuratively.

Families bear much of the brunt of book writing—the long hours, absences, and distracted responses. We are deeply grateful to ours for bearing with us.

Many thanks to Greta Ashworth for bringing her artistry to the figures, Bethany Ross and Linda Forlifer at the Johns Hopkins University Press for greasing skids and overseeing book production, Joanne Bowser and Meg Hanna of Aptara for making the dreaded copyedit process easy, and Jake Kawatski for the indexing.

Finally, a special thanks to Vincent Burke, our acquisitions editor at the Johns Hopkins University Press, for his seemingly bottomless well of patience, support, and good humor.

FOREST ECOSYSTEMS

1

Introduction

NO ORGANISM LIVES IN ISOLATION. To be sustained, life requires continual inputs of energy and matter. For most organisms the ultimate source of energy is sunlight, captured in matter by photosynthesis. Earth is in the truest sense of the word a spaceship—there is a fixed supply of matter that must sustain us for a very long journey. Hence any given molecule of water, atom of nitrogen, carbon, oxygen, or any other essential element must be reused many times—that is, it must be **cycled.** Cycling may take various forms. Water is cycled from ocean to land and back again, powered by sunlight. Carbon is passed from organism to organism through feeding and eventually released back into the atmosphere as carbon dioxide, where it is once again used by green plants to capture solar energy. Nitrogen, phosphorus, and other elements necessary to sustain life are also passed from organism to organism and in this way cycled from the old to the new, permitting life to be continually renewed. Without this renewal, evolution would not be possible, and life as we know it would not exist.

No single species is capable of capturing energy and at the same time cycling matter, and thus living things are always found grouped together in **communities** that include species capable of capturing energy—in most cases green plants—and species that obtain their energy and the matter that composes their bodies by consuming, in one fashion or another, the tissues of other organisms. Such a community is not simply an aggregation of bodies, like travelers thrown together on an airplane, but a coherent group tied together by interaction among all its members and their environment. A community of species interacting among themselves and with the physical environment is an **ecosystem.** Ecosystems have the following distinguishing characteristics:

1. *A web of interactions and interdependencies among the parts.* Animals and microbes require the energy supplied by plants, and plants cannot persist without animals and microbes to cycle nutrients and regulate ecosystem processes. The

interdependencies within ecosystems relate to **function:** there must be species that photosynthesize, species whose feeding results in nutrients being cycled, predators that keep populations of plant-eaters from growing too large, and so forth. Some system functions can be performed by more than one species (a property called **redundancy**); in other cases a single species plays a unique functional role (such species are called **keystones**).

2. *Synergy* is the "behavior of whole systems unpredicted by the behavior or integral characteristics of any of the parts of the system when the parts are considered only separately" (Fuller 1981). Synergy characterizes any system whose components are tied together through interaction and interdependence (the human home is an example of a synergistic system; in fact the word *ecology* is derived from the Greek word for "home").

> Synergy—which means "the whole is greater than the sum of the parts"—is a basic property of matter. Physicists tell us that two electrons circling an atom become so mutually intertwined that one cannot be understood without reference to the other. Gravity is a synergistic force that emerges from the interaction of two or more objects. Virtually all patterns that we see in nature emerge from interactions: between organisms and their environment, among different species, and among different levels of the global hierarchy. The result is synergy: an interplay between local and holistic forces that generates complexity, dynamic stability, and, to the observer at least, unpredictability.

3. *Stability* is a simple yet complicated concept that does not mean "no change" but rather is analogous to the balanced movement of a dancer or a bicycle rider (Mollison 1990). The processes of disturbance, growth, and decay produce continual change in nature. Stability means that (1) changes are maintained within certain bounds and (2) key processes (such as energy capture) and potentials (such as the productive potential of soil) are protected and maintained (chapters 20–22).
4. *Diffuse boundaries.* Unlike an organism, an ecosystem does not have a skin that clearly separates it from the external world. Ecosystems are defined by connectance, and connections extend through space and time, integrating every **local ecosystem** (one that is localized in time and space) within a network of larger and larger ecosystems that composes landscapes, regions, and eventually the entire earth. Any given forest both influences and is influenced by cities, oceans, deserts, the atmosphere, and forests elsewhere on the globe. Moreover, every local ecosystem produces patterns that propagate through time, communicating with and shaping the nature of future ecosystems. The interconnections among ecosystems that exist at many different spatial and temporal scales result in what is termed **hierarchical structure,** which simply means that each ecosystem that we can define in space comprises numerous smaller systems and at the same time is part of and in interaction with a hierarchy of larger systems (fig. 1.1)

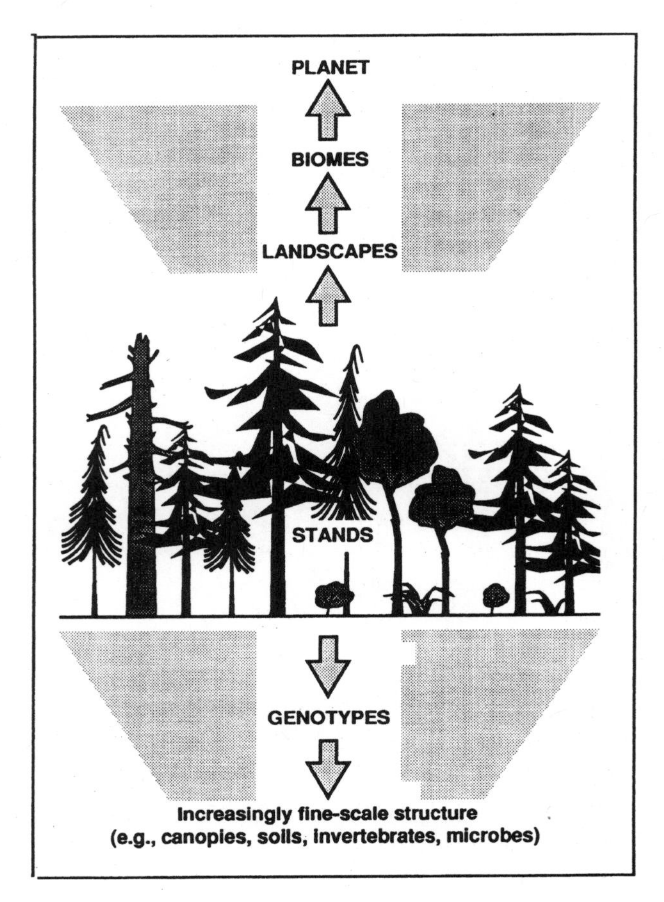

Figure 1.1. The hierarchy of nature. Every local ecosystem is part of a larger set of ecosystems that includes landscapes, regions, and ultimately the planet as a whole. Local ecosystems also comprise diversity at many scales, from individual plant and animal species and genotypes through microbes and the fine-scale structure of soils and canopies.

1.1 WHY STUDY ECOSYSTEMS?

> Management, to be successful, must be founded in a deep appreciation of the forest, of its origins and structure and of the complex interrelationships of all its component parts.
>
> JOHN T. HOLLOWAY 1954

We have already anticipated the answer to the question posed in this section: understanding ecosystems is prerequisite to keeping them healthy and productive. Let's pursue that point. There are at least two reasons why those who are interested in the management and conservation of forests should understand the nature of ecosystems. First, the growth of trees and the viability of all forest inhabitants are products of many interactions within the ecosystem, and forest management involves either direct or indirect manipulation of these interactions. Because all species within a community are linked with one another, it is not possible to manipulate the forest without influencing to some degree other organisms within the ecosystem. Removing even a single tree sends ripples through the system, like throwing a rock into a pond. Similarly, because of interconnectedness at the scale of landscapes, regions, and the globe, changes in one place are likely to have effects in other places as well (we shall see examples in later chapters). Such ripple effects are not necessarily bad and may even be good for the health of ecosystems. It is very important, however, to understand the consequences of various actions because without this understanding there is no way to predict effects on tree growth, species viability, or long-term stability.

As an example of the ripple effect, consider thinning, which is the common silvicultural practice of removing some trees within a stand to provide more growing space for those remaining. Thinning is a tool for shifting productivity from many trees to a few and, hence, producing large, commercially valuable individuals. But thinning may have numerous other effects on the ecosystem. Because more light reaches ground level, growth of understory shrubs and herbs is stimulated. These may compete with trees for water and nutrients, or they may increase the rate of nutrient cycling and therefore enhance tree growth; probably they will have both effects, and the net effect on trees will depend on environmental factors such as the abundance of water and nutrients. Understory growth after thinning provides food for ground-dwelling animals and perhaps nest sites for birds. Animals may prefer feeding on certain types of vegetation, thereby altering the species composition of the understory and perhaps the way in which the understory influences the trees. Spiders, ants, mice, and birds that find homes in the understory consume insects that feed on tree foliage and help prevent insect outbreaks. Because individual trees have more resources after thinning, they are better able to defend themselves against insects and pathogens. On the other hand, if trees or soils are damaged during thinning, the opposite could result.

The ripple effect also works in the opposite direction, and the effects of any given silvicultural practice may be tempered by what is going on in the surrounding landscape, or even globally. For example, although thinning may enable trees to better resist insects and pathogens, a thinned stand that exists as an island within a sea of unthinned stands may become vulnerable in the same way that a healthy person can become sick when thrust among sick people. Similarly, the benefits of thinning (or of any silvicultural practice) may be outweighed by stresses associated with pollution and/or global climate change.

The second reason for understanding ecosystems stems from the fact that forest management, once focused largely on producing wood fiber, is increasingly asked by society to balance multiple, often conflicting needs and values. It is abundantly clear that forests and other natural ecosystems provide numerous direct and indirect services to humans (Daily et al. 1997; Balmford et al. 2004). Anyone who has spent time with indigenous peoples has learned that forests are living pharmacies, containing many plants that produce valuable medicines. For example, the late David Forlines, a traditional healer, listed 30 species of forest plants used as medicines by indigenous people living on the Olympic Peninsula of Washington State (Forlines et al. 1992). Modern medicine has discovered compounds with anticancer properties in at least two forest species: western yew *(Taxus brevifolia)*, which grows only in old growth forests of the Pacific Northwest, and the tropical plant Madagascar periwinkle *(Catharanthus roseus)*. One drug extracted from the periwinkle has reduced mortality from lymphotic leukemia from 80 percent in 1960 to 20 percent in the 1980s (Fearnside 1989). Only a very small minority of forest plants have been examined for medicinal properties. Of course, medicines are not the only nonwood product of forests. A study in the Brazilian Amazon concluded that managing rain forest for fruit and rubber yielded twice the economic return of either intensive plantation forestry or conversion of the forest to pasture for livestock (Peters et al. 1989).

Some of the more important services that forests provide humans are quite intangible and difficult, if not impossible, to quantify in an economic analysis. Trees are a vitally important component of the global ecosystem. Though occupying less than 14 percent of the earth's surface, forests and savannas account for more than 40 percent of the total solar energy captured each year by green plants and contain the largest concentrations of organic material of any global ecosystem (Potter 1999; chapter 3). The low albedo of forests (i.e., their ability to absorb solar energy rather than reflect it back to space) greatly influences the earth's heat balance. The large amount of carbon dioxide cycled to and from the atmosphere by forests has a major influence on the composition of the atmosphere and, through this, on global climate.

Forests also form a critical link in global water and nutrient cycles. They are the major source of evaporation from land to atmosphere and, hence, play an important role in the distribution and quantity of rainfall. Because virtually all of the earth's major rivers originate in what are, or were, forests, the latter exert major control over floods and water quality. By regulating the movement of chemical elements from land to water, forests influence the productivity of streams, lakes, and oceans. One economist calculated that the value of Japan's forests for conserving water and soils for recreation exceeded their value when managed solely for wood products (Fukuoka 1985).

Much is left to be learned, but it seems likely that healthy forests are an important prerequisite to a healthy earth (Chivian 2003; chapter 3). Conversely, as parts of a strongly interactive global ecosystem, forests are influenced by events that originate far beyond their boundaries and will be increasingly so in the future.

Quite a different kind of forest value, but no less important than any other, is the aesthetic and spiritual value. Forests are among the last of the truly wild lands left on earth, and their diversity and beauty fill a universal human need to be connected to nature. "I am looking at trees," said the poet W. S. Merwin, "They have stood round my sleep and when it was forbidden to climb them they have carried me in their branches."

Spiritual identification with forests is deeply embedded in the human psyche. For Hawaiians, the forests are *wao akua*, the region of the gods. In his classic 1898 text on the roots of religion and folklore, *The Golden Bough,* James Frazer discusses spiritual identification with trees and forests (Frazer 1981):

> Tree-worship is well attested [throughout Europe]. Among the Celts the oak worship of the Druids is familiar to everyone. Sacred groves were [also] common among the ancient Germans, and tree-worship is hardly extinct amongst their descendants at the present day. At Upsala, the old religious capital of Sweden, there was a sacred grove in which every tree was regarded as divine. . . . Proofs of the prevalence of tree-worship in ancient Greece and Italy are abundant. . . . In the Forum, the busy center of Roman life, the sacred fig tree of Romulus was worshiped . . . and the withering of its trunk was enough to spread consternation throughout the City.

Such beliefs were by no means restricted to Europe and, though weakened in modern market-centered economies, can still be found. Thousands of sacred groves exist today scattered among many countries, and they are often refuges for biological diversity that has disappeared from the surrounding landscapes (Bhagwat and Rutte 2006).

[For indigenous peoples throughout the world,] trees considered as animate beings are credited with the power of making the rain to fall, the sun to shine, flocks and herds to multiply, and women to bring forth easily (Frazer 1981, 66). Superstition or wisdom? As we shall see a bit later, forests do indeed enhance rainfall, protect soils, and profoundly influence the livability of the earth. That the deep identification of humans with forests is still alive and well is attested by the vigor with which people throughout the World, from western North America to India, are fighting for the preservation of remaining forests (fig. 1.2).

Figure 1.2. A tree-sitter protests logging in an Oregon forest. (Courtesy of Thom O'Dell)

1.2 STATE OF THE WORLD'S FORESTS

> Even if some of the data are dubious, there seems little doubt that the situation of global forests now is substantially worse than it was in 1992. The FAO analysis of pre- and post-Rio patterns of forest loss indicates that all the international activities directed towards forest protection since the Rio conference have made virtually no difference to the worldwide rate of natural-forest loss.
>
> BRYAN 2002

As we enter the twenty-first century, many of the deeply troubling trends of the past decades continue and some even worsen. Deforestation, pollution, chronic insect and pathogen problems, and unnaturally severe fires are exacting a toll on natural forests throughout the world. The recently completed Millennium Ecosystem Assessment concluded that "Forests have essentially disappeared from 25 countries, with 9.4 million hectares being lost annually" (Mooney et al. 2005). Sixty percent of the ecosystem services evaluated by the Millennium Assessment were "either being degraded or being used unsustainably" (Mooney et al. 2005). Climate change triggered by industrial humans is without doubt underway, though the full range of its effects on forests is far from certain. However, there are also positive developments, and a great deal of uncertainty.

Figure 1.3. Twentieth-century Greece. Much of the Mediterranean basin was forested 2,500 years ago.

Humans have been deforesting the earth since we first settled down to become farmers and townspeople (Perlin [1989] gives a history of deforestation). Eight thousand years ago, a thriving agricultural culture existed in the western "fertile crescent," an area that encompassed much of the Near East. Around 5000 B.C., however, many of the villages were abandoned. Until recently archaeologists believed that this culture collapsed because of climate change, but recent excavations indicate that deforestation was the actual cause. Evidence gathered by Rollefson and Kohler-Rollefson (1990) suggests that the oak forests of the area were cut for housing, and that overgrazing by large goat herds prevented forest regrowth. Without tree cover, soils deteriorated and farming became impossible.

The pattern of deforestation, soil loss, and subsequent impoverishment was repeated many times through history (Perlin 1989; Maser 1990). In the *Critias*, Plato complained about deforestation and soil loss in Greece (fig. 1.3): "The mountains were [once] high and covered with soil . . . [and] bore thick forests. . . . [The] land enjoyed an annual rainfall from Zeus which was not lost to it, as now, when it quickly flows off from the thin, unreceptive ground into the sea; but the soil it then had was rich and deep, so the rain soaked into it and was stored up in the retentive loamy soil."

Forests throughout the Mediterranean basin were cut for shipbuilding and firewood, to fuel Roman silver smelters, and to provide farmland for growing populations. The once extensive native forests of the British Isles are virtually nonexistent, and most of the original deciduous forest that once covered eastern North America was converted to cities, suburbs, and farms (though new forests have reclaimed many abandoned farms). Many tropical forests were cut by Europeans during the colonial period and converted to plantations of one kind or another. As of 1996, it was estimated that 53 percent of the world's original forest cover remained (World Resources Institute 2004).

Although deforestation is not new, it has certainly reached new levels during the past 150 years, and especially the last 50 years. Houghton and Hackler (2004) calculate that land use change pulsed 156 billion tons of carbon to the atmosphere between 1850 and 2000, about 63 percent of which was from the tropics. Much of this has been since the mid-twentieth century—the result of tropical deforestation (figs. 1.4, 1.5).

Reliable information on changes in global forest cover is notoriously difficult to obtain. The United Nations Food and Agriculture Organization (FAO) makes a valiant attempt every 10 years, using information compiled separately by member countries and submitted to the FAO. In its report for 2000, FAO estimated that the rate of deforestation for the globe as a whole averaged 0.38 percent per year during the 1990s, while reforestation of degraded lands acted to reduce the loss of total global forest cover to an estimated 0.22 percent annually (FAO 2001). The FAO estimates that with the exception of Siberia, total forest cover in the boreal and temperate regions remained relatively constant during the 1990s and may even have increased slightly (FAO 2001; Balmford 2002).

Remote sensing has been used to estimate changes in forest cover for about three decades (Asner et al. 2005). Use of satellites avoids the problems associated with the FAO's reliance on reporting by individual countries but has limitations of its own and less than perfect reliability (see the discussion by Hansen and DeFries [2004]). Satellite estimates

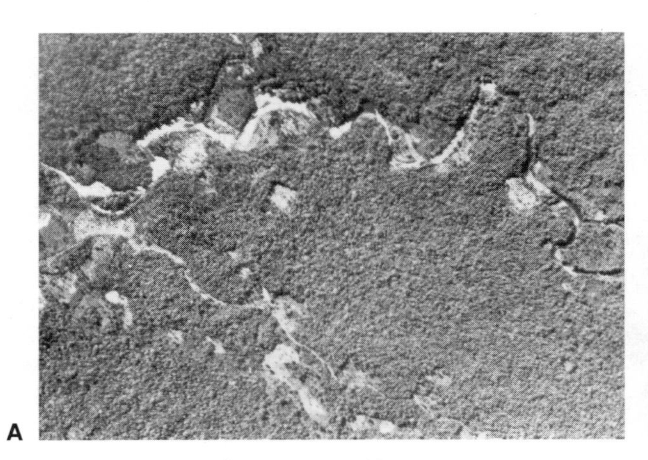

Figure 1.4. Change in forest cover between 1950 (A) and 1961 (B) in an area near La Selva Biological Station, Costa Rica. (Courtesy of Phil Sollins and Phil Robertson)

for the annual loss rate of tropical forests range between 0.5 and 1 percent. One recent study estimated that 8 million ha of humid tropical forests were lost or severely degraded annually between 1990 and 1997 (Achard et al. 2002; but see Kaiser 2002). Such estimates do not include areas subjected to selective logging (which could not be detected remotely until recently). Asner et al. (2005) used improved techniques to image selective logging in portions of the Brazilian Amazon and concluded that "selective logging doubles previous estimates of the total amount of forest degraded by human activities."[1]

[1]In the tropics, selective logging can result in considerable collateral damage to unlogged trees, as well opening the forest and increasing vulnerability to wildfires.

Hansen and DeFries (2004) compared satellite images from 1980 through the 1990s and concluded that deforestation had accelerated during the 1990s, primarily because of high rates of forest loss in South America and tropical Asia. During the 1990s, the annual carbon flux from land use change (mostly tropical deforestation) averaged 2.3 billion tons, slightly more than one-third the amount released by burning fossil fuels (Marland et al. 2003; Houghton and Hackler 2004).

Loss of tropical forests stems from various factors, but expanding agriculture plays a major role. Between 1960 and 2000, crop- and pasturelands increased from 38 percent to over 45 percent of the useable land area in the developing world, increasing the pressure on threatened species (Green

Figure 1.5. This site in Brazil once supported tropical rain forest with several hundred tree species on each hectare.

et al. 2005). Kaimowitz (2005a) summarized a report by Niesten et al. (2005) as follows:

> Since the early 1970s, the area devoted to oil palm, cocoa, and coffee in developing countries has doubled, from 50 million hectares to 100 million hectares. That is an area three times the size of Germany. Much of this growth came at the expense of forests. . . . Brazil grows thirteen million more hectares of soybean today than it did thirty years ago. . . . Most of that growth came in the wooded savanna areas of South-Central Brazil, but it has been gradually moving northward into the Amazon.[2] . . . Opening up new areas to produce cocoa has been the main cause of deforestation in Cote D'Ivoire, Ghana, and parts of Indonesia . . . Of the twenty-five global biodiversity hotspots identified by Conservation International, three-quarters are in major coffee growing areas.

Although deforestation during the latter twentieth century was concentrated in the tropics, it was not exclusive to the tropics. Hobson et al. (2002) compared current satellite imagery with historic records for the boreal transition zone in Saskatchewan and concluded that the annual deforestation rate had averaged 0.89 percent between the mid-1960s and mid-1990s, a rate almost three times greater than the global average.

Lepers et al. (2005) drew on various sources to synthesize information on land-cover change between 1981 and 2000. Deforestation, as evaluated from satellite imagery and expert opinion, is concentrated in certain hot spots, mostly in the tropics, but also in Siberia and even small areas in the central Rocky Mountains of the United States. Some caution is needed in interpreting these data. Forests may be lost by conversion to agriculture (as most commonly happens in the tropics), by suburbanization (a growing problem in the eastern United States), through logging, or because of disturbances such as wildfire or hurricanes. In the case of logging and natural disturbances, the loss of forest cover is usually temporary, although converting natural to managed forests reduces or eliminates some developmental stages (i.e., old-growth and extended early successions). Moreover, any disturbance repeated too frequently will eventually degrade a forest (chapter 20). Wildfire was a particular factor in many of the world's forests during the 1990s (chapter 7), possibly because of global warming. In Russia alone, 45 million hectares were burned by wildfires over a six-year period in the 1990s (Sukhinin et al. 2004). During the 1990s, western logging companies also established a presence in Siberia and account for some unknown proportion of forest loss.

Deforestation is often initiated by the construction of roads that allow access to loggers, farmers, and hunters, a dynamic that holds in any area of pristine forests but is a particular factor in the tropics. Tropical forests may eventually be clear-cut and converted more or less permanently to grazing or crops, or if not, the more open conditions and consequent drying associated with shifting cultivation and selective logging of valuable species set the stage for large, destructive wildfires (Laurance et al. 2001). Comparing satellite images between 1990 and 1997, Achard et al. (2002) calculated annual deforestation rates exceeding 3 percent for Madagascar, portions of Amazonia, and much of Southeast Asia. By one estimate, Indonesia's biologically rich lowland forests, composing 70 percent of that nation's land area, will be completely destroyed by 2010 (Jepson et al. 2001). El Salvador lost 46 percent of its remaining natural forests during the 1990s, and the Philippines lost 19 percent (World Resources Institute 2004). During the decade of the 1990s, forest cover declined at an average rate of 3 percent for 15 African countries and 2.6 percent for 8 Central American countries (Bryan 2002).

The scale of species extinctions during the twentieth century is on track to rank with five other periods of mass extinction in the earth's known geologic history (Thomas et al. 2004) and may exceed all others (Briggs 1991). Species are threatened for various reasons and in a variety of ecosystem types; however, loss or degradation of forested habitats plays a significant role, particularly in the tropics. Perhaps the most thorough documentation of species loss due to deforestation in the moist tropics was done for Singapore by Brook et al. (2003). Since 1819, more than 95 percent of Singapore's original forest cover has been cleared. Comparing historical with current species records for a variety of taxonomic groups (birds, butterflies, freshwater fish, mammals, etc.), Brook and colleagues estimated losses of up to 73 percent of all species in Singapore since the early 1800s. Of 91 forest bird species known in Singapore in 1923, and 168 species suspected to be present in 1819, only 31 remained by the late 1990s (Castelletta et al. 2000). Projecting the Singapore experience to Southeast Asia as a whole, Brook et al. estimate that if current rates of deforestation continue, from 13 to 42 percent of species will be regionally extinct by the end of the twenty-first century.

Changes in total forest cover tell only part of the story, however. Fragmentation of forests by various intensive land uses and permeation by roads diminish habitat quality and increase risk of disturbances such as fire (chapters 7, 20). Conversion of old-growth to younger stands, which has happened widely in temperate forests, retains forest cover but affects habitat for many species (chapter 22).

Bryant et al. (1997) define **frontier forests** as large, ecologically intact natural forests that are relatively undisturbed by humans and which are likely to survive indefinitely without human assistance.[3] They estimate that 40 percent of the

[2]Even when concentrated in the savannahs, soybean farming has a ripple effect in the rain forests. The large soybean farms displace small farmers, who move north to clear and farm in Amazonia.

[3]See Innes and Er (2002) for a critique of the frontier forest concept as a tool for prioritizing conservation efforts.

world's remaining forests qualify as frontier forest, these being concentrated in Amazonia and the boreal regions of Canada and Russia. As with most broad-scale measures of forest condition, considerable uncertainty exists about how many forests qualify as frontier forests. But there is also a more fundamental uncertainty about the degree to which forests can be directly or indirectly affected by humans and still retain ecological integrity (Innes and Er 2002). The latter is one of the more important issues facing conservation, because, as Soule and Kohm (1989) point out, "biologists are painfully aware that there are virtually no unpolluted, unperturbed sanctum sanctori left on the planet." In the age of global climate change and related stresses, there are serious questions about the ability of even the most remote forests to survive indefinitely.

While the world's remaining natural forests face a variety of threats, since the early 1990s increasing attention has been paid to protecting and restoring forest ecosystems. Victor and Ausebel (2000) summarize these trends.

> Much useful activity is already under way. Environmental NGO's around the globe have organized behind forest protection. All major forestry firms now participate in various activities to lessen the environmental harms of forestry. Multilateral development funders such as the World Bank have added the protection of forests and their role in alleviating human poverty to their agendas. The United Nations engages forestry issues through the FAO and the ongoing effort to implement commitments made at the 1992 Earth Summit in Rio de Janeiro (at which forestry policies were hotly contested). Since Rio, an alphabet soup of panels, forums, and task forces on forests have filled U.N. meeting rooms. This year, the U.N. launched an annual Forum on Forests to provide an outlet for the many clamoring voices. Forests do not suffer from a lack of attention in international politics.

Positive developments since the early 1990s include the following:

- The 1992 Earth Summit at Rio initiated an ongoing international effort to grapple with the environmental and social complexities associated with sustaining forests.
- The concept of managing ecosystems and landscapes to sustain a wide variety of values emerged as a direction for the future (Lindenmayer and Franklin 2002).
- The Northwest Forest Plan, aimed at protecting old-growth-associated species within the range of the northern spotted owl (western Oregon and Washington and northwestern California), set a standard for scientific rigor in regional ecosystem management strategies (FEMAT 1993). A variety of other initiatives emerged, such as Canada's Model Forest Program, with the goal of learning how to manage forest ecosystems sustainably (Natural Resources Canada 2007).
- The number of moist tropical forests designated as World Heritage sites grew to 33, for a total area of nearly 25 million ha under that special designation (Sayer et al. 2000).
- Green certification emerged as a force for sustainable management of commercial forests worldwide.
- In the United States, a national commission was established to promote and fund scientific research on sustainable forestry, where sustainability includes all the values delivered by forests (National Commission on Science for Sustainable Forestry, http://ncseonline.org/NCSSF/).

One of the most heartening developments of recent years is the growth of grassroots efforts to restore degraded forests and watersheds. For example, students at the Rishi Valley school in the eastern Ghat Mountains of India plant 20,000 trees and shrubs per year and distribute 100,000 more to local villages. Faculty at the school have documented the return of more than one hundred species of native birds to restored forests, including the rare Yellowthroated Bulbul (Kaplan 1997; Santharam and Rangaswami 2002). The Green Belt movement has organized local groups to plant 30 million trees across Kenya, and in 2004, the founder of the movement, Wangari Maathai, was awarded the Nobel Peace Prize. (To learn about forest restoration in Central America, see www.ecoworld.com/Projects/ReforestCentralAm/ReforestCentralAm_home.cfm).

While heightened concern over the fate of the earth's forests is a welcome development, the degree to which it makes a difference remains to be seen. Bryan (2002) concludes that the international activities directed towards forest protection since the Rio conference have been ineffective in reducing the worldwide rate of natural-forest loss.

Reclaiming degraded forests and developing practices that balance human needs with the preservation of natural capital will be a central task in the coming years. Success will depend on a variety of social as well as environmental factors—human uses can perhaps be managed, but people and forests cannot be separated. A clear understanding of the structure and functioning of natural forest ecosystems is not only helpful, but essential for determining the bounds of sustainable use.

1.3 THE STUDY OF NATURE: BALANCE AND FLUX

> Stones rot, only the chants remain.
>
> SAMOAN SAYING

Perhaps the single most distinguishing feature of the known universe is change. More than a "thing," nature is a dynamic process of growth, decay, birth, death, migrations, fluxes of energy, and cycles of elements. However, to simply say all is change greatly oversimplifies reality. Fluxes of energy and matter (including those packaged as organisms) occur at a variety of temporal and spatial scales. Consider a leaf. Elements move in and out, and biochemical processes occur over short time scales. These processes are packaged within a structure (the leaf), which is relatively stable, and necessarily so because without a stable context within which to operate, the faster processes could not persist. The leaf has a limited lifetime and will eventually return its matter and energy to the soil, but the tree persists and will make new leaves, while generations of little things are born and die. Similarly, trees mature and die, but the forest remains. Disturbances replace old forests with younger ones, but change at the scale of large landscapes and regions is usually slower and less drastic—if that were not so, old-growth-dependent species could not persist.

Understanding the nature of change, and especially the stabilizing factors that constrain and canalize change, is the key to sustainable use of the biosphere. On that score, nature has much to teach us. Although ecosystems are dynamic, they are not thrown about willy-nilly by uncontrollable forces. Healthy ecosystems maintain certain trajectories, and those trajectories are strongly influenced by spatial context, temporal connections (biological legacies), and balances—in species composition, species groups, landscape patterns, soil and atmospheric chemistry, global climate, and various other factors. The term *balance* is sometimes misunderstood as denoting something that doesn't change, but balances are dynamic, like a dance, rather than static, like a barstool. Understanding the interrelationship between balance and flux—or, to put it another way, between structure and process in nature—is the essence of modern ecology, land management, and conservation biology.

> The importance of understanding the processes of nature is well illustrated by what happened when wildfires were excluded from forests of the interior western United States. The U.S. Forest Service began fighting fire in the West in the early 1900s, with the objective of protecting the forest. We now know that fire played a crucial ecological role in these systems, and its removal set in motion a chain of events that wrecked the health of forests throughout the region, increasing their susceptibility to insects and pathogens, and making them vulnerable to fires that are much more destructive (and difficult to control) than those we fought to exclude. Avoiding similar disasters in the future will take more than good intentions (those early foresters had the best of intentions); it will require knowledge.

The basic message is this: if we are to maintain the integrity of any ecosystem—whether managed or a reserve—we must understand not only what it is, but what makes it what it is. From both a scientific and a management perspective, such understanding requires that we probe more deeply and ask what terms such as *healthy*, *integrity*, and *balance* mean in an ecosystem context, and how we measure them. This is not a simple task—nature is very complex (anyone who thinks ecology is easy best find another area of study)—but we ignore the processes of nature at our peril.

1.4 A BRIEF OVERVIEW OF THE BOOK

Chapters 2 and 3 essentially continue the introduction, setting the stage for what is to follow. Chapter 2 defines and discusses some basic terms and concepts necessary to understand the language of ecosystem science. In chapter 3 we briefly examine the global ecosystem and the role played by forests within it.

Chapters 4 through 8 deal with variability in space and change in time. In chapters 4 and 5 we discuss, respectively, global and local patterns in vegetation type, and the environmental factors that influence these patterns. In chapters 6 through 8 we turn to the dynamic nature of ecosystems in time, which results from disturbances of various kinds and the reaction of organisms and ecosystems to disturbance.

Beginning with chapter 9, we shall move from the larger-scale landscape patterns and processes to consider the structure and dynamics of local ecosystems. Chapters 9 through 13 explore the relationships between structure and function at the scale of landscapes and local ecosystems, including the relationships between landscape patterns and processes, the factors that create and maintain diversity, the interrelationships and interdependencies among species, and how relationships affect the workings of the ecosystem as a whole.

Chapter 14 focuses on the soil ecosystem, which is the product of a unique interrelationship between plants, animals, microbes, and minerals.

Chapters 15 through 19 deal with basic ecosystem-level processes: primary productivity (chapter 15), tree nutrition and nutrient cycling (chapters 16-18), and herbivory (chapter 19).

Finally, having learned about forest structure and processes at various scales, we can take up in chapters 20 through 22 what is probably the most holistic and integrated process that occurs in ecosystems: stability, or the mechanisms by which ecosystems persist in the face of changing environments and external disturbances.

Chapter 23 closes the book with an exploration of future directions for managing and conserving the earth's forests.

1.5 SUMMARY

An ecosystem is a community of interacting species, taken together with the physical environment within which it exists and with which the individuals composing the community also interact. Ecosystems are characterized by (1) a web of interaction and interdependencies among the parts, (2) synergy, (3) dynamic stability, and (4) diffuse boundaries. Understanding the structure and functioning of ecosystems is prerequisite to maintaining healthy forests and the multiple values that they produce for humans. Although actions aimed at protecting earth's ecosystems have gained momentum since the early 1990s, the pace of degradation has accelerated. Threats to tropical forests are particularly acute. The Millennium Ecosystem Assessment estimates that 60 percent of the services provided by natural ecosystems of all kinds are being degraded or used unsustainably. Conserving and restoring the ecosystems that provide vital services will be a central challenge of the coming decades.

2

Basic Terminology and Concepts

THIS CHAPTER HAS FOUR OBJECTIVES. First, to define and discuss some of the basic terminology associated with the study of ecosystems. Second, to briefly introduce the different general topics into which the study of ecosystems is divided. Third, to explore some things that ecosystems have in common with organisms and computers: all are **systems,** and as such have certain unique attributes. Understanding how an ecosystem works requires some knowledge of how any kind of system works.

The fourth and final objective spins out of the discussions that are associated with the first three objectives, and that is to introduce you to some of the issues that will be covered in more detail elsewhere in the book.

Some of the terms that you will learn in this chapter, and many others that appear throughout the book, are more in the nature of concepts than of definitions. This is an important point, one that will influence the way you learn and think about ecosystems; therefore, the first step is to clearly understand the distinction. The word *define* means "to draw boundaries around something." An automobile, the chair that you are sitting in, or a loblolly pine tree has certain characteristics that allow it to be clearly separated from other objects. Theoretically, at least, a space traveler who has never seen an automobile could take a list of its characteristics out on the street and find an automobile, with no danger of confusing it with a telephone pole or even a bicycle. Many commonly used terms are not so easily defined, however, because they do not have attributes that are readily measurable. In particular, boundaries cannot be drawn around abstract notions—or concepts—in the same sense that they can around objects. The space traveler who goes out on the street to find an example of *love* may be able to distinguish it from an automobile, a telephone pole, or a bicycle but may otherwise be very frustrated. Many of the terms commonly associated with ecosystems are concepts rather than objects, and hence rigorous definition is not possible. The best we can hope for is enough shared understanding to allow communication.

2.1 SOME BASIC CONCEPTS

2.1.1 Populations and Metapopulations

A **population** is a group of individuals of the same species living close enough together that they experience the same environmental conditions and interact with one another in some significant fashion. For sexually reproducing organisms, the critical interaction defining a population is exchange of genetic material. Because individuals within a population exist within the same environmental matrix and interbreed, they evolve similar sets of characteristics. It does not follow that all individuals within a population are genetically identical—far from it. This is a very important point for forestry that we explore in more detail in chapter 13.

Species that have extensive ranges, such as most temperate region trees, contain many different populations. Douglas-fir, for example, occurs from Canada to central Mexico, and from the Pacific Ocean to the edge of the Great Plains. Throughout this area the tree encounters vastly different environments, and each elicits a different set of adaptive responses, producing a mosaic of differing populations. Individuals within a population are not totally isolated from other populations. Trees separated by quite long distances may still interact through transfer of pollen, and individual animals frequently move from one population to another. A group of two or more populations that exchange genes is called a **metapopulation.**

> It is often difficult to draw a boundary around a population or a metapopulation. As we shall see in chapter 3, environmental factors may change significantly over very short distances, perhaps because of change in soils, elevation, or the direction in which a slope faces (this is called **aspect**). Hence trees growing close together and exchanging genes may experience very different selective pressures. On the other hand, in some areas, forests dominated by a few or even one tree species may stretch unbroken for hundreds of kilometers with little change in environment, while in moist tropical forests, where there are literally hundreds of different tree species, individuals of a given species may be separated by large distances. In each of these cases, it would be difficult to precisely delineate a population. Even with this practical limitation, however, the concepts of population and metapopulation are very important in ecology.

The fact that a given stand of trees constitutes a single population doesn't mean that other species living in that forest are single populations. Take, for example, a population of ponderosa pine on a mesa surrounded by desert. Each individual tree supports separate populations of many microbial, and probably some insect, species. On the other hand, a population of mobile animals, say ravens, may be spread over several stands on nearby mesas.

2.1.2 Communities and Metacommunities

A **community** is an aggregation of interacting species. To return to our mesa, the ponderosa pine trees, along with associated plant species, microbes, insects, birds, small rodents, large vertebrates, and so on form a community. Throughout the book, we use the terms *community* and *forest* interchangeably. In other words, a forest is more than just trees.

The term *community* is used both to describe an actual place on the ground—a forest, for example, that you can walk into and where you can hug the trees, listen to the birds, and turn over logs to find insects and fungi—and in more abstract terms to delineate certain groupings of communities that share similar characteristics. For example, the term *mixed conifer community* refers to a collection of communities in which various conifer species dominate the tree layer. **Community-type** is sometimes used to refer to the abstract community. The term *biocoenosis,* common in the European literature, is roughly equivalent to *community*.

A **metacommunity** is a set of local communities that are linked by dispersal of multiple potentially interacting species (Leibold et al. 2004).

2.1.3 Ecosystems and Metaecosystems

An **ecosystem** comprises a community along with the environmental factors and physical forces that act on it. Our ponderosa pine on the mesa, its soils and associated mineral elements, atmosphere, seasonal rainfall and temperature patterns, and normal disturbance events (for example, periodic fire) combine to make a coherent, interacting whole that is an ecosystem. Like community, *ecosystem* may be used concretely, to describe a particular place on the ground, or abstractly, to describe a type (e.g., the lodgepole pine ecosystem). The roughly equivalent European term for ecosystem is *biogeocoensis;* this is simply *biocoenosis* with a *geo* added to denote inclusion of physical factors. A metaecosystem is defined as a group of ecosystems interconnected by flows of energy, materials and organisms (Loreau et al. 2003).

A viable ecosystem must have the following:

- *A source of energy.* This is usually solar energy captured by green plants.
- *A supply of raw materials.* These include carbon, nitrogen, phosphorus, water, and so on.
- *Mechanisms for storing and recycling.* In most ecosystems raw materials are input periodically from outside the

system (e.g., rainfall), but often in insufficient amounts or at the wrong time to meet requirements. For example, in the mountains of western North America, most precipitation occurs as snow in the winter and is released as water during a very short period in the spring. To be available for summer growth, at least a portion of this water must be stored in the soil. In virtually all natural terrestrial ecosystems, one or more essential nutrients are input at very low rates, and productivity is sustained by some combination of recycling and utilization of stored reserves (chapters 15 through 18).

- *Mechanisms that allow it to persist.* Climatic fluctuations and periodic disturbance of one kind or another—in forests, often fire or wind—are common to almost all terrestrial environments, in fact they are so pervasive as to be considered legitimate components of ecosystems. If the ecosystem is to remain viable, its biological component—the community of organisms—must be able to "roll with the punches" and either resist these disturbances or recover efficiently from them (chapters 20 through 22). Furthermore, the community must have mechanisms that prevent it from self-destructing, which, for example, keep plant-eating animals from consuming all of the vegetation and thus cutting off the energy source (chapter 19). Successfully coping with disturbance does not mean that the character of the ecosystem is unchanged. On the contrary, species composition and physical characteristics may change radically after a major disturbance, slowly returning to the original state over many years. Hence an ecosystem is dynamic rather than static, and its boundaries stretch over time as well as space. This phenomenon of change initiated by disturbance, called **succession,** is of central importance to forestry, which by its nature disturbs (chapters 6 through 8).

> It should not be construed that ecosystems can absorb any insult and continue to function. **Foreign disturbances,** those that are not part of the history of a given system, may cause irreversible change. Ecosystems across the globe are currently subjected to various human-induced disturbances for which they have evolved no defense mechanisms. Air pollution and, particularly in tropical forests, improper land-use practices are examples of foreign disturbances that seriously threaten forests. The possibility of significant change in global climate because of the greenhouse effect has considerable implication for stability of all ecosystems. We shall return to these points at various places throughout the book.

It is worth reiterating a point made in the previous chapter: *an ecosystem is not just a set of organisms and molecules; like a friendship between two people, it is a complex of interactions.* This is not to say that every species in the system interacts with all others; a fungus on the root of a red maple is unlikely to directly influence or be influenced by a pileated woodpecker nesting in old dead oak. As components of the same forest, however, the fate of these organisms is to some degree intertwined. You may find it useful to think of an ecosystem as in some ways analogous to a waterbed. There are many molecules in the bed that will never bump up against one another; however, if you push down at one place on the bed, it will surely pop up at another. Despite no direct contact, a connectedness binds the molecules together. This kind of connectedness is the essence of an ecosystem.

> It is the focus on interaction rather than place that leads to the hierarchical nature of ecosystems. Because of the nature of connectedness it is impossible to draw firm boundaries around ecosystems. Our ponderosa pine forest on the mesa is an ecosystem. The stream that flows in the canyon of that mesa is also an ecosystem, but it is connected to the forest in numerous ways: the forest regulates its water supply and the nutrient elements that enter it, the forest inputs energy to the stream through leaf fall and logs that create pools so that organic and mineral inputs can be stored rather than being washed immediately downstream. The stream is an ecosystem, but it is also part of the larger forest-stream ecosystem. Songbirds that nest within the forest during the spring and summer, and help regulate foliage-feeding insects, winter in tropical forests of Central America. Hence, the fate of the forest on the mesa is tied to that of other forests far removed in space. The forest on the mesa receives rainfall from storms that rise in the Gulf of California, mineral elements in dust from the desert, and pollutants from Los Angeles. Its drought years may be triggered by interactions between the ocean and the atmosphere that originate thousands of kilometers away in the southern Pacific. The forest in turn regulates the amount of water, nutrients, and sediments entering rivers and oceans and influences rainfall at distant points by various mechanisms. As no organism is completely isolated, neither is any single ecosystem, and we must think in terms of metaecosystems. Ultimately, the earth as a whole is the metaecosystem that humans must learn to understand and live harmoniously within if we are to survive as a species.

2.1.4 Ecology

Ecology, which comes from the Greek *oikos,* for "home" (the same root that gives us *economics*), is defined in various ways. A common early definition was the study of the relationship between organisms and their environment.

Ecology has also been defined as the study of the factors controlling the growth and distribution of populations. Both of these are correct but incomplete because they focus on components of the ecosystem rather than the ecosystem itself (e.g., we don't call computer science the study of factors constraining the flow of electric current). The simplest, most comprehensive definition is that given by Odum in his classic text (1971) which boils down to the study of how nature works. As Odum stresses, *nature includes humans*. This is a particularly important point for resource managers, whose primary reason for learning about ecology is to understand how humans can extract values from ecosystems in a way that is harmonious with nature rather than disruptive.

2.2 THE SUBDISCIPLINES OF ECOLOGY

Ecology covers a wide range of phenomena, and, just as biologists group into physiologists, geneticists, taxonomists, and so on, different ecologists tend to concentrate on different aspects of ecosystems. In the following discussion, we delineate the types of things studied by the different subdisciplines of ecology and, in the process, preview some of the questions that we will concern ourselves with throughout the text. Although it is necessary for you to know what these subdisciplines are, remember that they are artificial distinctions, created by humans in order to aid understanding. No such neat divisions occur in nature. In fact, the trend in all natural sciences is increasingly integrative, especially at the level of ecosystems, landscapes, and the planet. Natural scientists still specialize—no single person can understand all—but the important questions facing society require specialists to talk and work together.

2.2.1 Physiological Ecology

Physiological ecology is the study of how environmental factors influence the physiology of organisms. On our mesa, we might ask how the pine trees regulate water loss during drought periods and what effect this has on their photosynthesis, or we may study the way in which a cold-blooded animal such as a lizard controls its body temperature in daytime heat and the cold desert night.

2.2.2 Population Ecology

Population ecology is the study of the dynamics, structure, and distribution of populations. For example, in an even-aged stand of slash pine, we may study the relationship between the stocking density of trees and individual tree sizes, or we may ask why the size of southern pine beetle populations within the stand suddenly increases sharply and then a few years later decreases perhaps even more sharply. These concerns lead us to ask how resources are divided among individuals, and what environmental factors regulate the size of populations. We may move from the level of interactions within an individual stand to ask what environmental factors influence the distribution of a species—why is red spruce found where it is and not found where it isn't? This leads to questions about **habitat,** which is the relationship between patterns in the ecosystem and the food and shelter requirements of organisms. When (as is often the case) we are concerned with the array of habitats provided rather than the habitat for a single species, we enter the realm of community ecology.

2.2.3 Community Ecology

Community ecology is the study of interactions among individuals and populations of different species. How does the shrub greenleaf manzanita influence the growth of white fir in the Sierra Nevada? What role does the fig wasp play in the reproductive cycle of the fig tree? How does drought influence the susceptibility of slash pine to the southern pine beetle? One of the primary interests of community ecologists is the factors that influence species diversity, which leads to questions about the diversity of habitats and how they change in space and time.

2.2.4 Evolutionary Ecology

Evolutionary ecology is linked closely to population ecology. The physical and biological environment acts as a filter that allows some individuals within a population to pass, and screens out others. Those who pass contribute genes to the next generation; thus, there is a continual interplay between the environment and the genetic composition of populations.

Because interactions among species are extremely important in shaping the genetic character of populations,

> Since the genetic character of populations is shaped by the environment, we might expect a single population, because all of its members experience much the same conditions, to be genetically homogeneous. Yet this is not what we find in nature. Populations of virtually all sexually reproducing organisms contain surprisingly large amounts of genetic variability. Perhaps the central question facing evolutionary ecologists is why populations contain so much. Is it there merely by accident? Or does it serve a purpose, that is, do individuals living within what we perceive as a uniform environment really experience diverse selective pressures? How much foresters can manipulate the genetic structure of natural forests without destabilizing the system depends on which of these alternatives is correct.

evolutionary ecology is also closely linked to community ecology. For example, plants have evolved numerous mechanisms to avoid being eaten by animals, and animals in turn evolve ways of counteracting these defenses. The result is a continuing interplay, a dance in which the moves of any given participant must be viewed as part of a larger pattern. In many cases, some of which are very important in forestry, interaction between two species benefits both, and sometimes the bond that evolves is so strong that one partner can't survive without the other (chapter 11).

2.2.5 Ecosystems Ecology

The subdisciplines described above focus on components of ecosystems. In physiological ecology the individual is the center of concern. In population and evolutionary ecology we step back a bit and consider the individual as part of a larger group of similar individuals, acquiring in the process concepts that have no meaning at the level of individuals: density, genetic variability, coevolution. If we step back yet again, we see that individuals and populations are embedded in a network of interaction. As evolutionary ecologists, for example, we learned that slash pine and the southern pine beetle are partners in a coevolutionary dance in which the moves of each are tempered by those of the other. If we enlarge the field of vision of our ecoscope,[1] we see numerous other dancers on the floor and realize that the moves of any one pair are part of a larger pattern. The relationship between the tree and the beetle may be tempered by the woodpeckers that eat the beetle, the tree's partnerships with certain fungi, the bacteria that cycle nutrients, the kinds of vegetation that neighbor the tree, the level stocking control imposed by the forester; all of these in turn respond to patterns of climate, soils, disturbance, and so on, which might be thought of as providing the beat to which the dancers respond.

At the ecosystem level we are interested in structural and functional attributes of the system as a whole:

- *The reciprocal influences between patterns and processes,* where patterns span scales from stands (e.g., the number of canopy layers) to landscapes (e.g., the distribution of community types or age classes across the landscape) to regions and the entire globe, and processes include all things that involve movement, change, or flux
- *Productivity*—the conversion of solar energy and nonliving chemicals to plant chemical energy and mass through photosynthesis (**primary productivity**), and conversion of the energy and mass in plants to energy and mass in animals and microbes (**secondary productivity**)
- *Food webs*—the way in which energy is distributed among the organisms of the system
- *Cycling of matter*
- *Stability* or the processes that allow the system to adapt to uncertain and often catastrophic change in the environment
- *Interactions between land, air, and water*

[1]The ecoscope is a magical device, powered by imagination, that allows one to focus back and forth in space from the very small to the very large, and in time from the very brief span to the very long.

2.2.6 Landscape Ecology

Just as ecosystems exist in a hierarchy from local systems to the globe, the study of ecosystems is also structured hierarchically. Stepping back from the level of the individual forest, we see that a particular landscape may comprise a variety of local ecosystems that differ in factors such as species composition, or that the landscape may have one dominant tree species, but of different ages. One of the most significant advances of recent years in the ecological sciences has been the recognition that what goes on in one area can profoundly influence other areas. In 1983, Forman introduced the new discipline of landscape ecology as follows (quoted in Turner et al. 2003a): "What do the following have in common? Dust bowl sediments from the western plains bury eastern prairies, introduced species run rampant through native ecosystems, habitat destruction upriver causes widespread flooding downriver, and acid rain originating from distant emissions wipes out Canadian fish. Or closer to home: a forest showers an adjacent pasture with seed, fire from a fire-prone ecosystem sweeps through a residential area, wetland drainage decimates nearby wildlife populations, and heat from a surrounding desert dessicates an oasis. In each case, two or more ecosystems are linked and interacting."

Landscape patterns influence numerous ecological processes, including the spread of disturbances, food web dynamics, gene flows, migration, and local climatic patterns, and many of the processes influence, in turn, changes in landscape patterns. Landscape ecologists study these reciprocal interactions between spatial patterns and ecological processes (Turner et al. 2003a). Though the term *landscape* is often used to denote our intuitive sense of what the word means—roughly an area humans can see when standing on a high point—for ecologists landscapes occur at a variety of scales; an eagle has one landscape, a ground squirrel another, a beetle yet another. Whatever the scale, the scientific focus is on linkages between spatial pattern and process.

2.2.7 Earth Systems Science

As John Lawton recently observed, "One of the key challenges of the twenty-first century is to forecast the future of planet earth" (Lawton 2001). Those who study the earth as a system—essentially global ecology—deal with biogeochemical interactions among land, water, and atmosphere. It is at this level that processes such as the earth's climate and the large-scale cycles of matter must be

pathogens, plants often respond by producing defensive chemicals, thus altering the growth rate of the animal or pathogen; the plant is not a passive prisoner of its environment but has the ability to influence its own destiny. All organisms receive feedback from their environment in terms of food supply and habitat quality and modify their behavior accordingly. (This doesn't mean that they consciously modify behavior, although they might.) Territoriality, for example, is a behavioral response to resource and habitat limitations. As we shall see in chapter 15, we also find what some refer to as **feed-forward mechanisms,** which are analogous to a thermostat detecting and responding to a drop in outside temperature before the room temperature drops (e.g., trees utilize feed-forward mechanisms to control water loss from leaves). Both feedback and feed-forward are, in effect, internal communications that permit a system to have a measure of control over its own destiny.

Two types of feedback occur in systems: positive and negative. Positive feedback is "more-making"—some of one thing leads to more of the same; negative feedback is the opposite—some of one thing leads to less of the same. *The productivity and stability of ecosystems result from complex interactions between positive and negative feedbacks.* Population growth is a good example of a process that is regulated by a combination of positive and negative feedback. When at relatively low levels, a population—say, of bark beetles—grows geometrically at a rate that is determined by factors such as the average size of a brood, the proportion of males to females, and survivorship to breeding age. If one female produces 10 females who survive to each produce 10 offspring, and those 100 each produce 10 offspring, the positive feedback of "more producing more" results in explosive population growth. A population of bark beetles (or anything else) cannot grow infinitely large, however; at some point resources become limited, perhaps coupled with increased prevalence of some pathogen or viral disease, and negative feedback comes into play to slow and eventually stop population growth. Ecologists use the term **density-dependent controls** to describe negative feedbacks on population growth.

Many processes within ecosystems exhibit positive feedback. Perhaps the most common type of relationship between species is **mutualism,** in which the different species benefit one another in some way (chapter 11). The mutual benefit accruing to each species from such a relationship represents positive feedback: if A helps B and B helps A, then more A leads to more B, which leads to more A, and so on. Trees (and most other plants) exist in a positive feedback relationship with soils: energy supplied by the tree creates biological, chemical, and physical conditions within the soil that lead to better tree growth (chapter 14). To give another example, tree crowns rake the air that moves through them for water and nutrients, resources that help the trees grow more tree crowns and thus do more raking. Positive feedback also appears in various forms at the level of the global ecosystem. For example, warmer temperatures resulting from a buildup of carbon dioxide and other greenhouse gases in the atmosphere accelerate the decomposition of soil organic matter, which releases more carbon dioxide to the atmosphere.

Like many aspects of nature, positive feedback has two faces: it is both immensely beneficial and—if it goes too far—potentially quite destructive (e.g., a nuclear reaction, or human population growth). In nature, the destructive potential of positive feedback is constrained by negative feedback, of which two types are most common: resource limitations, and predation.

It is the explosiveness of human population growth that is stressing the biosphere and that sooner or later will be brought under control by negative feedbacks from the environment. Amaranthus provides the following relevant example (personal communication). "Take a single bacterium and place it in a test tube with bacteria food. Every minute the population doubles (e.g., two bacteria at the end of one minute, four bacteria at the end of two minutes, eight bacteria at the end of three minutes) until at 60 minutes the tube is completely out of bacteria food.

At what point are the resources half depleted? At 59 minutes! At 58 minutes, 75 percent of the resources remain. At 55 minutes 97 percent of the food remains. If you were a concerned bacterium and at 55 minutes attempted to mobilize your fellow bacteria concerning the impending shortage, they would take away your flagella.

At 59 minutes, the plot thickens as concern grows. The best bacteria minds in the tube are brought together and engineer a remarkable breakthrough. Research and development are able to quadruple the current food in the tube. There is great joy in Bacterialand. And what did the quadrupling of the resource base buy? An additional 2 minutes!"

2.3.2 Pattern

One of the more common examples of the importance of pattern in nature is human language. Everyone who has worked on a computer has seen the message *syntax error* (*syntax* meaning "an orderly or systematic arrangement"). We have given the computer a pattern of letters that it doesn't recognize. Energy, the primary motivator in classical physics, plays no role here—the computer must have a certain arrangement of things or it doesn't respond.

As in language, the patterns generated within organisms and ecosystems transmit information. Genes, for example, do not function through transmitting energy like pool balls colliding, rather they deliver messages. If we rearrange the order of nucleic acids on a chromosome (something that commonly happens both in nature and the laboratory), the energy content is unaffected, but the message may be dramatically altered. The organism containing this altered chromosome may respond with what in effect is a syntax error message and die; or it may develop some new characteristic.

Information (we use *information, pattern, syntax,* and *structure* interchangeably) is important in many ways within ecosystems, occurring at all hierarchical levels from genes through landscapes. Many organisms secrete chemically encoded messages into the environment as a means of communication. For example, when a female mountain pine beetle lands on a susceptible lodgepole pine, she releases a particular chemical compound that spreads through the air and calls other beetles to the tree. When the beetle population reaches a certain level on that tree, another chemical is released that spreads a different message: stay away! (These two chemicals are acting, respectively, as positive and negative feedbacks on growth of a beetle population.) Such chemical messengers, called **pheromones** (a combination of the Greek *pherein,* meaning "to bear," and *hormone*), are in a very real sense a language. Each compound is like a word, carrying a highly specific message that depends on the arrangement of atoms composing the compound. Pheromones may be used as communication between species also. For example, many mycorrhizal fungi, which, as we discuss in chapter 11, live within tree roots and are essential for good tree growth, produce fruiting bodies belowground. For spores to be disseminated, these fruiting bodies must be dug up and eaten by animals (the spores pass intact through the gut and are spread by defecation). The fungus attracts animals by releasing pheromones. One particular species, common in Europe, is very good to eat, and sows have been used for years to find the fruiting body. Why are sows effective at this? Recently it was found that the fungus releases a pheromone that is identical to one produced by boars as a sexual attractant! (This particular pheromone also sexually stimulates humans.)

The chemical language of ecosystems includes a class of compounds called **allelochemicals,** secreted by many plants and some animals. Allelochemicals function as a form of defense against feeding animals and pathogens and also aid some plants in securing and holding space. A black walnut tree, for example, secretes a very powerful chemical that inhibits growth of other plants (including other black walnuts) in its vicinity. Allelochemicals are very common in nature. Because they influence the distribution of plants and the behavior of animals, they play an important role in determining both the structure and the dynamics of ecosystems.

Patterns in the structure of local ecosystems and their surrounding landscapes largely determine the numbers of species in the system. For example, bird species diversity is often closely related to the number of canopy layers in a forest. Large grazing animals such as elk require a certain pattern of open areas for grazing and forest for protection. Patterns in soil characteristics such as depth, texture, and nutrient content may produce corresponding patterns in the vegetation.

The importance of pattern in nature cannot be overemphasized. As we discussed earlier—and it is worth repeating—whereas in classical physics, effects are proportional to mass and energy, the same is not true in either physical or biological systems. A few spores of Dutch elm disease introduced into the United States created an effect somewhat akin to a pebble thrown into the ocean creating a tidal wave. Such phenomena simply can't be understood in terms of mass and energy but must be seen as products of information. This seemingly esoteric point has considerable relevance to forestry. Foresters are interested in capturing energy in wood, but the primary tool for doing this is the alteration of ecosystem structure, that is, patterns within the system. Thinning, fertilization, harvest, genetic selection, control of noncommercial vegetation, and virtually everything else we do in the forest alter, either directly or indirectly, the way the system is put together—its syntax. Hence, we devote considerable space in this book to understanding the nature of pattern in forests.

Chaos is the name given by scientists to a very special kind of pattern that occurs in both physical and biological systems characterized by strong positive feedback among system components. Chaotic systems, which include the earth's weather and growth cycles of some populations (e.g., Canadian lynx and African locust), are very sensitive to environmental conditions and, because of that sensitivity, highly unpredictable in their behavior. Some have speculated that whole ecosystems are chaotic (Schaffer and Kot 1985); however, this has not yet been established with certainty. It seems probable that a better understanding of the characteristics of chaotic systems will eventually lead to important insights into the trajectories followed by populations and perhaps whole ecosystems (Schaffer and Kot 1985; Pool 1989; Howlett 1990); however, a detailed discussion is beyond the scope of this book. For a very readable account, see Gleick (1987).

2.4 SUMMARY

Ecology, the study of the structure and function of nature, deals with a variety of subjects, including how organisms are affected by their environment, interactions among the different species that compose a community, the influence of environment on evolution, landscape patterns, and cycles of matter at scales ranging from local ecosystems to the globe.

Ecosystems have at least one thing in common with organisms and computers: all are systems and as such have certain unique attributes. The behavior of any kind of system is determined by two things: (1) its energy supply and (2) its information content, where information is anything that exerts an influence by virtue of its pattern rather than its energy content. (Stoplights, genes, and rock music are examples of things that act on the world strictly through their information content.) In forests, such things as species mixes, the number of canopy layers, and landscape patterns represent within-ecosystem information that regulates the course the system follows through time. Forestry alters patterns within forest ecosystems and landscapes and, hence, has both direct and indirect effects that ramify through the system.

3

Forests as Part of the Global Ecosystem

IN CHAPTER 1 WE SAW that every local ecosystem has a Janus-like character, facing both inward and outward. Inwardly, each ecosystem comprises many smaller systems that should we focus the ecoscope to a very fine scale, we would see are legitimate ecosystems in their own right—the community of organisms within the canopy of a single tree, for example. Outwardly, each ecosystem composes one part of a larger system, and with the ecoscope opened all the way, we see the global ecosystem itself: land, oceans, and air in dynamic interaction. In this chapter, we take a brief look at the global ecosystem and the importance of forests within it.

3.1 A BRIEF LOOK AT THE GLOBAL ECOSYSTEM

3.1.1 The Distinct Signatures of Life

Imagine that you are a space traveler entering our solar system for the first time. Turning your instruments to scan the atmospheres of the planets, it becomes immediately apparent that the third planet is quite different from the others. More detailed analysis shows why: it is the only one that supports abundant carbon-based life. How is it that a distant analysis of a planetary atmosphere can reveal the presence of life? The answer is simple: organisms imprint distinctive patterns on atmospheres. Compare, for example, the earth's atmosphere with that of Venus and Mars (table 3.1). Two gases, nitrogen and oxygen, make up 99 percent of the earth's atmosphere. In contrast, carbon dioxide composes well over 90 percent of the atmospheres of Venus and Mars—with nitrogen and oxygen occurring in very small amounts.

Life alters atmospheric chemistry because organisms take up, utilize, and cycle some elements more than others. Carbon dioxide occurs in small amounts in our

Table 3.1
The composition of the atmospheres of Venus, Mars, and Earth

	Percentage			Surface Temperature (°C)
	Carbon dioxide	Nitrogen	Oxygen	
Venus	98.00	1.9	trace	477
Mars	95.00	2.7	0.13	−53
Earth	0.03	79.0	21.00	13

Source: Adapted from Lovelock 1979.

atmosphere at least in part because carbon is fixed within living things or their residues (e.g., coal, oil, and soil organic matter). The same process that moves carbon from air to bodies (photosynthesis) also releases oxygen to the atmosphere. Respiration—the process by which organisms liberate the energy captured in carbon compounds to power the business of life—reverses photosynthesis, removing oxygen from the air to recombine with carbon and form carbon dioxide. Fire has the same effect; in fact, respiration is nothing more than controlled fire. The relative proportion of carbon dioxide and oxygen in our atmosphere depends on the balance between photosynthesis and respiration/combustion, which is, in turn, strongly affected by burial of organic matter in ocean sediments and other places where it is protected from recombination with oxygen. As every one is well aware, the buildup of atmospheric carbon dioxide and consequent global warming are due in large part to the fact that industrial societies have been busily tapping into carbon stored in coal, oil, and natural gas, not to mention cutting down forests that are also storehouses of carbon (we return to this point later in the chapter).

Many physical aspects of our planet result from the activities of life. Nitrogen is abundant in our atmosphere because a few types of microbes that live in soil, streams, and oceans utilize nitrogen compounds for energy, releasing nitrogen gas to the air in the process. The cycle of water is greatly affected by land plants, without which water that is transpired from leaves to the atmosphere would flow instead to the oceans. The chemistry of the atmosphere, in turn, profoundly influences factors that are of considerable consequence to life, such as global temperature and penetration of harmful shortwave radiation from the sun.

3.1.2 Natural Capital and Ecosystem Services

> The ecosystems-services concept makes it abundantly clear that the choice of "the environment versus the economy" is a false choice.
>
> COSTANZA 2006

> The global economic importance of forests will increasingly lie in their nontimber products and especially in their environmental functions.
>
> LESLIE 2005

The earth's natural and managed ecosystems provide humans with a variety of goods and services. While the value of goods such as food or fiber is readily captured in the marketplace, the value provided by ecosystem services, such as climate amelioration or watershed protection, is usually not. As Costanza et al. (1997) point out, "The economies of the earth would eventually grind to a halt without the services of ecological life-support systems, so in one sense their total value to the economy is infinite." Nevertheless, ecologists and ecological economists have devoted increasing attention to placing a monetary value on ecosystem services (Costanza et al. 1997; Daily et al. 1997, 2000; Daily 1999; Balmford et al. 2002). Integrating several studies, Costanza et al. (1997) estimated that the global value of ecosystem services averaged slightly more than $33 trillion annually, a value they considered a minimum estimate. For comparison, the average global gross national product was approximately $18 trillion per year at the time of the study. Oceans, especially coastal areas such as estuaries and reefs, accounted for about two-thirds of the ecosystem service values. Of the terrestrial contribution, forest and wetland services were of equal value.

An important research thrust in the coming decades will be to better understand the ecology of ecosystem services, so that ecosystems can be kept healthy and, where necessary, restored (Daily et al. 2000). Improved estimates of the value of ecosystem services, and the development of pricing and market mechanisms that allow better accounting of those services in policy decisions, will require ecologists and economists to work together. Market-based tools are already available to some degree for carbon markets and watersheds. For example, "an analysis of 27 U.S. water suppliers revealed that treatment costs for drinking water derived from watersheds covered at least 60% by forests were one-half the cost of treating water from watersheds with 30% forest cover, and one-third of the cost of treating water from watersheds with 10% forest cover" (Postel and Thompson 2005). Faced with the choice of building a new filtration system or protecting forested watersheds, New York City concluded the least expensive route was to commit money to protecting and restoring the city's watershed. (See Postel and Thompson [2005] for a detailed discussion of the New York City experience and other examples of investments in watershed protection.)

3.1.3 The Gaia Hypothesis

> In the real world the evolution of life and the evolution of the rocks, the oceans, and the air are tightly coupled.
>
> LOVELOCK 2004

> The earth system behaves as a single, self-regulating system comprised of physical, chemical, biological, and human components.
>
> THE AMSTERDAM DECLARATION 2001[1]

Gaia was the name given by the ancient Greeks to their goddess of the earth. James Lovelock and Lynn Margulis applied this name to their hypothesis that the sum of life on earth actively regulates the temperature and composition of the Earth's surface (including the atmosphere) (Margulis and Lovelock 1974, 1989). In this view, life on earth comprises a cybernetic system that regulates and maintains physical conditions within bounds that are suitable for life. In Lovelock's words (Lovelock 1988, 19):

> The Gaia hypothesis, when we first introduced it in the 1970's, supposed that the atmosphere, the oceans, the climate, and the crust of the Earth are regulated at a state comfortable for life because of the behavior of living organisms. Specifically, the Gaia hypothesis said that the temperature, oxidation state, acidity, and certain aspects of the rocks and waters are at any time kept constant, and that this homeostasis is maintained by active feedback processes operated automatically and unconsciously by the biota. . . . Life and its environment are so closely coupled that evolution concerns Gaia, not the organisms or the environment separately.

As already discussed, modern science leaves no question that life imprints distinctive patterns on the atmosphere, modifies the chemistry of water (chapter 17), and shapes both the chemistry and the physical structure of rocks (without life, soil would not exist; see chapter 14). When first introduced, however, the Gaia hypothesis was revolutionary and (like all revolutionary ideas) quite controversial. Today it is widely accepted by earth systems scientists (in fact earth systems science is in effect the active exploration of the Gaia hypothesis), and "ecology as taught in academic circles has become more Gaian or has faded away" (Margulis 2004). Controversy remains, especially among some evolutionary biologists (whose worldview contains no mechanisms by which global integration could evolve). Margulis (2004) attributes the continuing controversy in part to "confused definitions, incompatible belief systems of the scientific authors, and inconsistent terminology across the many affected disciplines." An excellent summary of current thinking has been presented by Schneider et al. (2004).

Margulis (1990) argues that "life does not 'adapt' to a passive physico-chemical environment. . . . Instead life actively produces and modifies its environment." The result is that the earth is a self-regulating system. As Lovelock (2004) points out, this is more than an hypothesis, it is a new way of organizing facts about life on earth. Throughout this book, we see examples where physical environments are indeed regulated by organisms at the scale of local ecosystems and regions. Plants, for example, devote a great deal of energy to creating and maintaining soil conditions that are favorable for plant growth. At the regional scale, the cycling of water by forests moderates temperatures and maintains rainfall that allow forests to grow. Models predict that should forests be cleared from the Amazon basin, rainfall would decline to the point that the same forests could no longer grow there (Shukla et al. 1990). Note that the effects of plants both on soils and on rainfall are examples of positive feedback between the biota and the physical environment.

In their systems analysis of feedbacks among plants and carbon dioxide, Beerling and Berner (2005) conclude "the coupled evolution of land plants, CO_2, and climate over the last half billion years has maintained atmospheric CO_2 concentrations within finite limits, indicating the involvement of a complex network of geophysiological feedbacks. But insight into this important regulatory network is extremely limited."

The earth is without doubt a highly coupled system. However, the emerging picture at the global level is not one of cybernetic control in the same sense that a thermostat acts through negative feedbacks to maintain room temperature within a narrow range, but of a complex mixture of positive and negative feedbacks between the biotic and abiotic worlds, resulting in abrupt changes and wide swings in climate that shape the evolution of the biota, which, in turn, reshapes biotic feedbacks to climate: the earth is not only coupled, it is coevolved (Beerling and Berner 2005).

The history of science is one of old paradigms grudgingly giving way to the new. The question of the reality of a self-regulating planet is typical of many others we face in ecology today. Whether it is the whole earth, a region, a landscape, or a single local ecosystem, understanding the nature of complex systems is exceedingly difficult, and the tendency of complex systems to change abruptly and sometimes irreversibly (chapter 20) makes tinkering with them a risky business.

3.1.4 Global Cycles of the Elements

Figure 3.1 shows the cycles of some important elements through the global ecosystem. Some 26 chemical elements are required by either plants, microbes, or animals (chapter 16). Moreover these are required in rather specific proportions: phytoplankton, for example, contain carbon, nitrogen, and phosphorus in the ratio 106:16:1, called the **Redfield ratio** (Redfield 1958). The chemical elements required by life are not distributed evenly over the

[1]The Amsterdam Declaration was issued by a joint meeting of the International Geosphere Biosphere Program, the International Human Dimension Program on Global Environmental Change, the World Climate Research Program, and the International Biodiversity Program, held in Amsterdam on July 13, 2001. The Declaration included a number of points, of which the quote given above was the first.

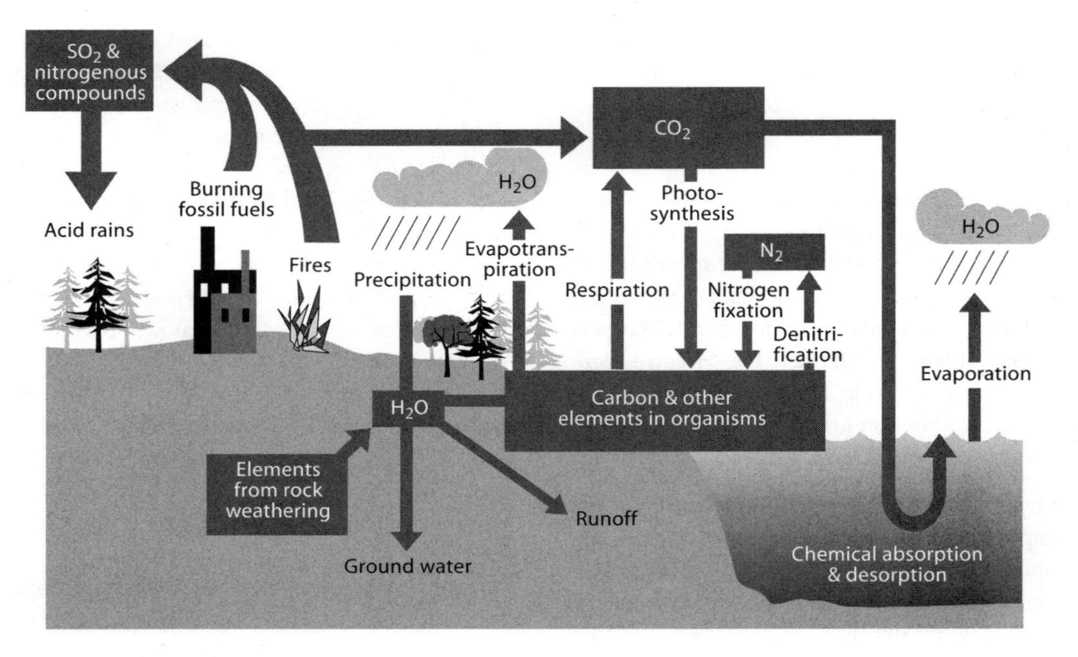

Figure 3.1. The global cycles of some biologically important elements. (NASA 1988)

globe. Water and carbon are concentrated in the oceans, molecular nitrogen and oxygen in the atmosphere, and other elements essential for life in rocks. Hence cycling from one portion of the biosphere to another is essential to maintain life everywhere. The study of the cycling of elements on the earth is called biogeochemistry (for a good review, see the book of that title by Schlesinger [1997]).

Two cycles can be distinguished: (1) the **gaseous cycle,** in which elements move through the atmosphere during at least some portion of their global cycle; and (2) the **sedimentary cycle,** in which elements move from land to water and then to sediments, where they remain until moved to land again by tectonic activity. Relatively few of the chemical compounds required by life exist as gases at normal temperatures and pressures; however, the ones that do—carbon, oxygen, nitrogen, sulfur, and water vapor—are required in relatively large amounts by organisms. (The nutrient elements essential for life are discussed in chapter 16.) All of these gas-phase elements also exist in solid form and, hence, move via water as well as via the atmosphere. Phosphorus, calcium, potassium, magnesium, iron, and a host of other sedimentary-cycle elements that originate from rocks and are required by life move primarily by water, although they may also be transported through the atmosphere as windblown dust (chapter 17).

The time scales of gaseous and sedimentary cycles are quite different. The former cycles on the scale of years, the later on the scale of eons. For example, the equivalent of the entire atmospheric content of carbon passes through the terrestrial biota every six years (Schlesinger 1997), while a molecule of phosphorus eroded from a hillside and moved by water to eventually reside in deep ocean sediments may not move to land again for millions of years. The essential nutrients that have a gas phase (C, N, H, O, S) also have a solid phase; hence, they too are subject to being buried in deep sediments. Life plays a critical role in determining the fate of these elements by converting (as a byproduct of metabolism) the organic forms to gases, thereby keeping these elements in circulation. We follow these points in a more detailed discussion of the global cycles.

3.1.4.1 Carbon

Carbon actively cycles among three major reservoirs on earth: the atmosphere, oceans, and land, the latter comprising vegetation and soils (fig. 3.2). Of these, the oceans store by far the largest amounts of carbon, with about 50 times more than is held in the atmosphere and 10 times more than is held in land vegetation and soils (Falkowski et al. 2004; Sabine et al. 2004a). The majority (97%) of carbon in the oceans is in dissolved inorganic forms (Falkowski et al. 2004).

The amounts of carbon stored in various global pools reflect differences between fluxes into and out of those pools. Carbon dioxide is soluble in water and moves between atmosphere and ocean along partial pressure gradients. In water, CO_2 combines with H_2O to form bicarbonates and carbonates, the predominant forms of carbon in oceans (only carbon dioxide exchanges with the atmosphere). Because carbon dioxide is more soluble in cold than in warm water, the oceanic sink strength is greater at high than at low latitudes; hence, oceans generally act as a net carbon sink at high latitudes and a net source to the atmosphere at low latitudes (Le Quere and Metzi 2004). Through

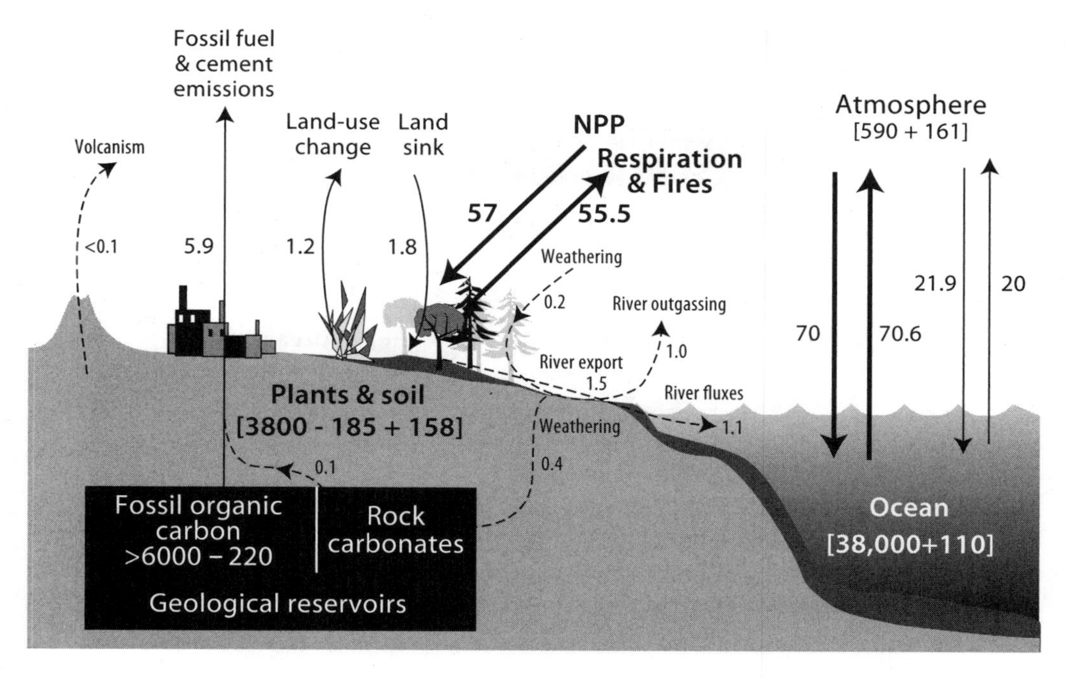

Figure 3.2. Global carbon pools and fluxes (pools in Pg and fluxes in Pg/y; 1 Pg = 10^{15} g). For pools, the first number represents the estimated preanthropogenic value, while additions or subtractions represent human perturbations. For example, 185 Pg is estimated to have been lost and 158 Pg added to the plant-soil pool due to human activities. For fluxes, dark arrows denote preanthropogenic and light arrows human perturbations. Of the total soil-plant pool, nearly 80 percent is in soils. Of the total flux due to heterotrophic respiration and natural fires, more than 90 percent is respiration. Values are averages over the 1980s and 1990s. (Estimates for most values may vary among authors). NPP, net primary productivity. (Adapted from Field and Raupach 2004)

photosynthesis and fixation in biomass, phytoplankton lower carbon dioxide partial pressure in the ocean, thereby strengthening the oceanic carbon sink. Some marine organisms form shells from calcium carbonate, which also lowers carbon dioxide partial pressure in the ocean.

While carbon storage in the oceans and carbon fluxes between the atmosphere and the oceans are primarily physicochemical processes in which phytoplankton play a significant regulatory role, movement from the atmosphere to land is almost exclusively a product of life, and movement from land to atmosphere is largely so. Most carbon fixed in terrestrial **net primary productivity** (NPP) is eventually returned to the atmosphere in heterotrophic respiration or fire, although the time required may range from months to centuries. In soils of any vegetation type, but particularly forests, a high proportion of NPP is stored in slowly cycling pools such as humus and (in forests and shrublands) dead wood, where it may be retained for centuries. Compared to heterotrophic respiration, fire is a relatively minor source of atmospheric carbon dioxide, but in some years it can add large amounts of carbon to the atmosphere. For example, in 1997/1998, huge forest fires in Indonesia are estimated to have pulsed an amount of carbon equal to 13 to 40 percent of the annual emissions from fossil fuels, resulting in a doubling of the growth rate of carbon dioxide in the atmosphere (Page et al. 2000; Schimel and Baker 2000).

> Two terms are particularly important in describing the fluxes associated with living, or once living, things. The first, *net primary productivity* (NPP) is the difference between carbon removed from the atmosphere in photosynthesis and carbon returned to the atmosphere in plant respiration. On land, the carbon represented by a positive NPP may be stored in biomass or soils, or exported to rivers and ultimately the ocean. Most carbon fixed through NPP is eventually returned to the atmosphere by heterotrophic respiration (i.e., animals and microbes), fire, or, in modern times, land use change (e.g., conversion of forest to agriculture). The difference between NPP and the sum of all natural processes other than plant respiration that cycle carbon from the earth's surface to atmosphere is termed *net ecosystem productivity*.

As of the early twenty-first century, humans have released about 400 petagrams (1 Pg = 10^{15} g) of carbon to the atmosphere during the Anthropocene (the industrial age beginning roughly 200 years ago), largely through burning fossil fuels and land use change (Sabine et al. 2004a). For perspective, atmospheric carbon dioxide content at the dawn of the industrial age was on the order of 590 Pg (Sabine et al. 2004a). Of the 400 Pg introduced by

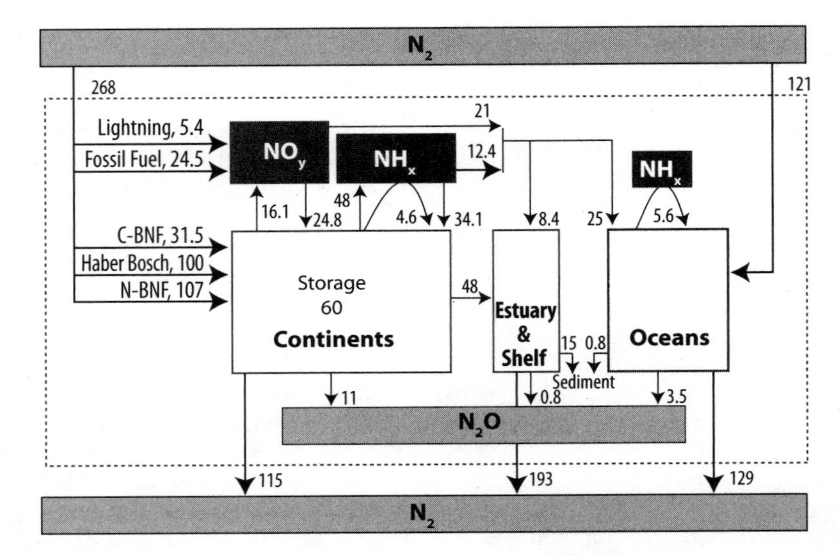

Figure 3.3. Components of the global nitrogen cycle for the early 1990s (Tg N/y). All shaded boxes represent reservoirs of nitrogen species in the atmosphere. Creation of reduced nitrogen (N_r) is depicted with bold arrows from the N_2 reservoir to the N_r reservoir (depicted by the dotted box). Denitrification, the creation of N_2 from N_r, is also shown with bold arrows. All arrows that do not leave the dotted box represent interreservoir exchanges of N_r. The dashed arrows within the dotted box associated with NH_x represent natural emissions of NH_3 that are redeposited on fast time scales to the oceans and continents. N-BNF, biological nitrogen fixation within natural ecosystems; C-BNF, biological nitrogen fixation within agroecosystems; Haber Bosch, an industrial process. (Adapted from Galloway et al. 2004, with kind permission of Springer Science and Business Media)

humans, 161 Pg stayed in the atmosphere and the rest was absorbed by either land or oceans. Sabine et al. (2004b) calculate that from 1800 to 1994, somewhere between 30 and 40 percent of total emissions was from land use change, mostly deforestation but also other factors such as conversion of native grasslands to agriculture (which accelerates decomposition of stored soil carbon). During that period, the land was a net source to the atmosphere (i.e., on a global basis emissions from land use change exceeded NPP), and oceans comprised the entire net carbon sink. The relative ocean-land sink strength began to shift in the 1990s. Although emissions from deforestation remained high during that period (especially in the tropics), the net terrestrial sink strengthened significantly, equaling and perhaps even exceeding that of the ocean (Sabine et al. 2004b).

The higher terrestrial sink in the late twentieth century has been attributed to increased NPP in the temperate and subboreal forests of northern hemisphere (Ciais et al. 1995); however, more recent work attributes it to tropical forests rather than those of the north (Stephens et al. 2007). Caspersen et al. (2000) found little evidence for enhanced growth rates in forests of the eastern United States. Integrated over large areas (e.g., North America), carbon fluxes between the land and atmosphere can vary widely over periods of a few years, reflecting climatic effects on the balance between NPP and respiration (chapter 15), as well as on the frequency of wildfires (Bousquet et al. 2000; chapter 7).

3.1.4.2 Nitrogen

Nitrogen is unique among the elements for several reasons. A constituent of proteins and nucleic acids, it is required in relatively large amounts by organisms (chapter 16). Diatomic nitrogen (N_2) is by far the most abundant gas in our atmosphere; however, unlike carbon dioxide, nitrogen in its gaseous form is unusable by all organisms except a few primitive bacteria capable of converting N_2 gas to ammonia (NH_3) by a process called **nitrogen fixation** (chapter 17). All life on earth is utterly dependent on the nitrogen-fixing capabilities of these primitive organisms.[2]

Figure 3.3 shows a diagram of global nitrogen fluxes as of the early 1990s. Rather large uncertainties are associated with most of these fluxes, in particular the values for biological fixation, lightning fixation, and denitrification. Atmospheric N_2 gas is converted to NH_3 (fixed) in three ways: (1) by microbes, (2) by humans through an industrial technique called the Haber-Bosch process (used to manufacture fertilizers), and (3) through a strictly chemical process catalyzed by lightning.

By far the majority of fixed nitrogen within ecosystems occurs in organic forms, either in living bodies or soil organic matter. However, small amounts of inorganic nitrogen compounds also occur, particularly NH_4^+ and NO_3^-.

[2]Nitrogen fixation involves the chemical process reduction; therefore, all fixed forms of nitrogen (primarily NH_4^+ and NO_3^-, are referred to as reduced nitrogen (denoted as N_r).

Both organic and inorganic nitrogen-containing compounds undergo various transformations within local ecosystems (chapter 18). One transformation of particular relevance in forests is the conversion of organic nitrogen to gaseous nitrogen during fire, a process that can result in the loss of a significant proportion of total aboveground nitrogen from burned forests. Some species of nitrogen-fixing plants have evolved special adaptations to recover quickly from fire; these plants are very important in replenishing ecosystem nitrogen stores following a burn (chapters 8 and 17).

Fixed nitrogen is unfixed and returned to the atmosphere as gas in several ways, but especially through the combustion of organic matter in fires (including burning fossil fuels) and through a process called **denitrification,** in which certain microbes that live in oxygen-poor (anaerobic) environments use NO_3^- rather than oxygen as a terminal electron acceptor during metabolism, releasing mostly N_2 but also N_2O to the atmosphere (chapter 18). Denitrifiers keep nitrogen in circulation; without these organisms the atmospheric pool of nitrogen would be eventually depleted through fixation and burial in sediments. Gutschick (1981) estimates that were denitrification in the ocean to be reduced substantially while the amount of nitrogen in runoff from land to sea remained at current levels, the total pool of atmospheric N would be depleted within 70 million years.

Most of the organic and inorganic nitrogen compounds that are carried by rivers from land to streams and oceans come from ecosystems that are disturbed or stressed in some way, for example areas with insufficient plant cover to hold soil against erosion, or systems with nitrogen inputs in excess of what can be retained (which may be due to heavy fertilization of agriculture fields or to nitrogenous pollutants). Healthy, intact terrestrial ecosystems retain nitrogen reasonably effectively, though not so effectively as once thought (chapter 17).

Galloway et al. (2004) estimate that humans more than tripled the rate of nitrogen input to terrestrial ecosystems between 1860 and the early 1990s, and predict a further doubling by 2050. While small additions may benefit ecosystems by stimulating plant growth, large additions create several potentially serious problems, including altering nutrient balances, generating acidity that leaches essential cations from soils, acidifying freshwater ecosystems, causing eutrophication and hypoxia in coastal ecosystems, facilitating invasion of exotic weeds, and triggering biodiversity loss in terrestrial and aquatic ecosystems (Vitousek et al. 1997; Galloway 1998; Cowling et al. 2001).

3.1.4.3 Sulfur

Sulfur, whose primary source is from rocks, is moved from the earth's surface to the atmosphere in three ways: (1) emitted from burning fossil fuels as sulfur dioxide (SO_2) (along with certain gaseous nitrogen compounds, the source of acid rain); (2) released by phytoplankton as the gas dimethylsulfide (DMS); and (3) emitted as hydrogen sulfide (H_2S) by microbes living in anaerobic environments such as bogs and sediments. As is the case with NO_3^-, oxidized sulfur compounds are used by certain bacteria as terminal electron acceptors during metabolism; these bacteria are abundant only where lack of oxygen excludes oxygen-requiring (or aerobic) microbes. Sulfur moves from the atmosphere to the surface in precipitation or dryfall.[3]

Burning of fossil fuels accounts for two-thirds to three-quarters of the gaseous sulfur transferred from the earth's surface to the atmosphere. Release of DMS provides most of the rest (Andreae 1986). DMS is produced by phytoplankton throughout the world's oceans; the familiar smell of the sea comes from DMS. Once in the atmosphere, much of the DMS is oxidized to sulfuric acid, which may combine with gaseous ammonia to form tiny particles that are eventually washed out by precipitation (Andreae 1986). DMS is believed by some researchers to be an important global source of the ice nuclei required to trigger rainfall.

> To fall from the sky as precipitation, water must coalesce into droplets that are large enough for gravity to overcome their tendency to float in the atmosphere. Formation of droplets (actually, ice crystals) requires a so-called nucleating agent, something for the small water drops to cluster around. Any number of small particles may serve as nucleating agents, including DMS, volatile organic compounds, and sodium iodide crystals. (The latter are used in cloud-seeding programs that attempt to induce rainfall.) Evidence suggests that soot from biomass burning acts as an important nucleating agent for rainfall in equatorial Africa (Cachier and Ducret 1991).

3.1.5 The Hydrologic Cycle

Storage reservoirs and fluxes of water are shown in fig. 3.4. By far the bulk of earth's water is contained either in oceans or in ice. Water is transferred from the surface to the atmosphere through evaporation and transpiration (both of which are powered by solar energy), and from the atmosphere to the surface in precipitation.

Figure 3.5 shows the fate of rain falling on a moist tropical rain forest in the Amazon basin: 74 percent is either transpired or evaporated from leaf surfaces, and 26 percent flows to streams. This is a typical pattern. Globally, approximately two-thirds of the precipitation falling on land in an average year is recycled back to the atmosphere by evapotranspiration, the remainder flowing ultimately to the oceans (Shukla and Mintz 1982). Forests play an important

[3]Dryfall refers to the transfer of atmospheric dust and gases to surfaces.

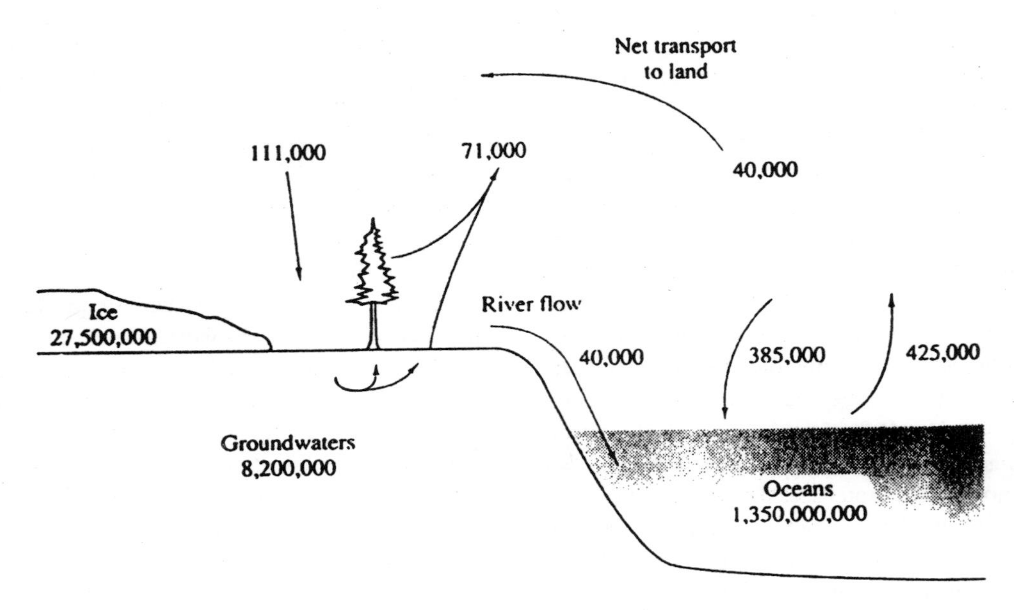

Figure 3.4. The global water cycle. Values are in cubic kilometers for pools and cubic kilometers per year for fluxes (1 km^3 = 10^{15} g). (Schelsinger 1991)

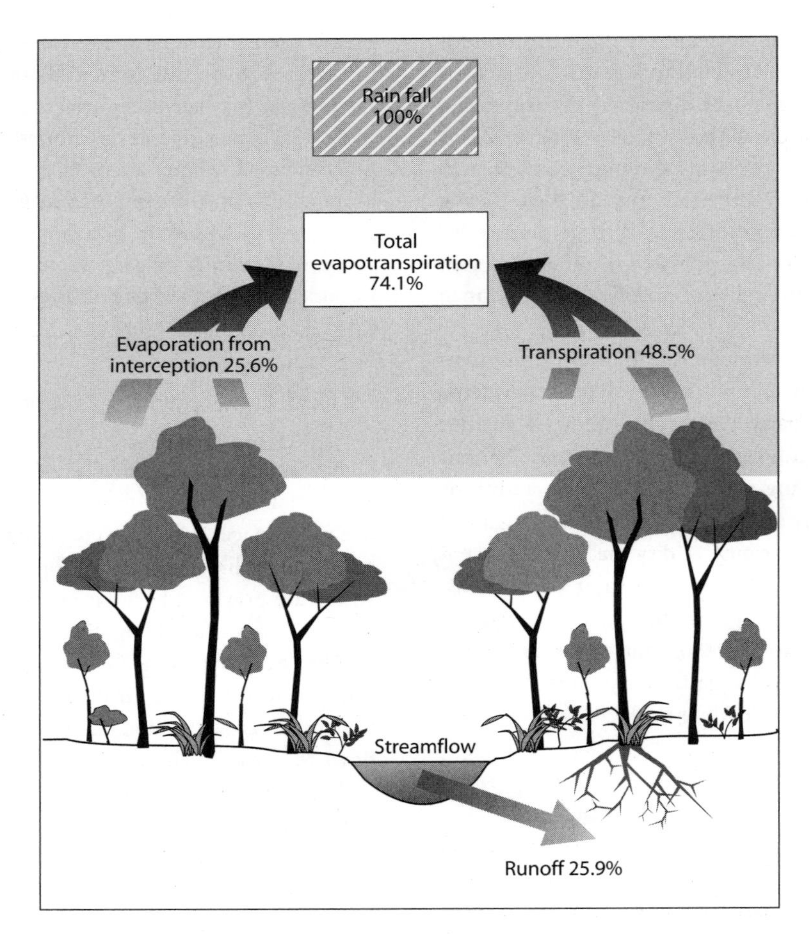

Figure 3.5. The water balance of an Amazon rain forest. (Salati 1987. Copyright © 1987. Reprinted by permission of John Wiley and Sons, Inc.)

role in the hydrologic cycle, something we discuss in more detail in section 3.2.5.

3.1.6 Linkages among Land, Sea, and Air

The various parts of the globe are linked through flows of air and water, and also through migratory animals such as birds that winter in the tropics and summer in the temperate and boreal zones, and anadromous fish that transport nutrients from the ocean to freshwater streams. Productivity of freshwater and marine organisms depends on nutrients from land, especially phosphorus, iron, and nitrogen (Fiadero 1983; Martin et al. 1989; Schlesinger 1997).[4] By weathering rocks and controlling nutrient loss from soils to streams and oceans, land plants modulate the nutrients that are available for the growth of aquatic organisms (Schwartzman and Volk 1989). Aquatic life, in turn, fixes atmospheric carbon and releases nitrogen and sulfur to the atmosphere, preventing these elements from being buried in deep ocean sediments and keeping them in circulation where they are available to other organisms (Fiadero 1983).

The oceans act through atmospheric teleconnections[5] to profoundly affect weather patterns on land. Periodic, rapid reorganizations of the ocean-atmosphere system, termed **climate modes,** are now known to occur commonly in all the earth's oceans. The best known is the El Niño–Southern Oscillation (ENSO), which redistributes rainfall throughout the world (Holmgren et al. 2001). Alterations between the El Niño and La Niña phases of the ENSO produce corresponding swings between relatively dry and relatively wet conditions on land, with different parts of the globe responding differently. The El Niño phase brings drought to Australia, Southeast Asia, India, the Amazon basin, Central America, southern Africa, and portions of North America, while the La Niña phase tends to make these areas wetter and other areas drier (Palmer and Brankovic 1989; Swetnam 1990; Swetnam and Betancourt 1989; Holmgren et al. 2001).[6] Through its effects on climate, the ENSO indirectly influences ecosystem processes such as primary productivity, herbivory, and fire (Holmgren et al. 2001). Fire cycles in particular are strongly affected (chapter 7).

The ENSO is not the only ocean-state oscillation with significant influences on global weather. The Pacific Decadal Oscillation (PDO) also seesaws between warm and cool ocean conditions, but with a longer wavelength and somewhat smaller temperature amplitude than the ENSO (roughly 1–2°C over 15–30 years for the PDO, as opposed to roughly 2–3°C over 1 year for the ENSO), and a much larger affected ocean area than the ENSO (UN Atlas of the Oceans, www.oceanatlas.com). The PDO's primary climatic effect is to alter the jet stream across North America, which, in turn, shifts storm tracks from north to south on a roughly two-decade cycle (Latif and Barnett 1994). The PDO appears to have entered its cool phase early in the twenty-first century, steering the jet stream northward and lowering precipitation in the southwestern United States.

The Atlantic and Indian oceans also exhibit distinct climate modes with strong effects on land. The North Atlantic Oscillation, a pressure differential that seesaws between the northern and subtropical Atlantic, affects the distribution of winter temperature and moisture throughout Europe (UN Atlas of the Oceans www.oceanatlas.com). The Indian Ocean Dipole, an ENSO-like oscillation discovered in 1999, interacts with the ENSO to influence the strength of the Indian summer monsoon (Ashok et al. 2001).

Conditions on land influence the global climate in a variety of ways (Pielke 2002). Terrestrial vegetation contributes to regulating water cycles, nutrient fluxes, and, through both albedo and influences on atmospheric composition, the earth's heat balance. Over geologic time, nutrient inputs to oceans are regulated by weathering and retention on land. Salinity is the product of freshwater losses through evaporation and freshwater inputs through rain and runoff from land. Salinity, in turn, drives large-scale ocean currents and heat transport from the tropical Atlantic to high latitudes, producing, among other things, temperate climates in subarctic areas of northern Europe. Historic droughts in Africa and tropical Mexico are related to increased precipitation and/or rapid glacial melt at high latitudes, which infuses freshwater into the North Atlantic and thereby alters the patterns of ocean circulation (Street-Perrot and Perrot 1990). Droughts in northern Africa have, in turn, been linked to decreased hurricane activity off the coast of the southeastern United States (William Gray, as reported by Kerr 1990).

Strong evidence indicates periods of rapid (<10 years), globally widespread, significant climate change in earth's history. These are almost certainly due to rapid reorganization of the ocean-atmosphere; however, the triggers for these events remain uncertain (NAS 2002; Wang et al. 2004). Global ecology—the study of the earth as an ecosystem—is in its infancy, and much remains to be learned about interactions among the different areas of the planet.

[4]Fanning (1989) provides evidence that the phosphorus limitations in the ocean are found downwind of the most populated areas of North America and East Asia. He suggests that nitrogen contained in acid rain has fertilized the oceans and thereby shifted them from nitrogen to phosphorus limitations. As we shall see in chapter 11, this is quite similar to a pattern that has been seen in at least some European forests, where acid rain has resulted in a shift from nitrogen to phosphorus limitations.

[5]*Teleconnections, or teleconnection patterns,* refers to recurring, persistent, large-scale patterns of atmospheric pressure and circulation that span large geographic areas. "Teleconnection patterns reflect large-scale changes in the atmospheric wave and jet stream patterns, and influence temperature, rainfall, storm tracks, and jet stream location/intensity over vast areas. Thus, they are often the culprit responsible for abnormal weather patterns occurring simultaneously over seemingly vast distances" (NOAA 2005).

[6]In India, at least, an El Niño does not always produce drought. Recent work shows that the occurrence of severe droughts during an El Niño depends on whether the maximum sea-surface temperature occurs in the central or eastern tropical Pacific (Kumar et al. 2006).

3.2 ECOSYSTEM SERVICES PROVIDED BY FORESTS

> Healthy ecosystems are central to the aspirations of humankind.
>
> THE MILLENNIUM ECOSYSTEM ASSESSMENT, 2005A

In his book *Collapse: How Societies Choose to Fail or Succeed* (2005a), Diamond assesses what led to the collapse of historic civilizations. He identifies five groups of interacting factors, with environmental degradation in general and deforestation in particular playing a prominent role in some cases. Diamond (2005b) contrasts historic societies that persisted with some that collapsed:

> Ninth-century New Guinea Highland villagers, 16th-century German landowners, and the Tokugawa shoguns of 17th-century Japan all recognized the deforestation spreading around them and solved the problem, either by developing scientific reforestation (Japan and Germany) or by transplanting tree seedlings (New Guinea). Conversely, the Maya, Mangarevans, and Easter Islanders failed to address their forestry problems and so collapsed.

An earth without forests would be quite different than the one we have today, and certainly less livable for humans (Color plates 1–8 show some examples of the earth's forests). In chapter 1 we briefly discussed the multiple values that forests provide humans. In the remainder of this chapter we explore in more detail how the earth's forests influence global heat balance and cycling of carbon and water, and the role of forests in moderating weather, controlling floods, and modulating inputs of nutrients and sediments to surface waters.

3.2.1. Forests Use Solar Energy to Drive Processes

Forests—indeed plant communities of all types—are systems that transform solar energy into work, which, in turn, manifests as various processes that support the internal functioning of the forest and connect it to the surrounding landscape and the globe as a whole. Two general types of processes may be distinguished: (1) those that directly affect organisms, such as processes providing food, habitat, medicinal plants, and wood; and (2) those that directly affect the movement of energy and elements. Transformation and cycling of elements is called **biogeochemical cycling.**

Although different, these two types of processes are also very much interrelated. Providing nourishing food, for example, requires that the biogeochemical cycle be regulated, while as we saw at the beginning of this book, cycling of elements requires the animals and microbes that are supported by plant energy. Moreover, both processes feed back to support the ability of forests to continue capturing energy in photosynthesis (fig. 3.6).

Figure 3.7 illustrates the ecosystem functions and services provided by forests, along with the factors that influence those outputs. Canopies clean the air stream by raking natural substances and pollutants from it; canopies also moderate climates by transpiring water and absorbing heat energy. By modulating the rate at which water moves through soil and

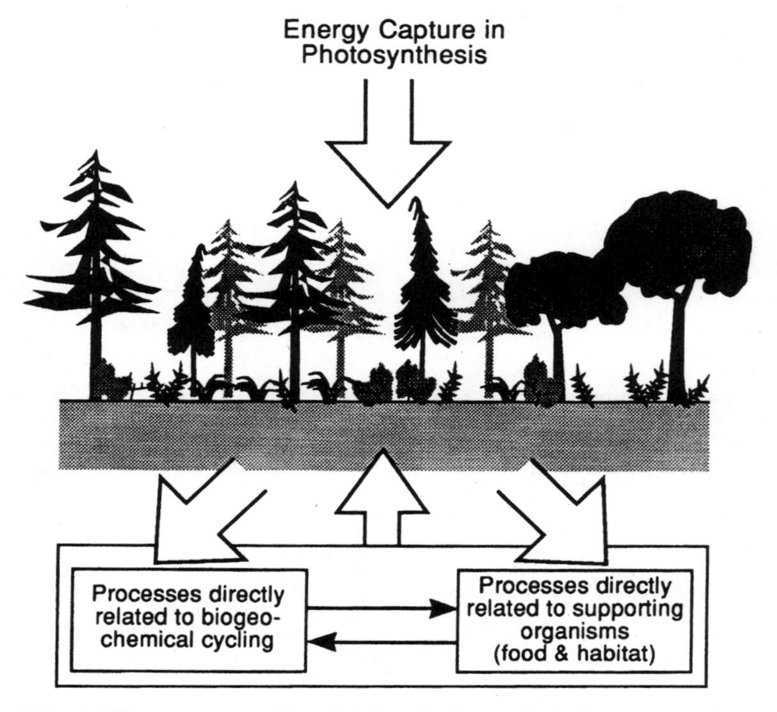

Figure 3.6. The energy captured by green plants in photosynthesis drives processes that feed back to support more photosynthesis.

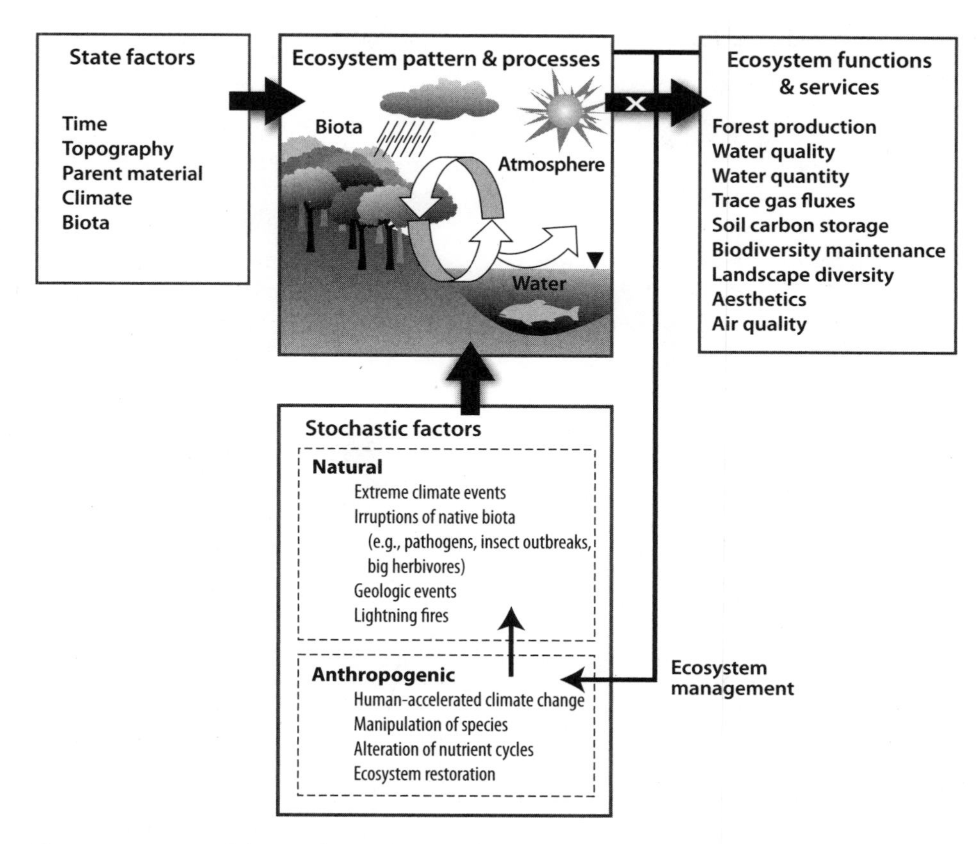

Figure 3.7. A conceptual diagram showing how state factors and stochastic factors influence functions and services provided by a northern hardwood forest, Hubbard Brook Experimental Forest, New Hampshire. (Adapted from Groffman et al. 2004)

modifying its chemistry, forests abate floods and clean the water flowing into streams and aquifers. Forests store carbon, emit trace gases, support a rich diversity of organisms, and provide significant aesthetic values. The degree to which a given forest performs these functions depends on **state factors,** such as topography, soils, climate, and composition of the biotic community; and **stochastic factors,** such as disturbances and human interventions (Groffman et al. 2004).

All types of plant communities provide ecosystem services to one degree or another, but forests are particularly important because of where they grow and what they are. The earth's forests are found only where there is a moderate to good supply of water; hence, they are particularly important in protecting watersheds and cycling water. In turn, water allows forests to array many leaves, and leaves clean the air, rake water from the air, cycle water to the atmosphere, and absorb and reflect solar energy. That trees have many leaves and are tall and woody means they capture and store more carbon than any other type of plant. We now examine the ecosystem services provided by forests in more detail.

3.2.2 Global Leaf Area

Including both those that are intact and those that are degraded, forests occupy roughly 30 percent of the earth's land surface (table 3.2). More relevant than land area in terms of global processes, however, is the leaf surface arrayed by plant communities and the amount of carbon they store. It is leaves that through photosynthesis pump carbon from the air into living matter, and leaves that transpire water from soils

Table 3.2

The land surface occupied by different types of vegetation

Vegetation type	Land area (10^6 km^2)	% of total
Forest		
Tropical closed	12.0	9
Open*	7.5	6
Temperate	9.5	7
Boreal	9.0	7
Total forest	38.0	29
Savanna	18.0	14
Grasslands	13.5	10
Tundra	9.5	7
Deserts and scrublands	27.0	21
Cultivated lands, wetlands, etc.	25.0	19
Total	131.0	100

Source: Adapted from Brown and Lugo 1984, Botkin 1984, and Watt 1982.
*Open forest is defined as having at least 10% tree cover but with sufficient canopy opening to permit understory growth.

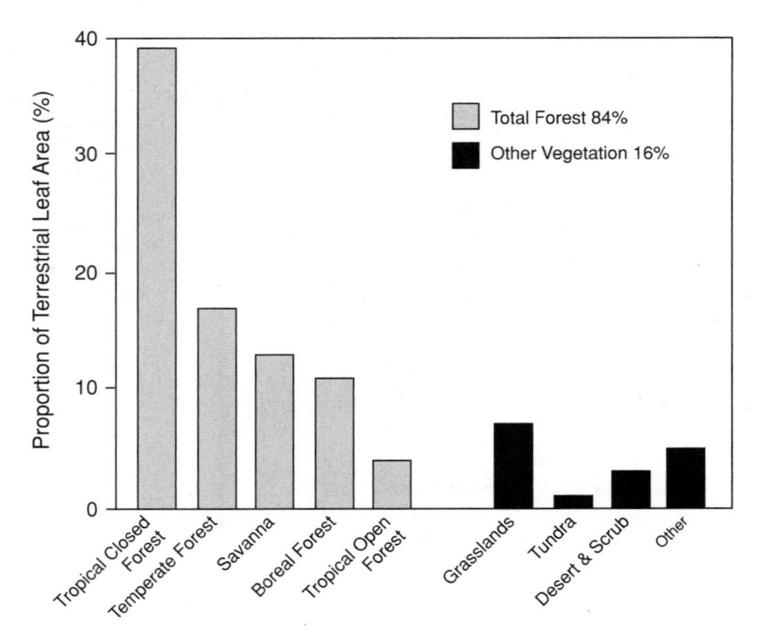

Figure 3.8. The estimated proportion of terrestrial leaf area accounted for by different types of vegetation.

back to the atmosphere. Leaves absorb a much higher proportion of the sun's energy than does bare ground. Hence, the amount and distribution of leaf area on land surfaces greatly affects the earth's heat balance.

As seen earlier, forests occur (or once occurred) in those areas that are relatively well watered (chapter 4), which permits forests to support more leaves and store more carbon per unit of ground area than is possible for deserts and grasslands. If we consider only healthy, undegraded ecosystems, forests in relatively moist environments support from 2 to perhaps 10 times more leaves per unit ground area than grasslands, and from 10 to perhaps 50 times more than deserts. Viewing this on a global basis, forests account for approximately 70 percent of the earth's leaf surface area (fig. 3.8).[7]

3.2.3 Global Heat Balance

All surfaces on earth reflect some of the solar energy that falls on them and absorb the rest. The proportion of incident sunlight reflected by a given surface is called its **albedo.** Light-colored surfaces, such as clouds, snowfields, and sand dunes, have a high albedo, which is to say they reflect a high proportion of the sunlight that falls on them. Dark surfaces, such as oceans and forest canopies, have a low albedo; they absorb a high proportion of the solar energy falling on them. A given surface has a lower albedo when wet than when dry, because liquid water readily absorbs solar energy. Old-growth forests generally have a lower albedo than young forests (plates 9, 10).

For a given level of incident solar radiation, the average temperature of the earth depends chiefly on two factors: (1) the amounts and types of greenhouse gases in the atmosphere, and (2) the collective albedo of its absorbing surfaces. The planetary albedo, in turn, depends on (1) the relative proportion of the surface covered by forests, grasslands, deserts, tundra, oceans, and snow and ice[8]; (2) cloud cover; and (3) the concentration of aerosols in the atmosphere that reflect solar energy (e.g., the common pollutant sulfur dioxide). Forests absorb a higher proportion of incident solar energy than other vegetation types, with albedos ranging from about 0.07 to 0.25 (table 3.3); that is, 75 to 93 percent of the solar radiation falling on forest canopies is absorbed. In general, less than 2 percent of the solar energy absorbed by canopies is used in photosynthesis (chapter 15)—most goes into evaporating water, the water vapor thus produced serving as a vehicle that carries heat into the atmosphere and distributes it from one place to another on the globe. Through

[7]To calculate the proportions shown in fig. 3.8, the following projected leaf areas per unit ground area were assumed (average yearly basis, i.e., a deciduous forest that has 6 m^2 of leaves per m^2 ground area for 5 months per year would have a yearly average leaf area of $5/12*6 = 2.5\ m^2/m^2$): tropical closed forest, 7; tropical open forest, 1.5; temperate forest, 4; boreal forest, 3; savanna, 1.5; grassland, 1; tundra, 0.2; desert and scrub, 0.25; other, 0.5. The actual values are very crude estimates. However the relative values among the different vegetation types probably have less error.

[8]To illustrate one way in which a planet might be self-regulating, James Lovelock, the father of the Gaia hypothesis, created via computer an imaginary world populated by two kinds of daisies: dark-colored ones and light-colored ones. The albedo of Daisyworld, hence its average temperature, depended on the relative proportion of the planet's surface that was covered by either dark or light daisies. Lovelock then varied the luminosity of Daisyworld's sun. To find out what happened, see Lovelock's book *The Ages of Gaia* (1988).

Table 3.3
Albedos of different vegetation types

Vegetation type	Albedo
Temperate forest	0.12 (summer)
	0.25 (winter)
Tropical forest	0.07
Savanna	0.16
Field, grassland	0.16 (summer)
	0.60 (winter)
Desert	0.35

Source: Adapted from Sagan et al. 1979.

this phenomenon, the planetary heat balance, hydrologic cycle, and climate become closely linked.

How important are forests to the global heat balance? Potter et al. (1975) used a simulation model to investigate how climate might be altered by changes in surface albedo due to removing the earth's tropical rain forests. They found that a world without rain forests was cooler and drier than at present. Sagan et al. (1979) calculated how global albedo had been affected by human land use patterns over the past few millennia. They concluded that by converting savannas to deserts (through overgrazing) and forests to fields, and through the salinization of agricultural fields (at least in some cases linked to deforestation, see Perlin [1989]), humans have contributed significantly to global climate changes during the past several millennia.

3.2.4 Carbon Storage

Although considerable uncertainty exists regarding the carbon budgets of terrestrial ecosystems, forests are clearly the largest sinks and greatest storehouses on land. Sabine et al. (2004b) estimate that forests account for 82 percent of the carbon stored in land plants, and 40 percent of the carbon stored in soils (excluding that in permafrost) (table 3.4). A little less than two-thirds of total forest carbon stores (plants and soils) is in the tropics, and about one-quarter in the temperate zone. Similarly, forests account for 52 percent of terrestrial net primary productivity, two-thirds of that in the tropics. An additional 24 percent of terrestrial net primary productivity occurs in tropical savannas and grasslands. Note that the poor database and rapid rate of deforestation across the globe create a great deal of uncertainty in these values. Carbon uptake by plants is most readily seen in the annual fluctuations of atmospheric carbon dioxide, which changes on the order of 3 percent from growing season to nongrowing season (Schneider and Londer 1984). D'Arrigo et al. (1987) estimate that seasonal growth of the boreal forests alone accounts for 30 percent of the annual atmospheric carbon dioxide fluctuation as measured at Mauna Loa, Hawai'i.

Table 3.4
Plant carbon, soil carbon, and net primary productivity in the world's major biomes

Ecosystems	Area (10^6 km^2)	NPP (Pg C y^{-1})	Plant C (Pg C)	Soil C (Pg C)
Tropical forests	17.5	20.1	340	692
Temperate forests	10.4	7.4	139	262
Boreal forests	13.7	2.4	57	150
Arctic tundra	5.6	0.5	2	144
Mediterranean shrublands	2.8	1.3	17	124
Crops	13.5	3.8	4	248
Tropical savannas and grasslands	27.6	13.7	79	345
Temperate grasslands	15	5.1	6	172
Deserts	27.7	3.2	10	208
Wetlands				450
Frozen soils				400
Total	174.8	57.5	652	3,194

Source: Adapted from Sabine et al. 2004.

3.2.4 Water Cycling

Hahai no ka ua i ka ulula'au. (The rain follows after the forest.)

HAWAIIAN SAYING

Shukla and Mintz (1982) quote the biography of Christopher Columbus, written by his son: "On Tuesday, July 22d, he departed for Jamaica. . . . The sky, air, and climate were just the same as in other places; every afternoon there was a rain squall that lasted for about an hour. The admiral writes that he attributes this to the great forests of that land; he knew from experience that formerly this also occurred in the Canary, Madeira, and Azore Islands, but since the removal of forests that once covered those islands they do not have so much mist and rain as before."

It is common wisdom among people who live close to the land that forests play a critical role in regulating the hydrologic cycle. For example, in Wilk's (2000) survey of villagers living within tropical catchments in India and Thailand, people commonly believed that trees increase infiltration and long-term water supply to streams, increase humidity and therefore precipitation, increase evaporation and therefore cloud formation, and increase rainfall by "capturing" passing clouds and extracting their moisture. Wilk summarized villagers' views as follows: "Water was pictured very strongly as being part of a cycle so that (whether) used by humans or trees (it) is not considered lost but only displaced to return again as rain."

Forest hydrologists disagree among themselves about the validity of some commonly held beliefs concerning trees and water. Skeptics find little hard evidence in support of a beneficial role of trees, at least for some hydrologic functions, and in some cases a demonstrable negative effect on streamflows (Andreassian [2004] and Bruijnzeel [2004] summarize the history of this controversy). Like many other issues in ecology, the large scale and high variability of hydrologic processes make generalizations that are both broad and accurate extremely difficult, expensive, and in some cases impossible, a situation that leads inevitably to scientific controversy. In the remainder of this section we look more closely at the hydrologic function of forests and the issues in this debate. The subject is too complex to explore in detail here. Interested readers should see recent reviews by Andreassian (2004), Bruijnzeel (2004), and Huston et al. (2004).

Land plants influence the water cycle in at least six ways:

1. Leaves transpire water back to the atmosphere, which leads to more rainfall.
2. Forests in particular produce canopies that have low albedo and are aerodynamically rough, both of which enhance cloud formation.[9]
3. Plants produce aerosols that enhance cloud formation. (Because of their high leaf surface areas, forests may be particularly important in this regard.)
4. Leaves facilitate water transfer from air to ground by providing surfaces upon which water vapor condenses. (Forests are also particularly important in this process.)
5. By creating a diverse structure within soils (chapter 14), plants greatly enhance the water-storage capacity of soils, modulating streamflows and reducing erosion from overland flow.
6. Through a process called **hydraulic redistribution,** deep-rooted plants pump water between deep storage reservoirs and surface soils (Burgess et al. 1998; Horton and Hart 1998; Lee et al. 2005). Downward hydraulic redistribution occurs primarily during periods of excess moisture and serves as a water conservation mechanism. Upward hydraulic redistribution, which is referred to as **hydraulic lift,** occurs primarily at night during dry periods. Because stomata are closed at night, water is released to the upper soil layers where it is available for transpiration by the entire plant community during the day. Dawson (1993) estimated that hydraulic lift could increase transpiration in temperate forests between 19 and 40 percent. Similarly, Lee et al. (2005) estimate that hydraulic lift increases dry-season transpiration in the Amazon by approximately 40 percent.

All plants influence water in one way or another. Because forests grow in areas with more rainfall than grasslands and deserts, however, they occupy key nodes within global and regional hydrologic cycles. In the United States (excluding Alaska), approximately two-thirds of water runoff comes from forests (14% of the total comes from National Forest lands) (Sedell et al. 2000). Moreover, as discussed earlier, forests support more leaves, hence more evaporative surface, than other plant communities. On a global basis, forests account for approximately 60 percent of evapotranspiration from land (Oki and Kanae 2006), and evapotranspiration, in turn, plays various key roles in regional and global climates.[10] Gordon et al. (2005) estimate that deforestation has decreased global flows of water vapor from land to the atmosphere by 4 percent, an amount nearly equal to the increased vapor flow due to irrigation.

3.2.4.1 Forests and Rainfall

As Columbus observed, keeping water vapor in circulation creates the potential for more rainfall. For example, about twice as much rain falls in the Amazon basin as can be accounted for by water vapor that moves onto land from the ocean: the excess is almost certainly generated by evapotranspiration (Salati 1987). Computer simulations suggest that if grasses were to replace forests as the major vegetation type in that region, the Amazon basin would become much drier—unable, in fact, to support tree growth (Lean and Warrilow 1989; Shukla et al. 1990). The forests create the conditions that allow the forests to grow. (We encounter other examples of self-sustaining positive feedback in ecosystem processes throughout the book.)

Where forests are allowed to regrow, hydrologic functions characteristic of forest begin to recover, and the disparity between primary forests and deforested areas is reduced (Bruijnzeel 2004). In areas where shifting cultivation or forestry is the dominant land use, trees usually do regrow on cleared areas. Where ranching or permanent cropland is the primary land use, however, they do not (pastures are typically burned to prevent recolonization by trees). In the Brazilian Amazon, Cardille and Foley (2003) estimated that of 25.32 million ha deforested between 1980 and 1995, 54 percent was in active pasture in 1995, 36 percent was regrowing forest, 7 percent was in cropland, and slightly less than 3 percent was degraded. These proportions indicate a longer-term regional shift in land-atmosphere interaction, with implications for cloud formation and precipitation (cf. Costa and Foley 2000).

Some tropical areas, particularly in Asia and the Pacific, have experienced long-term declines in rainfall during the twentieth century, a period of heavy deforestation (Bruijnzeel 2004); however, it has not always

[9]Bright surfaces (high albedo) absorb less solar energy than darker surfaces; hence, they have less energy to drive evaporation and transpiration.

[10]On a global basis, evapotranspiration from land supplies only 15 to 20% of the water evaporated to the atmosphere each year; the rest comes from oceans (Westall and Stumm 1980; Spiedel and Agnew 1982). However, less than 10% of the water evaporated from oceans moves over land surfaces; most falls back into the oceans as rain. Water vapor from oceans certainly contributes to precipitation on land, especially in coastal areas; however, on a global basis, roughly 65% of the water that falls onto land has come from land. Shukla and Mintz (1982) modeled an earth without land vegetation. They found that precipitation in Europe declined to virtually zero, while over much of North America it declined by 70 to 80%. Across the globe, only those coastal areas that received precipitation directly from the oceans did not experience a severe drop.

been straightforward to link deforestation within a given basin to changes in rainfall within the same basin. Complex interactions between ocean conditions and land conditions are likely to make such simple correlations problematic. Nevertheless, the emerging picture is that large-scale deforestation alters land-atmosphere processes in such a way as to influence climate across broad regions. To quote Bruijnzeel (2004), "there is cause for concern. Of late there is increasing observational (as opposed to purely model based) evidence that forest conversion over areas between 1000 and 10,000 km^2 causes feedbacks in the timing and spatial distribution of clouds."

Perhaps the clearest examples of deforestation in one area influencing climate in adjoining areas are in maritime-influenced tropical regions. In both Costa Rica and Puerto Rico, loss of forest in lowland areas has been linked to altered atmospheric processes and reduced cloud cover in mountain forests (Lawton et al. 2001; Van der Molen 2002). As montane cloud forests are very effective at raking water from clouds that pass through their canopies (Bruijnzeel 2004), reductions of cloud cover in such forests would almost certainly translate into lower water input to the system. Modeling indicates that large-scale alterations of land-atmosphere feedbacks also influence precipitation over continental areas (Costa and Foley 2000). Bruijnzeel (2004) speculates that deforestation may have a greater effect on larger-scale atmospheric circulation in continental than in maritime climates: "Larger-scale conversions (>100,000, >1,000,000 ha) may cause even more pronounced changes in atmospheric circulation, to the extent of actually affecting precipitation patterns, even under more continental conditions, such as Amazonia."

Simulation models predict that forests also influence climate (precipitation and temperature) at the scale of regions and the globe. For example, modeling predicts that converting forest to cropland in the eastern and central United States cooled the climate of that region, particularly at the edge of the major forest zone in the midwest (Bonan 1999). Models also predict that deforestation in the tropics can influence climate of extratropical regions throughout the globe (Chase et al. 1996, 2000; Zhao et al. 2001a; Werth and Avissar 2002). Lee et al. (2005) calculate that the increased transpiration supported by hydraulic lift reduces temperatures by up to 2°C throughout the water-stressed regions of the globe.

What can we conclude about forests and precipitation? The earth's climate emerges from complex interactions among oceans, atmosphere, and land. The oceans and atmosphere are in particularly close interaction, and the atmosphere communicates climatic signals between oceans and land through large-scale teleconnections. Vegetation both reacts to and shapes these atmospheric signals, acting primarily through albedo and evapotranspiration to influence climates at regional and global scales (Pielke 2002). Forests are unique among the earth's vegetation types in their relatively low albedo, high transpiration rates, and canopy roughness, all factors that tend to increase cloud formation and, therefore, influence not only precipitation but the global heat balance as well. In continental interiors, the extent of forest cover is almost certainly a primary driver of local climate (e.g., Werth and Avissar 2002).

The issues are complex, but the emerging picture is that the old wisdom had it right: rain follows the trees; however, one may have to look regionally and even globally to see that pattern.

3.2.4.2 Forests and Stream Hydrology

According to Huston et al. (2004), "the single most important factor regulating the hydrologic interaction of terrestrial and aquatic ecosystems is the amount of vegetation, including both living and dead plant material, covering the landscape." However, other factors come into play as well, including topography, soils, bedrock, seasonal distribution of precipitation, and, for forests, whether trees are evergreen or deciduous (Post and Jones 2001). That complexity can produce large variability in hydrologic regimes[11] among watersheds in different environmental settings, and even a single watershed can vary from year to year depending on climatic variability.

There are two issues to be dealt with regarding the effect of plants on stream hydrology. The first is the total amount of water flowing to streams (water yield), which over a year is a simple input-output function.

$$WY = P + CW - ET - \Delta S$$

where WY is water yield; P is precipitation; CW is inputs from canopy-raking of cloud water; ET is evapotranspiration; and ΔS is changes in storage (e.g., in soils).

The second issue relates to the timing of flows, which is strongly influenced by precipitation regime (e.g., seasonality) coupled with the ability of soils to absorb and store water. However, topography, bedrock characteristics, and impermeable surfaces (e.g., roads, compacted soils) also play a significant role.

Let's first consider total water yield irrespective of timing. Evapotranspiration from land modulates the flow of water to streams, rivers, lakes, and oceans. Because forests have more transpirational surface than other vegetation types, and in certain climatic zones may also supplement precipitation and counterbalance transpiration by raking water from fog and clouds, they play a special role in regulating the flow of water to streams. Early data regarding the effect of vegetation on total water yield from individual watersheds was summarized by Bosch and Hewlett (1982). In the studies they reviewed, water yield from pine and eucalypt forests increased an average of 40 mm for each

[11]Post and Jones (2001) define a hydrologic regime as the relationship between precipitation inputs and stream outflows in a basin.

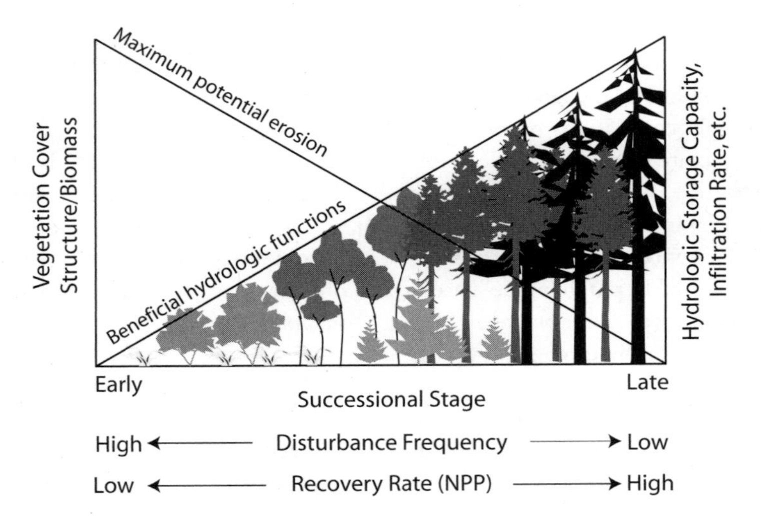

Figure 3.9. The influence of disturbance frequency on hydrologic function. NPP, net primary productivity. (Huston et al. 2004)

10 percent drop in leaf area, while water yield from deciduous hardwood forests increased an average of 25 mm for each 10 percent drop in leaf area. Fully stocked forests transpire about 50 percent of total site precipitation if trees are evergreen, and 25 to 40 percent if deciduous (depending on the length of time in leaf) (Post and Jones 2001). In some cases—most notably fast-growing eucalyptus plantations in relatively dry environments—the trees appear to transpire more water than is received in annual precipitation (Calder et al. 1997).

The effect of forest cover on streamflows can be seen most clearly in experimentally paired, small watersheds where one of the pair has been partially or totally clear-cut. Patterns generally follow that shown in fig. 3.9. Water yields increase following clearing (by human or natural disturbance), returning to predisturbance levels as the vegetation recovers (e.g., Jones and Post 2004). However, there are at least two exceptions to that pattern. One is where fog drip (cloud-raking by canopies) provides significant moisture, in which case streamflows have been seen to diminish after forests are cleared. The second relates to differences in water yield with forest age. In the Pacific Northwest of the United States and in southeastern Australia, older forests have been found to transpire less than younger ones (Vertessy et al. 2001; Moore et al. 2004). In mountain ash forests *(Eucalyptus regnans)* of Australia, "catchments covered with old-growth stands . . . yield almost twice the amount of water annually as those covered with regrowth stands aged 25 years" (Vertessy et al. 2001). This phenomenon is not due solely to differences in leaf area between old and young stands (though that may play a role), but to the fact that old trees use less water per unit of leaf area than young ones (Moore et al. 2004). Whether the same is true of other tree species is an open question (Andreassian 2004).

When thinking about changes in water yield related to changes in land use, it must be borne in mind that water transpired from one watershed falls as precipitation elsewhere, and vice versa; water not transpired means less rainfall somewhere. Hence, when viewed on a regional rather than a local scale, vegetation (especially forest) does not so much decrease streamflow as it distributes it in space. Bruijnzeel (2004) concludes, "The normally observed increase in streamflow totals after forest clearing at the local scale . . . may be moderated or even reversed at the largest scale because of the concomitant reduction in rainfall induced by atmospheric circulation feedbacks."

3.2.4.3 Timing of Flows

In regions with pronounced seasonality of precipitation, timing of flows is critically important. During wet seasons (including spring snow melt), hydrologic control by land vegetation decreases the probability of floods. In dry seasons, sustained streamflow depends on the ability of soils to store and release water, which, in turn, depends on the presence of a vigorous plant community (chapter 14).

Do forests help prevent floods and contribute to maintaining streamflow during dry periods? These have been points of controversy among forest hydrologists, in part because the answers are not simple and depend on the environmental setting. Gentry and Lopez-Parodi (1980) attributed increased flooding in the upper Amazon basin to deforestation. Myers (1986) discussed the situation in the Himalayas: "The Himalayan forests normally exert a sponge effect, soaking up abundant rainfall and storing it before releasing it in regular amounts over an extended period. . . . While forest cover remains intact, rivers not only run clear and clean, they flow throughout the year. When the forest is cleared, rivers turn muddy and swollen during the main wet season, before shrinking during the dry season."

Let's first consider flooding. Andreassian (2004) summarized long-term results from 137 paired watershed studies

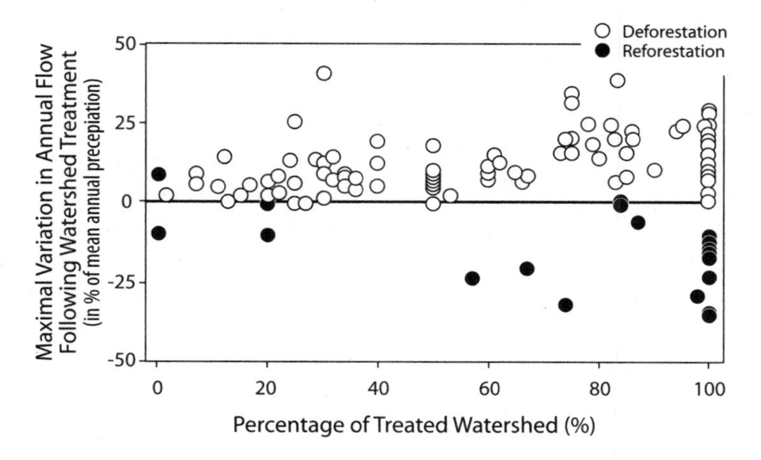

Figure 3.10. Maximum variation in annual flow (in percentage of mean annual rainfall) following watershed treatment as a function of percentage of basin subjected to treatment. (Andreassian 2004. Reprinted with permission from Elsevier. Copyright 2004)

from the Americas, Australia, New Zealand, Europe, Asia, and Africa (fig. 3.10), concluding that "deforestation generally increases flood peaks and flood volumes." The magnitude of the effect varies widely, however, depending on many factors, particularly the extent and rate of logging within a basin, how logging is done, the road network, the rate of regrowth, and the weather patterns in any particular year. Intensive site preparation, vegetation control, extensive roading (especially when poorly designed), and disruption of riparian forests have the greatest likelihood of increasing the magnitude and duration of peak flows (Reiter and Beschta 1995).

Studies in western Oregon show that clear-cutting and roads alter hydrology via different mechanisms that act synergistically. Clear-cutting alters evapotranspiration and snow accumulation and melt, resulting in increased deep soil water storage that persists until the leaf area of deep-rooted trees and shrubs has fully recovered (Harr 1976). Roads alter hillslope flows by converting subsurface to surface flow, hence providing pathways for deep soil water to reach the surface and flow rapidly to streams (Harr et al. 1975). On the H. J. Andrews Experimental Forest (in the Oregon Cascades), peak flows did not differ between a watershed that was 100 percent clear-cut without roads, and one that was 25 percent clear-cut with roads (Jones and Grant 1996). In both, average peak discharge increased by more than 50 percent (compared to unlogged forest) for the first 5 years after treatment, then began to decline, but 25 years after treatment still remained 25 to 40 percent higher than pretreatment peak flows.

As one moves from small to progressively larger drainage basins, only dramatic effects on hydrologic functions are statistically detectable in short-term studies, and decades-long records are often necessary to discern trends. In the western United States, records are just now becoming sufficiently long to allow trends in larger basins to be separated from noise. Jones and Grant (1996) examined 50- to 55-year records from three pairs of 60- to 600-km^2 basins in the western Cascades of Oregon (paired based on proximity and different rates of logging). Though variability was high, the trends in each basin pair were consistent and statistically significant; a 5 percent difference in cumulative area clear-cut translated to a difference in yearly peak flows ranging from 10 to 55 percent, depending on basin pair. None of the basins were heavily clear-cut (maximum 25% of total basin area), yet Jones and Grant (1996) estimated that peak discharges in the basins had increased by 50 to 250 percent compared to prelogging. When there is an extensive road network, as in the logged basins studied by Jones and Grant, the total amount logged underestimates total hydrologic effect.

3.2.4.4 Low Flows

In environments with extended dry periods, the ability to maintain streamflow is a function of two factors: (1) the capacity for soils and bedrock to absorb and store water, and (2) the portioning of stored water between evapotranspiration and streams. These two factors are closely linked, as the spongelike quality of soils depends on the presence of a healthy plant community. However, understanding the relation between forests and streamflow during dry periods (indeed, during any period) requires both a clear separation of the tree effect from the soil effect, and a clear understanding of how trees protect and build soils.

The most common pattern seen in paired catchments is that trees reduce low flows (Johnson 1998). Perhaps this is not surprising, because trees pump water to the atmosphere that otherwise would flow to streams. In contrast to the controlled studies, however, "reports of greatly diminished stream flows during the dry season after tropical forest clearance are numerous" (Bruijnzeel 2004). Bruijnzeel

(2004) discusses the probable reasons for this discrepancy between experiment and experience.

> The circumstances associated with controlled (short-term) catchment experiments may well differ from those of many "real world" situations *in the longer term.* To begin with, the continued exposure of bare soil after forest clearance to intense rainfall . . . , the compaction of topsoil by machinery . . . or overgrazing . . . , the gradual disappearance of soil faunal activity . . . , and the increases in area occupied by impervious surfaces such as roads and settlements . . . , all contribute to gradually reduced rainfall infiltration . . . in cleared areas. As a result, catchment response to rainfall becomes more pronounced and the increases in storm runoff during the rainy season may become so large as to seriously impair the recharging of the soil and groundwater reserves feeding springs and maintaining baseflow. In other words, the "sponge effect" is lost.

What can we conclude about the hydrologic functions of forests?

- Because they grow in areas with relatively high precipitation, forests play a special role in the global hydrologic cycle. Most of the world's streams and rivers arise in areas that are, or once were, forested.
- Streamflow is affected by a diversity of factors, and the role of forests within the complex of other influences may be highly variable, both between basins and within a given basin over time.
- Forests use more water than other types of plant communities, diverting it from streams and redistributing it as precipitation elsewhere. That has the effect of reducing flood peaks but may also reduce base streamflows when trees are planted in relatively dry environments.
- Through their protective effects on soils, forests help maintain low flows during drought periods. Effects of deforestation on low flows depend strongly on what replaces the forest, and particularly on factors that impair the soil's ability to absorb and store water. Paired basin experiments that do not reflect the full range of human effects on soil may not accurately predict what happens in the real world.
- Long-term effects of deforestation or afforestation on water may be quite different than short-term effects, both because the trees change in their water-use characteristics as they age, and because of the long-term soil-building capacity of forests. The same factors that produce high water use in trees—dense foliage and high net primary production—also enhance the ability of forests to build and maintain soil integrity.

3.2.5 Soil Stabilization

Forests, particularly those in mountains, regulate the inputs of nutrients and sediments to water, and through this, the productivity of freshwater and nearshore marine ecosystems,[12] the lifetime of reservoirs, and the navigability of rivers. Proper functioning of aquatic ecosystems requires a balanced input of nutrients: not too little, not too much. Too little nutrient input limits productivity; too much nutrient input results in **eutrophication,** a condition in which excessive growth of aquatic plants (algal blooms) depletes oxygen and thereby kills fish and other aquatic animals. According to Cherfas (1990), the mouths of most of the world's rivers appear to be eutrophying. The nutrients driving this eutrophication come from various sources, including fertilizers washed from agricultural fields, human wastes such as detergents, runoff from feedlots, and soil erosion from fallow agricultural fields and deforested hillsides.

As we discussed in chapter 1, erosion from deforested mountain slopes is a major problem in some parts of the world. In Nepal, for example, between 30 and 75 tons of soil are washed away annually from each hectare of deforested land (Myers 1986). Landslides and surface erosion from deforested hillsides create significant problems for both humans and natural ecosystems. For example, in November 1991, six thousand people were killed by landslides from deforested mountain slopes in the Philippines. Accumulations of silt from eroding hillsides have rendered several of the world's rivers unnavigable and are significantly decreasing the storage capacity of reservoirs (Myers 1988).

Sediment transport to streams can come from three sources: surface erosion, gully erosion, and landslides (also called mass wasting). Erosion from soil surfaces and gullies is negligible in undisturbed forests, in large part because of the protection afforded by canopies and, especially, litter layers. In his summary for the United States, Stednick (2000) concludes that "undisturbed forest watersheds usually have erosion rates from near 0 to 0.57 Mg per hectare per year. . . . Erosion rates have been estimated as <0.2 Mg per hectare per year for three-quarters of eastern and interior western forests." An exception is where environmental factors limit tree cover and bare soil is exposed.

Landslides are more complicated. In some mountain ranges, including the Coast Ranges of western North America and the Himalayas, a combination of factors including geology, steep topography, and high rainfall lead to landslides that originate in failures below the rooting zone. Forests do not protect against these, and in extreme cases (e.g., torrential rains from hurricanes) the weight of trees may even contribute to their occurrence (Bruijnzeel 2004). Such events are normal and, if not too frequent, in some cases can benefit long-term fish habitat by pulsing large wood that creates pools (Reeves et al. 2003). Forests, however, do protect against shallow landslides, that is, those originating at depths within the rooting zone (Bruijnzeel 2004).

[12] Although estuaries and other nearshore marine ecosystems compose only 10 percent of the oceans' surface, they account for 30 percent of its productivity (Cherfas 1990).

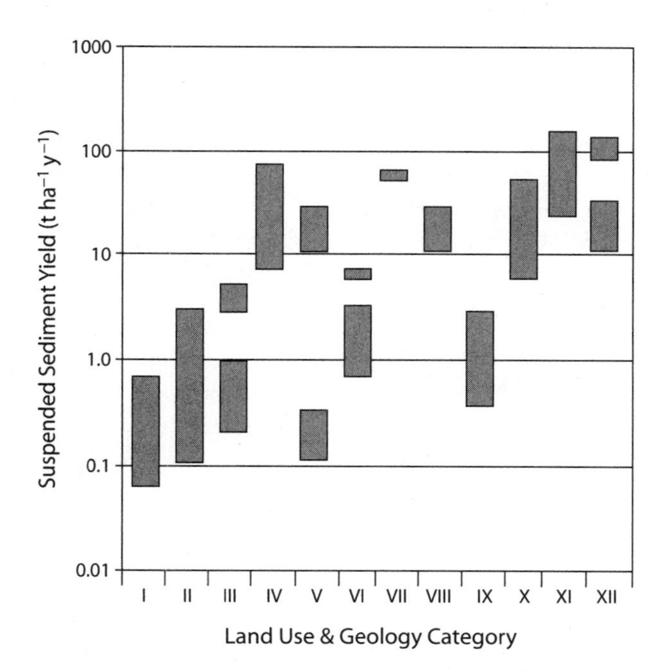

Figure 3.11. Ranges in reported catchment sediment yields in Southeast Asia as a function of geological substrate and land use. Categories: I, forest, granite; II, forest, sandstones/shales; III, forest, volcanics; IV, forest, marls; V, logged (*lower bar:* reduced-impact logging); VI, cleared, sedimentary rocks (*lower bar:* microcatchments); VII, cleared, volcanics; VIII, cleared, marls; IX, medium-large basins, mixed land use, granite; X, medium-large basins, mixed land use, volcanics; XI, medium-large basins, mixed land use, volcanics plus marls; XII, urbanized *(lower bar)*, mining and road building *(upper bar)*. (Bruijnzeel 2004. Reprinted with permission from Elsevier. Copyright 2004)

The effects of land use practices on sedimentation have been easier to demonstrate than effects on water flows. As with water, the best-documented studies have been conducted in experimental watersheds, although much clear evidence exists from other sources, such as comparisons of logged and unlogged watersheds and historical observations. Figure 3.11 shows the ranges of reported sediment yields from Southeast Asia according to land use and geology. The first four categories are forests on different geologic substrates, and the remainder either cleared basins, mixed land uses in large basins, or highly intensive land alteration (category XII, mining or urbanization). In most cases, sediment yields from forested basins are lower than from basins with human use, the exception being forests on marls, which are a highly erosive form of limestone. The lower bar in category V is an area of reduced-impact logging, which shows sediment yields comparable to those in undisturbed forests. Bruijnzeel comments that although high levels of erosion from roads were documented in the reduced-impact logging area, much of that was apparently blocked from reaching streams by down wood left behind in the logging operation.

Sediments associated with forestry come from four primary sources (Bennett 1982; Swanson et al. 1989): surface erosion from roads, surface erosion from clear-cuts, mass transport during slash burns, and landslides associated with either roads or clear-cuts. Studies in the Cascade Range of Oregon found that landslides, especially from poorly designed roads during major storms, pulsed large amounts of sediment in brief episodes, while surface erosion from roads and clear-cuts was more chronic and over time has accounted for roughly the same sediment delivery as landslides (Swanson et al. 1989).

As discussed above, the amount of erosion after clearing forests depends on the geological substrate; it also depends on slope steepness and roads (including skid roads). In the Pacific Northwest, the greatest increases in erosion following logging have been on slopes steeper than 35 percent, however the precise relation between slope steepness and erosion is likely to vary with geologic substrate (Swanson et al. 1989). Erosion is dramatically increased by roads on steep slopes. On the H. J. Andrews Experimental Forest in the Oregon Cascades, a small basin that was completely clear-cut but unroaded delivered greater than 20 times more sediment to its stream over a 10-year period than its unlogged control, while a nearby basin that was only 25 percent logged, but with roads, delivered over 40 times more (Swanson et al. 1989).

As with many issues in forestry, short-term studies may not accurately predict longer-term effects. In the case of sedimentation, sediment from eroding hillsides may be stored in upstream reaches and released slowly over time, or perhaps in delayed pulses triggered by large storms (Swanson et al. 1989; Bruijzneel 2004). Zeimer et al. (1991) modeled the long-term cumulative effects of harvesting on sedimentation for coastal northern California to central Oregon (an area with high landslide potential). Their simulations predicted that sediment produced during the first century following harvest would be initially stored in small tributaries, to be washed into larger streams during the second century. That prediction was supported by data from Casper Creek in northern California, which is still adjusting to logging-related sedimentation from the late nineteenth and early twentieth centuries (Ziemer et al. 1991).

3.3 FORESTS AND HUMAN HEALTH

How wild nature influences our health is a relatively new topic of research, but one that is likely to gain in importance. For example, deforestation in Mexico probably led a strain of encephalitis virus to switch mosquito hosts from one that required forests to one that did not (Bienen 2004, reporting research by Weaver et al.). In contrast to the former host, the new host mosquito prefers to feed on mammals, including humans.

Ultimately, anything that affects ecosystem health ramifies to one to degree or another to influence human health (fig. 3.12). Consider, for example, Hurricane Katrina, which devastated the U.S. Gulf Coast in 2005. Either directly or indirectly, people had destroyed the barrier islands, coastal marshes, and coastal forests that formerly buffered inland

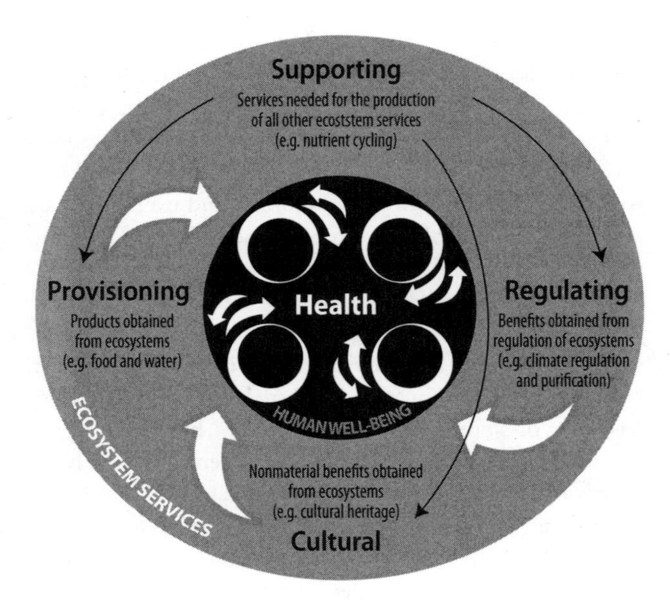

Figure 3.12. Interrelationship between ecosystems services, aspects of human well-being, and human health. (Millenium Ecosystem Assessment 2005a)

areas from the direct consequences of severe hurricanes. The result was a level of devastation that would not have occurred with a similar-sized hurricane one hundred years ago. Similarly, areas with intact mangrove forests suffered much less damage from the Indian Ocean tsunami of December 2004 than areas where mangroves had been removed (Danielsen et al. 2005).

The Millennium Ecosystem Assessment (2005b) put it this way: "Ecologically unsustainable use of ecosystem services raises the potential for serious and irreversible ecological change. Ecosystem changes may occur on such a large scale as to have a catastrophic effect upon the economic, social and political processes upon which social stability, human well-being and good health are dependent. This suggests that a precautionary approach to environmental protection is most likely to protect and enhance health. Unavoidable uncertainties about the impacts of global environmental changes on public health should not be an excuse for delaying policy decisions."

3.4 SUMMARY

The global ecosystem is tied together by cycles of water and nutrients and through the dynamics of the climate, all of which are significantly influenced by forests. Though occupying only about one-third of the earth's land area, forests account for over two-thirds of the leaf area of land plants and contain roughly 70 percent of the carbon contained in living things. Forest canopies have a lower albedo (reflect less solar energy) than other vegetation types, hence the extent of forest cover affects the earth's heat balance. Because of their high leaf area, forests account for most of the earth's photosynthesis, and the bulk of water evapotranspired from land to the atmosphere. By keeping water in circulation, forests effectively create rainfall and may effect climate regionally and globally. Forests play a vital role in regulating streamflow and water quality, though the details vary depending on environmental context. Deforestation has contributed to the greenhouse effect, led to decreased precipitation in some areas, and produced large accumulations of silt in rivers, lakes, and estuaries.

4

Major Forest Types and Their Climatic Controls

THE MAJORITY OF EARTH'S FORESTS lie in two broad bands, one circling the globe in the tropical regions and the other in the Northern Hemisphere roughly between latitude 35° N and the Artic Circle (plate 11). Except for oceans and areas cleared by humans, forests of the tropics and of the far north form relatively continuous bands around the globe, while those of the midlatitudes are interrupted by the grasslands and deserts of interior North America and Eurasia. Forests occur south of the tropics, particularly in Australia, New Zealand, and South America; however, their extent is limited by the lack of land masses in the south.

The type of vegetation that occurs in an area is determined by precipitation, temperature, soils, landforms, and historical factors (e.g., patterns of human use, or past glaciation). The distribution of major vegetation types—forest, desert, grassland—largely reflect global patterns of precipitation and temperature. Similarly, broad patterns in forest type are largely determined by temperature and precipitation, while local patterns are a function of soils, landform, and history. The former—broad patterns in forest type—are the subject of this chapter. Local variability is discussed in chapter 5.

The influence of precipitation and temperature on vegetation type is shown in fig. 4.1. Because trees grow tall, they have more leaves than other types of plants. Although this means more photosynthetic surface and therefore relatively high productivity, there are also costs associated with having many leaves, one of which is water use. Whenever stomates open to take in carbon dioxide, water is unavoidably lost. This is not necessarily waste; water is the conveyor belt that carries minerals from the roots to the crown, and evaporation of water is essential to keep leaves from overheating. Nevertheless, trees use a great deal of water and will not grow where precipitation is too low. Thus, forests are absent from large areas between about latitude 25° and 35° N (plate 11) because of insufficient precipitation; instead,

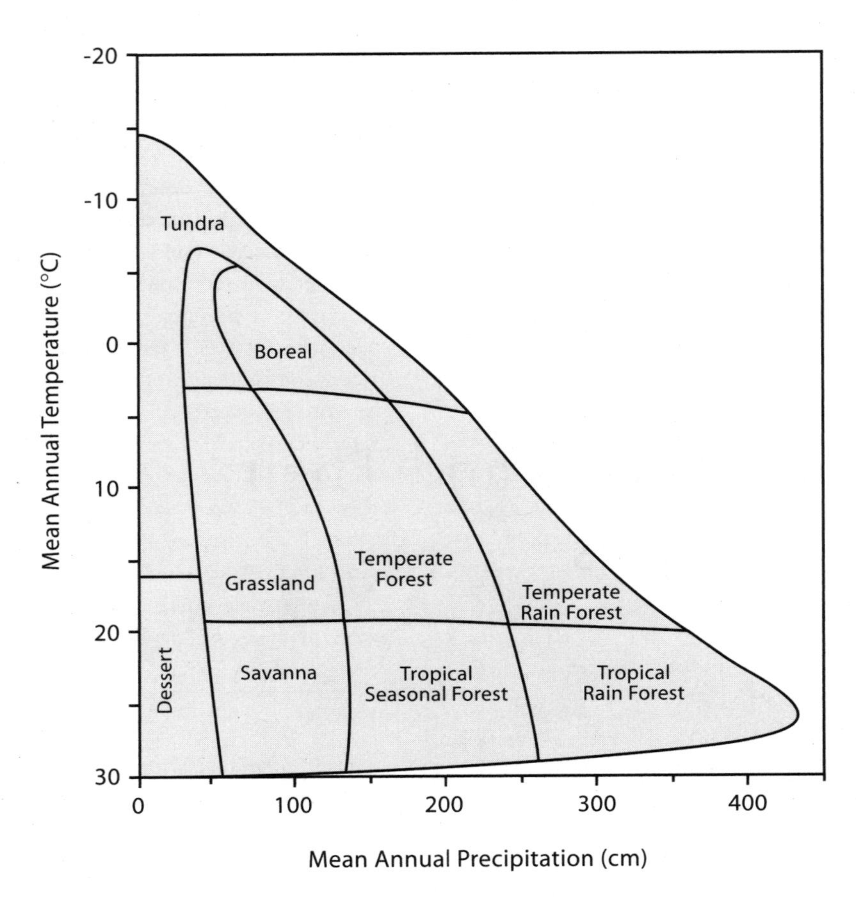

Figure 4.1. The influences of mean annual temperature and yearly precipitation on the type of vegetation. (Adapted from Whittaker 1975)

we find grasslands and deserts, although forests can occur in mountains at these latitudes. Similarly, because of low precipitation within the interior of the North American and Eurasian land masses, forests give way to the great grasslands of the North American plains and the Eurasian steppe.

The boundary between forest and grassland, however, at least in North America, appears to be controlled less by the total yearly precipitation than by its seasonal distribution. Winter precipitation is stored in deep soil reservoirs that are accessible to deep-rooted woody plants; hence, areas with relatively low total precipitation tend to support trees if sufficient precipitation falls in winter, and grasses if precipitation is concentrated in spring (Neilson et al. 1990). Areas with winter precipitation are also the only sources of year-round streamflow (Neilson et al. 1990). Fire has also been an important factor controlling the boundary between forest and grassland in North America, with frequent fire favoring grasses over trees.

In the far north and at high elevation, trees become limited by either winter low temperatures or short growing seasons and are replaced by tundra. The same factors limit the northern boundaries of certain bird species (Root 1988).

4.1 THE INFLUENCE OF CLIMATE ON FOREST TYPE

4.1.1 The Major Forest Biomes

Systems for classifying the earth's terrestrial vegetation include those developed by Walter (1985), Bailey (1995), and Olson et al. (2001).[1] All are similar in broad outline, differing primarily in detail. Of the earth's 14 terrestrial **biomes** as defined by Olson et al. (2001), 7 are forested:

- Tropical and subtropical moist broadleaf forests
- Tropical and subtropical dry broadleaf forests
- Tropical and subtropical coniferous forests
- Temperate broadleaf and mixed forests
- Temperate coniferous forests
- Boreal forests (also called taiga)
- Mangrove forests

One additional biome, Mediterranean forests/woodlands/scrub, includes substantial areas of forest and woodland. In the following, we briefly review climatic characteristics at the

[1] More information on the Olson et al. classification, including photographs of different ecoregions, can be found at www.worldwildlife.org/science/ecoregions/terrestrial.cfm.

coarser resolution of tropics/subtropics, temperate, and boreal regions, then move on to consider the major biomes within these latitudinal zones.

4.1.1.1 Tropical and Subtropical Forest Biomes

Tropical forests occur roughly within the area bounded by the Tropic of Cancer (latitude 23.5° N) and the Tropic of Capricorn (latitude 23.5° S). Climatically, according to Walter (1985, 183), "the tropics are taken to end where frost occurs or where, even in the absence of frost, the mean annual temperature is below 18 degrees C and cultivation of tropical plants such as coconuts, pineapple, coffee etc. is no longer worthwhile." Evans (1982) defines the tropics as that area where the average temperature of the three warmest and the three coldest months does not differ by more than 5°C. (Evans refers to *average monthly* temperatures; daily temperature fluctuations often exceed 5°C in the tropics.) Areas with temperatures too cool to be considered tropical, yet with a mean annual temperature equal to or greater than 10°C for five to eight months per year, are often termed *subtropical*. Examples include mountain forests within tropical latitudes, and some areas, such as the southeastern United States, that may be called either temperate or subtropical depending on the classification scheme. Whether a particular forest type is defined as subtropical or warm-temperate is not of great importance so long as its ecological characteristics are clear.

Even with no winter, many areas of the tropics nevertheless have two seasons—rain and no rain—and the seasonal distribution of rainfall determines the type of forest found in an area. The influence of precipitation pattern on forest type, important in temperate as well as tropical areas, is discussed in more detail in section 4.1.2.

4.1.1.2 Temperate Forests

Temperate forests—those of the midlatitudes—are characterized by a distinct winter, but not one that is too cold or long to support broad-leaved angiosperms (hardwoods). Although conifers may be a common component of these forests, with a few exceptions to be discussed later, they are dominated, or potentially dominated, by broad-leaved deciduous trees. (Bear in mind the qualifier *potentially dominated*, as we shall return to that later.) Examples include the forests of eastern North America, northeastern Asia, and western and central Europe (not including higher mountains). These regions have long growing seasons (four to six frost-free months) and precipitation either distributed evenly throughout the year or peaking in the summer (Walter 1985). Where precipitation is concentrated in the winter, such as the west coast of North America and the Mediterranean basin, temperate forests tend to be dominated by either conifers or evergreen broad-leaved species rather than deciduous broad-leaved species; an exception to this is south-central Chile, a winter-rain area where native forests comprise a high proportion of deciduous species in the genus *Nothofagus*. The few temperate areas with year-round precipitation and frost-free winters also support evergreen broad-leaved species. Higher mountains throughout the temperate zone are dominated by conifers. We shall discuss these special cases in more detail in section 4.1.2.

Earlier we mentioned that many temperate forests are dominated, or potentially dominated, by deciduous broad-leaved trees. In some instances, conifers are one phase in a sequence of vegetation types leading to hardwoods: this is the case in the lowlands of the southeastern and south-central United States, where pines are often the principal species. Given time and lack of disturbance (logging, fire), however, hardwoods would come to play a dominant role in these forests.[2] The northward extent of southern pines is apparently restricted by the frequency of hard frosts (Neilson et al. 1989).

The northward extent of temperate hardwood forests is limited primarily by low winter temperature; however, most

> At least two reasons exist for the greater sensitivity of most deciduous trees to short, cool summers. First, they must develop leaves fully before significant photosynthesis occurs, whereas conifers can begin a high level of photosynthesis relatively quickly in the spring (how quickly depends on species; some are very responsive and will even photosynthesize during warm periods in the winter; others, such as some spruces, break down chlorophyll during the winter and must resynthesize photosynthetic machinery each spring). A second reason relates to the different leaf structures of conifers and angiosperms and illustrates an important principle of physiological ecology. The broad, flat leaves of most deciduous angiosperms have a high surface area relative to their mass; hence, they lose heat more readily than a conifer needle, which has a relatively low surface area to mass. Hardwood trees of higher latitudes (willows, aspens, birches) tend to have relatively small leaves, thereby reducing the area from which heat may be lost. The ratio of surface area to mass is also important in the temperature relations of animals. Within a given species, mammals tend to be larger in the north than in the south (e.g., whitetail deer). This is because large bodies have less surface to mass than small bodies. (To see this, calculate the surface to volume ratio of two spheres, one with radius 10 cm, the other with radius 100 cm.)

[2]In both the temperate forests of northwestern North America and the boreal forest, the situation is often reversed: hardwoods may occupy sites for relatively brief periods, but conifers eventually come to dominate (change in species dominance over time is discussed in chapter 8). Note that dominance does not mean exclusion. In fact, conifers and hardwoods frequently coexist in temperate and the southern boreal forests.

deciduous trees also require longer, warmer summers than boreal and montane conifers. According to Walter (1985), the former must have a growing season of at least 120 days with mean daily temperature greater than 10°C, while the latter require these temperatures for only 30 days. (As an aside, it is important to realize that though generalizations such as the temperature relationship of Walter's are useful and you will see many in this book, nature is very diverse and creative, and there will be exceptions to almost any rule.)

4.1.1.3 Boreal Forests

The sensitivity of many hardwoods to low winter temperatures, coupled with their requirement for relatively long, warm summers, results in a broad transitional zone in North America and Eurasia in which cold-tolerant conifers gradually replace hardwoods as the major forest dominants. Walter (1985) considers the true boreal forest to begin where hardwoods are a relatively minor forest component. That occurs at roughly latitude 50° to 60° N, depending where you are, and is most precisely defined by a (roughly) east-west line, north of which is a high probability of winter temperatures falling below −40°C. The boreal forest extends north to the point where the growing season becomes too short and unpredictable for conifers to complete their yearly development (about latitude 70° N).

> The temperature −40°C is biologically important because it is the point at which pure water (i.e., containing no impurities for ice crystals to form around) freezes, something that temperate hardwoods cannot tolerate. Most deciduous hardwoods avoid freezing damage by something called *deep supercooling* (Burke et al. 1976). Water retained within the cells of these plants does not freeze at temperatures above −40°C because the concentration of solutes that could act as ice nuclei is maintained at a very low level. This strategy fails in very cold climates, where trees have adapted to tolerate extreme cell dessication during winter; in other words, cells of boreal trees do not freeze during winter because they contain virtually no water (Burke et al. 1976).

The southern and northern boundaries of the boreal forest roughly coincide with, respectively, the midsummer and midwinter boundaries between the cold, dry artic air mass and the warmer, moister Pacific air mass (Bonan and Shugart 1989). In winter, artic air pushes south, producing the cold temperatures that restrict temperate hardwoods; in summer, Pacific air pushes north, bringing with it sufficient warmth and moisture to allow trees to complete a growing season. As a result, the climate of the boreal forest is characterized by short, moist, and moderately warm summers, and winters that are long, cold, and dry. The majority of boreal forests are in Siberia and Canada, with lesser amounts in Alaska and Scandinavia (fig. 4.2).

> There are other examples in which the seasonal movements of large air masses control vegetation patterns. As we shall see later, the climate of the Pacific Northwest is characterized by wet winters and dry summers, and this results, in turn, in dominance by conifers rather than hardwoods. The wet winter–dry summer climate of that region results from the north-south movement of a large high pressure system called the Pacific High. In winter, the Pacific High moves south toward Hawaii, and the weather of the Pacific Northwest is dominated by a moist low pressure system called the Aleutian Low. In summer, the Pacific High moves north, shoving the Aleutian Low ahead of it and bringing dry weather to the Pacific Northwest (Norse 1990). Southeastern Alaska remains under the influence of the Aleutian Low and, hence, has rainy summers. The distribution of forest, grassland, and deserts across the southern tier of the United States is thought to be controlled by the long-term seasonal movements of a large high pressure system called the Bermuda High (Neilson et al. 1989). In general, the factors that influence movement of such large-scale air masses are poorly understood, although they are almost certainly a phenomenon of the global ecosystem.

4.1.2 Subdivisions of the Major Forest Types: Montane Forests

Forests of high mountains are called **montane.** The temperature change over 1,000 m in elevation is similar to that occurring over 5° latitude (this is called **Hopkins's biogeoclimatic law**); therefore, montane forests within any given latitudinal zone commonly have some characteristics of more northerly forests. Tropical and subtropical conifer forests occur in the mountains of central America, Mexico, and a few regions of tropical Asia (see www.worldwildlife.org/science/ecoregions/biomes.cfm for more detail). In North America some tree species common to boreal forests of Canada also occur in the Appalachian Mountains as far south as northern Georgia (e.g., balsam fir, red spruce) and in the western mountains to northern Mexico (e.g., Englemann spruce, white fir, lodgepole pine). Both the Alps and the Himalayas are characterized by broad-leaved dominance at low elevations and conifer dominance at higher elevations. Mountains influence precipitation as well as temperature, however, so vegetation change with elevation is not exactly like, and in some cases may be quite different than, changes with latitude.

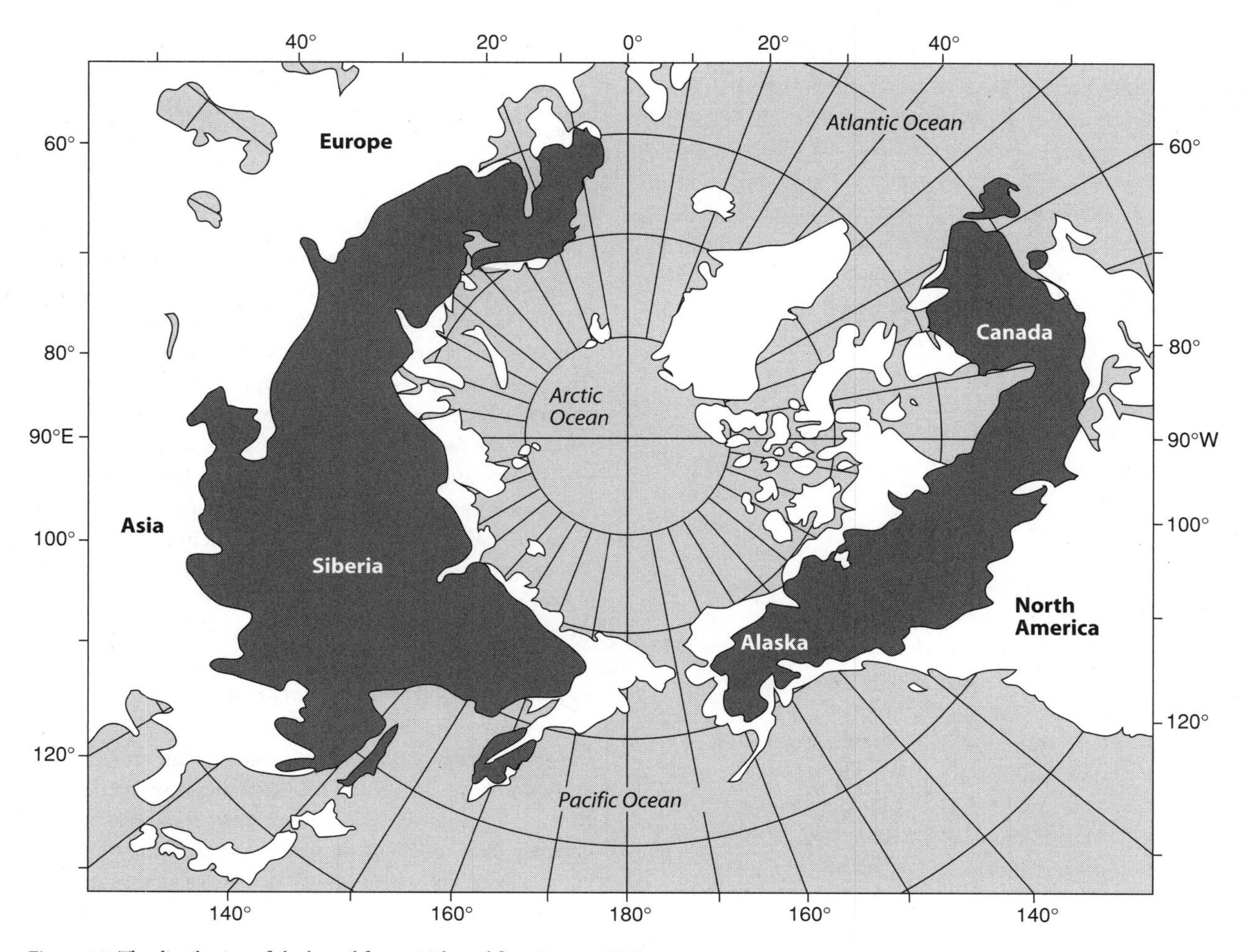

Figure 4.2. The distribution of the boreal forest. (Adapted from Larsen 1983)

4.1.3 Subdivisions of the Major Forest Types as Influenced by Precipitation Patterns

4.1.3.1 Tropical Forests

Both the total amount and the seasonal distribution of precipitation have important influence on forest type in the tropics. Many areas experience distinct dry seasons, ranging from a few weeks to several months. Where drought periods last long enough, trees shed leaves in order to conserve water, and tropical forests are differentiated according to the length of time during each year that leaves are retained. Walter (1985) distinguishes four types: (1) evergreen tropical rain forest, (2) seasonal rain forest, (3) tropical semievergreen forest, and (4) deciduous tropical forest (sometimes called monsoon forest). He describes the conditions under which the latter three types occur as follows (Walter 1985, 72):

> When a short dry period occurs in the very wet tropical region, the endogenous rhythm of tree species adapts itself to the climatic rhythm. The general character of the forest remains unchanged, but many trees lose their leaves at the same time, or sprout or flower simultaneously, so that the vegetation does in fact exhibit definite seasonal changes in appearance (seasonal rain forest).
>
> If the dry season becomes even longer, the type of forest changes. The upper tree story is made up of deciduous species . . . , whereas the lower stories are still evergreen, so that it can be termed tropical semievergreen forest. . . .
>
> With a further decrease in rainfall and a lengthening of the dry season, all of the arboreal species are deciduous, so that the forest is bare for long or short periods of time. These are the moist or dry deciduous tropical forests.

Where the dry season is on the order of 6 months, a fifth type of tropical forest occurs: a stable mixture of trees and grasslands called savanna (Jordan 1983). Figure 4.3 shows the relationship of annual rainfall and length of the dry season to the occurrence of these five forest types in India.

4.1.3.2 Temperate Forests

Although most temperate forests are characterized by the dominance (or potential dominance) of broad-leaved deciduous trees, four types are dominated by evergreens: (1) moist conifer forests, (2) evergreen broad-leaved sclerophyllus forests, (3) dry conifer forests, and (4) temperate broad-leaved rain forest.

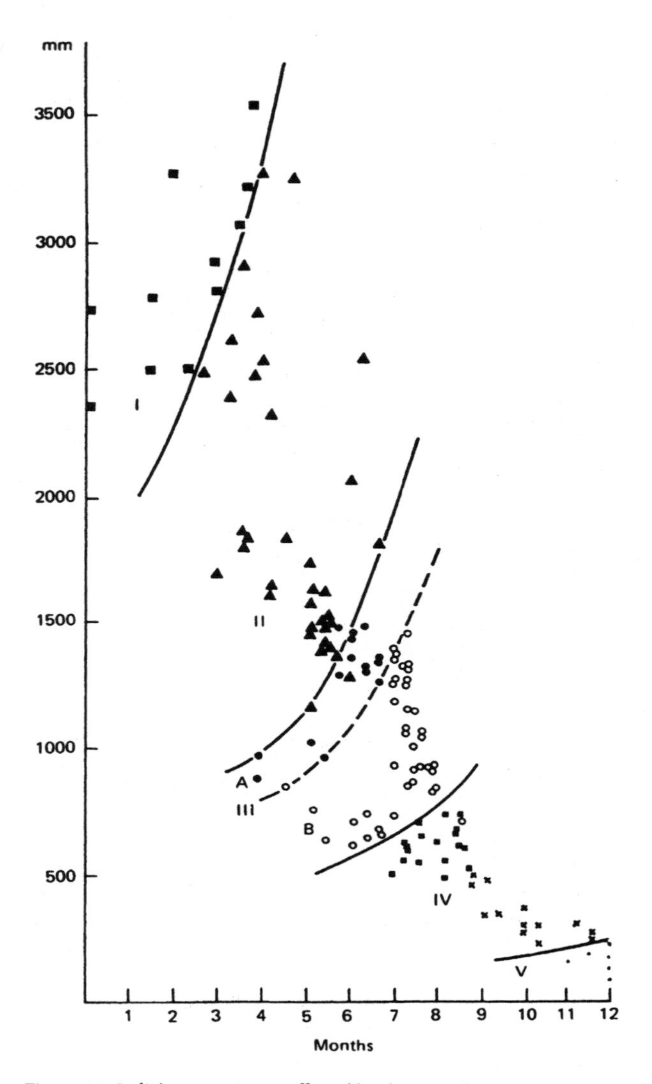

Figure 4.3. India's vegetation as affected by the annual precipitation (y axis) and length of the dry period (x axis). Zone 1, evergreen tropical rain forest; zone 2, semievergreen tropical rain forest; zone 3, monsoon forest (A, moist; B, dry); zone 4, savanna; zone 5, desert. (Walter 1985)

The first two types occur in areas where rainfall is concentrated in the winter months and, although frosts may occur, winters are relatively mild. Under these conditions photosynthesis, limited by lack of water in the summer, may occur at relatively high levels in the winter. Having no leaves during the winter, deciduous trees are less well adapted to this type of climate than are evergreens.

Climates in which precipitation is concentrated in the winter, and in which total precipitation is less than about 1,000 mm/y, are called Mediterranean. Where precipitation is sufficient to support trees, Mediterranean forests are composed of evergreen broad-leaved species and drought-tolerant conifers (especially pines). Leaves of broad-leaved trees in this type are **sclerophyllous** (they have hard, leathery or waxy surfaces that decrease water loss). In the Northern Hemisphere, temperate broad-leaved sclerophyllous forests occur in scattered patches in western and southeastern North America, in the Mediterranean basin, and over large areas of temperate Asia (Ovington 1983). Many species occur in the Northern Hemisphere's Mediterranean forests, but oaks predominate. It is in the Southern Hemisphere, however, in Australia, where sclerophyllous broad-leaved forests attain their most extensive continuous cover. There the dominant genera are *Eucalyptus*, represented by over 450 species, and *Acacia*, with over 600 species in Australia alone.

As one proceeds either west, toward the ocean, north from the sclerophyllous broad-leaved forests of northern California, or into the southern coastal regions of Chile, precipitation increases but remains concentrated in the winter. These are the regions of the giant conifers, which are among the most long-lived and massive forests on earth. In North America they stretch from northern California to southeastern Alaska and include the redwoods; the mixed conifer-hardwood forests of northern California and southern Oregon; the Douglas-fir–hemlock–true fir forests of Oregon, Washington, and British Columbia; and the Sitka spruce of southeastern Alaska. Their productivity—higher than many tropical rain forests when measured in bolewood production—is due to a combination of factors. As in Mediterranean climates, winters are mild enough to permit significant photosynthesis. In contrast to Mediterranean climates, however, yearly precipitation is quite high and, in fact, is comparable to that of the moist tropics (greater than 2,000 mm). Furthermore, soils of the region tend to be deep and capable of holding a great deal of water; hence, winter precipitation can be stored for use during the dry summer.

East of the moist conifer zone of the Pacific Northwest, annual precipitation drops rapidly, eventually resulting in the arid grasslands, shrublands, and deserts of the interior. Bordering these areas, where precipitation is barely sufficient to support trees, are forests and temperate savannas of drought-tolerant pines and junipers. Similar dry coniferous types, usually pines, occur throughout the Northern Hemisphere. In the boreal forests of western Canada, areas that are in the rain shadow of mountains are occupied by jack pine rather than the spruces that dominate moister sites (Bonan and Shugart 1989). In the southwestern United States and northern Mexico, forests and savannas of pine, juniper, and oak border the western edges of the Great Plains and the southwestern deserts. Farther south in Mexico, pines occupy the high mountains rising out of the desert. Pines are found throughout the Mediterranean basin, often mixed with or bordering broad-leaved sclerophyllus vegetation. Unlike the pine forests of the southeastern United States, the dry pine forests are generally not transitional to broad-leaved species, though pines and drought-tolerant hardwoods may occur together.

Temperate areas with mild, frost-free winters and relatively high precipitation (greater than about 1,500 mm) that is well distributed throughout the year support temperate broad-leaved rain forest. Small areas of this forest type occur in the southeastern United States and at various locales in Asia (Japan, southern China, the lower slopes of the Himalayas). Much of the area once covered by this type in

Asia has been cleared. Temperate broad-leaved rain forests reach their greatest extent in the Southern Hemisphere, particularly the southern coast of Chile, Tasmania, and New Zealand. New Zealand was covered by this type prior to human settlement (Ovington 1983). Although other species occur, those in the genus *Nothofagus* predominate in temperate rain forests of the Southern Hemisphere.

4.1.4 Ecoregions

Olson et al. (2001) use ecoregions to serve as a basis for identifying patterns of species diversity and setting conservation priorities. In their classification of the world's terrestrial ecosystem types, 14 biomes are further broken down into 867 ecoregions, which they define as "a large area of land or water that contains a geographically distinct assemblage of natural communities that

(a) share a large majority of their species and ecological dynamics;

(b) share similar environmental conditions, and;

(c) interact ecologically in ways that are critical for their long-term persistence."

4.2 LATITUDINAL GRADIENTS IN FOREST CHARACTERISTICS

As one proceeds higher in latitude from the equator toward the poles, climatic changes have important influences on the structure and processes of forest ecosystems. As we have seen, some areas in the tropics experience a regular annual drought season during which trees lose leaves and growth is much reduced, but tropical temperatures are suitable for year-round biological activity. Temperate and boreal forests, in contrast, occupy areas with wide seasonal changes in temperature, restricting most biological activity to spring and summer. Boreal and high-latitude montane forests, in particular, have a very short period in which organisms must complete their yearly cycle of growth and reproduction. For these and other reasons, some poorly understood, forests differ across climatic zones in aspects of their structure and the rates at which various processes occur. We shall discuss these differences at many points throughout the book. This section briefly introduces three patterns of particular relevance to ecology and forestry: species diversity, distribution and cycling of essential elements, and productivity.

4.2.1 Species Richness

Striking differences in species richness occur among the major forest types. The global pattern of angiosperm family richness is shown in plate 12. Highest diversity by far occurs in the moist tropics. Figure 4.4 shows the tree species occurring along a 60-m transect in Southeast Asia. Each letter represents a different family and each number a different species. For example, D1 is *Coylebolium melanoxylon*, which is

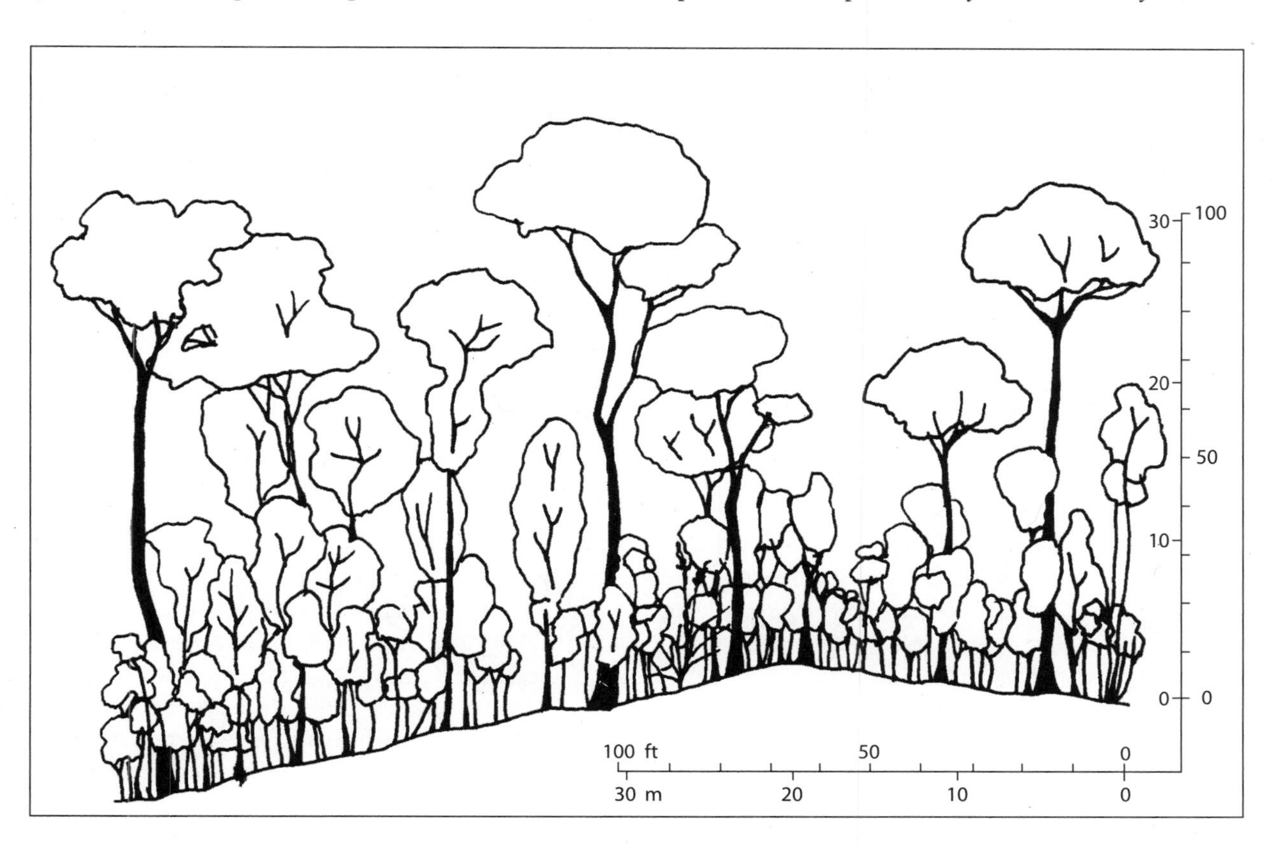

Figure 4.4. The forest profile along a 60-m × 8-m transect in Borneo. (Ashton 1964)

in the family Dipterocarpaceae; D4 is *Shorea acuta*, also in the Dipterocarpaceae; A5 is *Semicarpus ruforelutinus*, in the Anacardiaceae. Along this 60-m transect are 114 individuals, representing 53 species and 22 families. It is not uncommon to find anywhere from 50 to greater than 100 tree species per hectare in tropical forests. In contrast, two centers of North American temperate forest diversity—the southern Appalachians in North Carolina and Tennessee and the Klamath Mountains in Oregon and California—contain no more than 30 woody species (trees and shrubs). Up to 48 woody species occur in some Australian forests. Many temperate and most boreal and montane forests contain at the most three to six tree species at any one time, and monospecific stands are not uncommon. Note, however, that a natural forest comprising a single tree species is *not* genetically uniform. Most temperate and boreal tree species contain high levels of genetic diversity within stands (chapter 13), suggesting that if comparisons were made on the basis of genotypes rather than species, the diversity of northern forests might compare more favorably with that of the tropics. It would also compare more favorably if herbaceous plants were included, which tend to be quite diverse in some temperate forests. Moreover, richness—the total number of species—is only one of several measures of biological diversity (chapter 10); little is known about how other measures compare across the earth's biomes (Roy et al. 2004).

The structural diversity of species-rich forests (the variety of sizes, forms, and seasonal patterns exhibited by the trees) is greater than that of species-simple forests, even when the latter are genetically diverse. Since structure in the tree layer provides living and working space for the other occupants of the ecosystem, it is not too surprising that the diversity of animal species varies in the same manner as diversity of tree species. For example, there are 222 known species of ants in Brazil, 63 in Utah, and 7 in Alaska. Roughly 500 species of land birds nest in Central America, 150 in the temperate zones of North America, and less than 100 in Alaska (MacArthur 1972). (Roughly one-half of bird species nesting in temperate forests migrate to tropical forests during the winter.) Figure 4.5A shows the diversity of terrestrial birds in 206 grid cells distributed across the globe (note the

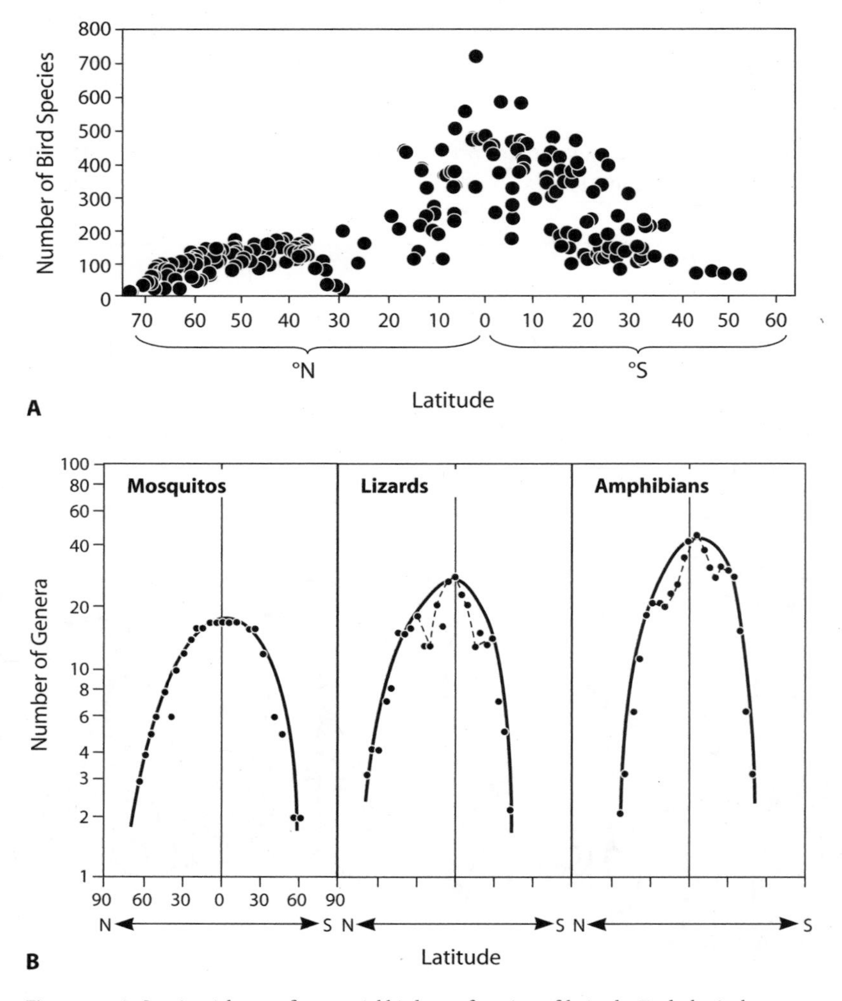

Figure 4.5. A. Species richness of terrestrial birds as a function of latitude. Each dot is the species richness in a 48,400 km^2 cell. (Turner and Hawkins 2004) B. The latitudinal patterns in genera of mosquitos, lizards, and amphibians. (Adapted from Stehli 1968)

high variability in the tropics). Figure 4.5B shows similar latitudinal patterns for the number of genera of mosquitoes, lizards, and amphibians (Fischer 1960). The number of species in each of these groups drops sharply as one proceeds either north or south from the equatorial regions.

The decrease in species richness as one goes from the equator toward the poles is one of the longest-recognized and most-discussed patterns in modern ecology and has been documented for numerous taxa in both marine and terrestrial ecosystems (recent reviews include those by Gaston [2000], Willig et al. [2003], and Turner and Hawkins [2004]). We revisit this phenomenon in more detail in chapter 10. Before moving on to the next section, however, it is worth noting that as with virtually all things ecological, there are exceptions. For instance, species that live within other species for at least part of their life cycle, either as parasites or as helping partners (mutualists), often do not follow the standard pattern (Willig et al. 2003). One group of mutualists that does follow the standard pattern and whose members are more diverse in the tropical than in higher latitude forests is the group of arbuscular mycorrhizal fungi, which participates in a cooperative living relationship with many higher plant species (Opik et al. 2006; chapter 11).

> In general, soil is a biologically rich, but poorly understood component of natural ecosystems (chapter 14), and it is quite possible that other groups of soil organisms are also more diverse in northern forests than in the tropics. Less than 2 percent of known terrestrial animal species are vertebrates (May 1988); the rest are arthropods, nematodes, worms of one kind or another, and other small creatures. Many of these live in the soil or high in canopies, places where they are seldom seen and measured. Microbes comprise another group of organisms whose diversity is largely unknown.

What explains the sharp latitudinal gradient in species richness? A wealth of hypotheses have been put forward, no one of which seems adequate in itself to explain the observed patterns. We return to this question in chapter 10.

4.2.2 Storage and Cycling of Essential Elements

Plants require chemical elements. Carbon is the primary structural element of all living organisms. Carbon skeletons are modified by the addition of various elements such as oxygen, hydrogen, nitrogen, and, because they are derived primarily from rocks, what are called the mineral elements: phosphorus, potassium, calcium, iron, and so on. The mineral elements, along with nitrogen, are often referred to as **nutrients** (from the Latin *nutrire,* meaning "to nourish"). Most nutrients are input to ecosystems in relatively small amounts, and productivity is closely tied to the rate at which they are cycled from dead tissues to new growth. A few nutrient elements, for example, potassium, are bound very loosely within the organic matrix of plant tissues, but others, such as nitrogen, phosphorus, sulfur, and calcium, are held tightly and released only when the carbon skeleton is dismantled. Decomposers—invertebrates and microbes that make their living by consuming in one fashion or another dead organic matter—are a critical link in this cycling of nutrients. Without the decomposers, essential nutrients remain tied up in dead tissues rather than becoming available to support new growth (chapter 18).

In terms of ecosystem function, there are both good and bad aspects to the retention of nutrients by dead organic matter. Once released by decomposition of organic compounds, three things can happen to a nutrient element: it can be taken up by a living organism (either a microbe or a plant), it can be held in the soil, or it can be lost from the ecosystem in various ways (most often carried by water to streams). Although there may be some 'leakiness' in undisturbed forests (chapter 17), loss of nutrients, in effect a decrease in the bank that is drawn on to support new growth, is relatively uncommon because the rate of nutrient uptake by living plants and microbes closely matches the rate of nutrient release through decomposition. Hence, in ecosystems with rapid decomposition, the majority of nutrients are found in living tissues, while in ecosystems with slow decomposition, the majority are associated with dead organic matter, either within or on the surface of the soil.

Like most organisms, the activity of decomposers is limited by environmental factors, including temperature and moisture. The warm and wet tropics are quite favorable for invertebrates and microbes, and decomposition occurs very rapidly. At higher latitudes, and at high elevations, cold temperatures limit decomposition during much of the year. In temperate or montane forests with summer drought (e.g., much of western North America), both temperature and moisture may be suitable for decomposition only during very short periods in the late spring or early fall.

Figure 4.6 shows the relation between decomposition rate and the accumulation of organic matter on the soil surface for various forest types. The turnover coefficient (the *y* axis in fig. 4.6) is a measure of the rapidity of decomposition. When the turnover coefficient equals 1, litter (e.g., leaves, branches) is decomposed within 1 year after it falls; when the turnover coefficient equals 0.1, decomposition takes on the average 10 years, and at 0.01 it takes 100 years. Note the positioning of forest types in this graph, and also note the log-log scale. Although considerable overlap occurs between types, highest turnover coefficients (approaching 1) and lowest accumulations of organic matter (generally below 1,000 g/m^2 of soil surface) are in tropical forests. Lowest turnover and highest soil organic matter are in coniferous forests (both temperate and boreal), while temperate broad-leaved forests have intermediate levels.

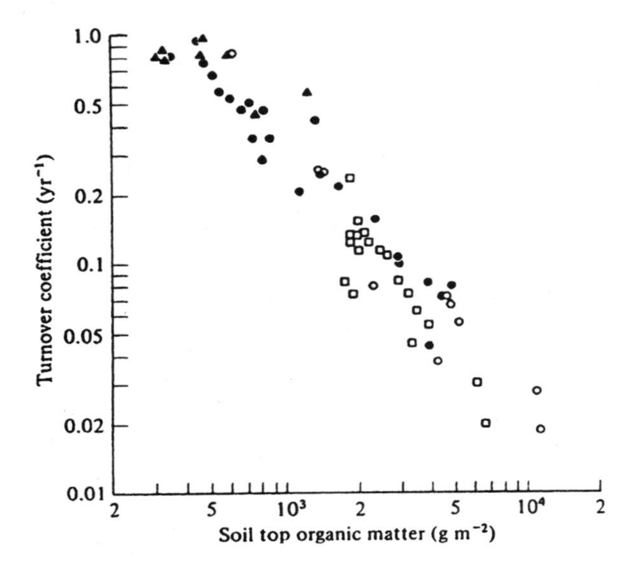

Figure 4.6. The relationship between decomposition rate and soil organic matter for different types of forest. (O'Neill and DeAngelis 1981. Reprinted with the permission of Cambridge University Press)

As discussed earlier, rapid decomposition in tropical forests is matched by rapid nutrient uptake. Consequently, compared to forests of higher latitudes, a high proportion of ecosystem carbon and associated nutrients is contained in the living vegetation of tropical forests (note that not all belowground organic matter is dead, a certain proportion—sometimes quite high—is living roots and microbes).

Rapid decomposition and concentration of ecosystem nutrients in aboveground biomass of tropical rain forests has important implications for tropical forest management. Silvicultural techniques such as clear-cutting and control of noncommercial vegetation can result in the loss of a much higher proportion of site nutrient capital in the tropics than in temperate and boreal forests. Nutrient losses from ecosystems are discussed in chapters 17 and 20.

4.2.3 Productivity

Forest productivity correlates only roughly with latitude (chapter 15). The highest net primary productivity (NPP) values tend to occur in moist tropical forests, and the lowest in boreal forests, with temperate forests having intermediate values (Kucharik et al. 2000) (plate 13). However, NPP may vary widely, especially within tropical and temperate forests, depending on water availability, temperature, and soil fertility (factors controlling productivity are discussed in chapter 15). The NPP of some temperate forests, such as those of the southern Appalachians and the Pacific Coast of North America, compare favorably with relatively high producing tropical forests. In general the database is poor, especially in the tropics and the boreal zone.

One interesting difference between tropical and extratropical forests relates to how growth is allocated among different tree parts. Jordan (1983) compiled a large amount of data showing that compared to temperate forests, those of the tropics allocate a higher proportion of their net growth to leaves and a lower proportion to wood. This is true not only for native forests of the tropics, but for plantations as well.

4.3 HOW WILL GLOBAL CLIMATE CHANGE AFFECT THE DISTRIBUTION OF FORESTS?

Most environmental scientists agree that the earth is warming due to human activities (Houghton et al. 2005). Widely different modeling approaches predict that average global temperatures in the decade from 2020 to 2030 will be from 0.3 to 1.1°C warmer than in the decade from 1990 to 2000 (Knutti et al. 2002; Stott and Kettleborough 2002). Uncertainty grows for periods after that with projected mean global temperatures in 2100 ranging from 1.2 to 6.9°C higher than in 1990 to 2000, depending on the intensity of fossil fuel use (Stott and Kettleborough 2002). The largest uncertainties at a continental scale are in North America and Europe (Stott et al. 2006). None of these scenarios include the potential effects of strong positive feedbacks to warming, such as the earth's forests switching from a net sink for carbon to a net source (as might occur with widespread fire or continued deforestation), or the release of methane (a strong greenhouse gas) stored in frozen soils. Models that do consider such feedbacks predict significantly greater warming (e.g., Cox et al. 2000).

Anthropogenic climate change will significantly affect all of the earth's ecosystems and, in fact, already is (IPCC 2007). In their review, Walther et al. (2002) conclude, "There is now ample evidence of the ecological impacts of recent climate change, from polar terrestrial to tropical marine environments. The responses of both flora and fauna span an array of ecosystems and organizational hierarchies, from the species to the community levels." Echoing this, Cotton (2003) states, "There is mounting evidence that global climate change has extended growing seasons, changed distribution patterns, and altered the phenology of flowering, breeding, and migration."

There is little doubt that protecting forests against adverse effects of climate change and easing the transition from one forest type to another will be a major task of foresters in the future. Therefore, it is very important to understand how climate change and its spin-offs, such as altered disturbance regimes, might affect forests. This is a subject we return to at various points throughout the book, particularly in chapters 7, 15, and 20. In the remainder of this section we briefly review how climate change might influence the distribution of forests.

Changes in the distribution of plants in response to global change are estimated by combining general circulation models (we also refer to these simply as *climate models*), which are used to predict future climate scenarios, with global vegetation models, which are biogeography models that relate the distribution of plants to climate (Aber et al. 2001). Dynamic global vegetation models, the latest generation of modeling related to global change, combine general circulation models, biogeography models, fire models, and biogeochemistry models (Aber et al. 2001; Bachelet et al. 2003). Fire models predict how changing climate will affect the frequency and severity of wildfires, while biogeochemistry models relate climate to element cycling. Both are critical to predicting carbon storage and fluxes and, therefore, the direction and strength of terrestrial feedbacks to global change.

4.3.1 Biogeography Models

There are two approaches to biogeography modeling. **Correlative models** use statistical techniques to associate the present distribution of vegetation to climatic variables. Sugar maple, for example, occurs within a certain range of temperature and moisture, which corresponds to particular topographic and edaphic settings within a given geographic area. As global change shifts climates, the range of temperature and moisture that sugar maple is adapted to will move somewhere else, and that "somewhere else" is where sugar maple has the potential to occur in the future. Whether it actually occurs there is a question we return to later.

Process models, on the other hand, simulate how environmental variables influence the ecophysiology of plants and from that deduce how future climates will affect distribution (Neilson et al. 1998). Process models generally work by calculating how factors such as water availability and temperature constrain the amount of leaf area that can be supported on a site, and from that deduce NPP, water use, and the general plant form corresponding to that leaf area (e.g., grasses, shrubs, trees). The more widely used process models also incorporate the effects of atmospheric carbon dioxide on plant NPP and water-use efficiency.

Each approach has strengths and weaknesses. Combined with general circulation models, correlative models can simulate how climate change may effect the distribution of individual tree species, while process models only predict effects on broad life-forms (e.g., grass, shrubs, broad-leaved trees, needle-leaved trees). On the other hand, correlative models are constructed using the climatic conditions that exist today, while process models have the flexibility to simulate new conditions, such as combinations of temperature and moisture that do not exist on the current landscape. Very importantly, process models allow incorporation of biogeochemistry (nutrient cycling, water and carbon fluxes) and, as mentioned before, can simulate how increased levels of atmospheric carbon dioxide affect plant growth and drought resistance (chapter 15).

Both correlative models and the process models developed in the 1990s are equilibrium in nature, which is to say they predict where climatic conditions may be suitable for a given species or life form in the future, but they cannot simulate the processes of migration and adaptation that will be required for plants to extend their ranges, nor do they simulate the disturbances likely to accompany a rapidly warming climate. Therefore, such models must be considered as highly optimistic scenarios. Given the rapidity with which climate is predicted to change in the coming decades, the reality might be quite different. The so-called transient biogeograpic models, which attempt to account for the ecological processes of change, present enormous challenges and are currently being developed (Aber et al. 2001; Bachelet et al. 2003).

4.3.2 Equilibrium Predictions

Despite their limitations, by translating the predictions of climate models into implications for the earth's biota, equilibrium models give useful information and, in particular, help identify potential problems. For example, if the climate suitable for a given species moves, say, 200 km north from its present location, the challenges and stresses faced by that species are quite different from those should the suitable climate shift by only a few tens of kilometers. Similarly, if process models predict a significant change of leaf area in a given biome, substantial changes are indicated in climatic factors that directly affect plant communities.

As of the early twenty-first century, about five equilibrium biogeography models and several general circulation models were in use; hence, several different combinations of the two model types are possible, each yielding more or less unique predictions and, in aggregate, producing a range of modeled futures. All general circulation models simulate a future climate under a fixed level of **radiative forcing** (heat absorption by the atmosphere) at or near that corresponding to twice the preindustrial concentration of carbon dioxide,[3] and all predict that warming will be unevenly distributed, with high-latitude areas generally warming more than tropical areas.

Plate 14 shows predictions of one widely used equilibrium process model, the Mapped Atmosphere-Plant-Soil System (MAPSS) (Neilson 1995), combined with a general circulation model from the Hadley Centre, in the United Kingdom. Among the various general circulation models, the Hadley model predicts in the midrange of temperature increase during the twenty-first century. MAPSS incorporates the potential physiological effects of increased atmospheric carbon dioxide. Higher levels of carbon dioxide may affect plants in various direct and indirect ways (Aber et al.

[3]Several gases in addition to carbon dioxide absorb heat radiated from the planet's surface and thus act as greenhouse gases. For simplicity, their effects are often translated to an equivalent carbon dioxide effect, and the whole package reported as *equivalent carbon dioxide forcing.*

2001), but two in particular have been incorporated into biogeographic models: (1) increased NPP (whether such an increase can be sustained is a matter of debate; see chapter 15), and (2) increased water-use efficiency (chapter 15). The latter has the greatest potential to influence plant distributions, as higher water-use efficiency allows plants to extend onto more droughty sites than would otherwise be possible.

Several patterns are worth noting in plate 14. Not surprisingly, closed boreal forest is predicted to expand northward into the broad, ecotonal area of open forest and tundra MAPSS refers to as *taiga/tundra*. In general, grasslands expand onto arid lands, savannas onto grasslands, and forests onto savannas, reflecting the combined effects of (1) a more active hydrologic cycle (including more precipitation) due to warmer temperatures, and (2) greater water-use efficiency resulting from higher levels of atmospheric carbon dioxide. Most other climate-vegetation model combinations yield qualitatively similar predictions, though quantitative details such as the amount of expansion or contraction of a particular biome differ. Table 4.1 shows the range of values for percentage change in biome types obtained from simulations with two biogeography models in combination with three climate models (six different combinations). More recently, IPCC (2007) makes a quite different prediction for eastern Amazonia, where their models show the region becoming more arid, rather than forest encroaching on savanna as in the models discussed above; drying soils due to higher temperatures result in just the opposite, forest being replaced by savanna (and current savannas being replaced by more arid vegetation).

Extensive correlative modeling has been done for the eastern United States, an area in which greater than average warming is predicted. Iverson and Prasad (2002) utilized a huge database (>100,000 plots) to model the relationships between environmental variables and the distribution of 76 tree species. In order to predict future shifts in the location of suitable habitat, they then applied those relationships to five different general circulation model predictions of climate toward the end of the twenty-first century (see www.fs.fed.us/ne/delaware/atlas for updates to the Iverson and Prasad model). Depending on which model was used, optimal climate moved north more than 20 km for 38 to 47 species, and more than 200 km for 8 to 27 species.

How much will optimal habitats shift on a global basis? Malcolm et al. (2002) approached this question using combinations of two process-based global vegetation models and two general circulation models. They laid a grid over each of the earth's terrestrial biomes and compared the geographic location of each grid cell in the current climate with its location in a climate with double the current amount of carbon dioxide (assumed to occur in 100 years). The straight-line distance between current and future cells divided by 100 years indicates the distance plants in a given biome would have to migrate each year in order to keep pace with the changing climate. Malcolm et al. termed this value the migration rate, although that does not imply plants will—or can—actually migrate at that rate. Figure 4.7 shows results from the Malcolm et al. model. On a global basis, 60 to 70 percent of biome cells moved less than 315 m/y, while roughly 25 percent of cells moved more than 1,000 m/y (fig. 4.7A). The required migration rate was not spread evenly across the globe, however (fig. 4.7B). The mean rate of migration required to keep pace with climate increased sharply at latitudes above 40° N, reflecting the greater warming predicted for the boreal and arctic regions.

Table 4.1
Range in model predictions for change in areal extent of various biome types under a two-times CO_2 climate

	Future area as percentage of current area	
Biome type	With CO_2 effect	Without CO_2 effect
Tundra	43–60%	43–60%
Taiga/tundra	56–64%	56–64%
Boreal forest	64–116%	68–87%
Temperate evergreen forest	104–137%	84–109%
Temperate mixed forest	139–199%	104–162%
Total temperate forest	137–158%	107–131%
Tropical broadleaf forest	120–138%	70–108%
Savanna, woodland	78–89%	100–115%
Shrub, steppe	70–136%	81–123%
Grassland	45–123%	120–136%
Total shrub, grassland	105–127%	111–126%
Arid lands	59–78%	83–120%

Source: Adapted from Neilson et al. 1998.

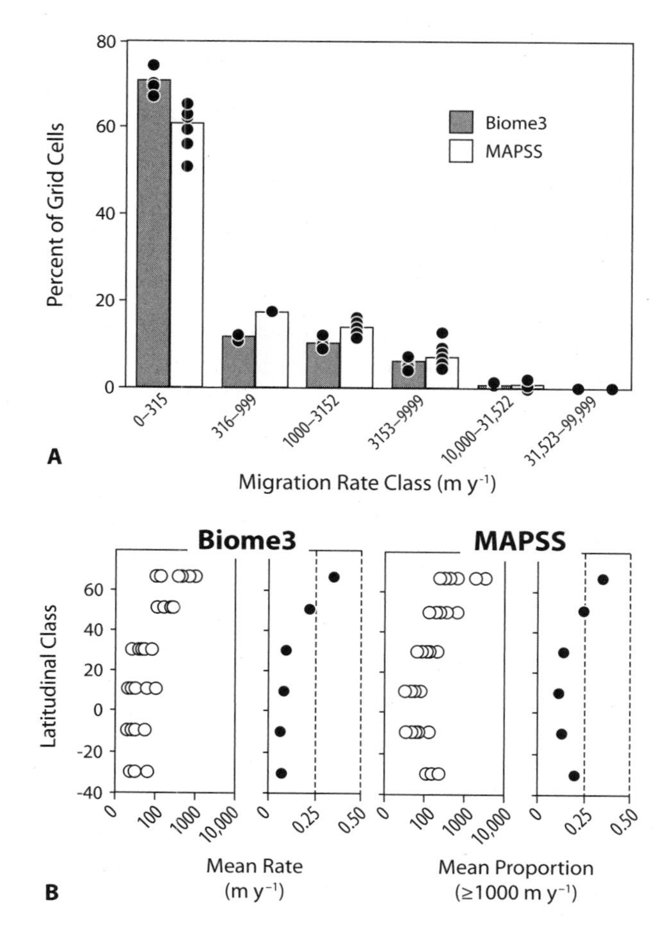

Figure 4.7. A. Average percentage of grid cells in various migration rate classes for two global vegetation models. Dots represent individual global circulation models. B. Mean migration rates and proportions of grid cells with rates greater than 1,000 m y for 20° latitudinal classes. Positions on the ordinate represent the mean latitudes of the cells within the classes. (Adapted from Malcolm et al. 2002)

4.3.3 Mountains

On the average, thermal zones in mountains of the United States are expected to shift upward from 650 m to 900 m in response to doubled carbon dioxide forcing (Neilson et al. 1989). Early equilbrium models predicted that, in consequence, the lower tree line in western mountains would be much higher than at present, the area occupied by high-elevation trees such as true firs greatly reduced, and alpine zones eliminated.

Figure 4.8 shows equilibrium predictions of how vegetation zones in the Cascade Range of northwestern North America would be altered by average temperature increases of either 2.5 or 5°C. With a 2.5°C change, alpine zones would disappear, and the area occupied by high-elevation tree species (mountain hemlock and silver fir) would shrink significantly. Sagebrush steppe, which does not now occur in the Cascades, would cover about 50 percent of the east slopes of the range (the east slopes are much drier than the west slopes). With a 5°C change in average temperature, about 50 percent of the west slopes would be covered by oak savanna and grasslands (neither of which now occur in the range). Seventy-five percent of the east slopes would be sagebrush. More recent transient models agree with some of these predictions and differ significantly in others (Bachelet 2002, 2003). There is general agreement that oak savannas will spread up the west slopes (in the southern Cascades nearly to the crest). However, rather than shrublands moving up the east slopes, the recent models show conifer savannas spreading to occupy former shrublands (this holds throughout the west). The contrasts between these two sets of predictions illustrates that temperature changes alone may not adequately characterize potential changes in the distribution of vegetation; nevertheless, however, the mountain vegetation may ultimately respond, it will undeniably be experiencing significantly warmer climatic conditions. We consider the physiological and ecological implications at various points throughout the book.

4.3.4 Global Change and Water

How global change will affect the hydrologic cycle is clear in some respects, and not so clear in others. Some feeling for the effect of temperature on available water is provided by the data of fig. 4.9, which shows the relation between temperature and runoff to rivers in the western United States. (The temperatures shown are not actual but are weighted according to the seasonal distribution of precipitation; see the figure legend.) Consider an area that receives 700 mm of precipitation per year, enough to support forests in the Rocky Mountains. An increase of 5°C in average yearly temperature, with no change in the amount or seasonable distribution of precipitation, reduces runoff by 30 to 40 percent. The drop in runoff reflects a greater proportion of water being evapotranspired at higher temperatures; in other words, forests and grasslands will demand a higher proportion of the water they receive. (The story is, in fact, more complex than this simple example; see chapter 15 for more detail.)

As we discussed in chapter 3, however, increased evapotranspiration results in increased cloudiness and, in at least some places on the globe, increased precipitation. In other words, warmer temperatures will not just evaporate more water, they will drive a more vigorous hydrologic cycle. But, while evapotranspiration tends to be distributed according to radiant energy, precipitation is a more complex phenomenon whose behavior is famously difficult to predict. Total global precipitation will increase with global warming, but how that increase will be distributed is less certain.

Adding to the complexity are the ecophysiologic responses of plants, whose transpiration supplies most water to the atmosphere from land. On the one hand, an increasing concentration of atmospheric carbon dioxide allows plants to use water more efficiently, thereby counteracting to some degree the increased evaporative demand

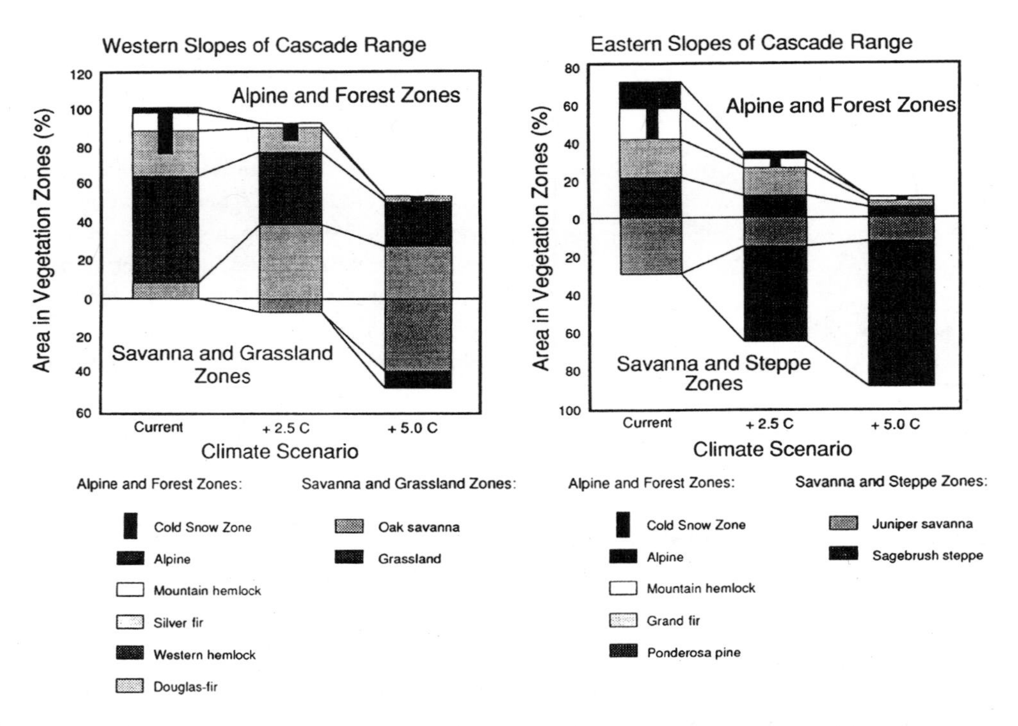

Figure 4.8. The percentage of area in major vegetation zones on the western and eastern slopes of the central Oregon Cascade Range under current climate and with temperature increases of 2.5 °C and 5 °C. (Franklin et al. 1992)

induced by warmer temperatures. On the other hand, higher temperatures require more water to be expended in keeping leaves from overheating, a process of particular importance in the tropics.

In biogeographic models, water balance on a given site is reflected by leaf area (square meter of leaf surface per square meter of ground surface). This is based on the well-established fact that except on very nutrient poor soils, leaf area is limited by available water (chapter 15). Therefore, the models assume that changes in available water will be followed by adjustments to site leaf area (which will, in turn, regulate water flow to streams). Changes in leaf area, in turn, provide a picture of how much the climate is predicted to change on a given piece of ground. In fact, it is change in leaf area that drives the changes in plant life-forms (e.g., shrubs to trees) predicted by process-driven biogeographic models.

Table 4.2 shows the percentages of different biomes predicted to gain or lose leaf area under a doubled carbon dioxide climate. The ranges encompass predictions from two biogeographic models applied to three climate models.

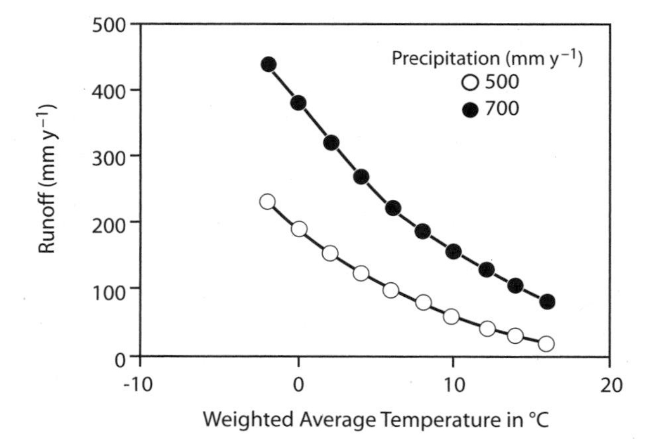

Figure 4.9. The average runoff to rivers in the western United States as affected by temperature. The weighted average yearly temperature (*x* axis) is the product of the average monthly temperature multiplied by the average precipitation for that month. Each curve represents sites with the same annual precipitation but different weighted average temperature. The sharp drop in runoff with increasing temperature reflects increasing evapotranspiration. (Adapted from Revelle and Waggoner 1989)

Table 4.2

Percentage of different biome types predicted to either gain or lose leaf area under a two-times CO_2 climate scenario

Biome type	% gaining leaf area	% losing leaf area
Tundra	20–58%	0–1%
Taiga/tundra	92–95%	1%
Boreal conifer forest	36–93%	0–20%
Temperate evergreen forest	46–67%	1–18%
Temperate mixed forest	50–91%	1–29%
Tropical broadleaf forest	16–87%	1–42%
Savanna, woodland	46–84%	7–17%
Shrub, steppe	64–80%	1–24%
Grassland	45–78%	5–46%
Arid land	53–80%	0–13%

Source: Adapted from Neilson et al. 1998.
Note: Ranges encompass predictions from two different biogeographic models (MAPSS and BIOME3) applied to three climate models. Values reflect increased plant water-use efficiency due to higher levels of atmospheric CO_2.

According to these scenarios, the majority of all biomes will become wetter, as reflected in higher leaf area. It is important to note, however, that the biogeographic models used to derive the predictions of table 4.2 incorporate effects of atmospheric carbon dioxide on plant water-use efficiency. The model predictions are quite different if a carbon dioxide effect is not included, particularly for temperate and tropical forests and for grasslands. Without enhanced water-use efficiency, a high proportion of temperate forests and grasslands become drier (lose leaf area), while tropical forests either lose leaf area or do not change.

Greater demand for water with no change in total precipitation will increase the probability that forests will become drought stressed during years of below-average precipitation. Increased droughtiness will, in turn, exacerbate the frequency and perhaps intensity of fires (Clark 1988, 1990a; Overpeck et al. 1990) and will also make trees more susceptible to insect pests. (The relation between climate and wildfire is discussed further in chapter 7, while the effects of drought on insect outbreaks are covered in chapter 19.)

4.3.5 Transient Biogeographic Models

At this juncture, humans are like travelers standing on the bank of a large river and looking at the other side. Equilibrium models give some idea of what kind of altered climate awaits on the other side but tell nothing about the process of getting from where we are now to where we will be in the future. The latest generation of models—called transient models—tackle the challenges of getting across the river. *Transient* literally means "of short duration," and transient global change models simulate the time-process of change, usually in yearly time steps. Coupling transient biogeographic models with transient climate models allows feedback between the two to be modeled, which is a significant step toward real-world dynamics. For example, if higher levels of atmospheric carbon dioxide fertilize land vegetation and, therefore, promote more carbon storage, coupled models adjust atmospheric carbon dioxide accordingly, which feedbacks to plant performance, and so on. In the same way, if warmer temperatures lower carbon storage through droughts, fire, or draw-down of soil carbon, coupled models increase levels of atmospheric carbon dioxide, which is a positive feedback to warming.

Constructing realistic transient models is a significant challenge because of the complexity of possible responses by land vegetation. In particular, the process of species migration is poorly understood and likely to be hampered by several factors unique to the modern world (see below). Changes in disturbance regimes (e.g., drought, fire, invasive species, insects, and pathogens; see chapter 7) will almost certainly have a strong effect on transient responses. As of the early twenty-first century, transient biogeographic models simulated drought, fire, and, to a lesser extent, phytophagous insects but did not deal with migration or invasive species.

Bachelet et al. (2003) modeled future vegetation in the lower 48 states of the United States using two transient biogeographic models coupled with two climate models, one simulating a moderate increase in temperature and the other a more substantive increase. Plate 15 shows results from two of the model combinations (one biogeographic with two climates) for 2095. For comparison, vegetation distributions as of 1990 are also shown (one based on Kuchler's vegetation map and the other simulated by the biogeographic model). Both future scenarios shown assume plant water-use efficiency correlates positively with increasing concentrations of atmospheric carbon dioxide. The two biogeographic models used by Bachelet et al. (2003) differed considerably in their predictions; however, there were also similarities. Both simulate a westward expansion of eastern forests, although that seems problematic since farms dominate the area forests would expand into. In the western United States, as we saw earlier both models predict shrinking deserts and expansion of savannas and woodlands into shrub-steppe (i.e., mixed grasslands and shrublands). A later model that includes the effects of fire predicts major changes in the distribution of vegetation in the eastern United States that do not occur in the absence of fire (Bachelet 2006). In particular, incorporating the effects of fire results in most of the eastern United States being occupied by xeric coniferous woodlands, oak savannas, and grasslands, with temperate deciduous and mixed forests remaining only in the northern Lake states and New England.

4.3.6 Migration

Although one certainly responds to the other, change in climate and change in vegetation are two different things (Maslin 2004), and exactly how the predicted changes in vegetation will play out on the ground is unclear. Trees will not simply hop on a bus and move until they find the right climate; they must migrate by seed dispersal.

What are the chances species can migrate fast enough to keep up with changing climate? Recall from earlier in the chapter that the models of Malcom et al. (2002) predicted a substantial need for migration rates of 1000 m/yr, especially north of 40° latitude (fig. 4.7B). For reference, the pollen record shows that trees migrated at rates of 200 to 400 m/yr in response to the climatic swings between glaciation and warming during the Quaternary, a period with no farms, roads, and cities to impede movement (Davis and Shaw 2001). The implication of this disparity is that a number of species are not going to be able to keep up with their preferred environment, especially in the north (the situation is less bleak in the tropics). Likely consequences include the following:

- Many tree species in the temperate and especially the boreal biomes will be increasingly maladapted to their environment, although genetic variability could buffer that to some degree (Davis and Shaw 2001).

- Maladaptation to their environment will diminish the ability of individuals to resist and recover from disturbances (especially those outside the range of historic variation).
- Plants that are adapted to disperse widely are most likely to keep pace with the changing climate, a fact that has led many researchers to conclude the world will become increasingly weedy in the future (e.g., Malcolm et al. 2002).

To complicate matters even more, rather than local climates simply shifting northward or up in elevation, it is possible and perhaps even likely that whole new combinations of climatic factors will emerge (Neilson et al. 1989). Consider, for example, seasonal changes. If summer indeed warms more than winter, trees whose northern boundaries are limited by winter cold, such as northern hardwoods and southern pines, will experience warmer summers than at present, even when they have adjusted their ranges northward to match the new climate. Based on winter temperatures, for example, northern hardwoods could eventually extend roughly 600 km north of their present boundary; however, their summer temperatures in this new locale will be hotter and perhaps drier. The upshot is that existing tree populations will increasingly experience temperature extremes and probably other climatic factors that have occurred rarely or not at all in their past. How forests might respond is anybody's guess and is complicated by various factors discussed in chapters 15 and 23.

4.3.7 Feedbacks and Uncertainties

It is important to understand that while the evidence in support of significant global warming during the twenty-first century is very strong, there are, nevertheless, many uncertainties about what may actually happen. Climate is a phenomenon of the global ecosystem, and numerous, poorly understood positive and negative feedbacks will come into play to influence the magnitude of greenhouse-related climate change. For example, as we saw in chapter 3, soils are a major storage reservoir for carbon. Warmer temperatures will increase the rate at which soil carbon is decomposed and released to the atmosphere as the greenhouse gases carbon dioxide and methane[4,5]; warmth creates conditions leading to more warmth, a positive feedback to the greenhouse effect. If trees grow faster under future climatic conditions, that would create a sink for atmospheric carbon, resulting in negative feedback to global warming. On the other hand, should warmer, drier conditions result in more drought mortality, insect kill, and forest fires, the consequent transfer of carbon from biomass to the atmosphere would produce a positive feedback to warming. The expansion of forests, as predicted in particular for the far north, will have contrasting effects that we discuss in more detail below.

4.3.8 Feedbacks between Land Use and Climate

Vegetation affects climate through both biogeophysical and biogeochemical processes. Biogeophysical effects include canopy roughness, which affects transpiration, air movement and cloud formation, and albedo, which affects the absorption and reflection of solar energy. The primary biogeochemical effects are uptake and release of carbon dioxide and release of other biotically mediated greenhouse gases, especially methane. Fire, a physical process that cycles carbon from vegetation to atmosphere, is a hybrid process spanning both categories.

Land use—especially converting land to or from forest—has opposing physical and chemical feedbacks to climate. Afforestation, including the natural expansion of forests into nonforested biomes, increases the strength of the carbon dioxide sink and creates a negative feedback to climate warming. But it also decreases albedo and increases canopy roughness, both of which feedback positively to climate warming. Deforestation has the opposite effects. Changes in albedo between forest and nonforest are particularly striking in temperate and boreal areas during winter, as illustrated in fig. 4.10.

Sitch et al. (2005) reviewed modeling studies on feedbacks between land use and global warming. The balance between physical and chemical effects varied depending on region. In the tropics, changes in carbon storage tended to outweigh physical effects, while in the north, where snow cover comes into play, physical effects played a more significant role. In the boreal zone, for example, expansion of forests onto tundra increases carbon storage, a negative feedback to global warming, while at the same time sharply lowering the albedo of afforested areas, a positive feedback to global warming. Modeling by Betts (2000) predicts that the decreased albedo would offset increases in carbon storage and perhaps even overwhelm it, to produce a net positive feedback.

While biogeophysical feedbacks are relatively straightforward and can be predicted with some accuracy, biogeochemical feedbacks are another matter, and models give mixed messages concerning the degree to which the earth's vegetation will store more carbon in a warmer climate. There are at least four large uncertainties:

1. The degree to which carbon dioxide fertilization will enhance growth of individual plants and stands
2. The effect of warmer temperature on plant and soil respiration
3. The magnitude of carbon losses due to fire, insects, drought, and other disturbances
4. The regional and global distribution of vegetation types, which depends both on land use patterns and rates of migration

[4]Methane (CH_4), a strong greenhouse gas, is produced when organic matter is decomposed under anaerobic conditions, such as in bogs or marine sediments.

[5]In some forests of the far north, the opposite could occur. By thawing frozen soils, warmer temperatures could increase soil water to the point that it inhibits decomposition and leads to a build-up of soil organic matter (Dunn et al. 2007).

Figure 4.10. Clear-cuts in the Washington Cascades.

These issues are covered in more detail at various places in the book: factors affecting carbon fluxes in chapter 15, disturbances in chapter 7, and rates of migration earlier in this chapter and again in chapter 20. We touch briefly on them in the remainder of this chapter.

4.3.9 Carbon Dioxide Fertilization and Ecosystem Respiration

If carbon is a limiting nutrient, higher atmospheric levels of carbon dioxide will act as a fertilizer and improve plant growth. Sitch et al. (2005) modeled global terrestrial NPP under a variety of future land use and climate change scenarios and concluded that carbon dioxide fertilization would make land vegetation a net carbon sink through the twenty-first century, regardless of future land use. However, experiments using open-topped chambers (which are large enough to accommodate small trees) show that when soil nutrients are limiting, such fertilization is a short-lived phenomenon (Oren et al. 2000; Finzi et al. 2002; Norby et al. 2004). Since soil nutrients are frequently limiting in forests (chapter 15), it seems doubtful that higher atmospheric carbon dioxide concentrations will produce a long-term carbon sink.

Respiratory losses of carbon due to higher temperatures might also tend to counteract a carbon dioxide fertilization effect. Soil respiration is particularly important for at least two reasons: first, soils are the largest carbon reservoir on land (chapter 1); and second, soils emit not only carbon dioxide, but methane (CH_4) and nitrous oxide (N_2O), gases with, respectively, 20 and 200 times the warming potential per molecule of carbon dioxide (Chapin et al. 2002). Both nitrous oxide and methane are produced as end products of decomposition when oxygen levels are low, and wetlands in particular produce large amounts of methane.

Soil respiration[6] increases exponentially with temperature (Chapin et al. 2002), while plant respiration increases linearly (Waring and Schlesinger 1985). Gross photosynthesis increases with temperature as well, but over a narrower range than respiration, resulting in declining NPP as temperatures climb beyond approximately 25°C (chapter 15). Modeling by Cox et al. (2000) predicts that because of respiratory losses associated with increasing temperatures, the land biota will switch from being a sink for carbon dioxide to being a strong source in about 2050. However, that prediction is also questionable. In a 10-year study of soil warming in a New England hardwood forest, increased soil respiration was of short duration because the supply of readily decomposable (labile) organic matter was quickly exhausted (Melillo et al. 2002). Moreover, the increased decomposition that did occur accelerated the release of nitrogen in plant-available forms (chapters 15, 18), stimulating tree growth and producing a counteracting carbon sink that outweighed the carbon loss associated with higher rates of decomposition. A similar result was obtained in a long-term soil-warming experiment with Norway spruce in Sweden (Stromgren and Linder 2002; Eliason et al. 2005).

Many questions remain concerning how the combined effects of higher atmospheric carbon dioxide and warmer temperatures will affect carbon effluxes from soil. Among the most important are how the quality, which is to say the decomposability, of soil carbon will be affected, and whether slow pools (those that for one reason or another

[6]Soil respiration results from at least three separate processes: (1) autotrophic respiration associated with the growth and maintenance of roots and mycorrhizal fungi; (2) respiration of microbes living in close association with roots (rhizophere microbes), which is tightly coupled to the supply of labile plant carbohydrates; and (3) respiration of heterotrophic decomposers (Pendall et al. 2004). Only the last represents a draw-down of soil organic matter.

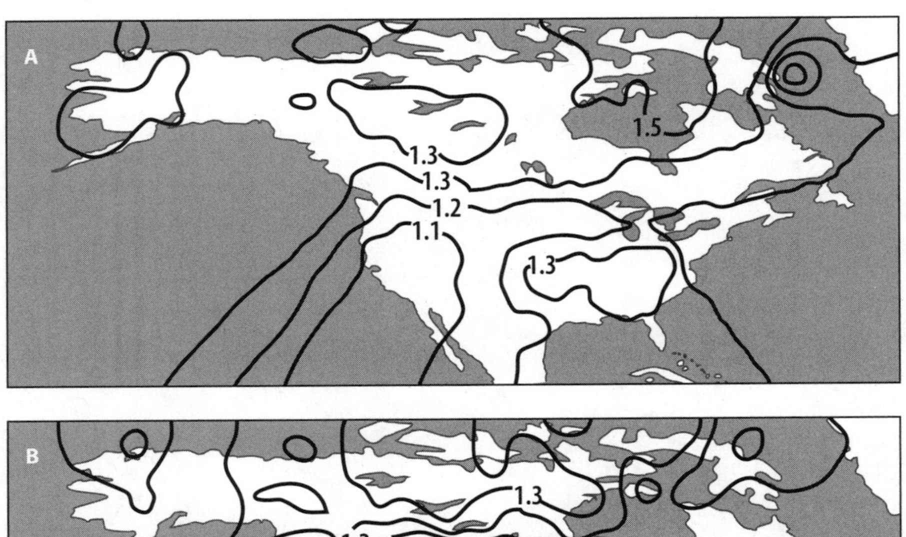

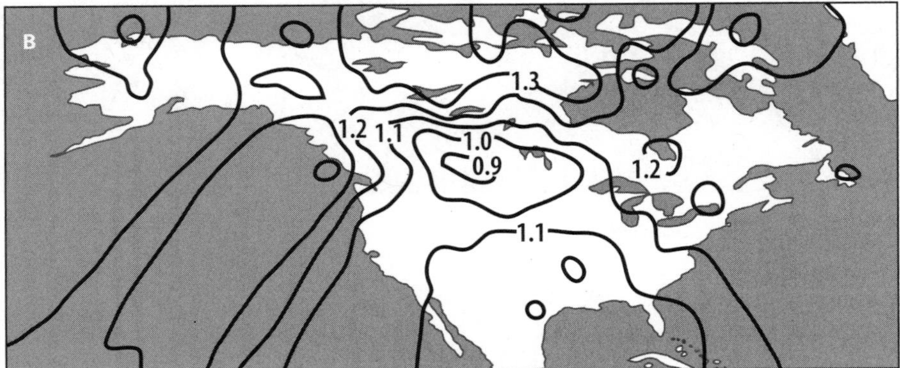

Figure 4.11. The ratio of the mean seasonal severity sating (SSR) estimated for the period 2055–2064 to the SSR averaged for the period 1985–1994. A value of 1.0 indicates no change, while values greater than 1.0 indicate increased SSR. SSR is a measure of fire weather severity and a rough indication of area burned. A. Canadian general circulation model. B. Hadley general circulation model and fire model of Flannigan et al. (1998). (Adapted from Dale et al. 2001)

turn over very slowly) will be primed to turn over more quickly (chapter 18). Scientific thinking about the nature and stability of soil organic matter is evolving, and some long-held ideas are now being questioned (Sollins et al. 2007, chapter 15).[7] The issue of priming is particularly important, because by far the majority of soil carbon exists in slow pools, with turnover times ranging from decades to millennia. Studies have shown that elevated carbon dioxide produces a priming effect in grasslands, but it is not clear whether the same is true of forests (Pendall et al. 2004).

In their review, Pendall et al. (2004) conclude: "The limited body of experimental evidence suggests that soil C cycling and decomposition may increase dramatically when warming and elevated CO_2 are combined. . . . With few exceptions, however, current experiments do not adequately capture responses of slow pool C to altered environmental conditions."

4.3.10 Disturbances

Disturbances and their functional roles in ecosystems are complex topics dealt with in some detail at other points in the book (chapters 7, 19, 20). Carbon sequestration is reduced, at least temporarily, by crown fires, windstorms, ice storms, drought, and outbreaks of insects, pathogens, and viruses. Additionally, fire pulses carbon from organic pools on land to the atmosphere, something that may happen to a lesser degree with other disturbances as well. At a global scale, cooling that results from higher albedo in disturbed areas may offset warming due to carbon pulsed to the atmosphere by the disturbance, especially during winter in northern forests (Randerson et al. 2006).

Disturbances either result directly from or are strongly influenced by climate, and significant climate change will almost certainly affect the extent, frequency, and severity of disturbances (Dale et al. 2001). If the new disturbance regime results in more carbon being sequestered on land, the feedback to greenhouse-related climate change will be negative; if it results in less, the feedback is positive and will reinforce global change.

Not surprisingly, model predictions of future disturbance regimes vary depending on what climate model is being used. Figure 4.11 shows predictions of altered fire severity using two climate models, the Canadian general circulation model and the Hadley general circulation model. Indices are the ratios of seasonal severity rating (SSR) in 2055 through 2064 to that in 1985 through 1994, so

[7]The paper by Sollins et al summarizes a number of papers delivered at a conference and published in Biogeochemistry, Volume 85 (1). Chapter 15 delves into this subject in more detail.

values greater than 1.0 indicate an increase in SSR, while values less than 1.0 indicate a decrease. The Canadian model, which predicts warmer temperatures than the Hadley, shows increased SSR over all of North America, with 30 percent increases in the Canadian boreal region and the southeastern United States. The Hadley also predicts increased fire severity over much of North America, but not so extensive or dramatic as the Canadian model.

Increased fire risk predicted for North America results from drought. When coupled with the Canadian climate model, two transient biogeographic models "suggest a rapid, drought-induced decline in southeastern forests" (Bachelet et al. 2003). On the other hand, the Canadian model predicts that the western United States will become somewhat wetter. In contrast, the Hadley climate model forecasts either unchanged or significantly higher average annual soil moisture for late-twenty-first-century North America, but strong drying trends in the Amazon basin, Central America, and Indonesia (plate 16). Should these tropical rain forests dry and burn, large amounts of stored carbon will be pulsed to the atmosphere, strongly reinforcing global warming.

4.3.11 Understanding Complex Responses: Where Are We?

At the midpoint of the twenty-first century's first decade, what do we know about potential feedbacks between terrestrial ecosystems and global change? In their review, Norby and Luo (2004) conclude: "Analyses of ecosystem responses to global change must embrace the reality of multiple, interacting environmental factors. There is a wealth of information on plant responses to CO_2 and temperature, but there have been few ecosystem-scale experiments investigating the combined or interactive effects of CO_2 enrichment and warming."

In other words, we have a dearth of knowledge at what might be considered the most fundamental level: interactions among plants, soils, and the atmosphere. When we move to higher levels of ecologic structure and organization—food-web dynamics, disturbances, migration—the uncertainty grows dramatically. We are indeed heading into unknown territory.

4.4 SUMMARY

The distribution of forests, grasslands, deserts, and tundra is determined by precipitation and temperature. Forests support more leaves than do other vegetation types; hence, they require more water. Forests also require a certain minimum growing season. As one moves from the tropics to high latitudes, changes in precipitation and temperature produce different vegetation zones, including three distinct forest types: tropical, temperate, and boreal.

Tropical forests occur roughly between the Tropics of Cancer and Capricorn and are dominated by broad-leaved angiosperms that may or may not be deciduous. Low rainfall between latitudes 25° and 35° produces deserts and grasslands. Temperate forests begin at about 25° (except in continental interiors, where low precipitation produces grasslands and deserts). Temperate forests are often dominated by deciduous broad-leaved angiosperms; however, notable exceptions exist, such as the conifer forests of northwestern North America and broadleaf evergreen forests that occur in the more southerly portions of the temperate zone. Boreal forests, dominated by cold-tolerant conifers, replace temperate hardwoods where winter temperatures fall below −40°C and extend north to about latitude 70° N, where trees are replaced by tundra.

Distinct forest types also occur within each of these broad groups. Tropical areas with a distinct dry season support deciduous rather than evergreen trees. In the temperate zone, regions that have wet winters and dry summers are dominated by conifers or evergreen hardwoods rather than deciduous hardwoods. Conifers also dominate higher-elevation forests within the temperate zone.

The major forest types differ in various respects other than species composition. Most notably, the diversity of plant and animal species drops sharply as one moves from the tropics to the higher latitudes, as does the rate at which nutrients are cycled. In general, productivity also drops from the tropics to the higher latitudes, although forests within each region differ widely in this regard, and some temperate forests are more productive than some tropical forests.

Global warming is already altering the earth's climate and affecting ecosystems. Present forest distributions will change, but poorly understood factors such as migration rates, altered disturbance regimes, and feedbacks between land and atmosphere create many uncertainties about the process of change. Significant stresses are likely. Stabilizing forests and easing the transition from one type to another will perhaps be the greatest challenge for future foresters.

5

Local Variation in Community Type

The Landscape Mosaic

5.1 A CASE HISTORY

The Big Horn Mountains rise like a massive island out of the arid Great Plains of north-central Wyoming, east of the main chain of the Rockies (fig. 5.1). When the range lifted during the Eocene geologic epoch (50 million years ago) it consisted of various layers of sediments—shales, limestones, sandstones—laid down over granite bedrock by the ancient seas that covered interior North America in the past. Erosion has since redistributed the sedimentary layers, exposing granitic bedrock in various places, and resulting in a patchwork of rock and soil types. This complex geology, combined with the sharp gradients of precipitation and temperature resulting from peaks that rise to 4,000 m, offers an excellent example of how the interacting effects of climate and soils influence local vegetation patterns. (The vegetation of the Big Horns was studied by DeSpain [1973].)

Walking from the eastern base of this range to the summit, one passes through a succession of forested and nonforested vegetation types. Figure 5.2 shows the influence of elevation and soil parent material on vegetation. At the lowest elevations, soil type (through its influence on water availability), determines whether a site is occupied by trees or grasses. Drought-tolerant ponderosa pine, the only tree species found at this elevation, occurs on alluvial fans (collections of material eroded from the upper slopes), which form a relatively coarse textured soil allowing water to penetrate to the deep tree roots. Soils formed from shale are fine textured, and though they hold more water than do coarse-textured soils, they do not allow this deep penetration. Hence shale-derived soils at this elevation are occupied by shallow-rooted grasses.

As we climb, the yearly precipitation increases. At about 1,800 m, stands of Douglas-fir appear, first on sheltered north-facing slopes, then eventually on gentle slopes of all

Figure 5.1. The Big Horn Mountains, Wyoming. A, Rising from the plains. B, Ponderosa pine at the prairie interface. C, Lodgepole pine forests at the midelevations. D, Spruce-fir forests intermingle with subalpine meadows at the higher elevations. E, At lower elevations on the dry west-slopes, Douglas-fir grows only along the steep canyon walls.

aspects. Around 2,000 m, lodgepole pine appears. It is, however, restricted to soils derived from granites and sandstones; Douglas-fir still occupies soils formed from limestone and dolomite. The differential preference of these two trees for soil type (seen throughout the Rocky Mountains) results partly from chemical characteristics of the parent material. Limestone- and dolomite-derived soils are generally higher in nutrients (particularly calcium) than are those formed from granites and sandstones. Along with many other pines, lodgepole is quite tolerant of low-nutrient soils.

Water availability still plays an important role in the distribution of vegetation at higher elevations. This is evident if we look at differences among aspects. Even on granitic soils, droughty south aspects tend to favor Douglas-fir more than lodgepole pine (the former being more drought tolerant than the latter). Southerly aspects with soils formed from sediments support grasses and drought-tolerant shrubs such as big sagebrush *(Artemisia tridentata)*.

Climbing higher, we encounter Engelmann spruce *(Picea engelmannii)* and subalpine fir *(Abies lasiocarpa)*. These

C

D

E

Figure 5.1. *Continued.*

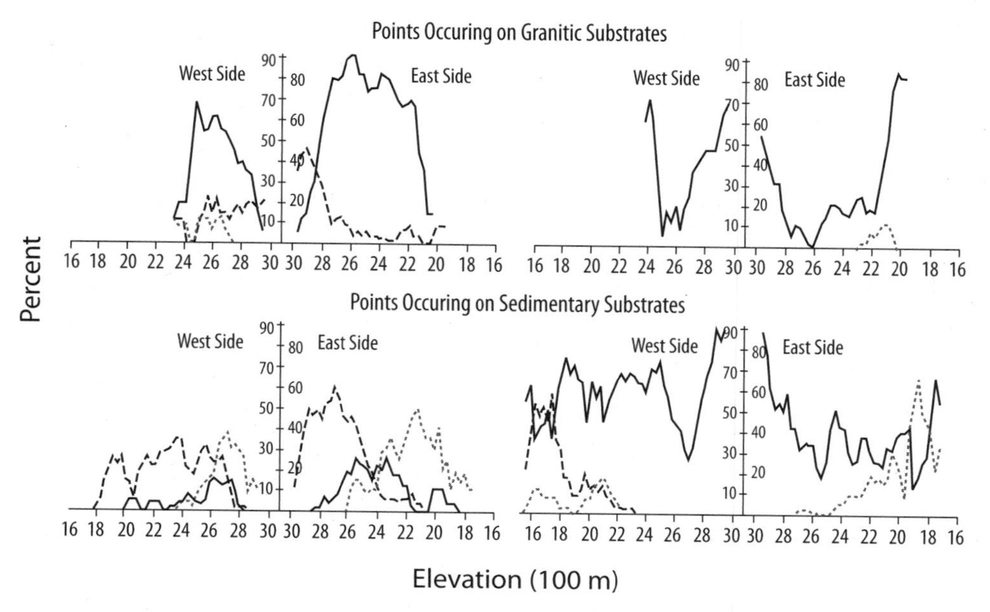

Figure 5.2. The influence of elevation and parent material on forest type in the Big Horn Mountains, Wyoming. The *y* axis is the percentage of cover. The left side of each graph represents forest patterns on the western slopes of the range, and the right side patterns on the eastern slopes. *Left,* the dashed line represents Englemann spruce / subalpine fir, the solid line lodgepole pine, and the dotted line Douglas-fir. *Right,* the dotted line represents ponderosa pine, the dashed line Rocky Mountain juniper, and the solid line nonforest. (Adapted from DeSpain 1973)

species, which require relatively moist sites, first appear in the understory of Douglas-fir and lodgepole pine stands at about 2,300 m. From 2,600 m to timberline (3,000 m), well-developed spruce-fir forests occur, particularly on north aspects and fine-textured, fertile soils derived from shales. On the relatively infertile soils formed from granites and sandstones, spruce-fir forests occupy only a thin band between lodgepole pine and timberline.

There are probably various reasons why trees rather than grasses occur on shale-derived soils at higher elevations in the Big Horns. Greater annual precipitation allows more water to reach lower soil levels; however, trees at this elevation also root higher in the soil profile than do trees in the more droughty sites at lower elevations and, hence, do not rely as heavily on water draining to lower soil levels. Greater annual precipitation also permits development of a closed forest canopy that inhibits grasses by intercepting light.

On passing over the alpine meadows of the mountain crest and descending the western slopes, we encounter a vegetation pattern that differs in many respects from that of the eastern slopes. Because summer rains in the Big Horns come from moist air moving north and west out of the Gulf of Mexico, the west side of the range is drier than the east side. This results in a higher proportion of grasses and shrubs, fewer spruce-fir and lodgepole pine forests, and a much broader band of Douglas-fir than occurred on the eastern slopes. For reasons that are presently unclear, no ponderosa pine occurs at the lowest elevations on the west side, as contrasted with the east. Rather, Douglas-fir forms the lower forest boundary, at the lowest elevations occupying only the steep, north-facing slopes of canyons that have been cut through the sedimentary rocks.

Two questions may be asked concerning the distribution of vegetation types in the Big Horns. First, are the various types distinct, or do they overlap? Second, is the distribution stable (i.e., if we return in five hundred years will we find the same patterns)? Although boundaries between vegetation types can be quite sharp, particularly where two distinct soil types meet, most are diffuse and include an area of variable size in which at least some vegetation from both types is found. Such a zone of transition is called an **ecotone.** In the Big Horns, ecotones exist between meadow and forest and between Douglas-fir and spruce-fir forests. In contrast, because of the distinct preference of the two species for different rock types, there is little overlap between Douglas-fir (growing on soils formed from sediments) and lodgepole pine (growing on soils formed from granites).

How stable are the various vegetation types? If we pay attention to tree composition in the understory of forests, we will have some idea of what the future forest would look like in the absence of major disturbance (such as wildfire) or significant change in climate. At lower elevations, for example, trees in the understory of Douglas-fir stands are also Douglas-fir, which tells us that barring some event that alters the species composition of tree seedlings, the next generation of trees on these sites will also be Douglas-fir.

With increasing elevation, however, the higher annual precipitation has two important effects. First, because the greater supply of water allows the forest to support more leaves, less light reaches the understory. Douglas-fir seedlings generally do not grow well in heavy shade (they are said to be moderately intolerant of shade); hence, fewer occur in the understory of higher-elevation stands. Second, increasing precipitation allows spruce and fir seedlings, which tolerate shade but not drought, to become established. As we climb, then, these species begin to predominate in the understory of Douglas-fir stands, indicating that Douglas-fir is transitional to spruce-fir on these sites. (This is the important ecological process called **succession,** which is covered in detail in chapter 8.)

Does it necessarily follow, however, that spruce-fir will in fact replace Douglas-fir on these sites? Ignoring the question of climate change, let us do some detective work that gives insight to stand history. Perhaps we notice old scars at the base of the trunk on some larger Douglas-fir and suspect that these were caused by past fires. Some digging turns up fragments of charcoal buried in the soil, confirming our suspicion: historically, periodic forest fires are likely to have destroyed understory spruce and fir seedlings, but not all of the large Douglas-fir (the scars are proof that some survive). Hence disturbance tends to perpetuate the Douglas-fir forest at this elevation, despite the fact that climatic factors favor spruce and fir. In fact, it is likely that fire suppression during the past decades has allowed more spruce and fir to become established than was normal.

A history of past disturbance is no longer a reliable guide to future stand composition, however. We now have global warming to contend with (chapter 4), not to mention air pollution and increased ultraviolet radiation due to thinning of the tropospheric ozone layer, all factors that could greatly influence the composition and overall health of forests. (The effects of air pollution on forests are discussed in chapters 7 and 20.) Focusing for the moment on climate change, how might vegetation patterns in the Big Horns be affected? Should warming, as predicted, push mountain vegetation zones upward by 650 to 900 m, spruce-fir forests would extend almost to the highest peaks, leaving virtually no alpine meadows. Other forest types would move correspondingly upward. The mosaic of ponderosa pine and grasslands presently at the lowest elevations would extend upward well into the zone now occupied by spruce-fir forests. This scenario does not account for predicted increases in the frequencies of drought, wildfire, and insect outbreaks. Drier conditions could limit spruce and fir to the most-protected north slopes at higher elevations and would probably also reduce the area of lodgepole pine forests relative to Douglas-fir and ponderosa pine. Engelmann spruce and subalpine fir do not tolerate fire and could be eliminated from the Big Horns (and other western mountains), should fires occur more frequently. All tree species would be significantly impacted by more frequent insect outbreaks.

At this point, let us review what we learned about vegetation patterns in our walk across the Big Horns.

- Plant species vary in their ecological tolerances and requirements for resources. In the Big Horns, as in all mountains, average yearly precipitation increases with elevation, but water availability is also influenced by aspect and soil type. Nutrient availability varies with soil type as well. Moreover, at higher elevations the frost-free season is short, which influences nutrient availability through its effects on cycling rate. The resulting patchwork of resources and environmental conditions produces a mosaic of differing vegetation types across the landscape.
- The mosaic of plant community types is not necessarily stable. Sufficient change in climate will alter vegetation patterns, but even in a stable climate, species composition and structure of communities usually change over time. In the Big Horns, this is seen in the high-elevation Douglas-fir stands, where Engelmann spruce and subalpine fir predominated in the understory and, presumably, in the absence of disturbance, would eventually dominate the forest.
- Disturbance influences vegetation pattern. In the Big Horns, Douglas-fir and lodgepole pine (both species that would be replaced by spruce and fir on at least some of the high-elevation sites) occur because of past fires. These species have evolved mechanisms that allow them to cope successfully with fire. Older Douglas-fir have thick bark that protects cambial tissue from heat and allows at least some trees to survive fire, as long as it does not spread into the crowns. The same is true for ponderosa pine. Lodgepole pine has evolved quite a different survival mechanism. Many (but not all) have what are called **serotinous cones,** which remain closed, holding their seed, until heated to 50 to 70°C. Fire usually kills the thin-barked trees, but enormous quantities of seed are released from cones that may have remained unopened on the tree for decades. Paradoxically, though trees are killed, perpetuation of the species on many sites is possible only with periodic fire. Disturbance in one form or another plays a role in all ecosystems, and all systems contain a set of species that have evolved mechanisms to cope with disturbances they have experienced in the past. (Foreign disturbances—those that a particular community has not previously experienced—are another matter; see chapter 20.)
- Boundaries between vegetation types (ecotones) may be quite sharp, as when one soil type abuts against another, or on the ridgeline separating a north from a south aspect. They also may be wide, with one type merging gradually into another. Later, we see that the nature of ecotones may be quite different among different climatic zones.

In summary, landscape patterns are determined by interactions among three factors: (1) the physical matrix of landform and soils; (2) disturbance, which modifies (usually temporarily, but not always) both the physical environment and the biota; and (3) relationships among the various species that compose the community. These factors do not act independently but are interrelated to one degree or another, a point we return to in section 5.3. Two levels of community pattern emerge from these interactions: a relatively stable pattern determined by topography and soils (called **topoedaphic**), and superimposed on the topoedaphic pattern, a dynamic mosaic of patches in differing stages of successional development. As we saw in chapter 2, landscape ecologists study the factors that influence the development and dynamics of spatial patterns, and how spatial patterns influence biotic and abiotic processes such as animal habitat and migration, water flows, and the propagation of pests, fire, and wind (e.g., Gutzwiller 2002; Liu and Taylor 2002; Turner et al. 2003a). These aspects of landscape pattern are of considerable practical interest to foresters and conservationists.

In the remainder of the chapter we discuss the relation between topoedaphic factors and local vegetation pattern in more detail, and how vegetation patterns can be used as indicators of the underlying physical environment. Patterns resulting from disturbance and succession are further discussed in chapters 6 through 8.

5.2 TOPOEDAPHIC INFLUENCES ON VEGETATION PATTERNS

5.2.1 Topography

Three topographic features interact to influence the physical environment and, therefore, the species composition and structure of communities. These are elevation, aspect (the direction in which a slope faces), and slope steepness.

In mountains, the most striking landscape patterns are often associated with a change in elevation. Average annual temperature decreases with higher elevations, while precipitation increases, at least up to the elevation at which clouds form. Vegetation types change correspondingly. Figure 5.3 shows a typical elevational transect in the Venezuelan Andes. Desert and scrub vegetation dominate below about 1,000 m. When annual precipitation reaches about 600 mm, species-rich deciduous forests appear, shedding leaves during the dry season. These forests then give way to semievergreen forest as precipitation continues to increase with elevation. At about 2,500 m, cool temperatures result in cloud formation (the elevation at which this occurs varies from one mountain range to another), and forests are characterized by abundant orchids, mosses, and lichens. These so-called **cloud forests** are the wettest ecosystems on the transect. Not only is precipitation high, but a great deal of water condenses directly from clouds onto leaves and other surfaces. Direct condensation of water vapor from clouds or fog onto leaves is a common phenomenon and can significantly affect the hydrology of forested basins (chapter 3).

Above the cloud forest, precipitation decreases, and in contrast to lower elevations (where vegetation types are determined primarily by water availability), low temperature begins to play an important role. The zone between cloud forest and tree line is dominated by conifers in the genus *Podocarpus*.

Boundaries between vegetation types along an elevational gradient can be rather sharp, but more often they are diffuse (i.e., broad ecotones). Figures 5.4 and 5.5 show this for two very different locales in the United States: mesic forests of the Great Smoky Mountains, and semiarid, predominantly oak forests of the Santa Catalina Mountains in southern Arizona. As readily seen in fig. 5.5, the broad zones of overlap between species along elevational transects often result in the most species diverse forests occurring at the midelevations of mountain ranges. Not only species, but different genotypes within a given species (called **ecotypes**) may occupy different elevational ranges, as is the case for *Fagus grandifolia* in the Great Smoky Mountains (fig. 5.4).

Aspect and slope steepness combine to influence vegetation patterns in the temperate and boreal zones through

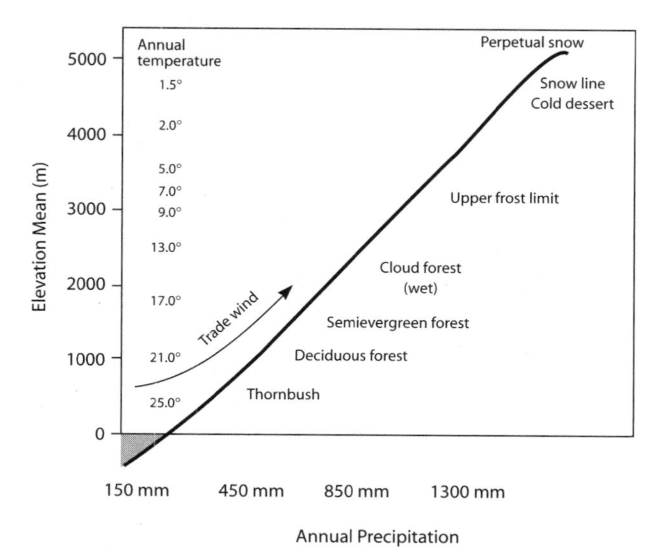

Figure 5.3. The change in vegetation with elevation on the western slopes of the Andes Mountains, Venezuela. (Adapted from Walter 1985)

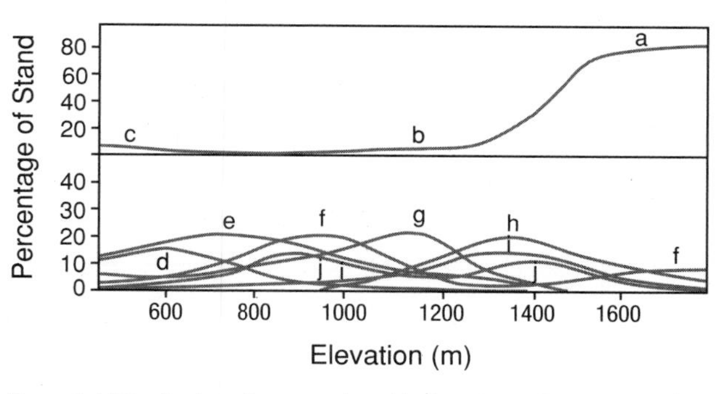

Figure 5.4. Distribution of tree species with elevation in the Great Smokey Mountains, Tennessee and North Carolina. *Top, Fagus grandifolia;* a, b, and c are different ecotypes. *Bottom,* d, *Acer rubrum;* e, *Tsuga canadensis;* f, *Halesia monticola* (bimodal distribution); g, *Tilia heterophlla;* h, *Acer spicatum;* i, *Aesculus octandra;* j, *Betula allegheniensis* (bimodal distribution). (Whittaker 1967. Reprinted with permission of Cambridge University Press)

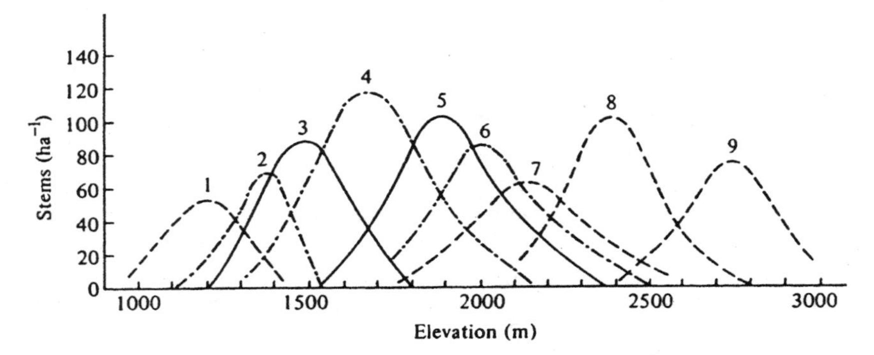

Figure 5.5. The elevational distribution of dicot trees in the Santa Catalina Mountains, Arizona. Species shown with solid lines are evergreen oaks of the black oak subgenus *Erythrobalanus*, those with dot-and-dash lines are evergreen oaks of the white oak subgenus *Lepidobalanus*, and those with dashed lines are other broad-leaved species. 1, *Vauquelina californica;* 2, *Quercus oblongifola;* 3, *emoryi;* 4, *arizonica;* 5, *hypoleucoides;* 6, *rugosa;* 7, *gambelli;* 8, *Acer grandidentatum;* 9, *glabrum*. (Whittaker 1969)

effects on the amount of solar radiation that is received. Figure 5.6 shows the influence of these factors on total daily direct radiation (without clouds) at latitude 45° N. Differences in the radiation received by different aspects vary with slope steepness, latitude, and season. At the time and latitude for which the calculations of fig. 5.6 were made (i.e., the equinoxes and latitude 45° N, respectively), this maximum occurs on south-facing, 45° slopes. During the winter, when the sun is lower in the sky, maximum radiation will occur on steeper slopes. At a given date, the slope steepness at which the maximum occurs declines with latitude. (For more detail, see Monteith 1975.)

Roughly 90 percent of solar radiation either evaporates water (including transpiration) or heats surfaces (Black 1977). Hence the amount of solar radiation received at the surface of canopies influences water balance and surface temperatures much more than it affects photosynthesis. In midlatitudes, the influence of temperature on vegetation usually manifests over elevational gradients, while aspect and slope steepness primarily influence water balance. Where water is sufficiently limiting, aspect-related differences in solar energy input can have profound effects on landscape patterns. For example, the Rocky Mountains often have forest on northern aspects and grasses on southern, with a very abrupt transition at the ridgeline (fig. 5.7). Analogous differences occur in many arid mountains and have important implications for forestry—especially during reforestation, when young seedlings are particularly susceptible to drought.

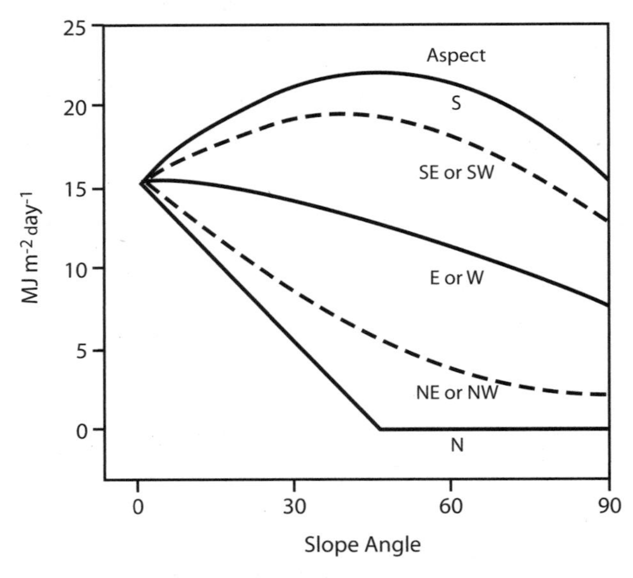

Figure 5.6. The amount of solar radiation incident on a slope varies with the aspect and steepness of the slope. Values are the daily integral of direct solar radiation at the equinoxes for latitude 45° N. (Gamier and Ohmura 1968)

In forests of the far north, water limitations are less important than cold soils and excessive water. At these high latitudes, low soil temperatures restrict nutrient cycling and allow permafrost to form, restricting the depth of roots. This is the case on north slopes, which tend to be occupied by black spruce, a species that grows slowly and tolerates low nutrient availability. The extra radiation received on southerly aspects prevents permafrost from forming and speeds nutrient cycling, allowing growth of relatively productive, nutrient-demanding species such as white spruce and quaking aspen (Bonan and Shugart 1989; Viereck et al. 1983) (fig. 5.8). Permafrost also develops on flat areas, even on southerly aspects, and water drains slowly from these areas. Such areas also collect cold air, which keeps soil temperatures lower than on surrounding slopes.

The collection of water and/or cold air in topographic depressions may influence forest patterns even within relatively flat terrain. An example of this is shown in fig. 5.9, which depicts landscape patterns in an area of the South African Kalahari region. During the rainy season, soils in low areas become saturated with water, restricting tree growth because of poor aeration (i.e., poor movement of atmospheric gases into soils), and perhaps because the deep-rooting trees have no advantage over shallow-rooted grasses. Slight changes in elevation are sufficient to allow water to drain from upper soil layers yet still permit the

Figure 5.7. Trees on the northern slopes and grasses on the southern slopes is a common pattern in arid and semiarid mountains.

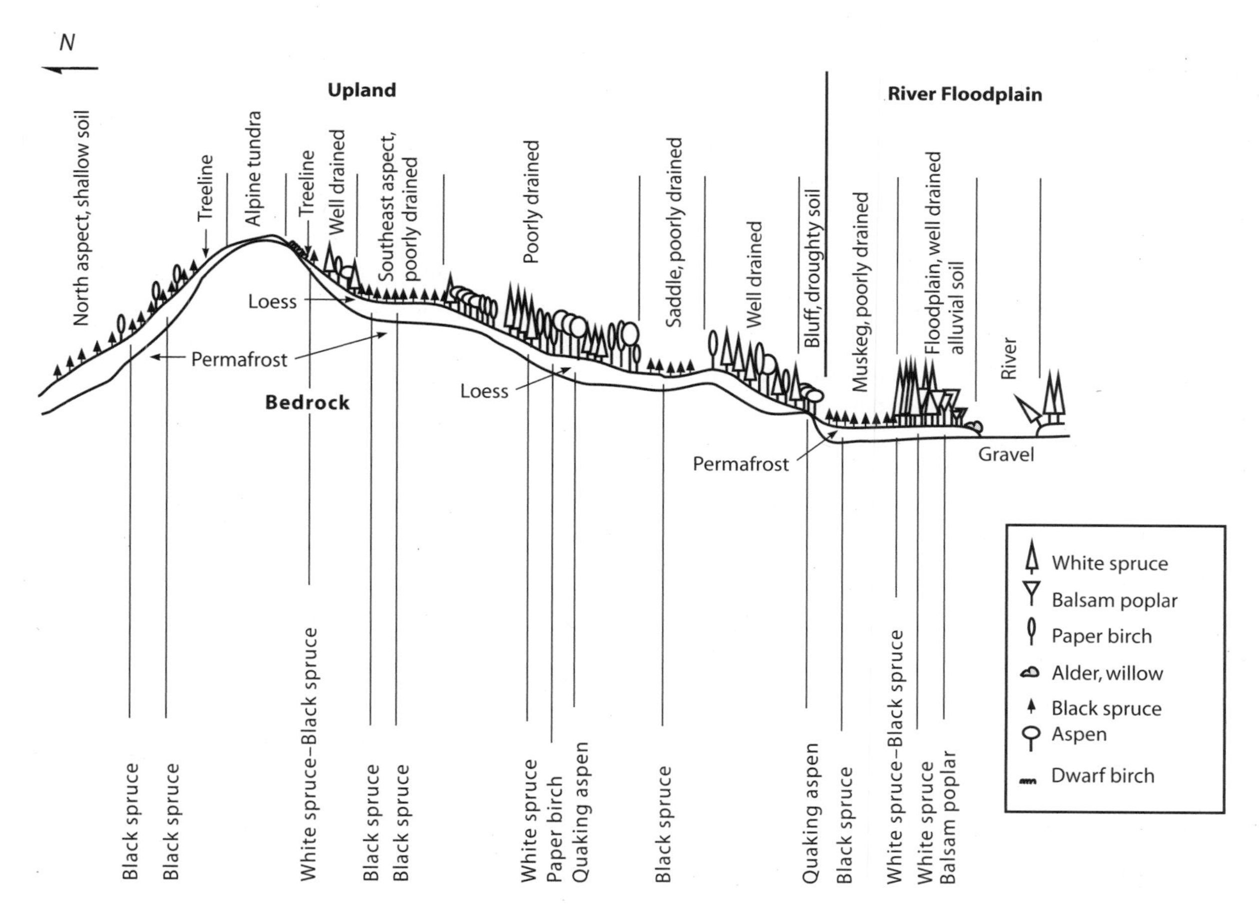

Figure 5.8. The distribution of trees across an Alaskan landscape (Fairbanks area). (Adapted from Viereck et al. 1983)

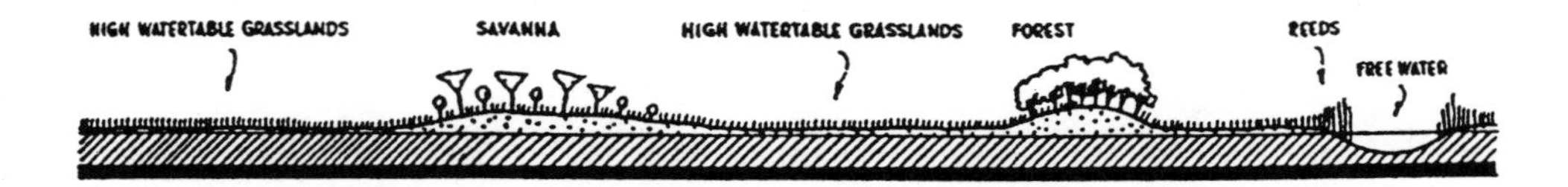

SAND FLATS AND RISES OVERLAYING A QUASI-HORIZONTAL PAN HORIZON

Figure 5.9. In the Kalahari region of South Africa, high water tables restrict trees to areas of slight elevational rise. (Adapted from Tinley 1982)

> A simple calculation demonstrates the importance of the effect of aspect on potential evapotranspiration. For the conditions of fig. 5.6, a north- and a south-facing slope, both at 45° steepness, differ by about 20 $MJ/m^2/d$ in the amount of direct solar radiation received. At 20°C, the latent heat of vaporization of water is 2.454 MJ/g. Therefore, a wet surface oriented perpendicular to the sun would lose nearly 10 times more water on the south than on the north aspect. It does not follow that transpiration would differ by this magnitude, because plants, through opening and closing stomata, modulate water loss (chapter 15). Nevertheless, a valid principle is illustrated by this simple example: differences in the input of direct solar energy create large differences in the demand for water at different points in the mountainous landscape. Plants may modulate transpiration, but only to a degree. Water must be evaporated in order to cool leaf surfaces, and water loss occurs when stomates are opened to admit carbon dioxide for photosynthesis.

deep-rooting trees to reach water. These areas are occupied by either savanna or deciduous woodland, depending on the height of the rise. The flat pine forests on the eastern slopes of the Cascades in Oregon are dominated primarily by ponderosa pine; however, lodgepole pine occupies topographic depressions. This is because lodgepole is better able to tolerate the frequent frosts resulting from cold air collecting in the depressions, and it is also better able to tolerate the occasional flooding that occurs when topographic depressions collect water.

5.2.2 Soils

Both physical and chemical characteristics of soils influence the type of vegetation that occurs on a site. Important physical characteristics include texture (i.e., the proportion of sand, silt, and clay), rock content, organic-matter content, aggregation, and depth to impermeable layers that restrict rooting or drainage. Chemical characteristics include pH and the relative amounts of various nutrients and other chemical compounds. These factors are not independent of one another, nor are they independent of the vegetation. (The nature of forest soils is discussed more fully in chapter 14).

5.2.2.1 Physical Properties

One of the more common ways that soil influences vegetation pattern is through its control over water. The physical structure of soil may effect water availability to plants in three general ways: (1) the total amount of water that can be retained within the soil profile; (2) how tightly water is held within the soil, which, in turn, determines the ease with which plants can obtain it; and (3) water distribution within the profile. In general, there is an inverse relationship between the ability of a soil to retain water and the ease with which that water is obtained by plants. Water retention correlates closely with pore size in soils, which is, in turn, a function of the relative composition of sand, silt, clay, and organic matter. Clay soils have small pores and retain water quite efficiently, but a relatively high proportion of that water is difficult for plants to extract. At the other extreme, plants readily extract water from the large pores of sands, but sands have relatively small water-holding capacity. Organic matter creates a heterogeneous pore-size distribution within soils and generally enhances water availability in both sands and clays, but too much organic matter can result in waterlogging. Through their roots and mycorrhizal fungi, plants themselves play a significant role in creating and maintaining the pore structure of soils.

A good example of how soils influence vegetation patterns through control over water is given by Brush (1982), who studied factors determining the distribution of forest types in Maryland. Her results from the Coastal Plain are shown in fig. 5.10. Paradoxically, from the perspective of a plant, the driest soils in this region are the tidal marshes. Water is of course abundant, but the high salt content makes it physiologically difficult for plants to obtain. Tidal marshes are occupied exclusively by loblolly pine.

The next driest substrates are fragipans, sands, and clays. (Fragipans have a layer close to the surface that is impervious to water; hence, their capacity to store water is low.) These soils are occupied by chestnut, post, and blackjack oaks, with willow oak–loblolly pine forests also occurring on clays. Intermediate moisture availability (a condition

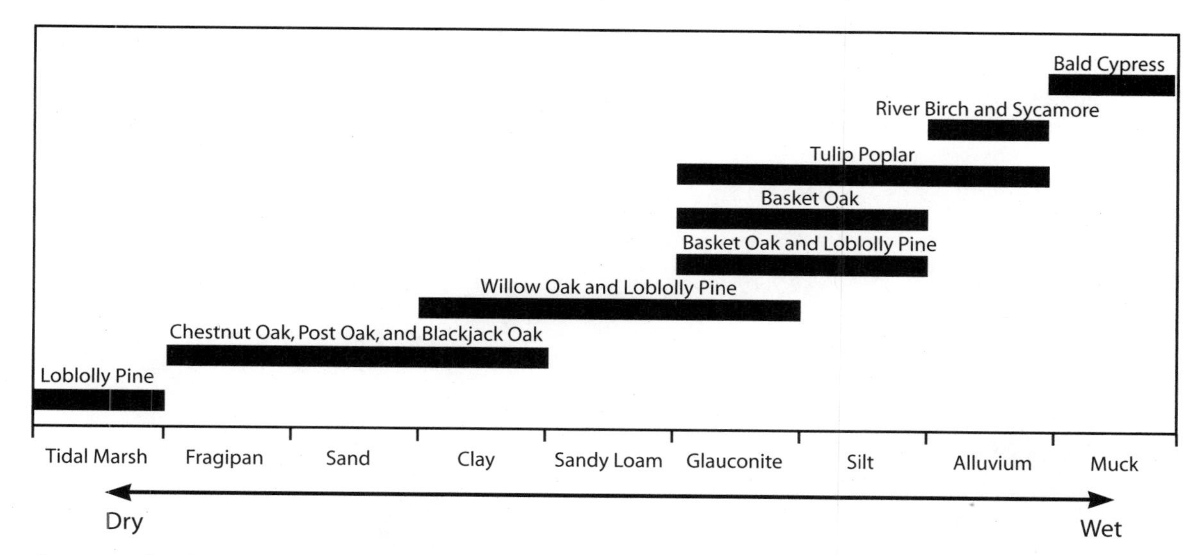

Figure 5.10. The influence of soil water on forest type in the coastal plain of Maryland. (Brush 1982)

referred to as **mesic**) occurs on sandy loams, glauconite sands (glauconite is a silicate of iron and potassium), and silts. These soil types store water for long periods but are sufficiently well drained to prevent waterlogging. Mesic soils within the Coastal Plain are occupied by willow oak–loblolly pine, basket oak–loblolly pine (basket oak replacing willow oak with increasing water availability), pure basket oak (north of the range of loblolly pine), and tulip poplar. Note the range of conditions under which loblolly pine is found. A species that grows in a diversity of habitats is said to have wide **ecological amplitude.**

The wettest soils in the Coastal Plain are those that are periodically flooded. These include alluvial soils (which were formed by deposition during floods) and mucks (which are soils that have accumulated large amounts of organic matter, because flooding reduces decomposition). River birch and sycamore occur on the former, and bald cypress on the latter.

One particularly important point about Brush's findings is that distinctive forest communities could be identified only in mature forests; the species composition of young forests did not correlate with soils. This common observation underscores the fact that landscape patterns are not determined solely by the physical environment. Rather, they are a product of biotic interactions modified by the physical environment. For example, in Brush's study the absence of mature basket oak forests on clay soils is not because this species cannot grow there, but because other species are better adapted and, therefore, dominate at the expense of basket oak. On the more mesic soils, this situation is reversed; basket oak is the better adapted, and over the course of forest maturation, it excludes other oaks.

Soil depth often influences composition of the plant community as well. Shallow soils over an impermeable bedrock have little water-holding capacity and favor drought-tolerant plants. On the other hand, if underlying bedrock has cracks and fissures, these act as water reservoirs for deep-rooting trees and shrubs. Walter (1985) describes an arid region of southwest Africa (185 mm rainfall per year); deep sands in this rainfall zone support only grasses. In Walter's example, however, 10 to 20 cm of sandy soils overlies a sandstone bedrock characterized by both large and small fissures. Water trapped in the fissures allows two species of woody shrubs to persist. One is small and obtains water from small fissures; the other is large and obtains water from large fissures. The distribution of shrubs corresponds to the structure of the underlying sandstone. Similar patterns occur wherever growing-season water is in short supply. Coarse-textured substrates favor deep-rooted woody plants, while fine-textured substrates favor shallow-rooted herbs and grasses. Figure 5.11 shows an example from El Malpais National Monument in New Mexico, where ponderosa pine establishes only on old lava flows, where water is stored in deep fissures.

In some circumstances chemical reactions within soils produce impermeable layers. The fragipans of the Maryland Coastal Plain are an example. Another common example is the formation of **lateritic**[1] layers in oxisols, which are old, highly weathered soils common to some tropical areas. Oxisols are characterized by high levels of iron oxides and low levels of silica (which has been weathered away). When allowed to dry sufficiently, the oxides form impermeable layers at variable distances below the soil surface. There has been some concern that clear-cutting tropical forests growing on oxisols would allow soils to dry and lateritic crusts to form, thereby restricting revegetation. Apparently, however, plants reoccupy logged areas rapidly enough to prevent the necessary soil drying (Jordan 1985). In more arid regions of the tropics, impermeable lateritic layers do form

[1]Laterite is derived from the Latin word *later*, which means "brick."

Figure 5.11. The boundary between ponderosa pine and meadow in this photograph corresponds to the extent of an old lava flow. Pines colonize the lava, where they can access water stored in fissures, but not the meadow, where water is used fully by the grasses.

in soils and can influence vegetation patterns. Figure 5.12 shows such a situation in southern Africa. Here the lateritic layer occurs at variable depths. Where it is near the surface, soils are flooded in the rainy season but have little storage capacity; therefore, they dry quickly in the drought season. These sites are occupied by grasses or savanna. Where the lateritic layer is relatively deep, not only are soils also deep, but the layer collects water draining from adjacent areas, permitting deciduous forests to grow.

5.2.2.2 Soil Chemistry

Soil is a product of vegetation acting on rocks; hence, it has characteristics of both. This is particularly true of soil chemistry, which depends to at least some extent on the type of vegetation growing on it. Therefore, one must be careful when ascribing differences in vegetation to differences in soil chemistry. Nevertheless, it is clear that chemical properties intrinsic to bedrock can influence the type of vegetation, and that at least in some cases, vegetation patterns reflect underlying geology. We have already seen that throughout the Rocky Mountains, lodgepole pine and Douglas-fir tend to segregate on granitic and sedimentary substrates, respectively. Numerous similar examples can be found. In Australia, for instance, the type of rock from which soils are derived determines the relative abundance of different eucalyptus species (Austin et al. 1990). As we shall discuss in chapter 16, species differ both in the total amounts of nutrients that they require, and in the ratios of different nutrient elements. Species that require relatively large amounts of nutrients may not tolerate nutrient-poor substrates such as granites, while other species have evolved

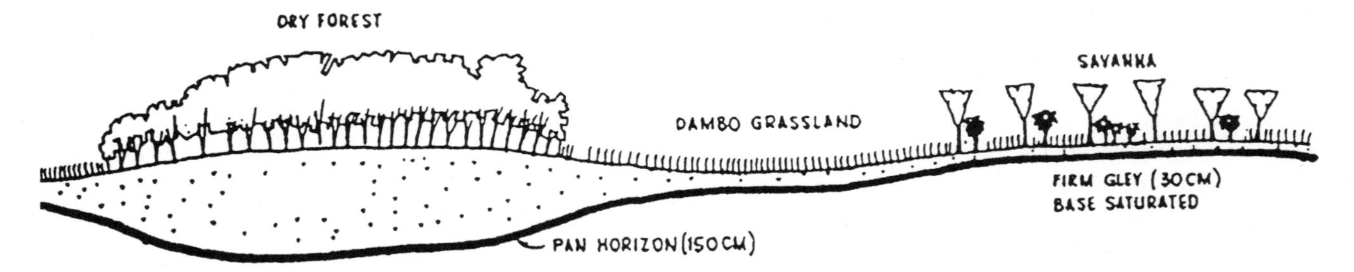

Figure 5.12. In soils that form impermeable layers, the type of vegetation may depend on the depth to that layer. This schematic drawing shows a South African lateritic soil with forest restricted to areas where the lateritic layer is relatively deep. (Adapted from Tinley 1982)

Table 5.1
The distribution of forest types in different climatic zones of eastern Australia as affected by soil fertility

Climatic region*	Latitudes	Climatic zone†	Soil Nutrient Status‡ Eutrophic	Mesotrophic	Oligotrophic
Tropical	11–18	Lowland (mesophyll vine forest)	Complex raingreen	Mixed	Sclerophyll simple evergreen
		Subtropical (notophyll vine forest)	Complex raingreen	Mixed	Simple evergreen
Subtropical	20–32	Lowland (notophyll vine forest)	Complex raingreen	Mixed	Simple evergreen
		Intermediate subtropical-submontane (notomicrophyll vine-fern forest)	Mixed	Simple evergreen	Simple evergreen
		Submontane (microphyll fern forest)	Evergreen	Evergreen	Evergreen
Intermediate tropical warm temperate	32–37	Lowland (notomicrophyll vine-fern forest)	Mixed	Simple evergreen	Simple evergreen
Warm temperate	37–39	Submontane (microphyll fern forest and thicket)	Evergreen	Evergreen	Evergreen
		Lowland (microphyll fern forest)	Evergreen	Evergreen	Evergreen
		Montane (nanophyll mossy thicket)	Evergreen	Evergreen	Evergreen
Cool temperate	c. 43	Nanophyll mossy forest	Evergreen	Evergreen	Evergreen

Source: Adapted from Webb 1968.
*Temperatures decrease from tropical to cool temperate.
†Zones receiving at least 100 cm of annual precipitation.
‡Eutrophic, high nutrient levels; mesotrophic, moderate nutrient level; oligotrophic, low nutrient levels.

various adaptations to cope with infertile soils—and sometimes to thrive on them.

In general, evergreen trees (whether conifers or angiosperms) seem to tolerate low-nutrient soils better than deciduous trees. For example, within a certain elevation range (600 to 1,100 m) in the Sierra Madre Occidental of northwestern Mexico, variation in bedrock chemistry produces a mosaic of evergreen-oak woodland surrounded by subtropical deciduous forest. The former occurs on acid, infertile soils derived from hydrothermally altered rocks; the latter occurs on less acid, more fertile soils derived from unaltered volcanic rocks (Goldberg 1982). However, this relationship depends on elevation: above 1,100 m evergreen oaks occur on both fertile and infertile soils. (The effect of rock type on soil characteristics is discussed in chapter 14.)

The influence of soil fertility on vegetation type has been particularly well studied in Australia (Beadle 1966; Webb 1968; Austin et al. 1990). Within the tropical and subtropical regions of eastern Australia, fertile soils support complex rain-green forests (i.e., trees shed leaves in the dry season), while infertile soils support tree species such as eucalypts that are evergreen and that also tend to have sclerophyllous and/or xeromorphic leaves (table 5.1). Sclerophylous leaves are hard and tough; xeromorphic leaves have a set of characteristics that confer drought resistance, including small size and thick, waxy cuticles. Tough leaves, which are typical of trees growing on nutrient poor soils, result from a high content of lignin, a compound that does not contain nutrients other than carbon, hydrogen, and oxygen. The low nutrient content of such leaves gives us some insight as to why evergreens are more successful on infertile soils than deciduous trees.

The relative proportions of rain-forest genera and xeromorphic genera within Australian forests correlate more closely with soil phosphorus content than with rainfall (fig. 5.13). One must be careful, however, about concluding that close correlation between two variables indicates that one causes the other. In the data of fig. 5.13, soil phosphorus may indicate some other nutrient or soil physical property

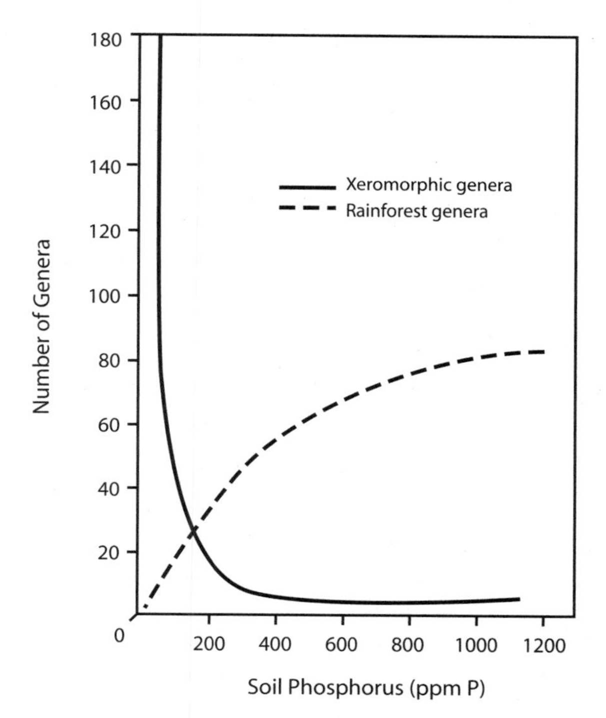

Figure 5.13. The relationship between levels of phosphorus in Australian forest soils and the occurrence of dicotyledonous plant species with genera typically occurring in rain forests, or having xeromorphic leaves. (Adapted from Beadle 1966)

In general, a given plant species tolerates conditions that are unfavorable for other species because of adaptations that allow it to do so. Trees that grow on infertile soils have various mechanisms permitting them to cope with a low nutrient supply. Nutrients may be stored in old foliage or wood and cycled to new growth when needed. For example, black spruce tolerates low nutrient availability at least in part because it retains needles for 25 to 30 years; hence, it can maintain a relatively large supply of nutrients "in-house" (Bonan and Shugart 1989). According to Grubb (1989), a major factor in the ability of some trees to grow well on infertile soils "seems to be the astonishingly complete withdrawal of certain nutrients, particularly phosphorus and potassium, from older wood, first emphasized by Beadle and White (1968) for eucalypts." (Trees growing on fertile soils also store and cycle nutrients from old to new tissues; this is discussed further in chapter 18.) Plants that tolerate infertile soils frequently have tough (sclerophyllous) leaves and chemical defenses that discourage herbivory and consequent loss of nutrients to animals (chapter 19).

Some species are able to cope with infertile soils because they are very efficient at gathering nutrients from the soil. Frequently this is related to a close association between plant roots and fungi that is called a **mycorrhiza** (from the Latin for "fungus-root"). We discuss this widespread and extremely important partnership between plants and fungi in chapter 11. Basically, however, the plant supplies carbohydrate to the fungus, and the fungus benefits the plant in numerous ways, including an enhanced uptake of some nutrient elements. The particular type of mycorrhiza formed by a plant can play an important role in where it is able to grow. The most striking example of this is in moist tropical forests, where highly infertile soils are most often occupied by tree species forming **ectomycorrhizae,** types of mycorrhizae distinguished by highly visible structures on the root surface (Gartlan et al. 1986; Ashton 1988). Tropical ecologists generally believe that a tree's ability to occupy highly infertile soils is due to the nutrient-gathering capabilities of its fungal partner (Gartlan et al. 1986; Janos 1987).

Some species that grow on infertile soils have inherently slow growth, and therefore, a low demand for nutrients. This is the case with black spruce in the boreal forest (Van Cleve et al. 1983) and also on certain types of highly infertile soils in the tropics and subtropics (Beadle 1966; Grubb 1989). Other trees, however, such as species of eucalyptus and in the widespread tropical family Dipterocarpaceae, may grow to a larger size on infertile soil than other species growing on nearby fertile soils (Grubb [1989] calls these **superplants**). It is not always clear what enables superplants to do this. Some, but not all, are able to form partnerships with nitrogen-fixing bacteria. Superplants are frequently species that form ectomycorrhizae (e.g., eucalypts, dipterocarps), and the nutrient-gathering power of the tree's fungal partner is almost certainly involved in their ability to tolerate infertility.

Trees and other plants also alter soil fertility, in some cases improving it and in others acting through their own chemical composition to reinforce and even exacerbate infertility. Fertility may be improved in various ways. Some species form partnerships with bacteria that convert atmospheric nitrogen into a chemical form usable by plants (this process, discussed further in chapters 11 and 17, is called **nitrogen fixation**). Tree crowns rake nutrients from the air (chapter 17), and the organic matter added to soils by trees and other plants serves as a nutrient storage reservoir. Rapid nutrient cycling (especially in tropical forests) allows trees to use a small amount of nutrients efficiently (chapters 14, 15, 18). On the other hand, the tough, nutrient-poor leaves produced by trees growing on particularly infertile soils are not easily decomposed; hence, soil infertility is reinforced by slow nutrient cycling. This is the case in some boreal and montane forest types, and in the highly infertile white sands in the tropics.

(e.g., water-holding capacity) that influences vegetation patterns; most often in nature a complex of interacting properties is involved. In Belize, for example, rain forests grow on relatively fertile soils, particularly with regard to magnesium, while infertile soils are occupied by pine savanna. However, Baillie (1989) points out that "the edaphic environments of the two ecosystems differ morphologically, hydrologically, and chemically. . . . It is not possible to select any one soil property as being the critical determinant of the distribution of the two ecosystems."

In at least some cases, the ratio (as opposed to the total amount) of various chemical elements in soils is quite important in determining vegetation type (Tilman 1986)—something that needs much more study in forests. Hogberg (1989) suggests that in the tropics, the ratio of nitrogen to phosphorus influences forest type. He points out that tree species that dominate dry savannas (the soils of which tend to have relatively low ratios of nitrogen to phosphorus) frequently form partnerships with nitrogen-fixing bacteria. Tree species that dominate on soils poor in both nitrogen and phosphorus form partnerships with ectomycorrhizal fungi, which are adept at gathering both elements. (Virtually all tree species form partnerships with at least one and commonly several types of microbes. These relationships are discussed in chapter 11).

One of the most striking examples of the influence of element ratios on vegetation occurs with soils formed from serpentinite, a rock type that is produced when magma

Figure 5.14. The sharp change in tree cover in the upper left of this photograph is caused solely by a change from sedimentary bedrock to serpentinite (low tree cover on the latter).

flowing out of rifts in the ocean floor absorbs seawater. Serpentinite is characterized by an unusually high ratio of magnesium to calcium, and soils forming from this rock generally have a distinctive flora. The result is often a dramatic change in both tree density and species composition at the boundary between serpentinite and other rock types. Figure 5.14 shows such a boundary in the Klamath Mountains of northern California and southern Oregon.

5.3 THE EMERGENT LANDSCAPE: INTEGRATION OF TOPOGRAPHY, SOILS, AND DISTURBANCE

Like most ecological phenomena, landscape patterns result from interactions among many different factors (fig. 5.15). Soils, and even topography, can be altered by disturbances such as volcanic eruptions, landslides, or abusive management practices (chapters 7, 20). Disturbance patterns may be strongly influenced by topography and by the vegetation mosaic itself. The intensity of fires, for example, tends to vary with factors such as slope steepness and aspect, and some community types are more flammable than others (chapter 7). Biotic factors that influence landscape patterns, such as competition, cooperation, herbivory, and pathogenesis (chapters 8, 11, 19), are, in turn, influenced by both the physical and biotic environment and by the history of disturbances.

Soils forming from a single rock type often differ depending on their location within the landscape. Erosion redistributes materials from steep slopes to flat areas below; hence, soils may be shallow and rocky on the former and deep and fertile on the latter. A great deal of topographically related soil variability may exist within quite small areas. For example, Pregitzer et al. (1983) found five different soil types within a 300-m horizontal distance in a Michigan watershed, each soil type with a distinctive forest type.

Erosion rate is at least to some extent controlled by plant cover, which, in turn, depends on water and nutrient availability. Hence, under some circumstances, droughty and infertile soils resulting from erosion on steep slopes decrease plant cover, and the emergent system reinforces erodibility. Because of the role of plant cover, erosion can differ with the aspect: droughty southern slopes have less plant cover, therefore more erosion and, consequently, a topography that differs from that of northern aspects.[2] Redistribution of material through erosion often produces patterns of rock type—and therefore soils—that correlate with topography. Earlier, we saw an example in the Big Horn Mountains, where the sediments composing the surface bedrock when the mountains formed had been redistributed to the valleys, leaving granitics exposed in the uplands. Because different rocks weather at different rates, topographic features depend to some extent on the distribution of rock types; hence, steep slopes may be associated with some easily erodible rock such as mudstone, while adjacent basalts (relatively resistant to weathering) have a different topography. A further complication is introduced in many forested areas because soils are

[2] Low plant cover does not always imply high erosion rates. Some soils, such as the serpentines of the Klamath Mountains (fig. 5.14), are relatively stable even though sparsely stocked.

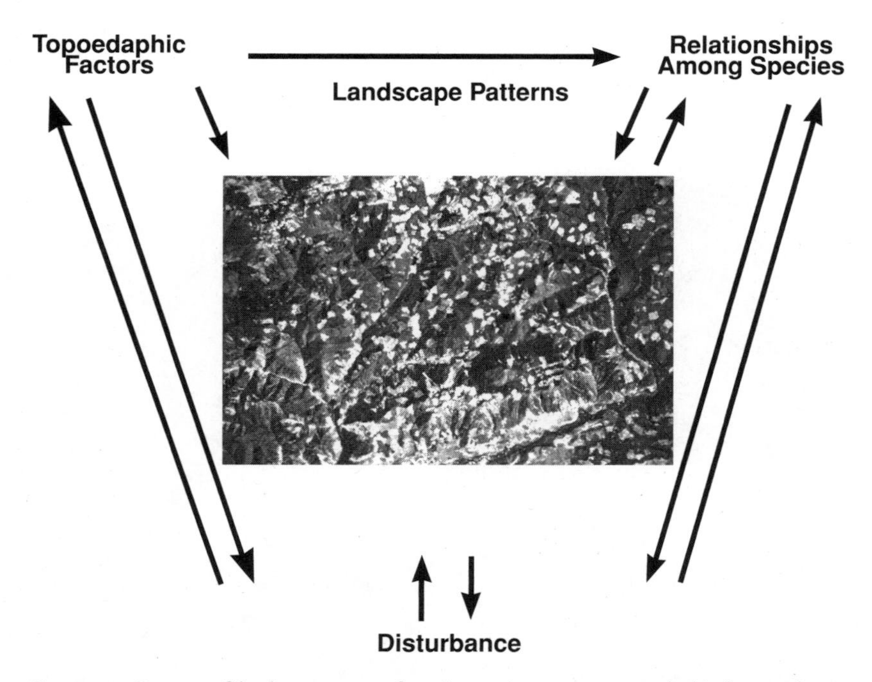

Figure 5.15. Patterns of landscape emerge from interactions among topoedaphic factors, disturbances, and relationships among species. The inset shows satellite imagery of a 2,600-km^2 area of the central Cascade Mountains. The change in this case is dark colors are old-growth forest, light colors are young plantations established by clearcutting. (Inset from Spies et al. 1994)

formed either partly or entirely from volcanic ash, which modifies or eliminates the influence of bedrock.

When disturbance is superimposed on all of this, the emergent landscape pattern may be quite complicated indeed. Earlier, for example, we saw that the distribution of rain forest and sclerophyllous forest in Australia was closely tied to soil phosphorus content. However rain-forest trees are quite sensitive to fire, whereas sclerophyllous vegetation tolerates fire. Therefore, areas that burn frequently do not support rain forest, regardless of soil type (Webb 1968). A very similar pattern occurs in Africa, where long-term experiments show that rain-forest trees invade dry miombo woodlands when fire is excluded (Trapnell 1959).

This chapter opened with a walk through a landscape mosaic of the Big Horn Mountains. We close this section with a more abbreviated look at the forest mosaic in two areas. This is to reinforce the general nature of landscapes: wherever we look, especially in mountains but also in lowland forests, we find complexity.

5.3.1 Sarawak, Borneo

Figure 5.16 shows a **catena** (i.e., a profile of soils and vegetation as related to topography) in the Mulu National Park of Sarawak, on the island of Borneo, in Malaysia. Because the interbedded layers of sedimentary bedrock angle downward, erosion has produced very different slopes on either side of the hill to the left. In the center of the figure, eroded material from the hill has combined with alluvium deposited during the Pleistocene geologic period (when the river on the right was much higher) to form a substrate composed of various rock sizes (indicated by the triangles). The most productive, species-rich forests in this catena occur where soils are well, but not excessively, drained (numbers 1, 5–7, 13, 15, 16, 18). Impeded drainage in zones 2–4, 8–12, 14, and 17 result in slow decomposition of organic matter (because of low oxygen concentration in flooded soils) and, therefore, bog formation. Forests in these zones are less productive and are composed of fewer species than those of adjacent zones.

5.3.2 The Siskiyou Mountains

Some of the more thorough analyses of the effect of environmental factors on local vegetation patterns in the United States were conducted by the ecologist Robert H. Whittaker, who carried out detailed studies of three mountain ranges, the Great Smokies of North Carolina and Tennessee, the Santa Catalinas of Arizona, and the Siskiyous of southern Oregon and northern California (Whittaker 1956; Whittaker and Niering 1965, 1968; Whittaker 1967). The Siskiyous are particularly interesting because of the complexity of their rock types. As with the Big Horns, a walk through the Siskiyous would take you through a forest mosaic whose patches correspond to interactions among elevation, aspect, and, especially, rock type. Consider the two rocks shown in fig. 5.17. On soils formed from quartz diorite (which are sandy and moderately productive), the drought-tolerant forests reach their highest elevation on ridges and southern slopes,

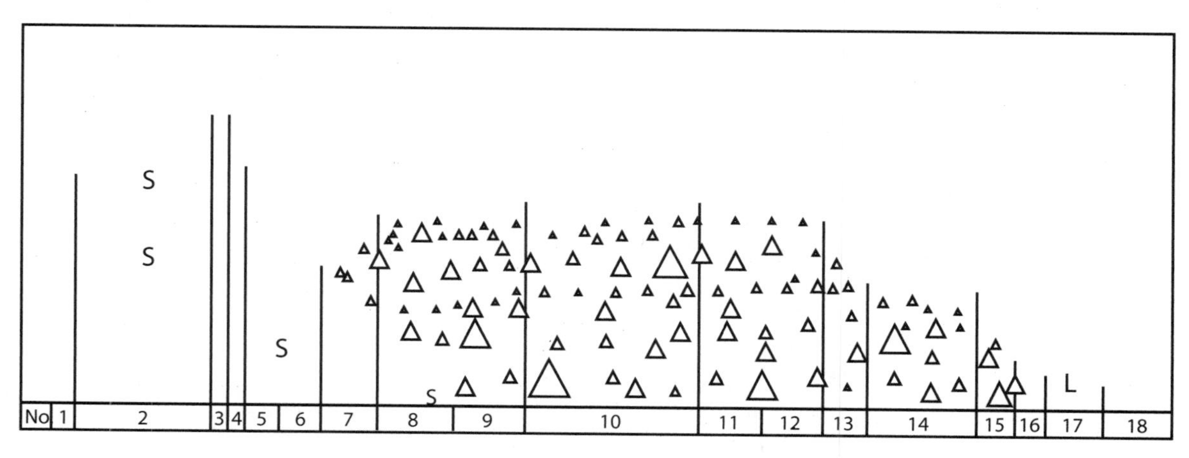

Figure 5.16. The relationship between substrate, topography, and forest type in an area of the Mulu National Park, Sarawk, Malaysia. The hill to the left is composed of interbedded layers of sedimentary rocks (S), which because of their orientation have eroded differently on either side of the hill. The hill in the center is composed of material that has eroded from the hill to the left, plus rocks of various sizes *(triangles)* that were deposited as alluvium in the geologic past, when the river at the right was much higher. Numbers along the bottom denote different types of forest. (Adapted from Brunig 1983)

The Siskiyou Mountains are one of the best examples of how events buried in the geologic past reach forward to influence the landscape patterns that we see today. For the forester, who must understand the characteristics of the landscape to make intelligent management decisions, awareness of geologic history can be an important tool. Like many mountain ranges along sea coasts, the Siskiyous were formed by colliding tectonic plates at the boundary between the continent and the ocean. The collision compressed old layers of sediment along the continental shelf, lifting them to form mountains. In addition, sediments—and occasionally volcanic rocks—from the ocean floor were lifted over the continental plate and added to the growing mountain range. Sections of ocean floor pushed beneath the continental plate during collisions metamorphosed in the earth's interior and sometimes rose to the surface as magma, forming a diversity of bedrock types, depending on the original seafloor composition and the conditions under which metamorphosis occurred. Rock types in the Siskiyous were further diversified by both volcanic eruptions on land and glaciation, resulting in "a geologic nightmare, a chaotic mixture of a wide variety of rocks originally formed at different times, in different ways, and at widely separated places all swept together into a hopelessly confused heap" (Alt and Hyndman 1978). The jumble of rock types produced in this way results in a strikingly diverse landscape, with different rock types producing soils that differ both chemically and physically. These factors, in turn, influence the composition of plant communities; the Siskiyous support the most diverse conifer forests in the world.

Not all mountains (not even all coast ranges) are as geologically diverse as the Siskiyous. For example, both the Cascades and the Andes were formed when seafloor plunging beneath the continental crust melted and returned to the surface during volcanic activity. The result is a rather uniform bedrock. This is a general phenomenon: wherever coastal ranges form from the collision of plates, they are paralleled by volcanic ranges about 150 km inland (Alt and Hyndman 1978).

but mountaintops are occupied by subalpine forests regardless of topographic position (fig. 5.17A). On the highly infertile serpentine soils, however, open pine woodlands extend to the summit on ridges and southern aspects, and even northern aspects at midelevations are occupied by an open forest-shrub complex. Subalpine forests on serpentine occur only on northern aspects (fig. 5.17B).

5.4 VEGETATION CLASSIFICATION

To make good management decisions, foresters must understand both the opportunities and constraints presented by the environments in which they work. For example, rotation lengths and harvest volumes depend on knowledge of potential site productivity. Also, regeneration techniques may vary depending on whether a site is mesic, droughty, or frost prone. Because it is usually too time-consuming and expensive to directly measure all the environmental variables that may be relevant to tree growth on a given site, foresters often use the plant community as an indicator of the physical environment.

Vegetation is widely used as an environmental indicator in western North America, where the term **habitat type** is

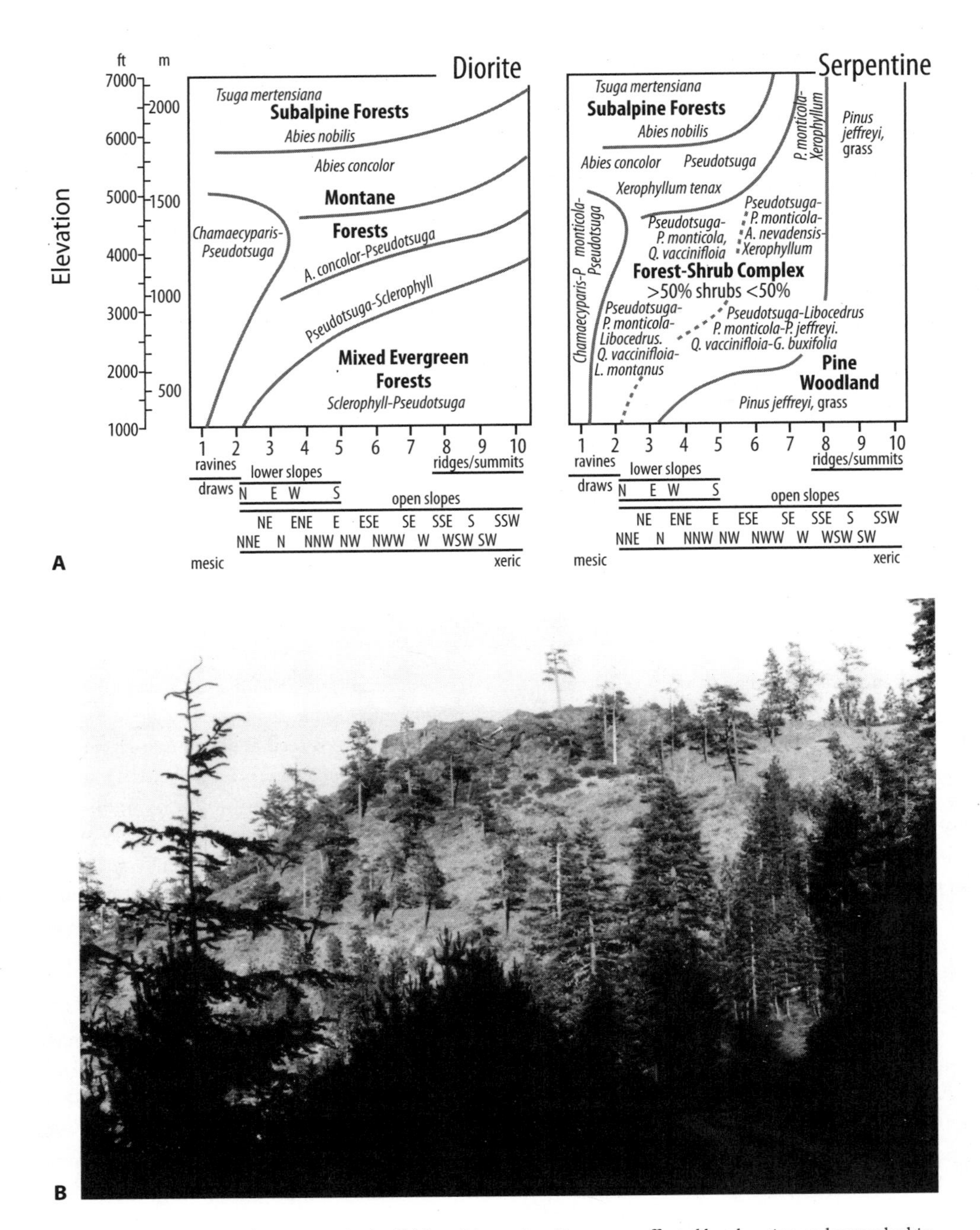

Figure 5.17. A, Patterns of vegetation in the Siskiyou Mountains, Oregon, as affected by elevation and topoedaphic factors. Soils formed from diorite are moderately productive; those formed from serpentine are unproductive because of imbalanced nutrient content. B, High-elevation forests on serpentine soils, Siskiyou Mountains, Oregon. (Whittaker 1960. Reprinted with permission from Ecological Society of America)

most frequently used to denote "an aggregation of all land areas potentially capable of producing similar plant communities" (Pfister et al. 1977). Two things in particular should be noted about this definition. First, habitat type refers to a particular land type and its accompanying environment, not to the vegetation; vegetation is merely an indicator (just as the rise of mercury in a glass tube is an indicator of temperature rather than the temperature itself). Second, the word *potential* means that the mix of plant species occupying a site at a given time is not necessarily indicative of the habitat type. This is because, as we have seen, landscape patterns are a function of two things: environment and disturbance. If we wish to use vegetation to classify environments, we must not let patterns that are induced by disturbance confuse the issue. Therefore *potential* refers to the vegetation that would occupy a site in the absence of disturbance. As we shall see in chapter 8, this is referred to as the **climax vegetation.**

In forests, a particular habitat type is identified by both the climax tree species and one or more understory species. For example, in northern Idaho, ponderosa pine occurs with the grass *Stipa comata* on very dry sites. The presence of these two species is used to denote the *Pinus ponderosa*/*Stipa comata*, or Pipo/Stco, habitat type. With increasing moisture, various other grasses and shrubs replace *Stipa comata*. For example, the presence of the shrub *Symphorocarpus albus* in association with ponderosa pine denotes the Pipo/Syal habitat type. As moisture increases further, probably coupled with increasing elevation and decreasing temperatures, ponderosa pine is replaced by climax Douglas-fir, perhaps in association with the grass *Calamogrostis rubescens*, in which case the habitat type is classed as Psme/Caru. Western hemlock *(Tsuga heterophylla)* is the climax tree on the most mesic sites in northern Idaho and always occurs with the shrub *Pachistima myrsinites*, hence the habitat type Tshe/Pamy. Douglas-fir is often the dominant tree on Tshe/Pamy habitat types, but this reflects past disturbance. A look at species composition of the understory tree layer shows that in the absence of disturbance, western hemlock will eventually replace Douglas-fir on these mesic sites (Douglas-fir is climax on drier sites).

Figure 5.18 and table 5.2 show some of the ways that habitat types are used to guide forest management in the northern Rocky Mountains. Figure 5.18 compares potential productivity (termed *yield capability*) among habitat types of northwestern Montana. The mesic Tshe/Clun (western hemlock/*Clintonia uniflora*), Thpl(western red cedar)/Clun, and Abgr(grand fir)/Clun habitat types are the most productive, while the dry ponderosa pine and climax Douglas-fir types and the cold (high-elevation) types dominated by subalpine fir (Abla) are least productive. Not all sites with subalpine fir are unproductive, however, and the importance of understory species in habitat classification

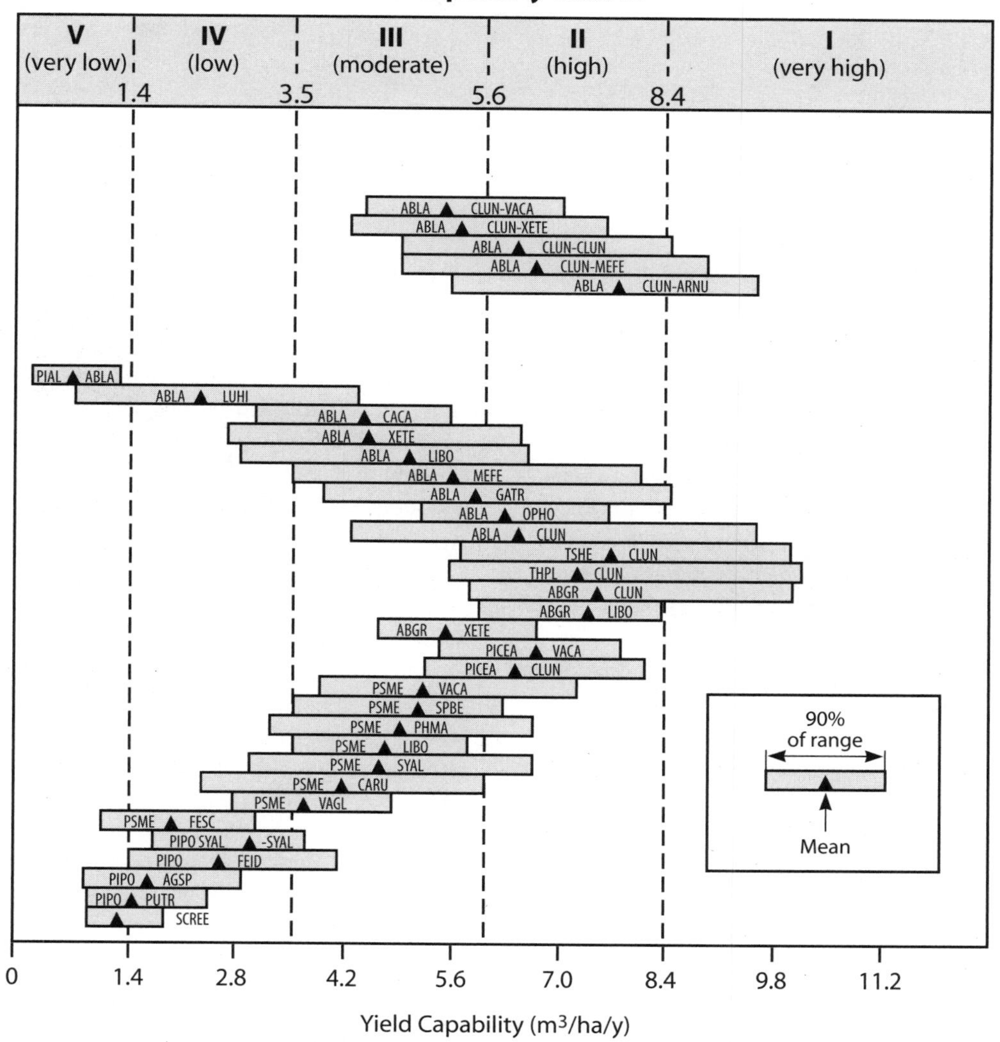

Figure 5.18. The yield capability classes of different habitat types in northwestern Montana. Yield capability refers to the potential wood production of fully stocked forests. (See text for explanation of the designations of habitat type.) (Pfister et al. 1977)

Table 5.2
The potential of northern Idaho habitat types to produce different resource types

	Resource potential ratings*				
	Timber	Big-game forage	Domestic-livestock forage	Water yield	Esthetic recovery rate
Pinus ponderosa/Stipa comata	2	1	6	1	1
Pinus ponderosa/Agropyron spicatum	2	2	10	1	3
Pinus ponderosa/Festuca idahoensis	3	3	10	1	3
Pinus ponderosa/Purshia tridentata	3	5	7	2	4
Pinus ponderosa/Symphoricarpos albus	5	4	8	2	5
Pinus ponderosa/Pbysocarpus malvaceus	6	7	7	3	7
Pseudotsuga menziesii/Symphoricarpos albus	6	5	7	3	6
Pseudotsuga menziesii/Physocarpus malvaceus	8	8	4	4	8
Pseudotsuga menziesii/Calamagrostis rubescens	6	4	5	4	3
Abies grandis/Pachistima myrsinites	9	10	3	5	10
Thuja plicata/Pachistima myrsinites	9	10	2	6	10
Thuja plicata/Athyriumjilix-femina	7	6	1	?	9
Thuja plicata/Oplopanax horridum	7	6	1	?	9
Tsuga heterophylla/Pachistitna myrs-mites	10	10	2	7	10
Abies lasiocarpa/Pachistima myrs-mites	7	10	2	8	9
Abies lasiocarpa/Xerophyllum tenax	4	4	1	6	4
Abies lasiocarpa/Menziesia ferruginea	6	6	1	10	8
Tsuga mertensiana/Xerophyllum tenax	3	4	1	10	8
Tsuga mertensiana/Menziesia ferruginea	5	6	1	10	8
Abies lasiocarpa/Vaccinium scoparium	4	4	3	8	2
Pinus albicaulis/Abies lasiocarpa	1	2	4	10	1

Source: Adapted from Pfister 1976.
*Ratings range from very low (1) to very high (10).

becomes apparent in comparing productivity of Abla/Luhi types with those of Abla/Clun.

Table 5.2 shows a ranking of northern Idaho habitat types with regard to their production of various values: timber, wildlife, domestic livestock, water, and aesthetics. Rankings such as these can be used to guide management strategies. For example, Tshe/Pamy types have high timber yields, produce good wildlife forage, and recover rapidly from disturbance. Other factors being equal, this habitat type might be managed for timber and forage. In contrast, Abla/Vasc types produce relatively little timber or forage but have high water yields. Moreover they do not recover quickly from disturbance. Silvicultural prescriptions in Abla/Vasc types would probably prioritize watershed protection above wood and forage. Regardless of management emphasis, however, modern forestry strives to protect and foster all values produced by forests, a sometimes difficult challenge that requires landscape and regional perspectives (Kohm and Franklin 1997; Lindenmayer and Franklin 2002). We discuss this issue in some depth in chapter 23.

The concept of habitat types—that certain plant species are reliable indicators of environment—is probably applicable to forests anywhere in the world. Similar systems are used throughout western North America and are gaining popularity elsewhere. Pregitzer and Barnes (1982), for example, have identified sets of understory plant species they call *ecological species groups* that reliably indicate soil drainage, texture, and fertility. It is important to realize, however, that habitat types identified in one area are unlikely to correspond to those occurring in other areas with different sets of environmental conditions.

5.5 SUMMARY

Landscape patterns are produced by a combination of disturbance, which usually initiates a temporal sequence of differing plant species, and local variation in topography and soils (topoedaphic), which determines the particular set of species most likely to grow to maturity. Topography influences vegetation type through changes in elevation, aspect, and slope steepness. Soils act on vegetation via their physical structure and chemistry. Topography, soils, disturbance, and plants interact in complex ways to create much of the pattern in nature. In the absence of disturbance, the species composition of plant communities in a particular area reflects underlying topoedaphic factors; therefore, plants are often used as indicators of particular environmental conditions.

Change in Time

An Overview

> No matter what forms we observe, but particularly in the organic, we shall find nowhere anything enduring, resting, completed, but rather that everything is in continuous motion.
>
> GOETHE 1790

The Chobe National Park in Botswana is typical of the dry forest savannas that cover much of eastern and southern Africa: open woodlands mixed with grasses, supporting what may be the most diverse vertebrate communities in the world. Although protected from exploitation by humans, the acacia woodlands that dominate the park are in a state of severe decline (Mosugelo et al. 2002). The proximate reason can be found within the ecosystem itself: high numbers of elephants and other herbivorous animals are overbrowsing the acacia trees, resulting in severe damage to mature trees and death of young seedlings. Without young seedlings to replace older trees, density of acacia within the park is dropping sharply.

Elephants and acacia trees have coexisted in Africa for eons, why does this coexistence seem to have gone awry? Has it really gone awry, or are we seeing one stage of a longer natural cycle? At least part of the answer lies in history. In the latter part of the last century a combination of drought and disease sharply reduced populations of elephants and other browsing animals within the park. Many of the elephants migrated in search of greener pastures. Without browsing pressure, acacia trees increased their cover, resulting in the extensive stands that existed when the park was established in the early 1900s.

During the twentieth century, populations of elephants and other browsing animals recovered in Chobe National Park, peaking about 1960. With this recovery, acacia began to decline again (fig. 6.1). In the future, drought, disease, or perhaps some other phenomenon is likely to again reduce populations of browsers and acacia will recover.

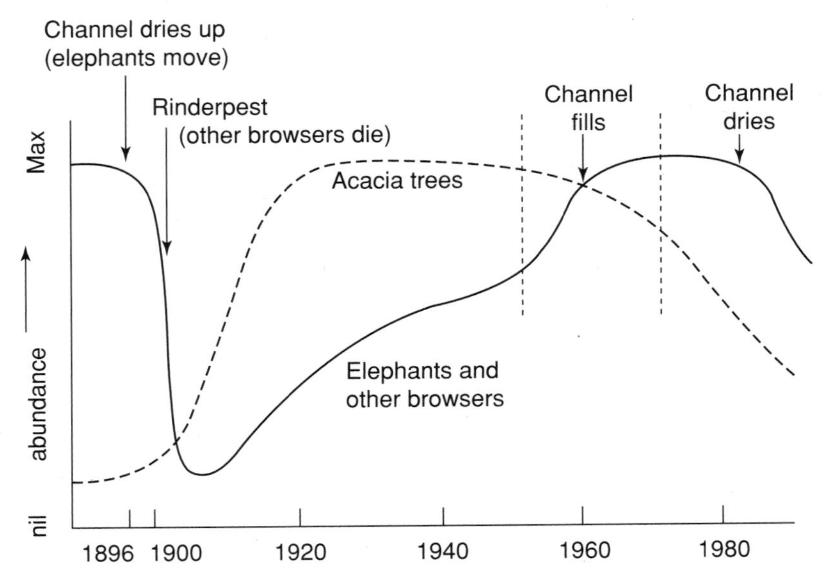

Figure 6.1. Change in the relative abundance of elephants and acacia trees in Chobe National Park, Botswana. (Lewin 1986)

However, the cycle is unlikely to repeat, at least in the same form as it took in the past. This is because the twentieth century has seen a phenomenal expansion in the numbers of an animal that completely excludes elephants and most other large grazers and browsers from much of their former territory, greatly reducing the ability of wildlife to follow natural historic migration routes (as elephants did in order to find feeding grounds). That animal is, of course, the human. The result is that a natural cycle that once played out over a very large landscape is now compressed into a much smaller one. Adding to that is the fact that like all places in the world, southern Africa faces a future climate regime likely to be quite different than any in the past several thousand years. The past is no longer a reliable guide to the future.

The history of elephants and acacia trees in Chobe National Park illustrates a pattern that many ecologists believe is common to most, if not all, ecosystems. Whereas local ecosystems were once thought to exist in **equilibrium**—that is, the populations that composed any given system remained constant unless disturbed by some outside influence—the prevailing view now is that the various species composing a local ecosystem system rarely exist in equilibrium with one another, but rather in a disequilibrium reflecting a combination of historical events, current environmental conditions, and interactions among the organisms themselves.

6.1 EARTH MUSIC

Virtually all ecosystems exist within a matrix of fluctuating environments punctuated by periodic disturbances ranging from mild to severe. Like a complex piece of music, the earth pulses with rhythms of varying strength and periodicity, a phenomenon that most commonly (but not exclusively) manifests as long- and short-term variations in climate, which, in turn, influence the dynamics of the biosphere. Seasonal patterns of temperature and precipitation shape evolutionary adaptations and constrain life histories. Moon cycles drive tides and elicit behavioral responses in organisms. Cyclic variations in solar intensity alter evaporation from the oceans, producing corresponding fluctuations in the monsoon rains that water large areas of the tropics (Wang et al. 2005). Glacial-interglacial cycles drive large-scale migrations, not only in the north, but in the tropics as well (Mayle et al. 2000). A recently discovered 62 million year cycle in extinctions is currently unexplained (Rohde and Muller 2005).

During the closing years of the twentieth century and continuing to the present, scientific knowledge (if not understanding) of planetary dynamics has increased exponentially. It is now known that periodic reorganizations of the ocean-atmosphere system produce distinct climate modes that ramify to alter conditions throughout the globe. The El Niño–Southern Oscillation, operating on a roughly three- to six-year cycle, has profound effects on ecosystems, inducing drought (and therefore fire), triggering shifts in vegetation types, and producing changes in net primary production that ramify through food chains (Holmgren et al. 2001). Basin-scale changes in the Atlantic Ocean drive multidecadal fluctuations in the summer climates of both Europe and North America (Sutton and Hodson 2005). Other ocean-atmosphere modes include the Pacific Decadal Oscillation, the North Atlantic Oscillation, the Arctic Oscillation, and the Antarctic Oscillation, each operating on a unique time scale but also interacting with the others to produce synchronized climate fluctuations in regions that may be far apart (Wang and Schimel 2003).

Local populations of one or more species commonly disappear for a time, but the species do not become extinct because persistence elsewhere provides the potential to return to former haunts once the conditions are right. Elephants left Chobe National Park for a period, but they came back; fire may replace a forest with a herbaceous community, but the site will return to forest at some point in the future. Long-term climatic cycles are commonly accompanied by regional shuffling of species.

But the game has now changed. Humans have preempted much of the stage on which nature's drama played out. Like the African elephant, many species are increasingly relegated to parks and other natural preserves—islands within a sea of farms, cities, and intensively managed forests. Whether nature's play can remain viable on a much smaller stage is a central question facing ecologists, conservation biologists, and land managers. Increasingly, natural scientists are coming to believe that the strategy of conserving biological diversity on preserves scattered within a highly managed, homogenized landscape is doomed to fail: natural diversity, in the emerging view, must be protected within the managed landscape, as well as in preserves. This is an issue of central importance to land managers, because it underlies at least in part the recent movement in the United States and elsewhere toward a more ecologically sensitive type of forest management (e.g., Kohm and Franklin 1997; Lindenmayer and Franklin 2002). Understanding the arguments behind this view requires considerable knowledge about how ecosystems work and will be deferred to the final chapters dealing with ecosystem stability and the future.

The following two chapters deal with two primary processes of change in natural ecosystems: disturbance and succession. Disturbance and succession are common elements of all terrestrial ecosystems, and the character of ecosystems results from interplay between these two forces of change. Understanding the role played by these natural forces is important for resource managers, who, to manage wisely, must understand the historic forces that have shaped ecosystems. Even in parks and wilderness areas, as illustrated in Chobe National Park, it is important to understand that stability may depend on periodic disturbance, and that preempting the natural disturbance regimes can in the long run produce unwanted change.

Disturbance occurs at many intensities and scales, from the death of a single, old dominant tree to a catastrophic forest fire to a volcanic eruption. Disturbance is not necessarily a negative factor in ecosystems; more often it is an agent of renewal and rejuvenation. However, under some circumstances, a disturbance can catalyze threshold changes in species composition that are not successional in nature but relatively permanent; this is especially true when the disturbance is one to which the species occupying the system are poorly adapted. This phenomenon, which amounts to a destabilization of the ecosystem, is discussed in chapter 20.

Many forest species persist only because of periodic disturbance. Without the windthrow that exposes mineral soil seedbeds, some northern forests convert to bogs. As we discuss further in chapter 10, some people believe that much of the rich diversity of moist tropical forests reflects historic patterns of shifting cultivation. Numerous forest types across the globe, including pine, eucalypt, and tropical woodland, are maintained by periodic fire.

Fire suppression in the ponderosa pine forests of western North America provides a classic case study of the importance of periodic disturbance to the stability of some ecosystems. Eliminating fire in these systems allowed true fir to establish under the ponderosa pine (historically, periodic fire had destroyed establishing fir seedlings). Many forest entomologists believe that fir, not as drought tolerant as ponderosa pine, becomes stressed on these sites and, hence, is susceptible to defoliating insects (the mechanisms by which trees defend themselves against insects are discussed in chapter 19). Not only are the fir defoliated, they may serve as foci for the spread of insect populations into habitats that would not otherwise have been successfully attacked. In this view, fire suppression destabilized not only the sites on which fire was suppressed, but a whole region.

By definition, disturbance initiates a change in community structure, though the degree of change varies widely depending on the nature and severity of the disturbance. Perhaps the most signature trademark of disturbances to forests is the death of dominant trees, which allows light to penetrate and encourages growth of herbs, shrubs, and subdominant trees. New plants may establish, especially annuals and light-demanding perennials, creating a diverse plant community composed of some newcomers and some old-timers that persist as legacies of the predisturbance forest. While that is the most common scenario, in other cases, such as the relatively gentle ground fires that characterize dry forests in interior North America, disturbances spare the dominant trees and kill small trees and shrubs, changing structure from below rather than above.

Naturalists have long observed that the change in community structure initiated by disturbance is not permanent. Trees grow and eventually reassert dominance. Depending on the density of the overstory and the shade tolerance of individual species, plants that dominated the postdisturbance community either drop out or persist in the understory. Eventually, composition of the tree layer may change from species that cannot reproduce in their own shade, to species whose seedlings can tolerate shade. This sequence of change is called **succession.**

Disturbance and succession are two sides of the same coin: both produce change in community properties but differ in the direction of change that each produces. Hence the clearest way to understand one is in terms of the other: succession imposes a certain pattern—a direction—on ecosystems; disturbance, which might just as well be termed *antisuccession,* is

Although in theory the concepts of disturbance, succession, and stability seem simple and well grounded in observations of how nature works, in practice they are quite elusive, and one must be suspicious of hard-and-fast definitions because they often turn out to be arbitrary and misleading. Bormann and Likens (1979) conclude, "Defining 'disturbance' itself can be something of a problem, because it is difficult to draw a line between . . . events that come within the scope of normal plant succession (for example, the fall of old trees), and . . . events that might be considered to deflect (succession)." Matters are no clearer with succession, which over the years has probably been the single most controversial topic in ecology. The situation is succinctly stated by West et al. (1981), "No two ecologists seem to have a unified concept of the details of succession."

One reason for this fuzziness is that change in ecosystems is too complex and variable to be adequately explained by a few simple concepts. As anyone who works with natural systems soon discovers, nature seldom lends itself to the level of clear definition that modern humans have come to expect: as Highwater (1981) points out, we live in a "multiverse" rather than a "universe." This may be the most important thing that students and resource managers have to learn.

an event or chain of events that pushes the system in the opposite direction. To clarify this, we shall first examine the nature of the pattern imposed during succession.

Envision a landscape of bare rocks exposed by a retreating glacier. This is an inhospitable environment for life. It has no soil, hence no water and nutrient storage capacity to support plants. Animals find little shelter and no food. Some life is adapted to such conditions, however. Lichens colonize the rocks, obtaining nitrogen directly from the air and other nutrients by releasing acids that break down the rock. They scrounge water from cracks in the rock. Lichens provide a food base for animals and, mixing their own organic matter with the products of rock weathering, slowly build a soil, which allows higher plants to establish. These, along with the set of animals and microbes that accompanies them, further modify the environment, casting shade and building litter layers, resulting in yet another set of plants and animals—adapted to the new conditions—coming to occupy the site.

This sequence represents what is termed **primary succession;** the term *primary* is applied because the sequence started in an environment—bare rock—with no prior biological influence.[1] It is an idealized picture—some higher plants, including trees, are quite capable of colonizing fields of bare rock and do not have to wait on lichens (fig. 6.2)—but nevertheless illustrates a common pattern in primary succession:

1. A past disturbance (glaciation in the example above, but it could be something else, e.g., volcanic eruption) wipes out most or all traces of life on a site.
2. A set of organisms adapted to survive and reproduce in these primary conditions becomes established, the colonizing plants often characterized by an ability to extract nitrogen directly from the atmosphere and other nutrient elements directly from rock (nutrient requirements of plants and nutrient cycling are discussed in detail in chapters 16 through 18).
3. Colonizing organisms modify the site, accumulating nutrients and building soil, thus creating the conditions that permit a second wave of organisms to establish.
4. Over many years, the site becomes increasingly modified by the biotic communities that occupy it. That combination of minerals, dead organic matter, complex biochemical molecules, and living organisms that we call soil continues to be built, litter accumulates, and—particularly in climates capable of supporting trees—the accumulation of leaf area increasingly shades and buffers the interior of the community from environmental extremes.
5. As one set of organisms modifies the site, it is replaced by another set better adapted to the new conditions. Nitrogen-fixing plants become supplanted by non-nitrogen-fixers. Plants that require high light levels to survive are replaced by plants that can establish in shade.
6. Barring another disturbance, a relatively persistent community eventually comes to occupy the site; in the case of forests the persistent community is often, but not always, dominated by tree species that are able to reproduce in shade. The qualifying term *relatively* must be taken seriously when applied to the persistence of late successional stages. As we saw in the example of Chobe National Park, an unchanging community is seldom, if ever, attained, at least in forest ecosystems.

This theme has many variations, but one feature is common to all primary successional sequences: primary succession involves a progressive imprinting of biological features onto a physical landscape. The first and most important biological imprint is the soil. Soil is the product of organisms acting on rocks and contains elements of both. Early colonizers that initiate soil formation pave the way for all that follow. As succession proceeds, accumulating biomass imprints a site in various ways. Leaves intercept light, thereby restricting establishment of new plants to those that can tolerate shade. The forest canopy also moderates temperature fluctuations within the stand. Soils accumulate increasing amounts of both living and dead organic matter, and in boreal, montane, and many temperate forests, slow decomposition results in accumulations of dead organic matter on

[1]This statement is not entirely correct; some types of rocks, particularly those formed from sediments, do carry the imprints of ancient life in the form of carbon, nitrogen, and such things as fossil shells.

the forest floor. This, in turn, decreases the nutrients available for new plant growth, because nutrients contained in leaves, branches, and fallen trees are not released through decomposition. Numerous other changes that may occur during succession are discussed in chapter 8.

Figure 6.2. Ponderosa pine, grasses, and lichens colonizing lava in El Malpais National Park, New Mexico.

Succession, then, is a progressive change in the composition of the community on the site, accompanied by an accumulation of biological imprints. Disturbance is an event, or a chain of events, that to one degree or another alters the biological imprints on a site (e.g., reducing dominance, transforming living to dead), and in doing so shifts the ecosystem from later to earlier stages of succession. Succession following a disturbance that removes some but not all of the biological imprints on a site is called **secondary succession.**

Disturbances that wipe the slate completely clean of biological imprints are uncommon. Even the catastrophic eruption of Mount St. Helens in 1980 left multiple biological legacies (chapter 8). However, disturbances that alter biotic influences sufficiently to initiate secondary succession are quite common; fig. 6.3 shows a typical successional sequence following stand-destroying wildfire in a western conifer forest. Crown fires, windthrow, insect epidemics, and clear-cutting all kill overstory trees and create the conditions that allow early successional plants to establish. Fires consume dead organic matter and thereby increase the availability of some nutrients. Landslides, erosion, and some kinds of logging practices remove soil—the first and most important of the biological imprints on a site—sometimes to the point of initiating primary succession. Floods and volcanoes deposit a layer of new minerals or soil rather than removing the old. Other disturbances, such as the fall of a single tree, may be quite gentle yet initiate pockets of early successional vegetation.

Disturbance and succession define a certain trajectory of change in time. Although species composition and other factors such as productivity differ at different points along this trajectory, the trajectory itself tends to remain stable. If a Douglas-fir forest is destroyed by a fire—and many have

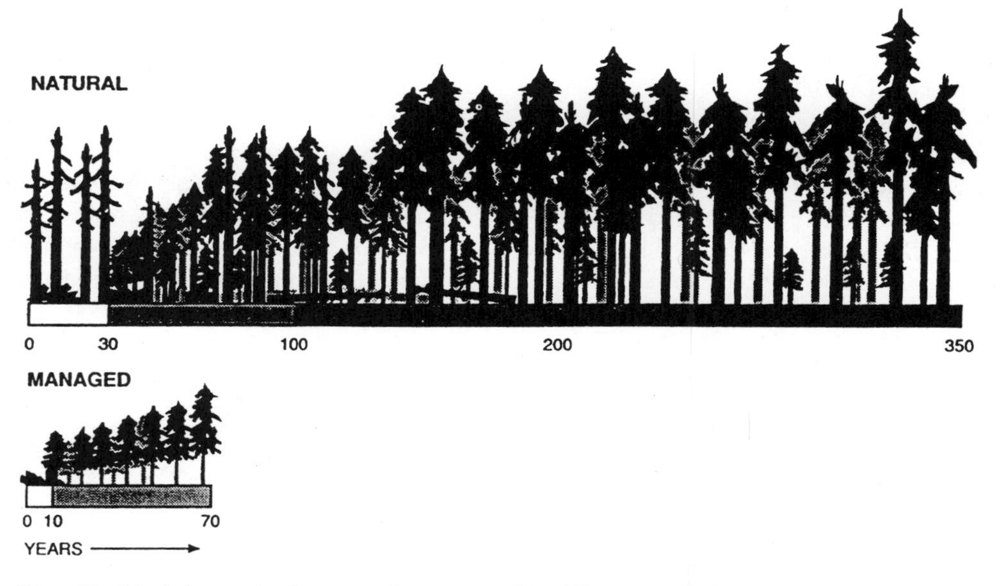

Figure 6.3. A typical successional sequence after a catastrophic wildfire in a conifer forest. The average age span of forests that are periodically destroyed by wildfire is determined by the average interval separating catastrophic wildfires, which ranges from approximately 100 years in some boreal and montane forests to over 1,000 years in some ancient forests of the Pacific Northwest of North America. The inset shows a typical age span for a forest managed primarily for wood production. (Franklin et al. 1989)

been in the past—we expect it to eventually return to Douglas-fir, not ponderosa pine or canyon live oak. This is an expectation of stability, that despite periodic disturbance a given ecosystem will retain the ability to reassemble the same set of species at the same level of productivity. Does such **trajectory stability,** as Orians (1975) has called it, really exist? The answer is, within limits, yes. Among the key questions taken up in chapters 20 through 23 are what determines the limits of stability, and what it takes to exceed them. With a significant change in climate, for example, can we really expect the disturbed Douglas-fir forest to return to Douglas-fir, or will disturbance be the trigger that initiates some different trajectory that comprises different sets of species?

This brings us to a final point: ecosystem changes also occur on the scale of hundreds and thousands of years, driven by changes in climate. Climate acts as a background variable that drives and conditions relationships within the biota and ultimately determines the ability of any given species to persist on a site. It was climate that through its impact on elephants, produced a resurgence of acacia in Chobe National Park, and climate that brought the elephants back. Whether this cycle repeats depends, among other things, on future climatic patterns.

We know that climate has fluctuated rather dramatically throughout geologic history, the ice ages being the prime example; and we also know that these fluctuations have significantly altered the extent and species composition of the earth's forests. Historic changes in pollen accumulation at a given site (the **pollen record**)—usually studied in lake sediments or bogs, because these are stable environments that preserve the temporal sequence of pollen accumulation—clearly show that climatic fluctuations can produce long-term, semipermanent alterations in the species occupying a given site (Delcourt et al. 1983; Mayle et al. 2000; Davis and Shaw 2001; Hughen et al. 2004).

Eighteen thousand years ago, the southern boundary of the continental ice sheet lay across what is now the northern United States, and boreal forests of spruce and jack pine reached well into the south. By 10,000 years before present (BP) the ice sheet had retreated, with jack pine rapidly occupying former tundra (Davis 1981), and mixed conifer-hardwood or pure hardwood forests replacing boreal forests over much of the east. A very warm and dry period from 8000 to 4000 BP produced major changes in vegetation patterns: hardwoods pushed far north, prairie far east, and pines dominated the southeast. The cooling trend beginning about 4000 BP readjusted these boundaries to roughly those we see today.

Similar changes occurred in western North America. Cwynar (1987) studied the sequence of forest types occurring in the northern Cascades since withdrawal of the most recent glaciers (somewhat over 12,000 years ago in that area). Three scales of change can be distinguished since withdrawal of the most recent glaciers. The first corresponds to relatively long-term fluctuations in temperature and precipitation, the second to a primary successional sequence initiated by glacial retreat, the third to secondary successional sequences arising from particular disturbance regimes. Within this milieu, a given community type persisted only so long as climatic fluctuations, site characteristics, and the disturbance regime—itself at least partially determined by climate—remained within the adaptational sphere of the species composing the community. The pioneering communities were open forests of lodgepole pine, cottonwood, Sitka alder, and mountain hemlock, with Sitka spruce, red alder, and subalpine fir appearing about 12,000 BP. A major warming trend beginning around 8000 BP was accompanied by both the appearance of Douglas-fir and an increase in fire frequency. With the return to a cooler, moister climate around 4000 BP, fire frequency declined, and community composition shifted to a combination of western hemlock, red cedar, and Douglas-fir, a forest type that persists today.

Milankovitch astronomic forcing, which is believed to be largely responsible for glacial-interglacial cycles at high latitudes, also influences ocean-atmosphere interactions that determine the boundaries between moist tropical forests and savannas or grasslands (Mayle et al. 2000; Hughen et al. 2004). Mayle et al. (2000) found that rain forests of the western Amazon have been expanding southward for the past 3,000 years, and that "their present-day limit represents the southernmost extent of Amazonian rainforest for at least the past 50,000 years."

Nature is dynamic, but it does not follow that ecological communities are tossed about willy-nilly by events out of their control. For example, migration of Amazonian rain forest during Northern Hemisphere deglaciation 10,000 to 15,000 years ago lagged behind climatic shifts by several decades, which Hughen et al. (2004) interpret as buffering capacity on the part of the forest. Any viable biological system must incorporate mechanisms that stabilize against external fluctuations; this is just as true of ecosystems as it is of organisms, although the open nature of ecosystems and their dynamically shifting species composition makes the issue of ecosystem stability much more complicated than that of organismic stability. The adaptational sphere—that set of environmental conditions to which the species that compose a given community is adapted—is usually quite broad. Unfortunately, however, the mechanisms by which systems stabilize themselves are poorly understood (chapter 20).

6.2 SUMMARY

Change is a common element of all forests. Disturbances that have occurred periodically in the past and to which species are adapted commonly renew and invigorate ecosystems, and by initiating changes in community composition enhance the diversity of landscapes. Disturbances such as wildfire and windthrow seldom engender permanent change in community composition; rather, they initiate changes in the structure and composition of ecological communities that eventually lead back toward some reasonable approximation of the original. This sequence is called succession. On the other hand, relatively long term changes can and do occur following some disturbances, and particularly as a result of climate change. In such cases the mechanisms by which a given set of successional communities stabilizes itself on a site fail, resulting in replacement by another set of communities. Because forestry involves disturbance, and because stability of the global ecosystem is tied to its forests, understanding the mechanisms of change and stability is critically important for foresters.

7

Disturbance in Forest Ecosystems

7.1 THE COMPLEX NATURE OF DISTURBANCE

Disturbances are often divided into two general types: exogenous, those originating outside the ecosystem (e.g., fire, wind, pollutants, introduced animals or pathogens), and endogenous, those originating from within the system (e.g., epidemics of native insects or pathogens). In practice it is usually difficult to separate the two. An endogenous factor, such as senescence of an old tree, which is often exacerbated by pathogens and insects, is likely to make that tree more vulnerable to external physical factors (e.g., wind). Conversely, endogenous disturbances are often triggered by exogenous disturbances. For example, drought or pollution stresses trees, lowering their resistance to native herbivores and pathogens (chapter 19). Abundant fuel (dead wood) produced by insect and pathogen outbreaks increases, in turn, the probability of a catastrophic fire. A more fundamental problem arises when we recall that ecosystems, unlike organisms, seldom have clearly identifiable boundaries: deciding whether something comes from inside or outside a system is not easy when where the inside stops and the outside begins is not clear. The modern view of earth as a single, large ecosystem implies that the only possible exogenous disturbances come from outer space.

The following points should be remembered when thinking about disturbance:

- The disturbance regime of a particular forest usually consists of a complex mixture of infrequent, large-scale events (e.g., a stand-destroying fire or a hurricane) and more-frequent, small-scale events (e.g., gentle ground fires or small gap formation).
- Any given disturbance may be the result of numerous, interconnected factors. Moreover, the disturbance regime of a particular ecosystem cannot be

Figure 7.1. An area in which mature conifers were killed by the root-rot *Phellinus weirii*. Hardwoods, resistant to this fungus, establish readily in such root-rot patches, with the net effect that the forest is diversified. Without a host plant, the root-rot eventually disappears from the patch, allowing conifers to reestablish.

understood in isolation but rather is tied to the systems to which it is linked and, ultimately, to events occurring within the global ecosystem.

- Natural disturbances, rather than being a bad thing to be avoided and prevented, often renew ecosystems and diversify landscapes (fig. 7.1). Foreign disturbances, those for which organisms within the system have no adaptation, are another matter.

This chapter deals with some disturbances that have been common to forests throughout history, including fire, wind, floods, and tectonic activity, and with one relatively new and highly disruptive one: invasive species. Insect and pathogen outbreaks are discussed in chapter 19, pollution in chapter 20, and disturbances associated with logging in various chapters. For several reasons, fire receives a disproportionate share of the attention in this chapter. On a global basis, fire is the most widespread and common disturbance to forests, with significant implications for both forest health and carbon cycling. Fire is a complex issue both environmentally and socially; depending on circumstances, it acts either as an essential agent of renewal or a highly disruptive force. In the United States, there is controversy about what forests are primed to burn outside of their "natural range of variability" and what, if any, active role managers should take in reducing risk of wildfire. Along these lines, it is important to note that the foreignness of a particular disturbance stems not only from the novelty of the disturbing agent (e.g., pollution vs. fire), but from changes in the frequency or intensity of common disturbance agents. For example, fires that burn more frequently or with higher severity than the historic norm may have significant consequences for ecosystem functioning, and excluding fires from fire-adapted forests definitely does (i.e., excluding a native disturbance may in itself constitute a disturbance). This point is explored in more detail in chapter 20.

7.2 FIRE

In the summer of 1986, 36,000 lightning strikes were recorded within a two-week period in a single Oregon county. Despite the high lightning activity, the summer had been dry, and one in a hundred strikes started a fire, many of which were in lodgepole pine stands that had recently experienced a mountain pine beetle outbreak. The combination of dry conditions and large amounts of fuel from beetle-killed timber resulted in extensive, catastrophic fires that burned over many thousands of hectares. This was only a prelude to what was to come, however. Dry conditions extended into the following winter and spring, the seasons in which western North America receives most of its precipitation. By late summer 1987, western forests were again tinder dry, and, on August 30, a series of dry lightning storms ignited more than 1,500 fires throughout California and southern Oregon (fig. 7.2). Fanned by low humidity and high winds, these fires burned over 300,000 ha before fall rains finally put them out. In 1988, a combination of

Figure 7.2. A, On the fire line, Siskiyou National Forest, September 1987. B, Fire patterns on the Siskiyou National Forest, following the 1987 wildfires. Lighter colors are dead trees, darker colors are living trees. In this area, the fire crowned out on southern slopes but not on northern slopes and draws and in riparian zones.

drought and high winds produced the most extensive fires to occur on the Yellowstone Plateau in at least the past 150 years, burning over 570,000 ha, including 45 percent of Yellowstone National Park (Christensen 1989). These were babies, however, compared to those that burned 7 million ha in Siberia and Mongolia in 1987, destroying many villages and killing numerous people, and those in 1982/1983 that damaged approximately 5 million ha of Borneo's forests (Wooster and Strub 2002).

The beat continued through the 1990s and into the new millennium. In 1997, more of the globe's forests burned than at any previous time in recorded history (Laurance and

Figure 7.2. *(Continued)* C, Fire pattern in an area of Yellowstone National Park after the1988 wildfires. Fire killed a swath of trees up the slope, leaving either side intact. (A, courtesy of Mike Amaranthus)

Fearnside 1999), and that may have been exceeded since. Much of what burned in 1997 and 1998 was in the tropics, especially Borneo, where fire once again covered several million hectares, volatilizing somewhere between 13 and 40 percent as much carbon dioxide as released in a typical year by burning fossil fuels (Page et al. 2002). During the same year, satellites documented nearly 45,000 separate fires in Amazonia, most associated with people clearing forest for crops and cows (Laurance and Fearnside 1999). In the early years of the new millennium, both Colorado and Oregon had their largest forest fires on record, and in Siberia, 22 million ha, an area roughly the size of the entire state of Oregon, burned in a single summer (Schiermeier 2005).

Throughout history, fire has been the single most pervasive disturbance in many of the world's forest types. Any forest that experiences periodic dry periods of sufficient length to allow fuels to dry will also burn periodically. While the Indonesian and Siberian fires mentioned above were by any measure huge, they must be put into context. In fire-prone regions, large numbers of wildland fires can, in aggregate, burn over immense areas. For example, between 1997 and 2006, an average of 74,832 fires burned 2.75 million ha of wildland each year in the United States (NIFC 2007). Before the advent of effective fire suppression in the mid-1900s, the numbers were significantly higher. It is important to understand, however, that total fire area can give a misleading impression of how much forest is actually destroyed. Fires commonly burn in a patchy fashion, leaving varying amounts of living trees, burning only through the understory in some areas and perhaps bypassing others altogether. In some forest types, ground fires rather than crown fires are the rule. As we discuss later, both weather and forest structure play critical roles in determining fire behavior.

The composition and structure of many forests is directly due to fire (fig. 7.3). Extensive even-aged stands of early successional, fire-adapted tree species are common in the boreal zone and temperate montane areas—lodgepole pine, jack pine, black spruce, to name a few. These result from **stand-replacement** fires that kill most or all of the trees, followed by successful natural regeneration within a short time period (producing a relatively even aged new forest). In some forest types, such as the Douglas-fir dominated forests of western North America and the montane ash forests of southeastern Australia, variable fire patterns combine (in some cases) with extended regeneration periods to produce a complex age structure at both the stand

"A *large* wildfire is one of great size; the word does not imply, in and of itself, catastrophe or damage of any kind. *High-intensity* and *low-intensity* fires define energy release rates: these are physical descriptors of the fire, not its ecological effects. *High-severity* and *low-severity* fire refer to the ecological effects of fires, usually on the dominant organisms of the ecosystem. Large, high-intensity fires are often fires of high-severity; large, low-intensity fires are usually, but not always, fires of low severity" (Agee 1997).

Figure 7.3. Some of the world's fire-dependent forests. A, Lodgepole pine, western North America. (The naturally regenerating seedlings are establishing beneath a fire-killed stand.) B, Miombo woodlands, eastern Africa.

and the landscape scale. Yet other forest types, such as ponderosa pine in western North America, longleaf pine in eastern North America, and miombo woodlands in Africa, experience frequent ground fires that kill few mature trees but are very important in shaping the character of the ecosystem. Without fire these forest types would not persist, and when fire is excluded (as it has been in the dry forests of western North America) their ecology is significantly affected.

In western North America, forests characterized by a mixed regime of both low- and high-intensity fires lie between the zones of infrequent, severe fires and frequent, gentle fires and may encompass more area than either. Relatively little is known about the mixed regimes beyond broad generalities. Their fire patterns (i.e., proportion of low- and high-intensity fires) are likely to be complex but follow a reasonably predictable gradient from the dry end (mostly low intensity) to the wet end (mostly high intensity). Their boundaries almost certainly shift with mid- and long-term climatic fluctuations.

Figure 7.3. *(Continued)* C, *Eucalyptus* spp., Australia. (A, courtesy of Gary Pitman; B, courtesy of Jummane Maghembe; C, courtesy of Michael Castellano)

Abundant charcoal in the soils of moist tropical forests indicate that even these high-rainfall ecosystems experience periodic fire that corresponds to the infrequent periods of drought (Sanford et al. 1985). The recent megafires in Borneo and the extensive fire activity in the Amazon, however, are modern phenomena related to humans. Aboriginal people used fire as a land management tool for millennia (e.g., Pyne 1995; Kimmerer and Lake 2001), but the current scale of human-related fire is almost certainly unique in history. We return to this point later.

Although often perceived as a disaster to be prevented, fire is usually a healthy thing in ecosystems where it has been common, renewing and maintaining the character of the ecosystem. Like most natural disturbances,[1] fire is a great diversifying agent. Even where stand-replacement fires burn, these are commonly very patchy, killing all trees in some areas and few or none in other areas. The result is a mosaic of vegetation in different stages of succession that greatly enhances landscape diversity and provides an array of habitats for different plants, animals, and microbes. Even gentle ground fires kill occasional mature trees and produce landscape diversity, though the patchiness is at the scale of meters rather than kilometers.

But we must be careful about oversimplification. Particularly in marginal habitats (e.g., cold, droughty, or infertile), wildfires that are too frequent or intense can reduce soil fertility and tip forests into relatively permanent shrublands or tundra. For example, wildfires in northern Quebec have favored the development of tundra at the expense of forest (Sirois and Payette 1991). In Zambia, simply shifting the season of annual burning from the wet to the dry part of the year eventually resulted in dry tropical forest being replaced by shrublands (Trapnell 1959). Moreover, the naturalness of a large, stand-destroying fire must be evaluated not only by the amount of forest it burns but in a regional context as well. No large fire that burns within a region that has been heavily modified by land use can be considered to be natural. For example, in 2002 the Biscuit fire in Oregon destroyed more than 50,000 ha of older forest habitat, an amount that two hundred years ago would have been a drop in the bucket of the vast old-growth forests of the Pacific Northwest. After, a century of logging, however, 50,000 ha represents much more than a drop in the bucket, and it has quite different implications for wide-ranging old-growth associates such as the northern spotted owl.

Excluding fires from forests where they have been common historically has produced quite unexpected and undesirable side effects. Perhaps the most notable example is in the dry ponderosa pine forests of western North America, where controlling the gentle ground fires that characterized these forests allowed a dense understory of small trees to establish. Because the younger trees act as **fire ladders,** carrying ground fires into the crowns of mature trees, the forests have actually become more susceptible to large, stand-destroying fires. Moreover, as we saw in an earlier chapter many of the newly established trees were *Abies* species (firs), which are more susceptible than pines to outbreaks of the defoliator spruce budworm. The consequent increase in available food permitted budworm populations to attain higher levels and persist longer than had been possible in the past (chapter 19).

[1]We use *natural* in this case to denote any disturbance common enough in the history of a forest type to allow species within the system to have become adapted to it. It does not necessarily imply lack of human influence.

Fire results from a conjunction of four factors: (1) an ignition source; (2) fuel, which is often closely tied to stand and landscape structure; (3) weather (especially drought and wind) that facilitates ignition and promotes fire spread; and (4) topography, which modifies both fuel and weather. Lightning and humans are the two major igniters of forest fire; historically throughout the world fire has been used to drive game, to clear land for cultivation, and to stimulate forage growth for domestic livestock. Once ignited, fire behavior depends on weather, fuels, and topography, which are traditionally referred to as the **fire behavior triangle** (Agee 1993).

7.2.1 Weather

Weather is a key determinant of fire intensity and therefore fire severity (Johnson and Larsen 1991; Meyn et al. 2007). Drought and high winds, in particular, set the stage for small fires to turn into big ones. Bonan and Shugart (1989) discuss boreal forests: "A few major fires that occur in extreme fire years account for the vast majority of forests burned. Sixty to 80 per cent of all fires in northwest Canada and Alaska are less than 5 ha in area . . . and 85 per cent of all fires in Canada between 1961 and 1967 were 4 ha or smaller. . . . Yet in severe years, individual fires can cover 50,000 to 200,000 ha." The total area burned by the sum of small fires can be quite large, however. Viereck (1973), for example, estimated that 400,000 ha per year burned in Alaska.

Historic records confirm that the most extensive wildfires in forests of the northern United States have occurred during extended dry periods (Clark 1990a; Whitlock et al. 2003; Pierce et al. 2004). Pierce et al. (2004) produced a particularly detailed record of fires in the northern Rocky Mountains over a time frame that encompassed both an unusually cold period (the Little Ice Age, lasting from about 650 years before present [BP] to 100 years BP) and an unusually warm, dry period (the Medieval Climate Anomaly, lasting from about 1050 years BP to 650 years BP; also referred to as the Medieval Warm Period). During the warm period, stand-replacing fires occurred periodically in both the montane lodgepole pine forests of the Yellowstone area and the drier ponderosa pine forests of central Idaho. The situation shifted with the advent of cooler, moister conditions during the Little Ice Age; Pierce et al. (2004) found no evidence of stand-replacing fires in either region during that period. In all likelihood, the ponderosa pine forests entered the regime of frequent ground fires encountered in the 1800s by the first Euro-American settlers, while the lodgepole pine forests, occupying a wetter and cooler environment, burned very lightly or not at all (Pierce et al. 2004). Clark (1990a) found a similar climatically related fire pattern in northern Minnesota: fires burned frequently during the warm, dry fifteenth and sixteenth centuries and infrequently during the Little Ice Age (fig. 7.4). The oldest age classes of trees in the once extensive old-growth Douglas-fir forests of Oregon and Washington were also established during the latter years of the Medieval Warm Period.

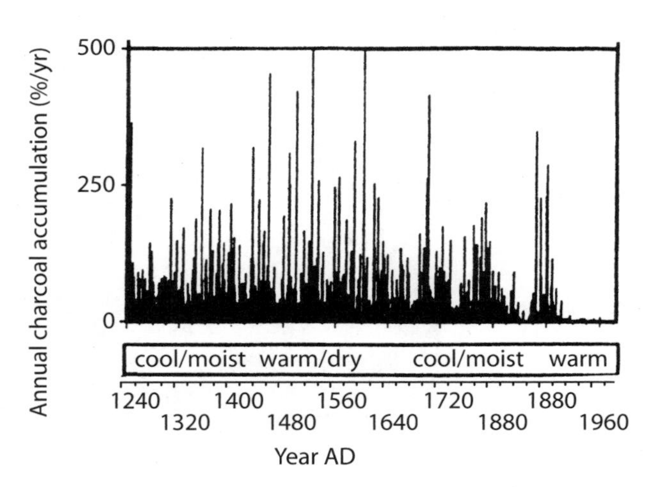

Figure 7.4. The frequency of historic fires in northwestern Minnesota as estimated by charcoal in lake sediments. Fires were most frequent during the relatively warm and dry 1400s. This was also a very active fire period in the Pacific Northwest. (Adapted from Clark 1990a)

The connection between fire and drought is different in the semiarid forests of the southwestern United States and northern Mexico. In contrast to the northern United States, the Little Ice Age was a period of relatively high fire activity in the southwest (Grissino-Mayer and Swetnam 2000); apparently, climatic cooling in that arid region moves forests into a moisture regime where fires are more, rather than less, likely. A similar pattern occurs in the dry mixed-conifer forests of Baja California, where fire frequency is higher on moist than on droughty sites (Minnich et al. 2000). In forest types where extended drought is normal, slow vegetative recovery following fire extends the time frame in which fuels have accumulated sufficiently to support another fire (Minnich et al. 2000). Additionally, shrubs typical of the understory on such sites tend to maintain high leaf-water content when young, making them relatively nonflammable (Barro and Conard 1991).

Because of the strong connection between fires and climate, and the strong connection between climate and dominant modes in the ocean-atmosphere system (chapter 3), it is not surprising that wildfire activity is linked through atmospheric teleconnections to events occurring globally. The El Niño–Southern Oscillation (ENSO) produces climatic effects that are particularly strong in the tropics but also ripple across the globe. In Indonesia and eastern Australia, La Niña years are accompanied by abundant rainfall, while El Niño years are drier than average. Millions of hectares burned in Australia and Indonesia during the particularly strong El Niño that occurred in 1982/1983 (Swetnam and Betancourt 1990), and the megafires that burned Borneo in 1997/1998 were once again during a very strong El Niño (Siegert et al. 2001). El Niños also bring drought and forest fires to northern Patagonia (Kitzberger et al. 2001). In the western United States, El Niños bring relatively dry winters to the northwest and relatively wet winters to the southwest, while La Niña produces the opposite

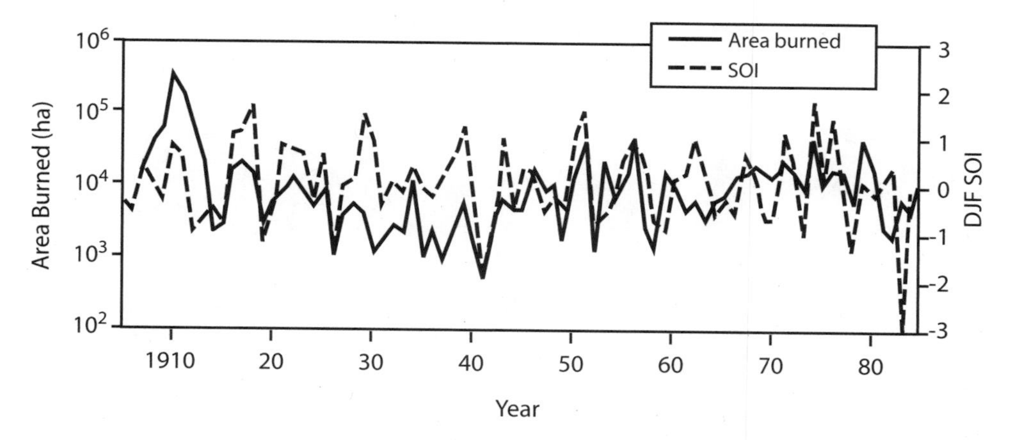

Figure 7.5. The correlation between El Niños and fire activity in ponderosa pine forests of the southwestern United States. The solid line represents percentage of trees scarred by fire in any given year; the dashed line corresponds to southern oscillation index (SOI). DJF denotes December, January, and February. Low DJF SOI values indicate an El Niño, high values indicate a La Niña. In the American Southwest, high fire activity is associated with La Niñas. This relationship does not hold everywhere, and in fact, in the Pacific Northwest it is quite the opposite. (Adapted from Swetnam and Betancourt 1990)

(fig. 7.5) (Swetnam and Betancourt 1990; Westerling and Swetnam 2003). However, unlike in more tropical regions there is no correlation between ENSO cycles and large fire years in the Pacific Northwest (Gedalof et al. 2005). In that region, winter precipitation is less important than spring and early summer precipitation in determining regionwide fire activity.

The ENSO is not the only ocean-atmosphere mode to influence fire behavior. In western North America, the effects of the ENSO on drought and fire are modified by the Pacific Decadal Oscillation (PDO).[2] Fires in northeastern California are most widespread when El Niño years coincide with the warm phase of the PDO (Norman and Taylor 2003). In the Lake Tahoe basin of Nevada, the most widespread fires occurred historically when the PDO shifted from its warm phase to its cool phase (Taylor and Beaty 2005). In general, the strong effects of ocean-atmosphere modes on wildfires stems not only from droughts, but from the characteristic oscillation between wet and dry periods produced by swings between dominant modes (e.g., El Niño to La Niña). Wet years stimulate vegetative growth that when dried during the drought phase, becomes fuel to feed fires (Norman and Taylor 2003; Taylor and Beaty 2005).

7.2.2 Fuels

The amount, distribution, and flammability of fuels (which include both living and dead biomass) are important determinants of fire severity; although in some types of forests (especially subalpine and boreal), extreme weather can trump the effect of fuels (e.g., Bessie and Johnson 1995; Agee 1997; Turner et al. 2003c). We return to this point in a later section.

Because they dry quickly, leaves and small branches (called **fine fuels**) frequently determine the rate of fire spread. These occur in two places within stands: as litter on the forest floor, and within crowns, the latter containing by far the greatest amount of fuel in nearly all cases. Two patterns of fuel distribution are relevant: vertical and horizontal. Suppose a discarded cigarette starts a ground fire within a forest stand. The amount and moisture content of fuels on the forest floor, along with wind speed, will determine the flame lengths that develop. For a given flame length, the probability of the fire moving from the ground into the crowns (called **torching**) will depend on the distribution of fuels in vertical space. If crowns of some or all trees are within reach of the flames, and if the foliage within the flame front is flammable (a function of moisture content and species), it will burn. If the vertical distribution of flammable crowns (including dead branches) is sufficiently continuous from the bottom into the main part of the canopy, torching will occur. Vertical fuels are often termed fire ladders or **ladder fuels.** In dry forests of western North America, understory trees create ladder fuels (fig. 7.6). In tropical forests, vines may play that role.

Once in the crowns, and conditioned by wind, the distribution of fuels horizontally through the stand determines tree-to-tree spread, or **active crown fire.**[3] For example, crowns are packed together in densely stocked conifer stands, producing a high degree of fuel continuity and favoring rapid fire spread. Crown fuels are commonly

[2]Recall from chapter 3 that the PDO is an oscillation in sea-surface temperatures similar to the ENSO, but also with significant differences. The PDO is centered in the north rather than the tropical Pacific, has less extreme temperature swings than the ENSO, covers a larger expanse of ocean than the ENSO, and has a cycle on the order 15 to 30 years rather than the ENSO's (roughly) 3 to 7 years.

[3]For detail on the conditions that produce crown fires, see work by Van Wagner (1977) and Rothermel (1991).

Figure 7.6. Ponderosa pine forest, Oregon. Understory trees create fire ladders that carry flames into the crowns of overstory trees.

quantified as **crown bulk density.** In stands of relatively uniformly sized trees, crown bulk density is usually calculated as foliar biomass per square meter of ground area divided by the average crown length (in meters), yielding a value expressed in biomass per cubic meter (biomass values are for dry foliage). Using average crown length in multistoried stands, however, results in significant overestimates of crown bulk density, hence of fire susceptibility (Perry et al. 2004). In multistoried, mixed-conifer stands in eastern Oregon, crown bulk density correlated closely with stand basal area,[4] and even more closely with leaf area index (Perry et al. 2004).

The degree to which fire is propagated through crowns varies greatly with plant species. Some burn readily and propagate fire; others are quite nonflammable and inhibit fire spread. Flammability of fuels is influenced by their moisture content and chemical composition. Oils, fats, and resins are quite flammable, while certain inorganic compounds inhibit combustion (Broido and Nelson 1964; Mutch 1970; Philpot 1970). Mutch (1970) hypothesized that fire-dependent species (e.g., lodgepole pine) have evolved chemical characteristics that facilitate fire. At least some ericaceous species that are frequent associates of boreal and montane conifers are also highly flammable, and chaparral species may or may not be. Some hardwood trees and shrubs appear to be quite nonflammable and, when intermixed with conifers, may actually act as "antifire ladders," preventing the fire from moving into conifer crowns rather than propagating it.

7.2.3 Fire Regimes: A Trade-off between Frequency and Severity

The historic fire regime of a given ecosystem can be characterized by three factors: (1) the average frequency with which fires have occurred, (2) their average severity or destructive power, and (3) how the first two factors have varied over centuries to millennia. Fire frequency and severity generally correlate negatively with one another. Where fires burn frequently, they are seldom highly destructive; infrequent burns, on the other hand, tend to be of the stand-replacement variety. There is a simple reason for this pattern: little fuel accumulates where fires are frequent, much where they are infrequent.

7.2.4 Fire Behavior: A Complex Function of Weather, Topography, and Structure

The severity of any disturbance depends on both the strength of the forcing factor, which is usually (but not

[4]Stand basal area is the sum of individual tree basal areas per unit area and is usually expressed as square meters per hectare. By convention, individual tree basal areas are measured at 1.37 m above the ground.

always) some aspect of climate, and the ability of the disturbed ecosystem to resist disruption, which is a function of its structure. For example, whether a house stands in a hurricane depends on both the strength of the hurricane and how solidly the house is built.

Topographic effects on fire patchiness are relatively predictable. A fire that is very severe on the slopes may burn riparian zones lightly or not at all. In the Northern Hemisphere, southerly aspects are drier than northerly aspects; hence, they tend to burn more frequently (Clark 1990a; Beaty and Taylor 2001; Heyerdahl et al. 2001). The steepness and complexity of topography affect fire patterns. In Oregon's Blue Mountains, different slope aspects burn differently where slopes are steep, but not where they are shallow (Heyerdahl et al. 2001). In Morrison and Swanson's (1990) study of fire history in two areas of the Oregon Cascades, the steeper and more highly dissected area had more frequent but lower-intensity fires than the area with gentler topography. Complex topography is associated with abrupt changes in aspect and perhaps soils, with numerous interbedded small streams and associated riparian vegetation. These factors modify the accumulation or flammability of fuels, hence the ability of fires to spread widely (Taylor and Skinner 2003).

The effects of species composition, stand age, and stand structure on fire severity are, in theory at least, predictable; however, at this point far too little is known about the influence of these factors on fire behavior. Because of large amounts of dead wood and the development of understory tree layers that act as fuel ladders, old forests were once thought to be especially vulnerable to catastrophic fire; however, as discussed later, this is not necessarily the case. As Van Wagner (1983) points out, the "rate of fire spread depends more on the quantity and arrangement of fine fuels than on the accumulation of downed logs or deep organic matter." Indeed, old decayed logs are spongelike in their ability to retain water and are likely to be among the least flammable fuels within a forest.[5] Once burning, however, they can smolder like cigarettes for long periods.

The relation between stand age and susceptibility to fire is complex and likely to vary depending on forest type; however, some generalizations are possible. Conifer forests tend to be most vulnerable to crown fires when young and again when senescent, and least vulnerable during their middle years (in long-lived conifers these middle years may encompass hundreds of years) (Van Wagner 1983; Perry 1988; Franklin et al. 1989; Stephens and Moghaddas 2005). Age is actually an indirect determinant of fire susceptibility, being important only in so far as it determines forest structure, particularly the distribution and flammability of fuels (fig. 7.7). Being close to the ground, crowns of young trees act as their own fire ladders, and if the stand is densely stocked, fire spreads readily once in the crowns. As stands age, crowns become higher, thereby reducing the vertical continuity of fuels, and with further aging, individual trees and their crowns become more widely spaced, thereby reducing the horizontal continuity of fuels. The latter stage, corresponding to what is often termed old growth, is the most fire-resistant stage of conifer forest development (Van Wagner 1983; Perry 1988; Franklin et al. 1989). As stands begin to break up (i.e., heavy mortality among overstory trees, generally coupled with the increased growth of flammable understory trees), susceptibility to catastrophic fire increases. The age at which that happens varies widely from one forest type to another. The giant conifer forests of northwestern North America may not break up until they are close to one thousand years old (or more), while boreal and montane conifers begin to break up when they are much younger. Dry ponderosa pine forests of western North America present a somewhat different situation. In these, the ability of flammable understory trees to establish (hence, an increased risk of stand-destroying fires) is not due to the breakup of the old stand, but to the exclusion of ground fires that historically had cleared the understory while sparing most mature trees.

In their study of patterns in two fires in the Siskiyou Mountains (Oregon and California), Alexander et al. (2006) found that in general, aspect was the most important factor determining fire severity (south aspects burning more severely); however, they also found that areas with large trees burned less severely than areas with small trees. Also in the Siskiyous, Thompson et al. (2007) studied what happened when the very large Biscuit fire of 2002 burned through areas burned by the Silver fire of 1987. They found that areas that were salvaged logged and replanted to conifers following the Silver fire burned more severely than areas left to recover naturally. The likely explanation for that pattern can be found in two factors: the high flammability of young conifer stands and the logging debris left by salvage. (Standing dead trees in unsalvaged areas were too large to carry fire readily).

As with older stands, the vulnerability of young conifer stands to wildfire depends on stand structure, the flammability of intermixed plant species, and (a particularly important factor in young plantations) the abundance of dead fuels left after logging the previous stand. The foliage of many conifers (and some other species such as eucalyptus) contains resins and oils that make it highly flammable. When conifers are widely spaced, however, it is the intervening fuels (both dead and living) that carry the fire. Weatherspoon and Skinner (1995) studied the relationships between fire severity and various environmental factors following the 1987 wildfires in the Shasta-Trinity National Forest in northern California. They found that dry logging slash, especially fine fuels (crowns), burns exactly as one would expect and increases fire hazard within young stands

[5]Following a stand-destroying fire in Oregon, down logs still averaged 157 percent moisture content, which is enough water to easily squeeze out moisture with your hands (Amaranthus et al. 1989).

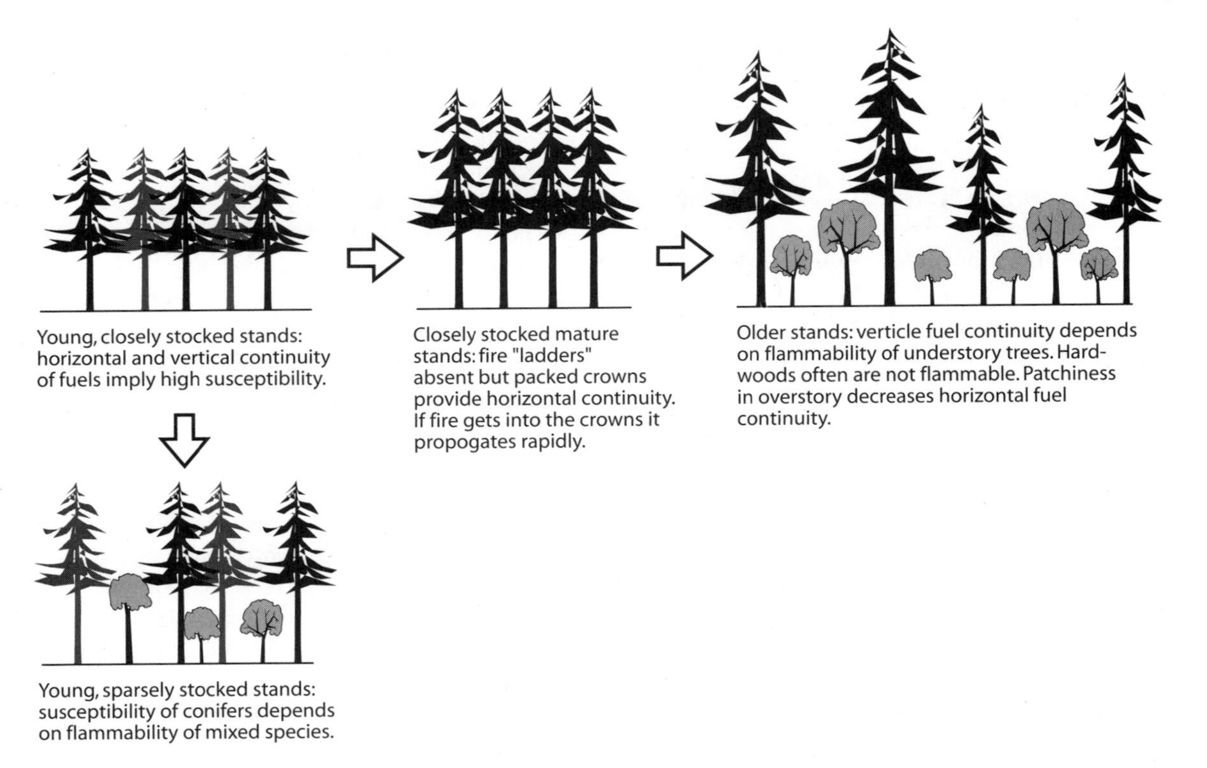

Figure 7.7. Some aspects of stand structure that influence susceptibility of conifer forests to crown fires.

(and also thinned or partially logged older stands). The hazard associated with logging slash can be reduced by various means, including controlled burning as it lay ("broadcast" burning), gathering in small piles that are burned, or fragmenting into smaller pieces that decompose more readily. The latter may be done either by hand or machine; when by hand it is commonly referred to as "lop and scatter."

The flammability of intermixed vegetation is a little-studied but potentially important factor influencing the susceptibility of both old and young conifer stands to fire. In their study, Weatherspoon and Skinner (1995) found that plantations with a grassy understory tended to burn more severely than those without. This pattern is likely to be generally true, because grasses tend to dry out in late summer and, hence, burn readily. On the other hand, Weatherspoon and Skinner found that fire severity correlated negatively with the presence of forbs. Whether forbs or grasses dominated the understory of plantations in the Weatherspoon and Skinner study depended, in turn, on how sites were prepared prior to planting. Grasses were more abundant where logging slash was piled with bulldozers (which often involves considerable soil disturbance), whereas broadcast-burned sites had more forbs.

Figure 7.8, a photo taken shortly after the 1987 wildfires in southwest Oregon, shows a striking example of hardwood trees protecting intermixed Douglas-fir. The former plantation in which this photo was taken was the site of a brush control study in which understory hardwoods had been removed from all but one area (the control plot). Fire consumed all Douglas-fir in this plantation except those in the control plot, which were apparently spared because surrounding hardwoods did not propagate the fire.

Finally, at least some of the patchiness with which fires burn is due to chance, driven by factors such as short-term fluctuations in weather (particularly humidity and wind), time of day (other factors being equal fires tend to lay down at night and grow during the day), and direction the fire is moving (fires tend to burn more intensely when moving up slopes than when backing down slopes; this is because fires moving upslope more effectively dry and ignite unburned fuels ahead of the fire front). Spatial patterns of fuel accumulation, which often trace back to previous fires, contribute significantly to patchiness of a given burn, a point we return to later.

7.2.5 Power-Law Behavior and Self-organization in Forest Fires

In the United States, Australia, Italy, and China, the frequency of wildfires of different sizes has been found to follow negative power laws (fig. 7.9) (Malamud et al. 1998, 2005; Ricotta et al. 1999; Song et al. 2001; Turcotte et al. 2002; Moritz et al. 2005).[6] Quoting from Turcotte et al. (2002): "Considering the many complexities of the initiation and propagation of forest fires . . . , it is remarkable that

[6]A negative power law has the form $y = cx^{-z}$, where c is a constant and z is a positive constant. On a log-log scale, the relationship becomes a straight line: $\log y = \log c - z(\log x)$.

Figure. 7.8. A, In the foreground is a former 15-year-old, Douglas-fir plantation completely destroyed by wildfire. In the midground is the only area in the plantation where intermixed hardwoods had not been herbicided. B, Inside that portion of the stand is where Douglas-fir was intermixed with hardwoods. The darker colors are live Douglas-fir saplings. The lighter colors are hardwoods whose foliage was killed by the fire but did not flame and hence did not propagate the fire.

the frequency-area statistics are very similar under a wide variety of environments. The proximity of combustible material varies widely. The behavior of a particular fire depends strongly on meteorological conditions. Firefighting efforts extinguish many fires. Despite these complexities, the application of power-law frequency-area distributions appears to be robust."

Complex systems theorists consider power-law behavior to be a characteristic of either self-organization (in natural systems) or purposeful design (in engineered systems).[7] A self-organizing system is one whose dynamics are strongly

[7]Power laws may arise from other dynamics also.

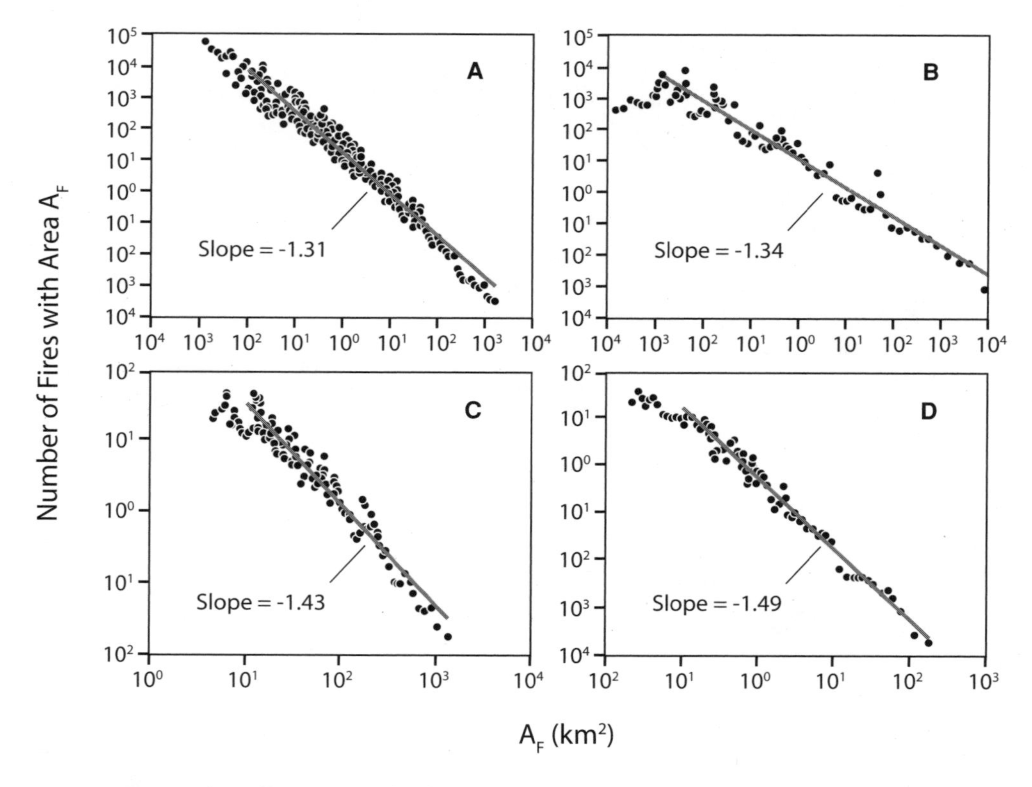

Figure 7.9. Noncumulative frequency-area distributions for actual wildfires in the United States and Australia: A, 4,284 fires on U.S. Fish and Wildlife Service lands (1986 to 1995). B, 120 fires in the western United States (1150 to 1960, as determined by tree rings). C, 164 fires in Alaskan boreal forests (1990 to1991). D, 298 fires in the Australian Capital Territory (1926 to 1991). Note log-log scale. See Malamud et al. (1998) for citations to original studies. (Adapted from Malamud et al. 1998)

influenced by interactions among components of the system. It is not free from outside influences, but external inputs are buffered and filtered by internal interactions. Organisms are quintessential examples of self-organizing systems; however, self-organization to one degree or another is probably common in nature (Perry et al. 1989a; Perry 1995; Sole et al. 1999; Turcotte and Rundle 2002). What does that mean in the context of ecosystems and disturbances? Disturbances either alter landscape patterns or act to reinforce existing patterns, in either case feeding back to influence the severity of future disturbances. In other words, disturbance effects structure and structure effects disturbance. The importance of climatic forcing not withstanding, power-law behavior of forest fires argues strongly for the importance of ecosystem and landscape structure in determining fire behavior.

7.2.6 Fire in Some Specific Forest Types

We close our discussion of fire with a brief look at characteristic fire patterns in some forest types that represent a broad range of fire histories. Keep in mind that fire regimes vary over centuries to millennia, depending on global climate. Particularly in mountainous terrain, fire regimes can also vary over relatively short distances.

7.2.6.1 Mesic Conifer Forests of Northwestern North America

Agee (1993) has an excellent, detailed review of fire histories for the variety of forest types found in the U.S. Pacific Northwest. The region west of the Cascades crest in Washington, Oregon, and British Columbia, which encompasses the west slopes of the Cascades and the Oregon Coast Range, is an area of moderate to high precipitation, relatively mild temperatures, and moderate to high productivity. The Siskiyou Mountains in southwestern Oregon and northwestern California are highly variable climatically and ecologically, with mesic and dry forests that in some cases are intermingled in close proximity. Infrequent crown fires typify the wetter, more northerly parts of the region (British Columbia, Washington, and the northern Coast Range in Oregon). For example, over the past one thousand years in Mount Rainier National Park, large, stand-replacing wildfires occurred on average every 434 years (Hemstrom 1979). Southward, in central and southern Oregon and northern California, fire regimes become increasingly patchy, with a mix of large, intermediate, and small fires (Agee 1993). In Douglas-fir forests of northern California, the driest portion of the region, fire-return intervals averaged 12 to 19 years (Taylor and Skinner 1998). Native American use of fire as a land management tool was probably common, especially

Two different models have been developed by complexity theorists to explain the power-law behavior of forest fires: self-organizing criticality (SOC), and highly optimized tolerance (HOT). In models, and presumably in the real world, self-organizing systems evolve to a critical state in which the frequency distribution of disturbances follows a negative power law. In the words of Sole et al. (1999), "large, far from equilibrium, complex systems, formed by many interacting parts, spontaneously evolve toward the critical point." As readily seen in the form of a negative power-law curve, self-organized criticality is characterized by many small and a few large disturbances, with the numbers between the two extremes increasing as size decreases from the largest to the smallest (a common way of describing this behavior is that disturbances occur at all size scales). Note that fractals are also described by negative power laws, so a self-organizing critical system is one with a fractal structure.[8]

In contrast to the spontaneous appearance of power-law behavior in self-organizing criticality, HOT models assume that either through purposeful design or self-organization, the structure of systems is tuned to minimize losses over the long term. Moritz et al. (2005) describe it thusly: "HOT emphasizes specialized configurations of interacting components, tuned via biological mechanisms or engineering design to be robust to fluctuations and variability in the environment. A key concept from HOT theory is that tuning for robustness involves trade-offs . . . and that the very mechanisms and interdependencies which increase robustness to common events also introduce new sensitivities or fragilities to rare or unanticipated disturbances." One result of these trade-offs is power-law behavior.

HOT models provide a somewhat better fit than SOC models to real forest fire patterns (Moritz et al. 2005), although both approaches boil down to robustness arising from self-organizing dynamics and so may tell us the same thing about ecosystems and fire. Moritz et al. (2005) discuss the application of HOT theory to forest fires as follows:

> Our working hypothesis is that the natural feedbacks associated with ecological processes over time have led to vegetation patterns that are well suited (if not completely "optimized") to the local fire regime, so that by modeling current conditions with reasonable fidelity, we may be capturing the net result of some underlying mechanism for ecosystem resilience. The most salient outcome of HOT models are [sic] the consequences associated with more resources being allocated to common events and less to rare ones. In terrestrial ecosystems, this tendency may lead to dominant patterns of species distributions and plant community structures that reflect the prevailing environmental conditions, whereas other patterns may reflect acclimation to marginal conditions or adaptations to infrequent extreme events.[9]

Similarity of real-world patterns with model predictions does not prove the two result from the same processes, and power laws can be produced by a variety of mechanisms (Newman 2000). For example, fractal patterns are widespread in nature (Mandlebrot 1983; Sole et al. 1999; Brown et al. 2002), including some structural features such as stream networks (Rodriguez-Iturbe and Rinaldo 1997) and forest gaps (Sole and Manrubia 1995), that influence the contiguity of fuels and therefore fire size.

in the lower-elevation forests of central and southern portions of the region (Carloni 2005).

Prior to logging, 60 percent or more of the forests were old growth (presently old growth occupies less than 20 percent). The majority of forests were dominated by Douglas-fir, with varying admixtures of western hemlock and western red cedar in mesic habitats, and various other tree species in dry or cold habitats. The mesic forests graded into subalpine forests at high elevations, and ponderosa pine or oaks at low elevations in the south and the interior valleys of the central portions of the region.

Stands were multiaged, with the oldest trees on the order of five to six hundred years old, and in some cases considerably older. Individuals and patches of various age classes existed within the matrix of older trees (fig. 7.10) (Morrison and Swanson 1990; Tappeiner et al. 1997; Sensinig 2002). The age class diversity probably resulted from a combination of two factors: an extended period of tree establishment following large crown fires, and a history of numerous small disturbances (i.e., events that killed from one up to several hundred hectares of trees) following the original burn. The smaller events were likely a mixture of wind, insects, pathogens, and so forth, but low- to moderate-intensity fires were probably the dominant factor.

As we discussed previously, conifer forests of the Pacific Northwest are most susceptible to stand-replacement fires

[8] A tree provides one of the best examples of fractal structure. Branching produces repeating patterns at smaller and smaller scales, producing a self-similar structure characteristic of fractals. A frequency diagram of branch size would yield a negative power law.

[9] "In ecological applications the word 'optimization' is perhaps more aptly replaced by 'organization,' reflecting the distinction between HOT and random, disorganized configurations, and highlighting the importance of structured interdependencies that evolve via feedback among and between biotic and abiotic variables" (Moritz et al. 2005).

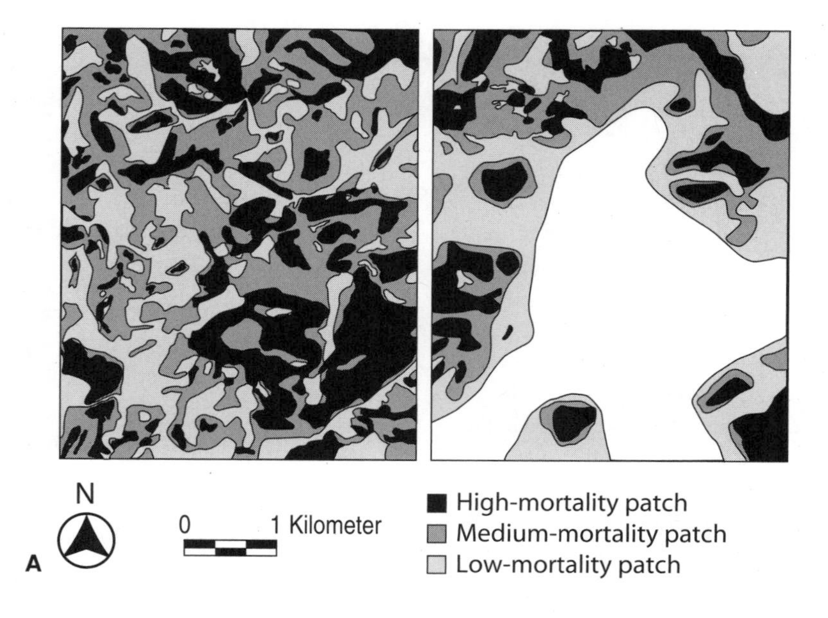

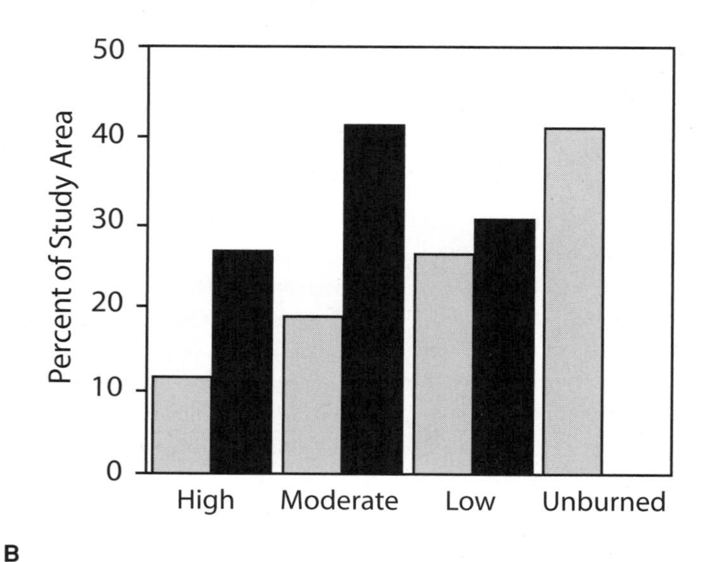

Figure 7.10. A, Wildfire patterns between 1800 and 1900 in two areas of the central Oregon Cascades. Note the complex mosaic created by fires of differing intensity. Black areas burned severely (70 to 100 percent mortality of overstory trees); dark gray areas burned moderately (30 to 70 percent overstory mortality); light gray areas burned lightly (<30 percent overstory mortality); and white areas did not burn. Each panel is approximately 4 km × 5 km. B, The proportion of each of the two areas of A that burned at different intensities *(dark bars, left side of A; light bars, right side of A)*. (Morrison and Swanson 1990)

during their first one hundred or so years (Franklin and Hemstrom 1981). The data shown in fig. 7.11 illustrate this pattern for the 1987 wildfires in southwestern Oregon.

7.2.6.2 Ponderosa Pine and Mixed Conifers

The most widely distributed pine in North America, ponderosa pine occupies relatively dry habitats from central Mexico into southern British Columbia. Ponderosa pine may grade into oaks, grasslands, or desert shrubland at lower elevations and form a component of mixed-conifer forests as elevations increase. At midelevations, ponderosa pine often forms a **fire climax,** which is to say it dominates areas that in the absence of fire, could be (and are being) occupied by other conifers. As we discuss below, alteration of the fire regime over the past century has shifted forest composition in such areas.

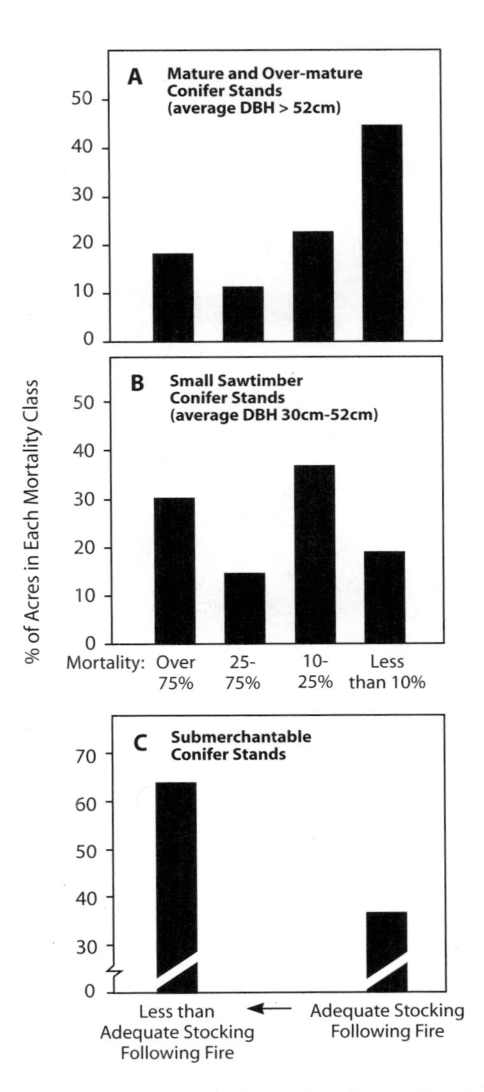

Figure 7.11. Fire severity by age of forest, Silver fire complex, Oregon. Older stands had the least amount of mortality and young stands the most. (Adapted from U.S. Forest Service 1988)

Historically, many ponderosa pine forests were characterized by frequent surface fires that kept dead fuels from accumulating and living fire ladders from becoming established, thereby increasing the chances that the next fire would also burn with low intensity (Agee 1993). Like some other tree species, ponderosa pine develops thick, fire-resistant bark that allows mature trees to survive fires that do not "crown out." Trees may, however, be scarred by fire, and these scars, preserved within the wood of old trees, provide a convenient historical record of fire frequency (fig. 7.12). Because of this, fire history is better understood in ponderosa pine than other forest types. Fire scars are not error free, however, and the validity of fire histories based on fire scars has been challenged (Baker and Ehle 2001).

As with other forest types, the historic frequency of fires in ponderosa pine forests varies with average annual precipitation. Bork (1985) used fire scars to reconstruct fire-return intervals in three different habitats, all dominated by ponderosa pine but spanning a rainfall gradient from relatively moist to quite dry. Over the previous five to six hundred years, fires burned most frequently in the middle of the moisture gradient, with a given 16-ha patch averaging one fire every 11 years, and least frequently on the driest site, averaging one fire every 24 years. The site with highest precipitation was intermediate, averaging one fire every 16 years (fig. 7.13). In all likelihood, fire frequency was limited by the amount of fuels at the driest site (which bordered desert), and by the moisture content of fuels at the wettest. As elevation increases in the western mountains, one moves into a mixed-conifer zone that may include ponderosa pine, Douglas-fir, and true firs (although ponderosa may have been a fire climax historically). The transition from dry ponderosa to mixed conifers includes sharp moisture gradients related both to elevation and aspect, producing fire regimes that may vary over relatively short distances from predominantly surface fires to mixed regimes that include both surface fires and crowning (e.g., Veblen et al. 2000; Schoennagel et al. 2004; Hessburg et al. 2005). The boundary between predominantly surface fires and mixed regimes was not static but varied depending on climatic drivers, and (as noted earlier) past shifts in global climate produced periods when crown fires were more common in at least some ponderosa pine forests than has been the case over the past few hundred years.

Cattle grazing[10] plus extirpation of Native Americans altered fire regimes in ponderosa pine forests beginning in the mid-1800s, and this trend accelerated with the advent of fire control in the early 1900s. As a result, fuels have built to abnormally high levels, including understory fire ladders that seldom or never occurred under the historic regime of frequent underburns. These forests are now vulnerable to more severe fires than in the era before humans began "protecting" them (Swetnam and Betancourt 1990; Agee 1993).[11] This situation is not unique to ponderosa pine forests. Pollen and charcoal analyses in southeastern Sweden show that fire was common until about two hundred years ago (the advent of forestry and fire control), and forests of the study area are more dense now than at any time in the past 2,500 years (Lindbladh et al. 2003). One consequence is that several beetle species that require open forests or other structural conditions created by fire are at risk of extinction.

[10]Cows grazing within open ponderosa pine forests reduce grass cover. Grass, in turn, is important in carrying surface fire, and reduced grass cover results in fewer and smaller surface fires.

[11]A long fire-free period does not always lead to greater flammability. Odion et al. (2004) studied patterns of burning during 1987 fires in the Klamath Mountains of Oregon and California. They found that forests with previous fires since 1920 burned more severely than forests that had not burned since 1920. Odion et al. speculate that previously unburned forests had developed a relatively fire-resistant understory of shrubs and small trees.

Figure 7.12. A. Fire-scarred ponderosa pine. B. Matching the position of fire scars on living trees with growth rings permits the dating of past fires. (Courtesy of Joyce Bork)

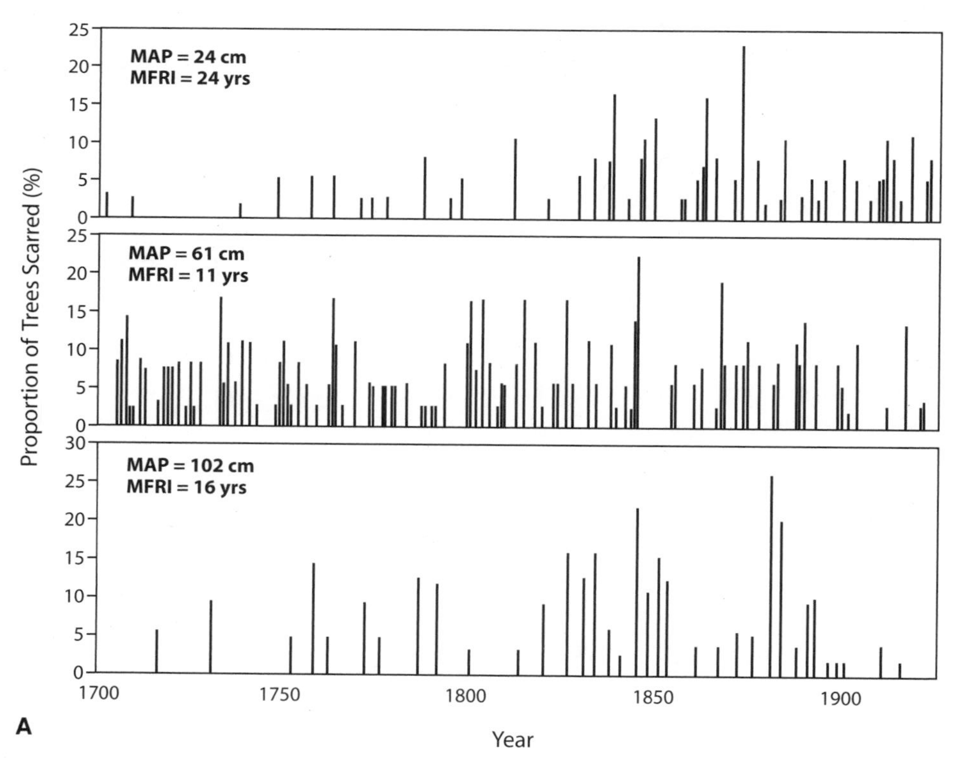

Figure 7.13. A, The history of fires between 1700 and 1920 in three ponderosa pine forests in central Oregon. The length of the bars represents the proportion of trees within 16-ha plots that were scarred during any given year. MAP, mean annual precipitation; MFRI, mean fire return interval (to 16-ha plots).

Figure 7.13. *(Continued)* B, Old-growth ponderosa pine in central Oregon.

7.2.6.3 Boreal and Northern Rocky Mountain Subalpine Forests

Productivity of boreal and subalpine forests is strongly influenced by cold temperatures, short growing seasons, and nutrient rather than water limitations; consequently, stands tend to be densely stocked with relatively short trees, which sets the stage for large crown fires (Schoennagel et al. 2004). Stand-destroying crown fires occur every one to two centuries in subalpine forests (Schoennagel et al. 2004), and in boreal forests at roughly 50- to 200-year intervals (in very moist sites up to 500-year intervals) (Bonan and Shugart 1989). Fire cycles become even longer in the forest-tundra boundary of Canada, where islands of spruce forest have persisted for as long as several thousand years and, in contrast to more southerly boreal forests, maintain a multiaged structure (Payette et al. 2001). Even large fires, however, do not produce "desolate areas of uniform devastation" (Turner et al. 2003c). The severe fires that burned Yellowstone National Park in 1998 "created a spatially complex mosaic of unburned and burned patches, encompassing a wide range of burn severities. . . . The majority of severely burned areas were within 50-200 m of unburned or lightly burned areas" (Turner et al. 2003c) (e.g., fig. 7.2C). Patchiness may be related to various factors, but natural firebreaks such as ridges, streams, and wetlands play a significant role. Hardwoods, which compose a significant component of many boreal forest landscapes, are less flammable than conifers and may reduce fire severity.

Fire severity is tied closely to weather in boreal and subalpine forests (Schoennagel et al. 2004). In the Rocky Mountains, infrequent severe droughts triggered by

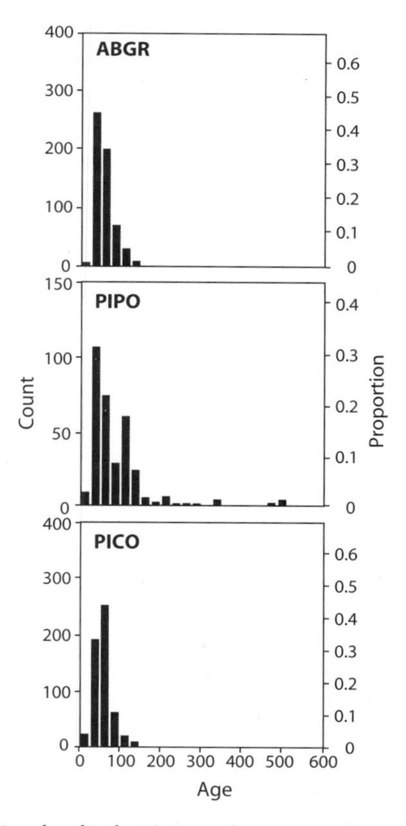

Figure 7.14. Age class distribution in ponderosa pine and mixed-conifer forests in central Oregon, by species. Stems greater than 5 cm DBH per 1.4 ha. ABGR, *Abies grandis* (grand fir); PICO, *Pinus contorta* (lodgepole pine); PIPO, *Pinus ponderosa* (ponderosa pine). (Adapted from Perry et al. 2004)

Figures 7.14 through 7.16 show an example of forest structure in dry forests on the east slopes of the Cascade Mountains in central Oregon, and implications for severe fire. Forests at the lowest elevations in that area are pure ponderosa pine, as sites are too droughty to allow other conifer species to establish in significant numbers. Midelevations are within a mixed-conifer zone, but the older trees are mostly ponderosa pine because a history of frequent surface fires prevented fire-sensitive tree species from establishing. Without frequent surface fires, however, the species composition of stands began to change dramatically. Figure 7.14 shows the age class distribution (as of 1995) by major species (note the different scales on the *y* axes). More than 80 percent of stems are younger than one hundred years. Younger age classes in the midelevations consist primarily of grand fir *(Abies grandis)* and lodgepole pine *(Pinus contorta)*, representing a major shift in species composition from young to old. Younger trees at the relatively arid lowest elevations are predominantly ponderosa pine.

Figure 7.15A shows the vertical distribution of crown bulk density averaged over 14 0.2-ha plots. Note the high proportion contributed by trees younger than one hundred years, especially below about 15 m. Stands with fewer than 40 old-growth trees per hectare average about 30 percent greater crown bulk density below 12 m height than stands with 40 or more old-growth trees per hectare (fig. 7.15B), reflecting the greater establishment and growth of small trees when less shade is cast by an overstory.

Figure 7.16 shows how stand basal area influences susceptibility to crown fires in these forests. (Basal area correlates closely with leaf area, which correlates, in turn, with crown bulk density.) Changes in basal area are achieved by simulated cutting of smaller trees, working up in size until the desired basal area is achieved (foresters call this approach *thinning from below*). Once stands fall below about 15 m^2/ha, extremely high winds are required to sustain a crown fire. Points labeled U are current conditions. The unthinned stands vary widely in susceptibility; however, all are above 15 m^2/ha.

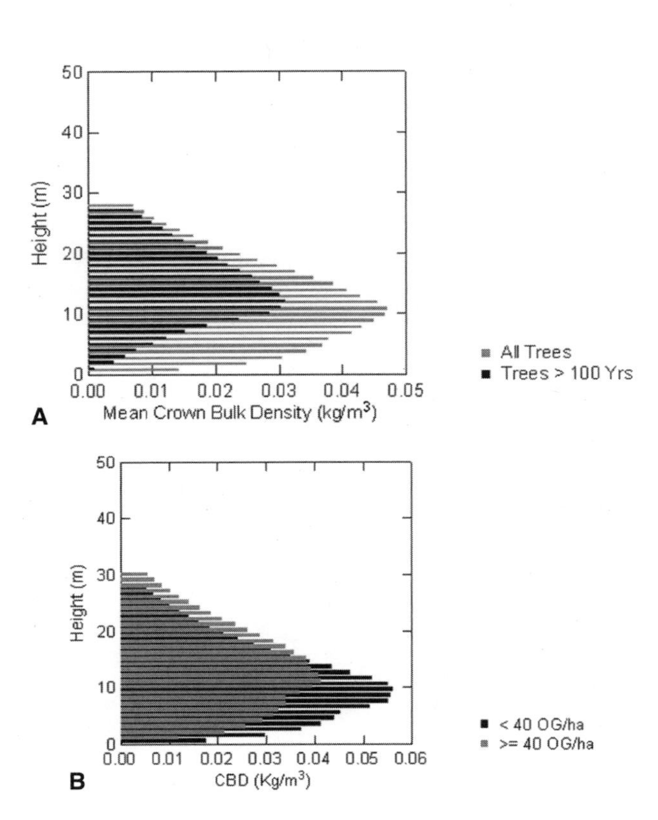

Figure 7.15. Some characteristics of crown bulk density in old-growth ponderosa pine and mixed conifer forests in central Oregon. A, Crown bulk density by height showing all trees and only that contributed by trees more than 100 years old. Values are means of 14 plots. B, Crown bulk density profiles as affected by the presence or absence of old-growth trees. (Adapted from Perry et al. 2004)

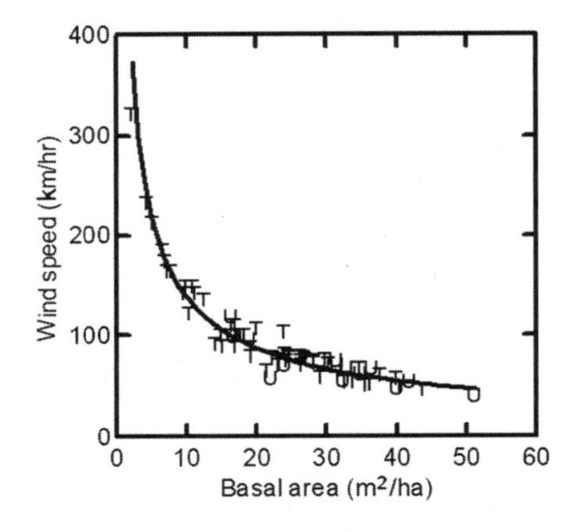

Figure 7.16. The relation between stand basal area and the wind speed necessary to initiate active crown fire in ponderosa pine and mixed-conifer forests, central Oregon. Models assume late-summer fuel moisture (i.e., very dry). U, stands as measured in the field; T, simulated thinning. All thinning is from below (i.e., only the relatively small trees in a stand are cut, working upward in size to the desired minimum diameter for leave trees). (From Perry et al. 2004)

high-pressure systems set the stage for large fires, especially when accompanied by high winds (Bessie and Johnson 1995; Turner et al. 2003c). Historically, these conditions tend to occur over wide areas and have been linked through atmospheric teleconnections to conditions in the northern Pacific and eastern North America (Paine et al. 1998).

As is the case with other forest types that throughout history have periodically burned, the influence of fire can be clearly seen in the adaptations of the trees. Referring to North America, Rowe and Scotter (1973) state:

A rash of large fires beginning in the late 1980s and continuing into the new millennium has sparked considerable discussion in the United States about "forest health" (see, e.g., Agee 1997; Covington 2000; Allen et al. 2002; NCSSF 2003; Turner et al. 2003c; Brown et al. 2004). Which forests are burning outside of their natural range of variability? What are the implications? What should be done about it? Central to the issue is an unnatural buildup of fuels over the past century due to fire exclusion. As we saw earlier, however, a warmer climate has also played a strong role (Westerling et al. 2006).

Which forest types have unnatural fuel buildups and which do not? What treatments, if any, are appropriate to reduce fire hazard? There is wide agreement on the following situation in the western United States:

- Forests characterized by a history of low-severity fires have an abnormal buildup of fuels and are primed to burn with severities not seen for at least the past several hundred years.[12] This applies especially to the low-elevation ponderosa pine forests.
- Forests with a history of high-severity fires have been little affected by fire exclusion and are not burning outside their historic ranges. This applies especially to the lodgepole pine forests of the northern Rockies.
- Midelevation forests between these two extremes present a more complicated and less well understood situation. They are often described as having a "mixed-severity" regime, that is, some combination of low- and high-severity fire. In reality, we find a gradient from low severity to high severity that follows elevational and topographic gradients in moisture. The gradient is not fixed but varies with climatic cycles that alter precipitation patterns.

Brown et al. (2004) provide the following guidelines for management actions to reduce the risk of abnormally severe fire (see also Allen et al. 2002).

- Place highest priority on historically low-severity fire regimes, with secondary priority on mixed-severity fire regimes, and lowest priority on high-severity fire regimes;
- be based on the ecology of the species and site, with presettlement conditions as a sustainable reference, recognizing different goals for a wide variety of landowners;
- be site specific and consider the landscape context, including watershed conditions and fish and wildlife habitat;[13]
- consider a range of restoration steps, rather than attempt complete restoration with a single treatment everywhere;
- maintain the most fire-resistant, large-tree component of the forest . . . ;
- consider mechanical thinning and prescribed fire as acceptable tools [to] reduce surface fuels, increase height to live crown, and decrease crown bulk density, while keeping big trees of resistant species;[14]
- where thinning is used, apply minimal impact harvesting techniques, and focus first on areas where road systems are largely complete;
- make a commitment to long-term monitoring and adaptive management so that we learn by doing.

Reducing risk of fire spread does not necessitate doing the same thing everywhere. Foresters have long employed fire breaks of one kind or another at strategic points on the landscape (e.g., Agee et al. 2000). Shaded fuel breaks, so called because a forest canopy is retained, are most common in forests (Agee et al. 2000), but strips of fire-resistant species such as alders have also been used. These are not intended to be stand-alone treatments, however, but components of a more widespread fuel-reduction strategy. Landscape-level strategies can be more sophisticated than firebreaks, at least in theory. Modeling by Loehle (2004) indicates that "arranging treated acres into a grid, analogous to bulkheads on a ship, drastically reduces the acreage that must be treated to achieve a fireproof condition" (Loehle 2004).

Of the ten commonest trees in the boreal forest—jack pine, lodgepole pine, aspen, balsam poplar, paper birch, tamarack, black spruce, white spruce, balsam fir, and alpine fir—the first seven can be classed as pioneers in their adaptations for rapid invasion of open areas. White spruce shows some pioneering characteristics too, although it is not so well equipped for dispersing seed at all seasons as pine and black spruce. Only balsam fir and its close relative alpine fir seem poorly adapted to reproduce immediately after fire. . . . Significantly, balsam fir is a usual constituent of old forests and is most common in the moister, less fire prone sections of the boreal zone.

[12]Forest types with low-severity regimes over the past millennium experienced high-severity fires during earlier periods (see text); hence, high-severity fire is not outside the range of historic variability in these forests. It is important to keep in mind, however, the very different landscape and regional contexts between now and a thousand years ago. After a century of cutting, the area of old-growth ponderosa pine forests is much diminished. Do we want to risk letting what is left burn up?

[13]For example, because closed forests accumulate fewer snowpacks at ground level than open forests, heavy thinning in snow zones can disrupt winter yarding and travel routes for elk and deer.

[14]By providing light for small trees to grow, cutting large, fire-resistant trees increases future fire hazard. For example, see fig. 7.6B.

Adaptations that facilitate colonization following stand-replacement fires include the ability to sprout from roots (aspen), small, readily dispersed seeds (poplar, tamarack), and retention of seeds in cones that require heat in order to fully open (jack pine, lodgepole pine, black spruce). The latter trait is called serotiny, from the Latin *serus*, meaning "late" or "delayed." Literally millions of seeds per hectare can be stored in serotinous cones that accumulate on trees for decades until a fire comes along to trigger their opening.

Not all individuals within species that bear serotinous cones are, in fact, serotinous. In most stands, or at least those of lodgepole pine, some proportion of trees are open coned, and a few bear both cone types. This fact provides evidence for a history of fires that have varied in intensity and also illustrates how environmental variability acts to maintain genetic diversity within populations.Serotiny is a genetic adaptation to stand-destroying fires; however, when fires are patchy, as they frequently are, living trees with nonserotinous cones are also able to disperse their seed, and hence, the open-coned genotype remains in the population (Perry and Lotan 1979). The proportion of serotinous trees within stands varies widely. In Yellowstone National Park, for example, lodegpole pine stands at higher elevations, where fire-return intervals are on the order of 300 years, contain fewer serotinous trees than stands at lower elevations, where fire-return intervals are around 180 years (Turner et al. 2003c).

7.2.6.4 Deciduous Forests of Eastern North America

Shelford (1963) recognized three large subdivisions of the eastern deciduous forest:

1. The mesic northern and upland regions, which occupy much of the northeastern United States and extend southward in the Appalachian and Cumberland Mountains to western North Carolina and eastern Tennessee (typified by shade-tolerant species such as sugar maple, yellow birch, and American beech, but with many variants)
2. The southern and lowland regions, occupying relatively dry sites in the northeastern United States and much of the lowlands of the southeastern and south-central regions (typified by oak/hickory)
3. Riparian forests that extend westward into the Great Plains

For our purposes it is sufficient to consider only the first two, keeping in mind that these represent two poles of the environment, and ecotonal forests exhibit properties of both to one degree or another. There is a distinct difference in the disturbance regimes between the mesic forests and the drier oak-hickory types (Lorimer 2001; Lorimer and White 2003). Fire is rare in the mesic forests (wind is the most common disturbance), while surface fires have been common historically in oak-hickory. Historically, oak forests in the eastern and midwestern United States experienced surface fires every 4 to 20 years (Shumway et al. 2001; Abrams 2003); in fact, early colonists described the northeastern oak forests as having widely spaced trees with grassy understories, almost certainly a result of periodic burning by Native Americans (Lorimer and White 2003). In New England, "in general, the hilly moraines with finer textured soils and more mesic forests supported lower fire occurrence than the flatter, sandier, and more droughty outwash plains, with their flammable vegetation of oak, scrub, oak, and huckleberry" (Foster et al. 2004).

While surface fires were common in the drier forests, crown fires in broad-leaved forests of any kind were (and are) rare or nonexistent (Foster et al. 2004). Summer rains are much more common than in the west, humidity within forests remains high, and the combination of warmth and moisture—something not common in the forests of western North America and the boreal zone—contributes to relatively rapid decomposition, hence little fuel accumulation (Bormann and Likens 1979). Growing as they do on mesic sites, hardwoods such as maples and beeches maintain high leaf moisture and are therefore relatively nonflammable (Foster et al. 2004). Mesic northern hardwoods did experience stand-replacing fires, probably during periods of abnormal drought or in the aftermath of severe blowdowns; however, the fire-return interval was one to three thousand years (Lorimer and White 2003).

7.2.6.5 Moist Tropical Forest

Charcoal in soils beneath moist tropical forests of Amazonia indicates that despite heavy rainfall and absence of a distinct dry season, fire has occurred in these ecosystems (Sanford et al. 1985; Saldarriaga and West 1986; Kauffman and Uhl 1990).[15] Fire has been infrequent, however, and probably coincident with periods of extended drought. Fire-return intervals in *tierra firme* (upland) forests of Amazonia have varied between about 400 years and 1,550 years.[16] Fire activity in all forest types varies with climate over long cycles, and the moist tropics are no different. Historic reconstructions in monsoonal forests of northeastern Cambodia (using buried charcoal) show that the most recent 2,500 years has had the lowest fire activity of the past 9,300 years (Maxwell 2004). The period of 8000 to 9300 BP was one of particularly high fire activity in those forests.

[15]Studies in the largest remaining blocks of undisturbed tropical rainforest (Amazonia, the Congo basin, and the Indo-Malay region) indicate once-thriving human settlements and extensive agriculture (Willis et al. 2004).

[16]In contrast to moist forests, many dry tropical forests, such as the miombo woodlands of Africa, have frequent underburns.

The occurrence of wildfire in moist tropical forests is limited by high fuel moisture. A study in the Rio Negro region of Venezuela, one of the wetter areas of Amazonia (3,500 mm precipitation annually), found abundant fuels that were quite flammable once dry (Kauffman et al. 1988). Fuel drying depends not only on rain-free periods, but also on relative humidity, which depends, in turn, on forest structure. In their study in the Rio Negro, Uhl et al. (1988) found that fuels dried sufficiently to burn only when relative humidity dropped below 65 percent. During rain-free periods (which seldom exceed a few days in these ecosystems) relative humidity does drop below this level in open-canopy forests, but this occurs only rarely in closed-canopy forests. (Forest type in this area depends on soil texture, nutrients, and drainage patterns.) Uhl et al. conclude that slight shifts in microclimate that result in lower relative humidity could make fire more commonplace in moist closed forests of Amazonia. Such shifts could, and have already, occurred through human activity and may also be triggered by changes in global climate patterns (Kauffman and Uhl 1990).

The influence of humans was clear in the Indonesian megafires of 1997/1998. Pulpwood plantations experienced heavy damage, with almost two-thirds destroyed (Siegert et al. 2001). Damage in other forests correlated with time since logging. Nearly 50 percent of forests that had recently been selectively logged were severely damaged, compared to 26 percent of forests that had been selectively logged 15 to 20 years prior to the fires and 17 percent of unlogged forests (Siegert et al. 2001). Selective logging in rain forests opens the canopy, leaves large volumes of flammable debris, and encourages the growth of small trees, all of which increase the probability of severe fire. Once burned, forests do not recover to fire-resistant rain forest, or at least do so slowly, a phenomenon seen in both Indonesia and Amazonia, leading researchers to conclude that selective logging initiates a positive feedback loop in which recurring fires maintain forests in a fire-prone condition (Cochrane et al. 1999; Siegert et al. 2001). Cochrane et al. (1999) documented patterns of burning during 1997 fires in Amazonia. They found that the probability of burning increased sharply with the number of previous fires: 23 percent of previously unburned forests burned in 1997, 39 percent of forests that had experienced one previous fire burned in 1997, 48 percent of forests that had experienced two previous fires burned in 1997, and 69 percent of forests that had experienced three or more previous fires burned in 1997. Cochrane et al. describe the situation as follows: "Forest fires create positive feedbacks in future fire susceptibility, fuel loading, and fire intensity. Unless current land use and fire use practices are changed, fire has the potential to transform large areas of tropical forest into scrub or savanna."

Human influences on fire in most tropical forests are complex and varied. Kaimowitz (2005b) discusses the findings of Dennis et al. (2005), who studied factors contributing the 1997/1998 fires in Indonesia:

> The authors found the fires causes varied greatly between sites. Large plantation companies and small farmers both used fire to clear land and as a weapon against each other in land tenure disputes. Companies cleared land for oil palm and timber plantations. Small farmers cleared it mostly for annual crops, coffee, or rubber. Villagers burnt undergrowth to make it easier to get into areas to fish and hunt. Many fires accidentally got out of control and burned areas they weren't intended to.
>
> Many factors encourage fires. Multiple government agencies allocate the same land to different groups and fail to recognize villagers´ rights. People have little incentive to make sure the fires they set don't escape and burn places they did not mean to. Burning may be the cheapest way to clear land, and companies willing to explore other options don't get much support. Logging creates flammable debris and makes forests drier and more susceptible to fires. Draining peat swamps for plantations makes them more flammable. The government has little capacity to regulate fires or teach people to manage them.
>
> These problems require major changes in land tenure and coming up with agricultural and forestry policies and approaches adapted to each region. That's why it is much easier for policymakers and donors to sit back and do nothing. That way everything will be fine until its not

Uhl et al. (1988) listed four ways in which human activity increases the susceptibility of moist tropical forest to fire: "(1) Human activity generally involves fire in one way or another and thus ignition potential increases; (2) forest cutting for any purpose leaves slash on the ground and thus increases the fuel load; (3) opening of the forest, by increasing the amount of radiation reaching the forest floor and decreasing relative humidity, allows fuels to quickly dry to the ignition point; (4) deforestation, at the basin-wide level, could change overall climatic patterns—decreasing evapotranspiration, total precipitation, and mean relative humidity—thereby increasing fire likelihood."

In general, removing or opening the canopy increases temperature and winds within a forest, and that, in turn, decreases relative humidity and accelerates the rate at which fuels dry and become flammable. Fifty percent reduction in canopy cover within a primary forest can increase average temperatures within the forest by 10°C and decrease relative humidity by 35 percent (Kauffman and Uhl 1990). Uhl and Kauffman (1990) studied the susceptibility of four vegetation types to fire in the eastern Amazon. There, primary forests stayed sufficiently cool and moist that they would not burn even after a 30-day dry period. In pastures, midday temperatures averaged 10°C hotter and relative humidity averaged 30 percent lower than in primary forest, and that resulted in the grasses drying sufficiently to burn within 24 hours after a rainfall. Openings in selectively logged forests

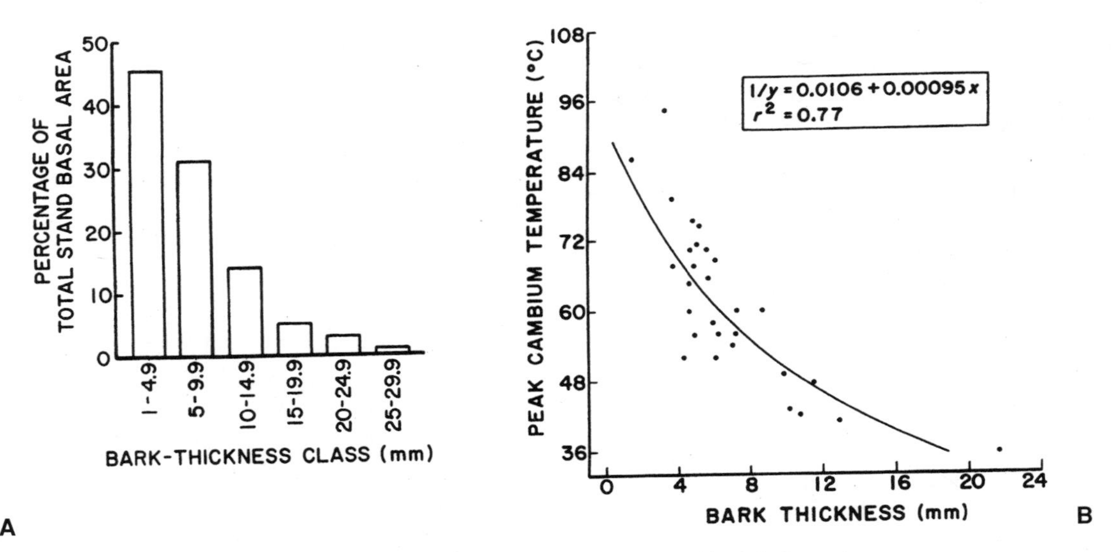

Figure 7.17. A, The distribution of stand basal area among trees of different classes of bark thickness, Amazon basin, Brazil. The stand basal area is the sum of the basal areas of individual trees, calculated at 1.4 m above ground level. B, The relationship between bark thickness and peak cambium temperature during low-intensity fire, Amazon basin, Brazil. Cambium temperatures greater than 6°C for over 60 seconds have a high probability of killing a tree. A low-intensity fire is likely to kill trees representing well over 50 percent of the basal area in rain forests of the type in A. (Uhl and Kaufman 1990. Copyright © 1990 by Ecological Society of America. Reprinted by permission)

became dry enough to burn after 5 to 6 rainless days, and secondary forest would burn after 8 to 10 rainless days.

These patterns have significant implications for conservation strategies. As the proportion of pastures and partially logged stands increases within a given landscape, the probability that fire will be propagated into intact primary forests also increases. Increased frequency of even low-intensity ground fires would be a disaster for moist tropical forests, whose trees are thin barked and unlikely to survive what in northern conifer forests would be a very gentle burn. Uhl and Kaufmann (1990) estimate that 98 percent of primary forest trees in Amazonia would be killed by a low-intensity surface fire (fig. 7.17). Therefore, primary forest reserves imbedded within a highly altered and fire-prone landscape are unlikely to be stable in the long term.

7.3 WIND

After humans and fire, wind is probably the most common disturbance to forests, with scales of damage that range from a single tree to many hectares. Intense continental windstorms (tornados and downbursts[17]) impact thousands of hectares of forest in the eastern and central North America each year (Peterson 2000). Downbursts may come in clusters (called derechos) associated with intense thunderstorms. In 1999, what was probably a derecho damaged nearly 200,000 ha of forest in northern Minnesota (Peterson 2000). In 1995, a derecho blew down 60 to 100 percent of the trees over 15,300 ha in upstate New York and damaged an additional 277,000 ha. Peterson (2000) gives other examples of severe wind disturbances to forests of eastern North America. Like fires, however, small events, creating openings on the order of a few tens of hectares or less, are much more common than very large ones, although infrequent large windstorms may account for the most dramatic effects, (e.g., Seymour et al. 2002; Lorimer and White 2003). Canham and Loucks (1984) used presettlement survey records in northeastern Wisconsin to calculate that between 1834 and 1873, an average of 52 patches of forest (a patch being defined as at least 1 ha in area) were blown down annually (this did not occur in southern Wisconsin, an area with fewer intense thunderstorms). The majority of these were less than approximately 30 ha in size. Similarly, 50 percent of the patches blown down by a severe derecho in the Adirondack Mountains in 1995 were less than 10 ha in size, while only 3 percent exceeded 100 ha (Lorimer and White 2003).

Historically, intervals between catastrophic continental windstorms were quite long, at least in the northeastern and north-central United States. Canham and Loucks estimated that small blowdowns in northern Wisconsin had average return cycles of many hundreds of years, and large ones perhaps several thousands. Large blowdowns in the northeastern United States occurred at intervals of well over one thousand years (Lorimer and White 2003). Because of the long periods free of stand-destroying disturbances, by far the majority of mesic forests of the northeastern United States were old growth at the time of European settlement (Lorimer and White 2003).

[17]Downbursts are intense downdrafts of varying sizes. Usually associated with thunderstorms, downbursts can generate winds at ground level well in excess of hurricane speed.

Coastal forests, and even those some distance inland, are subject to periodic hurricanes, of which there have been ample examples since the late 1980s. Two powerful hurricanes, Joan and Gilbert, swept through tropical forests in the Caribbean and Central America in 1988 (Boucher 1990). Joan, with 250-km/h winds, destroyed much of the canopy over one-half million hectares of primary rainforest in Nicaragua. Gilbert, up to that time the strongest hurricane ever recorded in the Western Hemisphere (Lynch 1991), defoliated dry tropical forests in the Yucatan peninsula of Mexico and blew down about 25 percent of the trees in montane forests of Jamaica. Another powerful hurricane (Hugo) passed directly over the largest rain forest in Puerto Rico in the fall of 1989, heavily impacting about 9,000 ha of the 11,000-ha Caribbean National Forest. Hugo then moved north along the east coast of the United States, coming ashore in South Carolina to heavily damage about one-half the sawtimber stands in the Francis Marion National Forest. At least six large hurricanes hit the Caribbean in the 1990s, some of these extending into the southeastern United States.

Hurricanes are a part of the scene in coastal forests, and even inland areas can be affected. For example, Fran (in 1996) damaged forests in the Duke University Forest in central North Carolina, and Opal's (in 1995) effects extended as far west as the Great Smokies in western North Carolina. Eight major hurricanes hit New England between 1635 and 1938, affecting forests well inland (Foster et al. 2004). As with large fires, hurricane damage is generally quite patchy, depending on topography, wind direction, and other factors (Lugo 2000; Boose et al. 2004; Foster et al. 2004). Hugo changed forests in the Luquillo Mountains of Puerto Rico from "mostly continuous canopies with occasional gaps to a structure of many gaps with occasional areas of intact, continuous canopies" (Brokaw and Grear 1991; Brokaw and Walker 1991). The influence of mountainous terrain on patterns of damage across landscapes is clearly seen in Puerto Rico. There, hurricane winds commonly come from the east, and south slopes along with pockets on north slopes are protected from the most damaging winds (Boose et al. 2004). (Wind movements through mountains can be complex and difficult to predict, however.)

Tree death may vary widely from one hurricane to another. For 23 hurricanes in the Caribbean and Southeast Asia, mortality varied from less than 5 to 100 percent, with most incidents falling between 10 and 80 percent (Lugo and Scatena 1996). Even 10 percent is a significantly high pulse of mortality for moist tropical forests, however, since background mortality (i.e., in the absence of a major disturbance) is on the order of 1 to 2 percent per year (Lugo and Scatena 1996). Not surprisingly, stronger hurricanes produce more damage. That fact is quantified using the Fujita Scale, in which F0 (sustained winds of 18–25 m/s) corresponds to leaves and fruits being blown off and branches broken, F1 (sustained winds of 26–35 m/s) to some trees being blown down, F2 (sustained winds of 36–47 m/s) to extensive blowdowns, and F3 (sustained winds of 48–62 m/s) to most trees down (Fujita 1987; Boose et al. 2004). Even hurricanes of roughly the same strength may produce quite different levels of tree mortality in different areas, however. According to Boose et al. (2004), "At the stand level, damage caused by winds of a given speed and duration varies as a function of stand composition and structure, which, in turn, are strongly influenced by land use and natural disturbance history". Conversion of lowland forests to farms, for example, may increase the exposure of upper-slope forests to damaging winds (Boose et al. 2004).

Throughout centuries hurricanes have shaped the composition of forests by favoring species that are adapted either to stand during high winds or to recover rapidly by sprouting from roots or efficient seed dispersal (Walker et al. 1991; Lugo and Scatena 1996). In Nicaragua, trees most likely to stand against Hurricane Joan were either very tall or very short. The so-called rain-forest emergents—trees that may be 20 m or so taller than the main canopy—stood best, while midsized trees were the most likely to be blown down (Boucher 1990). Emergents are generally slow growers with very dense wood that allows them to stand in high winds. Palms—generally short, shade-tolerant trees that are abundant in the understory of Central American and Caribbean rain-forests—also stood well. In South Carolina, trees native to the coastal plain, where high winds are relatively common, were less damaged by Hugo than trees whose range extended out of the coastal plain (Gresham et al. 1991).

Of the rain-forest trees that do not stand well against high winds, many recover quite rapidly. Initial surveys in Nicaragua indicated that more than three-fourths of the primary rain-forest trees blown down by Hurricane Joan had resprouted within four months (Boucher 1990), and by the fifth year the canopy, though shorter than the predisturbance forest, had completely closed (Vandermeer et al. 1996). By the late 1990s, the posthurricane forest was floristically very similar to the prehurricane forest, but with a greater representation of some tree species, especially sun lovers that benefited from the reduction in overstory canopy. Pioneering tree species that rely on seed dispersal to establish new seedlings were poorly represented in the recovery from Joan and are likely to spread back into devastated areas quite slowly for two reasons: (1) because of the distance that seed must travel from undamaged trees, and (2) because populations of birds that disperse seed of many tropical trees were severely affected by the hurricanes (seed dispersal by birds is covered in chapter 11). Pine forests that are blown down are not likely to recover as rapidly from hurricanes as the tropical forests. Pines do not sprout from roots but rather regenerate through seed (with a few exceptions, most notably redwood and sequoia, this is true of all conifers).

Mortality and recovery following Hugo followed a different pattern than occurred in Nicaragua following Joan (Scatena et al. 1996; Lugo 2000). In Puerto Rico, as in Nicaragua, forest biomass recovered rapidly, but pioneering tree species[18] were a more important component than in Nicaragua following Joan. Lugo (2000) attributes this difference to the fact that Puerto Rico has a history of more frequent hurricanes than Nicaragua (50 per century in Puerto Rico vs. 10 per century in Nicaragua). More frequent disturbances would maintain a greater presence of pioneer tree species in the forest and therefore a greater potential for seed dispersal following a large hurricane. This underscores an important point: history is a critical determinant of how ecosystems respond to disturbances (Boose et al. 2004), and generalizations from one region to another should be made cautiously.

Lugo (2000) discusses how hurricanes have shaped ecosystems on the Luquillo Experimental Forest, Puerto Rico. (These patterns may not apply to regions with different topographies, environments, land uses, and disturbance histories; e.g., see Boose et al. 2004).

A large hurricane results in the following:

- Massive tree mortality, not only at the time of the hurricane but also months to years after
- High species turnover, coupled with increased diversity of successional pathways; depending on the extent and type of damage (e.g., large gaps, small gaps, hurricane-triggered landslides), different sets of early successional species are favored, creating diversity across the landscape
- Enhanced diversity of age classes due to differential tree kill and recovery
- Accelerated turnover of biomass and nutrients
- Enhanced carbon sinks (e.g., from wood buried in landslides or transported to streams)

A history of frequent hurricanes has produced the following:

- Strong selection on organisms, especially with regard to life-history patterns (e.g., age to attain reproductive maturity)
- Convergence of community structure and organization among different forests, particularly with regard to canopy structure (smooth), aboveground biomass (low), and species dominance (high)

[18] As the name implies, pioneers are adapted to colonize disturbed areas. They generally produce abundant, widely dispersed seeds and are relatively short-lived. In the neotropics, trees in the genus *Cecropia* are common pioneers.

At least some bird species were heavily impacted by the 1988 and 1989 hurricanes (Boucher 1990; Askins and Ewert 1991; Lynch 1991; Waide 1991a, 1991b; Will 1991). In the tropical forests, populations of fruit- and nectar-eaters appear to have been sharply reduced (reflecting the loss of their food), including the rare Puerto Rican parrot. In South Carolina, Hurricane Hugo severely damaged 90 percent of the habitat of the endangered red-cockaded woodpecker. This bird excavates nest holes in large old pine trees, and many in the path of Hugo were blown down. The abundance and species richness of birds in the tropical forests recovered within 17 months after the hurricanes; however, there was a shift in the composition of bird communities. Not surprisingly, posthurricane forests had fewer deep-forest bird species than prehurricane forests, and more species that prefer open fields and forest edges (Waide 1991b).

The consequences of recent hurricanes on bird populations contain some important lessons for ecologists, foresters, and conservationists. The primary message is that species restricted to a few local habitats can be rapidly pushed to extinction when those habitats are lost; species with widespread habitats are buffered against that. Hurricanes have undoubtedly reduced bird populations many times in the past, but once the natural recovery of the trees has restored suitable habitat, intact bird populations from forests outside the hurricane track have served as a source of recolonizers, a "bank account" from which individuals can be drawn to replace losses elsewhere. As suitable habitat becomes restricted to fewer and fewer areas, however, the bank account shrinks to where no reserves remain to replace losses. A species that finds itself in this situation is surely doomed to extinction (at least in the wild). The vulnerability of isolated populations has much to tell us about strategies of conservation, and about the way forests must be managed if one of society's objectives is to prevent extinctions. The old strategy of conservation centered on saving species in reserves that were essentially islands in a sea of highly managed forests. The new strategy is to manage forests in such a way that habitat is maintained outside preserves as well as inside. We shall return to this point in some detail in chapters 22 and 23.

Forest management can increase the risk of windthrow by suddenly exposing trees that had matured within sheltered forest interiors. A high proportion of trees blown down in Pacific Northwest forests, for example, are along clear-cut edges (fig. 7.18). Trees left standing within clear-cuts may also be subject to windthrow, particularly in areas that experience high winds. If exposed slowly rather than suddenly, interior forest trees can adapt to withstand at least moderate winds.

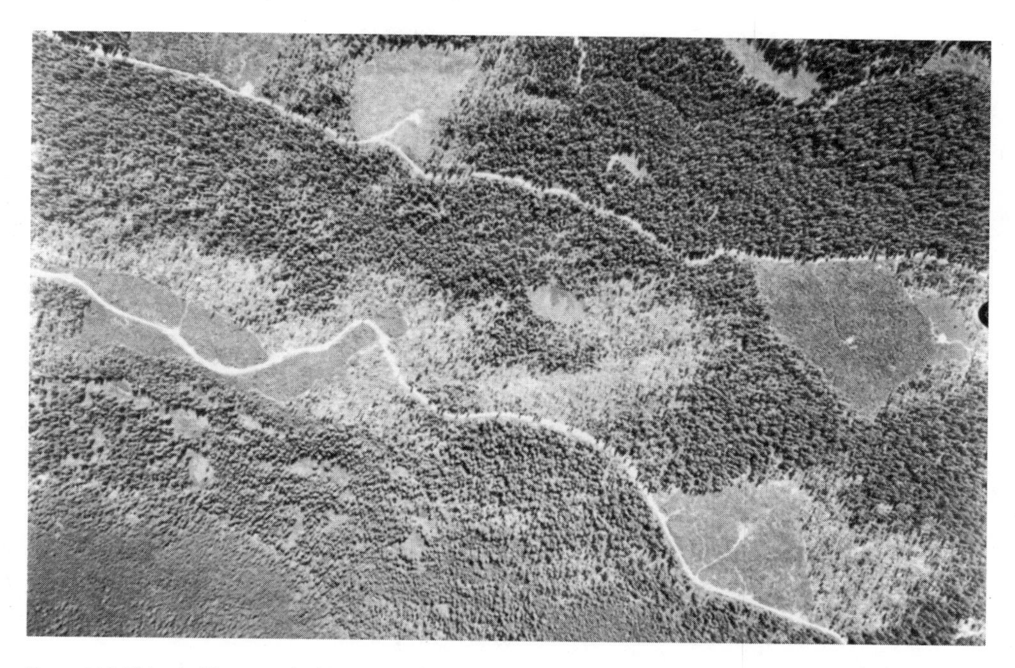

Figure 7.18. This windthrow on the Mount Hood National Forest (Oregon) resulted from a strong winter storm. Gray areas are blowndown trees. Sharp edges, such as occur along roads and clear-cut boundaries, are particularly vulnerable to windthrow. In areas of the Mount Hood National Forest that experience periodic strong winds, a few relatively small clear-cuts created a self-reinforcing dynamic in which blowdown along clear-cut edges created a larger amount of edge vulnerable to additional blowdown with the next storm. The subsequent unraveling of the landscape has proved difficult to stop and provides a good example of how relatively small changes in landscape pattern can sometimes trigger large changes in the dynamics of disturbance. (Courtesy of Mount Hood National Forest)

7.4 TECTONIC ACTIVITY

In areas with high levels of tectonic activity, volcanoes and earthquake-triggered landslides are significant sources of forest disturbance. The 1980 eruption of Mount St. Helens, for instance, was only one of a series that have destroyed (or at least perturbed) forests of the western United States throughout geologic history (fig. 7.19). Hawaii's Mauna Loa has erupted 39 times since 1832 (Hawaii Institute for Volcanology 2004), in aggregate covering 806 Km^2 of the volcano's surface. Lava flows destroy all forests in their path excepting small islands ("kipukas") that for one reason or another are bypassed.

Earthquakes are another form of tectonic disturbance. Earthquake-triggered landslides are common in the temperate forests of Chile, often occurring on sites destabilized by previous forest fires, as living roots contribute to soil stability in steep lands (Veblen 1985). Forest destruction by earthquake-triggered landslides has also been documented in tropical forests of Central America and New Guinea and is probably common elsewhere in the tropics (Garwood et al. 1979). Garwood et al. estimate that 38 percent of Indo-Malayan rain forest and 14 percent of the American lie in zones of high tectonic activity. Intense rainfall associated with hurricanes can also trigger landslides (Lugo 2000).

7.5 FLOODING

Streamside (or riparian) forests are often perturbed and even destroyed by flooding. In mountains, high runoff from snowmelt is an annual event sometimes severe enough to wipe out riparian forests. In flatlands, large areas can be affected by floods or outright change in a river's course. Using satellite imagery (Landsat), Salo et al. (1986) estimated that 12 percent of lowland forests in the Peruvian Amazon were in early successional stages along rivers, and a total of 26.6 percent of forests had characteristics of recent erosion or deposition triggered by meandering rivers (fig. 7.20).

7.6 INVASIVE SPECIES

An invasive species is one "that is not native to an ecosystem and whose introduction does or is likely to cause economic or environmental harm or harm to human health" (Executive Order No. 13112, 64 Fed. Reg. 6183–6186 [1999]). Invasive species are a serious and growing problem worldwide. In the United States alone, Pimentel et al. (2000) estimate that invasive species cause forest products losses exceeding 2 billion dollars per year. Worldwide, the consequences of invasives on native biological diversity and ecosystem health are immeasurable. In the United

Figure 7.19. Chile's Lonquimay volcano (background) erupted on Christmas Day 1989, covering forests that once stood in the foreground with a thick layer of ash, seen as black in the photo.

States, 57 percent of the imperiled plant species studied by Wilcove et al. (1998) were negatively affected by invasive species, making invasives second only to habitat loss and degradation as a cause of native species endangerment. In some areas, such as the islands of Oceania, invasive species represent the most the serious threat to native biological diversity. Vitousek et al. (1996a) argue that by breaking down the diversity-preserving barriers to large-scale species movements, human-mediated invasions threaten to homogenize the planet.

Invasive species have both direct and indirect effects on native species (see the review by Chornesky et al. 2005). Pathogens, pests, and other "feeders" of one kind or another have the most widespread direct effects. In the United States, introduced pathogens have virtually eliminated American chestnut from eastern forests and, in the west, greatly damaged white pines, larches, and Port Orford cedar. A recent introduction (first detected in 1994), the pathogen *Phytophtora ramorum* (sudden oak death) is spreading rapidly in California and Oregon and threatens oaks and other hardwoods throughout the United States. Various insects have reduced populations of true firs, eastern hemlock, and beech in the United States, and European gypsy moth *(Lymantria dispar)* has defoliated a wide variety of tree species across millions of hectares. Two new introductions to the United States, emerald ash borer *(Agrilus planpennis)* and the generalist feeder, lobate lac scale *(Paratarchardina lobata lobata)*, threaten, respectively, ash species nationwide and up to 150 other species of native trees and shrubs. West Nile virus has spread rapidly across North America, killing some humans and "untold numbers of birds, mammals, and reptiles" (Marra et al. 2004).

Introduced ungulates (hoofed mammals) are a particular problem on islands with no native grazing mammals (e.g., Hawai'i), hence no evolutionary pressure for plants to develop and maintain defenses against grazing. Invasive predatory insects such as yellowjackets can decimate native arthropods.

Invasive plants can significantly impact native plants. This is most obvious in the case of aggressive invaders that either form closed, monotypic stands that exclude native plants, or overgrow and smother native trees and other plants (climbing vines). The most prominent example of the latter in the continental United States is kudzu (Pueraria montana var. *lobata)*, a semiwoody vine native to Asia (introduced to the United States as a potential forage plant) that now covers 2 to 3 million ha in the southeastern United States. *Passiflora tarminiana*, an aggressive woody vine native to the Andes, has become a serious pest in various areas of Oceania and southern Africa. In Hawai'i, where it is called banana poka, *P. tarminiana* is rapidly invading native forests (fig. 7.21).

Invasive species can have profound system-level effects. Asner and Vitousek (2005) put it this way: "Some invaders alter the structure and/or functioning of the ecosystems in which they occur; these do not just compete with or consume organisms in their new habitats, they change the rules of the game under which all organisms exist". Such species are called **transformers** (D'Antonio et al. 2004). One widespread example is invasive grass. Several characteristics of

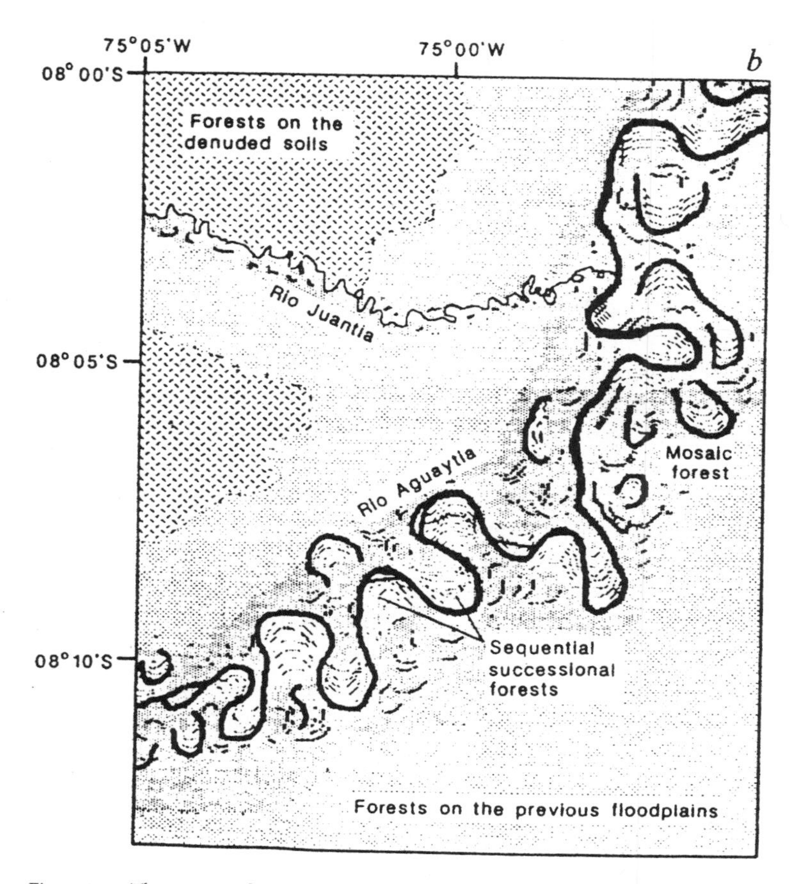

Figure 7.20. The pattern of successional forests created by a meandering river. Periodic changes in the river's course create new pockets of primary succession (sequential successional forests), resulting in a landscape mosaic of different types of forest (mosaic forest). Previous floodplains, occupied by later successional forests, are slightly higher in elevation than current floodplains (because of a buildup of past river depositions) and, hence, are not disturbed by the river in its current position. Denuded areas *(upper left corner)* are being eroded by the Rio Juanita at flood stage. (Salo et al. 1986. Reprinted with permission of Nature. Copyright © 1986 by Macmillan Magazines, Ltd.)

Figure 7.21. The exotic vine banana poka *(Passiflora mollissima)* covering native trees in Hawai'i.

grasses make them more flammable and quicker to recover from fire than woody vegetation.[19] Often purposely introduced for forage or as ornamentals, various species of African and Asian grasses have greatly increased the incidence of wildfire in both North and South America, Oceania, and Australia (D'Antonio and Vitousek 1992). D'Antonio and Vitousek (1992) describe the dynamic that develops when grasses invade forest landscapes: "Invasion can set in motion a grass/fire cycle where an alien grass colonizes an area and provides the fine fuel necessary for the initiation and propagation of fire. Fires then increase in frequency, area, and perhaps intensity. Following these grass-fueled fires, alien grasses recover more quickly than native species and cause a further increase in susceptibility to fire."

Some invaders alter soil nutrient balances in ways that give them an advantage over natives. In Hawai'i, the invasive nitrogen-fixing tree[20] *Myrica faya* increases biologically useable soil nitrogen up to fourfold in young lava substrates (Vitousek et al. 1996a; Asner and Vitousek 2005). Native plants and associated soil organisms that colonize lava are adapted to low nitrogen levels; hence, nitrogen enrichment benefits invaders at the expense of natives. Another invasive nitrogen-fixing tree, *Falcartia molluccana,* greatly increases rates of nutrient cycling in stands it invades, increasing supply rates of both nitrogen and phosphorus (Hughes and Denslow 2005). In contrast, an invasive understory herb reduces the nitrogen available to native overstory trees (Asner and Vitousek 2005), and fire spread by invasive grasses lowers site fertility by volatilizing nitrogen (Mack et al. 2001).

Belowground consequences go beyond nutrients. Wolfe and Klironomos (2005) conclude that "in numerous ecosystems, the invasion of exotic plant species has caused major shifts in the composition and function of soil communities." They give the example of garlic mustard *(Alliaria petolata)*, a European native that invades the understory of North American forests. Garlic mustard produces compounds that inhibit mycorrhizal fungi required by native trees (mycorrhizal fungi are discussed in chapter 11). Plants are not the only invaders that transform soil processes. In North America, invasive European and Asian earthworms alter nutrient availability and greatly affect native soil organisms (Bohlen et al. 2004a, 2004b).[21]

By altering ecosystem structure, invaders can transform the rules of engagement for native predators and prey. In Hawai'i, attempts to reintroduce the highly endangered Hawaiian crow to its forest habitats were unsuccessful, largely because of heavy predation by the Hawaiian hawks.[22] A major factor contributing to the vulnerability of the crow was invasive ungulates, which destroyed the rich understory of subcanopy trees and shrubs that once provided cover for the crow.

One invasive species can promote the establishment and spread of another; in fact, this is common. Invasive ungulates and birds, for example, transport the seeds of invasive plants, and the ungulates create seedbeds by disturbing the soil. In Hawai'i, water collected in wallows created by feral pigs provides breeding habitat for invasive mosquitoes, which, in turn, serve as vectors for two introduced diseases of native birds, avian pox and avian malaria. Native and exotic herbivores differ in their effects on plant invasions; native herbivores inhibit, while exotic herbivores promote the spread of invasive plants (Parker et al. 2006).

Invaders can have effects that ramify beyond the area of invasion. Hillsides denuded by invasive ungulates erode soil into streams and nearshore marine ecosystems (fig. 7.22). Fast-growing timber plantations planted onto grasslands or semiarid shrublands may use more water than the native vegetation, thereby lowering streamflows. This happened in Fiji, where pines planted on former grasslands reduced dry-season flows by 50 to 60 percent, and in South Africa, where forestry plantations of various exotic trees reduce streamflow by nearly 30 percent compared to native *fynbos* (i.e., shrublands) (Postel and Thompson 2005). In these cases the trees happen to be exotics, but trees native to a particular country could well have the same effect when planted outside their normal range.

A species does not necessarily have to cross a political boundary in order to become invasive, it is the ecological boundaries that count. Deer, for example, commonly overgraze native vegetation when introduced to new areas (e.g., Coomes et al. 2003), but they do the same in their native habitats when loss of predators or other ecological changes allow their populations to grow (chapters 9, 19). In their survey of invasive plant species in Ontario wetlands, Houlahan and Findlay (2004) concluded that "(e)xotic species were no more likely to dominate a wetland than native species. . . . The key to conservation of inland wetland biodiversity is to discourage the spread of community dominants, regardless of geographic origin." This is not to minimize the threat of exotic organisms, which in general are more likely than natives to create problems; however, the ecological root of invasive behavior is a breakdown in regulatory mechanisms (e.g., predation, restrictions to movement), which can happen with natives as well as exotics. (Regulation is discussed in chapters 9, 11, and 19).

[19]Because of their high surface-to-volume ratio, grasses dry quickly. In the tropics they may also produce relatively large amounts of biomass. When tops are killed by fire, the large, living root systems are capable of producing new top growth quickly.

[20]Nitrogen fixation is discussed in chapter 11. Basically, a few specialized bacterial species are capable of converting atmospheric nitrogen gas (N_2), which is unusable by other organisms, to useable NH_4. Some species of plants form partnerships with nitrogen-fixing bacteria, and these are called nitrogen-fixing plants.

[21]In many northern forests native earthworms have not recolonized since glaciation (Bohlen et al. 2004a, 2004b). Other northern forests have native earthworms, but without the transforming power of the introduced worms.

[22]The Hawaiian crow is now extinct in the wild. The Hawaiian hawk, which is also listed as endangered, is doing reasonably well.

Figure 7.22. Impacts of goats in Hawai'i. *Upper:* Steep, bare mountain slopes on Kauai. *Lower:* Eroding hillslopes on the island of Kaho'olawe produce a sediment plume that impacts nearshore ecosystems (gray band in the photo).

Throughout history, species have commonly moved around with past climate changes and hitchhiked along with humans in our many migrations. The modern era differs for several reasons. Humans have become highly mobile, and the accelerated global economy translates to more movement of forest, agricultural, and horticultural products. Rapid movement over long distances means that invasive organisms are likely to be transported without the complex of natural enemies that regulates their populations in their native environment (Torchin and Mitchell 2004). Moreover, ecosystems are disturbed by multiple stresses unique to the modern era, creating both literal and figurative openings for invaders to establish and spread. A study in Madagascar, for example, found that "logging was the overriding factor influencing establishment of nonnative plants" (Brown and Gurevitch 2004). In the northeastern United States, incursions of invasive shrubs into native shrublands and early successional forests correlate closely with the prevalence of both present-day and historic agricultural fields (Johnson et al. 2006). However, Sakai et al. (2001) conclude that "few communities are impenetrable to invasion by exotic species."

Not all introduced species become invasive. The so-called tens rule of Williamson (1996) says that approximately 10 percent of introduced exotics establish and approximately 10 percent of those spread and become problems. However, for vertebrates introduced from Europe to North America or vice versa, the proportions are closer to 50 percent at each step (Jeschke and Strayer 2005). Attempts to correlate invasive behavior with life-history characteristics have had mixed success (see reviews by Sakai et al. 2001 and Daehler 2003). Characteristics such as rapid population growth, fast-growing individuals, widely disseminated seeds, clonal reproduction, or tasty fruit facilitate the spread of species whether they are exotics or natives. In 79 comparisons between co-occurring native plants and invasive exotics, Daehler (2003) found no clear pattern of invasive species having intrinsically greater growth rates, competitive ability, or fecundity. Instead, the relative performance of natives and exotics depended on growing conditions, invasive exotics having the advantage when resources were relatively high and natives having the advantage when resources were low. As we saw earlier, some of the most successful invasive plants are the nitrogen fixers, which pack their own nutrition. Recent evidence shows a strong phylogenetic control over invasiveness. In California, the most highly invasive grasses are the least related to native grasses (Strauss et al. 2006). Strauss et al. conclude that "the presence or absence of multiple, closely related species and not just a single most closely related species may more effectively limit the success of an invader once it has become established." Exactly how relatedness influences invasiveness is unclear, however it seems likely that native herbivores would have a greater likelihood of being preadapted to close relatives of native plants, which would support the "escape from natural enemies" hypothesis (Strauss et al. 2006).

Once established, invasive species are difficult or impossible to get rid of, although biological control (importation of natural enemies from the invasives native habitat) has had some notable successes (chapter 19). The Global Invasive Species Database (2005) discusses biological control of banana poka in Hawai'i:

> Three biocontrol agents have been released in Hawaii. *Cyanotricha necryia*, a foliage-feeding moth, was released in 1988 but failed to establish. Another moth species, *Pyrausta perelegans*, was released in 1991. It feeds on the buds, leaves, fruit, and shoot tips of *P. tarminiana*. It has established but is not common. A leaf spot fungus, *Septoria passiflorae*, which was released in 1996, is now widespread and causing large disease epidemics. There have been *P. tarminiana* biomass reductions of 80–95% over more than 2000 ha, giving indications that the leaf spot fungus has great potential. Other agents that are being investigated include *Zapriotheca* nr. *nudiseta*, a fly that feeds on flower buds, as well as *Josia fluonia* and *J. ligata*, two species of defoliating moths. . . . In damp areas *P. tarminiana* may suffer from slug herbivory (Binggeli 1997).

Biological control is not risk free, however, because introduced control agents have the potential for spin-off effects on native species and ecosystems (see Simberloff and Stiling 1996). This was especially true with early attempts, which often suffered from poor understanding of the biology and ecology of the species involved. Modern biological control is much more sophisticated (e.g., Murdoch and Briggs 1996; Fagan et al. 2002; Taylor and Hastings 2005). Nevertheless, Simberloff and Stiling (1996) argue that "we must view biological control as risky. . . . Specific projects should not be assumed to be innocuous until substantial effort has been expended to support this assumption." In the final analysis, the most effective and safe control measure for invasive exotics is not letting them in.

In some cases invasive species can perform valuable ecological functions. For instance, native plants may be unable to reestablish on severely degraded land (either because of harsh environmental conditions or lack of seed sources), while invasive plants can. The invasives stabilize and rebuild soils and, at least in some cases, ameliorate environmental conditions sufficiently for natives to restablish. Lugo (2004) discusses how this occurs in Puerto Rico:

> Invasive alien tree species . . . often form monospecific stands on deforested lands that were previously used for agriculture and then abandoned. Most native pioneer species are incapable of colonizing these sites. . . . Alien trees may dominate sites for 30 to 40 years, but by that time native species begin to appear in the understory. By 60 to 80 years, unique communities comprising both alien and native species are found on these sites. . . . The invasion of a site and the formation of an alien-dominated forest serve important ecological functions, such as repairing soil structure and fertility, and restoring forest cover and biodiversity at degraded sites.

Lugo points out that this dynamic is in response to a new disturbance regime (agriculture) to which the native species are not adapted. "These examples," he goes on to say, "do not imply value to alien species invasions. They simply demonstrate that ecological phenomena underpin these invasions and that further study of these occurrences is needed, in order to gain an understanding of the causes and consequences of such events" (Lugo 2004).

7.7 SUMMARY

Disturbance of one kind or another is a pervasive feature of forests. Disturbances may be categorized as either exogenous, originating from outside the ecosystem (e.g., fire), or endogenous, originating from within the system (e.g., outbreaks of native insects). In practice, however, endogenous and exogenous factors often interact; for example, weather influences susceptibility to insects, insect infestations increase vulnerability to catastrophic fire. The disturbance regime of a given forest is usually a complex mixture of infrequent large events punctuated by frequent smaller events.

Fire is the most common disturbance in a wide variety of forest types, including coniferous forests of northern, western, and southern North America; eucalypts of Australia; and dry tropical forests such as the miombo woodlands of east Africa. Species common to these areas have adapted to fire in various ways, and their persistence depends on periodic fire. Other common disturbances are wind, tectonic activity, flooding, and insect infestation. Natural disturbances—whether exogenous or endogenous—are an integral part of ecosystems and, more often than not, are agents of renewal rather than destruction. On the other hand, foreign disturbances, those of a type or severity to which native species are maladapted, can be a serious threat to ecosystem integrity. These have become increasingly prevalent during the past several decades and include uncommonly severe fires, rapid climate change, and invasive species. The latter in particular represent a serious and growing threat to ecosystems worldwide.

8

Patterns and Mechanisms of Succession

> During more than a century of debate, many concepts of succession have been put forth, yet none of these appears to be comprehensive enough to adequately capture the dynamic and highly variable processes that occur during succession.
>
> CRISAFULLI ET AL. 2005

Succession involves change at three different levels: (1) compositional, or the change in the particular mix of species and life forms (e.g., herbs, grasses, shrubs, trees) that dominate the system at any point in time; (2) structural, or the spatial patterns in which the elements of the system occur (chapter 9); and (3) system-level properties such as species diversity, nutrient cycling, and productivity. These three levels, which in effect correspond to the dynamics of the parts (composition) versus the dynamics of the whole (structure, processes, functions), do not necessarily follow similar patterns during succession. Various combinations of species, for example, might produce the same overall species diversity or system productivity (e.g., Ewel et al. 1991). On the other hand, the three levels are also interrelated, and to understand the phenomenon of succession one must understand all.[1]

8.1 HISTORICAL NOTES

Among American ecologists, two names stand out in the development of successional ideas during the early years of the twentieth century: Frederick Clements and H. A. Gleason. Clements (1916) believed that communities were superorganisms, and that

[1]McIntosh (1981) and Finegan (1984) give relatively dispassionate accounts of the long-running controversy among ecologists about the nature of succession. It is debatable how much this argument has contributed to our understanding of natural processes; however, it does reveal something about the sociology and psychology of science.

succession was a maturation of the community toward its most mature state, which he called the **climax:** "Succession must then be regarded as the development or life-history of the climax formation. It is the basic organic process of vegetation, which results in the adult or final form of this complex organism. All the stages that precede the climax are stages of growth. They have the same essential relation to the final stable structure of the organisms that seedling and growing plant have to the adult individual." For Clements, the composition of the climax vegetation was uniquely determined by climate: "Such a climax is permanent because of its entire harmony with a stable habitat. It will persist just as long as the climate remains unchanged, always providing that migration does not bring a new dominant from another region."

Like Clements, Gleason recognized the importance of environment in determining the composition of plant communities; however, he rejected Clements's idea that a given community was a repeatable entity that established wherever a given set of environmental conditions occurred. In his 1926 paper, "The Individualistic Concept of the Plant Association," Gleason argued that two factors made each community distinct from every other. First was the independent nature of plant species: "Every species of plant is a law unto itself, the distribution of which in space depends upon its individual peculiarities of migration and environmental requirements. . . . The behavior of the plant offers in itself no reason at all for the segregation of definite communities" (Gleason 1926).

Gleason's second factor was **randomness,** which meant that the composition of a given community was not completely predetermined by environmental factors. Rather, any number of plant species might be able to occupy a given site but may or may not depending on whether their seeds were dispersed into disturbed areas. Gleason argued that community composition might be repeatable in regions with few species, simply because there were few alternative communities that might develop. But, as species diversity increased within a region, so too did the variety of community types that might develop on a given site during the course of succession.[2]

As with most of the polar issues that have consumed an undue amount of the time and energy of ecologists over the years, the "truth" (at least, as we understand it today) contains elements of both viewpoints but is adequately captured by neither. Most modern ecologists reject Clements's idea of communities as superorganisms. But, if a community is not an organism in the same sense as an oak tree or a swallowtail butterfly, neither is an individual plant a law unto itself, as Gleason suggests. Every organism is part of and in interaction with a larger community: feeding, being fed upon, competing, cooperating, and coexisting. As we discuss later in the chapter, ecologists increasingly recognize positive interactions among plants, leading Callaway to conclude that "the prospect that the distribution and abundance of any species in a plant community may be positively affected by . . . other species suggests that communities are organized by much more than the 'fluctuating and fortuitous immigration of plants and an equally fluctuating and variable environment' as stated by Henry Gleason."

The problem with highly polarized issues is that they present only stark choices: is a community an organism or merely a collection of individuals that are laws unto themselves? The correct answer is "none of the above." The proper questions to ask are: What type of a system is a community? What is nature of the interactions among its components? What is the relative degree of autonomy and interdependence among community members? What environmental and evolutionary forces shape the character of communities?

Gleason was on to something when he invoked the importance of chance, which is a powerful force in nature. Dale et al. (2005b) discuss the 1980 eruption of Mount St. Helens in Washington State:

> The timing of the eruption—that is, time of day, time of year, and the stage of development (succession) of plant and animal communities—strongly influenced patterns of survival and succession. The early morning eruption allowed nocturnal animals to be protected in their subterranean retreats. The early spring eruption meant that snow and ice created refuges and that plants had not yet broken winter dormancy at high elevations. The early successional stage of many recently harvested forest sites led to a profusion of wind-dispersed seeds of pioneer plant species. The importance of timing underscores the significance of chance in survival and successional pathways.

Successional patterns are influenced by the type, size, and season of disturbance, and by the availability of seed, all of which have at least some degree of stochasticity associated with them. Also, as Gleason predicted, the more species existing in a given climatic region, the more uncertainty in the initial composition of the successional community on a given piece of ground (see the discussion on neutral biodiversity models in chapter 10). On the other hand, the Clementsian concept of repeatable plant associations clearly holds in some areas; as we discussed in chapter 5, it appears in the western United States as the widely used habitat types—repeatable plant communities believed to indicate certain environmental conditions.

Even in species-rich areas, the importance of chance in community composition is lower than one might expect. Plants evolve strategies to reduce the uncertainty associated

[2]The same contrasting viewpoints developed in early twentieth-century Europe. The Russian Sukatchew and the Frenchman Braun-Blanquet argued that plant communities were repeatable entities, while the individualistic view was developed by the Russian Ramensky and the Frenchman Lenoble.

with randomness and to retain a presence, or a potential presence, on a site. Biological legacies, including (but not restricted to) buried seeds, live roots from which new tops sprout, and mycorrhizal fungi, are passed from the old community to the new, shaping the new community in the image of the old. In the last chapter we saw a good example of this from Nicaragua, where sprouting from roots after hurricanes tended to maintain the prehurricane community composition. In this chapter, we see that biological legacies are of widespread importance in maintaining community composition through disturbance. Moreover, we learn that despite the undeniable (and not surprising) fact that different species differ in their environmental requirements (as Gleason argued), the Clementsian view of mutual dependence among members of a community also has validity.

Based on tree responses to hurricanes in Jamaica, Bellingham et al. (1995) proposed four general types of population response to disturbance. **Resilient** species have high mortality but also high recruitment. **Usurpers** have low tree damage but also high recruitment and release of individuals in the understory. **Resistant** species have low mortality and little increase in recruitment. **Susceptible** species have high mortality that is not compensated for by release of understory individuals or increased recruitment. Following a hurricane that struck an old-growth mixed pine and hardwood forest in northern Florida, Batista and Platt (2003) classified tree species responses according these four syndromes:

> *Pinus glabra* exhibited the Resilient syndrome, with high tree mortality but massive recruitment after the hurricane. Two subcanopy species *(Carpinus caroliniana* and *Ostrya virginiana)* and three canopy species *(Liquidambar styraciflua, Quercus michauxii* and *Q. nigra)*, exhibited the Usurper syndrome, showing low tree damage, some release of understorey trees and saplings, and substantial recruitment. *Ilex opaca* (subcanopy), and *Fagus grandifolia* and *Nyssa sylvatica* (canopy), exhibited the Resistant syndrome, characterized by low tree damage and little increase in recruitment. The main canopy dominant, *Magnolia grandifolia*, exhibited the Susceptible syndrome; it had large reductions in growth and survival and no detectable release of understorey individuals.

Batista and Platt went on to conclude that "long-term persistence appeared to depend on periodic hurricanes (or equivalent large-scale disturbances) for Resilient and Usurper syndromes, but might become compromised by recurrent hurricanes for the Susceptible syndrome."

Clements, Gleason, and others who followed them focused on end points, of which Clements's climax was the archetype. Today, however, ecologists are more concerned with the mechanisms and processes that shape community dynamics (Pickett et al. 1987a, 1987b). The following two sections explore successional patterns and then consider the mechanisms behind those patterns.

8.2 COMPOSITIONAL AND STRUCTURAL CHANGE DURING SUCCESSION

8.2.1 The Stages of Succession

Gomez-Pompa and Vazquez-Yanes (1981) distinguish five successional[3] stages in evergreen rain forests of Mexico:

- A short period of dominance by ephemeral herbs (a few weeks or months)
- Dominance by secondary shrubs that eliminate pioneer herbs by shading (6 to 18 months)
- Dominance by secondary trees of low stature (3 to 10 years)
- Dominance by taller secondary trees (10 to more than 40 years)
- Dominance by tall primary trees (until destroyed by disturbance)

Although the number of distinct stages and timing will vary depending on environment and total species pool, the above sequence is a fair representation of succession in all forests. Sites are initially dominated by rapidly growing, often short-lived herbaceous species; these are followed by shrubs and pioneer trees, which, barring another disturbance, are eventually succeeded by late successional, usually shade-tolerant tree species. Figure 8.1 illustrates this pattern for a successional sequence initiated by a meandering river in Amazonia.

Many early successional plant species are intolerant of shade; hence, their seedlings do not survive and grow beneath an established canopy, and in the absence of disturbance these species do not persist (except in some cases through buried seed). Late successional plants, however, are at least moderately tolerant of shade. The earliest successional stages are relatively short in both time and stature of the dominant vegetation, the intermediate stages increasingly longer and taller, culminating in Clements's climax community, one that (in theory) persists indefinitely (but in fact rarely does).

It is important to keep in mind that the sequence discussed above refers to **dominance,** or the degree to which the site is occupied by canopies and roots, not to the presence or absence of a given life-form. As we have already discussed, shrubs and trees that sprout from roots or grow

[3]Another word for a successional sequence is "sere," from the Latin *serere* to join or connect. The adjective form is "seral."

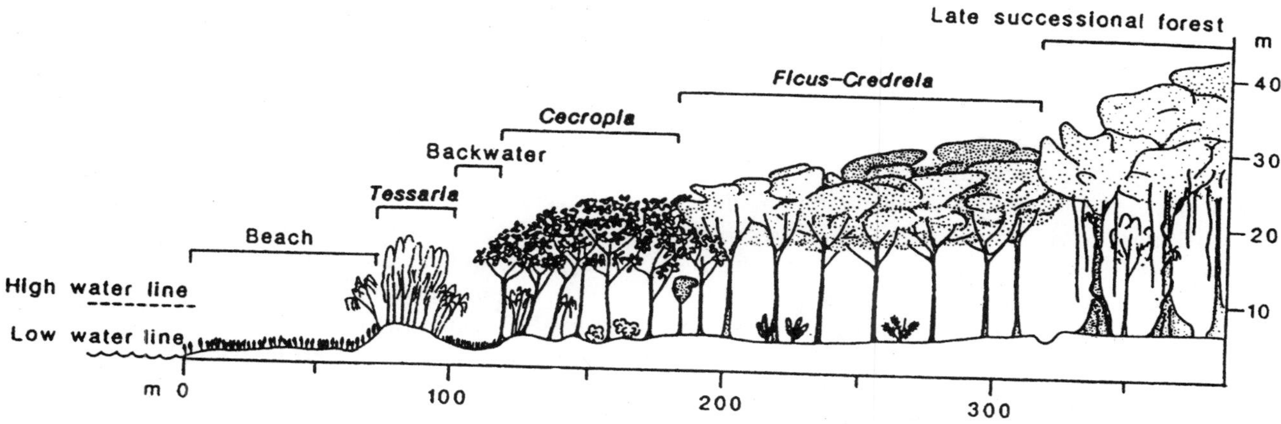

Figure 8.1. The successional sequence initiated by a meandering river in Amazonia. Created by fresh deposition of river sediment, the beach is colonized primarily by a single tree species, *Tessaria integrifolia*. *Tessaria* modifies the harsh beach environment, allowing other tree species to establish. Zones of older deposition of river sand (from previous meander positions) become dominated by *Cecropia tessnian nil.* As the *Cecropia* forest ages, gaps in the canopy are filled by various tree species, over time leading to dominance by trees in the genera *Ficus* and *Credeta,* If an area remains undisturbed by the river long enough, late successional forest will develop. The disturbance and succession created by river meandering result in a complex pattern of intermingled communities across the landscape. (Salo et al. 1986. Reprinted with permission from *Nature*. Copyright 1986 by Macmillan Magazines, Ltd.)

from buried seed (legacies of the old forest) are likely to be present during the earliest stages of succession (unless the disturbance is so catastrophic that surface soils are lost). In many cases these do grow fast enough or have sufficient density to dominate a site immediately after disturbance; however, sprouting clonal shrubs (e.g., salmonberry and salal in western North America) can attain high densities after disturbance (Tappeiner et al. 2001), as can some species that grow from buried seed (e.g., nitrogen-fixing shrubs in the genus *Ceanothus*).

Many species that come to dominate forests in late successional stages are also able to pioneer newly disturbed sites. Western hemlock, a shade-tolerant, late successional tree of the Pacific Northwest, has light, readily dispersed seed and frequently pioneers clear-cuts in relatively moist habitats. In New England forests, both early and late successional. tree species establish soon after disturbance, with fast-growing pin cherry dominating early in succession, and slower-growing species emerging to dominate later (Bicknell 1982; Hibbs 1983) (fig. 8.2). Throughout any given successional sequence, the community at any one point in time is likely to contain not only the dominants, but seedlings of future dominants. The period prior to complete canopy closure is in particular a time of great species richness, containing mixtures of herbs, shrubs, and tree seedlings that, in turn, create diverse habitat for animals. Where seed dispersal, grazing by ungulates, or other factors limit the density of establishing trees, the forb-shrub-tree stage can last for decades (e.g., Tappeiner et al. 1997; Poage and Tappeiner 2002). When forests comprise an overstory of early succesional trees and an understory of late successional trees, disturbances can in some cases have the paradoxical effect of moving succession forward rather than setting it back. For example, a severe windstorm in Minnesota blew down early successional overstory trees, allowing sites to be taken over by late successional hardwoods that had been growing in the forest understories (Arevalo et al. 2000).

8.2.2 The Early Successional Community

The particular mix of species forming the earliest successional stage (sometimes referred to as **pioneers or if they were present on site before the disturbance, legacy plants**) depends on multiple factors, including (1) the species pool (i.e., what propagules are available), (2) the type and

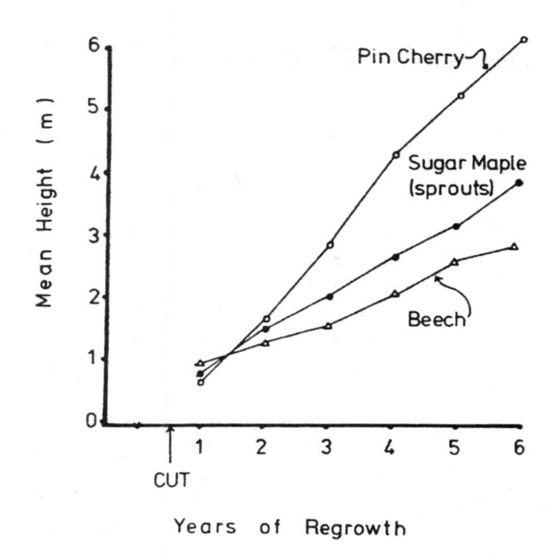

Figure 8.2. Early height growth of three different tree species in a New Hampshire clear-cut. (Adapted from Bicknell 1982)

severity of disturbance (e.g., windthrow, moderate-intensity fire, high-intensity fire), (3) the timing of disturbance, (4) the abiotic environment (e.g., microclimate, water and nutrient availability), and (5) the biotic environment (direct and indirect interactions among species). The first three factors, which are the subject of this section, determine the mix of species that is available for colonization, while the last two factors (the abiotic and biotic environment) determine which of the available species actually survive and establish. (Abiotic environmental factors are dealt with in section 8.2.3, and biotic factors in section 8.3).

There are three sources of early successional plants: (1) seed transported from elsewhere by wind or animals, (2) seed stored on site (e.g., in litter, soil, or serotinous cones), and (3) individual plants that survive the disturbance in some form (e.g., the aboveground parts of the plant may be destroyed, but the roots survive, sprout, and produce a new top). Primary successions, by definition, involve only the first of these. Secondary successions may involve one or all, depending on the particular set of adaptations present in the species pool and on the nature and timing of the disturbance that initiates the succession.

Gomez-Pompa and Vazquez-Yanes (1981) term plant species that produce mobile, widely dispersed seed **nomads.** Nomadic species are the sole colonizers in primary successions, and they are usually an important component of the pioneering community in secondary successions. Nomad seeds are either wind or animal dispersed. Many common herbaceous pioneers of burns and clearcuts have seeds that are light and readily dispersed by wind, as do some trees. Cottonwood is a classic example of a nomadic tree species. Within two years after the eruption of Mount St. Helens, black cottonwood seedlings blanketed large areas of mudflow along the Toutle river. Few, however, survived. It is sometimes a long jump from seed germination to seedling survival and establishment.

Plants that produce wind-dispersed seed are relatively common in dry habitats, while seed dispersal by animals tends to be more important in wetter habitats (Howe and Smallwood 1982). Animal dispersal is particularly important in moist tropical forests (Gomez-Pompa and Vazquez-Yanes 1981). According to Howe and Smallwood (1982), "at least 50% and often 75% or more of the tree species (in tropical forests) produce fleshy fruits adapted for bird or mammal consumption." Of 37 tropical American bird species examined by Trejo (1976), the digestive tracts of 33 species contained seeds of one or both of the nomadic trees *Cecropia obtusifolia* and *Trema mircrantha*. Bats are also very important seed dispersers in the tropics, and in tropical riparian forests, so are fish, which eat fruits that fall into the water during seasons when the forest is flooded, the defecated seeds germinating after the floodwaters withdraw (Howe and Smallwood 1982). Animals also disperse seeds of many temperate plants (Howe and Smallwood 1982), for example, pin cherry, an important pioneering tree in eastern North America (Bormann and Likens 1979a). Seed dispersal by animals has important implications for patterns of diversity within a forest, because composition of pioneering communities is less tied to nearby trees as seed sources than it is where seeds are primarily wind dispersed. For example, 65 percent of the seeds input to gaps in a Costa Rican rain forest were from species growing 50 m or more from the gap (Denslow and Diaz 1990). (Seed dispersal by animals is discussed further in chapter 11.)

In the previous chapter we saw that the ability of tropical trees to sprout from roots assured that the earliest pioneers after Hurricane Gilbert hit Nicaragua were species that had been present in the previous forest. This phenomenon is not restricted to the tropics: most, if not all, forest communities include species that are adapted to recover quickly from disturbances. For example, in the black spruce forests common to the interior of Alaska, burned sites are quickly occupied by sprouting grasses, shrubs, and small trees (willows, birch) and by numerous black spruce seedlings originating from seed stored in semiserotinous cones (Viereck 1973). In the Mediterranean region, "most plants that appear after fires come from survival organs such as rhizomes, lignotubers, bulbs, and seeds, which were present in the soil before the fire, or were dispersed immediately afterwards from nearby plants. . . . All the pre-fire species are present almost immediately after fire, even if the relative abundance . . . changes later on" (Trabaud 2003). In the Pacific Northwest, wildfire creates a mosaic of species that either survive the fire through heat-resistant bark, or regenerate through sprouting or from buried seed (fig. 8.3). The earliest colonizers in areas where the aboveground vegetation is killed are generally nomads that persist for a few months to a few years before being succeeded by former residents growing from sprouts or buried seed (Halpern 1988). The mix of species occupying a particular piece of ground depends on how hot the fire was at that point: hot fires tend to delay development of sprouters or kill them outright.

Literally millions of seeds per hectare representing numerous plant species may accumulate in forest soils and litter (Bormann and Likens 1979). Ewel et al. (1981) germinated 67 species from an 11-m^2 area of soil taken from a Costa Rican forest. Strickler and Edgerton (1976) germinated 38 shrub and herb species from soil and litter of a mixed conifer forest in Oregon. Buried seeds are the primary mode of regeneration for pin cherry, an important pioneering tree in New England forests (Marks 1974). In contrast to those of early successional plants, seeds of late successional species generally do not persist longer than a few years in soil.

Seeds stored in soil and litter are often those of nomads. In tropical rain forest, for example, "the most constant and abundant colonizing species of clearings and devegetated land combine wide dispersibility, prolonged survival of seeds in the soil, and specialized mechanisms for triggering

Figure 8.3. A, Sprouting hardwoods beneath a fire-killed stand in Oregon. B–D, This series of photos shows forest recovery following hurricane Hugo, which hit Puerto Rico on September 18, 1989. All were taken from the same spot. B, Four months after hurricane. *(Continued)*

germination" (Gomez-Pompa and Vazquez-Yanes 1981). However, in some ecosystems seeds of pioneers are not dispersed to the site from elsewhere but rather persist in soil from one disturbance event to another. This is common among fire-adapted, sclerophyllous shrubs of western North America. For example, clear-cutting and broadcast-burning old-growth Douglas-fir forests, some up to eight hundred years old, often leads to a thick cover of nitrogen-fixing shrubs in the genus *Ceanothus*. Seeds of these species have persisted in the soil since the previous early succession, hundreds of years ago, and are triggered to germinate by the heat of the broadcast burn (fig. 8.4). Seeds of many early successional plant species germinate only when triggered by some disturbance-related environmental factor

Figure 8.3. *(Continued)* C, One year after. D, Four years after. (B–D courtesy of R. B. Waide and the Luquillo LTER Project)

such as heat or change in the spectral composition of light (the ratio of red to far-red wavelengths differs depending on whether light has passed through a canopy or not).

Disturbance severity acts as a filter on the available species pool, modifying the composition of the early successional community. Disturbances that preserve soil but destroy aboveground parts favor sprouters or species with seed that is stored in soil. Species with serotinous cones are generally an exception. In cases where fires are so intense that serotinous cones are destroyed, sprouting hardwoods such as aspen are favored (Viereck 1973).

Fires that generate excessive heat in the soil either delay recovery by sprouting plants (Halpern 1988; Kauffman 1991) or kill the roots so no sprouting is possible. Tropical trees, many of which sprout prolifically following windthrow, are particularly vulnerable to roots being killed in fire (Kauffman 1991; Whigham et al. 1991). Without sprouters, the composition of the early successional community depends on seeds stored in the soil or input to the site following disturbance. Disturbances severe enough to destroy soil (e.g., landslides) generally initiate primary succession, and sites must be colonized by seeds from elsewhere; however, biological legacies

Figure 8.4. Douglas-fir saplings intermingled with the nitrogen-fixing shrub *Ceanothus yelutinus* (snowbrush) in an Oregon plantation. Seeds of snowbrush lie buried in soil for hundreds of years and are triggered to germinate by fire (in this case, a slashburn after the previous stand was clear-cut).

have a surprising ability to persist and shape early successional communities (see section 8.3.4).

Successional patterns following large, infrequent disturbances may or may not differ from those of smaller disturbances, depending on species adaptations (Romme et al. 1998). For example, tree species with abundant serotinous cones will disperse seeds into large burns as effectively as they do into small burns, while trees that must disperse seeds by wind will not (Romme et al. 1998). Both situations occurred in the 1988 fires in Yellowstone National Park (Turner et al. 2003c). Lodgepole pine stands in the lower elevations of the park had high levels of serotiny, while lodgepole stands in the higher elevations, where fires have been less frequent historically, had mostly open-coned (i.e., nonserotinous) trees. Five years after the 1988 fire, an area with prefire serotiny of 65 percent, had 211,000 lodgepole pine seedlings per hectare, while areas where prefire serotiny was less than 1 percent had only 600 lodgepole pine seedlings per hectare (Turner et al. 2003c).[4]

Where succession depends primarily on seed dispersed from living plants, recovery after large, infrequent disturbances is a nucleation process that builds around the landscape-level legacies represented by surviving patches of forest (Turner et al. 1998). Surviving patches are not as rare as some of the rhetoric around forest fires would suggest. Few if any fires destroy everything in their wake. Atmospheric turbulence generated by an intense fire is now believed to create coherent structures of variable fire intensity and, consequently, a physically driven pattern of unburned patches (Johnson et al. 2003). Moreover, as discussed in chapter 7, topography results in patchiness of surviving vegetation in both fires and windstorms. In the boreal forest of Canada, the amount of surviving forest as a proportion of burn area remains fairly constant regardless of burn size (although survivors tend to be grouped in larger patches in large than in small fires) (Johnson et al. 2003). Consequently, Green and Johnson (2000) calculate that the maximum dispersal distance from either the burn edge or surviving patches does not exceed 150 m, regardless of fire size.

Nevertheless, large, infrequent disturbances impose distinct patterns on succession. In addition to the nucleation discussed above, Turner et al. (1998) list the following trends with increasing distance from seed sources: (1) initial densities of organisms decline; (2) chance arrival becomes more important than competitive sorting[5]; (3) community composition is initially less predictable; and (4) community composition recovers to predisturbance levels more slowly. Despite the uncertainties, however, some aspects of recovery in large, infrequent disturbances are reasonably predictable. Species that sprout or have seeds either buried on site or dispersed for long distances compose a higher proportion of the recovering plant community as one moves further from seed

[4]Not all lodgepole pine trees produce serotinous seeds, and stands vary in the number of serotinous trees they contain.

[5]Competitive sorting implies that several species are present from the beginning, and these compete for resources to determine who survives.

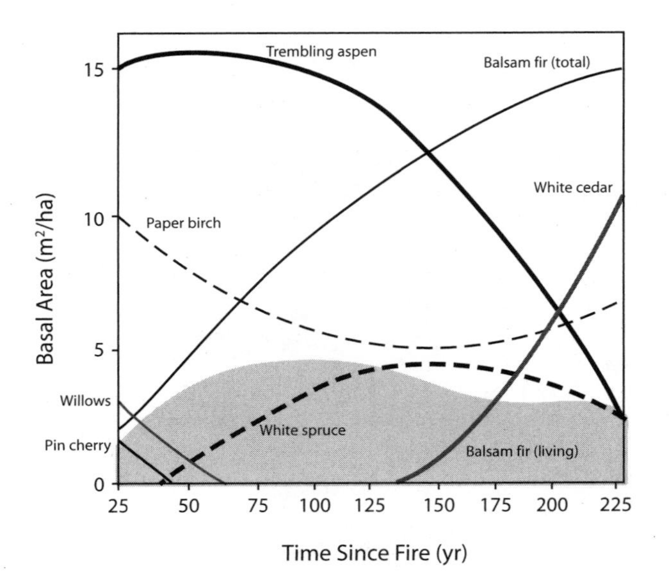

Figure 8.5. Accumulation of basal area of different tree species after fire in boreal forests of Quebec. All curves are statistically significant but with low explanation value, indicating high variability. Balsam fir total includes both living and dead (killed by spruce budworm); the stippled area denotes the living balsam fir. (Bergeron 2000, with permission of the Ecological Society of America)

sources. In boreal and temperate forests dominated by nonserotinous conifers, this means that the first trees to colonize large disturbances are sprouting or light-seeded hardwoods such as maples, aspen, and birch. Both conifer and hardwood seedlings established relatively early on the debris-avalanche deposit at Mount St. Helens; however, many of these died during a drought year. Only red alder was able to persist in significant numbers (because of its rapid early growth), and by 20 years posteruption that species was by far the most important of the early successional plants (Dale et al. 2005a).

The trends discussed above have at least two implications for successional patterns in large disturbances. First, the recovering landscape is heterogeneous. For example, 11 years after the 1988 fires in Yellowstone National Park, the density of lodgepole pine saplings varied over six orders of magnitude (0 to 535,000 per hectare), with areas of low pine density dominated by herbs and a single species of sprouting shrub (Turner et al. 2004). Variability in pine density correlated with levels of serotiny in prefire stands and variations in the severity of the fire. Secondly, slow recolonization by potentially dominant trees results in an extended period of species-diverse, complexly structured stands. Figure 8.5 shows this for postfire stands in southern boreal forests of Quebec (Bergeron 2000). Stands are dominated by aspen and birch for more than 50 years after fire, with smaller components of balsam fir, pin cherry, willows, and, eventually, white spruce. Balsam fir adds basal area quickly after 50 years, but large numbers of trees are killed by spruce budworm. As stands enter the old-growth phase (around 175 years after fire), they comprise roughly equal components of aspen, birch, spruce, and white cedar (which does not appear until well into the one hundredth year). Similar patterns have been documented in temperate conifer forests of western North America (but without the spruce budworm)[6] (Tappeiner et al. 1997; Poage and Tappeiner 2002; Sensinig 2002). Landscape and stand-level heterogeneity, in turn, significantly influence a variety of ecosystem processes, including the spread and severity of future fires (chapter 7), susceptibility to insects and pathogens (chapter 19), diversity of animals and microbes (chapter 10), productivity, and nutrient cycling (section 8.4).

Along with size and severity, timing of a disturbance also filters the available species pool. Composition of the early successional community often depends on coincidence between the time at which a disturbance occurs and the natural rhythms of species within the colonizing pool. For example, even conifers without serotinous cones can disperse large amounts of seed from fire-killed trees, but the fire must occur during a year with a good cone crop and at a time when seeds have matured but cones not yet opened. (Obviously the fire cannot be so severe as to consume the cones).

Three different time scales are important: time of year, the year itself, and the interval between disturbances. The first two relate to coincidence between disturbance and the availability of propagules (seeds or sprouts), the third to lifespan. Plant species vary in their seasonal rhythms (i.e., **phenology**); hence, a disturbance occurring at one time of year may select for quite a different set of early successional plants than one occurring at another time of year. The ability of some species to sprout following destruction of aboveground parts varies seasonally: destroyed at one time of year these recover vigorously, at another time not at all. Seeds of different species mature—and are available for colonization—at different times of the year. For example, in interior Alaska most wildfires occur during June and July, coincident with the ripening and dispersal of aspen and balsam poplar seed, but before seeds of white spruce and paper birch ripen (Viereck 1973). In the tropics, where yearly rhythms of the biota are not constrained by winter, species vary widely in phenology. Trees that produce animal-dispersed seeds tend to fruit year-round, while those producing wind-dispersed seeds fruit only during the dry season and, moreover, seeds are dispersed only on days with relatively low humidity (Gomez-Pompa and Vazquez-Yanes 1981). Other factors being equal, a disturbance coinciding with this dispersal would probably result in a relatively high proportion of wind-dispersed species in the pioneering community, while one that did not would have more animal-dispersed species.

Many trees produce seeds at intervals of several years. In Alaska, for example, birch produces heavy seed crops at least once every 4 years, but white spruce only once every 10 to 12 years (Viereck 1973). Hence the capacity of a given

[6]Spruce budworm is quite active in the montane forests of the intermountain West, but not in the temperate forests west of the crest of the Cascades (chapter 19).

species to deliver seed to a newly disturbed site depends on (among other things) whether the disturbance coincides with a good seed year. Such a coincidence is not totally random, however, because weather conditions (hot, dry springs) that increase the probability of wildfire also trigger seed production by white spruce (Viereck 1973).

Finally, the interval between disturbances can also influence composition of the pioneering community. Cattelino et al. (1979) give the example of an 80-year-old western hemlock forest in northern Montana that had established following a fire without being preceded by lodgepole pine, which is the usual early successional tree of that area. In this case the interval between fires had been sufficiently long that shade-intolerant lodgepole pine had dropped out of the forest, leaving only hemlock to colonize. Unlike many late successional trees, hemlock produces light, widely dispersed seed, so it can play the role of the pioneer. In chapter 7 we saw that patterns of succession following hurricanes differed between Puerto Rico and Nicaragua, the former having a history of more frequent hurricanes than the latter, and therefore more nomadic tree species in the prehurricane forest to deliver seeds to disturbed areas.

The timing of fire in a forest, hence the pattern of succession, is due partly to chance, such as a lightning strike in the right place and at the right time. However the dice are also loaded to some degree, because the probability that ignition turns into conflagration is much greater at some stages of stand development than at others. The trajectories of ecosystems through time usually result from just such a peculiar mix of randomness and determinism. A pattern that is random, but with a central tendency, is called *stochastic*, from the Greek *stochastikos*, which literally translates to "skillful in aiming" (e.g., the pattern of hits derived from firing arrows at a target).

8.2.3 Succession in Animals and Microbes

Communities of animals and microbes respond to successional changes in the composition and structure of plant communities. Some species of birds, for example, nest preferentially in early successional shrub-forb habitats, while other prefer mature or old-growth forests (chapter 10). Schowalter (1989) found that in forests of both the eastern and the western United States, the species composition and structure[7] of canopy insect communities differed significantly between old and young forests; old growth had a greater diversity of spiders and a much more equitable ratio of phytophages (foliage-feeders) to their insect predators (the ratios were approximately 1:1 in old growth and 8:1 in young stands; the predators were primarily spiders).

Soil microbial communities are commonly dominated by bacteria in the earliest stages of succession and become fungal dominated as trees or shrubs increase in abundance (e.g., Ohtonen et al. 1999). The response of mycorrhizal fungi[8] to disturbance and succession has been particularly widely studied. Some species of are more prevalent in early stages of succession, others are more prevalent in later stages, while yet others occur throughout the sere (Amaranthus et al. 1994; Smith and Read 1997; Goodman and Trofymo 1998; Jupponene et al. 1999; Smith et al. 2000, 2002; Griffiths and Swanson 2001; Allen et al. 2003; Bonet et al. 2004; Outerbridge and Trofymow 2004; Douglas et al. 2005). Species more common in older forests are often associated with large, decaying wood or deep litter layers. Similarly, some species of saprophytic fungi specialize in early successional habitats, especially recently burned areas (e.g., morels, prized by people for food) (Wurtz et al. 2005), while others are more common in later stages (e.g., wood decay fungi). Lichen and moss communities also vary with succession, some species being largely restricted to old-growth forests or younger forests that contain old-forest legacies such as large remnant trees (McCune 1993; Neitlich and McCune 1997; Boudreault et al. 2000). Some lichens typical of old-growth forests are very poor dispersers that take many years to rebuild populations following a disturbance (Sillett et al. 2000).

8.2.4 The Abiotic Environment Influences Early Successional Patterns

As we discussed in chapter 5, landscapes are mosaics of differing resource availabilities determined largely by topography and soils. Patterns of resource availability interact with species adaptations and biotic interactions to determine establishment success and dominance at any given time (Tilman 1982, 1984, 1985). The influence of the abiotic environment is most obvious in mountains, where any given disturbance is likely to contain numerous habitats with differing moisture and nutrient availability (related to slope steepness, aspect, and differences in rock type) that produce a corresponding mosaic of vegetation during early succession. Figure 8.6 shows the relationship between habitat type (chapter 5) and cover of various shrub and tree species on two 20-year-old clear-cuts in the Oregon Cascades. As we have already seen, the *Ceanothus* species are nitrogen-fixing, pioneer shrubs; Douglas-fir is an early and western hemlock a late successional species (i.e., hemlock is more tolerant of shade than Douglas-fir and, as succession proceeds, gradually replaces Douglas-fir as the dominant tree, a process that takes many hundreds of years); vine maple *(Acer circinatum)* occurs in all successional stages.

[7]When applied to the mix of species composing an ecosystem, *composition* refers to what species are present, and *structure* to their relative abundance. Both terms, but especially *structure*, may also have wider meanings (see chapter 9).

[8]Mycorrhizal fungi are discussed in chapter 11.

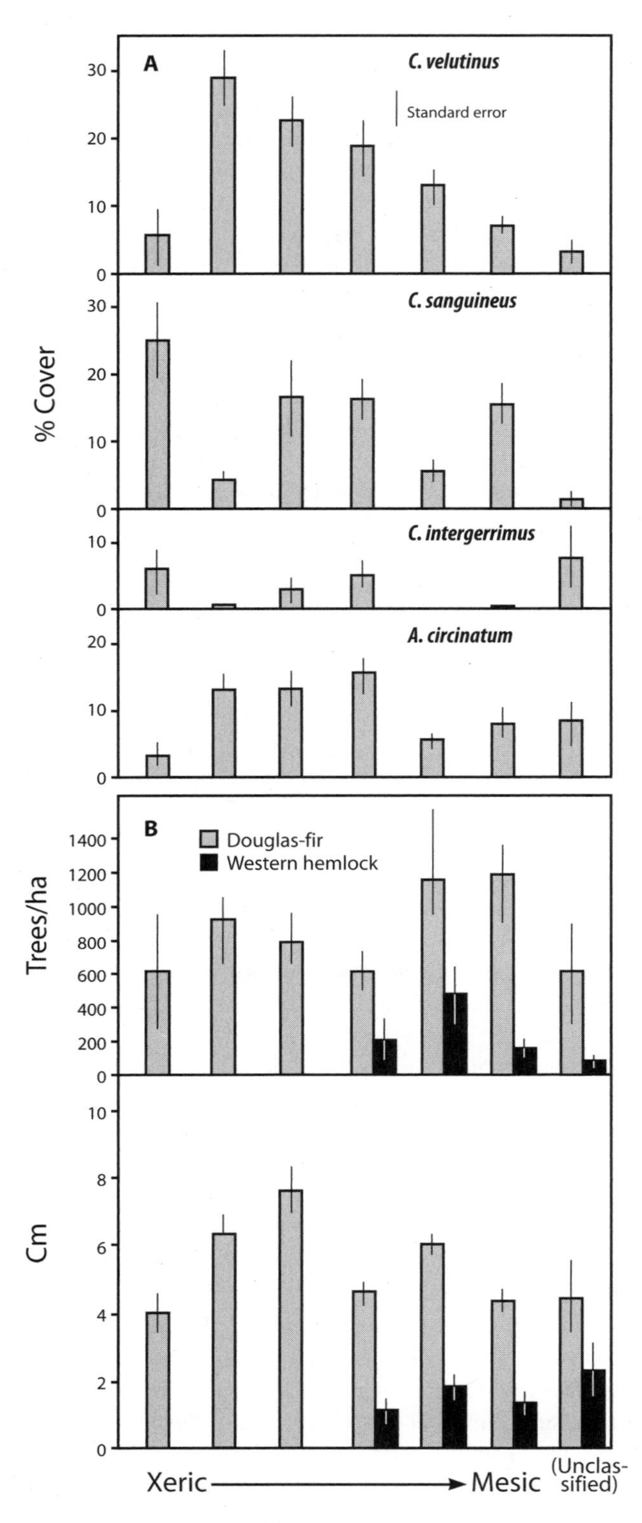

Figure 8.6. The relation between habitat type and cover of various shrub and tree species on two 20-year-old clear-cuts in the Oregon Cascades. Each bar (or pair of light and dark bars) represents a different habitat type. From left to right, habitat types become increasingly mesic. The far-right bar represents areas that could not be classified in any habitat type, primarily scree slopes. A, Cover of three *Ceanothus* species and vine maple *(Acer circinatum)*. B, Stocking density and average basal diameter of Douglas-fir and western hemlock. (Adapted from Egeland 1982)

Note that species composition varies from xeric (i.e., droughty) to moist habitats. The *Ceanothus* species attain highest cover on relatively dry sites and maple *(Acer)* in maple habitat types (i.e., where it occurred in the predisturbance forest), and hemlock successfully pioneers only on the relatively moist sites; Douglas-fir attains highest stocking on the moist sites but is well represented in all habitat types. (Early successional trees characteristically tolerate a wider range of environments than do late successional trees.) The result is a patchwork of early successional forest that corresponds to some degree with landform.

8.2.5 The Effect of Land Use History

The legacies of land use can profoundly influence successional patterns, with effects that may persist for centuries or even permanently (Fuller et al. 2004; Motzkin et al. 2004; Flinn and Vellend 2005). In their review of the research, Flinn and Vellend (2005) conclude: "From this body of work, the principal result is abundantly clear: land use history has a consistently strong influence on plant diversity and distributions. The impact of past land uses on vegetation patterns can equal or even override the effects of topography, soils, subsequent disturbance, and current management. . . In many landscapes, variation in plant community composition largely reflects the former distribution of woodlots and agricultural fields."

In chapter 7, we saw that native trees have difficulty colonizing abandoned agricultural fields in Puerto Rico. According to Motzkin et al. (2004) land uses that differ significantly from natural disturbances can alter successional pathways by "altering physical site conditions directly and thereby influencing subsequent plant performance and competitive interactions, or changing the suite of species that can potentially occupy a site; allowing for the establishment of species that are able to persist for long periods of time . . . on sites where they would, in the absence of disturbance, be unable to establish or compete effectively; or removing species that are extremely slow to recolonize." The latter category comprises species with limited dispersal capability, including some late successional forest herbs (Flinn and Vellend 2005), nonvascular plants, lichens, and flightless beetles, as well as species whose remaining seed sources are rare and widely dispersed.

In the United States, land-use legacies have been studied most extensively in New England. Large areas that were logged and converted to farms during the settlement period were eventually abandoned and reclaimed by forest. But the new forests were quite different from the old. Fuller et al. (2004) discuss how forest composition of that region has been altered and homogenized:

> The species that were most important at the time of European settlement and that declined most abruptly after European settlement are beech and hemlock, both

of which are susceptible to disturbances such as fire and the compaction of the soil surface by grazing animals. . . . These slow growing and shade tolerant species are slow to reinvade sites from which they have been removed by agricultural activity or fire. Relative to other species, both are slow to recover and dominate sites where they have been heavily cut. In contrast, the species that have increased across the region, including red maple, gray and paper birch, and the oaks, exhibit distinctly different behavior. Each of these responds well to disturbance by fire, cutting, and field abandonment as they sprout prolifically, grow relatively rapidly, and disperse and establish relatively easily in open sites. (Consequently) from the pre-European pattern in which hemlock and northern hardwood species dominated in the uplands and oaks and hickories dominated in the lowlands, the landscape has changed to one in which oak, red maple, and black birch are ubiquitous across the region, with lesser amounts of white pines and hemlock and very little beech.

Land use has also had long-lasting effects on nutrient cycling in New England. Available soil nitrogen increased with clearing and farming, and nearly a century after many farms in the region were abandoned is only now returning to preclearing levels (McLauchlan 2007). As we saw in Chapter 7, higher soil fertility can favor early successional and invasive species over late successional ones.

8.2.6 The Steady-State Forest

Once the forest becomes dominated by species that reproduce successfully under their own canopy, or in gaps created by the death of old trees, community composition may become relatively stable. This is Clements's climax and is also called the **steady-state** or **equilibrium community.** The dominant trees are called **climax species.** The steady-state forest is not static but composed of a dynamic mosaic of patches created by old trees dying and young trees filling the gaps that are left, a condition termed **shifting-mosaic steady state** by Bormann and Likens (1979b). Species composition may or may not change over time at any one point on the ground, but on the larger scale of the forest and landscape, and in the absence of major environmental change, it remains relatively constant. Age structure of the forest changes from the relatively even-aged condition of earlier successional stages to many-aged. Biomass accumulation levels out to zero, and total biomass remains relatively constant. (Bormann and Likens [1979a, 1979b] discuss the steady state.)

When thinking about steady states, it is important to distinguish between forests and forested landscapes. In theory, all forests attain a steady state as the end point of succession. In fact, however, many do not, or if they do, they do not stay there long because disturbance is always part of the scene. This is particularly true in forests that are prone to crown fires, lay along hurricane tracks, or grow in areas with a great deal of tectonic activity (such as Hawai'i or southern Chile). On the other hand, forested landscapes may maintain a relatively stable distribution of stands in different successional stages (the shifting mosaic), even where disturbances are frequent. The landscape area within which such a relative steady state occurs depends on the average scale of the disturbance: a regime dominated by small-scale disturbances, such as minor windthrow or low-intensity fire, produces a steady state within relatively small areas. This seems to be the case in moist tropical forests that are not on hurricane tracks, mixed conifer-hardwood forests of eastern and central North America, dry ponderosa pine forests of interior western North America, and dry miombo woodlands of southern Africa. In each of these, the steady state is characterized by frequent minor disturbances—death of old trees, minor wind throw, ground fires—that create space within which young trees can establish and grow. The steady state of ponderosa pine forests, longleaf pine forests, and miombo woodland depends on frequent ground fires, in the absence of which new species invade and the character of the forest changes. Where the disturbance regime is characterized by large events (e.g., high-intensity crown fires), a steady state is more elusive and may be found only within very large landscapes, or not at all. As we have seen, the disturbance regime of many forest types is characterized by relatively frequent, small-scale events punctuated by infrequent, large-scale events, in which cases the scale at which constancy is found on the landscape varies over time. Considering what has been learned in the past decade about the dynamic nature of the earth's climate system, and the profound influence of climate on ecosystem dynamics, it is more accurate to think about the steady state through time as comprising an envelope of possible landscape and regional patterns rather than some constant proportion. Even without significant change in climate, steady-state conditions may have a limited lifetime. Later in the chapter we see that in the absence of disturbance, many forests enter a decline phase.

Although many forests (as opposed to forested landscapes) never attain a steady state, or do so only temporarily, the concept of the steady state, or climax, is valid in all forests and, moreover, is an important tool for foresters. As Pickett and McDonnell (1989) point out, "The idea (of the climax) served as an important tool for bounding the study of succession. Without it, the great variety of specific trajectories and states that communities could exhibit would have been difficult to order." The steady-state forest, whether or not it is ever attained, is a relatively sensitive indicator of climate and soils. Hence, the presence of certain climax plant species tells the land manager something about a site, regardless of whether the climax vegetation ever attains dominance. This is the concept of the habitat type, which we discussed in chapter 5.

Remote imagery is an increasingly important tool with a variety of applications in ecology. Hall et al. (1991) used Landsat images to document the nature of the shifting mosaic in a 900-km^2 area of the Superior National Forest in northeastern Minnesota, a little less than one-half of which was in wilderness (no logging allowed). Using spectral characteristics (i.e., light reflected from the surface) of both visible light and near-infrared, Hall et al. were able to distinguish different tree species and also the degree of crown closure. Using that information, they identified five successional stages from Landsat photos. From early to late, these were (1) clearings, (2) areas of regeneration (cleared areas covered by low shrubs and young trees), (3) mature stands of deciduous trees, (4) mixed deciduous- conifer stands, and (5) closed-canopy, pure conifer stands. They then compared 1983 photos with 1973 photos to determine the rate of change from one type to another. The landscapes of both the wilderness and nonwilderness areas were very dynamic. Over the 10-year period, approximately 50 percent of the stands changed from one successional stage to another. Despite the dynamism at the stand level, however, the proportion of different successional stages across the landscape remained relatively constant.

8.2.7 Forestry and Succession

Because mature trees are harvested, forestry of necessity initiates succession. In fact, in its ecological aspects, silviculture[9] largely involves management of successional processes. This can take many forms, with different implications for successional and other ecological processes, a topic dealt with in detail in chapter 23. For economic reasons, virtually all approaches involve harvesting well before forests attain old-growth status, thereby truncating the later successional stages. By removing trees for sale, all approaches leave fewer biological legacies than natural disturbances. Even-aged forest management, which usually means plantations, alters early stages, as well, by (1) planting, (2) focusing on one or a narrow range of tree species, and (3) aiming for rapid site capture by crop trees, all of which act to shorten the period spent in the diverse shrub-forb-tree stage. The degree to which that occurs depends on the intensity of management, which depends, in turn, on the goals of the owners and the costs and benefits of managing intensively. At the landscape scale, forestry tends to leave a more even distribution of patch sizes than natural disturbances, which usually involve many small and a few large events (e.g., a power-law distribution; see chapter 7) (Perry 1998). Later in the chapter and at various points throughout the book we consider implications for ecosystem processes and biological diversity.

8.3 MECHANISMS OF SUCCESSION

To understand the process of shifting species dominance that ecologists call succession, we must begin by asking what biological and ecological mechanisms underlie the temporal dynamics—both change and constancy—of biological communities. Clearly, disturbance is one such factor that we have discussed in some detail. In this section we concern ourselves with the trajectories that are set in motion by disturbances, and why they are what they are.

The most important questions at this time center on the degree to which the species, or structures, that are present at any one stage of forest development influence the composition of succeeding stages, and the mechanisms by which that is accomplished. Over the past two or three decades, ecologists who study succession have focused most of their attention on community dynamics following establishment of the earliest successional stage, and in particular how species that dominate relatively early in the sere influence the transition to later seral species. Since the mid-1980s, an increasing amount of attention has been devoted to how biological legacies (i.e., patterns originating in the old forest) shape successional patterns in the new community. These might be viewed, respectively, as mechanisms of change and threads of continuity, the latter referring to mechanisms that constrain change and maintain a course leading back to the original community composition.

8.3.1 Mechanisms of Change: Species Interactions

Pickett (1977) argued that change of species in time was directly analogous to change of species in space. Because any one species can successfully adapt to a limited number of environments, changing environments (whether in space or time) result in different communities. However, change of environment in space is not the same as change in time. The former results from climate and topoedaphic factors, the latter from biological factors such as accumulating nutrients and soil organic matter, and shading. Through their effects on both the aboveground and belowground environment and their interactions with potential successors, species at each stage of succession participate in and even control to some degree the factors that lead to their replacement.

According to Lawton and Brown (1992), "There is growing evidence for priority effects in community assembly (either species A or species B can establish in the habitat; which one actually does depends upon which arrives first). Priority effects may then lock community development

[9] Silviculture is defined as the art and science of growing trees as a crop.

into alternative pathways, generating different end points or alternative states." We have already discussed the factors that influence who arrives first, and we shall return to that point in section 8.3.2. The question now is "Through what mechanisms do the earliest arrivals shape subsequent patterns of community development?"

In an influential paper, Connell and Slatyer (1977) proposed three ways in which a plant might influence a potential successor: facilitation, tolerance, and inhibition. These terms refer to the effect of environmental modification by early colonizers on subsequent establishment of later arrivals. In the facilitation model, only early successional species are able to colonize disturbed sites, and these modify the environment in such a way that it becomes *less* suitable for their own species and *more* suitable for others. In the tolerance and inhibition models, disturbed sites are potentially colonized by both early and late successional species (i.e., there is nothing inherent in the newly disturbed environment to prevent colonization by late successional species), which then modify the environment in such a way that new individuals of early successional species are unable to become established, with late successional species either unaffected by these modifications (tolerance model), or also inhibited (inhibition model).

Connell and Slatyer provided a valuable framework for thinking about species interactions during succession; however, except in a few cases, successional dynamics rarely fit neatly into one or another of the categories they proposed (Pickett et al. 1987a, 1987b). Interaction between individuals of two different plant species during succession often contains elements of both inhibition (e.g., competition for resources) and facilitation, with the net effect varying, depending on factors such as soil fertility, climate, and the relative stocking density of each species (Callaway 1997; Callaway et al. 2002; Bruno et al. 2003). The net effect may also vary over time, a relationship dominated by competition at one stage of stand development becoming predominantly facilitative later on, or vice versa. Moreover, successional dynamics can rarely be reduced to interactions between only two species: the nature of the relationship between any two individuals is conditioned by numerous other plants, animals, and microbes.

With these background comments, we now explore mechanisms of interaction in more detail. We begin with facilitation, and then move to inhibition.

8.3.1.1 Facilitation during Primary Succession

Primary successions probably always involve facilitation of one kind or another, most often related to soil building and nutrient accumulation. Soils do not develop without plants and associated microbes (chapter 14), and these pioneers facilitate establishment of their successors by weathering rocks, accumulating carbon and other nutrients, and providing the energy base that allows populations in the biological community to develop and grow.

Nitrogen accumulation is particularly important during primary succession. A key element in plant growth, nitrogen is abundant in the air but present in only minute amounts in rocks. Whereas green plants pump carbon from the atmosphere into the soil through photosynthesis, only a few species of bacteria and cyanobacteria (i.e., the nitrogen fixers) are capable of converting atmospheric nitrogen to a form usable by other living things. (Nitrogen fixation is discussed in some detail in chapters 11 and 17). Plants colonizing new substrates, such as rock or deep volcanic ash, must be either preceded by or associated with nitrogen-fixing microbes. Most often, colonization involves a close association between plants and nitrogen fixers, and for very good reason. Except for cyanobacteria, nitrogen-fixing microbes cannot photosynthesize and, hence, must have a source of carbon. The plant needs nitrogen, and the microbe needs carbon. Nature's solution to this problem is obvious: partnerships are formed. Lichens and mosses are two such partnerships that figure prominently in at least some early successions (Richards 1987; De Las Heras 1990). Some species of higher plants, such as alders and various legumes, have nitrogen-fixing bacteria living within their roots in visible swellings called nodules; alders are often the most abundant colonizers of rock exposed by withdrawing glaciers, avalanche slides, or volcanic debris flows. Nonnodulated plants such as conifers and grasses may have nitrogen-fixing bacteria living near their roots and on leaf surfaces, enabling them to colonize bare rock.

Nitrogen fixation requires large amounts of energy, however, and nodulated plants are generally intolerant of shade. Therefore, as succession proceeds, the deepening canopy of nonnodulated plants eventually shades out the nodulated ones. At least in this case, species change through time because changing resources (from nitrogen limitation to light limitation) favors different sets of adaptations. In what he calls the **resource ratio hypothesis,** Tilman (1984, 1985, 1986) argues that relative change in the availability of different resources is generally an important mechanism for species change during both primary and secondary successions.

Pioneers also facilitate the establishment of other plants during primary succession by establishing populations of beneficial soil microbes. Ehrenfeld (1990) discussed this in her review of primary succession on sand dunes (also see Rose 1988):

> The soil microflora interacts with plants in promoting soil development in dune ecosystems. Hyphae of both saprophytic and endomycorrhizal fungi help bind sand grains into aggregates through the excretion of amorphous polysaccharides, which in turn serve as substrate for colonization by bacteria, actinomycetes, and algae. The presence of a diverse microflora enhances the process of aggregation. The degree of soil aggregation increases [as succession proceeds]. . . . Soil aggregates (mg/kg soil) increase from 5 in the foredunes to 40 on

the mobile dune slope, 300 on the dune crest, and 1260 on young fixed dunes. The aggregates contain a variety of fungal species, including mycorrhizal species, and various bacteria . . . thought to be nitrogen fixing. There is an interactive effect between plant root growth and aggregate formation. . . . [The] total amount of aggregation, and concomitantly the abundance of all microfloral species . . . increases dramatically in the presence of roots.

Figure 8.7 shows how sand changes to soil during succession on dunes (Rose 1988).

Life requires a variety of nutrients in addition to carbon and nitrogen, most of which are contained in rocks (chapter 16). Plants and associated microbes play a vital role in facilitation by accelerating rock weathering and releasing these essential mineral nutrients. The profound effect of trees on rock weathering was demonstrated by Bormann et al. (1998), who compared elements released

A

B

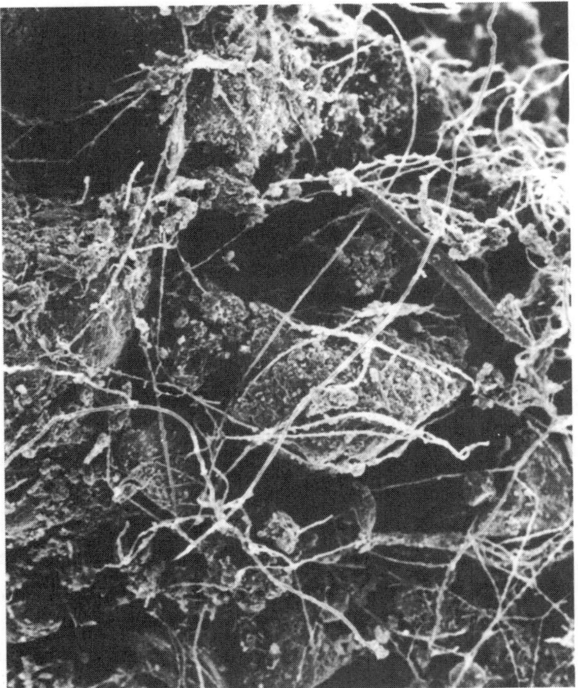

C

Figure 8.7. Electron micrographs of soil across a dune successional sequence. A, Sand from the beach (no vegetation) (scale, 150:1; area, 0.5 mm × 0.8 mm). B, Sand beneath the earliest successional community (various herb and grass species). Fungal hyphae (probably mycorrhizal) are beginning to bind individual grains of sand into loose aggregates (scale, 300:1; area, 0.25 mm × 0.40 mm). C, Sand—now soil—beneath forest (the most advanced stage of succession on the dune) (scale, 440:1; area, 0.20 mm × 0.25 mm). Hyphae and other forms of organic matter have bound sand grains into aggregates, creating physical and biologic structure within the soil (see chapter 14). Compared with the beach (A), this soil has 14-times-greater concentration of carbon, 1.6-times-greater concentration of nitrogen, and 8 times more microorganisms per gram of soil. (Courtesy of Sharon Rose, 1988)

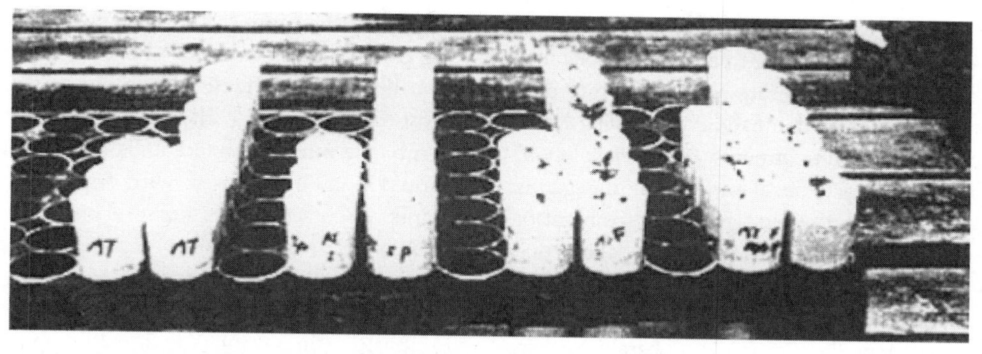

Figure 8.8. *Alnus tenufolia* growing in different soil media and with various combinations of microbial symbionts. The seedlings on the left are controls; the seedlings next to the controls are treated, from left to right, with (1) the mycorrhizal fungus *Alpova diplophloeus,* (2) the nitrogen fixing actinomycete *Frankia* sp., and (3) *A. diplophloeus* plus *Frankia*. Seedlings are grown only in perlite *(upper)*, or with perlite and ground basaltic rock *(lower)*. (Adapted from Yamanaka et al. 2003)

> Nitrogen-fixing bacteria are just one example of the close, mutually supportive relationships that exist between plants and some types of soil microbes (chapters 11 and 14). One of the more common partnerships found in nature is that between plants and numerous species of fungi, in which the fungus lives within the plant root, forming a structure called a mycorrhiza that is an intimate mixture of plant and fungal tissue (*mycorrhiza* literally means "fungus-root"). The fungus gets food from the plant and repays the plant in numerous ways (chapter 11). Plants also form working partnerships with some types of soil bacteria other than the nodule-formers.

from sand-filled lysimeters[10] ("sandboxes") that were either planted with red pine or kept mostly clear of vegetation. After 8 years, Bormann et al. calculated that 10 times more calcium and 18 times more magnesium were released from sandboxes planted with pine than from nonvegetated sandboxes. In a study that compared mineral elements in streams draining either a forested or a nonvegetated catchment in Japan, Asano et al. (2004) found approximately three times more bedrock-derived SiO_2 and Na^+ in the stream draining the forested than in the stream draining the nonvegetated catchment.

Much of the rock weathering accomplished by plants is probably due to fungi and bacteria living in association with roots. Many microorganisms are capable of weathering rock; however, species that form partnerships with algae, mosses, or higher plants are likely to be the most effective because they have a ready supply of energy in photosynthates (Hoffland et al. 2004). Lichens, a symbiosis between a fungus and an alga, can increase rates of rock weathering 10-fold (Stretch and Viles 2002). Mycorrhizal fungi are believed to actively weather rocks, as are certain soil bacteria that produce iron-complexing proteins called **siderophores** (literally, "iron-carriers"). Because rock contains little or no nitrogen, tripartite relationships between plants, mycorrhizal fungi, and nitrogen-fixing bacteria are probably the most effective system for weathering.

[10] A lysimeter is a boxlike device used for measuring nutrients released in solution. When the box is large enough, trees or other plants can be planted within, as was the case with Bormann et al.'s sandboxes.

This was demonstrated experimentally by Yamanaka et al. (2003), who grew the nitrogen-fixing tree *Alnus tenufolia* in a medium that contained either perlite alone or perlite plus ground basaltic rock. In each medium, the alder seedlings were given one of four inoculation treatments: (1) no inoculation, (2) inoculation with a mycorrhizal fungus (in the genus *Alpova*), (3) inoculation with the nitrogen-fixing symbiont of alder (an actinomycete in the genus *Frankia),* and (4) inoculation with both the mycorrhizal fungus and the nitrogen fixer. Uninoculated seedlings hardly grew, nor did inoculated seedlings planted in only perlite (fig. 8.8). With basaltic rock in the growing medium, adding the mycorrhizal fungus alone boosted growth a small amount, and adding *Frankia* boosted it quite a bit. Seedlings grew best when both microbes were added in the presence of rock (growth with both microbes averaged 1.35 times higher than with *Frankia* alone and 37 times higher than with *Alpova* alone). As discussed in later chapters, tripartite relationships are probably common in forest ecosystems. From the standpoint of facilitation during succession, tripartite associations work in a mutually supportive way to mobilize elements from rock, fix nitrogen and carbon from the atmosphere into organic forms, and create soils that serve as the productive base for the developing ecosystem.

8.3.1.2 Facilitation during Secondary Succession

Facilitation may occur during secondary as well as primary succession. For example, large amounts of nitrogen can be lost from ecosystems during fire (chapter 17), and nodulated plants are often among the earliest pioneers on burned sites. These are frequently rapid growers that initially compete with other trees and shrubs for water and nutrients. Foresters have often viewed them as undesirable competitors with crop trees; however, in the long run they facilitate growth of other plants in the ecosystem by restoring soil fertility (chapter 17).

Early successional plants may create certain structures or habitats that facilitate establishment of later successional species (Finegan 1984). Providing cover or perches for animals that disperse seeds is one way this happens: fruits are eaten and the seed defecated, and nuts are dispersed through the caching behavior of animals. Seeds buried by birds (particularly nutcrackers and jays) and mammals (e.g., squirrels, bears) are a primary avenue for establishing heavy-seeded species such as oaks, beeches, hickories, and some pines (e.g., whitebark, limber). Lanner (1985) estimates that in a good seed year, a single Clark's nutcracker may cache 100,000 whitebark and limber pine seeds. Johnson and Adkisson (1985) calculated that jays dispersed 150,000 nuts from a beech woodlot. Animals are exposed to predators when in the open and frequently constrain their movements (including seed-caching) to areas with cover. Jays, for example, avoid open fields when burying acorns (Bossema 1979). Hence, the cover provided by early successional trees and shrubs facilitates seed dispersal of late successional trees.

One of the more common examples of facilitation during early succession (at least in some environments) is the so-called **island effect:** tree or shrub seedlings establish most readily in the vicinity of an already-established tree or shrub (the **nurse plant**) (Padilla and Pugnaire 2006). This should not be confused with facilitation by nodulated plants, discussed above; nurse plants may or may not be nodulated. Tree seedlings invading savannas in Belize, for instance, establish preferentially near other trees (Kellman and Miyanishi 1982); the same is true for tree seedlings establishing in savannas in the Philippines and in abandoned pastures in Amazonia (Tupas and Sajise 1977; Nepstad et al. 1990). The island effect has been often noted in both forests and deserts of western North America. Both ponderosa and pinyon pines require nurse plants to establish on certain droughty or frosty sites (Phillips 1909). Live oak seedlings are strongly associated with some species of woody shrubs in central California (Callaway and D'Antonio 1991), and Douglas-fir seedlings establish preferentially beneath some species of oaks and ectomycorrhizal shrubs in northern California (Wilson 1982; Horton et al. 1999). One study of natural regeneration in Oregon found nearly five times more Douglas-fir seedlings beneath Pacific madrone trees than in the open (Amaranthus et al. 1990) (fig. 8.9). Not all trees and shrubs necessarily act as nurse

Figure 8.9. Douglas-fir seedlings growing beneath Pacific madrone.

plants on a given site, however. For example, while abundant Douglas-fir seedlings establish beneath canopies of Pacific madrone and some species of oaks, none establish beneath nearby Oregon white oak stands.

> Degraded forests often become dominated by woody shrubs that foresters have traditionally viewed as competitors to be removed before planting trees. However, experience shows this view can be wrong, or at least oversimplified, especially in environmentally harsh conditions. Gomez-Aparicio et al. (2004) discuss the situation in the Mediterranean region of Europe:
>
> > After a millenarian history of overexploitation, most forests in the Mediterranean Basin have disappeared, leaving many degraded landscapes that have been recolonized by early successional shrub-dominated communities. Common reforestation techniques treat these shrubs as competitors against newly planted tree seedlings; thus shrubs are cleared before tree plantation. However, empirical studies and theory governing plant–plant interactions suggest that, in stress-prone Mediterranean environments, shrubs can have a net positive effect on recruitment of other species. Between 1997 and 2001, we carried out experimental reforestations in the Sierra Nevada Protected Area (southeast Spain) with the aim of comparing the survival and growth of seedlings planted in open areas (the current reforestation technique) with seedlings planted under the canopy of preexisting shrub species. Over 18,000 seedlings of 11 woody species were planted under 16 different nurse shrubs throughout a broad geographical area. The facilitative effect was consistent in all environmental situations explored. However, there were differences in the magnitude of the interaction, depending on the seedling species planted as well as the nurse shrub species involved. Additionally, nurse shrubs had a stronger facilitative effect on seedling survival and growth at low altitudes and sunny, drier slopes than at high altitudes or shady, wetter slopes. Facilitation in the dry years proved higher than in the one wet year. Our results show that pioneer shrubs facilitate the establishment of woody, late successional Mediterranean species and thus can positively affect reforestation success in many different ecological settings.

Reasons for the island effect are not always clear, but there are at least three plausible mechanisms, any or all of which could be operating in a given situation. Nurse plants might (1) shelter seedlings from environmental extremes, (2) act as foci for seed inputs, and (3) provide enriched soil microsites. Shelter can significantly improve survival in droughty and also in cold environments. In droughty forests shade cast by early successional trees and shrubs may reduce water use by seedlings growing beneath them (i.e., less transpiration is needed to cool leaves). On high-elevation or other frosty sites, nurse plants provide a relatively warm nighttime environment by preventing excessive loss of radiant heat.

As discussed earlier, established trees and shrubs act as foci for seed inputs because they attract birds that often leave behind seeds (Nepstad et al. 1990; Callaway and D'Antonio 1991). Over one 6-month period in an abandoned pasture in Amazonia, nearly four hundred times more tree seeds fell beneath *Solanum crinitum* trees that were colonizing the pasture than fell in the open (Nepstad et al. 1990). Eighteen different tree species were represented in the seed rain beneath *Solanum*.

Plant islands may also facilitate establishment of later-arriving species during secondary succession through affects soil chemistry, biology, or structure. This is in some ways similar to (but in other ways quite different than) the facilitation that occurs through soil building during primary succession. Pioneers during secondary successions may restore soil carbon, nutrients, and organisms lost during disturbance. However, the most resilient members of the former community—generally species that are able to sprout from roots or grow from buried seeds—arrest soil degradation in the first place by preventing excessive nutrient loss and maintaining critical elements of soil biology and structure (Bormann et al. 1974; Bormann and Likens 1979; Perry et al. 1989).

Ecologists have known for some time that early successional plants accumulate nutrients and therefore prevent excessive nutrient loss after disturbance (chapters 18, 20). More recently it has become clear that islands of pioneering shrubs and trees (especially those that are legacies of the previous forest) also stabilize soil microbes that facilitate reestablishment of later-arriving plants. Survival and growth of tree seedlings establishing in disturbed areas may depend on their ability to quickly reestablish links with their belowground microbial partners, especially on infertile soils or in climatically stressful environments (Amaranthus et al. 1987; Perry et al. 1989). That would seem not to be a problem for plants that can sprout from roots, because they presumably never lose contact with belowground partners (e.g., Egerton-Warburton et al. 2005). It could, however, be a problem for trees whose seeds must be dispersed to disturbed sites from elsewhere. What happens to their microbial partners during the period the host plant is absent? One possibility is that the microbe simply goes dormant until its host plant reestablishes. Another possibility is that the microbe is flexible enough to utilize other food sources, perhaps soil organic matter or a pioneering plant. The latter seems to be the case in tropical and at least some temperate forests. The most common mycorrhiza-forming fungal species in tropical forests are widely shared among different tree species, as are some of the fungi that form mycorrhizae with temperate trees and shrubs. Generality may be more widespread early in succession than later. In the first 6 to 12 months following a wildfire in California chaparral, all recovering plants hosted the same types of mycorrhizal fungi, regardless of the plant's prefire affiliation (Egerton-Warburton et al. 2005).

In addition to ameliorating the physical environment and accumulating nutrients and carbon beneath its canopy, an early successional plant that supports microbes needed by later-arriving plants provides a critical biological boost to aid in establishment of the late arriver. Horton et al. (1999) provided a particularly good example in their study of invasion of California chaparral by Douglas-fir. The chaparral consists of patches of two different vegetation types: dominated by *Arctostaphylos* species, which form mycorrhizae with some of the same fungal partners as Douglas-fir; and dominated by *Adenostoma* species, which form mycorrhizae with different fungi. Douglas-fir successfully establishes within the *Arctostaphylos* but not within the *Adenostoma*. Horton et al. provided strong evidence that the ability of the tree to establish among one vegetation type but not the other was due to shared mycorrhizal fungi.

Because an established pioneer will also compete with new arrivals for light, water, and nutrients, the net interaction is potentially complex and the outcome variable depending on the environmental context and the time frame considered (Callaway 1997). In some cases, facilitation may be "banked" in the soil and not cashed in until the pioneers are reduced in density or have dropped out of the system altogether. For example, plants colonizing the forefronts of a receding glacier in Washington State do not benefit from growing near preestablished willows, but when planted into soils transferred from beneath willows into open areas, the prior presence of willows gives other plants a significant growth advantage (Jumpponen et al. 1998; also see Kytoviita et al. 2003).

> In southwestern Oregon and northwestern California, Douglas-fir and various hardwood trees and shrubs share some of the same mycorrhizal fungi. The hardwoods sprout from roots following disturbance, while Douglas-fir must reestablish from seeds. Douglas-fir seedlings tend to survive and grow better in the proximity of at least some hardwood species than in the open, and controlled studies indicate that the phenomenon is related to soil biology (Amaranthus and Perry 1989a, 1989b; Amaranthus et al. 1990; S. Borchers and Perry 1990) (fig. 8.10). The Douglas-fir are believed to "plug into" the network of hyphae extending from the hardwood mycorrhizae, which allows seedlings to rapidly develop their own water- and nutrient-gathering capacity (Horton et al. 1999). But the phenomenon is complex and appears to involve other factors as well, including nitrogen-fixing bacteria and perhaps bacteria that stimulate root-tip production by seedlings.[11] Nutrients also cycle faster in soils near hardwoods than in the open, a reflection of greater biological activity. The evidence amassed so far suggests that hardwoods of this area act as selective filters of soil biology, retaining beneficial soil microbes and inhibiting detrimental ones. Studies of unreforested clear-cuts have found that inability of seedlings to establish may be related to the buildup of certain types of microbes that inhibit seedlings and their mycorrhizal fungi (Friedman et al. 1990; Colinas 1992). In at least one instance, soils near sprouting hardwood islands within a clear-cut were relatively free of deleterious microbes (C. Y. Li unpublished, cited in S. Borchers and Perry 1990). At present, ecologists have only a rudimentary understanding of the complex relationships among plants and soil organisms and how these influence successional dynamics.

8.3.1.3 Inhibition

Any plant species that grows rapidly and occupies territory fully following disturbance has the potential to inhibit establishment and growth of other plants. The critical questions in terms of successional dynamics relate to what this implies for the successional trajectory and, in particular, for eventual return to the predisturbance community. How long lasting is the inhibition? Is its primary effect to lengthen the time sequence of changing species dominance without altering ultimate species composition, or does it lead to an entirely new community (perhaps dominated in perpetuity by the inhibitory plant)? Unfortunately, answers to these questions are seldom clear, largely because they require studies (or observations) that extend over many years, and until recently, these have been rare.

In ecology, what you see today is not necessarily what you get tomorrow, and judging long-term successional trajectories based on species dominance during early succession is particularly risky, especially when the underlying processes are poorly understood. As already discussed, many early successional species do not reproduce well under their own canopies; hence, they cannot dominate sites in the long-term unless continually reestablished by disturbances. For instance, many sites that had been clear-cut in the Oregon Cascades during the 1960s and early 1970s became dominated by nitrogen-fixing shrubs in the genus *Ceanothus*. Foresters and some scientists were concerned that *Ceanothus* might

[11]Some types of soil bacteria greatly stimulate root production by plants. One of these, in the genus *Agrobacterium*, is used by biotechnologists to enhance root production in plants that contain valuable medicines. *Agrobacterium* appears to accomplish that task by transferring genes to the plant (Pennisi 1992).

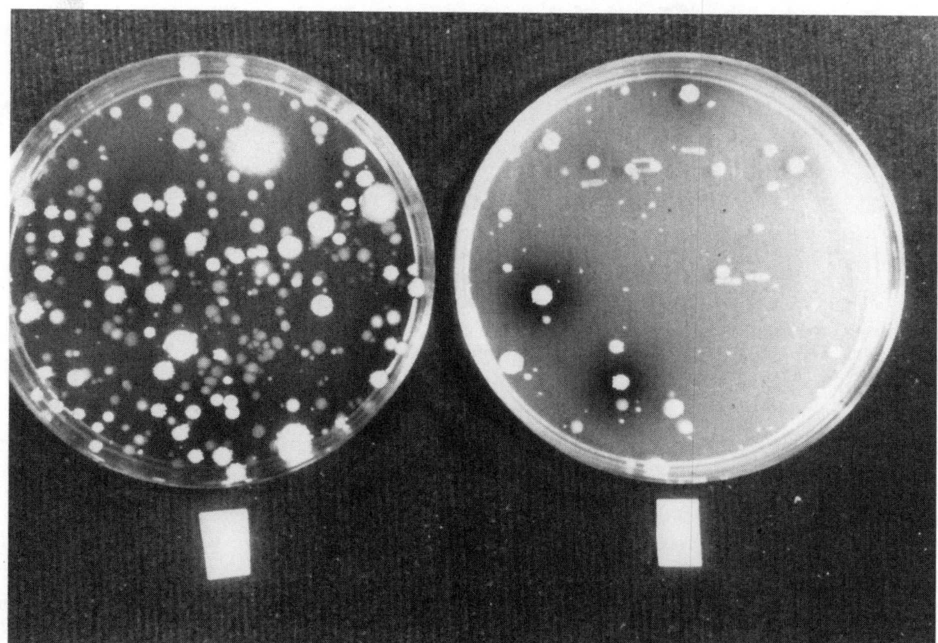

Figure 8.10. A, The growth of Douglas-fir seedlings in soils collected at different distances from hardwood trees sprouting in a 5-year-old clear-cut. From left to right, soils were collected less than 1, 2, 4, and 5 m from a hardwood (four species of hardwood, all had the same effect). The pots shown were randomly selected from experimental replications; patterns were highly significant statistically. The stimulating effect of soils collected near hardwoods appears to result from biologic factors rather than inherently different soil fertility, which suggests that it stems from patterns imposed on the soil by the hardwoods (as opposed to the hardwoods simply growing in nutrient-rich soil microsites). B, These Petri dishes contain actinomycetes isolated from the same soils in A. The dish on the left is from soils collected 1.5 m from a tanoak sprout; the dish on the right is from soils beneath the tanoak. Actinomycetes are filamentous bacteria that are known to allelopathically inhibit plants and other microorganisms (including mycorrhizal fungi). Low numbers of actinomycetes beneath hardwoods may contribute to greater mycorrhizal formation by and better growth of Douglas-fir seedlings in soils beneath hardwoods. (A, from Borchers and Perry 1990; B, courtesy of C. Y. Li)

exclude trees for many decades. Eventually, however, intermixed conifers began to grow above the shrub canopy, and most of those sites are now dominated by Douglas-fir (fig. 8.4). In one of the few situations where long-term records were kept, Douglas-fir was overtopped for 35 years by red alder before emerging above the alder canopy and eventually dominating the stand (Miller and Murray 1978). Inhibition was not permanent in these cases, because of the inherently different growth rhythms of the species: *Ceanothus* and alder grow fast and reach maximum heights at a relatively young age, while Douglas-fir grows more slowly but maintains growth, eventually becoming taller than the others.

Figure 8.11. This former plantation was destroyed by wildfire. In the foreground is an area seeded with annual grasses after the fire; the area in the background (not seeded with grasses) is occupied by native shrubs that sprouted from roots or buried seeds. The recovery of native shrubs completely failed in the grass-seeded area. Douglas-fir seedlings planted among the living grasses grew poorly, although seedlings planted one year later among dead grasses grew well (the dead grass thatch probably acted as a mulch to conserve soil water).

On the other hand, inhibition can be relatively long lasting when circumstances permit an aggressive early successional plant to form a dense, monospecific cover that effectively excludes other species. The pioneer could exclude other species by preempting site resources so fully and quickly that no other species can establish, or it might release chemicals that inhibit other plants and/or their mycorrhizal fungi. Biochemically mediated positive or negative interactions among plants and/or microbes (called allelopathy) appear to be relatively common (Rice 1984) and have influenced succession in at least some instances. In eastern North America, failures of trees to reestablish in old clear-cuts, abandoned agricultural fields, and areas burned by wildfire have been related to allelopathic inhibition of tree seedlings by herbs, ferns, and grasses (Horsley 1977; Fisher et al. 1978). In the Sierra Nevada of California, the herbaceous perennial *Wyethia mollis* has spread widely in old burns and allelopathically inhibits tree regeneration (Parker and Yoder-Williams 1989; Riegel et al. 2002). A similar situation occurs in northern Europe, where tree regeneration is allelopathically inhibited by the dwarf shrub *Empetrum hermaphroditum* (Nilsson and Zackrisson 1992). In these examples, the inhibitory plants are often natives that once had been present in relatively low numbers and were apparently triggered into a more aggressive mode by abnormal disturbance; in other words, a balance was disrupted. In Pennsylvania, tree seedlings were originally eliminated from recovering clear-cuts by forest fires and exceptionally high populations of deer[12] (Horsley 1977). In California, overgrazing allowed the unpalatable *Wyethia* to spread at the expense of more tasty plants (Parker and Yoder-Williams 1989). Although details vary, Nilsson's (1992) summation of the factors underlying the spread of *Empetrum hermaphroditum* in northern Europe describes a common dynamic of imbalances induced by the change from natural to anthropogenic disturbance regimes: "At the end of the nineteenth century active and passive forest fire elimination programs together with cutting operations started to change the vegetation patterns and processes over vast areas of the Boreal landscape."

Woody plants can have particular difficulty getting a foothold within established grass communities. In western North America, annual grasses are often deliberately sown in recently burned forests to stabilize soils. However, the grasses can completely inhibit the rapid recovery of native shrubs (Amaranthus et al. 1993) (fig. 8.11). In Central and South America, areas cleared of forest are frequently seeded to grasses to provide cattle pasture, then abandoned after a few years because they decline in productivity. Trees have

[12]Twentieth-century deer populations have been significantly higher than the previous norm in the eastern United States. The probable reasons lie in the extensive clearing of eastern forests since European settlement, which extirpated large predators (wolves and mountain lions), and the subsequent abandonment of many farms, which created an abundance of deer habitat.

great difficulty reinvading abandoned pastures. According to Nepstad et al. (1990), "Directly or indirectly, grasses present barriers to tree seedlings at every step of establishment in abandoned pastures with histories of intensive use. Seed dispersal into grass-dominated vegetation is low because grasses do not attract birds and bats that eat fleshy fruits of forest trees. Grasses provide food and shelter for large populations of rodents that consume tree seeds and seedlings. . . . The dense root systems of grasses produce severe soil moisture deficits in the dry season and compete for available soil nutrients. Finally, grasses favor fire so that tree seedlings that do surmount the numerous obstacles to establishment are periodically burned."

In at least some instances, allelopathy is involved in interactions between grasses and woody plants. Theodorou and Bowen (1971) showed that decomposing grass roots inhibited mycorrhiza formation by radiata pine planted in Australia and suggested that toxins released by plant roots were responsible. (Several different types of litter contain water-soluble compounds that inhibit, and in some cases stimulate, mycorrhizal fungi [Schoeneberger and Perry 1982; Rose et al. 1983.] Invasion of prairie by forest shrubs and trees in Oklahoma appears to be a two-step process in which certain species of woody shrubs establish first, aided by allelopathic inhibition of the prairie plants (Petranka and McPherson 1979). Established shrubs then provide islands that facilitate tree regeneration.

If a pioneer successfully excludes other plants and also is capable of reproducing under its own canopy, at least in theory it can hold a site indefinitely. Such is the case with the Pacific coast shrub, salmonberry *(Rubus spectabilis),* which produces pure stands of 30,000 or more stems per hectare following disturbance. By sprouting from basal buds and rhizomes, salmonberry quickly replaces old stems with new ones, thereby creating uneven-aged stands (Tappeiner et al. 1991). Once a pure stand (i.e., without intermixed tree seedlings) attains a sufficiently high density, plants such as salmonberry are likely to persist until weakened by pathogens or insects, or confronted with a disturbance to which they are not adapted. Research in grasslands suggests that long-term dominance by a single species is ultimately discouraged by the buildup of species-specific root pathogens (Klironomos 2002).

8.3.2 Facilitation and Inhibition by Microbes

In the past, ecologists thought of succession as a process driven primarily by plant-plant interactions. In recent years, however, it has become clear that many other elements of the ecosystem can either directly or indirectly influence successional trajectories, including in particular microbes, animals, and abiotic environmental factors. We will be dealing with this subject through the next several sections; here we briefly discuss the sometimes profound, but for the most part little studied, effects that soil microbes can have in shaping the composition of successional plant communities. The starting point for this discussion is balance. As we have already seen, soils contain microbes, such as mycorrhizal fungi and some types of bacteria that directly benefit plants. Soils also contain microbes that are pathogenic or otherwise inhibitory toward plants. Particular microbial species within those broad groups seldom affect all plant species equally (e.g., a given species of mycorrhizal fungus may benefit some plants and not others, and the same is true for the detrimental effect of pathogens). In some instances, a microbe that stimulates one plant species is pathogenic toward another (Chanway and Holl 1992). Because of the selectivity of their action, the composition of the microbial community on a site feeds back to affect the relative success of different plant species (Janos 1980; Allen and Allen 1990; Perry et al. 1990, 1992; Chanway et al. 1991; Klironomos 2002). The relationship is reciprocal, because a microbe that depends on living plants for food—whether it is a mycorrhizal fungus or a pathogen—presumably cannot persist indefinitely in the absence of a host. Therefore, feedback relationships develop between composition of the plant community and composition of the soil microbial community (Perry et al. 1989; Klironomos 2002).

The availability of beneficial microbes is known in at least some instances to determine whether their host plants can establish on a site, or how well they grow once established (Mikola 1970; Marx 1975; Perry et al. 1987; Horton et al. 1999; Allen et al. 2003). Mycorrhiza formation by plants may be reduced where host plants have been absent too long, on highly disturbed areas (e.g., where erosion is severe), and in some instances even with rather mild soil disturbance (Perry et al. 1982; Parke et al. 1983a; Perry and Rose 1985; Jasper et al. 1991, Dickie and Reich 2005). Inoculating seedlings either with mycorrhizal fungi or with forest soils or litter has significantly improved survival of trees planted in mine spoils, abandoned fields, old clear-cuts, and natural grasslands (Shemakhanova 1967; Marx 1975; Parke et al. 1983b; Valdes 1986; Amaranthus et al. 1987; Lobuglio and Wilcox 1988; Danielson and Visser 1989; Stemstrom et al. 1990; Helm and Carling 1992). Inoculating with certain types of rhizosphere[13] bacteria significantly improves growth of outplanted tree seedlings in Canada (Chanway 1992).

In order to understand how soil biology might influence succession, we must first ask how disturbance affects soil biology. Jones et al (2003) thoroughly reviewed what has been learned about how clear-cutting influences mycorrhiza formation on seedlings (mycorrhizae are discussed in some detail in chapter 11). Clear-cutting virtually always alters the species composition of mycorrhizae on seedling roots compared to what is formed in forest soils, perhaps toward those better adapted to the altered physical and chemical environment.[14] The proportion of root tips that are colonized is less affected, and depends a great deal on methods of site preparation. Techniques that destroy the

[13]The rhizosphere is the zone immediately surrounding and heavily influenced by plant roots.

[14]Mycorrhizal species more typical of the forest may still be present in clearcut soils, but tend not to form mycorrhizas as readily as in forest soils.

litter and duff layers (e.g. broadcast burning, bulldozer scarification) have a greater impact than those which protect these important habitats. Proximity to mature trees is an important determinant of what mycorrhizal morphotypes[15] form as well as their overall diversity (Simard et al. 1997, Cline et al. 2005, Dickie and Reich 2005). In clear-cuts, mycorrhizal diversity declines with distance from mature trees, either the forest edge (Hagerman et al. 1999) or individual trees retained within the clearcut (Cline et al. 2005). Do changes in mycorrhiza diversity and morphotype influence seedling success? There is no simple answer to this question, and mostly it has been little researched. Simard et al (1997) found that seedlings planted within a British Columbia forest and trenched away from contact with surrounding tree roots had both lower mycorrhiza diversity and lower photosynthesis than seedlings with access to surrounding roots. Earlier, we saw how improved growth and survival of seedlings in the proximity of established shrubs or trees (the island effect) was related in at least some cases to mycorrhiza formation. However, it would be premature to conclude that lower diversity and altered mycorrhiza types necessarily reduce seedling performance. If changes do indeed reflect adaptations to new conditions, the effect on seedling performance could be neutral or positive. Much more work is needed.

Inhibitory soil microbes include well-known pathogens, such as root rots and the so-called **damping-off** fungi, and less well known groups, sometimes called **exopathogens,** that can have sublethal inhibitory effects on plants and/or mycorrhizal fungi. Actinomycetes, a form of filamentous bacteria, have been implicated in reforestation failures in the Pacific Northwest (Freidman et al. 1989). *Streptomyces*, from which the antibiotic streptomycin is derived, is a genetically diverse soil actinomycete that has complex effects on other organisms. Depending on isolate,[16] *Streptomyces* are known to allelopathically inhibit plants, bacteria, and/or plant pathogens, and to either inhibit or to stimulate mycorrhizal fungi (Rose et al. 1980; Perry and Rose 1983; Freidman et al. 1989; Richter et al. 1989; Rojas Melo et al. 1992). Numbers of *Streptomyces* have been found to be higher in soils of unreforested clear-cuts than in forest soils, and, within clear-cuts, higher in soils between islands of sprouting trees and shrubs than in soils beneath the islands (Freidman et al. 1989; C.Y. Li, unpublished data).

Some of the more interesting research questions relating to successional dynamics have to do with the role of the belowground community. What triggers the buildup of inhibitory microbes in some disturbed areas, and how widespread is that phenomenon? How long can mycorrhizal fungi or beneficial rhizosphere bacteria persist without host plants? How does the composition of the early successional community influence the composition of the soil microbial community, and how does that feedback to influence the successional trajectory? In his review of ecological specificity between plants and microbes, Chanway (1991) concludes, "A major interdisciplinary research effort involving plant population biology and rhizosphere microbiology is required to further elucidate the role of rhizosphere micro-organisms . . . and evolved specificity on plant competition, species distribution, and community structure in different environments." Considerable progress has been made on understanding the role of soil microorganisms in structuring the plant communities of grasslands (e.g., Klironomos 2002), but a similar understanding of forests lags behind.

8.3.3 Higher-Order Interactions: Those Involving More Than Two Species

Community interactions invariably involve several species, not just two; hence, complex relationships may develop during succession. In the North Carolina Piedmont, for example, early successional pines inhibit the establishment of fast-growing hardwoods such as *Liquidambar*, which, in the long term, facilitates entry of slower-growing oaks and hickories (Peet and Christensen 1980). Animals often modulate plant-plant interactions during succession. In the Pacific Northwest, browsing elk and deer prefer hardwood shrubs over conifer seedlings, hence accelerating succession from shrubs to trees. This was demonstrated by Hanley and Taber (1980), who excluded elk and deer from a portion of a clear-cut in western Washington. In areas accessible to elk and deer there were 8.7 woody stems per square meter, 50 percent of which were Douglas-fir, while the area with animals excluded had 19 woody stems per square meter, 11 of which were salmonberry (a particularly aggressive competitor with conifer seedlings). In eastern Oregon, early successional communities from which deer, elk, and cattle are excluded are dominated by *Ceanothus* species. Where those animals are present, however, browsing limits height growth of the shrub, and sites are dominated relatively quickly by conifers (fig. 8.12). On the other hand, where trees are favored food, or (as is more often the case) when excessively high animal numbers result in a shortage of preferred food, animals will definitely retard succession and, as we saw with elephants and acacia trees, even jeopardize the existence of trees on a site. In chapter 9 we see that reintroducing predators such as wolves can alter that dynamic and indirectly benefit the health of the plant community.

The relationship between numbers of large grazing mammals and damage to trees is not always straightforward, however. In Oregon, for example, Weyerhaeuser Company, which runs cattle in its tree plantations in order to control grass and shrub densities, found that browsing on trees was reduced by increasing the number of cattle! Why this counterintuitive result? When grasses, the food

[15]A mycorrhizal morphotype has a unique color or structure that often but not always corresponds to a distinct fungal species.

[16]An isolate is a genetically uniform colony of some microbial species.

Figure 8.12. An illustration of how large herbivores can alter ecosystem structure (ponderosa pine plantations on Oregon State University's Hall Ranch, northeastern Oregon). A, Elk, deer, and cows were excluded from the area shown (the shrub is *Ceanothus velutinus*). B, Cows were excluded from the area shown, but elk and deer were not. *(Continued)*

preferred by cows, are grazed too lightly, they harden off and become unpalatable, and cows turn to trees. Heavier grazing keeps that from happening.

Plant species can be grouped into **guilds**[17] based on common interests in mycorrhizal fungi and perhaps other beneficial soil organisms (Perry et al. 1989).[18] One consequence is that facilitation, or at least some aspects of facilitation, can be selective; early colonizers during secondary succession facilitate subsequent colonization by members of the same guild by providing a legacy of mycorrhizal fungi (and perhaps other beneficial soil organisms) and inhibit colonization by members of other guilds because they provide

[17]The term *guild* is used in two ways in ecology. According to Webster's, *guild* is an "old term for a group of plants in some way dependent on other plants." Later, the term was applied to groups of species that have similar food requirements (e.g., the guild of insect-eating birds; see chapter 9). Our use here is more in the sense of the original, to indicate a group of species—plants, animals, and microbes—that are mutually interdependent.

[18]In chapter 11 we see that guild members can be physically linked through mycorrhizal hyphae, and that carbon and other nutrients pass between plants through those links.

Figure 8.12. *(Continued)* C, Elk, deer, and cows fed in the area shown. Elk and deer significantly reduce shrub density but have less impact on grasses. Cows reduce grass cover. (Courtesy of Bill Krueger)

no such legacy (e.g., the study by Horton et al. 1999, discussed in section 8.3.1). Members of a guild compete as well cooperate, with the net effect depending on which resources are most limiting to plant establishment and growth. Interactions are not always facilitative, or at least not immediately so (Kytoviita et al. 2003). As discussed earlier, facilitation between guild members may be immediate or banked, to be realized at some later time (e.g., Jumpponen et al. 1998).

> The guild concept may be extended to include animals that are tied into a relational network with plants and their mycorrhizal fungi. For example, truffles, the belowground fruiting bodies of some species of mycorrhizal fungi, are the primary food source for many mammals that spread spores of the fungi (Maser et al. 1978; Johnson 1996). In forests of the Pacific Northwest, the primary diet of the endangered northern spotted owl is the northern flying squirrel, whose primary diet is truffles. Hence the long-term welfare of both flying squirrels and spotted owls depends on successional trajectories that lead back to trees that support truffle-producing mycorrhizal fungi (fig. 8.13).

Ecologists have barely scratched the surface in studies of interactions during succession. Some flavor of the complex relationships that can occur is provided by the following two studies. The first relates to complex interactions among plant species and soil organisms. In southwestern Oregon, Douglas-fir seedlings were planted into adjacent areas, one of which had just been cleared of a dense stand of the woody shrub whiteleaf manzanita *(Arctostaphylos viscida),* while the other (an old, abandoned field) had been cleared of annual grasses (Amaranthus and Perry 1989; Amaranthus et al. 1990). A subset of the Douglas-fir seedlings in each area was given a small amount of soil collected beneath the hardwood tree Pacific madrone *(Arbutus menziesii).* Whether given madrone soil or not, seedlings survived and grew much better in soil previously occupied by manzanita than in the old field. Adding madrone soils greatly stimulated growth of seedlings planted in the former manzanita stand and increased the activity of nitrogen-fixing bacteria by more than fivefold in seedling rhizospheres. However, madrone soil given to seedlings growing in the old field did not improve their growth and actually decreased their rhizosphere nitrogen fixation by 80 percent. The effect of madrone soils was biological, because soils that had been pasteurized to reduce biological activity had no effect. Beyond that, however, the exact mechanisms that produced these responses—or why the effect of madrone soil should be so different in the old field than in the former manzanita stand—are unknown. The point is that one plant, through its associated microbes, can either facilitate or inhibit another plant, depending on the environmental context.

The second example deals with an unexpected effect of grazers. Bowers and Sacchi (1992) excluded small and large grazing mammals from certain areas within an early successional community in Virginia. In the first year of the study, purple clover *(Trifolium pratense)* occurred at greater densities where grazers were excluded than where they

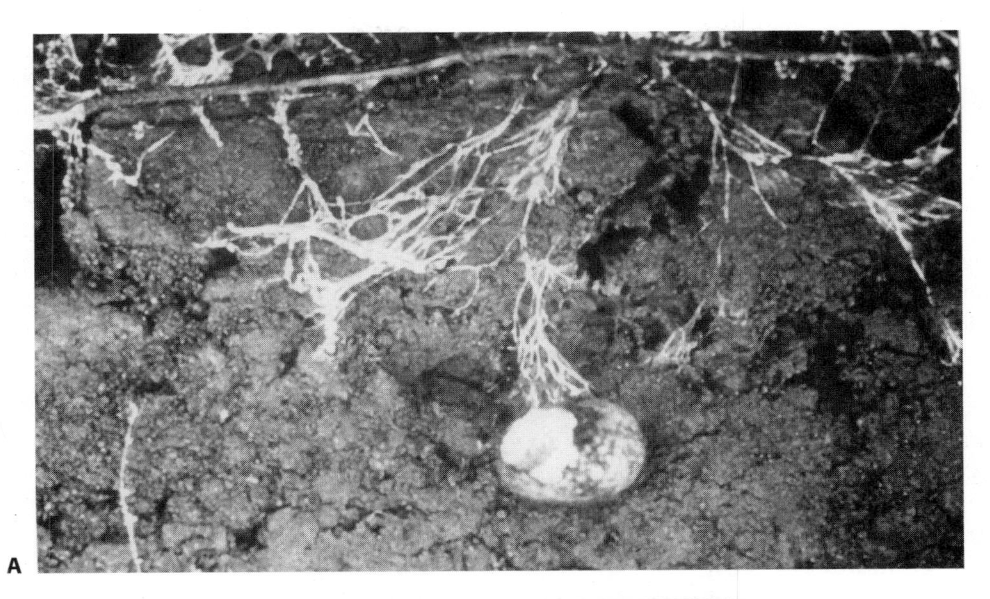

Figure 8.13. Truffles, which are the belowground fruiting bodies formed by some species of mycorrhizal fungus, are an important food source for mammals in many areas of the world, including forests and deserts (essentially, wherever the plant community contains species that support truffle-producing fungi). Defecating animals spread fungal spores (which pass unharmed through the digestive tract) along with nitrogen-fixing bacteria and other microbes living in the truffle. One hypothesis holds that logs facilitate the reinoculation of disturbed areas with truffle-producing mycorrhizal fungi, because the logs provide cover from predators for small mammals that disseminate spores (Maser et al. 1978). A, Truffles produced by an ectomycorrhizal fungus in the genus *Hysterangium*. B, A northern flying squirrel munching on a *Hysterangium* truffle. Northern flying squirrels compose a major part of the diet of northern spotted owls. (A, courtesy of Chris Walker; B, courtesy of Jim Grace)

were not. By the second year, however, the reverse was true. High numbers of clover plants in ungrazed plots led to the buildup of a fungal pathogen on the clover, causing the population to crash. In sum, the grazers had a short-term negative impact, but a longer-term positive effect on clover. Bowers and Sacchi's comment is appropriate for concluding this section: "Emphasis in community ecology (and in plant succession) should be on interactive networks rather than on species pairs." This leads us to the concept of biological legacies.

8.3.4 Threads of Continuity

I want to tell you how land shows the power of linking back, of re-connecting, of remembering.

BRIAN SWIMME

On a Sunday morning in May 1980, residents of Corvallis, Oregon (where two of us lived at the time), heard a distant and obviously powerful explosion. It was the eruption of Mount St. Helens, 250 km to the north. Franklin et al. (1985) describe it this way:

> The . . . eruption began . . . when a large earthquake triggered a massive avalanche of debris involving the entire upper portion of the mountain. Movement of this mass unroofed the core of the mountain, where super-heated ground water flashed to steam, unleashing a blast of steam and rock debris in a 180° arc to the north. Mud flows rampaged through the valley-bottom forests to the west and southeast. Volcanic ash rained from the sky northeast of the mountain from the morning of 18 May into the next day. In the early afternoon of 18 May and in the subsequent eruptions . . . [flows of hot gases and pumice] spilled northward out of the newly formed crater across the deposits of debris left by the avalanche.

Scientists studying the recovery of plants and animals at Mount St. Helens found some surprises (fig. 8.14). Conventional wisdom, which was strongly influenced by succession on abandoned agricultural fields, viewed succession as a wave of early pioneering species that prepared the way for the eventual arrival of late successional species. Recovery of Mount St. Helens was much more complex than that and was strongly shaped by both living and nonliving legacies of the predisturbance ecosystem (Dale et al. 2005c). Quoting from Franklin and MacMahon (2000):

> Biologists were confounded by the richly varied and often surprising circumstances that not only left organisms alive but also allowed processes and structures to persist. As a result, ecosystem recovery proceeded rapidly and by diverse pathways.
>
> Excepting only the sites of pyroclastic flows and lava extrusions, organisms survived almost everywhere as complete individuals, perennating parts (such as rhizomes and roots), seeds, and spores. Not all survivors persisted, but many did, including species representing all trophic levels, most life forms, and all successional stages—pioneer and late successional.

Three years after the eruption, 230 plant species—90 percent of those in preeruption communities—had been found within the area affected by the blast deposit and mudflows. Franklin and MacMahon (2000) continue:

> In the soil, animals such as pocket gophers and deer mice survived; pocket gophers proved important in mixing old

Figure 8.14. Early plant recovery following the eruption of Mount St. Helens. Some of these species sprouted from rootstocks; others originated from seeds that were either windblown or carried by animals (e.g., birds). The largest trees are cottonwoods, which have light, wind-dispersed seeds. Early recovery is best in erosional gullies for two reasons: (1) they concentrate water and seeds, and (2) they have less ash cover than uneroded areas do. Initial pioneers facilitate the subsequent establishment of plants by ameliorating the physical environment, building soils, and providing perches for birds that spread seeds.

soil with new volcanic ash and disseminating spores of mycorrhizal fungi. Late-lying snow banks melted away to reveal intact beds of tree saplings and shrubs that immediately resumed growth. Amphibians emerged from the sediments of lakes and ponds to reproduce. . . . Organisms did disperse from long distances into the devastated zone, and many established themselves. These included species considered to be pioneers as well as those characteristic of the late successional forest. . . . There were immense legacies of organic structures, most notably blast-flattened tree boles and standing dead trees (snags).

Webster's defines legacy as "anything handed down from . . . an ancestor." In an ecological context, legacies are anything handed down from a predisturbance ecosystem, including both the legacies of human land use (section 8.2.5) and the biological legacies that are a ubiquitous feature of natural disturbances. This section focuses on the latter, which include surviving propagules and organisms, such as buried seeds, seeds stored in serotinous cones, surviving roots and basal buds, mycorrhizal fungi and other soil microbes, invertebrates, mammals, living trees, dead wood, and certain aspects of soil chemistry and structure, such as soil organic matter, large soil aggregates, pH, and nutrient balances (Franklin et al. 2000; Lindenmayer and Franklin 2002).

Most, if not all legacies probably influence the successional trajectory of the recovering system to one degree or another. That is clearly the case with surviving plant propagules, which directly affect composition of the early successional community. Other legacies may shape successional patterns in more subtle ways, or perhaps not at all—this is a relatively new area of ecology that needs much more research. Despite the uncertainties, however, a variety of plausible mechanisms exist through which legacies might influence succession. For example:

- *Soil biology.* As we have already discussed, the composition of the soil biological community following disturbance is a legacy that potentially influences the relative success of different plant species during succession.
- *Dead wood.* Dead wood influences system recovery in several ways. At Mount St. Helens, snags and logs "influenced geomorphic processes, such as erosion and deposition of sediments (and) provided critical protective cover, habitat, and food and nutrient sources for a variety of organisms. Recolonization of some species depended on these structures; for example, western bluebirds and Oregon juncos require snags and logs, respectively, for nesting" (Franklin and MacMahon 2000). Standing dead snags mitigate environmental extremes within disturbed areas by shading and preventing excessive heat loss at night. Down logs within forests are centers of biological activity, including not only organisms of decay, but also roots, mycorrhizal hyphae, nitrogen-fixing bacteria, amphibians, and small mammals (chapter 14), and at least some of these are lifeboated through disturbances within the protective cover of the logs. After disturbance, down logs reduce erosion by acting as physical barriers to soil movement (Franklin et al. 1985) and provide cover for small mammals that disseminate mycorrhizal spores from intact forest into the disturbed area (Maser et al. 1978). The spongelike water-holding capacity of old decaying logs helps seedlings that are rooted in them survive drought (Harvey et al. 1987) (fig. 8.15).
- *Soil aggregates and soil organic matter.* In an earlier section we saw that plants and associated microbes literally glue minerals together to form soil aggregates, which are intimate mixtures of minerals, organisms, and nutrients. These aggregates are essentially little packages of mycorrhizal propagules, other microbes, and nutrients that are passed from the old forest to the new (Borchers and Perry 1992). Soil organic matter in general, whether contained in aggregates or not (most is), provides a legacy of nutrients for the new stand. Depending on its origin and stage of decay, soil organic matter can either stimulate or inhibit plant pathogens (Linderman 1989; Schisler and Linderman 1989).
- *Soil chemistry.* Different plant species may affect soil chemistry quite differently, either by the particular array of nutrients they accumulate, their effect on soil acidity, or allelochemicals they release. To the degree these chemical imprints persist after the plant is gone, they constitute legacies that in theory at least, could influence the composition of the early successional community.
- *Living trees.* As already discussed, living trees maintain links with the belowground ecosystem and thereby maintain islands of mature forest soil within disturbed areas. In Sweden, "large Scots pines usually survive fire, and a significant proportion of the organic layer, where the majority of mycorrhizas are located, is left intact below the uppermost charcoal layer" (Jonsson et al. 1999). Residual living trees also provide structural complexity in the aggrading ecosystem, which translates into habitat (chapter 10).
- *Living patches.* Patches of surviving forest provide benefits beyond those provided by individual trees. They maintain intact ecosystems for small things and serve as foci for recolonization of the broader landscape. One study found 13 species of truffle-forming mycorrhizal fungi in small fragments of mature forest that did not occur in surrounding plantations (Amaranthus et al. 1994). Eight of these were associated with large, decaying logs.

Legacies interact with one another, creating chains of direct and indirect influence. Earlier, we saw how pocket gophers, which survived belowground, facilitated establishment of plants at Mount St. Helens. The soil exposed by these digging animals contained nutrients and mycorrhizal spores, and its organic matter retained water during

Figure 8.15. Immediately after a crown fire that killed all trees in this stand, logs such as this one averaged 150 percent moisture content and contained living roots and mycorrhizal hyphae. Surviving logs, which probably shelter invertebrates and nonmycorrhizal microbes as well as mycorrhizal fungi, may act as foci for the recovery of some groups of soil microbes and invertebrates. They definitely act as dams that reduce soil erosion. Logs do not always survive fire, however; it depends on the intensity of the fire and the moisture of the log.

drought (Allen et al. 2005). Also as discussed previously, once the initial plants are established, they often become foci for the recovery of other plants.

8.3.6 Summary of Successional Mechanisms

Patterns of species establishment and changing dominance during succession are likely to be a complex mixture of the three models proposed by Connell and Slatyer (1977). Franklin and MacMahon (2000) discuss lessons learned from recent, large disturbances: "We find our predictive power is very limited. Models based on individual mechanisms such as facilitation, inhibition, and tolerance can contribute to our understanding of succession but are relevant only to specific locations in the disturbed landscape. Under the complex conditions of real disturbance, many processes are operating simultaneously, and some are the unpredictable consequence of stochastic events. In this context, reductionist approaches and an emphasis on the null hypothesis with its 'either/or' outcome can be very misleading."

To a certain degree, composition of the pioneer community is determined by which species arrive first, which is, in turn, a function of interactions between the nature and timing of disturbance, on the one hand, and the early successional environment, which filters colonizers according to their adaptations, on the other. Biological legacies play a crucial role in recovery and act to shape the new community in the image of the old. Dominance changes over time in part because developmental patterns differ among species—some are fast growing and short-lived, others are slow growing and long-lived, and yet others are somewhere in between—and in part because, for various reasons, some species establish more successfully in a preexisting community than in a newly disturbed site. Figure 8.16 illustrates this pattern for a hypothetical moist tropical forest.

Interactions among plant species during succession often include elements of both facilitation and inhibition and are influenced by complex interactions among climate, resource availability, and many nonplant species—animals, pathogens, mycorrhizal fungi and other soil microbes. Moreover, the nature of interaction may change with time, inhibition becoming facilitation or facilitation becoming inhibition. Hence, one must proceed cautiously when judging interactions among plants during succession, especially over short time frames.

8.4 ECOSYSTEM CHANGES DURING SUCCESSION

In chapter 6 we discussed succession as a progressive imprinting of biological influences onto a site. This process involves various changes in ecosystem properties, some that are quite clear and consistent across a wide variety of ecosystems, and some that are neither. In the former category are biomass accumulation, control over nutrient cycling, and, particularly

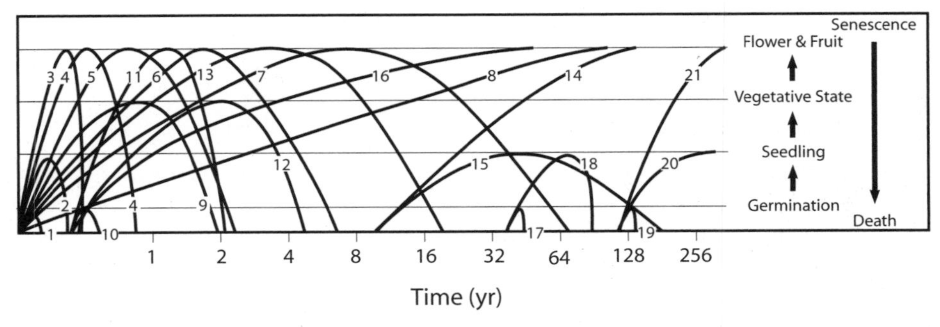

Figure 8.16. A hypothetic sequence of plant establishment during secondary succession in the moist tropics. Each number corresponds to a hypothetic plant species, and the *y* axis indicates the degree to which a given species completes its full life cycle from germination to seed set. For example, species 3 and 5 germinate and set seed in less than 1 year. Species 8 germinates early in succession (or perhaps sprouts from roots) but develops slowly, not producing seeds for several decades. Species 18 establishes several years after disturbance and persists for several years, but it never matures to the point of producing seeds. Species 8, 14, 16, and 21 represent what are called primary rain-forest species, or those that emerge to dominate the forest during later successional stages. (The terminology is potentially confusing; remember that *primary* as used here refers not to the first colonizers but to species of primary importance in the late successional forest). Some primary species become established only very late in succession (e.g., number 21), while others may be present from the early stages but grow slowly and not emerge as dominant until much later (e.g., number 8). (Adapted from Gomez-Pompa and Vazquez-Yanes 1981)

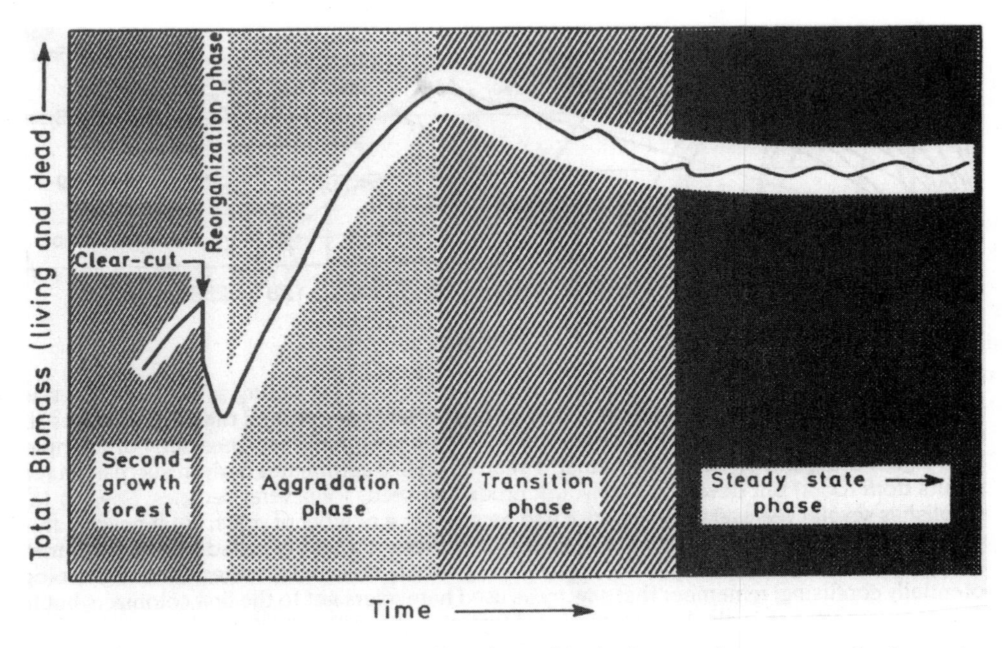

Figure 8.17. Bormann and Likens (1979) proposed four phases of the development of an ecosystem after clear cutting a northern hardwood forest: reorganization, aggradation, transition, and steady state. (Bormann and Likens 1979)

in primary successions, soil formation. In the latter are species diversity and the nature of interaction among species. All of these topics are covered in detail elsewhere in the book. At this point, we discuss them only briefly.

8.4.1 Biomass Accumulation

Forest succession is characterized by rapid accumulation of leaf area and biomass relatively early in the sere. As succession proceeds, the rate of accumulation progressively slows, eventually reaching a point at which new organic matter input through photosynthesis is balanced by its loss through respiration (including the respiration associated with decomposition of dead organic matter). This pattern is shown schematically in fig. 8.17. Based on the rate of biomass accumulation, Bormann and Likens (1979) define four phases of ecosystem development during secondary succession in northern hardwood forests:

- *Reorganization.* A period immediately following disturbance when pioneering plants are establishing

Table 8.1
Comparison of different successional classification schemes

Typical stand age (years)	Classification				
	Franklin et al. 2002	Oliver and Larson 1990	Spies and Franklin 1996	Carey and Curtis 1996	Bormann and Likens 1979
	Disturbance and legacy creation				
0					
	Cohort establishment	Stand initiation	Establishment phase	Ecosystem initiative	Reorganization phase
20					
	Canopy closure				
30		Stem exclusion	Thinning phase	Competitive exclusion	Aggradation phase
	Biomass accumulation/ competitive exclusion				
80		Understory reinitiation		Understory reinitiation	
	Maturation		Mature phase		Transition phase
		Old-growth		Botanically diverse	
150					
	Vertical diversification		Transition phase (early)	Niche diversification	
					Steady state
				Old-growth	
300					
	Horizontal diversification		Transition phase (late)		
800					
	Pioneer cohort loss				
1200			Shifting gap phase		

Source: Adapted from Franklin et al. 2002.
Note: "Typical" stand ages vary depending on community type and environmental conditions.

- *Aggradation.* A period of rapidly accumulating leaf area and biomass, often corresponding to the economic rotation of even-aged forest crops
- *Transition.* A period of some biomass decline from the peak reached during the aggradation phase
- *Steady state.* Constant biomass, except for slight fluctuations around the mean

Later we see that if the disturbance-free period is sufficiently long, many forests enter a decline phase.

Other classifications of forest development stages focus more on structural changes during secondary succession than on biomass accumulation (Oliver and Larson 1990; Carey and Curtis 1996; Spies 1998; Franklin et al. 2002) (table 8.1). Though details vary, all include (1) an early period of plant establishment; (2) a period when the canopy closes, understory establishment is limited, and increasing competition leads to density-dependent mortality (termed **self-thinning**); and (3) a period of diversification, when death of dominant trees (density-independent mortality) adds large dead wood (a critical habitat) and creates canopy gaps that allow understory development, creating both horizontal and vertical heterogeneity. Large dead wood, fine-scale patchiness, and vertical heterogeneity are major defining elements of old-growth forests. With a sufficiently long disturbance-free period, late successional trees reach the canopy and attain dominance. That rarely occurs in disturbance-prone systems.[19]

While recognizing the heuristic value of classifying forest development into distinct stages, Franklin et al. (2002) caution that these are oversimplifications: "Classifications of stand structural developmental stages are arbitrary. First, development is clearly continuous rather than a series of discreet stages. Second, many processes, such as those that create spatial heterogeneity, operate throughout the life of the stand. Third, individual stands may skip particular developmental stages."

The time frame for forest development varies greatly with environment and the nature of disturbance. Reorganization, a highly critical period in ecosystem recovery, usually occurs quickly; however, disturbances that are especially severe (e.g., involve soil loss) or foreign (i.e., those for which the native plants are not adapted) may result in an extended reorganization period (chapter 20). The period spent in the aggrading phase varies widely from one forest type to another. The long-lived conifer forests of the Pacific Northwest, for example, accumulate biomass (living and dead) for hundreds of years, but that would not be the case in forests dominated by shorter-lived trees. As in the reorganization phase, the rate of biomass accumulation depends on the severity and historic frequency of disturbances. In Amazonia, repeated clearing and burning slows biomass accumulation by as much as 50 percent (Zarin et al. 2005).

The period of aggradation also varies depending on what biomass component is considered. In its earliest stages, the aggrading forest preferentially allocates biomass

[19] Although, as discussed earlier, under some circumstances late successional trees can dominate relatively early in succession.

to organs necessary to secure resources: leaves and roots. Leaf area usually reaches a maximum in the aggrading ecosystem long before living biomass. Fully stocked, 20-year-old Douglas-fir plantations support as much leaf area as 500-year-old forests in the same habitat, but it takes 500 years to attain an equivalent living and dead biomass. By 6 years of age, a tropical rain forest dominated by *Cecropia* had developed nearly as much leaf area as a mature forest, and about 60 percent of the living biomass (Gomez-Pompa and Vazquez-Yanes 1981). Biomass accumulation in soils and on the forest floor depends on decomposition rate, which is, in turn, a function of environment and chemical characteristics of the biomass. Moist tropical forests build up virtually no forest floor organic layers during succession, while in boreal and temperate coniferous forests, forest floor aggradation may continue for hundreds of years. Organic matter also accumulates belowground, particularly during primary succession. Figure 8.18 shows aggradation of soil and forest floor organic matter during the first 180 years of succession on areas exposed by receding glaciers in Alaska. Primarily associated with nitrogen-fixing *Alnus tenufolia*, aggradation of soil organic matter was accompanied by accumulating nitrogen (fig. 8.18B) and decreasing bulk density (fig. 8.18C). In other words, rock was being turned into soil. Because of rapid turnover of roots and mycorrhizal hyphae, organic matter accumulates more slowly in soils than in living trees and forest floor. For example, over a 40-year period, loblolly pine planted on abandoned agricultural fields in South Carolina accumulated nearly five times more carbon aboveground (in living trees and forest floor) than belowground (living roots and soil carbon) (Richter and Markewitz 2001). Accretion in soils was especially slow, which probably reflects a high rate of decomposition and cycling of carbon back to the atmosphere (Richter and Markewitz 2001) (chapter 18). An important implication is that forest soils degraded through land use practices can be expected to recover very slowly.

Finally, in addition to temporal variability, the spatial scale of forest development varies widely also, depending on the scale of disturbances common to a given type. The importance of spatial scale is particularly relevant to the steady-state phase. The notion of a steady state has fallen out of favor in modern ecology, which views the "natural" state of nature as one of change rather than constancy. However, as we have discussed in earlier chapters, one might see either change or constancy within the same forested landscape, depending on how closely one looks. The northern hardwood forests studied by Bormann and Likens, for example, are characterized by gap-phase replacement: a tree establishes, grows, matures, senesces, and dies, making room for another tree. At the scale of one or a few trees, there is constant change. At the landscape scale, however, we might find that the rate of gap formation and replacement are fairly constant, resulting in a *relative* degree of constancy in species composition, stand structure, and biomass. Even that steady state is transitory, however, because it persists only within certain bounds of climate and disturbance that are likely to be exceeded sooner or later, whether by clearcuts, abnormal drought, an unusually destructive windstorm, virulent exotic pests, pollutants, long-term climate change, long-term soil change, or any of numerous other factors that might impose a fundamentally different pattern on the system.

8.4.2 Forest Decline

Some late successional forests cannot sustain levels of biomass in the absence of disturbance, and some cannot sustain themselves at all. Perhaps the clearest example of the latter is in the far north (Van Cleve and Viereck 1981). Because low temperatures lead to slow decomposition, dead organic material accumulates on forest floors. This negatively affects tree growth in two ways: it ties up site nutrients in unavailable forms, and it insulates and prevents the soil from warming during the brief summers. The latter initiates a series of events that reinforce one another and eventually make the site unsuitable for trees (fig. 8.19). Slow decomposition in the cold soils further reduces nutrient cycling and results in even greater accumulation of organic matter. Organic acids accumulate and lead to lower soil and forest floor pH. Because of its high water-holding capacity, increasing amounts of organic matter may lead to too much soil water. Permafrost (the condition in which soils never thaw) may develop or, if already present, its level may rise in the soil. This further

> In chapter 1 we learned that feedback is a distinguishing characteristic of systems. Degeneration of boreal forest in the absence of fire is an example of positive feedback, in which a process or group of processes interact in such a way as to increase their own activity: in the example given in the text, accumulation of organic matter cools soils and leads to further accumulation of organic matter, which favors development of mosses, which cool soils even more, and so on. Positive feedback is not necessarily a "bad" thing—various ecosystem processes incorporate positive feedback loops. For example, in the reciprocal interaction between plants and soils, plants create fertile soils, which, in turn, create better growing conditions for plants; population growth is a positive feedback mechanism: organisms beget more organisms and these beget yet more. However, positive feedback is also a potentially destabilizing force within ecosystems, leading, if not counterbalanced by other processes within the system, to significant alteration of system characteristics. Excessive growth of the human population during the last several decades is a good example of the disruptive power of a runaway process.

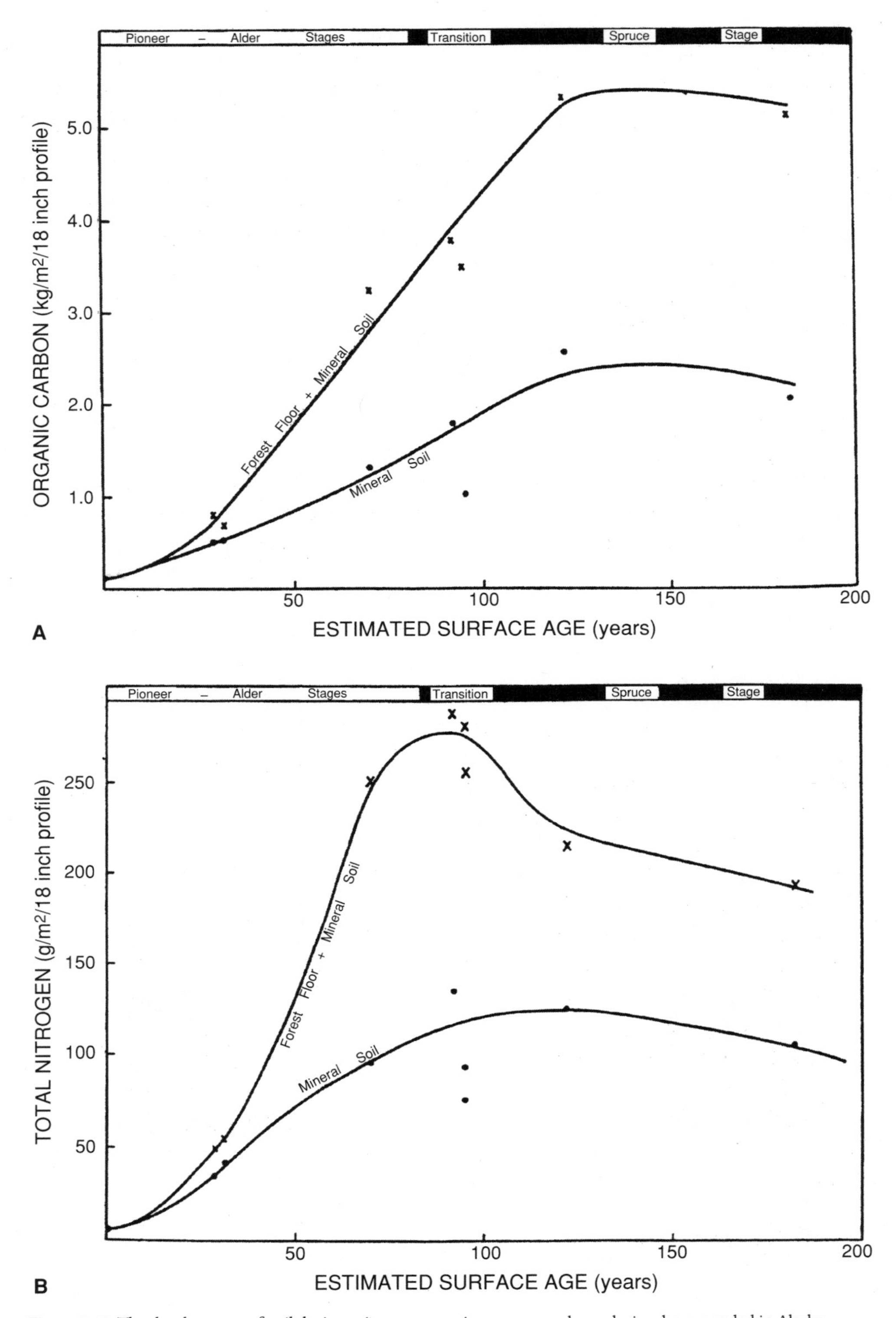

Figure 8.18. The development of soil during primary succession on areas where glaciers have receded in Alaska. A, Soil organic carbon. B, Soil nitrogen. *(Continued)*

restricts drainage and decreases the rooting volume available to trees. All of these changes—reduced nutrients, increased acidity, higher soil water—favor the survival and growth of mosses, which further insulate the soil, compete for nutrients, and generally reinforce the processes contributing to forest decline. The end result of this process is a **muskeg** (i.e., a bog) that supports few or no trees (Heinselman 1981). This eventuality is prevented by one of two things: fire, which destroys organic layers, warms the soil, and initiates a new round of succession, or windthrow, which exposes mineral soil and provides a seedbed upon which tree seedlings can establish, thus perpetuating the forest (Bormann 1989).

Forests of the far north are somewhat of a special case, but they are not unique. In the absence of disturbance, other forest types commonly enter a decline phase, or retrogressive succession, in which productivity and biomass drops off (Birks and Birks 2004; Wardle et al. 2004b). *Soil fatigue*, as this condition has been called, occurs widely, but is poorly understood. Wardle et al. (2004b) studied six forested chronosequences in which the later stages showed significant declines in tree basal area[20] (one each in Australia, Sweden, Alaska, and Hawai'i, and two in New Zealand). The decline phase in all was associated with increasing nutrient limitation, especially of phosphorus. In most of the chronosequences the decline phase also included changes in the decomposability of litter and the composition of the soil microbial community that are common responses to nutrient limitation. At least during primary succession, many forests appear to accumulate biomass at an unsustainable rate. Disturbances pulse nutrients into the system, either by accelerating decomposition (e.g., fire) or supplying new sources (e.g., through alluvial deposits or by exposing fresh rock to roots and mycorrhizal hyphae).

[20]Basal area, the sum of individual tree basal areas per unit area, is a commonly used proxy for potential site productivity.

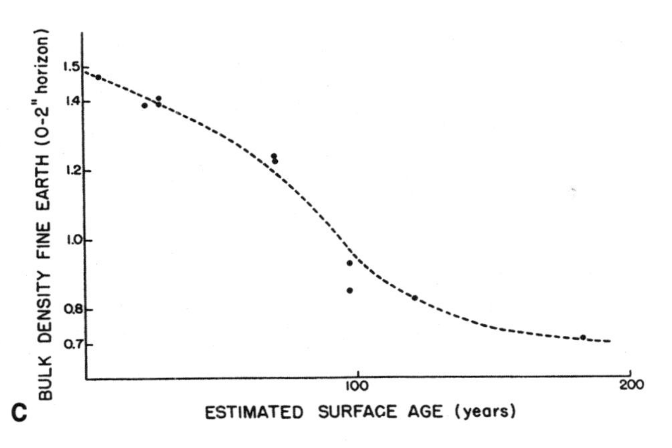

Figure 8.18. *(Continued)* C, Bulk density.

8.4.3 Nutrient Cycling

Biomass accumulation during succession is accompanied by transfer of nutrients from soil to biomass. This process is extremely important to ecosystem recovery following disturbance, particularly in high-rainfall zones or where soils have low nutrient-holding capacity. In such cases, nutrients not tied up in either plant or microbial biomass are quickly leached from soil to streams, and site productivity declines. (We return to this point in chapter 17.) As we saw in the previous section, nutrients can become increasingly limiting if forests go through long periods without disturbance.

8.4.4 Species Diversity

Species diversity (as opposed to composition) may change in one fashion or another during succession, but patterns are not as consistent as those related to aggradation and nutrient

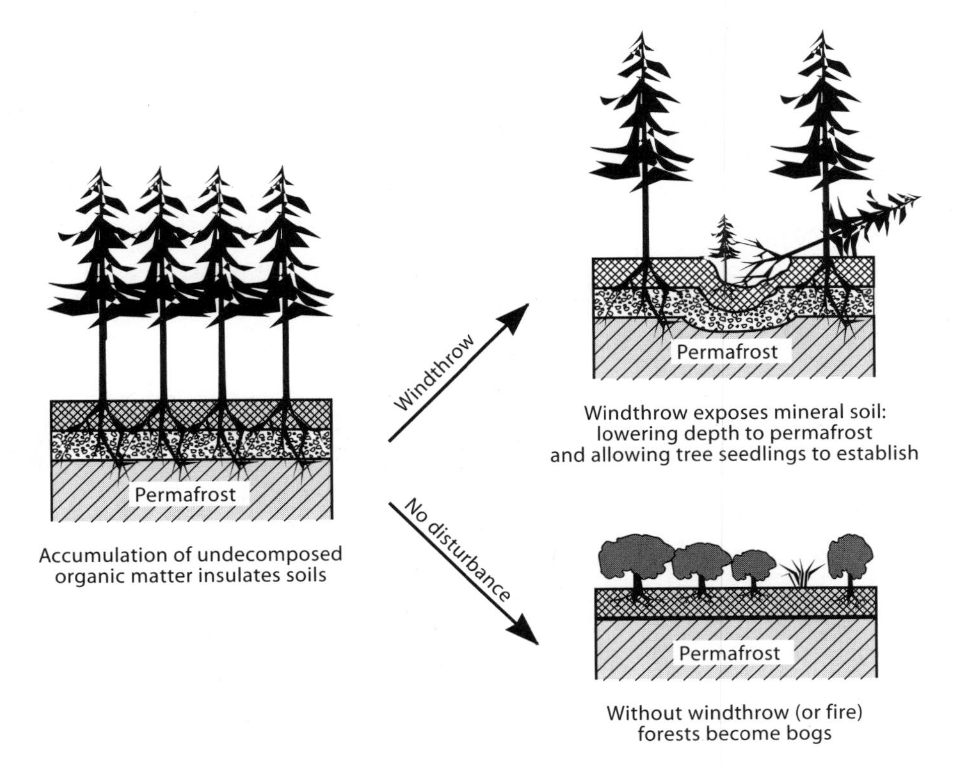

Figure 8.19. In the absence of disturbance, some northern forests are succeeded by bogs.

cycling and depend very much on the nature of disturbance (especially the amount of structural legacy that survives).

Species diversity has two components: (1) **richness,** or the number of species in the community, and (2) **evenness,** or the distribution of numbers (or biomass) among species that are potentially utilizing the same resources (also called **equitability**). Both correlate negatively with **dominance,** or the tendency of one or a few species to dominate the community. Species richness is low during the early stages of primary succession; few plants are adapted to colonize primary substrates; however, it is often relatively high during the early stages of secondary succession, when the reduced dominance of the overstory allows a variety of plants to flourish. Richness remains high and may even increase until early successional trees close their crowns and exert dominance, at which time pioneering plants decline in importance and diversity decreases. With time, growth of later successional tree species results first in an upturn in plant diversity (because of greater evenness), then, as they attain dominance, another decline. For example, pin cherry was the early dominant following destruction of New England white pine forests by hurricane in 1938. By 1948, however, three other species (red maple, white ash, and red oak) had grown into positions of codominance with the cherry, increasing evenness. By 1978, red oak and paper birch were exerting strong dominance, with consequent decline in evenness (Hibbs 1983). In forests with many tree species, diversity may fluctuate up and down as successive waves of species grow into the canopy and attain dominance. In forests with few tree species, initial canopy closure may be followed by a relatively long period of low diversity.

Animal species richness is generally tied closely to plant successional patterns. Early successional forbs, grasses, and shrubs provide green forage for mammals and abundant seeds for small mammals and birds. On the other hand, destruction of litter layers and removal of downed logs has a profound impact on soil animals. In the Oregon Cascades, clear-cuts that have been broadcast burned have 50 percent fewer species of soil invertebrates than forests (A. Moldenke, unpublished data).

In forests dominated by few tree species, the closed canopies and low structural diversity of midsuccessional stages provide little food and habitat diversity for vertebrates, although they are important as cover for large mammals and nesting sites for some birds. The buildup of litter during this stage probably enhances the diversity of soil invertebrates. Multilayered canopies, which are often associated with transitional stages of the sere, provide diverse habitats for birds. Figure 8.20 illustrates this pattern for pine-hardwood forests of the southeastern United States. In this area, early successional forests are often quickly dominated by fast-growing pines. Slower-growing hardwoods, primarily oaks and hickories, gradually increase in size, attaining codominance at 80 to 100 years and replacing the pine after about 150 years. Numbers of species and breeding pairs of birds increase during the earliest stages of succession, decrease as pines come to dominate

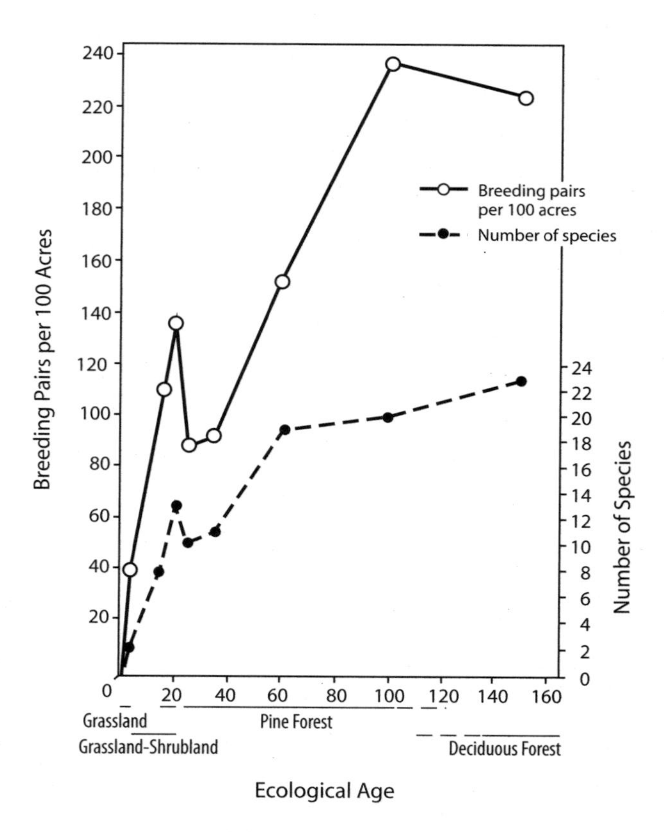

Figure 8.20. Successional changes in the total bird density and number of bird species in the Piedmont area of the southeastern United States. (Johnston and Odum 1956. Copyright © 1956 by Ecological Society of America. Reprinted by permission)

the site, then increase sharply as hardwoods become more important. In their study, Johnston and Odum found that the number of species of breeding birds doubled between year 35 and 60. However, when richness is calculated as species accumulation curves, which portray the cumulative increase in species as more plots are added (chapter 10), the story becomes more complicated. Viewed in that way, either young stands or older stands in the southeastern United States may support greater bird species richness, depending on which subregion is considered (Loehle et al. 2005a). As Loehle et al. point out, "reported relationships between forest age and bird richness are complex and variable." One reason is that forest age alone is an imperfect measure of the diversity of habitats it provides. A young pine plantation, for example, may be a densely stocked monoculture, a diverse forest supporting a variety of plant species, or (more likely) something in between. Because of the abundant biological legacies, early successional stands following natural disturbances are not easily characterized as "young" in the same sense as a plantation established on a clear-cut. In the Pacific Northwest, richness of vertebrate species differs little among successional stages resulting from natural disturbances (fire, wind), and the majority of species that have been studied occur throughout the sere (see the review by Hansen et al. 1991). Hansen et al. express the opinion of most ecologists in the region: "A likely explanation for the similarity in species distribution is that structural differences among these natural forest stages are insufficient to strongly influence most species of plants and animals. The natural disturbance regime and structural legacy in all . . . age-classes provide the resources and habitats required by many species. The important conclusion is that the canopy structures, snag densities, and levels of fallen trees found in unmanaged young, mature, and old-growth stands appear to make all three of these seral stages suitable habitat for most species of forest plants and vertebrate animals." Note that "most species" does mean "all species." Some are still restricted to one or another seral stage.

Snags and logs resulting from the death of large, old trees greatly enhance diversity by providing habitat for mammals, birds, insects, and fish. Thomas (1979) lists 39 bird and 23 mammal species in the Blue Mountains of eastern Oregon and Washington that use cavities in snags. In the same area, 179 species of amphibians, reptiles, birds, and mammals use downed logs. In the Cascade Mountains of western Oregon, snags or logs are essential habitat for 45 species of terrestrial vertebrates (Harris 1984). In some areas, large logs that fall into streams create pools, providing important habitat for aquatic organisms. Unless removed in salvage, logs and snags are abundant in the early successional stages following natural disturbance. Their abundance declines over time, reaching a low in the middle seral stages, then increases again as forests enter the old-growth stage.

To really know the state of biodiversity, one must go beyond questions of species richness and ask about the status of groups with different needs. Birds, for example, divide into guilds based on their nesting preferences; some prefer mature forests, some early successional stages, some pines, others hardwoods. In chapter 10 we will see that total richness of bird communities correlates positively with landscape heterogeneity. Moreover, and somewhat surprisingly, at least in some cases, so does species richness within a given nesting guild.

To recapitulate, species diversity is relatively high in the mix of grasses, forbs, shrubs, young trees, surviving old trees, snags, and logs that characterize early successional stages; it may or may not drop in midsuccessional stages (depending on the degree to which one or a few tree species come to dominate the site); then richness increases again with the increase in spatial and structural diversity as forests move into the old-growth phase. Because of this pattern, even-age, short-rotation forestry can have considerable impact on biological diversity (fig. 8.21). Clear-cutting reduces structural legacies, planting and control of noncommercial vegetation accelerates passage of the system through the early grass-forb-shrub-small tree stage, and short rotations completely truncate later successional stages. Forest managers in many areas attempt to preserve or replace key habitat. For many years, European foresters have used nest boxes, which some, but not all, cavity-nesting birds will use. Currently, approaches to ecosystem management are being developed that attempt to protect diversity by emulating at least some of the patterns of natural disturbances (Lindenmayer and Franklin 2002) (fig. 8.22). Ecosystem management is discussed in chapter 23.

8.4.5 Hydrologic Regulation

In chapter 3 we saw that regulation of streamflow by forests arises from interactions between the frequency of disturbance and plant growth rate (fig. 3.11). In this section, we

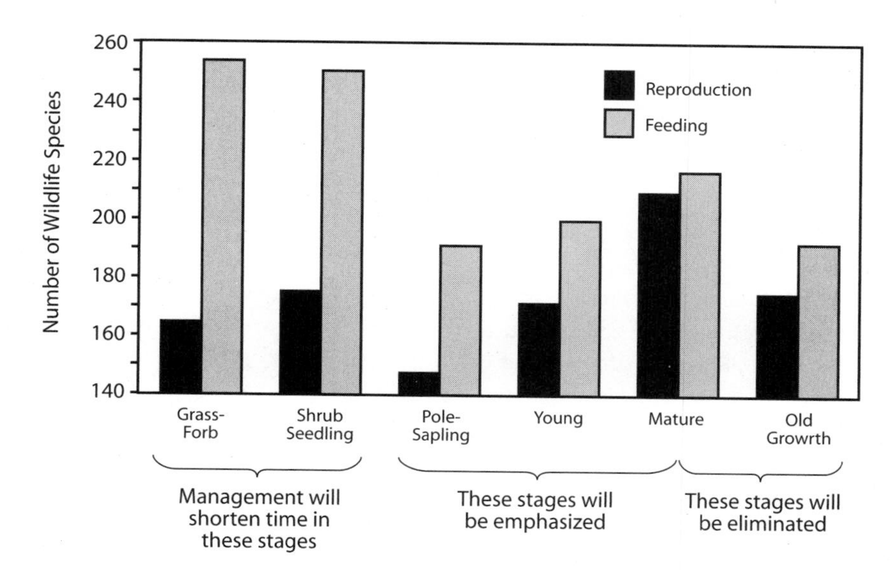

Figure 8.21. The numbers of bird and mammal species that reproduce and feed during different successional stages in the Blue Mountains, Oregon. Even-aged, short-rotation forestry eliminates the old-growth stage. Management often shortens the period that any one site spends in the early successional stage, because noncommercial vegetation is controlled to hasten the dominance of the site by crop trees. However, short rotations may actually increase the amount of early successional habitat at the landscape scale, because clear-cutting creates early successional habitat at a faster rate under short rotations than under long rotations. (Adapted from Thomas et al. 1979)

Figure 8.22. Living trees left in a clear-cut to provide a source of future snags and logs. H. J. Andrews Experimental Forest, Oregon.

circle back to look at some specific examples. Andreassian (2004) summarized results from several paired watershed studies distributed across a wide range of locales and forest types as follows: (1) peak flows are almost always greater from logged (i.e., early successional) than from unlogged basins (this does not appear to hold, however, for infrequent, extreme floods) and; (2) initiating early succession by clearing forests almost always increases and reforesting decreases low flows (however, see below). Jones and Post (2004) did a more detailed study of the time course of successional effects on streamflows for 14 logged and control basin pairs in the western and eastern United States (fig. 8.23; also see Swank et al. 2001). First, consider total yearly streamflow (fig. 8.23A). Sites vary widely in the time required for total yearly flows to return to predisturbance levels. In deciduous forests with seasonal snow, a strong initial response is followed by relatively rapid recovery (15 to 20 years). Conifer forests with seasonal snow also have a strong initial response but recover much more slowly; after 40 years streamflows remain well above predisturbance levels. After the fifteenth year, some of the early successional deciduous forests are using more water than their unlogged partner basin, reflecting high water use by densely stocked young forests.

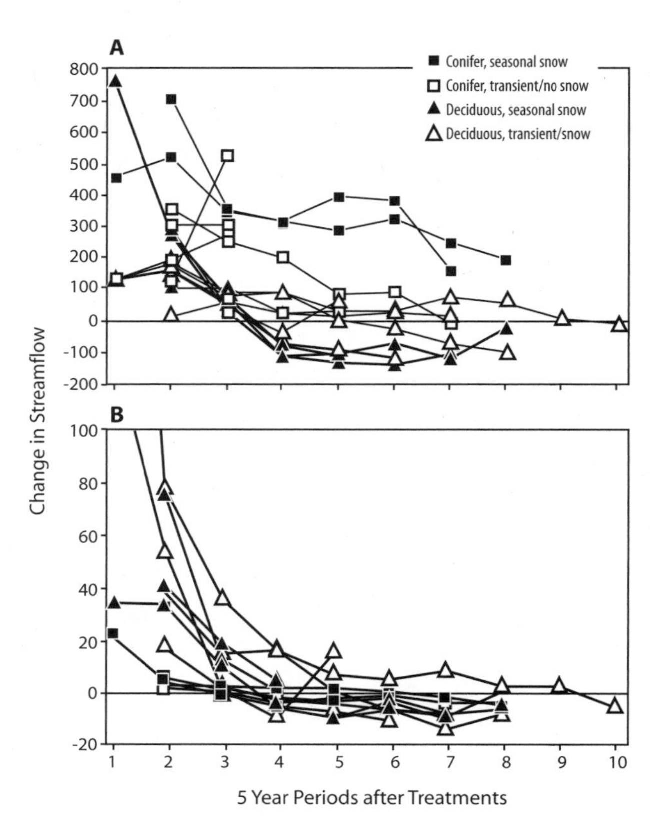

Figure 8.23. Effect of time since harvest on absolute changes in streamflow for 14 treated and control basin pairs in the United States. A, Entire year. B, August low flow. Values on the *x* axis are 5-year periods since treatment. Period 1, pretreatment; period 2, 1–5 years after treatment; and so forth. Values on the *y* axis are differences in yearly streamflow (mm) between treated and control basins. Sites represent both western coniferous and eastern deciduous forests. (Adapted from Jones and Post 2004)

Summer low flows tell a somewhat different story, especially in western North America (fig. 8.23B). All except the deciduous sites with transient snow cover had returned to baseline levels within 15 years of harvest, and in most sites early successional basins have lower late-summer flows than control basins (i.e., values on the *y* axis are less than zero). Comparing parts a and b of fig. 8.23, it is evident that the excess yearly streamflows from early successional conifer forests in the seasonal snow zone results from spring and early summer runoff, most or all of which originates from snowmelt.

In most of the 14 basin pairs studied by Jones and Post (2004), the early successional forests had less capacity than older forests to capture, store, and release water to streams during dry seasons. Similar patterns have been seen in two forest types in Australia (Cornish and Vertessey 2001; Vertessey et al. 2001). Jones and Post (2004) suggest that various factors may be involved, including altered interception capacity and the development of a canopy epiphyte community in older forests.

One consistent factor in paired basin studies is wide variability among sites, which reflects a variety of interrelated factors, including soils, climate, and differences in stand structure (Andreassian 2004; Jones and Post 2004). The latter includes species composition and the accumulation of leaf area, which determine, in turn, interception (particularly of snow), transpiration, and evaporation. In snow zones, coniferous and deciduous forests respond quite differently to clearing that initiates early succession. Mature conifer forests capture snowfall in their canopies, where it tends to sublimate back into the atmosphere. Consequently, in coniferous forest types, early successional communities accumulate much greater snowpacks than mature forests (which appears as peak flows snowmelt). Without winter foliage to collect snow, however, mature deciduous forests behave more like early successional vegetation in the degree to which snow collects on the ground, and clearing makes less difference.

8.5 THE EMERGENT LANDSCAPE REVISITED

Throughout this chapter and the last we have discussed patterns of disturbance and succession mostly from the viewpoint of a single piece of ground. In order to really grasp the dynamic nature of forest ecosystems, it is necessary to step back, enlarging our scope to include the landscape. As we saw in chapter 5, landscapes are often composed of numerous community types, corresponding to changes in elevation, aspect, and soils. Patterns of disturbance and succession overlay this with yet another pattern. Disturbances are often quite variable across the landscape. Fires ebb and flow in intensity, crowning out and killing mature trees in some areas, underburning in others, perhaps skipping others all together. Early successional patterns vary with local environments, intensity of disturbance, and proximity of seed sources, and probably by chance as well. Following the same disturbance, some patches of ground may become dominated by trees very quickly, while others spend an extended period of time in the shrub or small tree stage. Late successional trees appear relatively quickly within some areas of a midsuccessional forest, very slowly in others. The result of all this is a dynamic mosaic of varying plant species and stages of development, which, in turn, creates a rich diversity of habitats for animals and microbes.

8.6 SUMMARY

Succession involves change in both community composition and in ecosystem structure and processes. Plant species pioneering newly exposed substrates (i.e., primary succession) build soil through accumulation of organic matter, nitrogen fixation, and rock weathering, thereby creating conditions that facilitate survival of other species. Secondary succession—that which occurs following disturbances such as clearcutting, windthrow, or fire—is strongly shaped by legacies of the old stand, particularly large dead wood, surviving live trees, and plant species that regenerate by sprouting from roots or from seed stored in soil or serotinous cones.

Dominance during the typical forest sere shifts from grasses and herbs to shrubs to a sequence of tree species: first fast-growing shade intolerants and eventually slower-growing shade tolerants. At any one point in time, however, communities are likely to contain species of both current and future dominants. Eventually, at least in theory, a steady state is reached in which there is no further change in species. In practice, however, that condition is seldom attained at the stand level, though it may be at the landscape level. With a sufficiently long disturbance-free period, forests may enter a decline phase that is apparently related to increasing nutrient limitations.

Succession is characterized by complex interactions among plants, animals, and microbes. In some cases pioneer plants facilitate survival and growth of later comers: building and stabilizing soil physical structure, nutrients, and biotic communities, thereby providing key habitat for animals that spread seeds. In other cases early successional plants inhibit growth of later comers. Most frequently, interaction includes elements of both facilitation and inhibition, and the net effect may not be obvious for many years.

At the ecosystem level, succession involves a rapid accumulation of biomass and transfer of nutrients from soils to plant biomass. Species diversity is often high early in succession, drops as one or a few early successional trees come to dominate, then turns upward again during the transition from early to late successional tree species. Patterns are highly dependent on the nature of disturbance, particularly the amount of structural legacy retained. Regulation of streamflow varies through succession. Streamflow is higher in basins draining early successional stages, especially during periods of peak flow, but as rapidly growing young forests come to dominate, streamflow declines. Evidence indicates that as forests age, they become more efficient at trapping, storing, and releasing water to streams, especially during periods of low flow.

The temporal dynamics of forests cannot be grasped by looking at only a single point on the ground but must be understood at the scale of landscapes. The typical landscape comprises a mosaic of communities at different successional stages and occupying different habitats. The result is a rich, dynamic tapestry of vegetation providing an array of habitats for animals and microbes.

The Structure of Local Ecosystems

Patterns are defining characteristics of a system and often, therefore, indicators of essential underlying processes and structures.

GRIMM ET AL. 2005

Life is about diversity, heterogeneity, and novelty.

MARQUET ET AL. 2004

One of the more commonly accepted definitions of biodiversity is the *variety of life and its processes* (Keystone Center 1991), which covers almost everything but nevertheless tells us something important. Biodiversity is more than just species; it is life in all of its manifestations. Learning to live sustainably, however, requires knowing more. A good beginning is the concept of hierarchy. In chapter 1, we saw that the spatial patterns of nature are arranged hierarchically, which is to say that every entity that we choose to delineate (whether organism, ecosystem, landscape, or something else) has a dual aspect: it comprises smaller entities and at the same time is part of and interacts with larger entities (Allen and Starr 1982; O'Neill et al. 1986; Grene 1987). In the words of Noss (2002), "Every local area is a piece of a bigger ecological puzzle. Its . importance can be appreciated only in relation to a larger whole." Thus far, we have discussed higher levels of the hierarchy: patterns of ecosystem types at the scales of the globe and local landscapes. We have also discussed larger-scale processes: global cycles of the elements, disturbance, and succession. In this and the following chapters, we turn to those levels that managers deal with and that the public perceives most directly: the structure of local ecosystems, and the processes that create and maintain that structure.

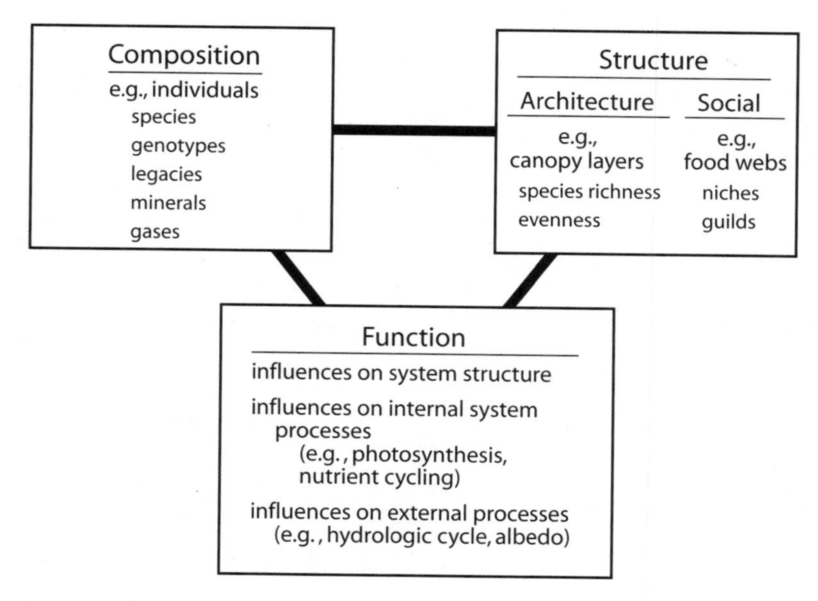

Figure 9.1. The components of biodiversity in local ecosystems.

Before proceeding it is necessary to clarify the descriptor *local.* Ricklefs (2004) argues that "ecologists should abandon circumscribed concepts of local communities. Except in cases of highly discrete resources or environments with sharp ecological boundaries, local communities do not exist. What ecologists have called communities in the past should be thought of as point estimates of overlapping regional species distributions. The extents of these distributions can be understood only by considering interactions within the region as a whole."

How much territory does such a "point" estimate comprise? There is no precise answer to that question, nor can there be. Imagine an inverted pyramid. When taking a landscape or regional perspective we start at the top and work down. When taking a local perspective we start at the bottom and work up. As with *landscape, local* is a subjective distinction and can be equated with what foresters and ecologists have traditionally called *stands*. Very roughly, *local* encompasses heterogeneity at scales of a few hundreds of meters or less (whereas landscapes encompass heterogeneity at larger scales) and includes components that do not occur at landscape scales, especially three-dimensional biotic factors such as vertical heterogeneity (e.g., crown layers). Local ecosystems are shaped by biotic and abiotic forces whose domain extends far up the pyramid and are linked to other local ecosystems in a network of **metacommunities** and **metaecosystems** (Holyoak et al. 2005; Loreau et al. 2005). Food webs, in particular, integrate spatial domains ranging over several orders of magnitude, as do cycles of the elements. This chapter is about local structure because we start at the tip. But as you will see, we also range well up into the body of the pyramid.

Franklin et al. (1981) divide the biodiversity of a given ecosystem into three components (fig. 9.1):

Composition. Ecosystems are composed of organisms, species, and groups of closely interacting species, genetic diversity within species, legacies of organisms (e.g., dead wood and soil organic matter), and various inorganic components (e.g., minerals and gases).

Structure. Ecosystem structure arises from the patterns in which these basic building blocks occur. It is useful to distinguish two aspects of system structure: architectural and social. *Architecture* denotes the physical aspects of structure and is most often applied to spatial patterns in the plant community: the number of canopy layers, for example, or patchiness in the distribution of species and age classes (fig. 9.2). *Social* refers to patterns in the way that individuals, species, or groups of species relate to one another and to the system as a whole: who eats whom, for example, who helps whom, or how species allocate resources. But the term *structure* is most frequently used to denote the physical (architectural) aspects, and to avoid confusion we adhere to that usage here unless otherwise noted. Also unless otherwise noted, we refer to the social aspects of system structure as the **social network.**

Although the least obvious to the casual observer, social relationships provide "rules" that shape the composition, structure, and function of local ecosystems. But nature has few one-way streets, and certainly none in this case: system architecture also provides rules that modify and shape the social network of a given community, producing an **interaction network** that includes relationships among all aspects of system structure (Price 2002).

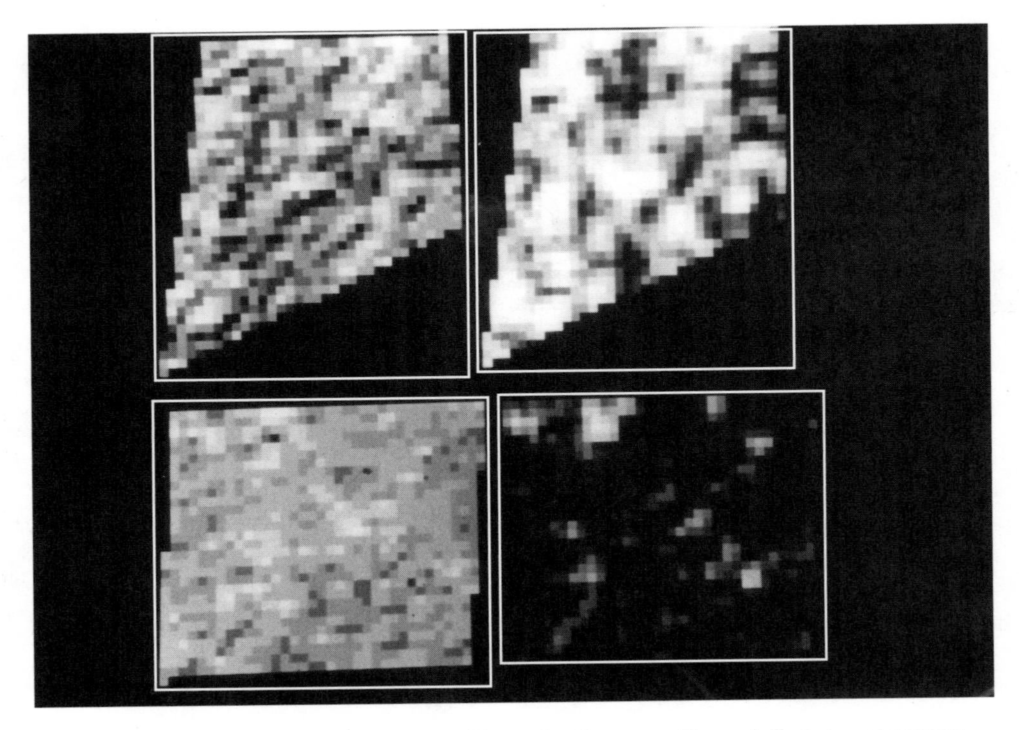

Figure 9.2. Remote imagery of canopy structure in old-growth and young conifer stands illustrates an important aspect of architectural structure in forests. The upper left shows an old-growth canopy, the lower left shows the canopy of a young stand. Note the difference in canopy texture: the older stand appears rough and the younger stand smooth. That difference is accentuated in the images to the right, in which a computer has been used to emphasize differences in canopy texture. Briefly, the computer compares the degree of light-dark contrast between adjacent 10 m × 10 m pixels within each of the images on the left (i.e., within each stand). A high degree of contrast (which equates to a high degree of within-stand canopy texture) appears as white in the transformed images (the old-growth stand), while low contrast appears as dark (the young stand). Older stands commonly have more complexly textured canopies than do fully stocked young stands, because crowns in the former tend to be more clumped horizontally and spread deeper vertically, while crowns of the latter form more of a monolayer. (Courtesy of Warren Cohen)

A forest with complex physical structure is like a house with many rooms. It provides the occupants greater opportunity to find a space providing protection from enemies, and it allows the evolution of unique niches that effectively reduce conflicts over resources. As we have seen in earlier chapters, ecosystem structure is dynamic rather than static, continually shaped and altered by interactions among the environment, disturbances, and the organisms themselves.

Function. *Webster's* defines function as "the normal or characteristic action of anything."[1] In forests, trees function as a food base and a habitat for animals and microbes, and animals and microbes function to cycle nutrients and regulate balance among populations. In these examples, we see two aspects of function: (1) influence on processes (e.g., photosynthesis, nutrient cycling, population growth), and (2) influence on system structure and social networks (e.g., balance among different populations). In addition to these internal functions, we can also identify external functions, which are influences of the community as a whole on its surroundings. Regulation of water and nutrient fluxes, stabilization of soils, and absorption and reflection of solar energy (i.e., albedo) are examples of external functions.

In general, functions can be thought of as actions that flow from some aspect of structure or the interaction web. For example, forest structure functions to provide habitat and regulate processes such as photosynthesis and evapotranspiration. The **food web** functions to cycle nutrients, maintain balance among populations, and shape evolutionary trajectories. In some cases, functions differ between local ecosystems with similar composition but different structure (e.g., young versus old forests), while in other cases systems with different composition but similar structure function similarly. The latter case was demonstrated by an elegant experiment that created a successional sequence similar to that of nearby rain forests, but with a different suite of plant species (Ewel et al. 1991). "Constructed" successional vegetation was as effective as natural rain forest successions in protecting and restoring soil fertility, whereas crop monocultures were not. (Note that forests perform

[1]Webster's New World Dictionary, 2nd college edition, David B. Guralnik, Editor in Chief, William Collins + World Publishing Company, Cleveland, 1978

Scientists of all kinds, from physicists to molecular biologists to ecologists, deal with complexity by seeking patterns in nature, then attempting to identify rules that govern the creation and maintenance of pattern. Patterns provide the basis for organizing (i.e., grouping things together in an orderly way), which reduces complexity by reducing the number of entities that must be dealt with. Rules provide the basis for predicting the conditions under which a given pattern or patterns will occur. The clearest example in biology is taxonomy: individuals are organized into species, species into genera, and so on, with evolutionary ecology providing the rules that govern taxonomic patterns. At another scale, we can easily recognize broad classes of community types—forest, grassland, desert—and we know that climate provides at least some of the rules that determine where these will be found.

Debate among ecologists about the patterns and pattern-generating rules within communities dates back at least to Clements and Gleason (chapter 8). Chase (2003) suggests that "understanding the patterns of community assembly . . . will be central to our understanding of the patterns of species composition and diversity that are at the heart of community ecology." He summarizes the current state of affairs as follows:

> The pattern of community assembly remains controversial. One view holds that there is a one-to-one match between environment and community. In this case, regardless of the historical order in which species invade, if all species (within a regional pool) have access to a given community, composition should converge towards a single configuration in localities with similar environmental conditions. The alternative view (known as multiple stable equilibria) holds that different historical sequences of species entering a locality can lead to different final community composition, even when the environment in each locality is similar and all species have access to the locality.

One good rule of thumb for those trying to understand the patterns of relationship in nature is to be skeptical of broad, general principles, which frequently reflect the human need to organize diverse observations more than they mimic reality. As the Hawai'ians say, *'A'ohe pau ka 'ike ka halau ho'okahi* ("There are many stories, and they all are true"). This is not to say, however, that such conceptualizations are worthless. On the contrary, without generalization, we would be overwhelmed by the rich diversity of nature. The key is to remember that our generalizations are just that; as a sage pointed out long ago, do not confuse the finger that points at the moon with the moon itself!

many different functions, and overlap between two systems in one function does not imply overlap in all).

Interwoven through composition, structure, and function are the **processes,** which, basically, describe anything that changes with time.

As the title suggests, this chapter focuses on physical structure and social networks, although both will be continuing threads throughout the book. Sections 9.1 and 9.2 introduce some of the basic aspects of forest structure and the concepts of habitat and niche, which are the primary interfaces between the physical aspects of structure and the organisms that compose the social network. Sections 9.3 through 9.5 delve further into the social network, including food webs, niche diversification, species richness and evenness, and trade-offs between dominance and diversity. Section 9.6 reaffirms the importance of viewing structure at different scales, from landscapes through microbes to genetic diversity within populations of a single species. In following chapters our discussion moves to consider functional aspects of biodiversity. Chapter 10 deals with how structure functions to create and maintain biodiversity. In chapters 11 through 13 we discuss in more detail the nature of relationships among organisms, and chapter 14 deals with a very special component of ecosystem biodiversity: the soil. Our exploration of the functional aspects of ecosystems continues in chapters 15 through 22, which deal with the critical system processes of primary productivity, nutrient cycling, herbivory, and stability.

9.1 FOREST STRUCTURE

Table 9.1 lists the various components of the physical structure of forests. In this section we touch on just a few of these: vertical and horizontal spatial patterns (both largely driven by canopy structure) and belowground patterns (the latter covered in more detail in chapter 14).

9.1.1 The Canopy

Along with roots and mycorrhizal hyphae, canopies are the primary interface between trees and the external environment. They are the site of light absorption and gas exchange with the atmosphere and provide habitats for a myriad of organisms. All of these factors are strongly influenced by both the vertical and horizontal structure of crowns within the stand. In the words of Solé et al. (2005), "canopy structure provides the forest with a three-dimensional structural matrix where a wide range of community functions must be accomplished. Thus, the comprehension of canopy structure characteristics . . . provides an appropriate framework to study the relationship between structure and function in ecosystems."

Despite its vital ecological importance, until recently, difficulty of access meant the crown layer was largely

Table 9.1
Components of forest structure

- Foliage
  - Leaf area
  - Vertical distribution
  - Leaf shape, density
  - Canopy gaps and horizontal pattern
- Tree crowns
  - Shape
  - Length
  - Life form (e.g., deciduous, coniferous)
  - Diameter, area, density
  - Position in stand
  - Branch characteristics
  - Cavities, breakage, decay
- Tree bark
  - Texture
  - Thickness
- Tree boles
  - Diameter
  - Height
  - Cavities, breakage, decay
  - Gaps and spatial pattern
  - Age distribution
- Wood tissues
  - Volume
  - Biomass
  - Type (e.g., sapwood, heartwood)
- Standing dead trees
  - Diameter
  - Height
  - Decay state
  - Volume, mass
  - Cavities
- Fallen trees
  - Diameter
  - Height
  - Decay state
  - Volume, mass
- Shrub, herb, and moss layers
  - Biomass, volume
  - Height
  - Life form
  - Spatial pattern
- Forest floor and organic layers
  - Depth
  - Decay state
- Pit and mound topography
  - Area
  - Height / depth
- Roots
  - Size
  - Density, decay state
  - Biomass
  - Spatial pattern
- Soil structure
  - Aggregations
  - Organic matter distribution
- Landscape structure
  - Stand / patch type distribution
  - Patch size
  - Patch shape
  - Habitat connectivity
  - Edge density

Source: Spies 1998.

uncharted territory. That has changed for various reasons, including, in particular, remote sensing, modern techniques for measuring gas exchange, and the increasing use of large cranes to provide direct access.[2]

The density[3] of any one canopy[4] layer determines the amount of light that penetrates beneath that layer (chapter 15), and therefore the density profile of the canopy both reflects and regulates the development of vegetation layers (i.e., vertical diversity) (fig. 9.3). Heterogeneity in the understory vegetation (i.e., small trees, shrubs, herbs and grasses, bryophytes) is primarily determined by two factors: (1) horizontal heterogeneity in the canopy profile (e.g., gaps), which is operative in all forest types with the potential to produce a closed canopy; and (2) light ground fires or other patchy disturbances that effect only the understory (or at most kill a few large trees), which is a well-known factor in dry forest types and probably occurs in many other types also. The understory, in turn, influences ecosystem processes in various ways[5] (e.g., Krebs et al. 2001a; Nilsson and Wardle 2005): its productivity can be surprisingly high (comparable to the trees in some boreal forests); it acts as a filter that through competition with tree seedlings, shapes the future forest; it is accessible food for animals that cannot climb or fly (and is often more palatable than tree foliage even for those that can); and it can strongly affect decomposition and nutrient cycling.[6]

Canopy closure is determined by tree species (through crown architecture), environment, disturbance history, and stand age. Leaf area, therefore canopy closure, is limited by shortages of water and, in extreme cases, nutrients (e.g., fig. 5.14). Frequent, relatively mild disturbances create a small patch (gap) mosaic, while infrequent, more severe disturbances initiate a larger-scale successional dynamic that includes not only changes in species composition and dominance but longer-term changes in forest structure independent of species (chapters 7 and 8). Mixed disturbance regimes (which are probably the norm in many, if not most, forest types) create a complex mix of spatial patterns. For example, although large-scale disturbances (i.e., fire and insects) have been commonly assumed to drive the structure of boreal forests, his review of the literature led McCarthy (2001) to conclude "evidence is mounting that indicates that, despite the ubiquitous nature of catastrophic disturbance in the boreal forest, such disturbances do not fully explain the range of observed structural and compositional patterns found in the boreal forest. Micro-scale gap dynamics seem to be a significant and often overlooked disturbance factor in boreal forests." Similarly, montane ash forests of Australia, long believed to be driven by large disturbances, are now known to experience many finer-scale disturbances that play an important role in shaping their structure (Lindenmayer et al. 2000). The same is true of many mesic temperate forests and tropical forests that lie along hurricane tracks.

9.1.2 Aboveground

Table 9.2 lists structural processes occurring during forest development from establishment to old growth.[7] Early successional forests may or may not close their

[2]As of 2005, 10 canopy cranes were operating in Europe, Asia, Australia, and the Americas (Pennisi 2005). *Forest Science* volume 50, issue 3 (2004) contains several articles reporting results from crane studies.

[3]Canopy density can be expressed in various ways, but usually as either the biomass of leaves per unit volume of space or the leaf-area index (square meters of leaf area per square meter of ground area). See chapter 15 for further information.

[4]*Canopy* includes all foliage from the ground upward (Franklin and Van Pelt 2004).

[5]Especially the shrub-forb-grass layer, which unlike tree seedlings holds a permanent rather than transitory place in the understory.

[6]Eviner and Chapin (2003) discuss the mechanisms by which plants in general affect ecosystem processes.

[7]Two quite different definitions have been given for *old-growth forest*. Oliver and Larson (1990) define it as the end stage in succession, when the forest is dominated by shade-tolerant species. A more common definition is based on structural characteristics, especially the presence of old trees, spatial heterogeneity (e.g., gaps), large dead wood, and (depending on forest type) multiple canopy layers (Franklin et al. 1981).

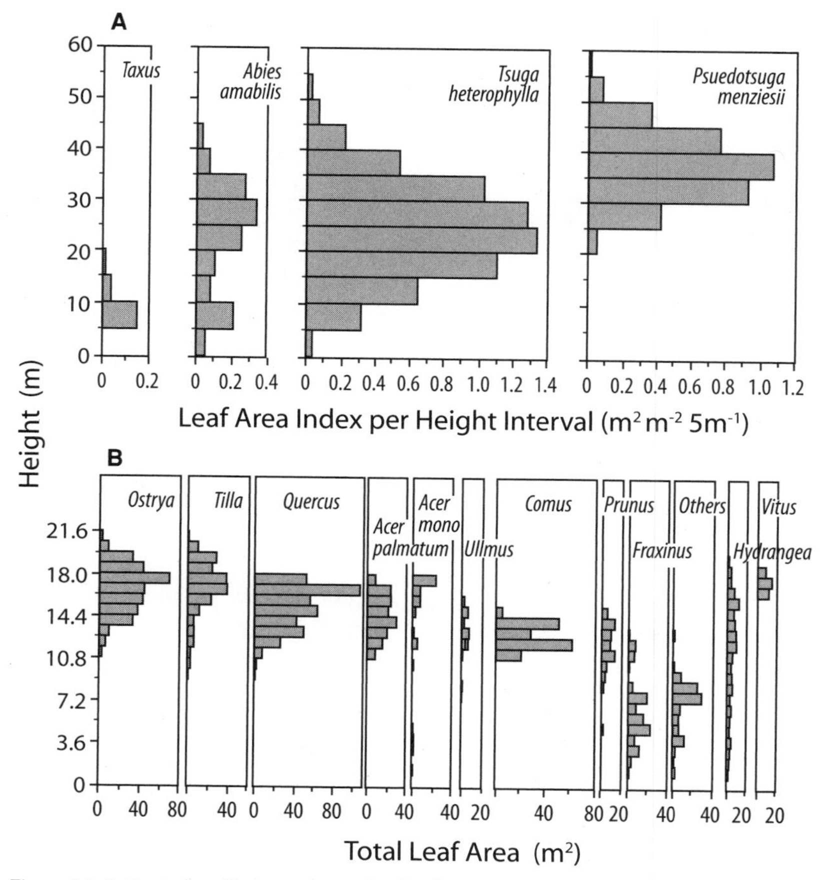

Figure 9.3. A. Vertical profile in 5 m intervals of leaf-area index for the dominant tree species in the Wind River old-growth forest, Washington State. (Adapted from Parker et al. 2004) B. Vertical profiles of leaf area inside a scaffolding tower (225 m^2 surface area, 22 m height) at Tomakomai Experimental Forest, Japan. *Ostrya japonica, Tilia maximowiczii Shirasawa, Quercus mongolica, Acer palmatum* Thunb., *Acer mono*, and *Ulmus japonica* Sargent are canopy tree species, while *Cornus controversa* Hemsl., *Prunus ssiori* Fr. Schm., *Fraxinus lanuginosa f. serrata* Murata, and others are understory tree species. *Hydrangea petiolaris* Sieb. et Zucc. is a vine. (Ishii et al. 2004)

upper canopies quickly (chapter 8); but if they do, maximum site leaf area is attained early and is concentrated in a monolayer. Tree mortality in such stands is driven primarily by competition in which some trees fall behind and succumb, while others survive and grow, the whole process following a very regular pattern (chapter 12).

As trees age and grow tall, stand leaf area is spread over longer crowns, and the canopy becomes increasingly three-dimensional and open (i.e., containing more open space), resulting in light penetrating deeper into the canopy and a consequent increase in understory development (e.g., shrub cover and tree regeneration) (Bailey and Tappeiner 1998; Kerns and Ohmann 2004; Lindenmayer et al. 2004) (fig. 9.4). The transition from mature to old growth involves a further internal restructuring of the canopy, but without a concomitant increase in height. Whereas the three canopy layers defined in figure 9.4B (euphotic, oligophotic, and open space) occur in distinct vertical layers in mature stands, in old growth (at least for conifers) the layers are more distinct in the horizontal than in the vertical; in other words, active photosynthesis extends deep into the canopy in old growth.

Variable-density thinning has been suggested as a way to introduce both horizontal and vertical heterogeneity into closed forests (Carey 2003). Thinning has variable and sometimes unpredictable effects on understory development. In some cases, thinning may promote development of herbs most commonly found in old growth (Lindh and Muir 2004) and of tree regeneration at levels comparable to old growth (Bailey and Tappeiner 1998). Thinning can also promote shrub cover, but composition of the shrub community may differ from that of old growth (Bailey et al. 1998). Understory responses to thinning may differ drastically between stands in the same watershed and similar environmental settings, with responses ranging from highly diverse understories to a monolayer of an aggressive clonal shrub (D. A. Perry, personal observation). Reasons for such diverse responses are not clear but may involve historic factors or a chaotic response in which small environmental differences have large community-level effects.

Table 9.2

Structural processes occurring during the successional development of forest stands, in approximate order of their first appearance

- Disturbance and legacy creation
- Establishment of a new cohort of trees or plants
- Canopy closure by tree layer
- Competitive exclusion (shading) of ground flora
- Lower tree canopy loss
  - Death and pruning of lower branch systems
- Biomass accumulation
- Density-dependent tree mortality
  - Mortality due to competition among tree life forms; thinning mortality
- Density-independent tree mortality
  - Mortality due to agents, such as wind, disease, or insects
- Canopy gap initiation and expansion
- Generation of coarse woody debris (snags and logs)
- Uprooting
  - Ground and soil disruption as well as creation of structures
- Understory re-development
  - Shrub and herb layers
- Establishment of shade-tolerant tree species
  - Assuming pioneer cohort is shade-intolerant species
- Shade patch (anti-gap) development
- Maturation of pioneer tree cohort
  - Achievement of maximum height and crown spread
- Canopy elaboration
  - Development of multilayered or continuous canopy through
    - Growth of shade-tolerant species into co-dominant canopy position
    - Re-establishment of lower branch systems on intolerant dominants
- Development of live tree decadence
  - Multiple tops, dead tops, hole and top rots, cavities, brooms
- Development of large branches and branch systems
- Associated development of rich epiphytic communities on large branches
- Pioneer cohort loss

Source: Franklin et al. 2002.

Older stands develop gaps for two reasons. First, dominant trees begin to die (rarely because of competition, as in young stands, but through the natural process of senescence that leaves trees vulnerable to insects and pathogens); that mortality is often aggregated, opening relatively large gaps (Franklin and Van Pelt 2004). Second the older a stand the greater the chance that it has experienced one or more intermediate-level disturbances (e.g., low- to moderate-intensity fires, windstorms) that kill isolated individuals or small to medium-sized groups of larger trees (Weisberg 2003; Wood 2004). The death of large trees results in accumulations of large dead wood, and rotting fungi become more prominent in living trees. Dead and moribund trees are a signature characteristic of old growth and support a rich food web comprising fungi and both invertebrate and vertebrate animals (chapter 10). Dead trees and their food webs also form the backbone of younger stands that are allowed to recover naturally from stand-initiating disturbances (e.g., fig. 9.5A). The abundance of mosses and epiphytic lichens generally increases as stands age, and the species composition may shift (in conifer forests that pattern is at least partly due to the increase in shrubs and hardwood trees that serve as primary substrates for some moss and lichen species) (e.g., Peck and McCune 1998; Price and Hochachka 2001). The presence of certain lichen species is used along with large dead wood and the density of large trees as old-growth indicators in the northeastern United States (Whitman and Hagan 2007). The details of the general developmental pattern discussed above vary (fig. 9.6); however, the basic outline remains relatively similar across forest types (e.g., McGee et al. 1999; Lindenmayer et al. 2000).

Structure builds upon structure. Consider the epiphytes (*epi* meaning "surface," and *phyte* meaning "plant"). These lichens, bryophytes, and vascular plants (ubiquitous in forests worldwide) live on the surface of other plants and perform a variety of important functions, providing habitat for invertebrates, food for both invertebrates and vertebrates, and influencing element cycles (Nadkarni 1994). A single large fern growing as an epiphyte on a tropical rainforest tree may support as many invertebrates as the entire rest of the tree crown (Ellwood and Foster 2004). Epiphytes build on and amplify existing structural patterns because most are selective about where they grow, and different substrates usually support different epiphyte communities. For example, in the mixed-conifer-hardwood forests of southern Oregon and northern California, heavy moss cover accumulates on the boles of black oaks, but not on white oaks, Pacific madrone, or any of the four conifer species growing in the same stands. Figure 9.7 shows how lichen communities stratify according to different oak species in the Missouri Ozarks (Peck et al. 2004). Lichen communities in all forest types partition the available substrates not only according to tree species, but among soil surfaces, rocks, and dead logs and within individual trees. In the boreal forests of Estonia, for example, 107 lichen species are found only on tree boles, 24 species only on branches, 24 species only on woody debris, and 18 species only on the ground (Lohmus 2004). Furthermore, species segregate on boles depending on height, and on branches depending on size.

The use of large railroad cranes to access canopies has greatly improved our knowledge of all aspects of crown structure, including the distribution of epiphytes and invertebrates. Using a crane at Wind River Experimental Forest (Washington), McCune et al. (2000) discovered that "more species of epiphytes showed a distinct association with the very tops of trees (within two m of the top) than any other single habitat in the forest." They speculate that this is due to birds landing on treetops and inadvertently spreading the propagules of lichens and bryophytes. Schowalter and Ganio (1998) found that tree species at Wind River differed significantly in the structure of canopy arthropod communities: western red cedar and western hemlock communities were dominated by sapsuckers, and grand fir by detritivores, while Douglas-fir had a relatively even distribution of predators, gall formers, detritivores, and sapsuckers (also see Schowalter 2006).

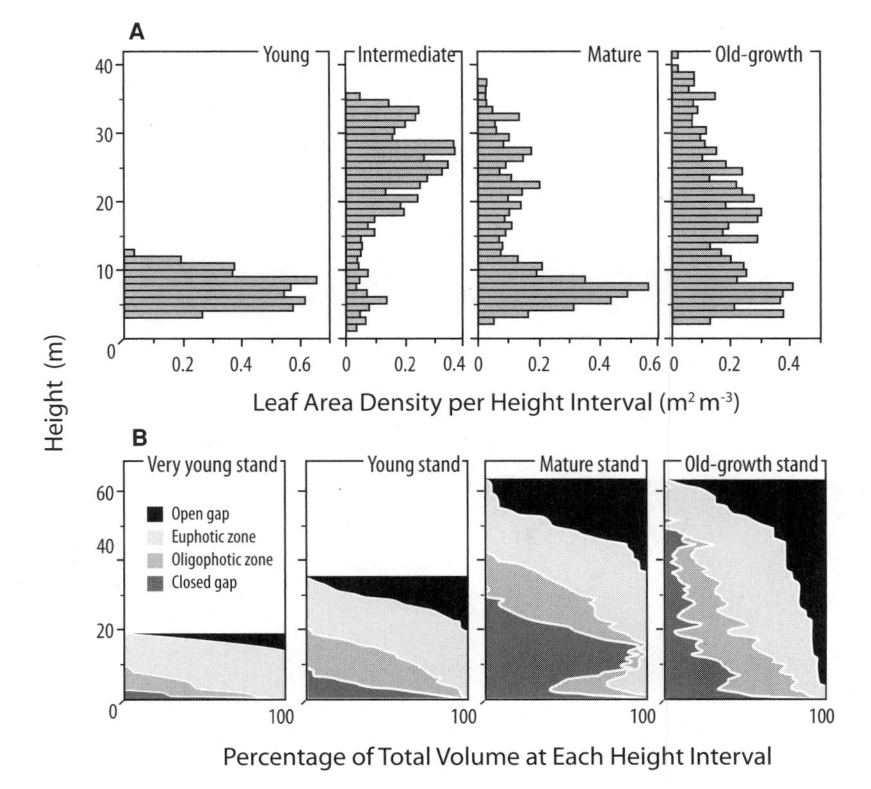

Figure 9.4. A. Vertical distribution of foliage area in the canopies of an age sequence of stands in the tulip poplar association, Mid-Atlantic region of the United States. Profiles determined using laser altimetry (LiDAR. (Parker and Russ 2004. Reprinted with permission from Elsevier. Copyright 2004) Lim et al. (2003) review the use of LiDAR in forestry. B. LiDAR determined canopy characteristics in four age classes of Douglas-fir western hemlock forests, Oregon. *Euphotic* and *Oligophotic* refer, respectively, to the zone in which most light energy is captured and the rest of the canopy. By convention, the cutoff point between the two zones is 65% of LiDAR signal returned. (Lefsky et al. 1999. Reprinted with permission from Elsevier. Copyright 1999)

A

Figure 9.5. A. This relatively young forest in New Brunswick is recovering from a spruce budworm infestation in the mid-1900s. It exhibits some of the structural characteristics commonly associated with old-growth forests, most notably dead wood and heterogeneous horizontal and vertical structure. *(Continued)*

Figure 9.5. *(Continued)* B. An unsalvaged stand following the Biscuit fire, Oregon. C. Old-growth montane ash forest, Victoria, Australia. *(Continued)*

Figure 9.5. (*Continued*) D. Old-growth dry forest, Big Island, Hawai'i. (Courtesy of Susan Cordell) E. A mature mixed conifer forest, Siskiyou Mountains, Oregon. (*Continued*)

Figure 9.5. *(Continued)* F. Old-growth Acadian Forest, New Brunswick.

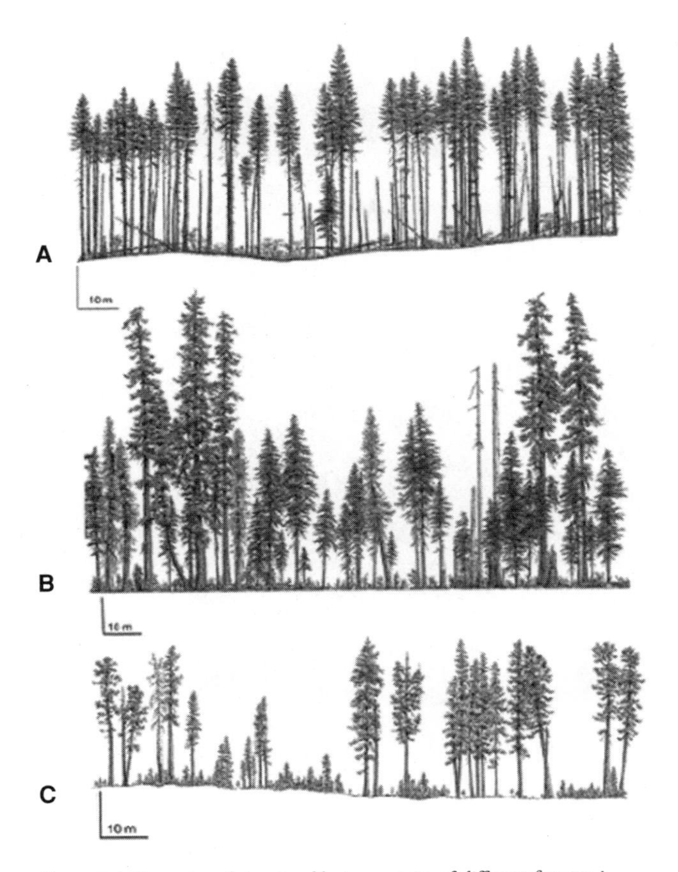

Figure 9.6. Examples of structural heterogeneity of different forests: A. Maturing even-aged Douglas-fir stand developed following a 1902 wildfire in the Cascade Range of southern Washington State. B. Old-growth, 650-year-old stand of Douglas-fir, western red cedar, and western hemlock in the Cascade Range of southern Washington State. C. Old-growth ponderosa pine stand in southeastern Oregon. (Franklin and Van Pelt 2004)

9.1.3 Belowground

As with canopies (but for different reasons), understanding the belowground presents unique challenges, and much remains to be learned.[8] Unlike canopies, however, the belowground is not readily accessible to our eyes and remains a black box until specific techniques are used to probe it.[9] These reveal a great deal of patterning, much of which can be attributed to plants or animals that feed on plants (see Wardle [2002] for a thorough discussion of links between the aboveground and belowground). In fact, a cocreative relationship exists between plants and soils, something that we touched on in chapter 8 and is a recurrent theme throughout this book. In this section we briefly look at three examples of belowground ecosystem structure.

9.1.3.1 Spatial Patterns of Mycorrhizal Fungi

Mycorrhizal fungi form a vital symbiotic relationship with plants, colonizing roots and providing numerous services in return for plant photosynthates (chapter 11). Because of technical difficulties, relatively few studies have documented the spatial structure of mycorrhizal fungi; however, new techniques show considerable promise.

[8]Vertical profiles of soil structure have been extensively studied are the central concept employed to classify soils. Horizontal heterogeneity is much less well understood.

[9]Chris Maser refers to the belowground as the *management unconsciousness* (personal communication).

Table 9.3
Ecological characteristics of old-growth forests

Development time	Structure	Patch size	Stability	Examples of forest types
Long	Low	Mixed	Enduring	Pinyon-juniper, subalpine white pine, dwarf black spruce
Intermediate	Moderate	Fine grained	Enduring	Balsam fir, black spruce
Short	Moderate	Coarse grained	Short-term transient	Aspen, red alder, jack pine, black spruce, sandpine
Intermediate	High	Coarse grained	Long-term transient	Douglas-fir, ponderosa pine, white pine, red pine, Atlantic white cedar
Intermediate	High	Fine grained	Surface fire dependent	Ponderosa pine, oak-hickory, longleaf pine, Douglas-fir–incense cedar
Intermediate	High	Fine grained	Enduring	Beech-maple, hemlock spruce

Source: Spies 2004.

The classification shown in table 9.3 illustrates the diversity of old-growth forests based on characteristics of space, time, and structure.

Development time. Long: centuries to millennia are required for development as a result of poor site conditions and very slow growth and establishment. Intermediate: centuries are required for development because of the long life spans of species, low site quality, or slow growth rates. Short: decades are required for development because of the short life spans of dominant tree species.

Structure. Low: forests comprise small trees with short life spans (<250 years), low-density stands, and low dead wood because of high decay rates. Moderate: forests comprise small to medium-sized trees with short life spans, and low to moderate accumulations of dead wood. High: forests comprise large to very large trees with long life spans (>250 years), and moderate to high accumulations of dead wood.

Patch size. Fine grained: dominant tree species typically regenerate in small canopy gaps and disturbance patches (<1 ha). Coarse grained: dominant trees typically regenerate in moderate to large stand-replacing disturbances (>1 ha). Mixed: a combination of both fine- and coarse-grained disturbances regenerate the dominant tree species.

Stability. Short-term transient: old-growth stages exist for decades before successional change or stand-replacement disturbance. Long-term transient: old-growth stages typically exist for one to several centuries before successional change or stand-replacement disturbance. Surface-fire dependent: stands can persist for centuries as long as surface fires burn with a frequency of decades or less. Enduring: old-growth stages exist for centuries to millennia because canopy trees can regenerate in canopy gaps.

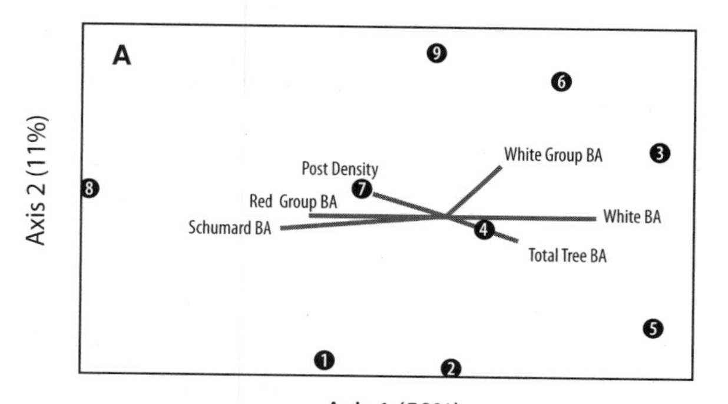

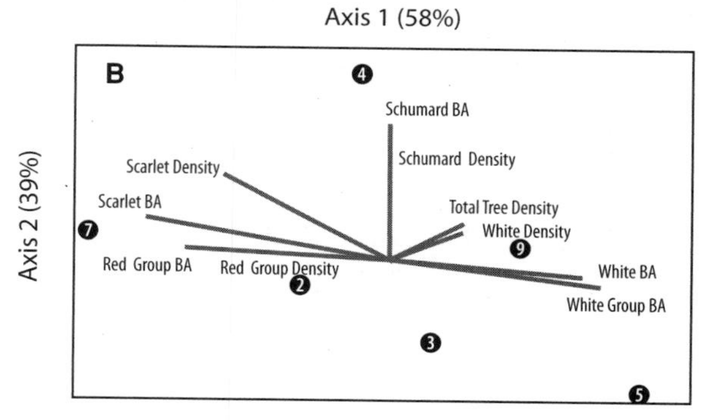

Figure 9.7. An ordination diagram showing the strong pattern of association of the density and basal area of the white and red oak groups and individual oak species (white oak, red oak, scarlet oak, post oak, and Schumard oak) with lichen communities of the midbole and canopy microhabitats. *Ordination* (literaly, "to order") is a commonly used statistical technique to simplify complex data sets. For example, if 40 lichen species are measured, ordination can be used to group these with regard to some measure of similarity (e.g., occurring in the same plots, or on boles or branches, or on white oaks or red oaks). Ordination creates a new set of axes (e.g., axis 1 and axis 2 in the figure), and in the resulting graphical space those points (e.g., lichen species) that are close to one another are more similar (e.g., tend to occur together) and points far away from one another less similar. In the figure, the degree of correlation with certain habitats is superimposed on the ordination of lichen species. The direction of lines associated with different oak species or species groups indicates similarity; for example, in B (canopy) the lichen communities of scarlet oak and white oak are quite different, while those of Schumard oak fall in between. The length of the lines is proportional to the magnitude of the correlation, for example, in A (midbole), total tree basal area is not as good a predictor of lichen communities as the basal area of white oaks. Numbers 1-9 represent different plots. Percentages of total variation explained by each axis are shown parenthetically. (Adapted from Peck et al. 2004)

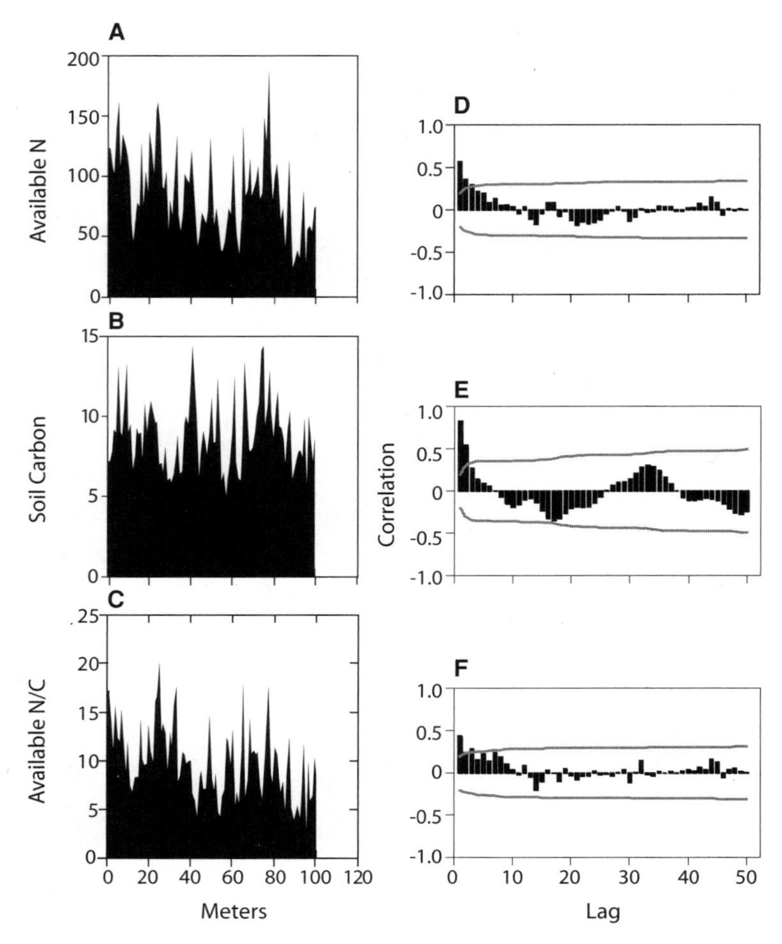

Figure 9.8. Patterns of soil nitrogen and carbon along a 100-m transect in old-growth Douglas-fir hemlock forest. A. Variation in available nitrogen (N, ppm) along the transect. B. Variation in soil carbon percent along the transect. C. Variation in the ratio of available nitrogen to soil carbon (N/C). D–F. Autocorrelation plots for available nitrogen, soil carbon, and the ratio of available nitrogen to soil carbon, respectively. An autocorrelation plot shows the average correlation between any given point along the transect and points at given distances from it. Lag represents the distance from the beginning point (i.e., Lag = 0 is the beginning point). Lines are 95% confidence intervals.

Dickie et al. (2002) used DNA analysis to show that these communities divided up vertical space in soils beneath a red pine plantation. Lileskov et al. (2004) examined the mycorrhizal composition of roots taken from soil cores and found that some species and communities[10] of mycorrhizal fungi formed patchy distributions, while others did not. Most patches of similar community composition were at scales of less than 3 m but ranged up to 25 m in some cases.

The so-called mat-forming mycorrhizal fungi create dense concentrations (i.e., mats) of rhizomorphs that significantly alter soil chemistry, perhaps by increasing rock weathering (Griffiths et al. 1994). Two of the primary species of mat formers in the Pacific Northwest occur in two levels of pattern (Griffiths et al. 1996). At the smallest scale, mats tend to occur in clusters within areas a few meters in width. The density of clusters decreases with distance from the nearest potential host tree, creating a second level of patterning whose scale depends on the density of host trees.

9.1.3.2 Spatial Patterns in Soil Organic Matter

Figure 9.8 shows patterns of soil carbon, available nitrogen,[11] and the ratio of available nitrogen to soil carbon[12] along a 100-m transect in an old-growth Douglas-fir–western hemlock forest. All three vary widely over a few meters or

[10]Many different species of MF may form on a single root, so patchiness is usually in the composition of the MF community rather than one species occurring in a patch with no others. The latter would be unusual.

[11]*Available nitrogen* is the amount of nitrogen as determined by an anaerobic incubation (related to but not the same as *total soil nitrogen* or *aerobically determined nitrogen mineralization*). This measure is assumed to reflect a labile nitrogen pool that is readily available to plants; this pool is thought to partially reflect microbial biomass.

[12]This ratio is related to the quality (i.e., decomposability) of soil organic matter. A higher ratio indicates a higher quality of organic matter as a source of nitrogen for plants.

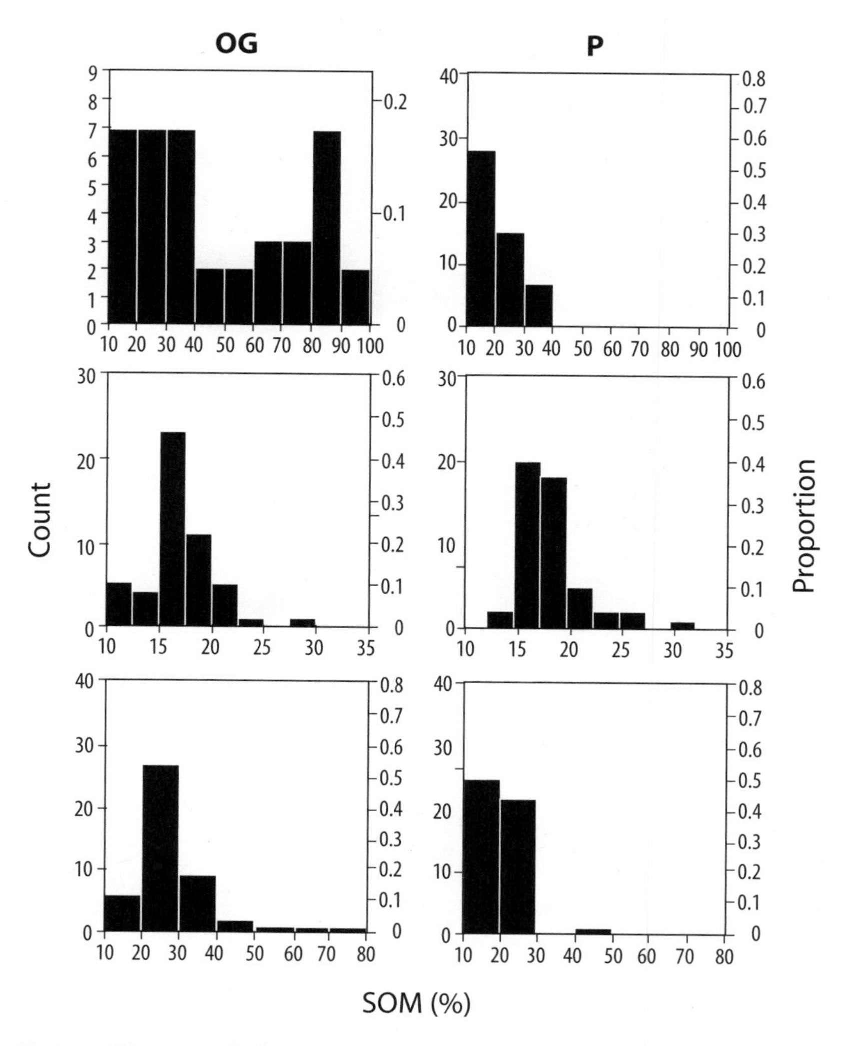

Figure 9.9. Histograms of soil organic matter (SOM, %) in three pairs of old-growth and mid-aged stands ("pole" stands) in the Oregon Cascades.

less. The autocorrelation plots tell us something about the scale of patchiness for each variable. Both carbon and available nitrogen are patchy (i.e., have similar values) at scales of 2 to 3 m, while the ratio of the two (an indication of the "quality" of soil organic matter, or how readily it delivers nitrogen to plants) occurs in patches of up to 10 m. There appears to be a second, larger pattern for soil carbon, which shows waves of positive and negative correlation, with peaks and troughs spanning 10 to 15 m. That is the approximate width of tree crowns in the forest; in other words, in all likelihood we are seeing broad spatial patterns imposed on soil by trees. As we discuss further in chapter 14, trees and other plants have profound effects on soil chemistry.

Figures 9.9 and 9.10 show histograms[13] of the distribution of soil organic matter (SOM) in 12 Oregon forests, three pairs of old-growth and nearby mid-aged stands ("pole" stands) in each of two regions, the Cascade Range (fig. 9.9) and the Siskiyou Mountains (fig. 9.10). At least three levels of pattern are evident in these figures: within stands, between stands in the same region, and between regions. Within most stands in both regions the distribution of SOM is skewed to the left, that is, most points have relatively low values. In the Cascades, two of the old-growth stands have long tails (a few points have exceptionally high values, probably representing old, buried logs); mid-aged stands may or may not have tails, but none are as long as in the old-growth stands (i.e., the mid-aged stands do not have points with exceptionally high SOM values). In the Siskiyous, which are drier and have a history of more frequent ground fires than the Cascades, old growth does not have the signal of old buried logs, and consequently the distribution of SOM is quite similar between old-growth and mid-aged stands.

9.1.3.3 Pattern Due to Soil Animals

Animals are a ubiquitous component of soils and affect their structure through various activities such as burrowing,

[13]A histogram shows how sample points are distributed among different measured values. The values are plotted on the x axis, and the number of points corresponding to a given value are plotted on the y axis.

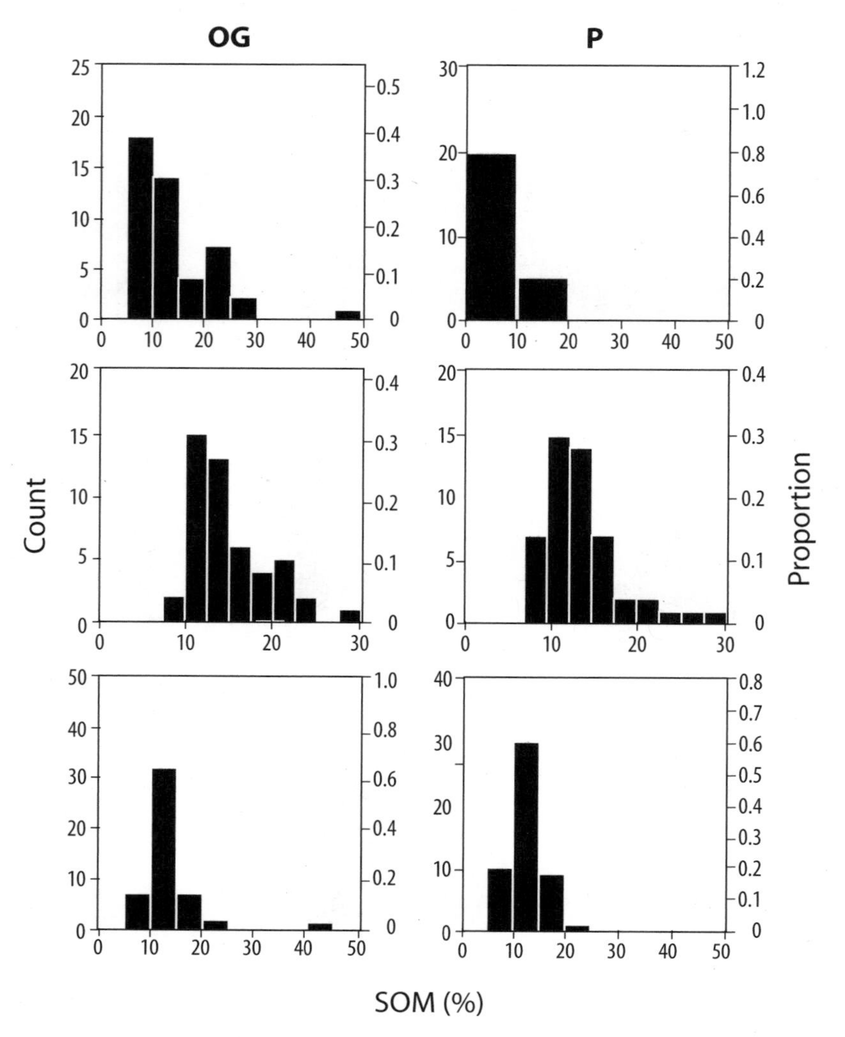

Figure 9.10. Histograms of soil organic matter (SOM, %) in three pairs of old-growth and mid-aged stands ("pole" stands) in Oregon's Siskiyou Mountains.

mixing, and constructing nests (chapter 14). For example, Amazonian soils are characterized by soft spots and chambers to a depth of 9 m. Chambers are the nests of leaf-cutter ants, and soft spots are probably the vestiges of old nests (de Olivera Carvalheiro and Nepstad 1996). Soft spots are chemically and biologically distinct from surrounding soil.

9.2. HABITAT AND NICHE

9.2.1 Habitat

To survive, every organism must have food, water, and protected space that provides refuge from enemies and the elements. In simplest terms, **habitat** is a place (or type of place) that provides these basic necessities. For a species to persist, enough habitats must exist to support a viable population (i.e., sufficient individuals to ensure genetic integrity and to buffer against losses due to local disasters) (chapter 22).

Habitats may be either extensive or localized. Broad habitats such as elevational ranges are composed of mosaics of smaller, more subtle habitats, some of which result from physical factors such as soil variability, topography, and streams, some of which result from environmental modifications by the biota themselves. Plants growing in a shaded forest understory occupy a different habitat (at least in terms of light) than found in the upper canopy. A seedling rooted in an old decaying log occupies a different habitat than one rooted in mineral soil. Local habitats are frequently used differently by different species. For example, southern red-backed voles *(Clethrionomys gapperi)* and various species of mice (*Peromyscus* spp.) occur in deciduous forests of the eastern United States, but voles are restricted to rock outcrops within the forest, while mice inhabit both forests and rock outcrops. Voles do not move into forest even when mice are experimentally removed (Wolff and Dueser 1986).

9.2.2 Niche

No concept in ecology has been more variously defined or more universally confused than "niche."

REAL AND LEVIN 1991

> Ecology's love-hate relationship with the niche concept has been long and not especially pretty.
>
> HAIRSTON 1995

Niche has been used in three senses by ecologists (Whittaker et al. 1973; Schoener 1989): (1) as referring to the functional role of a species (or species group) within a system, (2) as referring to the habitat of a species *(the place niche),* (3) as referring to a combination of the first two. The most common usage today is probably the third, as expressed in the simple definition[14] given by Chase and Leibold (2003): "the requirement of a species for existence in a given environment and its impacts on that environment." In other words, a niche is both: (1) an "opening" or opportunity for a species to find suitable living (and reproducing) conditions within a given system, and (2) the set of attributes that a given species brings to that opening. In this view, a niche is an interaction between an individual species and the ecosystem in which it is embedded. Prerequisite to this definition is that niches are flexible rather than rigid; though constrained by characteristics of the ecosystem, a niche also takes on the characteristics of the individual species that fills it. The death of a dominant tree, for example, creates an opening in local system structure that might be filled by various plant species as long as they are adapted to the abiotic and biotic conditions. The gap is quite literally a physical opening in the canopy that allows light to reach seedlings in the forest understory, but it is also an opening in the social fabric. Which seedlings succeed in the gap will depend on biotic as well as abiotic characteristics of the opening: competition for water and nutrients, the species of mycorrhizal fungi that are supported by surrounding trees, and the abundance and types of pathogens and herbivores. Conversely, each species that might pass the various biotic and abiotic filters and establish in the opening brings a set of unique attributes with it: how it interacts with surrounding trees, soil microbes, and herbivores, and its effects on soil chemistry and microclimate. A community is somewhat like a crowded dance floor at a rock concert: each pair of dancers has considerable freedom to express their individuality, and others will adjust to them—but only within limits. The movements of all are shaped by the beat of the music, which in our analogy equates with environmental "givens" such as climate, topography, and soil mineralogy.

The function of an organism is generally so intertwined with its morphology, physiology, and habitat that niche has become a kind of conceptual mulligan stew in which virtually everything that can be measured (and many things that cannot) are thrown into the pot. Certain aspects of the niche, such as habitat and (for animals) diet, are at least in theory readily measured; however, it is seldom if ever possible to completely specify the niche of an organism. Because of this difficulty, the Hutchinsonian niche is frequently divided into subcomponents such as the **habitat niche** or the **feeding niche.** Grubb (1986) distinguishes four general niche dimensions for plants:

1. The *habitat niche* encompasses a variety of factors related to environmental tolerances and the abiotic and biotic needs of plant species. Through its influence on nutrient and water availability, soil is a major component of the habitat niche. Requirement for light is also a part of the habitat niche; shade-tolerant species capable of growing in forest understories occupy a different habitat than those shade-intolerant species that require relatively high light levels. Requirements for mutualists (e.g., mycorrhizal fungi, pollinators) are part of the habitat niche.
2. The *life-form niche* refers to differences in inherent size (e.g., shrubs versus trees) or morphological characteristics (e.g., rooting depth, branching pattern).
3. The *phenological niche* refers to the seasonal pattern of growth and development. Within the constraints imposed by climate, plants within a given community may vary widely in phenology.
4. The *regeneration niche* includes "the requirements for effective seed set, characteristics of dispersal in space and time, and requirements for germination, establishment, and growth that have to do not only with shape and size [openings], but also with weather, pests, and diseases" (Grubb 1986). Different modes of regeneration, such as sprouting from rootstocks versus establishing from seed (and others discussed in chapter 6, also constitute part of the regeneration niche.

The utility of the niche concept has been challenged, especially by Hubble (2001) and other proponents of the neutral theory of biodiversity and biogeography (discussed in chapter 10), who assert that community organization can be adequately explained without invoking niches. In contrast, Chase and Leibold (2003) believe that the niche "has provided and will continue to provide the central conceptual foundation for ecological studies." For example, an old hypothesis in ecology (called the **competitive exclusion principle**) states that no two species share the same niche (Elton 1927). In practice, the competitive exclusion principle is impossible to prove, because the complete niche of an organism is not measurable. The idea that it expresses (i.e., species occupying the same trophic level should evolve to use resources differently), however, provides testable hypotheses about the nature of relationships among species and how these influence ecosystem structure. Note that the competitive exclusion principle does not say species cannot overlap in some aspects of their niches; as we shall see later, niche overlap is fairly common and extremely important for ecosystem function.

Various people have argued that the niche concept needs rethinking and revision (Hubble 2001; Chase and Leibold

[14]We refer to this definition as simple because Chase and Leibold (2003) also give more elaborate definitions; however, these do not differ fundamentally from the more simple one.

The term *niche* was introduced into the ecological literature by Grinnell (1914), who viewed it in the same sense as the common English usage: a recess in a wall. For him, a niche was an opening in community structure or function that was filled by a species. Later, Elton (1927) took the view that a species' niche was what it did in the environment, especially its role in the food web. In a very influential paper, Hutchinson (1957) included all interrelationships between organism and environment within the niche, including things such as habitat, environmental requirements and tolerances, feeding habits and position in the food web, rooting depth, diurnal activity, and life-history characteristics. The so-called Hutchinsonian niche is a multidimensional volume in hyperspace, with axes corresponding to these factors and any others that relate to the species' role in the ecosystem. Over time, a formal niche theory developed that combined aspects of both the hole-in-the-wall and the what-species-do views and was powered by the belief that all species interactions of any consequence are driven by competition for resources. Niche theory was a thriving cottage industry for ecologists for nearly 40 years, but in the late 1970s a backlash developed that eventually led to a bitter feud among some of the protagonists. The backlash was based on several things, most notably (1) the realization (among some) that competition alone was insufficient to explain species interactions, and (2) the lack of statistical rigor in many of the tests of niche theory. Work since has been aimed at addressing both of these issues. Unfortunately, there is an ecological analogue to the physicist's uncertainty principle[15]: the more precisely and rigorously one measures an ecological system, the more difficult it becomes to capture its essence. And so, as tightly controlled experiments were adopted to satisfy the requirements of rigorous science, the complex dynamics that characterize ecological systems became increasingly marginalized.[16] Currently, "some of the best studies combine observational and pattern analysis, theoretical predictions, and experimental approaches. This pluralism has allowed a much deeper understanding will depend on biotic as well as abiotic characteristics of the of pattern and process that is not possible with a singular approach" (Chase and Leibold 2003).

[15]One cannot precisely and jointly measure both position and momentum of an elementary particle.

[16]There is an old story about a man who, returning home in the wee hours, found his neighbor intently searching the ground beneath a street lamp. "Can I help you find something?" he asked. "I lost my key," replied the neighbor. "Do you know about where you dropped it?" "Yes," replied the neighbor, "over there," pointing to a dark corner of the street. "If you dropped it over there, why are you looking here?" asked the man. "Because this is where the light is," replied the neighbor. The task of ecologists is not to bring the search to where the light is, but to bring the light to where the key is.

2003). Chase and Leibold (2003) identify (and begin work on) several aspects that would bring more realism to niche studies, including (1) incorporating the regional, landscape, and ecosystem contexts within which local interactions occur (see Tilman 2004), and (2) distinguishing the separate roles of the two major niche components: habitat and function. For example, the habitat niche governs the establishment of a species within a local community, while the functional niche determines the effect of that species on the community (Chase and Leibold 2003).

9.3 FOOD WEBS: PATHWAYS OF ENERGY FLOW WITHIN ECOSYSTEMS

> Every living thing is a sort of an imperialist, seeking to transform as much of its environment as possible into itself.
>
> BERTRAND RUSSELL

> Managers are increasingly aware of the need for science to inform the stewardship of natural lands and resources. If ecologists are to address this need, we must increase the scope of our inferences, while maintaining sufficient resolution and realism to predict trajectories of specific populations or ecosystem variables. Food chain and simple food web models, used either as core or component hypotheses, can help us to meet this challenge.
>
> M. E. POWER 2001

The complex sets of interactions that define ecosystems have been called **interaction webs** (Price 2002). Interaction webs knit diverse species and structures into the integrated whole we call an ecosystem. Two processes are particularly important in this regard: (1) creation, which simply means that in the process of creating its own niche an individual of one species also creates conditions that individuals of other species come to rely on (i.e., becomes part of their habitat niche); and (2) feeding, a multifaceted process that always involves transfers of metabolic energy and nutrients and includes predation, herbivory, parasitism, and the many different kinds of mutualism found in nature. The two are intimately linked, as a creator must also eat: the pileated woodpecker foraging in an old log for insects will fly off to excavate a cavity that ultimately will be used by numerous other species, the beaver in his pond is food for wolves. It is virtually impossible to talk about ecology above the level of individuals without either directly or indirectly invoking one or both of these processes, and they will come up at many points throughout the book.

In this section we focus on the basic structure of food webs. The network of feeding relationships within ecosystems has numerous important consequences for system functioning. For example, tree growth is closely tied to the rate at which animals and microbes consume dead leaves and other tissues, thereby releasing essential nutrients for tree uptake (chapter 18). Feeding by predators combines with weather and plant defenses to maintain populations of **herbivores** (i.e., animals that eat living plants) below levels that would injure the plant community (chapter 19).

In figure 9.11, parts A, B, and C show, respectively, the major functional groups within a food web, the estimated quantities of energy flowing through the food web of a temperate hardwood forest in New Hampshire, and a simplified plant-vertebrate food web from a boreal forest in Canada. Parts B and C of figure 9.11 represent, respectively, a **flow web,** which associates values with the flows of energy, and a **topological web,** which simply shows who connects to whom with respect to feeding (i.e., without quantifying the amount of food that passes). (www.foodwebs.org/gallery_index.html has some beautiful topological food webs).

In figure 9.12, parts A–C show, respectively, the topology of a soil food web and (in B and C) **interaction webs** for a portion of the web of 9.12A. Interaction webs show reciprocal influences between species rather than the flow of energy. For example, in figure 9.12B, arrows run between omnivorous nematodes (predators) and bacteria (prey), and in figure 9.12C a longer reciprocal loop takes in the flagellates, which are predators on bacteria and prey for nematodes. In an interaction web, a loop is defined as any closed chain of trophic effects in which each species is visited only once (Neutel et al. 2002). The length of a loop is defined as the number of species (or species groups) it includes. The values associated with each arrow are estimates of the influence of one group on another. In an interaction web that depicts predator-prey relationships, top-down effects are always negative and bottom-up effects

Berlow et al. (2004) discuss the types of questions addressed using food web theory:

- *Questions about community-level patterns.* For example: (1) Are natural systems characterized by a predictable nonrandom patterning of interaction strengths? (2) What are the underlying mechanisms that generate nonrandom patterns of interaction strengths? (3) What are the consequences of this patterning for community-level functions, such as stability and persistence?
- *Questions about species-specific dynamics.* For example: (1) What is the minimum detail necessary to predict the effects of a change in abundance or extinction of one species on other species abundances or extinctions in the web? (2) Which particular species have disproportionately large effects on community structure and species' population dynamics? (3) Which species are particularly vulnerable to extinction?

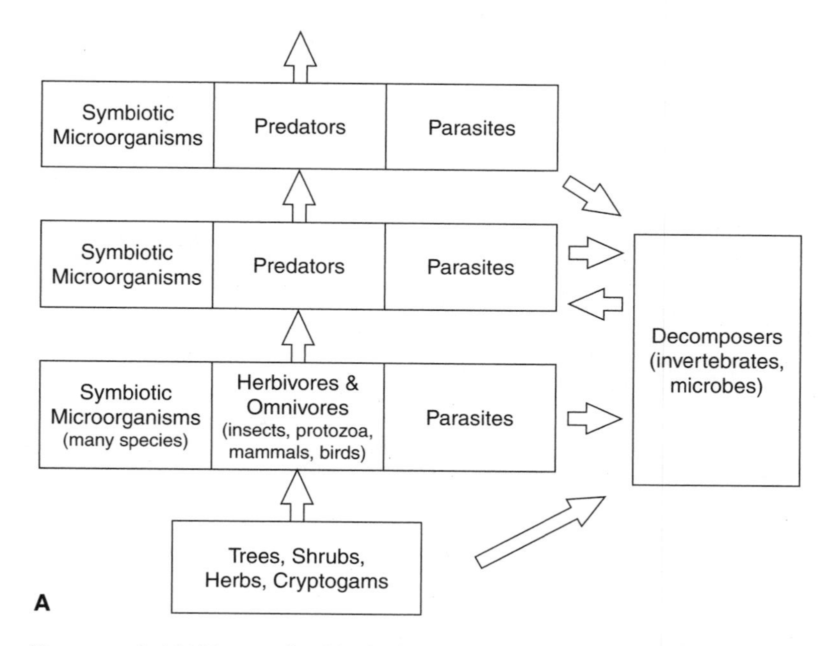

Figure 9.11. A. A highly generalized food web. Green plants compose the first level (or base) of the web. Matter and energy flow from plants to the second level, which comprises plant eaters (herbivores and omnivores), plant parasites (pathogens), and plant symbionts (mycorrhizal fungi). Matter and energy flow from the second level to predators, parasites (pathogens of animals), and symbionts (gut microflora) on the third and higher levels. (There are seldom more than six levels.) Feeding is not necessarily restricted to the level immediately below a given species; omnivores, for example, eat both plants and animals. Decomposers feed on all levels. *(Continued)*

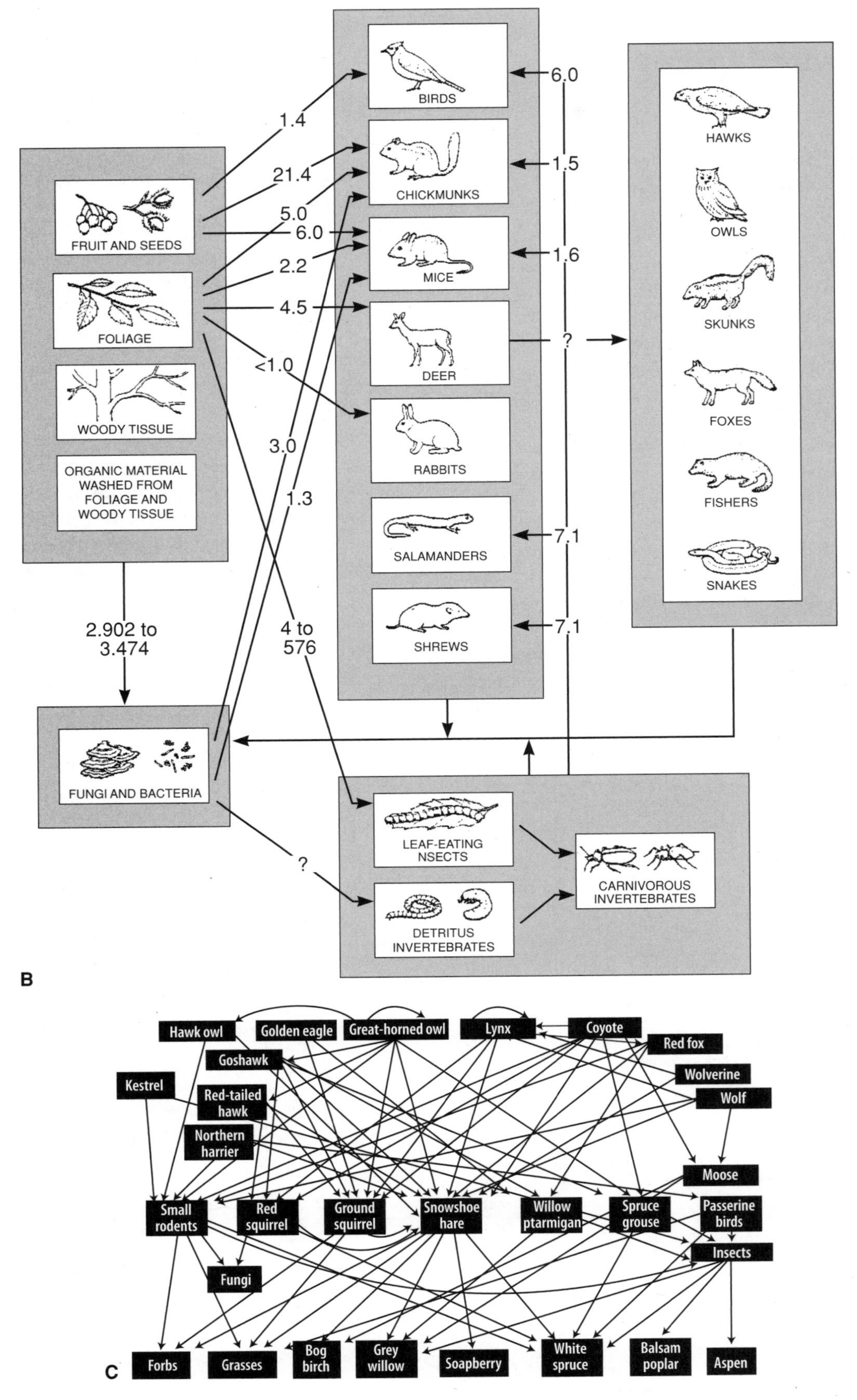

Figure 9.11. (*Continued*) B. Flow of energy through the food web of a northern hardwood forest ecosystem (numeric values are kilocalories per m^2 per year). As is typical of all forests, by far the most energy goes directly from plants to fungi and bacteria (decomposers and mutualists). Energy flow to leaf-eating insects varies widely because of periodic insect outbreaks. Amounts of energy flowing to the higher trophic levels (hawks, foxes, and so on) are unknown. (Adapted from Gosz et al. 1978) C. A simplified plant-vertebrate food web of a Canadian boreal forest, the Kluane Project (Krebs and Boonstra 2001. By permission of Oxford University Press, Oxford)

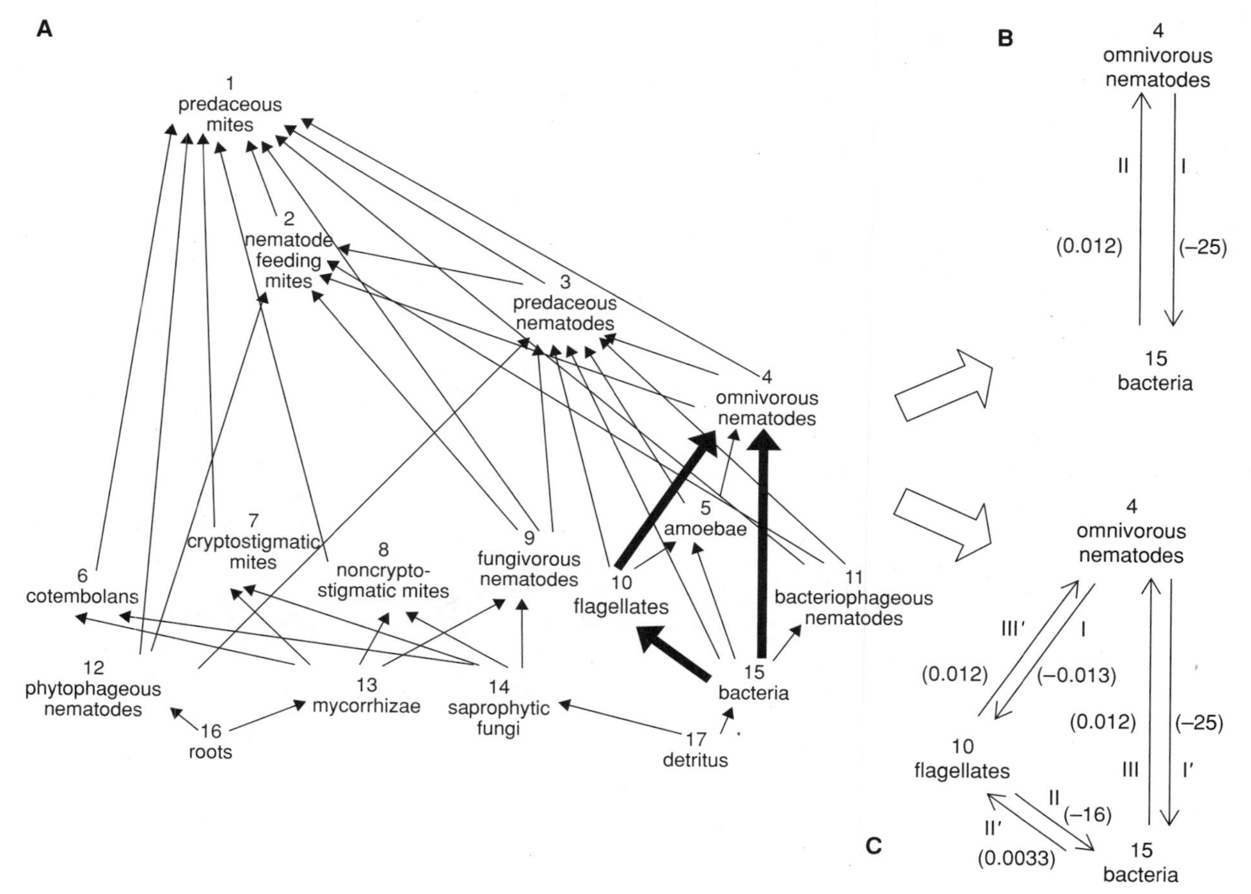

Figure 9.12. Feeding relations (A) and loops (B, C) in a soil food web. Values in parentheses in B and C show interaction strengths. B. A single predator-prey relation forms a loop of length 2, containing a bottom-up effect (prey on predator) and a top-down effect (predator on prey) (I, II). C. A three-species omnivorous relationship forms three loops of length 2 and two loops of length 3. The latter includes one loop containing two top-down effects and one bottom-up effect (I, II, III), and one containing two bottom-up effects and one top-down effect (I′, II′, III′). (Neutel et al. 2002. Reprinted with permission from AAAS)

are always positive (e.g., being their food, bacteria have positive effect on nematodes, and nematodes have a negative effect on bacteria).

At the most basic level, all food webs have two general types of patterns, one relating to energy and matter flow from one species to another (i.e., feeding or being fed on; sometimes termed **vertical relationships**), the other relating to allocation of energy and resources among species that potentially use the same resources (termed **horizontal relationships**). Vertically, food webs are divided into **trophic levels,** which represent each stage of energy transfer. The first (or basal) trophic level is occupied by green plants. The second trophic level comprises three distinct functional groups in terrestrial ecosystems (the groups are not necessarily taxonomically distinct; i.e., the same species might occupy more than one):

- **Herbivores** and **pathogens** are animals, microbes, and viruses that feed on living plants, usually (but not always) with no direct benefit to the plant and, in many cases, to the plant's detriment (chapter 19).
- **Plant mutualists** are animals and microbes that feed or otherwise obtain energy from living plants and that directly benefit the plant in some fashion (chapter 11).
- **Decomposers** are invertebrate animals and microbes that feed on dead tissue such as fallen leaves (invertebrates within this group are also referred to as **detritivores** and microbes as **saprophytes**) (chapter 18)

The third trophic level is occupied by predators, parasites, and mutualists of second-level organisms; the fourth level by predators, parasites and mutualists of third-level organisms; and so on.

By convention, energy is said to flow upward from lower to higher trophic levels. A **basal species** is one that eats no other species (e.g., most green plants). **Omnivores,** which include many invertebrates and vertebrates (e.g., bears, jaguars, wild turkeys, humans) are by definition species that feed on at least two, and sometimes more than two, trophic levels. Food webs dominated by insects and their predators and parasites (including those of forests) have more complex patterns of omnivory than those dominated by vertebrates (e.g., grasslands) (Pimm

Figure 9.13. *Darlingtonia californica,* the pitcher plant. Endemic to serpentine soils in the Klamath Mountains of Oregon and California, this plant traps and consumes insects.

1982). Omnivory is relatively common in soil food webs (Neutel et al. 2002), the aboveground web of neotropical forests (Reagan and Waide 1996), and most likely both soil and aboveground webs of all forest types. The presence of omnivorous invertebrates in soil food webs has been shown to increase the growth of tree seedlings (Setala 2002).

Decomposers are the ultimate omnivores, feeding on the fecal matter of insect grazers, the body of a dead jaguar, or on fallen leaves. The unique role of decomposers produces circularity, or loops, within flow-based food webs. (As we saw earlier, loops are an in-built feature of interaction webs because arrows run in two directions. The loops we refer to here are in a one-way flow). Lawton (1989) describes a celebration of this circularity in the "national anthem" of Yorkshire, the verses of which identify "the pathway from humans, worms, ducks, and back to humans." Note that the transfer of energy from one organism to another does not always involve death and consumption in the sense that we are familiar with when we think of eating. Parasites live on or in the body of their host, as do viruses. Mutualists such as mycorrhizal fungi, gut bacteria, and plant pollinators obtain energy from their hosts in payment for rendering valuable services. We discuss these relationships in more detail in chapter 11.

Although ecologists commonly use the term *complexity* in the same sense that most people understand it, "the condition of being difficult to analyze, understand, or solve,"[17] when applied to food webs, complexity refers specifically to the average number of trophic links per species (Montoya et al. 2006). This is unfortunate terminology because the average number of links is an inadequate characterization

[17]Encarta ON line Dictionary. http://encarta.msn.com/encnet/features/dictionary/DictionaryHome.aspx

Some energy transfers do not fit into the neat view of bottom-to-top energy flow. For instance, as mentioned earlier with regard to decomposers, food webs contain **circularity,** in which a species consumes another that occupies a higher trophic level. Some green plants, for example, trap and digest insects (fig. 9.13). (The plants, however, are seeking nutrients from the insects, not energy.) Reciprocal feeding among different life stages introduces circularity into food webs. For example, in a Puerto Rican forest, adult coqui frogs eat centipedes, wolf spiders, tarantulas, and scorpions, all of which return the favor by eating juvenile coquis (Reagan and Waide 1996). Lateral transfers of energy and nutrients are not only possible but in all likelihood common; studies have shown that carbon compounds move from one plant species to another, presumably through interlinking mycorrhizal hyphae or root grafts (chapter 11). In marine ecosystems, some plankton that feed on photosynthetic algae retain algal chloroplasts and photosynthesize with them (Stoecker et al. 1987).

of the organizational complexity of food webs, particularly biologically rich ones in which species vary widely in their number of links (i.e., mean values do not tell you much) (Montoya and Solé 2003). Our use of *complexity* in this book will be in the more general sense unless otherwise noted. An alternative (and, for us, preferred) term for the average number of trophic links per species is **linkage density.**

The **connectance** of the system is defined as the number of trophic links between species as a proportion of the total number possible (trophic links are those between trophic levels, i.e., who eats whom). Estimates of connectance are usually made for entire food webs and range from less than 5 percent up to around 30 percent, depending on the particular food web in question (e.g., Schoenly et al. 1991). As we discuss later, however, food webs contain subwebs within which connectances are significantly higher than in the web as a whole (Melián and Bascompte 2004).

Linkage strength is "the magnitude of the effect of one species on the growth rate of another" (Montoya et al. 2006), while **interaction strength** is a more general term that refers to either (1) the property of an individual link, or (2) the impact of a change in the properties of one link or of a set of links (e.g., all links to and from a given species) on the dynamics of other species or on the functioning of the whole system (Berlow et al. 2004) . In the first category, individual interaction strengths are treated as if they are independent of the network in which they are embedded, in the second category, they are, in theory, inseparable from their network context.

Intermediate trophic levels in forests are dominated by insects and microbes (especially decomposers and plant mutualists). Viruses are surprisingly important at the second and higher levels in aquatic ecosystems and probably in terrestrial systems as well. However, with a few (and growing number of) exceptions, food web studies have been heavily biased toward the larger, more easily measured plant and animal species, and toward aboveground as opposed to belowground food webs. Invertebrates and microbes (which account for most of the energy flow within ecosystems) have frequently been lumped together with no attempt to distinguish individual species or even major species groups, although beginning with the pioneering work of Ingham et al. (1989), Schoenly et al. (1991), and Moore et al. (1996) that situation began to change. Lafferty et al. (2006) found that including parasites in food webs increased connectance and greatly affected other factors such as length and linkage density. Bacteriophages (viruses that feed on bacteria) are ubiquitous and have a major effect on the turnover of bacteria, yet food webs do not include them. Similarly, the foliar endophytes of plants and the gut flora of vertebrates and invertebrates almost certainly play a role in energy transfers, probably a large one, yet are not accounted for in food webs. With these limitations in mind, following are general patterns that have been observed.

9.3.1 Who Eats Whom?

With rare exceptions, each species above the basal trophic level feeds on more than one species below it, and each species below the top is fed on by more than one species above it, creating food webs instead of food pipelines. Ecologists have devoted considerable effort to the search for fundamental principles that govern the structure of food webs (e.g., Dunne 2005; Bascompte et al. 2006; Montoya et al. 2006; Thompson 2006), a challenging task because few studies have documented web structure in a comprehensive way. One important aspect has to do with the distribution of strong and weak links, which tells how strongly pairs of species influence one another (Berlow et al. 2004; Montoya et al. 2006). A wide variety of food webs have a few strong and many weak links (Berlow et al. 2004). Weak links arise from: diffuse feeding relationships in which a predator (for example) has a variety of prey species, and the prey is fed upon by several predator species; and omnivory that produces long loops (i.e., closed trophic loops that include three or more trophic levels) (Neutel et al. 2002). Moreover, recent work shows that interactions in at least some food webs tend to be asymmetric, that is, if plant species P strongly depends on bird species B for seed dispersal, the strong dependence tends not to run the other direction; bird species B feeds on the fruits of many plants and has a relatively low dependence on plant species P (fig. 9.14). These structures—the prevalence of weak interactions (i.e., diffuse relationships) and asymmetry in relationships—are proving to have significant implications for ecosystem stability and dynamics (Thompson 2006). Quoting from Berlow et al. (2004; see their paper for original citations):

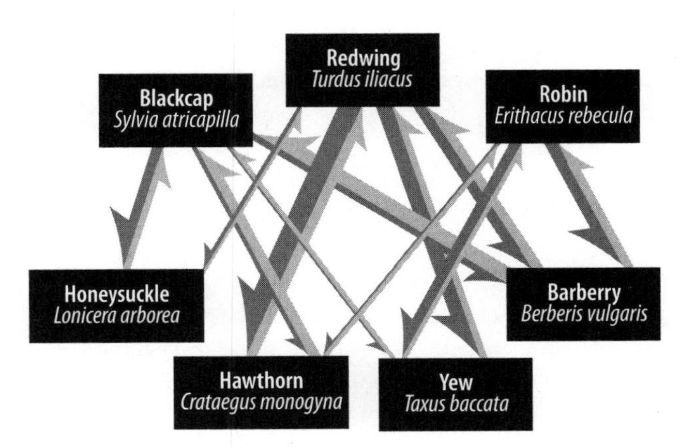

Figure 9.14. Part of an interaction web from a montane forest in southeastern Spain. As in all interaction webs, each interaction between frugivore and plant is represented by two arrows, one showing the dependence of the plant on frugivore *(dark arrows)* and one showing the dependence of the frugivore on the plant *(light arrows)*. Dependences are based on frequency of visits. The relative frequency of visits is shown by the thickness of the arrows. Note that in most cases the dependences between plant and frugivore are unequal, which introduces an asymmetry into the relationships. (Figure adapted from Thompson 2006. Food web data from Bascompte et al. 2006)

"Recent work suggests that non-random patterning of strong and weak links can be critical for the stability or persistence of theoretical and empirically observed complex communities. Additionally, not only stability but also ecosystem functions might be strongly mediated by the arrangement of interaction strengths."

Most patterns described above were derived from vertebrate food webs but may hold for many invertebrate webs as well. For example, herbivorous insects in a tropical forest in New Guinea were found to feed on several different tree species, resulting in a predominance of weak rather than strong interactions (Novotny et al. 2002). It should not be construed, however, that all food webs fit the above patterns, and there are many examples of specialization that leads to strong interactions. For example, both the native arthropods and native birds of the Hawai'ian Islands tend to be strongly specialized with regard to plant species, perhaps the result of evolution under isolated conditions. In many cases, insects specialize within plant taxonomic groups above the species level, but generalize among species within the group (Novotny et al. 2002; Prado and Lewinsohn 2004). Thus the strength of an interaction depends on the level of taxonomic resolution.

9.3.2 Food Webs Are Compact

The number of trophic levels varies depending on area (especially for isolated systems such as islands and lakes), history, and perhaps other factors (Post 2002); however, an analysis of the seven most comprehensive published food webs shows that 80 percent and 97 percent of species are, respectively, within two and three links from one another (Williams et al. 2002). As Dunne (2005) points out, this has profound implications for the existence of domino effects, or what ecologists refer to as **trophic cascades** (chapters 11, 20, and 21); however, the potential for trophic cascades depends not only on linkage distance but on the strength of the links as well.

9.3.3 Food Webs Contain Compartments, or Subwebs

These are portions of a larger web that are distinguished by relatively high levels of connectance (Melián and Bascompte 2004). All ecosystems have subwebs, the most obvious being the aboveground and belowground webs. As mentioned earlier, the soil food web is further compartmented into the live and dead webs that are based, respectively, on living roots and detritus (fig. 9.12A). In forests, the litter layer may have its own subweb, and each tree supports its own community of closely interacting microbes, epiphytes, and arthropods. Different tree, shrub, and herbaceous species may have their own subwebs. For example, in conifer-dominated forests the majority of Lepidoptera species (moths and butterflies) are associated with hardwood trees and shrubs (Hammond and Miller 1998). Within trees, the bole and the crown may support small subwebs, and each branch within the crown may as well. If we take a metaecosystems perspective, each local ecosystem is a subweb within the metaweb.

The emerging picture, then, is of a hierarchy of subwebs ranging from the very small (e.g., around an individual root) to the very large (landscapes and regions). What connects the subwebs? Higher trophic levels tend to integrate and connect lower-level compartments (e.g., a hawk eats both seed-eating and insect-eating songbirds, and a predatory soil mite probably eats both bacteria-feeding amoebae and fungal-feeding microarthropods); but not all predators are complete generalists. For example, three species of bats in the southeastern United States differ in the proportion of beetles, moths, and sucking insects in their diets (Carter et al. 2004). Recent research indicates that subwebs within local ecosystems commonly differ in their rates of energy processing, resulting in "fast" and "slow" energy channels (Rooney et al. 2006). Like the asymmetry in interactions, the asymmetry in flows is believed to play a significant role in stabilizing food webs (Rooney et al. 2006). We discuss this further in chapter 20.

Plants are the ultimate connectors between the above- and belowground webs, but some insects (e.g., beetles) commonly spend part of their life cycles in each subweb, existing as larvae belowground and adults aboveground. Migratory species connect regional webs. Birds are the obvious but not the only example; for instance, anadromous fish connect oceans with fresh waters and (through fish-feeders such as bears and eagles) watersheds.

9.3.4. Biomass Is Pyramid Shaped

The biomass of organisms within a given trophic level drops exponentially from the base to the top. For example, it takes about 1,000 ha of old-growth Douglas-fir forest to support one pair of spotted owls, which occupy the fourth trophic level. Old-growth trees on 1,000 ha contain approximately 450 million kg of carbon (including heartwood, which is not living), and a pair of owls about 2 kg. There are other species in the fourth trophic level of Douglas-fir forests (mostly insects), but their total biomass nevertheless remains an extremely small fraction of that of trees. An old rule of thumb is that 90 percent of energy is lost with each step up the trophic ladder; put another way, each trophic level supports roughly 10 percent as much biomass as the level immediately below it. Like all rules of thumb, this one should not be taken too literally—particularly the 10 percent figure; there are simply too many variables to allow accurate generalizations. Ten thousand kilograms of prey supports approximately 90 kg of vertebrate carnivores, a ratio of 0.9 percent (Carbone and Gittleman 2002). Vertebrate prey are usually mobile, and predators may have to expend considerable energy in the chase rather than devoting it to biomass production. Protozoa and invertebrates that feed on sedentary prey (i.e., bacteria and fungi) probably have much higher rates of transfer.

As a general rule, the biomass of animals (at least those that live aboveground) correlates positively with aboveground primary productivity (McNaughton et al. 1989). In forests, this pattern is especially evident when comparing tropical forests growing on highly infertile soils, with those on more fertile sites (McKey et al. 1978; Janzen 1983; Waterman et al. 1988). There are various reasons for this pattern. Low productivity by the trees produces less energy to pass up the food chain. Furthermore, the loss of nutrients to herbivores is a serious matter for trees growing where nutrients are in very short supply, and leaves on these sites tend to be heavily defended by a variety of chemicals that reduce their palatability (chapter 19). One of the better-documented examples of the latter is leaf-eating colobine monkeys, whose biomass within a given forest correlates closely with the food quality of leaves measured as the ratio of protein to defensive chemicals (Waterman et al. 1988). In the forests of the Appalachian Mountains of the eastern United States, the biomass of macroinvertebrates that feed in litter layers correlates positively with forest productivity, and so does the biomass of oven birds that feed on the macroinvertebrates (Seagle and Sturtevant 2005). On the other hand, a relatively high proportion of tree productivity on infertile or droughty sites occurs belowground in roots and mycorrhizal tissues rather than in leaves and flowers (chapter 15), and it is possible that belowground biomass is greatest and soil trophic chains longest on sites with intermediate plant productivity.

Biomass declines as one ascends the trophic levels for two reasons. The first applies only to the difference in biomass between the first and second levels and is particularly relevant to forests. A relatively high proportion of the biomass of land plants (especially woody ones) is not very edible and is stored as nonliving carbon rather than passed up the food chain. Soil humus (chapter 9) is a good example. Most ecologists argue that except during epidemic insect outbreaks, considerably less than 10 percent of the net energy fixed by trees in photosynthesis passes to organisms on the second trophic level (chapter 19). This may well be true, but we also know relatively little about the biomass of decomposers and mutualists in forest ecosystems, the vast majority of which are microbes and insects living belowground and difficult to study.

The second reason applies to all levels and all types of ecosystems. A relatively high proportion of the energy passed from one level to the next is lost as respiration at each step. The processes of living, such as hunting, feeding, courting, mating, playing, fighting, singing, staying warm (for warm-blooded animals), and so on, are all fueled by energy, and that energy is derived by respiring the carbon that originated in plants. This loss of energy at each step in the trophic ladder, coupled with the productivity of the plants and how it is allocated (e.g., wood versus leaves), is one factor limiting the number of trophic levels that can be supported in a given ecosystem.

9.3.5. Numbers Are Spindle-Shaped

The number of species increases as one moves from the basal to the intermediate trophic levels, then declines again from intermediate levels to the top vertebrate predators. This pattern has a straightforward reason (Lawton 1989). Plants are fed on mostly by insects, and any given plant (especially a tree) provides a structurally complex set of microhabitats (bark, leaves, flowers, big roots, little roots, top of the crown, bottom of the crown, and so on) that allow small-bodied animals (and microbes) to carve out a diversity of niches (Price 2002). The result is a great diversity of little creatures: the canopy of a single tree or the soil beneath 1 m of ground surface is likely to contain hundreds of species of invertebrates. As one moves up the food chain, predators require more territory. Hence, there are fewer options for niche diversification and, consequently, fewer species. Parasites, however, are an exception; these small-bodied creatures are quite diverse, even when feeding at the top of the food chain (Schoenly et al. 1991). Decomposers are another exception.

9.4 NICHE OVERLAP AND DIVERSIFICATION

The greatest potential for niche overlap, hence competition, exists among species that potentially use the same resources. Whether species actually compete (or have competed in the past) depends on how limiting the resource is. Plants, for example, usually fully use their basic resources (e.g., light, nutrients, and water) and, hence, potentially compete with one another for these factors (plants also cooperate). On the other hand, insect herbivores are frequently regulated by predators and climate, rather than food supply (chapter 19). Therefore, their food is not limiting, and different insect species can consume the same food without competing among themselves (excepting for periods when the regulating factors break down and insects outbreak).

Territoriality and habitat selection reduce or eliminate direct competition among some mammals and birds. In the tropical forests of Barrow Colorado Island, off the coast of Panama, four species of ant birds were found to maintain stable populations over eight seasons, despite fluctuations in their prey (Greenburg and Gradwohl 1986). Although the particular individuals occupying a given territory varied over time, territorial boundaries remained constant. This constancy in the space allotted to each breeding pair, not prey abundance, regulated numbers of ant birds. It does not follow that this is the case for all bird species; population levels of some vary widely with resource supply (we shall discuss an example below).

Whether a species is diurnal or nocturnal is a common form of niche diversification. For example, in the tropical forests of Barro Colorado Island, "a fig tree full of ripe fruit attracts bats by the hundred, and a few kinkajous and wooly opossums as well, while pacas, spiny rats and perhaps common opossums forage below for fallen fruit. By day this same tree attracts monkeys, especially howlers, guans (dark, long-necked, long-tailed birds with bare, bright red throats), toucans . . . and other birds, while coatis and agoutis dispute the fallen fruit below" (Leigh Jr. and Ziegler 2002).

9.4.1 Guilds

A **guild** is a group of species within a given community that share some common interest or behavior. The term is most often used by ecologists to denote species using the same resource in a similar way (Root 1967; Simberloff and Dayan 1991): the guild of foliage-gleaning predators, the guild of hummingbird-pollinated plants (the hummingbirds being the shared resource), or the guild of canopy-nesting birds. It may also be used to denote certain life-history characteristics, as with the guild of neotropical migrant birds. In broad terms, a guild is any group whose members overlap in one or more niche dimensions and, hence, either strongly interact or have the potential for strong interaction among them. Guild members may or may not be taxonomically similar; the guild of foliage gleaners, for example, includes birds, ants, and spiders. Any given species is likely to belong to several different guilds, resulting in a complex intertwining of relationships within communities.

Community ecologists (who until recently have been preoccupied with competition as the primary structuring force in nature) have devoted most of their attention to how members of a particular guild either compete or evolve to avoid competition. This view is now being broadened to include either indirect or direct interdependencies among guild members. The original application of the term *guild* in biology by Schimper (1903) referred to plants that relied on other plants. Atsatt and O'Dowd (1976) were the first modern ecologists to discuss cooperative interactions within guilds; they hypothesized the existence of **plant-defense guilds,** or groups of plants whose common interest in not being eaten by herbivores led them to evolve cooperative defense strategies. As we saw in chapter 8, the guild concept has also been extended to include sets of species defined by a shared mutualism (Perry et al. 1989b, 1991). For example, plant species that form mycorrhizae with one or more of the same species of fungi, along with the fungi themselves, form a guild; so do networks of plants that require the same pollinators, along with the pollinators themselves. There can certainly be competition within guilds defined by mutualisms (e.g., plants vying for the services of a given pollinator); however, mutual interdependencies also characterize such guilds. We saw one instance of this in chapter 8, in which mycorrhizal fungi required by conifers were stabilized by sprouting hardwoods following disturbance. The Brazil nut tree provides a good example of interdependence based on shared pollinators. Attempts to grow Brazil nuts in monospecific plantations have failed, not because the trees will not grow under such conditions but because they will not produce nuts. Why? Insects that pollinate the tree need sources of food during periods when Brazil nuts are not flowering, which requires forests with diverse plant hosts. Nature is pervaded by such indirect interactions and interdependencies (chapter 11).

Guild structure plays a critical role in ecosystem processes, especially regarding system stability. As we discuss in some detail in later chapters (11, 19, and 21), the ability of more than one species to perform the same function maintains functional integrity of the system even though a species may be lost from it. Because guild members overlap in their functional niche, guilds represent redundancies in ecosystems (overlap in functional niches is also referred to as **complementarity**). For example, the loss of a foliage-gleaning bird species from a community does not mean that foliage-eating insects are free of predators, so long as other species in the foliage-gleaning guild remain. On the other hand, where evolutionary pressures act to fragment guilds by reducing niche overlap—the niche diversification discussed below—the end result can be **keystones,** or species that perform some unique function. Keystones are points of vulnerability in ecosystems. When they are lost, so is their function. Determining the degree of functional niche overlap in ecosystems is not an easy task (chapter 21).

9.4.2 Niche Diversification

Evidence suggests that throughout evolution, conflicts over food have frequently been resolved through specialization. This is thought by some to have been a significant diversifying force in nature (Futuyma and Moreno 1988).

Examples of food-related niche diversification among animals are widespread in nature. In the neotropics, fruits compose the primary diet of 405 species of birds, 33 species of primates, and 96 species of bats (Fleming et al. 1987). But the diverse array of fruits produced by the numerous plant species of these areas permits **frugivorous** animals (i.e., animals that eat fruit) to specialize to

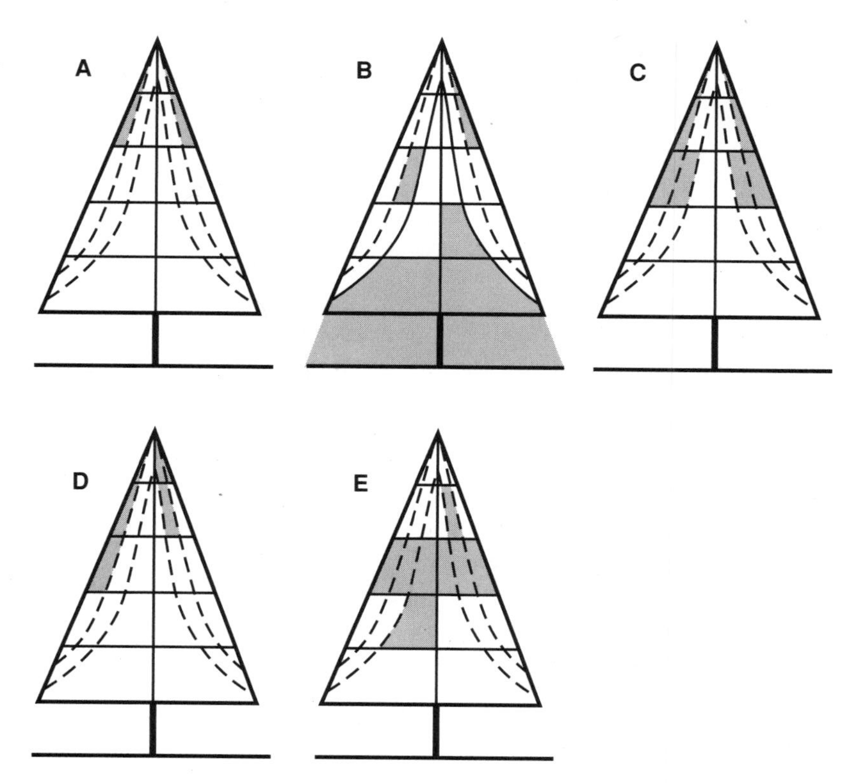

Figure 9.15. Spatial feeding patterns of five warbler species in conifer forests of New England. Shaded areas represent major feeding zones used by each species. A. Cape May warbler. B. Myrtle warbler. C. Black-throated green warbler. D. Blackburnian warbler. E. Bay-breasted warbler. (Adapted from MacArthur 1958)

For obvious reasons, hypotheses concerning forces that have acted in the distant past can be quite difficult to prove. There seems little question that evolutionary forces interacting with ecological factors have played a major role in shaping the structure of ecosystems. Many ideas about the exact nature of this dynamic (e.g., niche diversification because of food competition), however, have not been tested experimentally. Rather, they have grown from the interpretation of natural patterns in light of certain notions about how ecosystems "should" work (e.g., species "should" compete for food). Conclusions drawn in this way are not necessarily wrong, but neither are they necessarily right. Nature is full of surprises and frequently does not work in ways that conform to our preconceptions.

some degree, thus reducing competition or even avoiding it altogether. In the dry tropical forests around Monte Verde, Costa Rica, "only 13 of the 169 species of fruit eaten by birds are also eaten by bats" (Fleming et al. 1987). Similar studies throughout the tropics have found little overlap in the diets of fruit-eating bats, birds, and primates (Fleming et al. 1987).

Niche diversification may take numerous forms, including type of food, timing of feeding, place of feeding, and in some instances, the ability of a species to capitalize on food provided by periodic extraordinary events. Figure 9.15 shows spatial feeding patterns of five warbler species that coexist in conifer forests of New England. All are insect eaters and about the same size. These species feed in different positions in the canopy and in different manners, move in different directions through the trees, and have slightly different nesting dates. Commenting on these patterns, Krebs (1972) observed: "The feed zone differences seem sufficiently large to explain the coexistence of the blackburnian, black-throated green and bay-breasted warblers. The myrtle warbler is uncommon and less specialized than the other species. The Cape May warbler is different from these other species because it depends on occasional outbreaks of forest insects to provide superabundant food for its continued existence. During outbreaks of insects the Cape May warbler increases rapidly in numbers and obtains a temporary advantage over the others. During years between outbreaks they are reduced in numbers to low levels."

Numerous studies of niche diversification among animals have focused on the so-called **metric traits,** which are readily measurable attributes, such as body size, or in birds, the shape and size of the bill, that are assumed to reflect differences in diet. MacArthur (1972) discussed bird species diversity on the island of Puercos off the coast of Panama:

> [There are . . . four species of interior forest flycatchers on Puercos. The smallest, the beardless tyrannulet . . . has an average weight of 8 gm; the next smallest is the scrub flycatcher . . . with an average weight of 14.6 gm; then follows the short-crested fly catcher with a mean weight of 33.3 gm. Each of these is about double the weight of the previous one. Finally, the largest is the streaked flycatcher . . . that weighs an average of 44.5 gm. Other families do not seem to sort by size. For instance, there are two flower-feeding "honeycreepers," the banana quit . . . and the red-legged honeycreeper. They feed together among the flowers in the canopy and their mean weights are 10.7 and 12.8 gm, respectively. There is a plausible explanation [that flycatchers sort by size while honeycreepers do not]. Large flycatchers do eat larger foods than small ones . . . whereas there is no simple way that a large honeycreeper could eat . . . different food than a small one. . . . Rather the bills are of different shape, and it is very likely that these species eat nectar from different flowers or eat different insects while feeding on nectar.

On the other hand, Wiens (1984) documents cases where differences in body size among bird species seem to have no relation to diet. Therefore, one must conclude that the patterns of diversity in nature are themselves quite diverse.

Following this line of thought, it should not be construed from the preceding discussion that animal species always divide up the environment and avoid one another. In Hawai'i, for example, native bird species tend to flock together, often in areas where their insect prey are superabundant (Mountainspring and Scott 1985). Interestingly, this is not the case for native birds and exotics (i.e., introduced species), which tend to avoid each other. Of course, feeding is only one subset of many possible niche dimensions, and birds that flock together most certainly differ in other respects. If they did not, they would not be different species.

Niche differentiation is often encoded genetically—differences in metric traits being the most obvious example. This is not always the case, however. Niche differentiation can also result from factors such as behavioral modification (e.g., where or on what a species chooses to feed, rather than where or on what it is genetically capable of feeding), or because a particular species is excluded from a portion of its potential niche (either habitat or functional) by competitors, natural enemies, or accidents of history. Thus, ecologists talk about a **realized niche,** or that niche actually occupied by a species, and a **fundamental niche,** or that niche the species is genetically capable of occupying.

The Hawai'ian Islands offer what are probably the most outstanding examples of niche diversification on the planet (e.g., Price 2004; Price and Wagner 2004). The most isolated land masses on earth, they were colonized by relatively few species. In a veritable explosion of evolutionarily driven novelty, however, the few colonizers diversified into a myriad of species (fig. 9.16). The concept of niche diversification represents an intersection between two fundamental forces in nature: ecology and genetics (Eldredge 1986). Underlying the concept is the fact that genetic differences are driven in large part by ecological factors because "habitat provides the template on which evolution forges characteristic life-history strategies" (Southwood 1988). Not just habitat but all characteristics of the niche shape the genetics of organisms. Moreover, the relation is frequently reciprocal; the process of niche creation creates environmental conditions that can exert evolutionary forces not only on the niche creator but on associated species as well (Odling-Smee et al. 2003).

9.4.3 What Holds Guilds Together?

We have seen that species tend to overlap in at least some aspects of their functional niches, but if the overlap is in the use of some critical resource, there is evolutionary pressure to diversify. We have also seen that the degree of overlap in functional niches within a community has significant implications for ecosystem stability, because it determines which system processes are protected by redundancies and which are not.

If evolutionary pressures indeed push species either to diversify or to dominate a given niche by out-competing and excluding others, why do redundancies not eventually disappear, each critical function being performed by only one species? Posing that question is like walking into a minefield; there are many possible answers but rarely enough data to distinguish which may be correct. One factor that seems clearly established is that in some instances, species using the same resource are not limited by that resource and, hence, do not compete (i.e., there is plenty to go around). As mentioned earlier, this is the case with many insect species, which are controlled so effectively by predators or climate that they seldom run short of food (chapter 11). Note the close link between food web structure, guild structure, and ecosystem stability: by relieving competitive pressures, predators function to maintain the redundancies that are important for smooth functioning of the entire system. Another possible answer lies in the neutral theory of biodiversity (see section 9.1.2 and chapter 10), which postulates that diversity is driven by the combination of dispersal

Figure 9.16. Examples of adaptive radiation in Hawai'ian biota. A. Forty-five species of honeycreepers ('akohekohe) from a single colonizer (photo by Jack Jeffrey). B. Six endemic and one indigenous genera of lobelioids, comprising 110 species, from no more than five colonists. C. Eight hundred species of drosophilid flies from a single ancestor. D. More than one thousand endemic snails from about 15 colonists. (Snail photo by W. P. Mull)

limitations and random factors associated with establishment; in other words, niches play little or no role.

Species will also maintain overlapping niches whenever the evolutionary benefits outweigh the costs, which appears to be the case in guilds defined by shared mutualisms. Brazil nut trees cannot reproduce when taken outside of their guild, and Douglas-fir depends on periodically sprouting hardwoods to support its mycorrhizal fungi (chapter 8). Some species do evolve highly specific mutualistic relationships (i.e., their evolutionary path has taken them away from participation in a guild to go it alone). One of the better-known examples is the highly specific relationship between fig trees and fig wasps. By flowering year-round, fig trees avoid the need for other tree species to support its pollinator (the fig wasp); in turn, the fig wasp services no other tree. This is a vulnerable relationship: loss of one partner automatically means loss of the other. The evolutionary track taken by figs has also resulted in them becoming a critically important keystone resource in tropical forests. One of the few trees in tropical forests to fruit year-round, figs provide food during periods that otherwise would be very lean for fruit-eating animals (chapter 11).

9.5 THE TRADE-OFF BETWEEN DOMINANCE AND DIVERSITY

Recall from chapter 8 that species diversity has two components: richness, which is the number of species, and evenness, which is how equitably resources are distributed among species that occupy the same trophic level. By definition, evenness is low when one species dominates a given trophic level. Because every local population needs some minimum level of resources to persist, richness is usually also reduced in systems dominated by one or a few species.

The inverse relation between dominance and evenness can be displayed graphically, as in figure 9.17. Some measure of **presence** is chosen (e.g., biomass or leaf area). The x axis represents an ordering of species with regard to the presence value, while the y axis is the log of the **relative presence percentage,** or the proportion of the total presence measure that is accounted for by a given species. For example, if striped maple has a leaf area of 1 m in a forest with total leaf area of 6 m, the relative presence percentage of the maple is $(1/6)100 = 16.7$.

A sharply declining curve, such as that for the subalpine forest in figure 9.17, indicates that the community is strongly dominated by a few species and has a correspondingly low evenness. A relatively flat curve, such as that shown for the tropical wet forest, indicates a community with low dominance and high evenness. The patterns shown in figure 9.17 correspond to those discussed in chapter 4 and add some additional information: tropical forests are more diverse than temperate forests both in species richness and in evenness.

Most authors use the term *relative importance* in such a diagram rather than *relative presence*. This can be misleading, however, when importance is measured by characteristics such as biomass or leaf area, which may or may not accurately reflect the role played by a given species in ecosystem processes. In most cases, the species that dominate a particular trophic level also exert a major influence over ecosystem processes, but species with relatively low biomass may also be quite important. As we saw earlier, understory species may significantly influence nutrient cycling and, thereby, the growth of overstory species. Certain species are keystones that influence ecosystem processes far beyond what is indicated by their numbers or biomass, for example, nodulated nitrogen-fixing plants or species that provide unique foods or habitats. (Keystones are discussed further in chapter 11.)

Species with low presence most of the time may have the potential to expand their populations and become quite important at other times. Consider again, for example, the Cape May warbler. Most of the time, this species is a relatively minor component of the bird community in northern hardwood forests; however, it plays a very important regulatory role during periodic insect outbreaks. Similarly, many early successional plant species have a low presence over most of the life of a forest but, nevertheless, have great potential importance because of their ability to quickly occupy a site and to stabilize soils following a catastrophic disturbance (chapter 8).

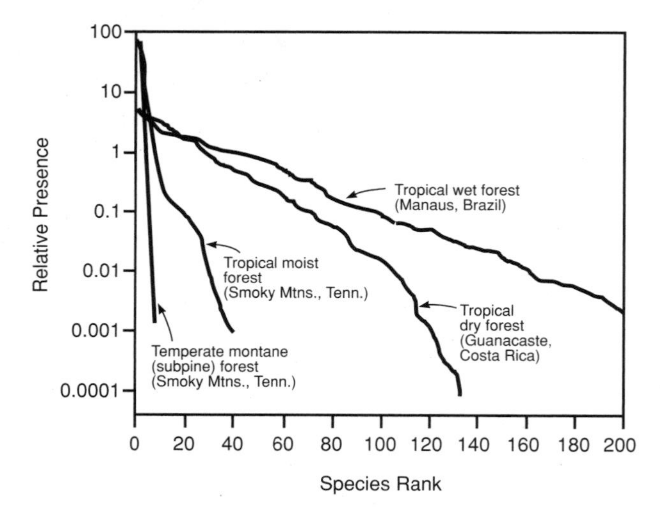

Figure 9.17. Dominance-diversity curves for four different forest types. (Adapted from Hubbell 1979)

9.6 SCALES OF DIVERSITY

So, Naturalists observe, a flea
Hath smaller fleas that on him prey;
And these have smaller fleas to bite 'em.
And so proceed ad infinitum.

SWIFT 1772

9.6.1 Within- and Between-Habitat Diversity

The diversity within a given area depends on (among other things) how large the area is and how much environmental variability it encompasses. Suppose, for example, that we install several plots to measure tree species diversity in the Big Horn Mountains of Wyoming (chapter 5). Many plots would contain only one tree species: ponderosa pine at low elevation, Douglas-fir or lodgepole pine at higher elevations. Some plots would contain as many as three species. Ecologists refer to the diversity measured on single plots as **alpha** or **within-habitat diversity.** This type of diversity is that which occurs within areas that have relatively homogeneous environments (e.g., soils, elevation, aspect) and no major discontinuities in successional stage (e.g., not encompassing both a clear-cut and an old-growth forest). On the other hand, when data from all plots are combined, the Big Horns as a whole contain five tree species. This is called **gamma diversity,** or that obtained when all of the plots within a given area are combined.

Gamma diversity reflects variability in environment and successional stage, but it is generally a rather open-ended concept with limited comparative value and uncertain biological meaning. In comparing, for example, the total number of tree species in the Big Horn Mountains with that in the Klamath Mountains of California and Oregon, we find the latter to be much more diverse. Is this simply because of the larger area that the Klamaths encompass, in which case the difference is of little interest ecologically, or does it reflect a greater complexity of environments and disturbance history within a given area of the Klamaths than in a similar-sized area of the Big Horns? To distinguish between these possibilities, ecologists use something called **beta diversity,** which may be thought of as the rate of change in species composition across any defined gradient. Whittaker (1972) used habitat gradients, but Bratton (1975) preferred the use of niche gradients to better reflect the fact that different species (and especially life-forms) have different habitat niches and therefore perceive different habitat gradients. Neither reflects the entire range of possibilities, as dispersal limitations could influence both alpha and beta diversity independently of habitats or niches (Hubbell 2001). In western Amazonia, beta diversity of two major plant groups is related to both habitat differences and dispersal limitation: habitat heterogeneity produces patchiness, and dispersal limitations result in more gradual changes in floristic composition at a regional scale (Tuomisto et al. 2003).

Beta diversity has been calculated in various ways, but most commonly as the quotient of gamma and alpha diversity (Whittaker 1972). For example, if five tree species occur in the Big Horns (i.e., gamma diversity), and the average plot contains 1.3 species (i.e., alpha diversity); the beta diversity of the tree component is 5 / 1.3, or 3.85. An average plot in the Klamaths would probably contain about three tree species; with a total of 18 species, beta diversity in these mountains is on the order of six. (Note that while alpha and gamma are measured in the same units [usually, number of species], beta is dimensionless.) However, a problem in comparing the beta diversity of regions using that approach arises from the fact that total species diversity (gamma) of a particular area depends on the size of the area (see below). Therefore, two regions could yield different beta diversities simply because one is larger than the other. To deal with that issue, Qian et al. (2005) calculated beta diversity as the slope of the relationship between the log of species similarity and either geographic distance or difference in climate. They used these measures to determine that beta diversity among angiosperms is greater in eastern Asia than in eastern North America (there are several reason why that might be so; see Qian et al. for a discussion).

Beta diversity measures horizontal and vertical spatial complexity. Across landscapes, it reflects topography, soils, and disturbance history (chapter 5). Mountainous landscapes tend to have high beta diversity (for obvious reasons). Wetlands, riparian zones, associated terrace-slope landforms, and other special habitats (e.g., rock outcrops) contribute significantly to beta diversity at the watershed scale (e.g., Goebel et al. 2003). Flat terrain can also have high beta diversity. Soils of the lowland tropics, for example, may vary widely over relatively short distances. Dispersal limitations combined with disturbances (e.g., periodic flooding and windthrow) can create a diverse forest pattern within topographically uniform landscapes.

Vertically, beta diversity measures the response of species to height in the canopy or depth in the soil. Earlier we saw that epiphyte communities vary as one goes from the ground upward, along with the height of trees. Butterfly communities in tropical forests also exhibit vertical beta diversity (Walla et al. 2004). Mycorrhizal communities may differ between litter layers and mineral soil.

Both the alpha diversity and beta diversity that are measured in a given region may vary depending on what part of the community is studied. Trees and herbs frequently exhibit differing patterns of diversity. In the Great Smoky Mountains of Tennessee and North Carolina, beta diversity of trees is relatively high at low elevations but declines as one ascends; beta diversity of herbs, on the other hand, remains high throughout the range (Bratton 1975). Beta diversity of understory plants is greater than that of trees in many temperate and boreal forests, a phenomenon that probably reflects the greater sensitivity of understory plants to small-scale environmental variation and is used in the habitat type classification systems discussed in chapter 5.

9.6.2 Species-Area and Species-Accumulation Curves

If one records the richness of species of a similar size (e.g., vascular plants, vertebrates) in increasingly larger areas, richness increases according to a power-law function. (If richness is recorded across size scales, for example vascular plants to nonvascular plants to invertebrates, it would increase with decreasing area sampled.)

The power-law relationship between species richness and the size of area sampled is one of the most venerable relationships in ecology and is an important tool in relating species loss to habitat fragmentation. The species-area

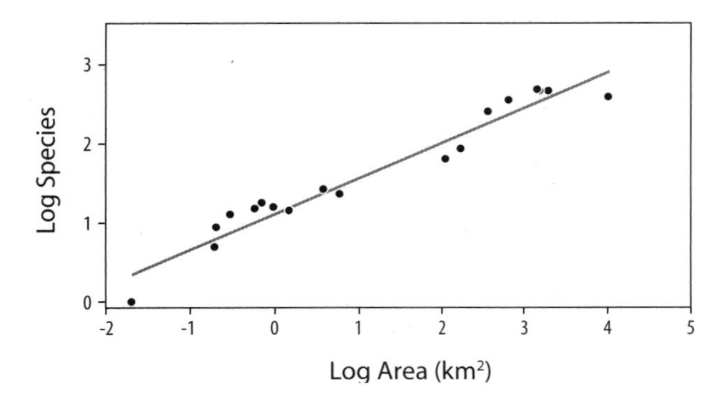

Figure 9.18. Species-area curves for plants in the Hawai'ian Islands. Points correspond to different islands. (Adapted from Price 2004).

relationship (SAR) is strikingly regular regardless of ecosystem or biome, uniformly following a power law of the form

$$S = cA^z$$

where c and z are parameters, S is the number of species, and A is area. On a log-log scale SAR becomes a straight line with a slope equal to z. Figure 9.18 shows species-area curves for the plants of the Hawai'ian islands, each point representing plant richness on a particular island (Price 2004). In some cases, species richness is graphed against the number of individuals sampled rather than the area (e.g., species richness of trees as a function of the number of trees samples) (Hubbell 2001).

Species-accumulation curves follow a form similar to that of SAR curves, but they relate richness to the number of samples (e.g., plots or soil samples) rather than increasing area within a defined locale. For example, Loehle et al. (2005a) used bird richness data from multiple plots scattered across the southeastern United States to calculate the rate at which richness accumulated as more plots were added (fig. 9.19). Note in figure 9.19 that different forest types exhibited either similar or different accumulation curves, depending on region. Accumulation curves for a variety of taxa are shown in figure 9.20. Note that unlike for other taxa shown in the figure, the genotypic diversity of east Amazonian soil bacteria shows no sign of leveling out with increasing sample size.

SAR reflects beta diversity over the area sampled. If the smallest area sampled corresponds to a uniform habitat, the intercept of the curve reflects alpha diversity. What accounts for the regularity of the species-area power function? Different researchers find different factors. Hubbell (2001) reproduced observed species-individual curves for a neotropical forest by assuming that dispersal limitation was the only factor operating. On the other hand, Martin and Goldenfeld (2006) show that the power-law SAR can be explained by the combined effects of species clustering and a predominance of rare species.

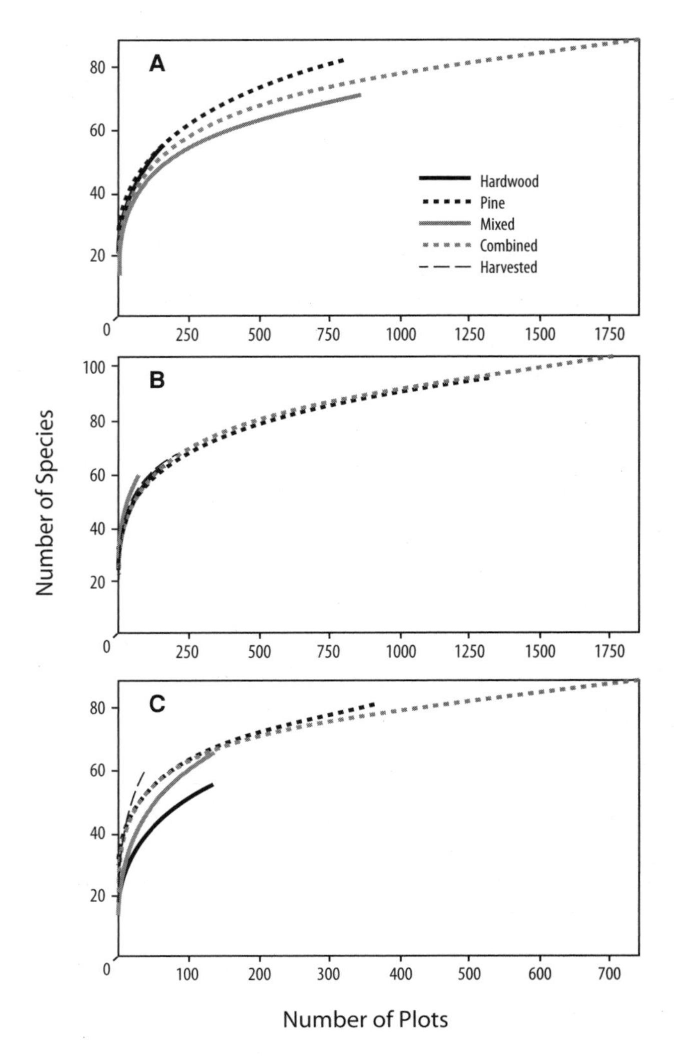

Figure 9.19. Species accumulation curves for bird communities within differe1nt forest types on study areas in Arkansas (A), two areas in South Carolina (B, C). (Loehle et al. 2005a. Reprinted with permission from Elsevier. Copyright 2005)

9.6.3 Hierarchies of Habitat and Species Diversity

Alpha and beta diversities represent a limited hierarchical classification: diversity at the scale of landscapes (i.e., beta) comprising an assembly of diversities within relatively homogeneous habitats (i.e., alpha). This classification is correct in so far as it goes; however, problems are encountered. What is relatively homogeneous habitat for one species or life-form may be a mosaic of differing habitats for others. As previously discussed, ecological diversity exists at many hierarchical levels.

If we use our imaginary ecoscope to look within what appears at one scale to be a uniform forest (i.e., a local ecosystem), we see a long series of habitats nested within habitats, including snags, rock outcrops, canopy gaps, different plant species, and different canopy layers. We find that insects are patterned in response to chemical differences

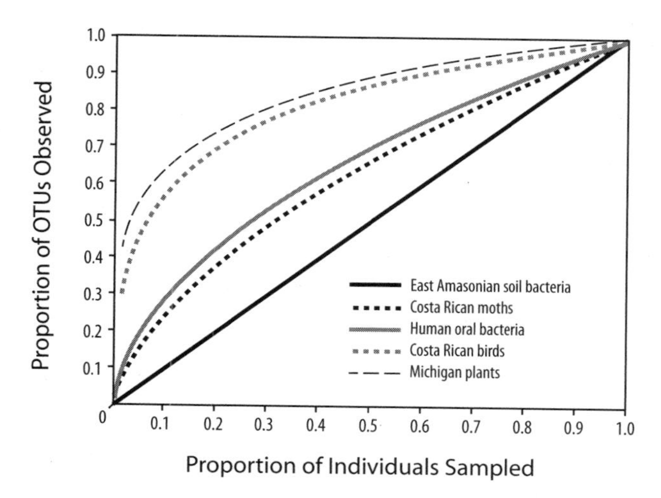

Figure 9.20. Accumulation curves for Michigan plants, Costa Rican birds (J. B. Hughes, unpublished data), human oral bacteria, Costa Rican moths, and East Amazonian soil bacteria. Estimates and are standardized for the number of individuals and species observed. OTU, operational taxonomic units. For birds, plants, and moths OTUs correspond to species. For bacteria, OTUs are measures that denote unique genotypes, such as such the number of unique terminal restriction fragments or number of 16S ribosomal DNA (rDNA) sequence similarity groups. (Adapted from Hughes et al. 2001; see Hughes et al. for original citations).

among individuals of the same tree species and between leaves growing on the same tree (chapter 19). On the forest floor, we find islands of logs in various stages of decay and surrounded by a sea of fine litter (e.g., leaves, twigs); within the soil, the ecoscope reveals a diversity of habitats, including zones that are strongly influenced by roots, aggregations of soil minerals and organic matter, termite and ant nests, and burrows of both small mammals and larger invertebrates. Each of these distinct structural features likely serves as habitat for various species of animals, microbes, or plants. From the viewpoint of a bacterium living on the surface of a root, landscape diversity is composed of surrounding roots.

Tuning the ecoscope to yet finer scales, we find that the animals living on plants serve as habitat for other animals. For example, a given species of plant-feeding insect is commonly parasitized by roughly 2 to 12 species of other insects (usually tiny wasps or flies), termed **parasitoids** (Hawkins and Lawton 1987). Insect herbivores and their parasitoids are estimated to compose over one-half of all known species of metazoans (Price 1980). Protozoa and microbes live within leaves, roots, and the guts of all animals. Diversity begets diversity.

9.6.4 Hierarchies of Genetic Diversity

9.6.4.1 Plants

It has long been known that any given plant species comprises genetically distinct populations (i.e., **ecotypes**) that correspond to the mosaic of habitats encompassed by the species. Trees of the same species may differ genetically depending on the soils, aspect, elevation, or latitude where they grow. In general, the more different the habitats of two individuals of the same species, the more likely they are to differ genetically. This phenomenon, which might be viewed as the within-species equivalent of regional and landscape diversity, is commonly used in forestry to derive seed-transfer guidelines—how far out of its native habitat a given seedling might be safely planted.

The genetic equivalent of within-habitat (i.e., alpha) diversity also exists within plant (and animal) species. With the exception of clonal species, each individual is genetically distinct from other individuals of the same species, even though they appear to occupy the same habitat. It is unclear how much of this diversity is adaptive (i.e., selected and maintained by subtle environmental variability) and how much is random (i.e., serves no particular adaptive function); but, despite the uncertainty regarding cause, it is abundantly clear that most species that have been studied maintain high levels of genetic diversity within local populations. Studies of temperate and boreal tree species have found that on the order of 75 to 90 percent of all specieswide genetic variability resides within populations, and only 10 to 25 percent between populations (Brown 1979; Hamrick et al. 1979; Loveless and Hamrick 1984). Conifers tend to have greater genetic differences within populations and less between populations than do temperate angiosperms (Loveless and Hamrick 1986).

The genetic structure of a given species—how much variability it contains and how that variability is distributed both within and between populations—depends on the rate of mutation, the amount of gene flow among individuals and populations, and the strength of selection either for or against some new gene. Gene flow, in turn, depends on the mating system (whether a species is outbreeding or inbreeding, animal or wind pollinated) and on the continuity of populations. Patchy distributions lead to more isolation and, in turn, greater differentiation among populations. Selection for or against a given gene depends on the fitness (or lack thereof) that it confers, which is a function of the biological and physical environment. Two populations of ponderosa pine growing adjacent to one another yet on very different soils (or aspects) would exchange many genes, but selection would quickly weed out any that produced a maladapted seedling.

Animal-pollinated (e.g., many angiosperms) and wind-pollinated (e.g., all conifers and some angiosperms) species differ in their genetic structure. Both contain similar amounts of within-population variation, but populations of animal-pollinated species tend to be more differentiated than those of wind-pollinated species (Brown and Moran 1981). The probable reason is energetics. Pollinators expend energy to fly from one tree to another and are likely to minimize this expenditure by not covering more territory than is necessary to meet their food needs; wind pollination (although quite wasteful of pollen grains) is energy free. This is a good example of linkages within ecosystems, the structure of one part (i.e., tree genetics) being influenced by processes occurring in another part.

Various forces act to maintain genetic diversity within populations:

1. Environmental variation favors different genotypes at different times or in different microenvironments. Similarly, heterozygous individuals are more fit in variable environments than are homozygous individuals. This is commonly called **adaptive variation.**
2. Heterozygosity improves individual fitness because it protects against **inbreeding depression,** which is a harmful or outright lethal combination of certain alleles. This is an adaptive value of heterozygosity having nothing to do with environmental variation.
3. Genes are maintained in the population because they have no adaptive role and, hence, are neither selected for nor against (i.e., so-called **neutral mutations**).
4. Genes may be at a selective disadvantage in a particular environment but are maintained in the population because they enter frequently enough to balance the rate at which they are lost through selection. An example would be pollen flow between stands of trees that are close enough to exchange large amounts of pollen but occupy environments that exert different selection pressures, hence eliciting different adaptations (e.g., different soil types, aspects, elevations).

The question of what forces maintain genetic diversity within populations is quite important in resource management. If diversity represents adaptations to past environmental variability, its loss would lower the capacity of populations to cope with environmental fluctuations. But what if it is nonadaptive (as in cases 2 and 3 above), does it then follow that genetic diversity can be reduced without threatening population stability? As we see in chapters 21 and 22, the answer to this question, at least for trees, is probably no. Whatever the mechanism(s) by which it is maintained, genetic diversity is probably important to keep.

9.6.4.2 Somatic Mutations in Plants

Unlike animals, individual plants are a collection of numerous different growing points (i.e., meristems), each bud and root tip producing its own set of mature tissues. At least initially, each of these growing points is genetically identical to the seed or root stock from which they spring; however, through time, long-lived plants such as trees may accumulate what are called **somatic mutations**—genetic changes that occur within vegetative cells rather than sexual

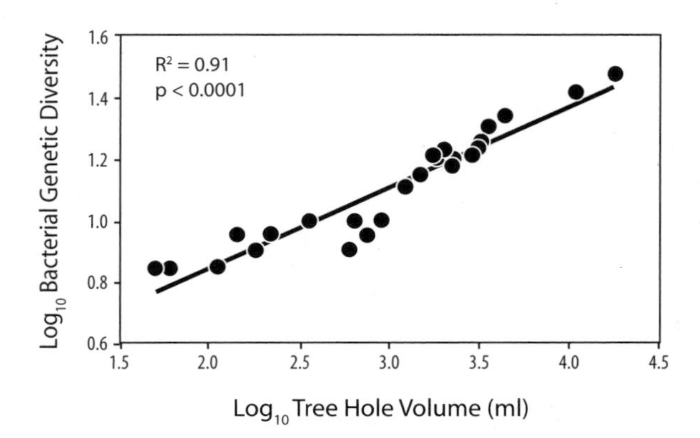

Figure 9.21. The species-area relationship for microbial communities in water-filled tree holes. Bacterial genotypic diversity increases with increasing island size (island = tree hole). Note log-log scale. (Adapted from Bell et al. 2005. Reprinted with permission from AAAS)

cells. This might happen, for example, when one or a few buds are hit by ionizing solar radiation that has managed to pass through the earth's protective ozone shield. All tissues and buds formed thereafter from such a mutation will carry the genetic imprint of the mutation rather than that of the original seed (unless they, in turn, somatically mutate).

In theory, long-lived plants could develop into mosaics of genetic variability; however, little is known about the extent of this phenomenon in nature or its ecological significance. Whitham and Slobodchikoff (1981) hypothesized that somatic mutations provided plants with flexibility in responding to environmental fluctuations and decreased the ability of pests and pathogens to evolve around plant defensive chemicals (chapter 19).

9.6.4.3 Animals

Like plants, animal species are genetically diverse. Vertebrates tend to be less variable than invertebrates (Nevo 1978). As with trees, habitat diversity probably plays a large role in this difference. Vertebrates are frequently mobile and exploit a variety of habitats; invertebrate species may be subdivided into several microhabitats within small areas. Edmonds and Alstad (1978) documented a particularly striking example of microhabitat differentiation in the black pine leaf scale, an insect that feeds on ponderosa pine by sucking juices out of needles. Each individual ponderosa pine tree in the stand that was studied supported a genetically distinct population of scale insects. Scales transferred from one tree to another survived poorly. In chapter 13 we see that the genetic structure of tree and other plant populations can profoundly influence animal communities and ecosystem processes.

9.6.4.4 Microbes

The advent of molecular genetics has had a profound effect on the ability to quantify genetic variation among microbes in field environments. A growing number of studies indicate that microbial taxa contain high levels of genetic diversity (Hughes et al. 2001), including soil bacteria (e.g., Borneman and Triplett 1997; Neufeld and Mohn 2005) and mycorrhizal fungi (Lanfranco et al. 1999; Sanders et al. 2003; Öpik et al. 2006). Arbuscular mycorrhizal fungi are genetically diverse even within single spores. The genotypic diversity of bacteria occupying water-filled holes formed by root buttressing of European beech follows a power-law SAR, with diversity increasing sharply with the size of the hole (Bell et al. 2005) (fig. 9.21).

9.7 SUMMARY

The patterns of nature can be divided into three components: composition, structure, and function. All occur in nested, interacting hierarchies. Local ecosystems are composed of organisms, species, groups of closely interacting species, genetic diversity within species, the legacies of organisms such as dead wood and soil organic matter, and various inorganic components (e.g., minerals, gases). Ecosystem structure arises from the patterns in which these basic building blocks occur and includes both the arrangement of things in space (i.e., physical structure or architecture), and the structure of social interactions (i.e., how species relate and interact). In forests, physical structure manifests horizontally, particularly with regard to canopy closure and the consequent influence on understory development, and vertically as layering in the canopy and soil.

Two fundamental concepts relating to social structure are *niche*, or the way species divide up habitats, resources, and ecosystem functions; and *interaction webs*. A niche is both an "opening," or opportunity for a species to find suitable living (and reproducing) conditions within a given system, and the set of attributes a given species brings to that opening. So, a niche is best viewed as the total set of interactions between an individual species and the ecosystem in which it is embedded.

Interaction webs include both food webs and the spin-off effects of niche construction (e.g., the array of habitats created by a beaver pond). The requirement of all organisms for metabolic energy and material resources is a major organizing force that integrates populations of individual species into the community and ecosystem as a whole. Two general types of food web patterning arise from this. One relates to energy and matter flow from one species to another (i.e., feeding or being fed on; sometimes termed *vertical relationships*); the other relates to allocation of energy and resources among species that potentially utilize the same resources (i.e., *horizontal relationships*). The network of feeding relationships within ecosystems has numerous important consequences for system functioning.

Species groups (called *guilds*) form an important component of the social structure of communities. Members of a given guild overlap in one or more niche dimensions; hence, they either strongly interact among themselves or have the potential for strong interaction. Guild structure plays a critical role in ecosystem processes, especially with regard to system stability. In cases where niche overlap creates conflict, species tend to reduce the overlap by diversify and specializing, or alternatively, one species may dominate and push others out. Niche diversification increases species diversity; dominance decreases it.

Alpha diversity refers to the species richness within a relatively homogeneous area, and within trophic levels is strongly related to the degree of dominance by one or a few species. *Beta diversity* refers to the rate of change in community composition across any defined spatial gradient and is related to habitat complexity and dispersal limitations. The relationship between area and species richness follows a well-defined power law called the *species-area curve*.

10

How Biodiversity Is Created and Maintained

> Empirical data reveal striking and often-repeated patterns in how species originate, persist, assemble in groups, and eventually go extinct. These regularities have always suggested . . . the existence of general causal mechanisms that shape both the evolution of biodiversity and the structure of ecological communities. But what are those mechanisms, and how exactly do they work?
>
> NORRIS 2003

Why are there so many species? This is a question that ecologists have struggled with for many years (e.g., Hutchinson 1959; MacArthur and MacArthur 1961; May 1988; Chesson 2000; Hubble 2001; Ricklefs 2004; Turner and Hawkins 2004). Actually, two questions are embedded within that one. First, in any given community type over evolutionary time, why has natural selection not produced a set of competitively superior superspecies, one dominating each trophic level? To ask that question in another way, through what mechanisms do species within the same trophic level coexist? The second question relates to the striking global patterns of species diversity discussed in chapter 3. Why do different environments vary so widely in the numbers of species they will support?

10.1 FORCES THAT GENERATE AND MAINTAIN DIVERSITY WITHIN COMMUNITIES[1]

Many hypotheses have been put forward to explain how species within communities coexist.[2] Most follow one of six general themes: niche theory, or niche assembly;

[1]*Community* as used here denotes the biota within an area that has no significant change in climatic factors that would influence diversity.

[2]These hypotheses refer to diversity within trophic levels, where the greatest challenges to understanding arise, and also where the vast majority of diversity resides.

neutral theory, or dispersal assembly; dispersal limitations; biotic interactions; self-organizing dynamics; or regional context and history. These are not mutually exclusive, and all potentially play a role in any given locale.

10.1.1 Niche Assembly

The variety of niches maintained in nature, coupled with the fact that species within a given trophic level specialize to one degree or another, prevents a single species from dominating.

One the most venerable ideas in ecology is that species avoid the costs and risks of competition by evolving niche specialization (chapter 9). It follows that species diversity is linked closely to niche diversity. Niches are diversified in both time and space. In time, the diversity of niches is greatly enhanced by both small- and large-scale disturbances that initiate succession and its attendant changes in species composition and forest structure (chapter 8). Spatially, changes in environmental factors (e.g., soils) coupled with mosaics of communities in differing successional stages not only enhance landscape diversity (chapters 5 and 8), they provide a reservoir of species for local ecosystems that are changing through time. A young, even-aged forest, for example, is poor habitat for the northern spotted owl. However, as long as forests that provide good owl habitat occupy a sufficient portion of the landscape, there will be a source of future owls to populate today's young forest when it ages and provides suitable habitat. What constitutes either "good" habitat or "sufficient" area is poorly understood for most species; this topic is pursued further in chapter 22.

10.1.2 Dispersal Assembly

Diversity arises from stochastic factors that determine which of the available pool of species actually occupies a vacant space within a community. Niche separation is not involved. Norris (2003) gives a simple example of the basic concept. "Imagine" he says, "a vast checkerboard of 'resource space.' Every square is occupied by an individual of some species, and the total number of squares can be counted and grouped to produce an abundance distribution. A few species will occupy many squares; most will occupy only a few.[3] As time passes, some squares become vacant (by an individual death) and are reoccupied by an individual of the same or another species. Each species has some chance of filling the vacancy, in direct proportion to its current share of the checkerboard."

The stochastic nature of this process results in random changes in species abundance through time on a given site, a phenomenon called **zero-sum ecological drift.** (The term *zero-sum* comes from the fact that total number of individuals in the community does not change.)

[3]This specification is a simplification the log-normal pattern of species distribution commonly seen in nature (chapter 9 in this volume).

All species are considered to be competitively equal and to have identical life-history characteristics (i.e., birth rates, death rates, and dispersal capabilities), from which it follows that relative species abundances are determined solely by chance draw from the available pool rather than specific ecological capabilities. The commonly used term **neutral theory** refers to the assumption of ecological equivalence among species. Neutral theorists do not claim that species really are identical ecologically, only that ecological differences play a minor role in determining patterns of diversity compared to stochastic factors.

Neutral theory in its current form was developed by Hubble (2001) as an outgrowth and elaboration of the highly influential theory of island biogeography (MacArthur and Wilson 1967). Neutral theory is heavily mathematical in its details and beyond the scope of this book to present in any but a cursory fashion. Norris (2003) and Bell (2001) provide good overviews.

10.1.3 Limited Dispersal

As we have seen, dispersal plays a central role in neutral theory. However ecologists recognized the importance of limited dispersal in shaping patterns of diversity long before the advent of neutral theory, and dispersal limitations may be incorporated into either niche or neutral models (Hurtt and Pacala 1995; Hubbell 2001).Limited dispersal results in patchy species presence, or greater beta diversity. Conversely, high dispersal rates lead to lower beta diversity, but higher alpha diversity (because of greater mixing) (Hubble 2001).

10.1.4 Biotic Interactions

Biotic interactions influence diversity in a variety of ways. Depending on species, individual trees may promote, discourage, or have no effect on alpha diversity of trees in their immediate vicinity. Wiegand et al. (2007), who showed this effect in two tropical forests, termed species with high diversity in their neighborhoods *accumulators* (these were more abundant in Sri Lanka) and those with low diversity in their neighborhoods *repellers* (more abundant in Panama). The effect did not extend beyond 20 m.

Animals promote plant and microbial diversity by distributing propagules from one suitable habitat to another; for example, birds and bats disperse seeds, small mammals disperse spores of mycorrhizal fungi. Mutualisms and other cooperative interactions among species enable members of the cooperating group to better resist factors that might otherwise exclude them from communities (e.g., cooperation among tree and shrub species during early succession; see chapter 8).

Higher trophic levels play a crucial role in regulating numbers of organisms in lower trophic levels (this is called **top-down regulation;** see chapter 11). Depending on the circumstances, top-down forces can increase diversity by

Ecologists debate the relative importance of niches versus randomness (i.e., neutral theory) in determining patterns of diversity. Because of the simplifying assumption of equality among species, neutral theory is more amenable to mathematical analysis than niche theory and, therefore, better able to generate quantitative predictions. Neutral theorists argue that the theory successfully reproduces commonly observed patterns in nature, such as species-area relationships and species-abundance distributions within communities[4] (Bell 2001). Critics reply, however, that the parameters contained within neutral theory allow the model to be tuned, in other words, that its output can be easily manipulated to fit field data. Moreover, at least with regard to species-abundance distributions, there is disagreement about how well the theory actually performs (McGill 2003).

Niche theory and neutral theory differ in their predictions of how community composition varies in space and time. Over the past several years a number of studies have compared the two with regard to how well they fit field data (including trees, herbs, birds, and small mammals, and in tropical, temperate, and boreal forests). Most find that real-world patterns are more consistent with the predictions of niche theory than those of neutral theory (Condit et al. 2002; Clark and McLachlan 2003; Gilbert and Lechowicz 2004; Graves and Rahbek 2005); however, neutral theorists have questioned the validity of at least some of these (Volkov et al. 2003). In support of neutral theory, patterns of tree diversity within gaps in both tropical and temperate forests are due more to chance than to niches (Brokaw and Busing 2000); however, survival of tree species within a given area is highly nonrandom in tropical forests, those that are most abundant to begin with having the highest rates of mortality (Wills et al. 2006). Such density-dependent mortality acts to increase diversity over time.

One approach to comparing the relative importance of the two mechanisms is to look at community inertia, or the tendency for communities to maintain a relatively stable composition over long periods. Inertia provides evidence that deterministic forces are at work (i.e., niches or biotic interactions reduce the role of randomness), though, depending on the degree of stability, a role for random forces is not necessarily ruled out. Using fossil assemblages, McGill et al. (2005) studied the inertia of small mammal communities in North America over a period spanning nearly one million years. They found "good evidence for inertia but . . . the results change . . . with taxonomic, spatial, and temporal scales." In other words, deterministic forces are at work, but so are random forces. In his study of contemporary small mammal communities in boreal forests, Morris (2005) also found that both deterministic and random forces were at work, a dynamic he referred to as "predictably stochastic."

As Ostling (2005) points out, "unless habitat and species change in lock-step, habitat effects do not rule out a simultaneous role for dispersal-assembly." It seems likely that both niche specificity and stochasticity play roles in determining patterns of species diversity, their relative importance varying as a function of environmental and ecological heterogeneity.

diminishing the ability of one or a few species to dominate the lower levels, or they can decrease diversity by doing just the opposite: allowing one or a few species to gain the upper hand. We saw examples of both situations in chapter 8: grazing by elk reduces shrub dominance and promotes diversity in early successional forests of the western United States, but overgrazing by cattle has allowed a single unpalatable plant species to exclude other plants from some areas.

Top-down regulation promotes diversity most effectively when species with the greatest potential to dominate are regulated more than those with lower dominance potential (termed **negative density** or **negative frequency dependence**[5]). Negative density dependence occurs in both tropical and temperate forests and is often associated with species-specific predators and pathogens (Pacala and Crawley 1992; Lambers et al. 2002; Marquis 2004).

10.1.5 Self-organizing Dynamics

First formulated by Hassell et al. (1994), the hypothesis of **self-organizing dynamics** represents a special form of diversity through species interactions. Models show that under some circumstances, interactions among two or more species produce distinct spatial patterning in species abundance that persist even in uniform environments. Self-organizing spatial patterns are neither niche driven nor random. One of the most commonly modeled situations is the interaction between herbivorous insects (e.g., moth larvae) and their parasitoids (tiny wasps and flies that parasitize the larger insects; see chapter 19), which produces elaborate spatial patterns, including spiral waves and latticelike concentrations of species abundance. Species coexistence is promoted if the spatial patterns resulting from self-organizing dynamics contribute to portioning of the landscape among different species.

[4] See chapter 9 for a discussion of these patterns.

[5] Wills et al. (1997) define *density dependence* as the correlation of life-history parameters (e.g., survival) with total biomass at the beginning of a period, and *frequency dependence* as correlation with total population numbers (e.g., numbers of trees) at the beginning of a period. Density and frequency are often related, but not always.

10.1.6 Regional Context and History

The diversity of any locale depends to some degree on the regional pool of species, which, in turn, depends not only on environment, but on historic factors such as immigration, emigration, speciation, extinction, and legacies of past land uses or other disturbances (e.g., Hillebrand and Blenckner 2002; Hawkins et al. 2003b; Rickleffs 2004; Oekland et al. 2005). Consider the Hawai'ian Islands—the most isolated land masses on earth. Relatively few species arrived and successfully colonized, but, of those that did, many adapted to the diverse environments of the mountainous islands by spinning off an astounding number of daughter species (termed **adaptive radiation**). Because of the low rates of immigration, the biota is species poor compared to those of similar continental environments (e.g., no native social insects, earthworms, or snakes, and the only native land mammal is a bat). However, high rates of speciation produced a very large proportion of endemic species (i.e., those that occur nowhere else). Both importation of new species and extinction rates of natives increased following the arrival of humans, a process that began with the Polynesians and accelerated with the arrival of other groups beginning in the nineteenth century. The diversity of the Hawai'ian Islands today is the product of these historic factors interacting with the island environment.

10.2 THE VARIATION OF SPECIES RICHNESS AMONG ENVIRONMENTS

> The oldest and one of the most fundamental patterns concerning life on earth is the increase in biological diversity from polar to equatorial regions.
>
> WILLIG ET AL. 2003

Over the years, more than 30 different hypotheses have been put forward to explain geographical variation in biodiversity, including but not restricted to the sharp increase in species richness as one moves equatorward from the poles. For forests in particular, the latitudinal effect is striking. Moist tropical forests commonly have up to 300 tree species per hectare, an order of magnitude more than the most species-diverse temperate forests (if shrubs and herbaceous plants are included, the disparity still exists but is less striking). Although attention often focuses on latitudinal gradients, the more general question applies to diversity differences across any environmental gradient (Currie et al. 2004).

Hypotheses group into two general categories (Hawkins et al. 2003a; Turner and Hawkins 2004): energy and climate, and history.

10.2.1 Energy, Climate, and Primary Productivity

Available energy (for photosynthesis and warmth, and closely correlated with climate) might influence diversity through effects on either **primary productivity** (the **productivity-richness hypothesis**) or the physiological need of organisms for certain temperature ranges (**the ambient energy hypothesis**). The two mechanisms together are often referred to as the **species-energy hypothesis.** Climate and productivity may also influence diversity through effects on mutation rates, a possibility referred to as the **climate-speciation hypothesis.**

There is abundant evidence that species richness correlates with primary productivity or some surrogate for productivity (such as water availability) (see reviews by Waide et al. 1999; Currie et al. 2004; Turner and Hawkins 2004). For example, Hawkins et al. (2003a) found that measures of energy and water best explained changes in species richness in 82 of 85 gradients exceeding 800 km in size and including measures of various taxonomic groups. What is the form of the relationship between richness and productivity? Waide et al. (1999) concluded that "no general consensus concerning the form of the pattern has emerged based on theoretical considerations or empirical findings." Hump-shaped (i.e., unimodal) patterns, in which richness peaks at intermediate levels of productivity, are most common, but monotonic increases of richness with productivity are also frequently seen (Mittelbach et al. 2001). The form of the relationship appears to depend on multiple factors, such as the trophic group studied (Mittelbach et al. 2001), the history of community assembly (i.e., the order in which species arrive [Fukami and Morin 2003]), how productivity is measured (Groner and Novoplansky 2003), and the scale encompassed by a particular study (Waide et al. 1999; Mittelbach et al. 2001; Chase and Liebold 2002).

Ecologists have put forward a "chaotic array of possible explanations" for the mechanisms by which productivity translates into species richness (Waide et al. 1999). A common starting point is the more-individuals hypothesis, which states that higher productivity translates into more individuals, which, through various mechanisms (e.g., more biotic interactions, lower probability of extinction[6]), translate into more species (see the review by Evans et al. [2005a]). There are limits, particularly for plant communities, for which high productivity is often accompanied by greater dominance of one or a few species and lower overall richness (see section 10.4.3).

Currie et al. (2004) and Evans et al. (2005b) evaluated various aspects of the productivity-richness hypothesis and concluded that without significant modification it could not account for observed patterns. One problem is that increases in richness are not spread evenly across species; that is, not only richness but community structure changes with productivity. Another problem is the emergence of

[6] More biotic interactions might generate pressure for niche specialization or perhaps increase top-down regulation. Lower extinction risk might result from greater population buffering, due to more individuals (chapter 22 in this volume).

dominance in productive plant communities. Explaining such changes in community structure requires more sophistication than simply saying more productivity makes more species. This is an active area of research, and more sophisticated models are undoubtedly forthcoming. What about the ambient energy hypothesis? Species richness in cold or dry environments is constrained by the inability of many species to cope physiologically. Currie et al. (2004) found this hypothesis "not entirely consistent with observation, although further tests are needed."

Rohde (1992) suggested that warmer climates and higher productivity in the tropics compared to higher latitudes would increase mutation rates, which provide the raw material for speciation. In support of Rohde's hypothesis, Wright et al. (2006) showed that "tropical plant species had more than twice the rate of molecular evolution as closely related temperate congeners."

Although climate and its surrogate, primary productivity, clearly influence species richness, the mechanisms are unclear. Waide et al. (1999) conclude that "rather than any one mechanism having hegemony, it may be the cumulative or interactive effect of (many) factors that determines the empirical pattern within a particular study." Relationships may vary, depending on environment. For example (and not surprisingly), Hawkins et al. (2003a) found that for animals, primary productivity (as measured by water availability) was the best predictor of richness in the tropics and temperate zones, but heat was best at latitudes higher than about 50°. Later in the chapter we see that correlations of diversity with climate may be confounded by history, both of past land use and of extinctions due to natural causes (e.g., climate change).

10.2.2 History

Historical factors almost certainly account for some of the differences in diversity seen today, especially different rates of speciation or extinction and the legacies of extreme disturbances such as glaciation. The issue is, however, controversial (Turner and Hawkins 2004). Various people have argued that the high diversity of the tropics can be traced to greater speciation than in nontropical areas, and slow dispersal out of the tropics because few species can adapt to the more stressful temperate and boreal environments (Turner and Hawkins 2004; Wiens and Donoghue 2004). In the words of Wiens and Donoghue (2004), "Many species and clades[7] are specialized for tropical climates, and the adaptations necessary to invade and persist in regions that experience freezing temperatures have evolved in only some. Tropical niche conservatism has helped maintain the disparity in species richness over time." In contrast to the prevailing view that speciation rates are greater in the tropics than at higher latitudes, Weir and Schutler (2007) found that both speciation and extinction rates of New World birds and mammals increased with latitude. In other words, species turnover rates follow a latitudinal gradient from low to high as one goes from the tropics poleward, and the higher turnover rates result in lower diversity.

> One explanation for why more species may have evolved in the tropics is a simple matter of space. Through much of earth's history, land masses were concentrated in tropical regions. "If much of the world was tropical for a long period before the present, then (all other things being equal) more extant clades should have originated in the tropics than in temperate regions" (Wiens and Donoghue 2004). Another explanation involves a positive feedback loop. Recent research indicates that rates of speciation in both plants and arthropods correlate positively with diversity[8] (Emerson and Kolm 2005). As Emerson and Kolm put it, "The answer to questions such as why are there so many species in the tropics might in part be because there are so many species in the tropics."

The remainder of the chapter examines some of the more important factors that create diversity (both local and large-scale) in some detail, but first a caveat. It is important to remember that no single factor is likely to explain either global or local patterns of diversity. High productivity and stable climates may set the stage for diverse ecosystems, but what actually plays out on that stage depends on niches being created through factors such as environmental complexity, disturbance, and species interactions. For example, well over one-half of the variation in vascular plant richness across the globe is explained by the combined effects of high energy inputs (measured by potential evapotranspiration), water availability, and environmental heterogeneity (Kreft and Jetz 2007). Moreover, the various factors discussed above are in some cases highly interrelated and may act in concert to either reinforce one another's effects on diversity, or cancel them. For example, where environmental factors are conducive to high species diversity, interactions among the species themselves may act to reinforce and even magnify that diversity. On the other hand, environmental factors conducive to high diversity may be overridden by contingencies, such as management that channels energy into producing a single crop rather than a variety of niches.

[7]A clade is a group of organisms that share a common ancestor and that includes all decendents of that ancestor. Clades may or may not correspond to taxonomic groups such as species.

[8]This is a largely overlooked prediction of island biogeography theory.

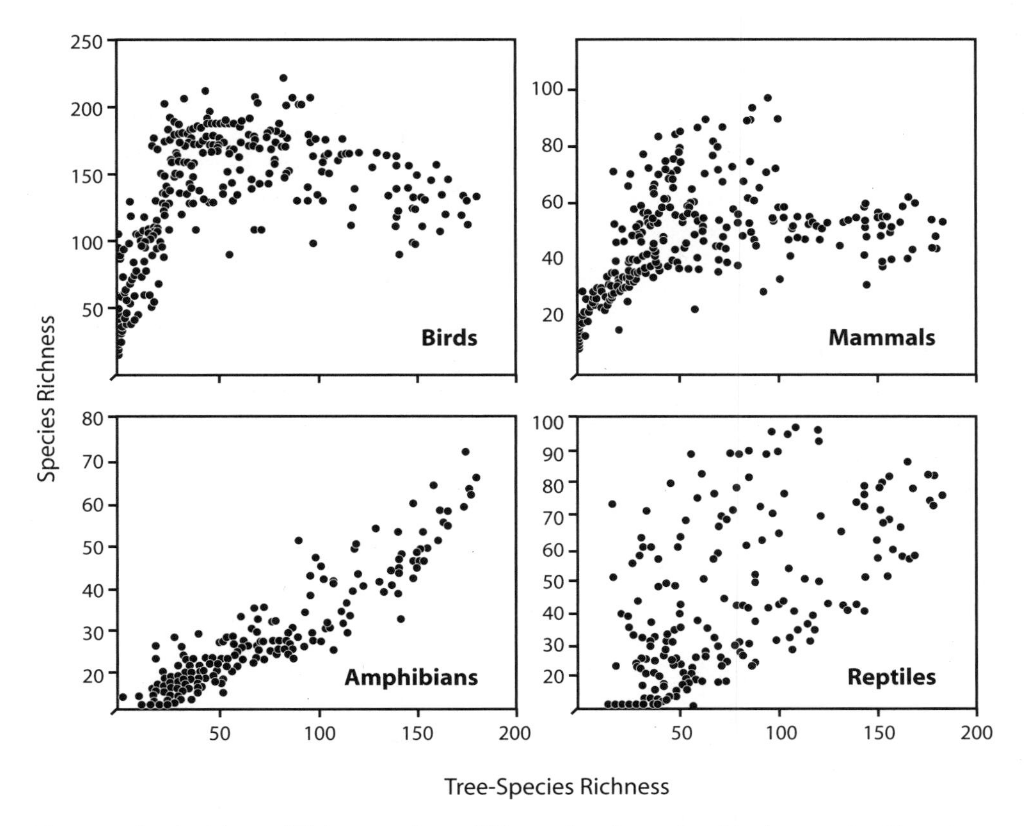

Figure 10.1. Species richness of birds, mammals, amphibians, and reptiles as a function of tree species richness in North America. Each point represents diversity within an area that may range from 60,000 to 125,000 km^2. (Adapted from Currie 1991)

10.3 RELATIONSHIPS BETWEEN FOREST STRUCTURE AND THE DIVERSITY OF ANIMALS AND MICROBES

The diversity of higher trophic levels is closely tied to the structure of forests and forested landscapes, a point of critical importance for forestry and conservation (chapters 19, 22, and 23). The most obvious element of forest structure is tree-species richness. Novotny et al. (2006) found little difference between temperate and tropical tree species in the numbers of folivorous insect species they support (respectively, 29.0 ± 2.2 and 23.5 ± 1.8 m^{-2}), suggesting the greater diversity of folivorous insects in the tropics stems from the greater tree richness there.

The relationship between numbers of vertebrate species and numbers of tree species depends on what scale one observes. As we saw in chapter 4, the sharp increase in tree species diversity as one goes from the high latitudes to the tropics is paralleled by an increase in vertebrate diversity, which is to say that numbers of tree species and numbers of vertebrate species correlate positively at the broad hemispheric scale (although correlation does not imply that one results from the other). At the continental scale, however, that is not necessarily the case. Currie (1991) divided North America, north of Mexico, into cells that ranged from about 60,000 to 125,000 km^2 in size, then studied factors associated with differences in species richness among the cells. Among the vertebrates, only numbers of amphibian species correlated strongly with tree-species richness (fig. 10.1).[9] Numbers of bird and mammal species did increase with numbers of tree species up to a point, after which adding more tree species made no difference. In Currie's study, diversity of terrestrial vertebrate species as a group (excluding bats) was most strongly associated with **potential evapotranspiration,** which is a measure of total solar energy received in an area: the more sunshine, the greater the numbers of vertebrate species. As we saw earlier, the positive correlation between vertebrate richness and some measure of solar energy input has been found by others.

Now let's step down in scale to a level that is more relevant to the individual resource manager: a local landscape (say, a 10,000- to 20,000-ha watershed) and the individual communities it comprises. And let's also broaden our view of structure to include not only plant-species richness, but stand-level factors such as canopy layers (which may or may

[9]It is probably misleading to compare animal diversity on a broad regional scale only to tree-species richness. For example, the center of vertebrate diversity in North America north of Mexico is the southwestern United States (Currie 1991), an area with relatively low numbers of tree species, but floristically rich deserts.

not be related to plant-species richness) and coarse woody debris. At this level, structural complexity of individual plant communities and diversity in landscape patterns translates into a complex of niches that enhance the diversity of animals and microbes.[10]

In order to understand the relationship between forest structure and animal (as well as microbial) diversity, we must ask how structure translates into niches. Much of the animal diversity in forests arises from the simple fact that trees are tall and, hence, have much more three-dimensional structure than does grassland. Vertical space within forests provides an array of feeding niches for animals. Harrison (1962) distinguished six mammal and bird communities in tropical lowland rain forests based on the canopy level and foods that they use; as Whitmore and Burnham (1975) describe, they are the following:

1. Above the canopy: insectivorous and carnivorous birds and bats
2. Top of canopy: birds and mammals feeding largely on leaves and fruits and to a minor extent on nectar and insects also
3. Middle-of-canopy, flying animals: mainly insectivorous birds and bats
4. Middle-of-canopy, nonflying animals: mixed-feeding animals, which range up and down tree trunks from crown to ground; also a few carnivores
5. Large ground animals: herbivores and attendant carnivores
6. Small ground or undergrowth animals: mammals and birds of varied diets taken from the forest floor, predominantly insectivorous or mixed feeders; plus some herbivores and carnivores

One need not look to complex tropical forests to find such patterns. In their study of the structurally rather simple oak woodlands of California and Mexico, Landres and MacMahon (1983) identified the following guilds of insect-eating birds:[11] (1) species that primarily gleaned foliage; (2) species that fed primarily from bark, either gleaning the surface or probing beneath (e.g., woodpeckers); (3) species feeding primarily on flying insects; and (4) species feeding primarily on ground-dwelling insects. Lindenmayer and Beaton (2001) describe how different species of possums and gliders[12] partition vertical space in Australia's Mountain Ash forests (the tallest flowering plants in the world): "The different vegetation layers in Mountain Ash forests have contributed to the evolution of various foraging patterns. . . . The Greater Glider is a canopy leaf-feeding specialist. The Yellow-bellied Glider often feeds on eucalypt sap high above the forest floor but below the canopy. The Common Ringtail Possum, Sugar Glider and Leadbetter's Possum are typically seen feeding in the understory or low shrub layer while the Mountain Brushtail Possum feeds both in the understory and on the forest floor." Figure 10.2 shows how various mammals and birds partition vertical space in the forests of Oregon's Coast Range.

Bird species richness has been shown to correlate positively with some measure of foliage height diversity in a variety of forest types: deciduous forests of eastern North America, mixed pine-hardwood woodlands in the southwestern United States, and Australian forests dominated by various sclerophylous species (MacArthur 1964; Balda 1969; Recher 1969) (fig. 10.3). This is not always the case, however, and for some birds, multiple canopy layers are a negative factor. One such species is the endangered red-cockaded woodpecker of the southern United States, which prefers old-growth pine stands with no hardwood understory (Lande 1988a). One must always be wary of generalizations regarding patterns in nature. The suite of species native to a given bioregion have presumably evolved to fit the habitat template (forest and landscape structure) that history has provided, and that usually differs in at least some respects from one forest type to another (Hansen et al. 1991).

Gaps created by the death of one to several canopy trees are a structural feature common to older forests of virtually every type (see section 10.4.5). Embedded as small, successional islands within a matrix of closed forest, gaps produce spatial heterogeneity at what might be considered a sublandscape scale. The resulting diversification of forest structure translates into diverse niches. Crome and Richards (1988) found the community of insectivorous bats within an Australian rain forest to consist of gap specialists, closed-forest specialists, and a third group that utilized both gaps and closed forest. In coniferous forests, gaps are often occupied by flowering plants (hardwood trees, shrubs, herbs, and/or grasses), which provide unique and important habitat. The majority of lepidopteran species (moths and butterflies) in both the eastern and western United States are associated with hardwood trees, shrubs, herbs, and grasses (Hammond and Miller 1998), while the abundance and diversity of birds within conifer-dominated forests correlates positively with the presence of hardwoods (reviewed by Hayes and Hagar 2002). In an earlier chapter, we saw that some lichen species in the northwestern United States also occur primarily on hardwoods growing within conifer-dominated forests. The same is true in spruce forests of central Norway, where Rolstad et al. (2001) found that "all (lichen) species used spruce trees as their main substrate, but *L. pulmonaria, L. scrobiculata,* and *Nephroma* spp. preferred scattered deciduous trees." As elsewhere, Rolstad et al. also found a preference among some lichens for older trees.

[10]See Southwood's (1988) discussion of the habitat template, and Ricklefs (1987) analysis of the interaction between local and regional diversity.

[11]*Guild* is used here to denote a set of species that have the same (or a similar) feeding niche.

[12]Possums and gliders are small marsupial mammals native to Australian eucalyptus forests.

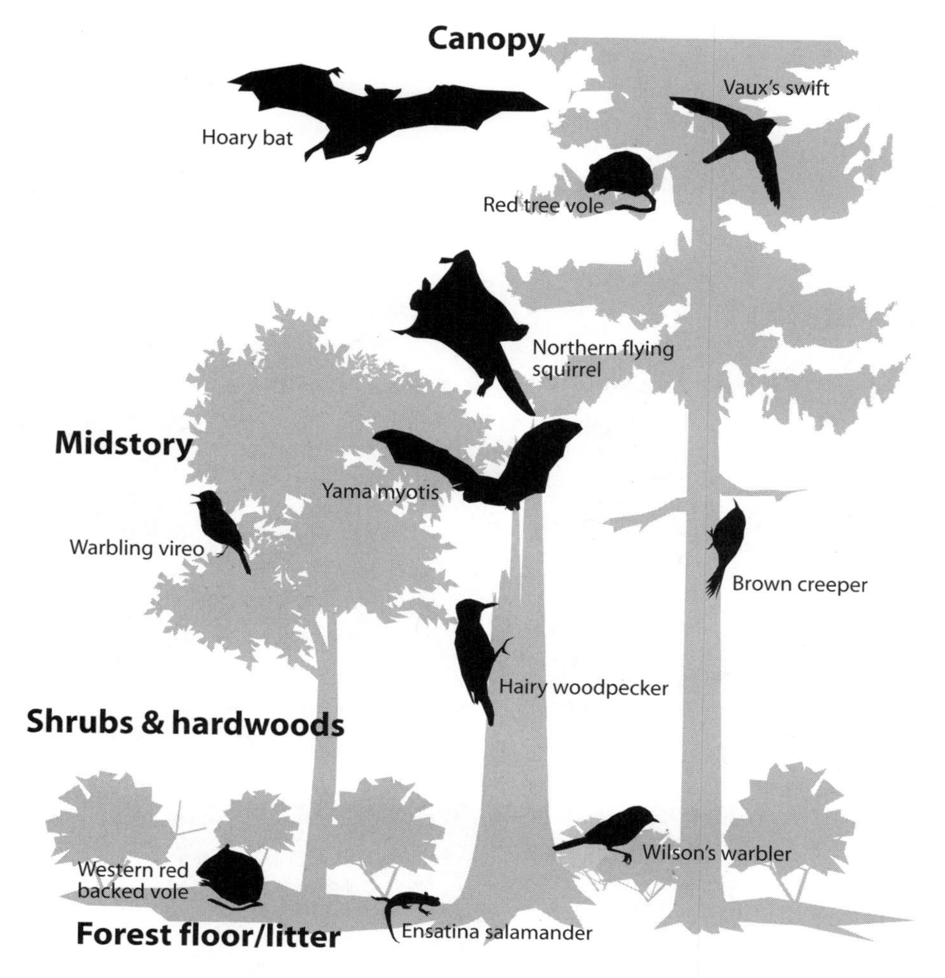

Figure 10.2. How various mammals and birds partition vertical space in the Oregon Coast Range. (Reprinted with permission from Hayes and Hagar 2002, published by Oregon State University Press. Copyright OSU Press 2002)

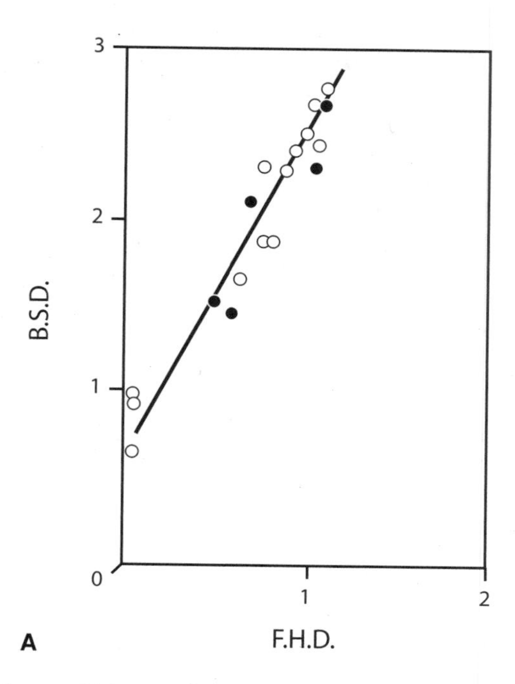

Figure 10.3. A. The relation between bird species diversity (BSD) and foliage height diversity (FHD) in Australia *(filled circles)* and North America *(open circles)*. The values shown for BSD and FHD are indices that account for both species richness and evenness (i.e., they are not simple counts of species numbers). MacArthur and MacArthur (1961) describe the method of calculation. (Recher 1969) *(Continued)*

Figure 10.3. *(Continued)* B. Vertical canopy structure in a tropical rain forest in Tanzania: vertical heterogeneity in many tropical and some temperate forests (e.g., northern hardwoods) results from the growth of young trees into gaps that are created by the death of one to a few dominant trees. (Courtesy of Jumanne Magembe) C. Vertical heterogeneity in this Oregon forest was produced by a wildfire that spared many overstory dominants, effectively thinning the stand so that understory tree and shrubs had sufficient light to grow. Fire-induced structural heterogeneity, such as that shown here, provides habitat for a diversity of bird species, including some that were formerly thought to be restricted to old-growth forests. *(Continued)*

Figure 10.3. *(Continued)* D. Vertical heterogeneity does not characterize all forests. Dry forest types such as this pinyon-juniper forest in Arizona may have little or none. E. Riparian forests such as this one in Montana provide critically important habitat for terrestrial animals, as well as greatly improving the quality of fish habitat by shading streams, stabilizing banks, and serving as a source of dead logs that fall into the stream and create pools. In the Blue Mountains of Oregon and Washington, "wildlife use riparian zones disproportionately more than any other type of habitat" (Thomas et al. 1979). *(Continued)*

Intuitively, one would expect nature to comprise many more niches for small creatures than for large: an elephant is likely to take little note of the fine-scale structure of an acacia crown, but insects and microbes may find a rich variety of habitats there. Unfortunately, the "little things that run the world," as Wilson (1987) aptly describes them, have received scant attention from ecologists until quite recently. The few existing studies show (not surprisingly) that forest structure has a profound influence on invertebrate diversity. Species assemblages vary among forest types, among different tree species within a given forest, and in some cases among different positions within a single tree. Schowalter and Ganio (1998) found that canopy insect communities differed significantly between different tree species within an old-growth forest in

Figure 10.3. *(Continued)* F. Epiphytes (plants that grow on tree branches) are an important component of animal habitat in moist tropical forests. This bromeliad (member of the genus *Bromelia*, part of the pineapple family) is growing in Reserva Poco Das Antos (Brazil), home to one of the world's most endangered primates, the golden lion tamarin. Insects are attracted by water that collects within the cuplike structure created by the bromeliad foliage. The insects attract birds, and bird's eggs form part of the diet of tamarins. Some biologists speculate that tamarins cannot persist in forests that do not have enough bromeliads.

Washington State. Assemblages of beetle species differ between plantations of different conifer species in Great Britain (Jukes et al. 2002), and among forest types in the rain forests of South America (Erwin 1988). Of a total of 1,080 beetle species detected by Erwin (1988) in four different forest types of Peru and Brazil, 900 occurred in only one type, and only 10 were found in all. In the Peruvian forest, a random sample of 1,000 individual beetles contained on the average 300 different species, and two forests separated by less than 50 m had only 9 percent of their total number of beetle species in common. In other words, both alpha and beta diversity of beetles were high in the forests studied by Erwin. The implication is that diversity is magnified significantly as one goes from trees to insects.

As is the case with vertebrates, and probably even more so, little things divide up vertical space within forests. Ectomycorrhizal fungi differ among soil layers; in a red pine plantation, the F layer, the H layer, and the B horizon[13] each supported distinct ectomycorrhizal fungi communities (Dickie et al. 2002). Both butterfly communities in neotropical rain forests and lichens in temperate forests exhibit vertical beta diversity (McCune et al. 1997; Walla et al. 2004). Canopy insect communities sometimes divide up vertical space in forests (Su and Woods 2001) and sometimes do not (Schowalter and Ganio 1998). (It is not surprising that different studies find different patterns. Nature is dauntingly complex and highly variable in space and time. Research has only scratched the surface, particularly with regard to little things.)

In chapter 8 we saw that that animal diversity changes with successional stage. In both the conifer forests of western Oregon and the deciduous forests of western North Carolina sucking herbivores such as aphids dominate young forests (see chapter 19 for the difference between sucking and chewing insects), but arthropod biomass of old forests is evenly split between defoliators and their predators (ants, mites, spiders, parasitic wasps) (Schowalter 1989). Not only are there more predators in old than in young forests, species richness and evenness of predatory insects are much higher. The pattern for Oregon forests is shown in tables 10.1 and 10.2. Too little is known about the forces that shape ecosystem structure to explain with certainty why such striking differences should occur in the arthropod communities of young and old forests, but it seems likely that at least part of the explanation lies in the greater number of niches resulting from the relatively high compositional and architectural diversity of old compared to young forests.

In forests, death and decay create life. Snags, logs, and living trees with decay provide important habitats for terrestrial and aquatic animals and for microbes.[14] These serve four general functions in forests (fig. 10.4):

1. They are the base of a terrestrial food chain that includes microbes, invertebrates, small mammals, and birds.
2. Snags and living trees with heart rot provide nest sites for cavity-nesting birds and dens for mammals. For example, 39 bird and 23 mammal species use snags for nesting or shelter in the Blue Mountains of Oregon and Washington (Thomas et al. 1979) (fig. 10.5). In Australia's mountain ash forests, a variety of arboreal marsupials require large trees with cavities for denning (Lindenmayer et al. 1993). Logs provide cover and even some nest sites and are especially important for a wide variety of invertebrates and microbes (including some species of mycorrhizal fungi).
3. Older, decayed logs serve as water reservoirs in forests with seasonal drought.

[13]Definitions of the various layers are given in chapter 14.

[14]Discussions of the habitat values of snags and logs include those by Thomas et al. (1979), Maser et al. (1979, 1988a, 1988b), McComb and Muller (1983), McComb et al. (1986a, b), Harmon et al. (1986), and Rosenberg et al. (1988). Additional articles are cited in these.

Table 10.1

Numbers of canopy arthropods per kilogram of foliage in old-growth and young conifer stands on the N. J. Andrews Experimental Forest, Oregon

		Old growth	
Anthropod group	Young Douglas-fir*	Douglas-fir	Western hemlock
Folivores			
Woolly aphids (*Adelges*, 2 spp.)	23,000 (26,000)	48 (57)	39 (40)
Aphids (*Cinara*)	100 (300)	0.4 (1.6)	0.0
Scales (4 spp.)	2 (8)	51 (41)	110 (100)
Budmoth (*Zeiraphera*)	0	2.8 (6.0)	0.0
Other defoliators (5 spp.)	0	0.9 (1.9)	0.9 (1.9)
Flower and seed predators			
Thrips (2 spp.)	0	3.0 (3.8)	0.7 (1.8)
Seed bugs (*Kleidocerys*)	0	1.1 (2.3)	17 (27)
Other seed predators (4 spp.)	0	0.4 (1.6)	0.9 (1.5)
Predators			
Ants (*Camponotus*, 2 spp.)	2 (8)	0.7 (1.8)	0.2 (0.8)
Aphid predators (3 spp.)	7 (14)	1.5 (3.8)	0.6 (1.3)
Parasitic flies	0	0.2 (0.8)	0.0
Parasitic wasps (>6 spp.)	6 (13)	1.5 (2.9)	1.7 (3.5)
Predaceous (*Hemiptera*, 3 spp.)	4 (11)	2.4 (4.7)	0.2 (0.8)
Neuroptera (*Semidalis*)	0	1.3 (2.6)	0.7 (1.8)
Predaceous mites (>3 spp.)	0	3.9 (5.7)	6.5 (6.4)
Spiders			
Metaphiddipus	6 (17)	1.3 (2.0)	3.3 (2.6)
Apollophanes	0	1.7 (3.3)	2.8 (3.7)
Philodromus (3 spp.)	0	1.7 (2.6)	1.3 (1.7)
Anyphaena	0	3.1 (3.9)	2.4 (2.8)
Other species (15)	6 (13)	2.8 (4.0)	5.7 (4.5)
Detritivores			
Springtails (*Collembola*)	0	0.7 (1.8)	0.2 (0.8)
Bark lice (*Psocoptera*)	0	0.6 (1.3)	17 (22)
Camisia carrolli	0	12 (16)	33 (22)
Other mites (5 spp.)	0	7.8 (8.9)	19 (14)
Migrant arthropods (6 spp.)	4 (11)	0.6 (1.3)	0.7 (1.4)

Source: Schowalter 1989.
Note: Values are expressed as means, with standard deviations in parentheses.
*Young stands are plantations averaging approximately 10 years of age.

Table 10.2

Biomass of canopy arthropods in old-growth and young conifer stands on the H. J. Andrews Experimental Forest, Oregon*

	Biomass (g/ha)	
Functional group	Old-growth forest	Young forest
Defoliating herbivores	180	0
Sucking herbivores	10	370
Predators	160	50
Others	30	0
Total	380	420

Source: Schowalter 1989.
*Young stands are plantations averaging approximately 10 years of age.

4. Logs dam streams, creating pools that are important fish habitat and protecting stream banks by reducing the erosive force of water; these are particularly important functions in mountain streams.

Different species of birds and mammals vary widely in the types of snags that they find suitable for foraging, nesting or dens. Snag size (height and diameter), degree of decay, and setting (e.g., within old forest or the shrub-forb stage) influence snag use. Studies in both western conifer and eastern deciduous forests of North America show that birds prefer snags greater than 20 cm DBH for foraging, though smaller snags are also used. According to Thomas et al. (1979), "Each cavity nesting species exhibits a decided preference for a specific height at which to build its nest." Bats prefer snags with loose bark still attached (they sleep beneath the bark). In eastern Oregon and Washington, bluebirds and wrens prefer to nest in snags standing within early successional grass-shrub-sapling stages, while the pileated woodpecker avoids these and chooses snags within closed forest (Thomas et al. 1979). (Note that these birds have evolved to segregate themselves along a niche dimension for nest-site preference).

A

Form of life	Uses of snags	Examples
Fungi, mosses, and lichens	Decayed wood serves as a growth substrate.	Fungus Moss Lichen
Invertebrates	Spaces under bark serve as cover and as places for feeding.	Pseudoscorpion Moth Beetle Ant
Birds	Cavities are used for nesting or roosting. Snags are used as perches and to support nests.	Flicker Nuthatch Pileated
Mammals	Cavities serve as dens of as resting or escape cover. Areas under loose bark are used by bats for roosting.	Bat Flying squirrel Marten

B

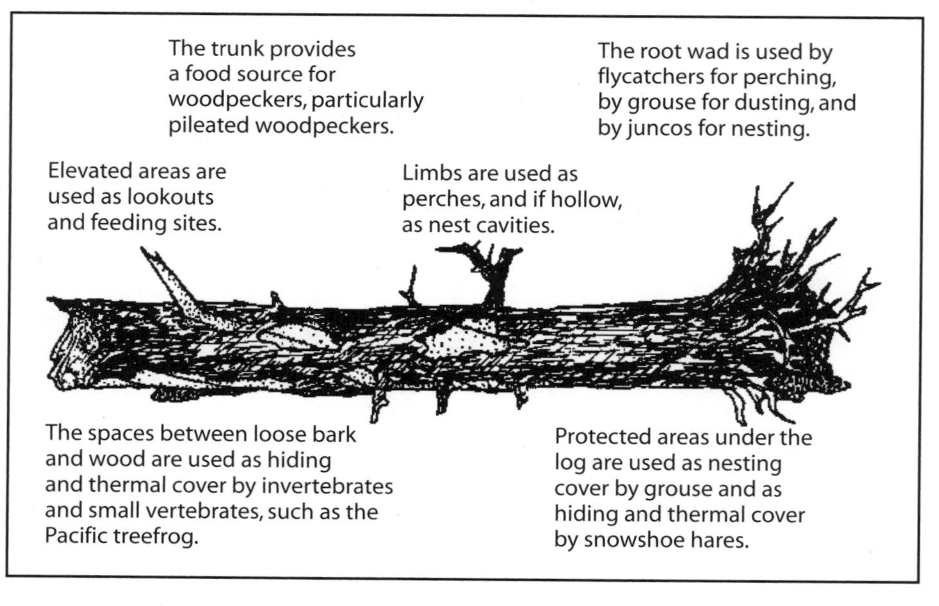

Figure 10.4. Some uses of snags and logs in forests. A. Snags. B. Logs. (Adapted from Thomas et al. 1979 and Maser et al. 1979)

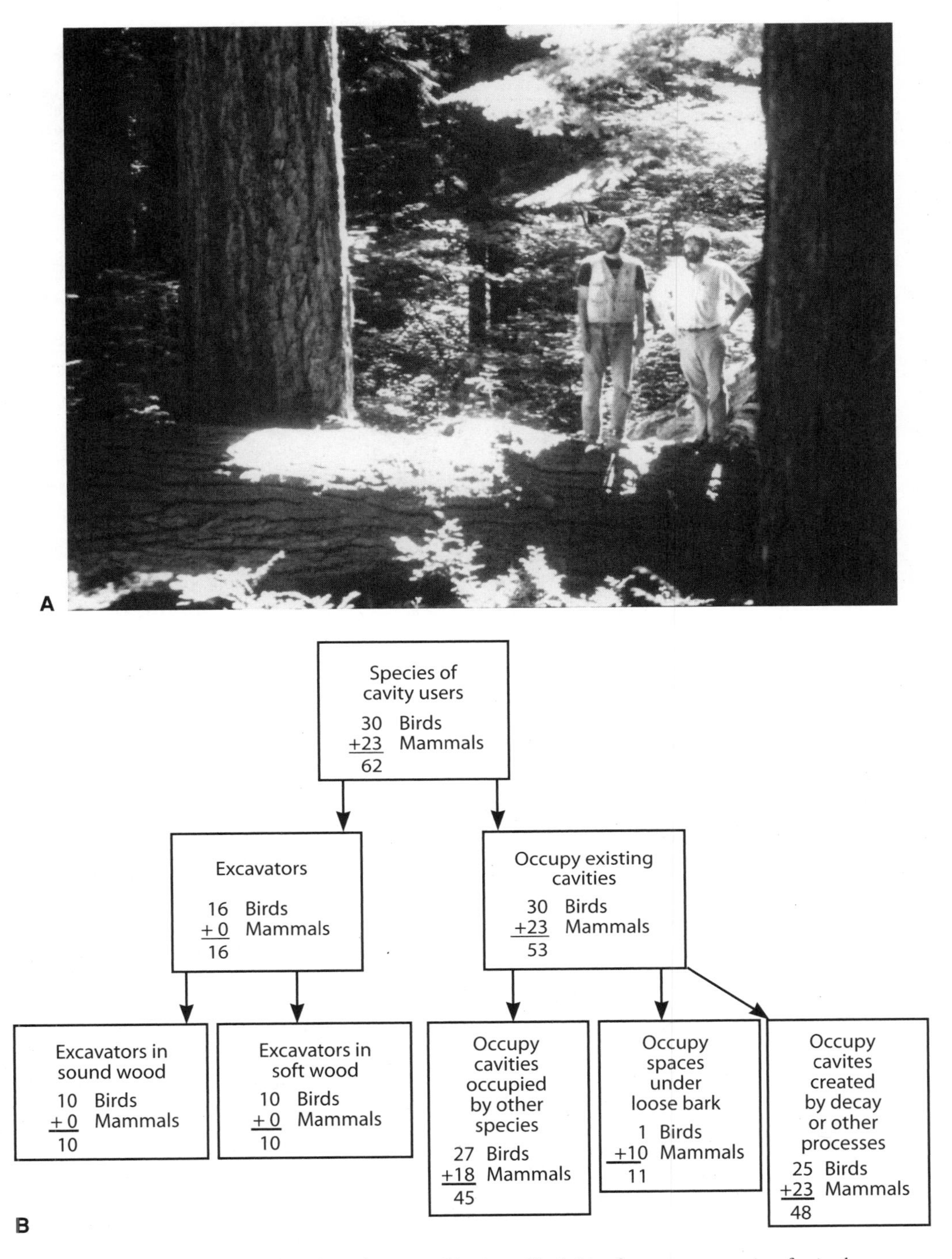

Figure 10.5. A. The large amount of dead wood (snags and logs) provides habitat for numerous species of animals and microbes. This downed log is in the Wind River Experimental Forest, Washington. B. Numbers of animal species that use snags in the Blue Mountains (eastern Oregon and Washington). (Thomas et al. 1979)

Woodpeckers, when present, are the primary excavators of cavities that may then serve as nests and dens for a succession of other species; although, if the snag is sufficiently soft, other birds such as chickadees may also excavate (Thomas et al. 1979). Some woodpeckers, such as the pileated woodpecker, can excavate relatively hard snags, but most primary excavators require soft snags. At least some degree of softening of the wood by heart-rot fungi is believed to be an essential prerequisite for any primary excavator.

On land, logs serve as hiding cover for small mammals and meet the total habitat needs of some species of amphibians, insects, and microbes. The Oregon slender salamander *(Batrachoseps wrightii)*, for example, makes its home within old, decaying logs (Carey 1989). A rich array of invertebrates and fungi depend on dead wood (these are termed **saproxylic**). According to Grove (2002), "The saproxylic habit includes representatives from all major insect orders (especially beetles and flies), and accounts for a large proportion of the insect fauna in any natural forest." The following sample of the 20 studies summarized by Grove (2002) illustrates the truly astounding species diversity associated with dead wood (see Grove for original citations): 339 saproxylic beetle species were recorded from a Queensland (Australia) rain forest; approximately 700 species of beetles collected in north Sulawesi (Indonesia) fed on either dead wood or fungi associated with dead wood; in Finland, 20 to 25 percent of all forest-dwelling species were saproxylic, including 800 beetle species (287 saproxylic beetle species were recorded from a single Finnish forest); in Canada, 257 species of saproxylic beetles were recorded from dead aspens in one forest type; in Washington State, more than 300 species of saproxylic beetles were recorded in Douglas-fir forests.

The water-holding capacity of old logs allows them to support an active microbial community during dry periods, including nitrogen-fixing bacteria and mycorrhizal fungi. Logs that fall into streams create pools that serve as important fish habitat and also provide food for aquatic insects (Sedell et al. 1988). The biomass of coho salmon in streams of coastal Oregon is directly related to pool volume, which depends, in turn, on stable log dams; logs are also used as protective cover by young coho, yearling steelhead, and older cutthroat trout (Sedell et al. 1988), and the structure imposed on streams by logs allows a greater number of fish species to coexist. The importance of logs to fish diversity was inadvertently demonstrated in the state of Washington following passage of a law (in 1976) that required debris to be cleared from streams at the time of logging. This resulted in fewer, smaller pools and a decline in the richness of salmonid species associated with increased dominance by a single species: steelhead trout (Bisson and Sedell 1984). Along the Pacific coast, the habitats provided by logs extends to the estuaries, where they provide perches for birds and convenient rafts for harbor seals; for reasons that are unclear, some fish of the open ocean, particularly tuna, congregate around floating logs (Gonor et al. 1988).

Snags are common in older unmanaged forests, where at least some trees die standing up, but snags are not unique to old growth. Unless salvaged, young forests established by fires have an abundance of snags and logs that are legacies from the previous forest. Figure 10.6 shows the pattern of accumulation of coarse woody debris in Douglas-fir stands. The rate at which logs decay, hence their abundance, varies widely among different forest types. Except for those in semiarid environments (e.g., ponderosa pine), conifer forests tend to have abundant logs in various stages of decay, while logs are less common in temperate deciduous and tropical forests.

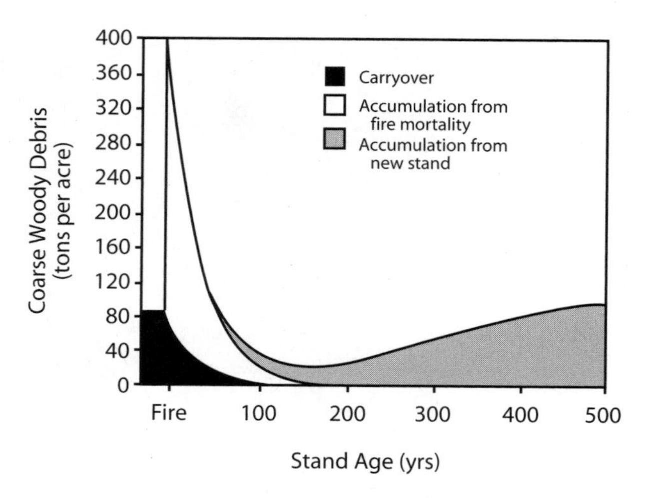

Figure 10.6. Patterns of accumulation of coarse woody debris after a stand-destroying wildfire in old-growth Douglas-fir. (Spies and Cline 1988)

10.3.1 Local Diversity Depends on Landscape and Regional Context

The diversity of a given locale depends not only on the habitats it provides, but on the landscape and regional context within which it is set. A homogeneous forest may support a diverse fauna if it is set within a heterogeneous landscape; and vice versa, animal diversity may be relatively low in a heterogeneous forest that exists as an isolated island within a sea of farms, suburbs, or forest monocultures. In general, local diversity is linearly related to regional diversity, a phenomenon that Lawton (1999) recognizes as one of the laws of macroecology.[15]

Landscape and regional patterns influence local diversity in various ways:

- Providing sources of migrants from the metacommunity (chapter 8) and, for local populations, genes from the metapopulation (chapter 22)
- Providing special habitats such as riparian zones and wetlands (chapter 22)
- Influencing species interactions that influence diversity, such as top-down regulation (chapter 19) and, for plants, pollination and seed-dispersal mutualisms (chapter 11)
- Influencing the frequency and severity of disturbances (chapter 7)
- Providing the template of habitat types and interconnections that determine the viability of populations and metapopulations

For a given species, heterogeneous landscapes may be either **fragmented** or **variegated.** Fragmented landscapes consist of islands of suitable habitat surrounded by nonhab-

[15]As the name suggests, *macroecology* is the ecology of large spatial scales and long time periods.

itat, whereas variegated landscapes consist of patch types that provide more or less of a continuum of suitable habitat (Hayes and Hagar 2002). The difference between variegated and fragmented landscapes depends on the species under consideration, as suitable habitat for one species may be unsuitable for another. As we discuss in more detail in chapter 22, forests that exist as fragments within a sea of farms or suburbs experience a variety of threats to deep-forest species. For example, birds that nest on or near the ground become vulnerable to dogs and cats, and nest parasites such as cowbirds[16] become more abundant (Wilcove 1985; Freemark and Collins 1992; Gates and Evans 1998; Donovan and Flather 2002). Permeation of forests with roads makes numerous deep-forest species more accessible to parasites and predators (including humans).

Early studies of how landscape patterns influence local diversity were focused on forest fragments within seas of farms or suburbs; however, research since the mid-1990s has turned increasingly to heterogeneous forested landscapes, where heterogeneity arises from varied successional stages and/or forest types, and usually includes areas of managed forest (e.g., McGarigal and McComb 1995; Rodewald and Yahner 2001; Hayes and Hagar 2002; Lichstein et al. 2002; MacFaden and Capen 2002; Loehle et al. 2005; Mitchell et al. 2005; and others cited in these articles). Several general patterns have emerged for birds (by far, the most-studied group):

1. Landscape patterns are often significant in determining the richness of bird species, but they generally account for less than 50 percent of total variability for a given species or guild (i.e., local habitat features are more important).
2. Total available habitat is more important than how the habitat is arranged.
3. Many species and guilds see the landscape as variegated rather than fragmented.
4. Many species, even some associated primarily with mature forests, respond positively to heterogeneity of stand ages and types.
5. Species and guilds differ in the landscape scale to which they respond.
6. A given species may respond to landscape patterns at more than one scale (e.g., within a 100-m radius and within a 1,000-m radius).
7. Nest parasites and predators are less of a problem in a managed forest matrix than in an agricultural one.

Various caveats go with these. The first and most obvious is that birds probably do not tell a lot about less-mobile organisms. Poor dispersers with specialized needs, such as flightless beetles, salamanders, some lichens and fungi, and vernal herbs, are of particular concern. Second, studies have been based on point counts,[17] which are widely used but have several weaknesses, including poor detection of rare species.[18] Third, studies deal with existing landscape patterns, which in many areas are so highly altered that sensitive species may have already disappeared or be too rare to detect in point counts. Such studies should be extrapolated with extreme care into more-pristine regions.

Table 10.3
A summary of the forces that create and maintain diversity in the tree layer

1. Fine-scale habitat diversity (i.e., a variety of niches) or broad transitions (ecotones) between coarse-scale habitats that tend to equalize the growth potential of different tree species.
2. Balanced resource limitations, which also equalize growth potential.
3. Biotic interactions of two general types: (1) *density-dependent* controls (i.e., factors—usually biotic (pathogens, herbivores)—that regulate numbers of a given species but do not exclude the species altogether); (2) mutualisms and other cooperative interactions that confer particular advantages on a species that would otherwise be at a competitive disadvantage. Mycorrhizal fungi are one example of a mutualism that allows particular plant species to persist in a community.
4. Disturbances of low to moderate intensity (including gap creation by death of one or a few trees). Disturbance is a strong diversifying force when it reduces the ability of one or a few species to dominate, and the disturbance is not so severe that it homogenizes landscapes rather than diversifying them.
5. Historic and regional context. History is the container for critical processes that influence regional diversity, including speciation, extinction, migrations, and disturbances. Local diversity depends, at least in part, on regional diversity. Historic disturbances such as glaciation or land use may leave legacies that influence patterns of diversity for centuries.
6. Stochastic factors and dispersal (dispersal assembly, or neutral theory). Composition is determined primarily by chance and dispersal, with niches playing a secondary role. Plant communities in which a sufficient number of species have limited dispersal capabilities will tend toward high beta diversity and low alpha diversity, while communities in which wide dispersal is common tend toward the opposite.

Note: Order does not indicate importance.

10.4 FORCES PRODUCING DIVERSITY IN TREES AND OTHER FOREST PLANTS

As we saw in the previous chapter, dominance mitigates against diversity, and the forces that create and maintain diverse forests are those that in some fashion reduce the ability of one or a few species to dominate the capture of resources (especially light). These diversifying forces can be grouped into six general categories shown in table 10.3. The remainder of the section considers these factors

[16]Brown-headed cowbirds lay eggs within the nests of other species. The nest parents feed the cowbird young, which then push the natural young out of the nest.

[17]Counting individuals detected by sight or sound at a series of points.

[18]See Ralph et al. 1995, available at no charge from PSW Distribution Center, 3825 East Mulberry, Fort Collins, CO 80524.

separately, but it must be remembered that they do not act separately, nor are they mutually exclusive. The diversity of any forest is likely to result from various of these factors.

10.4.1 Fine-Scale Niches

Heterogeneity on the scale of a few meters to a few tens of meters can influence regeneration patterns beneath intact canopies, thereby shaping the species composition of the future forest (Grubb 1977). Such heterogeneity may come from several sources, including small-scale topoedaphic variability, gaps created by the death of one or a few canopy trees, and decaying logs or other imprints of the forest.

10.4.1.1 Topoedaphic Factors

Both water and nutrients can influence plant community composition. Water appears to be particularly important, especially at local scales, and patterns of species occurrence reflect to some degree the underlying topoedaphic complexity and its effects on the channeling and storage of water and on evaporative demand. Riparian zones often support unique species assemblages in all forest types, and, as we saw in chapter 5, aspect differences can be a significant source of beta diversity in tree composition. Even in areas with high precipitation, the water-holding capacity of different soil types may influence which tree species occupy a site. In Malaysian rain forests, for example, tree-species composition varies between soils of low and moderate fertility, a phenomenon Palmiotto et al. (2004) showed to be related to soil moisture rather than nutrients. Sollins (1998) also found little evidence for a connection between tree species and soil nutrients in lowland tropical rain forest. We see in chapter 15, however, that nutrient availability can influence the ability of trees to gather water.

Reflecting the general debate about niches versus neutrality, ecologists disagree about the importance of niche differentiation in shaping diversity in tropical forests (this argument does not seem to play out to the same degree in temperate and boreal forests). In his review, Wright (2002) concluded that there is ample evidence for niche differences in tropical forests associated with microtopography, which causes "drainage, moisture, and possibly nutrients to vary from ridges to slopes and nearby streams often over just ten's of meters." However, commenting on a paper by Pitman et al. (1999) that shows many tropical tree species in Peru to be sparsely distributed as individuals, Rickleffs (2000) concluded that "the results are inconsistent with the conventional wisdom that tropical trees tend to be habitat specialists." The history of such polar controversies in ecology is that everyone is right in his or her own backyard. In other words, what one sees depends on where one looks. Some differences are related to the topoedaphic complexity of particular areas. Potts et al. (2004) compared tree communities of two areas in Malaysia that differed in their topographic complexity. Dispersal-assembly (neutral) theory did not fully explain patterns in either but performed especially poorly in the more heterogeneous environment. Moreover, the sheer numbers of species in tropical forests argue that both habitat specialists and habitat generalists exist, the latter being more prone to chance distribution. (The same is true for tree species in temperate and boreal forests.) Even where habitat specialization clearly occurs, forests commonly contain tree species that are habitat generalists. At the La Selva Biological Station in Costa Rica, Clark et al. (1999) estimate that a minimum of 30 percent of tree species are habitat specialists. On a 52-ha plot in Borneo, a much higher proportion of trees (73%) of tree species were aggregated on one of four soil types (Russo et al. 2005). In either case, there is leeway for generalists whose distributions are determined by dispersal, chance, or biotic interactions rather than by niche preferences.

10.4.1.2 Gaps, Logs, and Other Forest Imprints

Gaps created by the fall of one or a few canopy trees are major factors influencing the structure of tropical, temperate, and boreal forests.[19] (Good reviews of the very large literature on forest gaps include those by Brokaw and Busing [2000] and McCarthy [2001].) In the absence of canopy-opening disturbances such as crown fire or windthrow, most tree species require gaps in order to reach maturity; therefore, under such conditions the gap is very much the parent of the forest, one being shaped in the image of the other. Gaps are discussed in more detail later in the chapter.

Tree fall has effects other than creating gaps. In Alaskan forests, trees regenerate primarily on mounds of mineral soil that have been exposed by windthrow (Bormann 1989). Old, decaying logs serve as important regeneration substrates in various forest types, both in gaps and under closed canopies. The logs serve as water reservoirs during summer drought and perhaps as refugia from competition with other plants and from root pathogens (Harmon et al. 1986). Logs benefit small-seeded species to a greater extent than large-seeded ones (because the former are more susceptible to substrate drying) and, therefore, help shape species composition.

[19]Forests differ widely, however, in the dynamics of gap filling. Gray and Spies (1997) point out that whereas in the conifer forests of the Pacific Northwest "many gaps are devoid of tree saplings even 50 yr or more after gap formation . . . canopy closure may take as little as 2 yr in moist tropical forests . . . or 5 yr in temperate deciduous forests."

10.4.1.3 *Niches in Time*

Plants can also divide resources in time. Succession is the most obvious example, but differing growth rhythms may also allow some degree of resource sharing. A popular theory holds that diversity stems from demographic niches, or differences among trees in the rates of recruitment, growth, and mortality. In that view, trade-offs exist between strategies of fast growth and early mortality on the one hand, and slow growth and longevity on the other. However, in an extensive study of demography and species richness in tropical forests, Condit et al. (2006) concluded that tree richness could not be explained by demographic niches. They postulate tropical forest richness is better explained by the richness of regional source pools (which leaves the question of why regions differ in tree richness).

One version of resource sharing in time is called **storage theory,** which refers to the ability of a species to be "stored" until the environmental conditions are right for its growth (Kelly and Bowler 2002). Storage might be through buried seeds or long-lived individuals that can supply seed, and, in order to qualify as niche separation, the storage must lead to reduced competition with other species. A good example is the genus *Ceanothus*, which comprises several nitrogen-fixing shrubs native to western North America. *Ceanothus* seeds can be stored in the soil for centuries and are triggered to germinate by fire. The shrubs occupy sites in early succession (replenishing soil nitrogen in the bargain) and are eventually replaced by conifers. Another example was provided by Kelly and Bowler (2002), who studied patterns of tree establishment in a primary forest in Mexico. They concluded that different environmental conditions favored the recruitment of different tree species, thereby promoting coexistence.

10.4.2 Ecotones

Ecotones are areas in which the adaptations of individual species to environmental factors are relatively balanced, permitting coexistence. Patterns of species diversity in mountain ranges present some of the clearest examples. In chapter 5 we discussed the elevational distribution of tree species in two quite different environments, the Great Smoky Mountains of Tennessee and North Carolina, and the Santa Catalina Mountains of Arizona. In both, abundance of a given species tends to peak within a certain elevational range; however, the species is present over a much broader range. This pattern results in broad areas of overlap—hence diverse plant communities—in the midelevations. For example, both the highest and the lowest elevations in the Santa Catalina Mountains are dominated by a single tree species, while communities consisting of up to five species occur at midelevations. A given species occupies a broader elevational range in the Great Smokies than in the Santa Catalinas, probably because environmental gradients (especially precipitation) are less sharp in the more mesic environment of the former. The same general pattern, however, exists: the greatest species richness occurs in the midelevations.

It should not be construed from this that tree species diversity is always greatest at midelevations in a mountain range. Too many other factors come into play to permit such generalizations. With a few exceptions (e.g., in Sri Lanka), moist tropical forests are more diverse in the lowlands than on the slopes (Ashton 1988).

10.4.3 Productivity

Tree richness varies with primary productivity (and therefore climate and resource availability); however, the pattern is complex and must be generalized with caution (Waide et al. 1999; Mittelbach et al. 2001). Adams and Woodward (1989) showed that tree species richness within relatively large regions in the temperate zone (66,000–81,000 km^2) increased with net primary productivity (fig. 10.7). Currie and Paquin (1987) found essentially the same pattern in North America, where they showed that tree species richness within 60,000- to 125,000-km^2 regions correlated positively with actual evapotranspiration (i.e., numbers of tree species increased as climates got warmer and wetter). (Actual evapotranspiration integrates the amount of solar radiation received each year with the amount of precipitation,

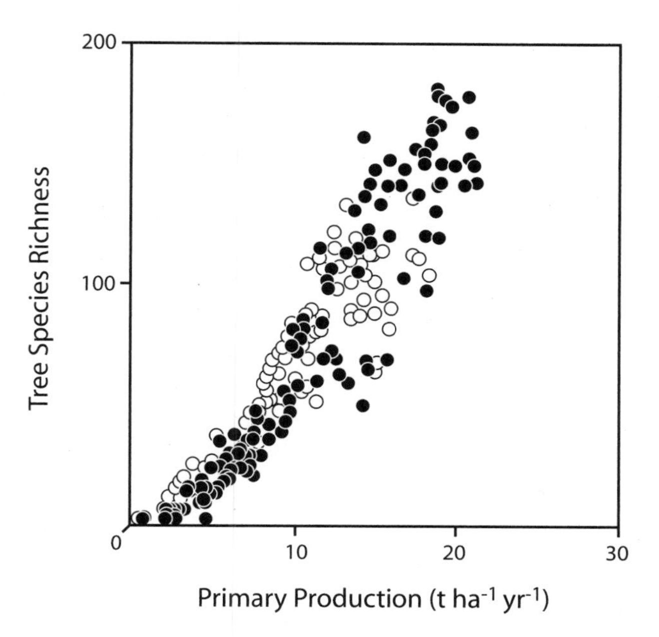

Figure 10.7. The relationship between tree species richness in Europe *(filled circles)* and North America *(open circles)* and net primary productivity. Individual points are based on regions varying from 66,000 to 81,000 km^2 in area. (Adapted from Adams and Woodward 1989)

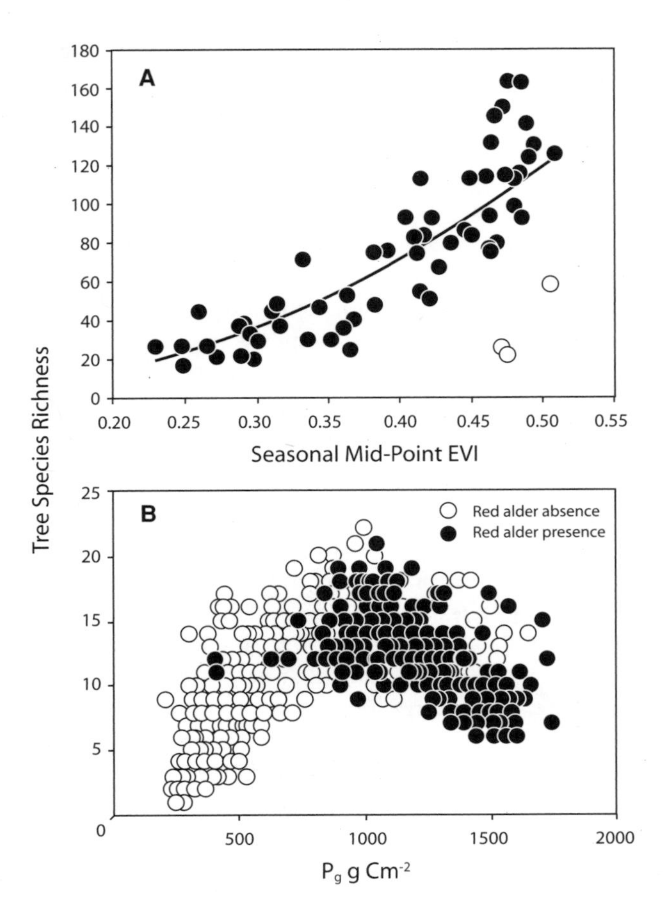

Figure 10.8. A. Relationship between tree species richness in the Pacific Northwest and the seasonal midpoint enhanced vegetation index (EVI). EVI provided the best fit to the data when three Pacific Northwest ecoregions *(open circles)* were excluded. The slope of the polynomial regression increases still more when the analysis is further restricted to those ecoregions with greater than 50% forest cover (r^2= 0.80). EVI is a measure of particular spectral reflections as sensed remotely. It is less sensitive to soil and atmospheric effects than the more widely used normalized difference vegetation index (NDVI) because it incorporates blue spectral wavelengths. As a result, EVI remains sensitive to increases in canopy density beyond where NDVI becomes saturated. (Adapted from Waring et al. 2006. Reprinted with permission from Elsevier. Copyright 2006) B. The relation between growing season (May–August) gross photosynthesis per square meter and tree-species richness in the northwestern United States. (Swenson and Waring 2006. With permission of the authors)

and is a reliable index of primary productivity.) In the United States, 75 percent of the variance in tree richness among ecoregions[20] is explained by the **enhanced vegetation index,** a satellite measurement that correlates closely with gross primary productivity (Waring et al. 2006) (fig. 10.8A).[21]

[20]Waring et al. (2006) used Level III Ecoregions as delineated by the U.S. Environmental Protection agency (see www.epa.gov/wed/pages/ecoregions.htm). There are 120 Level III Ecoregions in the continental United States, including Alaska.

[21]The 75 percent level of explaining power is achieved only when three of the ecoregions in the Pacific Northwest are excluded. These have only about one-half the number of tree species predicted by the enhanced vegetation index, probably because of extinctions during past climate changes (Waring et al. 2006).

Tropical forest diversity generally increases with precipitation. For example, Gentry (1982) found that the average number of tree species occupying 0.1-ha plots in the neotropics increased by 50 for each 100-cm increase in average yearly precipitation. This effect cannot be ascribed solely to precipitation, however, because soil nutrient availability within Central America generally declines as precipitation increases. Without experiments that manipulate one or more factors, natural patterns can rarely be related to variation in a single factor such as precipitation or soil nitrogen, because too many factors vary together.

The most comprehensive analysis to date was done by Francis and Currie (2003), who analyzed the relationship between climate (measured as water availability and heat) and the richness of angiosperm families across the globe. They summarized their significant findings as follows:

> Our most important result is that there is a globally consistent relationship between angiosperm richness and climate. Richness varies with climate within nearly every phytogeographic province and biome in very similar ways. Moreover, the richness of a given area can be predicted quite well using climate-richness models developed with data from other parts of the world, without the need to postulate other special circumstances for particular regions.
>
> Our second important result is that the form of the relationship between richness and heat depends on water availability, and the relationship between richness and water depends on heat. The best models accounting for richness patterns involve water deficit and either temperature or [potential evapotranspiration].
>
> Our third important result is that . . . the broadscale spatial relationships between richness and heat, and between richness and the climatic drivers of primary productivity, are not peaked. Richness increases monotonically with heat, provided that water is available. Similarly, on the global scale, richness increases as a positive monotonic function of the climate variables that control primary productivity (the simultaneous availability of water and heat).

The positive relationship between tree-species diversity and productivity (or its proxies, such as heat and water) does not hold up well when one compares local ecosystems within a given region. In many cases, local forest diversity is greatest where soil resources (either water or nutrients) are in intermediate supply. For example, two temperate regions with high forest diversity, the eucalyptus forests of Australia and the mixed conifer forests of western North America's Klamath Mountains, occur in moderately dry environments. Another center of temperate diversity, the Great Smoky Mountains, is relatively mesic, but the most diverse forests within these mountains occur in the midranges of the moisture gradient (Whittaker 1965). In the northwestern United

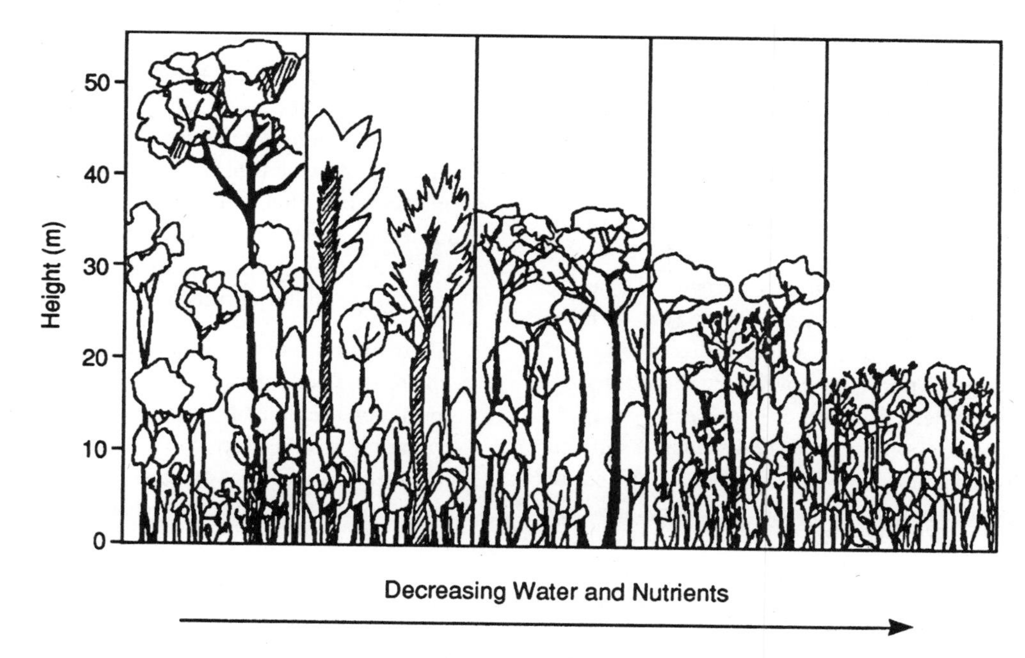

Figure 10.9. The relationship between resource availability (nutrients and water) and forest structure in Borneo. (Adapted from Brunig 1983)

States, the greatest tree richness occurs in areas of moderate productivity, especially where canopy closure does not exceed 70 percent (Waring et al. 2002; Swenson and Waring 2006) (fig. 10.8B). Speaking of Oregon (but probably applying equally well elsewhere), Waring et al. (2002) conclude that because of the strong dominance potential of a few fast-growing species, "some disturbance may be necessary to maintain high species richness on the most productive sites."

Tropical forests growing on soils with intermediate fertility tend to be more diverse than those occupying either nutrient-rich or nutrient-poor sites; however, one must look globally rather than regionally to discern this pattern. In Central America, where the young volcanic soils are more nutrient rich than in many other tropical areas, forest diversity tends to decrease with soil fertility (Huston 1980). Species richness declines with increasing availability of magnesium and phosphorus in Southeast Asian forests dominated by trees in the family Dipterocarpaceae, which includes several hundreds of species (Ashton 1988).

On the other hand, soils in the interior of South America, Africa, and parts of Southeast Asia tend to be very old, highly weathered, and moderately to highly infertile (chapter 14). Furthermore, low fertility in these soil types may be accompanied by relatively low soil water-holding capacity. In these regions, forest structure becomes markedly more simple as one moves across a gradient of decreasing fertility and water. This phenomenon is illustrated in figure 10.9 for forests of Borneo; similar patterns occur in Amazonia.

The peaked responses commonly seen at local and subregional scales (i.e., highest richness in the midranges of productivity) might result from parity in the availability of *different* resources, which allows species to compete on a more even basis. The role of resource parity in forest diversity is perhaps most obvious in the interplay between light and soil resources. Both the rate of canopy growth and the total leaf area supported on a given site are controlled by water and nutrient availability. Where water and nutrients are abundant, those species capable of the most rapid aboveground growth dominate light capture and, as a result, either exclude other species or inhibit their growth. Where soil resources are more limiting, this is less likely to occur; species that grow less rapidly aboveground but that are adept at gathering water or nutrients are on a more even footing, hence better able to maintain themselves within the community. These ideas have been developed in some detail by Huston (1979), Tilman (1988), and Smith and Huston (1989).

10.4.4 Animals and Microbes Enhance Tree Richness

Animals pollinate, disperse seeds, graze, and cycle nutrients; microbes also cycle nutrients and, in addition, may either help plants through mutualistic associations, or hinder plants through pathogenesis, allelopathy, or competition for nutrients. As long as these effects have some degree of selectivity, either because they are specific to certain plant species or are density dependent (e.g., acting more forcefully to limit a species whose numbers are high than one whose numbers are low), they will act to maintain plant diversity.

Suppose we could flick a switch, and all the herbivores would magically disappear from a given forest. Janzen (1983) speculates on the consequences for moist tropical rain forest:

> The absence of herbivores would lead to an immediate choking of the understory with seedlings from the many surviving seeds and from the enlarged seed crops. . . . Species composition of the understory should change, since some of the usual residents will not be able to cope with . . . competition from the sudden influx of invaders. . . . Species composition of the overstory should change rapidly as the numerical aspects of interspecific competition change; species whose seeds and seedlings suffered no mortality from animals will suddenly have to endure many competitive bouts with species that previously they only rarely encountered. . . . Species richness of trees at a site should decrease [because] the best competitors . . . will [no longer] be suffering from density-dependent disease and pest problems.

In chapter 11 we will see that Janzen's scenario proved accurate when herbivores were lost from newly created islands within a flooded landscape.

Marquis (2004) lists three ways in which top-down control by herbivores can influence plant diversity: (1) by "decimating plant populations, particularly at the seed and seedling stage"; (2) "by shifting competitive balance among plant species" (termed *density*, or *frequency dependence*); and (3) "by causing or reinforcing habitat specialization."

Janzen (1970) and Connell (1971, 1984) first suggested that pathogens and herbivores (primarily insects in forests) enhance tree diversity in tropical forests by inhibiting regeneration near the parent tree. This idea has come to be known as the **Janzen-Connell hypothesis,** and applies not only to parent trees and their seedlings, but to any concentration of individuals of the same species. Wright (2002) concluded that the "Janzen-Connell mechanism is likely to play an important role in the maintenance of plant diversity in tropical forests." Wright referred specifically to tropical forests, but recent research indicates the Janzen-Connell mechanism is also an important diversifying agent in temperate forests (e.g., Packer and Clay 2000; Lambers et al. 2002; also see below).

There seems little question that trees can hide from natural enemies in species-rich forests. One strong piece of evidence for this—and an important message for foresters—is that tree species occurring naturally as scattered individuals in tropical forest may experience considerable insect or pathogen problems when grown in plantations (Perry and Maghembe 1989). Herbivores and pathogens virtually always do best when their food plants are concentrated rather than dispersed (chapter 19). There are various reasons for this. The energy expended by herbivores in seeking out scattered individual food plants tends to negate the energy gain once an isolated host is found. Moreover the herbivore that is moving and searching may be more exposed to its own natural enemies than it is when not required to move about. Passively distributed pests and pathogens (e.g., spores of some fungal pathogens, some insects such as aphids and budworms) are likely to land on nonhost plants when dispersing within species-rich communities; hence, their survival rate is much lower than in communities where their host plants are concentrated.

For herbivores and pathogens to be effective diversifying agents, three conditions must be met:

1. There must be some degree of specificity between herbivores and pathogens on the one hand, and plant species on the other. If this were not so, herbivores and pathogens would affect all species equally and would not affect plant species diversity. Moreover there would be no enemy-free area to which a seedling could escape.
2. Populations of the pest or pathogen cannot be mobile enough to seek out and successfully attack escaping individuals.
3. The individual must escape its enemies without losing its friends. For example, plant species that depend on animals for pollination and/or seed dispersal must be able to be found by pollinators and seed dispersers. The majority of tree species that require mycorrhizal fungi to establish and grow must find the proper fungi in sites that are separated from others of the same tree species. (The terms *enemy* and *friend* should not be taken too literally; few interactions in nature are quite so black and white. Depending on conditions, enemies can also be friends, and friends can even be enemies on occasion.)

These conditions are frequently met. Janzen (1983) states, "One is struck by the extreme host-specificity of many of the insects that eat foliage in tropical forests. . . . At least half of the foliage-eating insects will have only one species of host plant and none will eat more than 10% of the species present (except leaf-cutter ants)."[22] Pathogens and insect herbivores have been shown to discriminate among different subgenera of *Eucalyptus* within a given forest (Burdon and Chilvers 1974). Six of seven temperate tree species studied by Lambers et al. (2002) experienced density-dependent mortality, most likely from species-specific seed predators or pathogens. Herbivores and pathogens that are not strictly species specific may nevertheless exhibit some degree of preference in the range of species they attack. Some root rots, which are common gap-creators in temperate forests, discriminate among tree species. Laminated root rot *(Phellinus weiiri)*, for example, readily kills Douglas-fir, attacks western hemlock less vigorously, and has no effect on maple and alder. At low to moderate levels, laminated root rot diversifies conifer forests by creating a niche for hardwoods.

---

[22]Leaf-cutter ants, which are common in moist tropical forests, gather foliage from a wide variety of tree species but do not consume the leaves directly. Rather, they are farmers who grow fungi on the leaves and graze the fungi.

How does a tree hide from its enemies but not its friends? At least part of the answer lies in the fact that plants evolve mechanisms to encourage friends on the one hand and to discourage enemies on the other. Fig trees, for example, which occur as widely scattered individuals in tropical forests, are thought by some researchers to emit a pheromone that attracts pollinating wasps (Janzen 1983). Pollen itself is a high-quality food, as are many fruits, while leaves frequently contain chemicals that make them difficult to digest or are outright toxic to herbivores (chapter 19). To survive on their hosts at all, herbivores and pathogens must manufacture detoxifying enzymes or utilize other energy-expensive ploys to defeat the plant defenses. In consequence, the leaf-eater gains far less nutrition per host plant than the pollinator or fruit-eater and, hence, cannot devote as much energy to the search. Old World monkeys provide a good example of how diet influences animal lifestyles: fruit-eaters such as macaques (which benefit plants by distributing seeds) range widely in search of nutritious food, while leaf-eaters such as colobines tend to be sluggish and rather sedentary (Altmann 1989).

Finally, predators and pathogens may cause or reinforce habitat specialization. This effect was demonstrated by Fine et al. (2004), who studied interactions among tree species, soil types, and insect herbivores in the Peruvian Amazon. In that area, relatively fertile clay soils and highly infertile white sands each support distinctive plant communities. Fine et al. reciprocally transplanted tree seedlings from both communities onto both soil types and either protected or did not protect them from herbivores. When protected, seedlings from clay-soil communities grew best on both soil types; however, when unprotected, seedlings from white-sand communities grew best on white sands and seedlings from clay-soil communities grew best on clay soils. Fine et al. conclude that "herbivores can be viewed as a diversifying force because they make existing abiotic gradients more divergent and cause finer-scale habitat specialization by magnifying the differences between habitats."

In chapter 11 we consider how mutualists (e.g., mycorrhizal fungi) might influence plant diversity.

10.4.5 Disturbance Maintains Diversity

In chapter 8 we briefly discussed the role of natural disturbances in diversifying forests through reducing dominance and initiating succession. This effect depends on the scale, frequency, and type of disturbance in relation to the historic norm in a given ecosystem. Disturbances of intermediate severity are a major diversifying force in forest ecosystems (termed the **intermediate disturbance hypothesis;** Connell 1978; Petraitis et al. 1989; Remmert 1991), including such things as low- to moderate-intensity fires, windthrow, shifting agriculture, and death of small groups of trees (or perhaps one large tree) through pathogens, insects, or simply old age. These create openings of varying sizes that are filled by the growth of young trees, thereby enhancing vertical and horizontal architectural diversity and (in most instances) enhancing tree-species diversity. Tree species growing into gaps may differ in maximum attainable height or crown shape, further enhancing structural diversity. Dead trees, now snags or downed logs, graduate to their next incarnation as habitats for animals and microbes. The result, which has been variously described as the *shifting mosaic* (Bormann and Likens 1979) or the *mosaic cycle* (Remmert 1991), is a dynamic landscape composed of diverse patches differing in size, age, and perhaps species composition.

Although the mosaic cycle concept applies to a variety of different forest types, the characteristic patch size varies from one forest type to another, depending on the average scale of disturbance (e.g., death of a single old tree, or death of groups of trees through fire, windthrow, insects, etc.). If trees become most susceptible to mortality at a given age, as is often the case, the patch structure of a forest becomes a legacy that persists through time; in other words, once an even-aged patch is initiated for any reason, it will tend to retain its identity as a patch through subsequent cycles of death and regrowth (Remmert 1991).

Another layer of complexity is added to the mosaic when landscapes comprise different topoedaphic conditions that are affected differently by disturbance. Consider, for example, pine forests on the coastal plain of the southeastern United States. Under natural conditions (i.e., before intensive forest management), many upland sites were dominated by mature or old-growth long-leaf pine stands, where frequent low-intensity fire maintained open understories (good habitat for red-cockaded woodpeckers, an endangered species). In contrast, the moister lowland forests, dominated by loblolly and slash pines, burned less frequently and were characterized by relatively dense stocking and understories of various hardwood species (good habitat for some migratory songbirds). Schowalter et al. (1981) hypothesized that fire and the southern pine beetle interacted to affect the structure of both upland and lowland forests, but in different ways. According to these authors, conditions in upland forests were such that trees were not overly susceptible to bark beetles unless scarred and weakened by fire, in which case beetles killed the trees. The result was a mosaic of relatively small, uneven-aged patches (a structure that is also characteristic of ponderosa pine stands in western North America). Overstory pines in lowland forests, in contrast, were more susceptible to bark beetles, which, by killing the pines, accelerated the conversion of mixed pine-hardwood forests to pure hardwoods. However, beetle-killed trees increased forest susceptibility to intense fires, which acted over the long term to maintain the mixed pine-hardwood forest.

Clearly, not all disturbances diversify forested landscapes. Acid rain can hardly be considered a diversifying agent, and even-aged, short-rotation forestry produces quite a different landscape pattern than natural disturbances, including elimination of some habitats (Perry 1998).[23] To understand the effect of disturbance on diversity of a given community, one must first ask what forces have shaped the community. Organisms are generally well adapted to their historic disturbance regime, and the adaptations of individuals and species may confer adaptability on the community as a whole, or at least on certain guilds within the community[24] (Perry et al. 1989b; chapter 8). Biodiversity is at least in part an important manifestation of this adaptation, reflecting a healthy response of the system to the shifting mosaic of environments created by disturbance. On the other hand, foreign disturbances (i.e., those to which the trees and other plants are not well adapted) may alter the structure of the system and reduce diversity. *Foreign* might mean an entirely new type of disturbance such as air pollution or short rotation forestry, or a shift in the intensity or frequency of a common disturbance, such as the change in fire regimes that has occurred in the dry forest types of western North America (chapter 7). We return to the question of foreign disturbances in chapter 20.

Earlier we saw that gaps created by death of one to a few canopy trees and successional regrowth into these are a fundamental structural component of many (but not all) forests (Pickett and White 1985; Denslow 1987; Brokaw and Busing 2000). Studies in a variety of forest types (i.e., temperate deciduous, moist tropical, and old-growth Douglas-fir) have found that gaps compose 3 to 25 percent of forest area (Runkle 1982; Brokaw 1985; Spies et al. 1990). Some forests have much more area in gaps. In the spruce-fir forests of New Hampshire, over 70 percent of the forest area at low elevation and 40 percent at high elevation consists of gaps in some stage of succession (Worrall and Harrington 1988). Most of these were created by the death of one or two trees, caused primarily by pathogens and insects at low elevation and windthrow at high elevation.

In both moist tropical forests of Central America and deciduous forests of eastern North America, an average 0.5 to 2 percent of forest area is converted to gaps each year. The structure of these forests is shaped by growth of trees into gaps. On the other hand, understory trees do not so readily fill one- to two-tree gaps in some forests, such as the conifers of western North America (Lertzman 1989; Spies et al. 1990); the structure of these forests is driven primarily by moderate- to large-scale fires and windthrows rather than by small gap creation.

[23]The effects of forestry on diversity are complex and can have positive as well as negative aspects. We discuss this in some detail in chapters 22 and 23.

[24]*Guild* is used here in the sense of an association for mutual support and benefit (chapter 8 in this volume).

A rather consistent general pattern of gap succession is observed in a variety of forests, particularly those without a pronounced dry season. Seedlings of species that are able to tolerate low light establish beneath intact canopies (commonly referred to as **advanced regeneration**). Although capable of surviving in low light, these understory seedlings generally do not grow rapidly in height until a gap is created, at which time the gap may be filled by advanced regeneration and species that may seed in after the gap is created. In forests with a pronounced dry season, understory seedlings do not establish as readily beneath intact canopies (presumably because of water stress), and gaps tend to be filled by pioneer species whose seeds arrive after the gap is created (Ashton 1988).

Species composition of the gap-successional community depends on a variety of factors. In forests composed of species with animal-dispersed seeds (predominantly angiosperm forests in mesic environments), the input of seed beneath any given point in the canopy contains a certain random element related to where an animal defecates and what it has been eating. As the number of tree species increases, the randomness of seed dispersal increases, becoming maximum in species-rich tropical forests.

The species that actually grow to fill a gap once it is created depend on numerous factors such as species adaptations, gap size, the degree of soil disturbance associated with gap creation, topographic position, height of the surrounding canopy, and effects of animals and pathogens (Popma et al. 1988). Depending on size and location (e.g., slope and aspect), gaps may favor different tree species with different resource requirements. The **gap-partitioning hypothesis** holds that species with different requirements establish in different parts of the same gap (e.g., edge vs. center). Light is often assumed to be the primary resource influencing patterns of colonization within gaps, but the situation is more complicated and varies depending on forest type. While "it has been well established that gap size is a major factor determining post-disturbance tree species composition" (McCarthy 2001), gap size (hence, light reaching the forest floor) is only one of several factors that potentially influence patterns of regeneration and tree growth within gaps. Brokaw and Busing (2000) reviewed studies of gap colonization in tropical and temperate forests, concluding that "gaps are filled mostly by chance occupants rather than the best adapted species. . . . This chance survival can slow competitive exclusion and maintain tree diversity."

In their study of experimentally created gaps in old-growth conifer forests of the Pacific Northwest, Gray and Spies (1997) found that "heterogeneity at the seedling scale ($<$10 cm) often overrode larger-scale environmental gradients ($>$2 m) associated with gap size and within-gap position." Factors that helped seedlings gather or conserve water made a greater difference to seedling survival than light, including shading by large logs and understory plants, and litter depth (litter dries relatively quickly). As seedlings, small-seeded tree species are more susceptible to water stress than those with large seeds (because the latter can put out deep roots more quickly); therefore, the presence or absence of microsite factors couple with the size and location of gaps to influence tree species composition within the gaps (Gray and Spies 1997). While especially critical in summer-dry forests such as those of western North America, microsite factors that influence seedling water balance are probably important in many forest types.

Other factors aside, light nevertheless does play a role in what species fill a gap. Relatively small gaps (e.g., produced by the death of a single large tree) in both moist tropical and moist temperate forests are most often filled by shade-tolerant species (Denslow 1987; Spies et al. 1990). Up to a certain point, increased gap size is accompanied by an increasing proportion of shade-intolerant species in the gap successional community (Denslow 1987). In large gaps, however, which are most common in forests that lie on hurricane tracks, the proportion of shade-intolerant pioneers within the gap successional community may decline because the pioneers are less able to disperse seeds into the interior portions of the disturbed areas (Vandermeer et al. 2000). In such situations, diversity may be greater in large than in intermediate-sized gaps, because the dominance potential of fast-growing pioneers is reduced (Vandermeer et al. 2000). Disturbance history is a critical determinant of gap dynamics and, consequently, beta diversity. Whereas forests along hurricane tracks have a relatively high proportion of shade-intolerant species, and large gaps are essential to maintaining tree species diversity (i.e., by keeping the pioneers from dominating the landscape), forests without a history of large disturbances have fewer shade-intolerant pioneering species, and tree diversity is highest in gaps of intermediate size (Molino and Sabatier 2001). The size of gap required to promote regeneration of shade intolerants depends on the height of surrounding trees. In old-growth forests of the Pacific Northwest, where dominant trees reach 80 to 100 m in height, even a several-tree gap will be colonized primarily by shade-tolerant western hemlock (Spies et al. 1990).

Topoedaphic factors may strongly influence both the characteristic size of gaps and the types of species that colonize them. For example, in the Monteverde Cloud Forest Reserve in Costa Rica, windswept ridge crests are occupied by "pygmy" forests in which the largest trees are only 5 to 10 m tall, while forests in the ravines—relatively protected from wind—attain heights of 15 to 23 m (Lawton 1990). The combination of severe winds and short trees on the ridges (shortness being an adaptation to high winds) results in large gaps that are relatively well lighted. These are colonized by slow-growing tree species. Gaps in the ravines are smaller and surrounded by tall trees and, hence, are not so well lighted. These are colonized by pioneers that grow very rapidly in height, a characteristic that Lawton (1990) suggests is an adaptation that allows pioneers to quickly escape the low light deep in the gap and reach the better light conditions higher in the gap.

Within certain limits, variable and unpredictable disturbances enhance diversity because they create a variety of niches related to establishment and growth. A tropical forest characterized by an array of gap sizes would contain a corresponding array of shade-tolerant and shade-intolerant tree species. Variability in the timing of disturbance interacts with differences among species in their seasonal rhythms to diversify forests. This is particularly true in the tropics, where the array of species producing viable seeds varies throughout the year and even from year to year. In chapter 8 we discussed the example of variable fire regimes in temperate forests that favor one kind of adaptation (say, sprouting from rootstocks) in one area and another adaptation (perhaps thick bark or buried seed) elsewhere.

Variable disturbances may maintain diversity within as well as between species. For example, lodgepole pine forests of the northern Rocky Mountains characteristically contain trees that produce only serotinous cones, that produce only nonserotinous cones, and that produce both. This diversity of cone types most likely reflects a history of alternating small and large fires that favor seed dispersal from one cone type at one time or place and dispersal from the other type at another time or place (Perry and Lotan 1979).

Forest diversity is enhanced if seedlings of a given species do not establish as successfully beneath their parent as do seedlings of other species, which means that the individuals filling a gap tend not to be the same species as the individual(s) that created the gap (i.e., species alteration). This phenomenon has been reported in both tropical and temperate forests. Of 27 tropical studies summarized by Clark and Clark (1984), 20 showed some evidence for species alternation and 7 did not. Some species may successfully establish beneath their parents, while other species in the same forest do not (Connell et al. 1984).

The situation seems much the same in the temperate zone as in the tropics. Species alternate through time in filling gaps in some temperate forests but not in others; within a given forest, some species alternate, while others replace themselves (Fox 1977; Woods and Whittaker 1981; Glitzenstein et al. 1986). A study by Forcier (1975) suggested a three-way species alteration in deciduous forests of the northeastern United States (fig. 10.10). At steady state, these forests are dominated by sugar maple, beech, and yellow birch in roughly equal proportions. Gaps created by the fall of large trees are usually colonized by yellow birch, which produces frequent crops of small, readily dispersed seeds. Birch is particularly successful where exposed mineral soil has created a favorable seedbed. For reasons that are unclear, sugar maple precedes beech in establishing beneath the birch, with beech seedlings finally establishing when the sugar maple reaches dominance. Mature beech trees produce shade-tolerant sprouts and,

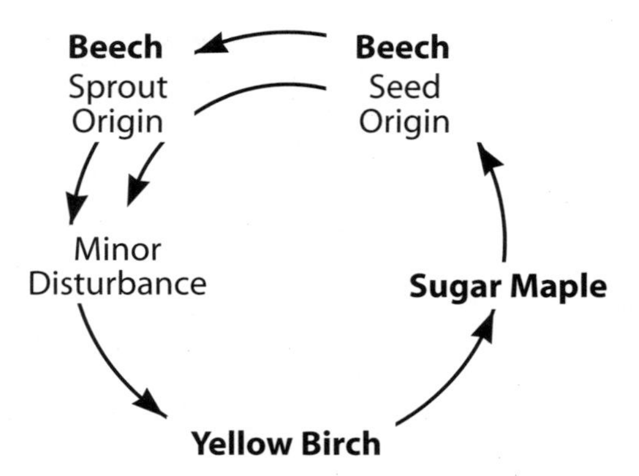

Figure 10.10. Hypothesized patterns of cyclic tree-species replacement following gap formation in deciduous forests of eastern North America. (Adapted from Forcier 1975)

hence, would seem ideally positioned to eventually dominate the forest. That this does not happen (at least in the slice of time that we observe) indicates that for some unknown reason, beech parents are frequently not succeeded by their sprouts.

What factors influence whether gaps are filled by trees of the same or of different species from that creating the gap? We have already discussed the role of natural enemies (i.e., the Janzen-Connell hypothesis; see section 10.4.4). In species-rich forests the probability of species alteration is also enhanced simply because of the large number of species in the available pool (i.e., ecological drift; see section 10.1.2). This is particularly true where seeds of most species are dispersed by animals and, hence, not tied to the proximity of parent trees. In most cases the reasons for alternation of species have not been studied, and definitive statements as to the cause are not possible.

10.4.6 History

> Ecology and climate . . . must act on evolutionary and biogeographical processes (e.g., speciation, dispersal and extinction) to determine patterns of species richness.
>
> WIENS AND DONOGHUE 2004

> Ecologists should raise regional and historical factors to equal footing with local determinism in their influence on the diversity-environment relationship and geographical patterns of diversity in general. No precedents exist that should compel ecologists to prefer one or the other type of explanation.
>
> RICKLEFS 2004

The diversity of ecosystems arises from some combination of contemporary and historical forces, the latter including factors that act primarily at regional scales, such as speciation, extinction, and migration, and factors that act both at regional and local scales, such as historic disturbances. Like rocks thrown in the pond of time, historical events produce ripples that persist long after the event itself, and even so-called antievents (e.g., long stable periods with no disturbance) shape the diversity we see today. For example, Hart et al. (1989) studied adjacent species-rich and species-poor forests of central Africa. They concluded that the "mixed-forest's greater diversity could not be explained by substrate differences, greater maturity, or greater predation on seeds or juveniles. . . . Monodominant tropical forests are widespread and may indicate areas that have not experienced large-scale disturbance for long periods." In other words, the explanation is lost in time.

In some cases the diversity of tree species within a given community reflects ongoing patterns of migration. For example, various lines of evidence suggest that species in the very large family Dipterocarpaceae (dipterocarps) have been slowly migrating from west to east in Asia since the Eocene, 40 million to 55 million years ago (Ashton 1988). At present, the center of dipterocarp species richness is Borneo, which has 287 species. Sulewai, just 80 km east of Borneo but across a stretch of water that acts as a migration barrier, has only 7 dipterocarp species. Some researchers believe that patterns of diversity that we see in today's temperate forests can be explained by differential migration of species northward following the retreat of glaciers 10,000 to 12,000 years ago. In the Pacific Northwest of the United States, "present climatic conditions are favorable to support a richer flora than is now present. During the late Pliocene, more than 40 angiosperm tree genera were represented in the flora of the Pacific Northwest, but massive extinction eliminated most of these under a colder and drier environment than exists today" (Waring et al. 2006).

Past glaciation has been invoked to explain the low diversity of European compared to North American and Asian temperate forests. Although Europe, North America, and northern Asia were all strongly influenced by glaciation during the Pleistocene era, fewer tree species survived this period in Europe than on the other continents. This is believed to stem in part at least from a quirk of geomorphology: European mountain ranges run east and west, hence their peaks presented a formidable barrier to species migrating southward in front of the glaciers. In contrast, the north-south orientation of mountains on the other continents facilitated southward migration through the valleys, allowing more species to survive. Alternatively, Adams and Woodward (1989) argue that differences in tree species richness among the northern continents are due to current climate rather than glaciation.

In chapter 8 we saw that historic land use, such as abandoned farms in the eastern United States, can influence patterns of diversity for centuries. Some researchers believe that historically, shifting agriculture was a powerful diversifying force in tropical forests. Connell (1978) hypothesizes that the relatively low diversity of tropical forests growing on infertile soils is not due directly to nutrient deficiency, but to the fact that these forests had no history of shifting agriculture. Similarly, in their study of tree diversity in central Africa, van Gemerden et al. (2003) concluded that "present-day tree species composition of a structurally complex and species-rich Central African rain forest still echoes historical disturbances, most probably caused by human land use between three and four centuries ago." (In recent years, however, intensification of human uses has tended to degrade rather than diversify tropical forests; see chapters 7, 20, and 22.)

Butterflies in Europe and Canada provide a good example of how patterns of diversity can vary between areas with different land use histories. In Canada, the richness of butterfly communities is closely tied to plant community patterns (Kerr et al. 2001). In western Europe, however, climate is the primary predictor of butterfly richness, and plant communities contribute relatively little (Hawkins and Porter 2003). This difference is not attributable to different climatic gradients between the two regions, and Hawkins and Porter (2003) speculate that it is related to the widely disparate histories of land use. Hypotheses about events buried in history are difficult and sometimes impossible to confirm or refute; however, that in itself makes them no less true.

10.5 SUMMARY

Numerous factors operating at different spatial and temporal scales combine to create and maintain biodiversity. At a very large scale (continental), differences in animal diversity among subregions (60,000–125,000 km^2) of North America correlate positively with total input of solar energy, while differences in tree diversity correlate positively with a combined measure of solar energy input and water availability. Different patterns emerge at the scale of regions, watersheds, and individual stands. Regionally, the greatest plant species diversity tends to occur in areas with intermediate levels of water and nutrient availability, with diversity falling off as resources become either more abundant or more limiting. At the scale of watersheds and individual communities, animal diversity is closely linked to the composition and architecture of the plant community, the legacies of trees such as snags and logs, and landscape patterns; plant diversity at this scale is linked to topoedaphic variability and disturbances, especially those of intermediate intensity. Competition among species within a given trophic level potentially reduces diversity; however, the tendency of one or a few species to dominate a given trophic level is reduced by five general mechanisms, all of which may operate to one degree or another in one place and time or another. First, different species adapt to different niches and are superior competitors within their own niche (i.e., niche assembly). Second, dispersal limitations combined with chance determine what species occupy a site (i.e., dispersal assembly). Third, population numbers in a given trophic level are regulated by higher trophic levels, thereby reducing the probability that one aggressive species can exclude others (i.e., the Janzen-Connell mechanism). Fourth, intermediate levels of resource availability, and in particular balance among the various resources, tend to keep one species from dominating and excluding others. Fifth, disturbances reduce the ability of one or a few species to dominate others. Overlaid on all these is the importance of history, which influences diversity in at least two ways. Historic patterns of land use and past natural disturbances may impart legacies that influence diversity for many centuries. Moreover, the diversity of any locale is tied to some degree to the diversity of the region within which it is embedded, and regional diversity arises from forces acting in evolutionary time.

11

The Biological Web

Interactions among Species

> If a local assemblage of organisms is to be regarded as a community with some degree of organization or structure then it is in the interactions between the organisms that we must look to provide this structure.
>
> CONNELL 1975

This chapter explores some of the more important relationships defining the structure of eosystems. The following section provides an overview (or perhaps more of a roadmap) of the structure of relationships within ecosystems. Later sections focus on the details of relationships within the overall system structure. Sections 11.2 through 11.4 deal with some common types of interaction between two species, especially mutualism and competition. In section 11.5 we consider higher-order interactions—those involving three or more species. Our discussion of interactions within ecosystems continues in chapter 12, in which we discuss some aspects of interaction among individuals of the same species (i.e., **conspecifics**), focusing in particular on how competition for growing space influences the structure of even-aged plant populations. In chapter 13, we consider some genetic and evolutionary implications of interactions among organisms.

11.1 THE STRUCTURE OF RELATIONSHIPS WITHIN COMMUNITIES

As we saw in chapter 1, any specific relationship among organisms within an ecosystem is embedded within and interacts with a hierarchy of interactions that encompass whole communities, landscapes, and ultimately most (or perhaps even all) living things on earth. It is no trivial matter to imagine such a complex web of interaction; this is one reason why reductionism is such a powerful metaphor in science; conceptually and experimentally it is simply easier to deal with than complex reality. Yet it is vitally important that we have a mental map of the system as a whole, no matter how crude that map may be. Those who study and manage ecosystems can hardly do so effectively without such a picture.

A long-running argument in ecology has to do with whether ecological communities are **organismic** (i.e., individuals and species are linked together as tightly as the parts of an organism) or **individualistic** (species are not linked together at all, except as sources of food or competitors for resources). Those contrasting views, for example, comprise the fundamental disagreement between the Clementsian and the Gleasonian views of succession discussed in chapter 8. Although a great deal remains to be learned about the structure of relationships in nature, at this point it seems that a community of organisms is neither. Rarely (as far as we know) are a pair of species within a community as tightly linked to one another as the parts of an individual organism; on the other hand, the rich tapestry of interaction among species in nature belies the individualistic view. An ecosystem is, in fact, a unique type of system neither solely organismic nor individualistic, but rather combining the attributes of both. Later, we shall see that instead of pulling the community apart, individuality—as expressed in the differing behaviors, life histories, and environmental requirements among the species composing an ecosystem—can be a force that binds species more closely together (see section 11.5).

How then do we think about the structure of relationships within ecological communities? What are these unique kinds of biological systems like?

11.1.1 Energy Flows in One Direction; Influences Flow in Many

To survive and reproduce, every organism must acquire chemical energy (usually in the form of reduced carbon compounds) and resources (i.e., nutrients and water). It must also have protected space (i.e., refuge from natural enemies and environmental extremes). The energy contained in photosynthates flows only one direction (from low to high trophic levels), but it performs work that feeds back to facilitate more photosynthesis, thereby maintaining the flow of energy through the system. Energy flow in a mutualistic relationship, for example, is returned as some service (e.g., nutrient acquisition, pollination), and it is the search for energy by decomposers that drives the nutrient cycle, without which green plants would soon cease to gather energy (chapter 18). Feedbacks such as these are what we shall call **influences** (i.e., factors that alter one or more aspects of system pattern). Influences flow both up and down the trophic ladder. Plants, for example, do much more than provide energy to the rest of the system; they influence the patterns of energy flow through higher trophic levels by providing structural and biochemical niches, and by influencing regional hydrology and climate (chapter 3). Food itself can be a force that influences patterns as well as providing energy. The chemistry of plant tissues, or the timing of their production, determines which herbivores can eat them and which cannot (chapter 19); some herbivorous insects utilize defensive chemicals produced by their food plants to defend themselves against their own predators.

Closely interacting sets of species are likely to influence the genetic structure of one another's populations, a process called **coevolution** (chapter 13). For example, predators act as selective agents that shape the genetic structure of their prey populations; in turn, the genetic characteristics of prey shape those of their predators. The relationship is closely interacting and reciprocal and may be mediated by microbial mutualists of both predator and prey.

Interactions among organisms in the intermediate trophic levels are mediated to one degree or another by organisms on higher and lower levels. By influencing the population densities of their prey, for example, predators alter competition among herbivores. Similarly, because they prefer some plant species more than others, herbivores alter competition among plants. Such higher-order interactions, in which the relationship between two species is modulated by others, pervade communities (Price 1991). They act not only to decrease competition (as in the examples discussed) but in a quite different vein also indirectly link organisms that have no direct relationship. For example, two widely separated trees of different species may nevertheless interact indirectly through common interests in the same pollinators or mycorrhizal fungi. An emerging view among some ecologists is that communities are composed of various subgroups that are defined by direct or indirect interdependence among group members, often centering around shared mutualisms (e.g., the guilds concept discussed in chapter 8; also see section 11.5).

11.1.2 Pattern and Repetition

The details of direct and indirect relationships among organisms are discussed in later sections. There are some further general aspects of internal system structure to deal with first. These relate to context and redundancy. **Context** refers to patterns of relationship in space or time. **Redundancy** denotes a kind of valuable repetition, such as the fact that more than one plant species in a community can support the same species of mycorrhizal fungi, seed dispersers, or pollinators (Lawton and Brown 1992).

To understand the importance of context and redundancy within the structure of community relationships, consider by analogy the structure of human language. The basic raw material of language is letters, just as the basic raw material of a community is the individual organisms, and the raw material of landscapes is the patches. Letters take on meaning only in a certain context (i.e., when they occur in certain patterns: words, sentences, phrases, chapters, books). This is also true with natural systems: the hierarchical pattern of relationships—among individuals within a given species, individuals of different species, and community types across the landscape—provides the context within which each individual organism must satisfy the requirements that allow it to survive long enough to pass on its genes. In nature, evolution serves as the "editor" who discards "meaningless" relationships (i.e., those that are not viable for one or more of the participants) and retains those that are viable.

Along with context, languages and ecosystems share the property of redundancy, which simply means that some rearrangement, substitution, addition, or even elimination of parts of the system may be possible without losing meaning from the whole. Consider, for example, the following sentences, which all say the same thing in different ways:

- Jack had rabbit to eat.
- Jack dined on rabbit.
- For his meal, Jack ate rabbit.

Different arrangements convey the same meaning. As any notetaker in a classroom knows, language contains enough redundancy that meaning is still conveyed even when certain parts are eliminated. On the other hand, some arrangements do not work at all, or if they do work, impart quite a different meaning. Consider the following sentences:

- Jack rabbit dined on.
- Rabbit had some Jack to eat.
- Hrsmoe dah kcakj.

Redundancies are important in ecosystems because they stabilize vital processes. Ecosystems cannot function without green plants that capture energy, as well as food web interactions that cycle nutrients and maintain balance between producers and consumers. These processes depend, in turn, on an array of partnerships (or mutualisms) among the species composing the system. These vital processes—energy capture, element cycling, maintenance of balance, maintenance of critical mutualisms—can be considered analogous to the meaning of a sentence or phrase: without them the system cannot function (e.g., Hrsmoe dah kcakj). As long as the critical processes are preserved, however, substitutions, deletions, or rearrangements of species do not significantly alter the functioning of the system as a whole.

The importance of context in nature (i.e., the pattern in which things occur) can be seen in such things as spatial arrangement at various hierarchical scales (i.e., species at the level of local ecosystems, patches at the level of landscapes). We saw in chapter 10 that susceptibility of trees to defoliating insects and pathogens may differ widely, depending on whether the species grows among others of its kind or is mixed with different species; this is an example of the importance of spatial pattern within local ecosystems. At another hierarchical level, consider the contrasting landscapes shown in figure 11.1. Although the same proportion of each simulated landscape has been cut over (49%), the patterns of cutting differ significantly and vary widely in their implications for animal habitat and movements, local climates, the susceptibility of stands to windthrow, and the propagation of fire, insect pests, and pathogens (Perry 1988b).

The importance of context also appears in the relationships among species, which in many cases are specific to particular sets of species. This is especially true in the case of mutualisms, such as those between plants and their pollinators or plants and their mycorrhizal fungi. A fig tree cannot be pollinated by just any animal, it takes a fig wasp; a fungus may form mycorrhizae with a number of tree and shrub species within a given community, but not with others, even though they are common to the same community.

Mutualisms are not the only relationships in nature in which context—the particular species mix—is important to ecosystem function. The shifting around of organisms by humans (both purposeful and accidental) provides numerous "experiments" on the introduction of species to communities where they have no prior history. Thus, the dogs, pigs, rats, and mongeese that followed humans to the Hawai'ian islands were much more than simply additions to the higher trophic levels, they devastated a bird community that had no prior experience with ground-based predators. Similarly, the implications of introducing the gypsy moth to North American forests have reached far beyond what could be predicted by the simple addition of one more defoliating moth species to a system already rich in defoliating moths. The gypsy moth (a relatively minor defoliator in its native forests) became a major pest when placed in a different ecological context. (As in language, some combinations in nature simply do not work very well.)

Like languages, redundancy within ecosystems allows relationships among the parts to be flexible rather than rigid. For instance, a given species that enters into a mutualistic relationship with another rarely puts all of its

Figure 11.1. An example of how the same level of cutting can produce quite different landscape patterns. Dark patches are cut and white patches uncut; 49% of the area is cut in each case. (Adapted from Li 1989)

proverbial eggs into one basket. Pollinators such as bees, moths, birds, and bats usually service several tree species; conversely, most tree species are pollinated by more than one animal. (This is not always the case; fig trees and fig wasps are an example of a pollination mutualism that is very specific to the two partners.) Mycorrhizal fungi are usually capable of forming mutualisms with at least two and usually many different plant species in a given community.

By stabilizing vital processes and providing flexibility in relationships, redundancy confers the property of adaptability, although not necessarily in the genetic sense. Environmental factors such as climate and disturbance patterns are generally variable (frequently unpredictably so), and individual species differ in the particular set of conditions (i.e., the environmental context) within which they are genetically adapted to best perform. Flexibility within the structure of communities permits the presence and activities of given species to vary over time without disrupting critical biotic links and ecosystem functions. Disturbance and succession provide the most common examples. As we saw in chapter 8, the composition of the early successional plant community in virtually all terrestrial ecosystems varies to some degree, depending on numerous factors. However, if various plants within the same system can stabilize soil physical structure and support the same mycorrhizal fungi (or pollinators), the maintenance of these critical links is not jeopardized by the loss (either temporary or permanent) of one species.

Flexibility in the composition of forests is apparent over very long time scales. Pollen preserved in bogs and lake sediments provides a window into history and reveals that the particular mix of tree species that compose many temperate forests has changed with shifting climates since the close of the last ice age (e.g., Foster et al. 1990, 2004; Brubaker 1991; Webb 1992). This individuality does not, however, imply a lack of common interest in soils, pollinators, seed dispersers, and probably other factors as well (see section 11.6, as well as chapters 20 and 21). This blend of independence and interdependence is perhaps the most characteristic feature of community organization.

The property of redundancy does not justify homogenizing of ecosystems. There are definite limits to the degree to which species composition of an ecosystem can be altered, and if these limits are exceeded the whole web of interconnection within the system can break down (chapter 20). Unfortunately, ecologists know far too little about what these limits are for a given system. As we discuss later in this chapter, some vital processes are known (or suspected) to be performed uniquely by only one species or a small group of species (i.e., the so-called keystones). In our limited knowledge of ecosystem structure, however, are we just seeing the tip of the iceberg of keystoneness? Some ecologists are skeptical of the whole idea of redundancy and argue that as far as we know, every organism plays a unique role in maintaining vital system processes. Lawton and Brown (1992) dubbed this latter view the **rivet hypothesis,** so named because Ehrlich and Ehrlich (1981) likened extinctions to pulling rivets out of an airplane. Our limited knowledge of the links between individual species and ecosystem processes is one of the more powerful arguments for maintaining biological diversity. To paraphrase the great forester and conservationist Aldo Leopold, the first rule of intelligent tinkering is to keep all the pieces.

11.2 INTERACTIONS BETWEEN TWO SPECIES: BASIC CONCEPTS

Interactions among individual organisms of different species can be broadly grouped into four categories: eating or being eaten, competition for resources, cooperation, and no direct interaction**.** The terminology used to describe these interactions is shown in table 11.1.

Symbiosis refers to individuals of two species living in intimate physical association with one another; the relationship could be either mutualistic or parasitic. Many mutualisms are also symbioses, for example, lichens (fungus and alga) or mycorrhizae (plant and fungus); however, many mutualisms are not symbioses, for example,plants and their pollinators.

Some interactions in nature do fit into the neat categories shown in table 11.1. Many others do not. In general, interactions between individuals of different species are complex and may change from one category to another depending on the circumstances. For example, certain moths prey on plants when larvae (a + / – interaction) but pollinate the very same plants when adults (a + / + interaction).

Table 11.1
Interactions between two species

Type of interaction	Species 1	Species 2	Nature of interaction
Predation	+	–	In both predation and
Parasitism	+	–	parasitism, species 1 benefits at the expense of species 2. The distinction is hazy, but predators tend to take large bites and parasites small bites. Parasites frequently invade the body of their hosts and eat from within; predators eat from outside.
Competition	–	–	Both species suffer.
Amensalism	–	0	One species is inhibited and the other unaffected.
Mutualism	+	+	Both species gain.
Commensalism	+	0	One species gains and the other is unaffected.
Neutralism	0	0	Neither species affects the other.

Many relationships between individuals of two or more species are best characterized as **limited partnership;** the individuals will cooperate as long as it is to their advantage to do so. Note that *advantage* does not necessarily imply an immediate reward. The mode in which a given species interacts with its environment is shaped by what confers fitness[1] over evolutionary time. Sacrifice may be of little consequence in the short run if it increases the probability of an individual passing on its genes in the long run (Thompson 1982; chapter 13).

The founders of modern ecology believed ecological communities were characterized by a mix of positive and negative interactions; however, in the 1960s this viewpoint shifted to one that saw competition for resources as the major force shaping the structure of ecosystems (Kareiva and Bertness 1997). The pendulum is now beginning to swing back, and ecologists are once again acknowledging the importance of positive interactions within communities[2] (Perry et al. 1989b; Perry 1995; Callaway 1997; Bruno et al. 2003). As Nowak observed, "Whenever nature achieves a major step, it involves cooperation (quoted in Klarreich 2002, p 60).

Species sharing the same trophic level may not compete for resources at all, or they may simultaneously compete and cooperate. For example, multispecies flocks of insectivorous birds commonly follow ant swarms in the tropics, feeding on insects flushed out by the ants. Bourliere (1983) lists two possible advantages to such mixed flocking: more efficient hunting in patchy environment, and more eyes and ears for protection against predators. Different species of monkeys also frequently forage together, even when they are quite similar in body size and diet. Once again, Bourliere (1983) relates that "the main advantage for mixed groups [of monkeys] appears to be to make easier the location of patchily distributed food sources and the detection of predators. A monospecific group of similar size might well offer the same benefits to its members, but at the cost of a stringent social hierarchy which would imply a greater energy expenditure for its enforcement."

In forestry, controlling noncrop vegetation usually increases the growth of crop trees, at least in the short term (e.g., Fleming et al. 2006), but there are environmentally dependent points of diminishing returns in which further control produces no additional growth increases. On the other hand, Hunter and Aarssen (1988) list the following ways in which coexisting plant species have been demonstrated to help one another: "improving the soil or microclimate, providing physical support, transferring nutrients, distracting or deterring predators or parasites, reducing the impact of other competitors, encouraging beneficial rhizosphere components or discouraging detrimental ones, and attracting pollinators or dispersal agents" (chapters 8, 15, 19, and 20).

[1]In genetic parlance, *fitness* refers to the ability of an individual pass its genes to future generations.

[2]Positive interactions may include either mutualism (both species benefit) or amensalism (one species benefits and the other is unaffected).

Environmental conditions may determine whether a relationship is facilitative or competitive (Holmgren et al. 1997). In an extensive study of plant-plant interactions in mountain ranges, Callaway et al. (2002) found that plants benefited from one another's presence in more stressful environments but competed in less stressful environments.

The net outcome of multifaceted interactions such as combined competition and cooperation may be difficult (or even impossible) to predict over the short times of most experiments. Moreover, while direct interactions between individuals of two species are the most obvious and better studied, they account for only a portion of the ecological web. Some of the more important work that lies ahead for ecologists will deal with indirect interactions between multiple species (see section 11.5).

> Much of the traditional ecological thinking about the role of competition in structuring communities has been shaped by mathematical models that treat two interacting species as if they were in a constant environment and isolated from other species in the system. Stone and Roberts (1991) developed a more realistic model that evaluates the interactions between any two species "within the framework of the community to which they belong." They refer to their approach as the *inverse method*. Note that the interactions they refer to are strictly *within* a given trophic level, not between trophic levels. In other words, they are interactions in which the participants potentially compete for resources. The criterion they use to determine whether a species benefits or suffers in interaction with another is population growth: in essence, they ask, "What happens to numbers of species A if numbers of species B increase?" If A increases, it benefits from B (at least within the range of increase of B that is modeled), if A decreases, it suffers from B. In their own words: "Remarkably, the 'inverse' method finds that generally a high proportion (20–40%) of interactions must be beneficial, or 'advantageous,' when not lifted out of the community context in which they actually occur. The contrary case, called here 'hypercompetitive,' in which each species suffers from every other species, can occur only if the environment is nearly constant, and the species closely akin to each other, with both of these conditions holding and persisting to a degree that must be considered implausible."

11.3 MUTUALISMS

Mutualisms may be the most common two-species interaction in nature. Whether this is true (it is an assertion that is in all likelihood unprovable), their ubiquity is undeniable. Plants, animals, and eukaryotic microbes that do *not* participate in at least one mutualism are rare, and in all likelihood nonexistent. For example, certain organelles within eukaryotic cells (notably mitochondria and chloroplasts) originated as bacteria that evolved a partnership with higher organisms (Sapp 1991), an old idea that was revived and championed by Lynn Margulis (Margulis 1981; Margulis and Sagan 1986). The development of life on land, which is a hostile environment compared to oceans, was almost certainly made possible by mutualisms between plants and certain types of fungi and bacteria (Pirozynski and Malloch 1975; Trappe 1987; Lewis 1991), and the continuation of terrestrial life as we know it remains dependent on these relationships. Both of the primary interfaces between plants and their environment (i.e., foliage and roots) are characterized by mutualistic symbioses with fungi (and probably associated bacteria as well), leading Pirozynski and Malloch (1975), and later Atsatt (1988), to suggest that plants are "inside-out lichens." (A lichen consists of a photosynthetic alga living within a fungal "house"; in a plant this relationship is reversed.)

Janzen (1985) distinguishes the following types of two-species mutualisms.

- Harvest mutualisms, or those related to gathering or processing resources
- Propagule dispersal mutualisms
- Protective mutualisms
- Pollination mutualisms
- Human agriculture and animal husbandry

We discuss these (except for the last) in more detail later in this section.

Mutualisms may be **pairwise** (i.e., a kind of monogamy involving only two species), or **diffuse** (i.e., each partner having other partners). Most are diffuse, which as we have already discussed provides redundancy that buffers the system against disturbance. In many cases the relationship is asymmetric, one partner being monogamous and the other polygamous, an arrangement that may permit networks of mutualists to maintain stably high levels of biodiversity (Bascompte et al. 2006).

Mutualisms may also be **obligate** (i.e., the partners are absolutely dependent on the relationship) or **facultative** (i.e., the partners are able to survive and reproduce outside the relationship). Some of the most important mutualisms in forests are commonly obligate, such as those between most trees and mycorrhizal fungi and between many flowering trees and pollinators. But many mutualistic interactions among species are probably facultative. Most obligate mutualisms are diffuse; for example, trees require mycorrhizae but can form the partnership with more than one fungal species. Mutualisms do exist that are both obligate and pairwise, but because the loss of one partner automatically means that the other will also be lost, species entering into such a relationship are considerably more vulnerable than those whose relationships are diffuse.

One unique example of a facultative forest mutualism is that between a small African bird called the honeyguide *(Indicator indicator)* and some honey-eating mammals such as badgers and humans. Isack and Reyer (1989) found that it takes humans a little over three hours on average to find a bee nest when guided by the bird, compared with almost nine hours when not guided. The bird benefits, because it generally cannot access a bee nest unless it is opened by a large mammal. Fires built by humans as a means of pacifying the bees also reduce the chances of the bird being stung. When embarking on a hunt, humans call a honeyguide by whistling; the bird may also solicit human help by flying close and emitting a characteristic double-noted call.

May (1989) points out that Isack and Reyer's study has implications that go beyond the relationship between humans and this species of bird. He notes: "In the tropical rainforests of Africa, Asia, and South America, about whose flora and fauna conventional science knows so little, a vast store of ecological knowledge resides among the diminishing groups of native people, who draw upon such knowledge for food-gathering, medicine, and other aspects of daily life. . . . [This knowledge holds] the promise of helping us answer important questions more quickly than will otherwise be possible. Yet these possibilities are disappearing, as the cultures of native people are being destroyed at an even faster rate than the forests and other places where they live."

11.3.1 Harvest Mutualisms

Those related to resource acquisition are the most ubiquitous and diverse of nature's mutualisms. Following is a list of some common harvest mutualisms in forests:

- Mycorrhizae, or the symbiotic mutualism between plant roots and fungi.
- Nonsymbiotic mutualisms between plants and an array of microbes and invertebrates that live in plant rhizospheres, where they perform a variety of services for the plant in return for food (chapters 14 and 18).
- Nitrogen-fixing mutualisms, which include a wide array of symbiotic and nonsymbiotic relationships between plants or animals and bacteria that are capable of "fixing" atmospheric nitrogen gas into a form that is usable by other organisms (chapter 17). In forests, nitrogen-fixing bacteria live within or in close proximity to the roots of many plant species, within or on the surface of leaves, and in the guts of some animals (e.g., termites and bark beetles).
- Digestive symbioses between animals and microbes that live within their guts; for example, protozoa living within the guts of termites provide the enzymes that break down cellulose. Eckburg et al. (2005) describe the intestinal microflora of humans as "an essential 'organ' in providing nourishment, regulating epithelial development, and instructing innate immunity."
- Various farming mutualisms between social insects and other organisms. Leaf-cutter ants, for example, are the principal consumers of vegetation in Central and South America (Wilson 1987), but rather than directly eating the leaves they are fed to fungi within ant nests, the fungi then grazed by the ants (fig. 11.2).[3] The ant farms also contain two other ancient and coevolved members, a fungal parasite that feeds on the food fungus and a bacterium that lives symbiotically with the ant and inhibits the parasite (Currie et al. 2006). The so-called higher termites (Macrotermitidaea), which are common throughout the dry tropics, have a similar mutualism with fungi. Some ants are herders rather than farmers, tending and protecting aphids in return for honeydew.

Figure 11.2. Leafcutter ants, common in moist tropical forests, are farmers. Leaves are used to grow fungi within ant nests, and the ants feed on the fungi. (Courtesy of Dave Janos)

11.3.1.1 Mycorrhizae

The word *mycorrhiza* comes from the Greek words *myco*, for "fungus," and *rhiza*, for "root," and so literally means *fungus-root* (fig. 11.3). A mycorrhiza is a structure in which root and fungal tissue combine in such a way that both plant and fungus benefit. The fungus receives food from the plant in the form of sugars and other organic molecules and performs numerous services in return (see reviews by Smith and Read [1997] and Jeffries et al. [2003]):

- Enhancing water and nutrient uptake, particularly nutrients such as phosphorus that do not readily move in

[3]Leaf-cutter ants are the most advanced species of a large group called the attine ants, all of which are fungus farmers.

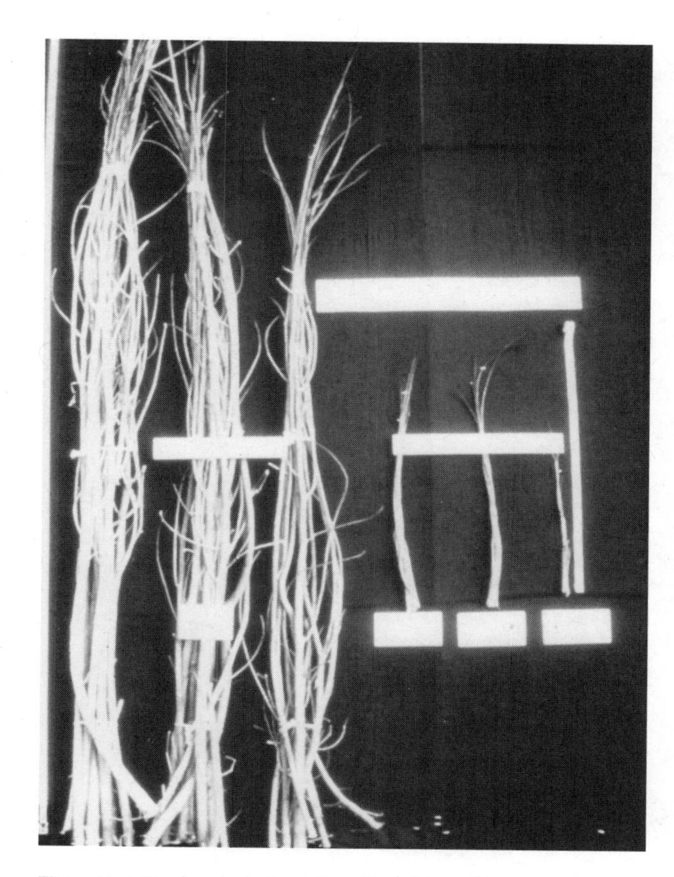

Figure 11.3. One-hundred-ninety-day-old seedlings of the tropical tree *Leucaena leacocephala*, grown under controlled conditions. Those on the left were inoculated with mycorrhizal fungi and those on the right were not. (Courtesy of C. Y. Li)

soils (Miyasaka and Habte 2001). Some mycorrhizal fungi decompose organic matter and cycle the nutrients contained therein directly back to their host plants (Read 1987, 1991b; chapter 18). Blum et al. (2002) proposed that mycorrhizal fungi weather rocks and transfer calcium directly from the rocks to their host (but see also Watmough and Dillon 2003).
- Protecting plants against root pathogens (Marx 1972; Smith and Read 1997).
- Extending the lifetime of small roots.

Mycorrhizal fungi also significantly affect community and ecosystem-level processes (Leake et al. 2004; Rillig 2004). They function in the following ways:

- Bind soil particles together into large aggregates, thereby producing a favorable structure for water retention and gas exchange (Rillig et al. 2002; chapter 14).
- Link individual plants of the same and different species together into common hyphal networks through which carbon and nutrients pass (see the review by Simard et al. 1997, 2003).
- Alter competition among plants, thereby influencing the structure and diversity of plant communities (Perry et al. 1989c, 1992; Smith and Read 1997; van der Heijden et al. 1998; Booth 2004).
- Alter the relationship between plant diversity and primary productivity (Klironomos et al. 2000).
- Contribute significantly to soil carbon and nitrogen pools (Rillig et al. 2002).
- Serve as an important food source for animals (Johnson 1996; Cazares et al. 1999).

Note that many of the functional aspects of mycorrhizal fungi, especially those that operate at the community or ecosystem levels (e.g., linking plants, altering competition), are well established in controlled studies but extremely difficult to study in intact natural ecosystems; hence, their significance in nature is rather poorly understood. As we discuss at various points throughout the book, this caveat holds for many (if not most) of the interactions in nature.

Many species of fungi form mycorrhizae, and not all perform each of the above functions equally. For example, some are proficient water gatherers, others are especially effective at protecting against pathogens, some operate primarily in organic soil layers, others prefer mineral soil, and so on. In all types, the fungal partner extends a network of hyphae from the mycorrhiza into the surrounding soil, increasing the water- and nutrient-gathering power of the plant manyfold (chapter 18).

Roughly 90 percent of the world's plant species form mycorrhizae with at least one (and frequently many) species of fungi (Molina et al. 1992). The numbers and types of mycorrhizae supported by trees varies depending on plant species and resource needs, the latter depending, in turn, on site and stand conditions. Fewer mycorrhizae are formed in fertile than in unfertile soils, and when greenhouse-grown seedlings are well watered and fertilized they may form none at all. Few if any forest soils are sufficiently fertile that trees require no mycorrhizae at all, and the symbiosis is probably obligate for most trees in the wild so long as soil resources are limiting to growth. Some studies find relatively low levels of root colonization in tropical and subtropical rain forests (Zangaro et al. 2000; Z-W Zhao et al. 2001; Gehring and Connell 2006), which is probably attributable to the low light levels beneath dense rain-forest canopies (Gehring and Connell 2006). Limited photosynthesis under such conditions decreases the demand for nutrients and water and increases the need to devote carbon to light-gathering rather than nutrient-gathering tissues (i.e., leaves rather than roots and mycorrhizal fungi).

The importance of mycorrhizae for pines and oaks was clearly demonstrated in numerous instances when those species were planted in the tropics or in grasslands where the particular types of fungi that they required were not present. Without their fungal partner, seedlings often did not survive; if they did, they did not grow well (Mikola 1970; Shemakanova 1967). Similarly, trees planted on highly disturbed areas, such as mine spoils or heavily eroded sites,

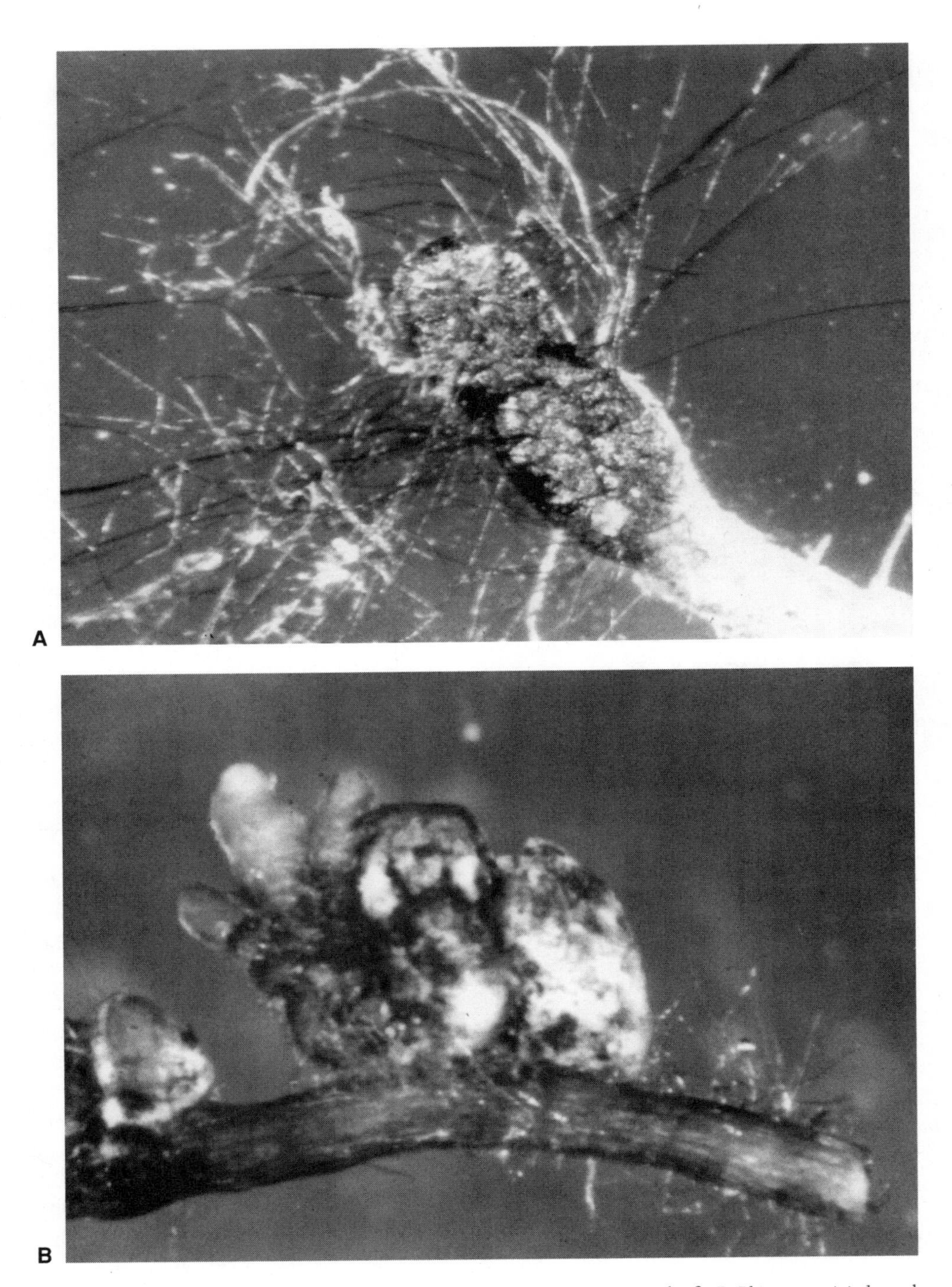

Figure 11.4. Some examples of ectomycorrhizae. A. *Cenoccocum geophilum* on Douglas-fir. B. *Rhizopogon vinicolor* and an unidentified brown mycorrhiza on adjacent Douglas-fir root tips. *(Continued)*

frequently do not succeed unless given the proper mycorrhizal fungus (e.g., Marx 1975). As we shall discuss in some detail in chapter 20, a weakened belowground community (including but not restricted to mycorrhizal fungi) is involved in both land degradation (a serious problem throughout the world) and the effects of pollution on trees.

Mycorrhizae are divided into several classes on the basis of various structural features such as whether hyphae penetrate plant cells or wrap around root surfaces (see Smith and Read 1997 for a recent classification). For our purposes it is convenient to focus on the three most common groups: the ectomycorrhizae (EM), the arbuscular mycorrhizae (AM), and the ericoid mycorrhizae. The AM may also be called vesicular-arbuscular mycorrhizae (VAM) or zygomycetous mycorrhizae. There is a distinct pattern in the occurrence of these groups (Smith and Read 1997). Heathlands are dominated by ericaceous mycorrhizal plants, grasslands by AM plants, boreal forests by EM trees, temperate forests by EM

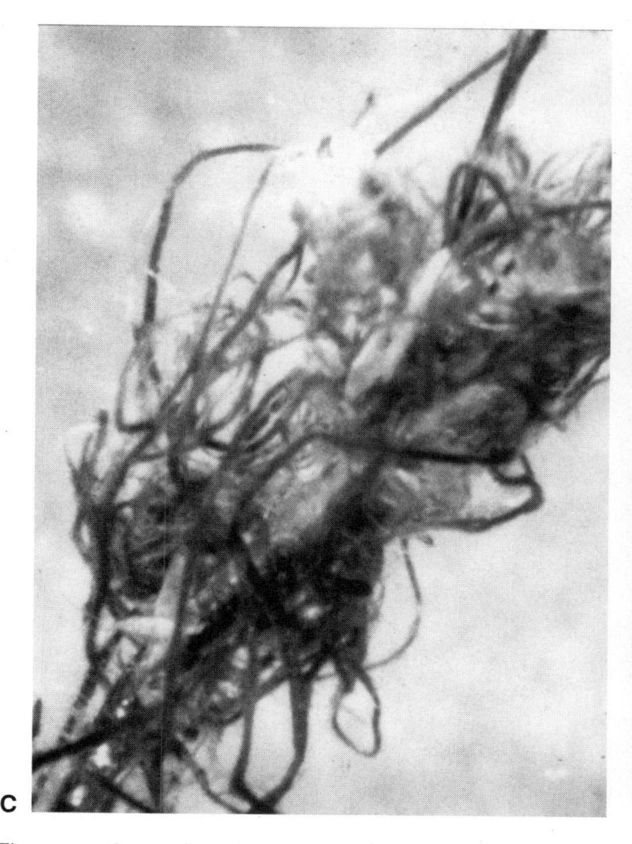

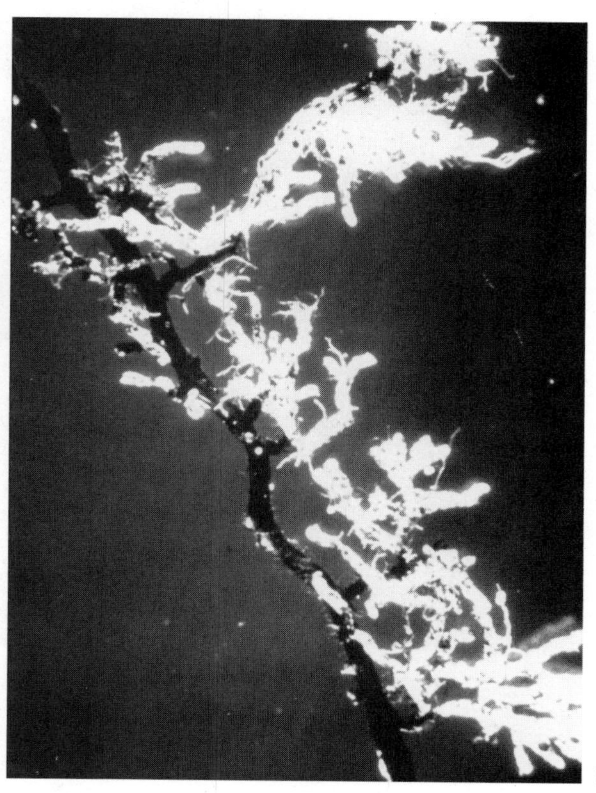

Figure 11.4. *(Continued)* C. *Rhizopogon vinicolor*. The ropelike strands are rhizomorphs, or organized clusters of hyphae specialized for transporting water and nutrients. D. Unidentified mycorrhiza on lodgepole pine. (A–C, courtesy of Sue Borchers; D, courtesy of B. Zak)

trees with significant components of AM trees, and most tropical forests by AM trees with significant components of EM trees; the extensive dipterocarp forests of Southeast Asia comprise EM trees. If all forest plants are considered rather than just trees, the major forest biomes differ little in the percentage of plants that form relationships with AM, averaging 64 percent for boreal, 56 percent for temperate, and 70 percent for tropical forests[4] (Treseder and Cross 2006).

The EM are distinguished by two morphological features. First, hyphae wrap around the surface of roots, forming what is called the **sheath** (also called the **mantle**). The sheath gives the root tip a swollen appearance and, depending on the species of fungus and the color of its hyphae, it usually imparts some color or combination of colors (e.g., white, black, blue, red, yellow, and brown are all common ectomycorrhizal colors). Also, EM may cause root tips to branch, sometimes repeatedly, although this does not always occur. The term *ecto,* meaning "external," is derived from the fact that these mycorrhizae modify the external appearance of the root (AM do not). Figure 11.4 shows some examples of EM.

The second feature used to distinguish EM is that they usually do not penetrate the root cells of the plant (AM do). Hyphae penetrate the intracellular spaces and wrap around root cells, forming what is called a **Hartig net** (fig. 11.5). In some cases—and particularly some plant species—hyphae do penetrate cells; these are referred to as *ectendomycorrhizae.* The different name reflects a structural difference and does not necessarily imply that the species of fungus differs from those forming EM. In some species of EM fungi, hyphae that extend into the soil are grouped together to form ropelike structures called **rhizomorphs,** which are specialized for long-distance transport of water and nutrients (AM fungi extend hyphae into the soil but do not form rhizomorphs).

Most plants that form EM are woody perennials. Many common temperate and boreal trees are ectomycorrhizal, including genera in the families Pinaceae, Fagaceae, Myrtaceae (which contain the eucalypts), and Betulaceae. Some important and widespread tropical trees also form EM, including species in the families Dipterocarpaceae and Caesalpinoidea.

Thousands of fungal species form EM, mostly in the subdivision *Basidiomycotina* (basidiomycetes) and *Ascomycotina* (ascomycetes). A given tree may have,10 or more EM types at any one time, and many more than this over its life. Mushrooms are the fruiting bodies of basidiomycetes, while the ascomycetes produce various types of fruiting bodies including the small, often inconspicuous cup-fungi and the edible morels. Some mycorrhiza-forming species within both the basidiomycetes and the ascomycetes produce

[4]These values should be considered provisional, as the sampling intensity varied widely among biomes (14 tropical stands, 3 boreal, and only 1 temperate).

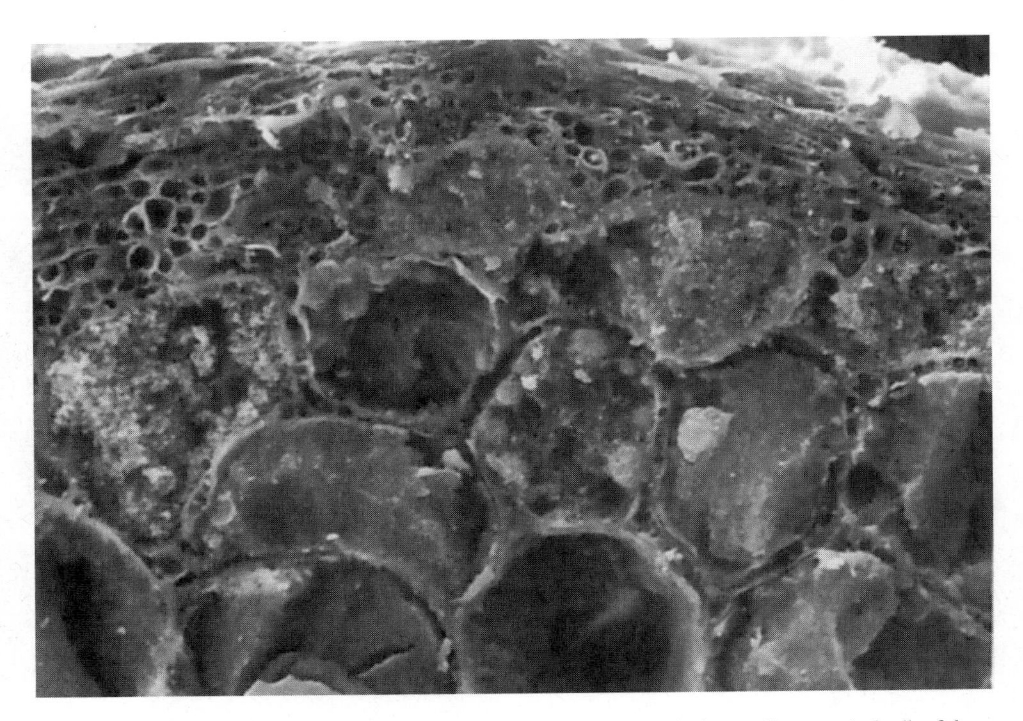

Figure 11.5. Hartig net and mantle formed by *Laccaria bicolor* on Scots pine. The large cells are cortical cells of the tree. Fungal tissue forms a mantle, or cortex, around the root surface, which shows in the photo as smaller cells with a honeycomb appearance. Fungal hyphae can also be seen encircling individual cortical cells, forming what is called the *Hartig net*. (Courtesy of Hughes Massicote)

belowground fruiting bodies called *truffles* or *false-truffles*. The spores of these belowground fruiters are spread by mammals; hence, the fruiting bodies are generally quite tasty. The European white and black truffles, considered to be delicacies, are the fruiting bodies of mycorrhizal fungi.

Several hundred fungal species in the subdivision *Zygomycotina* develop AM (quite a different group than those forming EM). Spores are always produced belowground. The vast majority of plant species form exclusively AM, including most trees of moist tropical forests, many temperate deciduous tree species, and some conifers (redwoods, cedars, and yews). Some tree genera have been known for some time to form both EM and AM (e.g., *Eucalyptus, Salix, Prunus, Acacia*). Some species in the family Pinaceae, formerly thought to exclusively form EM, are now known to also form AM (Cazares and Smith 1992, 1996).

Unlike EM, AM do not modify the external appearance of the root, and they characteristically penetrate the cells of the plant host. The name *vesicular-arbuscular* is derived from two kinds of structures formed by the fungus within the root. **Vesicles** are knoblike swellings that are formed within and outside plant cells by the hyphae of most (but not all) genera of VAM (fig. 11.6A); their function is unclear but probably relates to storage and perhaps propagation (Smith and Read 1997). **Arbuscles** are multiply branched, bushlike hyphae that usually occur only within root cells (fig. 11.6B). Because of their large surface area, arbuscles probably function to enhance the exchange of materials between fungus and plant.

Ericaceous plants, a group that includes many common forest shrubs and some small trees, may form EM, especially those in the genera *Arbutus* and *Arctostaphylous*. Other ericaceous genera have primarily ericoid mycorrhizae, in which the fungus penetrates plant cells. Formed primarily (or perhaps exclusively) by ascomycetes, ericoid mycorrhizal fungi are particularly adept at decomposing organic matter (Read 1991b).

There is a general tendency for trees that can form EM to be more common on sites that are very low in nutrients than trees that form only AM, and this is also true where climate limits the rate of nutrient cycling (e.g., boreal and montane forests) (Smith and Read 1997). For example, while most trees in the moist lowland forests of Cameroon form AM, soils that are particularly low in phosphorus are dominated by EM trees (Gartlan et al. 1986), a pattern that appears to hold throughout Africa (Hogberg 1986). Similarly, in the Asian tropics, trees in the EM family Dipterocarpaceae are most abundant on soils that are relatively low in phosphorus (Ashton 1988). It is not clear why low phosphorus availability seems to favor EM over AM trees. Both types of fungi are adept at gathering phosphorus; however, because of their greater hyphal development and particularly their ability to form rhizomorphs, EM fungi may be the better of the two types (Janos 1980; Ashton 1988). Another possible advantage of EM is their ability to store nutrients within their sheaths (Hogberg 1986). AM do not have sheaths.

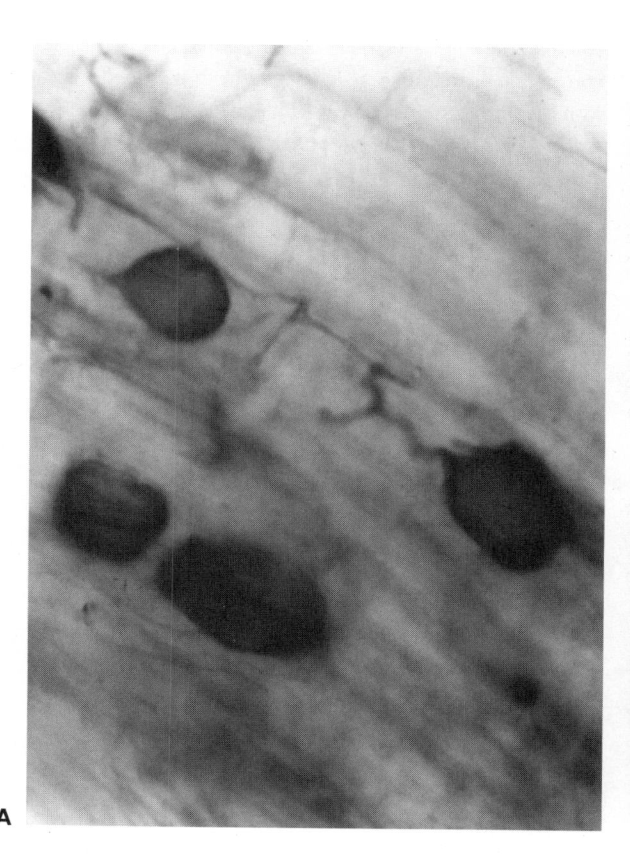

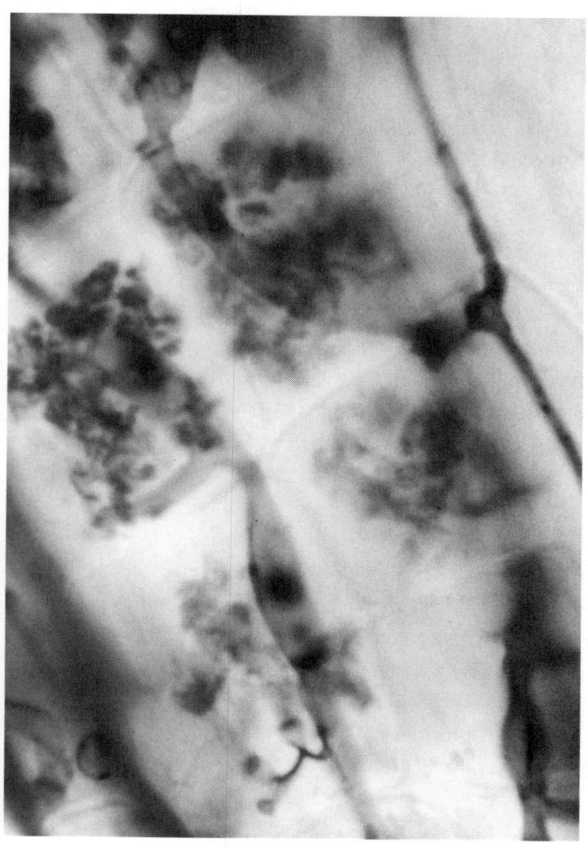

Figure 11.6. A. vesicles of a vesicular arbuscular mycorrhizal (VAM) fungus in a western hemlock root. Vesicles, which along with arbuscles, are produce only by VAM fungi, serve as storage organs and may mature into spores. Their presence is considered a definite sign that a root is colonized by VAM. (Courtesy of Efren Cazares) B. Arbuscles of a VAM fungus. Arbuscles are short-lived organs believed to function primarily in exchanging materials between fungi and plant cells. (Courtesy of Randy Molina)

Another possible difference between EM and AM fungi relates to their relative ability to decompose organic matter and directly cycle nutrients to their host. In boreal and subboreal forests, EM trees occur primarily on acid soils, while AM trees occur on more basic soils. The activity of decomposers tends to be lower on acid soils; hence, organic matter accumulates at the surface (chapter 18). At least some EM species commonly occur in the litter layers or at the interface between litter and mineral soil, and these are believed to cycle nutrients directly from litter to their hosts (Griffiths et al. 1994; Smith and Read 1997). The ability to decompose would be particularly advantageous to the tree-fungus partnership in environments where cold temperatures reduce the activity of saprophytes and, hence, slow litter decomposition and nutrient cycling (Janos 1980). As we saw above, ericoid mycorrhizal fungi are also capable of decomposing organic matter, and ericaceous plants are usually found where organic matter accumulates. (We return to the question of organic matter decomposition by mycorrhizal fungi in chapter 18.)

11.3.2 Pollination Mutualisms

Plants may be pollinated by wind, animals, or some combination of the two. In general, the importance of animal pollination declines from lower to higher latitudes, from lower to higher elevations, and from species-rich to species-poor forests (Regal 1982). This pattern reflects in part the dominance of conifers (which are always wind-pollinated) in boreal and montane forests, and in part the fact that temperate angiosperms are more likely to be wind-pollinated than are tropical angiosperms.[5]

Bawa (1990) estimates that at least 98 percent of the flowering plant species in moist tropical forests are pollinated by animals, as are most trees in the native forests of

[5]Wind-pollinated angiosperm genera include ash, oak, beech, elm, hazel, alder, birch, walnut, hickory, alder, chinkapin, tanoak, sycamore, and poplar (Regal 1982). Insects are the most common plant pollinators, especially bees (Janzen's "flying penises"). Birds and mammals (particularly bats) are also important pollinators in tropical forests and in temperate forests of Australia and South Africa, but pollination by vertebrates is rare in temperate forests of the Northern Hemisphere (Bawa 1990).

There are definite limits to the degree of generality in plant-animal mutualisms (i.e., pollination and seed dispersal), as evidenced by the fact that connectance decreases exponentially with species richness (Olesen and Jornado 2002). Accompanying the decline in connectance, a distinct structure emerges based on who partners with whom. The resulting organization manifests in at least two ways. First, connections are nested rather than random (Bascompte et al. 2003). What does that mean? Figure 11.7 compares idealized nested and random connections. If one orders plant species on the *y* axis from those having the fewest animal mutualists to those having the most (i.e., from the most specialized to the most generalized), and animal species on the *x* axis from those having the most to those having the fewest plant partners, a nested pattern shows a clear structure (fig. 11.7A). In simplest terms, for plants that have only one or a few animal partners, those animal partners are highly general in their choice of plant partners. The same holds for animals: the more specialized they are with regard to plant partners, the more general their plant partners. For example, consider plant species number 10 in figure 11.7A. It partners only with animal number 1; however, animal number 1 partners with all 10 plant species. Similarly, animal number 10 partners only with plant 1, which partners with all 10 animals species. Species within the core (outlined by the box) all partner with one another. A randomly structured mutualist network shows no such pattern (fig. 11.7B). An example of a real animal-plant mutualistic network is shown in figure 11.7C; it is not perfectly nested, but neither is it random.

Consumer-based food webs either do not exhibit nestedness to the same degree as the mutualisms (Bascompte et al. 2003), or they exhibit a complex, compound structure in which nestedness is embedded within compartments (Lewinsohn et al. 2006). Figure 11.8B shows a graphic array of nestedness values for real communities of seed dispersers, pollinators, and consumer-based food webs (each symbol is a separate community), and figure 11.8A shows summary statistics for the communities of figure 11.8B (1.0 is perfectly nested). Plant-pollinator and plant-seed disperser networks are similar to one another in their degree of nestedness, while consumer-based food webs are significantly lower.

The second kind of structure found in animal-plant mutualisms is a negative power-law relationship between *k*, the number of links per species (e.g., how many plant species a given pollinator services) and the proportion of species within a community having at least *k* mutualistic links (Jordano et al. 2003).[6] This phenomenon is shown in figure 11.9, where the number of links, *k*, is shown on the *x* axis, while the proportion of species with *k* or more links is shown on the *y* axis. Both the *x* and the *y* axes are scaled logarithmically, so that a straight line denotes a power-law function. The animal-plant mutualistic communities shown in the figure tend to follow truncated power laws, that is, they follow a straight line over certain ranges of *k* then begin to fall away. (In chapter 12 we see that even-aged stands of trees follow a very similar kind of relationship between tree size and stocking density.) Generalized food webs do not exhibit power-law structure (Dunne et al. 2002), a further indication that animal-plant mutualisms differ in fundamental ways from predator-prey relationships.

Australia. In North America, approximately 70 percent of tree species found north of the United States–Mexico border are animal pollinated, although wind-pollinated species such as the conifers and oaks account for the greatest proportion of forested area. There is one interesting example of combined animal and wind pollination: some tree species in the genus *Shorea* (which dominates many forests of Southeast Asia) are pollinated by tiny insects called *thrips* that are dispersed on the wind (Ashton 1988).

With few exceptions, plant-pollinator mutualisms are diffuse (i.e., general) rather than pairwise (i.e., specific) (Jornado 1987), and plant-animal mutualisms in general (i.e., pollination and seed dispersal) are highly asymmetric (Bascompte et al. 2006). Few trees flower year-round even in the tropics, and pollinators must generalize in order to maintain a continuous food supply. Many pollinators eat a variety of plant tissues and therefore do not depend exclusively on flowers. This is not true of bees, however, which subsist only on nectar and pollen and therefore depend on asynchronous flowering of their tree hosts throughout the year. Trees within forests that are especially rich in tree species have been found to suffer from pollination limitation, suggesting competition among trees for available pollinators (Vamosi et al. 2006).

Trees that flower year-round have the potential to develop very specific pollination mutualisms; such is the case with the various species of fig trees, each of which is pollinated by a single species of fig wasp. The one-to-one relationship that has evolved between fig trees and fig wasps is rare in pollination systems; however, in many cases the

[6] In chapter 7 we saw that forest fires also follow a negative power-law relationship between fire size and the number of fires of a given size. Recall that negative power laws can signify self-organizing behavior.

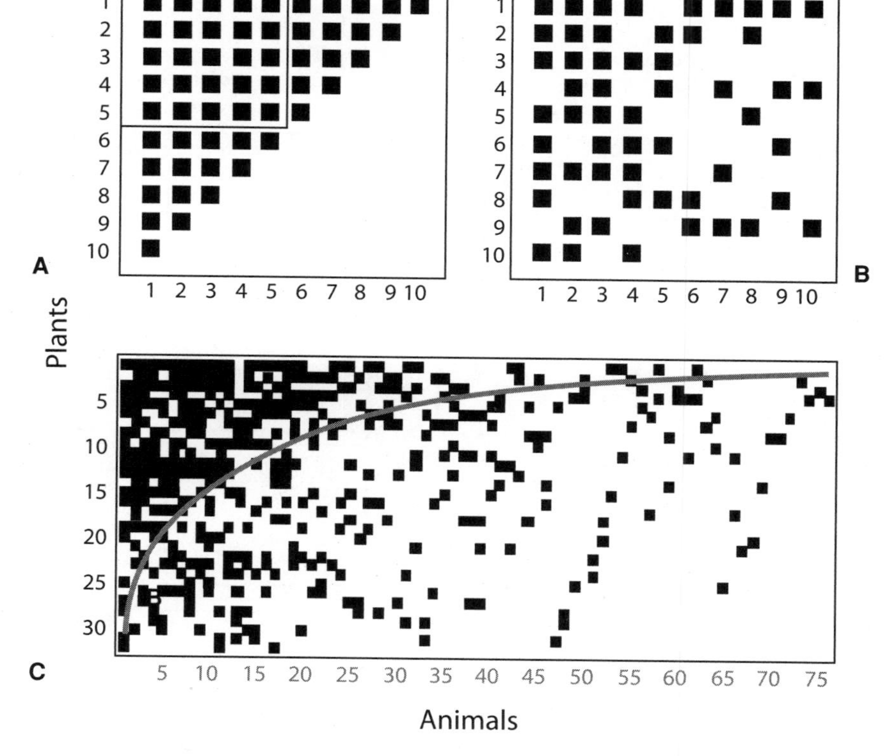

Figure 11.7. An illustration of plant-animal mutualistic interaction matrices. Numbers represent individual plant and animal species, ranked in decreasing number of interactions per species (i.e., species 1 in both groups has the most interactions, species 10 the fewest). A filled square indicates an observed interaction between plant i and animal j. A–C correspond to perfectly nested, random, and real mutualistic matrices (the plant-pollinator network of Zackenberg [J.M.O. and H. Elberling, unpublished work]), respectively. Values of nestedness are $N = 1$, $N = 0.55$, and $N = 0.742$ ($p < 0.01$) for A, B, and C, respectively. The box outlined in (A) represents the central core of the network, and the line in (C) represents the isocline of perfect nestedness. In a perfectly nested scenario, all interactions would lie left of the isocline. (Bascompte et al. 2003. Copyright 2003 National Academy of Sciences, U.S.A.)

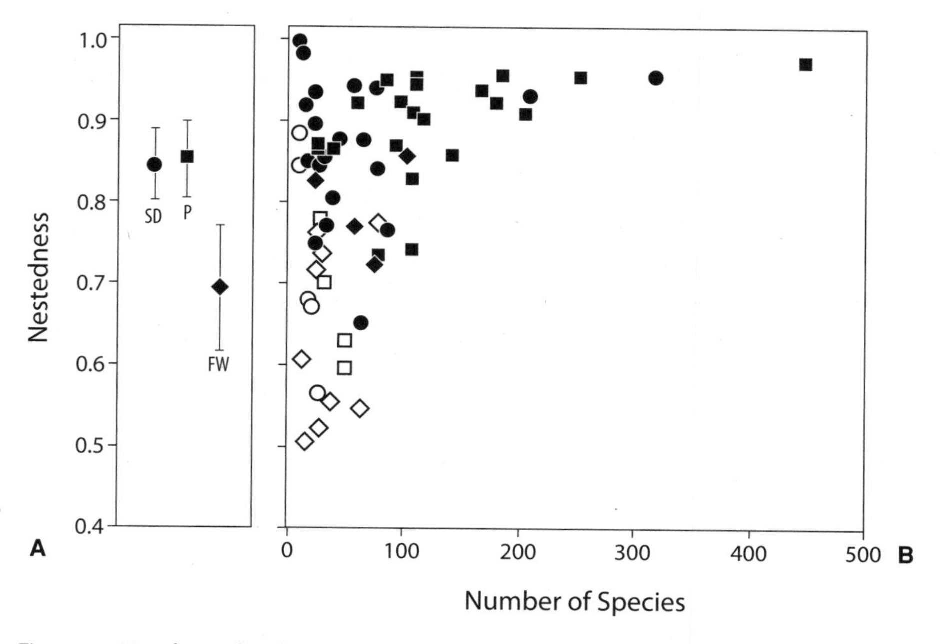

Figure 11.8. Nestedness values for seed dispersal (SD, *circles*), pollination (P, *squares*), and nonmutualistic food webs (FW, *diamonds*). A. Mean and SE of nestedness for the three types of networks. B. Nestedness versus species richness for all data sets. Each point corresponds to a specific community. Filled points indicate the level of nestedness is significant at the $p < 0.05$ level. (Bascompte et al. 2003. Copyright 2003 National Academy of Sciences, U.S.A.)

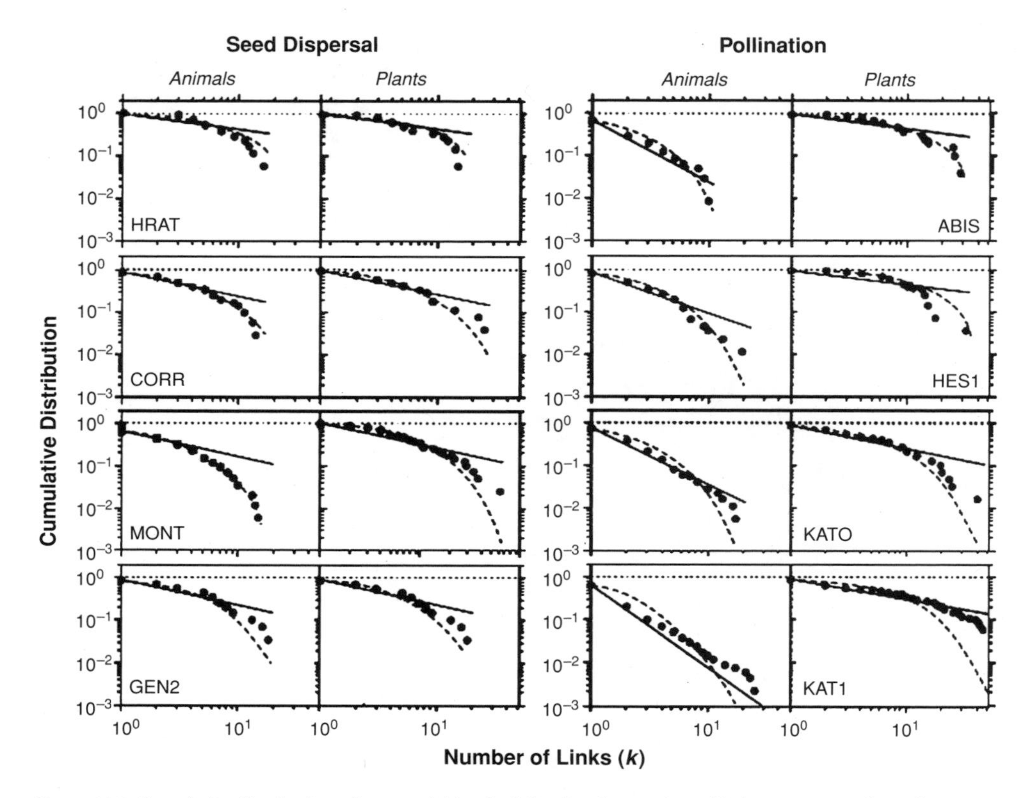

Figure 11.9. Cumulative distribution of connectivities (k, defined as the number of links per species) for pollination and seed-dispersal mutualisms. Panels show the log-log plots of the cumulative distributions of species with 1, 2, 3, . . . , k links *(dots)*. Solid lines denote power-law fits and dotted lines truncated power-law fits. Acronyms refer to locales of analyzed systems. (Jornado et al. 2003).

pollinators of a given plant species belong to the same taxonomic group (e.g., bees, beetles, bats) (Bawa 1990). Different animals are attracted by different colors: birds prefer green and red blossoms; bees prefer yellow and blue; flies, wasps, and beetles are attracted to brown or other drab colors (Faegri and van der Pijl 1971).

The structure of blossoms determines which animal can actually obtain pollen and nectar from a flower; hence, structure plays a very important role in the specificity of plant-pollinator mutualisms (Faegri and van der Pijl 1971). Open, "dish bowl" flowers are readily accessible and are pollinated by unspecialized animals such as beetles, wasps, flies, and bats. On the other hand, some types of flowers are not easy to access (e.g., those of the legumes) and are pollinated either by animals such as bees (whose size allows them to enter the flower interior), or by animals that can hover and probe with long tongues or beaks (e.g., moths, butterflies, birds). The fact that an animal of a given species visits a flower does not mean that animal is an effective pollinator (Bawa 1990). For example, of 10 different moth and beetle species visiting the flowers of *Calathea ovandensis,* a single moth species is responsible for 66 percent of the fruit that is set, and a second moth species accounts for another 14 percent (Schemske and Horvitz 1984).

Most studies of pollination have dealt with agricultural crops rather than forests; hence, much remains to be learned about this mutualism in forest ecosystems. Some of the more detailed work in forests has been done in the seasonal tropics, particularly Costa Rica (Frankie 1975). There, the predominant pollinators vary with season. Trees flowering in the dry season are most often pollinated by mid- to large-sized bees (roughly bumblebee size), while small bees and moths pollinate the majority of trees that flower in the wet season. In general, bees are relatively more important in drier forests and flies in wetter forests (Devoto et al. 2005).

In the aseasonal forests of Costa Rica (so called because they do not have a distinct dry season), about two-thirds of the overstory tree species appear to be adapted for pollination by bees, wasps, butterflies, or beetles (Frankie 1975). Bat-pollinated trees occur most commonly in the midcanopy layers, while hummingbird-pollinated plants (mainly vines, shrubs, and epiphytes) are most frequent in the understory.

11.3.3 Propagule Dispersal
11.3.3.1 Seeds

The seeds of over 60 percent of tree species in temperate forests are dispersed by animals, with approximately

50 percent of these producing nuts (i.e., hard-shelled fruits) and the other 50 percent producing fleshy fruits. Based on rates of postglacial tree migration, Wilkinson (1997) suggests that even predominantly wind-dispersed seeds are spread over large distances by birds.

More than two-thirds of canopy tree species in moist neotropical forests and from 35 to 46 percent of trees in tropical Asia and Africa produce fleshy fruits. The various species of palms are also important nut producers in tropical forests. In general, trees with animal-dispersed seeds are relatively more common in moist forests, and those with wind-dispersed seeds are more common in dry forests. (See reviews by Howe and Smallwood [1982], Lanner [1985], and Wheelwright 1988].) Seed-dispersal mutualisms are generally diffuse rather than pairwise, although in some cases at least one of the partners can be quite dependent on the other. Hutchins and Lanner (1982), for example, argue that whitebark pine relies solely on Clark's nutcracker to disperse its seeds.

Fleshy fruits and nuts are consumed by a wide variety of animals. Roughly one-third of bird species in both tropical and temperate forests eat fruit at one time or another (Wheelwright 1988); in the tropics, migrants from the temperate zone are important seed dispersers. According to Terborgh (1986), the most abundant warm-blooded animals in most tropical forests are birds and mammals that subsist primarily on fruits and nuts. Fish also eat fruits and disseminate seeds of some streamside trees in tropical riverine forests (Howe and Smallwood 1982).

Seeds within fleshy fruits may either be discarded or passed through animals unharmed and defecated along with a ready source of fertilizer; animals that cache nuts are frequently wasteful or forgetful. Birds and squirrels, for example, do not always retrieve their caches and, hence, become important disperses of tree species such as pines, oaks, and beeches. Lanner (1985) discusses the dispersal of pine nuts by birds:

> Nutcrackers and jays remove ripened seeds from cones. They fly to the ground, or to caching areas up to 22 km away, carrying one to over 100 seeds. . . . Seeds that are not eaten immediately are placed in the soil in groups of one to 15 or more, at 2–3 cm depth. In a mast year,[7] a single nutcracker may cache about 100,000 seeds. Seeds are retrieved and eaten throughout the fall, winter, and spring, and they are fed to newly hatched young. Unrecovered seeds, which in a mast year constitute a majority of those stored, often germinate and become established. Hutchins and Lanner (1982) showed that regeneration of whitebark pine *(P. albicaulis)* depends upon the services of Clark's nutcracker *(N. columbiana)*.

Jays and their relatives such as nutcrackers move copious amounts of nuts. Expandable throats and esophagi allow from three large nuts (e.g., white oak) to over 100 small nuts (e.g., pine) to be transported simultaneously. Individual birds cache literally thousands of acorns (Bossema 1979). In one study, 50 jays were estimated to transport and cache 150,000 acorns from 11 pin oak trees over 28 days (S. Darley-Hill, cited in Johnson and Adkisson 1986). Jays transport acorns and beechnuts up to several kilometers and bury them several centimeters beneath the surface in the vicinity of their nest sites (Bossema 1979; Johnson and Adkisson 1986). Nuts serve as winter food for birds that do not migrate, and spring food for returning migrants (nestlings are fed caterpillars that eat oak foliage).

Nuts are frequently buried in openings or at gap edges, probably because these areas are rather distinctive and serve as landmarks for locating the nuts later. Because of this behavior, seedlings germinating from nuts that are not retrieved by birds (or dug up by small mammals) have a good chance of emerging where there is enough light to survive and grow. The birds are very skilled at selecting only sound, viable nuts from trees. To the bird these are of course the most nutritious, but they are also the most likely to germinate and produce a healthy seedling if not retrieved and eaten.

11.3.3.2 Dispersal of Mycorrhizal Fungi and Associated Microorganisms

According to Gehring and Whitham (2002), spores of both AM and EM fungi "have been recorded in the feces of mammals from Australia, North America, Central America, and South America." Fungi whose spores occur only belowground, including the AM formers and some of the more common EM formers, are largely if not solely dispersed by animals. In North America, a wide variety of small rodents consume truffles, while marsupials play this role in Australia (Maser et al. 1978; Malajczuk et al. 1987). Small mammals are thought to be the primary dispersers of truffle-producing mycorrhizal fungi into disturbed areas; however, more research is needed on this point. Spores pass through the animal gut and are defecated unharmed, along with nitrogen-fixing bacteria and other microorganisms that live in association with the truffle (Li et al. 1986). Although most spores are passed within 24 hours, feces contain abundant viable spores for several days after an animal eats a truffle. Spores of truffles that are consumed immediately before hibernation are passed slowly over winter, many remaining in the gut to be excreted the following spring (Cork and Kenagy 1989).

11.3.4 Protective Mutualisms

Protective mutualisms include a variety of interactions in which one species protects another, with either food or shelter being the common payments. In most plant species,

[7]*Mast* is another term for *nuts*. Many tree species produce very heavy nut crops once every few years and few or none in between. Years of heavy crops are called *mast years*.

the two primary interfaces with the environment (leaves and roots) are characterized by symbioses with fungi, which in many cases function to protect their host plant from pathogens. Some types of nonmycorrhizal soil fungi and rhizosphere bacteria also inhibit plant pathogens. Ants form close mutualisms with some species of tropical trees, aggressively protecting the tree from defoliating insects and even from competing plants (chapter 19).

Fungal endophytes are a diverse array of fungi that live within the leaves and stems of every plant species examined to date (Stone et al. 2000). Endophytes of grasses have been extensively studied and are known to act as mutualists conferring an array of benefits on their hosts. Endophytes of woody plants are both much more diverse and much less studied than those of grasses. In the tropics, from 3 to 20 different species of fungal endophytes may occur within a single leaf (Lodge et al. 1996; Arnold 2002). Woody plant endophytes are morphologically similar to some plant pathogens, which has led some researchers to conclude they are unlikely to function as mutualists; however, genetic analysis of endophytes in white pine needles shows little genetic relatedness to white pine pathogens (Ganley et al. 2004).

Foliar endophytes have been implicated in tree resistance to pests in several studies, and in all likelihood this phenomenon occurs widely. Endophytes living within the bark of elm trees were found to inhibit the growth of bark beetles that spread Dutch elm disease (Weber 1981), and an endophyte of Douglas-fir foliage produces toxins that kill larvae of defoliating insects (Miller 1986; Carroll 1988). Barklund and Unestam (1988) have shown that the ability of a common pathogenic fungus (*Gremmeniella abietina*, known in the United States and Canada as Scleroderris canker) to infect Scots pine and Norway spruce correlates negatively with the presence of endophytes in foliage and twigs. Arnold et al. (2003) showed that foliar endophytes of the tropical tree *Theobroma caco* helped defend the tree from a major pathogen.

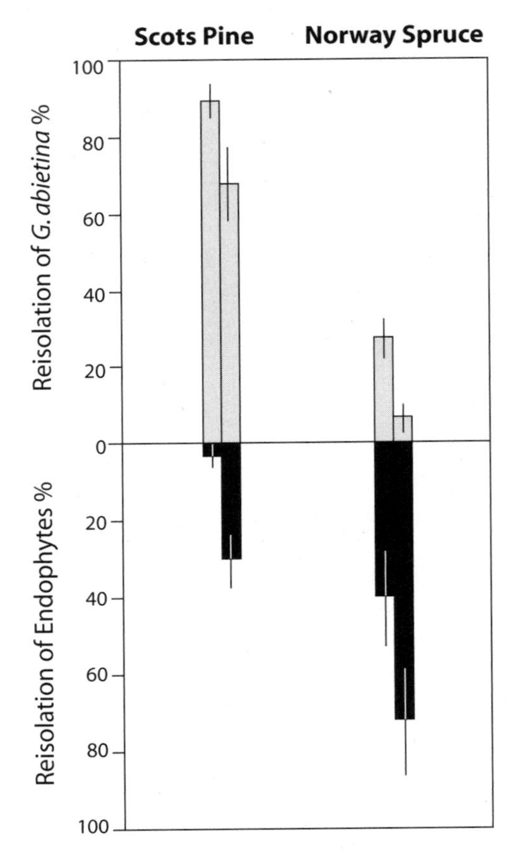

Figure 11.10. Correlation between the presence of the pathogenic fungus *Gremmeniella abietina* and foliar endophytes in the foliage of Scots pine and Norway spruce, showing the percentage of needles from which either the pathogen *(gray bars)* or the endophyte *(black bars)* were isolated. For each species, the left set of bars is from seedlings maintained at 0°C, while the right set of bars is from seedlings grown at 20°C. Note that conditions favoring the pathogen do not favor the endophyte, and vice versa. (Barklund and Unestam 1988)

The study by Barklund and Unestam is particularly interesting because it illustrates the importance of environmental factors in determining the outcome of species interactions. They found that endophytes grew best within conifer tissues when tree seedlings were grown at 20°C, while the pathogen grew best at 0°C (fig. 11.10). Moreover, when seedlings were sprayed with an acid mist in order to simulate the effects of acid rain, endophytes grew more poorly and the pathogen grew better. Barklund and Unestam suggest that pollution has increased the severity of the pathogen by inhibiting the protective endophytes. (Reports of Scleroderris damage to conifers in Europe and eastern North America increased dramatically during the 1970s and 1980s, coincident with increasing air pollution in these areas.)

11.4 COMPETITION

One of the oldest ideas in ecology is that individuals utilizing the same resource will compete if that resource is in short supply. For many years ecologists assumed that the sizes of all populations within a given community were ultimately limited by resources; hence, competition was believed to be an inevitable consequence of making a living, and the major determinant of community structure (i.e., the number of species and size of each population). Note, however, that constant combat for resources is not necessarily implied; over many generations, species may evolve ways to avoid competition (i.e., resource allocation; chapter 9). Nevertheless, the community is ultimately structured by competition, be it ongoing or, as a popular phrase puts it, the "ghost of competition past."

Many ecologists now agree that while competition for resources undoubtedly occurs (at least in some trophic levels), the notion that it is the major organizing force in nature is overly simplistic (Strong et al. 1984; Diamond and Case 1986; Hunter and Aarssen 1988; Callaway 1997; Bruno et al. 2003). The current view is best described as the recognition that ecological communities are complicated, variable, and with a structure that is shaped by

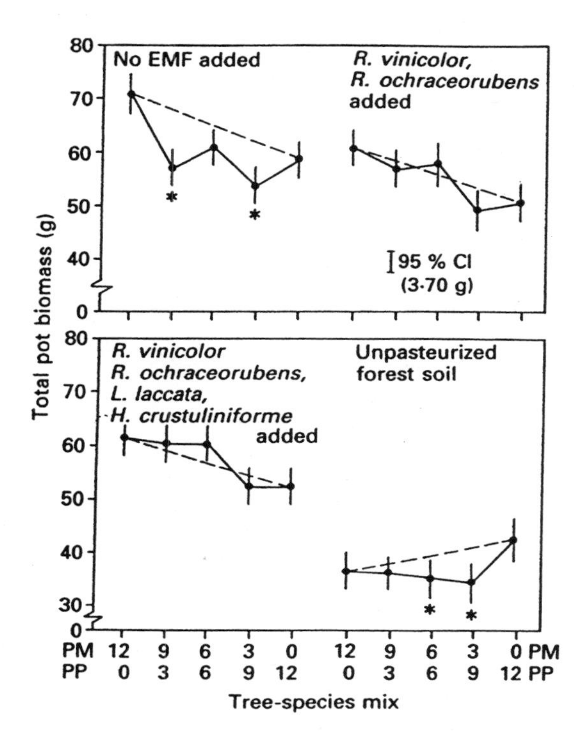

Figure 11.11. The influence of mycorrhizal fungi on competition between Douglas-fir and ponderosa pine seedlings. The *x* axis shows the number of seedlings of each species grown together in a given pot; the *y* axis shows the average total biomass per pot. The bars are 95% confidence intervals around each mean. Means with asterisks have a 95% probability of differing from the dashed line. When a starred mean is below the dashed line, seedlings of the two species mutually inhibit one another's growth. PM, Douglas-fir; PP, ponderosa pine. (Adapted from Perry et al. 1989c)

many interacting environmental and biotic factors in addition to resource competition.

At least two things come into play to modify the importance of competition within ecological communities. One is that species within a given trophic level may be limited by factors other than resources; hence, they seldom or never have to compete for food.[8] Predation, disturbance, or climatic fluctuations might all act to maintain species populations within a given trophic level below what their food supply would permit. This is frequently the case with herbivorous insects (e.g., Lawton and Strong 1981). As we saw in chapter 10, herbivores and pathogens can increase the species richness of plant communities by reducing the ability of any one species to competitively dominate others (see section 11.5). Higher-order interactions other than predation may come into play to reduce competition. Such is the case with at least some types of mycorrhizal fungi, which can convert a negative interaction between two plant species into one that is either neutral or mutually beneficial (Puga 1985; Perry et al. 1989c, 1992) (fig. 11.11).

How do mycorrhizal fungi mediate interactions among different tree species? More research is needed to answer this question definitively, however, hyphal linkages among plants probably play a role. Consider a forest in which different tree species form mycorrhizae with the same species of fungi (a common situation): a given fungus forms mycorrhizae with one tree, then extends hyphae into the soil and forms mycorrhizae with surrounding trees, thereby directly linking the trees to one another (fig. 11.12). There is no doubt that this phenomenon occurs within forests (and grasslands), and in fact it is probably the rule rather than the exception. Moreover, mycorrhizal fungi have been shown to facilitate the movement of carbon compounds and nutrients between individuals of different tree species, probably through the hyphal linkages that join them. Shaded trees, in particular, appear to obtain at least some carbohydrate from surrounding sunlit trees, although the ecological significance of such transfers is unclear at this time. We can conclude, however, that what appears as a stand of distinct, individual trees when viewed aboveground is much less individualistic when the belowground linkages are taken into account. (Simard et al. [2003] review studies of interplant transfer via mycorrhizal fungi and discuss the possible ecological significance.)

The other factor is that species may compete for the same resources and also benefit one another in ways that tends to dilute (or even negate) their negative effects. This probably occurs commonly in ecosystems but may not be readily apparent from casual observation—or even from experiments unless conducted over many years. In chapter 8, we discussed several examples of facilitation (i.e., one plant directly or indirectly helping another) during early succession. Atsatt and O'Dowd (1976) hypothesized that plant species participate in defense guilds (i.e., either directly or indirectly reduce herbivory, pathogenesis, or both within the community). For example, in British Columbia, Douglas-fir grows better in the absence of paper birch, but it is also more susceptible to Armillaria root disease (Baleshta et al. 2005). Flowering plants are common in young conifer forests, where they probably compete with the conifers for various resources; however, nectar produced by flowers of these plants is important in the diet of at least some insects that prey on defoliating insects. Hasson (1967) documented 148 species of parasitoids[9] associated with flowering plants in forests of northern Germany. What is the net effect of these plants on conifers? If they were not in the ecosystem, would faster conifer growth eventually be negated by root disease or

[8]Hairston et al. (1960) suggested that the importance of competition alternated up the trophic ladder: plants compete, herbivores are held down by predation and, hence, do not compete, and carnivores compete because they are at the top of the food chain and, therefore, have no predators. Evidence in support of this idea is equivocal (Schoener 1989b).

[9]Parasitoids are tiny wasps or flies that parasitize larger insects (chapter 19 in this volume).

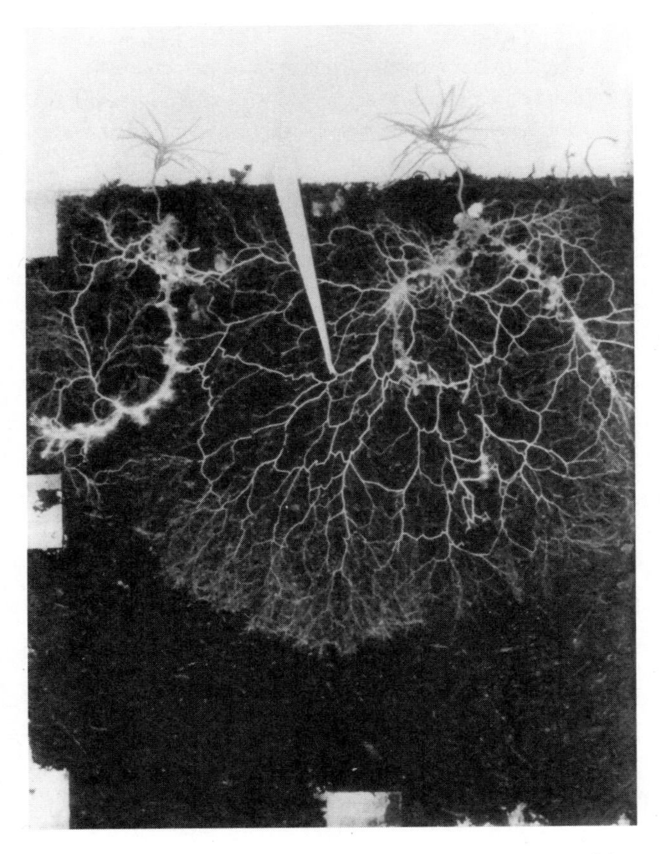

Figure 11.12. Mycorrhizal hyphae *(arrow)* linking the root systems of the two seedlings, one of which is Scots pine and the other lodgepole pine. (Courtesy of David Read)

larger populations of defoliating insects made possible by lower populations of their predators? Questions such as these are seldom entertained in competition studies.

In sum, long-term, diffuse interactions are the rule in ecological communities, rather than the exception (see section 11.5). Trees and other plants certainly compete among themselves for resources, but they may also depend on one another in numerous ways that are not readily apparent. Species that compete most of the time might benefit one another during certain critical periods.

> When one reflects on the multiplicity of indirect, diffuse, and subtle interactions that are possible in ecosystems, it is apparent that the experiments necessary to truly grasp the patterns of nature will be formidably difficult at best, and maybe even impossible (Holt 2006b). Like physicists, ecologists may have to accept an "uncertainty principle," in other words, we may never completely capture the richness of nature within the framework of scientific hypotheses and models. Recognition of this may have been why silviculture was originally defined as the art and science of managing forests.

11.5 HIGHER-ORDER INTERACTIONS

> Thou canst not stir a flower
> Without the troubling of a star
>
> FRANCIS THOMPSON[10]

Vandermeer et al. (1985) relate an old story about how spinsters saved the British empire.

> Darwin's observations of bumblebees and red clover led him to extrapolate that since field mice, who prey on bumblebee nests, were relatively scarce near villages, they could account for the presence of red clover there. The mice are presumably scarce there because of predation by domestic cats. A German scientist then continued to extrapolate that since cats were responsible for the presence of red clover, and since red clover was a staple food of cattle and since British sailors thrived on bully beef, one could conclude that Britain's dominant world position as a naval power was ultimately determined by the presence of cats. Thomas Huxley, tongue planted firmly in cheek, went on to note that old maids were the main protectors of cats, thus showing that the British empire owed its existence to the spinsters of England.

All of this after Darwin's initial observation was of course in the spirit of fun, but it nevertheless contains an element of truth. Because of the nature of connectance in ecosystems, interactions among organisms within a single species or between two species are in general greatly influenced by numerous other relationships within the ecosystem. Ultimately, any single effect is likely to emerge from a complex of interacting and interdependent causes.

It is not difficult to find real-world examples that show how profoundly ecosystem function depends on a web of higher-order interactions; in fact, this is almost certainly the rule rather than the exception (e.g., Gilbert 1980; Price et al. 1986; Perry et al. 1989b,1992; Carroll 1991; Krebs et al. 2001b). Subtle and easy to overlook in a healthy ecosystem, higher-order interactions usually get our attention only when some manipulation of one piece of the system ramifies through the interconnections to produce unforeseen results[11] (section 11.5.6).

Natural ecosystems are seldom, if ever, studied in sufficient detail to really understand the complex relationships that define the ecological web. It seems likely that only a small fraction of even the direct interactions among organisms in the wild have been documented. Certainly, the higher-order interactions are poorly understood and represent one of the frontiers of ecological research in the twenty-first century. The following example from a cultivated garden within the University of Michigan Botanical

[10]From *The Mistress of Visions,* 1913

[11]Vandermeer (1990) provides a mathematical model of how higher-order interactions influence community dynamics.

Gardens gives some hint of what might be expected in the wild (Vandermeer et al. 1985).

> (In) the experimental agriculture area . . . *Myzus persicae* [an aphid] attacks both beans and tomatoes. The . . . weed *Chenopodium album* attracts large numbers of *Aphis* sp. [another aphid], which attacks beans, which [has an indirect negative effect on tomatoes because beans] have a direct positive effect on tomatoes. The *Aphis* on *Chenopodium* attract large numbers of coccinelids which eat both *Myzus* and *Aphis*. . . . Thus *Chenopodium* has a direct negative effect on tomato through its competitive effect, an indirect negative effect [on tomato] through its competition with beans, an indirect negative effect [on tomato] by attracting *Aphis* which has a negative effect on beans, and an indirect positive effect [on tomato] by attracting *Aphis* which attracts coccinelids which eat *Myzus*.

Forests, grasslands, and deserts are considerably less amenable than experimental gardens to such observations; however ecologists are beginning to grapple with the role of higher-order interactions in the dynamics of wild ecosystems. As a general rule, interactions extend across space and ecosystem boundaries. In one of the better-documented examples of transboundary influences, Knight et al. (2005) described how fish contribute to plant reproduction. Dragonflies prey on the insect pollinators of certain plants; fish eat dragonflies, which relieves predation pressure on pollinators. Consequently, plants near ponds with fish are visited more often by pollinators and are less pollen limited than plants near fish-free ponds. This is an example of a **trophic cascade,** in which one trophic level influences nonadjacent trophic levels. We see other examples later in this chapter (e.g., section 11.5.1) and at various points throughout the book (e.g., chapter 19). Schmitz et al. (2000) reviewed studies dealing with how removing predators affects plants. They concluded that trophic cascades were common in terrestrial systems, but the effect varied, depending on the measure of plant performance used.

11.5.1 Isle Royale

Pastor et al. (1988) and Wilmers et al. (2006) discuss some of the multiple interconnections within Isle Royale

In the 1920s, the deer herd on the Kaibab Plateau, Arizona, apparently grew explosively, overgrazed its food source, and subsequently collapsed. Aldo Leopold attributed this overshoot in deer carrying capacity to the extirpation of large predators (especially wolves and mountain lions), which was official government policy in the late 1800s and early 1900s. Prior to 1970, the Kaibab story was commonly cited in ecology texts as an example of the importance of top-down regulation by large predators; however, questions arose about the validity of the data on deer populations, and alternative explanations were put forth to explain the explosion in deer numbers, if it indeed occurred. Subsequently, the story disappeared from texts.

The pendulum is now swinging back. A recent study by Binkley et al. (2006) provided strong evidence that the deer herd did increase on the Kaibab Plateau following extirpation of large predators, with significant impacts on plant communities. Recently Ripple and Beschta (2005) reviewed the history of extirpation of large predators in North America (especially wolves and mountain lions), and the subsequent effects on herbivore irruptions, and ultimately on plant communities: "In the late 1800's and early 1900's, wolves and other large predators in the western United States were besieged by widespread hunting trapping and poisoning efforts. . . . Ungulate irruptions, primarily of deer, began to occur following the occurrence of wolf extinctions. . . . In the eastern United States, the extirpation of large carnivores and subsequent deer irruptions occurred earlier."

Predictably, ungulate irruptions were followed by degraded plant communities. In Yellowstone National Park (YNP), overgrazing by elk after wolves were extirpated caused recruitment of aspen and cottonwood to drop dramatically. Ripple and Beschta (2005) note that "Results from YNP are consistent with other documented cases of trophic cascades in the Rocky Mountains, involving wolves, moose, willow, and birds in Grand Teton National Park (Berger et al. 2001) and wolves, elk, and aspen in The Canadian Rocky Mountains (White et al. 2003)."

Ripple and Beschta (2005) discuss the effects of reintroducing wolves into YNP: "Following the reintroduction of wolves, top-down trophic cascades have been observed, including altered patterns of ungulate herbivory, declining elk and coyote populations, new recruitment of woody browse species, and increases in the number of active beaver colonies on the northern range." Ripple and Beschta (2005) also observe that a single large predator may be insufficient to regulate lower trophic levels; an intact predator community may be required (e.g., wolves and bears).

Finally, it is important to bear in mind that multiple factors may combine to influence any ecological dynamic. As Jack Rumely, former professor of ecology at Montana State University, used to tell his students "never be a one-factor ecologist."

National Park in northern Michigan; among other things, the forests of Isle Royale are shaped by interactions among moose, wolves, beavers, early and late successional trees, microbes, climate, canine viruses, and disturbance patterns.

Isle Royale lies near the southern margin of the boreal zone. Late successional forests are dominated by three conifers: balsam fir, white spruce, and black spruce (*Abies balsamea, Picea glauca.* and *P. mariana,* respectively). Early successional forests include the fast-growing hardwood trees aspen (*Populus tremuloides*), balsam poplar (*P. balsamifera*), and birch (*Betula papyrifera*). Moose preferentially browse the hardwoods; they eat balsam fir only when nothing else is available, and spruce rarely (if at all). The direct effect of this is to hasten the return of disturbed areas to conifers. But there are also numerous interconnections with other parts of the system.

By feeding on the relatively nutrient-rich foliage of hardwoods in preference to conifers, moose determine the quality of food going to the soil invertebrates and microbes that decompose plant tissues; this, in turn, affects the rate of nutrient cycling. In the short run, nutrients are cycled more quickly, because the moose accelerate the transfer of easily decomposed organic matter (i.e., feces) to the soil. The longer-term effect, however, is quite the opposite. The secondary chemicals that make conifer foliage unpalatable to ungulates also reduce its decomposability, and decomposers are further inhibited by the acidity and low cation content of conifer foliage (see chapter 18 for factors controlling decomposition). Hence, feeding by moose eventually slows the nutrient cycle as sites become dominated by conifers.[12] Changes in the rate and pathways of nutrient cycling feedback in various ways to the relative growth of the various tree species. For example, defecating moose might spread spores from the belowground fruiting bodies of ectomycorrhizal fungi, and hyphae radiating out from conifer ectomycorrhizae may decompose litter, directly cycling nutrients from organic matter to their host trees.

By hastening the succession from palatable hardwoods to unpalatable conifers, moose negatively affect their own food source. However, wolves prey on moose, which combines with pathogens and plant defensive chemicals to prevent the moose from attaining population levels that cannot be sustained. Note that while wolves and other controlling factors have a direct negative impact on moose populations, their indirect, longer-term effects are positive. This can be seen in the recent history of Isle Royale (Andrewartha and Birch 1984).

Moose first arrived on Isle Royale about 1900 by swimming from the Canadian mainland. At this time, no wolves were on the island. Between 1915 and 1930, moose numbers increased from 200 to somewhere between 1,000 and 5,000. These numbers could not be sustained by the food supply, however, and by 1936 the population had declined to about 400 animals. A large forest fire that burned approximately 25 percent of the island in 1936 renewed the supply of favored moose food (i.e., early successional hardwoods), and by 1948 the population had rebounded to about 800. In the early 1950s, it crashed once again as hardwoods gave way to conifers. Wolves arrived on the scene in about 1948. From that point until 1980, moose populations were strongly regulated by wolves and only weakly by climate and food (Wilmers et al. 2006). In 1980, however, a severe outbreak of canine parvovirus (CPV) caused wolf populations to crash, a disaster from which they have never fully recovered. Wilmers et al. (2006) discuss the dynamic since 1980:

> Our study reveals that the release of moose from top-down control by wolves strengthens the contribution of climate to moose population dynamics on Isle Royale. The reduction in control of moose by biotic factors and the corresponding increase in abiotic climatic factors may erode the stability of this community. . . . Moose population dynamics prior to the outbreak of CPV in 1980, for instance, were characterized by a relatively slow and steady increase and decrease in the population. Since 1980, however, the moose population has displayed irruptive dynamics characterized by a steep increase in a population relatively free from significant predatory pressure, and dramatic declines in numbers when winters are severe.

The dynamic on Isle Royale is strongly influenced by landscape patterns resulting from disturbances of various types and scales. Moose require a variety of habitats within their 300- to 500-ha home ranges, including recently disturbed areas for browse, wetlands from which they obtain micronutrients (Pastor et al. 1988), and conifer forests for cover. Fire, logging, defoliating insects, and tree felling by beavers initiate early successional patches of different shapes and sizes that differ to one degree or another in their implications for moose. Events occurring at a global scale potentially alter fire patterns, tree susceptibility to insects, competition between tree species, and the food supply for moose, microbes, wolves, and ultimately all of the actors on the Isle Royale stage. For example, the severe winters that have exerted control over moose since the crash of the wolf population correlate with the North Atlantic Oscillation (chapter 3).

11.5.2 Generalized Food Web Interactions

In chapter 10, we introduced the concept of top-down and bottom-up regulation of ecosystem dynamics, where

[12]For recent reviews of linkages between above- and belowground biota, see Wardle et al. (2004b) and Wardle and Bardgett (2004).

top and *bottom* refer to positions within the food web. These are part of a larger array of possible regulatory interactions within food webs, which are described by Sinclair and Krebs (2001) as follows:

- **Bottom-up models** have an uninterrupted direction of control from lower to higher levels. They . . . predict that nutrient availability determines the biomass in the higher levels and thus assume that supply of nutrients is limiting. . . . In terrestrial systems this would apply to polar tundra, subarctic boreal forests with acidic soils, and nutrient-poor sclerophyll forests in Australia.[13]
- **Top-down models** have an uninterrupted direction of control from the top level downward. Thus, predators control the herbivore biomass, and herbivores control plant biomass. They . . . predict that nutrients are not limiting and community structure is determined by . . . predation. These models could be applicable to aquatic systems, particularly the more eutrophic lakes with many trophic levels. Larger fish species determine the presence or absence of small fish species, which, in turn, determine which zooplankton occur.
- In **herbivore-dominated models** the herbivore level controls both the predator and plant levels. Generally, such systems are grasslands.
- **Vegetation-dominated models** predict that soil nutrients are determined by plant biomass and that herbivores are limited by food supply. Such models would apply to forests where trees dominate the biomass. In particular, tropical forests absorb most of the nutrients in the soil, leaving the soil low in nutrients. At the same time, herbivores in these forests, largely insects, consume only a small fraction of the vegetation because plants are protected by structural and chemical defenses. Similar processes occur in temperate forests.
- **Dilution models** imply that both the lower and higher levels limit the intermediate trophic level. Thus there are both bottom-up and top-down effects. Such a process may be seen in aquatic systems, where both nutrient content of lakes and the predator community affect trophic dynamics. Insect-parasitoid-dominated systems may also be dilution systems, where parasitoids control herbivorous insects and nutrients control vegetation, which in turn affects herbivores.
- **Reciprocal models** suggest that there are two-way interactions between most of the trophic levels. These models . . . could apply to most ecosystems.

[13]This would also apply to tropical forests growing on extremely nutrient-poor soils.

Plants, particularly trees, have profound, indirect influences on other organisms that go beyond their function in the food web. As we saw in chapter 3, forests play a critical role in regulating the cycles of elements, especially carbon and water, and in determining planetary albedo.

Kennedy et al. (2006) suggest that rock weathering by early terrestrial plants and associated fungi set the stage for the appearance of higher animals. It is well known that for metabolic reasons, complex multicellular organisms (i.e., metazoans) could not evolve until atmospheric oxygen levels reached a certain level. What determines atmospheric oxygen levels? Plants release oxygen during photosynthesis, but virtually all of this is consumed in respiration, decomposition, or fires, all of which involve an oxidation process that exchanges atmospheric O_2 for CO_2. In order for O_2 to accumulate in the atmosphere, organic matter must be stored where it cannot be oxidized. Burial in deep-sea sediments is the primary way that happens, a process that largely depends on organic matter being sheltered from O_2 by clays (Kennedy et al. 2006). Organic matter produced on land enters the soil, where some proportion is protected by clays. Erosion eventually carries the organic matter–clay particles to rivers and ultimately to deep ocean sediments. The key to this process is rock weathering, from which clays are produced as a by-product. Kennedy et al. (2006) suggest that accelerated weathering by early land plants and their fungal associates (what Kennedy et al. call the **clay mineral factory**) accounts for a well-known stepwise increase in atmospheric O_2, which is widely believed to have been a necessary precursor to the appearance of metazoans.

Note that the terms *bottom-up* and *top-down* are commonly used in a less restrictive sense than in the definitions given above, being applied to any case in which a given trophic level regulates higher levels (i.e., bottom-up) or lower levels (i.e., top-down).[14] To avoid confusion, we refer to the dynamics defined by Krebs et al. (2001b) as *strict bottom-up* or *strict top-down*. Any other use of *bottom-up* or *top-down* will be in the less restrictive sense. Most natural systems probably combine elements of both top-down and bottom-up regulation, the relative importance of each likely to depend on a variety of poorly understood factors. Common sense (which is sometimes correct) argues that strongly resource-limited systems are more likely to be governed by bottom-up forces, but the evidence is mixed. For example, Gruner (2004)

[14]Similar to Hairston et al.'s suggestion that competition varies with trophic position (see note 8), Fretwell (1987) and Oksanen (1988) hypothesized that trophic levels "alternate between top-down and bottom-up regulation" (Krebs et al. 2001). This idea is known as the Fretwell-Oksanen hypothesis.

Figure 11.13. By damming streams and thereby creating pools and localized wetlands, beavers perform a key role in creating habitat for numerous other species.

manipulated bottom-up forces in a severely nitrogen-limited Hawai'ian forest by fertilizing and top-down forces by excluding birds from tree foliage. He found that canopy arthropods responded strongly to fertilization but were little affected by excluding their predators; in other words, he found strong bottom-up regulation. In contrast, an experiment in a nitrogen-limited boreal forest found strong top-down regulation (Krebs et al. 2001b). Both systems had mixtures of bottom-up and top-down control, but their relative importance differed between the two. As we discuss in chapter 19, herbivorous forest insects (e.g., defoliators, bark beetles) are regulated by both bottom-up and top-down forces (i.e., a dilution model).

11.5.3 Keystone Species

Individual species that are deemed to play some unique role in ecosystem structure or processes are called **keystone species.** The basic idea behind this concept is that some species are more important to community patterns and processes than others. The test of "keystoneness" is how different the community would be if the species in question were removed.

A species is frequently accorded keystone status when by feeding or other activities it significantly alters the structure (e.g., population densities) or dynamics (e.g., behavior)[15] of trophic levels below it, or when it provides unique food or habitat for other members of the community. Beaver are keystone species in northern forests because their ponds provide unique habitat for aquatic species, water birds, and amphibians (fig. 11.13). Through their effect on tree species composition, moose are a keystone species on Isle Royale. To the extent wolves control populations of beavers and moose, they would also be keystones because they regulate other keystones. In a unique and elegant experiment,[16] Krebs et al. (2001b) showed that snowshoe hares are a keystone species in North American boreal forests. Hares have far greater biomass than any other vertebrate; they consume more vegetation and are critical prey for the region's major predators (i.e., lynx, coyotes, and great horned owls). Krebs et al. (2001c) conclude that "if hares were eliminated, the boreal forest vertebrate community would largely collapse." Predators play a critical top-down regulatory role in the boreal ecosystem; however, the predator community contains a great deal of redundancy, and no single species qualifies as keystone (Krebs et al. 2001c).

Particular plant species that fulfill some irreplaceable function for higher trophic levels also qualify as keystones. Nodulated, nitrogen-fixing plants are one example. Fig trees are considered keystones in rain forests of South America and Borneo because they provide a reliable source of food for birds and mammals during periods when other trees are

[15]An influence that alters factors other than population density (e.g., behavior, phenology) is called a **trait-mediated effect.** Predators, for example, commonly alter the behavior of their prey and of other predators.

[16]Although inferences can be drawn from observation and modeling, the ultimate scientific test of whether a species is a keystone is to experimentally remove it. Difficult and expensive, selective removal of terrestrial vertebrates is rarely done on an experimental basis (though it has commonly been done during human colonizations). Krebs and colleagues did such an experiment. Their results are published in the book *Ecosystem Dynamics of the Boreal Forest* (Oxford University Press, 2001).

not fruiting or flowering (Leighton and Leighton 1983; Terborgh 1986). In the rain forests of Gabon, trees other than fig perform the keystone service of providing food during lean periods (Gautier-Hion and Michaloud 1989). Keystones do not have to be living; old dead wood is a keystone structure in many forests because of the unique habitat it provides (chapter 10).

But the plot thickens! If, for example, the fig tree is a keystone, then surely the fig wasp upon which the tree depends for pollination is also a keystone, and so there are keystone mutualisms. And how about the microbes upon which the keystone plants rely (either directly or indirectly), such as nitrogen-fixing bacteria and mycorrhizal fungi? Without these tiny and unapparent residents of the soil, most or all natural ecosystems on land (certainly all forests) would collapse!

There may also be keystone interactions (i.e., a keystone function arising from the actions of two or more species) that are not mutualisms. Helfield and Naiman (2006) give the example of salmon and bears in northern forests. Salmon returning from the ocean to spawn import nutrients (in their bodies) from the sea to freshwater streams. Salmon carcasses are then moved from streams into riparian forests by bears. Helfield and Naiman conclude that the joint action of salmon and bears may provide up to 24 percent of riparian forest nitrogen budgets.

We would hope that having read this chapter, it is clear to you that the concept of keystone species is a very knotty one at best. Certainly some species have inordinate effects on system structure and processes; the problem is that interrelationships are so complex and intertwined within ecosystems that numerous species and mutualisms might qualify as keystone. Speaking of keystone predators (but more widely applicable), Berlow et al. (2004) point out that "keystone predation is not a single strong predator–prey interaction, but rather a particular configuration or structural organization of strong and weak links: strong predation, relative to other predators of that guild, on a competitively dominant prey species. It demonstrates how a combined knowledge of both web structure and interaction strengths is a key to understanding how ecological communities function."

One critical factor underlying keystoneness is the concept of redundancy that we discussed at the beginning of the chapter: those species and structures whose functions within the system are not backed up by others are indeed keystone. Nothing else on Isle Royale dams like a beaver or eats like a moose; and past history suggests that though parasites and diseases may exert some control over moose populations, the Isle Royale dynamic would also be quite different without wolves.[17]

[17] A related concept is that of ecosystem engineers, species that modify the physical environment in some manner that creates niches for other species. Some keystones are also engineers (e.g., beavers), but many engineers are probably not keystones. *BioScience* 55 (7) 2006 contains several papers on ecosystem engineers.

11.5.4 Who's Your Buddy?

One simple rule of thumb that has been used to characterize higher-order interactions is that friends of friends are friends, enemies of enemies are friends, and friends of enemies are enemies (Post et al. 1985).

11.5.4.1 Friends of Friends Are Friends

This is most readily seen in diffuse mutualisms. Different plant species that depend on the same pollinators, mycorrhizal fungi, or seed dispersers become indirectly linked to one another; the welfare of one species is tied to that of another through a common partner. Like many interactions within ecological communities, those among the "friends of friends" are multifaceted: their relationships may be competitive (i.e., the trees vying for the services of their shared mutualist), cooperative, or combine elements of both cooperation and competition. Ultimately, however, species that share partners have a common interest in the welfare of that partner and, hence, come to indirectly depend on one another. This fact has significant implications for the structure of relationships within ecosystems, a point we shall take up in more detail in section 11.5.5.

11.5.4.2 Enemies of Enemies Are Friends

Perhaps the most common example of this type of indirect interaction is the role played by predators and parasites in controlling populations of herbivorous insects, which indirectly benefits the plants that are eaten by those insects (chapter 19). Many interactions among organisms that live in plant rhizospheres also indirectly benefit plants. As we saw earlier, mycorrhizal fungi protect plants against pathogens, as do certain rhizobacteria. Soil microbes may also compete with plants for nutrients, but soil insects and protozoa that graze on microbes indirectly benefit plants by releasing nutrients from microbial tissues (see chapters 14 and 18 for more on the rhizosphere).

As we have discussed previously, herbivores and pathogens can alter the competitive balance within a plant community and thereby indirectly benefit plants that might otherwise be dominated by faster-growing neighbors. As a rule, foliage and twigs of fast-growing, early successional trees and shrubs are more palatable than that of slower-growing species; hence, the faster growers are preferred by herbivores.[18] Large herbivorous mammals in particular have considerable influence on plant-plant interactions during early succession, although so might many small herbivores such as snowshoe hares. Earlier, we saw that moose preferentially browse early successional over late successional trees; the same is true of elephants (Mueller-Dombois 1972).

[18] Palatability of plant tissues to herbivores correlates positively with nutrient concentration (particularly nitrogen) and negatively with the concentration of the so-called secondary chemicals, such as tannins, resins, and alkaloids. (See chapter 19 for a more detailed discussion of plant secondary chemicals.)

11.5.4.3 Friends of Enemies Are Enemies

This one isn't too hard to figure out. Put yourself in the place of a strangler fig trying to take over a cecropia tree that is defended by ants!

11.5.5 Individualism Leads to Interdependence

Individualism on the part of different species that share the same mutualists serves to reduce competition for the services of those mutualists; paradoxically, it also ties the whole system more closely together. *Individualism* as we use it here may refer to any number of differences in species niches, such as different flowering times, mast years, responses to climate, or adaptations to disturbance.

To see how individualism can lead to interdependence, consider variation in the seasonal flowering patterns of different tree species that share the same pollinators. During periods when one tree species is not flowering, its pollinators are sustained by species that are in flower (Frankie 1975; Gilbert 1980). If we could flick a switch and remove all tree species that flowered from, say, March through April, species that flowered during other times—and that used the same pollinators—would be affected. Similarly, temperate nut-bearing tree species tend to produce heavy nut crops during different years. As is the case with staggered flowering, this helps to ensure a steady food supply for the birds and mammals that disperse seeds (Johnson and Adkisson 1986), reduces competition among nut-bearing trees for the services of seed dispersers, and indirectly links the various nut-bearing tree species (because they are "friends of friends").[19]

> Van Ommeren and Whitham (2002) give the example of juniper *(Juniperus monosperma)* and its associated mistletoe *(Phoradendron juniperinum)*. The mistletoe parasitizes the juniper; however, both use the same avian seed dispersers. Whereas berry crops in juniper vary widely a from year to year, those of mistletoe are quite consistent. Consequently, mistletoe provides a stable food resource for the birds that disperse juniper seeds. Van Ommeren and Whitham found two times more juniper seedlings in stands with high mistletoe density than in stands with little or no mistletoe. Evaluated solely as a two-way interaction between mistletoe and juniper, the mistletoe has a negative effect on the juniper; however, when viewed as a three-way interaction the effect is positive.

In the examples discussed, the differing life histories of different tree species provide a continuity in time in the food supply of pollinators and seed dispersers. Another kind of continuity in time (and another example of how diffuse mutualisms tie a community together) relates to the plant-mycorrhizal fungal community. If we move forward through the centuries in any forest, we would see the processes of disturbance, succession, and variable colonization producing periodic change in the plant species that occupy any given piece of ground (chapters 6, 7, 8, and 10). Within this shifting mosaic, each plant species that requires belowground mutualists will inevitably depend on some other species to support its friends during periods when it is absent. Conversely, the belowground mutualists will evolve generality in their host plants: the more species rich a forest, the greater the generality that we might expect in the partnership between plants and belowground mutualists. In fact, this is precisely what is seen in plant species-rich moist tropical forests, with their highly general AM fungi (section 11.3.1). As with trees and their pollinators, the individual responses on the part of plant species (in this case, to the timing and nature of disturbance) act to reinforce the indirect links among various plant species.

Note that all of the examples discussed in this section (and diffuse mutualisms in general) are redundancies within the structure of ecosystems (section 11.1.2) that serve to maintain processes even though the abundances of individual species are fluctuating. Such redundancies are a central feature of ecosystem stability (a point we discuss in some detail in chapter 21).

11.5.6 Interactions Link Biomes and Landscapes

Species interactions knit regions of space together in at least two ways.

1. Mobile links, such as anadromous fish and migratory birds, transport nutrients (and sometimes a multitude of other things).[20]
2. Top-down forces link any given piece of ground to a hierarchy of larger landscapes within which it is nested. This is because each succeeding trophic level as one moves up the food chain generally requires more space to make a living. The fruiting body of a mycorrhizal fungus requires only the tree it is partnered with; the squirrel that eats the fungus may require several hectares; the owl that eats the squirrel may require several thousand hectares.

One important consequence of this pattern is that top-down regulation by large predators is linked to landscape patterns. This was dramatically illustrated by a natural experiment in Venezuela, where damming a river to generate hydroelectric power created several islands within the new

[19] In contrast to most tropical and temperate forests, mast years are synchronized in those dipterocarp forests that grow on highly infertile "white sand" soils in Southeast Asia, perhaps to prevent populations of seed-eaters from building too high (Janzen 1974).

[20] Humans are the most active mobile link on the planet today and perhaps always have been.

lake. Terborgh et al. (2001) summarize the ecological consequences (see also Terborgh et al. 2006): "Limited area restricts the fauna of small (0.25 to 0.9 ha) islands to predators of invertebrates (birds, lizards, anurans, and spiders), seed predators (rodents), and herbivores (howler monkeys, iguanas, and leaf-cutter ants). Predators of vertebrates are absent, and densities of rodents, howler monkeys, iguanas, and leaf-cutter ants are 10 to 100 times that on the nearby mainland. The densities of seedlings and saplings of canopy trees are severely reduced on herbivore-affected islands, providing evidence of a trophic cascade unleashed in the absence of top-down regulation."

Although an unusually clear illustration, this is not a unique story. In many areas, encroachment by humans has reduced or eliminated the large areas of suitable habitat required by large predators, with cascading consequences discussed earlier.

In an article titled "Are predators good for your health?" Ostfeld and Holt (2004) review evidence for the role of top predators in reducing human diseases. Approximately 60 percent of human infectious diseases are zoonotic, that is, other animals serve as a reservoir for the pathogen. Most of these are mammals, predominantly rodents. Are predators good for our health, and have we encouraged disease by disrupting a balance? Ostfield and Holt summarize the state of knowledge:

> As is often the case in ecology, predicting the specific consequences of human-caused degradation of natural systems becomes difficult when we consider the complexity of the system. Here we have evaluated the assertion that that human health is protected when populations of rodent predators (largely carnivores and raptors) are unhindered by human activities, liberating them to regulate their rodent prey and thereby reducing the spread of pathogens from rodents to people. Beyond the fact that many human disease agents use rodents as a natural reservoir, this assertion remains largely untested. More information is needed on the importance of rodent density (vs. age structure, behavior, and other factors) for disease dynamics, and thus on human disease risk; on the impacts of different species of predators on rodent populations; and on the impacts of human activities on the various predatory species, as modified by their interactions with each other.

11.6 SUMMARY

Ecological communities are not as tightly linked as organisms, but neither are they simply collections of individuals. Rather, the community is a unique form of biological system in which the individuality of the parts (i.e., individuals and species) acts, paradoxically, to bind the system together. Multifaceted relationships characterize many interactions among species. The structure of ecological communities has two things in common with that of languages: context and redundancy. Context means that pattern is important (e.g., the arrangement of patches across the landscape, the particular mix of species in a local ecosystem). Redundancy refers to useful repetition, which implies that within limits, the species composition of a system can change without disrupting critical system functions such (e.g., the nutrient cycle).

Mutualisms, a type of interaction in which both partners benefit, are ubiquitous in nature. Among the many kinds that occur in forest ecosystems are those between fungi and plant roots (mycorrhizae), fungi and plant leaves and stems (endophytes), nitrogen-fixing bacteria and both plants and animals, and plants and the animals that pollinate flowers or disperse seeds. Diffuse mutualisms (i.e., each partner has several other partners) are much more common in nature than one-on-one mutualisms. Competition, an interaction in which each participants effects the other negatively, occurs when two species require the same limiting resource at the same place and time. Although species within the same trophic level will compete with one another, this does not always occur. Herbivores in particular are frequently limited by predators rather than resources and, hence, are seldom forced to compete. Plants compete but also cooperate in various ways, and their net effect on one another may be difficult to ascertain when viewed only over the short term.

Higher-order interactions, those involving three or more species, are the rule in ecosystems. Diffuse mutualisms are an important ecological binding force; different plant species, for example, become indirectly linked to one another through their common interest in mycorrhizal fungi, pollinators, and seed dispersers. Different species also become indirectly linked through the food web. Top down regulation by predators mediates interactions among prey species and, when the prey are herbivores, reduces feeding pressure on plants. Herbivores mediate interactions among plants. The complex of indirect interactions links systems in space and time and serves as a significant unifying and stabilizing force in nature.

12

Size-Density Relationships in Forests over Time and across Space

12.1 SELF-THINNING: AN ORDERLY PROCESS

Because they share what is essentially the same niche, others of the same species and age are generally the strongest potential competitors with a given individual. Intuitively, however, it seems clear that relationships among individuals within a given species must involve far more than competition for resources. In sexually reproducing species, a minimum population size is necessary to retain genetic viability (chapter 22); hence, any one individual has a clear evolutionary interest in the welfare of at least some others of its kind. Some hunters operate in teams or even armies, and the hunted frequently band together for protection. In what zoologists refer to as the *dear enemy* relationship, territorial animals that compete fiercely for space also band together to defend the group against predators or even share food. As the name implies, social insects have neither status nor much of a chance for survival as individuals.

Like interactions among individuals of different species, those among individuals of the same species are often a complex blend of cooperation and competition. Striking a proper balance between these two very different types of interaction involves trade-offs that are ultimately sorted through natural selection: what works survives, what does not work perishes.

This section focuses on how the struggle for growing space among individuals within even-aged tree populations results in a distinctive size structure within the population. In plant population biology, this phenomenon is referred to variously as the **self-thinning rule,** the **thinning law,** or the **negative 3/2 power law.** First, however, we discuss some more general aspects of even-aged plant community development.

12.1.1 Even-Aged Stand Development: Dominance and Death

Consider a newly established stand of a given tree species. Initially, the seedlings are small and grow without competing with one another, but eventually, the leaf area of the stand reaches a point where light and perhaps water and nutrients begin to limit growth (see chapter 15 for the effect of resource limitation on productivity). At this point, those trees that have initially grown best begin to suppress slower-growing individuals, and over time, stands often develop a bimodal size distribution (i.e., a group of relatively large trees dominating a group of small, suppressed trees, most or all of whom will eventually die). Note that positive feedback is involved here: the rich get richer and the poor get poorer. A similar phenomenon occurs within populations of some animal species (Huston et al. 1988; Schmid et al. 2000): fish, or invertebrates, for example, which like plants, do not grow to some genetically predetermined size but rather to a resource-determined size.

If for some reason a dominance hierarchy does not develop within even-aged plant populations (i.e., all individuals within a stand have the same growth potential) stands may stagnate, i.e., no trees grow well. Some lodgepole pine stands that establish after wildfire provide a good example of stagnation. Because of their serotinous cone habit, lodgepole pine can flood burned sites with millions of seeds per hectare. If early survival is high (which may be the case if climate is favorable in the growing season following the fire), seedling density can be so high, and the resources available to each seedling so low, that no individuals grow well enough to dominate others. Stands with several hundred thousand individuals per hectare have been documented, with maximum height less than 1 m at 80 years of age (Lotan and Perry 1983).

When a dominance hierarchy does develop, which individuals win the early race to secure space (and therefore resources) depends on complex interactions between genetic potential, early growth environment, and strictly random factors such as where a seed lands or a seedling is planted. The genetic potential for rapid early growth is certainly important; however, because the growth potential of a given genotype is likely to vary depending on an array of environmental variables, exactly which individuals have the genetic potential for rapid early growth may depend on factors such as the precipitation and temperature during the early years of stand development.[1] The random element of where a seedling begins its life is important in determining the early growth of individuals quite independently of their genetic capacity for growth. As we shall see in chapter 14, forest soils are in general highly variable; nutrient levels may vary severalfold over distances of less than 1 m. Large trees with extensive root and mycorrhizal systems integrate much of this variability, but small seedlings do not. A seedling that is genetically programmed to grow slower than other genotypes in the population may emerge as a dominant if it begins life in a nutrient-rich microsite; conversely, an inherently fast grower may fall behind if it starts in a rocky, infertile patch of soil.

Randomness, such as whether a seed falls in a fertile or infertile patch of soil, contributes to maintaining the genetic variability of tree populations and, serendipitously, is a very good thing for the persistence of the population. Consider, for example, a stand of trees that established during a period of exceptionally favorable weather, such as high rainfall during the growing season. If the emerging dominants in this stand were determined solely by genetic potential for early growth, most of the stand would eventually be composed of genotypes adapted to grow best when moisture was abundant. But this stand will almost certainly face drought years during its lifetime, and without individuals adapted to drought, the stand could be quite vulnerable to stress. At least in theory, the nongenetically based growth advantage gained by individuals that happen to start life in a favorable microsite should mitigate against such a narrowing of the gene pool.

12.2.2 The Self-Thinning Rule

So far, we have seen that within even-aged stands of trees (or any plant), some individuals for one reason or another exert dominance and others fall behind. In fully stocked stands of even-aged plants, this competition among conspecifics is reflected by a characteristic pattern of development that is variously called the self-thinning rule, the thinning law, or the negative 3/2 power law. The self-thinning rule is widely used by foresters as a guide to when and how much to thin stands.

Suppose we plot the relationship between stocking density and average seedling size for a newly established stand, planted at 1,000 trees per hectare (fig. 12.1). Two things about the graph shown in the figure 12.1 should be noted. First, we see the log-log scale, which will be explained later. Second, time does not appear explicitly but rather is implicit in the progression of the line that tracks changes in seedling size and density; this line is called the **time-trajectory of stand development** (or simply the **stand trajectory**).

During the early stages of stand development, competition among seedlings is not severe enough to cause mortality, and average seedling size increases with no decrease in stocking density. In figure 12.1, this is the portion of the trajectory that parallels the y axis, (stage 1). Eventually, the stand grows crowded enough that an increase in the average tree size cannot occur unless some trees die; at this point, the trajectory begins to curve left, denoting a reduction in stocking density (stage 2). Following its initial bend to the left, the trajectory asymptotically approaches and

[1] It should not be construed that fast early growth is always the best, either for the tree or the forester.

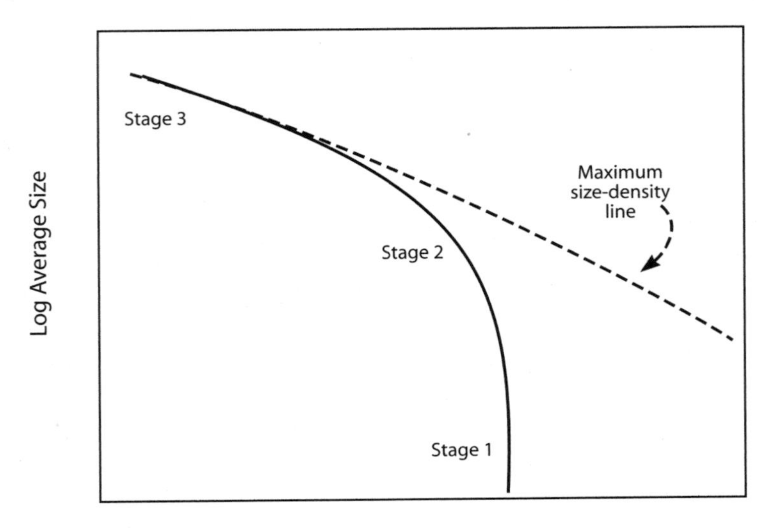

Figure 12.1. Schematic diagram of the self -thinning rule. In stage 1, trees increase in size with no corresponding decrease in density. Density-dependent mortality begins in stage 2, when an increment in average tree size is accompanied by a decrement in stand density. In stage 3, the relationship between increasing average tree size and decreasing stand density becomes linear on the log-log scale, following what is called *the maximum size-density line*. Note that the mortality referred to in the diagram is density dependent (i.e., it is related to competition between individual trees, which occurs only after stands attain a certain stocking density). Forests can also experience density-independent mortality, as in forest fires or hurricanes.

then follows quite closely along a straight line, which means that a given increase in average tree size is matched by a given decrease in stocking density (stage 3).

Note also that the final trajectory is a straight line only on the log-log scale. On a linear scale, the final trajectory would be curved, although the curve would still be quite regular in its behavior. Log-log scales are used in these types of graphs because linear relationships are easier to work with than those that are curvilinear.

The straight line that is approached asymptotically and then followed by the stand trajectory is called the **maximum size-density line** or, alternatively, the **self-thinning line.** For even-aged stands of a given plant species, combinations of average plant size and density always fall on or below the maximum size-density line of the species. The term *self-thinning rule* refers to the whole relationship depicted in figure 12.1, but in particular to the existence of the log-linear boundary between permissible and forbidden size-density combinations.

Virtually every plant species that has been studied follows a relationship of the type shown in figure 12.1 when growing in dense, even-aged monocultures; however, both the slope and the intercept of the self-thinning line may vary from one species to another (Westoby 1984; Lonsdale 1990). Figure 12.2 shows the maximum size-density line for several tree species grown in Great Britain (most of these species are not native to Great Britain). Note that the height of the line varies among the different species, which tells us that these species vary in the number of individuals of a given size that can be packed onto a given area. For example, when grown in Great Britain, a noble fir stand can contain as many as 740 trees per hectare that average 0.55 m^3 stem volume; while a European larch stand with trees that size will contain a maximum of only 366 trees per hectare. (As an exercise, what is the maximum number of trees of that size that a western hemlock stand will contain?)

Species also vary somewhat in the slope of their maximum size-density line, although the magnitude of this variation (generally within the range –1.2 to –1.8) is remarkably small. An early derivation based on plant geometry predicted that the slope of the line should be –1.5 (or –3/2, hence the alternative name, *negative 3/2 power law*). A more recent derivation based on plant metabolism rather than geometry predicts a theoretical slope of –4/3 (Enquist et al. 1998).

Why species vary in the heights and slopes of their self-thinning lines is not clear, but it certainly relates to the efficiency with which different species occupy space. This is related, in turn, to interactions between tree geometry (particularly crown width and the allocation of aboveground biomass to leaves vs. wood), and metabolism (particularly the efficiency with which trees use their leaves) (chapter 15).[2]

An example of how the self-thinning rule is used to determine when and how much to thin (in this case for

[2]Theoretical derivations of self-thinning characteristics include those by White (1981), Pickard (1983), Perry (1984, 1985), Hardwick (1987), Weller (1987), and Enquist et al. (1998).

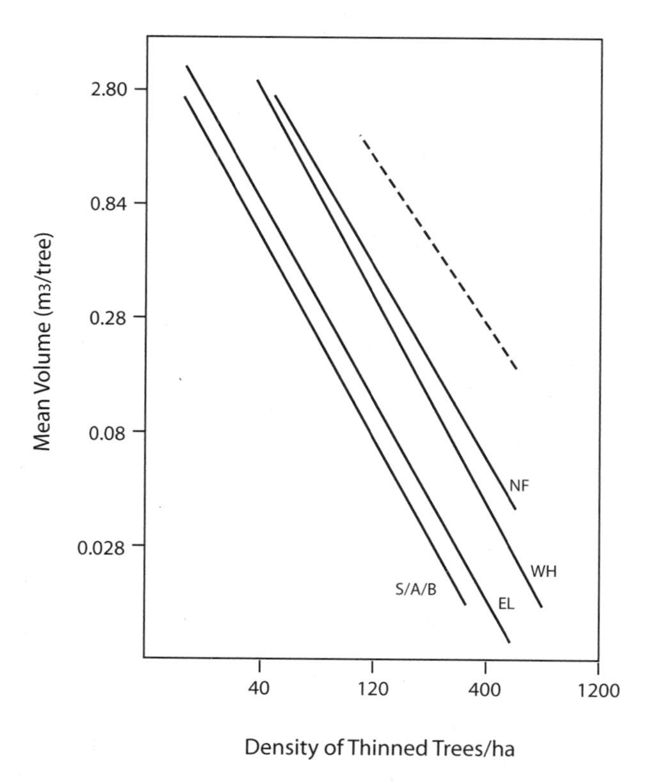

Figure 12.2. Self-thinning relationships for tree species planted in Great Britain. The dashed line has a slope of –2/3, and the line for each species includes stands that differ in site quality. EL, European larch; NF, noble fir; S/A/B, -sycamore, ash, and birch; WH western hemlock. (Adapted from Harper 1977)

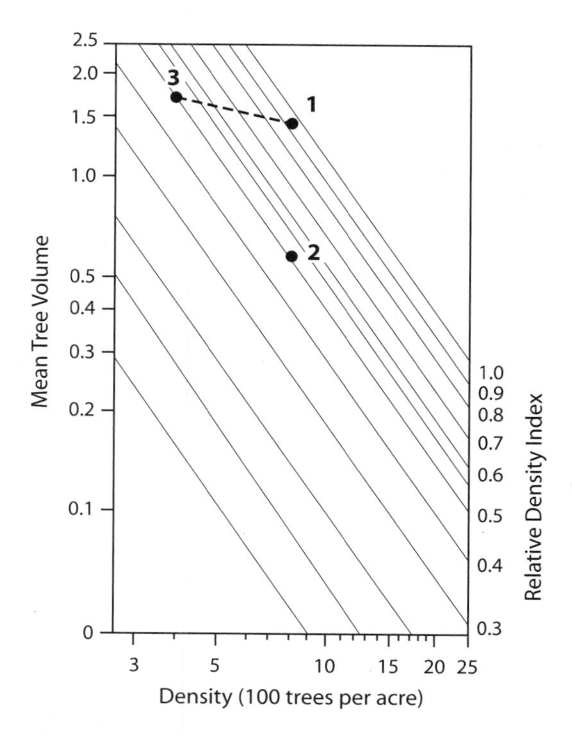

Figure 12.3. An example of how to use the self-thinning rule as a thinning guide. See text for explanation.

Douglas-fir) is shown in figure 12.3. A particular stand is plotted on the size-density diagram, and its **relative density** (RD) is calculated as the ratio between the actual density and the maximum allowable as defined by the maximum size-density line (Drew and Flewelling 1977, 1979). For example, the stand plotted at point 1 in figure 12.3 has 750 trees per hectare, with an average per tree stem volume of 1.48 m^3. For this average tree size, the maximum allowable density is 790 (from the maximum size-density line); hence, RD = 750/790, or 0.95. The second stand has the same number of trees as the first, but the smaller average tree size results in RD of 0.5.

As rough rules of thumb, crowns close somewhere around RD = 0.15 (i.e., 15% of the maximum number of allowable trees of a given average size), and mortality of suppressed trees begins after the RD exceeds 0.65 (Drew and Flewelling 1979). To prevent excessive competition on the one hand and understocking on the other, stands are often managed so as to maintain RD values between 0.55 on the upper end and 0.20 to 0.40 on the lower. Stand 1 is well past the point when it should have been thinned, while stand 2 does not need thinning unless extrawide spacing is a management objective. Suppose we wish to thin stand 1 to a RD of 0.5. Assuming that the thinning increases average tree size somewhat (because smaller than average trees are removed), the stand should be moved to point 3 on the RD = 0.5 line, which corresponds to 375 trees per hectare. By reducing the dominance of overstory trees, wide spacing (RD $<$ 0.3) is an effective way to promote diversity within stands.

The self-thinning rule is a particularly good tool for foresters, because, rather remarkably, the maximum size-density line for a given species is fairly insensitive to changes in site quality; therefore, the same rule can be applied across a wide variety of sites.

12.2 SIZE-DENSITY RELATIONSHIPS IN FORESTS: THE SPATIAL DIMENSION

Highly regular size-density relationships are not restricted to changes over time in even-aged forests. On a worldwide basis, forests follow a very regular pattern relating tree size to stocking density (Enquist and Niklas 2001). Figure 12.4 shows the relationships between average tree diameter and tree density (number per 0.1 ha) for forests worldwide (log-log scale). Note the relationship is very similar relationship to that of the self-thinning rule, which expresses change in time in any one stand rather than differences among stands. Figure 12.5 shows how the exponents of size-density relationships vary with latitude. Basically, the exponents vary widely among forests at similar latitudes; however, the general trend is of more negative exponents with decreasing latitude.

These relationships tell us two things. First, trees everywhere tend to divide up space in similar ways. The strong central tendency is for tree stocking density to scale as the

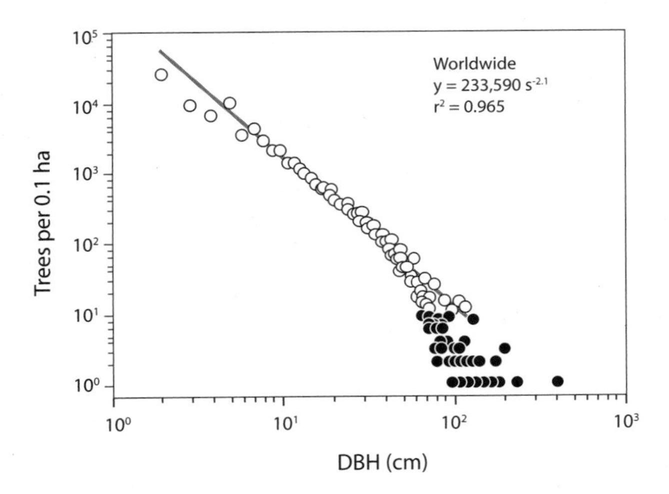

Figure 12.4. Log-log relationship between average tree size and tree stocking density for forests worldwide. Values on the *y* axis are trees per 0.1. ha. (Adapted by permission from Macmillan Publishers Ltd: Enquist and Niklas 2001, copyright 2001)

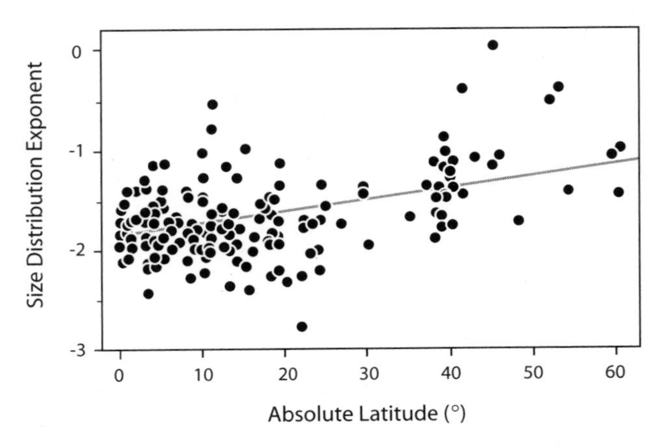

Figure 12.5. Latitudinal variation in the exponent of the size-density relationship in forests. (Adapted by permission from Macmillan Publishers Ltd: Enquist and Niklas 2001, copyright 2001)

–2 power of average basal stem diameter (which translates to the –3/4 power of aboveground biomass) (Enquist and Niklas 2001). As with the self-thinning curve, forests with relatively large trees and low stocking density fall away from the curve (i.e., the right end of the curve of fig. 12.4). The properties of this curve are reproduced by a simulation model in which individuals allocate biomass to leaf, stem, and reproduction and compete for space and light (Enquist and Niklas 2001). Secondly, despite the regularity at a global scale, the details of the relation (as expressed by the exponent of the size-density line) vary among forests.

12.3 SUMMARY

Forests follow regular patterns of development. This is expressed both with stands over time and in structural differences among forests across space. The struggle for space among individuals within even-aged plant populations results in a very distinctive pattern of stand structure. As plants grow and fill space, some become dominants and others fall behind to eventually die. Through time, fully stocked stands support fewer and fewer individuals of larger average sizes. When plotted on a log-log scale, the relation between average plant size and stocking density follows a straight line that depending on plant species, has a slope between –1.2 and –1.8. This relationship, called the *self-thinning rule*, is used by foresters to calculate levels of thinning. Similarly, comparing the size-density relationships among forests on a worldwide basis reveals a very regular relationship, with a strong central tendency for stocking density to scale as the –2 power of average tree diameter (or –3/4 power of forest biomass).

13

Genetic and Evolutionary Aspects of Species Interactions

> Ecological theory related to communities is incomplete if it does not account for the fact that ecological and evolutionary processes jointly affect community dynamics.
>
> COLLINS 2003

> The evolutionary analysis of community organization is considered a major frontier in biology.
>
> SHUSTER ET AL. 2006

The pattern of relationships within communities is ultimately determined by the fitness conferred on each participant; in other words, to the extent that we understand the mechanisms of natural selection, any lasting relationship will leave an imprint on the genetic structure of the participants that, in turn, will determine the characteristics of future interactions. Hence, the web of interaction both drives and is driven by the genetic structure of the participants. It then follows that understanding the mechanisms of evolution (i.e., the way genetic information is retained, edited, and either discarded or passed on) is a key aspect of understanding the structure and function of ecosystems. The degree to which evolution is influenced by cooperation, interconnectedness, and the complexity of relationships in nature is now receiving long overdue attention, and the wedding of ecology and evolution is becoming a reality. Long-standing and contentious issues such as the selection of groups are being looked at afresh. For the ecologist, these developments have already provided significant new insights into the organization of communities, and there is without doubt more to come. These new insights are perhaps even more important to the resource manager, who must decide how much to alter the composition of communities to produce a crop (or whether to alter the composition at all). Throwing away parts of a naturally selected group because they have no commercial value would be like throwing away the carburetor of an automobile because only the wheels carry us from one place to another.

Once believed to proceed slowly over many generations, evolution is now known in at least some cases to occur within the span of a few generations, a phenomenon that Stockwell et al. (2003) refer to as **contemporary evolution.** For example, minimum size limits on game fish have resulted in a genetic shift to sexual maturity at smaller sizes, resulting in evolution toward overall smaller fish (Stockwell et al. 2003; Zimmer 2003). On the Galapagos Island of Daphne Major, resident finches responded to the arrival of a new finch species in 1982 by evolving (over 22 years) a beak size that reduced niche overlap[1] with the new arrival (beak size determines what seeds can be utilized for food)[2] (Grant and Grant 2006). In later chapters we see that evolution is already occurring in response to global warming. Sudden environmental changes are believed to trigger contemporary evolution (Stockwell et al. 2003), in at least some cases by altering the nature of species interactions (Neuhauser et al. 2003); however, bacteria at least have been shown to actively evolve genetically distinct colonies even within constant environments (Maharjan et al. 2006).

The central aim of evolutionary ecology is "to understand the role of different ecological processes in producing patterns of macroevolutionary diversification" (Day and Young 2004). A full treatment is beyond the scope of this book; in this chapter we introduce three general aspects: (1) adaptations that are driven by interactions among individuals of different species, including coevolution, diffuse coevolution, and the role of microbial symbionts both in evolution and in enhancing the ability of their hosts to track environmental change; (2) the emerging field of community and ecosystem evolution, which includes, among other things, the evolutionary implications of niche construction and how genetic diversity within plant populations ramifies to effect structure and processes of entire ecosystems; and (3) group selection, or the possibility that natural selection acts on groups of organisms (e.g., populations, communities), as well as on individuals. Like most of the really interesting topics in science, these are at present rather poorly understood but hold the promise of becoming increasingly important to our understanding, in this case, of ecosystem dynamics.

13.1 THE ROLE OF BIOTIC INTERACTIONS IN EVOLUTION

In chapter 9 we saw that individuals within a given species adapt to abiotic environmental factors such as climate, soil parent material, and topography, thereby producing genetically distinct populations called *ecotypes*. The genetic composition of individuals and populations is also shaped by their biotic environment (i.e., through interactions with other organisms). For example, the types of flowers and fruits produced by individuals of a given plant species (as well as the seasonal timing of flower and fruit production) are selected at least in part through interactions with pollinators, seed dispersers, and seed predators (Snow 1971; McKey 1975; Howe and Smallwood 1982). Trees that are pollinated and whose seed are dispersed by bats hold their flowers and fruits away from the main crown, where they are presumably more accessible to a large flying animal (Whitmore and Burnham 1975). Lanner (1985) argues that through the types of trees from which they gather nuts, birds are responsible for the evolution of certain crown shapes in oaks, hickories, and some pines. On the other hand, plants evolve strategies to discourage seed predators. In an earlier chapter, we saw that large, infrequent seed crops produced by some tree species are believed to be an adaptation to seed predation, the idea being that populations of seed predators adjusted to low seed abundance cannot grow fast enough to consume the entire occasional large crop (Janzen 1974). (However, both American and Eurasian red squirrels were recently found to increase their reproduction in the summer *preceding* a large autumn cone crop [Boutin et al. 2006]). In addition, plants that participate in mutualisms must not only evolve ways to attract their mutualists, they must also evolve mechanisms that discourage cheaters—species that chisel on a mutual relationship by taking the inducement (e.g., pollen) without giving anything in return (Thompson 1982).

Certain types of animal behavior also might influence the genetic structure of plant populations, as in the case of nutcrackers, which most frequently cache pine nuts on southern aspects that remain relatively free of snow (Vander Wall and Balda 1977). Because southern slopes are also droughty during the growing season (in the Northern Hemisphere), pine seedlings that root deeply or possess some other adaptation to survive a dry spell are the most likely to survive until reproductive maturity, hence passing on their genes. The behavior of the bird sets the conditions to which the tree must adapt. Moreover, the nutcrackers' habit of collecting seeds from one or a few trees at a time and then caching these together produces a distinct genetic structure within stands of whitebark pine:

[1]Evolution of a trait that reduces niche overlap is called **character displacement.**

[2]Beak size determines the size of seeds that finches feed on.

the trees occur in clumps of closely related individuals (Furnier et al. 1987).

13.1.1 Coevolution

> [Coevolution] is one of the most important ecological and genetic processes organizing earth's biodiversity.
>
> THOMPSON 2005

Coevolution occurs when interaction between individuals of two species shapes the genetics of both participants (Ehrlich and Raven 1964; Thompson 1982, 2005; Futuyma and Slatkin 1983). Coevolution is very common between predators and prey. For example, plants evolve defenses against insects, and insects evolve responses to those defenses (chapter 19). The array of genetic responses available to each partner in this dance is constrained and shaped by all of the other abiotic and biotic factors impinging on the two. Considering the number of direct and indirect interactions that are possible among organisms, along with the adaptations necessary to persist in a frequently unpredictable abiotic environment, it is obvious that the intersecting forces of genetics and ecology can produce very complicated patterns. As with many of the issues in this book, information regarding the reciprocal influence of organisms on one another's genetic structure is far too scanty to make concrete statements regarding the prevalence of coevolution within ecosystems, and ecologists do not agree about its importance (Bernays and Graham 1988; Rausher 1988; Thompson 1988).

By definition, where interactions between two populations are sufficiently intense and of long enough duration to influence the fitness of both, coevolution will occur (fig. 13.1). The question is how commonly such interactions occur in nature. The two most obvious candidates (and the relationships to which ecologists and evolutionary biologists have directed most attention) are interactions between plants and plant-eaters, and mutualisms. Regarding the first, Ehrlich (1986)—who along with Raven coined the term coevolution in 1964—argues that "the coevolution of plants with herbivores . . . is responsible for the enormous diversity of biochemical compounds found in plants" (chapter 19). All of the psychoactive and medicinal drugs humans obtain from plants probably evolved as defenses against herbivores and pathogens.

Mutualists may also coevolve, but in this case, it is to facilitate rather than to inhibit interactions among the partners. Symbionts (i.e., mutualists living in close physical association, such as plants and nitrogen-fixing bacteria, or host-specific mycorrhizal fungi, animals, and the microbes that inhabit their guts) provide some of the more obvious examples of coevolution. Mutualists that do not live together also appear, in at least some cases, to have coevolved; Mori and Prance (1987) discuss pollination of some tree species in the Brazil nut family (Lecythidaceae): "The flower structures . . . tend to make their rewards accessible only to specialized pollinators. The androecial hood of many, but not all, of these species is tightly oppressed to the summit of the ovary, which limits entry to those bees strong enough . . . to force it open. The coiled flap makes nectar available only to species with long enough tongues to reach it."

13.1.2 Diffuse Coevolution

One source of skepticism about the importance of coevolution is that as discussed in chapter 11, relatively few interactions in nature are intensely one on one (although some certainly are). It then follows that the fitness of any one species emerges from numerous direct and indirect interactions (chapter 8). For example, plants evolve various defenses against herbivores, and herbivores evolve ways to defeat plant defenses; but this pairwise evolutionary sparring is shaped by various other factors, such as predation on the herbivores, the species composition of the plant community, and weather. Moreover, the herbivore not only may

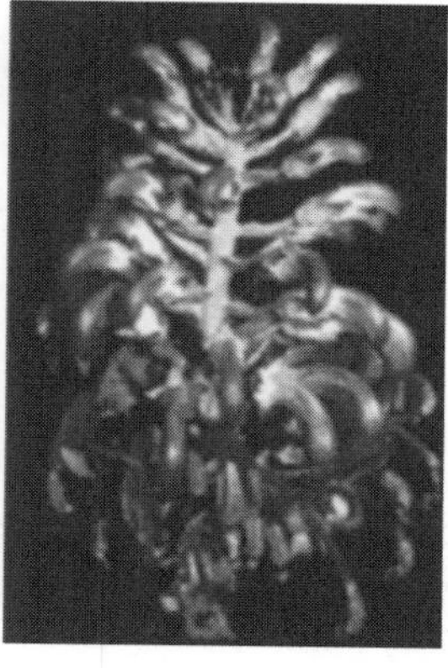

Figure 13.1. Coevolution between flower shapes in Hawai'ian *Lobelia* sp. and beak shape of their pollinator, the endemic I'iwi. (I'iwi photo by Jack Jeffrey)

evolve a way to defuse the plant's chemical defenses, it may actually use them to discourage its own natural enemies (chapter 19). The complexities of interaction within ecosystems have led some ecologists to argue that communities are characterized by **diffuse coevolution.** According to Fox (1988), "Diffuse coevolution implies that many species, on the same or different trophic levels, may simultaneously exert selective pressures on one another and be affected by changes in other component members." One implication of the complexities of diffuse coevolution is that two-way interactions may be modulated by different factors in different places, resulting in a **geographic mosaic** of coevolution (Thompson 2005).

One of the best examples of a geographic mosaic in coevolution has to do with North American conifers, squirrels, and various species of crossbills (*Loxia* spp.), birds that feed on conifer seeds (Benkman et al. 2001; Benkman et al. 2003; Thompson 2005). All crossbills have evolved a characteristic beak structure that allows them to pry open conifer cones, with different crossbill species specializing on different conifer species. Throughout much of the northern Rocky Mountains, red squirrels also feed on conifer seeds, and lodgepole pine has evolved cone traits that discourage squirrel predation (e.g., broad bases that make it difficult for squirrels to bite the cones off the trees). Crossbills that specialize on lodgepole pine in the northern Rockies have, in turn, evolved beaks that allow them to open broad-based cones, and the tree has apparently evolved no defenses against the bird. Red squirrels, however, do not occur in some peripheral areas, and there the crossbills exert significant evolutionary pressure on cone morphology. One isolated mountain range in Montana does not have squirrels but does have a moth species that preys on lodgepole cones. There, both crossbills and the moth exert evolutionary pressures on cone morphology (Siepielski and Benkman 2004).

A very similar pattern involving red squirrels, black spruce, and crossbills occurs in eastern Canada, where red squirrels are present on the mainland but absent in Newfoundland (Parchman and Benkman 2002).

In practice (i.e., in the complex real world), it is quite difficult to distinguish between diffuse coevolution, which by definition requires reciprocal influence among the participants, and simple adaptation to biotic and abiotic environmental factors, which does not. For example, predators of tree-eating insects likely influence the pairwise coevolution between tree and tree-eater, but are the genetics of the predators also altered by this process? Is there a reciprocal influence? In this particular example, the answer is probably yes, because chemical defenses produced by plants against herbivores frequently also make the herbivores more vulnerable to their predators (and in some cases the herbivores use a plant's chemical defenses in their own defense) (chapter 19). The result is a set of reciprocal direct and indirect interactions among three trophic levels that at least have the potential to alter the selection pressure on (and therefore the evolution of) each participant. Our understanding of the higher-order interactions in nature is so poor (chapter 11), however, that it is unclear just how far such arguments can be stretched when applied to community-level coevolution. Once again we are limited by our ignorance of the natural world.

13.1.3 Cooperation as a Source of Evolutionary Innovation

> Cooperation is needed for evolution to construct new levels of organization. Genomes, cells, multicellular organisms, social insects, and human society are all based on cooperation. . . . Thus, we might add "natural cooperation" as a third fundamental principle of evolution beside mutation and natural selection.
>
> NOWAK 2006

Cooperation may happen at various levels within and between species. As we saw in chapter 11, symbioses between microbes and both plants and animals are ubiquitous. For animals and plants that participate in such symbioses, evolutionary innovations are not tied to random mutations of single genes and sexual recombination; therefore, they do not necessarily occur slowly. Rather, the host can acquire whole new sets of genes—hence, whole new sets of traits—from its microbial partner(s). This possibility has profound implications for evolution. As Price (1991) summarizes, microbial "life has had billions of years . . . to radiate into every conceivable quarter of the earth, to solve every conceivable biochemical problem, and to surmount every obstacle of biotic interaction. By associating with these masters of intrigue and cunning, larger forms of life have thrived."

The most commonly accepted example of evolutionary innovation through symbiosis is the once widely rejected theory of **endosymbiosis;** this holds that organelles such as chloroplasts and mitochondria originated as free-living bacteria that entered the cells of higher organisms and evolved an obligate relationship (Margulis 1981, 1991). Various examples of other, less-advanced but nevertheless similar phenomena are now recognized. For example, many insects form symbioses with microbes that live in their guts and are passed from one insect generation to the next through mechanisms such as smearing feces on the surface of eggs (Schwemmler 1991). The relationship is frequently (but not always) obligate for the microbe (i.e., it cannot survive outside of the host). Depending on circumstances, the insect may or may not survive without the microbe, but its fitness (i.e., growth, fecundity) almost always benefits from the relationship (Nardon and Grenier 1991; Schwemmler 1991; Tiivel 1991). For example, symbiotic bacteria supply weevils with five vitamins that the weevil cannot synthesize with its own genes, enabling the insect to broaden its diet to include plants that do not contain those vitamins (Nardon and Grenier 1991). Similarly, insects that live on wood do so only with the help of various microbial symbionts with the genetic capability to digest cellulose and fix atmospheric nitrogen (Price 1988, 1991). In the plant world (as we discussed in chapter 11), bacteria in the genera *Rhizobium* and *Frankia* provide the genes that allow their plant partners to convert atmospheric nitrogen to usable form (also see chapter 17), and the genetic capabilities of mycorrhizal fungi to gather widely scattered nutrients and water may have been the key factor that enabled plants to colonize land.

In those cases that have been sufficiently studied, the genetics of the symbiosis frequently are not additive (i.e., it is not a simple matter of adding a package of microbial genes to a plant or an animal cell). Rather, the genomes of symbiont and host act in an integrated manner, producing a result that is more than the sum of its parts: nitrogen cannot be fixed without genetic instructions from both legume and *Rhizobium* (Postgate 1982); bacterial symbionts of insects may not produce certain amino acids without instructions from the insect genome (Tiivel 1991).

The examples cited deal with permanent evolutionary innovations that originated at some point in the evolutionary past. Microbial symbionts also play a somewhat different role by providing their hosts with genetic flexibility; rather than triggering a permanent evolutionary innovation, the microbe brings to the relationship a certain genetic fluidity that the host may not have by its self. For example, resistance by pea aphids to a parasitic wasp is not due to aphid genetics but, rather, is conferred by a bacterial symbiont of the aphid (Oliver et al. 2005). This possibility is largely unstudied in long-lived organisms such as trees, but it has the potential to greatly enhance the ability of an organism to genetically track environmental changes occurring at shorter intervals than its generation time. Although capable of somatic mutations (chapter 9), trees generally cannot evolve quickly enough to track climate fluctuations or to keep up with rapidly evolving pest populations. Six-hundred-year-old Douglas-fir trees established in a climate that is much different than the one they experience today and, without being able to change their genome (except for the possibility of somatic mutations), they have endured 600 or more generations of their insect pests, with all of the attendant novelty and problem-solving capabilities that six hundred generations can create. Enter the microbial partners of trees—mycorrhizal fungi, rhizobacteria, and fungal endophytes—which bring two types of genetic flexibility to their hosts. First, for certain traits, the host-symbiont combination presents a genetic mosaic rather than the single genome of the host. A single large tree, for example, may form mycorrhizae at any one time with scores of species and hundreds of genotypes of mycorrhizal fungi; not all of these fungi will necessarily have unique genetic capabilities, but many will. Second, the microbe brings to the partnership its rapid evolutionary capacity (Carroll 1988). Because at least some species of mycorrhizal fungi and foliar endophytes defend plants against pests and pathogens, these symbioses may help individual trees keep up in the "arms race" with insects and microbial pathogens. Similarly, the genetic composition of a given species of mycorrhizal fungus or fungal endophyte is likely to be quite fluid in response to changing environmental factors such as temperature and moisture, thereby providing host plants with considerable short-term genetic flexibility in rooting and resource gathering.

13.2 COMMUNITY AND ECOSYSTEM GENETICS

> The emerging field of community and ecosystem genetics seeks to understand the ecology and evolution of complex communities found in nature.
>
> WHITHAM ET AL. 2006

In 1986, Eldridge argued that a "balanced picture of nature sees the economic [ecological] and informational [genetic] as coexisting and utterly interdependent." Lewontin (2000) used the metaphor of the triple helix to describe the intimate interactions among genes, organisms, and environments. The term *community genetics* was coined by Collins (cited in Antonovics 1992) to describe the marriage of ecology and genetics, and recent research showing that different genotypes within a tree species[3] affect ecosystem processes differently has extended the concept to a genetics of ecosystems (Fisher et al. 2004; Schweitzer et al. 2004, 2005a, 2005b; Whitham et al. 2006).

Although discussed for two decades, community genetics did not come to the fore in ecology until early in the current century. In 2003, the journal *Ecology* published a special issue containing two seminal papers (Neuhauser et al. 2003; Whitham et al. 2003) along with a collection of short commentaries on those papers.[4] While both arguing for the importance of a community genetics perspective, Neuhauser et al. and Whitham et al. adopted quite different stances, and the various commentaries split along similar lines. Neuhauser et al. place themselves solidly in the Gleasonian school of community ecology, in which communities

---

[3]That different tree species affect ecosystem processes differently is old news. The new news is that diversity within species may have ecosystem-level implications.

[4]The commentaries can be found in *Ecology*, volume 84, pp. 574–601 (2003). They are an excellent introduction to the scope and breadth of thinking on the issue of community genetics.

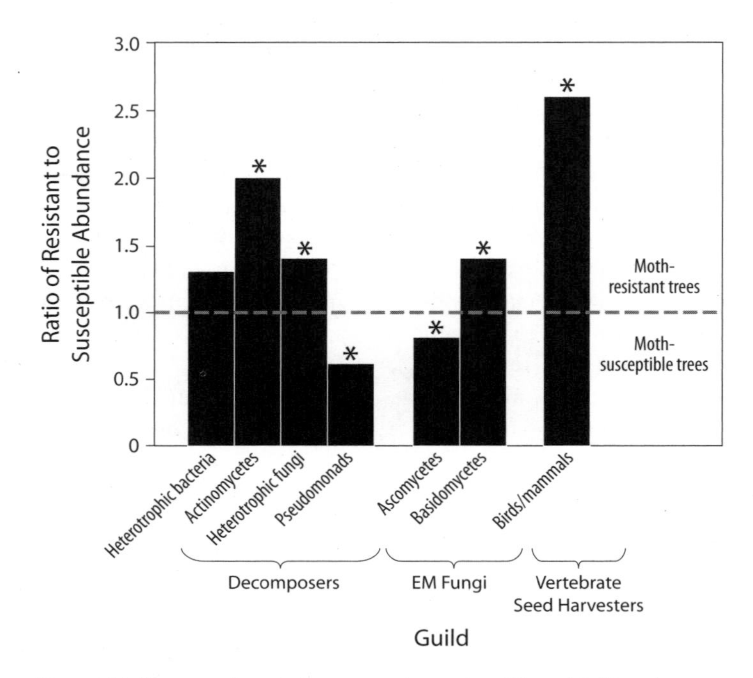

Figure 13.2. How genetic variation among pinyon pines *(Pinus edulis)* in resistance to a stem-boring moth *(Dioryctria albovittella)* affects associated bird, mammal, and rhizosphere microbe communities. Bars show the ratios of the abundance of each of seven species groups on resistant trees relative to susceptible trees. Bars with values >1 indicate greater abundance on resistant trees than on susceptible trees; bars with values <1 indicate greater abundance on susceptible trees than on resistant trees. An asterisk above the bar denotes a statistically significant difference at $p < 0.05$. EM, ectomycorrhizal. (Whitham et al. 2003. With permission of The Ecological Society of America. See Whitham et al. for citations to original data.)

are not entities in any kind of equilibrium but rather loose collections of species each seeking and responding to its own needs (chapter 8). Neuhauser et al. add a significant dimension to that picture by arguing that even in such nonequilibrium situations and over relatively short time periods, interactions among species can shape the genetics of the participants (i.e., coevolution happens). They give three examples in which that has been documented, and they argue that in today's world of strong anthropogenic forcing, ecosystems are being pushed into increasing levels of disequilibrium (e.g., by climate change, landscape fragmentation, genetic alteration of crops) whose implications require a community genetics perspective to understand.

Whitham et al. (2003) take a more holistic (and controversial) view. They argue that genetic effects stemming from a keystone species (chapter 11) ramify throughout the community and ecosystem to produce an **extended phenotype.** For example, figure 13.2 shows how differences among pinyon pines in resistance to a stem-boring moth alter the community of organisms associated with the pines Furthermore, Whitham and colleagues argue that the community phenotype is heritable, and in subsequent papers they calculate a formal heritability value associated with it (Shuster et al. 2006; Whitham et al. 2006).

At this point it is useful to briefly review some basic concepts. A **phenotype** is the face that an organism presents to the world and what selection directly operates on. Phenotypes are a product of both genetics and environment. For example, the basic shape of a tree crown may be (and often is) a product of tree genetics, while environment controls whether the leaves are adapted to sun or shade. **Heritability** is the proportion of phenotypic variation that is controlled by genetic variation. Whitham et al. (2006) defined **community heritability** as "the tendency for related individuals to support similar communities of organisms and ecosystem processes."

Do genetically related keystones (or foundation species such as dominant trees) support similar communities and processes? Whitham and colleagues addressed that question using replicated plots of *Populus* sp. of known genotype, and the answer was yes (Shuster et al. 2006; Whitham et al. 2006). In other words, if one knows the genotype of trees, the composition of the community and ecological processes associated with the trees can be predicted; they are "heritable." Not surprisingly, the degree to which the genetics of a keystone species produces a heritable community phenotype depends on the strength with which its genetically determined traits influence the system. In the case of *Populus,* differences among genotypes in the production of tannins were a driving factor. As we see in chapters 18 and 19, tannins exert strong control over leaf palatability and rates of decomposition and, therefore, over the structure of food webs and the cycling of nutrients.

As Whitham et al. (2006) point out, "for community and ecosystem phenotypes to have evolutionary significance, they must be heritable." This brings us to an old and still unresolved question in evolutionary ecology: Can groups evolve?

13.3 THE SELECTION OF COOPERATION WITHIN GROUPS

> There can be no doubt that a tribe including many members who were always ready to give aid to each other and to sacrifice themselves for the common good, would be victorious over other tribes; and this would be natural selection.
>
> CHARLES DARWIN 1871

Although altruistic behavior within populations is well documented, it is probably safe to say that most evolutionary biologists have been reluctant to believe that an individual might reduce its own reproductive potential to enhance the reproductive potential of others (although as suggested by the lead quotation of this section, Darwin did). The reasoning is straightforward. Because the winners in an evolutionary race are those who leave the most progeny, any behavior that limits the number of young that are produced carries the seeds of its own destruction; and the gene(s) coding for it will eventually be replaced by genes that code for selfish behavior.

Unfortunately, the straightforward reasoning cannot account for what we see in nature, and considerable effort has gone into developing plausible scenarios by which altruism can survive and even thrive. Nowak (2006) discusses the most common models:

- **Kin selection.** By saving kin, an individual also saves copies of its own genes. For example, siblings of most sexually reproducing species, on average, share 50 percent of their genes; an animal that dies to save three siblings therefore perpetuates 150 percent of its genome (Gould 1984). The concept of kin selection is attributed to the evolutionary biologist J. B. S. Haldane, who summed it up this way, "I will jump into the river to save two brothers or eight cousins" (quoted in Nowak 2006).
- **Direct, indirect, and network reciprocity.** Basically, this is "you scratch my back and I'll scratch yours." Reciprocity may be between two individuals (i.e., direct), or within groups where cooperating individuals don't encounter one another directly. If interactions among group members are random, that is termed *indirect reciprocity;* if structured in some way (e.g., cooperators tend to interact only among themselves), it is termed *network reciprocity.* Within human society at least, the selfish motivation that underpins indirect reciprocity may be the common (and fundamentally ecological) belief that "what goes round comes round."
- **Group or multilevel selection.** This can include both kin and unrelated reciprocators but differs in that selection can act on groups as well as individuals (Wynne-Edwards 1962; Wilson 1980). It is in the spirit of Darwin's argument quoted at the beginning of the section.

Nowak (2006) gives simple mathematical rules by which each of these mechanisms can persist, as well as discussing some variants on the three common themes.

One of the more thorough studies on the role of kinship in altruism was conducted by Emlen and Wrege (1988), who studied a bird called the white-fronted African bee eater. During breeding season and chick-rearing, mated pairs of this species commonly receive assistance from others in nest building and foraging. By very detailed study over several years, Emlen and Wrege were able to document who was genetically related to whom within one beekeeper colony. Hence, they were able to correlate helping behavior with relatedness. Of 174 cases in which paired birds were assisted by others, the helpers were kin in 154 (mostly adult sons and daughters of the nesting pair, although in some cases, parents assisted at the nests of their adult children). On the surface, this is compelling evidence for kin selection; however, this is not the complete story. One must still ask whether altruism is based in genetics or something else. To explore this question, Emlen and Wrege transferred 25 fledglings from the nest of their biological parents to another ("foster") home. When the fledglings matured, they assisted only their foster parents and treated their biological parents as nonkin. Thus, it seems that at least in this case, the choice of who to help was not based in the genes but in early imprinting, which suggests that reciprocal altruism (i.e., tit for tat) was a motivating force within this population. Vampire bats provide another example of reciprocal altruism within populations (Wilkinson 1990). Vampire bats, which feed on the blood of large mammals such as horses and cows, cannot survive more than two nights without a meal, yet the vagaries of hunting mean that any individual runs a fairly high probability of starving. To deal with this situation, bats have evolved what in essence is a "buddy" system: bats pair with one another (but not as mates), forming a small, mutual support group in which a member that has fed shares with one that has not. Bat buddies may or may not be related, so their cooperation cannot be explained by kin selection. The underlying cause of such cooperation is uncertainty in the environment (in this case food supply), which makes it worthwhile to share benefits as a means of spreading risks. In other words, to give an analogy from human affairs, an individual bat shares food as a form of insurance against the time when it will be without food. Like vampire bats, virtually all organisms that live in the wild face a great deal of uncertainty in one or more factors of their environment, such as food supply, weather, or the nature and timing of disturbance; therefore, cooperation as a form of risk-spreading might be widespread in nature, both within and between species.

While altruism was the early focus of group selection, thinking has expanded to focus on community-level interactions of any type that influence the genetic structure of individuals and produce traits (of individuals, communities, or ecosystems) that are subject to selection (Goodnight and Stevens 1997; Odling-Smee et al. 2003; Wade 2003; Whitham et al. 2003, 2005, 2006; Wilson and Swenson 2003). A growing body of experimental and modeling evidence points toward the reality of group selection. Mathematical analysis shows that interactions are a crucial component of heritability (Wade 2003), and their review of experiments led Goodnight and Stevens (1997) to conclude that "experiments clearly demonstrate that genetically based interactions among individuals are contributing to group differences and to the response to group and community selection. Thus, group selection is far more effective than predicted because it can act on genetic interactions both within and among species."

The term **multilevel selection** recognizes that selection may occur simultaneously across hierarchical levels spanning genes, molecules, organelles, organisms, populations, communities, ecosystems, and even metacommunities and metaecosystems (Wilson 1997). As with the related idea of diffuse coevolution, this concept emerges from the realization that although ecological communities are not as tightly knit as organisms, individuals nevertheless exist within a network of direct and indirect interactions that ramify through communities and metacommunities. It follows that the probability of any one individual passing on its genes cannot be neatly separated from the viability of at least some other residents of the same community or metacommunity, nor can it be separated from ecosystem-level processes that ultimately both reflect and drive community dynamics.

The degree to which group and multilevel selection are important forces in nature is an open question actively debated among evolutionary ecologists (e.g., see the commentaries cited in note 5). In the opinion of Traulsen and Nowak (2006) "group selection is an important organizing principle that permeates evolutionary processes from the emergence of the first cells to eusociality and the economics of nations"; however, that view is not shared by all (e.g., Collins 2003; Biernaskie and Tyerman 2005).

From the earlier section, it seems clear that obligate symbioses represent the merging of two species into one evolutionary unit. What other groups might qualify? Perhaps the network of species linked by shared mutualists (chapter 11)? Perhaps entire communities and metacommunities? Community level selection in microcosms (Swenson et al. 2000) and heritability of community phenotypes (Shuster et al. 2006) demonstrate potential. Wilson (1997) argues: "Natural communities frequently consist of a mosaic of semi-isolated patches . . . When many local communities exist that vary in their species and genetic composition, those that function well as a unit contribute differentially to the next generation of local communities. Traits can therefore spread, not by virtue of their advantage within local

Experiments have generally employed artificial selection of groups. By way of analogy, consider the common forestry practice of selecting superior-growing tree genotypes. Seed from fast-growing individuals identified in the field is collected, and progeny are grown in nurseries. These are eventually subjected to another round of selection, and seed collected from the best growers to used establish another generation, and so on. This is selection based on individuals, but the same basic techniques can be used for groups. One striking example involved selection for groups of chickens that got along with one another (and therefore laid more eggs) (Muir 1995). Selection that focused on individual chickens had resulted in superior egg producers who unfortunately were also quite aggressive, and when confined together in standard poultry farms spent more time fighting than laying eggs. Group selection solved the problem. Swenson et al. (2000) used the same technique to select for groups at the community level. Wilson and Swenson (2003) discuss that experiment: "[They] created soil and aquatic microcosms by inoculating with naturally occurring communities of microbes, measured them after a period of time for plant biomass [in the soil microcosms] and pH [in the aquatic microcosms] and selected from one end of the phenotypic distribution to inoculate a new set of micrcosms. The phenotypic distribution of the 'offspring' generations shifted in the direction of selection, demonstrating that ecosystem traits such as plant biomass production and fresh water pH can respond to community level selection."

Wilson and Swenson (2003) go on to discuss the mechanisms underlying community level selection: "In the case of community-level selection, evolution at the phenotypic level could be caused by genetic changes in the component species, changes in species composition of the community, or both. . . . The concept of 'community genetics' [or ecosystem genetics, insofar as communities are selected based on their ecosystem processes] should be expanded in certain contexts to include all changes in the composition of the community, between and within species."

communities, but by virtue of the advantage that they bestow on their local community, relative to other local communities."

With regard to community phenotypes, a key question relates to feedbacks.[5] Are the communities associated with a given *Populus* clone (for example) merely hitchhikers, or do they feedback to affect the genotype of the trees and of one another? Put another way, community-level selection can only be considered to occur when the fitnesses of all interacting species are in play (Whitham et al. 2003; Shuster et al. 2006).

The answers to questions about higher-level selection await a better understanding of interaction and interdependence within ecosystems. Goodnight and Stevens (1997) summarize the challenges ahead: "To explain the evolution of standing adaptations, one looks at patterns and attempts to infer process. Many potential evolutionary pathways for a particular trait exist. Some pathways involve individual selection, whereas others may involve higher levels of selection or selection acting simultaneously at multiple levels. Indeed, the possible pathways seem to be limited only by the biologist's imagination."

Despite (or perhaps because of) the rather daunting uncertainties concerning the workings of nature, ecologists and evolutionary biologists are now moving beyond the notion of natural selection acting only on individuals driven solely by struggle and competition (e.g., Day and Young 2004). One reason that scientific views are changing is simply that we are learning more. Modern techniques of genetic analysis have opened important new windows into the genetic structure of nature. Moreover, the closer we look at the social relationships within (and between) species, the more we see.

13.4 SUMMARY

The complex of direct and indirect interactions that characterize ecosystems plays a significant role in determining the fitness of individuals and, hence, shapes the evolution of all populations that reside and interact within the community. Flowering and fruiting patterns of trees, for example, have evolved to encourage mutualists and to discourage both parasites and predators. Coevolution occurs when an interaction modifies the evolution of both participants, as occurs between plants and plant-eaters. The idea of diffuse coevolution recognizes that the evolutionary trajectories of all permanent members of the community are likely to be shaped by a network of interactions. Perhaps the most powerful relationships regarding their influence on evolution are the ubiquitous symbioses between microbes and higher organisms. Symbiotic bacteria, fungi, and protozoa bring to their hosts entirely new sets of genes that permit rapid evolutionary innovation and flexibility in tracking fluctuating environments.

Although individuals are considered by most evolutionary biologists to be the basic units of selection, groups of interdependent individuals may also form units of selection (i.e., an individual may act altruistically to forego its own selfish gain for the good of the group). The theory of kin selection holds that such altruism occurs only within groups of related individuals, an altruistic act on the part of an individual being in fact a selfish act aimed at protecting copies of its own genes earned by its kin. The theory of group selection, on the other hand, holds that altruism occurs among unrelated individuals of the same or different species, probably because the altruistic act will be reciprocated in the future (i.e., tit for tat). Both kin selection and group selection appear to operate in nature, obligate symbioses being one example of the latter. The concept of group selection has been extended beyond altruism to encompass entire communities of interacting organisms, and its reality has been demonstrated experimentally. Keystone or foundation species (such as dominant trees) have been shown to produce heritable community phenotypes, where phenotype at the community level refers to the community of organisms associated with the keystone. The concept of multilevel selection holds that selection occurs across multiple hierarchical levels. Too little is known about the strength of interactions and interdependencies within ecosystems to predict how prevalent group and multilevel selection might be in nature; however, this field is developing rapidly.

[5]The concept of extended phenotype originated with Dawkins (1982), who defined it as "all effects of a gene on the world." Dawkins went on to point out that when the extended phenotype of a species feeds back to influence fitness (e.g., a beaver dam), it becomes interesting to the evolutionary biologist. Biernaskie and Tyerman (2005) argued that Whitham et al.'s (2003) community phenotype lacked a clear linkage to fitness. Whitham et al. (2005) respond with the argument that with or without feedbacks, the concept of extended phenotype has value because it places "community and ecosystem ecology within a genetic and evolutionary framework."

14

Soil

The Fundamental Resource

> Each soil is an individual body of nature, possessing its own character, life history, and powers to support plants and animals.
>
> HANS JENNY

The importance of soil to forest productivity and health cannot be overstated. Of the 27 elements required by plants, microorganisms, and animals, all but two (carbon and oxygen) are obtained either largely or solely from soil (the essential elements of plants are listed in chapter 16). Soil is important not only because of its immediate influence on productivity, but also because it is the repository of future productivity. In essence, soil is *the* fundamental resource of terrestrial ecosystems that ultimately provides the sustenance for all earthly organisms, yet it is a nonrenewable resource on an organismal time scale. Trees grow and die from natural causes or are harvested. Soil remains, providing a legacy that determines the performance of the next forest. But it is not a one-way interaction. Plants and soil exist in a tightly linked partnership. Just as plant growth cannot be understood without reference to soil, neither can soil structure or the processes occurring within soil be understood without reference to plants. In this chapter, we briefly review the composition and structure of soil, how that composition and structure are influenced by plants and other factors, and how soil influences, in turn, the resource supply to plants. A thorough review of forest soils is beyond the scope of this book. For a more complete introduction to forest soils, we suggest perusing a recent text by Fisher and Binkley (2000). Some older texts are still quite good, in particular those by Lutz and Chandler (1946), Wilde (1946, 1958), Remezov Pogrebnyak (1969; a Russian perspective), and Armson (1977; a Canadian perspective).

14.1 WHAT IS SOIL?

Soil is a multiphasic system (fig. 14.1). A typical, productive soil has roughly 50 percent pore space by volume, with about half of this pore space occupied by water and the other half occupied by air. The remaining 50 percent of the soil is solids; generally, a majority of these solids are in the mineral form, with organic materials comprising perhaps 5 percent of the total soil volume.[1] Water stored in pores and nutrients released from the mineral and organic components provide resources for plant growth. Gas transport though air-filled pores provides oxygen to the soil from the atmosphere, supporting plant root and aerobic respiration by soil organisms; in turn, gas movement from the soil to the atmosphere removes waste products of aerobic respiration (e.g., carbon dioxide) from the soil. Additionally, the surfaces and pores of soil provide habitat for a myriad of organisms that are essential for the cycling of nutrients required by plants.

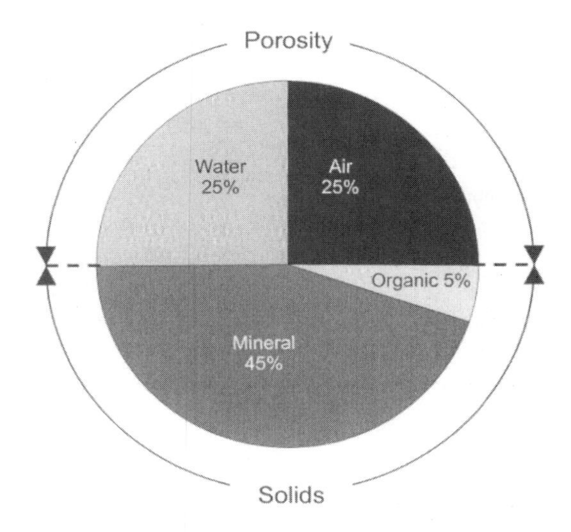

Figure 14.1. Soil is a multiphasic system that includes solids (mineral and organic material), gases (air), and liquid (water). An ideal volumetric distribution of these components for plant growth is shown for a typical agricultural surface soil with a loam texture. The relative amounts of water and air (the sum being soil porosity) vary inversely as the soil wets and dries. Surface forest soils typically contain more organic matter than agricultural soils. (Modified from Brady and Weil 2002)

Soil provides a variety of critical ecosystem services, including the support and nutrients for the production of food and fiber, cleansing the water that flows through the soil medium, imparting physical support for human structures (roads, buildings, etc.), and acting as waste repositories (municipal, industrial). But what actually is soil? It depends on your perspective. To a layman, it is the stuff you get on your pants (also known as dirt); to a civil engineer, it is the material you can excavate without blasting, and to a geologist it is material produced by weathering of the earth's surface (lithosphere). Soil scientists have generally taken two approaches for studying soils. The pedological approach (*pedon*, the Greek word meaning 'ground,' 'soil,' or 'earth') views soil as a natural body, much in the same way the astronomer views celestial bodies in the cosmos. This perspective focuses on the origin, description, and classification of soil. The information from this approach is as useful to an engineer as to a forester or farmer and does not necessarily focus on soil's immediate practical use. In contrast, the edaphological approach (*edaphos*, the Greek word meaning 'ground' or 'foundation') involves the study of soil as a medium for growth. This approach focuses on the causes for the variation in **soil productivity** (i.e., the capacity of a soil for producing a specified plant or sequence of plants under a specified management system, usually expressed in terms of yield) and the means for conserving and enhancing it (SSSA 1997).

Most of the research in soil science has been driven by the needs of agriculture. In North America, it was not until after World War II that forest soil science began to develop in its own right.[2] Forest soils are fundamentally different from agricultural soils not only because of the different plant communities that they support, but also because of the presence of a **forest floor,** an organic-rich layer that typically accumulates over the mineral soil. The organic nature of this layer, as well as the influence of the forest floor and the tree canopy on the soil microclimate, results in a different spectrum of soil organisms (typically more fungi relative to bacteria) than in agricultural soils. The organic acids produced by forest litter during decomposition also lead to the leaching of so-called base cations,[3] aluminum, and iron, which result in unique mineral layers or horizons (e.g., the E horizon) within the soil. While agricultural soils typically are devoid of larger particles (>2 mm in diameter, called *coarse fragments*) because of the selection of sites that lack such obstructions or because of selective removal of these materials from the soil, coarse fragments are generally abundant in forest soils. These fragments influence nutrient cycling, water storage and drainage, and soil temperature regimes. Finally, many forest soils are managed extensively, receiving little or no exogenous inputs such as fertilizers, herbicides, pesticides, or irrigation water. In some regions (e.g., the northwestern and southeastern United States and much of the Southern Hemisphere), however, forest soils are intensively managed much like most agricultural soils. Indeed, the differences in management intensity of forest and agricultural soils has become more blurred over time as forest practices intensify, and as more and more agricultural soils are managed in alternative, less intensive ways (e.g., no-till agriculture).

[1]On a dry mass basis, soil organic matter content is roughly the same. The organic matter content of forest soils is generally higher than that of agricultural soils disturbed by frequent tilling (which increases organic matter decay).

[2]For more on the history of forest soils, see Fisher and Binkley (2000).

[3]The cations Na^+, K^+, Ca^{2+}, and Mg^{2+} have traditionally been called *base cations*, but they are not true bases in the chemical sense (i.e., they do not combine with H^+). They are more appropriately called *nonacids* (Brady and Weil 2002).

14.2 THE SOIL PROFILE

Soil is the product of organisms and climate acting typically on rocks. It is a complex, intimate mixture of minerals, organic matter, and organisms. Many kinds of organisms (e.g., plants, microorganisms, and vertebrate and invertebrate animals) are part of the soil ecosystem, but plants are the ultimate source of organic carbon, which is a critical structural component of soil and the source of energy fueling the processes that occur within soils. The mineral component may be derived from igneous or metamorphic bedrock, volcanic flows or ash falls, or sediments that have formed in place (as in ancient ocean beds) or been moved from the ocean floor to land through tectonic activity. The initial mineral component of soils, regardless of its origin, is called **parent material;** parent materials frequently are quite heterogeneous within relatively small areas. Glaciers, water, wind, and gravity move material from one place to another, mixing rocks of various types in the process. In areas with past volcanic activity (e.g., the west coast of the Americas), soils receive periodic infusions of new minerals in the form of ash fall from eruptions. Furthermore, deposition of atmospheric dust from arid regions around the globe provides inputs of mineral materials that can have a significant impact on soil development and fertility.

Soil development, called **pedogenesis,** occurs on a temporal scale of tens to millions of years. Four major types of processes are involved in pedogenesis: (1) addition of organic and mineral materials to the soil as solids, liquids, and gases; (2) loss of these materials from the soil; (3) translocation of materials from one point to another in the soil; and (4) transformation of mineral and organic matter within the soil (Pritchett 1979). As we saw in chapter 8, addition of carbon and nitrogen to newly forming soils is a key component of primary succession. Transformation and translocation of materials involves the release of certain elements from parent material, generally through acid weathering that is mediated by the hydrogen ions in rainfall or organic acids released by plants and soil microorganisms; some of these elements move downward through the soil with water, or even out of the system to groundwater or streams.

The accumulation of organic matter both at the surface and in upper soil layers, coupled with the translocation of minerals and chemical elements downward during pedogenesis, produces a vertical structure within soils that is termed the **soil profile.** Pedogenic processes lead to the formation of different soil layers or horizons that comprise the soil profile. A **soil horizon**[4] is "a layer of soil, approximately parallel to the soil surface, with characteristics produced by soil forming processes" (Soil Survey Staff 1975). Soil horizons are distinguished in the field by a suite of soil properties, including organic matter content, color (dependent on chemical composition), texture (sand, silt, clay), structure, and pH (Soil Survey Division Staff 1993). There are six types of **master horizons** in soil, and they are denoted by the symbols O, A, E, B, C, and R (fig. 14.2). The presence and absence of these various horizons in large part determines the classification of the soils (see section 14.7).

The surface organic horizon (O horizon), often termed the *forest floor,*[5] consists primarily of plant litter in various stages of decay. Forest soil scientists have traditionally divided the forest floor into three layers (Pritchett 1979): L (the litter layer) is mainly plant material that is relatively fresh and still clearly recognizable in its original form (e.g., leaf, branch, insect body); F (the fragmented layer) is organic material that is partially decomposed (i.e., fragmented), but its origin is still recognizable to the naked eye; and H (the humus layer) is organic material in an advanced stage of decay such that the material is amorphous and its origin is not recognizable (see section 14.4.). These layers are roughly equivalent to the **subordinate horizons** Oi, Oe, and Oa, respectively, in the United States Department of Agriculture (USDA) Soil Taxonomy classification system (see section 14.7). Earlier versions of of the USDA's system divided the forest floor into just two layers (Buol et al. 1989): O1, still recognizable as to its origin, with some leaching of soluble components and discoloration (corresponding roughly to the L plus the F layers, or the Oi plus the Oe subordinate horizons); and O2, whose original form cannot be recognized (corresponding roughly to the H layer or Oa subordinate horizon).

Forests that produce readily decomposable litter generally support high numbers of large invertebrates (especially earthworms) that mix the more decomposed litter with mineral soil, creating what is called a **mull forest floor.** In contrast, little mixing of forest floor and mineral soil occurs in forests that produce recalcitrant litter, creating what is termed a **mor forest floor.** Mor forest floors tend to occur on infertile soils and beneath coniferous forests growing at high elevation or high latitudes; mulls are frequently associated with deciduous forests on soils with relatively high pH and abundant nutrients, particularly calcium (Lutz and Chandler 1946). Work in France indicates that the iron concentration of parent material may also play an important role in the type of forest floor layer (specifically the Oa horizon) that forms beneath forests (Toutain 1987). Most temperate forests (both coniferous and deciduous) have forest floors that are somewhere between mull and mor, and

[4]These *genetic horizons* are not equivalent to the *diagnostic horizons* used in the classification of soils in the USDA Soil Taxonomy classification system. "Designations of genetic horizons express a qualitative judgment about the kind of changes that are believed to have taken place. Diagnostic horizons are quantitatively defined features used to differentiate among taxa. Changes implied by genetic horizon designations may not be large enough to justify recognition of diagnostic criteria" (Soil Survey Division Staff 1993).

[5]Sometimes authors use the word *soil* exclusive of the O horizon, but certainly this important horizon is part of the *soil* just not the *mineral soil*. Furthermore, some authors use *litter* or *litter layer* for the entire O horizon regardless of the stage of decay.

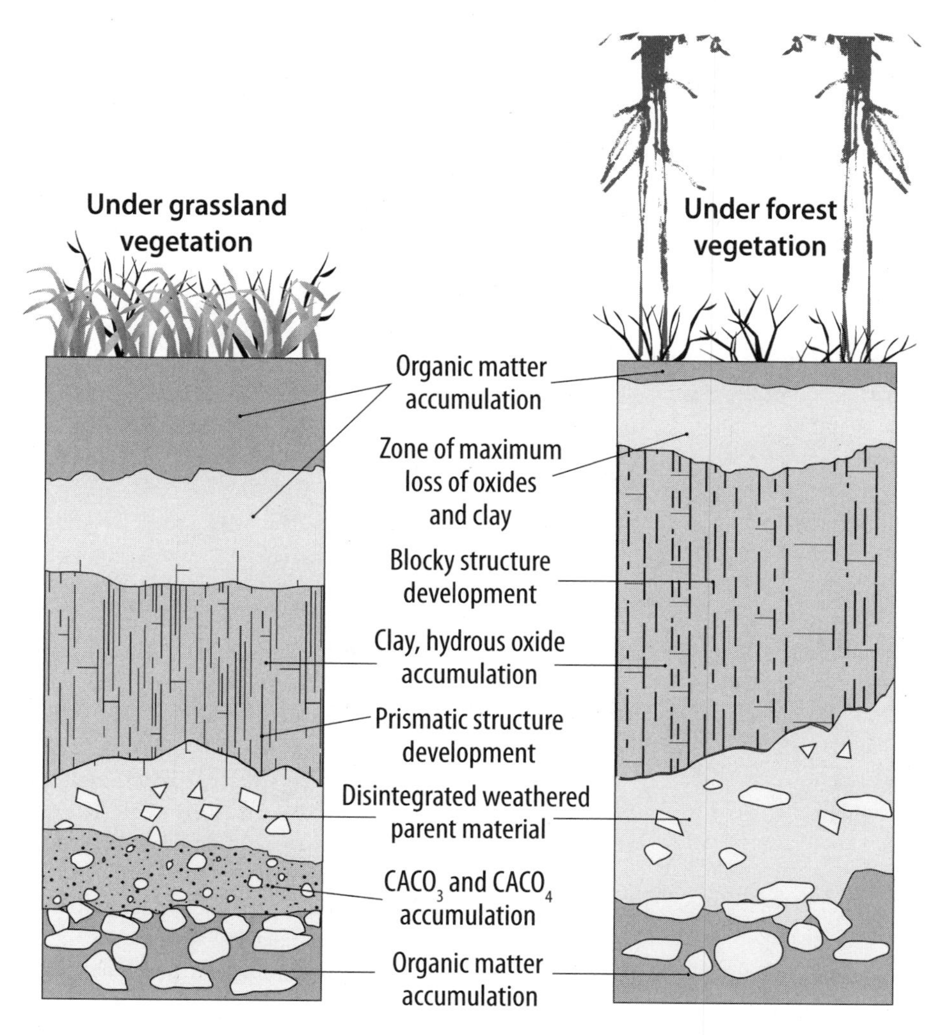

Figure 14.2. A soil profile is a vertical cross section of a soil showing the horizontal soil layers or horizons. Different soils have different types, thicknesses, and arrangements of soil horizons, and vegetation can have a profound influence on the type of soil eventually formed from the parent material (calcareous glacial till in this example). Here, soils developed under grassland and under deciduous forest are compared. The amount and distribution of organic matter differs considerably between the two soils. The forest soil exhibits an organic (O) horizon primarily from aboveground plant litter, in various stages of decomposition, along with a relatively thin, mineral A horizon where some of the surface litter has been incorporated and some fine roots have decomposed. In contrast, the A horizon that develops under the grassland soil is thick because of the high concentration, production, and turnover of fine roots near the surface under this vegetation type. Also, calcium carbonate has been solubilized and moved downward to lower soil horizons (C2k) under the grassland vegetation, while it has been completely removed from the profile in the more acidic and leached forest soil. Under both vegetation types, clay and iron oxides formed in the A horizon have been transported to the subsoil (forming the Bt horizon). Light-colored E horizons frequently form above B horizons under forest vegetation because organic matter input is more dispersed in the mineral soil and because organic acids produced during decomposition of tree litter remove the brown iron oxide coatings of soil particles in this horizon and transports them to the B horizon. (Modified from Brady and Weil 2002)

these are termed **duff mull** or **moder.** Tropical forests generally have very thin litter layers and no Oe and Oa layers (factors influencing the decomposition of organic matter are discussed in more detail in chapter 18).

Mineral soil layers are defined as having less than 12 to 18 percent organic carbon by mass, depending on the clay content (Buol et al. 1989). The top layer of mineral soil, or A horizon, is formed at the surface or below an O horizon where all or much of the original rock structure has been lost. It is further characterized by exhibiting an accumulation of humified organic matter mixed with the mineral fraction and is not dominated by properties characteristic of E or B horizons (defined below). Organic matter gives the A horizon a dark color. This surficial horizon is where weathering is most intense in the soil profile because it has relatively high water availability and biological activity, and it is the location where the greatest variation in soil temperature occurs. Similar to A horizons, E horizons are zones where organic matter accumulates and all or much of the original rock structure has been lost; however, the dominant feature of E horizons is the loss of silicate clay, iron, aluminum, or some combination of these (i.e., an **eluvial horizon**). These losses result in a concentration of silt- and sand-sized minerals that are resistant to weathering (e.g., quartz). Because less organic matter has accumulated, the E horizon is generally lighter in color than the A horizon.

The B horizon forms below an A, E, or O horizon, where again all or much of the original rock structure has been obliterated. B horizons are generally characterized by the accumulation of materials leached from overlying soil layers (i.e., an **illuvial horizon**). In humid areas (most forest soils), where precipitation exceeds evapotranspiration and results in significant leaching, the materials that tend to accumulate in the B horizon include silicate clay, acid cations (e.g., aluminum and iron), and humus. In these climates, much of the more soluble base cations (e.g., calcium, magnesium, potassium, and sodium) are leached from the surficial soil to deeper in the soil profile, or out of the soil altogether. In more arid regions (e.g., most grasslands) where precipitation is less than evapotranspiration and thus leaching is limited, soluble aluminum and iron are not produced in significant quantities or leached from the surface mineral horizons; as a result, the more soluble base cations accumulate in the B horizon in the form of salts of chlorides, sulfates, and carbonates (fig. 14.2). Although this horizon is dominated by translocated materials, some transformations (i.e., weathering and silicate clay formation) do occur in place.

C horizons are little affected by pedogenic processes and, hence, they lack the properties of the O, A, E, or B horizons. Generally, this horizon consists of minerals that have been physical weathered (i.e., broken into smaller pieces), but little change in the chemistry of the parent material has occurred. Frequently, some of the original rock structure is still apparent. This layer may also accumulate the more soluble cations leached from higher in the profile.

R layers are not true horizons in that they consist of hard bedrock (e.g., granite, basalt, quartzite, or indurated sandstone or limestone) that has been unaffected by pedogenesis. The bedrock may contain cracks, but these openings in the rock are too few and too small to allow much root penetration into the layer. Soil profile description is much more refined than exemplified in this brief discussion; more detail on this topic can be found by consulting Buol et al. (1989) and the Soil Survey Staff (1993)

14.3 PHYSICAL PROPERTIES OF SOILS

In the next three sections, we have separated the properties of soils into physical, chemical, and biological components to simplify our discussion of these concepts. However, the reader should be aware that such separation is totally artificial because the soil properties in these groupings interact in complex ways to influence soil processes.

In the USDA classification system**, soil separates** are individual mineral particles in soil that make up the **fine-earth fraction** (<2 mm in diameter). Mineral particles equal to or greater than 2 mm in diameter are called **coarse fragments** or **rock fragments.** These coarse fragments are described further, based on their shape and size of greatest diameter: **pebbles** or **gravel** if roughly round and 2 to 75 mm; **cobbles** if roughly round and 75 to 250 mm, or **flagstones** if flat; **stones** if 250 to 600 mm; and **boulders** if greater than 600 mm. Although the coarse fraction is frequently discarded for chemical and biological analyses of soil because of the relatively low surface-area-to-volume ratio, the coarse fraction has a significant impact on water and nutrient storage and the thermal properties of soil (Childs and Flint 1990). For instance, in water-limited forests where trees must rely on stored water, weathered bedrock (i.e., **saprolite**) underlying soils can supply 70 percent or more of the water used by trees during the growing season (Witty et al. 2003). Clearly, the coarse fraction of soils is an important forest soil component that deserves increased respect and attention.

Soil texture is the relative composition of primary mineral particles (i.e., soil separates) of the fine-earth fraction in the soil, and is divided into three broad groups in the USDA classification system based on effective diameter:[6] **sand**, 2 to 0.05 mm (gritty feel); **silt**, 0.05 to 0.002 mm (powdery feel); and clay, less than 0.002 mm (sticky feel). Because of the differential weathering rates of the **primary minerals** in parent material, and because soil solutes recombine to form new or **secondary minerals** in the smallest size fraction, there is a relative close correspondence between particle size and mineral composition: quartz dominates the sand and coarse silt fractions; primary silicates such as feldspars, hornblende, and mica are present in sand and in decreasing amounts in the silt fraction; secondary silicates dominate the fine clay; and other secondary minerals, such as oxides of iron and aluminum, are prominent in the fine silt and coarse clay fractions (fig. 14.3). Another reason soil texture is an important physical property is because as particle size decreases, the surface-to-volume ratio increases dramatically; this increases the surface area for chemical and physical interactions to take place between the solid and the other two phases of the soil (gases, water), and surfaces are also where the vast majority of soil microorganisms reside.[7]

Because of the large number of possible soil textures, soil texture is often grouped into textural classes. There are 12 of these classes in the USDA classification system, and these classes can be quickly determined with the use of a ternary diagram known as the **textural triangle** (fig. 14.4). Note that each class does not represent the same amount of area in the diagram because of the dominant effect of the clay-sized fraction on soil properties.

Soil separates are aggregated together to form larger (secondary) structures called **peds** or **aggregates.** The structure

[6]As measured by sedimentation, sieving, or micrometric methods.

[7]For instance, clay-sized particles may have more than 100,000 times the surface area per gram as sand-sized particles.

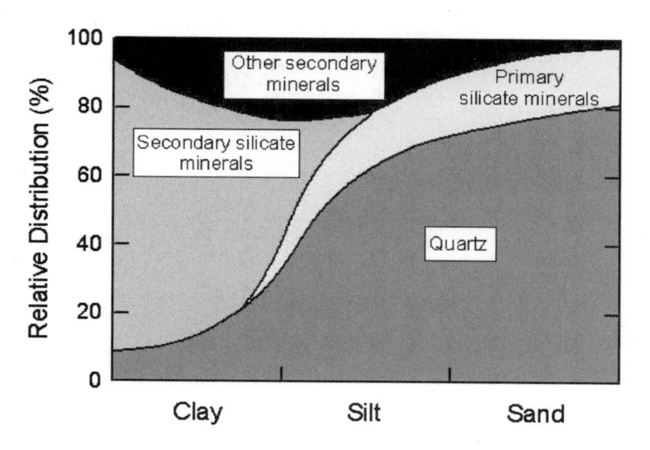

Figure 14.3. The size of soil particles covaries with the kinds of minerals present in soil to a significant degree. Quartz, being the most resistant to weathering, tends to dominant the sand-sized fraction, while primary minerals (present in the parent material), such as feldspars, hornblende, and micas, exhibit decreasing abundance from the sand- to silt-sized fractions. Secondary silicate minerals (those produced during soil development), such as smectites and kaolins, dominate the clay-sized fraction. Other secondary minerals, such as aluminum and iron oxides, tend to be most abundant in the intermediate to fine particle-size classes. (Modified from Brady and Weil 2002)

of a soil is defined as the arrangement of these peds within the soil profile. Unlike soil texture, **soil structure** can be observed only in place under field conditions and can be influenced by soil disturbances resulting from forest harvesting equipment, recreational vehicles, or other activities. Soil structure is an important determinant of water movement, heat transfer, and aeration. Soil structure is characterized on the basis of size (class), shape (type), and degree of distinctiveness (grade; fig. 14.5). Classes range from less than 10 mm (granular) to greater than 10 cm (prismatic or columnar). Grades are classified on the basis of inter- and intra-aggregate adhesion, cohesion, or stability: weak if difficult to see in a horizon; moderate if can be seen and most of material removed is aggregated; or strong if easily seen and most material removed is aggregated, and it can be handled without falling apart. Two types of soil lack structure (i.e., structureless soils) for quite different reasons. Massive soils or soil horizons contain so much clay that no planes of weakness are present. In contrast, single-grained soils do not have enough clay to hold the soil separates together.

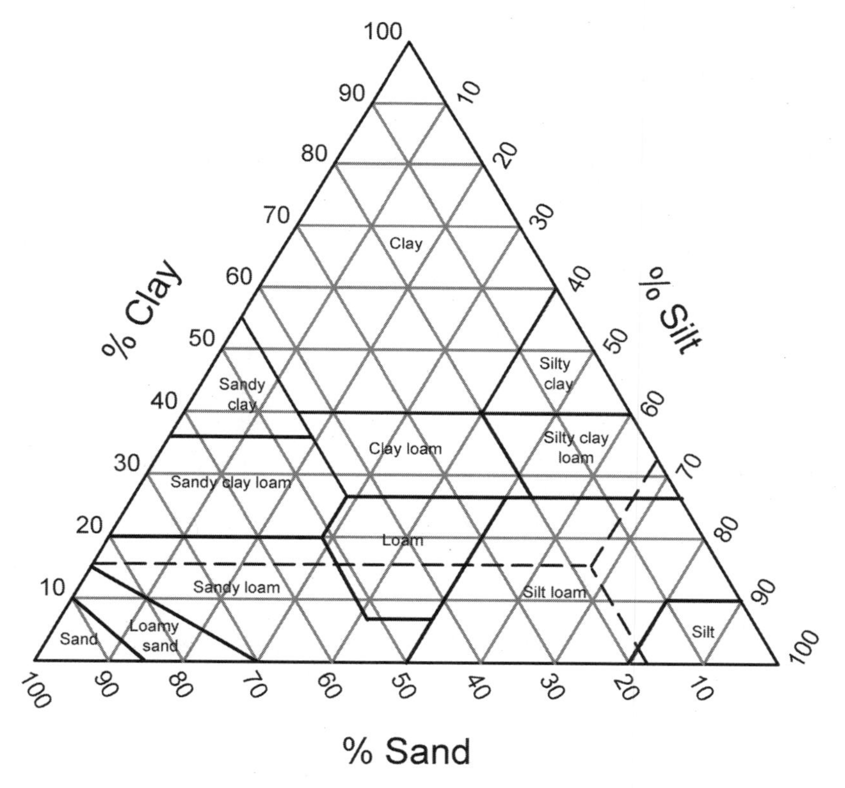

Figure 14.4. The textural triangle is a ternary diagram used to designate soil textural classes from information on the relative mass-abundance of sand-, silt-, and clay-sized particles. To use the diagram, find the appropriate clay percentage along the left leg of the triangle, and then draw a line from that location horizontally across the graph. Next, find the sand percentage along the base of the diagram (or silt along the right leg), and draw a line from that point inward across the diagram, parallel to the right leg of the triangle (or parallel to the left leg, if using percent silt). The zone of the diagram (denoted by bold lines) where these two lines intersect designates the soil textural class. Only two of the three values are needed because all three values must equal 100%. For instance, a soil sample that has 15% clay and 67% silt (and thus 18% sand) would be classified as a silt loam. The sizes of the zones for the 12 textural classes in the diagram are not equal because of the dominant impact of clay-sized particles on soil properties. (Modified from Brady and Weil 2002)

Granular
Structural units are approximately spherical. Characteristic of A horizons. The photograph to the right shows *strong fine* (1-2 mm diameter) and *medium* (2-5 mm diameter) *granular* peds.

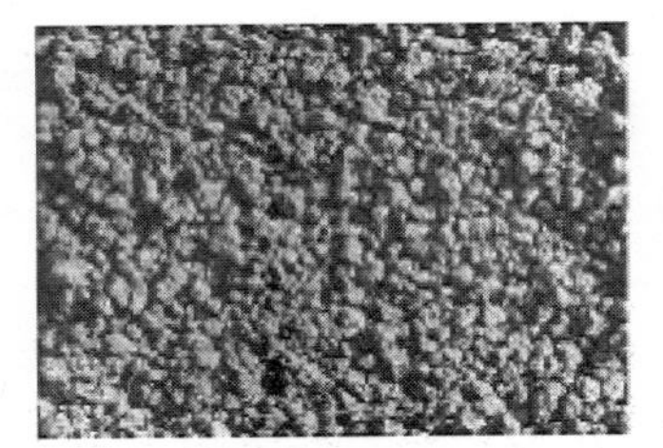

Platy
Structural units are flat and plate-like, and are generally oriented horizontally. May occur in any part of the soil profile but common in E horizons. Frequently caused by compaction or inherited from soil parent material. *Strong thin* (1-2 mm thick) *platy* structure is shown in the photograph to the right.

Blocky
Structural units are block-like with roughly equally sized faces. Two subtypes exist: if the vertices are relatively sharp the structural units are called *angular blocky*; and if the vertices mostly rounded the structural units are called *subangular blocky*. Common in B horizons especially in humid regions, but may also be found in A horizons. *Strong medium* (10-20 mm) and *coarse* (20-50 mm) *angular blocky* peds are shown in the photograph to the right.

Prismatic
Structural units have flat to rounded vertical faces with vertices that are angular or slightly rounded. Vertical dimensions of the units are distinctly greater than their horizontal dimensions. The tops of the 'prisms' are typically flat. Usually found in B horizons, especially in arid and semi-arid regions. The photograph to the right illustrates *strong medium* (20-50 mm in horizontal dimension) *prismatic* structure.

Columnar
Structural units are similar to prisms, with vertical dimensions distinctly greater than their horizontal dimensions. However, unlike prismatic structure, the tops of the prisms are rounded and generally more distinct. Usually found in B horizons, especially in arid and semi-arid regions where the sodium concentration in the soil solution is high. The photograph to the right shows a cluster of *strong medium* (each ped 20-50 mm in horizontal dimension) *columnar* peds. The entire cluster is about 135 mm across.

Figure 14.5. Soil structural types (shapes) found in mineral soils and their typical locations within the soil profile. (Photographs taken from Soil Survey Division Staff 1993)

The processes involved in the development of aggregates and soil structure are complex and not fully understood. They involve a combination of physical, chemical, and biological processes, with the physiochemical processes (mainly associated with clays) tending to be more important in the formation of smaller aggregates and the biological processes being more important in the formation of larger aggregates (Brady and Weil 2002). The most important of the physiochemical processes include the mutual attraction of clay particles (increased when divalent cations such as Ca^{2+} and Mg^{2+} are adsorbed on the clay surfaces; see section 14.4), and swelling and shrinking of clay masses coinciding with wet-dry and freeze-thaw cycles. The most significant biological processes involved in the development of soil structure include the burrowing and molding of earthworms (temperate zone) and termites (subtropical and tropical zones), the enmeshing of soil separates by root and fungal networks, and the binding of soil separates together by extracellular 'glues' (e.g., polysaccharides) secreted by soil microorganisms (Brady and Weil 2002). The restoration, maintenance, and enhancement of soil structure, particularly the larger aggregates that provide drainage and aeration (see below), are among the most important albeit some of the most difficult tasks of forest soil management, as these peds are readily destroyed via human activities such as tree harvesting, livestock grazing, and recreation.

Soil consistence is related to soil texture, structure, and the mineral composition of soils. It is defined as the ability

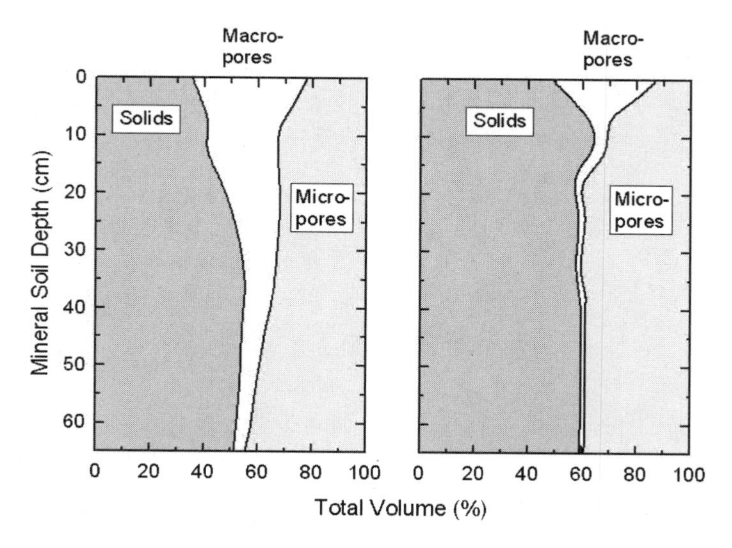

Figure 14.6. Soil porosity can be negatively impacted by management. The figure on the left shows the porosity of the upper 60 cm of mineral soil in an undisturbed, mixed hardwood forest in the South Carolina Piedmont, while the figure on the right shows the porosity of this same soil type (Typic Hapludult) in a nearby abandoned farmland. Capillary pores (also called micropores, approximately <75 μm in diameter) are so small that they retain water against gravitational forces, while noncapillary pores (also called macropores, approximately >75 μm in diameter) are large enough that they freely drain and fill with air following wetting. (Modified from Hoover 1949)

of a soil to resist deformation under stress. Soil consistence is measured in the field by feeling and manipulating soil by hand to determine the ease by which it can be reshaped or ruptured under different soil water contents. This soil physical property is important for soil stability (cohesion), erodibilty, and trafficability under various moisture conditions (McNabb et al. 2001).

Soil porosity is the total amount of pore space in a soil (volume not occupied by solids; fig. 14.1). It is important because it influences gas and water movement, which, in turn, affect root and soil organism activity. **Aeration porosity** is defined as the pore space occupied by air, while **water-filled porosity** (also known as **water-filled pore space**) is the amount of pore space occupied by water.

Soil porosity has two components: the total amount of pore space and the distribution of pore sizes that make up this pore space. In general, total porosity is less important than pore-size distribution. Both total porosity and pore-size distribution affect air and water movement and storage. Well-structured soils have a total porosity of about 50 percent, while an ideal pore-size distribution may be about half this volume in large pores (**macropores**, diameter >75 μm) and about half in small pores (**micropores**, diameter <75 μm). Macropores are important for drainage (water movement) and hence aeration (gas movement). In contrast, micropores are important primarily for water storage. In well-structured soils, macropores are generally found between peds, while micropores are found within the peds themselves (SSSA 1997). When soil structure is damaged, both total porosity and pore-size distribution may be affected, with generally a reduction in the total porosity and a shift in the distribution from macropores to micropores (fig. 14.6).

Bulk density of a soil is another important physical property that is closely related to total porosity. It is generally defined as the oven-dry mass[8] of the fine-earth fraction relative to the volume that this fraction occupies (including the solids and pores; sometimes called **fine soil bulk density**[9]). It is a function of soil porosity (which is influenced by soil structure and texture) and organic matter content. If one assumes a fixed particle density for mineral (D_p = 2.60–2.75) and organic materials (D_p = 0.90–1.30), porosity can be calculated from bulk density (D_b): percent porosity = $[1 - (D_b/D_p)] * 100\%$. The bulk density of forest soils varies widely, from 0.12 Mg/m^3 in some organic layers to almost 1.9 Mg/m^3 in coarse sands (McFee and Stone 1965; Fisher and Binkley 2000). Because of the lower particle density of organic matter and its tendency to improve soil aggregation, soils high in organic matter tend to have lower bulk densities than soils with lower organic matter contents.

The soil bulk density is important as a measure of both porosity and **soil strength**. In general, tree roots cannot

[8]Usually determined by heating soil between 100 and 110°C until a constant mass.

[9]Total soil bulk density is equal to the total soil material mass (i.e., fine-earth plus coarse fragments) divided by the total volume that these materials occupy (Childs and Flint 1990).

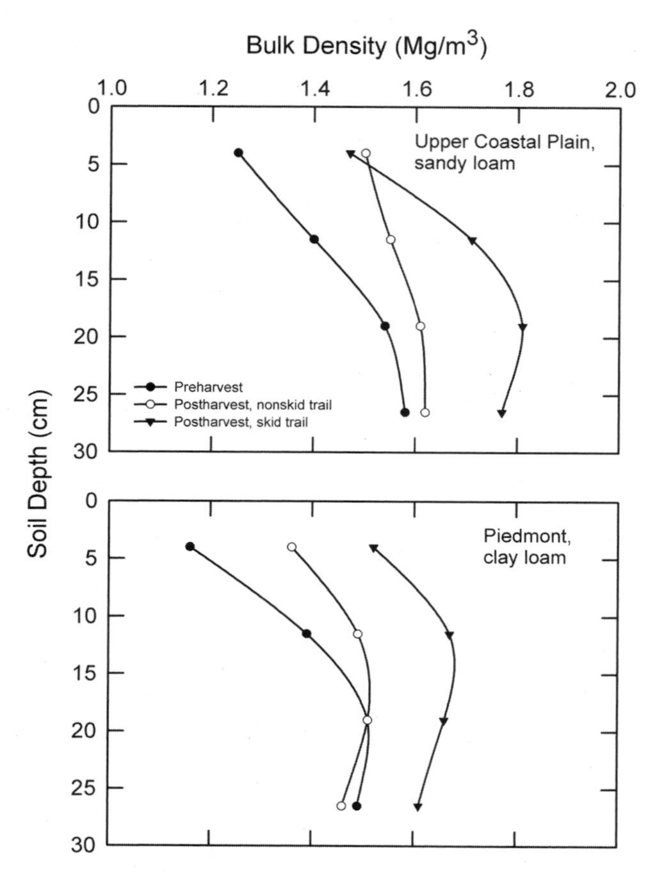

Figure 14.7. Forest harvesting operations can significantly impact soil physical properties, even at considerable depth. These two graphs show the effects of tree harvests using rubber-tired skidders on two soils of contrasting texture (Ultisols) in Georgia. Soil bulk densities increased at least to a 20-cm depth following harvesting, particularly on skid trails, where cut trees were dragged on the ground through the stand. (Data from Gent et al. 1984 and Gent and Morris 1986).

penetrate soil materials when $D_b > 1.55$ Mg/m^3 (if fine textured) to 1.75 Mg/m^3 (if coarse textured). The higher value for coarse-textured soil is because soil strength is lower at the same bulk density in coarse-textured soils as a result of lower cohesiveness; however, tree species vary considerably in their ability to penetrate soil material at different bulk densities (Pritchett and Fisher 1987). Tree harvesting frequently increases the bulk density of the soil, particularly in areas where tree boles have been dragged along the ground (so-called skid trails; fig. 14.7).

The gaseous composition of soil is fairly similar to that of the atmosphere. However, because of aerobic respiration by organisms in the soil (including plant roots) that produce carbon dioxide (CO_2) and consume oxygen (O_2), and because of the tortuous path that soil air has to make to exchange with the atmosphere (moving primarily through air-filled pores as the rate of diffusion of gases through air is about ten thousand times faster than through water), the soil air tends to be enriched in CO_2 and slightly depleted in O_2 relative to the atmosphere. **Soil aeration** is the gas exchange capacity of a soil and thus defines the degree of tortuosity of the gas transport pathway between the soil and the atmosphere. It is determined by total porosity, pore-size distribution, and the relative amount of pores filled with water.

The oxygen concentration of well-drained surface soils seldom falls much below atmospheric values of 21 percent (by volume), but in fine-textured soils with an overabundance of micropores, gas exchange can be greatly reduced. Under water-saturated (waterlogged) conditions, soil air may have CO_2 concentrations that rise to 5 to 6 percent, and O_2 concentrations may drop to as low as 1 to 2 percent (Fisher and Binkley 2000). Even under low O_2 concentrations, however, if water is moving through soil pores toward plant roots and other soil organisms, oxygen requirements for aerobic respiration can still be met. In contrast, under water-saturated conditions where water is stagnant, O_2 limitation for aerobic respiration can occur.

As O_2 concentrations decrease to low levels, root growth and aerobic microbial activity are reduced. The effects of poor aeration on trees tend to be a function of tree age: O_2 concentrations of soil air as low as 2 percent are usually not harmful to roots of mature trees for short periods, while O_2 concentrations below 10 percent may limit seedling root growth. Furthermore, some tree species are more tolerant of or even thrive under low soil O_2 concentrations, such as some species of *Alnus*, *Nyssa*, *Picea*, and *Taxodium* (Fisher and Binkley 2000). Not only do low O_2 concentrations affect aerobic respiration of plants directly, but low O_2 availability also results in increased production of compounds by anaerobic microorganisms that are toxic to the roots of many plants (e.g., ferrous iron and hydrogen sulfide; Paul and Clark 1996).

In some soils, even with high porosity, good overall pore-size distribution (e.g., half in macro- and half in micropores), and good drainage, the soil can have poor aeration if the surface soil structure (a few millimeters) is damaged by heavy rainfall or machinery. This is because all gas exchange with the atmosphere (both by gaseous diffusion along individual gas concentration gradients or by mass flow along total air-pressure gradients) occurs at the atmosphere interface with the soil. When the surface (upper few millimeters) structure is destroyed, water tends to accumulate on top of the soil; hence, these problem soils are often termed **puddled soils.**

Soils provide several signs of poor aeration besides those expressed through the slow growth of plants. For instance, under anaerobic conditions (when gaseous O_2 is depleted), iron is reduced from the ferric form (Fe^{3+}) to the more soluble (and hence mobile) ferrous form (Fe^{2+}). This reduction results in a change in the iron color[10] (hue) from yellow, red,

[10]Three components are used for defining soil color in the Munsell color system: hue (a measure of the chromatic composition of light that reaches the eye), value (denoting lightness and darkness, with lower values indicating darker shades), and chroma (color intensity, with higher values denoting 'brighter' colors).

or reddish brown to gray or blue green. This color change and the loss of oxidized iron coatings of soil minerals expose the gray colors of the underlying materials. The dull appearance or low chroma of the soil material under these conditions is called **gley.** Alternating aerobic and anaerobic conditions in the soil profile (such that occurs with a seasonal water table near the soil surface) can be indicated by soil mottling, where the fluctuating aeration levels cause the juxtaposition of oxidized and reduced forms of iron, giving mixed-colored patterns in the soil. The presence of gley soil or mottling in upper soil horizons is indicative of water-saturated conditions during at least part of the plant growing season and is used in delineating wetlands. The depth in the soil profile to which gley colors are found is frequently used to help define soil drainage classes, an interpretive classification of soils subjected to poor aeration (Brady and Weil 2002).

Water exists in three states in soil: as a vapor (gas), as a liquid, and as a solid (ice). Generally, soils have a very high (near 100%) relative humidity, except at or very near the surface, because of the tortuous nature of the gas exchange pathways of soil. By far, however, liquid water is the most important phase of water in soil. Water is not only required for plant growth, its availability also dictates mineral weathering rates and microbial activity that, in turn, influence nutrient cycling and availability to plants.

Water moves from areas of higher energy (potential) to areas of lower energy (potential). Soil water has three general types of energy that determine the total energy state of water (i.e., the **total soil water potential**). The **osmotic potential,** also called the **solute potential,** Ψ_o, is affected by the concentration of solutes in the soil solution; increasing the solute concentration in the soil water lowers the energy state of the water because these dissolved compounds enhance the ordering of water molecules. Hence, this component of the total soil water potential is negative in magnitude. The Ψ_o is usually not very important for water movement in soil except in deserts and saline soils, where solute concentrations become fairly high and considerable spatial heterogeneity in solute concentration can exist within the soil. The **gravitational potential,**[11] Ψ_g, relates to the potential energy stored in water relative to its position or elevation within the soil. The groundwater table is typically used as a reference point when evaluating the gravitational water potential component (i.e., $\Psi_g = 0$ at the groundwater surface), which makes this component positive in magnitude for soil water. The gravitational potential is generally responsible for movement of water down through the soil under saturated conditions (see below).

The **matric potential,** Ψ_m, is related to the reduction in the energy state of water as water molecules associate with soil surfaces (**adsorption**) or the interiors of pores

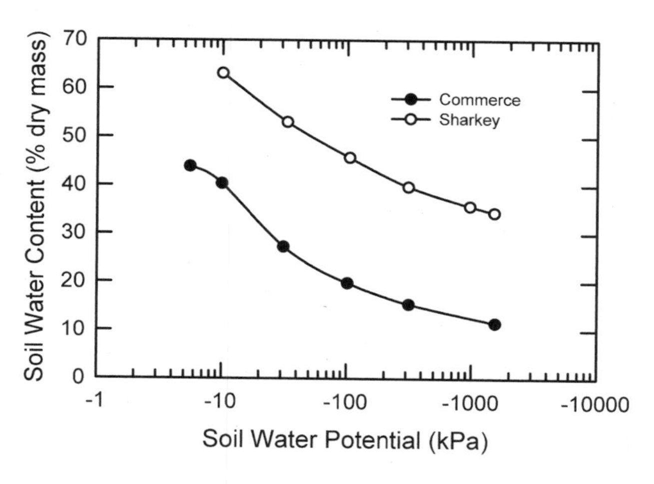

Figure 14.8. Soil water characteristic curves (also known as soil water release curves) for two soils with contrasting texture: a clay (Sharkey) and a silt-loam (Commerce). These curves describe the relationship between soil water content (volumetric or gravimetric) and soil water potential, and they are typically produced by slowly drying completely saturated soils. Note the log scale on the x axis. The clay holds more total water than the silt-loam, but it also holds the majority of this water more tightly than the silt-loam. From these soil water characteristic curves, the pore-size distribution of the soil can be estimated. (Modified from Bonner 1968)

(**capillarity**). The finer the soil particles (texture) and the smaller the pore sizes, the lower the Ψ_m at a given water content (fig. 14.8). Because these solid-liquid interactions reduce the energy state of water, this component is also negative in magnitude in the soil. The Ψ_m is primarily responsible for water movement under unsaturated conditions (when not all soil pores are filled with water), which is by far the most common soil water state.

The reference point for total soil water potential, Ψ_T, is zero (where $\Psi_T \approx \Psi_o + \Psi_g + \Psi_m$), based on pure water at its reference state and unaffected by soil solids. Water potential components are generally measured in terms of pressure (force per area).[12] After a torrential rain or following snowmelt, where water fills essentially all the soil pores (i.e., the soil is saturated and $\Psi_m \approx 0$), water drains from the larger to progressively smaller soil pores, until soil water is contained in pores of a size where $\Psi_g \approx \Psi_m$. The water that has drained from soil to this water state is often called **gravitational water,** and it is assumed that this water is essentially plant unavailable because it takes only two to three days following the cessation of water inputs for this soil water state to be reached. When the soil is at this water state (where micropores are still filled with water but macropores are filled with air), the soil is said to be at **field capacity** (FC; roughly corresponding to a $\Psi_m = -0.01$ to -0.03 MPa). Fine-textured soils continue to drain after this point due to differences in Ψ_m, but changes in soil water content occur much more slowly. When the remaining

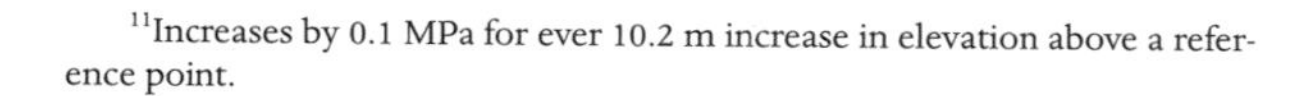
[11]Increases by 0.1 MPa for ever 10.2 m increase in elevation above a reference point.

[12]In the International System of Units (Système International, or SI), one Pascal (Pa) equals one Newton (N) of force acting over one square meter of area. An older unit of pressure, the bar, is equal to 0.1 MPa.

water in soil is contained in pores that are so small that the Ψ_m is lower than the water potential of plant roots, then the **permanent wilting point** (PWP) of plants has been reached. The Ψ_m at which the PWP is attained varies widely among plant species, but operationally it has been defined (based on a sunflower plant placed in a humid chamber) at a Ψ_m value equal to –1.5 MPa. Hence, the plant-available water storage capacity (PAWSC) of the soil can be calculated as

$$\text{PAWSC (m)} = \text{water content at FC } (m^3/m^3) - \text{water content at PWP } (m^3/m^3) * \text{soil depth (m)}$$

Soil texture, soil structure, and coarse fragment content greatly influence the PAWSC. For instance, a sandy loam soil at FC may contain as much plant-available water as clayey soil at FC and of the same soil depth, even though the sand contains only about half the amount of water at this water state (fig. 14.9).

Water movement in saturated soil is governed by Darcy's law, $Q/A = K * (H/L)$, where Q/A is the water flux density (volume flow per unit cross-sectional area perpendicular to the water flow per unit time), K is the saturated hydraulic conductivity (in meters per day), H is the hydraulic head (in meters), L is the length of flow (in meters); and H/L is the hydraulic gradient (unitless). Except during and right after a saturating rain or immediately after snowmelt, however, soils are in an unsaturated state. As soils dry out, water flow behaves differently because K changes as water flows through progressively smaller and

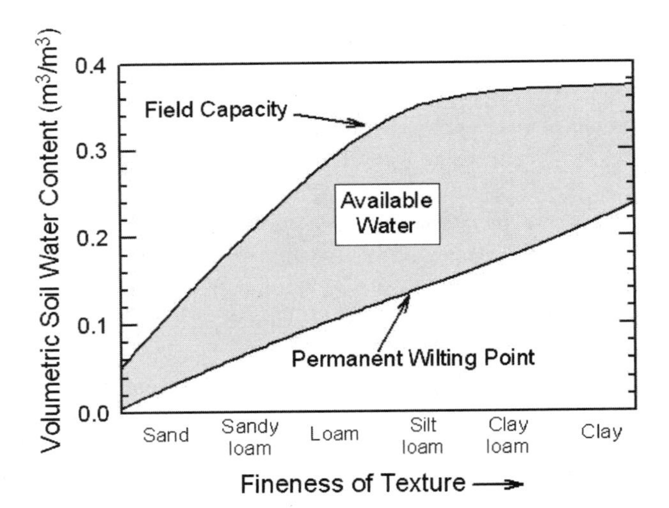

Figure 14.9. General relationships between soil water status and soil texture. The capacity of a soil to store water against gravitational forces (i.e., the field capacity) increases as the soil texture becomes finer up to silt-loams, and then levels off. In contrast, the wilting coefficient (also known as the permanent wilting point) continues to increase as soil become finer. As a result, plant-available water storage tends to be maximal for soils with silt-loam textures. Many other factors influence plant-available water storage besides texture, including coarse-fragment content and soil structure; hence, these patterns should be viewed as only approximate. (Modified from Brady and Weil 2002)

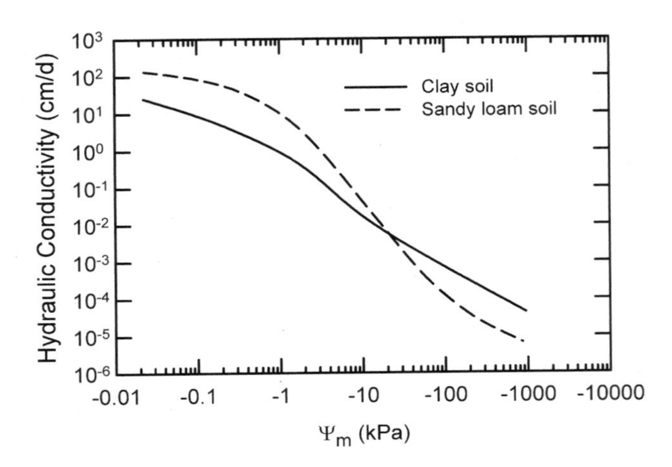

Figure 14.10. Idealized relationship between soil matric potential and hydraulic conductivity for two soils of contrasting texture. Saturated flow occurs at matric potentials near zero. Note that the hydraulic conductivity declines dramatically as the soil dries, and that coarse-textured soils can transmit water more rapidly than fine-textured soils near saturation, but that at low water contents fine-textured soils transport water faster. Most water movement in soil occurs under unsaturated conditions. (Modified from Brady and Weil 2002)

smaller pores, and the path of water flow becomes more tortuous as air fills pores and blocks water movement (embolisms). Hence, the rate of water flow declines in a highly nonlinear fashion as the soil dries (fig. 14.10). At saturation, water movement is slower in fine-textured soils than in coarse-textured soils because of the greater pore sizes in coarse-textured soils; however, as soils dry, flow rate decreases more gradually for fine-textured soil, so water transport remains appreciable in these soils even under dry conditions.

Soil temperature, like soil water content, is not a physical property of the soil per se, but it is largely influenced by soil physical properties coupled with the climatic regime. Soil temperature is important because it influences rates of mineral weathering and the activities of plant roots and other soil organisms and, therefore, affects rates of decomposition, nutrient mineralization, and nutrient and water uptake (chapters 15 and 18). In general, chemical and biological processes increase in an exponential manner with increasing soil temperature. This relationship has been frequently characterized as the **reaction quotient,** or **Q_{10} value:** $Q_{10} = R_{T+10°C}/R_T$, where R is the rate of a biological or chemical process at temperature T and at T + 10 °C. Typically, soil processes have Q_{10} values around 2 to 3; hence, changes in soil temperature of 10 °C result in a double or tripling of the process rate.

Soil temperature regimes (i.e., temporal and spatial [mainly depth] variation in temperature in soil) depend on the climate as well as soil thermal properties. Like water, heat moves from an area of higher concentration (denoted by the temperature of the material) to an area of lower concentration. Two important thermal properties characterize soils. The **volumetric heat capacity** is the amount of heat

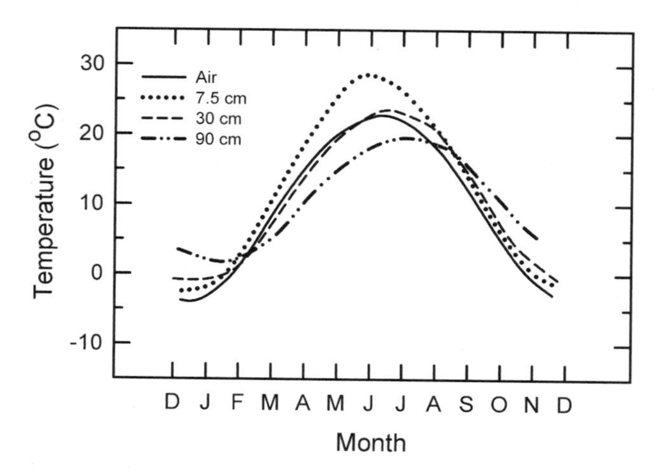

Figure 14.11. Mean monthly air and soil temperatures at different depths at Lincoln, Nebraska, over a 12-year period. This figure demonstrates two characteristics of soil temperature: as soil depth increases, the variation in temperature (in this case, monthly, but it is also true at other temporal scales) declines; and as soil depth increases, the lag in soil temperature (time difference between respective maximums or minimums) increases. (Modified from Brady and Weil 2002)

per unit volume of soil that needs to be added to the soil material to increase its temperature by 1 K. Water has the highest volumetric heat capacity of any soil material, while air has a very low volumetric heat capacity. As a soil wets up, the amount of radiant energy required to heat the soil increases. Hence, wet soil warms up and cools more slowly than dry soil. The other important thermal property of soil that alters soil temperatures in a given climate is the **thermal conductivity** of the soil. Thermal conductivity is the rate of heat movement through a given material for a given temperature difference (watts per meter-kelvin). Denser materials tend to have higher thermal conductivities because air has a very low thermal conductivity (i.e., it is a good insulator). Organic materials have lower thermal conductivities than inorganic materials, and thermal conductivities general decrease as particle size decreases (i.e., sands have greater thermal conductivities than clays).

These two thermal properties of soil materials result in two patterns that characterize soil temperature regimes. The first is that as you go deeper in the soil, the variation in soil temperature (within a day [diel], monthly, and seasonally) decreases. If you dig deep enough into a soil (typically several meters below the surface), you will reach a point where there is little temporal variation in soil temperature, and the soil temperature is very close to the mean annual air temperature. This is the physical basis for a wine cellar, a place to keep wine at a constant (and cool for temperate climates) temperature without the use of artificial heating and cooling. Second, as you go deeper into the soil, the time at which the soil temperature reaches a maximum or a minimum is delayed increasingly relative to the time that they occur at the soil surface (fig. 14.11).

One additional feature of water that influences soil temperatures is the large amount of heat energy consumed when water changes from a liquid (at its boiling point temperature) to a gas, called its **latent heat of vaporization** (2.26 MJ/kg), or this same amount of heat energy that is released when water condenses. Hence, surface soils cool during the day as water is evaporated from the soil surface (in the same way plant leaves are cooled by evaporating water from leaf stomata during transpiration [chapter 15]).

The presence of a forest floor (O horizon) can greatly modify the thermal regime of the underlying mineral soil (fig. 14.12). The low volumetric heat capacity and low thermal conductivity of this material result in the forest floor heating rapidly during the day and transmitting little of this heat to the underlying mineral soil. During the night, the presence of the forest floor also reduces heat loss from the mineral soil (much like a blanket), and the organic material itself cools rapidly. Hence, the presence of a forest floor reduces the temperature variation in the underlying mineral soil. This characteristic of the forest floor can negatively affect seedling regeneration because seedling stems surrounded by these materials can reach lethal temperatures during the day in warm summer months, thermally girdling and killing the seedlings. The lack of a forest floor overlying mineral soil of a moderate texture also can result in frost-heaving of seedlings during the spring in temperate climates. This is because these soils are fine enough to store some water so they can readily freeze at night, but not too much water, which, because of the high volumetric heat capacity of water, would tend to keep the soil warmer. When water freezes, it expands in volume, and the resulting induced forces can push material out of the soil. Without a forest floor, at night (particularly in an open area where trees do not reradiate heat back at the soil surface at night) these soils cool very rapidly, freeze, and then warm rapidly during the day. The resulting freeze-thaw action can push an entire seedling from the soil.

Other factors in forests also influence soil temperatures. For instance, the presence of snow can modify soil temperatures. Snow is a good insulator (low thermal conductivity) because of the air trapped within the snow crystals. Hence, when snow is on the ground during the winter months, it modifies the soil temperature by reducing heat loss (fig. 14.13). The indirect influence of snow cover on soil processes via modifying soil temperature regimes can be substantial given that, in some ecosystems (including some forests), 20 to 50 percent of annual carbon and nitrogen cycling and soil-atmosphere trace gas fluxes can occur during winter (Groffman et al. 2001). The presence of the forest canopy influences soil temperature regimes by reducing insolation during the day and reradiating radiant energy during the night back at the soil surface. This results in lower diel variation in temperature as well as lower soil temperatures during the growing season (fig. 14.14). In temperate forest nurseries, transparent plastic films are frequently used to cover the soil surface and

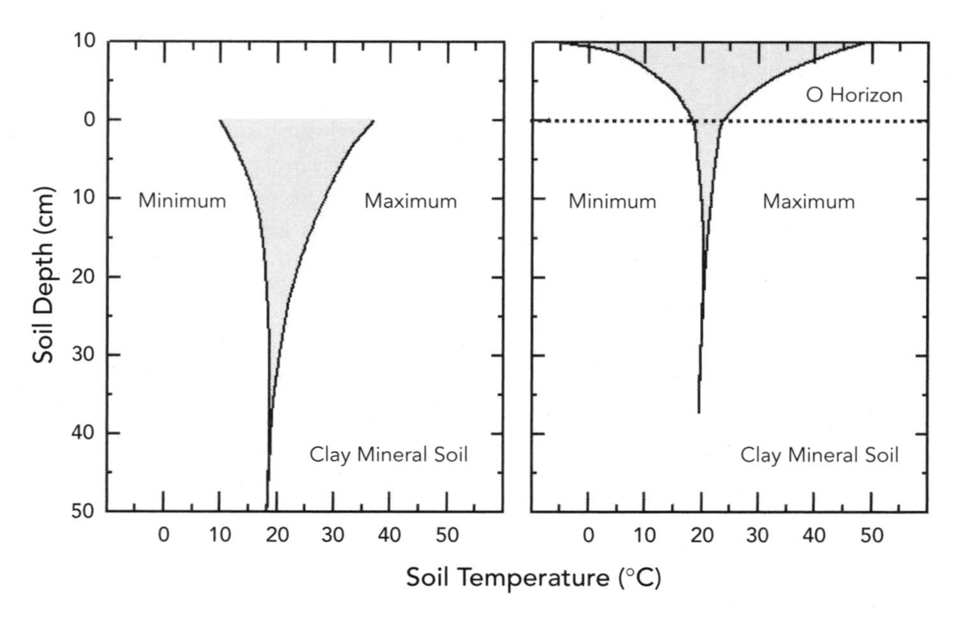

Figure 14.12. The presence of a forest floor (O horizon) can have a profound effect on underlying mineral soil temperature. These graphs show daily temperature variations with depth for a mineral soil lacking a forest floor (litter mulch, *left*) and a mineral soil covered by a forest floor (*right*). Because of the low thermal conductivity of the forest floor organic layer, little heat is transported to or from the underlying mineral soil, resulting in a decline in the diel temperature variation in the underlying soil. The low volumetric heat capacity of organic matter results in this layer showing large ranges in temperature compared to the underlying mineral soil. (Modified from Cochran 1969)

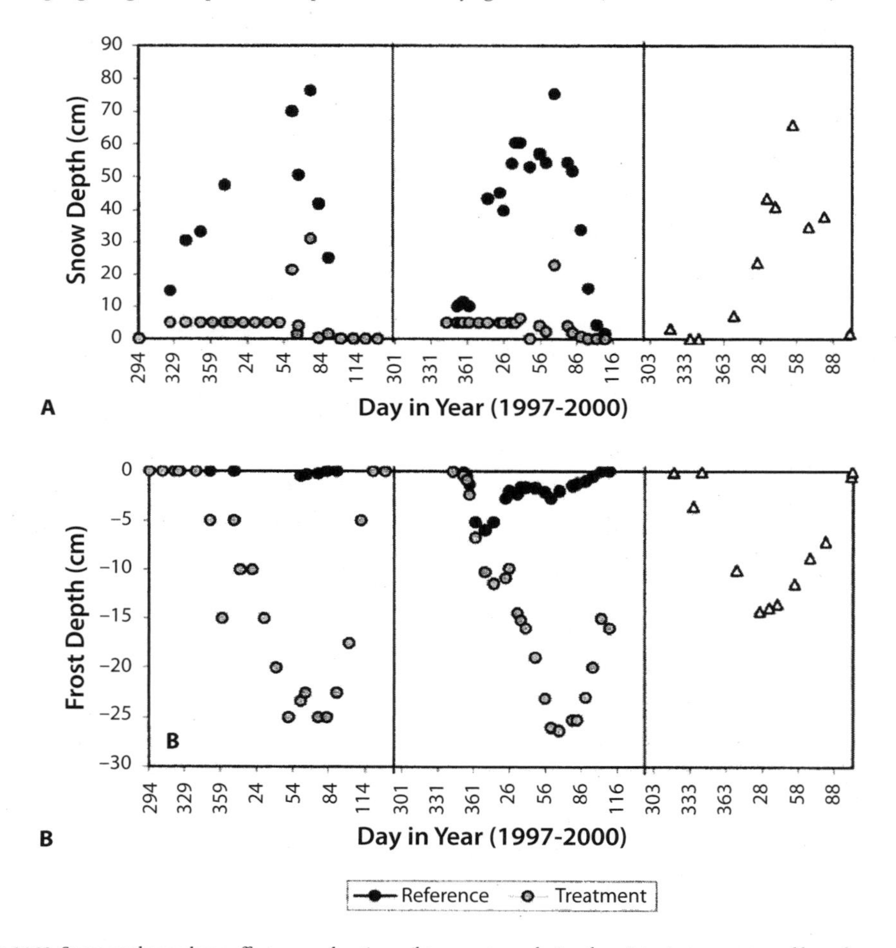

Figure 14.13. Snow can have a large effect on moderating soil temperatures during the winter in temperate and boreal climates. When snow (A) was removed from treated plots at the Hubbard Brook Experimental Forest in New Hampshire during two successive (and relatively mild) winters (1997–1998 and 1998–1999), the frost depth (B) increased substantially. As a result, soil biological processes such as organic matter decomposition can continue at a much higher rate under snow cover than when snow cover is low or absent in these climates. The recovery year (1999–2000), when snow was not manipulated in the treated plots, is denoted by triangles. (Hardy, J.P., Groffman, P.M., Fitzhugh, R.D., et al. Snow depth manipulation and its influence on soil frost and water dynamics in a northern hardwood forest. © 2001, *Biochemistry* 56:161, Fig. 2. Reprinted with kind permission of Springer Science and Business Media)

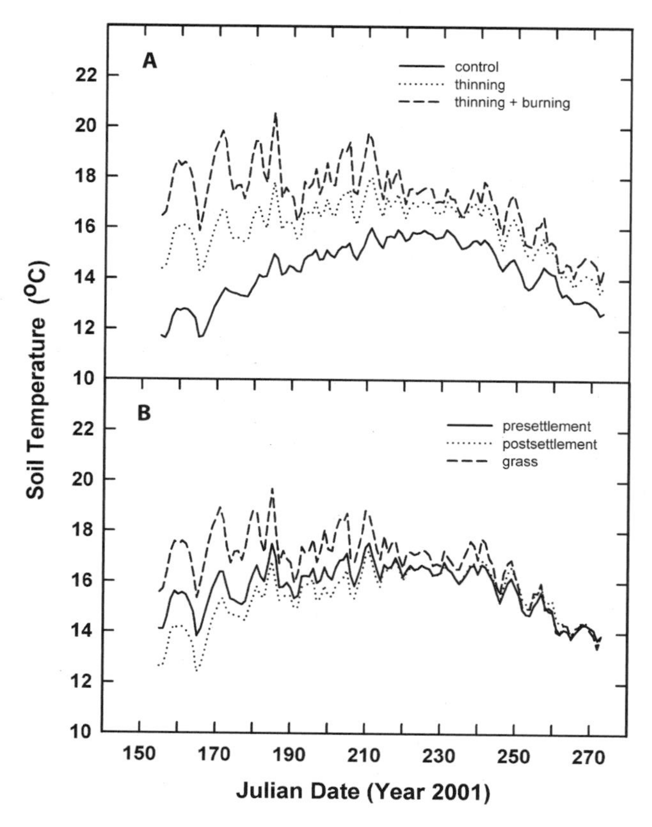

Figure 14.14. The presence of a forest canopy and a forest floor (O horizon) can modify mineral soil temperatures substantially. This figure shows the impact of ecological restoration treatments in a ponderosa pine (*Pinus ponderosa*) forest in the southwestern United States. Treatments (A) included thinning the forest to an open, savannalike forest structure similar to that found prior to Euro-American settlement of the area; thinning coupled with periodic (every four years) prescribed fire that also occurred in these forest prior to Euro-American settlement; and an unmanaged control. Thinning alone greatly increased insolation at the ground surface but had little effect on the mass of the forest floor, while thinning plus burning also resulted in a massive reduction in the forest floor mass. As a result, daily mean soil temperatures (7.5 cm mineral soil depth) are from 1 to 5°C higher in the thinned and also burned plots than the control plots during the growing season, with the soil temperatures in the thinned-only plots being intermediate. B. The impact of canopy type on soil temperatures in this study has been averaged across treatments. Note that soil temperatures are highest under grass canopies, lowest under postsettlement tree canopies, and intermediate under presettlement canopies (where tree canopy cover is lower but forest floor mass is higher). Such differences in soil temperature have a large impact on soil processes such as decomposition and nutrient cycling. (Reprinted with permission from Boyle et al. 2005)

increase the growing season by warming the soil earlier in the spring. This is achieved because the clear plastic transmits the sun's shortwave radiation during the day through the plastic film into the soil, while limiting heat loss during the night by reducing longwave radiation from the soil to the sky (the plastic also reduces convective loss of heat). A transparent plastic film that covers moist soil during warm, summer months can also be used prior to tree planting to reduce soil pathogens, sometimes being as effective as pasteurization and chemical sterilizing agents (Bihan et al. 1997).

14.4 CHEMICAL PROPERTIES OF SOILS

The chemical properties of soils are determined by the chemical composition of the parent material and the extent of weathering of this material, as well as the amount, composition, and decomposition of organic matter inputs from plants to the soil. As minerals contained in the parent rock (primary minerals) weather during pedogenesis, these minerals (the most common of which are feldspars, ferromagnesium minerals, and micas) are converted to new minerals called secondary minerals or **clay minerals.**[13] Not only does this transition result in a dramatic increase in the surface-to-volume ratio of the soil minerals (see above), it also generally results in an accumulation of a net electrical charge on the mineral surfaces. The electrical charge on soils, which results from both secondary minerals and soil organic matter, is critical to ecosystem functioning because it strongly influences nutrient retention and cycling (chapters 17 and 18).

Both physical and chemical processes are involved in transforming parent material to soil. **Physical weathering** processes include: changes in temperature (which result in expansion and contraction of the minerals contained in the rock); abrasion by solid materials contained in water, ice, and wind; and, to a lesser degree, expansion of plant roots in rock fissures and the activities of burrowing animals. The influence of changes in temperature on physical weathering rates is greatly enhanced if ice forms in surface cracks, owing to the considerable force applied to the rock when water expands during freezing. Physical weathering processes are usually greatest in very cold or very dry environments, and they result in the disintegration of parent material into smaller particles with little, if any, alteration in their chemical composition. However, the increase in surface area-to-volume ratio resulting from a reduction in particle size increases their susceptibility to chemical weathering processes. In contrast to physical weathering, **chemical weathering** is most intense when the climate is wet and warm, and results in chemical changes in the parent material as this material evolves to become soil material. Water is an important component in all chemical weathering processes, and frequently the activities of soil organisms provide the catalyst or accentuate chemical weathering rates.

A variety of different kinds of secondary minerals (or soil **colloids**[14], from the Greek word meaning 'glue') form in soils during chemical weathering of parent materials, determined by the types of primary minerals present, the weathering environment, and intensity of weathering (fig. 14.15). There are four major types of soil colloids (crystalline silicate clays, poorly crystalline silicate clays, aluminum and iron oxide clays, and humus; table 14.1). Crystalline

[13] So-called because they are usually found in the clay-sized fraction.

[14] Colloids are very small particles (typically 1 nm to 1 μm in diameter) of one substance dispersed in a different substance. Due in part to their very high surface area-to-volume ratio, soil colloids are responsible for most of the soil's unique chemical and physical properties.

There are seven basic types of chemical weathering reactions, all involving water and many influenced by biological processes. These various chemical weathering processes generally occur concurrently and are interdependent (Birkeland 1999; Brady and Weil 2002).

1. **Hydration.** This process involves the binding of intact water molecules to a mineral. It weakens the mineral structure, making it more susceptible to breakdown by the other chemical weathering processes. Hydrated oxides of iron (e.g., $Fe_{10}O_{15}\cdot 9H_2O$) and aluminum (e.g., $Al_2O_3\cdot 3H_2O$) exemplify common products of hydration reactions.
2. **Hydrolysis.** In hydrolysis reactions, water molecules split, with the hydrogen ions often replacing a cation from the mineral structure of the solid. For example, when hydrolysis occurs on a potassium containing feldspar (orthoclase), the feldspar may be weathered to the clay mineral kaolinite:

   $2KAlSi_3O_8 + 2H^+ + 9H_2O \rightarrow H_4Al_2Si_2O_9 + 4H_4SiO_4 + 2K^+$
   (orthoclase, solid) (aqueous) (liquid) (kaolinite, solid) (aqueous) (aqueous)

   The potassium released is soluble and subject to adsorption by soil colloids, uptake by plants or other soil organisms, or removal by leaching.
3. **Dissolution.** In this process, water hydrates the cations and anions in a solid until they become dissociated from each other and surrounded by water molecules, for example, halite (sodium chloride) and gypsum (hydrous calcium sulfate) dissolving in water:

   $NaCl \rightarrow Na^+ + Cl^-$
   (halite, solid) (aqueous) (aqueous)

   $CaSO_4\ 2H_2O \rightarrow Ca^{2+} + SO_4^{2-} + 2H_2O$
   (gypsum, solid) (aqueous) (aqueous) (liquid)
4. **Carbonation and other acid reactions.** Carbonic acid is a weak acid that is produced when carbon dioxide dissolves in water (a process enhanced by root and microbial respiration). Carbonic acid hastens the dissolution of calcite (calcium carbonate), a process known as *carbonation*:

   $CaCO_3 + CO_2 + H_2O \rightarrow Ca^{2+} + 2HCO_3^-$
   (calcite, solid) (gas) (liquid) (aqueous) (aqueous)

   Other sources of acidity exist in soil (see section 14.4), including hydrogen ions associated with soil colloids and organic acids, and the strong acids nitric acid (HNO_3) and sulfuric acid (H_2SO_4). All of these sources of acidity can be involved in similar chemical weathering processes as carbonation.
5. **Oxidation-reduction.** Minerals that contain iron, manganese, or sulfur are particularly susceptible to chemical weathering via oxidation-reduction reactions. For example, iron is usually present in primary minerals in the ferrous form (Fe^{2+}). When these minerals are exposed to air and water, oxidation of the iron to the ferric form (Fe^{3+}) occurs. This change disrupts the electroneutrality of the crystal, forcing other cations to leave the crystal lattice to maintain neutrality; the loss of these other cations either brings about the collapse of the lattice, or increases the susceptibility of the mineral to further chemical weathering by other processes. The chemical weathering of the mica biotite to the clay mineral vermiculite is an example of weathering due to oxidation.
6. **Complexation.** Organic acids, produced from the activities of soil organisms during decomposition (including oxalic, citric, and tartaric acids, as well as the much larger fulvic and humic acids) or leached from the plant canopy, increase chemical weathering rates because they provide hydrogen ions. They also can increase weathering rates by organic **complexation** reactions with Al^{3+} or other polyvalent metal cations held within the structure of silicate minerals. This process, also known as **chelation** (from the Greek word meaning 'claw'), removes the Al^{3+} from the mineral, weakening it, and making it more susceptible to other chemical weathering processes. In some situations, weathering by complexation might even exceed that brought about by hydrolysis.
7. **Ion exchange.** Sometimes, minerals can change from one type to another via the exchange of ions between the solution and the mineral. The ions that are most readily exchanged are those between silicate layers, such as calcium, potassium, and sodium (see section 14.4). In this form of chemical weathering, the basic structure of the mineral is unchanged, but the interlayer spacing may be altered due to differences in the sizes of the cations involved. For instance, cation exchange reactions result in the conversion of fine-grained mica (illite) to montmorillonite by replacing K^+ in the interlayer space with hydrated Mg^{2+}.

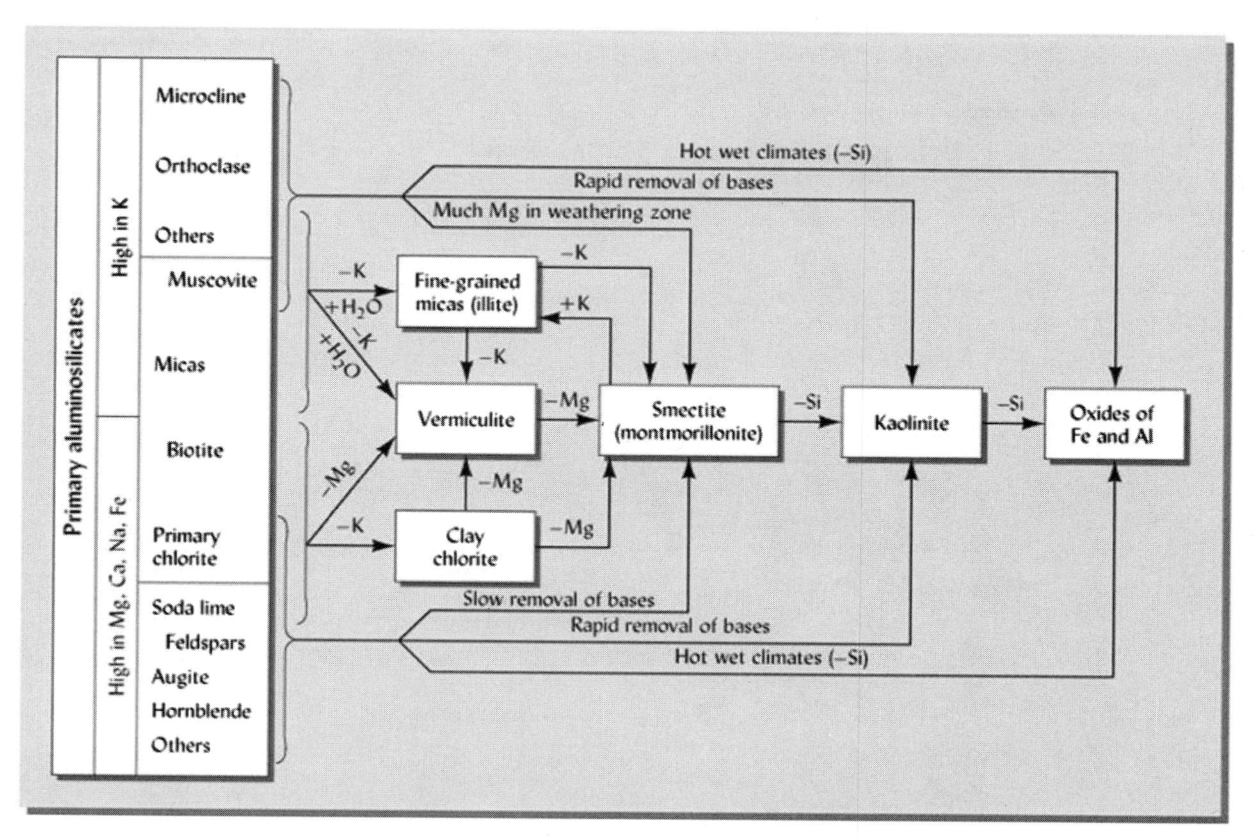

Figure 14.15. Factors that influence clay mineralogical development from primary aluminosilicate minerals in soil. Under conditions that result in mild weathering, fine-grained micas (illite), vermiculite, and chlorite dominate. As the extent of weathering increases, smectite (e.g., montmorillonite) formation is favored. At high levels of weathering, kaolinite (or other members of the kaolin group) and aluminum and iron oxides develop. The overall progression of mineral weathering of aluminosilicates is characterized by a loss of the so-called base cations (e.g., ions of calcium, magnesium, potassium, and sodium) and silicon to the soil solution, where these elements are eventually leached out of the soil. In contrast, most of the iron and aluminum remains in the solid phase. (Brady, N.C., and Weil, R.R. *The nature and properties of soils*. 13th ed. © 2002, p. 334. Reprinted by permission of Pearson Education Inc., Upper Saddle River, NJ)

Table 14.1

Major properties of selected soil colloids

Colloid	Type	Size (μm)	Shape	Surface area (m^2/g) External	Surface area (m^2/g) Internal	Interlayer spacing* (nm)	Net charge† ($cmol_c$/kg)
Humus	Organic	0.1–1.0	Amorphous	20–800‡	—§	—	−100 to −500
Vermiculite	2:1 silicate	0.1–0.5	Plates, flakes	70–120	600–700	1.0–1.5	−100 to −200
Smectite	2:1 silicate	0.01–1.0	Flakes	80–150	550–650	1.0–2.0	−80 to −150
Fine-grained mica	2:1 silicate	0.2–2.0	Flakes	70–175	—	1.0	−10 to −40
Chlorite	2:1 silicate‖	0.1–2.0	Variable	70–100	—	1.41	−10 to −40
Kaolinite	1:1 silicate	0.1–5.0	Hexagonal crystals	5–30	—	0.72	−1 to −15
Gibbsite	Al oxide	<0.1	Hexagonal crystals	80–200	—	0.48	+10 to −5
Goethite	Fe oxide	<0.1	Variable	100–300	—	0.42	+20 to −5
Allophane and imogolite	Poorly crystalline	<0.1	Hollow spheres or tubes	100–1000	—	—	+20 to −150

Source: Data from Brady and Weil 2002.

*From the top of one layer to the next similar layer; also called *c*-spacing.

†Centimoles of charge per kilogram of colloid; a measure of the ion exchange capacity.

‡It is very difficult to measure the surface area of humus, and different methods produce widely contrasting results.

§Denotes that there is no internal surface area for this colloid because it is either a nonexpandable (when hydrated) crystalline colloid or because it is essentially noncrystalline or amorphous (the latter being the case for humus, allophane, and imogolite).

‖Also known as a 2:2 type silicate because it has an interlayer sheet that has a structure similar to another octahedral layer.

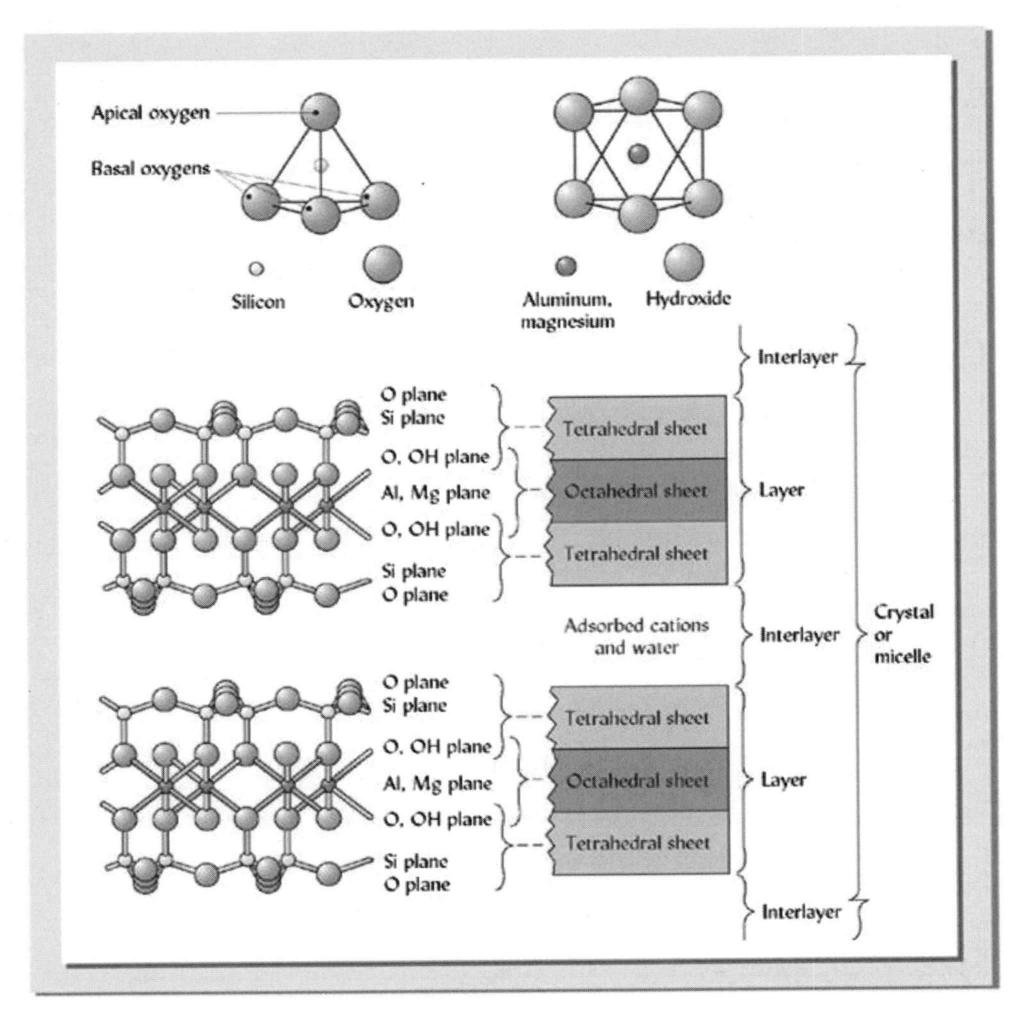

Figure 14.16. The molecular and structural components of silicate clays. The basic structures consist of a single, four-sided, tetrahedron composed of a coordinating silicon atom surrounded by four oxygen atoms, and a single eight-sided octahedron composed of a coordinating aluminum (or sometimes magnesium) atom surrounded by six hydroxyl groups or oxygen atoms. A silicate clay crystal consists of thousands of these tetrahedral and octahedral building blocks connected to give planes of silicon or aluminum (and magnesium) ions. These planes alternate with planes of oxygen atoms or hydroxyls, and apical oxygen atoms are shared between adjoining silicon and aluminum (or magnesium) planes. The silicon plane and associated oxygen-hydroxyls make up a tetrahedral sheet, while the aluminum (or magnesium) plane and associated oxygen-hydroxyls make up an octrahedral sheet. Different silicate clays contain different combinations of tetrahedral and octahedral sheets, and these groupings of sheets are called layers. Some silicate layers are separated by interlayers containing water and adsorbed cations. Each crystal or micelle (microcell) consists of many of these layers. (Brady N.C., Weil, R.R. *The nature and properties of soils,* 13th ed. © 2002, p. 322. Reprinted by permission of Pearson Education, Inc., Upper Saddle River, NJ)

silicate clays are the most abundant type of soil colloid in most soils (except in the Andisol, Oxisol, and Histosol orders of the USDA Soil Taxonomy system; see section 14.7). These clay minerals have a layered structure much like a stack of plates, and consist primarily of two kinds of sheets: a tetrahedral sheet dominated by silicon and oxygen, and an octahedral sheet dominated by aluminum, oxygen, and hydroxyl groups (fig. 14.16). The ratio of these two sheet types and the degree and location of substitution of ions for silicon and aluminum atoms in the mineral structure[15] largely define the crystalline silicate clay mineral (fig. 14.17). These minerals are predominantly negatively charged but show great diversity in the density of charge, plasticity and stickiness, and the ability to expand and contract when wetted and dried. Poorly crystalline silicate clay minerals, such as allophane and imogolite, also consist mainly of silicon, aluminum, and oxygen atoms, but they are not arranged into repeating units with long ranges.[16] These minerals usually form from volcanic ash and are dominant minerals found in the Andisol soil order. They have high amounts of both positive and negative charge (the net charge largely depending on pH) and strongly

[15]Called *isomorphous substitution* because only atoms of similar size can replace each other in a crystal lattice without disrupting or changing the crystal structure of the mineral.

[16]These minerals have been called *amorphous* or *noncrystalline* in the past, but they do typically have some repeating mineral structure; however, the repeating units have short ranges (Wada, 1985).

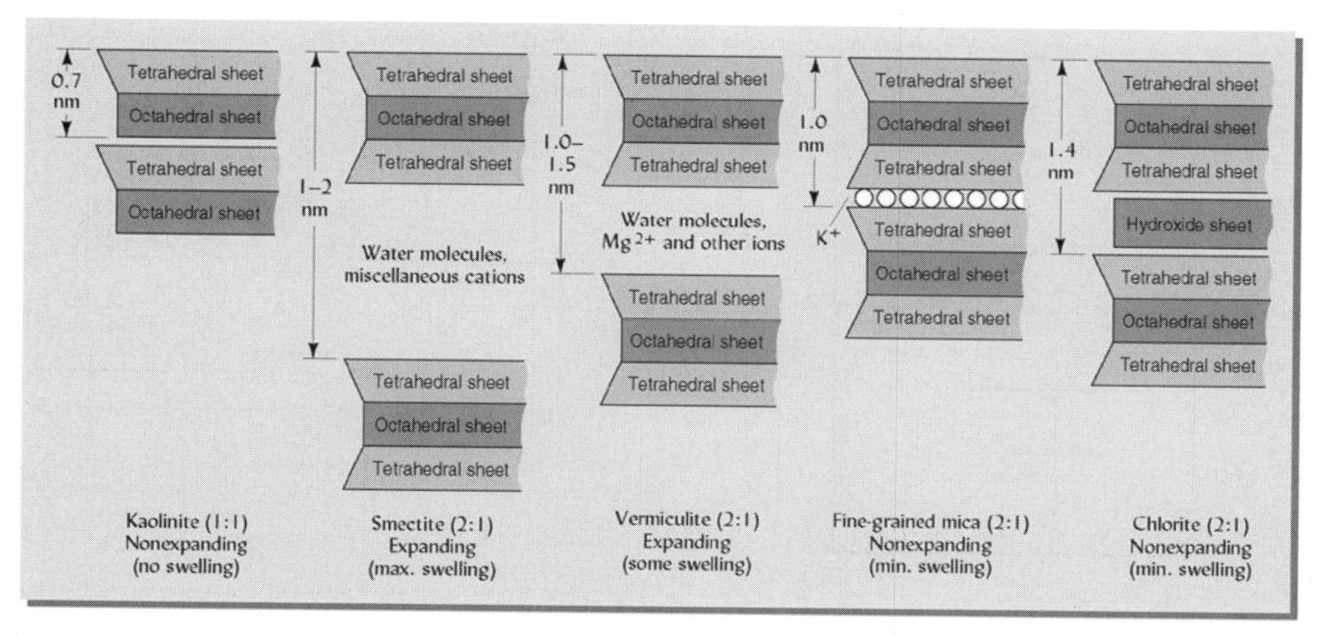

Fig. 14.17. Schematic diagrams of different types of common silicate clay minerals. Kaolinite is an example of one type of 1:1 (denoting the ratio of tetrahedral to octahedral sheets) silicate clay. The octahedral sheets in each of the 2:1 silicate clays can be either aluminum dominated (dioctahedral) or magnesium dominated (trioctahedral). In most chlorites, the trioctahedral sheets are dominant, while in most of the other 2:1 clays the dioctahedral dominates. Kaolinite does not expand or contract when hydrated or dehydrated, respectively, because the layers are held together by numerous hydrogen bounds. Maximum expansion/contraction is found with smectites (e.g., montmorillonite), while vermiculite has intermediate expansion/contraction because numerous Mg^{2+} ions in the interlayer act like electrostatic "glue" and limit expansion/contraction between adjacent 2:1 layers. Fine-grained mica does not expand/contract because K^+ ions in the interlayer bind the adjacent 2:1 layers tightly. Similarly, chlorite does not expand/contract because an additional octahedral-like sheet (consisting of hydroxides of aluminum, magnesium, and iron) tightly binds the adjacent 2:1 layers together. Because this interlayer sheet has a similar structure to the octahedral layer, frequently chlorites are called 2:2 layer silicate clays. (Brady N.C., and Weil, R.R. *The nature and properties of soils*, 13th ed. © 2002, p. 328. Reprinted by permission of Pearson Education, Inc., Upper Saddle River, NJ)

adsorb phosphate, especially under acidic conditions. They also tend to have high water-holding capacities and are plastic but not sticky when wet. Aluminum and iron oxide clay minerals are especially abundant in highly weathered soils (e.g., Ultisols and Oxisols) typically found in warm, humid regions. They consist both of crystalline forms and noncrystalline forms (the latter frequently coating other soil particles). Although commonly discussed as oxides, these clay minerals include iron and aluminum cations associated with both oxygen and hydrogen (i.e., hydroxides and oxyhydroxides). These minerals tend not to be very plastic or sticky, and have a range of net charge from slightly negative to moderately positive, depending on pH. Humus is not a secondary mineral, but an organic colloid formed after extensive decomposition of organic matter by soil microorganisms. It has high amounts of both negative and positive charge per unit mass, with the net charge always being negative but the degree depending on pH. It is an important colloid in almost all soils (particularly the organic soil order Histosols), especially in surface mineral horizons and in the forest floor (O horizon). Humus has an amorphous structure that consists of chains and rings of carbon atoms bonded covalently to hydrogen, oxygen, and nitrogen. Furthermore, humus, unlike clay minerals, has almost no cohesion. Humus has virtually no plasticity and stickiness but has a very high water-holding capacity.

The electrically charged surfaces on soil colloids are a critical chemical property of soil. In essence, these charged surfaces provide the 'parking spaces' that hold various cations and anions (many of which are plant nutrients) within the plant rooting zone and prevent them from being lost from the soil as water percolates through the soil profile. A measure of the ability of a soil to adsorb cations is called the **cation exchange capacity** (CEC) of the soil, and it is related to the type and amount of soil colloids present. The CEC is the amount of positively charged ions that can be electrostatically held on the negative charges on soil colloids. The term *exchange* means that these cations on the mineral surfaces can be replaced by other cations present in the soil solution. The exchange takes placed based on equivalent moles of charge and not on moles of mass. For instance, it takes two monovalent cations to exchange one divalent cation, but only one divalent cation to replace another divalent cation. Important plant nutrients such as calcium, magnesium, potassium, and ammonium are cations.

Most temperate and boreal soils have a net negative charge and, hence, tend to adsorb more cations than anions from the soil solution. Because of the net negative charge

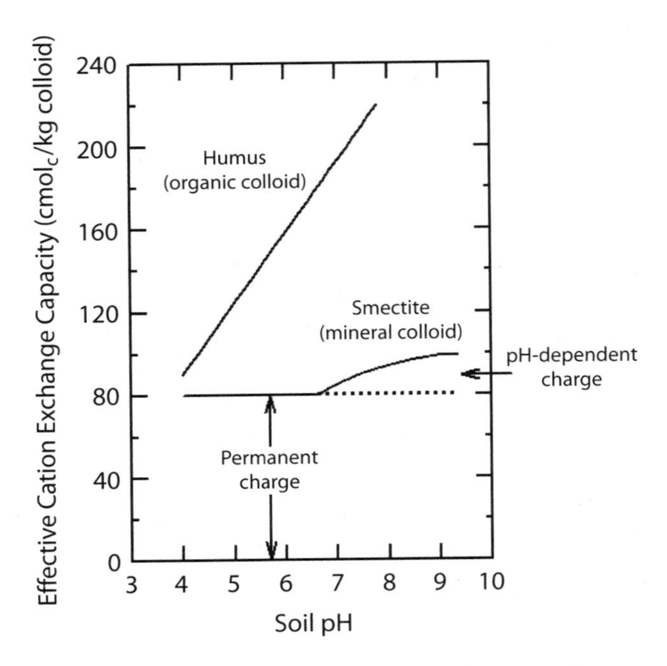

Figure 14.18. The cation exchange capacity of soil colloids changes with changes in pH. In humus, essentially all of the negative charges originate from the deprotonation of functional groups (e.g., carboxyl groups). For silicate clay minerals such as smectite, most of the negative charge is due to ionic substitution in the crystal (called isomorphous substitution), resulting in most of the mineral charge being permanent (i.e., unaffected by pH). However, all silicate clays have some pH dependent charge (usually at pH ≈ >6.0) because of deprotonation of hydroxyl groups at the edges of the crystal. (Modified from Brady and Weil 2002)

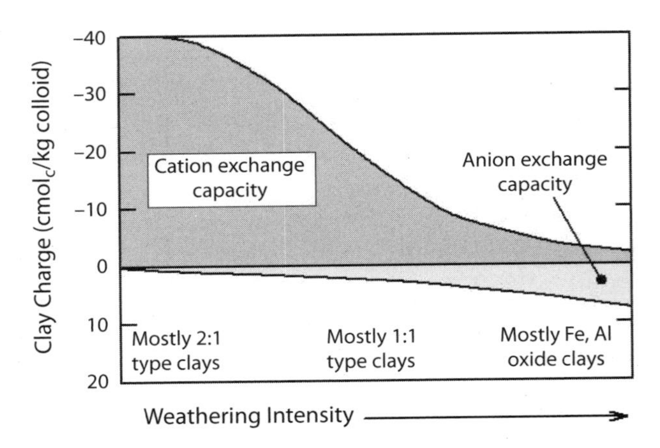

Figure 14.19. The cation and anion exchange capacities of soils (CEC and AEC, respectively) are a function of the types of clays present, which is a function of the degree of mineral weathering. High CEC and low AEC are found where 2:1 clays such as fine-grained mica (illite), vermiculites, and smectites are present, which occurs under mild weathering conditions. More intense weathering destroys these 2:1 clays (so-called base cations and silicon atoms are removed), leading to the formation of 1:1 clays (e.g., kaolinite) and then oxides of iron and aluminum. These clays have much lower CECs and higher AECs. (Modified from Brady and Weil 2002)

on soil colloids, anions (e.g., nitrate) in the soil solution generally move through soil pores more readily than cations (e.g., ammonium), which have their movement slowed by electrostatic attraction to the soil colloids.

The CEC depends to varying degrees on the soil pH because some of the soil colloids change their net charge as various mineral and organic compounds protonate (as pH decreases, reducing the net negative charge) or deprotonate (as the pH increases, increasing the net negative charge; fig. 14.18). Some secondary minerals (primarily aluminum and iron oxides and poorly crystalline silicates) may also have positive electrical charges on their surfaces and, therefore, an **anion exchange capacity** (AEC); however, the presence of positive charges on secondary minerals always depends on pH. Some soils (mostly in the tropics because of the dominance of highly weathered soils in this region, but also volcanic soils outside of those regions) have a variable net charge; in other words, depending on the pH, they may be dominated by either CEC or AEC (Sollins et al. 1988; fig. 14.19). Several plant nutrients are in anionic form that is readily exchangeable on anion exchange sites (e.g., nitrate and chloride).

Some ions are held on soil colloids much more strongly than by simple electrostatic attractions. These ions are not freely exchanged by other ions in the soil solution and are called **specifically adsorbed ions.** For instance, the phosphate anion (and to a lesser degree sulfate and molybdate anions) is frequently specifically adsorbed onto the surfaces of iron and aluminum oxides, where substitution of the anion for one or more water molecules (or hydroxyl groups) occurs at the mineral surface (known as **ligand exchange;** fig. 14.20). This process strongly affects the availability of phosphorus (as well as sulfur and molybdenum) to organisms in many soils (chapter 18). Other types of specific adsorption reactions occur in soil that result in the strong retention (and reduced plant and microbial availability) of some nutrient metals (e.g., cobalt and zinc) by oxide minerals and humus colloids (fig. 14.20).

Soil reaction is the activity of hydrogen ions (A_{H+}) in soil. The pH of forest soils is important because it affects the solubility of aluminum (which is toxic to many soil organisms, including plants), the weathering of soil minerals, the distribution of cations on the cation exchange complex, the availability of plant nutrients, and the activities of soil organisms (fig. 14.21; Fisher and Binkley 2000). Soil pH is measured on a logarithmic scale, where $pH = -(\log A_{H+})$. A difference in pH of one unit is equivalent to a 10-fold difference in A_{H+}. Pure water has a pH of 7 (at this pH, $A_{H+} = A_{OH^-}$[17]), which is considered neutral. At lower pH values, A_{H+} dominates and the soil is said to be acidic. At pH values greater than 7, A_{OH^-} dominates and the soil is said to be alkaline. Most mineral soils under oxidized soil conditions[18] have soil pH values

[17] A_{OH^-} is the activity of hydroxyls in the soil solution; $A_{H+} + A_{OH^-} \approx 10^{-14}$.

[18] When a given soil is exposed to reducing conditions due to high water content, the soil becomes more alkaline (pH increases).

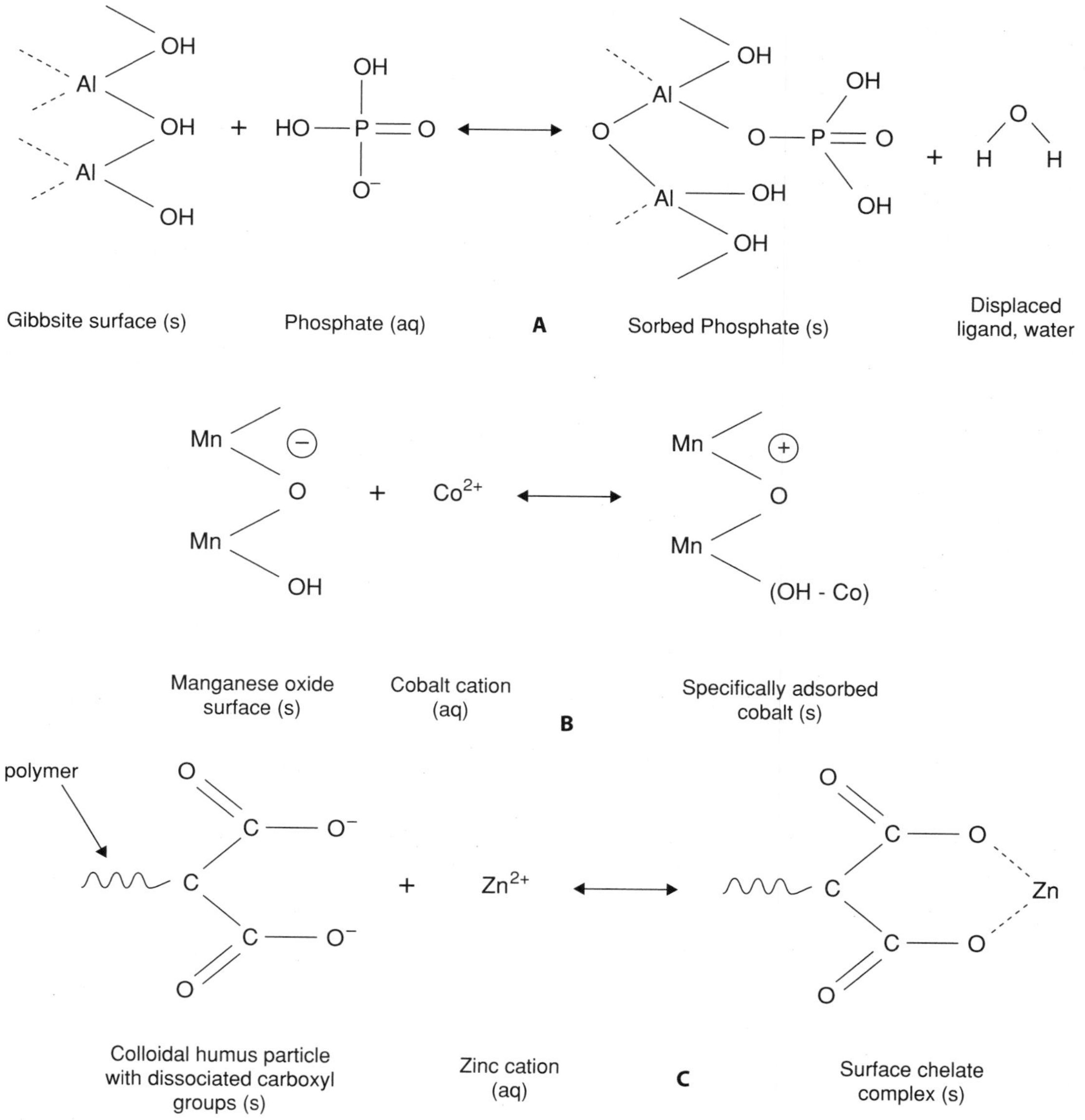

Figure 14.20. Specific adsorption of compounds onto soil surfaces is a much stronger interaction than simple cation exchange. Hence, specifically adsorbed compounds are not readily displaced by other compounds that do not undergo a similar type of surface interaction. A few of the several types of specific adsorption reactions are shown here. Of particular importance is the ligand exchange reaction (A) that the phosphate anion undergoes when interacting with aluminum and iron oxides (including hydroxides and oxyhydroxides). This reaction is responsible primarily for the low phosphorus availability in highly weathered soils dominated by these clay minerals. Many other types of specific adsorption reactions occur between elements/compounds and soil colloids. B. Adsorption of cobalt on a manganese oxide, and (C) adsorption of zinc on humus (surface chelation) are two examples of such reactions. (Modified from Singer and Munns 2006)

(measured in water) ranging between 3.5 and 8. In forest soils, soil pH values tend to be less than 4 only in organic horizons (Fisher and Binkley 2000). Soils that strongly resist a change in pH when hydrogen ions are added or removed are said to be well buffered. The buffering mechanisms of soils change as the pH of the soil changes (fig. 14.22). At pH values less than about 4.5, soils are buffered by the hydrolysis and precipitation of aluminum compounds (and iron compounds at pH values less than 3, but such low pH values are not conducive for plant growth), while at pH values greater than about 7.5, soils are buffered by the precipitation and dissolution of carbonates (primarily calcium carbonates). At intermediate pH values (between 4.5 and 7.5), most of the buffering occurs from cation exchange and protonation and deprotonation of pH-dependent exchange sites on clay and humus. Well-buffered soils tend to have high concentrations of organic matter and highly charged clays (Brady and Weil 2002).

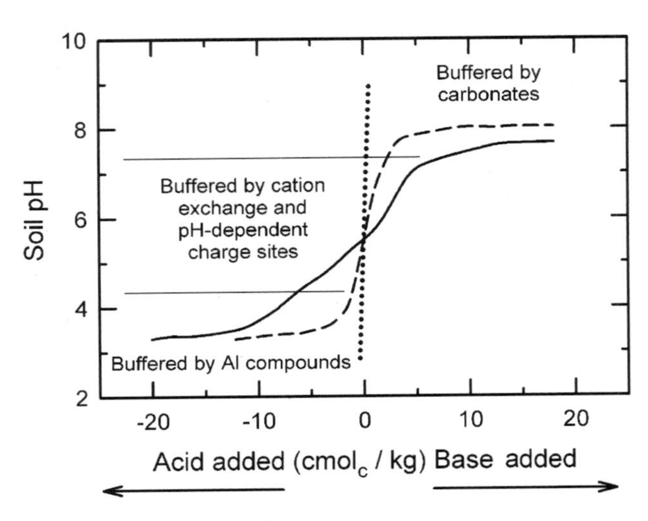

Figure 14.22. Soils are generally well buffered against changes in pH. The buffering capacity of a soil is evaluated by adding strong acids and bases to a soil sample and measuring the resulting pH. In this figure, unbuffered pure water (*dotted line,* A) is compared to a moderately buffered (*dashed line,* B) and a well-buffered (*solid line,* C) soil. Most soils are well buffered at low pH by the hydrolysis and precipitation of aluminum compounds and at high pH by the precipitation and dissolution of calcium carbonate. Most of the buffering at intermediate pH values is provided by cation exchange and protonation/deprotonation of pH-dependent exchange sites on clay and humus colloids. Generally, well-buffered soils have higher amounts of high-charged clay and humus than soils with poorer buffering capacities. (Modified from Brady and Weil 2002)

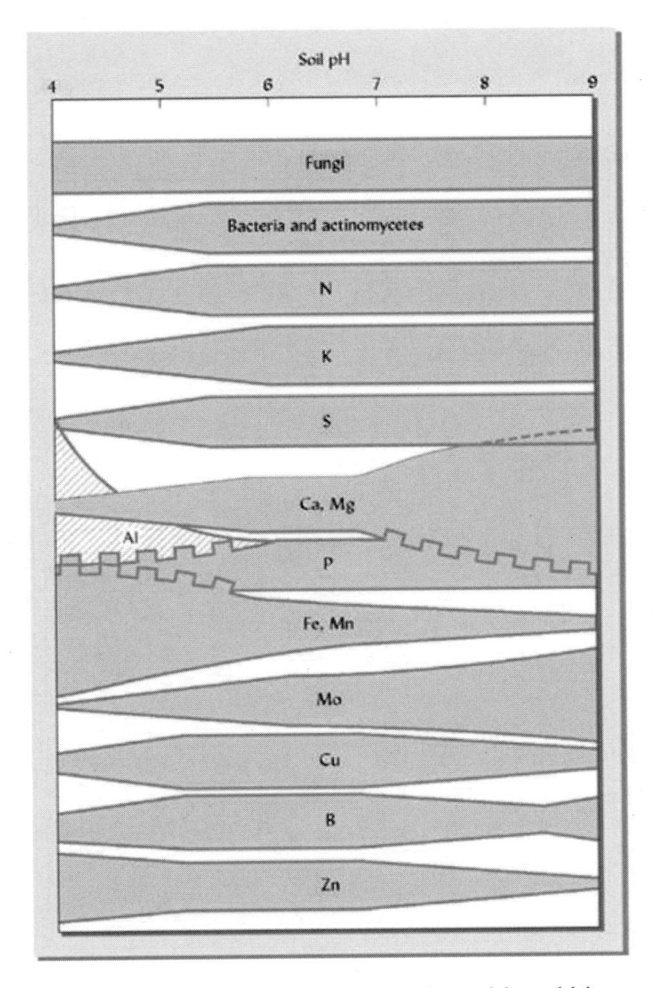

Figure 14.21. Generalized relationships between soil pH and the availability of nutrients to plants. These patterns as due primarily to the effect of pH on element solubility or the activities of soil microorganisms. The width of the band indicates the relative nutrient availability or microbial activity. The gear cogs between the P band and the bands for Ca, Al, and Fe represent the effect of these cations in reducing P availability. When nutrient availability is considered as a whole, soil pH values between 5.5 and 7.0 promote the greatest overall nutrient availability. This range in pH encompasses most forest soils, suggesting that differences in soil pH among soils or changes in soil pH at a given site do not substantially impact forest growth. (Brady N.C., and Weil, R.R. *The nature and properties of soils,* 13th ed. © 2002, p. 391. Reprinted by permission of Pearson Education, Inc., Upper Saddle River, NJ)

Three major pools of acidity in soil have been proposed (Brady and Weil 2002): **active acidity, salt-replaceable acidity** (also called **exchangeable acidity**), and **reserve acidity.** Active acidity is due to hydrogen (H^+) and aluminum (Al^{3+})[19] ions in the soil solution, and it is this acidity that is typically measured. Salt-replaceable acidity is the amount of acidity involving hydrogen and aluminum ions that is exchangeable by other cations using an unbuffered salt solution (such as 1 M KCl). It is typically 100 to 1,000 times greater than the active acidity pool (increasingly greater as the pH declines). Residual acidity is by far the largest pool of acidity in soil, being 1,000 times greater than active acidity in a sandy, low organic matter soil, to 50,000 or even 100,000 times greater in a clayey, high organic matter soil. This pool of acidity consists of hydrogen and aluminum ions (including aluminum hydroxyl ions) that are bound to humus and clays in nonexchangeable forms. Residual acidity is determined by titrating the soil with a strong base to a defined pH end point (often 8.2, based on carbonate equilibrium; Fisher and Binkley 2000).

The cation exchange complex in soil can be viewed as a weak acid, where the degree of dissociation (or deprotonation) depends on the nature of the acid[20] (i.e., whether the soil colloid is kaolinite, vermiculite, smectite, or humus), but also on the extent to which it has been titrated by addition of H^+ (or OH^-; Fisher and Binkley 2000). The degree of dissociation in soil science is referred to as the **base saturation**[21] of the soil and is defined as:

$$\text{percent base saturation} = \frac{\text{moles of charge of exchangeable } Ca^{2+} + Mg^{2+} + K^+ + Na^+ \star 100\%}{\text{moles of cation exchange capacity}}$$

The soil pH is often related to the soil base saturation, especially on a given soil type (Rhoades and Binkley 1996). Like a weak acid, the soil buffering capacity is generally greatest

[19] Aluminum acts like an acid by hydrolyzing water and combining with OH^- ions, leaving the H^+ to lower the pH of the soil solution.

[20] Known as *acid strength* (Fisher and Binkley 2000).

[21] As noted previously, these cations are not 'true' bases; hence, a more correct term to use is *percent acid saturation* (= 100%–% base saturation) when describing the degree of acidity on the soil cation exchange complex.

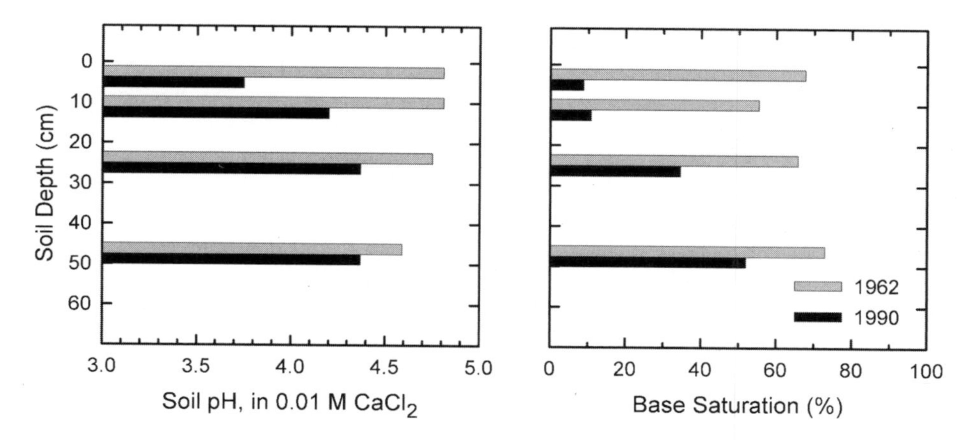

Figure 14.23. Changes in soil acidity and base saturation during development of a 28-year-old loblolly (*Pinus taeda*) forest in South Carolina. Trees were planted in 1956–1957 in an abandoned agricultural field (after approximately 150 years of cropping) that had been limed regularly as recently as 1954, and hence was less acidic than this soil would have been under natural conditions. The greatest acidification occurred in the surface mineral soil, with a concomitant reduction in base saturation. Soil was derived from a granite-gneiss parent material. (Data from Markewitz et al. 1998)

when the base saturation is about 50 percent (where base saturation approximates the degree of dissociation) and increases with soil CEC (where CEC is analogous to the amount of the acid). The fairly large buffering capacities of most forest soils make changing the soil pH by additions of acids or bases difficult.[22]

Generally, forest soils become more acidic over time (fig. 14.23) because the rate of production of H^+ exceeds the rate of its consumption. Numerous processes produce and consume H^+ in soil, however, soil acidification (as well as soil development) is generally favored in warm, well-drained, and high-rainfall areas, where the leaching of base cations from the soil proceeds more rapidly than the release of these cations and the consumption of H^+ during mineral weathering[23] (table 14.2). High soil biological activity also occurs under these conditions, favoring formation of carbonic acid due to rapid CO_2 production from aerobic respiration. Furthermore, soil acidification occurs more quickly in soil derived from parent materials low in base cations (e.g., granite and rhyolite)

As noted above, soil organic matter comprises a relatively small fraction of most forest soils. However, this soil

[22]For example, about 2 (sandy soils) to 10 (clay loam soils) Mg/ha of ground limestone ($CaCO_3$) would have to be added to a soil to raise the soil pH by one unit. This increase in pH typically lasts only a few years, necessitating a reapplication of the limestone (Brady and Weil 2002).

[23]Soils become acidic when H^+ ions added to the soil solution exchange with nonacid (so-called base) ions (e.g., Na^+, K^+, Mg^{2+}, Ca^{2+}) on clay and humus colloids. In humid regions, the nonacid cations are then removed from the soil in percolating water along with accompanying anions (e.g., HCO_3^-, NO_3^-, SO_4^{2-}). As a result, the exchange complex (and therefore the soil solution) becomes increasingly dominated by acid cations (H^+ and Al^{3+}). In more arid regions, the nonacid cations are mostly not removed from the soil by leaching and thus reexchange with the acid cations, minimizing or preventing a drop in pH (Brady and Weil 2002[0]).

Table 14.2
Acidifying and alkalinizing processes in soil

Acidifying (H^+-producing) processes	Alkalinizing (H^+-consuming) processes
Formation of carbonic acid from carbon dioxide: $CO_2 + H_2O \rightarrow H_2CO_3$	Input of carbonates or bicarbonates (e.g., $CO_3^{2-} + 2H^+ \rightarrow CO_2 + H_2O$)
Acid dissociation (e.g., $RCOOH \rightarrow RCOO^- + H^+$)	Anion protonation (e.g., $RCOO^- + H^+ \rightarrow RCOOH$)
Oxidation of N, S, Fe, and Mn compounds (e.g., $NH_4^+ + 2O_2 \rightarrow NO_3^- + 2H^+ + H_2O$)	Reduction of N, S, Fe, and Mn compounds (e.g., $SO_4^{2-} + 8H^+ + 6e^- \rightarrow S + 4H_2O$)
Atmospheric H_2SO_4 and HNO_3 deposition	Atmospheric deposition of Ca and Mg compounds
Cation uptake by plants	Anion uptake by plants
Accumulation of acidic organic matter (e.g., fulvic acids)	Specific (inner sphere) adsorption of anions (especially SO_4^{2-})
Cation precipitation (e.g., $Al^{3+} + 3H_2O \rightarrow 3H^+ + Al(OH)_3$; $SiO_2 + 2Al(OH)_3 + Ca^{2+} \rightarrow CaAl_2SiO_6 + 2H_2O + 2H^+$)	Cation weathering from minerals (e.g., $3H^+ + Al(OH)_3 \rightarrow Al^{3+} + 3H_2O$; $CaAl_2SiO_6 + 2H_2O + 2H^+ \rightarrow SiO_2 + 2Al(OH)_3 + Ca^{2+}$)
Deprotonation of pH-dependent charges	Protonation of pH-dependent charges

Source: Data from Brady and Weil 2002.
Note: Soil pH depends on the balance between processes that produce H^+ and those that consume H^+. Production of H^+ increases soil acidity and thus lowers soil pH, while consumption of H^+ increases soil alkalinity (reduces soil acidity) and thus increases soil pH.

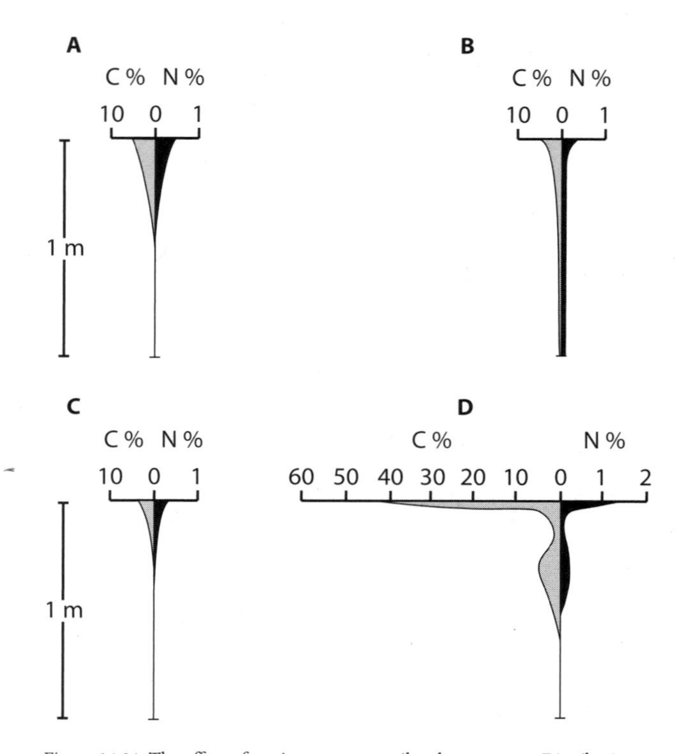

Figure 14.24. The effect of environment on soil carbon content. Distribution of carbon and nitrogen by depth in some representative forest soil profiles. A Mollisol (A) and an Alfisol (B), both of which are relatively young soils with high base status, have formed on stable surfaces in humid to subhumid temperate areas. Mollisols have greater accumulations of organic matter near the soil surface, generally owing to a significant herbaceous understory influence with a predominance of surface roots. Alfisols have a greater depth distribution of organic matter due to the deeper penetration of woody roots from trees. C. A highly weathered tropical soil with low base status (typical of Ultisols or Oxisols). Note the relatively low accumulations of carbon and nitrogen are concentrated in the surface soil layers, a phenomenon typical of Oxisols and Ultisols. D. A Spodosol, commonly known as a *podzol*. Spodosols are acid soils common in boreal and some montane forests and have a white surface mineral soil layer overlaying a darker layer that has accumulated relatively large amounts of carbon and nitrogen because cool temperatures, acidity, and organomineral complexes limit decomposition. (Modified from Parsons and Tinsley 1975)

fraction has a disproportionally large impact on the physical, chemical, and particularly the biological properties of soil. The organic matter content of soils ranges from almost zero in very young soils to over 80 percent by mass in some organic soils (Histosols). Most forest soils have somewhere between about 1 and 12 percent organic matter by mass[24] (Fisher and Binkley 2002) in the surface (i.e., A horizon), with lower amounts generally occurring in some dry tropical forests (where plant litter production is low relative to rates of decomposition; Jones 1989) and higher amounts in montane and boreal forests (where decomposition is slowed more by cold temperatures than plant litter production; Swift et al. 1979; figs. 14.24 and 14.25). Most soil organisms are **heterotrophic,** deriving their energy from the consumption of other living organisms (i.e., consumers) or the decomposition of nonliving soil organic matter (called **saprotrophs** or **detritivores**). The activities of these organisms, to a large degree, are responsible for nutrient transformations regulating plant nutrient availability (chapter 18).

Two major scientific reasons currently drive soil organic matter research: the putative role of soil organic matter in maintaining soil productivity (Dyck and Skinner 1990; Powers et al. 1990; Turner and Gessel 1990; Morris and Miller 1994; Powers 1999),[25] and the role of soil organic matter in the global carbon cycle (Kimble et al. 2003). The former has been a long-standing area of research, because soil organic matter influences so many properties and processes that affect soil productivity, including cation exchange and water-holding capacities, the formation and stabilization of soil aggregates, and the release of plant available nutrients (Brady and Weil 2002). More recently, the knowledge that soils contain about three times the amount of carbon than is contained in terrestrial vegetation has sparked interest in the role of soil as a source and a sink for atmospheric CO_2 (Kimble et al. 2003).

Organic matter in forest soils is found in a wide variety of forms. The soil organic fraction can be compartmentalized in may ways into various pools, including those based on fractions that can be physically separated (at least to some degree), and those that can be only conceptually separated based on turnover times of various pools (e.g., so-called active, slow, and passive soil organic matter fractions; Paustian et al. 1992). One physically based classification includes six different fractions:

1. *Live plant fine roots and their microbial symbionts*, including mycorrhizae, and mycorrhizal hyphae. The mean live fine root biomass ranges from 230 g/m^2 in boreal forests to 500 g/m^2 in temperate coniferous forests (Jackson et al. 1997). Far fewer estimates are available of the biomass of mycorrhizae and hyphae associated with the live fine roots in forests. Wallander et al. (2001) estimated that ectomycorrhizal external mycelia and mantles were between 70 and 90 g/m^2, or about one-third of the live fine root biomass in a Norway spruce *(Picea abies)* forest in southwestern Sweden, and Hart et al. (2005a) estimated that mycorrhizal biomass was about one-fifth of the live fine root biomass in surface

[24]Soil organic carbon may be roughly estimated from these numbers by dividing the organic matter values by 1.7 to 2.0.

[25]Interestingly, there are few well-documented examples showing a decrease in site productivity with losses of soil organic matter. Indeed, Fleming et al. (2006) recently reported that forest floor removal increased seedling growth in forests with Mediterranean-type climates but decreased seedling growth in warm-humid climates 5 years post-treatment. Such differential responses may be due to the varied effects that soil organic matter has on soil properties and processes. Note that long-term growth responses may be quite different.

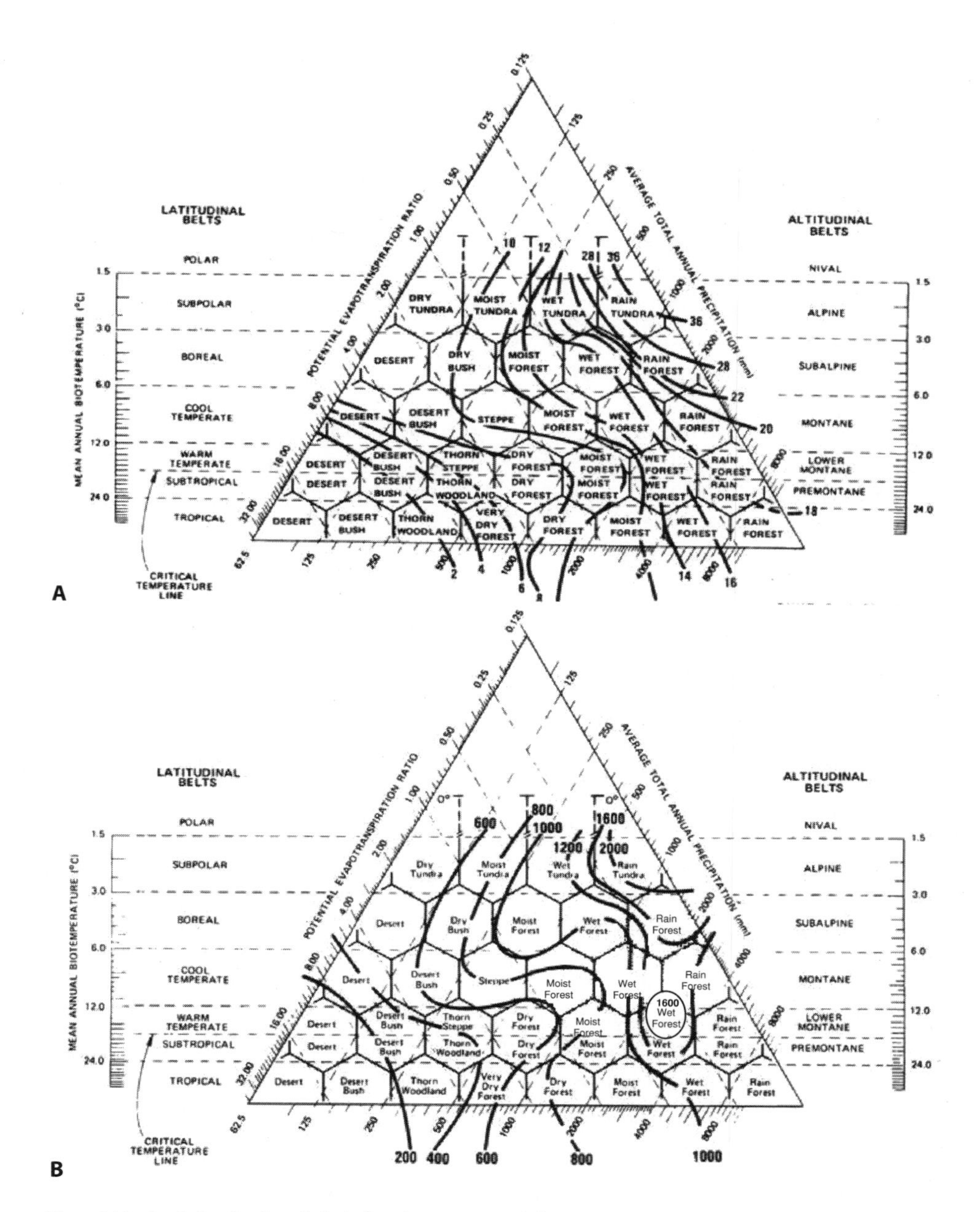

Figure 14.25. Isopleths of carbon (A, kg/m^3) and nitrogen (B, g/m^3) contents in soils from various Holdridge world life zones. Forest soils vary widely in their carbon (C) and nitrogen (N) storage, ranging from ~4 to >28 kg C/m^3 and from ~400 to >1600 g N/m^3. (Modified from Post et al. 1985)

soils of a ponderosa pine *(Pinus ponderosa)* forest in the southwestern United States. In contrast, Fogel and Hunt (1983) found that mycorrhizal biomass was four to eight times greater than total (live plus dead) fine root biomass (<5 mm diameter) in a young Douglas-fir *(Pseudotsuga menziesii)* forest in the U.S. Pacific Northwest.

2. *Soil biomass*, which comprises living organisms that are not part of a plant, including a wide variety of fungi, bacteria, protozoa, and invertebrates (see section 14.5). Accurate estimates of soil biomass are quite difficult; however, the microbial component has been estimated to compose from less than 1 percent to approximately 3 percent of total soil organic matter by mass (Smith and Paul 1990). Even this small proportion represents a large amount of living biomass. Brooks et al. (1985) estimated that the microbial biomass per hectare in a British wheat field was equivalent to the mass of one hundred sheep. Soil fauna generally comprise about one-tenth of the biomass of the soil microorganisms, with soil invertebrates accounting for the highest proportion, by far, of the total faunal biomass in forests (Zlotin 1985; Brady and Weil 2002). As mediators of many key processes, the importance of soil microflora and soil fauna in ecosystems far outweighs their total biomass (chapter 18).

Although at any one time soil organisms compose only a small fraction of the total soil organic matter, they turnover very rapidly (particularly the protists[26]). Therefore, much higher amounts of carbon pass through the biomass than are reflected in the standing crop. For example, Lousier and Parkinson (1984) studied the population dynamics of amoebae in the family Testacea in the organic layers of Canadian forest soils. They found that annual production of these protozoa was 250 to 300 times greater than their average biomass at any one time (mean residence time of the biomass of less than two days). The rate of turnover of soil microbial biomass ranges from less than one to several years, depending to a large degree on climate (faster in warm-humid climates, slower in cool-arid climates; Paul and Clark 1996).

3. *Dissolved organic soil organic matter* is produced and consumed by soil microorganisms or plants and can be stabilized (i.e., protected from decomposition) on or off of soil surfaces. It has an important role in the cycling of nutrients (i.e., carbon, nitrogen, phosphorus, and sulfur) within the soil, and the export of these nutrients from soils to stream and groundwaters (Herbert and Bertsch 1995). The amount of dissolved organic matter at any point in time is a very small component of the total organic matter pool in soils (roughly one-fifth or less the size of the microbial biomass; Hart et al. 1994), but the solubilization of soil organic matter into a dissolved form is believed to be the rate-limiting step in decomposition and plays an important role in the development and maintenance of soil structure (Chapin et al. 2002). Organic horizons are often a significant source of dissolved organic matter in soil. The biodegradability of dissolved organic matter varies widely among soils, apparently because of broad ranges in the chemical composition of these materials, which is a complex mixture of humic materials (see below), amino acids, carbohydrates, aliphatic and aromatic acids, and hydrocarbons (Herbert and Bertsch 1995). The sources of these constituents include compounds released by roots, mycorrhizal hyphae, and microorganisms into the surrounding soil or leached from living and dead plant tissues.
4. *Plant litter* includes both relatively undecomposed (i.e., identifiable) plant residues deposited either on the ground surface (mostly plant leaves, but also including twigs and plant reproductive parts) or underground (in the form of senescent plant roots and mycorrhizae). The litter[27] component of mineral soil can be separated by sieving along with the sand fraction (so-called coarse particulate organic matter) or by density fractionation (so-called light fraction, because this material is not associated with soil minerals [Gregorich and Janzen 1996; Sollins et al. 1999; see below]). These two fractions may not be equivalent (Sollins et al. 1999).
5. *Coarse woody debris*, or dead woody material larger than about 0.6 cm in diameter.[28] Old, decaying logs (i.e., tree boles) are a significant fraction of the soil organic matter in some forest ecosystems (Harmon et al. 1986). Decaying logs, which occur as distinct islands within the mineral soil, act as water reservoirs during dry periods and are sites of intense biological activity (Hart 1999).
6. *Humus*[29] (from the Latin word for 'soil') consists of dark-colored, amorphous organic material that is highly decomposed (i.e., the origin of the organic material is nonrecognizable). Although the formation of humus in soil (called *humification*) is not entirely understood, it includes both biochemical and abiotic processes that are not under enzymatic control (Paul and Clark 1996). It is synthesized from partially decayed (mostly resistant) plant and animal materials and microbial metabolites as a byproduct of organic matter decay. The resulting material is highly resistant to further decay, in part, because of its heterogeneous chemical structure (i.e., chemically protected from microbial decay). Colloidal humus, because of its high net surface charge (see table 14.1), chemically binds to secondary minerals, forming organomineral complexes that further increases its stability in soil because this interaction affords physically protection[30] from microbial decay (the so-called heavy fraction of soil organic matter; Sollins et al. 1999). Except in very young or arid soils, humus accounts for 80 percent to over 90 percent of the total soil organic matter. The chemical structure of humus is complex and incompletely understood, but it generally consists of aromatic structures and aliphatic chains that include carbohydrates and amino acids, linked together to form large, flexible molecules (Newman and Tate 1984; Stevenson 1994). Until fairly recently, soil humus was believed to be highly aromatic; however, recent studies using methods that do not require chemical extraction (e.g., ^{13}C-NMR) suggest that soil humus may be much more aliphatic than aromatic (Sollins et al. 2007).

Humus plays a very important role in soil fertility. It is relatively rich in organic forms of nitrogen, phosphorus,

[26]A heterogeneous group of organisms comprising those eukaryotes that are not animals, plants, or fungi. They are usually treated as the kingdom Protista or Protoctista.

[27]Many authors reserve the term *plant litter* just for plant residues deposited on the mineral soil surface.

[28]There are several other strategies for defining coarse and fine woody debris. For instance, coarse wood debris can be defined based on diameter (>10 cm at widest point) and length (>1 m long; Harmon et al. 1999); our definition is operationally defined in that grinding material >0.6 cm diameter for elemental analyses is difficult, and forest fuel censuses typically measure woody material using the planar intercept method that surveys surface woody material >0.6 cm in diameter (Brown 1974).

[29]Many textbooks and authors use the words *humus* and *soil organic matter* synonymously.

[30]Another kind of physical protection occurs when soil organic matter is contained in very small soil pores or inside soil aggregates that are relatively inaccessible to decomposers.

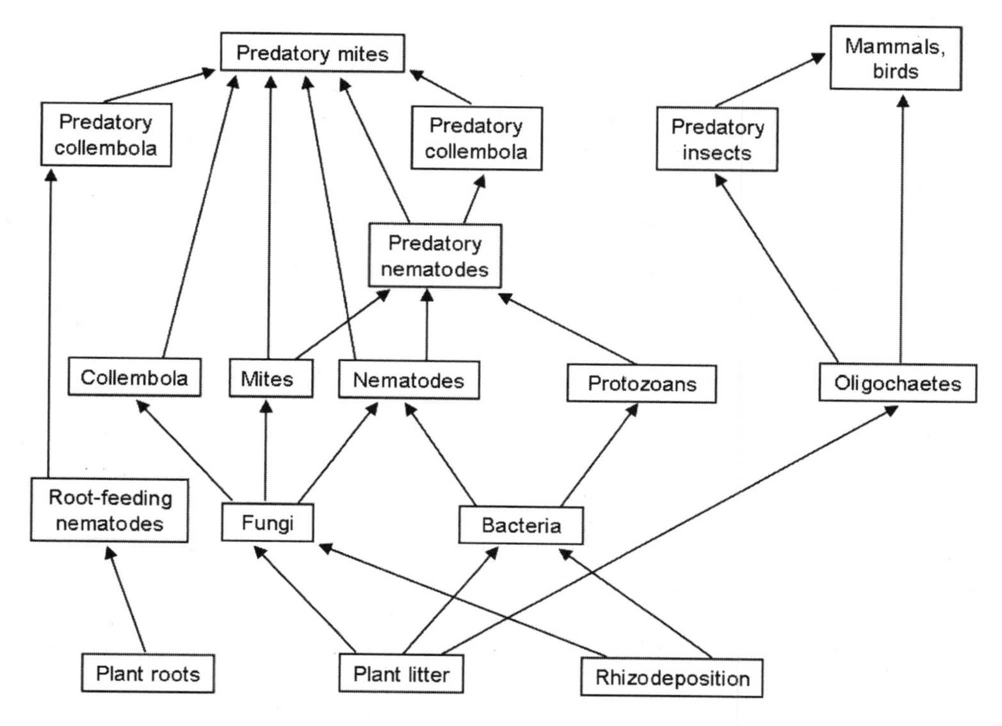

Figure 14.26. Possible trophic interactions in a simple food web in soil. Arrows denote transfer of energy and materials from lower to higher trophic levels. The varied feeding habits of many soil organisms results in them occupying multiple trophic levels, greatly increasing the complexity of the trophic interactions in the soil. Note that some 80–90% of the total metabolic activity in the food web is from fungi and bacteria (including the archaea and actinomycetes). (Modified from Bottomley 2005)

and sulfur, and, because of its stability against decomposition, serves as a reservoir of slowly available nutrients for plant roots, mycorrhizae, and the soil biomass. In addition to retaining nutrients, humus plays a critical role in balancing water retention and drainage within soils (soils must drain for air, and therefore oxygen, to diffuse in from the atmosphere). Humus absorbs large amounts of water but at the same time reacts with the mineral component to impose a structure on soils that facilitates free drainage and prevents waterlogging. Too much humus, however, can lead to waterlogging and anaerobiosis, inhibiting heterotrophic microbial activity.

The ultimate source of soil organic matter is carbon fixed by plants during photosynthesis. Plants divert a large proportion of photosynthate to the production of fine roots and mycorrhizal hyphae (chapter 15). Death of these parts, along with fallen leaves, twigs, and tree boles, creates the litter component. Dissolved organic matter leached from these materials, excreted from plant roots, microorganisms, or mycorrhizae, or biologically released during the decay process, provides a carbon source for the predominantly heterotrophic soil biomass. Humus is created as a byproduct of litter decomposition. Decomposition of litter is also the basis for nutrient cycling within ecosystems (chapter 18). Because soil organic matter is ultimately derived from surface litter and roots, it is most abundant in surface soil layers. Figure 14.24 shows the distribution of carbon and nitrogen by depth in several different soil types. With few exceptions (most notably peat), carbon and nitrogen (and hence soil organic matter) are concentrated in the surface 30 to 50 cm. This has considerable implications for forest management, because it is the surface layers of soil that are the most vulnerable to disruption during harvest and site preparation, and by recreational activities (i.e., off-road vehicular traffic).

14.5 BIOLOGICAL PROPERTIES OF SOILS

The soil biota comprises those organisms that spend all or part of their life cycle in the soil. A small amount of soil in the palm of the hand is likely to contain billions of individual organisms, with representatives of nearly every phylum in biology. The trophic relationships among these organisms are similar to those aboveground (fig. 14.26). In their role as regulators of ecosystem processes (chapters 11, 17, and 18), soil organisms perform several vital functions, including: (1) decomposing organic material; (2) serving as sources and sinks of key nutrients; (3) acting as catalysts of nutrient transformations; (4) converting atmospheric nitrogen into organic forms and reconverting organic nitrogen to gaseous nitrogen;[31] (5) directly interacting with living

[31]The activity of soil organisms also has a profound effect on the chemistry of the atmosphere because many soil organisms (primarily soil microorganisms) are involved in trace gas production and consumption (including carbon dioxide, methane, and nitric and nitrous oxides). These gases have a major influence on the radiative balance of the earth (Schlesinger 1997).

Table 14.3
General classification of some important soil organisms

Generalized grouping (width)	Major specific groups	Examples
Macroflora (>0.1 mm)		
Largely autotrophs	Vascular plants	Plant roots
	Bryophytes	Mosses
Microflora (<0.1 mm)		
Largely autotrophs	Vascular plants	Root hairs
	Algae	Greens, yellow-greens, diatoms
Largely heterotrophs	Fungi	Yeasts, mildews, molds, rusts, mushrooms
	Actinomycetes	*Streptomyces, Frankia*
Heterotrophs and autotrophs	Bacteria (and Archaea)*	Green bacteria, purple bacteria, flavobacteria (extreme halophiles, methanogens, extreme thermophiles)
	Cyanobacteria	*Nostoc, Anabaena, Microcoleus*
Macrofauna (>2 mm)	Vertebrates	Gophers, mice, moles
All heterotrophs, largely herbivores and detritivores	Arthropods	Ants, beetles and their larvae, centipedes, millipedes, grubs, maggots, spiders, termites, woodlice
	Annelids	Earthworms
	Mollusks	Snails, slugs
Mesofauna (0.1–2 mm)		
All heterotrophs, largely detritivores	Arthropods	Mites, collembola (springtails)
	Annelids	Enchytraeid (pot) worms
All heterotrophs, largely preadators	Arthropods	Mites, protura
Microfauna (<0.1 mm)		
Detritivores, predators, fungivores, bacterivores	Nematoda	Bacterial-feeders, fungal-feeders, omnivores, root-feeders
	Rotifera†	Bdelloid rotifers
	Protozoa†	Amoebae, ciliates, flagellates

Source: Data from Brady and Weil 2002.
*Traditionally grouped together in the kingdom *Monera,* both of these organism groups are prokaryotic but are classified separately in the domains Bacteria or Archaea based on differences in RNA. Aerobic, anaerobic, and facultative anaerobic types occur in both domains.
†Generally classified in the kingdom *Protista.*

plants and animals in various ways, ranging from mutualisms to pathogenesis; (6) synthesizing enzymes, vitamins, antibiotics, hormones, metal chelators, and allelochemicals that regulate populations and processes; and (7) acting as engineers and maintainers of soil structure.

Although there currently is no consensus on a system for taxonomic classification, the universal phylogenic tree for living organisms, based on comparative sequencing of 16S or 18S ribosomal RNA, proposes three domains: Eucarya (which includes all plants, animals, and fungi), Bacteria, and Archaea (Pace 1996). This classification system[32] clarifies that most of the biodiversity on earth, as well as most of the genetic diversity, is found in microorganisms.

A more traditional classification of soil organisms separates them into two main groups based on size and their historical grouping as plants or animals (table 14.3). The macroflora consist of vascular plant roots and bryophytes (e.g., mosses). These primary producers of organic matter provide much of the carbon and energy needed to support the soil food web (fig. 14.26). The microflora (<0.1 mm in body width) consist of bacteria (including archaea), actinomycetes, fungi, and algae (fig. 14.27). The soil microflora are responsible for the majority of the decomposition of and nutrient release from detritus in the soil, as well as being key players in the formation of soil structure. The soil fauna are generally grouped into three categories: macrofauna (>2 mm body width, >10 mm in length), mesofauna (0.1–2 mm in body width, 0.2-10 mm in length), and the microfauna (<0.1 mm in body width, <0.2 mm in length). Macrofauna consist primarily of snails and slugs (Mollusca), beetles (Coleoptera), earthworms (Megadrili), millipeds (Diplopoda), centipeds (Chilopoda), spiders (Araneida), woodlice (Isopoda), harvest spiders or harvestmen (Opiliones), sand or beach fleas (Amphipoda), and larger ants and termites (Isoptera). Some authors subdivide this group further to include the megafauna (>20 mm in body width), which consist of vertebrates (i.e., gophers, mice, moles) and larger invertebrates (for the most part, the lumbricids). These organisms create macropores that influence water infiltration and percolation and aeration (gas exchange), as well as mix the soil, altering horizonation and plant rooting patterns. The mesofauna include smaller ants and termites, pseudoscorpions (Chelonethi), pot worms (Enchytraeidae), garden centipedes (Symphyla), Diplurans (wingless arthropods with long antennae; Diplura), proturans (wingless arthropods lacking antennae and eyes; Protura), springtails (Collembola), and mites (Arcai). Arthropods in this group (e.g., mites and springtails) are sometimes referred to as microarthropods. These soil organisms fragment and ingest litter coated with microorganisms, producing large amounts of fecal material with greater

[32]In this classification system, viruses are excluded because they lack a cytoplasmic membrane with internal cytoplasm. Only when they are associated with another organism are they able to fulfill the basic life processes (Wollum 1999).

Table 14.4
Influences of soil organisms on nutrient cycling and soil structure as a function of organism size

Soil organism group	Soil function	
	Nutrient cycling	Soil structure
Microflora	Catabolize organic matter Mineralize and immobilize nutrients	Produce organic compounds that bind aggregates Hyphae entangle particles onto aggregates
Microfauna	Regulate bacterial and fungal populations Alter nutrient turnover	May affect aggregate structure through interactions with microflora
Mesofauna	Regulate fungal and microfaunal populations Alter nutrient turnover Fragment plant litter	Produce fecal pellets Create biopores Promote humification
Macrofauna	Fragment plant litter Stimulate microbial activity	Mix organic and mineral particles Redistribute organic matter and microorganisms Create biopores Promote humification Produce fecal pellets

Source: Data from Hendrix et al. 1990.

surface area and moisture-holding capacity than the original litter. The microfauna include rotifers (Rotifera), protozoans (Protozoa), and nematodes (Nematoda). These organisms are important predators on fungi and bacteria and are central to nutrient mineralization processes (conversion of organically bound nutrients to inorganic nutrients readily available for uptake by organisms; see fig. 14.26 and chapter 18; Fisher and Binkley 2000; Coleman et al. 2004).

Body width and length of the fauna is related to microhabitats and the impact that these soil organisms have on soil structure (fig. 14.27, table 14.4). Macrofauna are organisms large enough to create their own soil habitat and disrupt soil structure through their activity. The mesofauna are large enough to escape the surface tension of water and move freely in the soil, but small enough not to disrupt soil structure through their movement. The microfauna generally lack significant mobility because they are confined to water films on soil particles, and they have little impact on soil structure (Fisher and Binkley 2000; Coleman et al. 2004).

Judged by the sheer numbers of individuals and species, much of the life of forests (and indeed any terrestrial ecosystem) exists belowground (fig. 14.28). On a global basis, the majority of organisms are invertebrates, and many if not most of these spend at least a portion of their life cycle belowground (Wardle 2002). Over a million species of fungi and a similar number of species of nematodes may inhabit the planet, in contrast to perhaps a few thousand species of reptiles and birds. Indeed, in a single gram of surface soil,

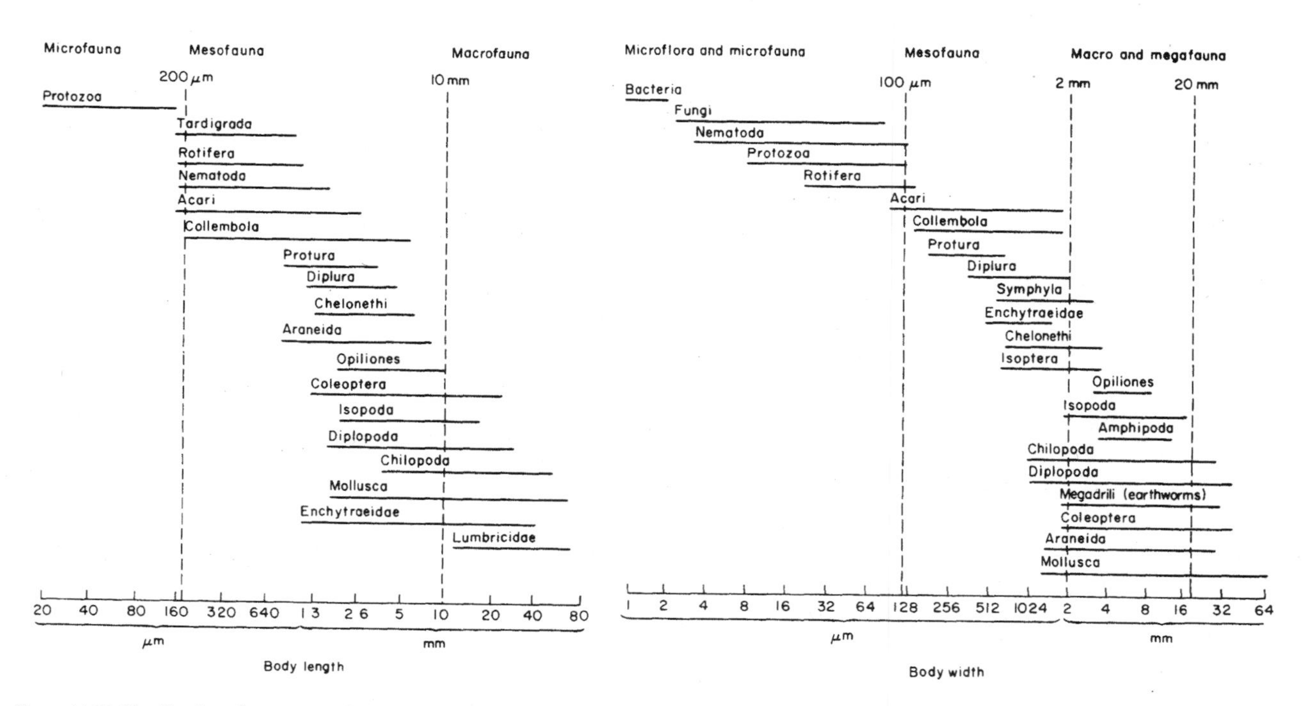

Figure 14.27. Classification of common soil organisms based on body length and width. (Swift M.J., Heal, O.W., and Anderson, J.M. *Decomposition in terrestrial ecosystem.* © 1979, Blackwell Publishing, Oxford)

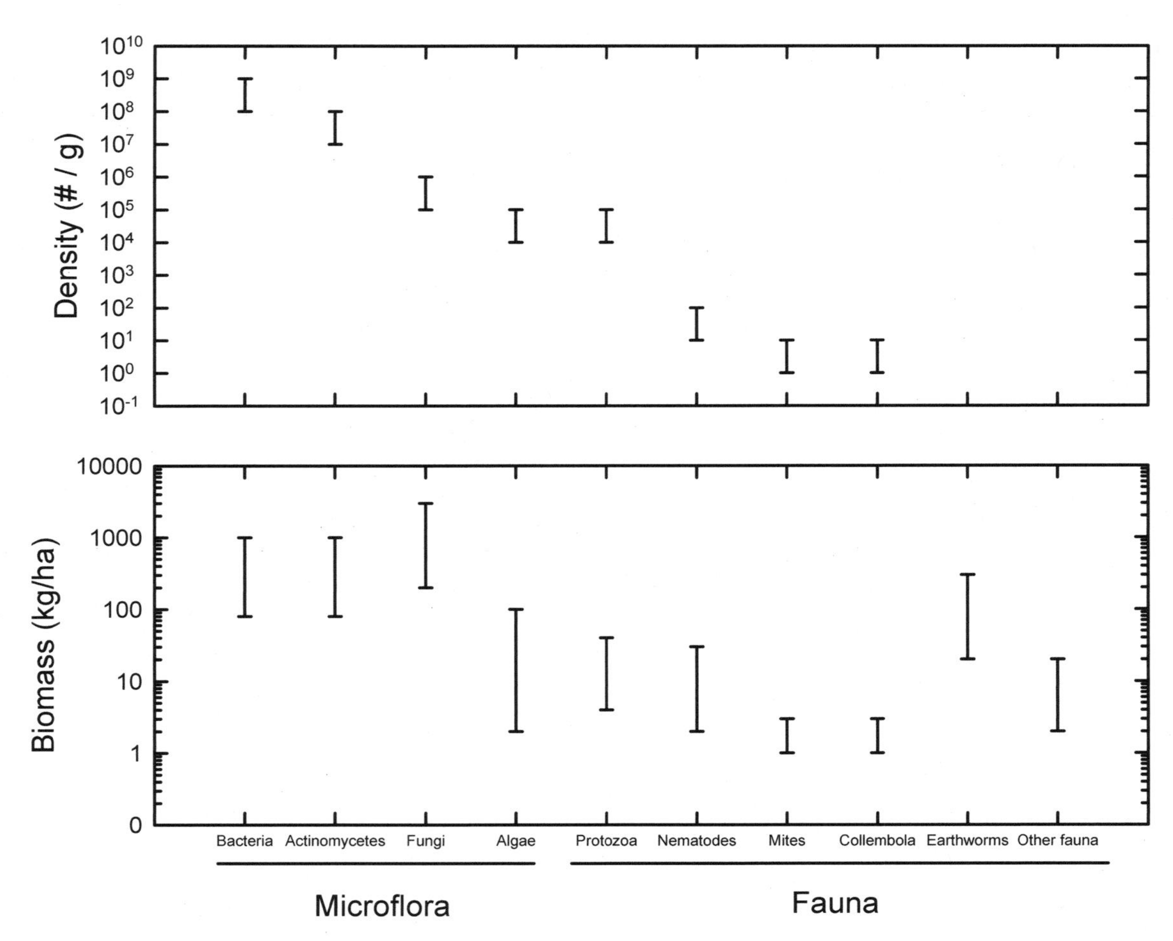

Fig. 14.28. Density and biomass (dry weight) of soil organisms in a typical A horizon (upper 15 cm). Because metabolic activity generally covaries with biomass, soil microorganisms (primarily the microflora) dominate life in the soil. No density data are shown for earthworms and other fauna because there is less than one individual per gram of soil. For perspective, the mean live fine-root biomass in forests ranges from 230 g/m^2 in boreal forests to 500 g/m^2 in temperate coniferous forests. (Jackson et al. 1997; data from Brady and Weil 2002)

several thousand bacterial species are likely present (Torsvik et al. 1994).

Species diversity and abundances of soil organisms vary depending on soil type, vegetation, and environment. However, the degree by which they vary depends on the organism group. For instance, soil microbial biomass (and presumably abundances as well) scales roughly with soil organic matter content (see section 14.4). Hence, those ecosystems with the greatest soil organic matter content (fig. 14.25) generally have the greatest soil microbial biomass. How the species diversity of soil microorganisms changes across forest and other terrestrial ecosystems is less clear. Fierer and Jackson (2006) recently found that soil bacterial diversity increased with soil pH across a wide range of terrestrial ecosystems; differences among soils were better explained by differences in soil pH than by vegetation type or other environmental factors such as soil organic carbon concentration, soil moisture deficit, or mean annual air temperature. Green et al. (2004) reported that despite high local diversity in ascomycete fungi across arid parts of the Australian continent, the diversity of this microbial group was similar across large geographic regions. Abundance of some of the larger fauna, in particular, varies considerably among different forest types. Large earthworms, for example, are 10 times more abundant in temperate deciduous than in conifer or tropical forests, while termites occur mostly in the tropics (table 14.5).

Table 14.5
Estimated biomass of soil fauna in different vegetation types

Vegetation type	Biomass (g dry weight per m^2)
Temperate deciduous forest	
Mull humus	8.0
Mor humus	3.5
Temperate coniferous forest	2.4
Tropical forest	1.8
Tundra	3.3
Temperate grassland	5.8
Tropical grassland	1.9

Source: Adapted from Petersen and Luxton 1982.

Below, we briefly describe some of the known functions of soil organisms. For a more comprehensive discussion of the classification and function of the soil biota, see work by Paul and Clark (1996), Coleman et al. (2004), and Sylvia et al. (2005).

14.5.1 Soil Macroorganisms

The macrofauna are all heterotrophs and largely herbivores and saprotrophs (table 14.3). They include such diverse organisms as moles, prairie dogs, earthworms, termites, ants, beetles, and millipedes. Although these organisms contribute only a very small fraction to the total soil metabolism (as indexed by the rate of CO_2 production), they can have a substantial impact on soil formation. Many of the organisms in this group have a profound effect on soil structure through their borrowing, pulverizing, mixing, and granulating activity on the soil matrix. These activities may move surface organic materials to the subsoil and provide large continuous pores that increase the movement of water and air through the soil. Ants are especially significant in temperate semiarid grasslands and boreal forests for their ability to break down woody materials and mix soil materials as they build their nests. In the drier tropics (<800 mm mean annual precipitation), termite activity dominates the soil fauna, while in wetter climates earthworms are the major movers and shakers of the soil fauna community. Indeed, earthworms are likely the most important macrofaunal species in soils. These organisms can ingest a mass of soil 2 to 30 times their own mass in a single day (Brady and Weil 2002). The burrows that remain provide continuous macropores, and the ingested soil remains that are expelled (called *casts*) provide water-stable aggregates that are also hot spots of microbial activity. Both of these processes result in improved soil structure for plant growth (see section 14.3). Earthworms can have rapid and dramatic effects on forests ecosystems, as has been recently documented with the invasion of northern temperate forests by exotic earthworm species (fig. 14.29; Bohlen et al. 2004a).

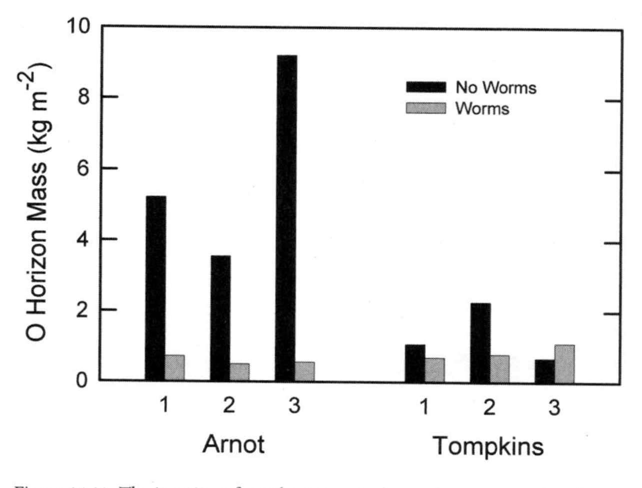

Figure 14.29. The invasion of north temperate forests by exotic species of earthworms can have a significant impact on soils. This figure shows forest floor (O horizon) mass in paired exotic earthworm–invaded and control forest stands at three locations in western (Arnot) and eastern (Tompkins) New York State. The substantial reduction in the forest floor mass following earthworm invasion dramatically altered the location and nature of nutrient cycling in soil profiles of invaded forests. (Modified from Bohlen et al. 2004a)

> Termites have a major effect on soil pattern in dry tropical forests, particularly species of so-called higher termites (Macrotermitinae), which like leaf-cutter ants in moist tropical forests are fungus farmers (Swift et al. 1979). Using aerial photographs, Jones (1990) found a rather consistent 200 Macrotermitinae mounds per square kilometer throughout Tanzania; this translates to an average spacing of 65 to 70 m between mounds, a figure that probably holds throughout the miombo woodlands of southern and eastern Africa, and perhaps in the dry forests of South America and Southeast Asia as well. Mounds may be three to four meters in height and cover as much as 30 percent of the soil surface (fig. 14.30B; Wood and Sands 1978; Golley 1983a). Termites are best known as wood-eaters, but they consume most any kind of litter, as well as living roots. The fungus-farming higher termites exhaustively glean the soil surface within their territory for organic debris, with the result that organic matter and nutrients become concentrated in mounds (Golley 1983a). Termite mounds may contain as much as 10 times more organic matter as surrounding mineral soil (Lee and Wood 1971). Jones (1990) hypothesizes that the extremely low organic matter content of dry tropical forest soils (often <0.5%) results from the thorough gathering and nearly complete digestion of carbon by the Macrotermitinae and their fungal partners. Mounds also have better water-holding capacity than surrounding soils, not only because of their relatively high organic matter content, but because in the process of constructing and repairing mounds termites bring up clays from lower soil depths to provide mound structure. In Africa, farmers frequently take advantage of the improved fertility and water-holding capacity by planting crops on termite mounds (Golley 1983a). Because of their relatively high clay content, mounded soils are also used to build houses (Fig. 14.30C).

The soil macroflora consist primarily of the roots of vascular plants in most soils, with bryophytes also important in some forests (e.g., feathermoss and sphagnum in boreal black spruce forests; Bisbee et al. 2001). Plant roots provide much of the carbon and energy needed by the soil

Figure 14.30. Macrofauna can have a profound affect on soil. A. The interior of an excavated leaf-cutter ant nest in Costa Rica. The gray mass in the center is fungi that have been growing on leaves gathered by the ants. (Courtesy of Dave Janos) B. A termite mound in an Australian forest. Large mounds can be several meters in height.

C

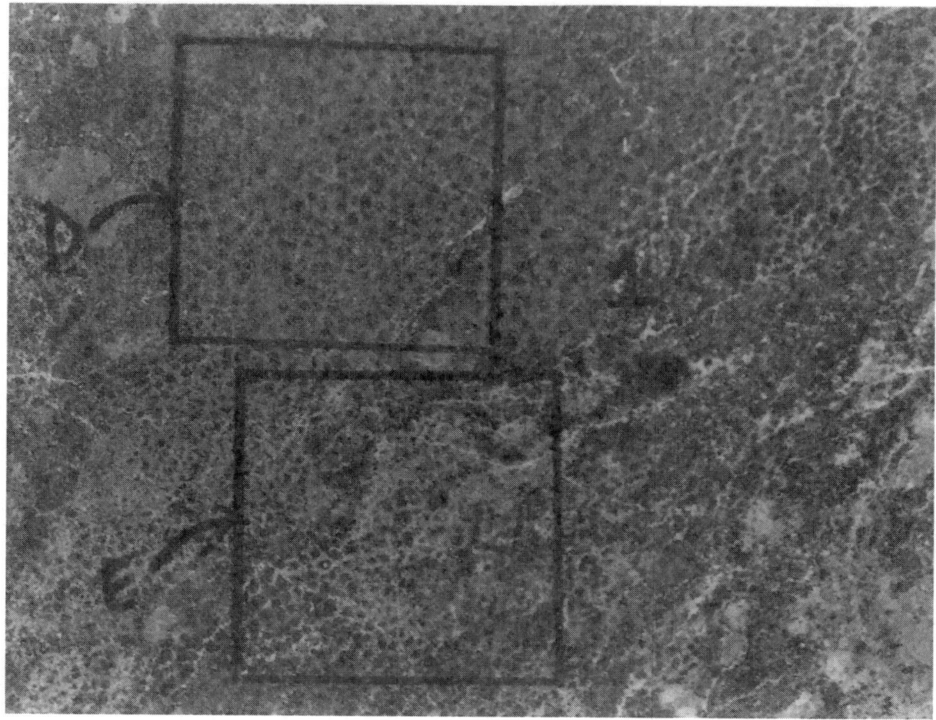

D

Figure 14.30. *(Continued)* C. A village in northern Zambia. The homes are made with clay from termite mounds. D. This aerial photograph shows the distribution of termite mounds in a dry tropical forest in Tanzania. Each square is 1 km^2. The dark "dimples" are termite mounds, which support vegetation distinct from the surrounding area. (Courtesy of Julia Jones; from Jones 1990)

heterotrophs[33] (i.e., soil fauna and most soil microflora). Root growth and death has similar positive effects on soil structure as the activities of the macrofauna.[34]

Most biological activity in mineral soil centers around roots, mycorrhizae, and mycorrhizal hyphae, where exudates and sloughed tissues support a rich food chain of microflora, microfauna, and small invertebrates. This zone of high biological activity is termed the **rhizosphere** if around a root, the **mycorrhizosphere** if around a mycorrhizal root, or the **mycosphere** if around a mycorrhizal hypha.[35]

Some of the processes occurring in the rhizosphere are illustrated in fig. 14.31. The chemical environment produced by both roots and rhizosphere microorganisms accelerates weathering and alters the mineralogy of secondary minerals (April and Keller 1990). Rhizosphere microorganisms release a wide array of compounds that inhibit pathogens, solubilize essential metals such as iron (the process of chelation), and glue mineral particles together into large soil aggregates (see section 14.3). Some groups of rhizosphere bacteria, other than the nodule-forming *Rhizobium* and *Frankia* that were discussed in chapter 11, are capable of fixing atmospheric nitrogen. This is also true of some nonrhizosphere bacteria; however, rhizosphere bacteria are likely to be more active nitrogen fixers simply because they have a more ready supply of energy than nonrhizosphere groups. Protozoa and invertebrates graze bacteria within the rhizosphere, in the process releasing nutrients for plant growth (chapter 18).

The rhizosphere is a complex powerhouse in which the energy contained within plant photosynthates fuels numerous processes that return benefits to the plant.[36] Specific strains of rhizosphere bacteria that enhance plant growth are called *plant-growth-promoting rhizobacteria*. Many different mechanisms are responsible for plant growth stimulation, including biocontrol of plant pathogens, biofertilization via asymbiotic nitrogen fixation, and phytostimulation (which directly increases plant growth through the production of plant hormones such as gibberellic and indoleacetic acids; Kennedy 2005). Some bacteria, mostly members of a group called fluorescent pseudomonads (so named because they are in the genus *Pseudomonas* and have the property of

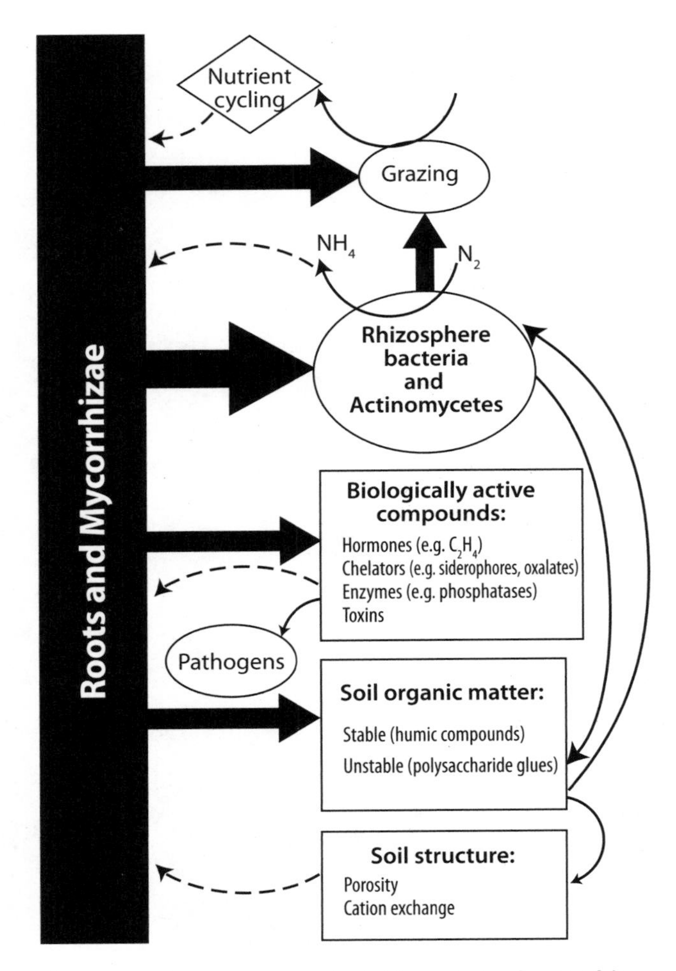

Figure 14.31. Rhizosphere processes. Dashed lines indicate influences of the rhizosphere on plants and mycorrhizal fungi. (Perry et al. 1987b).

fluorescing when placed under ultraviolet light), benefit plant growth by releasing a very strong iron-chelator called a *siderophore* (from the Greek for 'iron carrier') that binds iron so tightly that pathogens cannot get it (plants apparently can get the iron from siderophores; exactly how is not known). This particular form of defense against pathogens is shared by humans and other mammals, which have siderophore-like chelators in their blood. Mycorrhizal fungi also produce siderophores.

It should not be construed from the above that every rhizosphere organism benefits plants. Pathogens and other groups that inhibit plant growth are also present but (by definition) are kept in check within healthy rhizospheres. As mentioned above and discussed in some detail in chapter 18, food web dynamics (i.e., regulation of bacterial and fungal populations within rhizospheres by grazers) are critically important to maintaining plant health and, because plants are the energy source, the health of the rhizosphere community as a whole.

The influence of roots on microbial numbers is frequently expressed in terms of the **R/S ratio,** which is the

[33]In forests, about half of the net primary production is belowground (chapter 15 in this volume).

[34]Roots and associated mycorrhizal fungi (chapter 11 in this volume) stabilize soil aggregates. When coarse roots senesce and decay, they create large macropores that greatly increase water and gas transport throughout the soil. Root exudates also support soil microorganisms that secrete polysaccharides that further stabilize soil aggregates.

[35]Because of the interdependence and close physical relationship between plant roots and their mycorrhizal hyphae, we have included both as components of the rhizosphere.

[36]Foster (1986) distinguishes three zones within the rhizosphere: (1) the rhizoplane, within 10 μm of the root (or mycorrhiza) surface; (2) the inner rhizosphere, 10 to 400 μm from the surface; and (3) the outer rhizosphere, 400 to 3,000 μm from the surface. Both absolute numbers and species diversity of microorganisms are highest in the rhizoplane; however, microbial numbers are affected as much as 1,000 μm from the surface (the probable limit of diffusion of soluble organics), and nematodes can detect roots 3,000 μm away (Foster 1986).

ratio of microbial numbers per unit mass of rhizosphere soil to numbers per unit mass of nonrhizosphere soil (Richards 1987). The R/S ratio typically ranges from 10:1 to 50:1 for bacteria and 5:1 to 10:1 for fungi (Richards 1987). Not only numbers, but the types of microorganisms vary between rhizosphere and nonrhizosphere soils. Microorganisms of nonrhizosphere soil tend to be slower growing than rhizosphere groups and are better able to decompose complex organic molecules such as lignin and humic compounds (bacteria with these characteristics are termed **autochthonous**). Among the microorganisms that are more abundant in bulk soil than in rhizospheres are the actinomycetes (Neal et al. 1964; Friedman et al. 1989; in chapter 8, see fig. 8.10B).

Rhizospheres are typified by bacterial types that grow very rapidly on relatively simple substrates (termed **zymogenous**), and by groups that cannot synthesize their own amino acids, obtaining these essential building blocks of proteins from plant exudates. The composition of the rhizosphere community also differs between mycorrhizal and nonmycorrhizal roots, the former having fewer pathogens (Foster 1986). More detail on the differences between rhizosphere and nonrhizosphere microorganisms have been described by Starkey (1929, 1958), Katznelson et al. (1948), Richards (1987), and Kennedy (2005).

14.5.2 Soil Mesofauna and Microfauna

The mesofauna are all heterotrophs that are primarily saprotrophs or microbivores[37] (e.g., collembolans [springtails], some mites, enchytraieds [pot worms]) or largely predators (e.g., some mites; table 14.3). Collembolans and many mites are primarily fungivores[38] and may play a significant role as consumers of plant pathogenic fungi as well as in nutrient mineralization. Enchytraeids have been shown to have significant effects on organic matter dynamics and nutrient release (through their feeding on fungi and bacteria), and on soil physical structure by producing fecal pellets that may enhance aggregate stability. Proturan feeding habits are unknown, but there are some reports that they feed on mycorrhizae (Coleman et al. 2004).

Soil microfauna include rotifers, nematodes, and protozoa. This diverse group has a wide range in trophic habits, including saprotrophs, predators, and microbivores (table 14.3). Rotifers are not generally very abundant in soil compared to the other soil microfauna (fig. 14.28), as they require a significant amount of water in soil pores to be active (essentially being aquatic organisms, as are the rest of the microfauna). They generally are saprotrophs, although some rotifers consume unicellular algae (i.e., both primary consumers and microbivores). They are perhaps most abundant in moist, organic soils (Coleman et al. 2004).

[37]Heterotrophic organisms that feed on microorganisms.

[38]A subset of the microbivores; organisms that feed on fungi.

Nematodes are unsegmented roundworms found in almost all soils. In soils, they are restricted to water films around particles or between aggregates. Despite their narrow habitat, nematodes are the most abundant of the soil invertebrates, reaching population densities on the order of one million per square meter of soil surface (fig. 14.28; Richards 1987). Nematodes can survive dry periods by coiling up into a cryptobiotic (resting) state, where they show no detectable metabolic activity. They rapidly become active again following rainfall. Most nematodes are predators on other organisms including fungi, bacteria, algae, protozoa, insect larvae, as well as other nematodes. Nematode grazing on bacteria and fungi can have a marked effect on the abundances and activities of these microorganisms; in some ecosystems (e.g., grasslands), nematode grazing on bacteria can account for 30 to 40 percent of the nitrogen mineralized because of the higher nitrogen content of the bacteria (chapter 18). Some nematodes are also plant root parasites that lead to root attacks by secondary pathogens and reduced plant growth (Brady and Weil 2002).

Protozoa are mobile, unicellular organisms that generally thrive best in moist, well-drained soil. They are highly diverse (>350 species have been isolated from soils) and are the most numerous of the soil microfauna (fig. 14.28). They capture and engulf their food (in soils, primarily bacteria[39]), and, like the nematodes, they swim about the soil matrix in water-filled pores or in water films around soil particles. Under unfavorable environmental conditions (e.g., drought), they can also form cysts (a resting stage) and persist until conditions become more favorable. Also, like the nematodes, protozoa can have a significant influence on nutrient mineralization and organic matter decomposition through their effects on bacterial populations (Brady and Weil 2002; Coleman et al. 2004).

14.5.3 Soil Microflora

The soil microflora consist of root hairs and algae (largely autotrophs[40]), fungi and actinomycetes (largely heterotrophs), and bacteria and archaea (heterotrophs and autotrophs; table 14.3). Like higher plants, soil algae are photosynthetic eukaryotes and can be relatively abundant in surface soil layers where light availability is high (Zlotin 1985). Most grow best under moist to wet conditions (many are motile in soil water, like many of the soil microfauna), but some are also very important in hot or cold, arid environments (e.g., deserts). Some algae (as well as particular cyanobacteria, see below) also form symbiotic associations with fungi called lichens. Lichens are very important in

[39]Hence, protozoa are generally bacterivores, a subset of the microbivores.

[40]An autotroph (from the Greek *autos*, meaning 'self,' and *trophe*, meaning 'nutrition') is an organism that produces organic compounds from carbon dioxide as a carbon source, using either light or reactions of inorganic chemical compounds as a source of energy. An autotroph is a *producer* in a food web.

colonizing bare rock and other low-organic-matter environments and may contribute substantially to soil development (see section 14.6; Brady and Weil 2002).

Soil fungi are eukaryotic heterotrophic organisms with a high degree of diversity (tens of thousands of species have been found in soil). Due to their relatively large size and high abundance in soil, fungi typically compose the greatest fraction of the biomass in soil (fig. 14.28; Lynch 1983; Brady and Weil 2002). Fungi are generally aerobic organisms, although some can tolerate the low oxygen and high carbon dioxide concentrations found in poorly drained soils. Some soil fungi are not entirely microscopic; many species form large structures that can be seen easily with the naked eye (e.g., mushrooms).

As a means of simplifying the discussion of this diverse assemblage of organisms, sometimes fungi are broken into three main groups (yeasts, molds, and mushrooms fungi) that are not taxonomic units sensu stricto (Brady and Weil 2002). Yeasts[41] are single-celled organisms that reside primarily in waterlogged soils that have low oxygen concentrations (i.e., anaerobic). In contrast, molds and mushrooms are multicellular, filamentous fungi, characterized by long, threadlike, branching chains of cells. Fungal hyphae are individual filaments, while mycelia are filamentous networks of hyphae that resemble woven ropes. Rhizomorphs are more complex versions of mycelial strands, with a greater degree of tissue differentiation. These rhizomorphs are often seen as white or yellowish strands running through decaying organic matter in forests. Molds[42] develop extensively in acid surface soils of forests, where the other microflora (bacteria and actinomycetes) provide limited competition (see below). The ability of molds to tolerate (and even thrive) at low pH makes them particularly important saprotrophs in many forest soils. Mushroom fungi[43] are associated with forest and grass vegetation. The aboveground fruiting body of these fungi is a very small part of the entire biomass of the organism, the vast majority of which resides as an extensive network of mycelia belowground or within the organic substrate. Although not as widely distributed in soils as the molds, mushroom fungi are vital components of the soil biota because of their involvement in the breakdown of woody tissue and because some species (thousands) form ectomycorrhizae (chapter 11).

Soil fungi as a group are versatile decomposers of organic matter in forests. Some decompose primarily simple substrates such as proteins and sugars (so-called sugar fungi[44]), while others specialize in the decomposition of cellulose (so-called brown rot fungi because they leave behind the brownish lignin); still other fungi are capable of decomposing both cellulose and lignin (so-called white rot fungi because they leave behind a white-colored residue that is primarily cellulose; Paul and Clark 1996). Fungi play a major role in humus formation and soil structural development and typically tolerate lower soil water potentials (drier soils) than the soil bacteria (Paul and Clark 1996). In general, fungi are believed to have higher carbon-use efficiencies[45] than bacteria and decompose and release nutrients from complex organic materials after bacteria and actinomycetes cease to function. Some fungi are predators of soil nematodes, and many produce antibiotics that reduce competition from bacteria.

Not all fungi are beneficial to plants. For instance, many pathogenic fungi (e.g., *Fusarium*, *Phytophthora*, *Pythium*, and *Rhizoctonia*) cause damping off, which kills seedlings (especially severe in many forest nurseries) of most conifer and hardwood tree species. Furthermore, many root rot fungi (e.g., *Armillaria* and *Phytophthora*) kill older trees. Of particular concern in the coastal regions of the United States is the soil fungus *Phytophthora ramorum,* which is responsible for a phenomenon called sudden oak death. Since first reported in 1995 in central coastal California, tens of thousands of tanoaks *(Lithocarpus densiflorus),* coast live oaks *(Quercus agrifolia),* and California black oaks *(Quercus kelloggii)* have been killed.

Our discussion of bacteria will include archaea even though they are distinct phenotypically and evolutionary. This is because archaea have many functions in soil that are similar to those of the bacteria domain. More information on the differences between bacteria and archaea domains can be found in Paul and Clark (1996) and Alexander (2005).

Soil bacteria are very small[46] (fig. 14.27), single-celled prokaryotes and are the most numerous and perhaps the most taxonomically diverse group of organisms in soil (fig. 14.28). More than 400 named genera contain well over 10,000 species. This number will likely increase severalfold, as many of the rDNA sequence patterns of bacteria found in soil are not homologous with the named species (Paul and Clark 1996). They are second only to the fungi in terms of the amount of total biomass they represent in soil. Their abilities to form extremely resistant stages that survive dispersal by many mechanisms (including wind, water, and animal digestive tracts) and to rapidly proliferate when environmental conditions and food availability become favorable (with generation times typically of only a few

[41]This group consists primarily of members of the phylum Ascomycota (Morton 2005).

[42]This group comprises some of the members of the phylum Ascomycota (common genera in soil include *Aspergillus*, *Fusarium*, and *Penicillium*) and some of the members of the phylum Zygomycota (the common genus in soil is *Mucor*) (Morton 2005).

[43]This group comprises the phylum Basidiomycota (Morton 2005).

[44]Members of the phylum Zygomycota are generally considered sugar fungi (Paul and Clark 1996).

[45]Microbial carbon-use (or growth) efficiency is the ratio of the amount of carbon incorporated into new biomass (i.e., growth) to the total amount of carbon assimilated by the organism. In pure culture, fungi generally have higher carbon-use efficiencies than bacteria (30–70% compared to 20–50%); however, bacteria may have higher efficiencies in natural populations (Hart et al. 1994).

[46]Bacteria range in size from about 0.5 to 5 μm, or on the order of the size of a clay particle.

hours) have allowed them to flourish and spread to most soil environments (Paul and Clark 1996).

Soil bacteria are also extremely diverse trophically. Although most are saprotrophs (hence heterotrophs), photoautotrophs (e.g., cyanobacteria and the green and purple bacteria) and chemoautotrophs (e.g., nitrifying bacteria, sulfur-oxidizing bacteria, and hydrogen-oxidizing bacteria) also occur (Alexander 2005). Many of these chemoautotrophs are particularly important in nutrient-cycling processes (chapter 18).

Cyanobacteria were previously classified as blue-green algae, but these organisms are obligate phototrophs with oxygenic photosynthesis similar to that of higher plants. Cyanobacteria are often the primary colonizers on soil parent materials, either alone or as mutualistic associates of fungi in lichens. These organisms exhibit considerable tolerance of saline environments and are important in forming biological crusts (along with algae and other organisms); in arid soils, these crusts help stabilize the surface soil against wind and water erosion and increase soil fertility by contributing carbon and nitrogen obtained from the atmosphere (Belnap 2005).

Actinomycetes are a group of gram-positive[47] bacteria that historically were grouped with the fungi because they commonly form filaments that are often profusely branched. Many authors have renamed this group the *actinobacteria* because these organisms are prokaryotes like bacteria rather than eukaryotes like fungi. Generally, their numbers in soil are exceeded only by the other bacterial species as a whole (fig. 14.28). Most are aerobic heterotrophs that tend to be quite tolerant of environmental extremes. For instance, many can remain active at lower soil water potentials than most other bacteria and fungi. Actinomycetes are also tolerant of low osmotic potentials and, hence, are important in arid regions and in salt-affected soils. They are generally sensitive to acidic conditions and tend to develop best in soils with pH values between 6.0 and 7.5. Nearly all actinomycetes are saprotrophs capable of decomposing many resistant substances in soil, including lignin, chitin, pectin, keratin, complex aromatics, and humic acids (Paul and Clark 1996). Many actinomycete species produce antibiotics that kill other organisms.[48] In forests, some actinomycetes form mutualistic associations with plant roots that are involved in biological nitrogen fixation (e.g., the genus *Frankia*; chapter 17). Actinomycetes usually become dominant at the latter stages of litter decay when the more readily decomposable compounds have been already metabolized (Brady and Weil 2002).

Several studies have shown that the soil microbial community is relatively resistant to disturbances from forest management activities. For instance, 6 to 10 years after the complete removal of aboveground tree biomass and the forest floor and/or after severe compaction (≈40% increase in the bulk density) across multiple forest sites in the United States, the soil microbial community (biomass and structure) was relatively unaffected compared to results from more traditional, bole-only harvest methods (Shestak and Busse 2005; Busse et al. 2006). The impact of the bole-only harvest, however, did generally reduce the mineral soil microbial biomass relative to the undisturbed forests. Interestingly, reduction of competing vegetation (with herbicides) did reduce microbial biomass and alter its function (Li et al. 2004; Busse et al. 2006). This is consistent with the general hypothesis that soil microbial communities are tightly coupled to the dominant plant community (e.g., Wardle 2002; Hart et al. 2005b).

14.6 SOIL DEVELOPMENT

Soils develop from the forces of climate and living organisms, modified by topography, acting on parent rock over time. A useful mnemonic for these factors is *ClORPT*, where *Cl* is climate, *O* is organisms, *R* is relief or topography, *P* is parent material, and *T* is time (Jenny 1980). These factors of soil formation were first outlined in 1883 by the Russian Vasily Vasil'evich Dokuchaev, who laid the foundation of soil genesis and the roots of soil science as a discipline. Later, Swiss-born Hans Jenny took these factors and placed them into a quantitative paradigm of pedogenesis. In this paradigm, each factor is treated as if it were independent of the other factors. As Jenny himself realized, this is rarely the case. For instance, as we discussed in chapters 4 and 5, the regional climate (and to a lesser degree the chemical composition of the parent material) influences the kinds of vegetation (as well as other organisms) that can physiologically occupy a site. Hence, changes in regional climate invariably cause associated changes in the resident biota. Despite the lack of independence of these factors, Jenny's model still provides a useful construct for studying the complex interactions of soil development.

14.6.1 Climate

Given sufficient time, the regional climate[49] is the dominant factor of soil development. Indeed, an early soil

[47]In the nineteenth century, a Danish physician named Gram devised a test which separated bacteria into two broad groups based on their ability to retain or not retain a crystal violet stain. Those not retaining the stain are called *gram negative*; those that do retain it are called *gram positive*. Gram-negative bacteria include groups that are quite important in nutrient cycling (e.g., asymbiotic nitrogen fixers) and common rhizosphere inhabitants such as the fluorescent pseudomonads (see section 14.5.1 and chapter 17).

[48]About 90% of actinomycete isolates in soil are in the genus *Streptomyce*. Bacteria in this genus produce a suite of antibiotics that have been used in modern medicine, including erythromycin, neomycin, tetracycline, amphotericin, and streptomycin. S.A. Waksman was awarded the Nobel Prize in medicine in 1952 for the discovery of streptomycin as well as other antibiotics. The characteristic earthy odor emitted from organic-rich soils is a volatile organic called *geosmin* (a derivative of terpene) produced by streptomycetes.

[49]Regional climate is relatively unaffected by the soil-forming factors organisms and topography; however, local or microclimate is altered by these factors (chapter 5 in this volume).

classification system used in the United States classified soils into three groups based on the regional climate (Baldwin et al. 1938): zonal soils, soils with a well-developed soil profile and exhibiting properties that express the influence of the regional climate and to a lesser extent the dominant vegetation; intrazonal soils, soils with a more or less well-defined soil profile that reflect the dominant influence of some resident factor of relief or parent material over the zonal effects of climate and vegetation; and azonal soils, soils without well-developed characteristics due either to their youth or to some condition of relief or parent material that arrests soil development (for example, poor drainage at the bottom of a swale).

Figure 14.25 clearly demonstrates that carbon and nitrogen storage in soils is strongly influenced by regional climate. But many other soil properties are also influenced by climate. For instance, increases in mean annual precipitation result in increases in exchangeable acidity (see section 14.4) and associated decreases in pH in soils derived from both acid igneous and basic igneous parent materials (fig. 14.32). Furthermore, the amount of clay present in the soil profile increases with increases in mean annual air temperature at a given level of precipitation (fig. 14.33). The mean climate may, however, not be the best indicator of the influence of climate on soil properties. For instance, over long periods of time, the number of wet years dictates the depth of weathering and leaching more than does the mean annual precipitation (Jenny 1980).

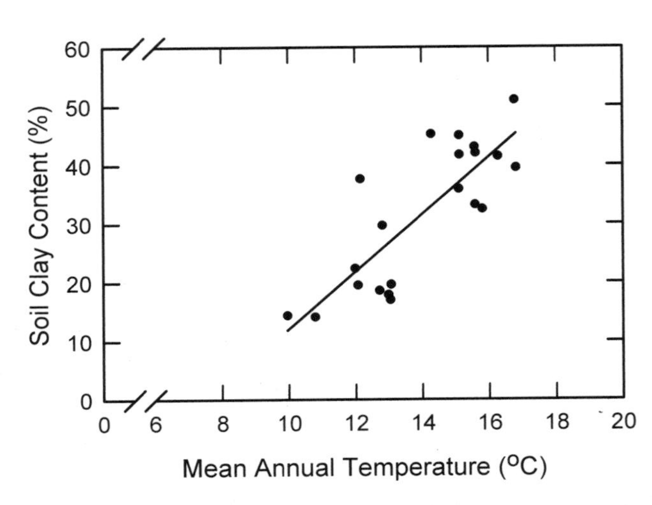

Figure 14.33. The clay content of soils increases with increasing mean annual air temperature (and hence soil temperature) due to accelerated rates of mineral weathering. This figure shows increases in average clay content to a depth of 1 m in soils derived from basic igneous rocks along the east coast of the United States. Changes in mean annual air temperature were obtained by varying the latitude where the soil was sampled. Each circle denotes a separate soil profile. (Modified from Jenny 1980)

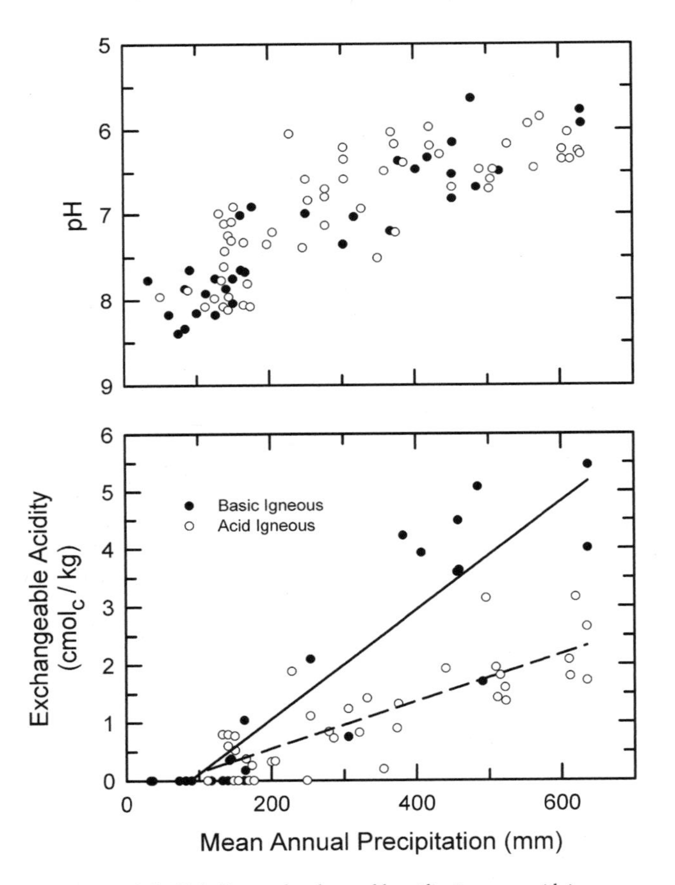

Figure 14.32. Soil pH declines and exchangeable acidity increases with increasing mean annual precipitation, where mean annual air temperature is held relatively constant. Soils derived from acid igneous rocks (e.g., rhyolite, granite) are denoted by open symbols, while those derived from basic igneous rocks (e.g., basalt, gabbro) are denoted by closed symbols. (Modified from Jenny 1980)

14.6.2 Organisms

The organisms present within a forest ecosystem influence soil development in a variety of complex ways. For instance, the soil biota influence the formation and characteristics of soils by: (1) "pumping" carbon and nitrogen from the atmosphere into soils (particularly near the soil surface) and, in the process, transforming these elements from inorganic to organic forms; (2) generating acidity that, in turn, accelerates rock weathering and the leaching of some chemical elements; (3) transforming mineral elements such as phosphorus and sulfur into organic forms; (4) altering the chemical balance and vertical and horizontal distributions by selectively cycling some elements (the nutrients), while allowing others such as sodium (which is a nonessential element for plant growth; chapter 16) to be leached from the soil; and (5) aggregating primary mineral soil particles (e.g., sand, silt, and clay) into organomineral structural elements (i.e., peds), thereby imposing a complex structure on soil that greatly influences the movement and retention of gases and water in the soil.

Although, Dokuchaev and Jenny viewed the biotic factor of soil development as a composite of all organisms present within the ecosystem, the influence of the plant community tends to dominate this factor of soil formation, and most studies of the biotic factor of soil formation have focused on differences due to dominant vegetation. For instance, in replicated vegetation patches in the southwestern United States, Welch and Klemmedson (1975) found that the amounts of carbon and nitrogen stored within the soil under perennial bunchgrasses were significantly higher than the amounts of these elements under ponderosa pine (the vegetation differences had been maintained for at least 50 years). Changes in other soil properties, such as clay and pedogenic $CaCO_3$ content, exchangeable cations, and pH, have also been compared across different vegetation types with marked differences occurring in many cases (Jenny 1980; Binkley and Menyailo 2005).

14.6.3 Topography

Topography influences pedogenesis largely through its effects on local (micro-) climate (e.g., insolation and evapotranspiration), patterns of subsurface water flow, and vegetation (chapter 5). The topographic factor includes slope, aspect, and position on the landscape (i.e., summit, shoulder, backslope, footslope, toeslope, or valley bottom; Brady and Weil 2002). Steeper (back) slopes undergo rapid erosion, which can expose new and relatively unweathered surfaces on slopes and deposit the older and more weathered surface layers in basins at the foot of the slope (fig. 14.34). Hence, soils found on steeper slopes tend to be thinner with coarser textures than soil developed on gentler slopes. Furthermore, soils on steep slopes also tend to be drier than soils on flat surfaces because of the downslope movement of water (either at the surface or subsurface) and the lower amount of precipitation intercepted at the soil surface compared to horizontal surfaces (fig. 14.34). Hence, the general movement of materials from backslopes to toeslopes and valley bottoms tends to result in more fertile soils occupying toeslope and valley-bottom topographic positions (Brady and Weil 2002).

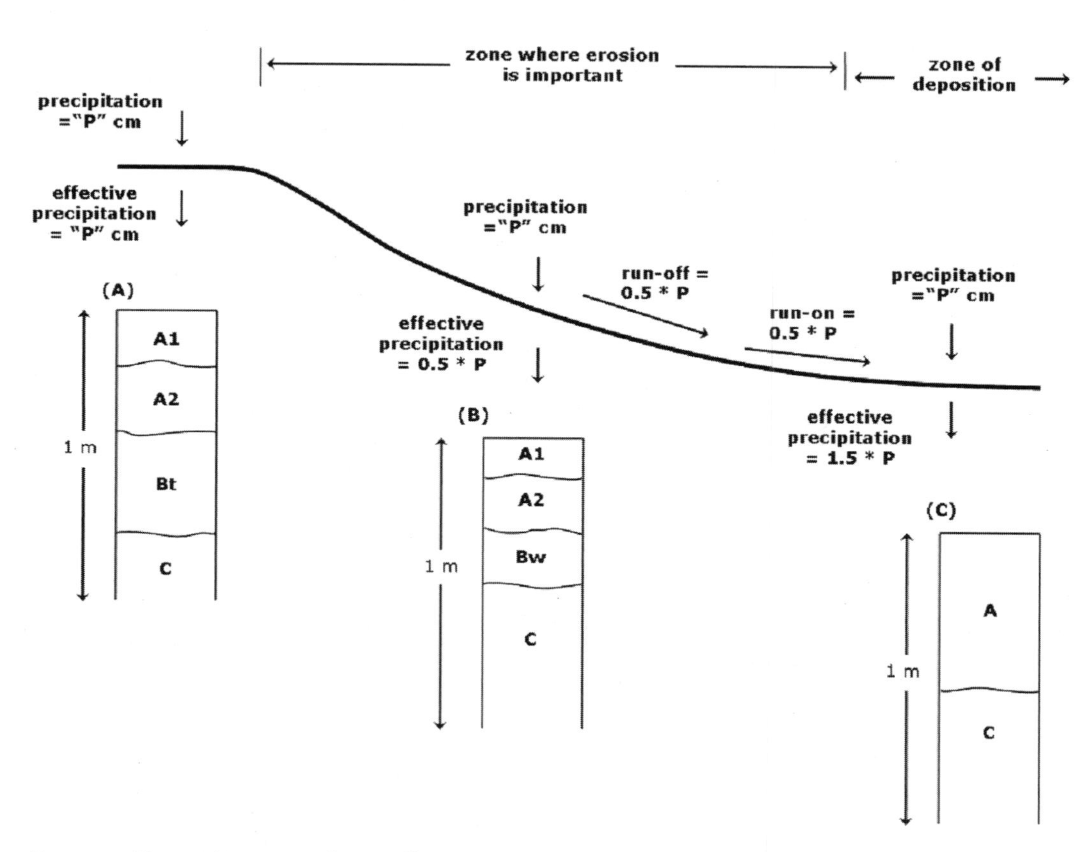

Figure 14.34. Topographic position influences effective precipitation (i.e., the amount of water infiltrating into the soil) which, in turn, influences soil formation. Effective precipitation and deposition of eroded material from upslope positions are greatest where runoff water from other locations accumulates (i.e., the toe slope or valley bottom); accumulation of this soil material coupled with decreased decomposition and clay formation because of prolonged soil wetness leads to thicker A horizons in cool, humid climates (*right profile*, C). Thinner A horizons and weakly developed B horizons (Bw) occur at midslide positions due to lower effective precipitation and greater surface erosion (*center profile*, B). Thicker A and better developed B horizons (containing more illuvial clay, called a Bt horizon) occur at flat, summit positions because of the stable land surface and moist, but not wet, conditions that promote clay formation and eluviation to the subsoil (*left profile*, A). (Modified from Singer and Munns 2006)

The aspect (the cardinal direction that the slope faces) also influences soil development through its effect on soil microclimate (chapter 5). Equatorward aspects (i.e., south-facing slopes in the Northern Hemisphere) tend to be drier and hotter than poleward aspects (i.e., north-facing slopes in the Northern Hemisphere). The warmer and drier soils on equatorward aspects reduces plant production (and frequently also changes the plant community), and this, in turn, reduces organic matter inputs to and accumulation in the soil.

14.6.4 Parent Material

Within a given climatic regime, the chemical composition and physical structure of parent materials determine the rate at which they are weathered, as well as many of the chemical and physical characteristics of the resulting soil. For instance, the grain size of parent materials determines to a large degree the texture of the soil (see section 14.3). Large grain sizes in parent materials (e.g., granite, diorite, gabbro, and arkose sandstone) generally lead to coarse-textured soils, while fine-grained parent materials (e.g., rhyolite, andesite, basalt, siltstones, shale/mudstone, and graywacke sandstone) generally lead to fine-textured soils. Soils derived from light-colored (felsic) rock (dominated by the minerals quartz and potassium-rich orthoclase feldspar) tend to be acidic, while soils derived from dark-colored (mafic) rock (dominated by the minerals pyroxine, calcium-rich plagioclase feldspar, olivine, and hornblende) tend to be more circumneutral or slightly alkaline. Sandstones cemented by calcium carbonate and limestone and dolomite parent materials tend to also produce fairly alkaline soils because of the abundance of carbonate released into the soil solution during weathering (Brady and Weil 2002).

The mode of origin of the parent material influences the texture and coarse-fragment content of soil, as well (Kimmins 2004). For instance, **residual soils** (soils derived from in situ weathering of bedrock) typically have finer textures near the surface and become less weathered and have higher coarse-fragment content with depth. In contrast, **transported soils** (those where the parent materials have been transported to the site of soil development) tend to have a more uniform texture with depth than residual soils; however, the texture and coarse-fragment content of the soil depends on the mode of transport of the parent material. For instance, parent materials deposited by lakes (**lacustrine**) or transported by wind (**aeolian** or **loess**[50] soils) generally are fine textured (i.e., clays, silts, and fine sands). Materials deposited in floodplains along rivers (**alluvial**) generally are fine to coarse textured (i.e., silts, sands, and gravels), depending on the distance from the channel (finer materials are deposited farther from the

[50]Aeolian (also spelled *eolian*) material is any wind-transported particle, while loess is wind-transported material that is mainly silt sized.

An old rule of thumb distinguishes four groups of parent material with regard to their influence on species composition and forest productivity. We have added two additional groups to this old classification of soil parent materials, which was based on European soils, to include geologically young and old soils. Most of the following descriptions are from Tamm (1921) and Wilde (1958); remember that these are only rough guidelines.

Group 1: Siliceous rocks. Includes sandstones, siliceous shales, conglomerates, and quartzites. These are all sedimentary parent materials with the exception of quartzite, which is a metamorphosed sandstone. Sandstones are composed primarily of quartz, conglomerates of gravel, and siliceous shales of silt- and clay-sized particles that are quartz-like in their chemical composition. The productivity of soils derived from sedimentary parent materials may also vary considerably depending on the cementing agent and other 'impurities' that may be present. For example, when cemented by calcareous materials, sandstones can form very productive soils, but when cemented by silica or iron with no other minerals present, sandstone soils are infertile. In general, soils developed from these parent materials have low nutrient availability and a low cation exchange capacity. Such soils are suitable primarily for less-nutrient-demanding tree species, particularly pines.

Group 2: Orthoclase-feldspathic rocks. Includes the igneous rocks granite, granitic porphyry syenite, and orthoclase felsites, and metamorphic rocks such as gneiss that are chemically similar to igneous rocks. Rocks of this group typically weather into coarse-textured soils (sandy loam to loam soils) that have high potassium availability, but low availabilities of calcium and magnesium. These soils are well suited for most commercially important tree species, with the exception of a few calcium-demanding species (e.g., aspen).

Group 3: Ferromagnesium rocks. Include the igneous rocks gabbro, diorite, diabase, basalt, and andesite, and the metamorphic rock hornblende gneiss. Enriched in ferromagnesium minerals such as augite, amphibole, or olivine, these rocks weather relatively rapidly to produce deep, fine-textured soils rich in calcium, magnesium, iron, and other nutrients (although potassium may be low). These soils generally support exceptionally productive forests of both deciduous and coniferous tree species. Basalts weather more slowly than other rocks in this group (especially flow basalts that lack vesicles), and young basaltic soils may be rocky.

Group 4: Calcareous rocks. Includes limestone, dolomite, chalk, and calcareous shales. The productivity of soils derived from these parent materials is extremely variable and is largely determined by climatic conditions, degree of weathering, and the clay content. Soils derived from clay-rich limestones in cold and humid regions tend to be deep soils and are among the most productive soils found in this climatic zone. These soils often support fast-growing stands of spruce and other so-called acidophilous conifers. In temperate regions, soils derived from deep, clay-rich limestones support calcium-demanding hardwoods and conifers such as black walnut, hickory, white and green ash, beech, white elm, black locust, red cedar, white cedar, ponderosa pine, Austrian pine, and Douglas-fir. Calcareous soils of a shallow depth, or those derived from relatively pure limestone or chalk, usually have low productive potential; in low-rainfall regions, these shallow soils are occupied by grass or woodlands.

Group 5: Volcanic ash. Parent materials of volcanic ash consist of tiny fragments of volcanic glass, quartz, feldspars, and ferromagnesium minerals (Buol et al. 1989). The most common ash deposits are relatively low in silica and high in base cations; these are found wherever volcanic activity occurs, principally around the Pacific basin in northwestern North America, Central America, western South America, Japan, Indonesia, New Zealand, and the Philippines (Buol et al. 1989). Many soils derived from volcanic ash are classified as Andisols in the USDA Soil Taxonomy system and as Andosols in the Food and Agriculture Organization (FAO) of the United Nations soil classification system (see section 14.7). Volcanic ash is carried for long distances following an eruption and frequently is mixed with and influences soils that were derived from other parent materials. Pumice, a type of 'frothy' volcanic material that is high in silica and low in nutrients, is found in limited areas of the northwestern United States, Central America, and northwestern South America.

Soils derived from volcanic ash and pumice generally show unique properties that are not found in soils derived from other parent materials under the same vegetation and climate (Wada 1985). The primary distinguishing characteristics of ash soils are: (1) large accumulations of soil organic matter; (2) the presence of allophane or imogolite, poorly crystalline secondary aluminosilicate minerals (these minerals complex very effectively with organic matter and probably account for the large accumulations of organic matter common to volcanic ash soils); (3) low bulk density, owing primarily to the high organic matter contents; and (4) high water- and cation-retention capacities (Buol et al. 1989). Except for some soils formed from pumice, Andisols generally are productive soils that have excellent **tilth** (the state of aggregation of soil and its condition for supporting plant growth) and high nutrient availabilities, with one exception: phosphorus availability tends to be low. Soils derived from volcanic ejecta tend to adsorb considerable amounts of phosphorus as phosphate (chapter 18), and, in many regions, these soils need to be fertilized with considerable amounts of phosphorus to maximize their productive potential (Buol et al. 1989).

Group 6: Previously weathered and transported parent materials. In the groups discussed earlier, rock weathering and nutrient release occur as ongoing processes within the plant rooting zone. Parent materials in this group are already highly weathered at the onset of soil development, far below the root zone of plants. These parent materials consist primarily of the most insoluble secondary minerals (usually aluminum and iron oxides) and contain little silica or base cations. Soils of this group occur primarily in the tropics and subtropics and often support species-rich rain forests.

Despite highly weathered and nutrient-poor parent materials, nutrients are retained very efficiently within the vegetation and the biological component of these soils, and forests growing on them can be quite productive (Jordan and Herrera 1981). The strong biological influence on soil fertility has very important implications for forest management on these soil types. Clear-cutting (particularly on short rotations) can break the biological chain that maintains nutrients within the rooting zone and can lead to sharp declines in soil fertility (chapters 17 and 18).

In the first four groups described, increasing fertility correlates with increasing calcium content of the parent material. The association between calcium and soil fertility is widely recognized among soil scientists and ecologists (Lutz and Chandler 1946; Jordan and Herrera 1981; Oades 1988), although the reasons for this connection are not totally clear. Calcium is required in rather high amounts by woody plants and many soil invertebrates (chapter 16). Calcium-rich rocks also tend to weather rapidly and produce more clay minerals than do acidic igneous rocks, and calcium is also an important determinant of the type of humus that forms in soil. Moreover, the basic igneous rocks that are high in calcium generally are also rich in other nutrient elements that exert control over site productivity. This is particularly true of phosphorus, which is believed to be an important determinant of carbon and nitrogen accumulation and nitrogen cycling in soils (McGill and Cole 1981; Tate and Salcedo 1988; Vitousek 2004). Pastor et al. (1984), studying a series of forests growing on different parent materials in Wisconsin, found that productivity was closely linked to nitrogen cycling rate, which, in turn, correlated with phosphorus availability. In general, however, the contribution of parent material to productivity is complex and multifaceted and cannot be attributed to any single nutrient element or physical characteristic.

channel). Materials transported and deposited by glacial ice (**glacial till**) are poorly sorted,[51] with materials that range in size from clay to boulders (with sharper edges), depending on the geological composition of the parent material. Glacial meltwaters (**glaciofluvial**) normally deposit very coarse (i.e., gravel and stones with rounded edges) materials. The texture of soils derived from parent material transported by gravity (**colluvial** materials) depends on the nature of the material prior to downslope movement.

The influence of parent material as a factor of soil formation tends to be greatest in relatively young soils. Over time, climatic factors come to dominate pedogenesis. Many forest soils are geologically young, however, and the influence of parent material can clearly be seen in soil characteristics, species composition, and the productivity of many forests (chapter 5).

14.6.5 Age

To recapitulate, weathering of parent material and soil development are accompanied by: (1) the progressive alteration of primary minerals to secondary minerals; (2) the differential loss of chemical elements; and (3) the progressive conversion of nutrients from mineral to organic forms. These factors simultaneously both create and destroy soil fertility. Initially, the forces of creation dominate, but as parent material becomes increasingly depleted of nutrients over time, the potential fertility of soils declines (Birkeland 1999; Vitousek 2004). The actual fertility does not necessarily decline, however, because nutrients are protected from leaching through their incorporation either within living organisms or dead organic matter. The net effect is that as a soil becomes older, its fertility becomes increasingly tied to the health and integrity of the community of organisms that it supports.

Relatively young soils occur in: (1) areas that are tectonically active, frequently the margins of continents where colliding plates produce volcanic activity or shove fresh ocean floor onto land (e.g., the west coast of the Americas); (2) areas that were glaciated during the Pleistocene (northern North America and Eurasia, parts of Australia); and (3) mountainous areas where erosion acts to keep 'fresh' rock within the plant rooting zone (Fyfe et al. 1983). Old, highly weathered soils predominate on stable land surfaces in the interior of South America and Africa, both areas that have experienced neither tectonic activity nor glaciation in recent geological history. These soils are weathered far below the rooting zone, and fertility is retained by retention and tight cycling within the biological community (Jordan and Herrera 1981; Fyfe et al. 1983).

[51]Parent material is considered sorted when similar-sized particles are grouped into depositional layers; water-transported parent materials are usually well sorted, while ice-transported materials are not.

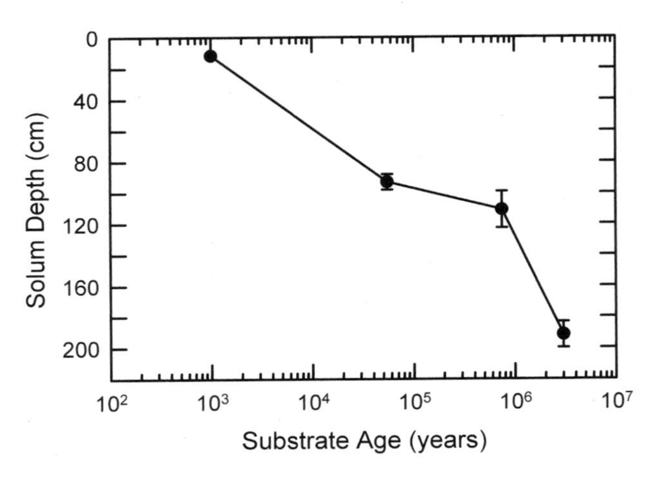

Figure 14.35. Soil depth increases with the length of time the parent material has been exposed to soil forming processes. The figure shows increases in solum depth (i.e., the depth to which pedogenic processes have modified the parent material, or the thickness of the A + B horizons) over three million years of soil development in piñon-juniper woodlands across the San Francisco Volcanic Field (near Flagstaff, Arizona). Increases in soil depth are associated with increases in the accumulation of carbon and nitrogen from the atmosphere, and the rates of their transformations in soil. In contrast, available phosphorus declines with substrate ("soil") age. Vertical bars denote ±1 SEM. (G. Newman, A. Kowler, and S. Hart, unpublished data, and Selmants 2007)

During soil development, soils become deeper (fig. 14.35) and soil texture becomes finer (fig. 14.36). These two factors provide for greater water storage capacity and greater exploitation of soil nutrient pools by plant roots and mycorrhizae. Besides these changes in physical characteristics, chemical changes also occur. For instance, carbon and nitrogen content increases over time, as these materials are ultimately derived from the atmosphere and not the soil parent material (chapter 17). In contrast, base cations and phosphorus, which are derived primarily from the parent material, are continuously lost from the soil due to leaching and surface erosion. Hence, we would expect that nitrogen availability would limit tree growth early on in soil development, while phosphorus and base cations would be more limiting to productivity later in soil development. Indeed, this model of ecosystem development has been substantiated using soil chronosequences in humid (Vitousek 2004) and semiarid regions (Selmants 2007) alike; although, in semiarid areas, increases in water storage and availability that accompany soil development also play an important role in regulating plant productivity (Selmants 2007). Furthermore, it appears that atmospheric deposition of dust, even from thousands of kilometers distant, may help sustain the productivity in highly weathered soils where all of the weatherable base cations and phosphorus should have long been lost from the soil or converted to plant-unavailable forms (Chadwick et al. 1999; see chapter 17).

14.7 SOIL CLASSIFICATION

Organisms are classified into groups (from kingdoms to species or subspecies) to allow more efficient communication of information, and the same is true for the classification of soils. Soils are different than biological organisms in that it is hard to define the individual. Because of the more gradual horizontal changes that occur in soils, soil scientists have developed the construct of the **pedon,** which is a three-dimensional body with lateral dimensions large enough to permit study of horizon shapes and relations. In practice, the minimum dimensions of a pedon are about one meter by one meter by one meter or more deep. The soil unit that is classified (i.e., the soil individual) is defined as the **polypedon,** which is a group of contiguous similar pedons (SSSA 1997).

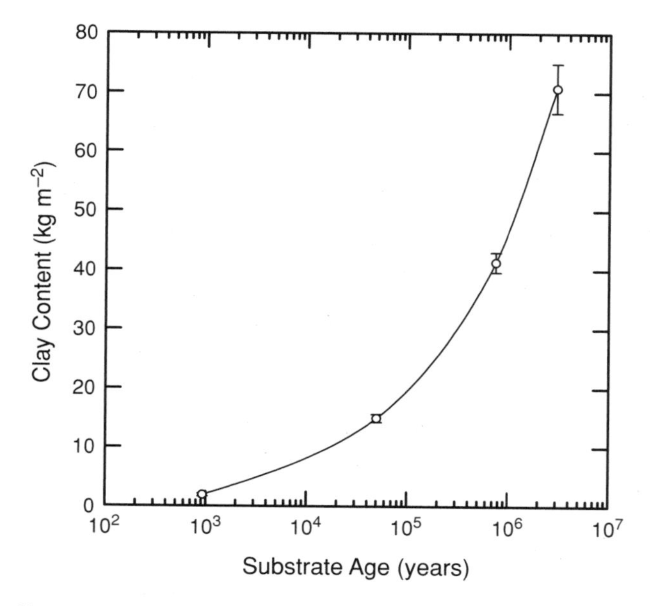

Figure 14.36. Increase in clay content (upper 15 cm of mineral soil) over three million years of soil development in piñon-juniper woodlands across the San Francisco Volcanic Field (near Flagstaff, Arizona). As with all soil chronosequences, the other state factors (i.e., climate: mean annual air temperature 10 °C and mean annual precipitation 360 mm; topography: $<5\%$ slope; organisms: dominant plants *Pinus edulis* and *Juniperus monosperma*; and parent material: basaltic cinders) were kept constant across the gradient. Increases in clay content have a dramatic effect on the rate of weathering, nutrient cycling, and the storage of plant-available water, which is a key limiting soil resource in this semiarid climate. Vertical bars denote ± 1 SEM. (Selmants 2007)

Many different soil classification systems (for distinguishing 'soil types') are used in the world today. Some are comprehensive systems (like those of the USDA and FAO), where any soil can be classified, while others are provincial, designed to be applied only to soils in that region (usually a country, for example, the Canadian soil classification system). Here, we briefly describe the soil classification system developed by the USDA. More information on this and other classification systems, and soil classification in general, can be found in work by Buol et al. (1989) and Soil Survey Staff (1999).

The USDA Soil Taxonomy system is a hierarchal classification system with six categories or levels (fig. 14.37). In this regard, it is similar to the Linnaean system used for classification of organisms (Brady and Weil 2002). This system uses observed soil properties of the soil individual (i.e., polypedon) to classify soils. At each lower category in the hierarchical system, more specific information about the soil is conveyed.

The USDA Soil Taxonomy system currently has 12 soil orders (the highest or most general category). At the lowest

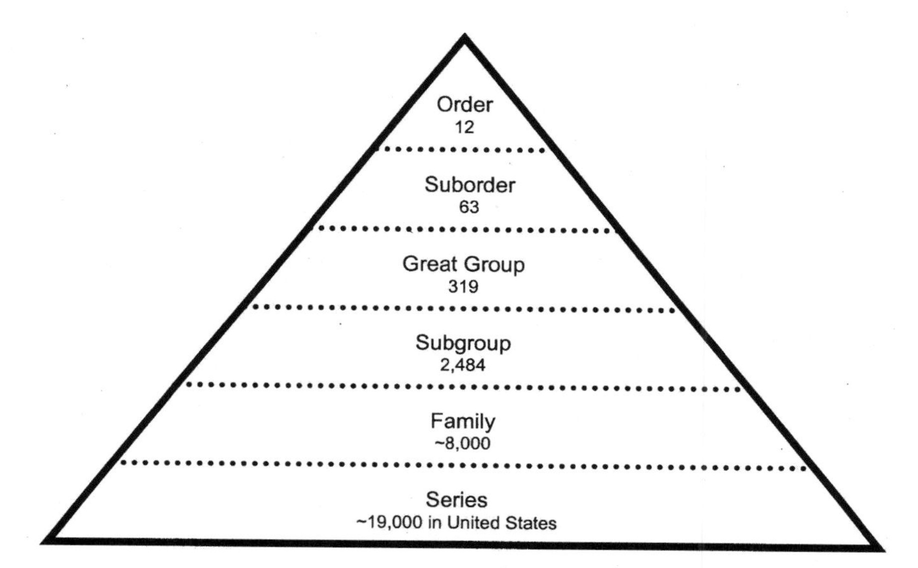

Figure 14.37. United States Department of Agriculture Soil Taxonomy system (Soil Survey Staff 1999) is a comprehensive and hierarchical classification system with six categories. Each lower category has a higher degree of specificity in regards to the range of soil properties contained within each taxon. The approximate number of taxa in each category is shown. (Modified from Brady and Weil 2002)

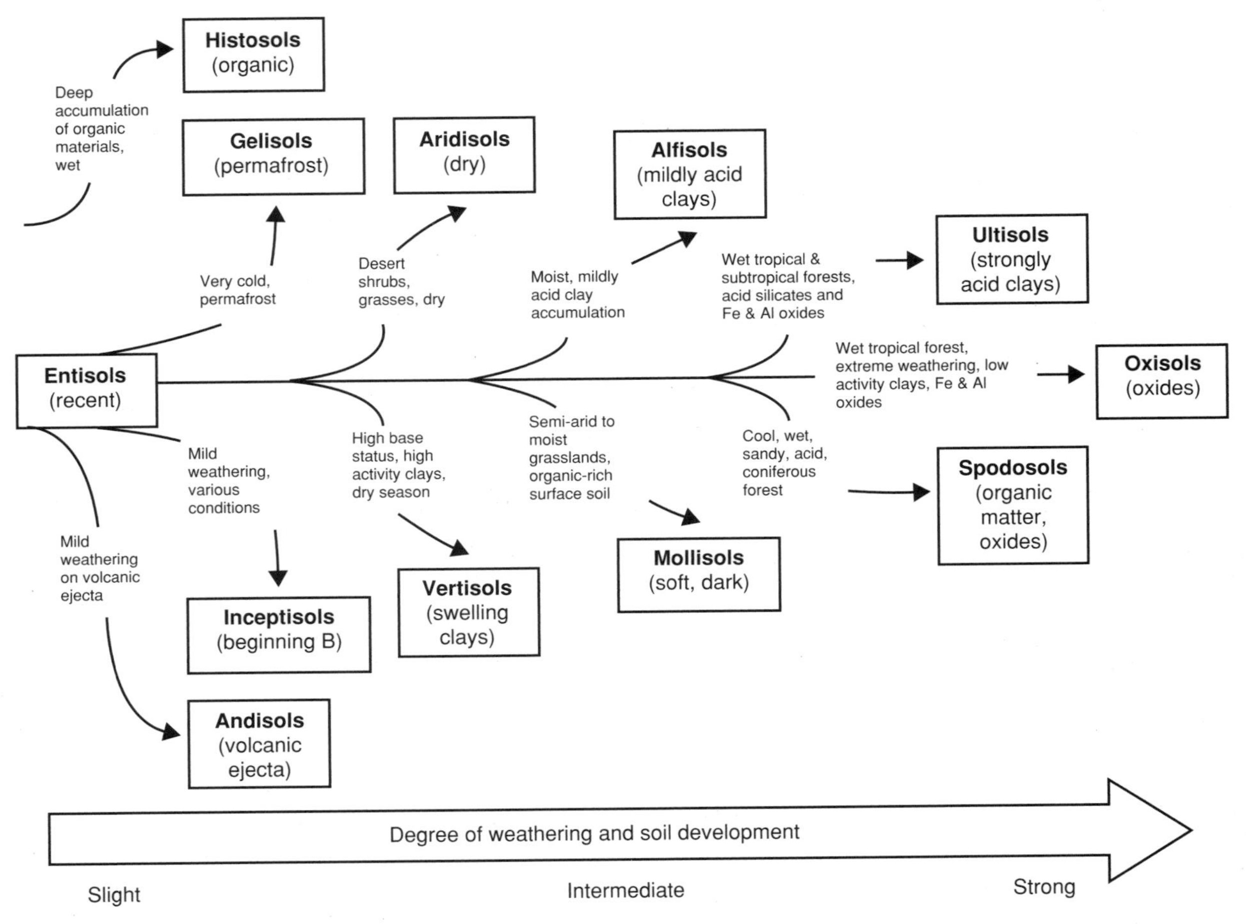

Figure 14.38. General relationships in the United States Department of Agriculture Soil Taxonomy classification system among the 12 soil orders along a soil developmental (i.e., weathering) sequence. General climatic and vegetative conditions under which each soil type typically forms is also shown. (Modified from Brady and Weil 2002)

category (soil series), which has the highest specificity, there are currently more than 19,000 taxa in the United States alone (Brady and Weil 2002). Although not officially a part of Soil Taxonomy, in the most detailed soil surveys, soils are broken down even further into soil phases, which contain specific information pertinent to the management of the soils (e.g., degree of slope where the series is found, or the amount and type of coarse fragments that are present in the soil profile).

Soil Taxonomy has a unique nomenclature (using a combination of nonsense and Latin- and Greek-based prefixes, suffixes, and word roots called *formative elements*) where the classification name at each lower catergory contains all the nomenclature of the higher category (except at the lowest category, the soil series, which usually bears the name of the location where that soil type was first described). For instance, the Holland soil series is a member of the soil family Ultic Haploxeralfs, fine loamy, mixed, mesic, and is classified in the following categories: order, Alfisols; suborder, Xeralfs; great group, Haploxeralfs; and subgroup, Ultic Haploxeralfs. The three letters *alf* identify each of the lower categories as being part of the Alfisol order. Unlike the Linnaean biological classification system, the nomenclature used in the Soil Taxonomy system (from order to family categories) allows the user to immediately know something about properties of that individual soil.[52]

Of the 12 orders, 11 are considered mineral soils (although they may have an organic horizon within the soil profile), and one is considered an organic soil (the Histosols). Histosols must contain at least 20 percent organic matter by mass, or commonly more than 80 percent by volume, in the soil profile. These soils are usually found in wet areas such as swamps, bogs, and marshes, where very high water potentials limit decomposition. Although the system is based on soil properties rather than presumed mechanisms of soil formation (Brady and Weil 2002), the 11 mineral soil orders can be envisioned along a pedogenic (i.e.,

[52]For instance, this soil family name conveys that this is a typical forest soil, with a clay-enriched and fertile subsoil that is found in a Mediterranean-type climate (summer-dry period). It has a medium-fine and fairly uniform texture with depth, a mixed-clay mineralogy (i.e., no single clay mineral dominates), and a mean annual soil temperature between 8 and 15°C at a 50-cm depth in the mineral soil. Furthermore, the soil is weathering toward a less fertile soil order, the Ultisols. Information contained within the soil series would include such specific information as the arrangement and thicknesses of soil horizons (Soil Survey Staff 1999).

Table 14.6
Common Soil Taxonomy orders (i.e., "soil types") found in major forest types

Major forest types	Most common soil orders
Boreal forests	Gelisols, Spodosols, Histosols, Inceptisols
Temperate conifer, mixed, hardwood, and montane forests	Alfisols, Inceptisols, Ultisols, Spodosols, Entisols, Mollisols
Tropical rain forests, monsoon forests, dry forests	Ultisols, Inceptisols, Oxisols, Andisols

Source: Data from Fisher and Binkley 2000.

degree of weathering) sequence (fig. 14.38). Entisols are the least-developed soil order and are found on parent materials recently exposed to weathering processes (e.g., recent floodplains, recent volcanic deposits, retreating glaciers exposing fresh till, etc.). Oxisols represent the other extreme in this soil development sequence; they are found in environments where extreme weathering of parent materials has occurred for extended periods of time (e.g., wet tropical forests), leading to soils dominated by low-CEC silicate clay minerals (e.g., kaolinite) and iron and aluminum oxides. In general, the most fertile soil types are those that are at an intermediate stage of development (e.g., Alfisols and Mollisols; Brady and Weil 2002).

As is the case for other plant communities (Brady and Weil 2002), there is some degree of correspondence between major forest types and soil types (table 14.6). This relationship is reflective of the importance of the biotic factor in soil development, the coupled feedback of soil in shaping plant communities, and the dominating influence of climate on the distribution of both forest and soil types.

14.8 SUMMARY

Soil is critical to the health and productivity of ecosystems. Of the 27 chemical elements required by one or another of the life-forms within ecosystems (plants require 18), all but two are obtained from soil. The electrical properties of soil, which are attributable to both clay minerals and organic matter, retain water-soluble nutrients against loss by leaching. Soil is a major thread of continuity in ecosystems. Organisms are born and die but soil remains, serving as a repository of site fertility and a primary 'communication' link between succeeding generations.

Six steps can be distinguished in soil formation:

1. Breakdown of rocks (i.e., weathering), primarily by acids contained in precipitation or released by organisms, accompanied by differential leaching or retention of chemical elements contained in the rocks.
2. Alteration of primary minerals contained in the rock to secondary forms (referred to as clay minerals), many of which have extremely high surface areas relative to their volume and a high density of surface electrical charge.
3. Addition of organic matter by plants and other primary producers (e.g., algae) and nitrogen by organisms that are fueled by photosynthates.
4. Development of a complex food web of microflora and fauna (both invertebrate and vertebrate), with photosynthate from plants and other primary producers at its base.
5. Transformation of photosynthates by soil organisms to unique, highly stable forms of organic matter called humus; like clay minerals, humus is characterized by very high surface area to volume ratios and a high density of electrical charge.
6. Aggregation of minerals and organic matter into intimate organomineral complexes tied together through a combination of electrical and physical forces, resulting in a structure characterized by a mosaic of solids interspersed with water- or air-filled pores of widely varying sizes.

Although these steps occur somewhat sequentially during soil formation (e.g., primary producers must be present before a food web can develop and humus can form), they also occur simultaneously. Soil formation cannot be divided into discrete steps but is a dynamic ongoing process that continually creates and maintains a unique entity in nature, something that cannot be neatly classified as 'animal, vegetable, or mineral' but, rather, is an intimate combination of all of these.

The characteristics of a particular soil vary according to five state factors: (1) regional climate (particularly temperature and precipitation); (2) parent material (rocks or other mineral components such as volcanic ash); (3) the characteristics of the biological community occupying the soil; (4) topography; and (5) time (i.e., how long the soil has been developing). Parent material influences forest productivity through the nutrient elements that it contains, the rate at which it weathers, and the nature of the secondary minerals produced by weathering. Rocks with high calcium content are frequently associated with more productive soils, though calcium probably is not the whole reason for this. Time has a nonlinear effect on soil productivity. Initially, weathering releases nutrients and provides the basis for plant growth. Eventually, however, weathering depletes rocks of their nutrients, and soil fertility increasingly depends on retention and cycling of nutrients within the biological community. This is the case in the rain forests in continental interiors within the tropical zone, where soils are both quite old and heavily weathered because of high precipitation and warm, relatively constant temperatures. Soils are usually younger and more fertile along continental margins, where tectonic activity periodically infuses fresh minerals.

If soils are critical to the health of plants, plants are equally critical to the health of soils. Plants are the source of organic matter that feeds soil organisms, accelerates weathering, and glues soil particles together to form aggregates. Most of the biological activity in soils occurs in rhizospheres (i.e., the zone around roots, mycorrhizae, and mycorrhizal hyphae). The diversity of pore sizes created by aggregation maintains a balance between soil water and soil air and also provides living and hiding space for soil organisms. Much of the rather large amounts of energy diverted by perennial plants to the soil represents an investment in future growth of the plant; plants and soils are joined in a 'dance of mutual creation.'

15

Primary Productivity

PRODUCTIVITY IS THE ACCRUAL of matter and energy in biomass. The first step in this process (termed **primary productivity**) is performed by green plants, which are the only organisms capable of capturing the electromagnetic energy of the sun and converting it to the chemical energy of reduced carbon compounds (i.e., **photosynthates**). **Secondary productivity** results when heterotrophic organisms consume plant tissues and convert some proportion of that matter and energy to their own biomass. Secondary producers, which are associated with the detrital and the grazing energy transfer pathways (chapter 9), compose a small proportion of total forest productivity, but are critically important regulators of ecosystem processes, particularly nutrient cycling (chapters 18 and 19). Gosz et al. (1978) give a relatively thorough balance sheet for energy transfers in a temperate deciduous forest.

The fundamental relationship governing matter and energy transfer in living systems is

$$6CO_2 + 6H_2O + \text{energy} \leftrightarrow C_6H_{12}O_6 + 6O_2 \text{ (green plants)}$$

$$\downarrow$$

$$C_6H_{12}O_6 \quad \text{(heterotrophs)}$$

Photosynthesis, which is the process of fixing energy in matter, proceeds from left to right in this diagram. **Respiration** (*R*), which is the process of extracting that fixed energy from matter, proceeds from right to left.

These simple transitions are mediated and shaped by numerous physiological, genetic, ecological, and social factors that require resources other than just carbon, H_2O, and sunlight. Enzymes, membranes, energy-transport molecules, and genes contain an array of chemical elements (i.e., nutrients) that must be gathered from the environment. Suitable habitat is needed for safe living and reproduction. Social organization integrates the individual into a population of similar organisms, and ecological organization integrates the individual into a community. A complex physiological and ecological structure evolves to sustain the acquisition of energy, and the energy, in turn, supports the structure.

Energy that is consumed during the business of living is reflected in the release of carbon dioxide from organisms to the environment (i.e., respiration, from right to left in the diagram), and the accumulation of matter and energy that we see as biomass increment is only a portion of that actually fixed in photosynthesis. The latter is termed **net primary productivity** (NPP), the former—which is equal to NPP plus *R*—**gross primary productivity** (GPP). NPP is traditionally measured as dry mass (i.e., with as much water removed as practical). The living biomass present on a site at any given time is called the **standing crop.**

NPP of a forest over a given period is composed of increment in tree tissues, including boles, branches, leaves, roots, mycorrhizae and other symbionts; tissues that are shed from the tree as litter (chapter 18); tissues that are consumed by heterotrophs; and trees that die. In practice, it is difficult or even impossible to measure all of these things. The task becomes much easier, however, if we focus on aboveground NPP, and most estimates of forest productivity deal only with the aboveground. During the 1980s, ecologists began devoting an increasing amount of effort to understanding belowground productivity and processes. Much remains to be learned, but it is now clear that energy allocated belowground by trees (as we shall see, a substantial amount) represents a critically important investment in the health and productivity of the entire ecosystem. Moreover, carbon diverted to belowground storage pools is a major sink for atmospheric carbon (chapter 3), a function that takes on considerable practical and economic significance in the modern world.

Gross primary productivity is generally estimated in one of two ways: based on canopy photosynthesis models that account for many of the factors that affect photosynthesis, or based on measured NPP and estimated *R*. Respiration is typically estimated from empirical equations that account for the sensitivity of respiration to temperature and the amount of living matter in the biomass, or as some proportion of NPP. Accurate measurements of NPP plus estimates of respiration should result in GPP values that are the same as photosynthesis of the entire tree crown or forest canopy. In the following sections we will first look at the process of light gathering in canopies, and the conversion of solar energy to chemical energy in the process of photosynthesis. We then look at respiration and the carbon balance remaining after respiration is taken out of photosynthesis (i.e., NPP). We will discuss how NPP differs among different forest types, across local landscapes, and between mixed stands and monocultures and move on to discuss resources and other environmental factors that control NPP and its allocation to production of different organs and substances. The chapter closes by discussing the possible effects of global warming on NPP. (The role of forests in global NPP is discussed in chapter 3).

15.1 LIGHT CAPTURE AND GAS EXCHANGE IN CANOPIES

15.1.1 Light Extinction in Canopies

In theory, canopy photosynthesis, and its equivalent, GPP, should level out as leaf area index (LAI; the average area of leaves covering the ground area in a forest) increases to a value above which most or all incoming light is intercepted. The LAI at which that occurs is not the same in all stands due to differences in the structure of canopies. To understand these points, it is necessary to discuss two things: the relation between light levels and photosynthesis of individual leaves, and the nature of light transmission through canopies.

15.1.1.1 Light Levels and Photosynthesis of Individual Leaves

For C_3 plants, photosynthesis of individual leaves increases sharply as light intensity increases from deep shade up to approximately 25 to 50 percent of full sunlight, then levels out and remains constant as light is increased further, at which point the leaf is said to become **light saturated.** In contrast, leaves of C_4 species do not become light saturated, continuing to increase photosynthetic rate up to full sunlight. This difference is related to the respective abilities of C_3 and C_4 plants to utilize carbon dioxide (Larcher 2003).

At low light levels, the ability of a leaf to maintain positive net photosynthesis depends on the respiratory costs associated with the leaf. Net photosynthesis of a leaf is equal to its gross photosynthesis less respiration of the leaf and its supporting structures (e.g., the twig). The point at which gross photosynthesis drops to a level equal to respiration (hence net photosynthesis equals zero) is called the **light compensation point.** This point varies among species, and one reason that shade-tolerant species are shade tolerant is because they have low respiration rates, thereby reducing the compensation point.

15.1.1.2 Photosynthesis of Canopies

The absorption of light during its passage through a canopy is described by the Beer-Lambert law (Monsi and Saeki 1953):

$$t(x) = \exp[-K^{\star}\mathrm{LAI}(x)]$$

where *t* is the proportion of photosynthetically active radiation (PAR) incident at the top of the canopy that is

Table 15.1
Extinction coefficients and leaf area at which 95% of incident light is intercepted by the canopy

Species	K*	L95†	Study
Coniferous			
Pinus resinosa	0.40	7.5	Waggoner and Turner 1971
Pinus resinosa/strobus	0.45	6.7	Mukammal 1971
Pinus radiata	0.51	5.9	Whitehead, unpublished data
Pinus sylvestris	0.62	4.8	Witehead and Jarvis, unpublished data
Picea sitchensis	0.53	5.7	Norman and Jarvis 1974
Picea abies	0.28	10.7	Baumgartner, unpublished data
Pseudotsuga menziesii	0.48	6.2	Ungs 1981
Broad-leaved			
Eucalyptus maculata	0.57	5.3	Dunin and Reyenga, unpublished data
Liriodendron tulipfera	0.29	10.3	Hutchison and Matt 1976
Fagus crenata	0.65	4.6	Kira et al. 1969
Castanopsis cuspidata	0.50	6.0	Kira et al. 1969
Quercus robur	0.39	7.7	Rauner 1976
Populus tremula	0.39	7.7	Rauner 1976

Source: Adapted from Jarvis and Leverenz 1983.
*The extinction coefficient is for the maximum solar elevation possible at the side of the given study.
†The leaf area index where, for a given *K*, 95% of incident light is intercepted by the canopy.

transmitted to a given point *x* within the canopy, LAI(*x*) is the leaf area index *above* point *x*; and *K* is a constant called the **extinction coefficient.** This equation simply says that light intensity falls off exponentially as it passes through each layer of leaves in the canopy. The rate at which it falls is given by the extinction coefficient (*K*). This description is of course an oversimplified view of reality; for one thing, leaves seldom occur in distinct layers. Nevertheless, the equation provides a reasonably good picture of what actually occurs within a canopy.

Table 15.1 shows *K* values that have been measured for various temperate forests. The column labeled L95 is the LAI where for a given *K*, 95 percent of incident light would be intercepted by the canopy. Interception of total light and of PAR are sufficiently similar that data on one can be extrapolated to the other without a great deal of error (Jarvis and Leverenz 1983).

K values range from 0.28 for *Picea abies* to 0.65 for *Fagus crenata*. These, respectively, correspond to 95 percent light absorption at LAIs of 10.7 and 4.6. There are no consistent differences between *K* values of conifers and broad-leaved species shown in table 15.1. For all species together, *K* averages 0.47, corresponding to full light interception at an LAI of about 7. In view of the above it is apparent why temperate deciduous forests seldom attain LAIs greater than 6 to 8. What is not readily apparent is (1) how conifers and some tropical forests are able to maintain very high LAIs without low-canopy leaves falling below the light compensation point and becoming a serious cost to tree carbon; and (2) what function is served by maintaining more leaves than those required for optimal light interception. To get at what may be going on here, we first explore tree adaptations that facilitate light transmission and capture, then take up the question of leaf functions other than photosynthesis.

15.1.2 Adaptation for Light Transmission and Capture

It should be evident from the above discussion that leaves within a canopy face a very heterogeneous light environment. Leaves in full sunlight at the top of the canopy have more light available than they can use, while if the LAI is sufficiently high, leaves toward the bottom are potentially light limited. In fact, however, canopy structure facilitates light transmission downward through the canopy, and leaves adapt to capture PAR efficiently even when light is low (Shinozaki and Kira 1977; Jarvis and Leverenz 1983; Kira and Kumura 1985; Stenberg 1998).

The leaves of a stand of trees are not distributed evenly in discrete layers but, rather, are organized into a spatially complex hierarchical grouping consisting of the crowns of individual trees, branch whorls within crowns, branches within whorls, and shoots within branches. The resulting spatial clustering of foliage facilitates light transmission downward through the canopy.

The orientation of leaves in relation to incident light varies with leaf position in the crown. Jarvis and Leverenz (1983) cite several studies showing that "leaf inclination varies from predominantly vertical in the upper canopy to predominantly horizontal in the lower canopy." In other words, leaves at the top allow a considerable amount of light to pass, and leaves at the bottom intercept as much as possible. This phenomenon is demonstrated in fig. 15.1 for a stand of sweet chestnut in Scotland; note that most leaves at the top of the canopy

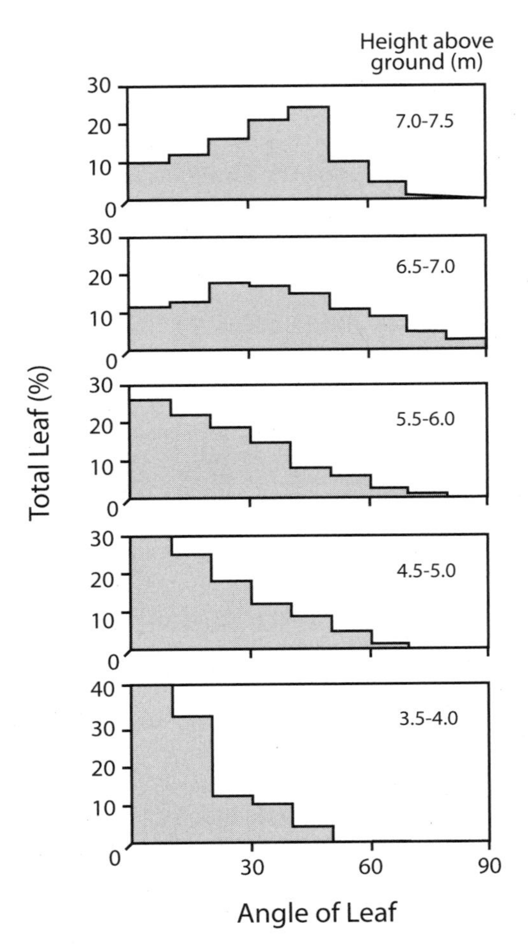

Figure 15.1. The distribution of leaf angles (measured from the horizontal) at different heights in a sweet chestnut canopy. (Ford and Newbould 1971)

are oriented at 30° to 45° from the horizontal, and there is a progressive flattening of leaf angle with depth in the canopy until the bottom leaves are finally oriented at 0° to 10° from horizontal.[1]

Leaves low in the canopy are characterized by a set of physiological and structural alterations that are subsumed under the heading **shade leaf.** Shade leaves differ from **sun leaves** in numerous ways that allow them to stay above the compensation point at low light levels. Adaptations include more chlorophyll per unit surface area, and lower respiration than sun leaves (Larcher 2003). Research shows that pea plants adjust the proportion of different light-gathering pigments within their foliage to match changing light conditions (Chow et al. 1990).

Shade leaves often have greater surface area per unit weight (or **specific leaf area**) than sun leaves, which increases their light gathering surface per unit of respiratory biomass. Figure 15.2 shows the change in specific leaf area and dark respiration with height in a *Castanopsis cuspidata* canopy. A similar pattern exists in other tree species.

[1]Calculating the leaf orientation that produces optimal photosynthesis is quite complicated. Leaves intercept the maximum amount of direct light when their surface is oriented at right angles to the incoming beam, but as we saw above, maximum light interception is not necessarily required to produce optimal photosynthesis. Moreover, the position of the sun in the sky varies both seasonally and with latitude; therefore, what is an optimal orientation in one place and time will not be optimal in another (Monsi et al. 1973).

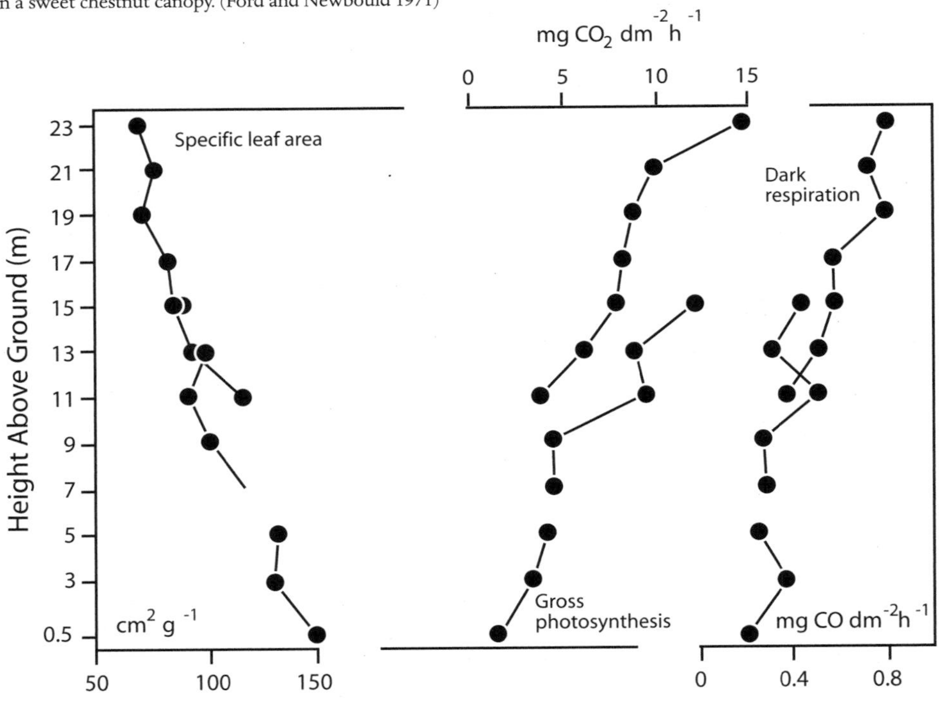

Figure 15.2. The change in specific leaf area, gross respiration, and dark respiration with height in a *Castanopsis cuspidata* canopy. Specific leaf area (SLA) is the surface area of a leaf divided by its weight; thin, flat leaves have high SLA, and thick or cyclindric leaves relatively low SLA. (Adapted from Kira 1975)

The eddy covariance technique provides noninvasive, ecosystem-scale land-atmosphere flux measurements. The concept is based on fluid mechanics principles applied to the vertical transport of mass and energy in the turbulent air overlying the land surface. For illustration purposes we focus here on the exchange of carbon dioxide between the land and the atmosphere. From a dimensional analysis point of view it is clear that the product of a gas concentration (c) with dimensions of mass per volume, and the vertical velocity of the transporting air (w) with dimensions of length per time, will yield a term (f_c) with units of mass per area per time, which is called a flux density (i.e., the amount of mass passing a unit area per time unit). Now, if the landscape is horizontally uniform, we make use of a powerful assumption that the flux density measured at the location of an instrumented tower is representative of the fluxes over some larger "footprint" area in the surrounding ecosystem (Schmid 1994). Because of the random nature of turbulent motions, the focus is on the fluctuations of the concentrations from their time-averaged values, and the product of these fluctuations (i.e., the covariance) is computed over a suitable averaging period: $f_c = \overline{w'c'}$, where the prime symbol denotes a deviation from the average, and the overbar represents a time average. To capture the full range of the turbulent motions responsible for the transport, the fluctuations must be measured at a high frequency (e.g., 10 times per second). Sonic anemometers have long been available for measuring w, but now with the widespread availability of open-path infrared gas analyzers for measuring carbon dioxide concentrations (c), there is a proliferation of eddy covariance instrumented towers around the world for measuring the net flux of carbon dioxide between terrestrial ecosystems and the atmosphere (Baldocchi et al. 2001). In fact, a search of the ISI Web of Science database lists more than two thousand published articles involving the eddy covariance (or its older name eddy correlation) technique. An important qualification on the interpretation of these measurements is that they apply to the *net* flux between the land and the atmosphere. In other words, if the instruments are properly placed well above a forest, then the measurement is of the rate at which carbon dioxide is crossing a horizontal plane at the measurement height. It is necessary to add ancillary measurements if one wishes to decompose this net flux into its main constituents, such as photosynthesis (which is extracting carbon dioxide from the atmosphere) and respiration (which is releasing carbon dioxide to the atmosphere) (e.g., Trumbore 2006). Another concern of note is that since this technique depends on a vigorous turbulent flow for the transport, it can be prone to underestimation errors in the nighttime, when the turbulence is weak. These errors can accumulate when measuring the total long-term (e.g., annual) fluxes of carbon, since the photosynthetic uptake is restricted to daytime, while the respiration continues through the night as well. Therefore, estimates of annual total exchange rates may be biased due to an underestimation of the total respiration. The research community has been focused on this problem, however, and has developed ways to correct the estimates (Baldocchi 2003).

Contributed by John Albertson, Duke University

15.1.3 Gas Exchange in Canopies

Canopy structure influences air movements both within the canopy and between it and the surrounding atmosphere. The consequent effects on GPP are quite complex. When air exchanges slowly between actively photosynthesizing canopies and the atmosphere, carbon dioxide supply may not keep up with consumption, and trees consequently become carbon dioxide limited. This commonly happens in dense stands of agronomic crops, in which leaves are packed into a relatively narrow vertical space that restricts the exchange of air between canopies and the atmosphere (Kira and Kumura 1983). The diffusion of air may be similarly limited in densely stocked young stands of trees; however, as trees grow taller, air exchange with the atmosphere is facilitated by the increasingly loose packing of leaves in vertical space. In natural stands, canopy roughness resulting from variation in tree height and gaps increases wind turbulence, hence water and gas exchange between canopy and atmosphere (Brunig 1970). A developing set of methods for estimating gas exchange in canopies is based on measurements of mass (water and carbon dioxide) and energy exchange between forests and the atmosphere above their canopies. These methods, collectively referred to as **eddy covariance,** couple physical models based on fluid dynamics theory with biophysical and biochemical models of leaf gas exchange to estimate GPP and ecosystem respiration (Baldocchi 2003). The ultimate goal is to estimate the difference between these two carbon fluxes (which is equivalent to net ecosystem carbon exchange [NEE]) and its dependence on ecosystem properties and climate and soil conditions. Additional information obtained by using stable isotopes of carbon and oxygen might help separate ecosystem respiration into its two components: plant autotrophic respiration and microbial heterotrophic respiration.

15.1.4 The Effect of Crown Shape

Individuals of most conifers and some angiosperm species have narrow crowns because their apical dominance allows crowns to grow upward more than outward. In contrast, many angiosperm species do not have strong apical dominance; hence the form of individual trees might be characterized as spreading rather than reaching. The short branches produced under the influence of apical dominance reduce the length of water transport path to leaves in comparison to the path under weak apical dominance. Shorter transport path improves the supply of water to the leaves (Mencuccini 2003), allowing high or more active photosynthetic surface (leaves) relative to respiratory volume (branches). This should increase the difference between individual tree GPP and respiration (i.e., NPP).

Similar to differences among species, individual genotypes within both conifer and angiosperm species vary in crown form. In both Norway spruce and Scots pine, the difference between a narrow and spreading crown is determined by a single gene (Karki and Tigerstedt 1985). Studies to date indicate that narrow-crowned genotypes are more productive on an area basis than trees with broad crowns (Brunig 1983; Cannell et al. 1983; Ford 1985; Karki and Tigerstedt 1985; Rook et al. 1985).

15.1.5 Nonphotosynthetic Functions of Leaves

Deciduous trees utilize woody tissues for nutrient and energy storage. In contrast, evergreens store nutrients and energy (starch) during nongrowing seasons in their leaves. (Nutrient storage by foliage is further discussed in chapter 18.) Hence, even where there is little or no gain in productivity at LAI values above 6 to 10, the "extra" leaves in evergreen forests may nonetheless fulfill an important function (Oren and Schulze 1989). Regardless of leaf longevity, the upper range of LAI is occupied by forests dominated by shade-tolerant species. Maintaining high LAI in these forests checks back the regeneration of more-light-demanding species, thus reducing competition for resources with established individuals and facilitating the recruitment of new shade-tolerant individuals into the population (Cattelino et al. 1979).

15.2 RESPIRATION BY TREES AND ECOSYSTEMS

In most forests, respiration of trees results in a carbon dioxide flux that is second only to photosynthesis (GPP). As a result, NPP (the balance between photosynthesis and autotrophic respiration) is typically positive; however, because of its magnitude, small relative changes in ecosystem *R* (which combines both autotrophic and heterotrophic respiration) may translate to large relative changes in **net ecosystem productivity** (NEP = GPP – ecosystem *R*). For example, certain disturbances (e.g., ice storms, hurricanes, clear-cutting, intensive partial cutting) can drastically reduce LAI by breaking tree tops, or altogether eliminate the canopy, and increase the biomass in detrital pools. The ensuing enhanced decomposition sometimes increases ecosystem *R* above the diminished GPP, resulting in negative NEP. Similar to GPP, tree and ecosystem respiration are affected by climatic variations and nutrient supplies that affect the rate of biochemical reactions in plants and microorganisms. In the following section, we discuss respiration by different components of the ecosystem, beginning with leaves and closing with heterotrophs.

15.2.1 Respiration in Photosynthetic and Nonphotosynthetic Organs

There are two kinds of respiration: light and dark. **Light respiration** occurs in leaves and is closely coupled to the photosynthetic process. C_3 plants lose large amounts of fixed carbon through light respiration, but C_4 plants do not. **Dark respiration** occurs in all living tissues and reflects the energy costs of synthesis and metabolism, the latter closely tied to forest structure (i.e., the biomass of stems, branches, roots, and leaves). Thus, plant respiration is often considered as two separate processes, that which is involved in growth activities, commonly termed **growth respiration,** and that which is involved in the activities of existing cells, commonly termed **maintenance respiration.** Growth respiration is a fairly constant proportion of NPP (i.e., about 1.25 g carbon is required to produce a tissue containing 1 g carbon) (Ryan 1991). Like any biochemical process, dark respiration increases with temperature and, hence, is coupled rather directly to the environment. Forests lose roughly between 50 and 80 percent of GPP as respiration (e.g., Kira and Kimura 1983), with higher losses occurring in warm climates and older stands within any given climate, older stands having a greater proportion of structural tissue relative to leaf area, and less-efficient foliage due to hydraulic limitation of water transport.

Whereas light respiration is a genetically determined factor that is not readily manipulated, the close coupling of dark respiration to stand structure means that it is influenced by common silvicultural practices. For example, trees respond to thinning by maintaining relatively large crowns with a high proportion of branches relative to leaves. Growth response to thinning therefore results from the trade-off between the increased resource availability to individual leaves and increased respiration due to more branches (and probably roots).

Both respiration at a given temperature and the sensitivity of respiration to changing temperature are different in different tissues, and in the same tissue in different species. Some tissues contain more living cells (e.g., leaves and fine roots vs. wood), as reflected in higher nitrogen and phosphorus concentrations, and respire at a higher rate. Leaves of some species contain more nutrients than leaves of other species, accompanied with higher photosynthetic and respiration rates. And fertilization can increase the nutrient concentrations in tissues, with a similar outcome.

15.2.2 Respiration in the Rooting Zone

Most of the microbial activity in forests takes place in the relatively moist forest floor and soil, and mostly in locations

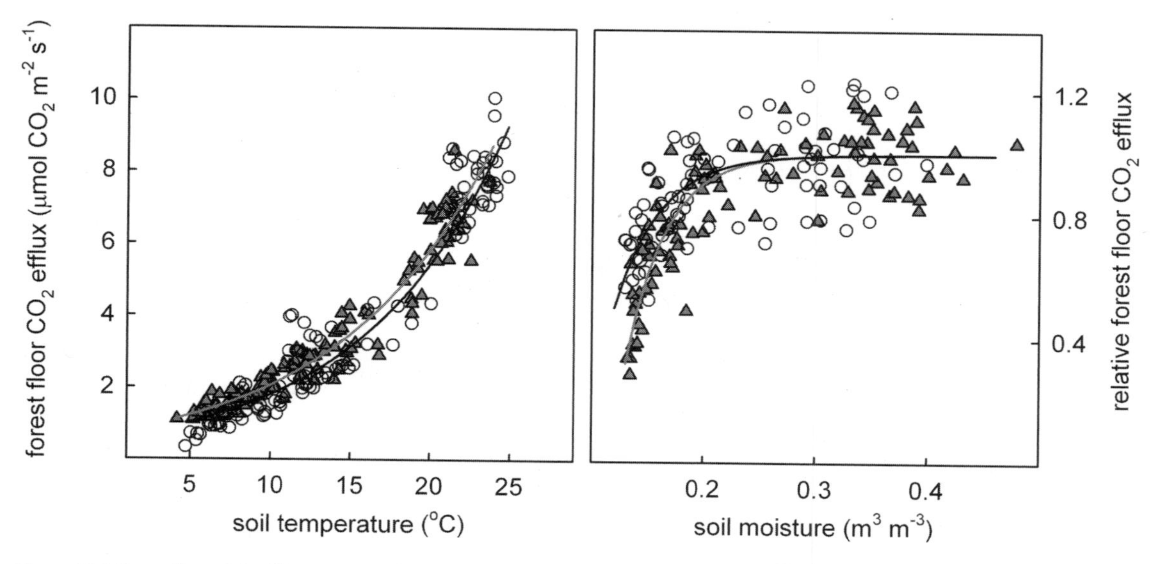

Figure 15.3. Forest floor CO_2 efflux, a proxy of respiration by plant roots and microorganisms, in a deciduous oak-hickory (circles) and a loblolly pine (triangles) forest in the southeastern United States in relation to soil temperature and moisture. (Data from Palmroth et al. 2005)

occupied by fine roots. This intimate colocation makes it difficult to separate carbon dioxide respired by fine roots from the carbon dioxide that is the product of heterotrophic respiration. As we show later, much of the carbon allocated to belowground is respired and emitted as carbon dioxide from the forest floor back to the atmosphere. Recently produced photosynthates allocated belowground are either respired immediately by roots, their associated mycorrhizae, or other microorganisms utilizing the simple carbon compounds exuded from fine roots, or later as microorganisms decompose dead leaf and root tissues.

Both root and heterotrophic respiration depend on temperature and soil moisture, increasing nearly exponentially with temperature, but increasing to an asymptote with soil moisture (fig. 15.3). The function most commonly used to describe the temperature response of respiration is the "Q_{10}" function, which essentially assumes that the rate of respiration increases by a constant proportion with each 10°C increase in temperature:

$$R = R_{10} \star Q_{10}{}^{\frac{T-10}{10}}$$

where in this specific formulation of the Q_{10} function, R_{10} is used as the baseline respiration at soil temperature of 10°C.

Like canopy photosynthesis, it is not possible to measure the respiration of entire trees or ecosystems, although ecosystem respiration at night can be estimated under windy conditions with eddy covariance. Respiration is thus measured with gas-exchange chambers set to enclose a few leaves or small sections of branches and stems, or to cover a small surface of the forest floor. In addition to respiration, the chamber temperature and, in soil chambers, the moisture in the soil are measured and used to parameterize relationships between respiration and temperature (e.g., Q_{10}) and soil moisture (e.g., fig. 15.3). These relationships are employed to estimate respiration per unit of soil area or stem surface area (or biomass). The estimates are then scaled to the ecosystem based on additional estimates of the soil or stem surface area in the forest. Such studies show that total forest floor–soil respiration (combining plants and microorganisms) is about 10 percent higher in temperate deciduous broadleaf than in adjacent coniferous forests (Palmroth et al. 2005). A possible explanation is that broadleaf forests transpire less than conifers, thus keeping the soil wetter, and their leaves decompose faster, producing less insulation on the soil surface and permitting higher soil temperature; both these factors impel higher respiration during the growing season.

15.3. NET PRIMARY PRODUCTIVITY

In the following sections we will first take at close look at the process of light gathering in canopies by linking LAI with NPP. We then discuss how NPP differs among different forest types, across local landscapes, and between mixed stands and monocultures.

15.3.1 The Interplay between Leaf Area and NPP

The NPP for a given area of ground is a function of two attributes of the plant community: (1) its leaf area, and (2) the efficiency with which leaves are used ($E(l)$ or $E(p)$). In other words,

$$\text{NPP} = (\text{leaf area})E(l)$$

Because LAI is equal to (leaf area)/(ground area), this relation can also be written

$$\text{NPP} = \text{LAI} \star E(l)c$$

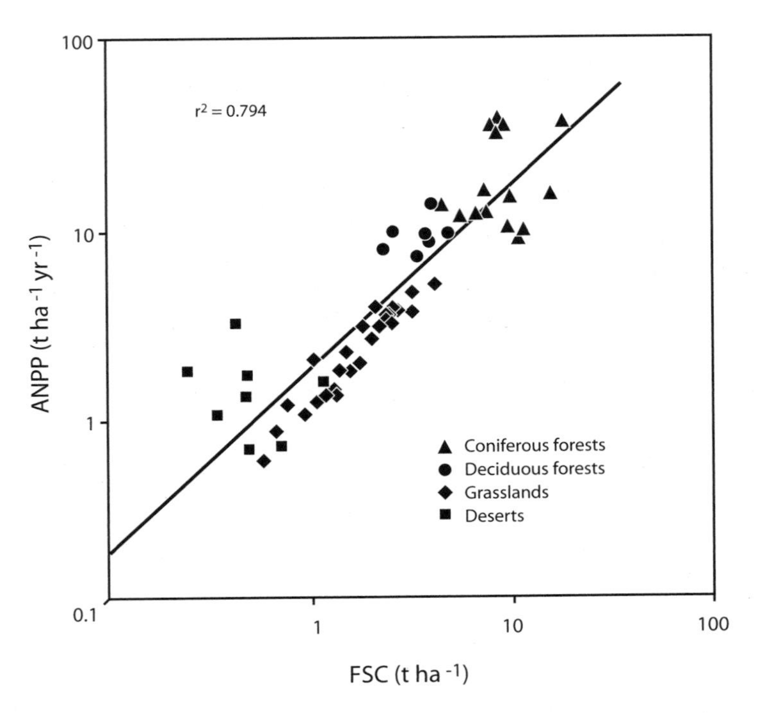

Figure 15.4. The relationship between foliar standing crop (FSC) and aboveground net primary productivity (ANPP) for North American vegetation types. (Webb et al. 1983)

Efficiency is a general term denoting the ratio between the output of some product and the input required to produce it. The output we are interested in here is NPP; however, there are various ways to calculate the efficiency of NPP depending on what input we choose. Three basic types of efficiency are most often used in ecology: (1) that based on the input of some resource, most commonly light, water, or nitrogen; (2) that based on land area, the idea being that the land surface a plant community occupies integrates all the various inputs; and (3) that based on the plant itself, usually either leaf area or leaf biomass, but sometimes total biomass.

Photosynthetic efficiency (sometimes referred to as **solar energy efficiency**) is the ratio of the energy content of NPP to the incident light energy (Kira and Kumura 1983). (This should not be confused with the efficiency of the photosynthetic process itself, which deals strictly with the immediate conversion of solar energy to chemical energy and does not include dark respiration.) Various measures of light energy are used in calculating photosynthetic efficiency, including total radiation, visible light, or only that portion of the spectrum that is used in photosynthesis (i.e., PAR, or wavelengths between 380 and 710 nm). PAR and visible light occupy essentially the same region of the spectrum. PAR composes a little less than 50 percent of the total solar energy reaching the earth's surface, so efficiency values calculated using PAR are about twice those calculated using total radiation.

The photosynthetic efficiency of most forests based on *aboveground* NPP and PAR is between 1 and 2 percent, with values up to 3 percent for some rapidly growing, young plantations (Kira and Kumura 1983). Photosynthetic efficiencies of other wild plants are within the same range, but those for some intensively cultivated crops are higher. Taking into account the best current estimates of belowground productivity, somewhere between 4 and 7 percent of the PAR incident on forest canopies ends up as NPP. This may seem very low; however, recall from your plant physiology course that the biochemistry of photosynthesis is such that only 20 to 30 percent of PAR is converted to carbohydrate, and some portion of this is quickly lost in respiration. In C_3 plants, about 20 percent of the fixed energy is immediately lost in light respiration (Larcher 2003) Given these physiological limits on efficiency, the ecologist should not expect too much.

Efficiencies based on land area or plant variables (designated by the letter *E*) include

$E(a)$ = NPP/(land area)
$E(b)$ = NPP/(total plant biomass)
$E(l)$ = NPP/(leaf area)
$E(p)$ = NPP/(leaf biomass)

Because they support a large leaf-surface per unit land area (i.e., LAI) (chapter 7), forests have high $E(a)$ relative to other vegetation types. On the other hand, the $E(b)$ of forests is relatively low, reflecting the large amount of nonphotosynthesizing structural tissues required to support the leaves. In other words, forests pay for their leaves with carbon—the currency of ecosystems.

where c is a constant that depends on the unit of ground area being used. Forests have a higher NPP per unit of ground area than other vegetation types because they support more leaves than do grasslands, deserts, and aquatic communities. Figure 15.4 illustrates the close coupling between NPP and photosynthetic surface for some North American vegetation types. Across the deserts, grasslands, and forests of the United States, a doubling of leaf biomass multiplies NPP by 1.9. The overriding importance of leaf area in determining NPP was further confirmed by Runyon et al. (1994), who measured *aboveground* NPP across an Oregon transect from arid juniper savanna to temperate rain forest. Despite total aboveground NPPs ranging from less than 2 to 17 Mg/ha/yr, the efficiency with which captured light was converted to aboveground biomass remained relatively constant at 1 g biomass per megajoule (MJ) of captured light. As we shall see later in this section, however, the tight relationship between leaf area and NPP breaks down after leaf biomass exceeds a certain level.

Leaf area determines the capacity of a canopy to absorb PAR, while the efficiency with which leaves are used, $E(l)$, reflects the conversion of PAR to biomass. When LAI is relatively low, $E(l)$ is independent of LAI, and NPP of a plant community is tied closely to the leaves it can display. This is the pattern we see in fig. 15.5. Once LAI attains a certain level, however, leaves begin to compete among themselves for scarce resources; in particular, leaves low in the canopy become light limited, thereby reducing overall canopy $E(l)$. As LAI increases beyond this point, $E(l)$ declines roughly in proportion to 1/LAI (fig. 15.5).

As a result of the inverse relationship between leaf area and leaf-use efficiency, NPP reaches a point of diminishing returns as more leaves are packed onto a site (i.e., after some point, adding more leaves gains little or nothing in productivity; to see this, substitute 1/LAI for $E(l)$ in the above equation). Figure 15.5 shows the relationship between aboveground NPP and LAI for major forest types.

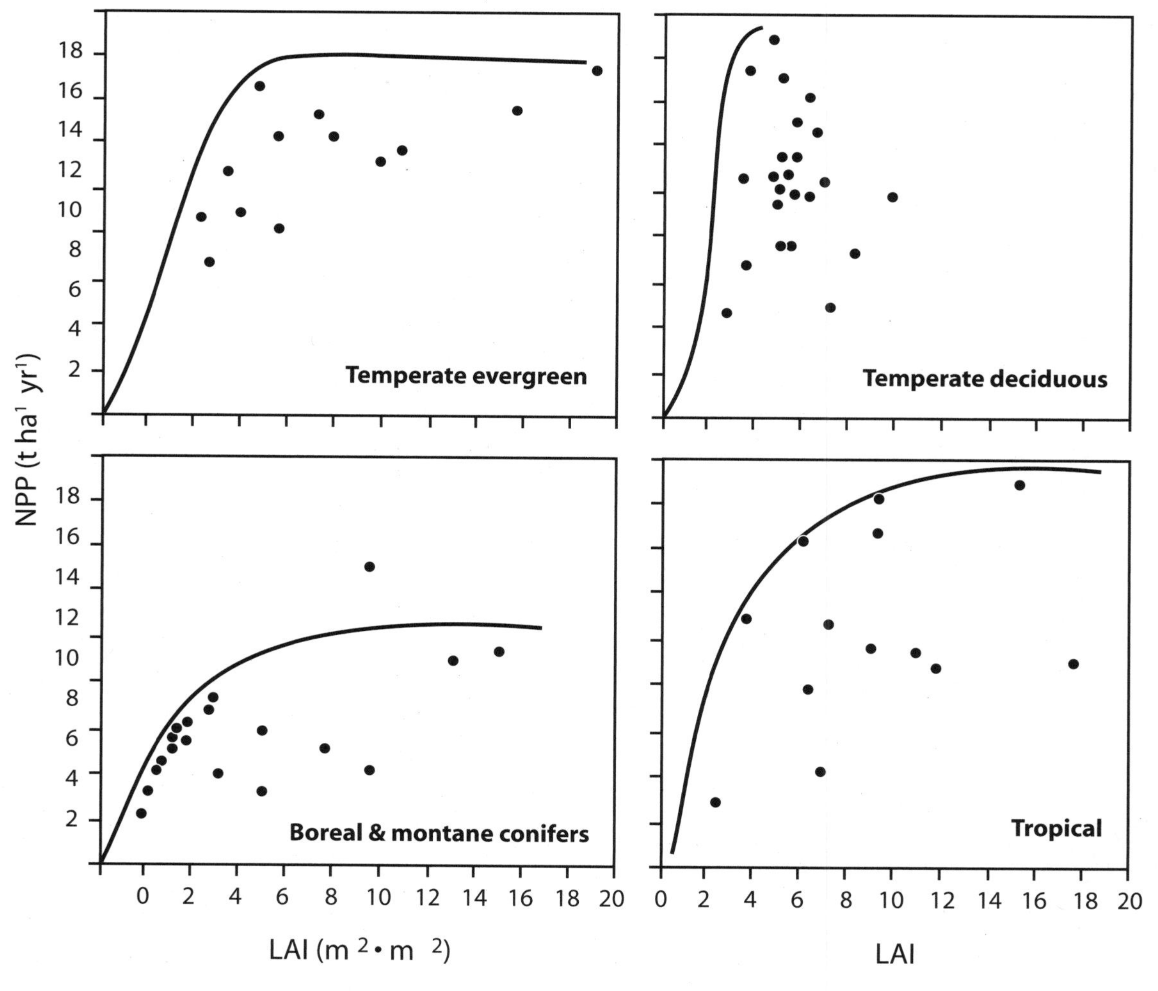

Figure 15.5. The relationship between aboveground net primary productivity (NPP) and leaf area index (LAI) for the major types of forest. (Adapted from Reichle 1981)

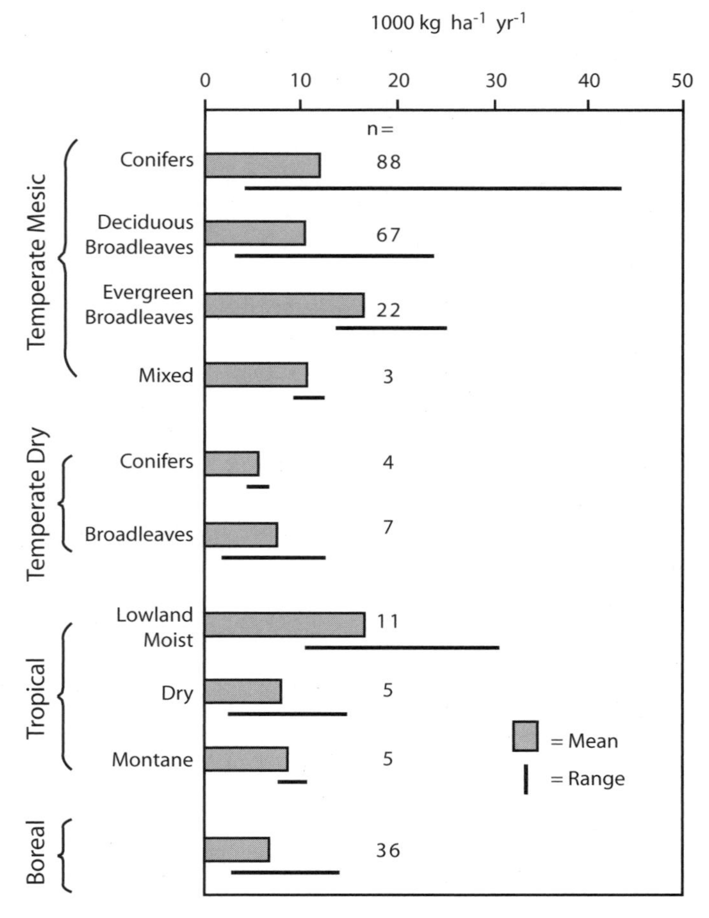

Figure 15.6. The aboveground net primary productivity (dry weight) for different forest types. Both natural and managed stands are included; however, data does not include fertilized stands, clonal plantations, or species planted outside their natural range. n, number of stands. (Data from Cannell 1982)

The data come from natural stands and plantations occupying a variety of sites throughout the world; therefore, there is a great deal of variability within a given type.

In all major forest types the sharpest gains in NPP occur as LAI increases up to about 6. After this point, the light-gathering capacity of canopies begins to saturate, and the rate of increase in aboveground NPP generally slows down or levels out (however, this is not always the case). Although evergreen forests (including moist tropical) attain leaf areas considerably higher than 6 (up to 18 or so), few of the temperate deciduous forests that have been studied exceed an LAI of 7, and none exceed 10. This reflects in part at least basic physiological and ecological differences between evergreens and deciduous species.[2]

[2]In forestry, the plateau in productivity over a wide range of leaf areas (or stocking densities) is known as the Laengsaeter relationship (named for a Danish forest biologist). The Laengsaeter relationship plays a very important role in silviculture, providing the forester with guidelines for thinning stands in such a way that individual tree growth is maximized without sacrificing overall stand productivity.

15.3.2 Aboveground NPP of Major Forest Types

Figure 15.6 shows average *aboveground* NPP (tons/ha/yr) of major forest types, along with the ranges of values that have been reported. Summarized from those compiled by Cannell (1982), these data include both natural and managed stands of various ages, but they do not include fertilized stands, clonal plantations, or species planted outside their natural range. The large variability within a given type reflects variation in factors such as soils, elevation, stand age, and the thoroughness with which investigators documented the various NPP components.

As we discussed in the previous section, differences among forest types largely reflect two factors: (1) the amount of energy-capturing surface (foliage), in turn, a function of water and nutrient availability, and (2) the length of the growing season. The highest average NPP has been reported for moist lowland tropical forests and evergreen broadleaf species occupying mesic (i.e., relatively warm and moist) temperate environments in Japan.

Some points need clarification regarding the patterns shown in fig. 15.5. While the range of LAIs shown in the figure encompass those reported for both natural and managed stands, the NPP values that are shown do not. As fig. 15.6 shows, higher aboveground NPPs have been reported for all of the forest types; in other words, some species, in some situations, continue to increase aboveground NPP at LAIs well above that at which most stands reach a plateau. These stands are either extraordinarily efficient in their use of leaves, allocate a relatively high proportion of total NPP to aboveground tissues, or both. In general, the curves shown in the figure are idealized and may or may not reflect what happens in any given situation.

As we shall see in section 15.5, maximum attainable leaf area is determined by the availability of water and nutrients, and on droughty sites or infertile soils, stands may never attain an LAI high enough to saturate the light-gathering capacity of their canopies. Even where high LAIs are attainable, the pattern of NPP increase with LAI may not follow the curves shown. This is the case with Douglas-fir plantations in the Oregon Cascades (Velasquez-Martinez et al. 1992): at relatively low LAIs, their aboveground NPP is lower than predicted by the curve in fig. 15.5, but at high LAIs their aboveground NPP falls on the curve. As a result, the relationship between LAI and aboveground NPP in these stands is linear rather than exhibiting a plateau. Why this pattern? One possible explanation is that stands with low LAIs have abundant understory plants whose productivity is not reflected when only tree growth is measured (i.e., aboveground NPP is fully accounted for in dense stands without understory, but not in open stands with understories).

Conifer and deciduous forests of the temperate zone (excluding dry sites) have similar NPPs; however, much higher maximums have been reported for the former. A 17-year-old Japanese cedar plantation *(Cryptomeria japonica)* produced a net of 43.7 tons/ha/yr (Kira 1975), and a 22-year-old western hemlock *(Tsuga heterophylla)* stand produced 32.2 tons/ha/y aboveground (Fujimori 1971). The lowest NPPs occur in boreal, montane, and dry forests, where the yearly growth period is limited by either low temperature or drought.

15.3.3 Net Primary Productivity at the Scale of Landscapes

The NPP within any given major forest type commonly varies across landscapes, which may be viewed as mosaics composed of differing levels of available water and nutrients, gradients in atmospheric carbon dioxide, and varying temperatures. This is most obvious in mountains, where available resources may vary over distances of a few meters. In the northern temperate zone, for example, relatively high solar radiation on south aspects results in a greater demand for water than on adjacent northern slopes; if water is a limiting resource, differences in NPP will result. This appears to be the case in the central Oregon Cascades, where (other factors being equal) the rate of leaf area expansion in plantations decreases from northern to southern aspects, shallow to steep slopes, and relatively rock-free to rocky soils (Velasquez-Martinez 1990). In many boreal forests, however, low temperatures are more likely than low water availability to limit growth, because low temperature leads to the development of permafrost layers in soil and limits decomposition and nutrient cycling (Van Cleve et al. 1981). Hence, in those systems, the greater solar radiation on southern aspects is a potential stimulus to NPP rather than a detriment to NPP.

Other factors being equal (*ceteris paribus*, in Latin) is an important qualifier to remember in all instances in which the influence of one or a few environmental factors over ecosystem processes is discussed. Processes are influenced by many factors, and the effect of any one (e.g., aspect) may vary depending on the others. In the young stands studied by Velasquez-Martinez, for example, the rate of growth in leaf area depended not only on aspect, but on foliar nutrients, elevation, and stocking density. Hence, the degree to which any two stands differed in their growth rates depended on the combined effect of all these factors and not any one. You might think of an ecological process (e.g., NPP) as a symphony, and the factors controlling that process as individual instruments. Following one of the instruments (e.g., base line) helps you appreciate what goes into making the symphony what it is, but does not tell you much about the symphony as a whole.

Changes in elevation affect atmospheric carbon dioxide concentrations, temperature, and precipitation, all factors that directly influence NPP (section 15.5). Temperature and moisture also indirectly influence NPP through their controls over the rate at which organic matter decomposes, which is a key factor in nutrient availability. As we saw in the last chapter, the cooler average temperatures that accompany increases in elevation slow decomposition, which, in turn, slows the nutrient cycle. Few studies have explored how changes in nutrient cycling with elevation influence NPP; however, Vitousek et al. (1988) showed that foliar nutrient concentrations drop with elevation in Hawaiian forests, a sign that NPP might be affected. On the other hand, trees growing at high elevations compensate at least somewhat for lower nutrient availability by diverting a relatively high proportion of their NPP to roots and mycorrhizal fungi (Vogt et al. 1989).

Soils can vary widely even in relatively flat terrain or on uniform slopes (chapter 14), and subtle variations in topography influence drainage and soil aeration. Forests adjacent to streams may receive fresh infusions of nutrients from periodic flooding. Historic disturbances such as landslides, fire, and windthrow alter soil properties and may either increase or decrease productivity, depending on specific effects and resource limitations at a given site. For example, areas that for some reason have experienced very hot or frequent fires may have lower soil-nutrient stores, hence lower NPP, than nearby areas with a milder burn pattern. On the other hand, disturbances such as windthrow that expose fresh rock can enhance NPP by accelerating nutrient release through weathering. In all forest types, both NPP and species composition vary widely, depending on landform and proximity to streams (Brunig 1983).

> A study by Espinosa-Bancalari and Perry (1987) illustrates the wide variation in NPP of a single species that can be encountered within small areas that appear relatively uniform. This work was conducted in a single 22-year-old Douglas-fir plantation occupying gentle terrain with no sharp breaks in aspect or slope steepness. Three quite distinct areas of NPP could be distinguished within this plantation, the most productive having 66 percent more LAI and nearly twice the total biomass as the least productive area. Why? The most likely explanation relates to subtle differences in soils. Because of their position on the landscape, soils beneath the slow-growing portion of the stand do not drain water readily, therefore they become waterlogged during the rainy season (i.e., winter and spring). This restricts rooting depth, leading paradoxically to drought stress during the late summer. Soils beneath the rest of the stand are well drained, silty clay loams overlying highly weathered sandstone. The difference between the fast- and intermediate-growing portions is associated with depth to the weathered sandstone, which is closer to the surface in the fast-growing portion. We cannot be sure, but rapid early growth of the fast portion of the stand may have stemmed from the ability of trees to access water stored in the sandy subsoil layer.

15.3.4 Net Primary Productivity in Mixed-Species Forests versus Monocultures

The NPP on any given piece of ground can vary widely depending on the plant community that occupies it, which may result either from differences in the ability of individual species to gather and to convert resources (section 15.3), or from community-level effects (i.e., direct and indirect interactions among species that affect NPP). The topic of this section—NPP of mixed species stands versus monocultures—is an aspect of community-level effects with important implications for the design of silvicultural systems.

Mixed-species silviculture has a long history in Europe, but unfortunately, little work has been done elsewhere. Lavrinenko (1972) summarized the extensive experience in the former Soviet Union, while Assmann (1970) and Kerr et al. (1992) have discussed the use of species mixtures in Germany and Great Britain, respectively. In short, European foresters have found that NPP can be greater in mixture than in monoculture, although this is not always the case. Much depends on the environmental context, and on the particular mix of species: some plants have good combining ability, and others do not. Time frame is also an important factor, especially when the primary advantage of the mixture is greater stability.

Productivity will be greater in mixture than in monoculture if one or both of the following conditions hold. First, more resources are available to the mixture, which may result from different species utilizing the environment more fully than a single species, or from one or more species in the mixture enhancing the availability of some limiting resource (e.g., nitrogen-fixing plants). Second, the mixture is more stable than a monoculture against climatic fluctuations, pests and pathogens, or other stresses.

15.3.4.1 Resource Availability in Mixture versus Monoculture

Perhaps the clearest example of enhanced resource availability in mixture is where one or more of the plants in the system is a nitrogen fixer. Given the importance of nitrogen to productivity (section 15.5), and its limited sources in many forests, it is not too surprising that nitrogen-fixing plants enhance the growth of associated trees and (in some cases) total ecosystem NPP (Chatarpaul and Carlisle 1983, Dawson 1983; Binkley 1986a, 1992). On nitrogen-deficient soils in the Pacific Northwest, mixtures of Douglas-fir and nitrogen-fixing alder species produce up to twice the biomass of Douglas-fir monocultures (Binkley and Greene 1983). Predictably, this is not the case on soils rich in nitrogen.

Mixtures might be more productive than monocultures if different adaptations among species within the mixture results in more complete use of resources than would be possible by one species. Consider light-gathering by canopies. Mixing shade-tolerant with shade-intolerant species frequently enhances productivity over that of the shade-intolerant species alone, although the mixture is not always more productive than the shade-tolerant species alone. In Germany, mixtures of shade-tolerant beech with various light-demanding species (e.g., oak, pine, larch) produce from 4 to 52 percent more dry matter than the oak or pine alone (Assmann 1970). Different species in a mixture may also root in different parts of the soil, thereby giving the mixed stand greater access to water and nutrients than would be possible with a single species. This is the case in at least some mixtures of beech and pine in Germany (Assmann 1970), and also with mixtures of Norway spruce and Scots pine in England (Brown 1992). Trees and shrubs may exhibit hydraulic lift, in which water taken up by deep roots is released by shallow roots. Dawson (1993) showed that shallow-rooted plants growing near sugar maples used a significant proportion of the water pumped to the surface by the maples.

One of the more striking (and thoroughly studied) examples of how rooting pattern can affect growth of tree mixtures comes from Scotland and Ireland. There, Sitka spruce grow much better on heavily organic soils (e.g., peats) when mixed with lodgepole pine or Japanese larch than when in monoculture (Kerr et al. 1992; Morgan et al. 1992). Reviewing the work on this phenomenon, Morgan et al. (1992) pieced the following story together. The peat soils tend to become waterlogged, which inhibits microbial activity, hence nutrient cycling; it also restricts the rooting depth of spruce because of low oxygen levels in waterlogged soils. The roots of pine and larch are less sensitive to anaerobic conditions than those of spruce and send roots into deeper soil layers, where they reduce waterlogging by pumping the water out to the atmosphere via their crowns. The greater aeration that results from less water stimulates microbial activity and therefore nitrogen cycling, which improves the nutrition of all the trees in the system. (There is also some evidence that pine mycorrhizae directly decompose organic matter.) Note that though species mixtures are clearly more productive than pure spruce on peat soils, it does not necessarily follow that the mixtures are more productive than pure pine or larch.

In some cases, nutrient cycling is accelerated by one or more of the species in a mixture. This has most commonly been noted when certain hardwood species are mixed with conifers; pines, spruces, and some oaks tend to produce litter that decomposes slowly and may even contain chemicals that inhibit the activity of saprophytes (Brown 1992; chapter 18 in this volume). According to Lavrinenko (1972), the growth of pine is significantly improved when mixed with certain deciduous species, probably because of accelerated nutrient cycling. In the former Soviet Union, mixing birch with pine and oak nearly triples the numbers of saprophytic bacteria in soils: similar patterns have been found in at least some mixtures of beech and spruce, the presence of beech stimulating the nutrient cycle and thereby improving growth of the spruce (Assman 1970). Soil collected in the vicinity of some common hardwood species of Oregon and California (not symbiotic nitrogen fixers) enhances growth of Douglas-fir seedlings, probably by stimulating the activity of beneficial soil microbes (Amaranthus and Perry 1989a; Amaranthus et al. 1990a; Borchers and Perry 1990).

Plant species can also have synergistic effects on soil organisms. Such is the case in mixtures of Scots pine and Norway spruce in England, which compared to monocultures of either species have significantly higher populations of soil beetles (Collembola) and of a particular type of worm (enchytreids) that are important decomposers (Brown 1992). As we discussed earlier, mixtures of these two trees are also more productive than either species alone. Finally, recall from chapter 11 that the NPP of tree species growing in mixture can be greatly affected by mycorrhizal fungi, some types of which can convert a negative interaction between trees into one that is either neutral or synergistic (Perry et al. 1989c, 1992). How the fungi do that is not totally clear, but they appear to mediate a more equitable distribution of nutrients between the tree species than occurs without the fungi.

Bear in mind that patterns discussed in this section likely depend on soil type and the species that are involved and should be extrapolated with caution. Some studies have produced contradictory results. In western Oregon, for example, maples have been found to both increase and decrease available soil nitrogen (Perry et al. 1987a; Fried et al. 1990). As we discussed in chapter 11, much research is needed on the nature of species interactions in forest ecosystems.

15.3.4.2 Stability of Mixtures versus Monocultures

The relationship between species diversity and community stability is a complex topic that we shall consider in some detail in chapters 19, 20, and 21. The point to be emphasized here is that the stabilizing effects of mixtures, if they occur at all, may be manifested only during certain critical periods (e.g., wildfire, insect outbreak, extremes of wind or drought). Because of this, studying the stabilizing effect of mixtures is considerably more difficult than is the case with nutrient enhancement and similar factors discussed in the previous section. When evaluating the benefits of mixtures,

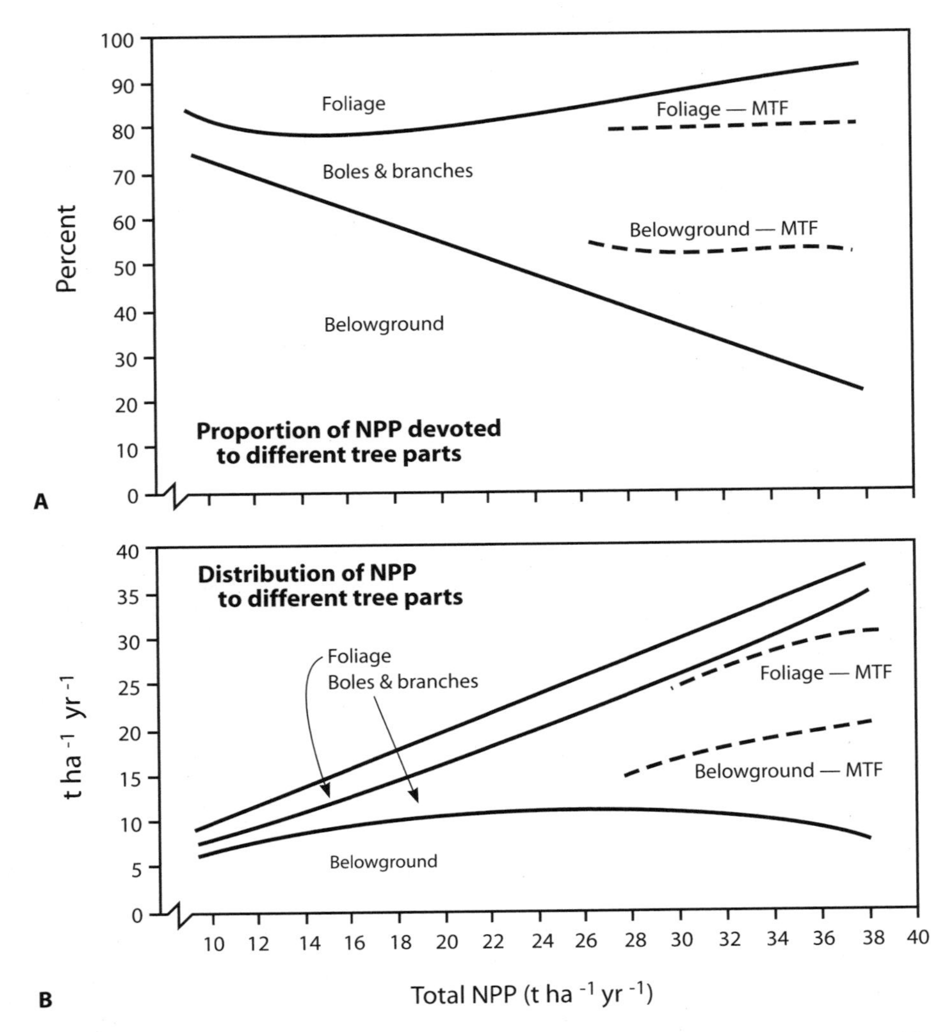

Figure 15.7. The estimated relationships between site productivity and the allocation of productivity to different tree parts. MTF, moist tropical forest; NPP, net primary productivity.

remember that what you see today is not necessarily what you will get in the future. We discussed a good example of this in chapter 7, in which an experiment in Oregon found that Douglas-fir trees grew better when mixed hardwoods were removed; however, after a wildfire swept through the experimental area, the only surviving Douglas-fir were those mixed with hardwoods. In this case, the short-term NPP of the conifer was somewhat better in the pure stand, but long-term NPP will be better in the mixture.

15.4 CARBON ALLOCATION IN DIFFERENT ENVIRONMENTS

So far we have concentrated on how environment influences total NPP, but environment also strongly influences how that NPP is distributed among the various tree parts. Trees allocate fixed carbon in varying proportions among foliage, reproductive structures (e.g., pollen, fruits), branches, boles, roots, and mutualists (e.g., mycorrhizal fungi, leaf fungi, rhizosphere bacteria, pollinators; see chapter 11 for a discussion of mutualisms). Carbon is also allocated between the so-called **primary** and **secondary chemicals,** the former comprising compounds such as sugars, starches, proteins, and others that are directly involved in physiological processes, and the latter including a diverse array of compounds such as phenolics, alkaloids, and terpenes whose function is more ecological than physiological (chapter 19).

15.4.1 Allocation to Foliage, Wood, and Belowground

Figure 15.7 shows a rough estimate of allocation to belowground, foliage, and wood (i.e., boles and branches) across a productivity gradient.[3] Figure 15.7A shows

[3]Values in fig. 15.7 are derived from three sources. The relation between aboveground NPP and allocation to wood is taken from Cannell (1985), and for moist tropical forests, from Jordan (1983). Belowground NPP for all types is estimated from the relation between aboveground fine litterfall and belowground C inputs given by Raich and Nadelhoffer (1989). To draw the curves, we assumed that all aboveground NPP not diverted to wood went to foliage, and that fine litterfall was 1.25 times the current year's foliar biomass.

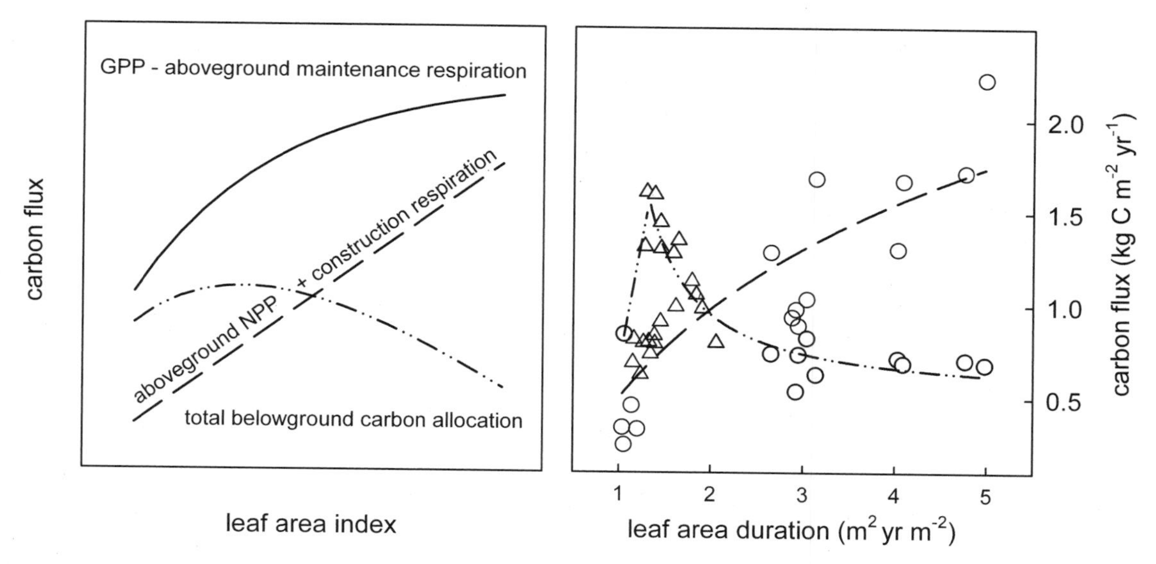

Figure 15.8. A schematic representation of the changes in net canopy photosynthesis (GPP; discounted by the maintenance respiration aboveground), aboveground net primary production (NPP; increased by 25% to include construction respiration), and the balance between these two quantities, the total carbon allocation to belowground, with leaf area index (left). Data from three stands of deciduous species and one of pine, based on leaf area duration in place of index (to account for differences in the length of growing seasons), show that carbon allocation above- and belowground roughly followed the schematic representation. It is possible that data would follow the schematic pattern more closely had there been sufficient data from any one site. (After Palmroth et al., 2006)

proportional allocation and 15.7B shows absolute amounts; note that the *x* axis is total NPP, not just aboveground. The patterns shown in fig. 15.7 reflect the fact that trees accord a higher priority for carbon to resource gathering tissues (i.e., foliage, fine roots, and mycorrhizae) than to boles. One consequence is that most change in NPP of forests across environmental gradients results from branches and boles (Cannell 1985). This is one reason why forests are such important sinks for atmospheric carbon dioxide: they stash it away in a form (i.e., wood) that is relatively stable and turns over slowly.

Excluding for the moment moist tropical forests, the proportion of carbon that is allocated belowground drops sharply as one moves from less to more productive environments. Although some evidence suggests that absolute amounts of carbon diverted belowground may vary relatively little among sites, or perhaps increase with total NPP (Nadelhoffer and Raich 1992), a more recent synthesis suggests that at least in stands of shade-intolerant species, total carbon allocation belowground decreases with LAI, and thus with NPP (Palmroth et al. 2006). Remember that stand-level photosynthesis (or GPP) increases rapidly as newly established stands form their canopies, but that the efficiency of additional foliage decreases once much of the PAR is absorbed (fig. 15.8). Stands of shade-intolerant species keep foliage from a few months to less than two years, making foliage production an important component of *aboveground* NPP. In such stands, aboveground NPP tends to increase more linearly with LAI. Total carbon allocation belowground is the difference between GPP and the sum of aboveground NPP and aboveground respiration, the latter also tending to increase with NPP. Thus, as LAI increases from zero, belowground carbon allocation should first increase as GPP increases, and then decrease as the increase in GPP slows. Palmroth et al. (2006) found the expected patterns in both aboveground NPP and total belowground carbon allocation in data compiled from four stands, three of broadleaf species and one dominated my loblolly pine (fig. 15.8). Because the stands came from environments that support growing seasons of very different length, they used leaf area duration (LAD) (section 15.5) instead of LAI in the synthesis.

Accurately measuring belowground NPP is fraught with difficulty, and the patterns in fig. 15.7 are rough estimates that may well change as more data becomes available. The patterns shown in fig. 15.8 replace belowground NPP with total belowground carbon allocation, which is the sum of belowground NPP and respiration, and can be estimated with a smaller relative error.

As with belowground NPP, foliage production varies relatively little over a wide range of sites. (For both roots and foliage, it is important to distinguish between production and standing crop; as we have seen in section 15.3, foliar standing crop varies widely, with the same probably true of roots.) The balance between foliage and belowground is quite sensitive to environmental factors. Moving

Plants use the carbon fixed in photosynthesis to build new tissue, but also for metabolism, transport, reproduction, and defense. The use and retention of this carbon is called **carbon allocation.** Knowledge of carbon allocation is important for fiber and food production (the useful parts generally get only a small fraction of photosynthesis), and for understanding and modeling the role of terrestrial ecosystems in the earth's carbon balance.

The first step in thinking about carbon allocation is to separate carbon use and carbon retention. Carbon use represents the total carbon flows from photosynthesis to a component (such as leaves or wood) or process (such as respiration). Carbon retention results from flows minus losses. Retention is the storage of carbon and is what we use and see. It is the easiest to measure but probably does not indicate carbon flows (Litton et al., 2007). In a mature forest, wood represents more than 90 percent of the biomass but receives only about 20 percent of the carbon fixed in photosynthesis every year (Litton et al., 2007). For wood, flows are low but so are losses, so retention is high. In contrast, leaves and fine roots turn over rapidly, so retention (and biomass) is low, but flow can exceed that to wood. Because use and retention differ among plants and parts, Litton et al. (2007) suggested separating different facets of allocation: (1) *biomass*, the amount of material; (2) *flux*, the annual flow of carbon to a component (wood, leave, roots) or use (production, respiration); and (3) *partitioning*, annual flux as a fraction of GPP—useful for modeling and for comparing ecosystems.

Progress in understanding carbon allocation has come from studying all the fluxes and components of the ecosystem (fig. 15.9) in concert (Litton et al., 2007; Palmroth et al., 2006). Using this approach, Litton et al. (2007) showed that in forests, flux was unrelated to biomass, so that easy-to-make measurements of biomass cannot serve as a proxy for flux as they can for annual plants. Litton et al. (2007) also showed that fluxes to all components rise as GPP increases, but that the rate of increase with GPP varies among components. Partitioning to respiration and foliage production in forests is fairly conservative—it varies little as GPP changes with species, climate and nutrition. Partitioning to belowground decreases and to wood production increases as GPP increases. These patterns seem to occur whether changes in GPP are driven by climate, species, nutrition, defoliation, or carbon dioxide level and are similar both within and across sites (Palmroth et al., 2006; Litton et al., 2007). If these patterns hold true, we will have a powerful new tool for predicting the response of ecosystem carbon cycles to changes in climate and carbon dioxide levels (Palmroth et al., 2006).

Contributed by Michael G. Ryan, USDA Forest Service, RMRS

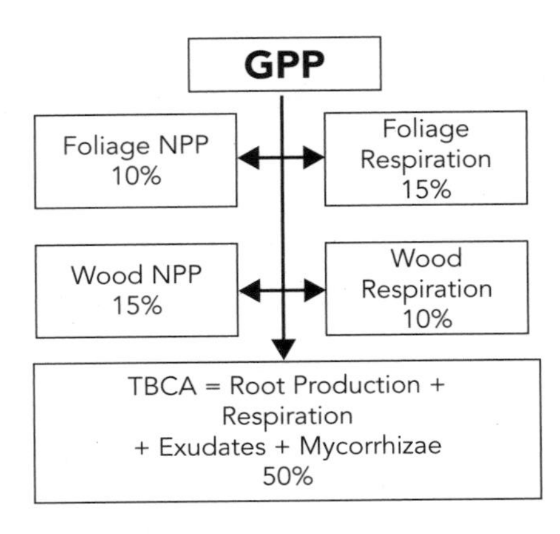

Figure 15.9. The primary components and processes receiving carbon fixed in photosynthesis and the percent of annual gross primary productivity (GPP).

from less to more productive sites in the temperate and boreal zones, the proportion of carbon allocated belowground drops more sharply than the proportion allocated to foliage, resulting in an increasing ratio of foliage to roots and mycorrhizae. This reflects the greater availability of water and nutrients on more productive sites (i.e., fewer roots and mycorrhizae are required to gather the resources needed to support a given amount of leaves).

Compared with temperate forests of similar productivity, moist tropical forests allocate a lower proportion of carbon to wood and a higher proportion to foliage and belowground. Moist tropical forests growing on highly weathered soils (i.e., Oxisols, Ultisols) (chapter 14) are productive despite growing on extremely infertile soils, because trees invest heavily in roots and mycorrhizae, thereby ensuring efficient nutrient capture and cycling. In Amazonia, for example, mineral soils are overlain by a mat of roots and mycorrhizal hyphae that efficiently absorbs the nutrients contained in litterfall and precipitation (Stark and Jordan 1978). Root mats are less common in flood plain forests that receive periodic infusion of fresh nutrients in flood water.

15.4.2 Allocation to Secondary Chemicals

Forests growing on infertile sites or at high elevation devote a relatively high proportion of carbon to secondary chemicals, particularly polyphenols (Janzen 1974; Perry and Pitman 1983). Secondary compounds mediate numerous interactions between plants and their environment: protecting against pests and pathogens (chapter 19), influencing the rate of nutrient cycling (chapter 18), and in the case of polyphenols, screening foliage from harmful ultraviolet radiation (Waring and Schlesinger 1985). Trees growing on infertile soils are thought to produce more secondary compounds as a mechanism to reduce nutrient loss, while protection from ultraviolet radiation is particularly important at high elevation.

15.4.3 Allocation Priorities

Trees on any one site alter their allocation priorities depending on growing conditions. Wood production is particularly sensitive to environmental fluctuations, and reduced diameter growth is one of the first signs of tree stress (Waring 1985). Wood production is also reduced in some species when large quantities of carbon are allocated to the production of reproductive organs.

Trees also readily adjust the relative proportion of leaves to roots and mycorrhizae, depending on nutrient and water availability. Unfertilized Scots pine, for example, produce on the average 95 percent as much fine root biomass as shoot biomass during a growing season, but fertilized trees produce only 30 percent as much root as shoot weight. Those that are both fertilized and irrigated produce only 20 percent as much (Axelsson 1981). For wild plants in general, "a 100-fold drop in the availability of a limiting nutrient causes a 1.5- to 12-fold increase in root:shoot ratio, depending upon species and initial growth conditions" (Chapin 1980). It is important to bear in mind, though, that the root:shoot ratio is not a measure of relative productivity, but of relative standing crop. Stands with high root turnover (death and replacement) could have a low root standing crop but high root productivity. Another example is loblolly pine growing on sandy soils with low moisture-holding capacity versus clay-loam soils with high capacity in sites that are otherwise very similar. The same genotypes maintain more than five times higher fine-root area (and biomass) per unit of leaf area (and biomass) when growing on the sandy than on the more fine textured soils (Hacke et al. 2000); both stands are able to extract about 85 percent of available soil moisture. Although production and standing biomass are not the same thing, the standing biomass of relatively short-lived organs can be used as a rough proxy for their production. The results demonstrate that some species can drastically alter the balance among organs through changes in carbon allocation, thus permitting efficient gathering of resources.

15.5 THE LIMITING FACTORS OF THE ENVIRONMENT

In the following sections we explore in more detail the influence of water, nutrients, temperature, and carbon dioxide on NPP. It is important to remember three general principles:

1. Resource supply is dynamic in both space and time. Limiting factors generally vary seasonally (e.g., nutrients in the rainy season, water in the dry season) and at different stages of forest development. The particular resource that limits productivity at any one time may differ across the landscape.
2. The various environmental controls over LAI and $E(l)$ do not operate independently but rather interact and condition their respective availabilities.
3. Trees are not passive prisoners of the environment but rather enhance their own growing conditions in various ways. Crowns rake water and nutrients from air columns. Temperature is ameliorated beneath canopies. Energy diverted belowground improves soil fertility and water-holding capacity, and transpiration increases regional rainfall.

A good starting point for discussing the general principles of environmental control over NPP is the simple relationship introduced earlier: NPP = GPP − R. In areas with a great deal of cloudiness (e.g., the moist tropics), GPP may be limited at times by low irradiance; however, in most cases the primary environmental controls over GPP are those that affect light-harvesting capacity and the conversion of that light to carbohydrate:

- *Water and nutrients.* Both influence the amount of leaves that can be supported and the efficiency with which light and carbon dioxide are converted to carbohydrate.
- *Temperature.* Correlates with the proportion of each year that plants are photosynthetically active and also directly affects the rate of photosynthesis and respiration.
- *Carbon dioxide.* A complex story better dealt with later in the chapter.

Regional differences in GPP of forests correlate closely with what Kira (1975) calls **leaf area duration,** which is the product of leaf area (determined by water and nutrient availability) and the length of the growing season (in months). This relationship is shown in fig. 15.10 for various broad-leaved and needle-leaved forests.

The dashed line in fig. 15.10 shows the approximate course of NPP, which does not increase linearly with leaf area duration but rather tends to fall at very high values. This occurs for two reasons. First, the warm temperatures associated with long growing seasons increase respiration more than they do gross photosynthesis. Second, as we saw earlier, more productive forests tend to maintain a higher proportion of biomass as wood, which also increases respiration costs.

It is important to realize that photosynthesis is not necessarily restricted to what is traditionally reckoned as the "growing season" (i.e., the period when plant tissues are expanding). When temperatures permit, conifers and other evergreens photosynthesize outside of the growing season, thereby increasing their productivity both because of a longer period for carbon fixation and because a smaller proportion of fixed carbon is lost as respiration during cool seasons than during the summer. In areas with dry summers and wet winters (e.g., the west coasts of both Americas), summer is the least favorable season for water and (perhaps) nutrient availability (Waring and Franklin 1979).

Figure 15.11 shows the influence of season on GPP and NPP of an evergreen-oak forest in Japan. Note that GPP closely follows temperature and irradiance, both of which are highest in summer and lowest in winter. (Irradiance is

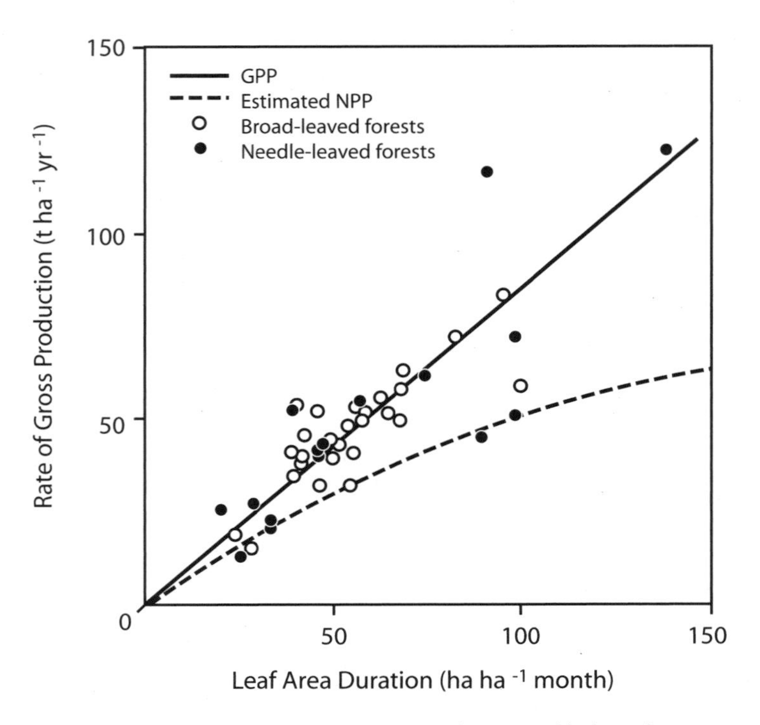

Figure 15.10. The relationship between primary productivity and leaf area duration for broad-leaved and needle-leaved forests. Data points and the solid line represent aboveground gross primary productivity (GPP). The dashed line is our crude estimate of how net primary productivity (NPP) might drop off across this gradient. (Adapted from Kira 1975)

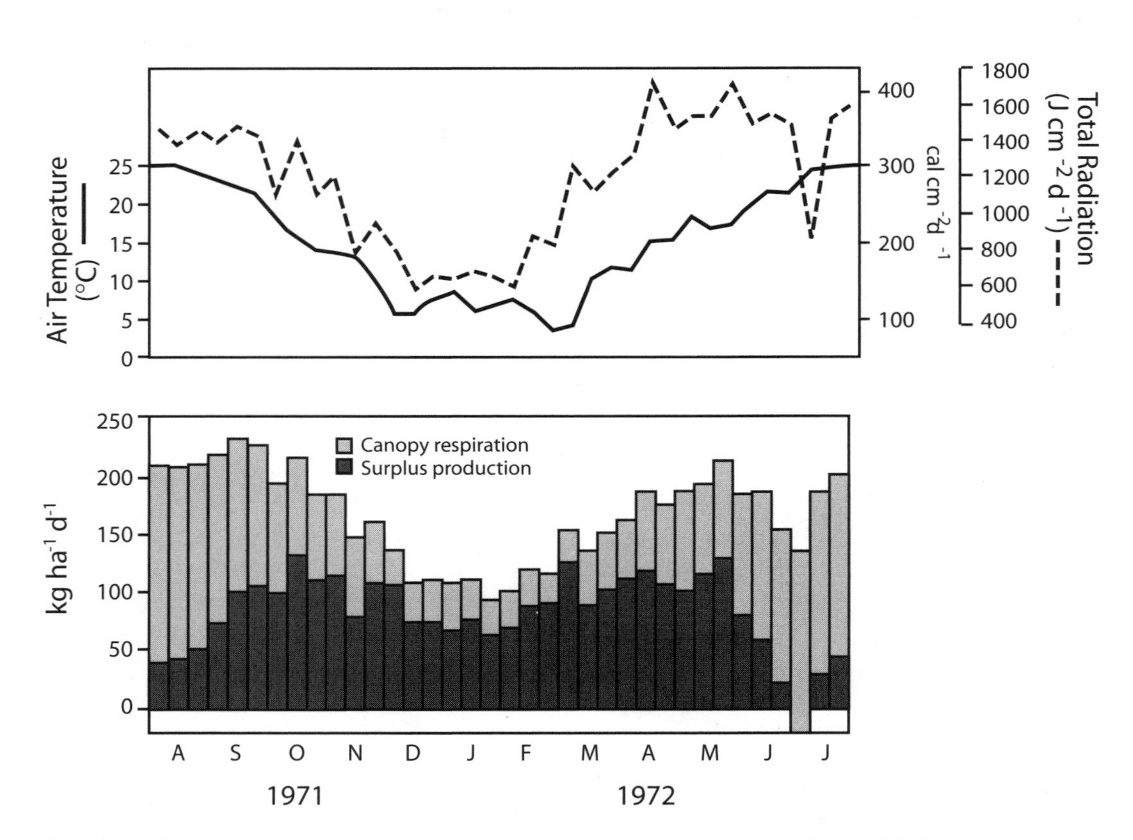

Figure 15.11. The seasonal changes in gross primary productivity (GPP) and net primary productivity (NPP) in an evergreen-oak forest in Japan. In the upper graph, the solid line is air temperature and the dashed line is total radiation. In the lower graph, the total height of bars is GPP, light shading is canopy respiration, and dark shading is "surplus" left over after canopy respiration. Note that the bar for June-July 1972 is not misprinted; respiration was estimated to exceed GPP in this period. (Adapted from Kira 1975)

low in winter because of cloudiness.) Well over 50 percent of GPP is lost through canopy respiration during summer; consequently NPP is higher in winter than in summer. Later we will discuss implications for forest productivity as the greenhouse effect warms the earth.

15.5.1 Water

Compared with other resources, water is frequently the most irregular and unpredictable in its supply (Harper 1977). Periodic water deficits occur in virtually every forest type, and seasonal drought is a regular feature throughout western and central North America, western South America, portions of Australia, and rather extensive areas of the tropics (chapter 4). Even trees growing in areas with abundant rainfall that is evenly distributed throughout the year may experience short periods of water deficit during most years and severe droughts periodically (Grubb 1977; Robichaux et al. 1984; Oren et al. 1998). During a dry summer in the oak-hickory forests of the eastern United States, an area in which rainfall is generally well distributed throughout the growing season, sugar maple kept its stomates closed 50 percent of the time, and oaks 20 to 30 percent of the time, in order to conserve water (Hinkley 1978). In the southeastern United States, past agriculture practices produced a layer that is impermeable to fine roots where soils contain high proportion of clay. As a result of the restricted rooting depth, loblolly pine forests growing on abandoned old fields reduce stomatal opening to about a third of maximum after two midsummer weeks without precipitation when soil moisture at the beginning of such rainless period is average, sufficient for maximum conductance (Oren et al. 1998).

Diffusion of water vapor from plant surfaces to the atmosphere is called **transpiration.** Although some water vapor diffuses through leaf cuticle, the major transpiration pathway is through open stomates. Since open stomates are the only port of entry for carbon dioxide into leaves, transpiration is an unavoidable consequence of photosynthesis. For various reasons, water diffuses out of stomates much more readily than carbon dioxide diffuses in (Larcher 2003). Temperate trees transpire from 200 to 350 L of water for every kilogram of aboveground dry matter produced, and tropical trees transpire more than twice this amount, or from 600 to 900 L (Larcher 2003). Applying these numbers to some of the productivity values that we discussed earlier, it can be seen that trees cycle an appreciable proportion of the rainfall that they receive back to the atmosphere. Transpiration is far from a total loss to the plant, however. Rather, it is essential for proper physiological functioning. The upward flow created by transpiration serves as a pipeline to deliver both water and nutrients to the leaf, and evaporation of water from leaf surfaces also keeps the leaf from overheating, a particularly important function in the tropics.

Plants must maintain at least 75 percent water content in functional cells (Bradford and Hsiao 1982). They accomplish this through a combination of uptake and control over losses by partial or complete stomatal closure. During dry periods uptake cannot keep up with loss, and cell turgor is maintained by closing stomates (with consequent reduction in carbon dioxide uptake). Stocker has summarized the situation quite succinctly as one of "tacking between thirst and starvation" (cited in Larcher 2003).

The rate of water loss from leaves is determined by three factors: (1) **evaporative demand;** (2) **boundary layer conductance,** or the rate at which water vapor is conducted from the leaf surface to the atmosphere through the boundary layer; and (3) **stomatal conductance,** which is the rate at which water passes from the leaf interior to the air immediately surrounding the leaf (i.e., the inner surface of the boundary layer). Evaporative demand determines the maximum potential transpiration; boundary layer conductance and stomatal conductance combine to modify this potential. (Transfer of water vapor through stomates and boundary layer can also be described in terms of **resistances;** boundary layer resistance and stomatal resistance are simply the reciprocals of their respective conductances).

Evaporative demand is driven by precisely the same factors that cause water to evaporate from an open pan of water: solar radiation (which provides the energy necessary to vaporize liquid water), temperature, and humidity. Temperature and humidity combine to create an **atmospheric saturation deficit** (also called **vapor pressure deficit** [VPD]), which is simply a measure of how much additional water vapor air can hold. Because warm air holds more water than cold air, at a given relative humidity the atmospheric saturation deficit increases with temperature.

When there is little change in atmospheric saturation deficit as one moves away from the leaf surface, the boundary layer is insignificant; boundary layer conductance is high, or conversely, resistance to water transfer is low. In this event, the leaf is said to be closely coupled to the atmosphere (Jarvis 1985). At a given radiation load, temperature, and humidity, transpiration of a closely coupled leaf (or canopy) is controlled largely by stomatal conductance.

The degree to which a leaf (or a canopy) is coupled to the atmosphere depends on leaf width and the turbulence of air around the leaf surface (or within and adjacent to the canopy). Narrow leaves have smaller boundary layers (i.e., are more coupled to the atmosphere) than broad leaves (Kramer 1983). Turbulence, which acts to couple leaves more closely to the atmosphere, results when wind moves across a surface and is accentuated if the surface is rough. It does not take much wind (approximately 2 m/s) to effectively wipe out a boundary layer (Larcher 2003). The upshot is that leaves near the surface of a forest canopy are coupled closely to the atmosphere, and stomates bear the burden of reducing water loss (Jarvis 1985). The smaller the leaves the closer the coupling. On the other hand, trees and other plants in forest understories are sheltered from wind and therefore are less coupled to the atmosphere.

The processes that control stomatal opening and closing are not completely understood; however, at least three different factors are involved: (1) carbon dioxide concentration in intercellular spaces, (2) water potential in the soil-plant continuum, and (3) humidity (Larcher 2003; Schulze et al. 1987; Oren et al. 1999). Stomates open in response to reduced carbon dioxide concentration in intercellular space; carbon dioxide concentration is reduced, in turn, by light, which provides the energy required to convert carbon dioxide to carbohydrate.

Water status of soils and plants is expressed in terms of potential energy. The water potential is defined as "the difference in free energy per unit volume between matrically bound, or pressurized, or osmotically constrained water and that of pure water" (Larcher 2003). The dimensions of water potential are energy per unit mass or per unit volume, which turns out to be the same as force per unit area. Hence, water potential can be expressed in the same dimensions as atmospheric pressure. Water potential was most commonly expressed as bars in earlier literature, but in the 1980s megapascals (MPa) came into use. One MPa equals 10 bars. By convention, the pure water reference state is taken as zero. Negative potentials indicate water that moves less freely than free water; the more negative the potential of soil or cell water, the more tightly it is held. Water always flows from areas of less negative to areas of more negative potential.

Leaf water potential is most commonly measured using a device called a pressure chamber, which consists of an enclosed chamber attached to a source of compressed gas (Scholander et al. 1965). A leaf (or a shoot) is placed in the chamber with the cut end (usually the petiole) protruding through a tight seal. Gas is introduced into the chamber, raising pressure until water is forced from the xylem. The pressure at which water can be seen oozing from the cut end, read from an attached pressure gauge, corresponds to the potential at which water is held in the leaf xylem. The degree of stomatal opening is determined with a device called a porometer, which measures resistance to movement of gas into the leaf. Kramer (1983) describes these instruments in more detail.

When humidity, soil water potential, or leaf water potential drop below a certain level, the carbon dioxide control is overridden and stomates close. Most work through the years has focused on the role of leaf water potential in triggering stomatal closure; however, work with annual plants suggests that stomates may actually be more closely coupled to soil water status than leaf water. Apparently, low soil water triggers roots to send chemical signals to leaves that cause stomates to close (see the review by Schulze et al. [1987]).

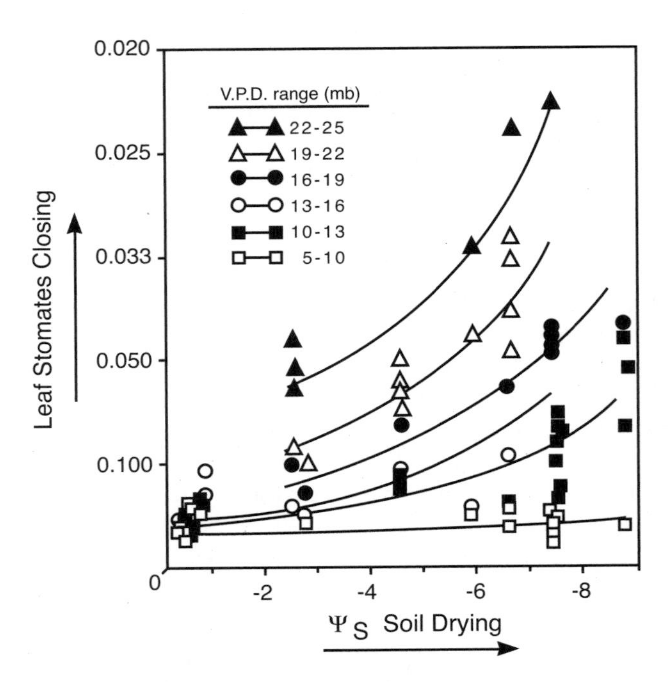

Figure 15.12. The relationship between stomatal conductance (expressed as cm/s; y axis) and soil moisture (expressed as water potential in bars; *x* axis) for different ranges of vapor pressure deficit (VPD). (Adapted from Tan and Black 1976)

Figure 15.12 shows the combined influence of soil water potential and VPD on stomatal resistance of a Douglas-fir canopy. Recall that VPD (or atmospheric saturation deficit) is a function of humidity and temperature, and stomatal resistance is the reciprocal of stomatal conductance. In other words, resistance increases with increasing stomatal closure. When VPD is low (because humidity is high), declining soil water potential has little effect on the stomates; however, increases in VPD trigger at least partial stomatal closure, even when soil water potential is relatively high (i.e., abundant soil water). The combined effect of drying soils and high VPD (i.e., low humidity and high temperature), which is a typical condition during the dry seasons experienced by many forest types, cause a sharp increase in stomatal resistance to water transfer.

The pattern shown in fig. 15.12 is likely to hold at least in a general way for most tree species. Trees control water loss from leaves very closely, closing stomates in response to changes in humidity and soil water in order to head off excessive loss from leaves well before it occurs (Larcher 2003). Depending on species and (perhaps) environment, trees begin to close their stomates when leaf water potentials are between −5 and −10 bars, with complete closure between −10 and −30 bars (Larcher 2003), although some tropical species continue transpiring at leaf water potentials below −40 bars (Medina 1983). The sensitivity of stomatal

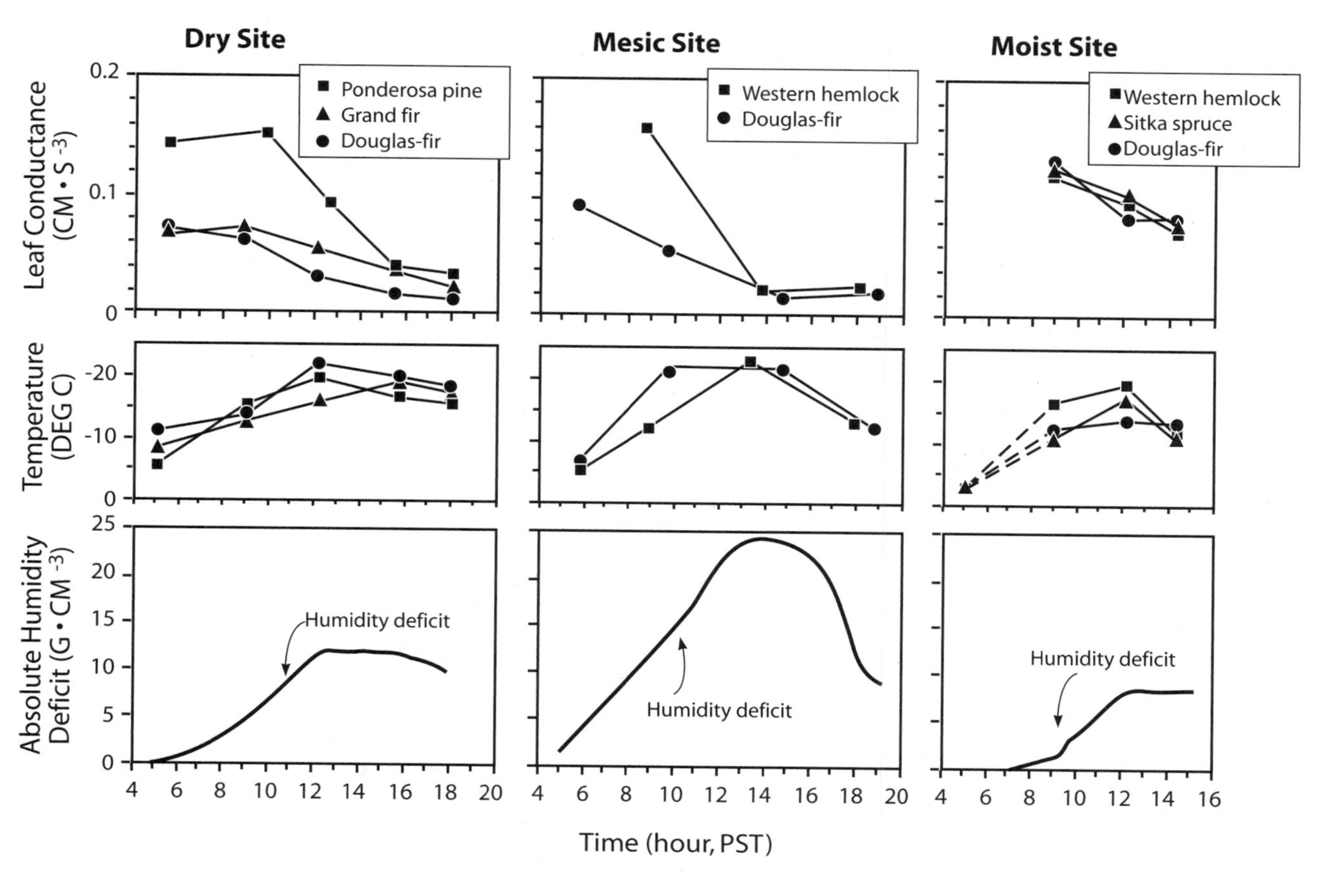

Figure 15.13. Diurnal changes in stomatal conductance and leaf water potential for trees in three habitats in Oregon. (Adapted from Running 1976)

conductance to VPD is related to the ability of plants to withstand low water potential without suffering excessive cavitation in xylem elements and permanent loss of hydraulic function (see references in Sperry et al. 2002). Such a loss may be catastrophic because stomates will remain nearly shut most of the time. Xylem comes in different types (i.e., diffuse porous, ring porous, nonporous), each constructed with various abilities to withstand low water potential. The ability to withstand low water potential comes with cost: such xylem typically cannot transport water very fast when water is amply available. As can be expected, xylem with low vulnerability to cavitation is likely to be found mostly in trees growing in xeric situations. Trees growing in situations in which soil moisture is ample and VPD is low typically produce xylem with high capacity for water transport, but they cannot and most of the time need not withstand low water potential (Hacke et al. 2001).

Temperate trees exhibit a distinctive diurnal pattern in stomatal resistance during the growing season. Early in the morning, photosynthesis reduces intercellular carbon dioxide concentration, triggering stomates to open. As temperature increases and humidity consequently drops, the carbon dioxide control is overridden and stomates begin to close, in certain situations closing completely (usually around midday), potentially remaining that way for the balance of the daylight hours (they may open at night when leaf water potential recovers; W.K. Ferrell, personal communication). Figure 15.13 illustrates this pattern for conifers occupying three sites across an east-west moisture gradient in Oregon (i.e., very low precipitation in the east to very high on the westerly coastal site). The data were collected in late August, when even the higher precipitation sites on the coast and the western slopes of the Cascades are relatively dry.

Species in all three environments of fig. 15.13 exhibit similar trends, but differences also occur. On the dry site, trees begin the morning with more negative leaf water potential than those on the moister site; however, by early to midafternoon, trees on all sites have reached potentials of −15 to −20 bars. Species within each site differ in their daily cycle. On the dry site, ponderosa pine keeps its stomates more open than do other species until midafternoon and at the same time maintains leaf water potential within the range of the other species. The pine is probably able to do this because it tends to root deeper than the other species, hence maintaining a better water supply to its leaves, and may have more xylem area supplying each unit of leaf area (Maherali and DeLucia 2001). On the western slopes of the Cascades, western hemlock behaves somewhat like the pine does on the drier site, maintaining its stomates more open (but leaves better supplied with water) than those of Douglas-fir; however, this is not the case on the coastal site. Compared with the Douglas-fir on the two other sites, trees of that species occupying the moist coastal environment maintain lower leaf resistance throughout the day. In Georgia, longleaf pine stands growing in areas with ample precipitation on a xeric site (xeric because of sandy soil) had similar leaf water potential and stomatal conductance as those on a more mesic site. The pines managed this by keeping density and, thus, leaf area index lower, allowing more soil volume and roots to supply each unit of leaf area, and by staying short, reducing the resistance to water flow to the leaves (Addington et al. 2006). Thus, adjustments at the stand and whole tree level maintained the physiological function of leaves at very similar levels in these stands.

Tropical trees also exhibit a diurnal pattern in stomatal opening and leaf water potential. In the tropics, however, leaf temperature can quickly climb to lethal levels without the cooling effects of transpiration. The low leaf water potentials maintained by some tropical trees probably reflect the high rates of transpiration that are necessary to cool leaves, and a xylem that is not very vulnerable to cavitation or is well suited to recover from cavitation nightly. The behavior of stomata in response to diurnally changing VPD and periodically changing soil moisture are consistent with stomatal regulation of plant water potential aimed at preserving the capacity of the xylem to transport water to leaves. Consistent with this role, stomates that have very high conductance when VPD is low (i.e., when soil moisture and PAR are high, and the plant's hydraulics support high water transport to leaves) are proportionally more sensitive to the diurnal increase in VPD (Oren et al. 1999). The same behavior is observed in trees and other plants in all biomes.

The problem of uncertain water supply (or certain drought) is circumvented to some degree by water storage in ecosystems. Water is stored both in the soil and in the tree. As we discussed in chapter 14, reserves of plant-available water in soils depend on pore structure and are enhanced by organic matter and the soil-aggregating effect of both roots and mycorrhizal fungi. Old, decaying logs store large amounts of water (Amaranthus et al. 1989c), and their abundance in some forest types (particularly in the Pacific Northwest and high-elevation, interior forests of North America) suggests that they could be major sources of water for trees during drought (Harvey et al. 1979; Harmon et al. 1986).

Trees (particularly conifers) store water in sapwood. Old-growth Douglas-fir trees absorb enough water in sapwood to equal the transpiration demand in 10 midsummer days (Waring and Running 1976, 1978). Pines maintain a relatively large proportion of sapwood to heartwood and, therefore, have large internal storage capacity. Along with deep rooting, water stored in sapwood probably contributes to the ability of ponderosa pine to keep its stomates more open than those of associated species on droughty sites (fig. 15.13).

Trees may also cope with water shortages by investing in large amounts of water-gathering surface (i.e., roots and mycorrhizal fungi) per unit of leaf area. At least some species of mycorrhizal fungi produce long, ropelike collections of hyphae called **rhizomorphs** that transport water over relatively long distances (Duddridge et al. 1980). Parke et al. (1983b) compared the effect of mycorrhizal fungi on photosynthesis and growth of water-stressed Douglas-fir seedlings; net photosynthesis per unit leaf area was 10 times greater, and leaf area almost 3 times greater, in mycorrhizal than in nonmycorrhizal seedlings.

15.5.2 Nutrients

In addition to carbon, oxygen, and hydrogen, which are the basic constituents of carbohydrate, plants require 13 to 16 chemical elements (depending on species), shortages of any of which can limit productivity. Nutrients are divided into **macronutrients,** which are required in relatively high levels by plants, and **micronutrients,** which are required in relatively low levels. Macronutrients (excluding carbon, oxygen, and hydrogen) are nitrogen, calcium, potassium, phosphorus, magnesium, and sulfur. Of these, nitrogen, calcium, and potassium are required in the largest amounts by trees. Micronutrients include (in no particular order of uptake) iron, boron, manganese, copper, zinc, molybdenum, and chlorine. With some exceptions, nutrients are input to ecosystems at rates far below tree-growth requirements, and productivity is closely tied to **nutrient cycling,** or the rate at which nutrients are released from dead organic matter by the microbes and animals composing the detrital food chain (chapter 9). (Chapter 16 covers the physiological functions of the various nutrients, while chapter 17 deals with nutrient inputs to ecosystems and chapter 18 with nutrient cycling).

15.5.2.1 Patterns of Nutrient Limitation

In reading this section, it is important to understand that our knowledge of nutrient limitations in forests (or indeed any natural ecosystem) is very rudimentary. By far, most research has centered on nitrogen and phosphorus. Only recently have we begun to learn that many other nutrient elements may either directly or indirectly influence forest productivity.

In unpolluted forests, nutrient limitations to productivity are best understood in terms of soils and the presence or absence of nodulated nitrogen-fixing plants. Most northern forests grow on young soils (dating from the last glaciation or, e.g., in the Pacific Northwest, from recent volcanic activity) that still have abundant weatherable minerals within the rooting zone. Nodulated nitrogen-fixing plants (when they occur) are pioneers that add nitrogen to soils at

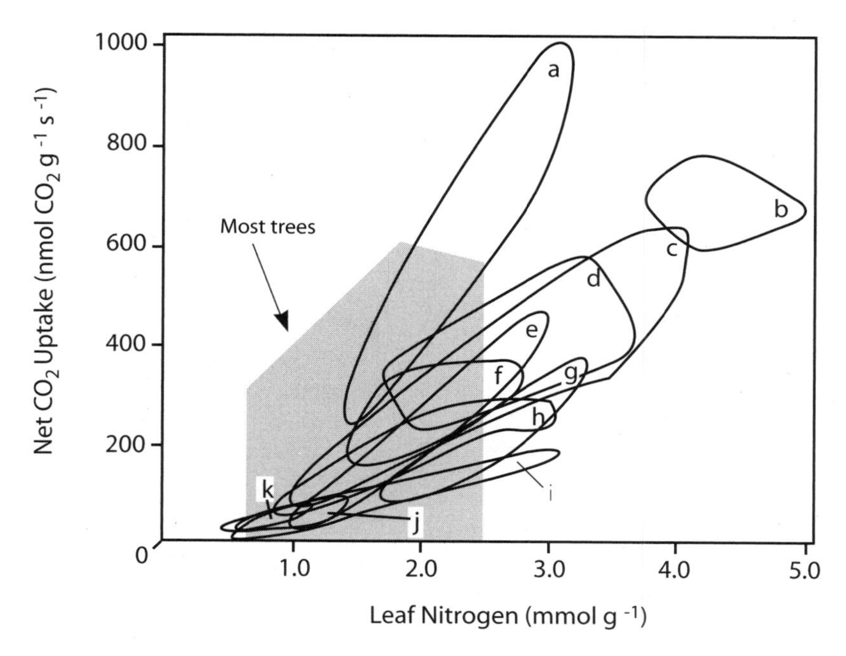

Figure 15.14. The relationship between leaf nitrogen concentration and net photosynthetic rate for different plant species (1 mmol/g = 1.4%). Letters represent different plant species; a is C_4, and all others are C_3. The shaded area denotes the range of leaf nitrogen concentrations within which most trees would be found. (Adapted from Chapin et al. 1987).

The historic focus on nitrogen reflects its central role in physiology and, in particular, the degree to which it is supplied through natural processes. Nitrogen is an essential component of protein, and proteins (as enzymes) regulate all physiological processes. Trees require more nitrogen than any other nutrient except (in some species) calcium and potassium. Roughly 75 percent of foliar nitrogen is contained in enzymes that are present within chloroplasts and function directly in photosynthesis (Chapin et al. 1987); approximately 20 percent is within a single enzyme, **RUBISCO** (ribulose-1,5-biphosphate carboxylase-oxygenase), which is the primary carbon dioxide–fixing enzyme of all plants. Figure 15.14 shows the relation between leaf nitrogen and net photosynthesis for several different wild plant species. Trees occupy the lower end of this diagram, with leaf nitrogen generally ranging from slightly less than 1.0 percent to about 2.5 percent.

Adding to the historic interest in nitrogen is the fact that it is input to many systems in very low amounts. Nitrogen differs from other nutrients in that it is not present in most rocks and, hence, is not supplied through weathering. Although most of our atmosphere is composed of N_2 gas (chapter 11), only a few types of bacteria and cyanobacteria possess the enzyme necessary to convert N_2 to ammonia.

infrequent intervals following major disturbances. These factors combine to make N limitations common in unpolluted northern forests.

Conifers commonly increase leaf area when fertilized with nitrogen. Figure 15.15 shows this pattern for Norway spruce and Scots pine growing in Sweden. These boreal forests have quite low LAIs, ranging between 1 and 2 m^2/m^2 (Albrektson et al. 1977). The fertilization experiment illustrated consisted of three levels of nitrogen with a single level of phosphorus, one level of nitrogen (intermediate) without phosphorus, and an unfertilized control. Each fertilizer treatment was applied to two plots in the spruce forest and four plots in the pine forest. All fertilizer treatments increased LAI of both species, up to nearly 6 in some plots of the spruce stand and 3 in the pine. Note, however, the considerable variation among plots of a single treatment. This kind of highly variable response to nutrients within a single forest is not uncommon and illustrates the extreme complexity of nutrient relations in forests.

Depending on factors such as soil type, disturbance history, and stocking density, nutrients other than nitrogen may also limit productivity of northern forests. For example, various studies have found potassium, magnesium, or sulfur limitations in Douglas-fir stands (Edmonds and Hsiang 1987; Velasquez-Martinez et al. 1992). Moreover, the availability of nitrogen in Douglas-fir forest soils correlates closely with levels of soil calcium (Velasquez-Martinez 1990).

Quite a different situation holds for forests growing on the highly weathered soils (i.e., Oxisols and Ultisols in the U.S. system) that are characteristic of the southeastern United States, parts of Australia, and tropical forests in the

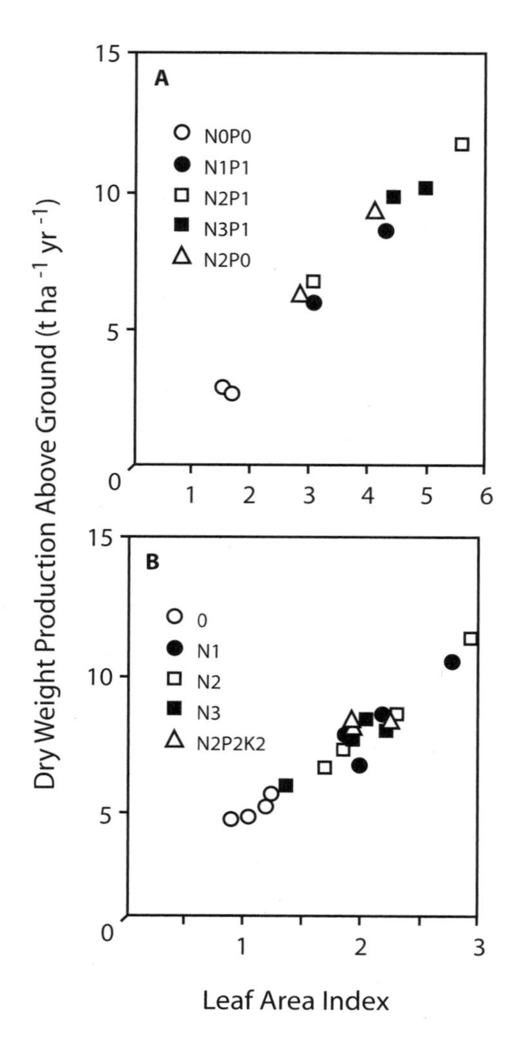

Figure 15.15. Leaf area and aboveground net primary productivity of fertilized and unfertilized Norway spruce (A) and Scots pine (B) in Sweden. N0P0 or 0, no fertilization; N, P, K, nitrogen, phosphorus, or potassium fertilization, respectively; 1, 2, 3,, increasing fertilizer amounts. (Adapted from Albrektson et al. 1977)

interior of South America, Africa, and Southeast Asia. Weatherable minerals have long disappeared from the rooting zone of these soils, leaving them depleted in basic cations (calcium, potassium, and magnesium)[4] and enriched in aluminum and iron. Aluminum and iron bind strongly with phosphorus, reducing phosphorus availability to plants (chapter 18). Phosphorus deficiency is common in parts of Australia and the southeastern United States that have old, sandy soils. The NPP of lowland tropical forests growing on Oxisols and Ultisols is also believed to be limited by phosphorus (Vitousek 1984b; Denslow et al. 1987) and in some cases by calcium and magnesium (Vitousek 1984b; Cuevas and Medina 1986, 1988). On the other hand, lowland tropical forests on Spodosols (periodically flooded sands) and at least some montane tropical forests are apparently nitrogen limited (Vitousek 1984b; Cuevas and Medina 1986, 1988). In the case of the Spodosols, periodic flooding may replenish cations that have been lost by leaching and thereby alleviate their deficiencies. The degree to which nitrogen is limiting in tropical forests is also related to the abundance of leguminous trees. Phosphorus-limited forests growing on Ultisols in Amazonia are dominated by legumes (Cuevas and Medina 1986), whereas nitrogen-limited montane forests are not. The presence of legumes, however, does not guarantee abundant nitrogen; shortages of other nutrients such as iron, molybdenum, and calcium may restrict the ability of these plants to fix nitrogen (chapter 17).

Forests of Europe and eastern North America have been heavily fertilized by acidic nitrogen- and sulfur-containing pollutants over the past few decades. In the Netherlands, Douglas-fir plantations that were previously nitrogen limited are now phosphorus limited (Mohren et al. 1986). Deficiencies of calcium and magnesium (both elements that are readily leached from soils by acidity) (chapter 18) became progressively more prevalent in Germany (Matzner et al. 1986; Rehfuess 1987; Oren et al. 1989). Declining growth of polluted German forests results from rather complicated interactions between nitrogen, magnesium, and calcium exacerbated (and perhaps triggered) by poor fine root development and mycorrhiza formation by trees, possibly the effect of high aluminum (Oren and Schulze 1989; Schulze 1989). (We return to this in more detail in chapter 20.)

15.5.2.2 Micronutrient Deficiencies

Few studies have searched for micronutrient deficiencies in forests; however, those that have frequently find them. Iron deficiencies have been reported from various areas in the Pacific Northwest and may be exacerbated by fire (Perry et al. 1984; Ballard 1986). Boron deficiency is apparently widespread, having been reported from numerous tropical, temperate, and boreal forests (Gonzalez et al. 1983; Carter et al. 1983). Boron deficiency (which in radiata pine tends to occur on volcanic soils, particularly where there has been some erosion) causes dieback of leaders and distorted growth (Turner and Lambert 1986b). Mitchell et al. (a) found that shortleaf pine seedlings did not respond to boron unless they were mycorrhizal; what this means is not clear. Mycorrhizal fungi might be critical to uptake of boron by shortleaf pine, or alternatively, nonmycorrhizal seedlings may be unresponsive to boron because they were limited by some other factor.

15.5.2.3 Nutrient Balance is Important

Except in cases of extreme deficiency of a single nutrient, tree growth is likely to be influenced by interactions among multiple nutrients, and **nutrient balance** must be considered as well as total amounts. Nutrients may either complement or interfere with one another. Two types of complementation can be distinguished: physiological and ecological.

[4]The term *basic* refers here to the tendency of these cations to serve as bases for acids; bases and acids combine to form salts.

Physiological complementation results from the fact that plant growth is the product of multiple physiological and structural factors that are mediated by different nutrients. These nutrients must be present in the correct proportions in order for the plant to function smoothly as a whole. For example, adequate calcium is essential for good root growth (Lyle and Adams 1971), hence a shortage of calcium can induce additional deficiencies by hampering the ability of trees to gather other nutrients and water. Another type of physiological complementation occurs at the molecular level. All enzymes must have nitrogen, but they also require one or perhaps several other nutrient elements for proper functioning. In parts of the Pacific Northwest, deficiencies of sulfur, which is required for some enzymes, limits the ability of trees to respond to nitrogen fertilizer (Turner et al. 1977).

Ecological complementation occurs when shortages of one element limit the input or cycling rates of others. For example, the microbes that fix nitrogen as well as the plants that host nitrogen-fixing microbes require adequate supplies of other nutrients to function properly, and nitrogen accumulation in soils has been found to be limited by one or another nutrient element (and by water). (Factors limiting nitrogen fixation are discussed further in chapter 17.) Another type of ecological complementation occurs when one or more elements limit the cycling rate of others. For example, although plants do not require sodium, animals do (Needham 1965). Inadequate sodium could, therefore, indirectly limit plants by limiting animals that are essential to the nutrient cycle.

Nutrient elements that are similar chemically can interfere with one another's uptake or physiological functioning. This is the case for the basic cations (chapter 18) and also for iron and manganese. Lafond and Laflamme (1968) provided a good example of interference between iron and manganese that through differential effects on tree species influenced the course of succession. In northeastern North America, jack pine seedlings regenerate well following fire that destroys humus layers but in the absence of fire are replaced by black spruce. Lafond and Laflamme showed that jack pine growth was inhibited by low concentrations of iron relative to manganese in humus layers, but black spruce seedlings were not. Fire provided a mineral soil seedbed with iron:manganese ratios more favorable to the pine.

15.5.2.4 Plants Adapt to Low Nutrients

As the previous example suggests, tree species frequently differ in their response to nutrients or nutrient balances. In particular, trees native to infertile sites adapt to low nutrient supply by evolving to grow slowly, hence they may not respond at all to fertilization. This is the case, for example, with black spruce (Van Cleve et al. 1983). Are black spruce forests nutrient limited? In evolutionary time, yes, but in ecological time, no. Plants adapt to low nutrient supply by evolving low inherent growth potential, which leads to the seemingly paradoxical conclusion that plant communities growing on soils with intermediate fertility may actually respond more to fertilization than communities on highly infertile sites (Chapin et al. 1986).

15.5.2.5 Unanswered Questions

Nutrients are more difficult to study than water for several reasons. First, there is no single, simple indicator of nutrient deficiency similar to stomatal closure for water. Second, the sheer number of nutrient elements makes measurement quite expensive. In addition, this expense is compounded because results obtained on one soil type or in a given environment may not apply elsewhere, thereby necessitating extensive field experimentation to characterize nutrient limitations over wide areas. Third, even if nutrient levels are determined (e.g., in soils or foliage) the problem remains of interpreting what that means for tree growth.

With a few exceptions, measures of soil nutrients are notoriously unreliable indicators of whether plant needs are being met. Determining nutrient levels in plant tissues (e.g., foliage or fine roots) is better, but still requires considerable experimentation to determine the relationship between a given level of tissue nutrient and productivity. The situation is greatly confounded because the amount of a given nutrient plants require varies depending on the presence of other nutrients and availability of light, water, and carbon dioxide. Returning to our symphonic analogy, the demand for different nutrients must be harmonized with the production and function of different tissues, which changes seasonally and with the availability of all nonnutrient growth resources (Oren et al. 1996). (The various methods employed to determine nutrient deficiencies are discussed in chapter 16.)

15.5.3 Temperature

Figure 15.16 shows the temperature dependence of net photosynthesis for several temperate and tropical tree species. Net photosynthesis of most temperate and boreal

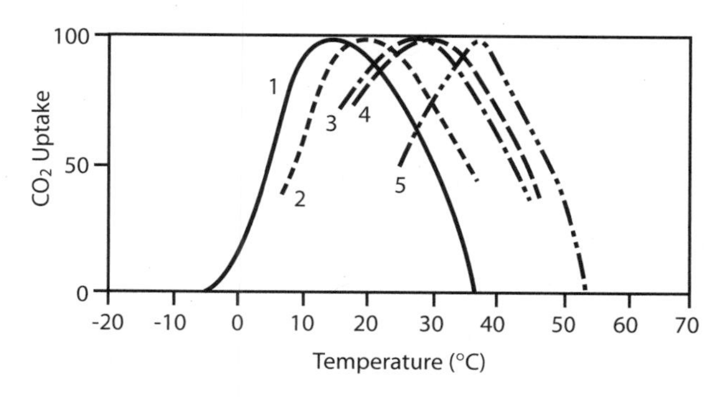

Figure 15.16. The temperature dependence of net photosynthesis for several temperate and tropical tree species. 1. *Pinus cembra*, Alps Mountains. 2. *Fagus sylvatica*, central Europe. 3. *Cassia fistula*, Java. 4. *Ficus retusa*, southern Japan. 5. *Acacia craspedoocarpa*, western Australia.

tree species peaks somewhere between 15 and 25°C and falls off rather sharply at both lower and higher temperatures.

Net photosynthesis declines at high temperatures because the enzymes mediating carbon fixation become less effective (and eventually completely inhibited) while respiration continues to increase exponentially. At the high end of the temperature range, net photosynthesis of temperate and boreal trees drops to zero between 35 and 40°C, while at the low end, photosynthesis in some species continues until temperatures reach –2 to –5°C (Larcher 2003). Evergreens that occupy areas with relatively mild winters photosynthesize during the winter.

Tropical trees also have a sharp peak in temperature dependence of photosynthesis; however, maximums are attained well above 25°C. The points of zero net photosynthesis are shifted correspondingly upward on both the low and high ends of the temperature scale (fig. 15.16). (Remember that most tropical trees have not been studied, and values such as these are based on a very small fraction of the total.)

Temperature also affects growth of new tissues independently of photosynthesis. According to Larcher (2003), the maximum rate of shoot expansion occurs between 30 and 40°C for tropical plants and 15 and 30°C for others. Root growth is much less sensitive to low temperature than shoot growth, which results in an asynchrony between above- and belowground growth. Roots of many species elongate in the spring (before shoots commence growth) and in the fall and early winter (after aboveground growth has ceased). In general, asynchrony in growth of different organs allows the supply of carbohydrates from photosynthesis to keep up with the maximum growth demand exerted by each organ (e.g., reproductive tissues, leaves, roots), thus allowing it to reach maturity as quickly as possible. With the exception of the bole, growing organs are more prone to damage from both biotic agents and abiotic factors than mature organs, so hastening maturation reduces risk.

15.5.4 Carbon Dioxide

The increasing carbon dioxide content of the atmosphere and consequent global warming have created considerable interest in the degree to which carbon dioxide limits NPP (Norby et al. 2005). If primary productivity is carbon dioxide limited, plants should respond to higher atmospheric carbon dioxide by increasing growth. In doing so, they soak up and fix in biomass some portion of the carbon dioxide being released to the atmosphere through burning fossil fuels and deforestation. The response of forests to global climatic change is a complex issue, however, involving not only atmospheric carbon dioxide but (among other things) temperature, water, nutrients, changes in seasonal climatic patterns, and altered disturbance regimes. (We will deal with this in a later section of this chapter and again in chapter 20).

When grown under controlled conditions, seedlings of several different tree species produce more biomass as atmospheric carbon dioxide concentration is increased (Rogers et al. 1983; Tolley and Strain 1984; O'Neill et al. 1987a, 1987b). Exceptions are few and include Douglas-fir, which showed no response (Hollinger 1987), and ponderosa pine, which reduced growth in response to a higher than ambient atmospheric carbon dioxide concentration (Woodman 1989). Among those species that increase growth, the magnitude of response varies widely (Bazzaz et al. 1990). In contrast to studies on seedlings, a recent synthesis of the response of NPP to elevated carbon dioxide levels (~550 ppm) in four free-air carbon dioxide enrichment experiments on young forest stands (<20 years old) composed of shade-intolerant species, shows a response that is highly conserved across a broad range of productivity, with a stimulation of approximately 23 percent (Norby et al. 2005). At low LAI, the response is mostly attributable to increased light absorption, but as LAI increased, the response is dominated by increased $E(l)$. The observed enrichment response is less than that of previous syntheses of tree growth responses, ranging from 29 percent in an analysis dominated by short-term seedling studies to 55 percent for field-grown trees. The difference may reflect generally higher carbon dioxide concentrations used in seedlings studies, and that often only aboveground biomass increment was measured in older studies.

The relevant question regarding carbon dioxide (and indeed regarding any single resource) is how limiting it is *in relation to other limiting factors*. Trees with plenty of water, nutrients, and light may increase their growth with more carbon dioxide, but trees that are inadequately supplied with one or more of these factors will not. In many experiments, improved growth in response to higher-than-ambient carbon dioxide concentration is transitory, probably because other limiting factors come into play (Kramer 1981). In one experiment, for example, dry weight of white oak seedlings increased 85 percent in response to an approximate doubling of atmospheric carbon dioxide concentration from 362 to 690 ppm, while foliar nitrogen concentration dropped well below levels considered to be limiting (Norby et al. 1986). The ability of seedlings to gather carbon was outstripping their ability to gather nitrogen, an imbalance that cannot be maintained. Similar response has been found in a pine forest where very strong initial response to elevated carbon dioxide was reversed after three years of continuous enrichment (fig. 15.17). The response resumed six years later, after an ice storm severely damaged the forest under ambient carbon dioxide concentration but only slightly damaged the plots subjected to elevated carbon dioxide (McCarthy et al. 2006). Twenty-year-old loblolly pine trees in southeastern United States and Norway spruce trees in Sweden growing on sandy soils that are even poorer in nitrogen did not respond at all to elevated carbon dioxide unless provided with nitrogen as well. Adding nitrogen to a section of the plot shown in fig. 15.17 reversed the decline in the response by the following year. Nitrogen limitation also changes the allocation of the extra NPP produced under elevated carbon dioxide.

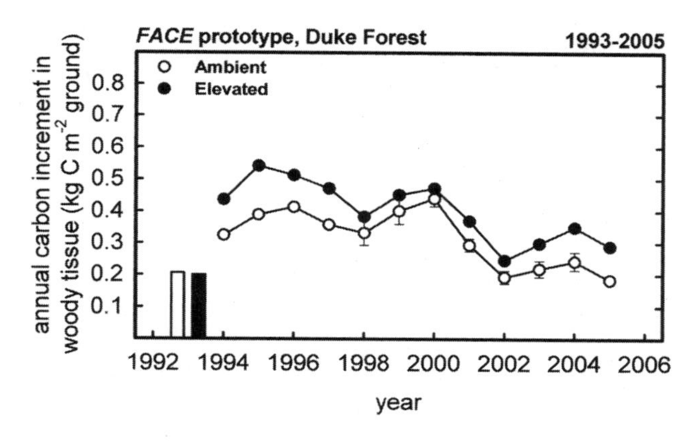

Figure 15.17. The response of stemwood production in loblolly pine forest to increased concentration of CO_2 (+200 ppm). Bars represent the average growth over the 10-year period since planting, before the commencement of enrichment. The noticeable decline in response after the first three years was reversed in a fertilized section the following year (not shown). A partial recovery of the response is apparent after the December 2002 ice storm (Data from Oren et al. 2001; McCarthy and Oren, unpublished)

For example, after a brief increase in stemwood NPP in a sweetgum plantation, the increase disappeared only to reappear as a more sustained increase in fine root NPP, likely reflecting an attempt by trees to overcome nitrogen limitation (Norby et al. 2004). Thus, other resources (e.g., nitrogen) not only affect the response of NPP, but also of carbon allocation to different organs.

On the other hand, trees (and plants in general) have a remarkable ability to adjust to changing conditions and maintain balance in their resource supply (Eamus and Jarvis 1989). Seedlings of several different species have been found to respond to elevated carbon dioxide concentrations by increasing allocation of carbon to roots and (in some cases) forming more mycorrhizae and exuding more carbon into the rhizosphere (Rogers et al. 1983; Norby et al. 1987; O'Neill et al. 1987a,b; Schäfer et al. 2003). Such adjustments keep the acquisition of belowground resources in balance with the carbon dioxide supply. Moreover, a relatively high level of atmospheric carbon dioxide allows plants to use water more efficiently, because less stomatal opening (hence, less water loss) is required to take in the same amount of carbon dioxide. There is clear evidence that trees and other plants have reduced the density of stomata produced on their leaves since the early 1800s, a common response to elevated carbon dioxide. Woodward (1987) counted stomata on leaves of seven temperate tree species and one shrub species that had been periodically collected and stored in the University of Cambridge herbarium over the past 200 years; stomatal density of all species declined over time by an average of 40 percent.

In the field, productivity is most likely to be limited by atmospheric carbon dioxide at high elevation. Although the concentration of carbon dioxide in air remains constant over a wide elevational range, the thinner air at high elevation results in lower absolute amounts of carbon dioxide. For example, in July at latitude 30° N, the density of air declines from 1.159 kg/m^3 at sea level to 0.835 kg/m^3 at 3,500 m elevation, corresponding to a 28 percent drop in carbon dioxide per unit atmospheric volume (LaMarche et al. 1984). Productivity, however, does not necessarily decline in proportion to the decline in atmospheric carbon dioxide with ascending elevation. Water availability generally increases with elevation, as does cloudiness; therefore, trees likely keep stomates open for a greater proportion of their growing season at high than at low elevation, thereby compensating to at least some degree for lower atmospheric carbon dioxide concentrations. Nevertheless, Tranquilini (1979) estimates that low carbon dioxide levels reduce photosynthesis of trees at timberline in the Alps by 10 to 20 percent below that of trees at lower elevation. As we discuss later, the rising concentration of atmospheric carbon dioxide during the last century has been implicated in the increased growth of at least some high-elevation forests.

15.5.5 Resource Interactions

We have seen that atmospheric levels of carbon dioxide influence the efficiency of both water and nutrient use. Similar interactions occur between other resources as well. For example, fertilization and irrigation together can produce greater increases in LAI and NPP than either alone; this is illustrated in fig. 15.18, which shows the results of an experiment in an Australian Monterey pine plantation (*Pinus radiata,* also commonly called radiata pine). In this experiment, net LAI increment in fertilized and irrigated plots was 50 percent greater than that of plots receiving only fertilizer, and more than double that of plots irrigated without fertilization.

Why the interaction between irrigation and fertilization in this example? Water certainly influences nutrient cycling; however, in this instance we are most likely seeing a seasonal effect. In the pine plantations of Australia, as in other evergreen forests with growing-season drought, growth of new foliage occurs early in the growing season, when soil moisture is still adequate and nutrients are most likely to be limiting. Later in the season, drought controls the amount of older foliage that is shed. Hence, net foliar increment over the course of a growing season is maximized when both these resources are abundant.

Nutrient cycling is influenced by both soil water availability and soil temperature; the latter two are also interrelated (fig. 15.3). Hence, as water and temperature vary seasonally, controls over nutrient availability also vary. The dynamics of the interplay between nutrients, water, and temperature vary considerably between different forest types. Consider temperate (or subtropical) forests that have seasonal drought. The rate of nutrient cycling is limited by low temperatures early in the growing season, when there is abundant water, but becomes limited by low water later in the season. In Australia, for example, soil respiration

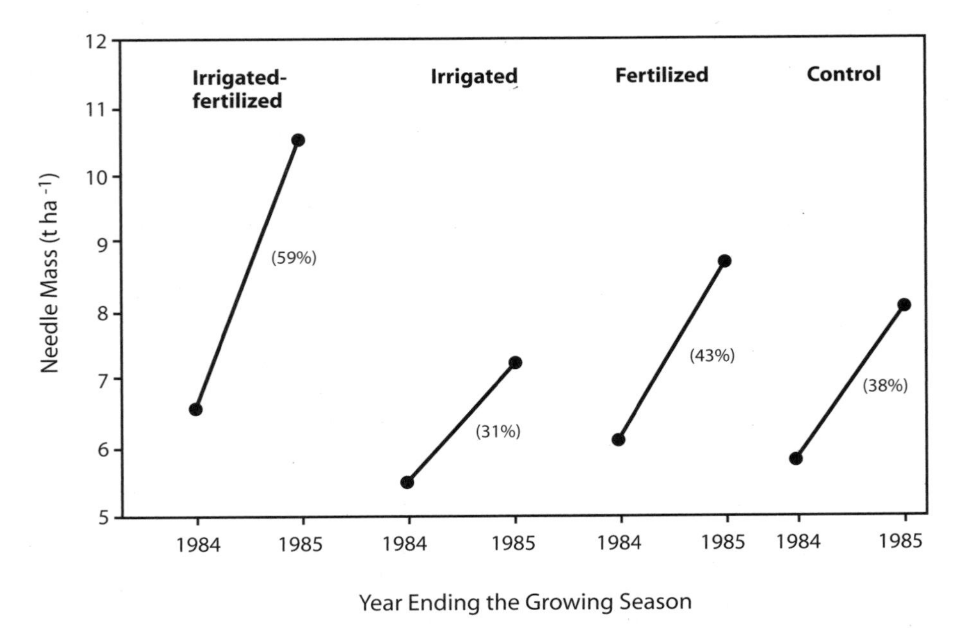

Figure 15.18. The effects of fertilization and irrigation on the growth of needle biomass in radiata pine plantations, Australia. (Data from Linder et al. 1987)

(which measures the activity of roots, mycorrhizal fungi, and other soil biota) varies closely with temperature when soil moisture is above 12.5 percent but depended more on water when soils were drier than this (Carlyle and Than 1988). (Do not consider 12.5 percent a "magic" number; the exact level at which soil moisture becomes limiting varies widely depending on soil type and other environmental factors.) Because the supply of nutrients to trees is closely tied to the activity of soil organisms, we might predict that tree nutrition is influenced by temperature at some times during the year and by water at others.

The situation differs in boreal forests. There, soil temperature exerts by far the major control over nutrient cycling and NPP (Van Cleve et al. 1983; Bonan and Shugart 1989), and too much rather than too little water becomes a problem. Low radiant energy in the far north reduces evapotranspiration, leading to excessive moisture in soils that do not drain rapidly. Too much water, in turn, greatly inhibits nutrient cycling by reinforcing low temperatures and limiting oxygen diffusion into soils.

15.6 TREES ARE NOT PRISONERS OF THE ENVIRONMENT

Forests modify the environment and influence resource supply in various ways, some of which are passive and some of which involve the direct investment of energy in the creation and maintenance of favorable growing conditions. For example, forests interact with and influence the within-stand atmosphere. Canopies are very effective rakes that increase both nutrient and water input to ecosystems by gleaning these elements from passing air masses (chapter 17). Evapotranspiration both cools the forest and cycles water that otherwise would escape to streams, creating the potential at least for short-distance cycling of precipitation. As chapter 3 showed, up to 50 percent of the rain falling in Amazonia originates from evapotranspiration by the rain forest itself (Salati 1987). As forest cover declines, water that would have been transpired back to the atmosphere instead goes to streams. The likely consequence is lower rainfall throughout the Amazon basin and subsequent stress on remaining forests of that region.

Forests also modify their growing conditions through influences on soils and nutrient cycling. Carbon and nitrogen pumped belowground by trees and their symbionts improve soil fertility and water-holding capacity. Rapid nutrient cycling by the forest prevents nutrient loss through leaching and is an especially important nutrient-conservation mechanism in both temperate deciduous and moist tropical forests (chapter 18). Because available nutrients are derived primarily from the decomposition of shed tissues, trees can actively regulate their own nutrient supply through the chemical composition (and therefore decomposability) of their litter. As we shall discuss in detail in chapter 18, nutrient content of litter correlates positively with nutrient availability, while decomposition correlates positively with the nutrient content of litter. If this sounds circular, it is. The upshot is that different tree species can produce quite different nutrient cycling patterns on the same site, a fact that may play a role in forest response to global warming.

Work by Nadelhoffer et al. (1985) illustrates how different tree species can produce quite different nutrient cycling characteristics, even on similar soils. They studied nitrogen cycling in stands of eight tree species in the University of Wisconsin arboretum (fig. 15.19). The rate of nitrogen

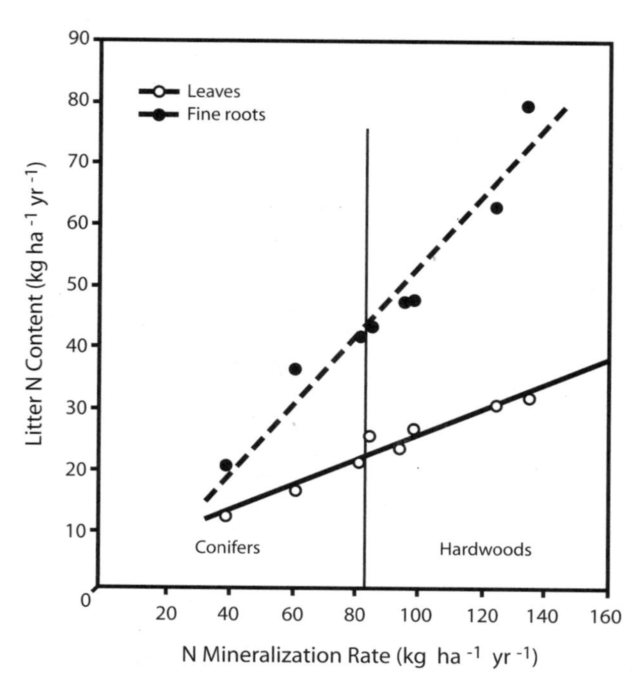

Figure 15.19. The relationship between nitrogen (N) mineralization rate and N content of litter for different tree species grown in the University of Wisconsin arboreturm. (Adapted from Nadelhoffer et al. 1985)

mineralization (a measure of nitrogen available for uptake) was lower under conifer than under deciduous forest (a pattern that is generally valid); note the strong relationship between nitrogen mineralization and litter nitrogen content.

The strong positive influence of forests on site productivity has important implications for ecosystem stability and ecologically sound forest management. We return to this point later in this chapter, and it is a central focus of chapter 20.

15.7 PRODUCTIVITY IN THE TWENTY-FIRST CENTURY

Pollution has significantly altered forest productivity, and evidence suggests that greenhouse effects have as well. In the absence of strict environmental regulations by industrialized countries, these trends will likely to continue and intensify in the coming decades. Pollution has rather clear negative effects on forest health and NPP, stressing trees and their mutualists through a complex of interacting effects that vary with the type of pollutant, soils, and (perhaps) tree species. (We discuss the effects of pollution on forests in some detail in chapter 20.)

In contrast to pollution, greenhouse effects on NPP likely vary a great deal among forests occupying different sites. Moreover, because of the multiplicity of possible factors and interactions among these factors, predicting effects on any given site is not straightforward. (Expected climate changes were discussed in chapter 4.)

Tree rings indicate that except for periods of extended drought (such as occurred in the 1930s), growth of at least some high-elevation and high-latitude forests has trended upward since the mid-1800s to early 1900s (Garfinkle and Brubaker 1980; LaMarche et al. 1984; Bork 1985; Graumlich et al. 1989; Peterson et al. 1990), a period that coincides with rising carbon dioxide levels and warming global temperatures. In the southwestern United States, this apparently results from the combined effect of warmer temperatures and increased carbon dioxide, with the latter predominating since 1970 (LaMarche et al. 1984). In the Pacific Northwest, warmth rather than atmospheric carbon dioxide appears to be the main factor (Graumlich et al. 1989); however, neither temperature nor precipitation explains the increasing growth trend of conifers in the Sierra Nevada (Peterson et al. 1990). Not all stands have been growing faster, and some may be stressed by the relatively warm temperatures of the twentieth century. Hamburg and Cogbill (1988) attribute declining growth of red spruce in the northeastern United States to warming that inhibits seedling establishment and favors other tree species. The ability to distinguish either positive or negative effects of climate change on the forests of Europe and eastern North America is greatly hampered by the high levels of pollution in those areas.

As the above studies suggest, the response to changing climate is likely to vary widely among different forest types, depending on (among other things) the limiting factors at any given site. It seems reasonable that high-elevation, high-latitude forests in particular should benefit from longer growing seasons and (perhaps) more carbon dioxide. If there were no accompanying decline in moisture availability, forests of dry sites should also benefit from greater water-use efficiency. Such predictions must be made very carefully, however. Unforeseen secondary interactions are possible and perhaps even probable, and uncertainty in future precipitation regimes is a formidable wild card.

The importance of site factors in determining forest response to climate change was illustrated by Kienast and Luxmoore (1988), who found that tree growth had increased since 1950 at 8 of 34 sites throughout the Northern Hemisphere. All eight of these sites had moderate drought or temperature stress. Bork (1985) found that ponderosa pine stands occupying either end of a precipitation gradient in Oregon have been growing faster since 1950 than at any time in the past several hundred years; however, this is not true of stands growing in the middle of the precipitation gradient. (Bork's data are shown in fig. 15.20; the curves in the figure are yearly ring growth adjusted for changes in tree size, and indexed to the average stand growth, which is represented by 1.00 point on the *y* axis.) The differing behavior of ponderosa pine across the precipitation gradient may or may not be related to their moisture regime. Although no higher in elevation, the high-precipitation stand has a heavier snow load than the other

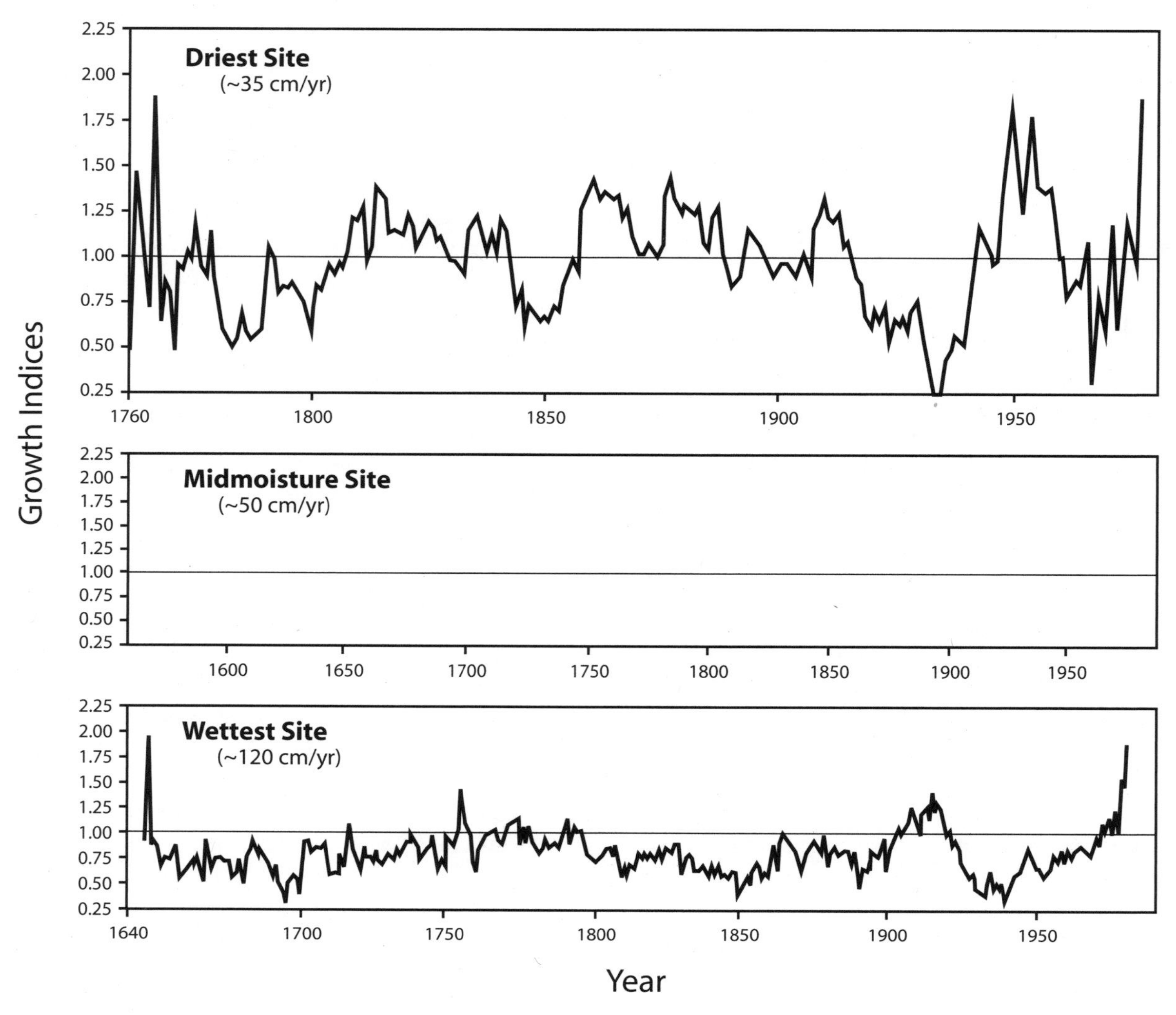

Figure 15.20. Growth indices, calculated from yearly ring growth, for three ponderosa pine forests across a moisture gradient in Oregon. Curves are ring growth adjusted to account for changes in tree diameter with age; 1.00 on the y axis represents average growth over the period of record at each site. Other values on the y axis represent proportions of the average. Note that the period of record varies among sites. All sites are at comparable elevation. (Bork 1985)

two, and this may translate into a shorter growing season. The dry-site stand might also be benefiting from the enhanced water-use efficiency associated with higher levels of atmospheric carbon dioxide. Note the large variability over the past few hundred years in the growth of all stands shown in fig. 15.20. This variability (most of which is associated with cycles of precipitation, such as the drought years of the 1930s) illustrates the difficulty of separating possible greenhouse-related growth trends from natural fluctuations.

Climate change is likely to affect forests in multiple direct and indirect ways. On any given site both positive and negative influences on NPP can be expected, with the net effect difficult to predict and possibly varying over time. On the positive side, warmer temperatures extend the growing season, and higher atmospheric carbon dioxide levels improve the efficiency of water and nitrogen use. Contingent on water availability, warmer temperatures will also increase the rate of nutrient cycling (chapter 18), although subsequent effects on productivity are far from clear.

Various factors could counteract these beneficial effects. Warmer temperatures increase respiration losses, a particularly significant factor in forests because of their large proportion of nonphotosynthesizing but respiring biomass. Kellomaki et al. (1988) suggest that pest management will be the greatest challenge for silviculturists in the coming years, a prediction borne out by the ongoing outbreak of bark beetles at an unprecedented scale in western Canada (Natural Resources Canada 2007b). How climate change might affect the ability of trees to defend themselves against pests and pathogens is unclear, but as we see in chapter 19, water stress makes trees more vulnerable to insects. Therefore, climate change could either increase or decrease tree resistance, depending on what happens to water. Plants that grow in high–carbon dioxide atmospheres invariably have lower foliar nitrogen concentrations than when growing at ambient carbon dioxide levels (i.e., they are less nutritious); insects respond to this by eating more (to get the nitrogen they need), but growing less (Bazzaz 1990). Bazzaz (1990) suggests that slower-growing insects would be more vulnerable to their natural enemies, which if true, might counteract to some degree the stimulating effect of warmer temperatures on insect population growth.

Shifting climate is likely to alter the seasonal cycles of resource availability and tree growth patterns, with uncertain effects on NPP. The yearly developmental cycle of trees is keyed to day length and temperature, and warmer springs and falls could paradoxically make trees more susceptible to late spring or early fall frosts (Cannell and Smith 1986; Kellomaki et al. 1988). Some tree species (e.g., Douglas-fir) have what is called a **chilling requirement,** which means the tree must have a certain period of cold temperatures in order to develop properly in the spring (Lavender 1984; Leverenz and Lev 1987). No one knows how field trees might respond if the chilling requirement isn't met.

The key role that precipitation changes will play is illustrated by a model from Pastor and Post (1988). The model predicts that with no change in soil water availability, a doubling of atmospheric carbon dioxide would increase NPP of forests in northeast and northcentral North America, with hardwood forests pushing north at the expense of spruce and fir. Model results were quite different, however, when moisture availability was allowed to decline. In this case, less productive (but more drought-resistant) pine and oak forests were most successful.

A central component of the Pastor and Post model is positive feedback between tree species and site conditions (section 15.6). In their simulation, the replacement of spruce and fir by northern hardwoods enhanced productivity beyond that resulting from climate alone, because the nutrient-rich, easily decomposable hardwood litter resulted in faster nutrient cycling. Just the opposite happened if spruce and fir were replaced by oak and pine, which tend to produce slowly decomposable litter.

Two aspects of the Pastor and Post model are generalizable to other areas. First, as discussed in chapters 4 and 5, the forest composition on a given site will change as new species migrate in and old species migrate out; if an altered climate shifts species dominance within existing communities, change will occur even before significant migration occurs (Bazzaz 1990). Second, feedback relations between plants and soils will strongly influence the course of community change.

Feedback will not be restricted to nutrient cycling. Belowground mutualists (i.e., mycorrhizal fungi and probably various rhizosphere bacteria) could play a key role. The transition from one community type to another will at least in theory be much smoother if at least some mutualists are shared between the old and the new plant species (Perry et al. 1990). This is probably the case throughout much of western North America, where at least some overlap exists between the mycorrhizal fungi that are supported by the various conifers (and even many of the hardwood species).

The story is not likely to be that simple, however, because the rate at which climate is predicted to change exceeds the capacity of woody plants to migrate (chapter 21). Bazzaz (1990) summarizes the potential outcome: "This rapid change [in climate] would likely result in the death of many individual plants and their replacement with early successional species. . . . Regenerating ecosystems may be the dominant ones over much of the landscape in a high CO_2 world."

In other words, the world is likely to become more weedy. If in the future a significant proportion of mature forests are indeed replaced by early successional vegetation, terrestrial carbon stores will drop sharply, creating a strong, positive feedback to the greenhouse effect. For example, pastures in Amazonia store only about 3 percent as much carbon in aboveground living biomass as the mature forest they replace (Kauffman et al. 1992). Harmon et al. (1990b) estimate that even rapidly growing young forests would take at least 200 years to replace the carbon held in old-growth Douglas-fir.

Another, more ominous dimension is added if one considers the possibility that early successional communities will not return to forest (or will do so very slowly). Should this happen, carbon stores on disturbed landscapes will not be recouped quickly—if at all. There is no doubt that under certain situations natural recovery mechanisms are disrupted and early successional plants persist indefinitely; as we saw in chapter 8, this has already occurred in some areas because of abusive land use practices (20). The potential for this to occur has been greatly exacerbated in the modern world, because exotic weeds of one kind or another are now widely spread. These plants are accomplished migrants, and at least some are considerably more aggressive than the native plants in a given region. Once annual weeds dominate a site, feedback relationships with soils could make it much more difficult for migrating woody plants to gain a foothold (Perry et al. 1990). Studies indicate that in at least some cases, dominance by exotic annuals can lead to complex changes in soil structure and biology that significantly impair the ability of tree seedlings to establish (Amaranthus and Perry 1987, 1989b; Perry et al. 1990).

Foresters can play an important role in easing the transitions among forest types that will occur. Current harvest and site preparation techniques must be critically reviewed in light of environmental changes. For example, clear-cutting exposes young tree seedlings (which are already more vulnerable to stresses than mature trees) to environmental extremes. When the silvics of a given species permit, clear-cutting, in our judgment, should be abandoned and replaced by other silvicultural tools that allow harvest without removing all trees, including shelterwoods, group selection, and perhaps single tree selection (chapter 23).

15.8 SUMMARY

Productivity is the accrual of matter and energy in biomass. Green plants are the primary producers in an ecosystem; animals and microbes that consume plant tissues are secondary producers. In forests, from 20 to 80 percent of the carbon fixed during photosynthesis is respired. The productivity that appears as biomass increment (termed net primary production [NPP]) equals gross primary productivity [GPP] minus respiration. NPP during any given period includes increment in tree tissues and tree symbionts, tissues that are shed (e.g., fallen leaves, fine roots), tissues that are consumed, and trees that die.

Forests are the most productive ecosystems on earth, covering a bit less than 50 percent of land surface area, but accounting for roughly 80 percent of terrestrial NPP (54% of total global NPP). Global forest productivity is declining because of the rapid clearing of tropical forests.

The NPP of any plant community is closely tied to its leaf area and the duration of photosynthetic activity each year. Forests are more productive than other vegetation types because they support more leaves per unit land area. The most productive forests are evergreens of the moist tropics and mesic temperate areas that photosynthesize year-round. Productivity within a given forest type may vary over relatively short distances across both mountainous landscapes and flat terrain.

Much of the increase in NPP as one goes from unproductive to more productive forests appears as wood (i.e., branches and boles). Allocation of NPP to foliage and belowground varies much less than that to wood. Current estimates of belowground productivity range from 20 percent to nearly 80 percent of total NPP, depending on the site.

NPP results from the interplay between leaf area per unit ground area (LAI) and the efficiency with which leaves are used, $E(l)$. As LAI increases beyond some critical point, the availability of light, water, and nutrients to a single leaf declines, and $E(l)$ is reduced. The structure of forest canopies facilitates transmission of light to leaves lower in the canopy, but $E(l)$ nevertheless declines in inverse relation to LAI. As a result, maximum NPP of most forests is attained at LAIs somewhere between 6 and 10 m^2/m^2; conifers and forests of the moist tropics may attain LAI values quite a bit higher than this. Although the "extra" leaves may not contribute greatly to NPP of these forests, they function in other ways such as nutrient and carbon storage.

Environmental controls over NPP are complex and understood only in broadest outline. Moreover, plants are not passive prisoners of the environment but act in numerous ways to modify the environments in which they grow. Hence, the productive potential of a given piece of ground emerges from the interaction between environmental factors and organisms. Depending on site and season, NPP may be limited by water, nutrients, temperature, light, or atmospheric carbon dioxide concentration. On any given site, different resources likely are limiting at different times during the year, for example, moisture during the dry season and nutrients or temperature during the wet season.

Water and nutrients influence both LAI and $E(l)$. Warmth is generally associated with longer growing seasons and therefore higher NPP, but warmth also speeds respiration and (because leaves are cooled by transpiration) increases plant demand for water. Resources interact in various ways to condition one another's availability. For example, to take in carbon dioxide, leaves must lose water; hence, atmospheric carbon dioxide concentration influences water use efficiency. The rate at which nutrients are cycled in ecosystems, which determines nutrient availability to plants, is tied closely to water availability.

Climate change will influence the NPP of forests; however, effects are likely to be complex and difficult to predict. Accelerated growth of some high-elevation, high-latitude forests during the last 100 years or so is apparently linked to greenhouse effects. Faster growth should not be expected in all forests; some may already be showing signs of greenhouse-related stress. Pest and pathogen problems may also increase. Effects of climate change on precipitation will play a key role in future forest productivity.

16

Forest Nutrition

LIFE AND ITS PROCESSES CERTAINLY can't be reduced to chemistry, but neither can they be understood without knowing something about the physiological and ecological functions of certain key elements—the nutrients. Nutrients carry energy, provide the basic structure of organic matter, and regulate key ecological and physiological processes. In chapter 15, we saw the importance of nutrition in primary productivity. Nitrogen fixation also depends on the availability of an array of nutrient elements other than nitrogen, and plant regulation of the hydrologic cycle depends on the nutrients that support root growth and regulate stomata. The relative abundance of different nutrients influences food web dynamics and therefore secondary productivity and, through effects on food webs, the rate at which the nutrients themselves are cycled (chapters 18 and 19). Ecosystem health is critically dependent on the ability of the system to retain nutrients through disturbances, and experience has taught us that the long-term productivity of managed forests can be reduced if system nutrient capital isn't adequately protected (chapter 20).

The next three chapters deal with nutrients and nutrient cycling. The first step in understanding the role nutrients play in ecosystem function is to ask what chemical elements are needed by life and why (i.e., what is their function?). This chapter addresses those questions. Section 16.1 surveys the essential elements and their role in tree physiology and introduces some aspects of their cycling within ecosystems (cycling is taken up in detail in chapter 18). Section 16.2 compares nutrient requirements among various tree species and forest types, and section 16.3 discusses methods of diagnosing nutrient deficiencies. (Readers who have had a recent course in plant physiology can probably skip this chapter.)

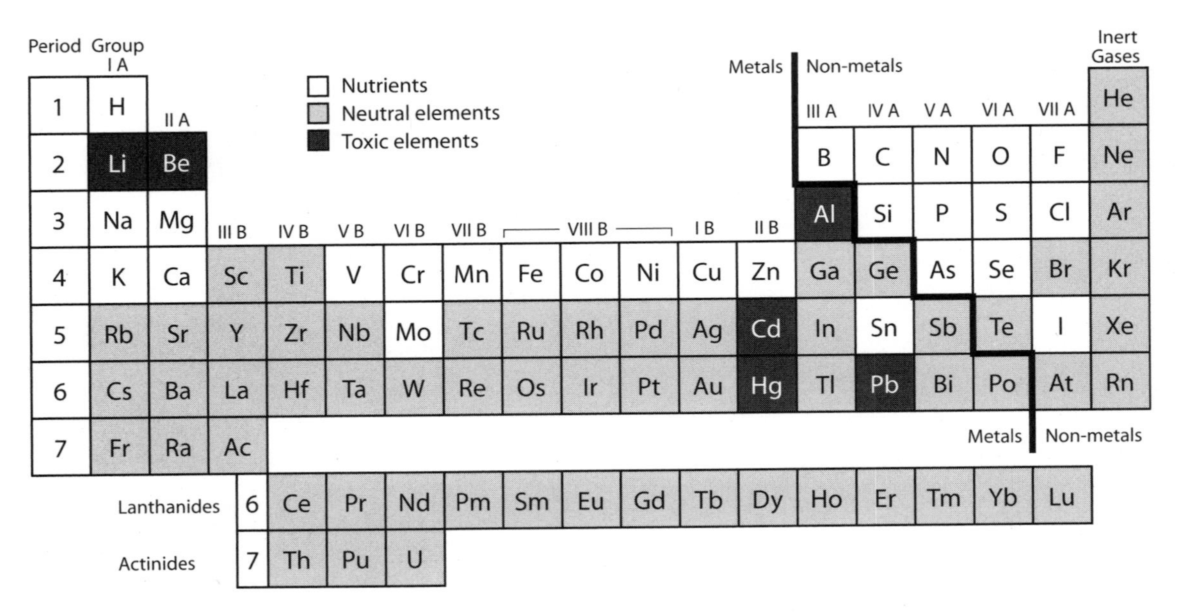

Figure 16.1. The table of the elements, showing essential nutrients (white), toxic elements (black), and "neutral" elements that have no known positive or negative effect on organisms. (Adapted from Bohn et al. 1979)

16.1 THE ESSENTIAL NUTRIENTS AND THEIR PHYSIOLOGICAL ROLES

16.1.1 The Essential Elements

Figure 16.1 distinguishes the chemical elements as to whether they are essential, toxic (at low concentration), or neutral to one or more life forms. Nutrients that are essential for plants and microorganisms are divided into the **macronutrients,** or those required in large amounts (i.e., hydrogen, carbon, oxygen, nitrogen, calcium, potassium, phosphorus, magnesium, and sulfur),[1] and the **micronutrients,** or those required in small amounts (i.e., boron, chlorine, manganese, iron, copper, zinc, molybdenum, nickel, and additionally vanadium for microorganisms). Animals require all of these—with chlorine as a macronutrient—plus sodium as a macronutrient and several additional micronutrients (fluorine, silicon, chromium, nickel, cobalt, arsenic, selenium, tin, and iodine) (Bohn et al. 1979; Bowen 1979). The so-called **beneficial elements** promote growth of many plant species but are not essential to complete the life cycle; these include sodium, silicon, and cobalt (Mengel and Kirkby 2001; Barak 2003). The list of essential micronutrients has expanded as we have become more aware of the biochemical details of plant and animal physiology and more species have been investigated (nickel was not added to the list until 1992).

The distinction between macronutrients and micronutrients is somewhat arbitrary. While levels of carbon, oxygen, nitrogen, calcium, and potassium in plant tissues are usually far above those of any micronutrient, concentrations of magnesium, and sulfur may not be much higher than those of iron and manganese. Some nutrients, for example manganese and chlorine, may be contained in plant tissues in much higher amounts than known requirements for physiological processes. Many tree species accumulate high levels of sodium, although in most cases it has no clear physiological function. (This is fortunate for animals, however, which require a source of sodium in their diet.)

In sufficiently high concentrations, virtually any chemical can be toxic to life, but lithium, beryllium, cadmium, mercury, aluminum, and lead are toxic at relatively low concentrations (they disrupt the functioning of certain enzymes). Nickel, which is an essential micronutrient for all plants, also becomes toxic if the levels are too high. Solubility of metals such as aluminum, mercury, and lead increases with decreasing pH: aluminum can reach harmful levels in soils affected by acid rain (Khanna and Ulrich 1984), and mine spoils frequently have high levels of lead and mercury.

As with many other aspects of life, it is important to remember that organisms are quite diverse and (within certain bounds) highly adaptive in their nutrient requirements and particularly in their tolerances. Plants growing on mine spoils or acid soils may evolve physiological adaptations that allow them to tolerate high levels of heavy metals (Larcher 2003). Some plants take up and concentrate nonessential and even potentially toxic elements such as aluminum and lead. Species in the family Ericaceae, for example, which are common associates of conifers throughout western North America, concentrate aluminum (Bowen 1979). In the montane conifer forests of northwestern North America, mountain hemlock concentrates large amounts of aluminum in

[1]Carbon, oxygen, and hydrogen are the only nonminerals among the essential elements.

aboveground tissues, while Pacific silver fir avoids this by sloughing roots that have accumulated aluminum (Vogt et al. 1987). Some species of fungi accumulate mercury and cadmium, and mycorrhizae may play an important role in protecting plants growing on acid soils from excessive uptake of heavy metals (Smith and Read 1997; Read 1987). Low to moderate levels of aluminum are reported to actually improve root growth of trees in the genera *Camellia* and *Eucalyptus* (Bowen 1979; Silva et al. 2004), and to enhance phosphorus uptake in *Pinus radiata* (Turner and Lambert 1986b). In *Eucalyptus,* aluminum toxicity is avoided by complexation with malic acid (Silva et al. 2004). Whether the accumulation of toxic elements by one species has ecological significance (i.e., whether it mediates the exposure of other species in the community to toxic chemicals) is not known.

16.1.2 Physiological Functions and Cycling Characteristics of the Nutrients

Four groups of nutrients can be distinguished regarding their role in plant physiology and their cycling characteristics within ecosystems. The following information is adapted and modified from Mengel and Kirkby (2001), Bowen (1979), and Clarkson and Hanson (1980). These authors discuss the biochemical aspects of nutrient function in some detail. (Larcher 2003 gives a somewhat different grouping). For simplicity, we have not included nutrients such as sodium, which are insignificant in plant physiology. The following is also briefly summarized in table 16.1.

16.1.2.1 Group 1: Carbon, Hydrogen, Oxygen

These nonmineral elements form the basic structure of organic matter. Carbon serves as the skeleton of organic molecules and composes between 40 and 60 percent of the dry weight of living tissue. Hydrogen and oxygen bond covalently to carbon, and together, the three compose common molecular types such as hydrocarbons (e.g., sugars, starches, cellulose), fatty acids (e.g., lipids, waxes), terpenes, and phenols (e.g., lignin, tannin). The numerous molecules formed solely by carbon, oxygen, and hydrogen function in a variety of ways. Most serve as either structural elements (e.g., cellulose, lignin) or energy-storage compounds (e.g., sugars, starches); however, some (most notably the terpenes and various phenolic compounds) protect plants against grazing animals, pathogens, and ultraviolet radiation.

Carbon and oxygen are taken up primarily as CO_2 from the atmosphere, and hydrogen is obtained when the water molecule is split during photosynthesis. Because they are released from organic forms as gases and therefore escape to the atmosphere (e.g., the evolution of CO_2 during respiration and O_2 during photosynthesis), these elements are not cycled to any large degree within local ecosystems. Some local cycling does occur;for example, organic molecules released during decomposition of dead tissues can be taken up by plants, and CO_2 or O_2 evolved during metabolism can be taken up by the same or different organisms. The acidity generated by hydrogen ions (H^+) significantly influences several soil processes (chapters 14 and 18).

16.1.2.2 Group 2: Nitrogen, Sulfur, Phosphorus

These nutrients form strong covalent bonds with carbon (nitrogen and sulfur) or oxygen (sulfur and phosphorus). (As we discuss in chapter 18, whether an element is bonded to carbon or to oxygen has important implications for the manner in which it is cycled in the ecosystem.) Combination of one or more of these elements with carbohydrate skeletons produces the basic structure of essential organic molecules such as proteins (nitrogen and sulfur), phospholipids (phosphorus), and nucleic acids (nitrogen and phosphorus). The addition of reduced nitrogen (NH_3) to a

Table 16.1
An abbreviated summary of physiologic functions and cycling characteristics of the most important plant nutrients

Group	Nutrients	Physiologic function	Cycling
1	C, O, H	Form the basic skeleton of all organic molecules. Sole constituents of polysaccharides, phenols, terpenes.	Cycled primarily at the global level. Little within local ecosystems.
2	N, P, S	Constituents of protein (N, S), nucleic acids (N, P), phospholipids (P), and the ATP-ADP complex.	Frequently contained in parent material at levels far below biologic demand (no N in parent material). Closely associated with soil organic matter and cycled very tightly. Most likely to limit productivity.
3	Ca, K+, Mg	Involved in numerous physiologic processes including osmotic regulation, stomatal behavior (K), photosynthetic transport (K), membrane integrity (Ca), photosynthesis (Mg). Essential for activation of numerous enzymes (especially K and Mg).	Usually abundant in parent material, although this varies. Not abundant in highly weathered soils. Cycling is also very tight for these elements and is tied to the soil's cation exchange complex.
4	Fe, Mn, Co, Zn, Mo, B	The micronutrients required in relatively small amounts. Numerous roles as activators and mediators of physiologic processes.	For the metals, solubility and therefore availability to plants and microbes depends on strong bonding to specialized organic molecules called chelates.

carbohydrate molecule containing an acid group (COOH) produces the basic structure of amino acids, which are, in turn, the basic building blocks of proteins. Sulfur is a constituent of various organic compounds, but 90 percent of plant sulfur is contained in the amino acids cysteine, cystine, and methionine (Stevenson 1986). Proteins are sources of energy and, in animals, also play a structural role. However, their most important function is as enzymes, the catalysts that govern the processes of life. Phosphorus is an essential constituent of ribosomes (the "protein assemblers" of cells), the molecules ATP and ADP (the energy transport system of cells), and nucleic acids (which combine to form DNA and RNA).

Like the elements of group 1, nitrogen and sulfur can occur as gases under normal environmental temperatures (roughly 80% of the earth's atmosphere is N_2, but nitrogen in this form can be used only by nitrogen-fixing microorganisms), but phosphorus does not. Unlike carbon, oxygen, and hydrogen, the elements of this group are cycled tightly within ecosystems. In soils, these elements are associated closely with organic matter, especially nitrogen. Phosphorus and sulfur also occur as negatively charged, inorganic phosphates and sulfates that combine with various metallic cations. Phosphorus in particular tends to form insoluble complexes with iron, aluminum, and calcium,[2] thereby becoming increasingly unavailable to plants as soils age (Crews et al. 1995). Compared with the other major nutrients, phosphorus is by far the least mobile and therefore least available to plants in most soil conditions (Hinsinger 2001). Plants counter that by devoting considerable energy to increasing phosphorus availability in their rhizospheres (e.g., by altering pH and releasing organic molecules that solubilize phosphorus) and arraying abundant gathering surface (primarily mycorrhizal) (Hinsinger 2001). Organic forms of phosphorus may be quite important to plant nutrition but are not measured in standard soil assays (McGrath et al. 2001).

16.1.2.3 Group 3: K^+, Ca^{2+}, Mg^{2+}

These monovalent or divalent cations play numerous roles in plant physiology. They may occur as free ions, salts of organic acids, or components of macromolecules. From 15 to 20 percent of magnesium in the cell is contained within the chlorophyll molecule, and calcium is a structural component of the middle lamella. All are involved at one level or another in enzyme function. Over 60 different enzyme systems require a monovalent cation for activation, and of the possibilities (K^+, Na^+, NH_4^+), K^+ is the most effective. Magnesium is essential to energy transfers involving ATP and ADP, serving as a bridge linking enzymes with the ATP or ADP molecules. Calcium plays numerous critical roles in plants. According to McLaughlin and Wimmer (1999), calcium plays an important regulatory role in "phosphorylation of nuclear proteins; cell division; cell wall and membrane synthesis and function; intra- and intercellular signaling; protein synthesis; responses to environmental stimuli, including low temperatures, gravity, insects and disease; stomatal regulation; and carbohydrate metabolism." Calcium is also a keystone element serving as a "messenger in many growth and development processes and in plant responses to biotic and abiotic stresses" (Reddy 2001).

As salts of organic acids, the elements of this group preserve electrical neutrality and act as osmotic agents that maintain cell turgor. As free ions, on the other hand, these nutrients maintain an electrical gradient across membranes, thereby allowing cells to convert electrical potential to work. The latter mechanism is the basis of nerve action in animals; in plants it mediates such things as stomatal opening and closure, which results from turgor changes in guard cells triggered by influx and efflux of K^+ (other cations don't have the same effect). Many of the processes mediated by these elements require a certain ion; others will not do. On the other hand, one cation can substitute for another in some physiological processes (NH_4^+, Na^+, Al^{3+}, and cations discussed below in group 4). Such substitutions frequently produce subpar results, however, and nutrient imbalances within tissues can disrupt physiological function. For example, virtually any cation can compete with calcium ions for binding sites on membranes, but none is as effective in maintaining membrane integrity. When (as sometimes occurs) there is an imbalance between calcium ions and other cations, a high proportion of binding sites become occupied by relatively ineffective elements, and membranes become leaky.

From a cycling standpoint, these elements differ from those of group 2 in several respects. They are usually more abundant in parent material. They are not closely associated with soil organic matter in relatively young soils (although their cycling depends critically on cation exchange sites provided by soil organic matter and clays). However, as soils age, minerals become depleted of these elements, and increasingly they are either fixed in biomass or leached to streams. In consequence, organic matter serves as a primary reservoir for these elements in highly weathered tropical soils (Silver et al. 2000). Nutrients in this group are quite soluble, enhancing both their availability to plants and their susceptibility to leaching from the system to streams. As we saw in chapter 15, excessive leaching of calcium and magnesium has led to deficiencies of these elements in stands affected by acid rain (Oren and Schulze 1989; also see chapter 20 in this volume). Calcium losses in particular have raised concerns about effects on both individual tree and ecosystem health (McLaughlin and Wimmer 1999; Schaberg et al. 2001). McLaughlin and Wimmer (1999) hypothesize calcium in particular exerts a significant control on both the structure and the function of forest ecosystems. That is supported by recent research showing that "tree species rich in calcium (are) associated

[2]Which metal phosphorus complexes with is determined by pH. At the acid pH values of most forest soils, complexes are with iron and aluminum.

with increased native earthworm abundance and diversity, as well as increased soil pH, exchangeable calcium, percent base saturation and forest floor turnover rate" (Reich et al. 2005).

16.1.2.4 Group 4: Iron, Manganese, Copper, Zinc, Molybdenum, Boron, Chlorine, Nickel

These are the most universally important plant micronutrients. All except boron (and perhaps chlorine) are essential for the activity of one or more enzymes. Manganese can substitute for magnesium to some extent as a bridge between enzymes and ATP or ADP. In addition to their role in enzyme activity, iron and manganese, which readily change valency state, mediate electron transfer (i.e., oxidation-reduction, or **redox** reactions) within cells and in the soil. Iron is an essential constituent of the cytochrome system and ferridoxin, which is involved in electron transport during photosynthesis. Manganese is also involved in electron transfer and, additionally, with oxygen evolution during photosynthesis. Chlorine plays a role in oxygen evolution during photosynthesis as well. The function of boron was a mystery for many years, but evidence now points to roles in membrane function, metabolic pathways, and protection against aluminum toxicity (Blevins and Lukaszewski 1998).

Except for iron, these elements occur in very small quantities within parent material; however, only small amounts are required by plants. Micronutrient deficiencies do occur and may be more widespread than now realized (see the following section on nutrient limitations). The metals in this group (iron, manganese, copper, zinc, molybdenum, and nickel) are insoluble to one degree or another at the pH values normally found in forest soils; hence, their abundance is not a good measure of their availability to plants. Solubility of the metals increases through a particular kind of organic combination called **chelation** (from the Greek *chele,* for "crab's claw"). In chelation, the metal ion is enveloped within a large organic molecule (the **chelate**), forming a soluble **chelate complex** (fig. 16.2). Chelates, which include humic and fulvic acids and a wide variety of compounds released into the soil by roots, mycorrhizae, and microbes, are essential to the iron, copper, and zinc nutrition of plants and microbes (Bohn et al. 1979).

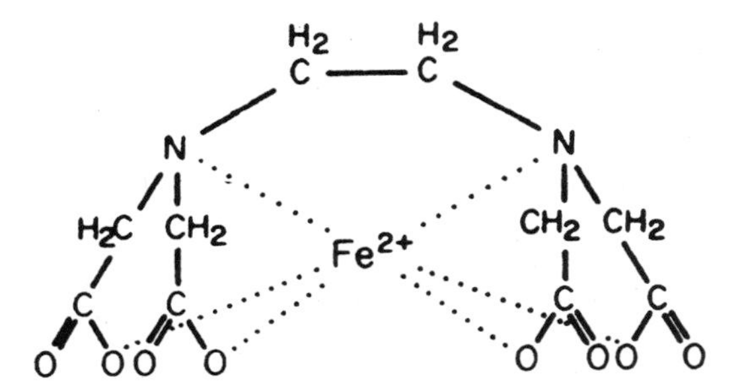

Figure 16.2. A chelate complex. The organic molecule is chelating the Fe^{2+} molecule. (Adapted from Bohn et al. 1972)

16.2 NUTRIENT REQUIREMENTS AND LIMITATIONS

16.2.1 Basic Nutrient Requirements of Organisms

Metabolically active plant tissues (i.e., excluding dead tissues such as heartwood) are approximately 70 percent water, 27 percent organic matter, and 3 percent minerals (Mengel and Kirkby 2001). Table 16.2 shows the approximate ranges of nutrient concentrations (other than carbon, oxygen, and hydrogen) in tree foliage. (Actual concentrations vary among species and with site.) On a dry weight basis (i.e., organic matter plus minerals), over 90 percent of tree foliar mass consists of carbon, oxygen, and hydrogen. The remaining nutrients compose only a small fraction of biomass, because their primary function in living systems is as regulators and mediators of processes rather than components of structure. Nutrients other than carbon, hydrogen, and oxygen occur in greatest concentrations in living tissues (foliage, cambial layers, and growing root tips) and in lowest concentrations in stem wood. As a rough guide, element concentrations in small branches and small roots are 50 to 70 percent those in foliage, in large branches and roots 20 to 30 percent those in foliage, and in stems 10 to 20 percent those in foliage. (Remember that concentration—amount per unit mass—is not the same as content.) In evergreens, foliar concentration of most nutrients declines with leaf age; calcium is an exception, tending to increase with leaf age.

Table 16.2
The ranges of nutrient concentrations reported in the foliage of forest trees

Element	Range of forest trees	
	Deficient	Healthy
Nitrogen	<1.0*	1.1 to 4.0
Calcium	<0.12*	0.1 to 0.7
Potassium	<0.50*	0.5 to 1.6
Magnesium	<0.08*	0.1 to 0.4
Phosphorus	<0.10*	0.1 to 0.3
Sulfur	<0.15*	0.2 to 0.3
Iron	<30†	50 to 100
Manganese	<0.01	0.01 to 0.5
Chlorine		0.001 to 0.3
Zinc	<5†	10 to 125
Copper	<3†	4 to 12
Boron	<7†	10 to 100
Molybdenum	?	0.05 to 0.25

Source: Adapted from Powers 1974.
*Values are percentage of dry weight.
†Value in parts per million.

Table 16.3
Some representative nutrient concentrations in different life forms

Life form	Ash, %	C, %	H, %	N, %	O, %	S, %
Tree foliage	4.4	46.4	5.5	1.6	42.0	0.14
Bacteria	8.0	48.5	7.4	10.7	24.8	0.61
Fungi	5.0	49.4	7.0	5.1	33.1	0.40
Invertebrates	7.4	41.4	4.5	8.7	35.6	1.2
Fish	4.3	47.5	6.8	11.4	29.0	1.0
Mammals	12.0	48.4	6.6	12.8	18.6	1.6

Source: Adapted from Bowen 1979; Rodin and Bazilivich 1967; and Ovington 1956.

Because nutrient cycling involves uptake and release of elements by plants, microbes, and animals, the relative concentration of various elements in tissues of these different life forms is an important determinant of cycling characteristics (chapter 18). Animals, richer in protein than plants, contain two to three times higher concentrations of nitrogen and sulfur than tree foliage (table 16.3). Bacteria are also relatively high in nitrogen and sulfur, while values for fungi fall between plants and bacteria. The ash content (nutrients other than carbon, hydrogen, oxygen,nitrogen and sulfur) of bacteria, invertebrates, and mammals is also higher than that of tree foliage. The values given in table 16.3 should be taken as rough representations only: chemical composition of species, and even individuals within each of these groups will vary to one degree or another, especially the trees and the

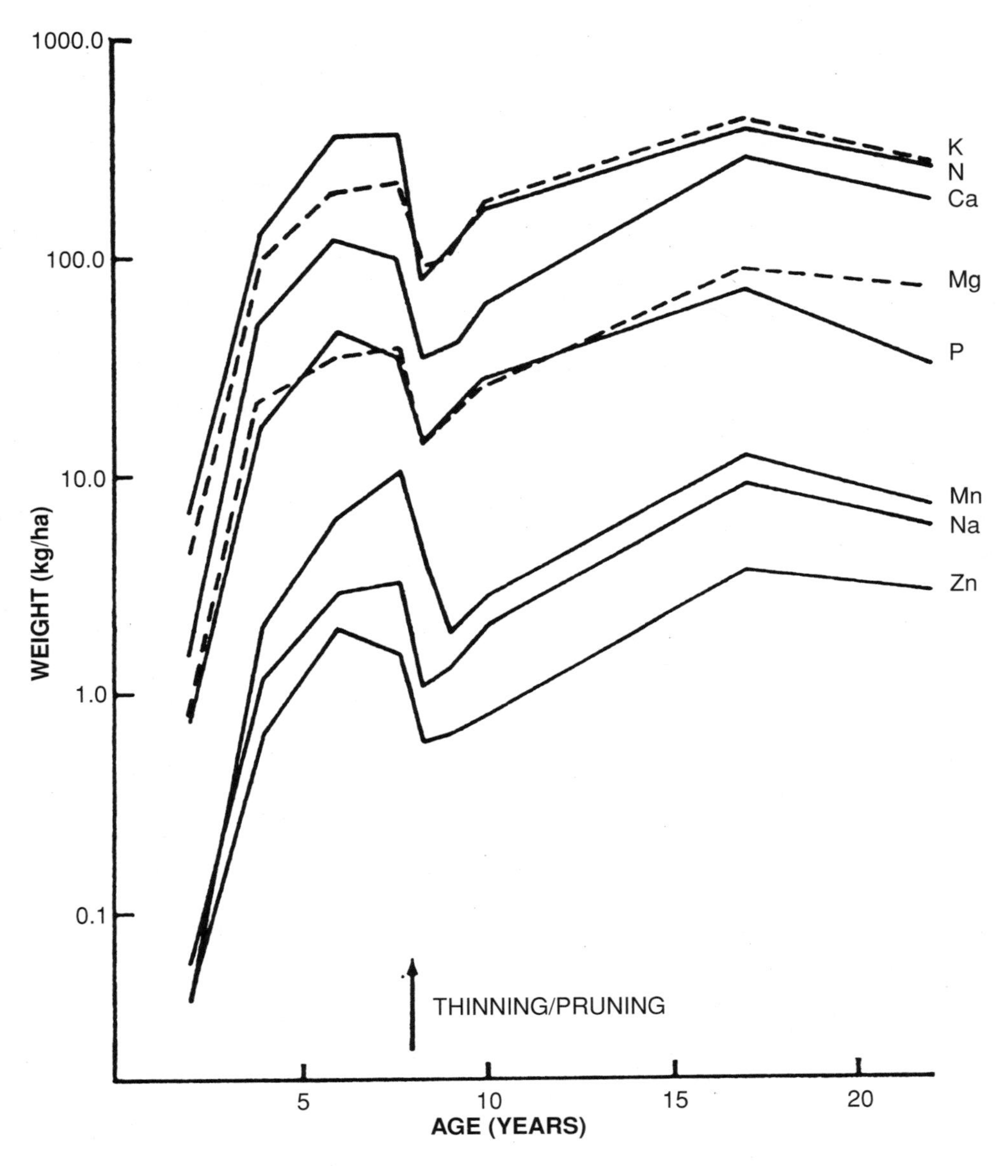

Figure 16.3. The accumulation of nutrients by stands of radiata pine in New Zealand. Note the log scale on the *y* axis. (Madgwick et al. 1977)

invertebrates. Nevertheless, the main point remains valid: plant tissues are lower than those of other life forms in all elements except oxygen.

So long as they are growing well, young, even-aged stands accumulate nutrients very rapidly, reflecting the rapid growth of foliar biomass early in stand life. Figure 16.3 shows this pattern for radiata pine stands in New Zealand. Although the demand for different nutrient elements varies widely among species and stands of the same species growing on different sites, fig. 16.3 exemplifies a common pattern with regard to the relative demand for different elements. Nitrogen, potassium, and calcium are accumulated in highest amounts by tree species, followed by phosphorus, sulfur (not shown), and magnesium. According to Mengel and Kirkby (2001), in all species of higher plants "the N and K content of green plant material is about 10 times higher than that of phosphorus and magnesium which is in turn about 10–1000 times higher than the content of the micronutrients."

16.2.2 Variations among Tree Species and Forest Types

Table 16.4 compares average foliar nutrient concentrations (percent of dry weight) among conifers, temperate angiosperm trees (all deciduous), and tropical angiosperm trees. The first thing to note is the wide variation among species within any given forest type; ignoring that for the moment, some general patterns can be seen. Temperate deciduous trees average the highest foliar nutrient concentrations, particularly when compared to conifers. However, it does not follow that conifers require fewer nutrients on a per-tree basis, because they frequently support more leaf biomass than angiosperms. Maintaining relatively low foliar nutrient concentrations appears to be a strategy of allocation for at least some conifers. Douglas-fir, for example, preferentially use nutrients supplied in fertilizers to produce additional foliage rather than increase the nutrient concentration of existing leaves (Velasquez-Martinez et al. 1992). The relatively low nutrient concentrations of evergreen trees are often seen in evergreen shrubs also. Ericaceous plants, for example, which commonly associate with conifers in many temperate and boreal forests, also tend to have low levels of foliar nutrients. Among the conifers, shade-tolerant species such as spruce and fir average somewhat higher foliar nitrogen and sometimes cations than shade-intolerant species.

The primary thing to note about the foliar nutrients of tropical trees shown in table 16.4 is the especially wide variation among species (seen in the ranges of nutrient concentrations), which makes the average values rather meaningless. Species differences within any of the broad forest types may reflect differing nutrient requirements, differing environments (e.g., fertile vs. infertile soils), or differences in the way the data was collected. Much of the variability in nutrient concentration among tropical trees reflects the wide variations in soil fertility that occur in the tropics. The link between the chemistry of soils and the chemistry of foliage in the tropics is illustrated in table 16.5; note these values are averages for the community rather than of a single species. The moist Panamanian forest grows on calcium- and magnesium-rich soils formed from sediments, while the premontane wet forest grows on soils formed from basalts and rich in micronutrients. In contrast, the Amazonian forest types shown occur on old, highly leached and very infertile "bleached white sands" and have very low foliar concentrations of phosphorus, calcium, and potassium.[3] The relatively high nutrient concentrations of dry tropical species also reflects, at least in part, the soils on which they grow, which are less heavily leached than those of the moist tropical forests. The phenomenon illustrated in table 16.5 is not restricted to the tropics. Soil fertility varies widely in all forest types, and that variation is usually reflected in foliar nutrients.

Another factor that must be accounted for when comparing nutrients among species is that nutrient concentration (i.e., weight of nutrients per dry weight of foliage) may not accurately reflect the physiological requirement for nutrients. The demand for nutrients involved in photosynthesis is greatest in cells nearest the surface of leaves, where

Table 16.4
Comparison of foliar nutrient concentrations among conifers, temperate angiosperm trees, and tropical angiosperm trees

		Angiosperms, % dry weight	
	Conifers, % dry weight	Temperate	Tropical
N	1.37% (1.07–2.16)	2.20% (1.36–2.81)	1.21% (.61–2.43)
P	0.14 (.07–.21)	0.26 (.10–.60)	0.10 (.01–.51)
Ca	0.57 (.38–1.03)	1.57 (.73–3.0)	0.87 (.11–3.46)
K	0.56 (.09–.20)	1.40 (.73–1.88)	0.88 (.19–2.80)
Mg	0.15 (.09–.21)	0.36 (.23–.67)	0.30 (.06–.78)
S	0.06 (.03–.07)	0.15 (.01–.37)	0.20 (.02–.40)

Source: Rodin and Bazilivich 1967; Ovington 1956; Klinge 1985.
Note: Values in parentheses are the ranges of the data (min and max). The number of tree species used to derive these values varies depending on the particular nutrient. For conifers, P, Ca, K, and Mg are based on 13 species, N on 12, and S on 4. Conifer genera include *Pinus, Picea, Larix, Tsuga, Pseudotsuga, Abies,* and *Juniperus.* For temperate angiosperms, Ca, K, Mg, and S are based on 8 species, and N on 7. All temperate angiosperms are deciduous and include species in *Quercus, Tilia, Fagus, Fraxinus, Ulmus, Betula,* and *Populus.* For tropical trees, P, Ca, K, and Mg are based on 33 species in 33 genera, N on 30 species in 30 genera, and S on 16 species in 16 genera.

[3]The different types of forest forming on bleached white sands result largely from differing water regimes. Tall or "high" caatinga occurs on relatively fine grained sands that are not flooded annually but do have high water tables (within 100 cm of the surface). Igapo, which consists of low vegetation, is flooded during high water each year. Campina is composed of low sclerophyllous vegetation assumed to be an early stage of succession to high forest (Jordan 1985).

Table 16.5
Foliar nutrient concentration in neotropical forests

Forest type	Nutrient %								Study
	Nitrogen	Phosphorus	Calcium	Potassium	Magnesium	Iron	Manganese	Molybdenum	
Tropical moist, Panama	1.40	0.1200	1.780	1.100	0.200	0.0176	0.0063	0.0004	Golley 1983b
Premontane wet	1.50	0.1000	1.050	0.830	0.270	0.0511	0.0352	0.0007	Golley 1983b
Amazonian Forests on Infertile Bleached Sands									
Campina	1.11	0.0480	0.370	0.658	0.264	—	—	—	Klinge 1985
Tall caatinga	1.08	0.0560	0.528	0.583	0.356	—	—	—	Herrera 1979
Tall caatinga	1.16	0.0730	—	—	—	—	—	—	Medina 1984
Igapo	1.73	0.0618	0.251	0.632	0.122	—	—	—	Klinge et al. 1983
Dry forests									
Deciduous sp.	2.10	0.1580	1.580	1.700	0.360	—	—	—	Medina 1984
Evergreen sp.	1.18	0.0710	—	—	—	—	—	—	Medina 1984

Note: Values are averages of various species in each community.

chloroplasts are concentrated; hence, expressing nutrients as per unit leaf area rather than per unit weight may better reflect the tree's status with regard to many nutrients (Smith et al. 1981). For example, compared to deciduous species, evergreens have relatively low surface area for a given leaf weight (Medina 1984), and the leaves of early successional species frequently have less surface per unit weight than those of late successional species. (Leaf area per unit weight is called **specific leaf area** [SLA]. Foliage with high SLA is relatively flat and thin, while foliage with low SLA is thicker and sometimes cylindrical.)

When foliar nutrients are expressed per unit leaf area rather than per unit leaf weight, differences between species with differing leaf morphologies may be narrowed or even reversed. This phenomenon is well illustrated by Medina's (1984) comparison of deciduous and evergreen species growing in dry tropical savannas. There, nitrogen and phosphorus concentration (per unit weight) were far lower in the evergreens than in the deciduous species (table 16.6). When expressed per unit leaf area, however, quite a different picture emerges. Evergreens average 4.3 g of nitrogen and 0.23 g of phosphorus per m^2 of leaf area, compared to only 1.6 g of nitrogen and 0.12 g of phosphorus for the deciduous species. Note in table 16.6 the similarity in both SLA and foliar nutrients between the tropical evergreens and the temperate conifer, Douglas-fir.

For the most part, it is very difficult to separate the degree to which environment on the one hand and genetic adaptation on the other contribute to differences in the nutrient requirements of trees growing on different sites. Trees growing on infertile soils are usually (and probably always) genetically adapted to conserve and utilize nutrients with high efficiency (Harrington et al. 2001; chapter 15 in this volume); hence, the differences between tropical moist forest growing on nutrient-rich soils in Panama and tall caatinga on bleached sands in Amazonia are likely to reflect (at least in part) genetic predisposition to low or high nutrient availability.

Although on a broad scale the chemistry of vegetation reflects the chemistry of soils, this is not necessarily true on a local scale, and species growing on the same soil can differ significantly from one another in both foliar nutrient concentration and total nutrient uptake. Note, for example, the difference between deciduous and evergreen species of the tropical dry site (table 16.6)—a pattern similar to that seen in the temperate zone and related to differing leaf morphology. Figure 16.4 shows biomass and nutrient content in stands of four different tree species growing on the same soil in Minnesota. Total tree biomass was greatest in

Table 16.6
Foliar nutrients of some species of deciduous and evergreen trees

	Specific leaf area (m^2/kg)	Nitrogen		Phosphorus		Source
		%	g/m^2	%	g/m^2	
Dry tropical forests						
Deciduous	14.18	2.21	1.6	0.17	0.12	Medina 1984
Evergreen	3.24	1.18	4.3	0.07	0.23	Medina 1984
Oregon Douglas-fir forests*						
High nitrogen fertility	2.35	1.10	4.5	—	—	Smith et al. 1981
Moderate nitrogen fertility	2.86	0.95	3.0	—	—	Smith et al. 1981

*Average growing season values of old and fully expanded new foliage.

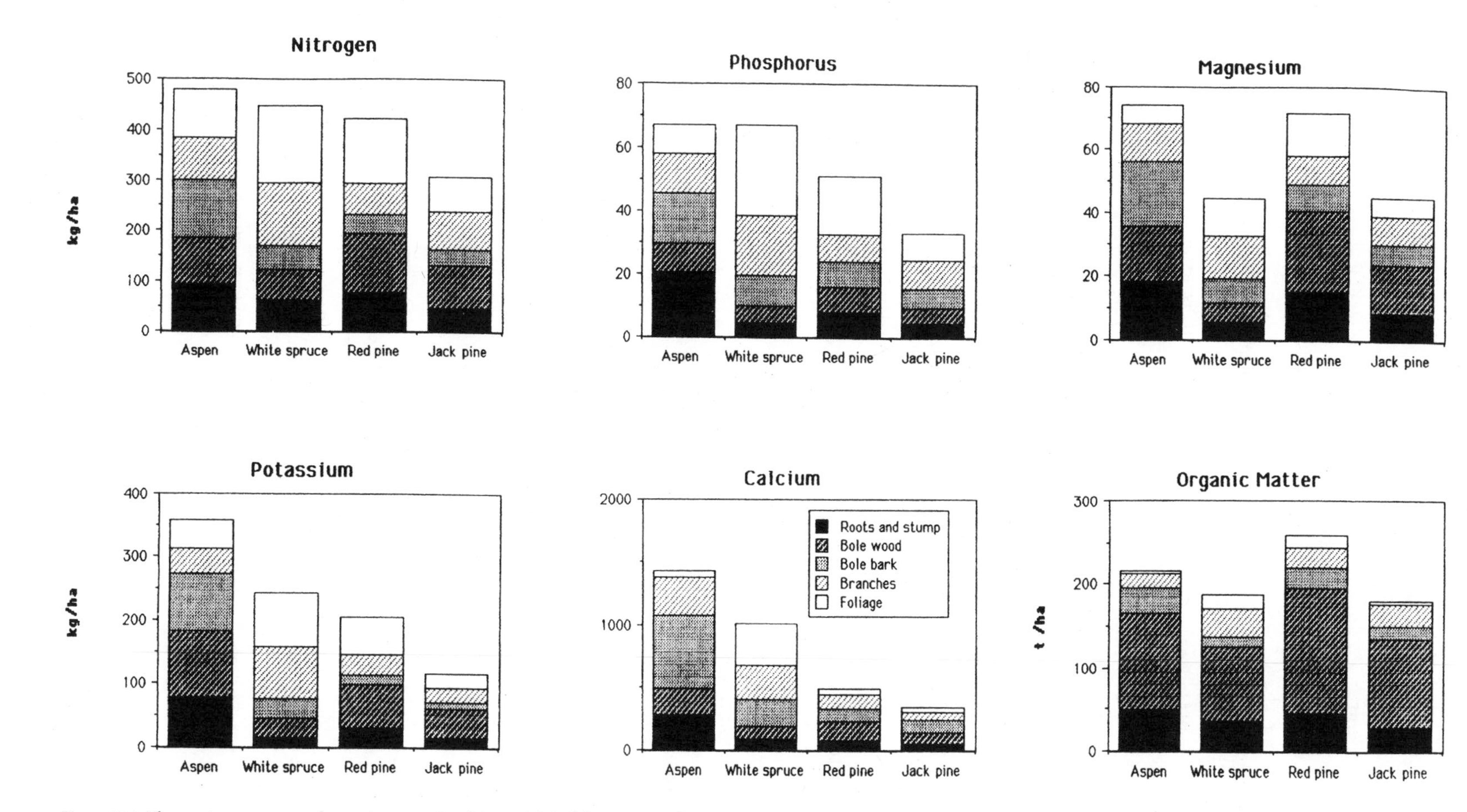

Figure 16.4. The nutrient content and organic matter (total dry weight) of four species of trees growing on the same soil type in Minnesota. (Adapted from Alban et al. 1978)

stands of red pine and differed little among stands of the other three species; however, total nutrient uptake did not follow the same pattern. Aspen stands accumulated relatively high levels of all nutrients (note the high nutrient levels in aspen bark), and jack pine stands accumulated relatively low levels. Note also the difference between shade-tolerant white spruce and the shade-intolerant pines in nutrient distribution among the various tree components. The majority of aboveground nutrients in the spruce stand are contained in crowns (branches plus foliage), while in the pine stands, which because of their intolerance to shade have smaller crowns for a given tree size, nutrients are distributed more evenly among boles and crowns. As we discuss further in chapter 17, differences in nutrient distribution among tree components have important implications for the impact of whole-tree harvest on site fertility.

16.3 DIAGNOSING NUTRIENT DEFICIENCIES

Deficiency of one or more nutrients is likely to limit the productivity of many, if not most, forest ecosystems (chapter 15). This statement must be interpreted with some caution. It does not necessarily follow that ecosystems on infertile sites are more stressed than those on fertile sites. Communities are usually well adapted to the conditions of their environment. "Success" in nature resides in the ability of individuals to survive and reproduce, which may or may not be related to biomass production. Nevertheless, the forester is usually interested in productivity, and it has been amply demonstrated that tree growth is linked to site fertility.

Determining which (if any) nutrients limit growth of a particular species on a particular site is not straightforward. It is beyond the scope of this book to discuss the problem and the various approaches to its solution in detail (see Tamm 1964; Van Den Driessche 1974; Keeney 1980; Powers 1983; Binkley 1986a ; Walworth and Sumner 1987; Binkley and Hart 1989; Scott et al. 2005). Briefly, two steps are involved:

1. A reliable and relatively easily measured indicator of nutrient deficiency must be derived. Reliability simply means that there must be a reasonable correspondence between the values that can be measured for the indicator and the nutrient status of the plant. Three types of indicators are traditionally used: nutrient concentrations or content of plant tissues (usually foliage, but sometimes fine roots), visual symptoms of deficiency (e.g., yellow foliage), and some measure of soil nutrient status.
2. The indicator must be calibrated (i.e., the range of values that can be measured or, in the case of visual symptoms, observed, must be associated with tree growth or some other measure of vigor). For example, suppose we measure 1.0 percent nitrogen in new foliage of ponderosa pine. Does that indicate nitrogen deficiency? What about 1.5 percent, or 0.5 percent? Calibration of indicators must be done through experimentation or extensive field sampling. The time and, particularly, expense involved in this are major limitations on our understanding of forest nutrition.

16.3.1 Tissue Concentration or Content

Foliar nutrient concentrations are the most widely used indices of tree nutrient status (e.g., Van Den Driessche 1974). Usually, living foliage is used, but nitrogen concentration of shed needles was found to be good indicator of nutrient status of Corsican pine (Miller and Miller 1976) and Douglas-fir (Van Den Driessche and Webber 1977). In conifers, nutrients are withdrawn from older needles to support new needle growth in spring, and the older needles recharged with nutrients from soil later in the growing season (chapter 18). Because of this pattern, old needles are believed to better reflect than new needles shortages in the supply of nutrients from soil. Powers (1984) recommends use of one-year-old rather than current-year needles to detect nutrient deficiencies in conifers.

Root nutrient concentrations have also been used to indicate tree nutrient status and in some cases are more responsive to fertilization than foliage concentrations (Dighton and Harrison 1983; Adams et al. 1987). Van Den Driessche and Webber (1977) found that the level of soluble nitrogen in Douglas-fir roots (as opposed to total nitrogen) differed significantly with level of fertilization. Stark et al. (1985) have also used the nutrient content of xylem sap to assess tree nutrient status.

Figure 16.5 shows a generalized relationship between tree growth and foliar nutrient concentration. In the **deficiency range,** tree growth increases linearly with foliar concentration. As foliar concentration increases, the **critical range** is entered, in which growth response (although still present) declines. The critical level is conventionally defined as "the foliar nutrient concentration at which yield attains 90% of the possible maximum" (Van Den Driessche 1974). In the **luxury range,** nutrients are present in excess of growth requirements, while in the **toxicity range** excessive nutrients inhibit growth.

The actual concentrations corresponding to each of these ranges vary for the different nutrient elements. Concentrations corresponding to the ranges are also likely to vary for any given element depending on tree species, tree age, stand density, and the specific measure of growth used as a response variable. For example, Miller et al. (1981) studied the relation between foliar nitrogen concentration and growth of Corsican pine of various ages and stand densities. They found that during the period prior to canopy closure, "optimum" nitrogen concentration (roughly corresponding to the transition between the critical and luxury ranges) declined logarithmically from 3.3 percent in young seedlings to 1.5 percent in trees 2 to 2.5 m tall. Following canopy closure, optimum increased to about 2.0 percent. At all stages concentrations producing maximum height

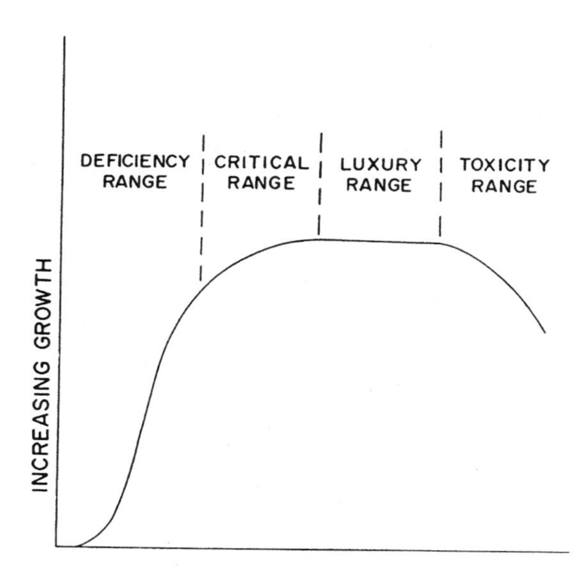

Figure 16.5. A generalized relationship between growth and foliar nutrient contents. (Van Den Driessche 1974)

growth were less than those producing maximum diameter and volume growth. Hoyle and Mader (1964) found that height growth of red pine correlated with foliar calcium concentration, while basal area growth correlated with potassium concentration.

Table 16.7 shows foliar nutrient concentrations corresponding to various levels of deficiency in several conifers native to western North America. In general, differences among the species are relatively minor and probably within the error bounds of the estimates. Calcium, required by lodgepole pine and western hemlock in relatively low amounts, is an exception. Table 16.8 shows estimates of micronutrient deficiency ranges for conifers (not species specific).

Interpretation of tree response to fertilization (or other change in nutrition, such as different soils) is often facilitated by analyzing differences in both foliar nutrient concentration and content, the latter being the product of concentration and leaf weight. This is because the most consistent long-term response of trees to improved nutrition is an increase in total leaf weight and leaf content of the limiting nutrient, and this may or may not be accompanied by an increase in concentration of the limiting nutrient. Various researchers have used a graphical technique to study the combined changes in leaf weight, nutrient concentration, and nutrient content resulting from fertilization, thinning, or other manipulations that might influence tree nutrition (Weetman and Algar 1974; Timmer and Stone 1978; Oren et al. 1988; Velasquez-Martinez et al. 1992). Binkley (1986a) provides a good description of the technique.

16.3.2 Nutrient Ratios

In general, nutrient ratios within foliage better indicate plant nutrient status than nutrient concentrations (Ingestad 1979; Powers 1984; Agren 2004; Gusewell 2004). Use of ratios rather than absolute concentrations is the basis of the diagnosis and recommendation integrated system (DRIS), a diagnostic approach first developed by agronomists (Walworth and Sumner 1987). Comerford and Fisher (1984) found that the ratio of nitrogen to phosphorus in foliage

Table 16.7
Foliar macronutrient concentrations believed to indicate a deficiency or adequacy in five conifer species

Element	Foliar concentration, %*					Deficiency
	Douglas-fir	Lodgepole pine	Western hemlock	White spruce	Western redcedar	
Nitrogen	1.05	1.05	0.95	1.05	1.15	Severe
	1.30	1.20	1.20	1.30	1.50	Slight to moderate
	1.45	1.55	1.45	1.55	1.65	Adequate
Phosphorus	0.08	0.09	0.11	0.10	0.10	Moderate
	0.10	0.12	0.15	0.14	0.13	Slight
	0.15	0.15	0.35	0.16	0.16	Adequate
Potassium	0.35	0.35	0.40	0.25	0.35	Moderate to severe
	0.45	0.40	0.45	0.30	0.40	Adequate
Calcium	0.10	0.05	0.05	0.07	0.07	Moderate to severe
	0.15	0.06	0.06	0.10	0.10	Slight to moderate possible
	0.20	0.08	0.08	0.15	0.20	Little (if any)
	0.25	0.10	0.10	0.20	0.25	Adequate
Magnesium	0.06	0.06	0.06	0.05	0.05	Moderate to severe
	0.08	0.07	0.07	0.06	0.06	Slight to moderate possible
	0.10	0.09	0.09	0.10	0.12	Little (if any)
	0.12	0.10	0.10	0.12	0.14	Adequate

Source: Adapted from Ballard and Carter 1985 and Edmonds 1989.
*Dry-mass basis.

Table 16.8
Foliar micronutrient concentrations believed to indicate a deficiency or adequacy in five conifer species

Element	Foliar concentration (ppm*)	Deficiency
Manganese	4.0	Probable
	15.0	Possible or near
	25.0	Adequate
Iron	25.0	Possible
	50.0	Unlikely
Zinc	10.0	Possible
	15.0	None
Copper	1.0	Possible
	2.0	Possibly somewhat
	2.6	Possibly slight
	4.0	None
Boron	10.0	Possible; NID[†] unlikely; if $N < 1.5$,
	15.0	NID possible, and if $N > 1.5$, NID unlikely
	20.0	None
Molybdenum	0.1	None

Source: Adapted from Ballard and Carter 1985 and Edmonds 1989.
*Dry-mass basis.
[†]NID, deficiency inducible by nitrogen fertilizer application. N, percentage nitrogen concentration.

predicted loblolly pine response to nitrogen fertilization better than nitrogen concentration alone (they concluded that neither measure was accurate enough for operational use). Ingestad (1979) studied the relationship between relative nutrient content and growth of seedlings of 10 plant species (seven conifers, one hardwood [birch], and one ericaceous shrub [*Vaccinium* sp.]). Depending on species, best growth occurred when seedlings contained the following proportions of nutrients relative to nitrogen: 45 to 70 percent as much potassium, 13 to 30 percent as much phosphorus, 4 to 8 percent as much calcium, and 4 to 8 percent as much magnesium.

Agren (2004) reasoned from biochemical considerations that nitrogen:phosphorus ratios should be maximum at intermediate growth rates and decline with either slower or faster growth. That is indeed the case with *Betula pendula* seedlings, in which foliar nitrogen:phosphorus ratios were approximately 12:1 at intermediate growth rates and 8:1 with both slower and faster growth (Agren 2004). However, critical ratios of young plants may not be a good indication of those for older plants (Gusewell 2004).

Nutrient ratios are better predictors of plant nutrient status than absolute concentrations for two reasons. First, nutrient concentration is itself a ratio of a given nutrient to dry mass, which is composed primarily of carbon, oxygen, and hydrogen. Because much of the foliar carbon, oxygen, and hydrogen are in structural rather than physiologically active tissues, nutrient concentration varies widely depending on the amount of structural tissue. This is the primary reason why nutrient concentrations vary with foliage age, among different tree components, and between evergreen and deciduous trees. When a ratio is made of two nutrients involved primarily in physiology rather than structure (e.g., nitrogen and phosphorus), the dilution effect of structural tissues is largely canceled, and a more accurate picture of the nutrient status of physiologically active tissues is obtained (Walworth and Sumner 1987).

The second reason relates more directly to physiology. Plant nutrition is a function not only of the concentration of a given nutrient element, but of the *balance* among the various elements. For example, nitrogen is a basic structural component of enzymes; however, to properly function, enzymes may need any of a variety of other nutrient elements. Sulfur is also a structural component of some proteins and must be present in the proper ratio with nitrogen or these proteins will not function. Recall that over 60 different enzyme systems require a monovalent cation (preferably K^+) for activation; most micronutrients also serve as cofactors for one or another protein. As with sulfur, deficiencies of these elements limit the ability of plants to utilize nitrogen fully. Finally, protein manufacture depends on ribosomes, which are phosphorus rich.

Some physiological processes of plants are quite sensitive to the relative proportions in which cations occur, because when imbalanced, these elements tend to interfere with one another's function. For example, calcium is readily replaced by other cations in membranes; however, these others do not maintain membrane integrity as well. Since membrane permeability modulates numerous physiological processes, imbalance between calcium and other cations may result in various physiological dysfunctions. Plants with leaky membranes may be unable to regulate the supply of starches and sugars to enzymes; hence, respiration increases. Plant defensive compounds that are normally sequestered within vacuoles (e.g., tannins and other polyphenols) can escape through leaky membranes into the cell proper, where they have the potential to interfere with physiological processes. A common example of this phenomenon is brownspot "disease" of fruit, which results from an imbalance of calcium and magnesium that diminishes membrane integrity and allows phenolic compounds to escape from vacuoles. Imbalance between calcium and other cations can also disrupt growth of new tissues, particularly roots. Taproot elongation in loblolly pine depends on the ratio of calcium to magnesium in soil (Lyle and Adams 1971). Similarly, root-tip formation by European spruce is inhibited by high ratios of aluminum to calcium in soils of acid-rain-affected stands. Aluminum has been shown to inhibit calcium and magnesium uptake by several coniferous species, but to stimulate potassium uptake (Ryan et al. 1986; Truman et al. 1986).

16.3.3 Visual Symptoms of Nutrient Deficiency

Nutrient deficiency is eventually accompanied by symptoms that are readily apparent, although correlating a given set of symptoms with deficiency of a specific nutrient isn't always straightforward. Epstein (1972) lists the following limitations on use of visual symptoms alone to diagnose deficiencies:

- Symptoms produced by deficiency of a given element may differ widely among different plant species.
- Deficiencies of different elements may produce similar or identical symptoms.
- Frequently plants may be deficient in two or more elements, complicating symptoms.
- Deficiencies may be severe enough to reduce yields well before visual symptoms are readily apparent. (This presents more of a problem in annual crops than in trees.)

Nevertheless, as long as the limitations are kept in mind, visual symptoms are useful as part of a suite of diagnostic tools, particularly for the field forester who does not have ready access to more sophisticated analytical techniques.[4]

16.3.4 Soil Measures

With a few exceptions, measures of soil nutrient levels have not proved to be good indicators of deficiency, probably because levels of soil nutrients measured in laboratory tests do not correspond well to the levels that are actually available to plants. In Douglas-fir forests of the Oregon Cascades, for example, most standard measures of soil nutrient availability correlate poorly with foliar nutrient levels (Velasquez-Martinez et al. 1992).

One type of soil measure that does correlate with tree growth, at least in some cases, is a determination of **mineralizable nitrogen.** This measures the amount of NH_4^+ and (depending on the way the test is done) NO_3^- released from soil samples during a given time period (called the **incubation period,** or simply **incubation**). Two types of mineralizable nitrogen measures are used: **aerobic,** in which the sample is incubated with a normal supply of oxygen; and **anaerobic,** in which the soil sample is waterlogged to exclude oxygen. Both NH_4^+ and NO_3^- are produced during aerobic incubation, while only NH_4^+ is produced during anaerobic incubation (bacteria that convert NH_4^+ to NO_3^- require oxygen and do not function in the anaerobic test).

Aerobic incubation can be done in either the laboratory or the field. When done in the field, it is a reasonably good measure of the rate at which nitrogen is converted from organic to inorganic forms during decomposition.

The anaerobic test (which must be done in the laboratory) is specifically designed to inhibit the activity of microbes that require oxygen and, hence, does not accurately reflect decomposition rates in the field. For example, anaerobic mineralization rates did not correlate well with aerobic rates in hardwood plantations of the southeastern United States (Scott et al. 2005). Nevertheless, the anaerobic test is simple, quick, and cheap and has proved to be a good indicator of nitrogen deficiency in some conifer stands (Powers 1984). Why the anaerobic test is a good indicator despite its "unreality" is not totally clear but is probably related to the fact that it is a good index of total microbial biomass in soil (Paul 1984; Myrold 1987), and microbial biomass represents a large proportion of labile (i.e., readily decomposed) organic matter (chapter 18).

In at least one instance, anaerobically determined mineralizable nitrogen proved to be superior to foliar nitrogen in predicting tree growth. Velasquez-Martinez et al. (1992) found that at a given stand density, growth efficiency of Douglas-fir correlated more closely with anaerobic nitrogen mineralization than with foliar nitrogen content. Why this should be so is not clear, but it may tell us that anaerobic incubations reflect various factors other than nitrogen supply that are important to trees. For example in the stands studied by Velasquez-Martinez et al. (1992), anaerobic mineralizable nitrogen correlated positively with total soil nitrogen, exchangeable soil calcium, and an adjusted measure of aspect (i.e., mineralization was higher on south than on north aspects), and negatively with soil rock content and slope steepness (Velasquez-Martinez and Perry 1997).

16.4 THE CONCEPT OF RELATIVE ADDITION RATE

Most of what is known about the effects of nutrient deficiencies on plants is derived from fertilizer experiments. Fertilization as commonly practiced, however, has little in common with plant nutrition under natural conditions. Under natural conditions, nutrients are mineralized and made available in relatively small increments, and there is a close correspondence between the availability of nutrients, their uptake by roots, and their incorporation into new biomass that tends to maintain constant nutrient concentration at root surfaces and in plant tissues.

In contrast, most fertilizer experiments add amounts far in excess of the immediate needs of the plant, confounding interpretation of plant response in two ways (Ingestad et al. 1981; Ingestad 1982). Initially, nutrient levels at the root are very high, disrupting uptake mechanisms (e.g., by one element interfering with uptake of another) and perhaps producing toxic effects or inducing drought stress through salt accumulations. Declining concentrations over time further confound the interpretation of fertilizer experiments, because it is difficult or impossible to determine exactly what concentration of a given element the plant has perceived over the life of an experiment.

To circumvent these problems, Ingestad and coworkers at the Swedish University of Plant Sciences developed procedures by which relatively small amounts of nutrients are supplied to plants at constant rates. Plant requirements for

[4] See Lyle (1969) for a guide to diagnosing nutrient deficiencies through foliar characteristics.

a given nutrient are determined by correlating plant growth rate with the rate of nutrient supply. Striking differences occur between results obtained by this procedure and those from experiments using more traditional approaches. For example, Ingestad and Lund (1979) studied growth of birch seedlings under two conditions of nutrient supply: seedlings fertilized with a solution containing a high concentration of nutrients, and seedlings fertilized with a solution containing low levels of nutrients. In the former case, seedlings required a nitrogen concentration of 400 to 900 $mmol/m^3$ of nutrient solution to maintain a relative growth rate of 0.15 g per g per day. In the latter case, the same growth rate was maintained with only 3.5 $mmol/m^3$. One clear implication is that much of the past research on plant nutrition may have little relevance to nature, and our understanding of plant nutrition in ecosystems is still at a rudimentary level.

16.5 SUMMARY

Plants require the chemical elements hydrogen, carbon, oxygen, nitrogen, calcium, potassium, phosphorus, magnesium, and sulfur in relatively large amounts (macronutrients), and boron, chlorine, manganese, iron, copper, zinc, nickel, and molybdenum in smaller amounts (micronutrients). Animals and microorganisms also require these, plus a few additional elements. Essential nutrients can be placed in four groups according to their physiological roles and cycling characteristics. Group 1 consists of carbon, hydrogen, and oxygen, which form the basic skeleton of all organic molecules and together compose over 90 percent of the dry mass of tissues. These elements have no mineral phase and are not cycled to any great degree within local ecosystems. Group 2 consists of nitrogen, sulfur, and phosphorus, which are components of various compounds such as proteins (nitrogen and sulfur), ATP (phosphorus), and DNA (nitrogen and phosphorus). One or more of these elements frequently limits productivity, and they are cycled quite tightly within local ecosystems. Group 3 is composed of the base cations K^+, Ca^{2+}, and Mg^{2+}. These elements play numerous roles in plant physiology. They are more abundant in parent material than other elements but are quite soluble and therefore subject to leaching. Retention and cycling of group 3 elements is tied closely to the cation exchange capacity in soils. Finally, group 4 consists of the micronutrients, which are primarily involved in enzyme function or electron transfer. Metals in this group (iron, manganese, copper, zinc, molybdenum, and nickel) tend to be insoluble at normal soil pH values, and their availability to plants is mediated by a class of organic compounds called chelators.

Tree species and forest types frequently differ in the nutrient content of their biomass. This is often related to site rather than to species differences; however, species growing on the same soil type may also differ from one another. Such species-related differences in nutrition are often due to morphological differences. The foliage of evergreens, for example, tends to contain more structural tissues (primarily carbon, oxygen, and hydrogen) than deciduous species. Nutrients other than carbon, oxygen, and hydrogen occur primarily in physiologically active tissues such as foliage and fine roots, and species that maintain large amounts of foliage (or fine roots) relative to their total biomass (e.g., shade-tolerant species) require relatively high levels of nutrients.

Various approaches have been used to detect nutrient deficiencies in forests. Foliar nutrient concentrations and nutrient content (i.e., concentration times mass) are the most widely used indicators of deficiency. Ratios of different nutrients in tissues promise to be better indicators of plant nutrient status than absolute concentrations. Visual symptoms can also be useful to detect deficiencies in the field, although these can be difficult to interpret. With some exceptions, analysis of soil nutrients is unsatisfactory because it is difficult to relate nutrient levels in soil to amounts that are available to trees. The most important exceptions to this are measures of mineralizable nitrogen, which have been found to successfully predict nitrogen deficiency and growth efficiency of some tree species.

Much past research on plant nutrition has been conducted under conditions that are quite different from those occurring naturally, and this research has probably given a distorted picture of plant nutrient requirements. More work is needed on the nutrition of wild plants under field conditions.

17

Biogeochemical Cycling

Nutrient Inputs to and Losses from Local Ecosystems

BIOGEOCHEMICAL CYCLING REFERS TO the cycling of biologically significant chemical elements among the biosphere and the various components of the geosphere (i.e., rocks, water, and air). The biogeochemical cycle actually consists of at least three cycles embedded hierarchically within one another: (1) cycling of nutrients within organisms, or the biochemical cycle (Switzer and Nelson 1972), an especially significant feature of perennial plants; (2) cycling of nutrients within local ecosystems, usually but not always mediated by the soil; and (3) cycling of nutrients between ecosystems and either the atmosphere or the hydrosphere (i.e., streams, lakes, oceans). Various factors may influence the rate at which these cycles proceed; however, biological activity exerts a major control.

This chapter focuses on the transfer of nutrients into and out of local ecosystems. Cycling within local ecosystems is covered in chapter 18. Both chapters deal primarily with essential elements other than oxygen and hydrogen.

17.1 AN OVERVIEW OF NUTRIENT INPUTS TO LOCAL ECOSYSTEMS

Although many of the nutrients taken up by vegetation are derived from the intrasystem cycle, inputs from external sources are important for two reasons: (1) depending on nutrient and locale, they can contribute significantly to annual tree requirements (e.g., Swank and Henderson 1976), and (2) they replenish nutrients that are lost from the system. The major losses of nutrients occur during disturbances such as wildfire or clear-cutting, and to evaluate the effects of management practices on site fertility, it is necessary to have some understanding of the sources and magnitudes of nutrient replenishment.

Nutrients are input to ecosystems via three primary pathways:

1. Atmospheric deposition, including precipitation, raking of clouds and fog by crowns, and dry deposition (all nutrients).
2. Weathering of primary minerals (all nutrient elements except nitrogen; although 98 percent of the earth's total store of nitrogen is contained within rocks, concentrations are so low that only negligible amounts are released through weathering).
3. Biological conversion, or "fixation," of atmospheric nitrogen to organic forms, an enzymatic process of which only a few bacteria, cyanobacteria, and actinomycetes are capable (chapter 11); despite the limited number of organisms capable of fixing nitrogen, historically this has probably been the most important source of nitrogen for terrestrial and probably also marine ecosystems.

These pathways are not independent of one another. The rate of biological nitrogen fixation is determined by a suite of environmental factors influencing the health and activity of nitrogen-fixing organisms, including (along with moisture and temperature) the availability of other nutrient elements required by **nitrogen fixers** (a more technical term is **diazotrophs**). Nitrogen inputs, in turn, help fuel the biological activity that contributes to rock weathering (chapter 14). Carbon fixed in photosynthesis provides the energy required to fix nitrogen and generates acidity that contributes to mineral weathering. The structure that develops as forests grow provides habitat for nitrogen fixers (e.g., decaying logs, foliage, rhizospheres, and mycorrhizospheres) and enhances atmospheric inputs through interception of dust and fog by canopies.

17.2 ATMOSPHERIC INPUTS

Excluding photosynthesis and nitrogen fixation, there are three basic pathways by which nutrients and other chemical elements are input to ecosystems from the atmosphere:

1. In precipitation (i.e., rain and snow)
2. Through dry deposition, including the absorption of dust and gases on canopy surfaces and by the soil
3. Through deposition of water vapor and associated chemicals onto canopy surfaces from clouds or fog

Both the total nutrient input from the atmosphere and the proportion of that input represented by dry deposition vary depending on the amounts and chemical forms of elements in the atmosphere. This, in turn, is strongly influenced by local sources and long-distance transport of dust, sea spray, and pollution (Swank and Henderson 1976; Bowen 1979), by the degree to which canopies are immersed in clouds or fog, and also by larger-scale patterns of air movement across the globe (Prospero et al. 1981).

Air masses moving onto continents from the ocean carry elements contained in sea spray, primarily chlorine, sodium, potassium, and magnesium (Ulrich et al. 1981; Chadwick et al. 1999). Air moving across continental interiors picks up a wide variety of chemicals from various sources: dust and gases from surface soils, emissions from fires and volcanoes, pollen and other organic emissions from vegetation, ammonia volatilized from fertilizers and animal wastes, and a variety of chemicals (especially nitrogen, sulfur, and heavy metals) released by industry and automobiles. This produces characteristic patterns of nutrient input to temperate forests. In the United States, for example, forests of the west coast get relatively few nutrients in precipitation (despite having high rainfall), because the rain moves in directly from the ocean. Eastern forests, on the other hand, get less precipitation but more nutrients in that precipitation (Cole and Rapp 1981). As we saw in chapter 15, areas of Europe and North America affected by acid rain have especially high levels of nitrogen and sulfur in precipitation. In Hawai'i, dust transported from Asia is an important source of phosphorus on old, highly weathered soils (Chadwick et al. 1999).[1]

Bruijnzeel (1991) summarized studies of nutrient inputs to tropical forests via precipitation. Although generally within the ranges found in temperate forests, amounts vary greatly among different forests. As in temperate forests, what is washed out of the air in the tropics depends on what is in the air. Industrial pollution is not a significant factor in the tropics; however, fires are common and dust from agricultural lands and deserts also contribute nutrients to the atmosphere. Dust can move long distances; soil blown off the Sahara desert in northern Africa is transported across the Atlantic Ocean and deposited in the Amazon basin, delivering approximately 1 kg/ha/y of phosphate and an unknown amount of other nutrients to the nutrient-deficient Amazon rain forest (Prospero et al. 1981; Garstang et al. 1991).

Because of the importance of plant canopies as depositional surfaces, measurements of precipitation taken in open areas or above the canopy significantly underestimate the amount of nutrients added to forests from the atmosphere. Actual inputs are quite difficult to measure accurately. To understand why, consider what occurs when a precipitation collector is moved from an open area to a point beneath the canopy. (Precipitation collected beneath the canopy is called **throughfall.**) The nutrient content of precipitation is altered in three ways as it passes through a canopy:

1. Nutrients contained in dust and gases that are present on leaf surfaces are soluble and become part of the throughfall. A heterogeneous mixture of gases and aerosols are either raked from horizontally moving air masses (especially clouds and fog) by the canopy or (in the case of larger aerosols) settle onto canopy surfaces

[1]Most Asian dist was transported during the last glaciation, when Asia was less vegetated and winds were higher than at present.

Much of the atmospheric input of sulfur and nitrogen is derived from gases, nitrogen from nitrous oxides and ammonia (or N_2 in the case of biological fixation), and sulfur from sulfur dioxide. These react chemically with foliar or soil surfaces in a process that is greatly facilitated if the gases are dissolved in water. Nitrogen and sulfur are transferred from the atmosphere to foliar surfaces with particular efficiency in forests that are frequently enshrouded in clouds or fog (Lovett et al. 1982), and from the atmosphere to surface soil layers when they are wet (Hanawalt 1969a, 1969b). Malo and Purvis (1964) estimated that 20 to 60 kg/ha/y of nitrogen was absorbed as ammonia (NH_3) in the New Jersey soils they studied, and Young (1964) also found that NH_3 was absorbed in Pacific Northwest soils. These values do not represent a net nitrogen increment to the ecosystem, however, because NH_3 is also released from canopies and soils to the atmosphere. Whether a given forest (or any vegetation type) absorbs more NH_3 than it releases (i.e., is a sink for atmospheric NH_3) or releases more than it absorbs (i.e., is a source of atmospheric NH_3) depends at least in part on atmospheric concentrations of the element. Work by Langford and Freshened (1992) in montane forests of Colorado shows that when atmospheric NH_3 concentrations were low, forest canopies released more NH_3 to the atmosphere than they took up. When atmospheric concentrations were high, net transfer was in the opposite direction (i.e., atmosphere to forest). In that area, atmospheric concentrations were high enough to produce net transfer to forests only when winds were from agricultural areas. (NH_3 is volatilized from animal wastes and fertilizers; agriculture is believed to be a significant source of acid rain in Europe.)

In addition to gases, nutrients are input from the atmosphere to forests as constituents of **aerosols,** particles ranging from 5 nm to 20 pm in size. At least 77 different elements have been detected at one level or another in atmospheric dust, including all of the essential plant and animal nutrients. (Bowen 1979 presents data on the aerosol composition of the atmosphere at various points across the globe.) Aerosols are readily washed out of the atmosphere by precipitation and may also be deposited on canopy surfaces by condensation, which is particularly important in cloud forests and in areas with fog (fig. 17.1), or, in the case of the heavier particles (which compose most of the aerosol mass in the atmosphere) settle out under the influence of gravity. Deposition onto canopies via clouds and fog increases with elevation and varies somewhat between conifer and deciduous forests (Weathers et al. 2006). According to Bowen (1979), a given aerosol stays in the atmosphere from 10 to 30 days; hence, much of the aerosol input to ecosystems is probably from local soils or local pollution sources (Swank and Henderson 1976; Bowen 1979). However, aerosols may also be transported long distances. Pollution from various sources adds large amounts of both gases and aerosols to the atmosphere.

Finally, deposition of organic molecules from the atmosphere is a seldom-quantified but potentially important source of nitrogen to ecosystems. Neff et al. (2002) summarize data from 41 sites showing that organic nitrogen consistently accounts for about one-third of total atmospheric nitrogen inputs.

under the influence of gravity; collectively referred to as **dry deposition,** these methods represent a significant pathway by which nutrients and other elements are added to local ecosystems. In forests that are periodically immersed in clouds or fog, elements are additionally deposited on canopy surfaces along with water droplets.

2. Nutrients that are leached from the interior of foliage rather than from leaf surfaces are also added to precipitation as it passes through the canopy. These are not new inputs to the system, however, but part of the system's internal cycle.
3. Some nutrients are directly taken up by leaves and epiphytes and do not appear as part of throughfall.

Nutrients that are washed from the canopy to the ground may take one of two routes: falling directly to the surface or running along branches and stems to the ground (called **stem flow**). The former would be captured in a standard collector placed beneath the canopy; however, measuring the latter requires collectors to be placed along stems. Even if all of the routes that water might take from canopies to the ground are measured, we still do not know what proportion of the nutrients in that water represent fresh inputs from the atmosphere rather than leaching from foliage, nor do we know what proportion of total nutrient deposition (both dry and wet) is absorbed directly by canopies and never appears in collectors.

Pollution adds large amounts of nitrogen and sulfur to some ecosystems (Holland et al. 2005; Weathers et al. 2006), a phenomenon that underlies acid rain and that has also generated concerns about nitrogen saturation and consequences for ecosystem health (e.g., Aber et al. 1997; chapters 3 and 20 in this volume).[2] Precipitation and other forms

[2]Nitrogen deposition has also been linked to the expansion of forests into grasslands in the Canadian Great Plains (Kochy and Wilson 2001).

Figure 17.1. The deposition of elements onto canopy surfaces via fog and clouds may add significant amounts of nutrients to some mountain or coastal forests. A. Eucalyptus plantation on the coast of Chile. B. Douglas-fir plantation in the Oregon Cascades.

of wet deposition accounted for about two-thirds of total inorganic nitrogen deposition across the United States between 2002 and 2004, with highest overall levels in the heavily industrialized and urbanized east. (See http://nadp.sws.uiuc.edu/amaps2/ for animated maps of trends in nitrogen deposition across the United States.) In the mountains of the eastern United States, deposition increases with elevation and, though highly variable, tends to be greater in coniferous than in deciduous forests (Weathers et al. 2006) (fig. 17.2).

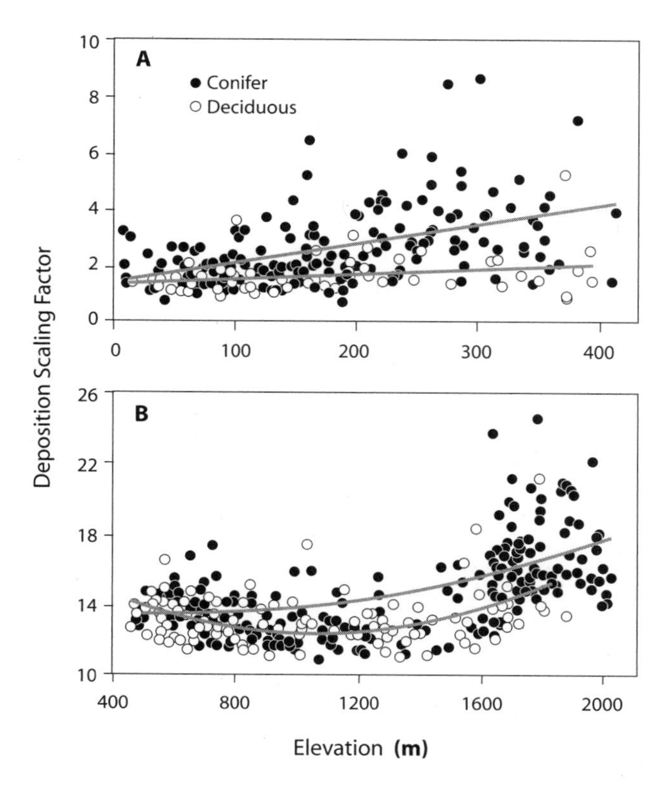

Figure 17.2. Scaling factors (unitless values that show relative deposition as a function of elevation and vegetation type for Acadia National Park (Maine) (A) and Great Smokey Mountains National Park (Tennessee/North Carolina) (B). Conifers are represented by filled circles, deciduous trees by open circles. (Weathers et al. 2006. With permission of the Ecological Society of America)

Dry deposition in particular is difficult to measure, and estimates may contain considerable error (Holland et al. 2005). In the northeastern United States, wet deposition of nitrogen and sulfur declines from west to east, reflecting concentrations of industrial activity to the west, while dry deposition declines from south to north, reflecting sources of aerosols in the heavily urbanized areas to the south (Aber et al. 2004). Significant amounts of some elements may also be added to ecosystems through deposition from deserts or agricultural lands.

Calcium, potassium, and magnesium are major constituents of dry deposition in both temperate and tropical forests (Swank and Henderson 1976; Ulrich et al. 1981; Jordan 1982; Chadwick et al. 1999). Between 1969 and 1972 in one German beech forest, the proportion of total atmospheric inputs accounted for by dryfall ranged from 0 percent for phosphorus to 78 percent for potassium and 89 percent for manganese (Ulrich et al. 1981). Proportions of dry input for other nutrients were generally on the order of 30 to 50 percent. In contrast, dry deposition accounted for 20 percent of the atmospheric phosphorus inputs to a Caribbean pine forest in Belize, but only 8 percent and 9 percent of the potassium and calcium, respectively (Kellman and Carty 1986).

Nutrients deposited on canopies may be absorbed before they are washed to the ground (particularly elements that are in short supply). In a mixed hardwood forest in the eastern United States, dry deposition was estimated to account for 56 percent of the total atmospheric inputs of sulfate and inorganic nitrogen (i.e., nitrate and ammonia), 59 percent of the potassium, and 67 percent of the calcium, (Lindberg et al. 1986). In that study, 70 percent of the deposited nitrogen was retained in the canopy, probably absorbed directly by foliage or epiphytes growing on the foliage. Most of the acidity (i.e., H^+) accompanying the nitric and sulfuric acids that composed a high proportion of the dry deposition to that site was also absorbed by foliage, with an accompanying release of calcium and potassium ions from foliage. Direct absorption of nutrients by leaves and canopy epiphytes has been shown to be quite important on nutrient-poor sites in Amazonia (Jordan et al. 1979); there, throughfall actually contains less calcium, sulfur, and phosphorus than is present in precipitation above the canopy.

17.3 INPUTS FROM WEATHERING OF PRIMARY MINERALS

In any system with fresh rock in the rooting zone, weathering constitutes the major source of all nutrients except nitrogen (or sulfur where air pollution is a factor). That includes areas glaciated during the Pleistocene; with recent volcanic activity; or with unstable soils that periodically expose fresh rock. At the Hubbard Brook Experimental Forest in New Hampshire, weathering accounts for from 85 to 100 percent of the yearly inputs of potassium, calcium, magnesium, and iron (Bormann and Likens 1979b). Weathering accounts for only 4 percent of the sulfur inputs to this forest, however, because of the large amount of sulfur in acid rain.

In the absence of mechanisms that bring fresh rock to the surface, the weathering zone moves downward and eventually beyond the reach of roots. On old soils in tectonically stable areas (e.g., Amazonia), little or no unweathered rock remains within the rooting zone, and fresh nutrient inputs are restricted to those from the atmosphere. Similarly, atmospheric deposition is more important than rock weathering for a temperate forest in Chile, even though nutrients in the marine air moving into the forest come predominantly from sea salts rather than the richer sources represented by atmospheric dust (Kennedy et al. 2002). Most fine roots in the Chilean forest are in organic layers, where sea salts are retained and apparently tightly cycled to trees and back. Except for a single deep-rooted tree species, the forest is decoupled from nutrients released by rock weathering, and consequently from its function as a regulator of stream chemistry. However, even in areas with old soils, many landscapes are patchworks of stable (i.e., flat) and erosionally active areas. On four- to five-million-year-old Kauai Island (Hawai'i) only the stable uplands are nutrient poor, the bulk of the landscape is on steep slopes where erosion brings fresh rock and therefore fresh nutrient sources to the surface (Porder et al. 2005).

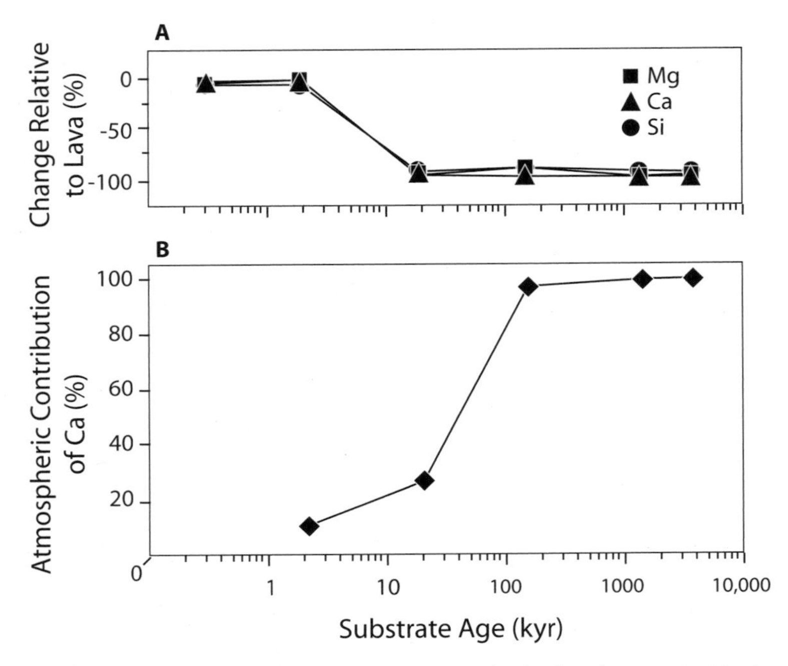

Figure 17.3. Changes in elements across chronosequences of soils in the Hawaiian Islands. A. Soil concentrations of magnesium, calcium, and silicon derived from lava relative to concentrations in unweathered lava. B. Proportional contribution of atmospherically derived calcium to total calcium pool in soil. (Adapted from Chadwick et al. 1999)

Changes in weathering rates over time have been studied in detail across a chronosequences of soil ages in the Hawaiian Islands (Chadwick et al. 1999). As soils are built up from lava, the concentrations of calcium, magnesium, and silicon derived from weathered lava (rather than atmospheric deposition) remain similar to those in unweathered lava for about two thousand years, then drop sharply to very low levels by 11 thousand years (fig. 17.3A). Over the same time, the proportional contribution of atmospherically derived nutrients to the total soil pool climbs sharply, and by a soil age of one hundred thousand years approaches 100 percent for calcium (fig. 17.3B).

Weathering rates are strongly influenced by climate (Dessert et al. 2003; Oliva et al. 2003). On Hawai'i island, there is a sharp transition in weathering rates between sites above and below a mean annual precipitation of 140 cm (Stewart et al. 2001). In areas with higher precipitation (but the same age of soils) weathering is estimated to have proceeded very rapidly in the early stages but dropped sharply over time, reflecting the depletion of weatherable rock (fig. 17.4). In areas with lower precipitation, weathering rates were lower initially but have remained relatively constant over time, so that current rates are highest in lower-rainfall areas.

Weathering is also strongly influenced by plants and microbes. In chapter 8 we saw that silicon and sodium fluxes from bedrock were three times larger in a forested catchment in Japan than in a geologically similar one with bare ground (Asanao et al. 2004). Similarly, weathering rates of calcium and magnesium were 10 and 18 times greater (respectively) in lysimeters planted with red pine than in

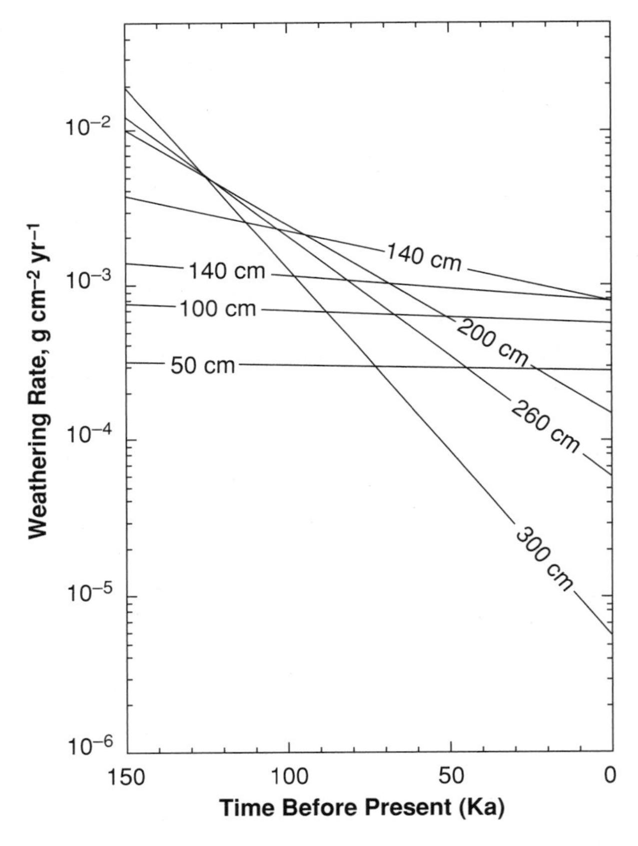

Figure 17.4. The effect of precipitation on estimated weathering rates of Hawai'i lavas. Curve labels are mean annual precipitation. (The two curves for 140 cm represent two estimates.) (Stewart et al. 2001. With permission from Elsevier)

Weathering rates are typically calculated using a simple **mass balance** accounting approach, although execution of this technique is not nearly as simple as the concept behind it (Clayton 1979). To illustrate this approach, suppose we want to quantify the amount of calcium released by rock weathering on a given watershed. The following equation is used for a starting point:

$$Ca(rw) + Ca(atm) = delCa(biomass) + delCa(soil) + Ca\ efflux$$

where Ca(rw) is the annual calcium release from weathering of primary minerals, Ca efflux the annual calcium loss to stream water, Ca(atm) the annual calcium inputs from the atmosphere, delCa(biomass) the change in calcium contained in biomass, and dclCa (soil) the change in calcium contained in abiotic soil components (i.e., other than primary minerals).

This equation simply states that calcium inputs into a site from the atmosphere and rock weathering must appear in one of three places: increment in biomass, increment in soils, or efflux from the site in stream water. DelCa(biomass) and delCa(soil) may also take negative values, that is, appear as decrements rather than increments; for example, delCa(biomass) would be negative if mass balance measurements encompassed a period when a site was clear-cut or destroyed by wildfire.

The equation can be rearranged thusly:

$$Ca(rw) = [Ca\ efflux - Ca(atm)] + delCa(biomass) + delCa(soil)$$

Calcium inputs from weathering are then determined by carefully measuring everything on the right hand side of the equation. This same procedure is used to calculate the release of other elements contained within rocks as well.

Accurately measuring the quantities on the right-hand side of this equation is generally expensive, time-consuming, and difficult. Soils are usually assumed to be in steady state during mass balance determinations; hence, change in soil storage, for example, delCa(soil), is set equal to zero. This assumption is almost certainly invalid for elements such as phosphorus, iron, and aluminum, which tend to be immobilized in secondary minerals (chapter 14). Because hydrogen ions tend to replace other cations both on exchange complexes and in secondary minerals, the assumption of steady-state soils is also invalid for ecosystems that are subject to acid rain.

Changes in the nutrient content of biomass, for example, delCa(biomass) are also frequently set equal to zero. This is probably seldom (if ever) the case, and Likens et al. (1977) showed that erroneous results are produced when changes in biomass are ignored in mass balance calculations.

Nutrient efflux can be estimated in two ways: by installing gauges on streams, or by placing lysimeters below the root zone. Gauging streams is by far the best approach, because it provides a relatively direct, integrated measure of what is leaving a given watershed, thereby eliminating the large variability that virtually always is associated with measurements at specific points on the ground. However, stream gauging also has its shortcomings. Depending on geology, nutrients and water can leach below the rooting zone yet never appear in the stream, being retained in very deep soil layers or passing to storage in underground aquifers. A further complication arises in determining where the nutrients in water came from—rocks or the atmosphere. (As we saw above, nutrients added in dryfall are particularly difficult to measure accurately.)

An increasing use of isotopes since the late 1980s has greatly improved the ability to distinguish nutrient sources and therefore calculate accurate weathering rates. Strontium is particularly useful in this regard for two reasons. First, different strontium sources have distinct isotopic signatures, making it possible to distinguish rock-derived from dust-derived sources on a particular site. Second, it is chemically quite similar to calcium (and to a lesser extent magnesium and potassium), which means it can serve as an acceptable proxy for the dynamics of these important plant nutrients. Isotopes of neodymium have also been used as tracers for phosphorus (Chadwick et al. 1999).

vegetation-free lysimeters (Bormann et al. 1998). Bacteria and both saprophytic and mutualistic fungi greatly accelerate weathering (Hoffland et al. 2004). Various mechanisms account for the biotic influence on weathering rates. Carbon dioxide released in respiration combines with soil water to form weak acids that contribute to dissolving rocks. Some bacteria and fungi also produce complex chemicals that break down mineral bonds (chapter 18). Fungi and to a lesser extent fine roots contribute to physical weathering by growing into interstices within rocks and forcing ruptures.

17.4 BIOLOGICAL NITROGEN FIXATION[3]

As discussed previously, nitrogen (like carbon) differs from other essential nutrients in that inputs to ecosystems come

[3]Springer Publishing has produced a series of edited volumes covering biological nitrogen fixation in detail. These include *Nitrogen-Fixing Leguminous Symbioses* (M.J. Dilworth et al., editors, 2007); *Nitrogen-Fixing Actinorhizal Symbioses* (K. Pawlowski and W.E. Newton, editors, 2006); and *Associative and Endophytic Nitrogen-Fixing Bacteria and Cyanobacterial Associations* (C. Elmerich et al., editors, 2006).

almost exclusively from the atmosphere. Nitrogen differs from carbon because all green plants are capable of transforming atmospheric carbon (carbon dioxide) into organic forms, while only a few microorganisms can convert atmospheric diatomic nitrogen gas to organic forms. Nothing lives without nitrogen, and life on earth in a very real sense is built on an edifice of primitive, one-celled organisms. Earlier in the chapter, we saw that some atmospheric nitrogen is converted to forms that are usable by organisms through lightning, and a great deal is also converted through combustion engines (the primary source of nitrous oxides in acid rain). However, biological conversion is still the major input throughout the tropics, as well as in higher-latitude forests not yet affected by acid rain (Chatarpaul and Carlisle 1983; Dawson 1983).

The most abundant gas in our atmosphere, N_2, is usable only by some species of bacteria, cyanobacteria, and actinomycetes (all of which are prokaryotes). These organisms possess the enzyme nitrogenase, which reduces N_2 gas to NH_3 (i.e., fixation), a form that is usable by higher organisms. As chapter 11 discussed, microorganisms capable of fixing N_2 are referred to simply as nitrogen fixers. Biological nitrogen fixation is common and widespread, including approximately 20 genera of nonphotosynthetic bacteria (aerobic and anaerobic) and 15 genera of photosynthetic cyanobacteria such as *Anabaena* and *Nostoc* (Hubbell and Kidder 2003). Nitrogen-fixing bacteria can be divided into three groups based on their ecology (taxonomically, there is at least some overlap between the groups, i.e., species in a single genus may occur in more than one group).

Symbiotic nitrogen fixers. These live in symbiotic association with another organism that provides energy (i.e., reduced carbon), other necessary nutrients and water, protection, and in some cases, biochemicals that are necessary for the proper functioning of the nitrogenase enzyme. Host organisms include the following:

- Some species of vascular plants, with the diazotroph most commonly living within root tissue, where it forms visible swellings called **nodules** (the plants that host nitrogen-fixing microorganisms in root nodules may also be called *nodulated nitrogen fixers*)
- Certain species of lichens and nonvascular plants
- Some fungal fruiting bodies
- Some animal species, particularly termites and bark beetles

The most important nitrogen-fixing microorganisms in forest ecosystems are bacteria in the genus *Rhizobium* (symbiotic with legumes); actinomycetes in the genus *Frankia* (symbiotic with various woody nonlegumes; these symbioses are called *actinorhizal*); and certain species of cyanobacteria (symbiotic with lichens, mosses, ferns, and some higher-plant species). Nitrogen-fixing microorganisms may also occur as endophytes of vascular plants, in other words, within plant tissues but not forming the nodules that are characteristic of rhizobial and actinorhizal symbioses.

Associative nitrogen fixers. These live in close association with plant roots, mycorrhizae, mycorrhizal hyphae, and also on the surfaces of leaves and even boles. They are distinguished from the symbiotic nitrogen fixers in that they are not known to live within the cells of host plants (although they may occur in the intracellular spaces of roots and mycorrhizae).

Free-living nitrogen fixers. These include the cyanobacteria (formerly called **blue-green algae**), which are capable of photosynthesis and hence provide their own energy source; various species of heterotrophic bacteria that use wood, litter, or soil organic matter for energy; and some chemotrophs. Most species of associative nitrogen fixers are probably also capable of living free, although more study is needed on this point.

Table 17.1 shows estimated rates of biological nitrogen fixation (BNF) for the earth's major ecosystem types (Cleveland et al. 1999). (By way of context, mature forests may contain from a few hundred to a few thousand kilograms per hectare of nitrogen in living and dead biomass. Depending on the stage of forest development, the environment, and other limiting resources, annual growth requirements range from less than 20 to over 200 kg N/ha/yr). In forested ecosystems, estimated average inputs through BNF range from less than 2 kg/ha/yr in boreal forests to 36 kg/ha/yr in tropical floodplains; the majority of BNF in all systems is provided by nodulated plants. Two caveats in particular go with these estimates, both arising from the fact that accurately measuring total BNF in natural ecosystems is very difficult (Vitousek et al. 2002). First, estimates are averages over time; BNF during early successional stages is often considerably higher than in later stages, particularly in temperate (and to a lesser extent boreal) forests. Second, the data on which estimates are based may not have captured all sources of BNF (and in fact probably did not), so should be considered underestimates. For example, recent research found that feather moss (*Pleurozium schreberi*), which is ubiquitous in boreal forests, forms a symbiosis with a nitrogen-fixing cyanobacteria in the genus *Nostoc* that alone accounts for 1.5 to 2 kgN/ha/yr (DeLuca et al. 2002). Later we discuss the possibility of large amounts of associative BNF in tree rhizospheres.

Rates of BNF shown in table 17.1 correlated positively with evapotranspiration and net primary productivity (Cleveland et al. 1999). Because nitrogen fixation is an energy-intensive process, it is not surprising that rates of symbiotic nitrogen fixation are tied closely to the ability of the host plant to fix energy in photosynthesis. Highest fixation rates are found in fully stocked stands of symbiotic nitrogen-fixing trees growing on sites that support high leaf areas (e.g., various species of alders). Lower rates of fixation occur on droughty or infertile sites. Any limitation on host photosynthesis will probably reduce the supply of photosynthates to nodules and therefore lower rates of nitrogen fixation; in fact, nitrogen fixers may have a greater

Table 17.1
Estimated rates of biological nitrogen fixation in different vegetation types

Potsdam* Vegetation Type	Area × 10^3 ha	Mean estimate, kg N ha^{-1} yr^{-1}	Range of estimates, kg N ha^{-1}	
			Symbiotic	Nonsymbiotic
Polar desert/alpine tundra	5.37	6.13	4.90	0.42–13.00
Moist tundra	5.36	6.13	4.90	0.42–3.00
Boreal forest	12.70	1.77	0.30	0.50–2.80
Boreal woodland	6.60	1.77	0.30	0.50–2.80
Temperate mixed forest	5.30	16.04	1.00–160.00	0.00–2.80
Temperate coniferous forest	2.51	16.04	1.00–160.00	0.00–2.80
Temperate deciduous forest	3.69	16.04	1.00–160.00	0.00–2.80
Tall/medium grassland	3.65	2.70	0.10–10.00	0.10–21.00
Short grassland	4.75	2.70	0.10–10.00	0.10–21.00
Tropical savanna	13.90	30.20	3.00–90.00	3.00–30.00
Arid shrubland	14.80	22.04	30.00–97.50	0.50–13.00[4]
Tropical evergreen forest (rain forest)	17.80	25.40	16.00	0.50–60.00
Xeromorphic forest	6.86	21.68	7.50–30.00	3.31
Tropical forested floodplain	0.18	36.45	243.00	NA
Desert	11.70	7.81	0.67–29.50	0.50–13.00
Tropical nonforested floodplain	0.36	34.80	28.50	6.30
Temperate forested floodplain	0.11	16.04	1.00–160.00	0.00–2.80
Temperate nonforested floodplain	0.10	2.70	0.10–10.00	0.10–21.00
Wet savanna	0.16	30.20	3.00–90.00	3.00–30.00
Temperate savanna	6.89	2.70	0.10–10.00	0.10–21.00
Temperate broadleafed evergreen forest	3.35	16.04	1.00–160.00	0.00–2.80
Mediterranean shrubland	1.47	2.51	0.10–6.90	1.00
Tropical deciduous forest	4.70	21.68	7.50–30.00	3.31
Global total	132.28			

Source: Adapted from Cleveland et al. 1999. Their use of *symbiotic* and *nonsymbiotic* differs somewhat from ours. In the table, *symbiotic* refers to nodulated plants only.
**Potsdam* refers to a particular classification system.

demand than nonfixers for nutrients such as phosphorus (Vitousek et al. 2002). Moreover, nodulation and nitrogen fixation, at least by legumes, requires calcium, cobalt, copper, iron, boron, and molybdenum (O'Hara et al. 1988) and can be limited by shortages of any of these elements.

In general, it is important to realize that ecologists have an incomplete understanding of the magnitude of nitrogen fluxes both to and from ecosystems. Several researchers have documented relatively large (>50 kg/ha/y) and so far unexplained accretions of nitrogen in litter and soils of coniferous forests (Binkley et al. 2000 review early studies; see also Bormann et al. 1993). We discuss this phenomenon further in section 17.4.8; the remainder of this section examines the various sources of biologically fixed anitrogen in more detail.

17.4.1 Nodule Formers

Of the various groups of nitrogen-fixing microorganisms, the bacteria that nodulate vascular plants are the best understood and (as far as is known at present) most important in terms of the quantity of fixed nitrogen per unit of area. Nodule formers have two advantages over associative and free-living nitrogen fixers. The first relates to energy supply. The N_2 molecule, in which the two nitrogen atoms are triple bonded to one another, is very stable; industrial manufacture of NH_3 from N_2 is accomplished only at very high temperatures and pressures. Nitrogenase can perform this at normal temperature and pressure, but large amounts of energy are required. Nodule formers have ready access to energy in the form of host photosynthates.

The second advantage relates to protection from oxygen, which irreversibly destroys the nitrogenase enzyme. Oxygen presents somewhat of a Catch-22 situation for biological nitrogen fixation. Although it destroys nitrogenase, conversion of reduced carbon to the energy that is needed to power nitrogen fixation is very inefficient in the absence of oxygen; hence, rates of nitrogen fixation are highest where there is some oxygen but not too much. This seems to be the case in nodules, where the problem is at least partially solved by separating the energy-generating components and the nitrogenase enzyme into different compartments within the nodule. Legumes also control oxygen levels within nodules by producing a special oxygen-binding molecule, **leghaemoglobin** (which is related to the haemoglobin contained in the blood of mammals).

The nodule formers occur in three taxonomic groups:

1. Bacteria in the genus *Rhizobium*, which are symbiotic with numerous species of legumes and also with some plant species in the genus *Parsaponia* (native to Southeast Asia) (fig. 17.5). Nodules are usually formed on

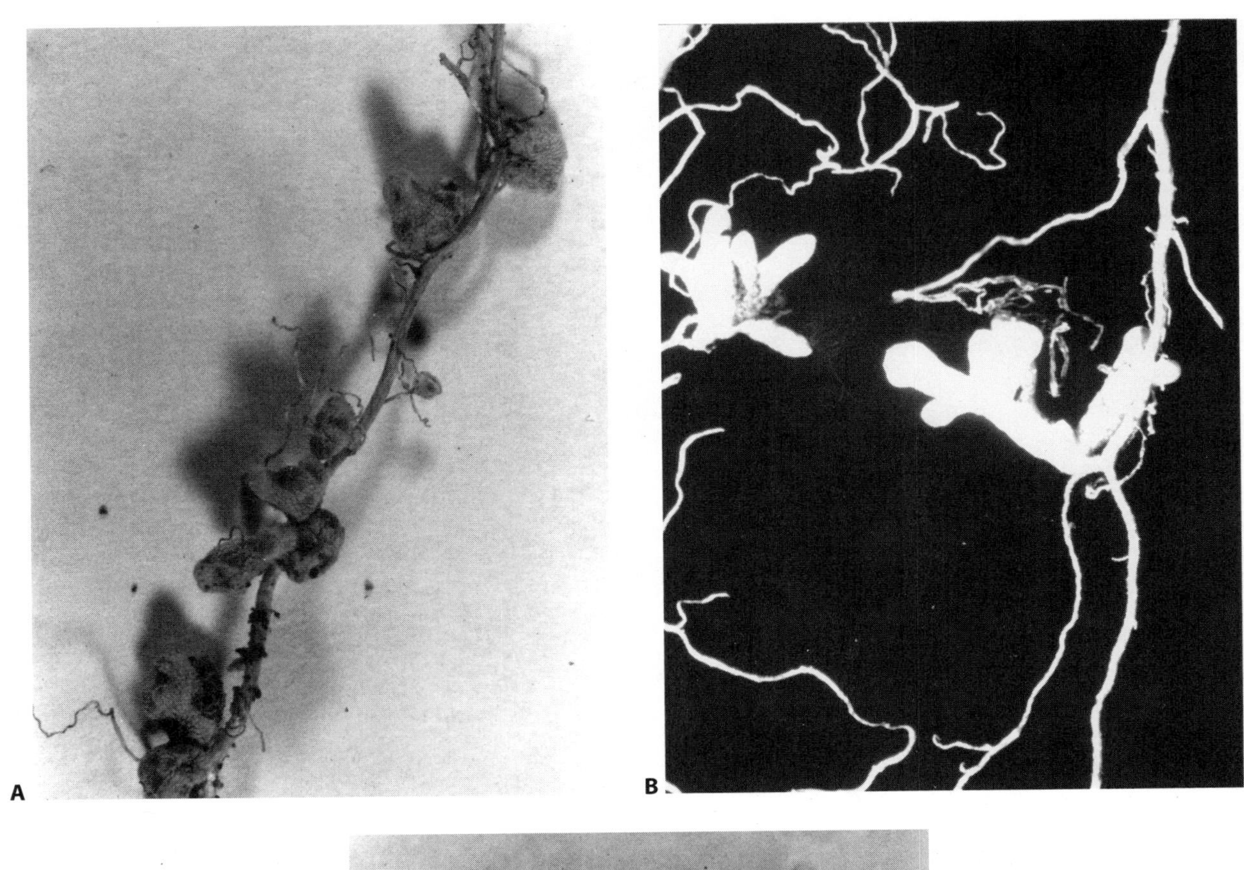

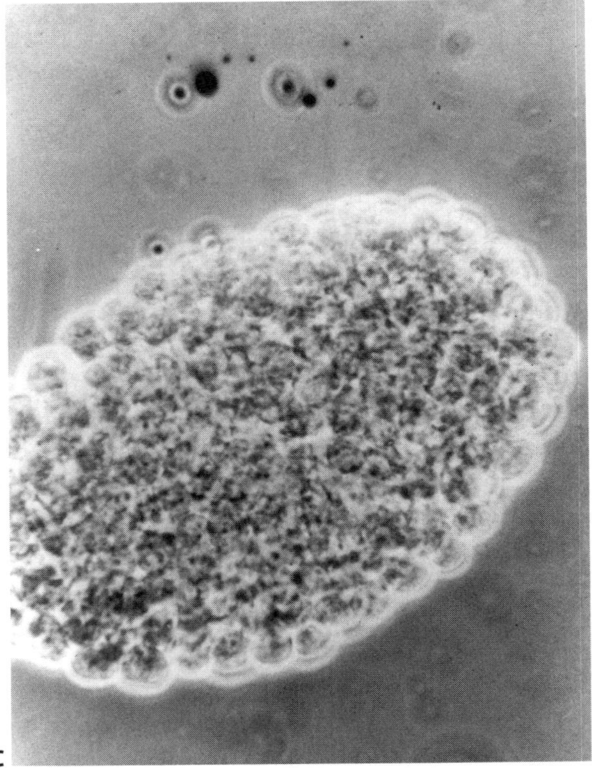

Figure 17.5. A. Nodules formed by *Rhizobium* sp. on the leguminous tree *Acacia senegal*. B. Nodules formed by *Frankia* sp. on *Ceanothus velutinus*. C. The interior of a *Frankia* nodule on *Alnus rubra*. (A, courtesy of Farouq El Hadi; B and C, courtesy of C.Y. Li)

roots but in very moist environments may also occur on stems.
2. Actinomycetes in the genus *Frankia*. Plants that form nodules with *Frankia* (called **actinorhizal**) are widespread and diverse, including species spread through eight plant families (fig. 17.6).
3. Some species of cyanobacteria that form nodules with plants in the family Cycadaceae and with plants in the genus *Gunnera* (Akkermans 1979).

The first two of these groups is quite important in forest ecosystems, and the host plants warrant further discussion.

17.4.2 Legumes

The Fabaceae (formerly Leguminoseae) is a very large plant family containing 16,000 to 19,000 species, many of which have not been investigated for nodules. Temperate-zone legumes include only a few tree genera (e.g., *Robinia, Prosopsis*); most are herbs. However, leguminous trees are common in tropical forests throughout the world and often dominate dry tropical forests and savannas (Hogberg 1986). In Amazonia, the abundance of leguminous trees follows a soil fertility gradient that stretches from the Guyana Shield (low soil fertility and abundance of legumes) to southwestern Amazonia (higher soil fertility and few legumes) (ter Steege et al. 2006). Abundant legumes may also reflect a history of small gap-forming disturbances, as legumes produce large seeds and consequently robust seedlings that may have a survival advantage under the relatively low light conditions that occur in small gaps (ter Steege et al. 2006).

Herbaceous legumes are common in some temperate forest ecosystems, where under the proper circumstances they can contribute significant amounts of nitrogen. In the interior of western North America, herbaceous legumes may be the principal (and in some cases only) nodulated, nitrogen-fixing plants in forested ecosystems (Fahey 1983; Hendrickson and Burgess 1989).

The Fabaceae is divided into three subfamilies:

1. Mimosoideae (trees, shrubs, woody vines, and a few perennial herbs). Ninety percent of the species studied in this subfamily have been found to be nodulated. Genera that are important in forestry or agroforestry occur primarily in the tropics and subtropics and include (to name just a few) *Acacia* (800 to 900 species distributed throughout the tropics and subtropics), *Albizia* (150 species, occurring mainly in savannas), *Leucaena* (50 species of trees and shrubs with pantropical distribution), and *Inga* (150 to 200 species in moist tropical forests).
2. Caesalpiniodeae (primarily trees and shrubs native to the tropics and subtropics). The dominant trees of miombo woodland in southern Africa are members of this subfamily, and members of the Caesalpiniodeae are common in Asia and South America as well. Species within the Caesalpiniodeae differ from those in the other subfamilies in that only approximately 40 percent of the investigated species have been found to have nodules, although it does not necessarily follow that species without nodules never nodulate (Hogberg 1986). Some species in this subfamily also tend to be ectomycorrhizal, whereas species in the other subfamilies are generally vesicular-arbuscular mycorrhizal (chapter 11).
3. Papilionaceae (trees, shrubs, and annual and perennial herbs). This subfamily includes the common wild and domestic herbaceous legumes of temperate areas (e.g., lupines, vetch, alfalfa, clover, and soybean). Ninety-eight percent of the species that have been studied are nodulated.

Legumes have numerous uses. Trees in the Mimosoideae and Caesalpiniodeae are used for timber, dyes, resins, medicines, and fibers, and trees and herbs in all of the subfamilies produce high-protein food for livestock and (in some cases) humans. In the tropics, leguminous trees are a principal component of agroforestry systems (Felker and Bandurski 1979; Rachie 1983).

One of the greatest values of the nitrogen-fixing, nodulated plants (both legumes and actinorhizal species) is in soil improvement (Virginia 1986). For legumes, this is particularly important in the tropics, but herbaceous legumes have also been employed as natural fertilizing agents in temperate forests. Herbaceous legumes have been intercropped with trees in Europe and the southeastern United States, but with variable success. Acid forest soils frequently must be fertilized with calcium, potassium, and phosphorus for legumes to establish and fix nitrogen (Haines et al. 1979; Rehfuess 1979). In Australia, *Lupinus arborous* establishes beneath thinned pine stands on infertile sands and is estimated to add a significant proportion of the nitrogen that the pines take up over an entire rotation (Sprent and Silvester 1973). (Some examples of productivity in mixtures of nitrogen-fixing and non-nitrogen-fixing plants were given in chapter 15.)

Many temperate legumes are most easily established in thinned forests or young plantations before crown closure. Under some circumstances, legumes interplanted with young trees compete for water or other resources, thereby negating the benefits of the nitrogen they add. Smethurst et al. (1986) found that both *L. arboreus* (a perennial) and *L. angustifiolius* (an annual) doubled soil nitrogen levels when intercropped with pine seedlings on infertile sands in Australia. Pines growing with the perennial, however, had high mortality because of competition for water. This problem did not occur with the annual legume, because it completed growth and senesced before the summer drought.

In Hawai'i, stands consisting of a leguminous tree (*Falcaltaria moluccana*) intercropped with *Eucalyptus saligna* produced nearly twice as much biomass as either tree grown alone (Binkley et al. 2003). The two tree species acted synergistically to improve site nutrition, the legume adding nitrogen through fixation and the eucalypt accessing sources of phosphorus not available to the legume when growing alone.

17.4.3 Actinorhizal Plants

Actinorhizal plants (i.e., symbiotic with *Frankia*) are the most important nodulated plants in both temperate and boreal forest ecosystems, although they also occur in the tropics. Two-hundred-twenty species distributed among 25 genera and 8 families are known to host *Frankia*. (Table 17.2 lists some of the common genera known to contain at least one actinorhizal species). Actinorhizal plants are taxonomically quite diverse but nevertheless have some characteristics in common. All are angiosperms and most are woody shrubs and trees. Actinorhizal plants are common early pioneers during primary and secondary succession, where they play a critically important ecological role by adding nitrogen and carbon to newly forming soils or (in secondary succession) by replacing nitrogen and carbon that were lost during disturbance. In chapter 8 we discussed examples of alder pioneering newly exposed rock in Alaska and of *Ceanothus* sp. occupying burned sites in the Pacific Northwest.

As with legumes, the amount of nitrogen fixed by actinorhizal plants varies depending on species, site, stocking density, and age of the plant, but it is generally a significant proportion of the total nitrogen inputs to the systems in which these species grow. Accretions of ecosystem nitrogen up to 300 kg/ha/y have been measured in stands of *Alnus* sp. (Tarrant and Trappe 1971) and up to 200 kg/ha/y in those of *Casuarina*, *Coriaria*, and *Hippophae* sp. (Haines and DeBell 1979). On the other hand, accretions as low as 1 to 35 kg/ha/y have been measured in stands of *Alnus* sp. growing in mixture with nonnodulated plant species or on soils poor in nutrients other than nitrogen (Bormann and DeBell 1981; Binkley 1982; Malcolm et al. 1985; Huss-Danell 1986). Accretions of 0 to 100 kg/ha/y have been measured for *Ceanothus* sp., which occupy droughtier sites than *Alnus* (Conard et al. 1985). In the interior of western North America, where many sites are droughty and cold, nodulated plants (if they are present) fix relatively small amounts of nitrogen. For example, *Sheperdia canadensis* was estimated to add slightly less than 1 kg/ha/y to a regenerating montane forest in British Columbia, and the legume *Lupinus arcticus* added another 2 kg/ha/y (Hendrickson and Burgess 1989). Despite being fairly low, fixation rates of this magnitude account for a significant proportion of total nitrogen inputs to North American montane forests (Fahey 1983).

Actinorhizal plants have been used in agroforestry systems and as soil improvers in Asia, Europe, and Central America. Some (e.g., *Alnus* sp.) are used commercially for pulp and lumber. As we discussed in chapter 15, nitrogen-fixing plants generally enhance productivity on sites where growth is nitrogen limited, although improved growth may not appear until trees are several decades old. Enhancement generally does not occur on sites with abundant nitrogen, or where some other resource (e.g., water, phosphorus) limits growth more than nitrogen.

Table 17.2
Genera that contain actinorhizal plants

Genus	Family	Geographic distribution
Alnus	Betulaceae	Europe, Siberia, North America, Japan, Andes
Casuarina	Casuarinaceae	Australia, tropical Asia, Pacific Islands (also introduced in other areas)
Ceanothus	Rhamnaceae	Western North America
Cercocarpus	Rosasceae	Western United States, Mexico
Chamaebatia	Rosasceae	Sierra Nevada of California
Colletia	Rhamnareae	South America
Comptonia	Myricaceae	North America
Coriaria	Coriariaceae	Southern Europe, Asia, New Zealand, Chile, Mexico
Cowenia	Rosasceae	Western North America
Datisca	Datiscaceae	North America (especially California), Himalayas
Discaria	Rhamnaceae	Andes, Brazil, New Zealand, Australia
Dryas	Rosasceae	Alaska, Canada, North Eurasia (circumpolar)
Eleagnus	Eleagnaceae	Asia, Europe, North America
Hippophae	Eleagrzaceae	Europe, Asia (Himalayas to Arctic Circle)
Myrica	Myricaceae	Widespread, tropics to Arctic Circle
Purshia	Rosasceae	Western North America
Rubus	Rosasceae	Widespread in north temperate zones; also Indonesia
Sheperdia	Eleagnaceae	North America
Trevoa	Thamnaceae	Pacific coast of South America

Source: Dawson 1983.

Fixation rates may be influenced by soil chemistry and microorganisms (other than the nitrogen fixers) in complicated ways. For example, Rojas et al. (2002) grew seedlings of *Alnus rubra* (red alder) and *Ceanothus velutinus* (snowbrush) in pasteurized clear-cut soils to which they sequentially added *Frankia* plus macronutrients, micronutrients, mycorrhizal fungi, and bacteria in the genus *Pseudomonas* (which are known to improve iron availability). They then compared amended soils to unpasteurized soils (i.e., clear-cut soils with their normal complement of microorganisms and nutrients). For snowbrush, nitrogenase activity was enhanced by adding *Pseudomonas,* while no other amendment had an effect, indicating that the clear-cut soils had adequate levels of *Frankia* and the arbuscular mycorrhizae required by snowbrush. Red alder exhibited a different and more complex pattern. Per-plant nitrogen fixation by that species was doubled by adding *Frankia*, but adding micronutrients negated the positive effects of *Frankia*. When mycorrhizal fungi were added (ectomycorrhizae in the case of alder) to the *Frankia* and micronutrient mix, the positive effects of *Frankia* were restored and even enhanced. Why did the two nodulated plant species respond so differently? They definitely use different types of mycorrhizal fungi (arbuscular mycorrhizae vs. ectomycorrhizae), and perhaps different strains of *Frankia*. However, a principle reason may relate to the fact that red alder is a much faster grower and therefore puts greater demands on soil resources. At the same age, red alder seedlings were 1.5 to nearly 3 times larger than snowbrush seedlings and were fixing 3 to 10 times more nitrogen per plant. Therefore, it is perhaps not surprising that alder seedlings were straining the soil's ability to supply *Frankia* and mycorrhizal fungi. The negative response of alder to micronutrients was unexpected and somewhat of a mystery (since nitrogen fixation requires micronutrients). The ability of mycorrhizal fungi to reverse the inhibitory effect suggests either nutrients or pathogens were involved. Perhaps the micronutrients stimulated pathogenic or competitive microorganisms. Similarly, the positive response of snowbrush to *Pseudomonas* is unexplained. Even though *Pseudomonas* are known to increase iron availability (chapter 18), the response of snowbrush probably does not reflect an iron deficiency because iron was added in the micronutrient mix (with no response by snowbrush and a negative response by alder). One inescapable conclusion is that the soil ecosystem is complex and poorly understood.

17.4.4 Nodulation of *Gunnera* sp. by Cyanobacteria

Gunnera sp. (i.e., herbs in the family *Haloragaceae*) occurs in the forests of New Zealand and Southeast Asia. Cyanobacteria in the genus *Nostoc* colonize plant cells at the base of leaves and form dense nodules. In New Zealand, Silvester and Smith (1969) estimate that *G. dentata* fixes 72 kg/ha/y of nitrogen, which is comparable to amounts fixed by legumes and actinorhizal plants. Becking (1976) estimated that 12 to 21 kg/ha/y of nitrogen were fixed by *G. macrophylla* in an Indonesian forest.

17.4.5 Other Sources of Biologically Fixed Nitrogen

17.4.5.1 Soil Wood

Nitrogen fixation occurs in decaying logs and large branches. This coarse woody debris (CWD) is a good habitat for free-living nitrogen fixers because of its abundant carbon (although it may not be readily decomposable), and CWD tends to retain water long after mineral soil has dried, hence oxygen levels are not too high. Nitrogen-fixing microorganisms within CWD are incapable of decomposing cellulose or lignin and probably live in close association with wood-decaying fungi that provide the bacteria with labile carbon in exchange for nitrogen (Jurgensen et al. 1987). CWD is frequently colonized by roots and mycorrhizal hyphae, and these may facilitate nitrogen fixation by importing nutrients that are in short supply in the wood.

Sollins et al. (1987), estimate that about 1 kg/ha/y of nitrogen is fixed within CWD in the Pacific Northwest. Estimates for the northern Rocky Mountains range from 0.16 kg/ha/y on dry forest types to 1.5 kg/ha/y on moist types (Jurgensen et al. 1987) and, for northeastern U.S. hardwoods, on the order of 2 kg ha/y (Roskoski 1980). In tropical forests, termites that consume dead wood are themselves hosts of nitrogen-fixing microorganisms (Breznak et al. 1973). Hence, the end result is the same as in temperate forests; a portion of the carbon within dead wood is used to power nitrogen fixation.

17.4.5.2 Associative Nitrogen Fixers

One of the most intriguing mysteries of the nitrogen budget is the tendency (noted by researchers for many years) of some plant communities to accumulate large amounts of soil nitrogen in the absence of known nitrogen-fixing plants. Stands of both grasses and nonnodulated trees (particularly conifers) have been found to accrete from 50 to over 150 kg/ha/y of nitrogen, which is an amount comparable to that fixed by some nodulated plants. Binkley et al. (2000), who reviewed several twentieth-century studies, refer to this as *occult* nitrogen fixation (we prefer the less judgmental term, *unexplained* nitrogen fixation) and conclude that many

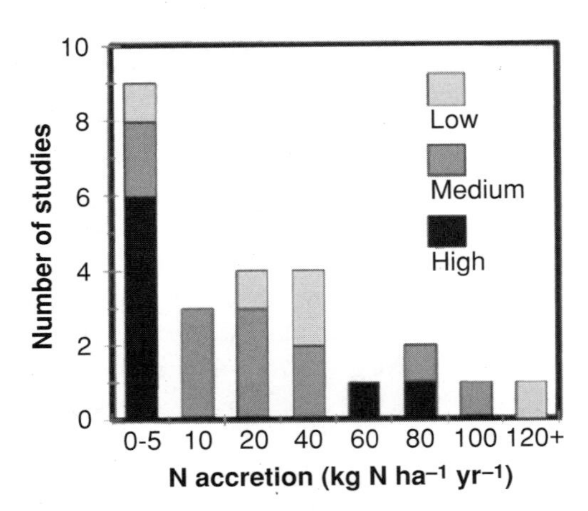

Figure 17.6. Histogram of reported rates of forest nitrogen accretion, with level of confidence warranted by the study. Levels of confidence assigned by Binkley et al. (2000) were based on their evaluation of the strength of the experimental design for each study. (Binkley et al. 2000. With kind permission of Springer Science and Business Media)

(but not all) reports are difficult to interpret unambiguously because of methodological problems (e.g., poor experimental design) (fig. 17.6). Nevertheless, it seems clear that under some circumstances associative fixers can be quite active. That is best documented in some members of the grass family (Elmerich et al. 2006) but is also true with trees. For example, in the so-called sandbox experiment, trees of various species were planted into large, sand-filled pits with known nitrogen content (Bormann et al. 1993, 2002). After five to six growing seasons, sandboxes planted to pitch pine and red pine had accumulated an average of anywhere from approximately 50 to nearly 150 kg/ha/yr of nitrogen[5] that could not be explained by precipitation inputs. Nitrogen fixation was measured in the rhizospheres of both species.

The debate about unexplained nitrogen inputs concerns amounts, not whether associative nitrogen fixers occur in tree rhizospheres; they do and their activity can easily be measured. Associative nitrogen fixers may occur either within close proximity to or on the surface of roots and mycorrhizae, as well as within the tissues of mycorrhizae and mycorrhizal fruiting bodies (Silvester and Bennet 1973; Malajczuk 1979; Florence and Cook 1984; Li and Castellano 1987; Li and Hung 1987; Amaranthus et al. 1990a). Nitrogen-fixing microorganisms also occur within the feces of animals that have been eating mushrooms or truffles (Li et al. 1986)—a sort of extended associative fixation. Rhizospheres are more favorable habitats for nitrogen-fixing microorganisms than nonrhizosphere soil for several reasons. Root and mycorrhizal exudates provide abundant labile carbon, and the resultant high biological activity reduces oxygen levels. Organic acids exuded into the rhizosphere by roots accelerate the weathering of primary minerals and release the nutrients they contain. Chelators produced by mycorrhizal fungi and rhizosphere bacteria (e.g., *Pseudomonas* sp.) solubilize iron and other metal nutrient elements and, by binding iron, probably increase the solubility of phosphorus as well (chapters 14 and 18).

17.4.5.3 Heterotrophic (Free-Living) Nitrogen Fixers in Soil and Litter

There have been a few reports of high rates of nitrogen fixation in soils and litter. However, because a high proportion of forest soil and litter likely is occupied by either roots or mycorrhizal hyphae, distinguishing free-living from associative fixation is virtually impossible in field studies unless roots and hyphae are experimentally excluded. Soils and litter are most likely to be favorable for nitrogen fixers when nitrogen limits decomposition (thereby giving fixers a competitive advantage over nonfixers), which may be common (Vitousek et al. 2002; chapter 18 in this volume), especially in certain microsites such as decaying wood. In Hawai'i, heterotrophic fixation is higher in litter with high lignin content than in litter with low lignin content and is closely related to the degree to which added nitrogen stimulates decomposition (Hobbie and Vitousek 2000; Vitousek and Hobbie 2000). Free-living fixation is also enhanced in wetland soils, where high moisture content limits oxygen concentrations (Vitousek et al. 2002).

Various studies of conifer stands have found that less than 1 kg/ha/y of nitrogen is added through biological nitrogen fixation in soils and/or litter (Granhall and Lindberg 1978; Baker and Attiwill 1984; Skujins et al. 1987; Heath et al. 1988); however, significantly higher levels (i.e., 7.5 to over 8 kg/ha/y) have been found in the soil and litter of deciduous forests in the eastern United States (Todd et al. 1978; Jones and Bangs 1985). This difference probably results from the more favorable summer rainfall regimes and higher litter nutrient content of eastern deciduous forests compared with conifer forests (chapter 18).

Ecosystems dominated by different tree species may differ with respect to sites of active nitrogenase activity. In Australia, estimated rates of nitrogen fixation were four times higher in litter than in soil beneath *Pinus radiata* (0.52 vs. 0.12 kg/ha/y, respectively), but this relationship was reversed beneath *Eucalyptus obliqua*, where rates in litter were only 20 percent of those in soil (0.19 vs. 0.89 kg/ha/y, respectively) (Baker and Attiwill 1984). Such differences may reflect differences between tree species in litter chemistry (i.e., nutrients, allelochemicals), patterns of root and mycorrhizal activity (if fixation is associative), or water balance in litter versus soils.

Under some circumstances, free-living cyanobacteria (which are photosynthetic) form mats on the surface of soils, where they fix nitrogen during periods of favorable temperature and moisture. These have been little studied in forest ecosystems; however, Isichei (1980) estimates that cyanobacteria add from 3 to 9 kg/ha/y of nitrogen to Nigerian savannas.

17.4.5.4 Lichens

Lichens that contain nitrogen-fixing cyanobacteria occur in the canopies and forest floors of a wide variety of temperate, tropical, and boreal forests (Millbank 1985). In the Pacific

[5] The large range of values results from different estimation techniques.

Northwest, various species of lichens that live in forest canopies (especially old growth) host nitrogen-fixing cyanobacteria. The most abundant of these, *Lobaria oregana*, is estimated to fix approximately 3 kg/ha/y of nitrogen (Denison 1979); in Amazonia, canopy lichens are estimated to fix 1.5 to 9 kg/ha/y (Forman 1975). The ground lichen *Sterocaulon paschale* was estimated to fix as much as 38 kg/ha/yr in a Swedish pine forest (Kallio and Kallio 1975). Note that accurate measurements under field conditions are difficult at best, and even estimates made by very careful investigators must be considered only rough approximations. (This holds for all types of nitrogen fixation.)

17.4.5.5 Microorganisms Occurring in Canopies

In addition to lichens, a variety of bacteria and cyanobacteria grow on or within tree leaves and branches, including at least some nitrogen fixers. Because high rainfall and humidity tend to prevent their desiccation, the moist tropics are a particularly favorable environment for canopy microorganisms. Bryophytes (e.g., liverworts) living on leaf surfaces retain moisture and enhance the activity of associated nitrogen fixers (Bentley and Carpenter 1980). Bentley and Carpenter (1984) estimate that 10 to 25 percent of the leaf nitrogen in the palm *Wolfia georgii* is obtained by direct transfer to the leaf of nitrogen fixed by cyanobacteria living on the leaf surface. Although free-living nitrogen fixers can be readily isolated from the surfaces of temperate trees, they apparently fix very little nitrogen (Sucoff 1979; Jones 1982).

17.4.6 Successional Patterns of Biological Nitrogen Fixation

Patterns of biological nitrogen fixation vary widely throughout succession. In both temperate and boreal forests, biological nitrogen inputs are generally high early in a sere, low in midsuccessional stages, and rise again somewhat during the old-growth stages. In the Pacific Northwest, for example, early successional actinorhizal plants fix large amounts of nitrogen; as these are shaded out and replaced by conifers, nitrogen fixation is restricted to soil, litter, logs, and rhizospheres, each of which is believed to contribute relatively small amounts of nitrogen. During these midsuccessional stages, conifer nitrogen requirements are presumably satisfied through nitrogen that was accumulated when actinorhizal plants occupied the site. As forests move into the structurally diverse old-growth stages, fixation increases again, although not to the levels of early succession. Nitrogen-fixing lichens, which are not abundant in midsuccessional stages, appear in old growth, and accumulations of CWD provide an energy source for free-living nitrogen fixers. In a deciduous forest of the eastern United States (which did not have significant components of nodulated plants), nitrogen fixation correlated with the amount of CWD, which was highest either early or late in succession (Roskoski 1980). In both logged and burned lodgepole pine forests, the highest activity of free-living fixers was in dead roots, followed by stumps, advanced or medium decay stems, litter, humus, early decay stems, and mineral soil (Wei and Kimmins 1998). In forests with few (or no) early successional nodulated plants, fixation in dead wood takes on considerable significance in replenishing nitrogen lost during disturbances. Wei and Kimmins (1998) conclude that nitrogen fixation in dead roots may "have an important, but largely unrecognized, ecological role in the long-term site productivity of lodgepole pine forests."

17.4.7 Limitations to Biological Nitrogen Fixation

The amounts of nitrogen added to ecosystems through biological fixation are limited primarily by the energy and nutrients available to nitrogen-fixing bacteria; however, low pH, drought, and low temperatures also limit nitrogen fixation.

17.4.7.1 Energy

Even symbiotic and associative nitrogen fixers become energy limited when environmental factors limit photosynthesis by their hosts (Whiting et al. 1986), which is one reason why many nodulated plants are intolerant of shade. Energy supply (i.e., the availability of reduced carbon derived ultimately from photosynthesis) is generally believed to be the most limiting factor for free-living nitrogen fixers (Stevenson 1986); however, the evidence does not support this in forest soils. In two separate studies, Brouzes et al. (1969) and Petelle (1984) both found that while adding sugar (i.e., an energy source) consistently increased rates of nitrogen fixation in agricultural soils, it had no consistent effect on nitrogen fixation in forest soils. These results imply that factors other than energy limited nitrogen fixation in the forest soils that were studied, which is not too surprising considering the large amounts of carbon input to forest soils via litter and root exudates (chapters 15 and 18).

17.4.7.2 Nutrients

Molybdenum and iron are components of nitrogenase, and magnesium and phosphorus are required in the energy transfers that fuel nitrogenase activity. Shortages of these elements will limit both free-living and symbiotic nitrogen fixation. Molybdenum in particular limits free-living nitrogen fixation in the Pacific Northwest (Silvester 1989); however, calcium has also been found to enhance free-living nitrogen fixation (perhaps an effect of pH).

Various other nutrients influence nitrogen fixation by nodulated plants. In legumes, multiplication of rhizobia within the rhizosphere is limited by low calcium availability, nodule initiation by low cobalt availability, and nodule development by low availability of copper and iron (O'Hara et al. 1988). Nitrogenase activity within nodules is also limited by any nutrient deficiency that reduces photosynthesis by the host plant. Enhanced nutrition of both nitrogen-fixing bacteria and host is the primary reason why nodulated plants fix more nitrogen when they are mycorrhizal than when they are not. (The three-way symbiosis among a host plant, nitrogen fixer, and mycorrhizal fungus is called a **tripartite association**) (Daft et al. 1985).

On the other hand, nitrogen fixation is reduced by too much soil nitrogen. Activity of the nitrogenase enzyme is inhibited by high levels of NH_3 in soil, and both nodulation and the numbers of free-living bacteria in soil can be reduced by nitrogen fertilization (Postgate 1982; Kolb and Martin 1988).

17.4.7.3 pH

Free-living nitrogen-fixing bacteria frequently are limited by the relatively low pH typical of many forest soil and litter layers. Jones and Bangs (1985) doubled rates of nitrogen fixation in the soil of an English oak forest by adding lime (CaOH), thereby raising soil pH from 4.5 to 6.0. Calcium fertilization has also been found to increase nodulation and nitrogen fixation by the leguminous tree *Leucaena leucocephala* (Hansen and Munns 1988).

17.4.7.4 Temperature and Moisture

In very dry or cold environments, plants may nodulate poorly, and nitrogenase activity may be low in nodules that do form. In the Himalaya Mountains, for example, nodulation and nitrogen fixation by *Alnus nepalensis* colonizing landslide tracks is highest on sites between 1,500 m and 2,000 m in elevation (fig. 17.7). Above 2,000 m, nitrogen fixation is limited by low soil temperatures, but low precipitation is probably the principal limitation below 1,500 m. Not surprisingly, alder seedlings do not survive well in areas where their nodulation is poor. Alders require moister environments than many other actinorhizal plants and often attain their greatest cover in either riparian zones or moist northern aspects.

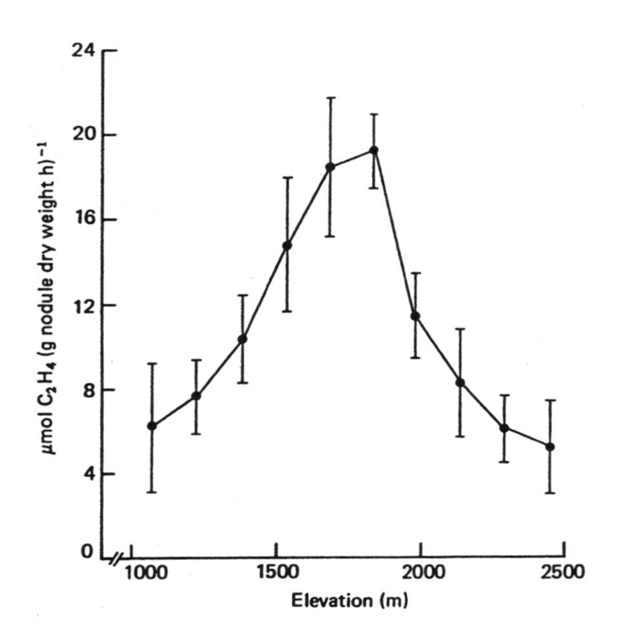

Figure 17.7. The influence of elevation on nitrogenase activity of nodules formed by Himalayan alder (*Alnus nepalensis*). Nitrogenase activity is measured in this case by the rate at which nodules reduce acetylene (C_2H_2) to ethylene (C_2H_4). The y axis denotes micromoles of C_2H_4 produced per gram dry weight nodule per hour. (Sharma 1988)

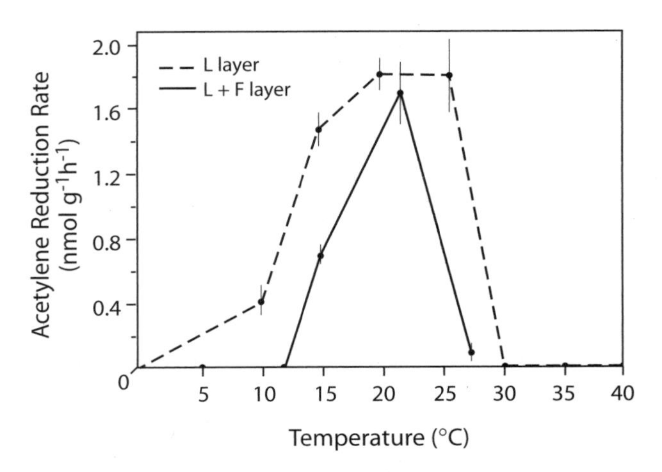

Figure 17.8. The effect of temperature on nitrogenase activity (as measured by acetylene reduction) of free-living diazotrophs in litter from Pacific Northwest conifer forests. (Adapted from Heath et al. 1988)

Nitrogenase activity in soils and litter also increases with increasing moisture content (up to about 250%) and with increasing temperature up to 20 to 25°C, after which it declines sharply (fig. 17.8) (Baker and Attiwill 1984; Heath et al. 1988). In temperate ecosystems where summer (when temperatures are favorable) is also a period of low rainfall (e.g., west coast of the Americas), no time of year presents optimum conditions for nitrogen-fixing microorganisms.

17.5 NUTRIENT LOSSES FROM UNDISTURBED FORESTS

The magnitude of nutrient leaching from soils depends on two factors: the amount of water moving through soils to streams, and the concentration of nutrient elements in the water. The latter, in turn, depends on both the chemical characteristic of and biological demand for a particular nutrient. Forests exert considerable control over some aspects but not others. Transpiration reduces the amount of water moving to streams, and a combination of abiotic chemical reactions and biological demand (by vegetation and soil microbes) reduces the concentration in soil water (Knight et al. 1985; Zak et al. 1990). Mobile (i.e., soluble) elements present in excess of biological requirements or in forms not used by biota leach below the rooting zone and often (but not always) to streams. For example, calcium and magnesium are rapidly lost from nitrogen-limited forests growing on young lavas in Hawai'i, while phosphorus is retained because it forms highly insoluble complexes with iron and aluminum (Chadwick et al. 1999). At least some biologically useable limiting nutrients also leach to streams, providing an important nutrient source for stream biota (e.g., Kaushal and Lewis 2005).

In their study of 31 mostly tropical watersheds, Lewis et al. (1999) found that stream nitrogen increased at about two-thirds the rate of runoff, that is, nutrient concentration in runoff declined as runoff increased. Other studies find the opposite: dissolved inorganic nitrogen concentrations increasing as soil moisture increases (Magill et al. 2004).

Since nitrogen is the most limiting nutrient in most (unpolluted) forests, ecologists have largely focused on the factors controlling its retention and loss. Until recently, nitrogen losses to streams have been understood strictly in terms of the nutrient retention model (Vitousek and Reiners 1975), which deals with inorganic nitrogen (ammonium and nitrate ions) and assumes that leaching losses of nitrogen are regulated by microbial and plant demand. Vitousek and Reiners (1975) proposed that nitrogen retention (in nitrogen-limited ecosystems) correlates with the ability of the system to fix carbon. For example, nitrogen would be held less tightly in slow-growing old than in fast-growing young and mid-aged forests. That hypothesis has rarely been tested but was true in the White Mountains of New Hampshire, where stream nitrate concentrations in old-growth watersheds were four times higher than in watersheds that had been logged and burned 80 to 110 years previously (Goodale et al. 2000). In contrast, stream phosphate concentrations are higher in watersheds dominated by young forests than in watersheds dominated by old forests (Binkley et al. 2004).

An increasing body of work shows that dissolved organic nitrogen is at least and possibly more important than inorganic nitrogen in nitrogen fluxes (Hedin et al. 1995; Campbell et al. 2000; Goodale et al. 2000; Neff et al. 2003; Rastetter et al. 2005; Goller et al. 2006). Dissolved organic nitrogen is a "structurally complex mixture of materials, ranging from simple compounds that are readily used by plants and microbes to polyphenols or tannins that are not easily metabolized" (Neff et al. 2003). As with inorganic nitrogen, the loss of simple organic compounds such as amino acids is potentially regulated via uptake by plants and microbes. More complex compounds have no such regulation and may represent a long-term leakage of nitrogen even from actively growing forests (Hedin et al. 1995; Neff et al. 2003; Rastetter et al. 2005). Dissolved organic nitrogen composes a high percentage of total dissolved nitrogen in forested soils that have been measured; for example, 55 percent in a montane forest in Ecuador (Goller et al. 2006); 28 to 87 percent in the White Mountains, New Hampshire (Goodale et al. 2000); approximately 40 percent in a subtropical Australian forest (Chen et al. 2005); and an average 50 percent of nonparticulate nitrogen in tropical American streams (Lewis et al. 1999).

Stream nitrogen concentrations are highly variable among forested watersheds across the United States (see the review by Binkley et al. [2004]). Some of the variability correlates with tree species composition, topoedaphic factors (e.g., geology, elevation), and precipitation inputs; however, much of it is unexplained. To some degree nitrate leaching reflects abundance, as losses are highest in the northeast, which also has the highest input of nitrogen in precipitation. In their analysis of stream chemistry for over three hundred forested watersheds in the United States, Binkley et al. (2004) found that nitrate dominated the nitrogen loads of streams draining hardwood forests, while dissolved organic nitrogen dominated streams draining conifer forests. That pattern is confounded by various factors (e.g., differing topoedaphic positions of conifers and hardwoods) that make interpretation difficult. The only experimental comparison (on the Coweeta Experimental forest, North Carolina) showed greater nitrate leaching from conifer forests (Swank and Vose 1997). Not all hardwoods affect stream chemistry similarly. In the Catskill Mountains (New York State), Lovett et al. (2002) found that stream nitrate was lower in watersheds dominated by oaks and beech than in watershed dominated by other hardwoods, and Lewis and Likens (2000) found a similar pattern in the Allegany Mountains (Pennsylvania and West Virginia). In general, tree species (e.g., oaks, conifers) with higher foliar carbon-to-nitrogen ratios retain nitrate better than those with low carbon-to-nitrogen ratios (e.g., northern hardwoods), which is consistent with the nutrient-retention hypothesis.

17.6 NUTRIENT LOSSES FROM DISTURBED FORESTS

Most nutrient loss from forest ecosystems occurs during disturbance and in the early stages of recovery from disturbance. Nutrients may be lost in any of several ways:

- Excessive nutrient inputs (e.g., pollution) may saturate the system and overwhelm its retention capacities, as well as causing increased leaching of other nutrients from soil pools.
- When trees are killed or weakened, their control over nutrients declines accordingly, and nutrients may be lost through leaching and erosion.
- Fire volatilizes some nutrients, especially nitrogen and sulfur; nutrients that are not readily volatilized may be lost as fly ash.
- Warming the forest floor (due to loss of canopy cover) accelerates decomposition of surface organic matter, releasing nutrients into solution, where they are vulnerable to loss via leaching or, in the case of nitrogen, through microbial conversion to gases.
- Nutrients are lost in association with logging. Some are removed along with harvested biomass; others may be lost during preparation of sites for planting.

In the following subsections we discuss these processes separately; however, it should be borne in mind that they do not occur separately in nature. Clear-cutting, for example, removes nutrients in biomass, usually causes some increased export to streams, and if logging residues are burned (as they frequently are), volatililzes some carbon, nitrogen, and sulfur (and even phosphorus and potassium if the fire is sufficiently hot).[6]

17.6.1 Nutrient Loss to Streams

Nutrients may be lost to streams either as dissolved substances (i.e., in solution) or as **particulates** (i.e., undissolved organic or mineral particles transported either by soil water or overland with eroding soil). Solution losses consist primarily of soluble anions and cations (e.g., nitrate, calcium), whereas particulates contain a high proportion of the less-soluble minerals (e.g., iron, phosphorus) and organically bound nitrogen, phosphorus, and sulfur. The relative importance of solution losses versus particulates varies among forest types and is greatly affected by the nature of disturbance. Dissolved substances are the major avenue of nutrient leaching in forests that are relatively rich in nitrogen (e.g., temperate deciduous) and in forests that are affected by acid rain. In unpolluted conifer forests, most nutrient leaching following disturbance is associated with organic particulates (Sollins et al. 1981). Landslides and erosion transport both mineral and organic particulates to streams and are of greatest significance in those areas with steep slopes and inherently unstable soils.

When a stand is destroyed by a natural disturbance such as wildfire, early successional vegetation quickly damps the nutrient loss to streams by reestablishing transpiration, immobilizing nutrients in biomass, stabilizing surface soil, and maintaining the nutrient sinks provided by mycorrhizal fungi and other soil organisms that rely on plants (e.g., Goodale et al. 2000; Belanger et al. 2004; Binkley et al. 2004; Egerton-Warburton et al. 2005; Smithwick et al. 2005). Additionally, bacterial populations bloom and shift toward predominance by saprophytes that subsist on the pulse of dead organic matter and serve as sinks for limiting nutrients (Egerton-Warburton et al. 2005).[7]

17.6.1.1 The Anion Mobility Model

The availability of mobile anions such as nitrate exerts a major control over cation loss from systems, a relationship referred to as the **anion mobility model** (Johnson and Cole 1980; Vitousek 1984b). The reason for this is straightforward. Cations always pair up with some anion, of which there are three possibilities: the soil exchange complex, immobile anions (e.g., calcium pairing with phosphates, which is uncommon at the pH of forest soils), and mobile anions. The more of the latter that are present in soil solution, the greater the potential for cation loss to streams. As we shall see, this has considerable significance for the impacts of acid rain on nutrient loss.

Three inorganic anions can be present in forest soils in sufficient quantities to influence cation movement: nitrate; sulfate; and bicarbonate (HCO_3^-), which is produced by dissolution of CO_2 in water. Sulfate is the least mobile and in unpolluted forests is also the least abundant. Soluble organic molecules (e.g., fulvic acids) can also serve as mobile anions (Schoenau and Bettany 1987). Fulvic acids transport nitrogen, phosphorus, and sulfur contained within the organic molecule, but little is known about their role in cation leaching.

In *unpolluted* forest soils, the relative abundance of bicarbonate and nitrate depends on soil pH and the various factors that influence nitrification and nitrate uptake by plants and microbes (chapter 18). Nitrogen-limited forests usually have low levels of nitrate in soil water, and bicarbonate and organic anions are the dominant mobile anion. Even bicarbonate is not abundant in many of these forests, however, because the chemistry of its formation is such that very little is produced at pH values below 5.0 (Bohn et al. 1979). Nitrate is produced in nitrogen-limited forests but tightly cycled within the microbial community. Where nitrogen is abundant, nitrate becomes more important as a leaching agent.

17.6.1.2 Acid Rain and Nutrient Leaching

Depending on the source of pollution, acid rain contains varying proportions of sulfate and nitrate, mobile anions that increase leaching of basic cations such as calcium, magnesium, and potassium. Increased cation leaching has been linked to acid rain in Europe (Abrahamsen 1983; Ulrich 1983), and Johnson and Todd (1987) estimated that sulphate deposition had increased cation leaching by 200 to 300 percent in eastern Tennessee forests. Measurements during the 1980s showed high levels of calcium leaching throughout the eastern United States that were probably related to

> Johnson et al. (1991) concluded that forest soils in North America, central Europe, Sweden, the United Kingdom, and Australia were acidifying with surprising rapidity. Some of this (but not all) can he attributed to acid rain. Johnson et al. concluded that in some cases, acidification can be traced to internal ecosystem processes such as uptake and storage of base cations by trees and nitrification associated with excessive nitrogen fixation.

[6] Chanasyk et al. (2003) review potential affects of wildfire and harvesting on soils in the boreal region of Canada.

[7] Nutrient losses to lakes following fire or logging in Canadian forests are discussed by Prepas et al. (2003).

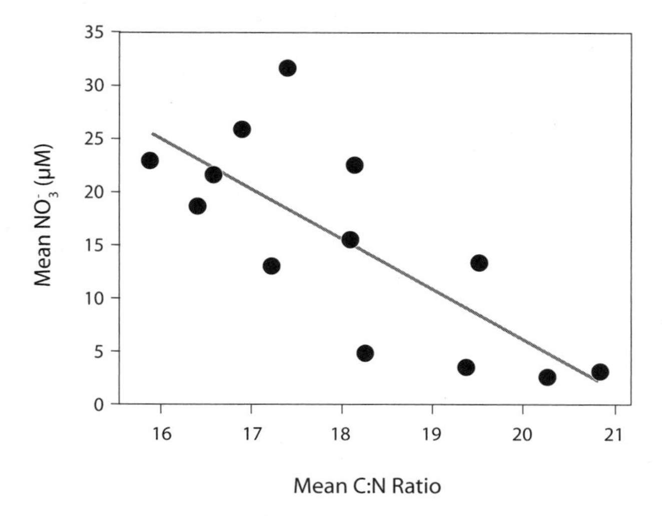

Figure 17.9. Mean stream NO_3 concentration versus mean carbon-to-nitrogen (C:N) ratio in soil organic horizons. The points represent 13 forested watersheds in the Catskill Mountains, New York. The line is the best fit regression line ($r^2 = 0.57$, $p = 0.0027$). (Lovett et al. 2002. With kind permission of Springer Science and Business Media)

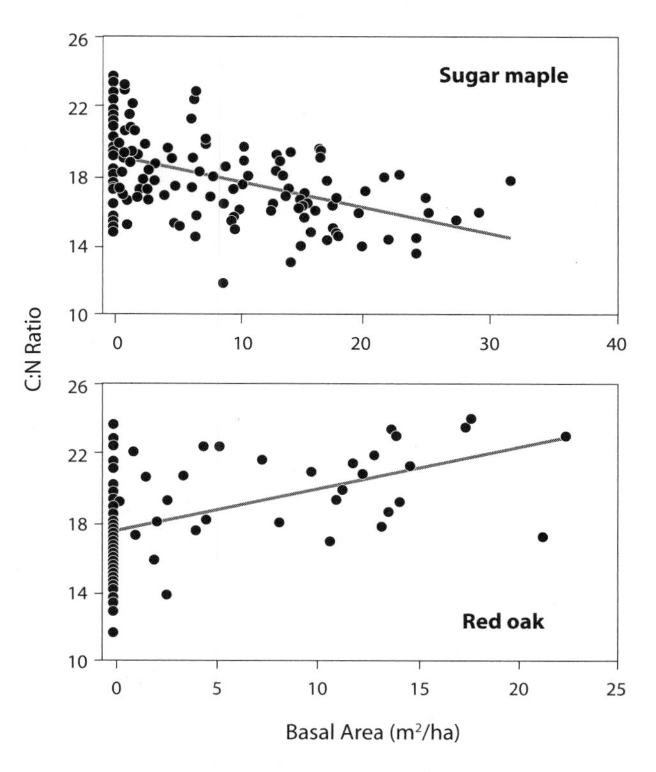

Figure 17.10. Organic horizon carbon-to-nitrogen (C:N) ratio versus basal area of sugar maple ($r^2 = 0.22$, $p = 0.0001$) and red oak ($r^2 = 0.20$, $p = 0.0001$) in mixed-species stands. (Lovett et al. 2002. With kind permission of Springer Science and Business Media)

sulphate and nitrate in acid rain (Federer et al. 1988). In the United States, sulfates in precipitation declined by 28 percent between 1985 and 2001, while nitrates declined in the northeast but increased elsewhere in the nation, and ammonium uniformly increased.[8]

Recent concern regarding ecological impacts has focused on nitrogenous pollutants and the potential for ecosystems to become saturated with nitrogen, especially the potential negative effects stemming from excessive inputs of NO_3^- (Aber and Magill 2004). Increased nitrate leaching is considered an early sign of nitrogen saturation (as discussed later, it is also an early sign of any kind of stress that weakens biotic sinks). Although elevated nitrate leaching is seen, to date only a small proportion of the excessive nitrogen inputs to forests of the northeastern United States has appeared in streams (Goodale et al. 2002). What happens to it? Different studies have produced different answers to that question. Van Breemen et al. (2002) estimated that about half of nitrogen inputs to the northeastern United States were returned to the atmosphere as gases, largely through denitrification (chapter 18), while 20 percent went to streams and 18 percent was stored in soil organic matter and wood (9% in each). In contrast, Goodale et al. (2002) found that 73 percent of nitrogen deposition in 16 watersheds of the eastern United States was fixed in woody biomass, with soil storage playing a relatively minor role. Other studies, however, have found that soil sinks are very important. On the Harvard Forest (Massachusetts), 70 percent of experimentally added nitrogen was retained, almost all of that in mineral soil (Magill et al. 2004). Lovett et al. (2002) showed that stream nitrate in the Catskill Mountains (New York State) correlated negatively with the carbon-to-nitrogen ratio of soil organic layers (fig. 17.9), which, in turn, depended on what tree species dominated the watershed (fig. 17.10). Berntson and Aber (2000) cite other studies from Europe and North America showing that "long-term immobilization of N in soils plays an essential role" (see their paper for citations to original studies in Europe and North America.). The most robust conclusion in the face of these conflicting data is that excess nitrogen is retained through various processes whose relative importance varies depending on local conditions.

Whether ecosystems will continue to store excessive nitrogen inputs is debatable. Polluted forests are characterized by elevated tree mortality, weakening a primary sink. That effect has been seen most dramatically in Europe but is appearing more frequently in the northeastern United States (Magill et al. 2004; chapter 20 in this volume). Since 1988, the Long Term Ecological Research program at Harvard Forest has maintained a controlled experiment involving large nitrogen additions to two different forest types, a red pine plantation established in 1928 and a mixed hardwood forest (Aber and Magill 2004). Most added nitrogen has been retained in mineral soil, but tree mortality has been high, particularly in the red pine plantation, and tree sinks are declining (Nadelhoffer et al. 2004). Moreover, fungal biomass has declined with high nitrogen inputs, and the microbial sink for nitrogen is weakening (Compton et al. 2004; Frey et al. 2004).

[8]Mark Nilles, U.S. Geological Survey, http://bqs.usgs.gov/acidrain.

17.6.1.3 Leaching Losses following Disturbance

The concentration of nitrate in soil water almost always increases following disturbances that kill or weaken trees, although the magnitude of increase varies widely among different forest types (Vitousek and Melilo 1979). Studies have focused on nitrate leaching associated with clear-cutting or wildfire, but insect defoliation and pollution stress can have similar effects (Swank et al. 1981; Ulrich 1983). Increases in nitrate following disturbance are well predicted by rates before disturbance; they will be highest in those ecosystems where soil microbes are least limited by nitrogen. This includes temperate deciduous forests, tropical forests, and forests with high inputs of nitrogenous pollutants.

Early successional vegetation exerts a major control over nutrient leaching, a fact that was clearly demonstrated by a simple but classic experiment conducted on the Hubbard Brook Experimental Forest in New Hampshire (Likens et al. 1978). The Hubbard Brook study consisted of clear-cutting followed by herbicide treatment that killed early successional vegetation. Figure 17.11 shows nutrients exported to streams from cut and uncut watersheds. Over 400 kg/ha/y of nitrogen was leached as nitrate from devegetated watersheds into streams, a yearly loss equivalent to 20 years of nitrogen inputs to this system. Losses dropped sharply after plants were allowed to regrow, beginning in 1969. Dissolved calcium and magnesium (carried by the nitrate) followed a similar pattern, as did particulates, most of which had been transported to streams through erosion from devegetated plots.

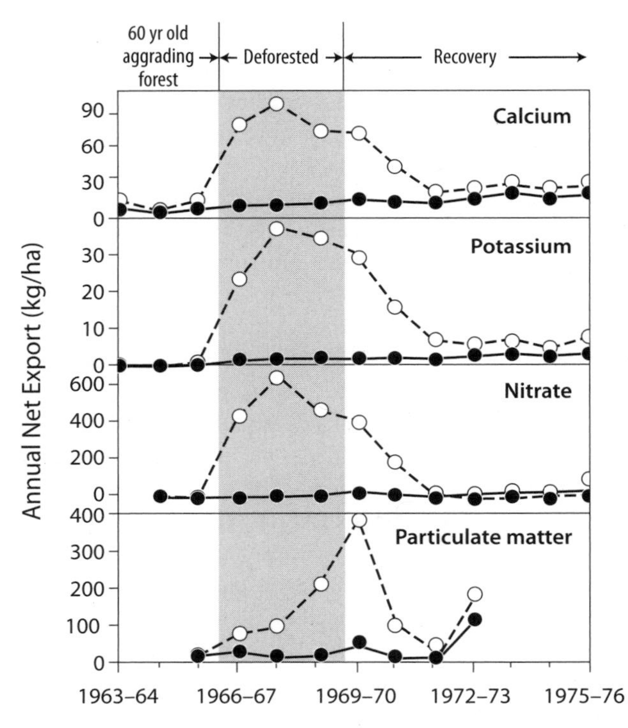

Figure 17.11. The loss of cation and nitrate to streams as affected by clear-cutting and treating with herbicide, Hubbard Brook Experimental Forest, New Hampshire. Open circles represent a clear-cut and herbicided watershed; closed circles represent an adjacent uncut watershed. Plants were allowed to regrow beginning in 1969. (Bormann and Likens 1979a)

In conifer forests, soil microbes are frequently nitrogen limited and immobilize a significant proportion of the nitrogen pulsed into the system by disturbance. Vitousek and Matson (1985), for example, showed that microbes sequestered nitrogen after a loblolly pine forest was clear-cut and therefore prevented leaching loss following removal of the plants. The capacity of microbes to immobilize nitrogen (or any other nutrient) directly relates to their having a ready source of carbon (without which they cannot grow). In the short run, microbes can draw on carbon in soil organic matter, dead roots, dead mycorrhizal hyphae, and surface litter; however, fresh carbon inputs from early successional vegetation are also important and (in the long run) absolutely necessary for the microbes. In a clear-cut loblolly pine site, both the removal of surface organic matter and herbiciding (common techniques used to prepare sites for planting) reduced the ability of microbes to immobilize nitrogen, as evidenced by greatly increased levels of nitrate in solution (Vitousek and Matson 1985).

In moist tropical forests, rapid mineralization, relatively low carbon-to-nitrogen ratio in organic matter, and heavy rainfall set the stage for large nutrient losses in leaching and erosion from disturbed sites (Jordan 1985). Nitrate levels in solution are generally high, and from 500 to 2,000 kg/ha of nitrogen have been reported to disappear from surface soils following clearing (not including the nitrogen lost in slash burning) (Sanchez et al. 1985; Matson et al. 1987). Erosion exports poorly soluble nutrients such as phosphorus and probably mycorrhizal spores as well (Habte et al. 1988).

17.6.2 Nutrient Losses Associated with Fire

Table 17.3 shows the temperatures at which various nutrients are volatilized. Temperatures may reach 1,100°C during hot fires and 60°C during cooler fires (J.B. Kauffman, personal communication). Nitrogen, organic phosphorus, and sulfur are volatilized to some degree even in relatively cool fires, while inorganic phosphorus, potassium, and sodium are volatilized in hot fires. All of the nitrogen and approximately 60 percent of the phosphorus and potassium content of organic matter are converted to gases during complete combustion (DeBano 1990). Calcium and magnesium are unlikely to be volatilized even in very hot fires, although they may be lost from the ecosystem as fly ash.

Figure 17.12 shows a flow chart of nitrogen losses following wildfire. It is primarily volatile nutrients in aboveground biomass that are lost from the ecosystem as gases during fire. Nitrogen is volatilized as readily as carbon, and nitrogen losses during wildfire or slash burning are highly correlated with fuel consumption (Little and Ohmann 1988; Wan et al. 2001) (fig. 17.13). Broad-leaved forests tend to have greater

Table 17.3
The temperatures at which various nutrients volatilize

Element	Temperature, C*	Source
Nitrogen	<200	White et al. 1972; Weast 1982
Sulfur	444	Weast 1982
Phosphorus		
Inorganic	774	Raison et al. 1985
Organic	360	Raison et al. 1985
Potassium	760	Weast 1982; Wright and Bailey 1982
Sodium	880	Wright and Bailey 1982
Magnesium	1107	Wright and Bailey 1982
Calcium	1240–1484	Weast 1982; Wright and Bailey 1982; Raison et al. 1985
Manganese	1962	Weast 1982

Source: Kauffman 1990.
*Temperatures may reach 600°C in cool fires and 1100°C in hot fires.

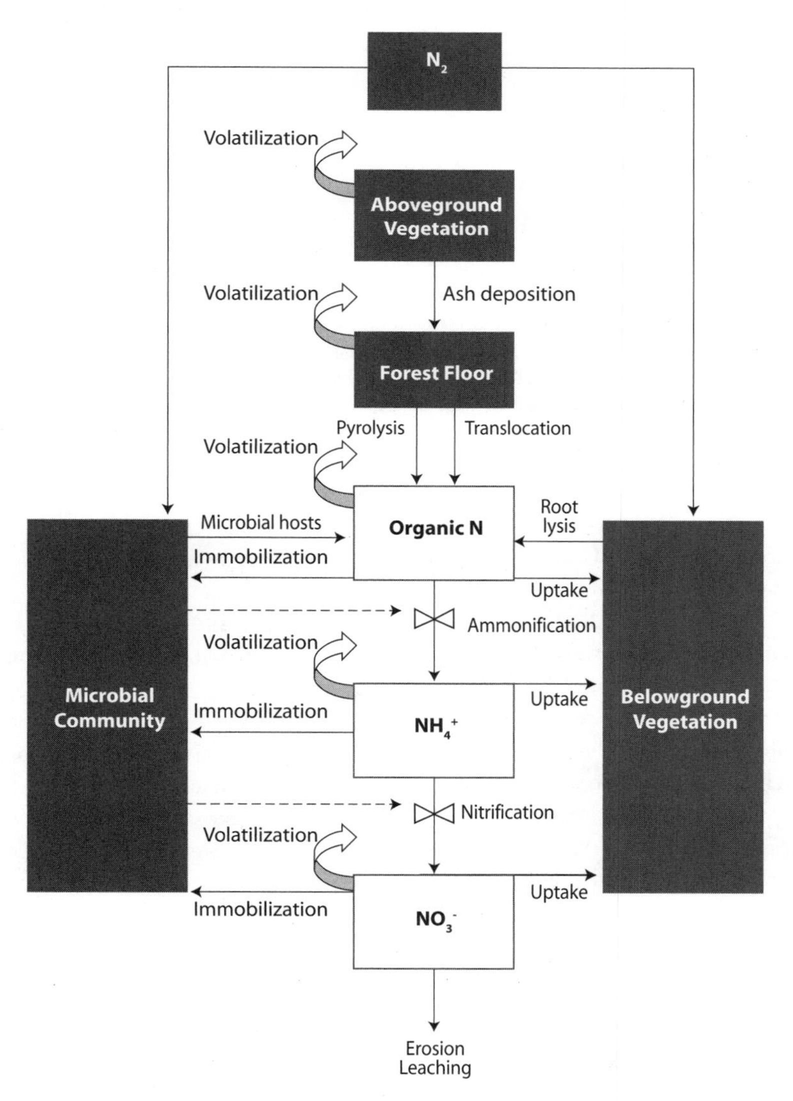

Figure 17.12. A flow chart of nitrogen fluxes following wildfire. (Smithwick et al. 2005. With kind permission of Springer Science and Business Media)

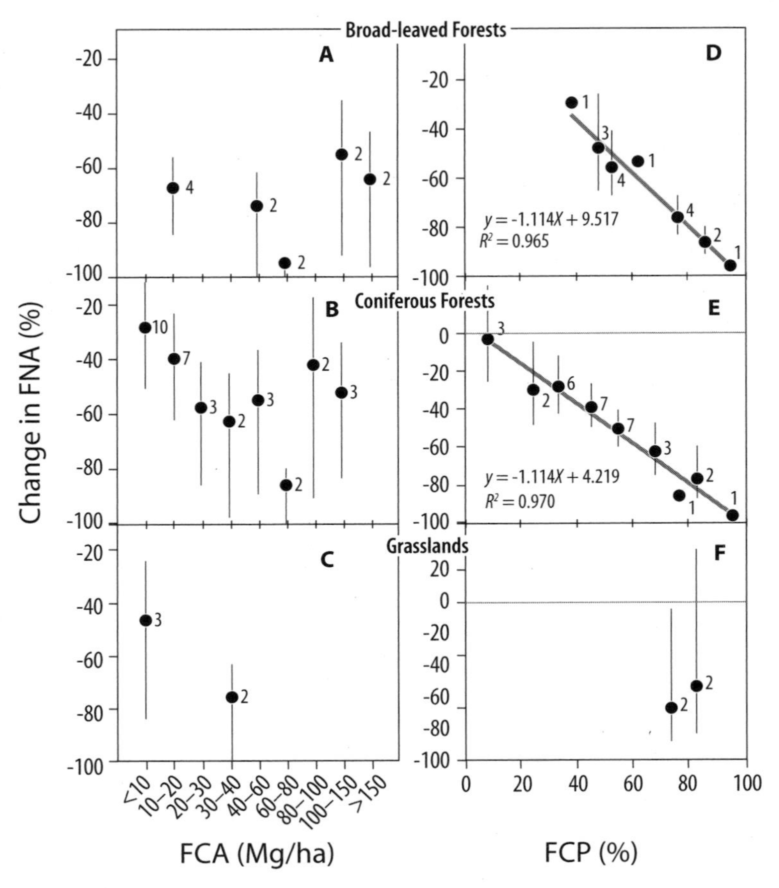

Figure 17.13. The effects of fuel consumption amount (FCA) (A–C) and fuel consumption percentage (FCP) (D–F) on the responses of fuel nitrogen amount (FNA) to fire in broad-leaved forests (A, D), coniferous forests (B, E), and grasslands (C, F). Open circles represent means. Vertical bars are 95% confidence intervals. Numbers adjacent to circles denote the number of studies from which means and confidence intervals are derived. (Wan et al. 2001. With permission of the Ecological Society of America)

nitrogen losses for a given level of fuel consumption than conifers (Wan et al. 2001)[9] (fig. 17.14). Nitrogen losses may amount to hundreds of kilograms per hectare in hot wild fires or slash burns (Feller 1982; Feller and Kimmins 1984; MacAdam 1987; Little and Ohmann 1988; McNabb and Cromack 1990; Kauffman et al. 1992; Wan et al. 2001), or from 10 percent to more than 50 percent of the total site nitrogen capital.[10] It is perhaps not surprising that nitrogen-fixing plants are often part of the early successional communities in ecosystems where wildfire is common. Fire can produce large losses of sulfur as well; slash burning following the harvest of a young secondary forest in Costa Rica volatilized 71 percent of the sulfur in residues (Ewel et al. 1981).

Soils are generally well insulated, and the maximum temperatures attained during fire decline rapidly with depth. For example, temperatures at the surface exceeded 60°C during a slash burn in Costa Rica but dropped to 10°C at 4 mm below the surface (Ewel et al. 1981). Under some circumstances, however, slash fires can produce high soil temperatures. This may occur when logging residues are piled before burning.

17.6.3 Gaseous Nitrogen Losses Resulting from Biological Processes

Some microbes convert nitrate to the gases N_2 and N_2O (chapter 18). More nitrates in soils following disturbance provides the necessary substrate, and studies have indeed found N_2O emissions to increase following clear-cutting and wildfire (Bowden and Bormann 1986; Anderson et al. 1988). Such emissions may be significant in terms of the atmospheric N_2O pool (hence ozone destruction and greenhouse effects), but they are only a small fraction of the nitrogen lost through leaching or volatilized by fire.

17.6.4 Harvest and Site Preparation

The amount of nutrients exported during harvest and site preparation depends on what biomass components are

[9] Broadleaved species in the data examined by Wan et al. (2001) included a high proportion of eucalypt and moist tropical forests.

[10] In their meta-analysis, Wan et al. (2001) found the following relationship: nitrogen loss (kg/ha) = 6.037*fuel consumption (Mg/ha) + 41.76; $r^2 = 0.80$, $p = 0.001$.

harvested and how residues and forest floor are handled after harvest. Recall from chapter 16 that nutrient concentrations in foliage and small branches are considerably higher than in wood; hence, whole-tree harvest removes a disproportionate amount of site nutrient capital relative to the gain in usable biomass. This has been demonstrated repeatedly in various forest types (Kimmins 1977; Johnson et al. 1982; Johnson and Todd 1987).

Table 17.4 shows a few examples of the increased nutrient removal when whole trees rather than only boles are harvested. Increased nutrient drain with whole-tree harvest ranges from 50 percent to over 300 percent, the highest values associated with species that have large crowns (in table 17.4, spruce, fir, and oak; note that oaks were harvested after leaf fall). Although highest nutrient concentrations generally occur in foliage, some species store large amounts of nutrients in bark. *Eucalyptus grandis*, for example, averages nearly 2 percent calcium in its bark (Turner and Lambert 1983); in cases such as this, removing only boles could eventually deplete the site of some nutrients.

In terms of soil effects, in their meta-analysis Johnson and Curtis (2001) found that whole-tree harvest reduced A horizon soil carbon and nitrogen by about 10 percent, while harvest that took only boles and left residues on site increased soil carbon and nitrogen in conifer forests and had no effect in hardwood or mixed forests. In general, forest soils appear to be very resilient, at least in the short term. A large study that was replicated over 19 sites across five locations in North America included treatments that removed varying levels of aboveground organic matter (Powers 2006). After five years, no effects on mineral soil carbon or nitrogen were detected, even in the most severe treatment (whole-tree harvest plus litter layers removed) (Sanchez et al. 2006) (but see below for effects on seedling growth).

Nutrients may also be lost during site preparation. As already discussed, broadcast burning of logging residues volatilizes several nutrient elements, especially nitrogen. Another common technique that is used where slopes are sufficiently gentle is to push residues into large piles using tractors or bulldozers, variously called **windrowing, piling and burning,** or **root-raking;** this practice redistributes nutrients contained in residues, forest floor, and often top soil into small areas, leaving most of the site depleted. Soils are also compacted. Numerous studies have found that windrowing reduces subsequent tree growth (chapter 20), two examples of which are shown in fig. 17.14, one from New Zealand (Dyck et al. 1989) and one from California (Atzet et al. 1989). In the New Zealand example (i.e., 17-year-old radiata pine), trees adjacent to windrows are growing nearly as well as those in unwindrowed area, but trees growing between windrows have produced only slightly over one-half as much volume as trees on unwindrowed area (fig. 17.14A). A similar pattern can be seen in the California case (i.e., 26-year-old ponderosa pine), where trees more than 3 m from a windrow average less than 40 percent of the volume of trees adjacent to windrows (fig. 17.14B). Unfortunately, this highly destructive practice continues throughout the world.

17.6.5 Do Harvest and Site Preparation Reduce Future Productivity?

This complex issue involves much more than nutrient removal (Grigal 2000). As Tamm (1979) pointed out, "We have to think in ecosystem terms when we try to judge the impacts of management operations on forest ecosystems." We return to this question in chapter 20; however, we anticipate the conclusions of that chapter with four generalizations:

1. High-impact management such as windrowing, whole-tree harvest, or excessively hot slash fires likely reduce site productivity, especially on nutrient-poor soils such as sands (Fox 2000).
2. Even relatively gentle treatments eventually reduce productivity if they result in more nutrients being lost than are replaced over the course of a rotation.

Table 17.4
The percentage increase in biomass and nutrient removal in whole-tree versus boles-only havesting

	Nitrogen	Phosphorus	Potassium	Calcium	Biomass	Study
Douglas-fir						
High site	52	71	45	—	23	Cole 1988
Low site	102	107	73	—	15	Cole 1988
Hemlock/cedar	165	117	77	95	43	Kimmins and Krumlick 1976
Lodgepole pine	53	54	14	15	15	Kimmins and Krumlick 1976
Loblolly pine	80	90	82	64	27	Johnson 1983
Oak*	186	214	233	166	159	Johnson et al. 1982
Spruce	288	367	236	179	99	Weetman and Weber 1972
Spruce and fir	232	283	103	97	38	Smith et al. 1986

Note: The percentage increase is calculated as (whole tree − boles only)/(boles only) × 100.

*Only sawlogs were harvested from the boles-only treatment, whereas all stems were taken during whole-tree harvest. This accounts for the large increase in biomass removed with whole trees. Trees were harvested after leaf fall, so whole-tree removal does not include foliage.

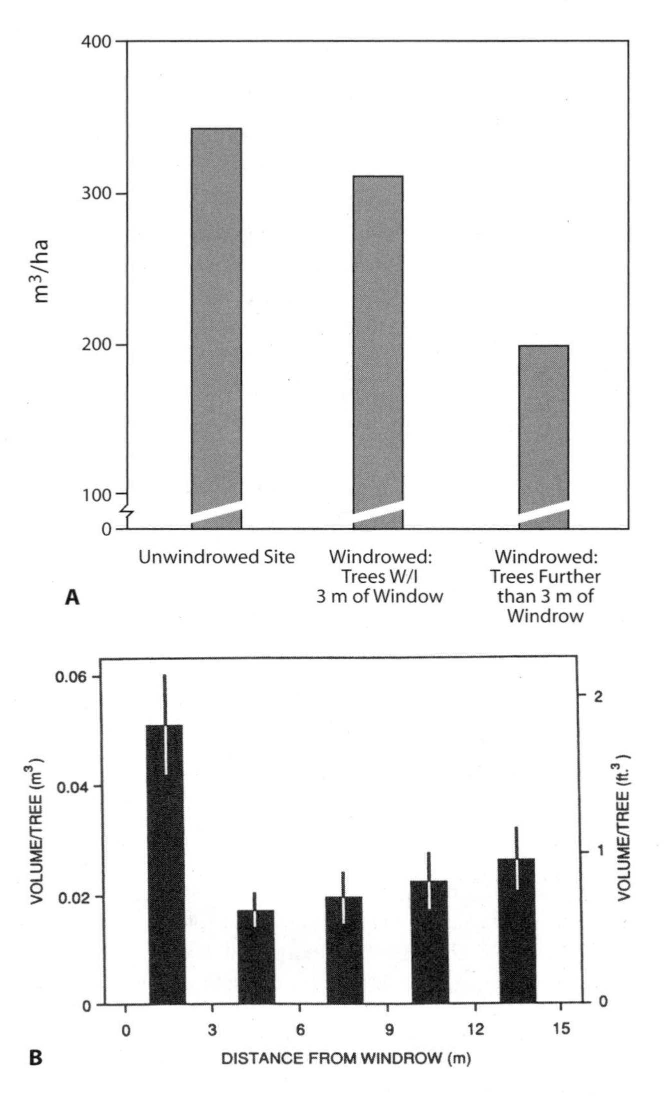

Figure 17.14. A. The stem volume per hectare in 17-year-old radiata pine plantations where slash had been either windrowed or not windrowed after the harvest of the previous stand. B. The average volume (per tree) of 26-year-old ponderosa pine growing at different distances from windrow in northern California. Bars represent standard errors. (A, adapted from Dyck et al. 1989; B, from R. F. Powers, unpublished, reprinted in Atzet et al. 1989)

3. Long-term impacts of any management practice depend on harvest frequency. Rotations that are shorter than the time required for sites to recover accumulate over time to produce large effects.
4. To keep managed ecosystems healthy, we must understand how they work. Whether a given set of management practices is high or low impact varies among forest types, depending (among other things) on environment and the natural disturbance regime. For example, a widely replicated experiment found that removing forest floor improved fifth-year seedling survival and growth in Mediterranean climates but reduced growth on productive, warm-humid, nutrient-limited sites (Fleming et al. 2006). In the latter case, seedling growth may have been more tightly linked to the nutrients released by rapid decomposition; if so, growth impacts may simply be delayed on the Mediterranean sites (Fleming et al. 2006). Species adaptations to natural disturbance influence system resilience following harvest. What is low impact for a temperate conifer forest that has evolved in a regime of periodic, stand-destroying wildfires may be high impact for a moist tropical forest. However, it should be clearly understood that no matter how practiced, forestry has unique aspects (removing products, building roads, etc.) and is not ecologically equivalent to a natural disturbance.

Short-term growth responses may tell little about the longer term. As Fleming et al. (2006) point out with regard to their fifth-year findings, "Results to date suggest that early seedling performance is often dominated by microclimate- and competition-related effects on resource availability and physiological stress. . . . As stand development proceeds and the expanding canopies place greater demands on nutritional reserves while moderating microclimatic conditions, productivity limitations associated with [organic matter] removal may become increasingly evident."

In an experiment in New Zealand, removal of forest floor and logging slash reduced nutrient supply to radiata pine seedlings growing on sandy soils; however, effects on growth did not appear until the eleventh year (C.T. Smith et al. 2000) Similarly, whole-tree harvest significantly reduced growth of Scots pine in Sweden; however, this did not manifest until the fifteenth year following planting (Egnell and Vallinger 2003).

17.7 SUMMARY

Nutrients are input to local ecosystems through three primary pathways: atmospheric deposition (all nutrients), rock weathering (all nutrients except nitrogen), and biological nitrogen fixation. Numerous elements are transferred to the atmosphere in sea spray, dust, fires, and pollution and are returned to the surface in precipitation or as dryfall. Because of their large surface area, tree crowns are particularly effective at raking elements from air masses moving through the crowns. Nutrient transfer from the atmosphere to leaf surfaces is facilitated when crowns are immersed in clouds or fog. Quantifying nutrient inputs from the atmosphere is difficult for various reasons, but it is clear that atmospheric inputs of nitrogen and sulfur are quite high in areas affected by acid rain. The total array of possible atmospheric inputs has not been quantified in most ecosystems.

Although relatively few microorganisms are capable of nitrogen fixation, these occur in a variety of forms and habitats. The greatest levels of nitrogen fixation are associated with bacteria and actinomycetes that form symbioses with some plants, most notably legumes and a variety of nonleguminous angiosperms called actinorhizal plants. Thousands of species of legumes are distributed throughout the world, including (particularly in the tropics) numerous leguminous trees. Not all legumes that have been studied host nitrogen-fixing bacteria, but most do, and leguminous trees are a primary source of nitrogen in some tropical forests. Actinorhizal plants include species in at least 19 genera (all but one woody) scattered through nine different families. Actinorhizal plants are common pioneers during primary and secondary succession in some temperate and boreal forests. They generally add large amounts of nitrogen to soils but are shade intolerant and eventually replaced by other species. Nitrogen-fixing microorganisms also occur in symbioses with lichens, with mosses, in rhizospheres and within mycorrhizal tissues, on the surface of leaves, within decaying wood, litter, animal feces, mineral soil, and the guts of some animals (e.g., termites and bark beetles). Rates of fixation in these habitats are commonly believed to be small compared with those associated with nodulated plants, although there is evidence that large amounts of nitrogen are fixed in the rhizospheres of some plant species. Even where nitrogen fixation is dominated by various small sources, these can represent in aggregate an important nitrogen input to forests. The magnitude of biological nitrogen fixation on a given site depends on the energy and nutrient supply to nitrogen fixers, temperature, moisture, and soil pH. As with other pathways of nutrient input, there are many uncertainties about the magnitude and sources of nitrogen fixation in ecosystems.

Limiting nutrients are generally retained tightly within undisturbed ecosystems; however, organic nitrogen can leak from undisturbed forests even when nitrogen is scarce. Excessive amounts of nutrients can be lost in hot wildfires and through management practices such as whole-tree harvest, piling logging slash with heavy equipment, and hot slash burns. In some forest types, especially tropical forests on old soils and forests growing on steep slopes, clear-cutting may result in large nutrient losses from leaching or erosion.

18

Biogeochemical Cycling

The Intrasystem Cycle

You shall ask
What good are dead leaves?
And I will tell you
They nourish the sore earth.

NANCY WOOD

In the early stages of primary succession, nutrients required for plant growth are derived exclusively from rock weathering, biological nitrogen fixation, and atmospheric deposition. These inputs are generally retained very tightly within the maturing ecosystem, and as succession proceeds, an increasing proportion of nutrients taken up by vegetation are not newly input but rather have been recycled from a previous existence in the biotic component of the ecosystem. Disturbances that initiate secondary succession usually result in some loss of nutrient capital, and the system again draws on fresh inputs to replenish these. However, the overall trend toward increased reliance on nutrient cycling continues. Eventually, by far the greatest proportion of nutrients required for new growth is derived from cycling.

The nutrient cycle occupies a key position in ecosystem processes. Primary production is regulated in large part by the rate at which nutrients are cycled (chapter 15), and the ability of the system both to store nutrients and to regulate their release provides stability against environmental fluctuations and facilitates recovery from disturbance (chapter 20). Different plant species and guilds may regulate nutrient cycling in such a way as to facilitate their own persistence in a community, thereby creating a self-reinforcing, positive feedback loop between ecosystem structure and processes (Pastor et al. 1987; Pastor and Post 1988). Given the characteristics and importance of the nutrient cycle, it is not surprising that nutrient-cycling research has been a central focus of ecosystem scientists, who over the years have produced a staggering number of printed pages on the subject.

18.1 OVERVIEW OF THE INTRASYSTEM NUTRIENT CYCLE

We begin our discussion of the nutrient cycle by taking a very general, "soil's eye" view of nutrient fluxes within the soil (fig. 18.1), then move to a more detailed look at reservoirs and transfers. Soil water is the conveyer belt that distributes soluble organic matter and nutrients throughout the soil matrix. Nutrients are exchanged between the soil solution and several reservoirs. Plants both take up and release nutrients to solution, and nutrients in litter and humus are incorporated into the bodies of microbes and soil animals and eventually released into the soil solution (where they are available for uptake once again). Nutrients are exchanged between solution and electrical charges on the surfaces of clays and organic matter (i.e., the exchange complex; chapter 14). Some nutrients, such as iron and phosphorus, form insoluble combinations and precipitate out of solution, entering again very slowly or through the action of special chemical compounds produced by plants and microorganisms. Nutrients may enter soil solution directly from the atmosphere, and some are released again into the atmosphere as gases. A few are leached from the local ecosystem to streams, but with some exceptions such nutrient loss is relatively minor in undisturbed forests (chapter 17).

The components of the intrasystem cycle are shown in fig. 18.2. The cycle can be viewed as a series of nutrient transfers (the arrows in fig. 18.2) among various biomass "compartments" that may be either living or dead (the boxes and circle). Compartments denoted by boxes turnover fairly rapidly (i.e., from a few hours to a few years), while humus and organomineral complexes (in the circle) are more persistent, with turnover times ranging from a few years to centuries. The diamonds in figure 18.2 represent losses from the local system.

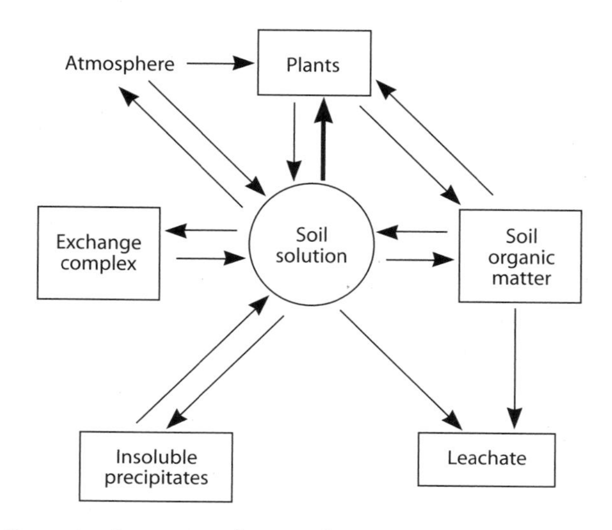

Figure 18.1. An overview of nutrient fluxes in local ecosystems. (Swift et al. 1979)

Remembering that any breakdown of ecosystem components is artificial to one degree or another, we consider the following compartments (most of these were discussed in detail in chapter 14):

- Aboveground plant tissues, along with associated epiphytes and microbes.
- Animals that graze on aboveground tissues (chapter 19).
- Belowground plant tissues and plant mutualists, including roots, mycorrhizae, mycorrhizal hyphae, and obligate rhizosphere bacteria.
- Detritus, which is any organic matter that is not contained within a living body (or cell) or is not incorporated into humus or organomineral complexes. There are three quite distinct forms: litter, dissolved organic matter, and exudates. A large proportion of detritus derives from plant litter (e.g., leaves, branches, roots, mycorrhizae); litter also includes dead consumers (e.g., microbes, protozoa, invertebrates, vertebrates) and both invertebrate and vertebrate feces. Dissolved organic matter is leached from living and decaying plant tissues and can compose a significant fraction of total litterfall carbon (e.g., Cleveland et al. 2004). Exudates are organic compounds that are released into the soil by roots, mycorrhizae, mycorrhizal hyphae, and microbes. A wide variety of compounds are exuded from living cells, including simple sugars, organic acids, and enzymes.
- Humus and organomineral complexes. These are relatively stable pools that are produced by microbial transformation of detritus to humus and the intimate association of organic matter with soil minerals. Nutrient turnover through these pools takes from several decades to several hundred years; hence they function primarily as a nutrient reservoir and a buffer against nutrient loss during disturbance.
- Decomposers, or animals and microbes that obtain energy and nutrients from detritus but that are not obligate rhizobacteria. These are the lowest-level consumers in the **detrital food chain** (chapter 9)
- Belowground grazers, including both animals and protists, that obtain energy and nutrients by feeding on decomposers, roots, mycorrhizal fungi, and rhizobacteria.
- Throughfall and stem flow, (i.e., soluble nutrients that are leached from living foliage or canopy epiphytes).

> Moorhead and Sinsabaugh (2006) restrict the term *decomposers* to that group of microbes that degrade cellulose and lignoceullose, and specify two other microbial groups active in the decay process: (1) *opportunists,* which utilize readily decomposable compounds present in the earliest stages of decay, and (2) *miners,* which degrade humified organic matter and are the primary actors in late stages of decay. For simplicity (and consistency with common usage), we refer to all of these groups as decomposers.

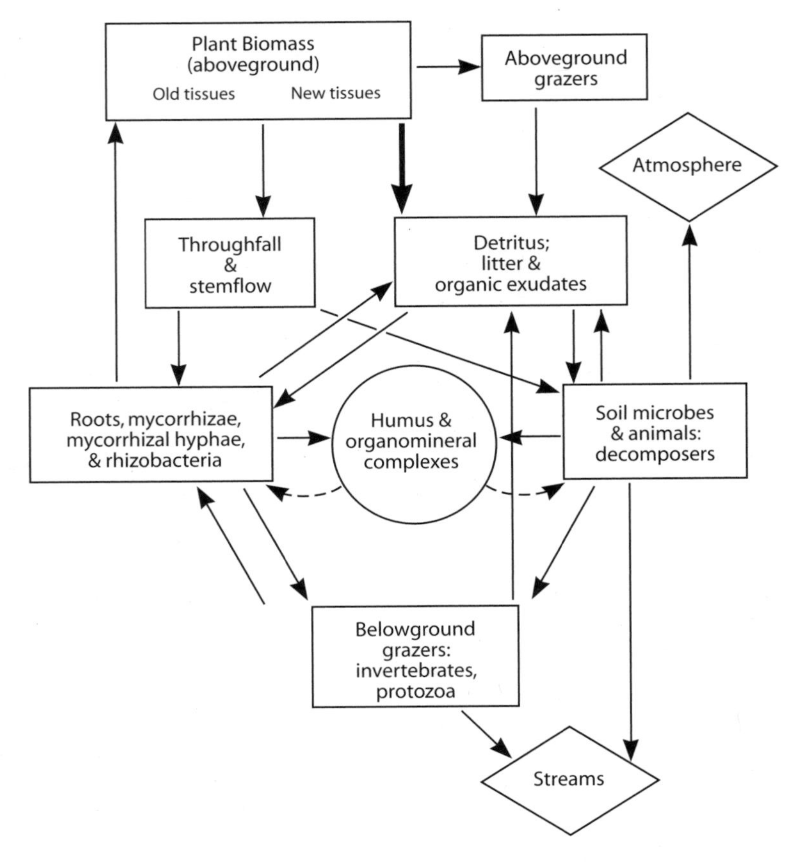

Figure 18.2. The nutrient storage compartments and fluxes between compartments in local ecosystems. Compartments denoted by boxes generally turn over in relatively short time scales (a few hours to a few years), while humus and organomineral complexes may persist from a few years to a few centuries. There are exceptions: nutrients in large wood may last centuries, and some in organomineral complexes may turn over more quickly. Triangles denote losses from the local system. Nutrients contained in unweathered bedrock are not shown, because they do not form part of the local cycle until released by weathering. (They are, however, an important source of nutrients to the cycle.)

Transfers among these compartments fall into four categories:

1. Living to dead, or the senescence and death of tissues or organisms.
2. Transfers through food webs, in other words, transfers that are driven by a search for both energy (e.g., reduced carbon) and matter (i.e., structural carbon and nutrients).
3. Transfers driven by the search for nutrients. These involve primarily plants, their mutualists (i.e., mycorrhizae), and associates (i.e., rhizosphere microbes) that have access to plant photosynthates as an energy and carbon source but obtain nutrients from detritus and soil solution.
4. Release of nutrients or reduced carbon compounds from living tissues (e.g., throughfall, stem flow, root and mycorrhizal exudates).

Nutrient transfers also occur within as well as among compartments. Trees withdraw nutrients from aging and dead tissues and use these to support new growth (i.e., **internal cycling**). Invertebrates recycle nutrients by eating their own feces (called **coprophagy,** from the Greek *copros,* for "dung"). Decomposers decompose other decomposers, and grazers feed on other grazers. In general, belowground food chains are longer and more complex than those aboveground (chapter 9).

To give some flavor of the magnitude of transfers among the various compartments, figure 18.3 shows nitrogen fluxes in an old-growth Douglas-fir forest in Oregon. Typical of the Pacific Northwest, the amount of nitrogen input to this forest in precipitation is quite small, and roughly equivalent to that leached from the system. Trees take up 42 kg/ha yearly (20 times more than precipitation inputs) and return slightly more than this in litter and as solutes in throughfall and stem flow. (Biomass of this old forest is either not changing or declining slightly; a young forest that is accreting biomass would return fewer nutrients to the soil than it takes up.) Typical of forests everywhere, relatively large amounts of nitrogen (and most other nutrients) are cycled internally.

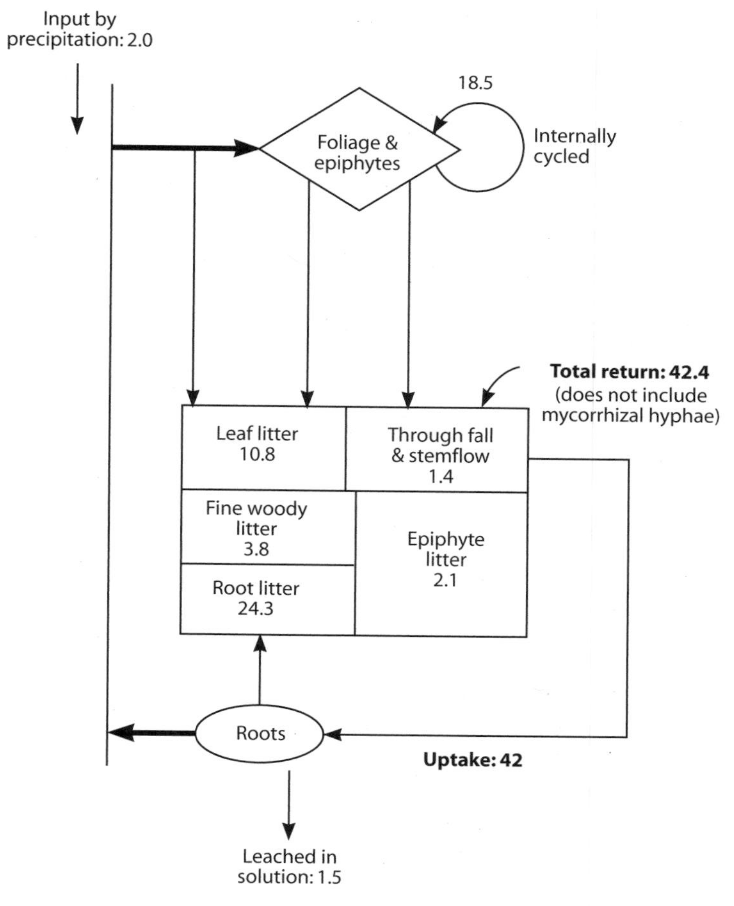

Figure 18.3. The nitrogen fluxes in an old-growth, Douglas-fir forest on the H.J. Andrews Experimental Forest, central Oregon Cascades. Values are in kilograms per hectare per year. (Adapted from Sollins et al. 1980)

18.2 THE CONTRIBUTION OF NUTRIENT CYCLING TO PRIMARY PRODUCTIVITY

Foresters have known for some time that tree growth is closely linked to the release of nutrients from decomposing organic matter. This relationship was first discovered in Europe because of the common practice of removing litter from forests to use as cattle bedding. Forest growth declined as a result, and in 1876, the German scientist Ebermayer linked this to removal of essential nutrients in the litter (Tamm 1979).

We now know that with the possible exception of the early stages of rapid biomass accumulation, by far the greatest proportion of a forest's yearly nutrient requirement is satisfied by elements that are cycled both externally (e.g., in litter) and internally (e.g., within the tree). Figure 18.4 shows the relationship between tree uptake (upper line) and release of nutrients in decomposition (lower line, shaded portion) for an age sequence of Scots pine plantations in Great Britain (Ovington 1959). Stands vary (because of site factors or management practices rather than age), but in most cases, at least 50 percent of yearly uptake is matched by decomposition.

The degree to which nutrient requirements are met by cycling external to the tree is estimated by the **recycling coefficient,** or K_m, which is simply the fraction of total yearly nutrient uptake that is released as yearly detritus, throughfall, or stem flow (Ulrich 1969; Larcher 2003). K_m is relatively low in young, rapidly growing forests, because a high proportion of nutrients is retained in the biomass increment (especially foliage); K_m increases as maximum site leaf area index is attained. Use of the term *recycling* for K_m expresses the idea that nutrients released in detritus will be recycled for future use by the tree. This is not always the case, however. In many ecosystems, decomposition and nutrient release proceed quite slowly, and large amounts of detritus build up, their nutrient content unavailable to the tree. Factors controlling the rate of decomposition are discussed in section 18.8.

18.3 DETRITUS

18.3.1 The Components of Detritus

Litter generally composes the largest fraction of inputs to the detrital pool, although under some circumstances (particularly during drought) exudates from living roots

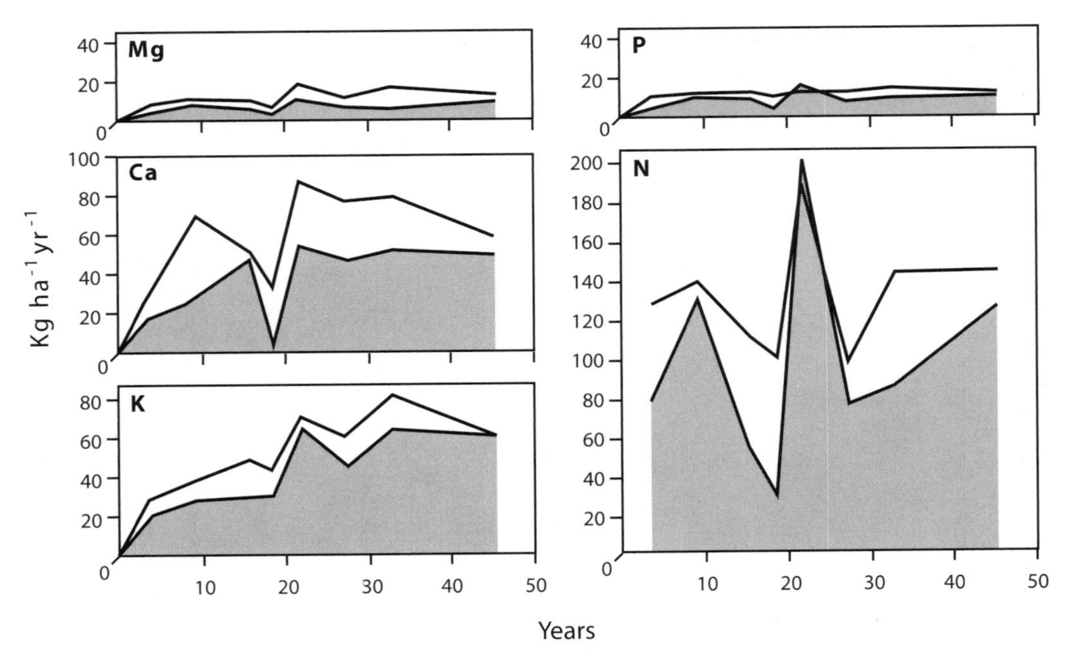

Figure 18.4. The nutrient uptake by trees *(top line)* and release in decomposition *(shaded portion)* in an age series of Scots pine plantations in Great Britain. (Adapted from Ovington 1959)

and mycorrhizae can attain surprisingly high levels. Eighty percent to well over 90 percent of litter is dead leaves, branches, soluble organic compounds, roots, mycorrhizae and mycorrhizal hyphae, boles, and so on, with the balance composed of a potpourri collectively termed **microlitter** and including things such as pollen, insect frass and bodies, cells that are sloughed from roots as they grow, lichens, feathers, and monkey hair. The feces of soil animals are a major component of belowground microlitter. Although microlitter generally accounts for only a small proportion of litter mass, it is usually nutrient rich and contributes proportionately more to litter nutrients than its mass alone would suggest. Under some circumstances (e.g., during insect outbreaks), microlitter can attain quite high levels. Cells sloughed from growing roots probably also attain high levels but are difficult to measure accurately.

18.3.1.1 Belowground versus Aboveground Litter

In mature forests worldwide, belowground litter inputs are estimated (based on soil respiration) to average two times higher than aboveground inputs (Davidson et al. 2002). Research in pine forests indicates that root decay is more likely to contribute to stable soil organic matter than foliar decay (Bird and Torn 2006), perhaps because ectomycorrhizal roots decay significantly more slowly than nonmycorrhizal roots (Langley et al. 2006). In contrast to aboveground litter, directly measuring belowground inputs is extremely difficult and can easily underestimate actual inputs. Cells sloughed from growing roots are not accounted for; moreover, very fine roots (unlike leaves) may have more than one period of death and regrowth during a year. Figure 18.5 shows average root turnover rates for different biomes, vegetation types, and (for forests) fine-root size classes. Root turnover increases exponentially from the boreal region to the tropics (i.e., with mean annual temperature), although different forests may differ widely from one another even at the same temperature (Gill and Jackson 2000).[1] On average, the pool of roots 10 mm in diameter or smaller (the "fine" roots of fig. 18.5A)[2] take longer than a year to turn over; however, the average lifetime of very small roots (1 mm or less in diameter) is less than a year (fig. 18.5B). Hence, periodic sampling of root biomass may miss periods of active death and regrowth of the smallest roots unless the sampling is conducted frequently. (To see this, imagine calculating yearly gas consumption in an automobile by measuring the level in the tank just twice a year.) Use of root observation chambers (i.e., rhizotons) and isotopic labels has added important information.

Average fine-root biomass (2 mm and smaller) varies from 0.57 to 0.82 kg/m^2 of soil surface, depending on biome (table 18.1) This is roughly one-third or less the biomass of annual leaf fall in forests (fig. 18.7); however, because some proportion of the fine roots turns over more than once a year, the annual root input to detritus is more comparable. In a Venezuelan forest, 25 percent of the

[1] A study in hardwood forests of the eastern United States indicates that there are actually at least two pools within the population of fine roots (<2 mm), one that turns over quickly and another slowly (Joslin et al. 2006).

[2] The term *fine roots* is not used consistently. Frequently it refers to roots 2 mm in diameter or smaller, but sometimes larger sizes are included (e.g., Gill and Jackson 2000).

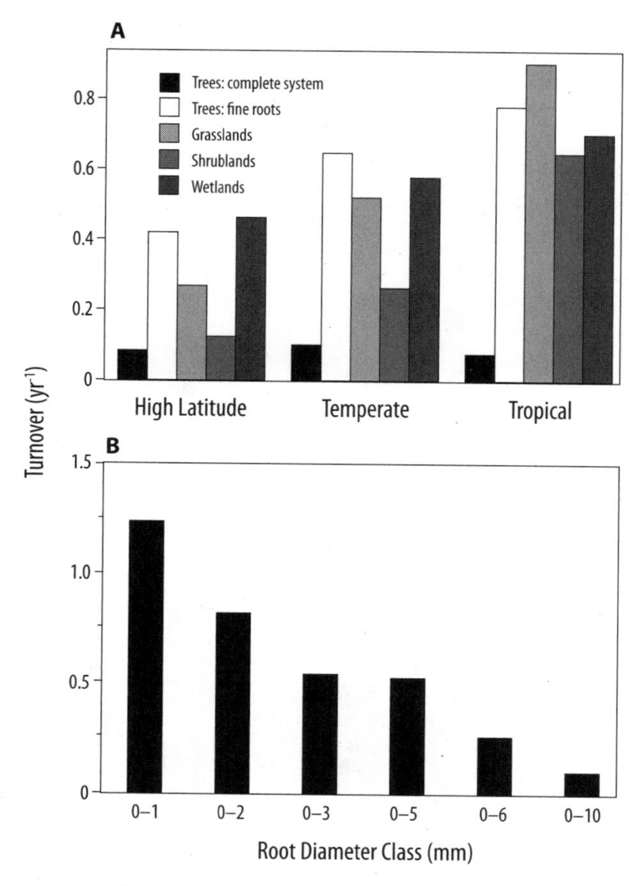

Figure 18.5. Turnover (per year) of roots. A. By biome for different vegetation types. *Complete system* refers to entire root system; fine roots are less than or equal to 10 mm. B. Effect of size on turnover of forest fine roots. (Adapted from Gill and Jackson 2000)

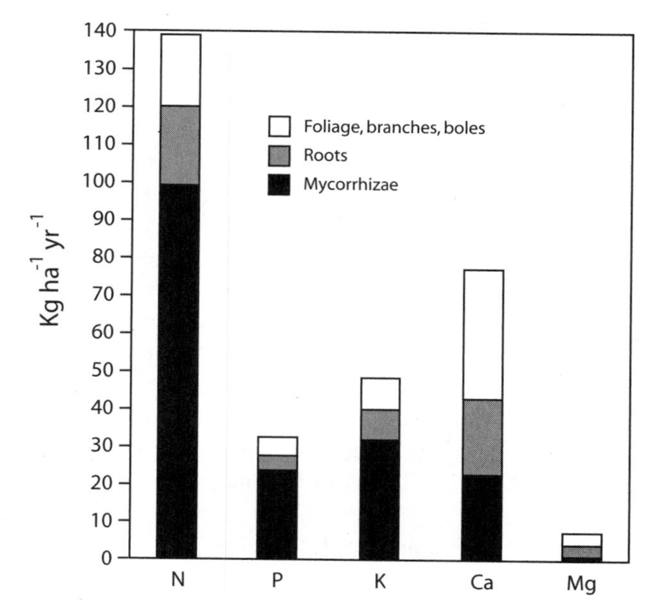

Figure 18.6. The nutrients cycled from trees to decomposers in aboveground and belowground litter of a Douglas-fir forest on a per-year basis. (Adapted from Fogel and Hunt 1983)

standing crop of fine roots turned over every month, producing nearly 6 times more litter nitrogen per year than did fine litterfall (343 kg/ha vs. 61 kg/ha, respectively) and 14 times more litter phosphorus (11.0 kg/ha vs. 0.8 kg/ha, respectively) (Vitousek and Sanford 1986). Fogel and Hunt (1983) estimated that turnover of roots and mycorrhizae[3] contributed from 77 to 87 percent of 1 year's total tree litter inputs of nitrogen, phosphorus, and potassium in a Douglas-fir forest, and about 50 percent of calcium and magnesium return (fig. 18.6). On the order of three-quarters of the nitrogen, phosphorus, and potassium in tree litter from that study was contained in mycorrhizae. Similar values have been reported for other forests. In a yellow poplar *(Liriodendron tulipifera)* stand, roots and mycorrhizae (not including external hyphae) accounted for 65 percent of the nitrogen and 93 percent of the potassium in annual tree-litter inputs (Harris et al. 1979). Vogt et al. (1982) studied nutrient cycling through roots and mycorrhizae in two Pacific silver fir *(Abies amabilis)* forests, one 23 years old and the other 180 years old; their data are shown in table 18.2. Belowground litter (i.e., fine roots, mycorrhizae, the reproductive structures of mycorrhizal fungi) accounted for 80 percent or more of all litter nutrients except calcium. More nutrients were cycled in the old than in the young stand, but the relative contribution of above- and belowground components remained roughly the same.

Belowground ecologists have long believed that hyphae extending out into the soils from the mycorrhizal root (variously termed *extraradical mycelia* or *extraradical hyphae*[4]) play a significant role in ecosystem processes but until recently we haven't had the ability to verify that. Advances in technique have greatly enlarged the tool kit[5] and are

[3]Recall from chapter 11 that a mycorrhiza is the structure formed by a root plus fungal tissue within and immediately surrounding the root. External hyphae grow from the mycorrhiza but are not part of it.

[4]Mycelia are coherent collections of hyphae.

[5]New techniques being applied to extraradical mycelia include isotopic labeling, biochemical markers, and advances in microscopy (Leake et al. 2004).

Table 18.1

Average fine root biomass for different biomes

Biome	Total fine root biomass (kg m^{-2})
Boreal forest	0.60 (0.13, 5)*
Desert	0.27 (0.10, 4)
Sclerophyllous shrubs and trees	0.52 (0.13, 6)
Temperate coniferous forest	0.82 (0.14, 10)
Temperate deciduous forest	0.78 (0.092, 14)
Temperate grassland	1.51 (0.12, 21)
Tropical deciduous forest	0.57 (0.098, 6)
Tropical evergreen forest	0.57 (0.069, 12)
Tropical grassland/savanna	0.99 (0.24, 5)
Tundra	0.96 (0.22, 5)

Source: Adapted from Jackson et al. 1997.

*Standard error and number of studies in parentheses.

leading to significant increases in our understanding (Leake et al. 2004). Wallander et al. (2001) found that the biomass of extraradical hyphae in a Swedish conifer forest was comparable to that of fine roots. Reports of arbuscular mycorrhizae extraradical hyphal biomass range from 0.03 to 0.5 mg per g of soil, which Zhu and Miller (2003) translate to 0.0054 to 0.09 kg per m^2 of soil surface.[6] Although those values are considerably smaller than those for fine-root biomass, detrital inputs depend on rates of turnover, about which there is considerable uncertainty (Leake et al. 2004). One study concluded that arbuscular mycorrhizae hyphae live only five to six days (Staddon et al. 2003); however, Leake et al. (2004) believe that estimate is erroneous and that turnover rates are significantly slower. This is an active area of research and new insights will undoubtedly be forthcoming.

18.3.1.2 Components of Litterfall in Three Forest Types: An Example

Figure 18.7 compares the mass and nutrient content of litterfall (excluding coarse woody debris) among three quite different forests: (1) old-growth Douglas-fir–hemlock in Oregon (Sollins et al. 1980), (2) moist evergreen tropical in Tanzania (Lundgren 1978), and (3) mixed deciduous-conifer in New Hampshire (Gosz et al. 1972). As always, comparisons among different studies must be made cautiously, but despite the very different forest types and investigators, these three systems illustrate some general patterns that are probably common to most forest ecosystems. (Recall that litterfall refers to aboveground litter only.)

Not surprisingly, leaves compose the largest proportion of both the mass and nutrient content of litterfall. Figure 18.7 shows that the proportion of leaves is quite similar among the three forests, ranging from 60 to 65 percent, while twigs and small branches compose a higher proportion of fine litter in the conifer forest. In general, wood accounts for a higher proportion of litter in conifer forests than in angiosperm forests (O'Neill and DeAngelis 1981). The distribution of litterfall nutrients among leaves, fine branches, and microlitter is very similar among the three forests, particularly for nitrogen.

The stands differ in the total dry weight of litterfall and even more so in the nitrogen content of litter. For example, while the total dry weight of litterfall is 30 and 159 percent greater in the temperate deciduous and tropical forests, respectively, than in the conifer forest, the former two are releasing nearly three and eight times more nitrogen in litterfall. On the other hand, phosphorus concentration is lowest in tropical litter, reflecting the greater phosphorus limitations in tropical than temperate broad-leaved and coniferous forests (McGroddy et al. 2004; section 18.9.1 in this volume).

[6] To arrive at those estimates, Zhu and Miller assume a 30-cm soil depth with bulk density of 1.2.

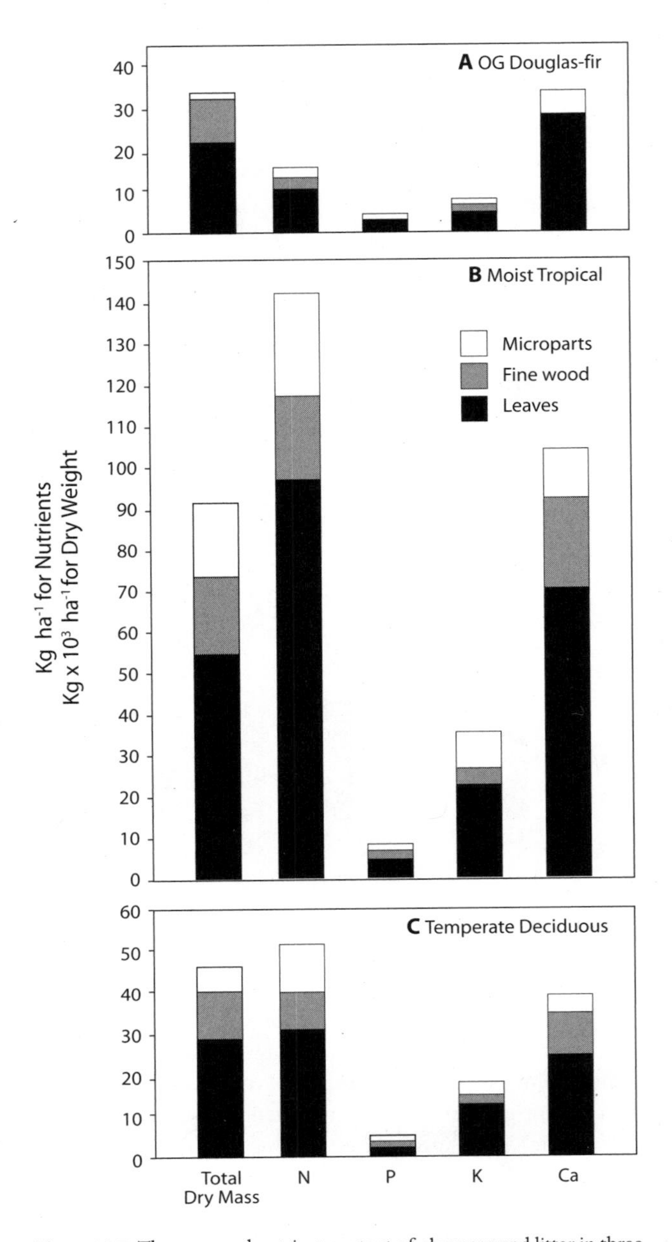

Figure 18.7. The mass and nutrient content of aboveground litter in three forests. Note that dry weight and nutrients are plotted on different scales. OG, old growth. (A, data from Sollins et al. 1980; B, data from Lundgren 1978; C, data from Gosz et al. 1972)

Microlitter plays a significant role in these forests, contributing from 20 to 25 percent of aboveground litter nitrogen in all sites and from 17 to 22 percent of phosphorus and potassium in the tropical and temperate deciduous forest. It is impossible to say whether the low proportion of microlitter in the conifer forest reflects a general pattern or is an artifact of the method or timing of sampling. Secondary production by aboveground food chains likely is greater in the two angiosperm forests than in the conifer (O'Neill and DeAngelis 1981). The relatively nutrient-rich foliage of angiosperms (chapter 16) probably supports more insect grazers than that of conifers (except during insect outbreaks),

Table 18.2
The annual nutrient turnover by litter components in stands of Pacific silver fir

	Nitrogen		Phosphorus		Potassium		Calcium		Magnesium	
	M	%	*M*	%	*M*	%	*M*	%	*M*	%
23-year-old forest										
Aboveground litterfall	14	14	1	7	2	8	11	24	1	8
Fine roots and mycorrhizae	60	60	10	67	20	77	30	67	10	86
Mycorrhizal fungal reproductive structures	27	26	4	26	4	15	4	9	1	8
Total	101	—	15	—	26	—	45	—	12	—
180-year-old stand										
Aboveground litterfall	20	12	3	10	3	9	17	33	3	20
Fine roots and mycorrhizae	110	64	20	67	20	62	30	59	10	67
Mycorrhizal fungal reproductive structures	41	24	7	13	9	29	4	8	2	13
Total	171	—	30	—	32	—	51	—	15	—

Source: Adapted from Vogt et al. 1982.

and angiosperm flowers are usually animal rather than wind pollinated. It would not be too surprising if more insect frass and bodies fell from angiosperm canopies than from conifer canopies.

18.3.2 Coarse Woody Debris

Coarse woody debris (i.e., logs, large branches, large roots) (CWD) contributes significant amounts to litterfall mass (Harmon et al. 1986). During the period that the temperate deciduous forest of figure 18.7 was studied, boles and large branches contributed 25 percent as much mass as all other aboveground detrital components, while in the conifer forest these components contributed twice as much. Values from short-term studies like these are unlikely to accurately reflect the true long-term input of CWD, however, because the fall of boles and large branches may be highly episodic (e.g., associated with windstorms). In the unlogged forest, the input of CWD over long time periods must equal the production of boles, large branches, and large roots, less any amounts that may be lost from the site (e.g., consumed in fires). One of the distinguishing characteristics of old growth in many forest types is the large accumulation of CWD.

The nutrient concentration of CWD is very low, and it plays quite a different role in the nutrient cycle than foliar or fine root detritus. In the studies of northern conifer forests summarized by Laiho and Prescott (2004), CWD contributed between 3 and 73 percent of aboveground litter input, but only 1 to 18 percent of nitrogen, 3 to 12 percent of phosphorus, and in the single study that documented calcium and potassium, 15 and 18 percent, respectively, of those elements (table 18.3). In forests where CWD persists in soil (chapter 14), it serves primarily as a water reservoir and habitat for a variety of organisms, including small vertebrates (e.g., salamanders, small rodents) and numerous species of saproxylic[7] insects and microbes (including nitrogen-fixing bacteria).

18.3.3 Dissolved Organic Matter

Soluble organic compounds leached from living tissues (e.g., foliage) and from decaying solid litter can compose from 1 to 19 percent of carbon inputs to soil microbes (Neff and Asner 2001). The importance of these compounds to soil microorganisms was illustrated by Jandl and Sollins (1997), who found that raking litter from the forest floor of a Douglas-fir plantation reduced soil respiration by one-third. Soluble organics comprise a variety of different compounds, ranging from small amino acids to complex organic acids and (correspondingly) readily bioavailable to recalcitrant fractions (Jandl and Sollins 1997). In their study of organic compounds leached from various litter species and age classes, Cleveland et al. (2004) found a large proportion to be highly biodegradable. Although ecologists are increasingly aware of the importance of soluble organic compounds in the nutrient cycle, they have been mostly neglected in nutrient cycling models. One exception is Currie and Aber (1997), who included soluble organics as a central driving force in their model of decomposition in forests of the northeastern United States

18.3.4 Exudates

The final detrital component to be considered is exudates. Roots, mycorrhizal fungi, and indeed all soil microbes release organic molecules from their cells into the surrounding soil. These include a variety of compounds,

[7] *Saproxylic* means living on dead wood. For insects, the actual food is probably not wood but the microbes that inhabit the wood (chapter 10).

Table 18.3
Mass and nutrient content of litterfall components in northern coniferous forests

Species	Stand age (years)	Boles	FWD	NWD	GV	CWD%	Reference
Mass (g m^{-2} $year^{-1}$)							
Abies balsamea	50–60	170	305	375		20	Sprugel 1984
	Mortality period	1,780	350	300		73	Sprugel 1984
	47	25					Lang 1985
	81	82					Lang 1985
Abies lastocarpa, Picea engelmannii	>350	48	49	103	5	24	Laiho and Prescott 1999
Picea glauca	120	7	80	124	26	3	Laiho and Prescott 1999
Picea contorta	90	68	90	150	56	22	Laiho and Prescott 1999
Pseudotsuga menziesii	600–100	135–392					Sollins 1982
	450	480–960*	130–180	170–280		56–66	Grier and Logan 1977
Pseudotsuga menziesii + others	450	266–395					Sollins 1982
Tsuga heterophylla, Picea sitchensis	105	274					Sollins 1982
Nitrogen (g m^{-2} $year^{-1}$)							
Abies balsamea	50–60	0.284					Sprugel 1984
Abies lastocarpa, Picea engelmannii	>350	0.067	0.296	0.874	0.071	5	Laiho and Prescott 1999
Picea glauca	120	0.006	0.330	0.830	0.318	1	Laiho and Prescott 1999
Pinus contorta	90	0.049	0.151	1.026	0.637	4	Laiho and Prescott 1999
Pseudotsuga menziesii	450	0.39	0.38	1.41		18	Sollins et al. 1980
Phosphorus (g m^{-2} $year^{-1}$)							
Abies lastocarpa, Picea engelmannii	>350	0.002	0.024	0.059	0.005	2	Laiho and Prescott 1999
Picea glauca	120	0.001	0.028	0.121	0.067	1	Laiho and Prescott 1999
Pinus contorta	90	0.003	0.014	0.089	0.105	3	Laiho and Prescott 1999
Pseudotsuga menziesii	450	0.06	0.05	0.38		12	Sollins et al. 1980
Potassium (g m^{-2} $year^{-1}$)							
Abies lastocarpa	50–60	0.120					Sprugel 1984
Pseudotsuga menziesii	450	0.15	0.12	0.58		18	Sollins et al. 1980
Calcium (g m^{-2} $year^{-1}$)							
Abies balsamea	50–60	0.372					Sprugel 1984
Pseudotsuga menziesii	450	0.64	0.56	2.99		15	Sollins et al. 1980

Source: Laiho and Prescott 2004 (see their paper for original citations).
Note: FWD, twigs, branches, bark fragments, reproductive tissues; NWD, needles, leaves; GV, ground vegetation; CWD%, proportion of CWD of total (aboveground) tree stand litter input.
*Including branches.

including organic acids, enzymes, amino acids, proteins, and simple sugars. The composition of exudates differs between mycorrhizal and nonmycorrhizal roots. Estimates of the amount of material released in this fashion range from less than 10 percent to more than 80 percent of the photosynthate that is shipped to roots (Reid and Mexal 1977; Whipps and Lynch 1986), with the higher values occurring during drought.

Like litter, exudates serve as a source of carbon and (depending on the type of molecule) nutrients and even physiologically active compounds (e.g., amino acids) for soil heterotrophs. Arbuscular mycorrhizal fungi release large amounts of the protein glomalin, which is apparently quite stable in soils and is believed to play a significant role in both carbon storage and soil aggregation (Steinberg and Rillig 2003). Exudates differ from litter in a qualitative sense; rather than being a waste product, they are actively excreted into the environment to perform some work (section 18.11.3).

18.3.5 Species and Environmental Effects on Litter Production

As figure 18.7 indicates, total yearly litter inputs vary among forest types, but there is also a great deal of variation within a given type. In general, annual litterfall correlates with aboveground net primary productivity (NPP) (fig. 18.8A). The relationship is not linear, because, as we discussed in earlier chapters, forests that occupy more productive environments divert greater amounts of NPP to wood. Over the long run, however, as boles and large branches die and fall to the forest floor, the total litterfall in forests that are not logged correlates more closely with NPP. Reflecting changes in NPP, litterfall decreases from the equator northward (fig. 18.8B).

The influence of environment on belowground litter production is uncertain, reflecting uncertainty in the relative production of aboveground and belowground tree parts in different environments (chapter 15). In their

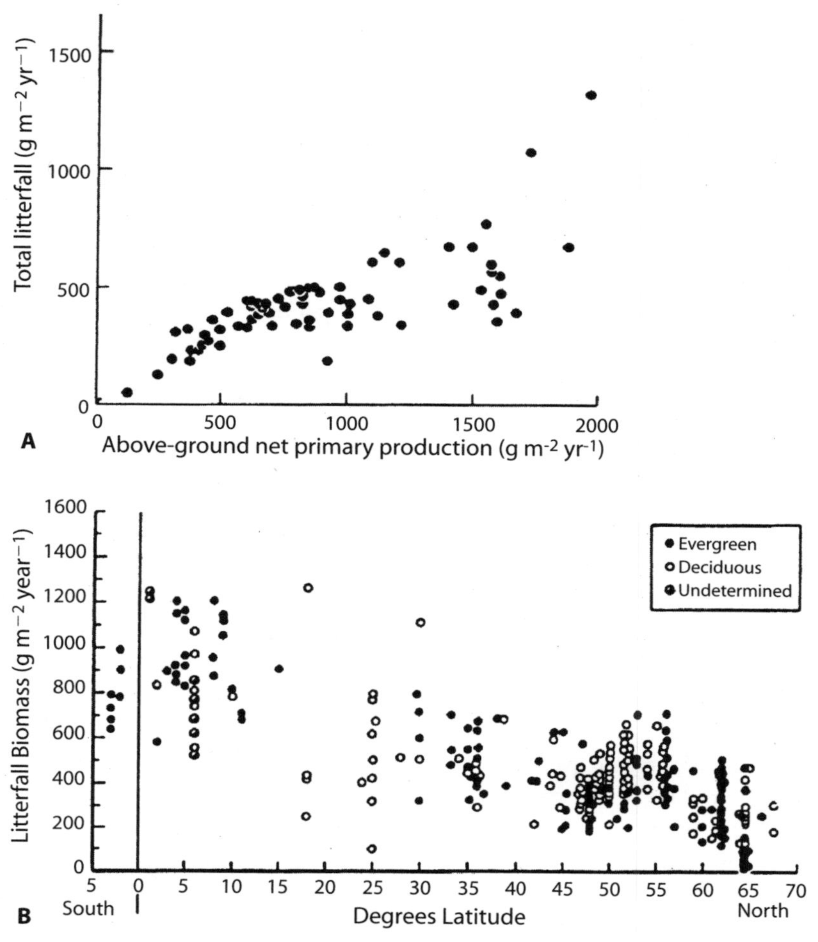

Figure 18.8. A. The relationship between aboveground net primary production and annual litterfall (aboveground litter) for the International Biological Program's woodlands data set. B. The biomass of aboveground forest litter as affected by latitude. (A, adapted from O'Neill and DeAngelis 1981; B, from Van Cleve et al. 1983)

summary of the literature, Vogt et al. (1986) showed a close, negative relationship between litterfall nitrogen and root turnover for both broad-leaved and needle-leaved forests (fig. 18.9). Powers et al. (2005) found fine-root biomass in Costa Rican forests to correlate negatively with soil nitrogen and phosphorus, which probably means that it also correlates negatively with aboveground NPP. These patterns suggest that as more nutrients are cycled aboveground, fewer are cycled belowground, and vice versa.

In general, the amount of a given nutrient contained in annual litterfall correlates closely with its availability in the ecosystem. Availability, in turn, is a function not only of environment but of the vegetation itself and, in particular, of the interaction between vegetation and environment. This point is illustrated by the work of Gosz et al. (1972), who compared the nutrient content of aboveground litterfall between a mixed hardwood and a spruce-fir forest on the Hubbard Brook Experimental Forest in New Hampshire (table 18.4). The lesser amounts of nitrogen, calcium, and potassium in litter from the spruce-fir forest reflect two things: (1) slower decomposition and therefore lower availability of these

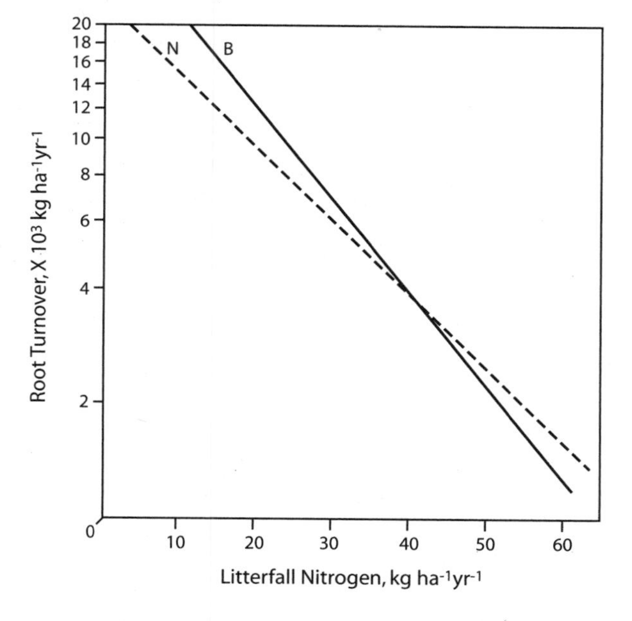

Figure 18.9. The relationship between the nitrogen content of aboveground litter and root turnover in broad-leaved (B) and needle-leaved (N) forests. Note the log scale on the *y* axis. (Adapted from Vogt et al. 1986)

Table 18.4
The nutrient content of aboveground litter from a spruce-fir and a northern hardwood forest

Element	Spruce fir forest*	Hardwood forest†
Nitrogen	44.90	55.80
Calcium	33.30	45.20
Potassium	11.70	21.10
Manganese	14.10	7.40
Magnesium	5.30	6.40
Sulfur	5.50	6.10
Phosphorus	4.00	3.70
Zinc	0.82	0.69
Iron	0.53	0.39
Sodium	0.11	0.09
Copper	0.03	0.04
Total	120.29	146.91

Source: Adapted from Gosz et al. 1972.
*750 to 790 m elevation.
†Hubbard Brook Experimental Forest, NH; 545 to 700 m elevation.

elements at higher elevations, and (2) the physiological responses of trees to this lower nutrient availability. Amounts of some nutrients, however, increased in the litterfall at higher elevation, in part at least because of changing species composition. Increases in litterfall zinc and iron, for example, resulted from higher proportions of birch (a zinc accumulator) and red spruce (an iron accumulator). Litterfall phosphorus and manganese increased in all species at higher elevation (even those that also occurred at lower elevation). Although this indicates greater availability of these elements at higher elevation, the greater availability may have been caused by the higher proportions of conifers rather than by inherent soil differences (Gosz et al. 1972).

18.4 THE INTRATREE NUTRIENT CYCLE

The content of most nutrients declines as tissues age (calcium is an exception), and this almost certainly reflects withdrawal for eventual transfer elsewhere in the tree.

From less than 10 to over 90 percent of the nitrogen, phosphorus, potassium, and magnesium required by new growth may be translocated from other tissues. Switzer and Nelson (1972) coined the term **biochemical cycle** to describe this withdrawal of nutrients from one tissue and their eventual translocation to support new growth.

Table 18.5 shows estimates of the proportion of aboveground growth met by internal transfers for various species. Calculations such as those shown in table 18.5 are made by comparing the nutrient requirements of new growth with the nutrient content of tissues that are shed (or for conifers, nutrient content of older needles still on the tree). This method is accurate only if there is no *net* biomass increment from one year to another and, hence, works reasonably well for older forests but not well at all for young forests.

The degree to which a given element is cycled biochemically depends on its solubility, and its availability in soils. Only nutrients that occur in soluble compounds within tissues, or can be easily transformed to soluble compounds, are cycled internally. Among the macronutrients, this includes nitrogen, phosphorus, potassium, and magnesium. Little or no calcium is retranslocated, and in fact, calcium content usually increases rather than decreases in leaves and twigs as they age, reflecting its importance in structural tissue and membranes. This does not appear to be the case in fine roots, whose calcium concentration has been found to decline with increasing root diameter (Nambiar 1987).

The relationship between site fertility and biochemical cycling is complex. For many years, researchers believed nutrients that are in short supply in soils are cycled more strongly within trees. The story is not quite this simple, however. The degree to which a given nutrient is cycled internally depends less on its abundance in soils than on two other factors: (1) its abundance *relative* to other nutrients (or other limiting factors such as water), and (2) whether periods of soil nutrient availability correspond to times when the tree is growing (Nambiar and Fife 1987; Oren and Schulze 1989). Forests with an abundance of nitrogen, either because of fertilization or acid rain, depend

Table 18.5
The estimated proportion of nutrients required for new foliage supplied by translocation from other tissue

	Nutrients (%)					
Species	Nitrogen	Phosphorus	Potassium	Calcium	Magnesium	Study
Douglas-fir	56	31	30	−180	−38*	Solins et al. 1980
Douglas-fir (age sequence)	43–47	46–95	12–25	—	—	Cole et al. 1977
Mixed hardwoods	33	—	—	—	—	Bormann et al. 1977
Loblolly pine	39	60	22	0	6	Switzer and Nelson 1972
Radiata pine	48	86	39	—	—	Fife and Nambiar 1982
Sclerophyll forests, Australia	—	40–60	—	—	—	Attiwill 1981
Deciduous species, Alaska	62	31	40	0	2	Van Cleve et al. 1983
Conifers, Alaska	27	23	42	0	7	Van Cleve et al. 1983
IBP data set, deciduous	26	14	0	0	0	Cole and Rapp 1981

IBP, International Biological Program.
*A negative value indicates an increase in nutrient concentration with foliage age.

quite heavily on internal cycles of other nutrients to match the nitrogen supply (Oren and Schulze 1989). Regarding growth cycles, trees in temperate and boreal climates grow new leaves and twigs early in the spring, when decomposers are still limited by low temperatures. Nutrients required for the early growth flush are transferred from old tissues and replenished later in the season, when nutrients released during decomposition exceed growth requirements.

Biochemical cycling can take two forms. In deciduous species, nutrients are withdrawn before leaf fall, stored in woody tissues, and then recycled to new leaves as they grow. The same thing occurs to some degree in evergreen species; however, here the older foliage acts as a source of nutrients for younger foliage. A survey of worldwide forests found carbon:nutrient ratios to be consistently higher in litter than in foliage, suggesting that nutrient withdrawal prior to leaf-drop is common and widespread (McGroddy et al. 2004).

In conifers, nutrient transfer from old to new results in a decline in nutrient concentration and content with needle age. Figure 18.10 illustrates this for ponderosa pine growing on two sites in California. In general, foliar concentrations of nitrogen, phosphorus, and potassium decline by 40 to 60 percent between first- and fifth-year conifer needles (Fife and Nambiar 1982). As figure 18.10 indicates, the decline in nutrient content is greater than changing concentrations would suggest.

Not only new foliage benefits from internal cycling; most (if not all) growing tissues probably draw to one degree or another on nutrients shipped from elsewhere in the tree. The greatest requirement for those elements that are involved in synthesis and energy transfers (e.g., nitrogen and phosphorus) occurs during the active growth phase, which in different tissues takes place at different rimes during the growing season. Melillo and Gosz (1983) suggest that the same nutrient capital may be shunted from one growing point to another, thereby doing multiple duty during a single growing season.

Whether fine roots retranslocate nutrients before death has long been debated, in large part because measurements are considerably more difficult in roots than in leaves. In their review of the literature, Gordon and Jackson (2000) found evidence for retranslocation of phosphorus and potassium, and hints that nitrogen may also be withdrawn before fine-root death. They stress, however, that more work is needed.

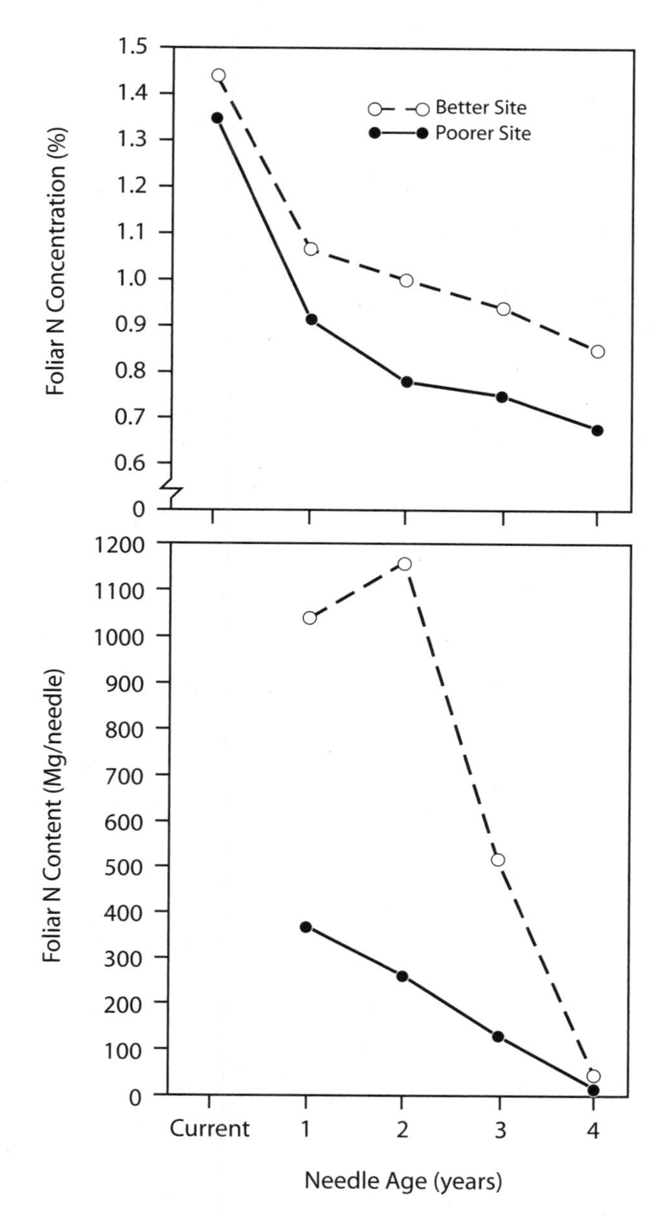

Figure 18.10. The change in the concentration and content of nitrogen as ponderosa pine needles age. Changes in nitrogen concentration are quite similar between the better and the poorer sites; however, changes in nitrogen content, which reflect the needle weight, are much greater on the better site. (Adapted from Powers 1984)

18.5 THROUGHFALL AND STEM FLOW

Throughfall and **stem flow** refer to nutrients in water that has passed through the canopy. As we discussed in chapter 17, the nutrient content of precipitation may either decrease or increase on its way through the canopy. Decreases reflect direct absorption by foliage or foliar epiphytes (including lichens, fungi, and bacteria); increases are more commonly reported than decreases and result from (1) nutrients that have collected on foliage through dry fall, and (2) nutrients leaching from foliage or foliar epiphytes. The former is an input to the system, while the latter is part of the intrasystem cycle. It is very difficult to distinguish between these two sources.

Table 18.6 shows the change in nutrient content of precipitation on passing through the canopy for several different forest types. Positive values indicate gain, and negative values loss. The temperate and tropical forests that are growing on relatively fertile soils mostly add nutrients to the precipitation, while the two tropical forests growing on highly infertile soils scrounge cations and phosphorus (the

Table 18.6
The nutrients in throughfall and stemflow in several forest ecosystems

	Nutrients, kg ha^{-1} y^{-1}						
	Nitrogen	Phosphorus	Potassium	Calcium	Magnesium	Sulfur	Study
Douglas-fir	1.40	0.90	12.90	4.20	—	—	Sollins et al. 1980
Mixed temperate hardwoods	0.00	0.00	24.00	11.00	16.70	—	Duvignead and DeSmet 1970
Moist tropical	0.90	0.60	17.00	7.10	6.20	6.80	Duvigneaud and DeSmet 1970
Infertile soils	3.75	−19.34	−6.34	−23.40	−2.05	27.68	Jordon et al. 1979
	11.56		4.30	−21.75	0.45	27.00	Jordon et al. 1979
Fertile soils	—	—	12.10	4.70	2.30	—	Brasell and Sinclair 1983
Moderately fertile	—	—	9.30	5.30	1.80	—	Brasell and Sinclair 1983

Note: Negative values indicate direct uptake of nutrients contained in precipation by canopies.

most limiting nutrients in those systems) from rainfall as it passes through the canopy.

As table 18.6 indicates, any nutrient may be leached from canopies by rainfall as long as it is in soluble form; however, the basic cations are usually more prevalent in throughfall and stem flow than nitrogen and phosphorus. This is particularly true for potassium, which is not known to occur in organic forms within tissues. Golley (1983b) summarized studies from both tropical and temperate forests, showing that on average, the potassium content of precipitation was increased 10-fold by passage through the canopy.

Intracycle nutrients in throughfall may come primarily from canopy components other than leaves. Reiners and Olson (1984) studied nutrient fluxes to and from various parts of balsam-fir canopies and found that dead twigs and lichens growing on twigs added well more than 10 times greater amounts of sulfate and potassium to precipitation water than were added by foliage. On the other hand, lichens were a strong sink for ammonia and nitrate in precipitation. Active nutrient release by dead twigs to precipitation water may explain why calcium (which is quite immobile in living tissues) sometimes occurs in high concentrations in throughfall and stem flow.

18.6 DECOMPOSITION AND NUTRIENT CYCLING: SOME BASIC CONCEPTS

> The perception of what is small is the secret of clear-sightedness.
>
> LAO TZU

Litter is decomposed by a complex of bacteria, fungi, and invertebrates that use litter carbon as an energy source and to build the basic structure of their bodies. These are called **saprophages,** from the Greek words for "rotten" (*sapros*) and "to eat" *(phagein)*. Decomposer microbes are also called **saprophytes** (*phyte* meaning "plant," a hangover from the days when everything not an animal was considered a plant). Saprophages are preyed on by various soil protozoa and invertebrates. Among the macronutrients, the availability of nitrogen, phosphorus, and sulfur are most strongly linked to decomposition, because in contrast to the cations, they are integral components of organic matter (chapter 16). This is especially true of nitrogen, which occurs almost exclusively (:99%) in organic forms within ecosystems.

Decomposition influences nutrient cycling in various ways other than the release of these three nutrient elements, however. Humic compounds, which provide much of the cation exchange capacity in soils and determine both aeration and water-holding characteristics (chapter 14), are synthesized by microbes during decomposition. Soil microbes produce an array of chelators that increase the solubility of some nutrients; particularly the heavy metals (section 18.11). Acidity generated by decomposition also influences the solubility of various nutrients. In general, detrital inputs from plants to soils fuel numerous processes that produce and maintain soil fertility.

One of three things may happen to litter nutrients during decomposition:

- They may be immobilized through incorporation into decomposer bodies, recalcitrant humic compounds, or organomineral complexes.
- They may be mineralized, or released to the soil solution, from which they can be taken up by plants or other organisms, bound on exchange sites, form insoluble compounds and precipitate out of solution, or leave the site through leaching or volatilization. Although the term *mineralization* implies conversion from organic to mineral forms (e.g., nitrogen contained in protein to NH_3) small organic molecules such as amino acids and even small proteins are also released to soil solution during decomposition.
- When the decomposer is a mycorrhizal fungus, they may be transferred directly to the plant.

Whether a given nutrient is immobilized or mineralized during decomposition depends on the chemical composition of the litter in which it occurs and, in particular, on its content relative to that of carbon and other nutrients in the litter. Saprophages and their predators effectively act as a series of nutrient filters, immobilizing some nutrients and mineralizing others, depending on their own requirements relative to litter composition. Nutrients that are immobilized at one step in the detrital food chain frequently are mineralized at a later stage and vice versa. If the content of a given nutrient in litter is inadequate for saprophages, they will draw on other sources within the system; hence, the effects of decomposition ripple out and affect nutrient availability throughout the soil. (These topics are explored in more detail later in the chapter.)

Litter may also be attacked by soil organisms that are searching primarily for nutrients rather than energy. This is the case with those species of mycorrhizal fungi with the ability to decompose, which have a ready energy source in plant photosynthate. It may also be the case for some rhizosphere organisms. Even decomposers may "decompose selectively," using enzymes to cleave phosphorus and sulfur from organic molecules without disrupting the carbon skeleton. This is not possible for nitrogen, however, which always occurs bound within the carbon skeleton.

As we discussed in chapter 15, tree growth usually correlates positively with the rate at which one or more nutrients are mineralized; however, it does not follow that mineralization is "good" and immobilization "bad." On the contrary, immobilization plays a positive role within the ecosystem. While nutrients that are immobilized are not immediately available for plant growth, neither are they subject to leaching loss during periods when plants are not active; immobilization is a critically important nutrient-conservation mechanism following a major disturbance such as wildfire or clear-cutting.

18.7 BROAD PATTERNS OF DECOMPOSITION: THE *k* VALUE

Decomposition of litter does not necessarily keep up with the rate at which it is input, and varying amounts of undecomposed and partially decomposed material accumulate in many forests. As we saw in chapter 14, accumulations on the mineral soil surface compose the *forest floor*, which varies from almost absent in some tropical forests to greater than living tree biomass in some northern forests.

Ecologists commonly express the rate of litter decay in a given ecosystem with a value designated as *k*, which is derived by dividing the weight of annual litterfall by the weight of dead organic matter in forest floor and soils:

k = (annual litter production / dead organic matter accumulation)

Suppose that 2 kg/m of litter were produced each year in a given forest, and 10 kg/m (or 5 years of litter production) had accumulated. If the total organic matter is in steady state, $k = 0.2$ simply states that 20 percent of each year's litter is decomposed each year. If one makes certain simplifying assumptions, it can be shown mathematically that the values $3/k$ and $5/k$ equal the time taken for 95 and 99 percent, respectively, of the standing crop of organic matter to turn over (Olsen 1966). The degree to which such calculations reflect real-world dynamics, however, should be viewed with skepticism.

Various approaches have been used to circumvent the problems associated with measuring belowground litter production in calculations of the *k* value. In some cases, the belowground is ignored and *k* calculated as the quotient of aboveground litter and forest floor accumulation (Olsen 1966). Swift et al. (1979) substituted estimates of NPP for annual litter production, assuming that the two are roughly equivalent in ecosystems that have attained maximum foliar and fine-root biomass; this approach is shown in figure 18.11, which provides the range of *k* values for various ecosystem types. The *k* values calculated by Swift et al. ranged from about 0.05 to 0.25 for conifer forests (i.e., from 20 to 4 years of litter production accumulated), 0.25 to 1.00 for temperate deciduous forests, and approximately 1.4 to 4.0 in tropical forests (i.e., less than 1 year of litter production accumulates in tropical forests).

Regardless of the approach, calculating *k* values for entire ecosystems requires oversimplified and usually unverified assumptions (e.g., the assumption of steady state), and this value should be taken as only a rough indication of average decomposition rate. The fact that *k* values are averages is important to keep in mind; some substrates will decompose much faster than *k* indicates and some much more slowly. Every forest ecosystem (including those of the moist tropics) contains organic matter in humic compounds and organomineral complexes that persist for decades and even centuries. Temperate, montane, and boreal forests accumulate coarse woody debris that also persists for decades and centuries (Harmon et al. 1986) and that ecologists rarely include in calculations of litter input and decay. On the other hand, *k* values such as those shown in figure 18.10 accurately portray broad trends; for example, the trend toward smaller *k* values as one moves from the tropics toward the poles reflects an increasing amount of dead organic matter accumulation relative to annual litter production. Put another way, the decomposition rate is much slower in boreal forest and tundra than in tropical forest. Another trend worth noting is that decomposition is generally slower in conifer than in temperate deciduous forests.

18.8 FACTORS CONTROLLING THE RATE OF DECOMPOSITION

Differing rates of decomposition in different ecosystems result from the interactions between environmental factors (e.g., climate, soil texture, pH, aeration) and resource quality (detritus being the resource). Quality is determined by nutrient concentration (i.e., high nutrients equal high quality) and the content of compounds such as lignin and tannins (which

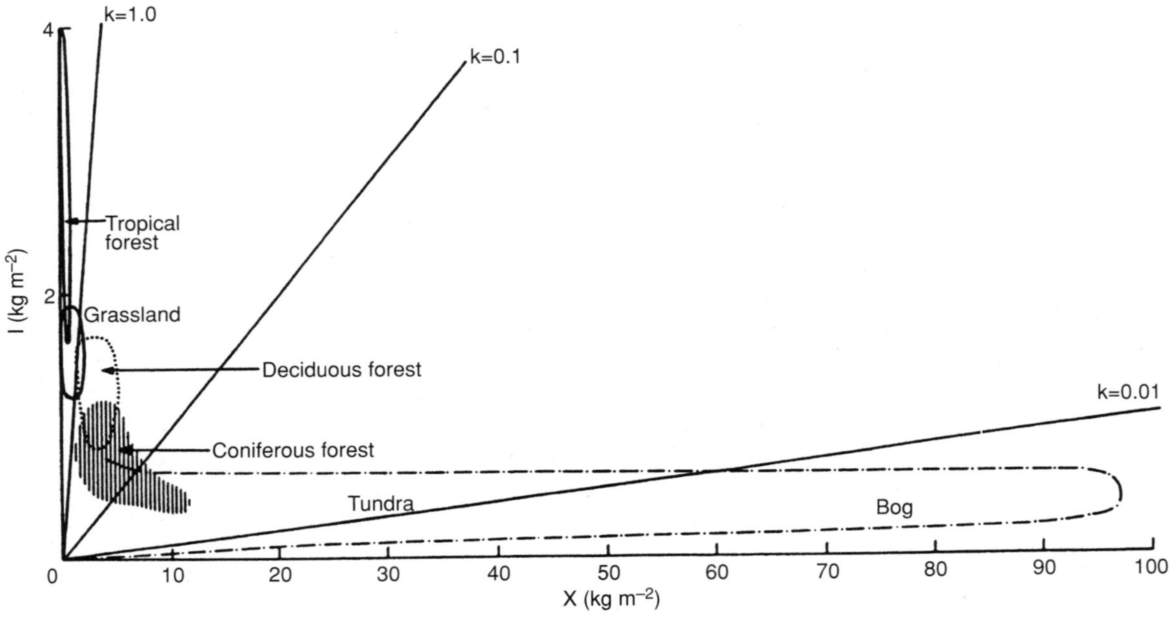

Figure 18.11. The yearly input (I) and total accumulation (X) of organic matter in different types of ecosystem. The decomposition rate is indexed by the value *k*, which equals I/X (Swift et al. 1979)

lower decomposability). Some clear general patterns have emerged from studies comparing the relative effects of climate and litter quality on decomposition. Climate has the strongest effects when comparing biomes at a global scale, where the best predictor of first-year leaf decomposition rates is actual evapotranspiration[8] (Aerts 1997). At continental scales and smaller, climate becomes less important and litter chemistry more so, especially in the tropics (Aerts 1997). Across Canada, three-year decomposition rates of forest litter were explained by the combined effects of mean annual temperature, mean annual precipitation, and litter chemistry[9] (Moore et al. 1999). In contrast to leaf litter, root decomposition is driven more strongly by root chemistry than climate, even when comparing across biomes at a global scale (Silver and Miya 2001), possibly reflecting the more buffered belowground environment in which roots decay.

Climate and litter chemistry are not independent of one another. Aerts (1997) describes the interactions among climate, litter chemistry, and decomposition as a triangular relationship. Resource quality in particular is strongly influenced by environmental factors. For example, trees in environments where decomposition is slow produce litter that is difficult to decompose, and vice versa.

18.8.1 Climate

With global warming looming, how climate influences decomposition is a topic of critical importance and much debate (Davidson and Janssens 2006). Increased decomposition would provide a positive feedback to global warming by releasing more carbon dioxide and methane[10] to the atmosphere. On the other hand, a counteracting and even stronger negative feedback could be induced if limiting nutrients released during decomposition led to increased plant growth.[11]

There is no debate that microbial activity is temperature and moisture dependent. A widely accepted rule of thumb holds that the rate at which soil organic matter is decomposed doubles for every 10°C rise in temperature; however, as Davidson and Janssens (2006) point out, "the origin of this rule-of-thumb . . . and the limits to its validity are less well known." Controversy arises because of the complexity of both the organic molecules being decomposed and the soil system within which decomposition takes place. Several factors may (and often do) modify the effects of temperature. Quoting again from Davidson and Janssens (2006), "Soils contain a 'veritable soup' of thousands of different organic-C compounds, each with its own inherent kinetic properties." In other words, how temperature affects decomposition rates depends on the compound being decomposed. Furthermore, relatively labile organic matter can, and often is, shielded chemically by more

[8] Actual evapotranspiration correlates positively with warmth and moisture.

[9] We discuss details of litter chemistry in a later section.

[10] The strong greenhouse gas methane is produced when decomposition occurs in a low-oxygen environment, such as wetlands.

[11] Because wood has a higher carbon:nutrient ratio than soil organic matter, the transfer of growth-limiting nutrients from soil organic matter to trees would result in more carbon being fixed in photosynthesis than was released in decomposition.

recalcitrant compounds or physically by soil minerals from the microbial enzymes that do the work of decomposing.[12] Finally, even at favorable temperatures, microbial activity can be limited by water. A meta-analysis of experimental warming studies in cold biomes found little effect of warming on decomposition, primarily because heating induced moisture limitations (Aerts 2006). Litter chemistry was the primary driver in those experiments.

One experimental approach has been to transfer litter or whole soils among different environments to determine effects on decomposition. Meentemeyer and Berg (1986) transferred Scots pine needles collected from a single stand to several different environments in Sweden. First-year weight loss of needles placed in pine and birch forests correlated closely with actual evapotranspiration, which reflected, in turn, changes in both precipitation and temperature. For unclear reasons, this pattern did not hold for needles placed in Norway spruce stands. The Long-Term Intersite Decomposition Experiment (www.lternet.edu/collaborations/syn_04.html) involves reciprocally transferring litter of 27 species among 28 sites in North and Central America. Gholz et al. (2000) analyzed five-year results for two tree species with contrasting chemistry, a tropical hardwood and a North American pine. Climate strongly affected decomposition but had good explaining power only when both temperature and moisture were considered. Chemical factors came into play also, as hardwood litter decomposed more rapidly than pine litter, especially in hardwood habitats (Gholz et al. 2000 refer to that phenomenon as *home field advantage*). Hart and Perry (1999) transferred whole soils between low- and high-elevation stands on the H. J. Andrews Experimental Forest in the Oregon Cascades. Results of that study are shown in figure 18.12. Moving soil down in elevation 800 m nearly tripled the rate of nitrogen mineralization over a six-month period (October 27 to June 4), while moving soil 800 m up in elevation reduced nitrogen mineralization by 87 percent. Hart (2006) performed a similar experiment in northern Arizona with essentially the same results.

Several soil warming experiments have been established in forests, grasslands, and tundra (all but one in North America and Europe). A meta-analysis of two- to nine-year results from these found a great deal of variation among sites, with an average 20 percent increase in soil respiration,[13] 46 percent increase in net nitrogen mineralization, and 19 percent increase in NPP (Rustad et al. 2001). Although researchers have speculated that soil warming would increase decomposition most at high latitudes, Rustad et al. (2001) found the opposite: the greatest increases in

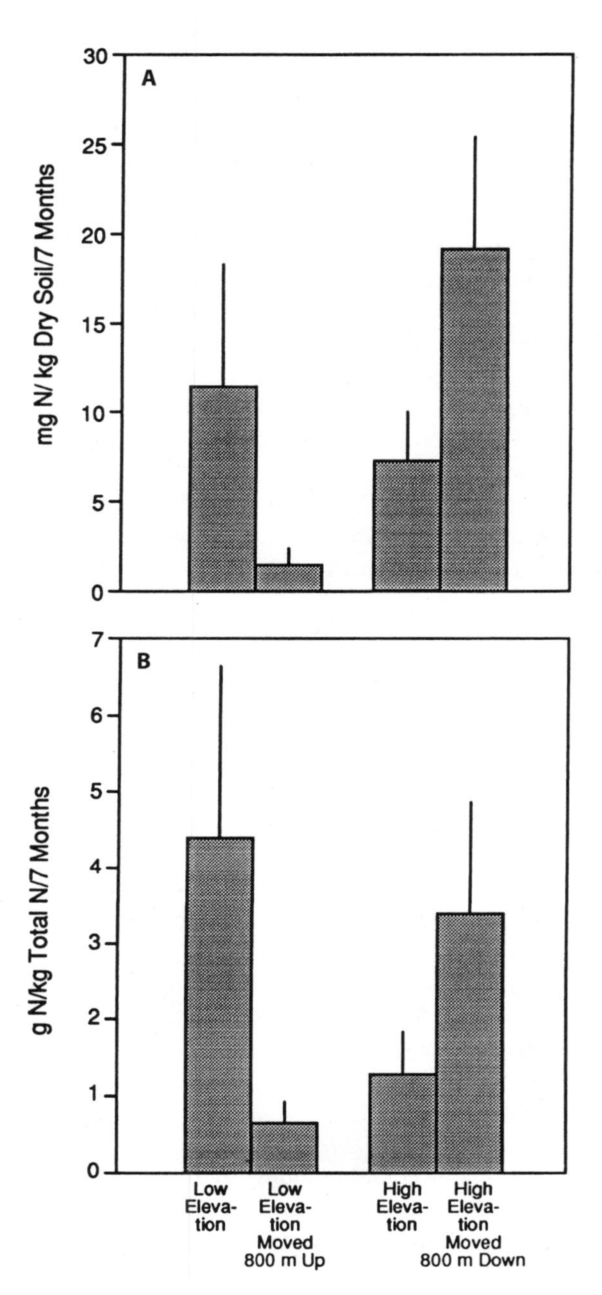

Figure 18.12. The effect of transferring soils across elevation bands on rates of nitrogen mineralization. Samples were from the top 15 cm of mineral soils in two old-growth conifer stands on the H J. Andrews Experimental Forest, Oregon Cascades. The low-elevation stand (490 m) is dominated by Douglas-fir with an understory of western hemlock and western red cedar. The high-elevation stand (1,310 m) is dominated by Pacific silver fir and Douglas-fir. Samples were either incubated in the stand of origin or transferred and incubated reciprocally between stands. Bars indicate standard errors. A. The mineralization rate per kilogram of dry soil. B. The mineralization rate per kilogram of total nitrogen. (Data from Perry et al. 1992; Hart and Perry 1999)

soil respiration in the network of soil-warming experiments occurred in warmer ecosystems. There was some indication in Rustad et al.'s meta-analysis that increases in respiration declined with time at all sites with a sufficient time record, and in the soil-warming experiment at Harvard

[12]Most decomposition is done by extracellular enzymes released by microbes into the environment.

[13]Soil respiration comprises two components, autotrophic (roots and mycorrhizal fungi) and heterotrophic (decomposers and predators). Only the latter reflects decomposition; however, it is very difficult to distinguish between the two.

Forest (Massachusetts) responses were definitely transitory, reflecting the depletion of a labile carbon pool (Melillo et al. 2002).[14]

In boreal, montane, and temperate forests with wet winters and dry summers, decomposition is slowed because increasing temperatures in summer are accompanied by decreasing moisture availability. However, soil organisms show considerable ability to adapt to low temperatures, and studies have shown a significant proportion of yearly decomposition may occur during winter (Brooks et al. 2005). In an Alberta aspen stand, for example, litter lost more weight in winter than in any other season (Lousier and Parkinson 1976). This was accompanied by active fungal colonization of litter (Visser and Parkinson 1975) and considerable activity by bacterial-feeding amoebae (Lousier and Parkinson 1984). In these studies, decomposers were active at soil temperatures above −3°C, and both litter moisture content and fungal activity declined with increasing summer temperatures. In such systems, snow insulates soil and keeps it from freezing, and water is available from small amounts of melting at the snow–forest floor interface. Heat from microbial respiration likely increases melting; therefore, the decomposers play a role in creating their own favorable microhabitat.

Finally, to better reflect the real world it is important to consider not only uniformly high or low temperatures, but how varying temperatures might influence decomposition. Toljander et al. (2006) found that temperature fluctuations promoted coexistence among different species of wood-decaying fungi. Highest rates of decomposition in their study occurred with fluctuating temperatures and intermediate levels of fungal diversity.

18.8.2 Soil

Soil has a profound (although indirect) influence on decomposition through its effect on the nutrient composition of plant material and therefore on litter. Soil also directly affects decomposition through its regulation of water and oxygen, its chemical composition, and its ability to physically protect organic matter.

Soil texture influences decomposition in various ways. In sandy or rocky soils, decomposition may be limited by low water availability, while in clayey or heavily organic soils, poor aeration becomes a problem. Silver et al. (2005) found that roots decayed more quickly in clay soils than in sandy loam soils in the lowland tropics, though water availability did not seem to be a factor in that difference. The decay of larger litter essentially transforms plant carbon into microbial cells and to a lesser extent free organic molecules, and these may become either physically or chemically[15] protected from decomposer enzymes. In fine-textured soils, a significant proportion of organic matter can be physically protected within soil aggregates (Borchers and Perry 1992).

Organic matter can also be protected by chemical interactions with soil minerals, including sorption onto mineral surfaces or complexation with iron and aluminum hydroxides. In their study of 10 acid subsoils from tropical, temperate hardwood, and conifer forests, Mikutta et al. (2006) found that from 59 to 96 percent of soil organic matter was stabilized through association with soil minerals. Kleber et al. (In press) hypothesize that organic matter sorption onto mineral surfaces occurs in three distinct zones, an inner contact layer, which is bound very tightly, a hydrophobic layer, which is bound somewhat less tightly and in which molecules may exchange with soil water to a limited degree, and an outer kinetic layer in which exchange with soil water occurs relatively freely.

Acidity, which correlates negatively with the abundance of basic cations, affects both total microbial activity and the composition of the decomposer community. Fungi are relatively more abundant at low pH than are bacteria, probably because fungi are better able to regulate their internal pH (Swift et al. 1979). In general, macroinvertebrate saprophages such as earthworms are more abundant in soils with pH values above 4.5 than in soils with pH values below this, although acid-tolerant species are not uncommon (Swift et al. 1979; Spiers et al. 1986).

18.8.3 Resource Quality

Litter is composed of various compounds that because of their chemical structure or nutrient content decompose at different rates. Litter may also contain certain compounds (called **modifiers**) that inhibit decomposers or their enzymes. Modifiers include some tannins and terpenes. The resulting resource quality (which might be thought of as the palatability of litter to decomposers) produces what can be a strong, self-reinforcing feedback loop within the intrasystem nutrient cycle. Slowly decomposing litter slows the nutrient cycle, thereby resulting in nutrient-poor tissues that will eventually become difficult-to-decompose litter; fast-decomposing litter speeds the nutrient cycle and produces more fast-decomposing litter.

Weight loss from litter occurs in stages. Initially there is a relatively rapid loss associated with both decomposition of simple substrates and leaching of soluble compounds;

[14]The Harvard Forest experiment also found that soil warming increased the supply of nitrogen to plants, supporting the hypothesis that warming will increase NPP in nitrogen-limited ecosystems (at least until limitations by other resources come into play).

[15]Physical protection differs from chemical protection in the following way. When physically but not chemically protected, physical disruption of the soil (e.g., by sonication) exposes organic matter to decay. When chemically protected, chemical rather than physical means are required to expose organic matter to decay. Note that chemical protection by sorption or complexation with soil minerals is not the same as recalcitrance that arises from the molecular structure of the organic matter itself.

this is followed by a much slower loss associated with decomposition of the more recalcitrant compounds (Berg 1984; McClaugherty et al. 1984, 1985, McClaugherty and Berg 1987). According to Oades (1988): "Rates of decomposition, even for simple substrates such as glucose, vary widely due to differences in water content, temperature, pH, and the availability of nutrients such as phosphorus and nitrogen for the microbial biomass. However, we can generalize and state that natural monomers from carbohydrates, proteins and many polyphenolic materials are decomposed in soil within weeks. Polymers are decomposed more slowly and resistance to decomposition increases with complexity."

Figure 18.13 shows the rate of disappearance of different compounds from forest litter (note that the *y* axis is a logarithmic scale). Sugars, which composed 15 percent of the dry mass of litter studied, almost completely disappeared within 1 year, while only 50 percent of the lignin had disappeared. It should be noted that soluble organic compounds are readily leached from litter; hence, disappearance does not necessarily imply decomposition.

Weight loss during the early stages of decomposition depends on (1) climate, particularly precipitation through its effect on leaching and temperature through its effect on volatile modifiers (Horner et al. 1988); (2) concentration of limiting nutrients in the litter; and (3) the initial concentrations of modifiers in the litter. During the next stage, decomposition is controlled by the chemical structure and nutrient content of the compounds remaining after stage 1 is complete. These are largely nutrient-poor structural compounds such as cellulose and lignin. Nitrogen is most often the nutrient that limits decomposition in unpolluted temperate and boreal forests, although as discussed earlier, the availability of other nutrients also comes into play (Vitousek et al. 1988). Whether drawn from soil or litter, decomposers must obtain the proper balance of all required elements.

The ability of insects and fungi to move nutrients around in systems (the latter through their hyphae) gives them an advantage in the decomposition of nutrient-poor litter; they can compensate for deficits by importing nutrients from elsewhere in the system. Increases in the absolute amounts of nitrogen in decaying conifer litter are commonly seen (i.e., the litter gains nitrogen). Some of this may result from free-living nitrogen fixation, but much of it comes from translocation of nitrogen from mineral soil to litter by fungi (Fahey and Knight 1986; Hart and Firestone 1991). Absolute amounts of both nitrogen and phosphorus increase during decay of Douglas-fir logs (Sollins et al. 1987). In this case, the phosphorus comes from precipitation, litterfall, and importation via roots and hyphae of both mycorrhizal and saprophytic fungi (Wells and Boddy 1990; Wells et al. 1990). The nitrogen also comes from these sources, but nitrogen-fixing microorganisms within the logs probably account for a large proportion of nitrogen accumulation.

In some cases, decomposition rate during later stages can be accurately predicted from carbon:nitrogen ratio of the litter; a higher carbon:nitrogen ratio implies slower decomposition (Taylor et al. 1989). For tree litter, however, either lignin content or the ratio of lignin to nitrogen is generally a better predictor of weight loss, but, the relationship with nitrogen is complex and variable over time. In middle stages more lignin to nitrogen implies slower decomposition (Fogel and Cromack 1977; Berg and Staaf 1980; Melillo and Aber 1984; Harmon et al. 1990). Figure 18.14 shows the strong, negative relationship between the canopy lignin content and soil-mineralizable nitrogen from several Wisconsin forests (Wessman et al. 1988). Nitrogen mineralization is

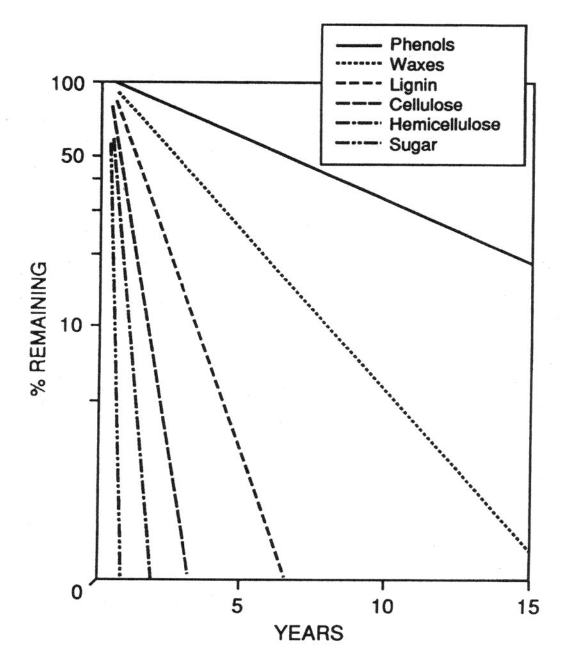

Figure 18.13. The rate at which different compounds are degraded during the decomposition of litter. Note the log scale on the *y* axis. (Adapted from Minderman 1956)

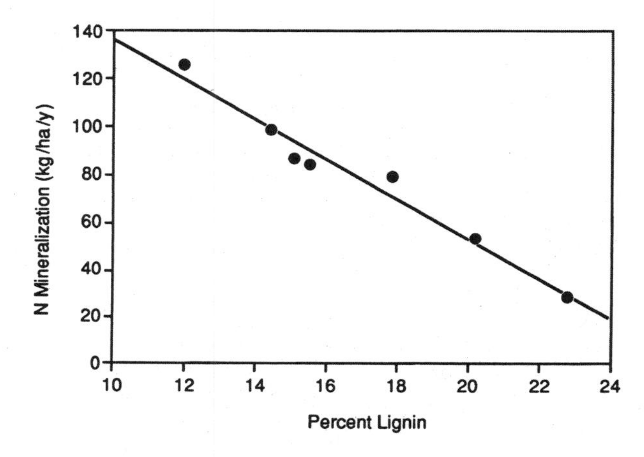

Figure 18.14. The relationship between nitrogen mineralization and canopy lignin in seven Wisconsin forests. (Wessman et al. 1988)

usually a reliable indicator of decomposition rate, but this depends on the carbon:nitrogen ratio of the substrate (section 18.9.1).

In later stages of decay the role of nitrogen changes dramatically, and higher concentrations inhibit rather than promote lignin decomposition. Berg and Meentemeyer (2002) discuss that effect: "The retardation of the decomposition rate in late stages may be so strong that decomposition reaches a limit value at which total mass losses virtually stop. . . . For no less than 106 long-term studies on litter decomposition, encompassing 21 litter types, limit values were significantly and negatively related to N concentration, meaning that the higher the N concentration in the newly shed litter (the lower the carbon:nitrogen ratio) the more litter was left when it reached its limit value."

Nutrients other than nitrogen may strongly influence litter decay. In some cases, especially phosphorus in tropical forests, this reflects the nutritional requirements of microbes (section 18.9). In other cases certain nutrients are required as cofactors for degradative enzymes, and in yet other cases the specific roles are not clear. Both manganese and calcium are emerging as important players in decomposition dynamics. Manganese is essential for the enzyme manganese peroxidase, which is involved in lignin degradation. Manganese oxides, which coat the surfaces of many soil bacteria, oxidize polyphenols and other complex organic compounds. In their study of 14 Norway spruce stands in Sweden, Berg et al. (2000) found two groups, one with a strong negative relationship between lignin content and year 2 to 5 litter decay (fig. 18.15A), and one in which litter decay was unaffected by lignin but positively related to manganese concentration (fig. 18.15B).

European foresters have long recognized that decomposition proceeds more rapidly in soils derived from basic rocks than in soils derived from acid rocks, and that mull humus (chapter 14) is more likely to develop on the former than on the latter (Lutz and Chandler 1946). Much of this difference has been associated with calcium, which according to Lutz and Chandler (1946) "is of outstanding importance in its influence on decomposition." In that subset of Norway spruce stands studied by Berg et al. (2000) in which lignin influenced decomposition rates, the lignin effect varied with litter calcium content; higher calcium:lignin ratios implied faster decomposition. In their study of global patterns of root decay, Silver and Miya (2001) found root calcium content to be an important determining factor. In soils of the central Cascades in Oregon, the rate of anaerobic nitrogen mineralization (which reflects microbial biomass and other forms of labile nitrogen) correlates closely with exchangeable soil calcium (Velasquez-Martinez and Perry 1997). As we saw in chapter 16, calcium plays multiple important roles in ecosystem processes (Reich et al. 2005). Some soil animals (e.g., earthworms) require calcium in relatively high amounts, providing a direct link to the decomposition process.

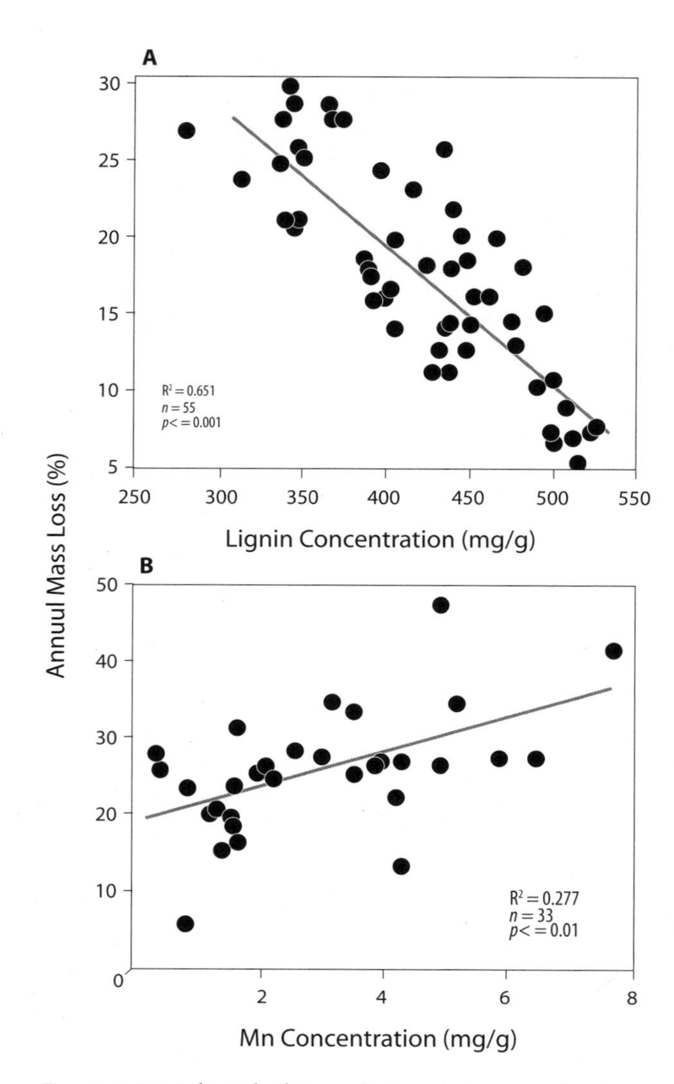

Figure 18.15. A. Relationship between lignin content and mass loss of litter for Norway spruce stands in which lignin was found to play a significant regulatory role. B. Relationship between manganese (Mn) concentration and mass loss of litter in Norway spruce stands where lignin does not play a regulatory role. (Berg et al. 2000)

18.8.4 Modifiers

Modifiers are compounds that "inhibit or stimulate decomposer activity by their chemical structure and are often (though not always) active at relatively low concentrations" (Swift et al. 1979). The last phrase of this definition (i.e., active at low concentrations) is the key to understanding modifiers. They are part of the broad class of system regulators discussed at various points in earlier chapters. Regulators are the "traffic lights" of a system, having far more influence on system trajectory than their abundance and own energy consumption would suggest.

Potentially stimulatory modifiers include various vitamins required by microorganisms, although there is little evidence that these play an important role in decomposition

Figure 18.16. The confluence of a blackwater river and a whitewater river in Amazonia. Whitewater rivers in the Amazon basin rise in the Andes and carry high sediment loads. Blackwater rivers rise in the lowland forest and are dark because of high concentrations of phenols.

(Swift et al. 1979). Inhibitory modifiers include various polyphenolic compounds, monomeric phenols, terpenes and resins, and waxes. Some microorganisms, particularly actinomycetes and some fungi, produce antibiotics, which are chemicals that inhibit growth of other microorganisms. (Earlier in the book, we saw that a commonly used antibiotic in medicine, streptomycin, is produced by actinomycetes in the genera *Streptomyces*, which are the most abundant actinomycete in soil.) The inhibitory modifiers fall within the class of so-called **secondary compounds** that were discussed in chapter 15, and that will appear again in chapter 19 as major regulators of herbivory in ecosystems.

Most work on modifiers has dealt with polyphenolic compounds, which have the property of binding to proteins and to some complex carbohydrates, such as cellulose and starch (see the review by Hattenschwiler and Vitousek [2000]). When bound with proteins or carbohydrates within litter, polyphenols protect these compounds from decomposer enzymes; when bound with decomposer enzymes, polyphenols render them ineffective (Swift et al. 1979; Homer et al. 1988). Polyphenols also interact with soil biota; however, effects are too diverse to allow simple generalizations (Hattenschwiler and Vitousek. 2000). Polyphenols are highly diverse,[16] and different molecular species may have contrasting effects on nutrient cycling. Moreover, polyphenols may inhibit cycling of nitrogen while promoting availability of phosphorus and perhaps basic cations (Hattenschwiler and Vitousek 2000).

Polyphenols are produced in greatest diversity and largest amounts by plant species on acid, nutrient-poor soils (Swift et al. 1979), that is, they are most active in systems where nutrient conservation is most important. For example, the so-called black water rivers of the moist tropics, which drain some of the most infertile soils anywhere in the world (Janzen 1974), run black because of the very high load of polyphenolics and other organic molecules (fig. 18.16). Some rivers of the Atlantic coastal plain of the United States run dark from the high load of phenols they carry (which are leached from the coastal swamp forests). In the past, water from these rivers was used to replenish ships' stores because of its great purity (i.e., freedom from bacterial contamination).

The strength with which tannins bind proteins declines with increasing pH, and over a range of pHs, binding is much weaker in the presence of potassium or sodium ions than in the presence of calcium or magnesium (Martin et al. 1985). The seeming contradiction between this and the fact that calcium increases rates of decomposition illustrates that soil is a complex chemical and biological medium whose dynamics are not likely to be fully captured in laboratory experiments.

[16]Polyphenols are the most common plant secondary metabolite. Several thousand different polyphenolic compounds have been identified.

18.9 EFFECTS OF FOOD-CHAIN INTERACTIONS ON DECOMPOSITION, IMMOBILIZATION, AND MINERALIZATION

18.9.1 How Carbon:Nutrient Ratios Affect Mineralization-Immobilization

Whether a given nutrient element is immobilized or mineralized at any point in the detrital food chain depends on its abundance relative to that of carbon and other nutrients in the litter. (Exceptions, particularly phosphorus, are discussed later.) Plant tissues in general (and particularly those of trees) have a relatively high proportion of complex carbohydrates, or compounds that contain only carbon, oxygen, and hydrogen (e.g., cellulose, lignin, tannins) (McGroddy et al. 2004). In contrast, the compounds that compose microbial cell walls and insect exoskeletons frequently contain nitrogen (e.g., the nitrogen-containing compound chitin is a common constituent of both fungal cell walls and insect exoskeletons). Hence, relative to their needs, saprophages find abundant carbon in plant detritus, but at least some nutrients likely are limiting (especially in woody-plant communities). Because of this, nutrients that are in short supply within a given ecosystem tend to be immobilized rather than mineralized during the initial stages of decomposition.

Table 18.7 shows representative ratios of carbon, nitrogen, and phosphorus in senesced litter and in saprophages and their predators. Ratios in fine-root litter are similar to those in leaf litter (Silver and Miya 2001), while carbon:nutrient ratios widen as both roots and aboveground tissues become woodier. Note that although carbon:nutrient ratios of fungi and bacteria are always more narrow than those of leaf litter, these organisms also exhibit considerable adaptability to the resource they grow on. For example, the carbon:nitrogen ratio of fungal mycelium is 14:1 when on leaves and 37:1 when on wood. Bacteria are considerably higher in nitrogen than fungi but also exhibit the same adaptability to the nutrient content of their resource.

Table 18.7
Representative nutrient ratios* for living foliage, litter, decomposers, and predators

		C:N	C:P	N:P
Foliage				
	Temperate broadleaf	35.1	922	28.2
	Coniferous	59.5	1,232	21.7
	Tropical	35.5	2,457	43.4
Senesced litter				
	Temperate broadleaf	54.8	1,702	29.1
	Coniferous	87.8	2,353	26.0
	Tropical	60.3	4,116	62.7
Fungal mycelia				
	On leaves	14	204	15
	On wood	37	544	15
Bacteria				
	In culture	3	8	4
	On leaves	12	53	4
Nematodes		4–10	—	—
Earthworms		4	40	10
Insects		5	6–80	12–16

Source: Values for foliage and litter are means from McGroddy et al. (2004). See their paper for standard errors and sample number. All other except nematodes from Swift et al. (1979), assuming C contents as given in Bowen (1979). Nematode data from Bowen (1979), Anderson et al. (1983), and Reiners (1986).
*Actual ratios may differ from those shown depending on species and environment.

To see how carbon:nutrient ratios combined with food web interactions determine immobilization-mineralization reactions, consider fungal decomposition of leaf litter from conifer forests. Litter averages approximately 47 percent carbon (on a dry weight basis), so 1,000 g of litter would contain about 470 g of carbon. At a carbon:nitrogen ratio of 88:1, conifer litter would contain approximately 5 g of nitrogen. Suppose that the fungus is 50 percent efficient in its conversion of litter carbon to fungal carbon (i.e., 50% of the litter carbon is converted to fungal biomass and the other 50% respired). This means that 235 g of carbon is available to produce more mycelia. At a carbon:nitrogen ratio of 14:1, this amount of mycelia would immobilize 17 g of nitrogen. The conifer litter has 12 g less nitrogen than the fungus requires to utilize the available carbon, so none is mineralized. The fungus may deal with the nitrogen deficiency it encounters in the conifer litter in several ways, which we discuss later. Now suppose the fungus is grazed by a fungus-feeding mite. In this case we assume 20 percent efficiency,[17] which means that 235 g of fungus would produce about 50 g of mites. At a carbon:nitrogen ratio of 5:1, the mites would require 10 g of nitrogen, leaving 7 of the total 17 g in the fungal mycelia free to enter the environment and be utilized by plants or other soil biota. It is that "free" portion that we refer to as being *mineralized*.

The belowground food chain and the decomposability of organic matter are more complex than this simple example indicates (Aber et al. 1991; Anderson 1991; Bosatta and Agren 1991). Nevertheless, the illustrated points are generally valid. These may be summarized as follows:

- *Nutrients that are in short supply within the system are immobilized; those in abundant supply are mineralized.* Thus, in conifer forests, nitrogen tends to be more strongly immobilized than phosphorus, while in many moist tropical forests, phosphorus tends to be immobilized most strongly (Vitousek 1984b). Fertilization with a limiting nutrient releases a constraint on microbial growth and results in increased immobilization of other nutrients (Robertson 1984).

[17] These efficiency values are for illustrative purposes and should not be taken too literally. However the arthropod is likely to be less efficient than the fungus by virtue of greater metabolic demands related to maintaining a more complex body and active life style. Invertebrates convert substrate into biomass with an efficiency of only 10 to 40% (Coleman et al. 1983).

- *As substrate passes up the food chain, an increasing proportion of the original carbon is lost in respiration, with consequent narrowing of carbon:nutrient ratios.* Coupled with the fact that carbon:nutrient ratios of microbes and animals are relatively similar, this results in increasing nutrient mineralization at each step up the food chain.

Before discussing how this rather simple view can be improved and refined, we first want to introduce the **critical ratio,** which is the carbon:nutrient ratio below which the nutrient is mineralized and above which it is immobilized. The critical ratio for nitrogen has traditionally been taken as somewhere between 25:1 and 30:1. This is calculated by essentially the same procedure we went through above (the carbon:nitrogen ratio of tropical litter in that example, from which a small amount of nitrogen was mineralized, was 25:1).

In fact, however, the critical carbon:nitrogen ratio varies widely (Haynes 1986). In some cases, nitrogen is mineralized from substrate with surprisingly wide carbon:nitrogen ratios; for example, the critical carbon:nitrogen ratio for Scots pine needles is 68:1 (Staaf and Berg 1977) and for Scots pine cones 167:1 (Berg and Staaf 1981). In other instances, relatively little nitrogen is mineralized from material with a carbon:nitrogen ratio far below 25:1. This is the case for organomineral complexes, in which the proportion of total nitrogen that is mineralized increases rather than decreases as the carbon:nitrogen ratio widens from 10:1 to 25:1 (Sollins et al. 1984).

From these examples, it is obvious that if one is to understand mineralization-immobilization reactions, it is necessary to go beyond carbon:nutrient ratios. Any factor limiting decomposition will effectively decrease nutrient availability and therefore also limit mineralization. Thus, material with abundant nitrogen, for example, may nevertheless be quite stable because it is shielded from microbial enzymes by lignin or modifiers, or protected within organomineral complexes, or because decomposers are limited for other reasons. One major factor influencing nutrient mineralization is the diversion by microbes of carbon, nitrogen, phosphorus, and sulfur to humic compounds. These are stable both because of their chemistry (i.e., amino acids being shielded from enzymes by polyphenols) and because they react with soil minerals that provide further protection. Nutrients are mineralized from this pool much more slowly than the carbon:nutrient ratio would suggest.

Organic matter is bound to clays by polyvalent cations, primarily Ca^{2+} and Mg^{2+} in neutral to alkaline soils and Fe^{+} and Al^{+} in acid soils. These compounds attach their positive charges to the negative charges of the clays and organic compounds, thereby bridging between the two (Oades 1988). Protection of nitrogen by clays is illustrated by the data of figure 18.17, which compares the rate of nitrogen mineralized from intact soils with that mineralized from the same soils with their aggregates disrupted. In the loamy soil (15% clay content), disrupting the aggregates effectively removed organic matter from the protection of clays, as manifested by the significantly increased rate of nitrogen

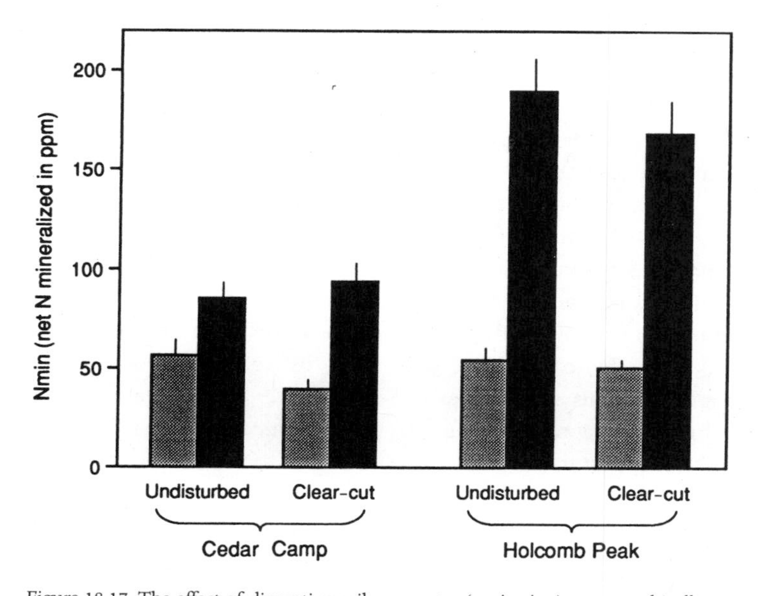

Figure 18.17. The effect of disrupting soil aggregates (sonication) on anaerobically determined, mineralizable nitrogen in two forest soils. Gray bars show intact soils, black bars sonicated soils. Cedar Camp soil has 8 percent clay and Holcombe Peak soil 15 percent clay. The difference in mineralizable nitrogen between sonicated and unsonicated soils is a measure of the amounts of labile, organically bound nitrogen physically protected within soil aggregates. Bars indicate standard errors. (Borchers and Perry 1992)

mineralization. This occurred to a lesser degree in the sandy soil (8% clay), in which decomposition was probably controlled primarily by chemical composition of the soil organic matter.

18.9.2 Interactions between Fauna and Microbes

Numerous studies have shown soil fauna to be critically important in the decomposition and mineralization of nutrients, particularly nitrogen and cations (Edwards et al. 1970; Hole 1981; Anderson et al. 1983, 1985; Coleman et al. 1983; Seasteadt 1984; Verhoef and De Goede 1985; Setala et al. 1988; Hunter et al. 2003b; Hattenschwiler and Gasser 2005; Lensing and Wise 2006). In one study that experimentally manipulated the presence of soil fauna, birch seedlings *(Betula pendula)* growing in soils with fauna were 57 percent heavier than seedlings growing in soils without fauna (Setala and Huhta 1991). Setala and Huhta attribute this result to better nitrogen and phosphorus nutrition as well as improved soil water where fauna were present. Apparently, Lutz and Chandler (1946) had it right when they admonished silviculturists to "whenever possible . . . give consideration to the creation and maintenance of forest conditions favorable to an active soil fauna."

Recall that the larger invertebrate fauna (e.g., earthworms, millipedes, isopods) are referred to as **macrofauna,** the small arthropods (e.g., mites and springtails) as **mesofauna,** and nematodes and protozoa as **microfauna** (chapter 14). Although they are usually classified as saprophages, the macro- and mesofauna are really omnivorous, consuming litter along with the fungi and bacteria it contains, consuming and recycling their own feces, and sometimes even consuming one another. The microfauna largely feed on fungi and bacteria, although some are omnivores.

Initially, decomposition proceeds rather slowly after a leaf hits the ground. Fungi are frequently the first decomposers to become active, some invading from surrounding soil and litter and some already present in the leaf when it is shed. Soil animals may be inhibited by tannins and other modifiers during this period, and they may also prefer some fungi in their diet. Oribatid mites, for example, will not feed on litter until it is colonized by saprophytic fungi (Hartenstein 1962). Calcium-demanding animals such as earthworms and oribatid mites may have a taste for fungi because the latter accumulate calcium in the form of calcium oxalates, which can be decomposed by microorganisms that live in the animal gut (Cromack et al. 1977).

Most modifiers (including tannins and other phenols) are water soluble and rather quickly leached from foliage in humid environments. Others (such as terpenes) are volatilized from litter, particularly in warm and dry environments (Homer et al. 1988). Once the concentration of modifiers falls below inhibitory levels, litter of both temperate and boreal forests is attacked and fragmented by various macro- and mesofauna, particularly earthworms, millipedes, isopods (e.g., sow bugs, pill bugs), oribatid mites, and collembolans (i.e., springtails). The activity of these animals generally increases the abundance and activity of microbes (Mikola et al. 2002). According to Edwards et al. (1970), "Litter decomposes at a rate directly related to the number of invertebrate animals in the litter and underlying soil."

Among temperate and boreal forests, macrofauna are most abundant in systems with high litter quality (e.g., temperate deciduous forests), while mesofauna dominate in conifer and other forests that produce acidic, nutrient-poor litter (Heal and Dighton 1985). However, earthworms can be abundant in conifer forests (particularly those in moist environments) and also in other forest types with acidic, nutrient-poor litter (Spiers et al. 1986). Most species in these systems are restricted to the forest floor, whereas those in deciduous forests move between the forest floor and mineral soil.

Macrofauna ingest an enormous amount of material, as much as 90 percent of which is passed through their guts rather than retained (Hole 1981). Spiers et al. (1986) estimated that earthworm casts composed 60 percent of the total volume of solids in the organic horizons of a British Columbia conifer forest. These may be effectively recycled many times, the fauna indiscriminately feeding on the mixture of fresh litter, their own fecal pellets, and admixed fungi and bacteria. In deciduous forests with high earthworm populations, the "annual nitrogen flux through the worm population may be several times the 30–70 kg/ha/y contained in leaf fall" (Anderson et al. 1985); nitrogen mineralization from oak litter is increased more than 10-fold by millipedes (Anderson et al. 1985).

The favorable effect of macro- and mesofauna on decomposition and nutrient mineralization results from at least five factors:

1. Fragmentation and increase in the surface area of litter during its passage through the animal gut (called **comminution**). This was once thought to be the major effect of macro- and mesofauna on the nutrient cycle (Edwards et al. 1970); but Kitchell et al. (1979) have argued that while litter surface area is indeed increased by smaller macroinvertebrates such as springtails and mites, it is decreased by passage through the guts of millipedes and earthworms.
2. Mutualistic associations between macro- and mesofauna and bacteria or protozoa that dwell in their gut (LaVelle et al. 1995). The fauna provide a warm, moist, food-rich environment within their digestive tract (movable feasts, as it were) for the gut residents. In exchange, the fauna receive essential enzymes that are produced by the bacteria or protozoa (Ineson and Anderson 1985; Gunnarson and Tunlid 1986). Interestingly, bacteria that occupy faunal guts include the gram-negative types that are also common in

rhizospheres. LaVelle et al. (1995) termed these mutualistic associations between microorganisms (i.e., plants and invertebrates) and microorganisms "biological systems of regulation," because of their importance in regulating the nutrient cycle.

3. By feeding over a wide area, the fauna integrate substrates with various nutrient compositions in their diet, which smooths out local nutrient deficiencies that might limit a less-mobile organism. This is certainly the case in forests with mull humus layers, where earthworms feed on both litter and mineral soil.
4. Physical changes due to faunal activity improve soil water-holding capacity (Setala and Huhta 1991).
5. The macro- and mesofauna restructure belowground food webs in such a way that more energy and nutrients flow to bacteria and the bacterial-feeding microfauna (Anderson et al. 1983).

The last point requires further explanation. Fungal hyphae generally do not survive passage through the gut, while invertebrate feces have two hundred to three hundred times more bacteria than the litter that is consumed (Anderson et al. 1983). Hence, fauna reduce fungi while favoring bacteria. This does not increase nitrogen mineralization in itself, but the shift to bacteria brings the bacterial grazers (primarily nematodes and various protozoa) into play. These have carbon:nutrient ratios similar to the bacteria they feed on, and their efficiency of conversion is relatively low. Therefore, they have the potential to mineralize large amounts of nutrients.

The role of bacterial-feeding microfauna (i.e., nematodes and amoebae) in the nutrient cycle has been studied primarily in grasslands. This work has found that nematodes consume up to 800 kg/ha/y of bacteria, containing from 20 to 130 kg/ha of nitrogen (Coleman et al.. 1984). Much of this nitrogen is released by the nematodes as NH_3 and small organic molecules (i.e., amino acids) that can be taken up by plants. Forest soils average approximately 50 percent of the numbers of bacteria and microfauna as in grasslands, but the relative proportions of microfauna to bacteria are similar in the two systems (Swift et al. 1979). Therefore, the bacterial-feeding microfauna likely are just as important to the mineralization of nutrients in forest soil as they are in grasslands, although the absolute amounts of nutrients that are mineralized through their grazing are probably less in forests than in grasslands by roughly one-half.

Lousier and Parkinson's (1984) research on bacterial consumption by protozoa in the forest floor of an Alberta aspen stand suggests that microfauna are indeed major actors in the nutrient cycle of forests. They studied a single group of protozoa, the testate amoebae. Annually, the amoebae consumed an amount of bacteria equivalent to 60 times the average bacterial standing crop; in other words, there was a very rapid cycling of bacteria through the amoebae. Nitrogen mineralization was not measured in this study, but on the order of 10 kg/ha/y of nitrogen was probably mineralized during the consumption of bacteria by amoebae, an amount roughly equivalent to 20 percent of the total tree requirements. Lousier and Parkinson suggest that soil protozoa respired from 10 to 15 percent of the total annual carbon input to the soil in this system.

Although the increase in bacteria resulting from feeding by macro- and mesofauna usually increases nutrient mineralization, overall decomposition drops and mineralization declines if faunal populations become so high that fungi are effectively overgrazed (Verhoef and De Goede 1985). Hence, animals that prey on the faunal saprophages (including predatory invertebrates, small mammals, and birds) are important to keep the system in balance. Predatory macrofauna such as ants and spiders have been shown to alter litter chemistry, presumably through top-down effects on the decomposer community (Hunter et al. 2003).

> Aboveground herbivores can significantly affect litter chemistry and therefore alter decomposition rates; however, depending on specific chemical changes, decay may be either speeded or slowed. For example, both scale insects and a stem-boring moth increase nitrogen and lignin:nitrogen ratios and increased first-year decomposition rates of pinyon pine litter (Chapman et al. 2003). However, a leaf-galling aphid resulted in higher concentrations of lignin and tannins in and slower decomposition of cottonwood litter (Schweitzer et al. 2005b). In chapter 19 we see that increased concentrations of secondary compounds in response to leaf herbivores is a common defensive strategy in many plants.

18.10 BIODIVERSITY AFFECTS DECOMPOSITION

Species mixtures have been shown in some instances to have nonadditive effects on decomposition and nutrient cycling, in other words, a given species may behave differently in mixture than in monoculture (Hattenschwiler et al. 2005). For example, Douglas-fir produces more nitrogen-rich litterfall and cycles more nitrogen aboveground when mixed with hardwoods than when in pure stands (Perry et al. 1987a). When hardwoods are removed, nitrogen content of Douglas-fir needle litter drops by approximately 10 kg/ha/y, and a corresponding amount is deposited as belowground litter. In northern hardwoods, native species litter decomposes faster in stands that are invaded by exotic plants than in noninvaded stands (Ashton et al. 2005).

Despite the fact that mixed plant species are the rule in nature rather than the exception, the majority of litter decay studies have dealt with single species. Of the 30 mixed-species studies reviewed by Gartner and Cardon (2004) (most dealing with temperate forests), about half found higher mass loss in mixture than would have been predicted from simple additive effects of decay rates in monocultures of the species involved, while approximately 30 percent showed only additive effects and 20 percent exhibited antagonistic effects (i.e., decay rates less than additive). Effects of mixtures on nitrogen mineralization were even more striking, 76 percent of studies showing synergistic effects (Gartner and Cardon 2004). Mass loss and nitrogen mineralization correlated poorly with one another, suggesting different mechanisms at work for carbon and nitrogen.

How does one plant species alter the decomposition and cycling characteristics of another? In some cases the answer is straightforward. Mixtures with nitrogen-fixing species benefit from the inputs of that element. Mixing eucalyptus with nitrogen-fixing species produces a synergistic effect, the latter adding nitrogen to the system and the former increasing phosphorus availability, resulting in the litter of both species having a more balanced nutrient content than when growing alone (Forrester et al. 2006). The belowground community plays a significant role in mediating how litter diversity influences decomposition and nutrient cycling. One way that happens is through smoothing the distribution of nutrients among different plant species or litter types. For example, certain types of mycorrhizal fungi act to balance the foliar nutrient contents of Douglas-fir and ponderosa pine when growing in mixture (Perry et al. 1989c). Saprophytic fungi may smooth the distribution of nutrients among different litter types (Hattenschwiler et al. 2005). However, differences in nutrients among litter types is not the whole story, as synergistic effects have been seen in litter mixtures comprising species with similar nutrient contents (Hattenschwiler et al. 2005). In some cases effects on microclimate are involved. For example, even though feather moss is slowly decomposing, it has a high water-holding capacity and through that enhances the decay rate of associated litters (Wardle et al. 2003). Litter-feeding soil macrofauna have been shown to either increase (millipedes) or decrease (earthworms) the synergistic effects of litter mixtures (Hattenschwiler and Gasser 2005). Hattenschwiler et al. (2005) conclude that "feedback effects between the composition and richness of litter and soil fauna appear to be important mechanisms for the understanding of how decomposition is influenced by litter diversity." However, they go on to point out that "litter-diversity effects on macrofauna feeding behavior and performance and the consequences for decomposition remained largely unexplored."

What about diversity in the decomposer community? Two things at least are clear. First, all trophic levels must be present for proper functioning. Second, there are keystone species, particularly some macrofaunal litter processors such as earthworms and millipedes. The role played by species richness within trophic levels is less clear and for the most part little studied. In the introduction to their study of how fungal diversity influences decomposition of conifer humus, Setala and McLean (2004) noted "there is virtually no knowledge as to how the diversity of decomposer microbes influences the decomposition rate of soil organic matter." They found diversity to increase decomposition only at the species-poor end of the fungal diversity gradient and concluded that there is a great deal of redundancy among saprophytic fungi. Toljander et al. (2006) found that wood decomposition was highest at intermediate levels of fungal diversity *and* a fluctuating temperature regime, suggesting niche differentiation among the fungi. On the other hand, Tiunov and Scheu (2005) found that while the decomposition rates of both forest soil and an artificial homogenized media were enhanced by saprotrophic fungal richness, the effect was stronger in the homogeneous media, where there would seem to be no options for niche differentiation (the environment was held constant). Tiunov and Scheu concluded that "facilitative interactions are more important than resource partitioning." (Later we see that some species of mycorrhizal fungi actually retard decomposition).

In studies of how the diversity of soil animals affects decomposition and nutrient cycling, only a few have found positive effects. Mikola et al. (2002) attribute this to two factors: (1) the range in diversity in most studies is too narrow to detect effects, and (2) simply counting species numbers is inadequate to capture the complex interactions that characterize the belowground ecosystem.

18.11 A CLOSER LOOK AT NITROGEN, PHOSPHORUS, AND SULFUR CYCLES

The transformations of nitrogen, phosphorus, and sulfur within soil are important to understand, because the forms in which they occur determine their availability to plants and (particularly with nitrogen) their potential for loss from the system to stream water or the atmosphere. Along with carbon, oxygen, and hydrogen, the nutrients nitrogen, phosphorus, and sulfur are most closely tied to movement of organic matter through the detrital food chain. In one sense, the fates of these three nutrients are intertwined both because they are integral components of organic molecules and because the ratios in which they occur relative to one another determine whether they will be mineralized or immobilized during decomposition. These nutrients also cycle independently of one another for a variety of reasons that we will explore in this section (McGill and Cole 1981; Stevenson 1986).

18.11.1 Nitrogen

Ecologists' understanding of the nitrogen cycle changed significantly during the 1990s and continues to evolve (Schimel and Bennett 2004). In the traditional view (fig. 18.18A) ammonia (NH_3) was central. Release of NH_3 from

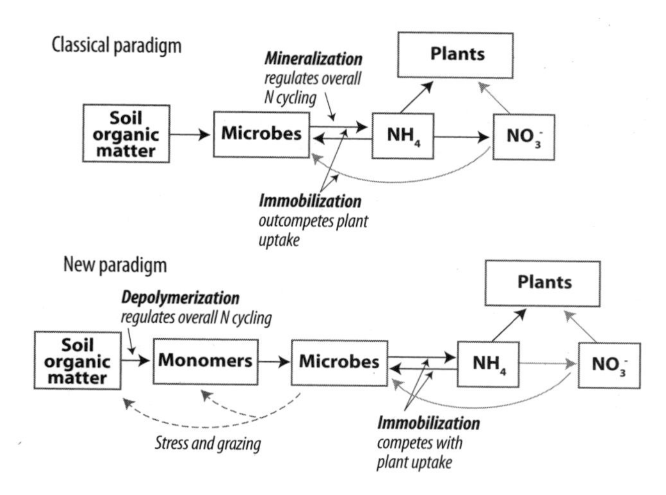

Figure 18.18. Comparison of the classical and the new paradigms of nitrogen cycling and uptake by plants. Gray arrows originating in the "nitrification" box indicate that nitrification is not important in all systems. (Schimel and Bennett 2004. With permission of the Ecological Society of America)

When considering nitrogen mineralization in ecosystems, it is necessary to clearly distinguish between gross and net rates, because in at least some cases these two measures give quite different information. Net mineralization rates are estimated by "incubating" soil or litter for some period (without plants), then calculating the amount of ammonia and nitrate that accumulates in solution. This method measures net amounts because it does not account for what is taken up by microbes, fixed in clays, or converted to gas during the incubation period. To measure the actual, or gross rates, one must account for all possible pathways, something that can be done only by labeling individual nitrogen molecules (with the heavy isotope ^{15}N) and then following their fate (Hart et al. 1992). This approach not only yields information about gross mineralization rates, it allows the various pathways that mineralized nitrogen might follow to be quantified to a degree that is not possible without labeling. Labeling studies show that microbes are a very strong sink for both NH_4^+ and NO_3^- (Vitousek and Andariese 1986; Schimel and Firestone 1989; Davidson et al. 1991, 1992; Hart and Firestone 1991; Zou et al. 1992; Hart et al. 1994). This is not surprising in nitrogen-limited ecosystems, which is what most conifer forests are assumed to be. Microbes are less of a sink for nitrogen in phosphorus-limited tropical forests (Matson et al. 1987; Vitousek and Matson 1988). One implication of the studies using ^{15}N is that measures of net mineralization greatly underestimate the actual rate of nitrogen cycling within nitrogen-limited forest soils. For example, Davidson et al. (1992) found that gross rates of mineralization in mixed-conifer forests of California were 60 to 80 times greater than net rates. Note that measures that look only at ammonia and nitrate do not account for that component of the nitrogen cycle represented by organic nitrogen forms.

organic molecules was the rate-limiting step in the nitrogen cycle, and demand for NH_3 by heterotrophic microbes regulated its availability to both plants and the chemotrophic bacteria that extracted energy by converting NH_3 to nitrate (*nitrifiers*). Except during brief periods following disturbance, only very small amounts of nitrate could be found, especially in forest soils (arid land soils differed in that respect), leading researchers to conclude that nitrification was severely limited by the availability of NH_3 (the substrate for nitrifiers). Two things in particular came together to change that view (Schimel and Bennett 2004): (1) the use of ^{15}N-labeled nitrogen to trace nitrogen movements and to separate gross from net mineralization, and (2) the fact that plants were actively growing even in situations where theoretically no nitrogen should be available to them.

The new view of the nitrogen cycle differs from the old in three major ways (fig. 18.18B). First, dissolved organic forms play an important role. Exoenzymes released by microbes cleave complex organic forms into simpler monomers[18] that may be directly utilized both by heterotrophic microbes and plants or their mycorrhizal fungi. As we saw in chapter 14, the cleavage of polymers into monomers is now considered by many to be the rate-limiting step in the nitrogen cycle (Chapin et al. 2002). Second, plants and their mycorrhizal fungi compete for nitrogen on an equal footing rather than getting only the leftovers (how they do that is uncertain and will be discussed in a later section). Third, rates of gross nitrification are quite high; little net nitrification is found because the nitrate is heavily utilized.

Most nitrogen within litter occurs in peptides (i.e., two or more amino acids joined together), proteins (i.e., a big peptide), DNA, or RNA. In all of these, the nitrogen atom is bound directly to carbon. Therefore, the enzyme that releases nitrogen from organic matter must cleave a carbon bond. We can also turn this around and say that during the process of cleaving carbon bonds, enzymes release amino acids and small peptides, and hence, nitrogen monomers may be generated by organisms with no interest in the nitrogen and that, rather, are searching for energy. Regardless of whether the search is for energy, nitrogen, or more likely, both energy and nitrogen, the release of carbon and the release of nitrogen from organic molecules go hand in hand.

The release of NH_3 from organic molecules is called **mineralization** or, alternatively, **ammonification.** (The

[18]A monomer is any simple organic compound that can bond with other monomers to form a polymer. Amino acids are one example.

term *mineralization* may also be extended to include the enzymatic cleaving of organic polymers to monomers). The common wisdom sees mineralization as microbially controlled; however, increasing evidence indicates that plants play an important regulatory role in at least two ways (Chapman et al. 2006): the nitrogen content of their litter (a function of uptake and internal recycling), and their mycorrhizae.

NH_3 is volatile and potentially lost to the atmosphere as gas. At the pH values common in forest soils, virtually every NH_3 molecule reacts with water to produce ammonium ions (NH_4^+) (Bowden 1986).

Several things may happen to NH_4^+ (which is not volatile).

- It may be *directly taken up* by plants, animals, or heterotrophic microorganisms.
- It may be *nitrified,* which involves two oxidations: the first from NH_4^+ to nitrite (NO_2^-), and the second from NO_2^- to nitrate (NO_3^-). Nitrification is mostly performed by a class of microbes called **chemoautotrophs,** which use the NH_4^+ (or NO_2^-) as an energy source; some heterotrophic microorganisms will also convert NH_4^+ to NO_3^-
- It may be *held on the cation exchange complex.*
- It may be *fixed nonexchangably* within the structure of some types of clay minerals (e.g., vermiculite, illite, montmorillonite) (Stevenson 1986).
- It may be *leached to streams.* This seldom occurs with NH_4^+. If an organism does not take it up, it usually will be retained on the cation exchange complex.[19]

Under certain conditions that we discuss later, nitrite and nitrate are converted to gaseous nitrogen compounds—primarily nitrous oxide (N_2O) and molecular nitrogen (N_2)—which are readily lost to the atmosphere.

18.11.1.1 Nitrification

Processes controlling nitrification are of particular interest to ecologists and foresters for at least two reasons. First, because nitrate is not held on the cation exchange complex it is susceptible to leaching loss when biotic sinks are weakened. Second, as mentioned above under certain conditions nitrate can be converted to gaseous forms of nitrogen and lost to the atmosphere. Chemoautotrophic nitrification occurs in two steps. Bacteria in the genus *Nitrosomonas* first oxidize NH_4^+ to NO_2^-, which is then oxidized to NO_3^- by bacteria in the genus *Nitrobacter*. The second step occurs very rapidly; little NO_2^- is ever found in natural soils. Both genera of bacteria use the nitrogen compounds rather than organic carbon as an energy source. Various heterotrophic microbes (i.e., those using reduced carbon as an energy source) are also capable of nitrification, including fungi, bacteria, and actinomycetes that are common in forest soils (Stevenson 1986).

Rates of net nitrification are generally low in soils of undisturbed temperate and boreal forests, but amounts vary widely among forest types. Highest rates have been measured in moist tropical forests growing on fertile soils and beneath stands of nitrogen-fixing trees such as red alder, intermediate rates in temperate deciduous forests, and low rates in conifers and tropical forests on infertile soils (Lamb 1980; Robertson 1984; Gosz and White 1986). Across forested sites in North America, rates of net nitrification decline strongly as lignin:nitrogen ratios in litter increase (Scott and Binkley 1997). In a study that encompassed a variety of North American forest types, Vitousek et al. (1982) concluded that the best predictor of net nitrification was the total amount of nitrogen circulating in a system: the more nitrogen that is cycling, the more nitrification occurring. At least in part because nitrogen uptake by trees is reduced or eliminated, disturbances that kill or weaken trees result in increased nitrate levels in soil solution. This phenomenon has led some to suggest that the amounts of nitrate in soil solution and stream water indicate the state of ecosystem health (chapter 20).

18.11.1.2 Production of Gaseous Nitrogen Compounds in Soils

Evolution of gaseous nitrogen compounds during the nitrogen cycle is significant not only because it represents a pathway of nitrogen loss from ecosystems but because its products (i.e., N_2O and NO) are among those chemicals that catalyze destruction of atmospheric ozone and act as greenhouse gases. Gaseous nitrogen compounds may be produced either biologically (i.e., by soil microbes) or through strictly chemical processes. Chemodenitrification occurs only when NO_2^- accumulates (Bowden 1986), something that seldom (if ever) occurs in natural soils. (We deal only with biological denitrification here.)

Microorganisms produce gaseous nitrogen through enzyme-mediated reduction of NO_3^- and NO_2^-. The wide variety of soil microbes performing this fall into two general groups (Bowden 1986; Sahrawat and Keeney 1986):

- Anaerobic organisms that use NO_3^- rather than O_2 as a terminal electron acceptor during metabolism, thereby producing N_2O, and N_2; this is called **denitrification.**
- Aerobes that produce N_2O, N_2, and probably NO from either nitrate or nitrite during the process of nitrification.

Denitrification occurs primarily under conditions of low oxygen availability and is particularly high in wetlands,

[19] As we saw in chapter 17, both NO_3^- and dissolved organic nitrogen are susceptible to leaching. Microbial uptake functions as a biological dam for NO_3^- and the more readily available organic compounds, providing the microbes have sufficient carbon or are not limited by other nutrients. However, some dissolved organic nitrogen is not utilized by microbes and may represent an important pathway of nitrogen loss from undisturbed ecosystems.

marshes, and periodically flooded riparian zones. Denitrification can also occur in microsites that develop low O_2 tensions within otherwise well-aerated soils, such as rhizospheres or the interior of aggregates. A variety of microbes produce gaseous nitrogen compounds under aerobic conditions, including *Nitrosomonas, Rhizobium*, and several other heterotrophic bacteria and fungi. Sahrawat and Keeney (1986) conclude that "N_2O production involves a complex ecological niche in the soil nitrogen cycle." N_2O is apparently produced by aerobic microorganisms via the enzyme nitrate reductase; the primary function of which is to reduce NO_3^- to NH_3 for incorporation into amino acids. It is not clear whether the evolution of N_2O by this enzyme is merely accidental or serves some physiological purpose.

The magnitude of gaseous nitrogen emissions from soils varies widely both among forest types and within a given site. Most studies of forest soils have found emissions on the order of a few kilograms per hectare per year or less, which is a relatively small proportion of the nitrogen in circulation. In some cases, however, emissions can be quite large; N_2O losses alone on the order of 20 to 40 kg/ha/y have been reported for some temperate hardwoods and moist tropical forests (Keller et al. 1983; Robertson and Tiedje 1984). Higher maximums of total gaseous nitrogen production have been reported for deciduous and tropical forests than for conifers, but the proportion of N_2O to N_2 that is produced increases with decreasing pH, therefore tending to be higher in conifer forests than in deciduous forests (Bowden 1986).

18.11.2 Phosphorus

The chemistry of soil phosphorous is quite complex, but a few general principles can provide good insights into the ecology of the phosphorus cycle. Cycling of phosphorus differs from that of nitrogen in various ways, of which the following are especially significant for plant nutrition:

- Generally at least one-half of soil phosphorus (excluding that in primary minerals) occurs in inorganic forms, a much higher proportion than that of nitrogen. Of the organic forms, 20 to 30 percent is contained in microbial biomass, again a much higher proportion than for nitrogen (about 2%) (Chapin et al. 2002).
- Inorganic soil phosphorus tends to form highly insoluble chemical combinations or to be strongly absorbed to clays, so it may not be readily available to plants even when large amounts are present. Because of this, mycorrhizal fungi are particularly important in the phosphorus nutrition of plants.
- Except for a few special compounds produced by some protozoa, phosphorus within organic compounds is not attached directly to carbon (as is the case with nitrogen) but rather occurs in ester bonds (C-O-P). This means that whereas nitrogen cycling is tied directly to the decomposition of carbon compounds, phosphorus can be cleaved from organic compounds without breaking a carbon bond and therefore is not directly tied to decomposition. (However, decomposition may make ester bonds more accessible to enzymes.)

The chemistry of soil phosphorus differs between those soils with a permanent cation exchange capacity (i.e., permanent charge; chapter 14), and variable-charge soils (i.e., having either a cation or anion exchange complex depending on pH) (Sollins et al. 1988). In the former, which include most temperate and boreal zone soils, inorganic phosphorus occurs primarily as **orthophosphate,** which is a salt formed when some metallic cation (in soils, primarily either iron, aluminum, or calcium) substitutes for the H^+ in phosphoric acid (H_3PO_4). Iron, aluminum, and calcium phosphates are highly insoluble; and because solubility of an element plays a major role in its availability to plants and microbes, phosphorus nutrition is strongly influenced by the degree to which it is bound to these elements, which, in turn, is controlled by pH. Figure 18.19 shows the relation between pH and availability (as measured by solubility) of inorganic phosphorus in permanent-charge soils. Within the pH range common in most forest soils (i.e., 4 to 6); a high proportion of inorganic phosphorus is precipitated as insoluble iron and aluminum phosphates. In variable-charge soils (including in particular the highly weathered soils of the humid tropics), phosphates bind to iron on clays rather than precipitating as insoluble complexes (Sollins et al. 1988). Although the end result is much the same in permanent-charge and variable-charge soils (i.e., low phosphorus availability) the differing chemical pathways by which this occurs have significant implications for phosphorus nutrition. While adjusting pH (e.g., by liming) frequently alleviates phosphorus deficiencies in temperate zone soils, it may have little or no effect in highly weathered tropical soils (Sollins et al. 1988). Mycorrhizal fungi are an important aid to plant phosphorus nutrition in both situations (section 11.18.2).

Organic phosphorus occurs in a variety of forms in soils, including microbial and plant biomass and humus. Biomass

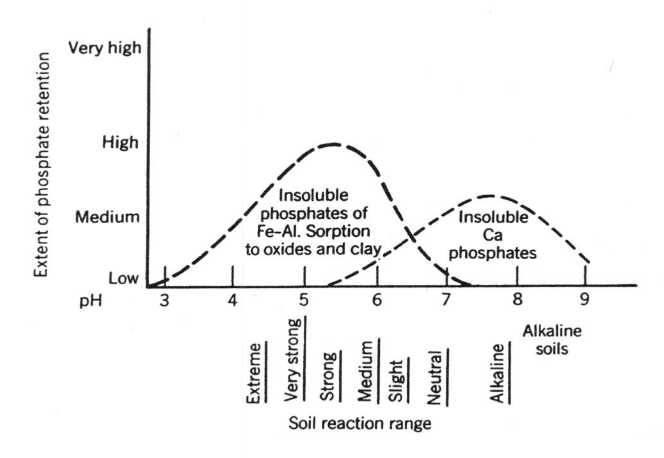

Figure 18.19. The effect of pH on the solubility of inorganic phosphorus in permanent-charge soils. In variable-charge soils pH may have little effect on the solubility of phosphorus. (Adapted from Stevenson 1986)

phosphorus is considered to be the most active fraction in terms of cycling rate (Stewart and Tiessen 1987). Levels of soil organic phosphorus correlate positively with soil carbon and nitrogen; however, the relation is not nearly as tight as that between carbon and nitrogen. Unlike nitrogen and sulfur, phosphorus contained in humus does not occur within the structural matrix of humic compounds. Fulvic acids, which contain a higher proportion of phosphorus than humic acids (Stevenson 1986), are more soluble than the larger humic acids, which may facilitate plant uptake of organic phosphorus but also could increase leaching of phosphorus from upper soil layers.

18.11.3 Sulfur

The sulfur cycle in forests received little attention before acid rain began dumping large amounts of sulfur on forests. Sulfur has some of the cycling characteristics of both nitrogen and phosphorus. In soils, it occurs in both organic and inorganic forms, the latter primarily sulfates (SO_4^{-2}). In unpolluted forests, most soil sulfur is in organic forms (predominantly amino acids or proteins in which sulfur is bonded directly to carbon) and compounds with ester-bonded sulfates.

As with phosphorus, inorganic sulfur may combine with iron and aluminum in low-pH soils to form insoluble complexes. However, sulfur is not adsorbed as strongly in soils as phosphorus and is displaced from adsorption sites by phosphorus (Vitousek et al. 1988). Sulfur adsorption by soils is inhibited by organic matter, and surface soils (which generally are high in organic matter) retain little inorganic sulfur (Johnson 1984; Stevenson 1986;). Lower soil horizons, on the other hand, may adsorb sulfates, an important ecosystem process because it prevents leaching losses.

18.12 PLANT UPTAKE

18.12.1 Some General Principles

Plant nutrient uptake involves two things: (1) movement of the nutrient to the plant's absorbing surface (or conversely, movement of the absorbing surface to the nutrient), and (2) absorption. Plant absorbing surfaces include nonmycorrhizal root tips, mycorrhizal root tips, and mycorrhizal hyphae. Nutrients move to absorbing surfaces either by mass flow within the soil water column or diffusion. Nutrients are also transported with moving soil and carried around by soil fauna. The importance of fauna in moving nutrients to plant absorbing surfaces is not known, but it could be appreciable. These organisms are quite mobile and frequently visit rhizospheres to graze on bacteria and fungi.

The degree to which a nutrient moves by mass flow versus diffusion depends on numerous factors, including its solubility in water (which is, in turn, pH dependent), its concentration in the soil, soil texture and the type of clay mineral, soil organic matter, and water availability. The effects of these factors can be briefly summarized by saying that if a nutrient is soluble at a given pH and not held on the exchange complex or trapped irreversibly within clay minerals, and if transpiration is pulling a water column toward a root or other absorbing surface, the nutrient has a good chance of being carried toward the absorbing surface in mass flow. As we shall see, these conditions are seldom met for most nutrients.

Mobile nutrients are those such as NO_3^- that are highly soluble and move readily in the water column. Insoluble compounds such as iron and aluminum phosphates are **immobile,** moving primarily by diffusion. Realizing that many factors influence nutrient mobility in soils, a rough ranking for a loamy soil with a pH below 6 is as follows (this ranking applies only to soil; mobility in plant tissues may be quite different):

- Highly mobile: NO_3^-
- Moderately mobile: Ca^{2+}, Mg^{2+}
- Moderately immobile: NH_4^+, K^+, Fe^{2+}, Mn^{2+}
- Highly immobile: Fe^{3+}, Mn^{3+}, phosphorus in combination with iron and aluminum

Mobility of the cations is reduced at least in part because they are held on the exchange complex; mobility of anions would be reduced in soils with variable charge (positive and negative exchange sites). Solubility of the micronutrients is highly dependent on their valency state, the more reduced forms (fewer positive charges) being the most soluble. Valency state, in turn, depends on pH, and all micronutrients except molybdenum become more soluble as pH declines (down to about 5.0) (Stevenson 1986).

Understanding the degree to which nutrients are supplied by mass flow versus diffusion is valuable, because it provides insights into how much energy a plant must invest in absorbing surface to maintain adequate nutrition. Mass flow can be important for nutrients that are abundant or required in small amounts (Chapin et al. 2002), but for the most part seems relatively unimportant in supplying nutrients to plants (Chapin 1980). McColl (1973), for example, found no relation between transpiration and nutrient uptake by ponderosa pine seedlings. With the possible exception of NO_3^-, calcium, and magnesium, nutrients further away than about 1 cm cannot diffuse to an absorbing surface fast enough to keep up with plant uptake (Nye and Tinker 1977).[20] The volume of soil surrounding an absorbing surface is quickly depleted of available nutrients, One exception is when nutrients are particularly abundant, in which case a high proportion of exchange sites within the soil are occupied, meaning that relatively few vacant negative charges exist that would pull a cation out of solution.

Although some nutrients may be adequately supplied through mass flow some of the time, other nutrients never

[20]Chapin et al. (2002) discuss factors affecting nutrient diffusion.

are under natural conditions. Hence, in all forests (and particularly in those growing on infertile soils), trees must produce extensive and well-distributed absorbing surfaces within the soil. The energy cost of this is potentially very large but can be reduced if absorbing surfaces have a high ratio of surface area to mass. Trees do maintain a higher root:shoot ratio on infertile than on fertile sites (chapter 15). However, the high surface area per unit of mass associated with long, skinny, absorbing surfaces (i.e., mycorrhizal hyphae) compared with short, fat ones (i.e., roots) dictates that trees invest much of their belowground energy in mycorrhizal fungi.

18.12.2 The Role of Mycorrhizal Fungi in Plant Nutrition

By far, the majority of the nutrient-absorbing surface of trees in natural forests is mycorrhizal (fig. 18.20). It does not follow that all absorbing surface is mycorrhizal, however; even heavily mycorrhizal trees are likely to have some nonmycorrhizal roots. Moreover, trees may alter the proportion of mycorrhizal versus nonmycorrhizal roots according to nutrient availability (Fahey 1992).

Recall from chapter 11 that a mycorrhiza is a combination of fungal and root tissue; however, the primary interface between mycorrhizal plants and the soil is not the mycorrhiza itself, but the clusters of hyphae (called *mycelia*) that radiate from the mycorrhiza. As we saw earlier, the biomass represented by these extraradical hyphae can be quite large; however, length is a better indicator of absorbing surface than biomass. Using special chambers that allowed mycelial development to be recorded, Read and Boyd (1986) measured from 10 to 80 m of mycorrhizal hyphae per 1 cm of fine-root length (Scots pine seedlings), or conservatively, approximately 200 m of hyphae per 1 g of dry weight of soil—a value two hundred thousand times greater than fine-root lengths measured in spruce and pine plantations (Read 1991a). Friend et al. (1987) found similar hyphal densities using quite a different approach; they measured the ingrowth of mycorrhizal hyphae into vermiculite cores placed in soils of Douglas-fir stands, finding from 7 to 200 m of hyphae per 1 g of vermiculite (the smaller values were in stands that had been fertilized). The degree to which such values can be extrapolated to real soils in real forests is unclear, but it seems safe to say that the lengths of mycorrhizal hyphae in forest soils are several orders of magnitude greater than the lengths of fine roots.

Mycorrhizal fungi influence plant nutrition in various ways: increasing absorbing surface, releasing chelators that solubilize metal nutrients, altering the absorbance characteristics of roots, increasing the lifetime of roots, decomposing organic matter and directly transferring nutrients from the organic matter to the plant, transferring nutrients among plants through hyphal linkages, protecting plants from root pathogens, and improving soil structure.

That mycorrhizal hyphae act as absorbing surfaces for the plant has been demonstrated many times. Mycorrhizae

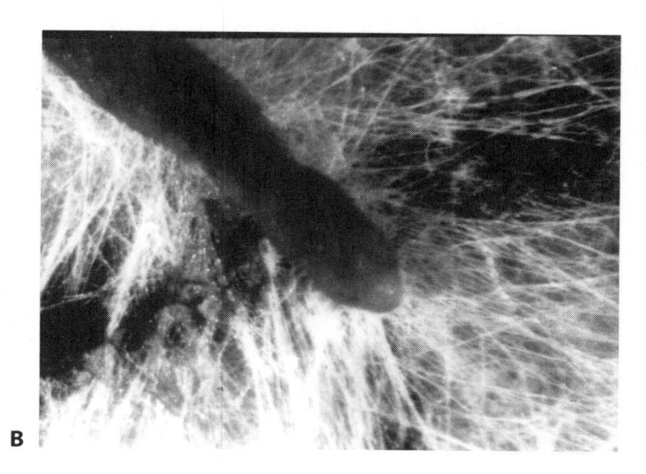

Figure 18.20. A. An autoradiograph of the roots and mycorrhizal hyphae of a Scots pine seedling. Roots are bold white, hyphae are finer white. The autoradiograph is taken by feeding the pine seedling radioactive carbon, then registering emissions on a photographic plate. B. Mycorrhizal hyphae emanating from the root tip of a ponderosa pine. (A, courtesy of David Read; B, courtesy of Dan Durall)

have been shown to increase plant uptake of phosphorus, nitrogen (NH_4^+, NO_3^-, and organic nitrogen), potassium, calcium, sulfur, iron, zinc, and copper and also to enhance rates of nitrogen fixation by nodulated plants (see the review by Smith and Read 1997). Mycorrhizae do not always enhance uptake of these nutrients, however; this depends on environmental conditions, mycorrhizal species, and plant species. In particular, while pot studies show that vesicular-arbuscular mycorrhizae enhance plant phosphorus uptake, demonstrating that in the field has been difficult. Smith and Read (1997) speculate that field results may differ from those of controlled studies because of other resource limitations (e.g., water) or grazing on vesicular-arbuscular mycorrhizae hyphae by soil animals in field environments.

To get some appreciation of the surface area gained by investing in mycorrhizal fungi as opposed to roots, consider the following. The diameter of an absorbing root is roughly 1 mm and that of a mycorrhizal hypha approximately 0.003 mm (Hunt and Fogel 1983). Because the ratio of surface area to volume of a cylinder increases as 1/radius, a hyphae has approximately 300 times more surface area per unit of volume than a small root. Ignoring factors such as different respiration rates and turnover between fine roots and hyphae, it can be roughly calculated that a net investment of 1 kg of carbon would yield about 8 m^2 of surface in fine roots, and approximately 2,400 m^2 in hyphae. (The surface area of a human lung is about 100 m^2.)

As an exercise, make a crude calculation of mycorrhizal hyphal area index, or the surface area of mycorrhizal hyphae per unit of surface area of ground (in analogy with leaf area index; chapter 15). Assume that beneath 1 m of forest soil surface, to a depth of 15 cm (generally the zone of most intense rooting), there is an average of 200 m of mycorrhizal hyphae per gram of dry weight of soil (i.e., the most conservative estimate of Read and Boyd). At an average hyphal diameter of 0.003 mm and a soil bulk density of 0.25, that works out to a hyphal surface area of 70 m^2 per m^2 ground surface! Going through the same exercise with Friend et al.'s lowest estimate of hyphal lengths (i.e., 7 m per 1 gram of dry soil in a fertilized stand) yields a hyphal surface area of 2.5 m^2 per m^2 of ground surface. In comparison, all-sided leaf areas range from approximately 3 to perhaps 15 m per 1 m in forests. Although crude, these calculations show that hyphal surface is at least comparable to (and perhaps significantly greater than) leaf area.

Ectomycorrhizae, with their extensive hyphal mantle that covers the root, store nutrients, doling them out to the host slowly over time; this has been shown for phosphorus and potassium (Smith and Read 1997). The net effect is that ectomycorrhizae enhance the ability of plants to take up and store at least some nutrients in excess of their immediate needs, thereby smoothing seasonal variations in availability.

Uptake and storage of phosphorus provides another example of how interactions among different nutrient elements influence the nutrient cycle. It has long been known that divalent and polyvalent cations (e.g., calcium, magnesium, aluminum) enhance phosphorus uptake by both mycorrhizal and nonmycorrhizal roots. As Strullu et al. (1986) explain, both the rate of phosphorus uptake and the total amount taken up is influenced by the concentration of soluble phosphorus within cells of the root and mycorrhizal hyphae. It is a simple principle of diffusion; the more of some molecule there is inside; the slower the net movement of molecules of the same type in from outside. In combination with calcium, magnesium, or aluminum, however, phosphorus forms insoluble granules within cells, effectively reducing the internal phosphorus concentration and thereby increasing the amount that is taken up. This same dynamic holds in both mycorrhizal and nonmycorrhizal roots, the primary benefit of ectomycorrhizae probably being that the sheath greatly increases the number of cells available for storage.

18.12.2.1 Production of Chelators

Recall from chapter 14 that a variety of organic compounds released into the soil by plants, mycorrhizal fungi, lichens, and soil microorganisms bind strongly to divalent and polyvalent cations. The binding is called **chelation** (from the Greek word for "claw," so named because of the manner in which the organic compound surrounds the cation). The organic molecule is termed a **chelator,** and the chelator together with its bound cation is called a **chelate complex.** Some of the more active chelators are associated in particular with mycorrhizal fungi and rhizosphere microbes.

Chelation is quite important in the nutrient cycle for three reasons in particular:

1. Solubility (and therefore availability) of heavy metal cations such as iron, aluminum, zinc, and manganese is increased by chelation.
2. Chelation of iron and aluminum releases phosphorus from insoluble complexes with heavy metals, thereby increasing phosphorus availability to plants, and chelation also speeds the weathering of primary minerals (Pohlman and McColl 1988). Considerable evidence

suggests that mycorrhizal plants can exploit sources of soil phosphorus that are unavailable to nonmycorrhizal plants, including phosphorus contained within the primary mineral apatite (Smith and Read 1997). It seems likely that this results at least in part from iron and aluminum chelators produced by the fungi.

3. Some iron chelators act as a sort of immune system within rhizospheres, defending plants against invading pathogens.

Two chelators in particular have received considerable attention regarding their effects on the nutrient cycle: oxalic acid and the very strong, iron-chelating proteins called **siderophores,** which literally means "iron carriers." Although oxalic acid is produced by plants, fungi appear to be the major source in soil. As far as is known, siderophores are produced only by mycorrhizal fungi and other microbes.

Oxalic acid combines with calcium to a form a crystalline molecule called **calcium oxalate.** Figure 18.21 shows calcium oxalate crystals coating the hyphae of a mycorrhizal fungus; in the Douglas-fir forest where the photo was taken, 50 percent of the exchangeable calcium in the soil was contained in oxalate crystals (Cromack et al. 1979). Oxalic acid has been found to enhance calcium uptake by tree seedlings (Arp and Ouimet 1986), and the sparingly soluble oxalate crystals act to prevent calcium loss from the system through leaching (Graustein et al. 1977). Oxalic acid also binds strongly to iron and aluminum, and soil minerals in the vicinity of oxalate-producing mycorrhizal mats weather more quickly than minerals that are not near mats (Cromack et al. 1979).

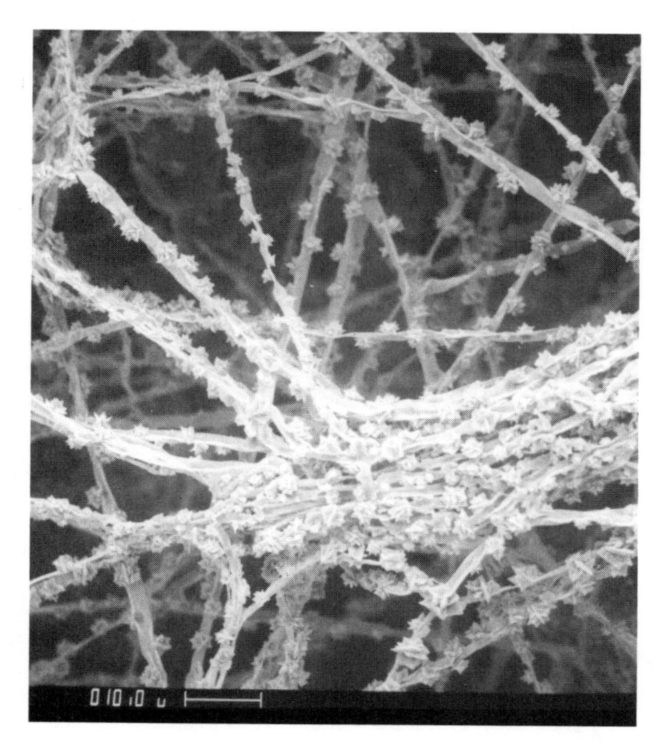

Figure 18.21. Calcium oxalate crystals coating the surface of mycorrhizal hyphae. (Courtesy of Kermit Cromack; scanning electron micrograph by Alan Pooley)

Siderophores are remarkably strong Fe^{3+} chelators (but not Fe^{2+}); they have been reported to pull iron out of Pyrex glassware in laboratories. As such, they undoubtedly are important in weathering minerals and increasing the availability of iron and phosphorus. Siderophores also function in plant defense, binding iron so tightly that root pathogens are unable to satisfy their requirements. (A siderophorelike molecule that is present in blood performs a similar defensive function in mammals.) A variety of different siderophores with different functions are produced by microorganisms; Bossier et al. (1988) reviewed the ecological significance of siderophores in soil.

Numerous microorganisms produce siderophores, but evidence suggests that mycorrhizal fungi and rhizosphere bacteria are the primary sources. Thus, plant-derived energy is the ultimate indirect source of siderophores in soil, and in fact, siderophore concentrations were found to be considerably lower in 8 of 10 Oregon clear-cuts than in adjacent forest soils (Perry et al. 1984). Again, we see a mutualistic relationship between plants and soil microbes. Plants supply energy in the form of reduced carbon compounds, and mycorrhizal fungi and other siderophore-producing microbes supply iron to the plants. Microbes are probably essential for the iron nutrition of dicots but not monocots (e.g., grasses), which apparently can gather iron on their own (Bossier et al. 1988).

18.12.2.2 Direct Cycling

That mycorrhizal fungi might decompose organic matter and transfer nutrients directly to the plant host was first suggested by Frank in 1894. Frank, who coined the term **mykorrhizen** in 1885, noticed that mycorrhizae proliferated in the presence of organic matter (an observation that has been substantiated many times since). In what came to be known as the **organic nitrogen theory,** he proposed that the fungus transferred organic nitrogen compounds directly from litter to host.

Frank studied ectomycorrhizae in the temperate forests of Europe. Much later, when the study of mycorrhizae was extended into the tropics, Went and Stark (1968) noted that vesicular-arbuscular mycorrhizae in moist tropical forests ramified throughout the thin litter layers and suggested that these mycorrhizae also decomposed litter and shipped the nutrients to hosts. (The term **direct cycle** was first used by these researchers.)

The idea of direct cycling has been quite controversial, particularly in the case of nitrogen, mainly because of skepticism about the ability of mycorrhizal fungi to produce the enzymes necessary to break carbon bonds and release nitrogen. (As discussed earlier, phosphorus is bound to organic matter through ester bonds, and both mycorrhizal fungi and plants produce the phosphatase enzymes necessary to cleave phosphorus directly from organic matter.) However, studies leave little doubt that at least some of the fungi that form ecto- and ericoid mycorrhizae take up and transfer proteins to host plants, something that roots of nonmycorrhizal plants cannot do (Smith and Read 1997). Whether litter-degrading enzymes are produced by the mycorrhizal fungi or the fungi stimulate the activity of saprophytes has been uncertain. Strong evidence indicates that mycorrhizae inhibit rather than stimulate saprophytes, perhaps by reducing water availability (Koide and Wu 2003). Moreover, it now seems clear that some ectomycorrhizal fungi produce the enzymes necessary to decompose organic matter (Smith and Read 1997). In contrast to ecto- and ericoid mycorrizae, there no evidence that vesicular-arbuscular mycorrhizae can decompose litter (Smith and Read 1997).

> The ability of mycorrhizal fungi (or associated bacteria) to break down litter and extract its nitrogen might explain some puzzling observations that have been made over the years. One is the **Gadgil effect,** first described by Gadgil and Gadgil (1971): when mycorrhizal fungi are present, saprophytic organisms do not decompose litter very well. The Gadgils suggested that this is because the mycorrhizae extract nitrogen from the litter, leaving the saprophytes nitrogen limited. On the other hand, Dighton et al. (1987) found the reverse of the Gadgil effect in their study: roots and mycorrhizae enhanced rather than retarded litter decomposition. The moral to this story is that nature is highly variable, and one should be very careful about extrapolating an observation too far.

18.12.2.3 Diversity in the Mycorrhizal Fungal Community

In general, species of mycorrhizal fungi differ in the benefits they confer on their host plant, including their effects on nutrient uptake. Considering that several thousand species of fungi form ectomycorrhizae and several hundred species vesicular-arbuscular mycorrhizae, and that considerable genetic diversity likely exists within any given species, it seems probable that mycorrhizal fungi in any given forest ecosystem occupy a large variety of functional niches (Allen et al. 1995). As we discussed in chapter 11, this diversity is likely to be of considerable benefit to the plant, because it magnifies the ways that the plant can respond to and deal with its environment. Relatively few experiments have addressed the issue, particularly with trees. Arbuscular mycorrhizal diversity has been shown to increase plant productivity in grasslands (van der Heijden et al. 1998). Jonsson et al. (2001) and Baxter and Dighton (2001) found that ectomycorrhizal diversity improved one or another aspect of growth or phosphorus nutrition of birch species. In their study, Baxter and Dighton were able to show that the effect stemmed from diversity per se rather than the inclusion of one or a few high-performing ectomycorrhizal species as diversity was increased.

Some ectomycorrhizal species form extensive hyphal networks (called **mats**) that frequently occur at the interface between the F and H layers (chapter 14). Mycorrhizal mats are easily seen as dense gray or whitish hyphae that are revealed by pulling the L and F layers hack from the H. Mycorrhizal mats are very efficient at extracting nutrients from water that passes through them. According to Alexander (1985), the nitrogen concentration of the soil solution drops by approximately 70 percent in its passage through heavily mycorrhizal surface horizons. Mycorrhizal mats impose many unique chemical and biological patterns on soil; studies in Oregon forests show that compared with nonmat soils, those occupied by mats have lower microbial biomass, higher levels of nitrogen fixation, faster litter decomposition, greater amounts of labile carbon, and more available phosphorus, potassium, and magnesium (Griffiths et al. 1990; Entry et al. 1991; Aguilera et al. 1993). On the other hand, pH and denitrification rates are lower in mat than in nonmat soils, as is labile nitrogen. Aguilera et al. (1993) hypothesize that labile nitrogen is lower in mat soils because it is more rapidly taken up and shipped to trees.

18.12.3 Rhizospheres

Recall from chapter 14 that the rhizosphere is the area surrounding a root that is influenced by that root. There are also mycorrhizospheres around mycorrhizal roots and mycospheres around mycorrhizal hyphae (e.g.,mycorrhizal mats are essentially all mycosphere). For simplicity, we refer to all of these as the **rhizosphere** in the following discussion. All are characterized by exudation of organic compounds and sloughing of tissues that create a unique chemical and biological environment that, in turn, profoundly influences plant nutrition. The rhizosphere is a factory, powered by plant energy and doing work that benefits the plant. It is also a filter, mediating and shaping interactions between the plant absorbing surface and the bulk soil. Most rhizosphere research has been performed on crop plants, and much of what can be said about tree rhizospheres must be extrapolated from this work.

18.12.3.1 Rhizosphere pH

Plants alter the pH of soils within their rhizospheres. This is largely a spin-off effect of the need to maintain charge balance within plant tissues (i.e., if a cation is taken up, a cation must be released into the soil; if an anion is taken up, an anion must be released). The cation that is released is primarily H^+ attached to an organic acid, although others may also be released, such as K^+ in exchange for NH_4^+ (Bledsoe and Rygiewicz 1986). OH^- and HCO_3^- are released as anions. Mycorrhizal and nonmycorrhizal roots differ in the types and amounts of ions that they release into the rhizosphere, which is not a surprise as they also differ in uptake characteristics. Bledsoe and Rygiewicz (1986) found that compared with nonmycorrhizal seedlings, mycorrhizal Douglas-fir released less K^+ and more H^+ and HCO_3^-. Whether mycorrhizal or not, trees usually take up more cations than anions, in the process releasing more H^+ than OH^{-a} and thereby acidifying the rhizosphere relative to bulk soil (Nilsson et al. 1982). The strength of this effect depends on the relative utilization of NH_4^+ versus NO_3^- Acidification in rhizospheres is increased by the high levels of biological activity, and hence respired carbon dioxide which combines with soil water to produce carbonic acid (chapter 14).

Up to a point, increased acidity in the rhizosphere benefits plant nutrition. As we discussed earlier, solubility of most metals increases as pH declines. Thus, acidifying the rhizosphere has the twofold effect of increasing availability of metal nutrients and freeing phosphorus from insoluble complexes with iron and aluminum. Consistent with their relatively high ability to acidify rhizospheres, mycorrhizal roots appear to be much more effective metal solubilizers than are nonmycorrhizal roots (Cairney and Ashford 1989). Another benefit of acidification relates to electrical fields: the proton gradient created between the inside and outside of the plant absorbing surface results in an electrical potential that is utilized by the plant to transport nutrients across membranes (Uribe and Luttge 1984).

Acidifying the rhizosphere can also affect plants adversely by increasing leaching losses of soluble cations and bringing toxic levels of aluminum and other heavy metals into solution. Production and release of oxalic acid (with its strong tendency to bind with calcium) is probably an adaptation of plants and mycorrhizal fungi to retain calcium within the acidic environment of the rhizosphere. Regarding heavy-metal toxicity, it is perhaps not surprising that at least some plant species occupying acidic soils are relatively tolerant of aluminum. This is true for ericaceous plants, montane conifers, and eucalypts (chapter 16).

18.12.3.2 Rhizosphere Communities

As we discussed in chapter 14, most microbial activity within soil occurs in rhizospheres and particularly mycorrhizospheres (i.e., zones around mycorrhizal roots), which may support from 10 to 100 times more bacteria than nonmycorrhizal roots (Foster 1986). Mycorrhizal mats are apparently an exception to this; Aguilera et al. (1993) found that mats had fewer bacteria than nonmat soils. On the other hand, mycorrhizal roots have fewer pathogens than nonmycorrhizal roots.

Apparently, many microbial types exist dormant within the soil until a root or mycorrhizal hypha comes by, at which time they become active. Synergistic relationships develop between mycorrhizal fungi and some groups of bacteria, the bacteria stimulating mycorrhiza formation, and the mycorrhiza, in turn, shaping the composition of the rhizosphere community through the types of organic compounds they exude (Rambelli 1973; Bowen and Theodorou 1979; Meyer and Linderman 1986; Garbaye and Bowen 1987; Chanway and Holl 1992; Strzelczyk et al. 1993).

Rhizosphere microbes have a profound effect on plant nutrition (Klein et al. 1990; chapter 14 in this volume). They release chelators (e.g., siderophores), enzymes, and hormones; fix nitrogen; and decompose organic matter. Work with both domestic and wild grasses shows that rhizosphere bacteria draw on their ready source of plant-derived energy to out compete saprophytes for nitrogen in litter. The nitrogen thus extracted is first incorporated into the biomass of rhizobacteria and then released and made available to the plant through grazing by protozoa (Coleman et al. 1984; Clarholm 1985). These same patterns likely hold in tree rhizospheres as well.

It should be remembered that not all rhizosphere microbes benefit plants; some are pathogens, for example. Nor does any one microbial type necessarily benefit the plant all of the time. Like all interactions among organisms within ecosystems, those in the rhizosphere are dynamic, complex, and poorly understood.

The ability of microbial communities to grow rapidly provides an important nutrient-conservation mechanism in ecosystems. Vitousek and Matson (1985) showed that soil microbes sequestered nitrogen after a loblolly pine forest was clear-cut, thereby acting as a biological dam that prevented leaching loss following removal of the plants. Even in intact forests, microbes may be important in retaining nutrients during periods when tree uptake is low. In hardwood forests of the northern United States, for example, uptake by soil microbes reduces nitrogen loss during early spring, when nitrogen uptake by trees is still low (Zak et al. 1990).

> The ability of microbes to sequester nutrients depends on their having an adequate source of carbon. Might trees in northern hardwood forests facilitate nutrient conservation during early spring by feeding the microbes carbon (via root exudates)? It would not be a bad strategy. Putting carbon into foliage would be risky at this time because of the chance of frost. The low carbon:nitrogen ratios of bacteria would allow much more nitrogen to be sequestered per unit of carbon invested than would be the case in woody tissues.

In environments with a distinct dry season, microbial biomass may expand during drought and contract with the arrival of rains—a seasonal pattern opposite that of plants.

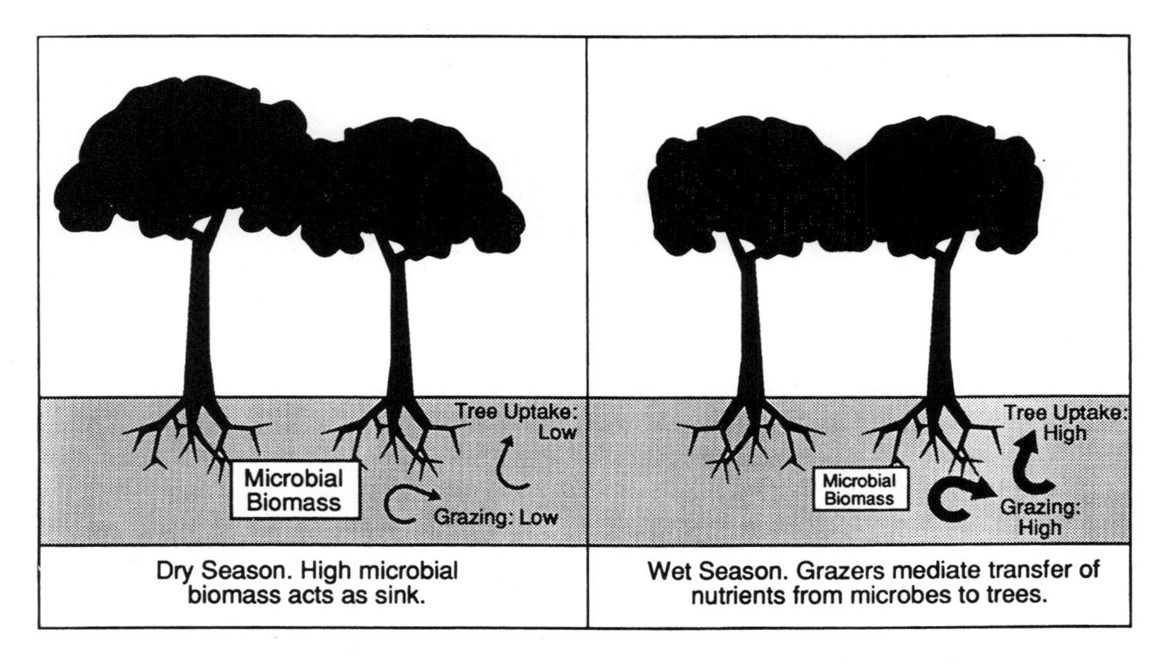

Figure 18.22. An illustration of how climate might interact with the belowground food chain to alter nutrient fluxes (as hypothesized by Singh et al. [1989] for seasonally dry tropical forests in India). During the dry season, microinvertebrate grazers (e.g., protozoa) cannot move freely through the soil; hence, microbial biomass and nutrient content grow. Rains provide the water films that allow grazers to move, and grazing, in turn, reduces microbial biomass and frees nutrients for uptake by trees.

Singh et al. (1989) documented this phenomenon in dry tropical forests and savann as of India and offered the following hypothesis. Microbes are less sensitive to drought than either plants or protozoa; hence nutrients released through death of fine roots and mycorrhizal hyphae during the drought are sequestered by the expanding microbial populations and protected because the protozoa that graze microbes are relatively inactive. Grazers become more active with the commencement of rains, thereby mineralizing nutrients that are subsequently taken up by the plants (which are also initiating growth during the rainy season) (fig. 18.22). Why would microbes be less sensitive to drought than plants and protozoa? The answer lies in the relationship between the size of soil pores and the tenacity with which they retain water (chapter 14). Bacteria are small enough to survive within the small pores that serve as water reservoirs during drought. While protozoa may be able to enter these small pores, their ability to move from pore to pore in the search for food requires a continuous water layer that does not occur during drought. Mycorrhizal hyphae can probably enter relatively small pores to obtain water for their plant hosts, but at some point during soil drying, even hyphae will be unable to access water that is still available to bacteria.

18.13 NUTRIENT CYCLING THROUGH SUCCESSION

Van Cleve et al. (1991) identified a sequence of what they termed **turning points** during succession in boreal forests of Alaska. This term is "to emphasize that in relatively short time intervals (less than 50 years) critical changes in ecosystem structure are accompanied by functional changes that have far-reaching effects on ecosystem development" (Van Cleve et al. 1991). The functional changes they identified were related to element cycling. Details varied depending on site and severity of disturbance, but the same general pattern could be identified in forests on both north and south slopes:

- Fire creates the potential for large leaching losses, but the flush of available nutrients from the fire allows early successional plants to rapidly grow and regulate nutrient loss. A hardwood tree cover develops, resulting in substantially more nutrients being stored in living biomass, as well as an increased rate of intrasystem cycling and the development of a forest floor. The combination of accumulating organic matter and increased biological activity reduces soil pH (which increases after fire).
- Transition from hardwoods to conifers allows a moss layer to develop on the forest floor (mosses cannot grow through heavy hardwood litter). Mosses insulate the soil, thereby greatly reducing nutrient cycling and (on north slopes) leading to the development of a permafrost layer. Probability of fire increases because of the accumulating dead biomass and increasing dominance of conifers.

There are many possible variations on this theme. On floodplains in Alaska, for example, the initial turning point is related to physical rather than biotic factors; sediments deposited by periodic flooding raise the height of river terraces, thereby allowing alder to establish and initiate the process of soil building by fixing nitrogen (Van Cleve et al. 1991). The next turning point, a shift from alder to balsam poplar, is initiated by an inhibition of alder nitrogen fixation and general slowing of the nitrogen cycle, both effects resulting from secondary chemicals (tannins and other phenolics) produced by the establishing poplar (Schimel et al. 1998). What is a critical turning point in one forest type may be of no consequence in another; however. For instance, reduced nutrient cycling because of cold soils is hardly a factor in tropical and subtropical forests, but rapid nutrient cycling because of warmth and moisture is. Embedded within all the variability, however, at least two critical points are common to all secondary successions:

- *Earliest stages of succession: damming the nutrient flow and replenishing capital.* In the initial stages following disturbance, rapid growth of microbes and early successional plants immobilizes nutrients, protecting them from loss. In some ecosystems, nitrogen-fixing plants replenish nitrogen capital. The ability of microbes to immobilize nutrients while plants are recovering depends critically on the availability of carbon (chapter 17).
- *Early to midsuccession: reestablishing the cycle.* During this phase, plants increasingly depend on the nutrient cycle to supply their nutrient needs.

How nutrient cycling may change as forests enter the late successional phase is a matter of some debate and probably varies widely from one forest type to another. Odum (1969) hypothesized that ecosystems should retain nutrients more effectively as they mature from early to late successional stages. But Vitousek and Reiners (1975) argued that losses were regulated by uptake in biomass and, therefore, would be lowest in the rapidly growing middle stages of succession. However, patterns are far too variable to fit into a simple successional model (Gorham et al. 1979). Regardless of successional stage, healthy systems retain essential nutrients both through tight coupling between decomposition and uptake, and through diversion of some nutrient capital to slowly available pools (e.g., humus), but not so much that trees become nutrient stressed. (As we saw earlier, complex organic molecules and the nutrients they contain may leak from even healthy forests.)

Net biomass increment is often assumed to slow as forests age; however, that is not always the case. For example, the NPP of older forests in the northern Rocky Mountains is 50% to 100% higher than predicted by models (Carey et al. 2001), and recent work in China found that old-growth forests actively accumulated soil carbon (Zhou et al. 2006). But regardless of how fast old-growth trees are growing, inputs of coarse woody debris to soils increase significantly in old-growth stages, rivaling litterfall in both temperate conifers and hardwoods (Sollins et al. 1980; Fisk et al. 2002). Because of its high carbon:nutrient ratio, coarse woody debris provides abundant carbon but insufficient nutrients for saprophytes, resulting in nutrient transfer from soil pools into the microbes feeding on the wood. Consequently, as forests age the major nutrient sinks (at least nitrogen and probably phosphorus and sulfur) shift from living trees to soil microbes (Fisk et al. 2002). Studies in both conifers and hardwoods have shown that old-growth forests cycle nitrogen more quickly than younger forests (Davidson et al. 1992; Aguilera et al. 1993; Fisk et al. 2002). In conifer forests, the increasing abundance of mycorrhizal mats as forests age is accompanied by an increase in labile soil carbon (Aguilera et al. 1993).

Finally, new questions are raised by the fact that in forests without excessive inputs of nitrogenous pollutants, most nitrogen loss is now known to be via dissolved organic forms that are probably unavailable to organisms (chapter 17). In these cases, a major regulating factor may be the ability of saprophytes and mycorrhizal fungi to produce the enzymes to convert unavailable to available forms (i.e., soluble polymers to monomers).

Earlier, we saw that transitional phases tend to be especially rich biologically, a phenomenon that extends to the soil ecosystem as well. Spatial patterning of nutrient cycling and other soil processes becomes more complex as forests age, reflecting various structural changes in the ecosystem such as accumulating litter and coarse woody debris, increasing diversity in the structure and composition of the biotic community, and increasing presence of mycorrhizal mats. Old-growth forests of the Pacific Northwest, for example, are dominated by an overstory of Douglas-fir but also have patches of maples, cedars, hemlocks, yews, and various understory shrubs. Each of these species has a distinct chemical signature (Killsgaard et al. 1987): Douglas-fir accumulates potassium and phosphorus, cedars and maples accumulate calcium, some other hardwood species accumulate manganese, and hemlock produces relatively acidic litter. The differing chemical signatures of plant species almost certainly influence soil organisms and therefore create patchiness in nutrient cycling. For example, calcium is associated with relatively high decomposition rates, low pH favors fungi over bacteria, and manganese inhibits the allelopathic effect of streptomycetes. Oxides of both manganese and iron can oxidize phenolic acids, which Lehmann et al. (1987) suggest may prevent phenols from reaching phytotoxic levels in soils of older forests. In forests characterized by gap-phase replacement (moist tropical, northern hardwoods; chapter 8), small-scale patchiness in tree species composition is a permanent feature of the landscape and likely to produce corresponding patchiness in nutrient cycling characteristics.. For example, in deciduous forests of eastern North America, sugar maple produces litter with uniquely low carbon:nitrogen ratios and is associated with high rates of nitrification (Lovett and Mitchell 2004).

Accumulations of litter and coarse woody debris in old-growth temperate and boreal forests also likely have complex effects on soil organisms and the nutrient cycle. Direct cycling by mycorrhizal fungi may become more important during this phase of forest development (Griffiths et al. 1990; Dighton 1991). Accumulations of high carbon:nitrogen litter probably enable populations of nonrhizosphere microbes to grow and immobilize nutrients, thereby altering the relative abundance of microbial types in the system. Secondary chemicals produced by litter inhibit some types of mycorrhizal fungi (Schoenberger and Perry 1982; Rose et al. 1983) and reduce net nitrogen mineralization and nitrification (Olson and Reiners 1983; White l986a, 1986b; White and Gosz 1987), although it is not clear in the latter case whether gross rates are also reduced. (Differences between gross and net mineralization rates were discussed in section 18.10.1.)

Trees occupying environments where nutrient cycling proceeds especially slowly may be genetically programmed to demand few nutrients. This is the case with black spruce in Alaska (Van Cleve et al. 1983). In at least some instances, however, trees become stressed by excessive buildup of organic matter, a condition well known in Europe, where it is called **soil sickness** (Florence 1965). Pastor et al. (1987) argued that recalcitrant litter produced by spruce forests would eventually lead to stress and perhaps predispose trees to insect attack. A soil-sickness type of phenomenon occurs in some redwood forests in the United States, where it appears to be relieved by ground fires that mineralize organic matter and rejuvenate microbes (Florence 1965). In chapter 8 we saw that the decline phenomenon seen worldwide in some older forests (called **soil fatigue**) is associated with nutrient limitations, especially phosphorus.

18.14 GLOBAL CHANGE AND NUTRIENT CYCLING

Earlier we discussed how warmer temperatures might affect decomposition and nutrient cycling. Another issue related to global change is atmospheric carbon dioxide. Ecological theory suggests that higher atmospheric carbon dioxide should increase the carbon:nitrogen ratios of litter, leading to greater sequestration of soil nitrogen by microbes and therefore limiting nitrogen supply to plants. Zak et al. (2003) summarized results from three free-air carbon dioxide–enrichment experiments (Duke Forest, Oak Ridge, and Rhinelander). Although higher atmospheric carbon dioxide increased both above- and belowground litter production, no effect on nitrogen cycling had been detected at the time of their study. However, a later study dealing only with the Duke site found that nitrogen mineralization in the high CO_2 treatment was only 50 percent of that in controls, suggesting that negative feedbacks to plant growth could occur in high–carbon dioxide atmospheres (Billings and Ziegler 2005).

18.15 SUMMARY

Nutrients are retained tightly within ecosystems and cycled from one organism to another (i.e., external cycling) and from one tissue to another within individual plants (i.e., internal cycling). With few exceptions, the majority of nutrients used to support primary productivity of all forest types are obtained from external and internal cycling. A very simplified version of intrasystem cycling is as follows. Trees and other plants shed senescent leaves, twigs, fine roots, and mycorrhizal hyphae. In the case of leaves and twigs (but perhaps not roots), at least some nutrients are withdrawn before shedding, stored elsewhere within the tree, and then cycled to new tissues when required. Shed tissues (collectively called *litter*) are consumed by insects and microbes that obtain both energy (i.e., reduced carbon) and nutrients from the litter. These organisms, in turn, may be eaten by protozoa or arthropods or may die and become food for other decomposers.

One of three things may happen to a given nutrient element at each step of this detrital food chain. It may be (1) incorporated into a body or microbial cell (i.e., immobilized), (2) released into the surrounding soil environment (i.e., mineralized), or (3) sequestered in stable humic compounds. Factors controlling humus formation are poorly understood. Whether a nutrient is immobilized or mineralized at a given step depends on its concentration relative to that of carbon in both the food and the feeder. In general, shed plant tissues have abundant carbon and low nutrient concentrations relative to decomposers; hence, nutrients contained within plant litter are immobilized during decomposition. However, decomposers and the organisms that graze on decomposers are similar in their carbon nutrient ratios; therefore, nutrients tend to be mineralized as litter passes up the detrital food chain. Once mineralized, nutrients are available for uptake again by growing plants, thereby completing the cycle. Most nutrient uptake by trees and other plants is through mycorrhizal fungi, which among other things enhance plant nutrient-gathering capacity by greatly increasing the belowground absorbing surface of trees. Some species of mycorrhizal fungi can also decompose organic matter and may cycle nutrients directly from litter to their host plant.

Many details of the intrasystem cycle are omitted from this summary. For example, at least some rhizosphere bacteria live on exudates from plant roots and mycorrhizal hyphae rather than on litter, and these are important in various aspects of the nutrient cycle. Other groups of soil bacteria obtain energy from reduced nitrogen and sulfur compounds rather than from reduced carbon; these chemotrophic bacteria are particularly important in the nitrogen cycle. Organic nitrogen- and phosphorus-containing molecules are increasingly recognized as being important in the nutrient cycle.

Higher temperatures associated with global warming may increase decomposition rates and therefore nutrient release, however that depends on two factors: (i) moisture availability, and (ii) the effect of higher atmospheric CO_2 on the carbon:nutrient ratios of leaf litter.

Plate 1. Old-growth temperate conifer forest, western Oregon.

Plate 2. Temperate deciduous forest, northeastern China.

Plate 3. Dry temperate forest, northern Mexico.

Plate 4. Riverine forest, Rondonia State, Brazilian Amazon.

Plate 5. Dry tropical forest, Zambia.

Plate 6. Temperate *Auracaria*/*Nothofagus* forest, Chile.

Plate 7. Eucalyptus forest, Australia. (Courtesy Michael Castellano)

Plate 8. Montane forest, Glacier National Park, Montana.

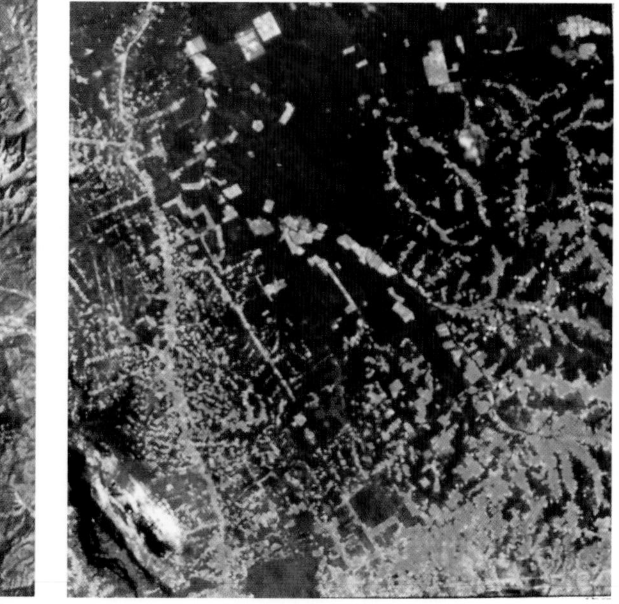

Plates 9 and 10. These two Landsat images show how different features of the earth's surface reflect or absorb light. Dark indicates little reflectance (high absorption), bright indicates high reflectance. *Left:* Western Washington State. The very dark portions on the left are the Pacific Ocean and Puget Sound. Bright red is young forests, mostly plantations established after logging of old growth. The dark portion in the lower right is old-growth forest in Mount Rainier National Park (note the straight line that separates the park from logged areas outside the park). *Right:* A portion of the Amazon Basin, Para State, Brazil. Dark red is primary rain forest. Blue is pastures and croplands established by clearing the forest. Note how forest clearing follows roads. (Reproduced by permission of Earth Observation Satellite Company, Lanham, MD, USA)

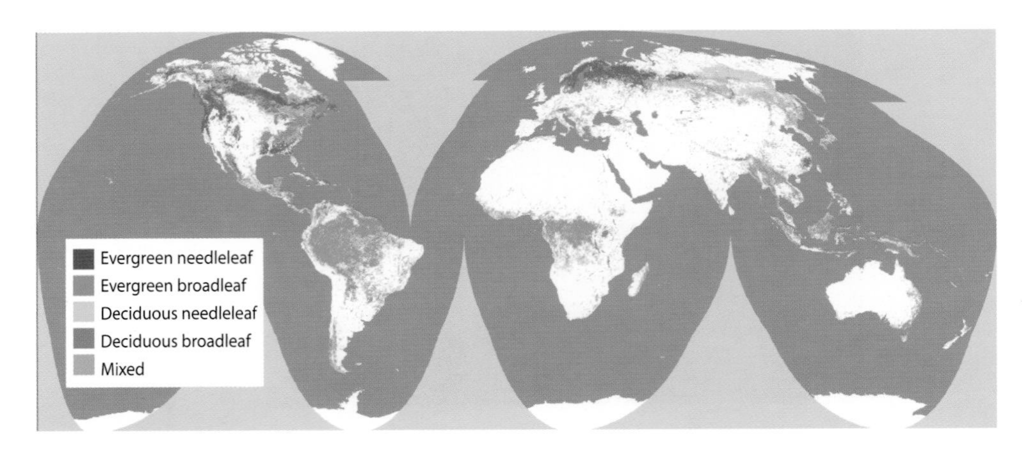

Plate 11. Distribution of the earth's forests. (NASA)

Plate 12. A map of the global variation in the number of angiosperm families per 30,000-km^2 quadrat. (From Francis and Currie 2003; copyright University of Chicago Press, reprinted with permission of the University of Chicago Press)

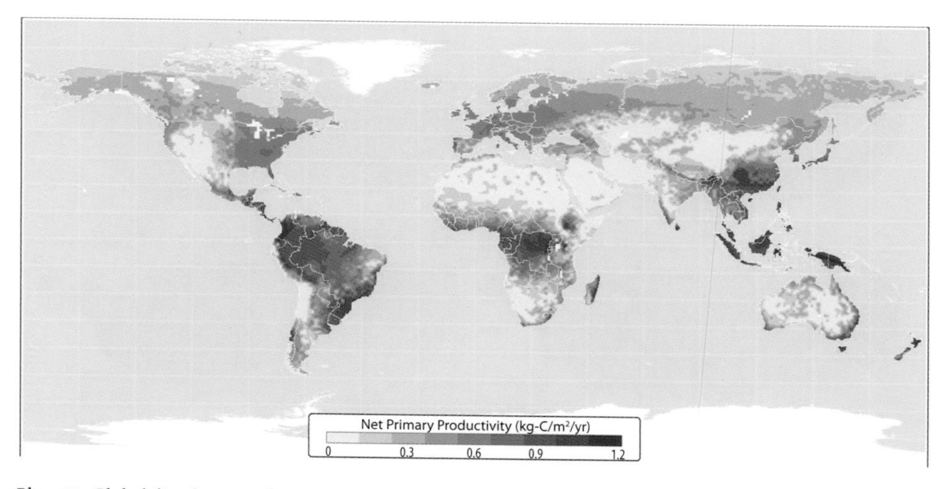

Plate 13. Global distribution of net primary productivity. (Adapted from Kucharik et al. 2000)

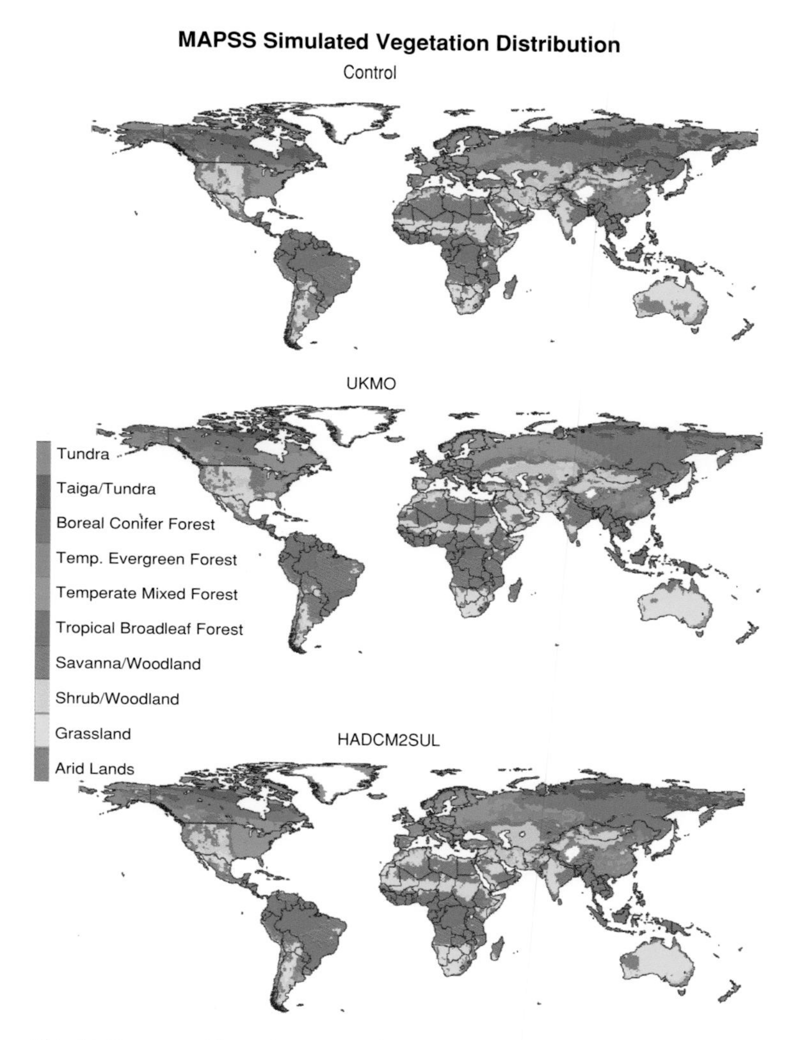

Plate 14. Current and future (2070–2100) distribution of global vegetation as predicted by the MAPSS model. Future prediction uses climate projections from the United Kingdom Met Office's (UKMO) Hadley Centre. The particular model used here is HADCM2SUL (Hadley Climate Model 2, including the effects of increased atmospheric CO_2 and sulfate aerosols). Among the various general circulation models, the HADCM2SUL predicts in the midrange of temperature increase during the twenty-first century. The version of MAPSS shown incorporates the potential physiological effects of increased atmospheric CO_2. (Adapted from Nielson and Drapek 1998)

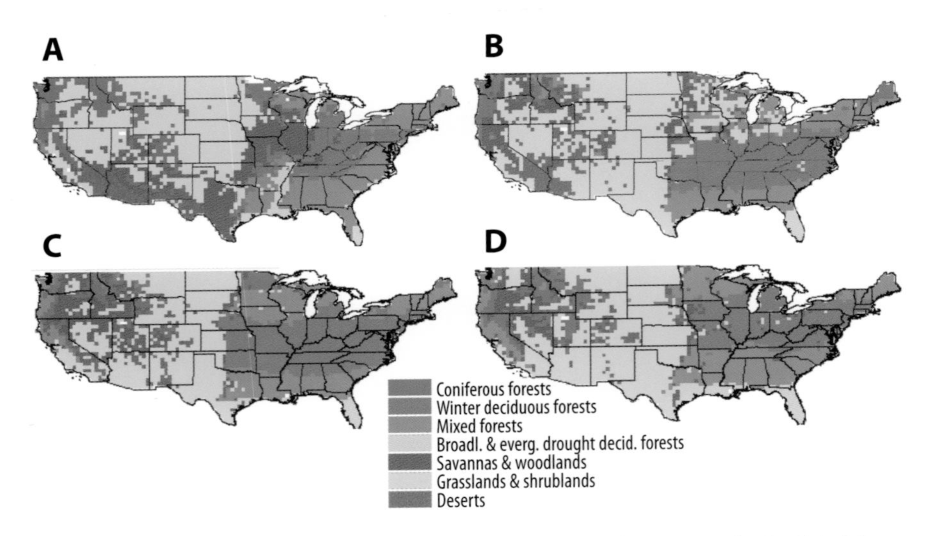

Plate 15. Current and simulated distribution of aggregated vegetation classes in 2095 for the United States. A. Current distribution according to the Kuchler classification. B. Current distribution as simulated by the dynamic vegetation model, MC1. C. Distribution in 2095 as simulated by MC1 using a moderately warm scenario (up to 2.8°C increase in average U.S. temperature by 2100) predicted by the HADCM2SUL General Circulation Model of the Hadley Climate Centre. D. Distribution in 2095 as simulated by MC1 using a warmer scenario (up to 5.8°C increase in average U.S. temperature by 2100) predicted by the CGCM1 general circulation model of the Canadian Climate Center. Both general circulation models assume a continuous increase in the atmospheric CO_2 concentration from 295 ppm in 1895 to 712 ppm in 2100. (Adapted from Bachelet et al. 2003)

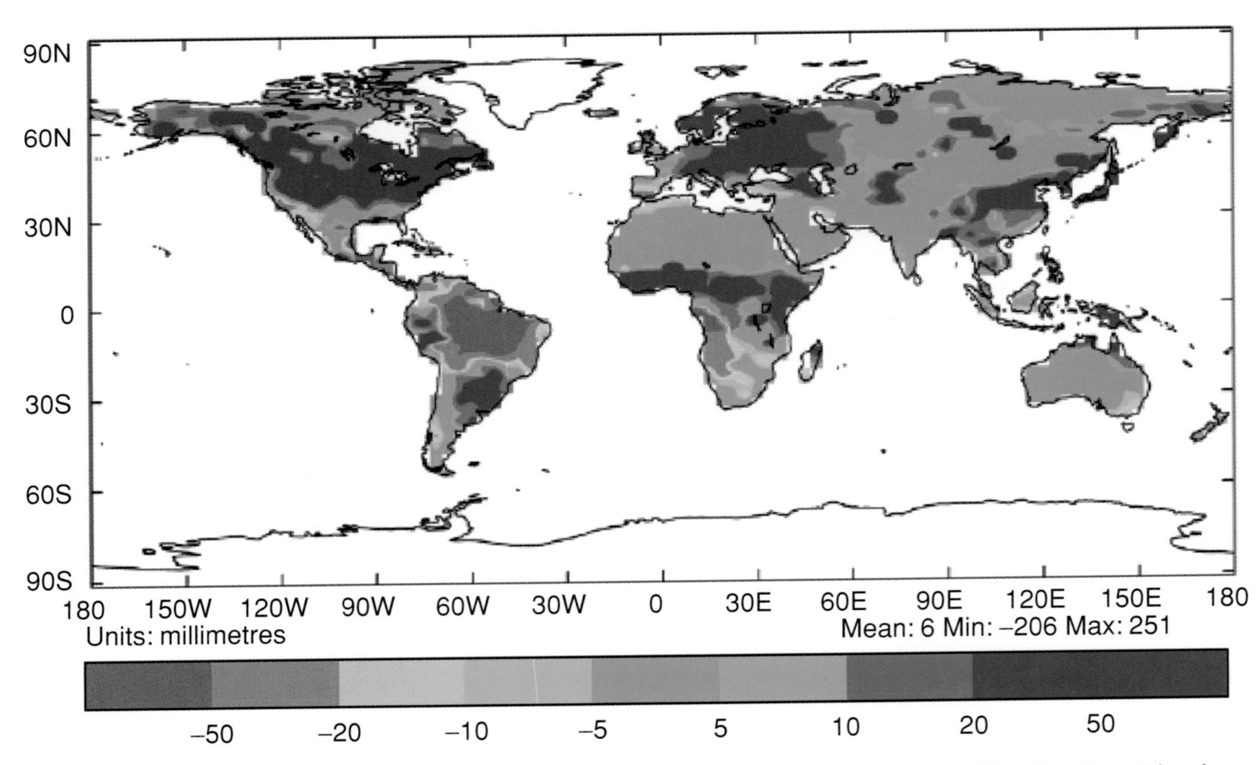

Plate 16. Change in average soil moisture content from 1960–1990 to 2070–2100, as predicted by the Hadley Centre model, HadCM3 IS92a. (Adapted from Hadley Centre for Climate Prediction and Research, The Met Office, United Kingdom, www.metoffice.gov.uk/research/hadleycentre/models/modeldata.html)

19

Herbivores in Forest Ecosystems

HERBIVORES ARE THE **PRIMARY CONSUMERS** in ecosystems, or those animals whose major dietary component is living plant tissue. But herbivory is about more than who eats plants. It is a story of how top-down and bottom-up forces, climate, species assemblages, landscape patterns, and even sunspot cycles come together to influence a key aspect of ecosystem function.

Animals eat plants to obtain energy, structural chemicals, and (at least in some cases) for the plant's medicinal properties.[1] In turn, they significantly influence ecosystem structure (e.g., canopy closure, relative abundance of different plant species) and processes (e.g., nutrient cycling) (Huntly 1991; Schowalter and Lowman 1999). Both vertebrates and invertebrates may be herbivores; however, in most terrestrial ecosystems invertebrates are the major primary consumers. The food web comprising herbivorous insects, their natural enemies, and their plant hosts may account for as much as 75 percent of earth's species (van Veen et al. 2006). Plant pathogens (fungi, bacteria, and viruses) consume or otherwise obtain energy and nutrients from living plant tissues; hence, by virtue of their trophic position they are also herbivores but are seldom referred to as such. The different herbivorous groups specialize to one degree or another on different plant tissues; for instance, **folivores** (or **phytophages,** a commonly used term for leaf-eating insects) eat leaves, **frugivores** eat fruits, and other animals may consume primarily nuts, phloem, roots, or pollen. Even species that eat only one type of tissue may eat it in different ways. For example, some leaf-eating insects are chewers that consume a leaf much like a human eats a salad, while others such as aphids probe into the phloem stream and extract sugars and amino acids without directly disturbing the leaf cells.

[1]Anecdotal evidence indicates that primates, elephants, black bears, and birds eat plants for their medicinal value (Cowen 1990).

Specialization often leads to interesting interactions between different groups. For example, in the African savanna, baboons extract the seeds from fruits of acacia trees and discard the fruit pulp, which is then eaten by elephants. In the process of gathering within the tree canopy, baboon troops knock down many whole fruits, which are also eaten by elephants. The seeds, however, are passed through the elephant digestive tract and are later retrieved by the baboons (who glean piles of elephant dung for this purpose). Similarly, in Coast Rica, fruit of the tree *Enterolobium cyclocarpum* is eaten by horses and cows, but the seed is passed through the digestive tract. *Liomys salvini,* a small rodent, uses odor cues to seek out piles of horse and cow dung at night, which it gleans for *E. cyclocarpum* seeds (Janzen 1983).

The type of plant tissue eaten and the mode of its consumption are two levels of specialization among herbivores. A third level is the species of plant that is eaten. A particular herbivore species is called a **specialist** if its diet consists of one plant species, genus, or family, and a **generalist** if its tastes are more cosmopolitan. (The term **monophagous** is often used for specialists and **polyphagous** for generalists).

The prevailing opinion has been that most phytophagous insects are host specialists (Jaenike 1990). Janzen (1983) speculated that at least one-half of the foliage-eating insects in tropical forests consume only one species of plant, and none, except for leaf-cutter ants, eat more than 10 percent of the available plant species. However, more recent studies suggest that a significant proportion of herbivorous insects in tropical forests consume more than one plant species, though many are restricted to a single plant genus, a similar pattern to that in temperate and boreal forests (Novotny and Basset 2005).

In both temperate and boreal forests, which have many fewer tree species than forests of the tropics, defoliating insects often have distinct preferences for certain tree species, but they will eat nonpreferred species as well. For example, western spruce budworm *(Choristoneura occidentalis)*, a defoliating moth larva, prefers true firs and Douglas-fir but will eat pines or larches as a last resort. Its close relative, the eastern spruce budworm *(C. fumiferana)*, prefers balsam fir, white spruce, and red spruce, in that order, but infrequently attacks black spruce and rarely attacks white pine (Albert and Jerrett 1981). In deciduous forests of the eastern United States, the various defoliating insects prefer oaks but will defoliate maples and birches (Stephens 1981). In both eastern and western forests of the United States, by far the majority of lepidopteran species (i.e., moths and butterflies) feed on hardwood trees, shrubs, or grasses rather than conifers (Hammond and Miller 1998). This is not surprising in the eastern forests that were studied, as conifers were not abundant. However, conifers dominate the western forests, which means the insects were highly selective and not influenced by what plants were most abundant.

Many insect herbivores of higher-latitude forests are quite species specific. For example, various species of sawflies, which are nonsocial hymenoptera (related to bees), are adapted to single tree species (Kuerer and Atwood 1973), and some species of bark beetles specialize on one or (at most) two tree species. Among the Lepidoptera (moths) of temperate regions, those species that feed on woody plants tend to be greater generalists than those that feed on herbaceous plants (Futuyama 1976); however, there are exceptions.

Tree species are far from genetically uniform but rather are composed of populations that differ in response to environmental and historical factors, and (with a few exceptions) populations are, in turn, composed of genetically distinct individuals. This is no less true of herbivores. An herbivorous insect species that is a generalist throughout its range may be a specialist in any given forest community, a phenomenon that Fox and Morrow (1981) call **local specialization.** Local specialization within a generalist insect species may be due to numerous factors, not the least of which is simply what food is available on a given site. Once subpopulations of an insect species begin feeding on different plants, genetic differentiation is likely to follow, sometimes quite rapidly. For example, the fall cankerworm (*Alsophila pometaria*), a defoliator of various deciduous trees in eastern North America, is composed of genetically distinct populations, which may occur no more than 100 m from each other (Mitter et al. 1979). The genetic characteristics of a particular cankerworm population apparently depend on the tree species on which it feeds. Individuals within a population also vary in their adaptations and may even take specialization to a new level, preferring only certain individuals within a plant species. Populations of the butterfly *Euphydryas editha,* for instance, contain some females that seek out only certain individuals of their host plant, *Pedicularis semibarbata,* while other females are not so finicky (Ng 1988).

Insects that spend many generations on a single host plant may become so closely adapted to that individual that they perform relatively poorly on adjacent plants of the same species. Edmunds and Alstad (1978) showed this for the black pineleaf scale *(Nuculaspis californica)* when they transferred scale insects between individual ponderosa pines and (as a check) between branches on the same tree. Survival of insects nine months after transfer is shown in figure 19.1. Note that survival was high for scale insects transferred within the same tree but low for insects transferred between trees. (Also see Karban 1989).

Because plant species vary genetically throughout their range, and genetic variation in plants may produce genetic variation in the herbivores feeding on them, we might predict that plant species with a wide geographic distribution host a greater diversity of herbivores than plant species with a limited geographic distribution. Strong (1979) reviewed several studies showing that this seems to be the case, at least

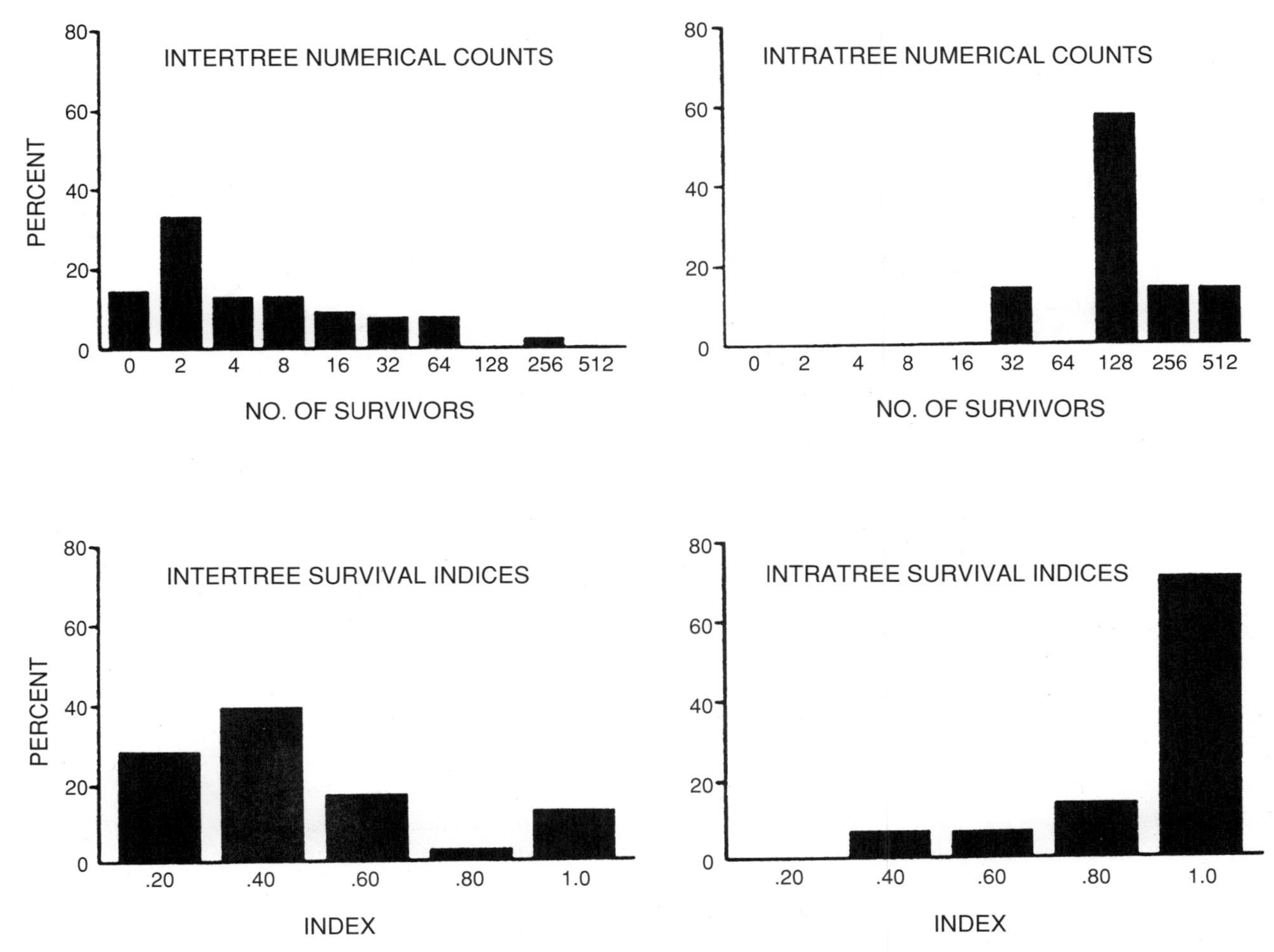

Figure 19.1. The survival of black pineleaf scales transferred between (intertree) and within (intratree) individual ponderosa pines. (Edmunds and Alstad 1978)

for insect herbivores. Southwood (1960, 1961) noted that conifers in Russia, which are widely distributed, have many more insect herbivores than conifers in Britain, which are more narrowly distributed, and that "local and rare" tree species in Hawaii support fewer insect species than widely distributed tree species. Figure 19.2 shows the relation between range size and number of insect species feeding on British trees (note the log-log scale). A tree species occupying an area 10 times larger than that of another tree species has roughly seven times more insect herbivores than the second species. Note that this relationship does not imply that the more widely distributed tree is fed on more heavily than the more narrowly distributed one, nor that at any given site one species hosts seven times more insect species than the other; it merely implies that throughout their respective ranges a greater variety of insects have adapted to feed on the more widely distributed tree.

As we discussed in chapter 9, animals compose only a very small proportion of the total biomass of forest ecosystems; however, this is not a good measure of their importance (chapter 11). Just as enzymes are a minor but absolutely essential component of organisms, animals in forest ecosystems are important because of the role they play in mediating and regulating certain processes. In previous chapters we saw that animals are important in maintaining the spatial structure of forests through their roles in pollination and seed dispersal, their influence on competition between tree species, and the grazing pressures they exert on clumped tree species (chapters 10 and 11). We also saw that animals influence (and in some cases control) the direction of succession (chapters 6 and 8), the health of the plant community via top-down regulation (chapter 11), and the nutrient cycle (chapter 18). The balance of this chapter deals with herbivores as consumers, exploring their effect on tree growth, the mechanisms by which they are regulated, why these mechanisms occasionally break down, and how forest management can affect levels of herbivory.

19.1 EFFECTS OF HERBIVORY ON PRIMARY PRODUCTIVITY

19.1.1 Herbivores at Endemic Levels

Consumption by forest herbivores may vary enormously over time, primarily because of fluctuations in

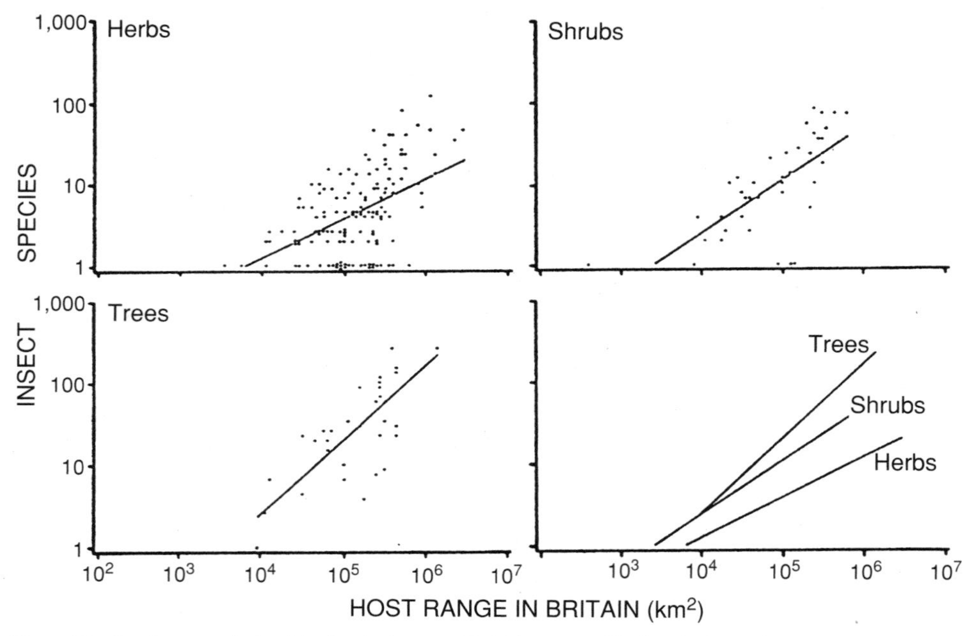

Figure 19.2. Species-area relationships for the insects of British genera of herbs, shrubs, and trees. In the lower right, the regression lines of each individual relationship are shown. Analysis of covariance finds that the regression lines cannot be discriminated from parallel and that the mean number of insect species differs significantly among the growth forms. (Strong 1979, © 1979 by Annual Reviews, Inc.)

insect populations. Therefore, when discussing effects of herbivory, it is necessary to distinguish periods when herbivores are at endemic levels from periods when they are at outbreak (i.e., epidemic) levels. At endemic levels herbivores eat only a small proportion of primary production. Table 19.1 shows the percent of net primary productivity consumed in various terrestrial and aquatic ecosystems. Consumption rarely exceeds 50 percent and in forests is generally less than 10 percent. Figure 19.3 shows the cumulative consumption of foliage during one growing season by insects in a Swedish Scots pine (*Pinus sylvestris* L.) forest (Larsson and Tenow 1980). Most folivores prefer young leaves, so total consumption is low during the spring and early summer, when new leaves are small, and increases rapidly during the later summer, when new leaves are fully expanded. Total consumption (15.5 kg/ha) amounted to 0.7 percent of total needle biomass, and 2.5 percent of the needles produced during the year (Larsson and Tenow 1980).

Table 19.1
Percentage of net primary production (by foliage) consumed by herbivores

Ecosystem	Organic input consumed (%)	Authority
Tropical forest	8.5	Misra 1968
Tropical forest	7	Odum and Ruiz-Reyes 1968
Temperate forest	1.5–2.5	Bray 1964
Temperate forest	1.5	Reichle and Crossley 1967
Temperate forest	3.4–9.2	Kazmarek 1967
Temperate forest	40	Andrezejewski 1967

Source: Golley 1972 (see Golley's publication for original citations).

Although total consumption is similar, the pattern of grazing in tropical forests is somewhat different than that in temperate forests. In the tropics, and particularly in seasonally deciduous forests, total defoliation of a given tree species commonly occurs. At any one time, however, only one or a few of the scores of tree species present are defoliated, and these replace lost leaves within two to four weeks (Janzen 1983).

19.1.2 Herbivores at Epidemic Levels

Insect populations undoubtedly fluctuate to some degree in all forest ecosystems, even the relatively stable tropics. Golley (1983b) gives the example of herbivory between 1977 and 1979 in Santa Rosa National Park, Costa Rica. The rainy seasons of 1977 and 1978 "were marked by extremely high general levels of herbivory and many species suffered total defoliation. In the rainy season of 1979, only one of the many previously defoliated species was again defoliated . . . and overall caterpillar densities were easily less than one tenth that of 1977 and 1978."

The most severe fluctuations in insect populations occur in forests of higher latitudes. A few of many possible examples of insect species that periodically outbreak from endemic to epidemic levels are the eastern spruce budworm (*Choristoneura fumifera*) in northeastern North America; the western spruce budworm (*C. occidentalis*), Douglas-fir tussock moth (*Orgyia pseudotsugata*), and mountain pine beetle

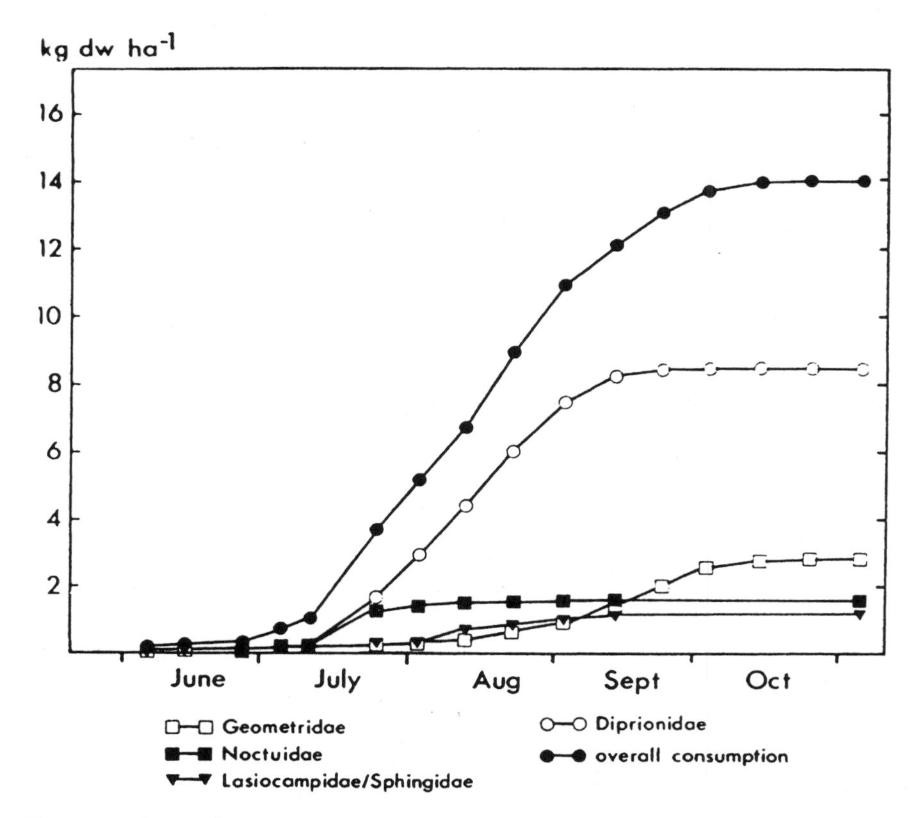

Figure 19.3. The cumulative consumption by needle-eating insect larvae in a stand of mature Scots pine *(Pinus sylvestris)*. (Larsson and Tenow 1980)

(Dendroctonus ponderosae) in western North America; and the gypsy moth *(Ocneria dispar* L.) and various sawflies in both Europe and North America. Outbreak species compose a relatively small proportion of herbivorous insects in forests (e.g., only 1 or 2% of forest Lepidoptera [moths and butterflies]) (Myers 1993). Because of the vast number of insect species, however, that still translates into a large number. Roughly 80 species of both European and North American Lepidoptera are outbreakers, at least 18 of which are cyclic, including tussock moth and spruce budworm (Myers 1993).

Outbreaks generally occur very rapidly. In phytophagous insects, consumption may increase to levels well over 50 percent of the total foliage present within two to three years. For example, figure 19.4 shows the change in density of Douglas-fir tussock moth larvae during two outbreaks, one in Arizona and one in the Blue Mountains of northeastern Oregon. (Note that the *y* axis [i.e., density of larvae] is logarithmic.) Before outbreak in 1967, there was slightly more than one larva per 0.65 m^2 (1,000 in^2) of branch area in the Arizona stand; by the time the outbreak had reached its peak two years later, there were about 15 larvae per 0.65 cm^2 of branch area. In the Blue Mountains, endemic levels of tussock moth larvae were lower than in Arizona (between 0.1 and 0.5 larvae per 0.65 cm^2 of branch area), and population growth during the epidemic phase was more dramatic, ranging from 150- to 450-fold increases over a three-year period. However, population increase during tussock moth outbreaks is modest compared to that of some other insects. In three outbreaks between 1949 and 1975 in the Engadine Valley of Switzerland, larch budmoth *(Zeiraphera diniana)* increased an average of 18,000-fold within a one- to two-year period (Baltensweiler et al. 1977). Larch budmoth outbreaks in the European Alps spread in highly organized waves in an east-northeasterly direction at an average rate of 220 km/year (Bjornstad et al. 2002).

Wickman et al. (1973) separate Douglas-fir tussock moth outbreaks into four phases, each lasting about one year (fig. 19.5). First is a release phase, when insect numbers increase past the point where they can be effectively controlled by the natural mechanisms that normally keep them in balance. The release phase is followed by a peak phase, during which the population increases exponentially to its maximum level. In the third year a decline phase begins with a drop in insect numbers that is even sharper than the previous year's increase. The rate of decline drops off in the fourth year, which Mason and Luck (1978a) call the postdecline phase. Outbreak and decline occur more rapidly in tussock moth populations than in many other insect populations. Nevertheless, the same general pattern holds for all outbreak species: numbers suddenly climb past a threshold beyond which natural controls can no longer effectively operate and increase to supra-abundance; then, because either the food runs out or populations of their pathogens reach high levels (or both), the population collapses.

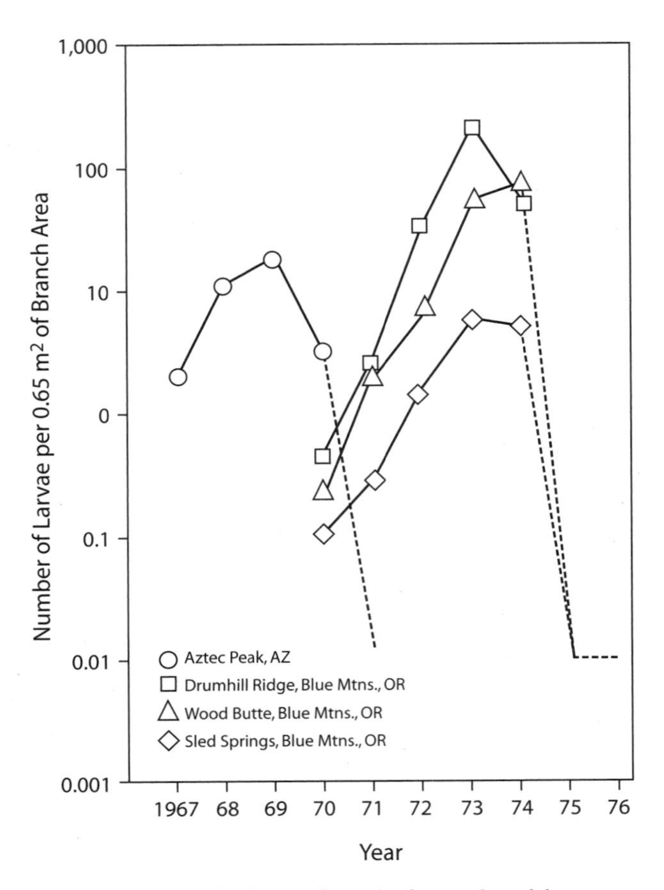

Figure 19.4. Trends in the density of Douglas-fir tussock moth larvae over two complete out breaks. (Mason and Luck 1978)

Intervals between epidemics vary among insect species, geographical regions, and over time in the same region, In some instances outbreaks occur with considerable regularity. In the Engadine Valley, 15 outbreaks of the larch budmoth were recorded between 1850 and the early 1970s. The interval between outbreaks was 8.64 ± 0.29 years, and the average length of an outbreak was 3.0 ± 0.22 years (Baltensweiler et al. 1977). Highly regular population fluctuations such as this are called *cyclic*.

> A somewhat different kind of cyclic behavior is exhibited by periodical cicadas (*Magicicada* sp.), which are among the most abundant insects in deciduous forests of eastern North America (reaching densities of 2.6 million/ha). They spend either 13 or 17 years belowground, feeding on the roots of broad-leaved trees, then emerge en masse and ascend to the canopy to mate, oviposit, and die within weeks (Koenig and Liebhold 2003).

The behavior of many other outbreak species is less predictable. Figure 19.6 shows the population variation of the tussock moth between 1900 and the mid-l970s. Beginning in the early l920s, outbreaks have occurred every 8 to 10 years in British Columbia, Washington, and Idaho. In other states, however, population cycles have been more irregular.

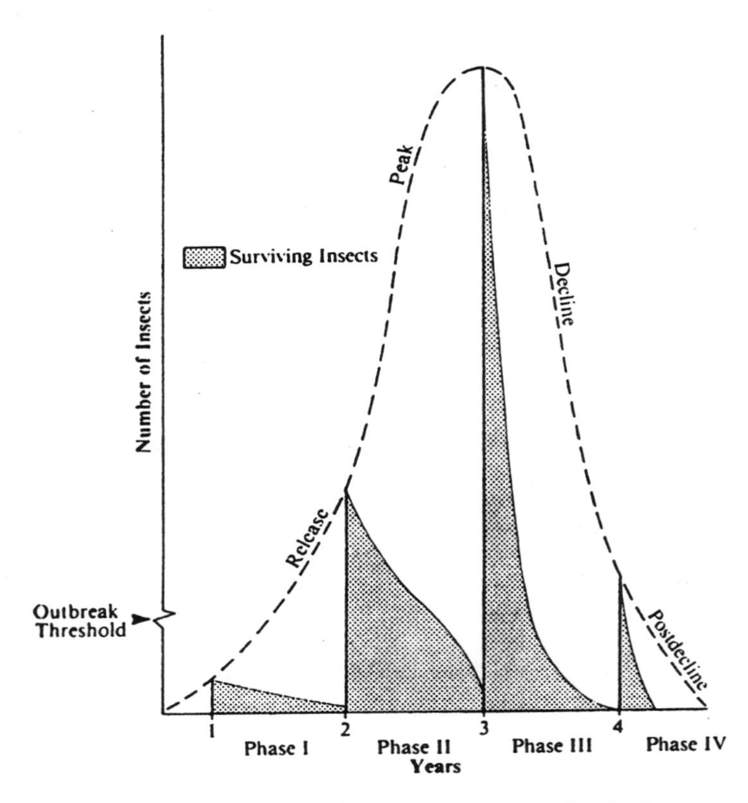

Figure 19.5. A schematic representation of a Douglas-fir tussock moth outbreak sequence with generalized patterns of within-generation survivorship. (Mason and Luck 1978)

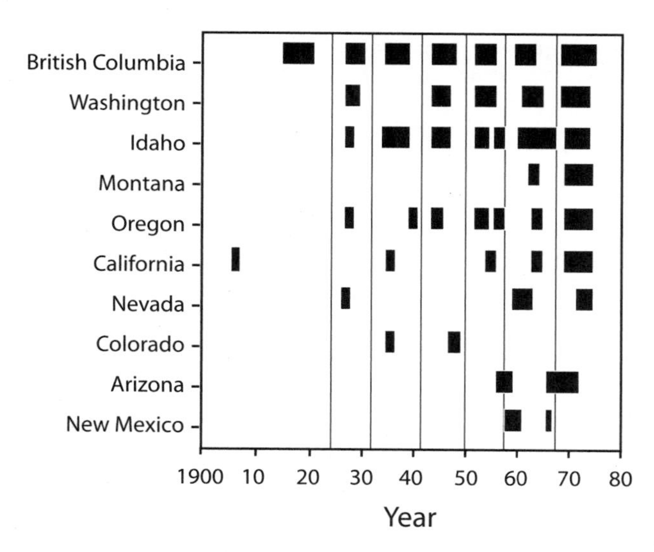

Figure 19.6. The years of mass outbreaks of Douglas-fir tussock moth by state and province. Vertical lines separate sequential, large-scale outbreaks. (Clendenen, unpublished data, 1975; Mason and Luck 1978)

Outbreaks may be irregular in individual forests but regular at the regional level; such is the case for both the spruce budworm in Colorado's San Juan Mountains (Ryerson et al. 2003) and the Pandora moth (a defoliator of ponderosa pine) (Speer et al. 2001).

Populations that fluctuate erratically are called **eruptive.** Some insects that exhibit cyclic or eruptive population fluctuations in one area have relatively stable populations in other areas. For example, the larch budmoth does not outbreak at all in some areas of Switzerland, and several species that are serious outbreak pests in the interior forests of western North America maintain stable endemic populations in forests west of the crest of the Cascade Range. Many outbreaking insects have quasi-cyclical population fluctuations. For example, when viewed over centuries, the Pandora moth exhibits relatively strong 20- and 40-year cycles, but the timing (phase) of the cycles shifts over time (Speer et al. 2001). Moreover, 20-year cycles dominate in some time periods and 40- year cycles in others (perhaps related to climate; section 19.2.6).

The behavior of outbreak insects may be altered by changes in environmental factors, a phenomenon clearly illustrated in North America by both the eastern and western spruce budworms. Whereas, during the nineteenth century there were 9 outbreaks of eastern spruce budworm in Canada, during the first 80 years of the twentieth century there were 21 outbreaks (Blais 1983). Moreover, nineteenth-century outbreaks were restricted to specific regions; twentieth-century outbreaks have coalesced and spread from region to region. Outbreaks of the western spruce budworm have also become more widespread and have lasted longer in the twentieth century than previously, although that has not happened in all regions (Anderson et al. 1987; Swetnam and Lynch 1989; Wickman et al. 1994; Ryerson et al. 2003). Other tree-eating insects have exhibited similar changes in behavior (McCullough et al. 1998).

The same reasons appear to underlie the change in insect behavior in both eastern and western North America (McCullough et al. 1998): a combination of logging practices and fire suppression has favored the spread of tree species that are preferred food. With more abundant food, populations grow faster and spread more widely. In interior western North America, for example, early loggers took mostly ponderosa pine and western larch, which are not preferred tree species of budworm or tussock moth. Stands consequently became dominated by various species of firs, which are preferred food. This situation was exacerbated by fire control, because the frequent low-intensity ground fires common to these forest types had killed young firs before they became established, without harming the large old ponderosa pine (chapter 7). In the absence of fire many more fir trees reached maturity than previously, thereby further expanding the food base.

In at least some instances, application of pesticides has prolonged outbreaks of eastern spruce budworm (Blais 1983). Without pesticides, insect populations grew rapidly, consumed the available food, and crashed. Pesticides slowed the growth of insect populations and kept more trees alive; however, that simply prolonged the food supply for those budworm that survived the chemicals. The net effect was to convert what had been boom and bust cycles on the part of the insect to persistent, chronic infestations. As we discuss later in this chapter, pesticides may also exacerbate long-term insect problems by killing predatory insects that feed on herbivorous insects.

19.1.3 Effect of Herbivores on Primary Productivity

Defoliation usually does not cause mortality, unless it is repeated too frequently for trees to recharge carbohydrate reserves. If photosynthetic area is significantly reduced, however, tree growth, fruiting, and seed production may be affected. Defoliation influences growth in various ways, not all of them negative. Figure 19.7, from Ericsson et al. (1980), shows the connections among defoliation, carbohydrate dynamics, and growth. Reduction of needle biomass decreases photosynthetic surface and may also remove starch reserves, both of which result in decreased growth. Transpiring surface also is decreased, however, which improves tree water balance and thus photosynthetic capacity. Lower levels of assimilates (especially starch) may actually stimulate photosynthesis in the short run as well.

Kulman (1971) reviewed 174 studies of growth loss in trees due to defoliation and concluded that in almost all, losses were directly proportional to the quantity of leaves removed. Response of pines and deciduous trees to defoliation was immediate if it occurred before seasonal growth

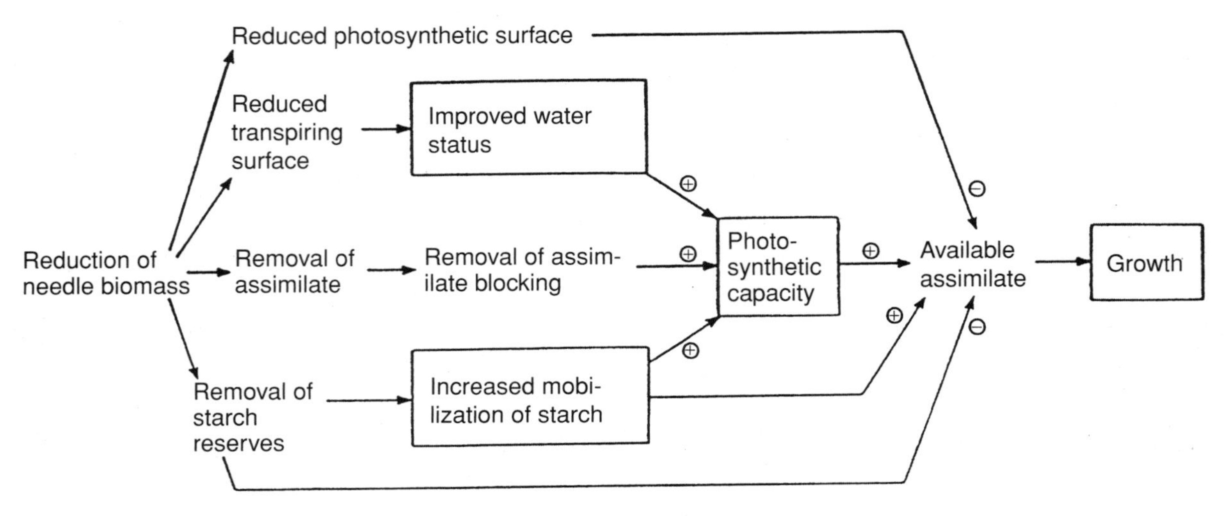

Figure 19.7. A schematic representation of hypothetic connections among defoliation, carbohydrate dynamics, and growth. + indicates a positive and – a negative effect on growth. (Adapted from Ericsson et al. 1980)

was complete, while that of spruce and firs was delayed for several years, perhaps because of greater amounts of stored carbohydrate in these species.

Effects of defoliation on growth and mortality depend on what foliage is lost (i.e., age class, position in crown) and in which part of the growing season it is removed. Consumption of new foliage in the early summer (before it has a chance to contribute significantly to whole-tree photosynthesis) may have little or no effect on tree growth; this was demonstrated by Ericsson et al. (1980) for young Scots pine *(Pinus sylvestris)* and is shown in figure 19.8. Early-summer defoliation in 1976 of up to 100 percent of new foliage had no effect on stem growth in 1977, while late-summer defoliation of new needles decreased stem growth in proportion to the amount of foliage removed. If one-year-old needles were also removed, growth was reduced regardless of whether defoliation occurred in early or late summer, although the effect was more severe for late-summer removal. Effects of defoliation may also vary from one tree to another. Piene (1989a, 1989b) suggests that balsam fir trees differ genetically in their ability to recover from defoliation by spruce budworm. In his studies, destruction of buds triggered some trees to produce epicormic shoots that permitted quick refoliation and survival. Trees that did not produce epicormic shoots died. Persistent defoliation would eventually drain carbohydrate reserves and limit the ability of trees to recover in this manner.

Tree mortality does occur during outbreaks of defoliating insects. For example, trees containing at least 300 million board feet of lumber were killed in the Pacific Northwest during a tussock moth outbreak in 1929 and 1930 (Wickman 1963). Often, however, mortality is concentrated in intermediate and suppressed size classes, so that the end result may, in fact, benefit the stand. Alfaro et al. (1982) studied an outbreak of western spruce budworm in British Columbia during the 1970s and found that mortality approached 100 percent in the smallest size classes and declined sharply as tree size increased (fig. 19.9). This result changed stand structure in much the same way as some kinds of thinning. That is, the insect acted as a natural silviculturist. In their analysis of historic budworm outbreaks in Colorado's San Juan Mountains, Ryerson et al. (2003) found that infestations produced barely detectable reductions in ring growth (such an analysis can only look at surviving trees). On the other hand, the budworm outbreak that persisted in the Blue Mountains of Oregon and Washington from the early 1980s through 1993 killed most host trees in some stands. Wickman et al. (1994) conclude that this "level of insect population and resulting damage is highly unstable and probably new to the ecological history of the Blue Mountains." The change in insect behavior can in all likelihood be traced to the large-scale landscape conversion and spread of host trees discussed earlier.

In contrast to defoliators, the so-called primary bark beetles usually kill living trees.[2] Many of these have a mutualistic relationship with fungi (commonly termed *bluestain fungi*) that act in concert with the beetle to kill trees (Paine et al. 1997).

The impact of herbivores on forests is often more extensive and subtle than growth losses and mortality. Herbivores that consume buds or mine leaders alter hormonal relationships that influence tree form. Examples of this are the Sitka spruce weevil, which colonizes terminals and causes trees to produce multiple leaders, and elephants, which browse trees in such a way that erratic branch growth may occur (Mueller-Dombois 1972). Seed and cone eaters consume little energy but may have a profound effect on tree fitness. One study found that defoliation reduced mycorrhiza formation in pinyon pine (Gehring and Whitham 1992), which would reduce the tree's ability to gather water and nutrients and defend itself against root pathogens (chapters 11 and 18).

[2]Secondary bark beetles generally colonize only weakened or recently dead trees.

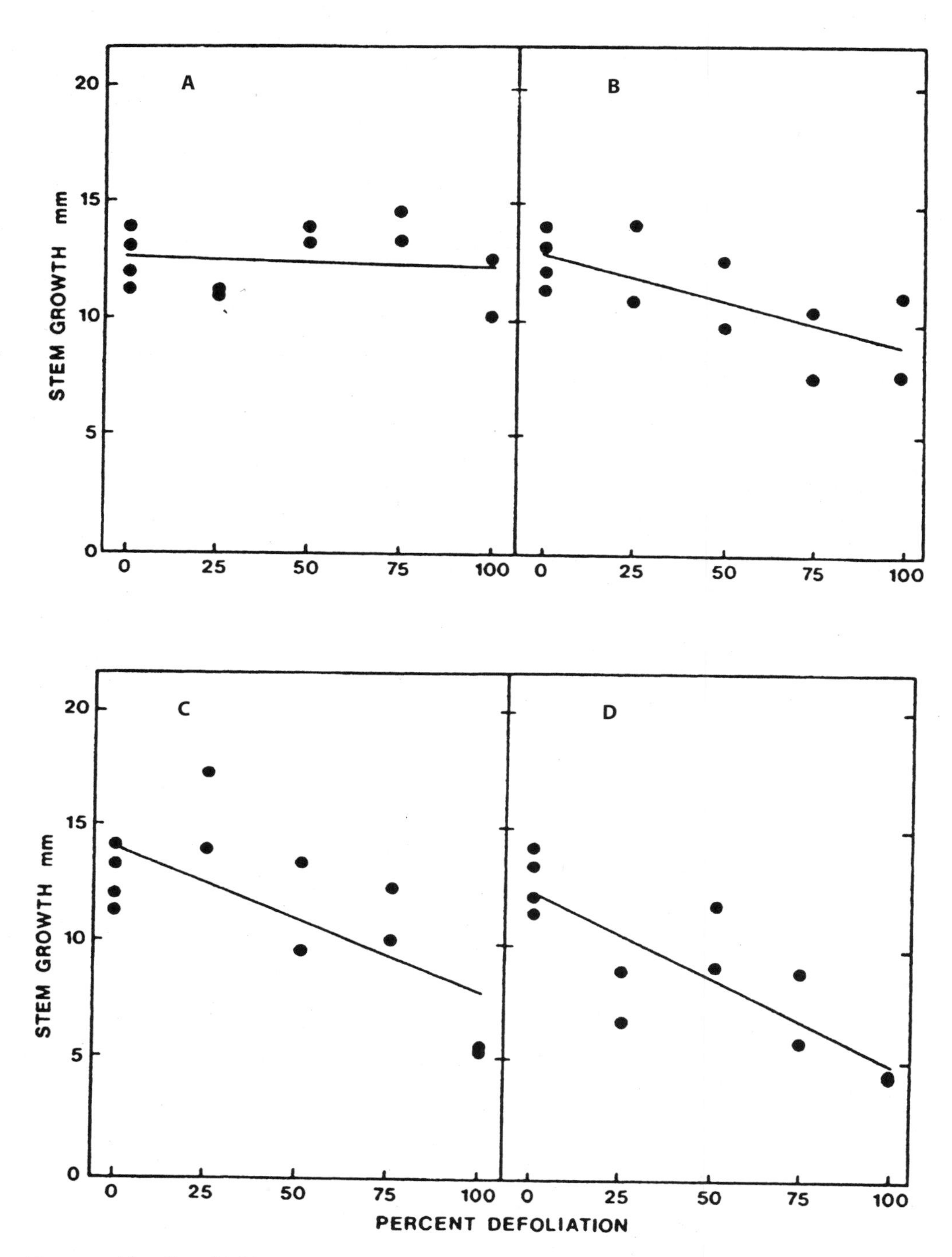

Figure 19.8. The effect of defoliation in the summer of 1976 on the growth of stem girth in 1977. A. Early summer defoliation of 1-year-old needles. B. Early summer defoliation of both 1-year-old and 2-year-old needles. C. Late summer defoliation of 1-year-old needles. D. Late summer defoliation of both 1-year-old and 2-year-old needles. (Adapted from Ericsson et al. 1980)

Trees have evolved a large variety of chemical compounds that act to inhibit herbivory, which is discussed in some detail in the following section. The energy cost of these defensive compounds (along with the cost of producing vacuoles to sequester them from the tree's own metabolic machinery) is difficult to measure but is likely to be considerable.

19.2 FACTORS CONTROLLING HERBIVORES

As discussed earlier, except during outbreak periods, herbivores in terrestrial ecosystems consume only a small proportion of the available food. Factors acting to keep herbivore populations at relatively low levels can be broadly

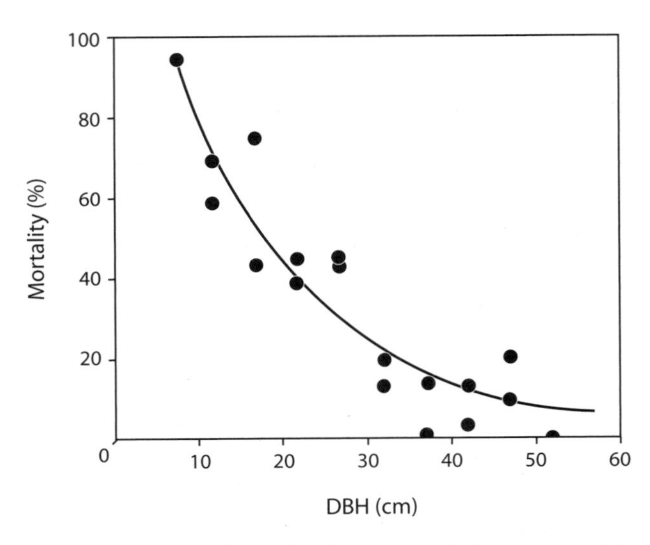

Figure 19.9. Tree mortality resulting from a spruce budworm outbreak in British Columbia, by tree diameter class. (Alfaro et al. 1982)

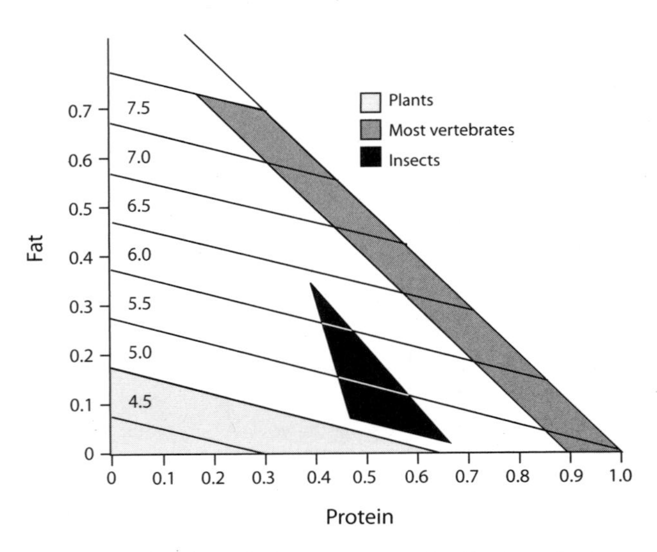

Figure 19.10. The relationship between composition (fractions of unit dry weight) and energy content (Kcal/g dry weight) of various groups of living organisms. (Adapted from Southwood 1972 and Morowitz 1968)

grouped into those that are properties of the individual plant, such as food quality, and those that are properties of the ecosystem, such as plant species diversity, the complex of predators and pathogens attacking herbivores, and weather. Individual-plant and ecosystem-level factors are not independent of one another. As we see later, the palatability of plants to herbivores can be profoundly influenced by weather and by surrounding plants. Plants also have various mechanisms that increase levels of predation on herbivores. In any given system, both individual-plant and ecosystem-level factors operate to control herbivore populations, although their relative importance varies over time and among different forest types. Understanding the mechanisms by which herbivore populations are maintained at low levels is very important for foresters, because such understanding provides insights into the causes of (and possible ways of avoiding) herbivore outbreaks.

19.2.1 Individual Plant Chemistry: Food Quality

As we saw in chapter 17, plants and animals differ in the basic composition of their tissues. Plant tissue is composed primarily of carbon-based structures such as cellulose and lignin, whereas animal tissue has a high proportion of protein and fat (fig. 19.10). While the nitrogen concentration of mature foliage rarely exceeds 2 percent in conifers and 3 percent in deciduous trees, the bodies of insect larvae may contain more than 10 percent nitrogen (Elser et al. 2000). This is shown in table 19.2 for gypsy moth feeding on various deciduous tree species in Russia (Rafes 1971). Concentrations of phosphorus and potassium (as well as nitrogen) are higher in larvae than in foliage, which suggests that herbivores are limited by the nutrient rather than the energy content of their food (this is also true in aquatic systems; Urabe 1993).

Because of nutrient limitations, the efficiency with which herbivores convert plant tissue into their own tissue is quite low (Elser et al. 2000). Figure 19.11 shows this effect for various invertebrate herbivores. For nitrogen concentrations in the range contained in most tree tissues (i.e., <3%), the efficiency of conversion of ingested food (ECI), defined as the ratio of herbivore growth to ingested food (Waldbauer 1968), is less than 20 percent. In contrast, the ECI values of predatory invertebrates such as spiders and parasitic wasps are around 50 percent (Mattson 1980). The low ECI values of invertebrate herbivores means that they must consume an enormous amount of food. With ECI values between 10 and 30 percent, herbivores must consume 3 to 10 g of plant tissue to produce 1 g of their own mass (Mattson 1980). Xylem sap-sucking insects ingest one hundred to one thousand times and chewing arthropods five to six times their body weight each day (Horsfield 1977; Mattson 1980).

Although herbivores can compensate somewhat for low nutrient concentrations in their diet by eating a large quantity, the energy and time costs that are associated with high

Table 19.2

Mean nutrient concentrations (%) of tree leaves and of frass and tissues of gypsy moth larvae feeding in the leaves

	N	P	K	Ca	Mg
Oak	2.59	0.05	1.04	1.15	0.13
Ash	1.79	0.18	1.50	2.12	0.33
Maple	1.33	0.14	1.38	1.87	0.38
Lime	2.65	0.11	1.79	1.65	0.21
Hazel	1.38	0.114	1.49	1.92	0.38
Frass	3.26	0.20	3.21	1.87	0.76
Larvae	9.62	1.53	2.98	0.39	0.27

Source: Rafes 1971.

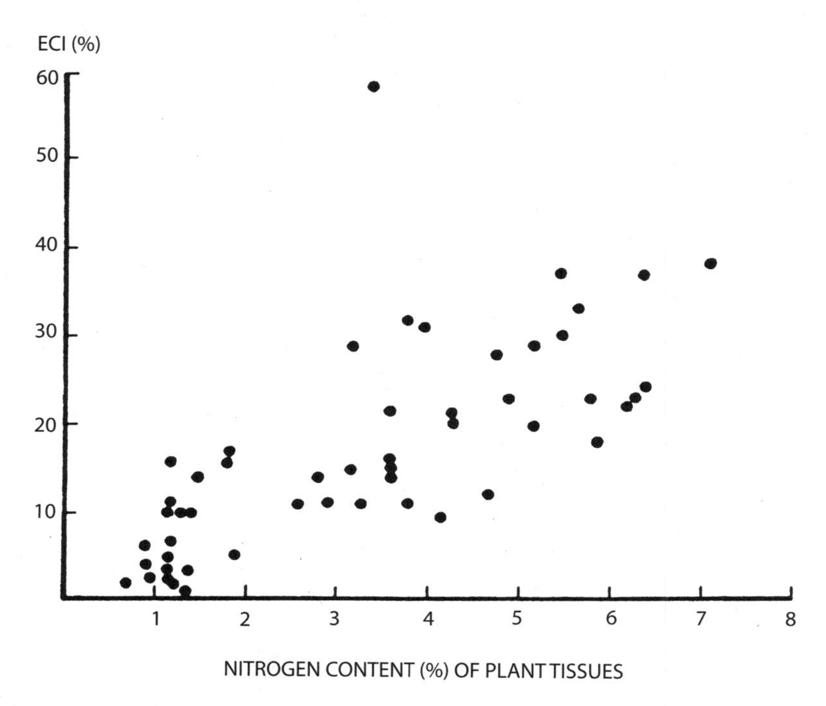

Figure 19.11. The efficiency of the conversion of ingested food (ECI) by various invertebrate herbivores in relationship to the nitrogen concentration of the food. (Mattson 1980)

levels of consumption result in lowered growth rates. Figure 19.12 shows how the ratio of energy to protein in the diet affects weight of eastern spruce budworm pupae (Shaw et al. 1978); as the proportion of protein in the diet declines, pupal weight drops. White (1984) hypothesized that the primary factor maintaining herbivore populations at low levels relative to their apparent food supply is shortage of nitrogen in their diet, and that outbreaks are triggered by increased levels of soluble nitrogen in plant tissues during drought periods. (Free amino acids in foliage often increase during drought, at least partially because they act as osmotic agents.)

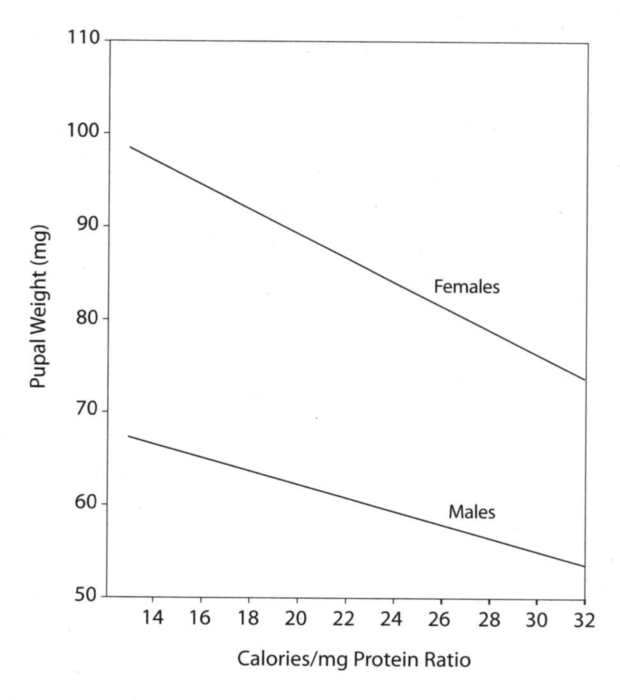

Figure 19.12. The regression of pupal weight against the calories per milligram of protein ratio in maturing, current-year needles of balsam-fir trees treated with different fertilizer prescriptions. (Shaw et al. 1978)

Herbivores have evolved various ways to increase nutrients in their diet. Many prefer tissues that are relatively high in nutrient content, such as swelling buds and newly emerged leaves. For insect herbivores, this means that emergence of larvae must be synchronized with production of new leaves. (Later we see that this can be an important factor in insect population dynamics.) Koalas tend to feed only on eucalyptus trees with relatively high foliar nitrogen (unless those trees have high levels of defensive chemicals; see the following section) (Moore and Foley 2005).

Insects such as termites and bark beetles (which feed on extremely nitrogen-poor tissues) have evolved symbiotic relationships with nitrogen-fixing bacteria. One particularly intriguing example of a possible association between nitrogen-fixing bacteria and a mammalian herbivore is that of a tribe in Southeast Asia that exists almost exclusively on sweet potatoes, a very nitrogen-poor food. Some years ago anthropologists discovered that members of the tribe excreted more nitrogen than they consumed; the implication is that nitrogen-fixing organisms live within the digestive systems of tribe members.

Although protein appears to be a major limiting factor in herbivore diets, other nutritional requirements of animals may be in short supply (or missing altogether) in plant tissues. Methionine, an essential amino acid for insects, and the water-soluble B vitamins are relatively scarce in plants (Lord 1968; Southwood 1972).

19.2.2 Individual Plant Chemistry: Defensive Chemicals

Plants, with the help of mutualistic foliar endophytes (chapter 11), further decrease their palatability to herbivores through a variety of compounds that because they are apparently not important in primary metabolic pathways, are called **secondary chemicals.** There are a bewildering array of different secondary chemicals in the plant world; thousands of different compounds have been identified, and undoubtedly many others remain to be discovered. Primary chemicals (e.g., sugars, starches, proteins) are the major actors in the physiology of plants, whereas secondary chemicals, although possibly playing some physiological role, appear to function primarily as mediators between plants and other organisms. In other words, their role is ecological rather than physiological, and a major part of their ecological function is in defense against herbivores and pathogens. The medicinal and psychoactive value of plants to humans and other animals stems from secondary chemicals.

All organisms share the same basic primary compounds (although in different proportions); however, the diversity of secondary compounds within the plant kingdom is enormous. Rhoades (1979) lists the following secondary substances that have been shown to deter herbivore grazing or reduce herbivore fitness (i.e., survival, growth, and reproduction): "alkaloids, pyrethrins, rotenoids, long-chain unsaturated isobutylamides, cyanogenic glycosides, phytoecdysones and juvenile hormone analogs, cardenolides and saponins, sesquiterpene lactones, nonprotein amino acids, glucosinolates and isothiocynates, oxalates, protoanemonin, hypericin, fluoro fatty acids, selenoamino acids, 6-methoxybenzoxazoline, gossypol, condensed tannin, phenolic resin and associated phenol oxidase and proteinase inhibitors of the soybean trypsin inhibitor type, . . . phytohemagglutinin, . . . and chromenes." Compounds within each of these groups are generally not homogeneous; for example, thousands of different alkaloids have been identified. Some of the secondary chemicals toxic to insects are produced by endophytic fungi living within plant leaves rather than by the plant, a phenomenon that has been documented in such widely different plant groups as conifers and grasses (Clark et al. 1989; Calhoun et al. 1992).

Secondary defensive compounds can act either directly by affecting herbivore physiology, or indirectly by signaling the herbivore's natural enemies (i.e., a call for help). Some act both directly and indirectly by slowing an herbivore's development and thereby making it more vulnerable to natural enemies. Some herbivorous insects turn plant defenses to their advantage by sequestering secondary compounds and using them as defense against their natural enemies (Nishida 2002).

Secondary chemicals that act directly on herbivore physiology can be grouped into two classes according to their mode of action. **Toxins** are relatively small molecules that disrupt metabolic or developmental processes of animals. Alkaloids are the most widespread toxic secondary chemical. Others include proteinase inhibitors, which inhibit the proteinase enzyme; nonprotein amino acids, which substitute for the proper amino acids and thus render proteins inactive; and juvenile hormone analogs, which disrupt insect development. Toxins are sometimes referred to as **qualitative defensive chemicals** because they are dosage independent (i.e., they are effective at very low concentrations) (Feeny 1975). Toxins are the primary mode of plant defense against mammalian grazers (Bryant et al. 1991).

Digestibility reducers are the second general class of plant defensive chemicals. These are relatively large molecules that reduce an animal's ability to assimilate food. Digestibility reducers are often called **quantitative defensive chemicals** because they are dosage dependent (i.e., unlike toxins, their effectiveness depends on their concentration) (Feeny 1975). The most widespread groups of quantitative defensive chemicals are resins and the various polyphenolics. Of the latter group, the most common are the tannins, a generic name for compounds that are able to bind protein (the name comes from their use in tanning leather). Anyone who drinks tea has had first-hand experience with a plant defensive compound: the astringent taste is due to precipitation of salival enzymes by tannins. In some cases tannins may have a toxic effect, particularly on mammals and birds, and their astringent taste may act as a feeding deterrent (Feeny 1970). Their primary mode of action, however, is to complex with carbohydrates and proteins, rendering them indigestible, and with animal enzymes, making them ineffective. The result is slower growth and decreased reproductive capacity of the herbivore rather than outright mortality. Feeny (1970) found that winter moth, a common defoliator of oak, had lower larval growth rate and reduced pupal weights when fed a diet containing as little as 1 percent oak-leaf tannins. In Africa, the phenolic content of vegetation influences the feeding pattern of colubus monkeys (*Colubus* sp.). Where foliage is high in phenols (including tannins), leaves make up only 37 percent of the colubus diet, but where vegetation is low in phenols, leaves make up 75 percent (McKey et al. 1978).

Tannin-protein complexes are unstable at high pH levels and the midgut pH of many insects is greater than 8, which means that in at least some cases tannins may be relatively ineffective digestibility reducers (Berenbaum 1980). Rhoades and Cates (1976) suggested that the low pH of conifer foliage is an adaptation to reduce the gut pH of herbivorous insects and thus stabilize tannin-protein complexes. However, the effectiveness of tannins as defensive chemicals is not always clear even in conifers. In Douglas-fir, terpene content of the foliage appears to be more important than tannins as a defense against western spruce budworm (Cates et al. 1983). Thus, although tannins and other polyphenolics are certainly the most widespread secondary chemicals of woody plants, there is much to learn about their efficacy as herbivore defenses. Nevertheless, the large amounts of energy diverted to the production and maintenance of polyphenolics argues persuasively that they do play an important role in tree fitness.

19.2.3 Distribution of Defensive Chemicals

> Plants have evolved an array of chemical and physical defenses against a diverse set of enemies and under variable abiotic conditions. In the development of a theory of plant defense, we seek explanations that take all of that into account.
>
> STAMP 2003

The distribution of secondary chemicals is highly heterogeneous, varying among plant species, with site quality, with geography, among trees within the same population, among leaves on the same tree, and over time in any single leaf. Differences among species are distinguished by two major patterns, one related to life history characteristics and another to availability of resources. As with most things in ecology, these factors are related to one another.

19.2.3.1 Life-History Characteristics

Herbs and fast-growing, woody pioneers generally have fewer and different types of chemical defenses than slow-growing, late-successional woody plants. Herbaceous plants contain primarily qualitative defensive chemicals (toxins), while woody plants contain a mix of both qualitative and quantitative types. For example, roughly 80 percent of woody dicotyledonous plant species, but only 15 percent of herbaceous species, have tannins (Rhoades 1979). Futuyma (1976) studied the secondary chemical constituents of 13 primarily woody families and 13 primarily herbaceous plant families of the northeastern United States and found that 11 woody families and 5 herbaceous families had phenolics. Only 3 woody families as opposed to 10 herbaceous families contained alkaloids. Among trees, early-successional species are grazed more heavily than late-successional species growing on the same site, suggesting that the former produce fewer defensive chemicals than the latter (Coley 1980, 1988; Coley et al. 1985) (table 19.3).

Table 19.3
Grazing rates (% leaf area/d) on pioneer and persistent plant species

	Mature leaves	Young leaves
Persistent species		
Mean	0.041	0.631
Standard error	0.024	0.151
Species, *n*	16	16
Leaves, *n*	452	509
Pioneer species		
Mean	0.466	0.663
Standard error	0.162	0.161
Species, *n*	11	11
Leaves, *n*	386	398

Source: Adapted from Coley 1980.

19.2.3.2 Resource Availability

Infertile soils are generally characterized by vegetation with high secondary-chemical content and low palatability to herbivores, a pattern especially evident in some parts of the tropics (McKey et al. 1978; Bryant et al. 1989). Chapter 18 showed one of the more striking examples: the highly leached white sands that Janzen (1974) calls "probably the most nutrient-poor soils in the world." These areas, which occur primarily in parts of tropical Asia, are distinctive not only because of their extreme infertility, but because they produce "blackwater" rivers, which are brown or black in color due to high concentrations of humic acids that originate from secondary chemicals in the vegetation. Though not as striking as blackwater river areas, similar differences between site quality and either secondary-chemical concentration or levels of herbivory occur elsewhere. In boreal forests, fast-growing trees that colonize recently disturbed areas are grazed more heavily than slow-growing trees in adjacent mature forests (Bryant and Kuropat 1980). Douglas-fir from the mesic, productive habitats west of the crest of the Cascade Range apparently have little or no foliar defenses against spruce budworm, while in the relatively harsh continental climates of the intermountain region, foliar defenses may be quite high (Perry and Pitman 1983) (fig. 19.13).

19.2.3.3 Why Do Plant Species Differ?

Stamp (2003) reviews the various hypotheses that have been invoked to explain differences among plants in the levels of chemical defense they employ (see also Wise and Abrahamson 2005). Four in particular have shaped the thinking about this issue over the past 30 years:

1. *Optimal defense hypothesis* (Feeny 1976; Rhodes and Cates 1976). Optimal defense holds that because

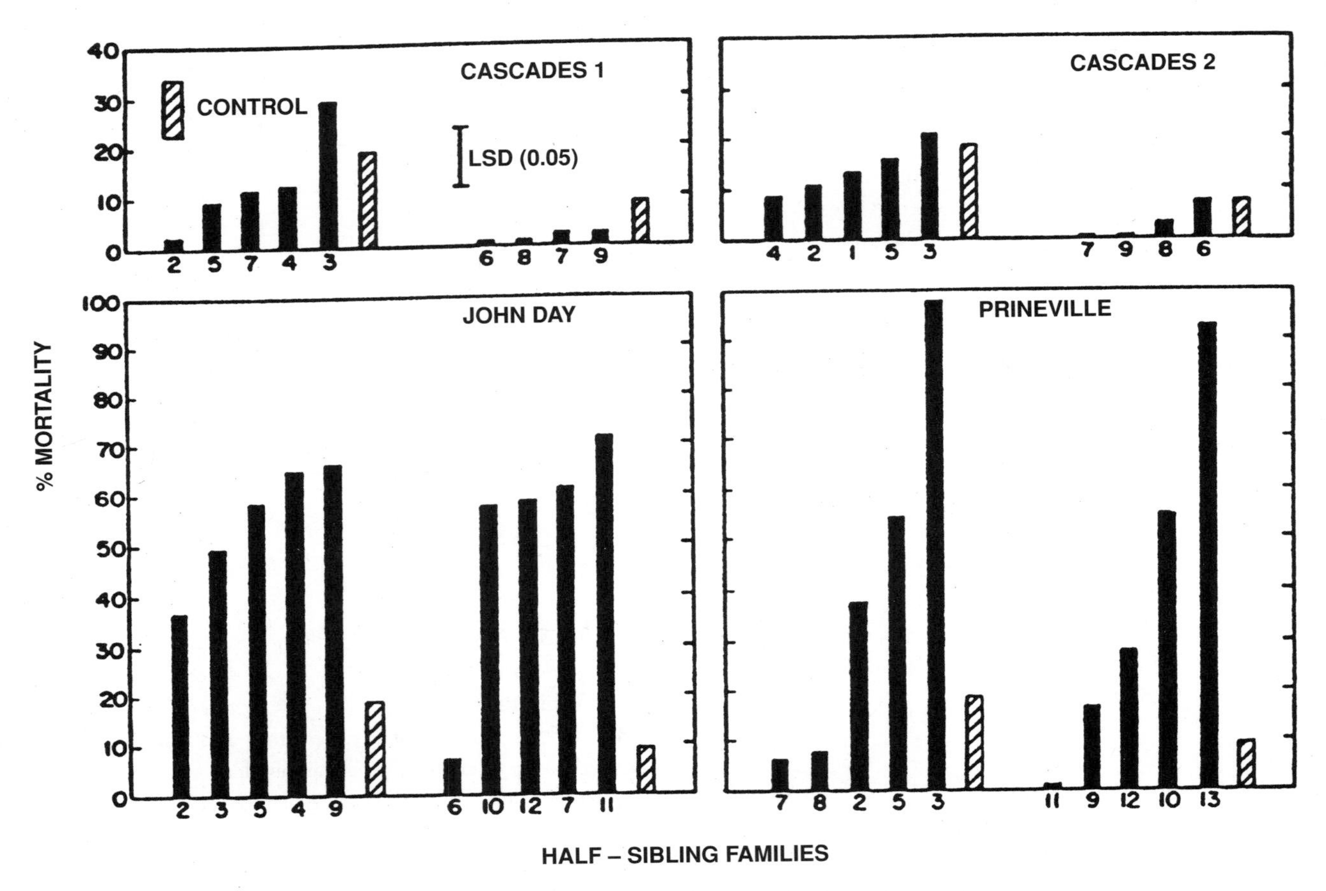

Figure 19.13. The mortality of western spruce budworm larvae fed an artificial diet mixed with 1- and 2-year-old Douglas-fir foliage. Bars represent mortality produced by seedlings from a single one-half sibling family. Cascades 1 and 2 are *Pseudotsuga menziesii* var. *menziesii* populations from west of the crest of the Cascade Range (a relatively mild, oceanic climate). John Day and Prineville are *Psme.* var. *glauca* populations from east of the Cascade Crest (a relatively harsh, continental climate). Control is the diet with no foliage added. Seedlings were grown in a common garden. (Adapted from Perry and Pitman 1983)

defensive chemicals are energetically costly to produce, plants should evolve to manufacture only what they need to prevent excessive loss to herbivores. Thus, for example, plants that are apparent to herbivores (i.e., easily found, such as long-lived trees) should defend themselves more heavily than unapparent plants such as ephemeral herbs (this aspect is often referred to as the apparency hypothesis).

2. *Carbon: nutrient balance hypothesis* (also called the environmental constraint hypothesis) (Bryant et al. 1983; Tuomi et al. 1988). This hypothesis postulates that the ability of a plant to produce defensive chemicals depends on the availability of carbon and other essential nutrients.
3. *Growth-rate hypothesis* (also called the resource availability hypothesis) (Coley et al. 1985; Coley 1987). In this view, fast-growing plants invest relatively little energy in expensive secondary chemicals because they are better able to tolerate herbivory than slow growers. The compensatory-continuum hypothesis and the limiting resource model (Wise and Abrahamson 2005) are basically elaborations on the growth-rate hypothesis.
4. *Growth-differentiation balance hypothesis* (Loomis 1932; Herms and Mattson 1992). This hypothesis holds that any factor that slows growth more than it does photosynthesis (e.g., shortages of water and some nutrients) (Herms and Mattson 1992) will produce an excess of carbohydrate that can be used to produce defensive chemicals. As Stamp (2003) explains it, "[Herms and Mattson (1992)] proposed a growth-differentiation continuum reflecting the allocation of resources to these processes. Competition in resource-rich environments selects for a growth-dominated strategy, whereas stress of resource-poor environments selects for differentiation-dominated strategy."

These hypotheses are not mutually exclusive and even overlap to one degree or another, particularly the last three, which all deal in one way or another with resource limitations. All have strengths and weaknesses, and no one is sufficient to fully explain the patterns seen in nature.

Comparing the optimal defense and the growth rate hypotheses provides a good illustration of the types of reasoning that scientists use to explain the dynamics of nature and also lays the groundwork for our later discussion of the importance of ecosystem-level defenses.

First, it is necessary to consider the trade-offs between costs and benefits of secondary chemicals. Costs are incurred because energy diverted to defense is (presumably) energy that cannot be used by plants for growth. Virtually all theories of plant defense are economic in the sense that they focus on cost-benefit analysis. The basic idea is that individuals within a given species evolve to optimize their allocation of energy to defensive chemicals, producing no more and no less than is required to balance the growth that is lost from herbivory against the growth that is foregone in the production of defenses. Because they are small molecules and present in low concentrations, qualitative chemicals are relatively inexpensive for plants to produce. However, precisely because they are small and therefore coded by relatively few genes, and because they have such a dramatic effect on herbivore fitness, qualitative compounds are thought to be especially vulnerable to the evolution of detoxification mechanisms on the part of herbivores. Herbivores can and do evolve resistance to simple defensive systems in a very short time. This has been especially evident in the rapidity with which insects evolve resistance to pesticides, and in the experience of agronomic plant breeders who "have a difficult time finding resistant plant genotypes that will remain resistant for the span of a resistance breeder's career" (Gould 1983). In contrast, quantitative compounds are expensive for plants to manufacture and maintain but are thought to be less vulnerable to coevolutionary response on the part of the herbivore.

Feeny, Rhodes, and Cates hypothesized that plants, or plant tissues, that were difficult for herbivores to find (*unapparent* in Feeny's terminology), either because they grew in species-rich communities or were ephemeral, would be defended primarily by qualitative-type compounds. A plant that is difficult to find will have relatively low herbivore pressure; thus, because the rapidity with which detoxification mechanisms are evolved in an herbivore population is likely to depend on how frequently the herbivore encounters the toxin, the evolutionary lifetime of toxins will be much longer in an unapparent than in an apparent plant or plant tissue. In contrast, both tissues that are not ephemeral (e.g., old needles on conifers) and apparent plants (i.e., those growing in species-poor systems or dominating species-rich systems) are, in Feeny's words, "bound to be found." Toxic defenses in these situations would be rapidly countered by herbivore populations; therefore, more stable, quantitative-type defenses are elaborated despite the fact that they are energetically more costly. An argument that is sometimes used in support of this hypothesis is that young leaves of woody plants, which are ephemeral in the sense that they soon become old, are low in defensive chemicals, particularly those that are quantitative. Feeny (1970), for example, found only 0.66 percent tannins in April foliage of *Quercus robur*, as opposed to 5.5 percent in September foliage.

Coley et al. (1985) also employ cost-benefit analysis to support the resource availability hypothesis, but, in contrast to the apparency hypothesis (which focuses on the pressures applied by the herbivore on the plant), Coley et al. believe the primary factor determining investment in defenses is plant growth rate. They give three reasons why slow-growing and fast-growing plants should differ in their production of defensive chemicals (Coley et al. 1985):

1. Slow growers usually occur in resource-limited environments, where replacing resources (e.g., nitrogen) lost to herbivores is more problematic than in relatively productive environments;
2. A given rate of herbivory removes a greater fraction of the net production of slow-growers than of fast-growers;
3. Fast-growing species that devoted the same proportion of their energy to defenses as slow-growers would experience a greater absolute reduction in growth.

There is strong evidence supporting the resource availability hypothesis. In general, phenol concentration is inversely related to nitrogen concentrations in tree foliage, suggesting (but not proving) that nitrogen is heavily defended when it is in short supply. As we discussed above, plants growing on infertile soils are often less palatable to herbivores than plants growing on fertile soils, and slow-growing, late-successional trees are generally less palatable to herbivores than fast growers that occupy the same site. On the other hand, shrubs and forbs that grow in forest understories in Alaska are more palatable to deer than shrubs and forbs that grow in open areas, even though the latter grow faster (Van Horne et al. 1988).

As with many competing theories in ecology, there are probably elements of truth in both the apparency and the resource availability hypotheses, but neither is likely to capture the whole truth.

19.2.3.4 Variation among Trees in the Same Population

The few studies addressing the issue have found that trees within the same population vary considerably in their defensive chemistry. Of the four Douglas-fir populations in figure 19.13, one had high levels of variability among half-sibling[3] families in their toxicity to western spruce budworm, and another had moderate levels of variability. Neighboring parent-offspring pairs and half-siblings of the neotropical tree *Quararibea asterolepis* vary more among themselves in leaf secondary chemistry than they do from unrelated individuals located at distance (Brenes-Arquedes and Coley 2005). Ruusila et al. (2005) found that different instars[4] of the moth *Epirrita atumnata* preferred different trees within birch stands. In other words, no single tree provides nutritious food throughout the larval development stage, which reduces the chance of the moth adapting to the tree's defenses (i.e., larvae on a single tree have a moving target to adapt to) (Ruusila et al. 2005). Note that what humans see as a monoculture, herbivores may perceive as a heterogeneous mixture.

Genetic variability within a host population could be an important factor stabilizing herbivore populations, particularly among plants that have predominantly qualitative-type defenses. For example, suppose a genetically homogeneous plant population contained a single alkaloid type. Evolution of a detoxification mechanism within an herbivore population would render the entire plant population vulnerable and permit the new herbivore genotype to proliferate rapidly. If the plant population contained 10 different alkaloid types, however, evolution of the ability to detoxify one of these would put only part of the population at risk. Furthermore, the new herbivore genotype would not increase rapidly for two reasons: (1) its food base (the now susceptible plant genotype) would be relatively limited and, (2) a proportion of its progeny would distribute to nonsusceptible genotypes and be lost.

The importance of plant genetic diversity in herbivore population dynamics was demonstrated by Dollinger et al. (1973), who studied herbivory on three species of lupines in Colorado, two of which were fairly homogeneous and one of which was heterogeneous with respect to alkaloid types. The heterogeneous species was the least grazed of the three. Susceptibility to herbivores has rarely been studied at that level in trees; however, McIntyre and Whitham (2003) found that different genotypes of hybrid poplar varied widely in their susceptibility to poplar bud gall mite *(Aceria parapopuli)*.

Sedentary insects (i.e., those spending many generations on a single tree) become highly adapted to the chemistry and probably the phenology of that tree. This was shown in the previously mentioned study by Alstad and Edmunds (1978), who transferred black pineleaf scale insects *(Nuculaspis californica)* among and within individual ponderosa pines. They found that scales survived much better when transferred within their "home" tree than when transferred to a "strange" tree (fig. 19.1).

19.2.3.5 Variation among Leaves of the Same Tree

Defensive chemistry and leaf palatability within the crowns of individual trees may vary considerably. For

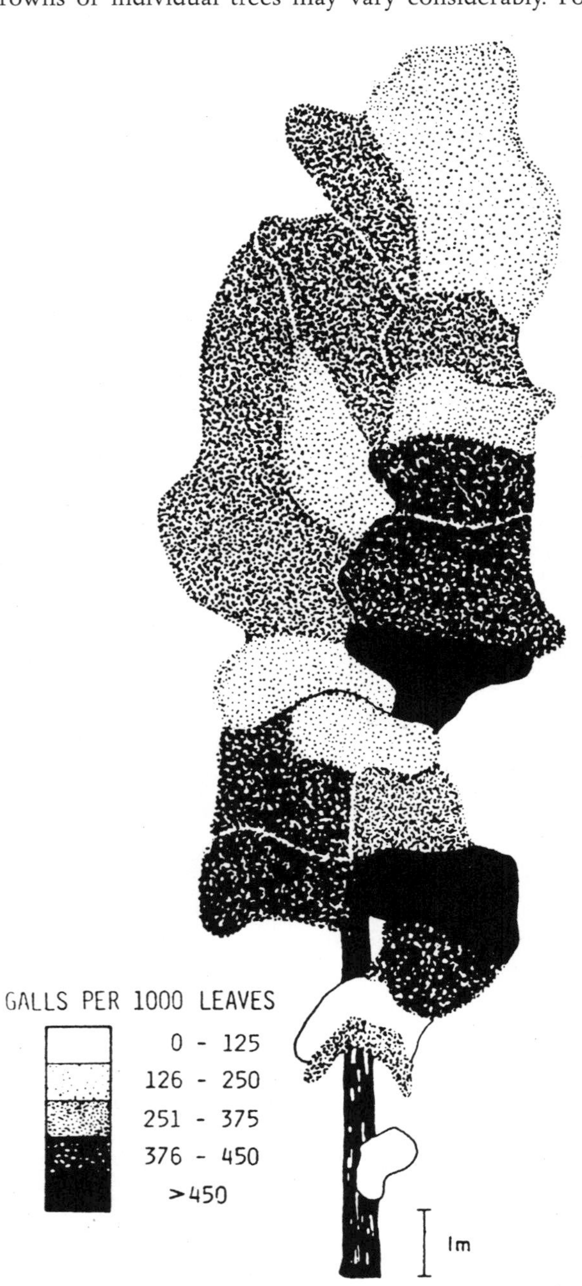

Figure 19.14. This schematic drawing of a 20.1-m *Populus angustifolia* tree shows how an estimated 53,000 galls (containing approximately 2,100,000 aphids) are distributed over 20 branches; the mosaic pattern of aphid settling is nonrandom and reflects an underlying mosaic pattern of host resistance. (Whitham 1983)

[3]A half-sibling family is a group of individuals that have only one parent in common. Seedlings grown from seed collected from a single tree will, of course, have the same mother but are likely to have different fathers and are therefore half-siblings.

[4]The different stages of development of an insect are called *instars*. The number of larval instars before attaining adulthood depends on species. For leaf-eating insects, it is the larval stages that are herbivorous.

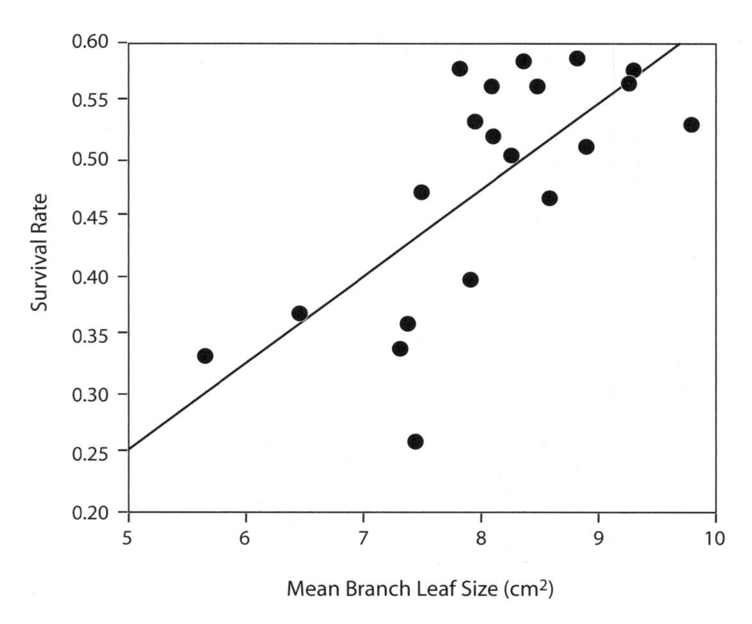

Figure 19.15. The survival of aphids during the colonizing period is highest on large-leaved branches. The survival of colonizing stem mothers per branch is plotted as a function of the mean leaf size of the host branch occupied ($r = 0.71$; $p < 0.01$; $n = 20$). All 20 branches are from the same tree. (Whitham 1983)

example, leaves of different ages usually differ in palatability, a phenomenon that we discuss in the following section. Palatability of leaves may also vary among individual branches on a single tree. Figure 19.14 shows the distribution of aphid *(Pemphigus betae)* galls within the crown of a cottonwood tree *(Populus angustifolia)*. Aphid densities vary not only with tree height but between adjacent branches and are correlated with identifiable leaf characteristics. Individual branches differ in the size of leaf that they produce, and as figure 19.15 shows, aphid survival is closely correlated with leaf size. Larger leaves have both lower levels of phenolics and higher survival of aphids. Variation in leaf size between branches remains consistent from year to year, and despite a complex life cycle that involves transferring from trees to an alternate, herbaceous host, colonizing aphids have the remarkable ability to discriminate fairly accurately between susceptible and nonsusceptible branches. Similarly, measures of larval performance for the oak leaf miner *(Tischeria ekebladella)* vary more within the crowns of oak trees than between trees (Roslin et al. 2006).

In some trees, leaves of the same age on a single branch differ in palatability. Figure 19.16 shows relative tannic acid values of individual leaves on single branches of yellow birch *(Betula lutea)* and sugar maple *(Acer saccharum)* (Schultz 1983). Horizontal axes represent mean values for all leaves on the branch. Vertical bars represent deviation of individual leaves from the mean, with bars above the horizontal axis indicating lower than average tannic acid content. In yellow birch, the youngest leaves are at the left of the graph, while in sugar maple, all leaves are the same age. The two youngest birch leaves (both 10 days old) have very different tannin contents, and sugar maple leaves within a single cluster differ as well.

Why would trees evolve to have variable levels of defenses within a given crown? The answer may lie in simple economics. Secondary chemicals are energy expensive, and moderate levels of herbivory may cost the tree less energy than producing enough chemicals to defend every leaf. In addition, forest ecosystems may be both more productive and more stable when moderate levels of herbivory occur (we discuss why later in the chapter).

19.2.3.6 Variation over Time

Levels of secondary chemicals (and therefore palatability) may change over time for various reasons, including age, environmental stresses that influence tree vigor, or in response to grazing in other parts of the tree or in nearby trees.

In many tree species, foliage becomes less palatable as it ages. Feeny (1970) showed that oak *(Quercus robur)* leaf tannins increased from 0.66 percent in April to 5.5 percent in September and calculated that winter moth larvae would have to consume four times more leaf material in September than in April to maintain the same growth rate. Similar changes in leaf quality with age are common in other species as well. Raupp and Denno (1983) compiled data from 30 studies comparing insect performance on old and new tree foliage; insects fared better on old foliage in only three studies. There are notable exceptions to this trend. For example, the various sawflies prefer old rather than new foliage of conifers. New foliage of jack pine, a common

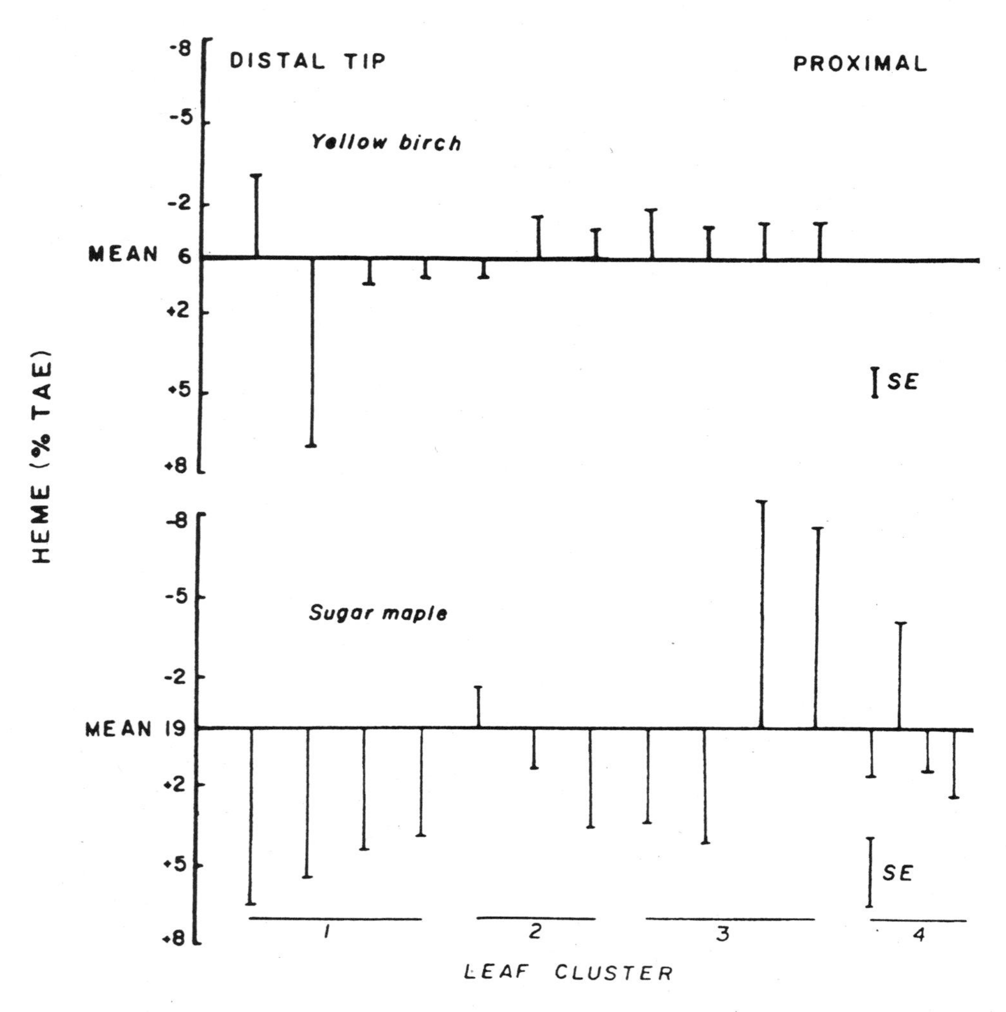

Figure 19.16. The value in terms of digestibility of individual leaves on individual sprays of yellow birch *(Betula alleghaniensis)* and sugar maple *(Acer saccharum)*. Horizontal axes are mean values as percentage tannic acid equivalents (TAE) of all leaves on the branch. Each vertical bar represents one leaf expressed in units of deviation from the branch mean. Upward-deviating bars indicate more-digestible, higher-quality leaves; downward deviations represent less-digestible leaves. In both cases, branch terminus is at left; for yellow birch, this means that the two left-most leaves are less than 10 days old. Sugar maple leaves are all evenly aged and grow in clusters as indicated. One standard error (SE) of the branch mean is shown at lower right. HEME, hemoglobin assay. (Schultz 1983)

host of several sawfly species, contains diterpene resins that inhibit sawfly feeding and are not present in old foliage (Ikeda et al. 1977). Similarly, *Chrysophtharia agricola,* a defoliator of eucalyptus species, prefers adult to juvenile foliage (Lawrence et al. 2003).

It is not surprising that the ability of insect larvae to detoxify defensive chemicals increases as the larvae age. In particular, the activity of mixed-function oxidases (i.e., enzymes that catalyze oxidation of various substrates and are thought to be the most important detoxification mechanism in herbivores) is more readily induced in late than in early larval instars (Chan et al. 1978; Gould and Hodgson 1980). Thus, as foliage and larvae age together, stronger defense on the part of the plant is countered by more effective circumvention of that defense by the herbivore.

Because secondary chemicals (especially quantitative types) are energy expensive, their production may be curtailed when trees are stressed. Poor growing conditions do not always lead to reduced chemical defenses. As discussed earlier, highest phenol levels are usually found on the most nutrient-poor sites. Nevertheless, insect outbreaks are often associated with drought, nutrient deficiency, or the stress that results from overcrowding in densely stocked stands (Rhoades 1983); various factors might contribute to that, including lowered chemical defenses (Mattson and Haack 1987; Ayres and Lombardero 2000). Thinning overstocked stands has been found to decrease the severity of bark beetle outbreaks (e.g., Mitchell et al. 1983), but Australian eucalypts are at greatest risk to defoliating insects when standing alone in pastures (Lowman and Heatwole 1992).

Plant chemical defenses are in many cases **inducible.** Chemical defenses that are always present are called **constitutive;** however, inducible defenses are produced in response to herbivores. Inducibility takes a variety of forms (see reviews by Karban and Baldwin 1997; Baldwin et al. 2006). Insect feeding (or attack by pathogens) may induce defenses within the attacked leaf, and also within unattacked leaves on the same plant; foliage that is produced in the year following defoliation may produce higher-than-normal levels of secondary chemicals. Haukioja (1990) terms this **delayed induced resistance** (as opposed to the **rapid induced resistance** that occurs within leaves of the attacked generation).

It is commonly observed that when at endemic levels, phytophagous (i.e., leaf-eating) insects consume only a small proportion of a given leaf. One explanation is that leaves respond to wounding by increasing levels of defensive chemicals. This is in fact the case. A meta-analysis[5] of 68 studies done on woody plants between 1982 and 2000 revealed a pattern of increased leaf phenolics and decreased larval weights following wounding (Nykanen and Koricheva 2004). However, plant responses varied with species (deciduous trees are more responsive than conifers) and stage of leaf development (wounding young leaves elicits a greater response than wounding old leaves does). A more recent study found similar responses; artificially damaging individuals of three plant species in a Mexican tropical dry forest resulted in lower late-season herbivore damage than in individuals protected from damage (Boege 2004). Two of the species had increased levels of total foliar phenolics and condensed tannins.

In a remarkable phenomenon dubbed **talking trees** by the press, unattacked plants may raise their levels of chemical defenses when a neighboring plant (of the same or a different species) is attacked (see the review by Baldwin et al. 2006). Evidence for this was first presented by Rhoades (1983), and shortly after, Baldwin and Schultz (1983) showed that damage to maple and poplar trees resulted in increased phenolics in undamaged trees in the same airspace (but not in root contact). Farmer and Ryan (1990) found that the volatile chemical, methyl jasmonate, a common plant secondary compound, induced formation of proteinase inhibitors in three plant species from two different families. When sagebrush (known to produce methyl jasmonate) was grown in an enclosed chamber with tomato plants, proteinase inhibitors accumulated in leaves of the latter (Farmer and Ryan 1990). Of course sagebrush and tomatoes do not grow together in nature, but the important point is that one plant species can elicit a defensive response in another species that is not even a close kin, which argues that the phenomenon could be quite general. The significance of interplant communication in nature is unknown; however, this is a rapidly evolving field of inquiry (Baldwin et al. 2006).[6]

Virtually all research on the detailed biochemistry of inducible defenses has been done on cultivated or wild herbaceous plants (e.g., tomato, potato, wild tobacco) (Green and Ryan 1972; Ryan 1983; Kessler and Baldwin 2002; Kessler et al. 2004; Chen et al. 2005; Felton 2005). Scores of plant genes are turned on by herbivory, although the function of most of these is unknown. Secondary chemicals known to be involved include volatile compounds (e.g., terpenoids) and various chemicals that disrupt the ability of herbivores to digest food. Volatiles act as signals that attract natural enemies of herbivores (Dicke and van Loon 2000) and stimulate induced defenses in nearby plants (Baldwin et al. 2006). Proteinase inhibitors are proteins that bind tightly with and thus inhibit proteinases, the major food-digesting enzymes of microorganisms and animals (including insects). Interestingly, when plants are cooked, proteinase inhibitors are converted from toxic to nutritious proteins, a fact that led Leopold and Ardrey (1972) to suggest that the invention of cooking may have played a large role in the success of primitive humans

19.2.4 Natural Enemies

Why is the world green? For many years, ecologists have wondered what forces kept herbivores from eating all the plants. In 1960, Nelson Hairston, Frederick Smith, and Lawrence Slobodkin published their **green world hypothesis,** which holds that the world is green because of predators. In chapter 11 we discussed the strong evidence for trophic cascades in which feeding by predators acts to benefit plants.

Herbivores are preyed upon by a complex of natural enemies, including pathogens, viruses, predators, and parasitoids. Parasitoids differ from predators in the same sense that a tapeworm differs from a Bengal tiger: the parasitoid is associated more intimately, often completing much of its life cycle within the host, and its feeding consists of many small bites rather than a few large ones.[7] Parasitoids are currently believed to constitute from 8 to 25 percent of all insect species (however, that proportion may be much higher if host specificity has been underestimated) (Smith et al. 2006). Douglas-fir tussock moth is preyed upon by at least 60 different species of parasitoids, many of which are wasps and flies that lay eggs within tussock-moth eggs, larvae, or cocoons. Mortality from cocoon parasitization can be greater than 50 percent (Torgersen and Dahlsten 1978). Parasitism of Lepidoptera larvae (i.e., caterpillars) exhibits highly complex, three-trophic-level patterns. In Canadian forests, the degree to which caterpillars of a given species

[5]Meta-analysis is a statistical technique for assessing general trends common to different studies on the same topic.

[6]Perry and Pitman (1983) found that a high proportion of Douglas-fir trees in an Oregon stand altered their defensive chemistry coincident with the buildup of spruce budworm in the area but before infestation of the stand itself. These results must be interpreted carefully, however, as intertree communication is not the only possible explanation.

[7]Hassell (2000) reviews host-parasitoid interactions.

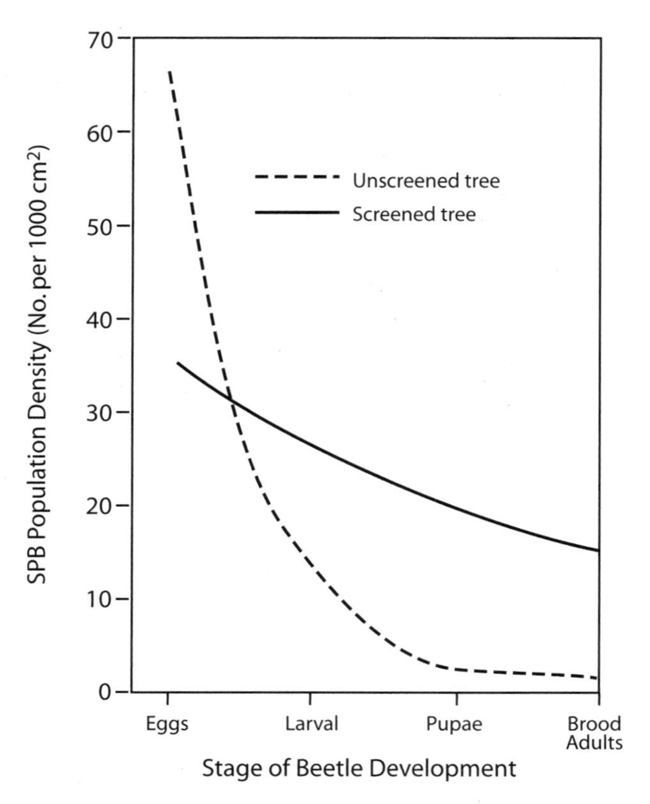

Figure 19.17. The survival of southern pine beetle (SPB) both in trees screened to prevent woodpecker predation and in unscreened trees. (Kroll, unpublished data)

are parasitized depends on both host tree and parasitoid species, resulting in a complex interactive landscape in which caterpillars of a given species find relative safety on some tree species and relative vulnerability on others, while caterpillars of other species also find areas of relative safety or vulnerability, but on different tree species (Lill et al. 2002). Similar patterns have been seen in forests of the United States (Le Corff et al. 2000; Barbosa et al. 2001). One implication is that parasitoids play a significant role in determining the host range of Lepidoptera (Lill et al. 2002).

Predators of herbivorous insects include both predaceous insects, such as spiders and ants, and vertebrates (primarily birds, but mice and other mammals as well). Considerable evidence indicates that predators (along with parasitoids) are significant regulators of herbivore populations, particularly when the latter are at low levels.[8] In one study in Japan, 97 percent of the scale insects in a heavily infested pine stand were killed by a single species of predatory beetle within one 4-week period (McClure 1986). Kroll (unpublished, see Thatcher et al. 1980) screened trees infested by the southern pine beetle to determine the effect of woodpecker feeding on the beetle populations. He found significantly greater beetle survival on screened trees than on trees accessible to woodpeckers (fig. 19.17). Thirty species of birds are known or suspected to feed on Douglas-fir tussock moth; with predation rates ranging from 3 to 30 percent of eggs, approximately 19 percent of larvae, and 19 to 49 percent of pupae (Mason et al. 1983; Torgersen et al. 1990). Torgersen et al. (1990) studied mortality of western spruce budworm larvae *(Choristoneura occidentalis)* on trees from which either birds, ants, or both birds and ants had been excluded. When budworm was at low densities, 10 to 15 times more budworm survived on trees from which both birds and ants were excluded than on trees open to birds and ants. Natural enemies were less effective at high budworm densities but nevertheless reduced budworm populations by about one-half. Crawford and Jennings (1989) found similar patterns for birds feeding on eastern spruce budworm *(C. fumiferana)*. When insect populations were low, birds consumed 84 percent of larvae and pupae, but that dropped to 22 percent when insect populations were at intermediate levels. Crawford and Jennings point out that the forest stands they studied "were dense, even-aged, [and] with little understory," factors generally not conducive to good bird habitat (chapter 10). They suggest that silvicultural operations (e.g., thinning) would increase numbers of birds in those stands and concluded that birds "are capable of dampening the seriousness of spruce budworm infestations when habitats are suitable for supporting adequate populations of these effective predators." Marquis and Whelan (1994) went a step further and documented how screening birds away affected growth of white oak *(Quercus alba)* saplings. Screened trees had twice as many herbivorous insects and approximately one-third lower biomass production than unscreened trees.

Earlier we saw that natural enemies use volatile chemicals released from damaged leaves to locate their prey. Some plants encourage predators to remain in the vicinity by offering a reward. The most well-documented example is the mutualistic relationship between species of swollen-thorn acacias (*Acacia* sp.) and associated ants in central America (Janzen 1966). Certain species of ants within the genus *Pseudomyrmex* nest only within the swollen-hollow thorns characteristic of some acacia species. Ant-acacias have various characteristics, such as enlarged foliar nectaries and year-round leaf production, that differ from other acacia species and undoubtedly have evolved to encourage ant colonization. The ants, in turn, protect the tree, not only from insect herbivores but from encroaching vines as well. Interestingly, ant-acacias do not have the chemical defenses common to other acacia species, presumably because they are adequately protected by ants. Other mutualistic associations can influence herbivory either directly or indirectly. In chapter 8 we saw that fungal endophytes (i.e., microfungi living symbiotically within leaves and bark) protect their hosts against herbivory. Mycorrhizae influence both the feeding success of herbivorous insects (Gehring and Whitham 2002) and the susceptibility of the herbivores to natural enemies (Gange et al. 2003).

[8]A rich literature concerns the effect of natural enemies on agricultural pests (e.g., *Ecological Applications* 9[2], 1998).

Although most ecologists acknowledge that natural enemies play a role in herbivore population dynamics, controversy exists over how effective they are when herbivore numbers are rapidly expanding. The key questions are whether individual predators and parasitoids can increase their consumption rate (i.e., functional response) and their population numbers (i.e., numerical response) quickly enough to keep up with the explosive growth that can occur within insect herbivore populations. Evidence is mixed. F.P. Hain (unpublished) found no correlation between numbers of natural enemies and levels of southern pine beetle at three locations in North Carolina (fig. 19.18); however, Kroll and Fleet (1979) showed strong correlation between southern pine beetle density and numbers of woodpeckers, which were up to 50 times more numerous in beetle-infested than in uninfested stands. Although the abundance of woodpeckers in beetle-infested stands was technically a numerical response, it most likely reflected a convergence of birds from the surrounding region on an area with abundant food rather than a dramatic increase in regional woodpecker populations (but this is a fine point, as the response was effective whatever its source). Parasitoids also responded quickly to increased beetle density in Louisiana (Goyer and Finger 1980) (fig. 19.19). Since the generation time of parasitoids is tied quite directly to that of their insects hosts (the host body being the incubation chamber for parasitoid larvae), their numerical response probably reflects real population growth.

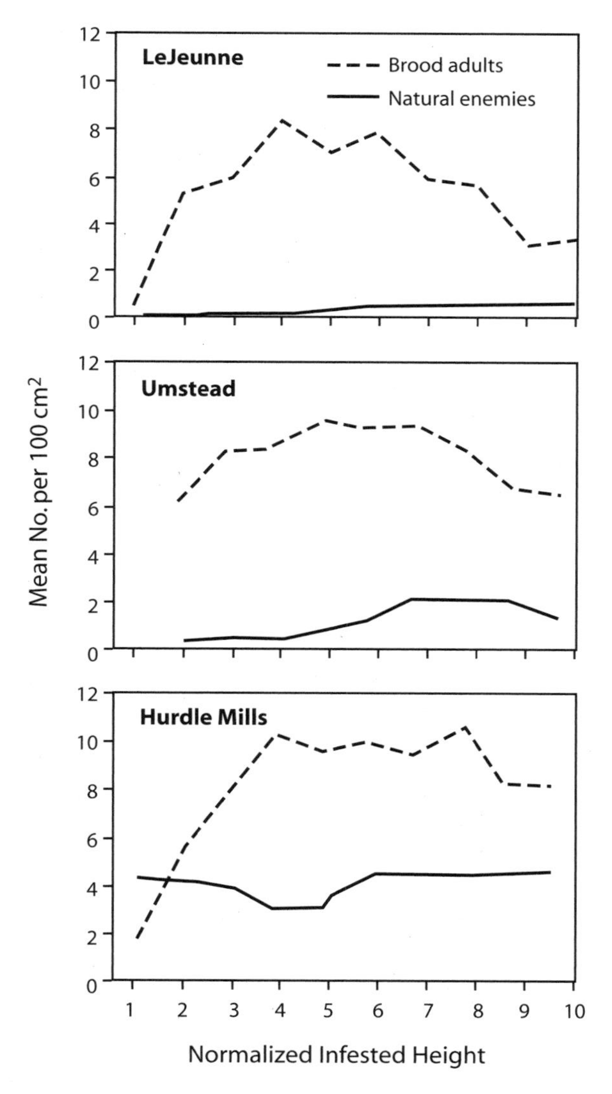

Figure 19.18. The numbers of natural enemies (including parasitoids) relative to numbers of southern pine beetle brood adults at three normalized infested height locations in North Carolina. (Hain 1987)

Holling (1959) studied predation on the European pine sawfly *(Neodiprion sertifer)* by three species of small mammals, the masked shrew (*Sorex cinereus cinereus*), the short-tailed shrew *(Blarina brevicauda talpoides)*, and the deer mouse *(Peromyscus maniculatus bairdii)*. Larvae of this insect drop from trees in early June, spin cocoons in duff, and remain there until September, when they emerge as adults. In Holling's study, percent predation of cocoons rose sharply with increases in cocoon density up to a certain point, then dropped off. However, the three rodent species responded differently (fig. 19.20). The rapid increase in percent consumption by *Blarina* was strictly functional (i.e., individuals ate more), whereas the response of the other species was both functional and numerical. The peak in percentage consumption shown in figure 19.20 can be thought of as a threshold: if sawfly cocoons surpass the level corresponding to the peak, they are beyond the control of these small mammal predators. Holling stresses that multiple species of predators are extremely important in regulating populations, not only because total consumption is increased but because the differing response pattern of the rodents produces a broader peak, which makes it more difficult for the sawfly to escape regulation.

In summary, natural enemies consume large numbers of insect pests, and this almost certainly contributes to preventing outbreaks. In 1972, for instance, Douglas-fir tussock moth populations began to build on the Eldorado National Forest (California), but populations crashed in 1979, before attaining levels that would cause visible tree defoliation. Mason et al. (1983), who followed that population cycle closely, attributed the crash to predators and parasitoids. Once pest populations climb beyond some threshold, however, predators are unlikely to damp an outbreak by themselves, because they do not have the capacity to keep up numerically with a rapidly growing pest population. Parasitoids, on the other hand, probably can keep up numerically, as can infectious diseases and viruses. Insects are attacked by highly species-specific baculoviruses, which Myers (1993) argued were instrumental in controlling population cycles of outbreaking moths and butterflies (i.e., Lepidoptera). However, the natural enemy complex must be considered in its totality. Models that incorporate *only* species-specific pathogens (e.g., viruses) or *only* predators do not fully capture the population dynamics of outbreaking

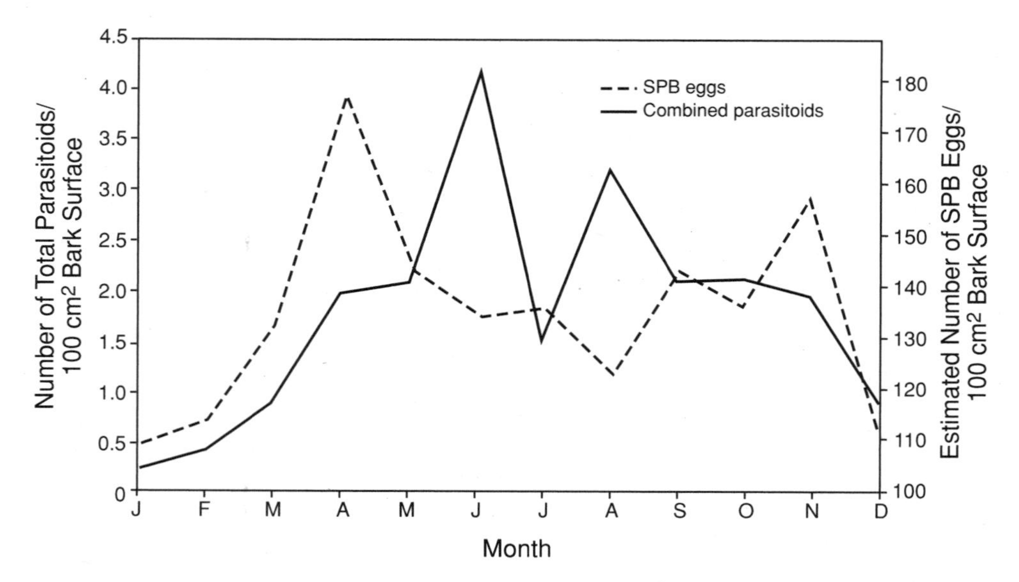

Figure 19.19. The seasonal abundance of parasitoids relative to numbers of southern pine beetle (SPB) eggs in SPB-infested trees in Louisiana. (Goyer and Finger 1980)

insects (Stone 2004). One model that incorporates both species-specific pathogens and generalist predators successfully reproduces the population cycles of gypsy moth (Dwyer et al. 2004). The dynamics of nature are seldom explained by single factors.

Two lines of indirect evidence support the importance of natural enemies of all kinds in helping to prevent pest outbreaks. First, there are numerous examples of sharp increases in insect herbivore numbers after predaceous insects have been removed by chemical sprays (Ripper 1956; McClure 1977, 1986). Second, herbivores that are relatively innocuous in their native habitat often become highly destructive when introduced to new areas, and in at least some cases such exotic pests can be controlled by introducing natural enemies from the pest's home territory (termed **biological control**). One of the more striking examples is the larch casebearer, a native of Europe. Not a serious problem in stands of its native host tree *(Larix decidua)*, larch

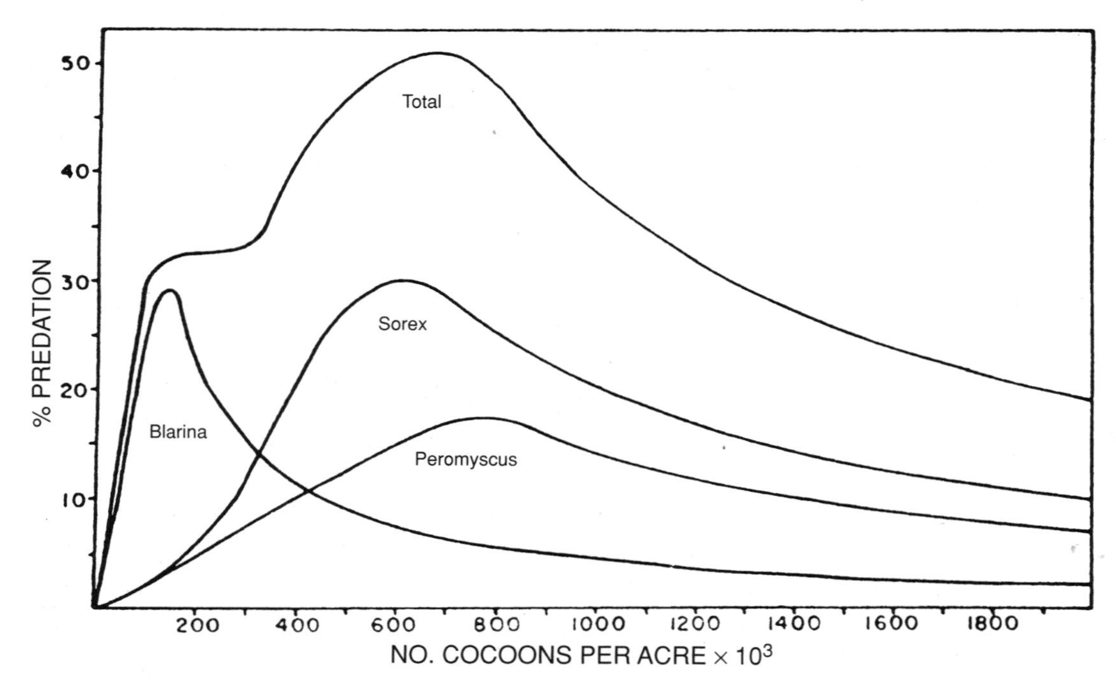

Figure 19.20. Functional and numeric responses combined to show the relationship between the density of European pine sawfly cocoons and the percentage predation by *Blarina brevicauda* (short-tailed shrew), *Sorex cinareus* (masked shrew), and *Peromyscus maniculatus* (deer mouse). (Holling 1959)

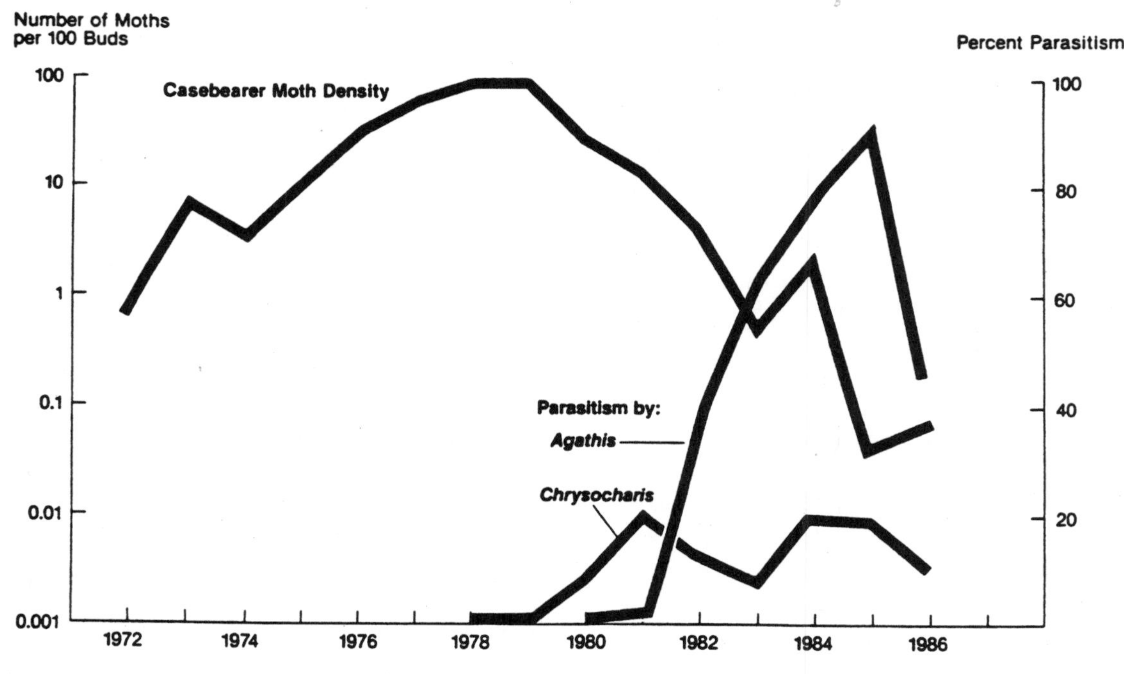

Figure 19.21. An example of biological control. In 1978, the exotic pest larch casebearer had attained densities of approximately 100 moths per 100 buds on larch trees in the Blue Mountains, Oregon. In 1978 and 1980, two parasitic wasp species from the casebearer's native locale (Europe) were introduced to infested stands, and by 1986, casebearer densities had dropped to only 0.1 moths per 100 buds. Note the log scale on the *y* axis. (Ryan et al. 1987. Reprinted from the journal of Forestry, published by the Society of American Foresters, 5400 Grosvenor Lane, Bethesda, MD 20814-2198. Not for further reproduction)

casebearer became a significant pest in North America following its first appearance in New York in 1886. Collection and release of parasitic wasps from the casebearer's native locale brought the pest under control in eastern North America and appears to be working in the west as well (Ryan et al. 1987) (fig. 19.21). Ryan (1987) gives an overview of biological control in forestry. (Biological control is not without risks, however. As we saw in chapter 7, natural enemies introduced to control invasive pests have in some instances significantly impacted native species.)

19.2.5 Stand and Landscape Diversity

> Cropping systems are highly pest prone for many reasons, of which simplicity (low biodiversity) and spatial uniformity are key.
>
> COYLE ET AL. 2005

There is widespread belief that as Vouté (1964) states, "mixed forests are less susceptible [than monocultures] to outbreaks," a phenomenon termed *associational resistance.* Gibson and Jones (1977) argued that monocultures have often led to an increase in the severity of insect and disease problems and set the stage for "the establishment and spread of pests of all kinds" (Gibson et al. 1982). However, others have concluded that associational resistance is not a general phenomenon but rather depends on a variety of factors including the tree species and herbivores that are involved (Watt 1992, Koricheva et al. 2006). There are at least three reasons why a clear, general statement cannot be made about this issue. First, different situations may yield quite different results. Second, as discussed earlier, stands composed of single tree species may nevertheless present a great deal of diversity to herbivores, both among trees and within individual trees, so what humans see as a monoculture herbivores may not. Third, outbreaks are phenomena of both stands and landscapes, and the importance of stand structure may be overwhelmed by its landscape context. For example, spruce budworm kill more trees in extensive pure stands of balsam fir than they do in pure stands that exist as islands in a sea of deciduous trees (Cappuccino et al. 1998).

There are numerous examples of lower susceptibility in diverse stands than in monocultures. For example, mahogany *(Khaya nyasica)*, a tree that grows naturally in species-rich forests, becomes quite susceptible to insects when grown in monoculture (Perry and Maghembe 1989). Vasechko (1983) reported that mixtures of tree species planted in the Russian steppes were less affected by insects than were monocultures. Densely stocked stands of loblolly and shortleaf pines are more heavily attacked by southern pine beetle when admixed hardwoods have been removed than when the hardwoods are present (Schowalter and Turchin 1993), and loblolly pines have higher levels of fusiform rust when herbaceous weeds are controlled than when they are not (Zutter et al. 1987). Balsam fir is less susceptible to the eastern spruce budworm when mixed with hardwoods than when in pure stands (Su et al. 1996, Needham et al. 1999). In Quebec, removing hardwood shrubs makes coniferous plantations more susceptible to some types of insects (Robert Gagnon, Ministère Des Forêts Du Québec, personal communication). Similarly, infection of Douglas-fir by Armillaria root disease increases as the density of admixed paper birch decreases (Baleshta et al. 2005).

But there are also examples in which diversity leads to greater susceptibility (i.e. *associational susceptibility*). For example, depending on the tree species involved, diverse stands can be more susceptible than monocultures to the virulent pathogen, sudden oak death (which attacks numerous tree species), because species that can survive infection serve as "Typhoid Mary's," providing reservoirs of inocula and thereby increasing the infection rate of species that are usually killed (Rizzo and Garbelotto 2003).

Two recent meta-analyses compared herbivory between mixed species forests and monocultures. Vehvilainen et al. (2007) analyzed results from seven long-term experiments involving three tree species (silver birch, black alder, and sessile oak) in the boreal and temperate zones of western Europe. Results depended on tree species and insect feeding guild. Leaf miners were lower in mixture than monoculture for all tree species, as was overall herbivory on birch. However, overall herbivory on alder and oak was higher in mixture than in monoculture. Vehvilainen et al. concluded that their results "suggest that the mechanisms behind the effects of tree species diversity on herbivores may be complex and most probably vary both between forests consisting of different tree species as well as between tree species within a forest."

One significant (and not surprising) factor influencing herbivory in mixtures is what a herbivore will eat in relation to what tree species are in the mixture. In their analysis of 119 studies distributed world-wide, Jactel and Brockerhoff (2007) found that herbivory by oligophagous species (i.e. specialists that feed on a single tree genus or family) was "virtually always reduced" in diverse forests compared to monocultures. As some of the previous examples illustrate, the protective effect was particularly strong in mixtures of conifers and hardwoods. In contrast, cases in which all tree species in a mixture were palatable to a polyphagous herbivore (i.e. generalists that feed on more than one family) could result in associational susceptibility, especially when one species is more palatable than others. For example, gypsy moths eat both oaks and pines, but prefer oaks. White pines growing with oaks are more heavily attacked by gypsy moths than when growing in pure stands, because the relatively large populations of moth larvae on the oaks serve as source populations to colonize the pines (Brown et al. 1988).

Exotic tree plantations present a special and somewhat complex case. In some cases, exotic plantations have experienced significant pest problems (Perry and Maghembe 1989). In Scotland, plantations of the introduced lodgepole pine are heavily attacked by pine beauty moth, while adjacent plantations of the indigenous Scots pine are untouched (Watt 1992). *Pinus radiata* plantations have been severely affected by Dothistroma needle blight in Brazil, Kenya, Zimbabwe, New Zealand, and South Africa (*P. radiata* and Dothistroma are both exotics in these countries) (Woods et al. 2005). Nevertheless, considering the vast areas over which *Pinus, Eucalyptus,* and *Acacia* have been planted in the Southern Hemisphere, they have been relatively free of insect and pathogen problems (Wingfield 2003). In fact, the impressive growth of these species as exotics is attributed in part to loss of their natural enemies.[9] However, "the artificial barrier between exotic and plantation species and pathogens is . . . crumbling. Despite intensive efforts to exclude pests and pathogens from countries dependent on plantations of exotic trees, new and seriously damaging incursions are occurring with increasing frequency" (Wingfield 2003). There is evidence that pathogens native to the area in which exotic trees are planted (once considered to be host specific and no threat to introduced trees) are adapting to the foreign trees (Wingfield 2003).[10] The worst of both worlds occurs when a given tree species is planted outside of its normal range (or habitat) and yet is still findable by its herbivores. A dramatic example occurred on the Cumberland Plateau in Tennessee, where large areas of native hardwood forests were cleared for loblolly pine plantations. Although the area is somewhat north of loblolly's natural range, the southern pine beetle located and heavily infested the plantations. The epidemic was so severe that plantations were abandoned, with an economic loss estimated in excess of one hundred million dollars (Evans 2003).

If we are ultimately to understand how species diversity affects outbreaks of insects and pathogens, we must ask questions about mechanisms. What factors control pests and pathogens, and how are these factors influenced by plant diversity? Three factors in particular are likely to be influenced by the diversity of stands and landscapes: resource concentration, the abundance of natural enemies, and cooperative interactions among plants.

[9] Wingfield (2003) points out that these tree species are behaving like exotic, invasive weeds.

[10] These developments call to mind Feeney's phrase "bound to be found."

19.2.5.1 Resource Concentration

One of the most important factors influencing the growth of herbivore populations is the concentration of host plants within stands and across landscapes, termed by Root (1973) the **resource concentration hypothesis.** Herbivores that encounter mixtures of host and nonhost species have less food per unit area than in monoculture; consequently, they spend more time searching, which is both energy expensive and increases exposure to predators (Schowalter and Lohman 1999). In the Jactel and Brokerhoff (2007) study, the protective effect of tree species mixtures increased with the proportion on non-host trees, indicating that resource concentration played a significant role.

Many pest problems in forestry have arisen because commercially valuable trees were planted not only as stand monocultures, but as landscape monocultures. For example, before colonization by Euro-Americans, loblolly and slash pines were restricted to bottomlands along streams in the southeastern United States. Perkins and Matlack (2002) estimate that stands of these species were separated from one another by 610 to 950 m of upland forest and savanna. With the expansion of plantations, the average separation of clusters of loblolly and slash pine has shrunk to only 20 to 50 m. Perkins and Matlack (2002) conclude that "these changes have increased the connectivity of the landscape as perceived by the southern pine beetle and fusiform rust, potentially facilitating their spread between previously separated watersheds."

19.2.5.2 The Enemies Hypothesis

Diverse plant communities may support higher numbers of natural enemies, a well known phenomenon in agroecosystems that is referred to as the "enemies hypothesis". Does it hold in forests? We saw in chapter 10 that bird species numbers are often positively associated with the number of canopy layers in forests; similarly, Vouté (1964) states that the ant community of a forest depends on density of the shrub layer. Many species of parasitoids require nectar or pollen as an energy source during their adult phase (DeBach 1974), which raises the possibility that flowering plants are particularly important in maintaining populations of these predatory insects within conifer-dominated forests (Vouté 1964).

In order to discern the ecological dynamics, it is usually (and probably always) necessary to go beyond simple measures of diversity and ask questions about the functioning of specific community assemblages. A good example was provided by Riihimaki et al. (2005), who studied predation and parasitism of the autumnal moth *(Epirrita autumnata)* when its host, silver birch, was grown in monoculture or in mixture with either black alder, Norway spruce, or Scots pine. The only significant difference occurred with ants in the birch-pine mixture. Apparently the ants were attracted to aphids on the pines, and from the pines radiated out to prey on moth larvae on the birches. Neither larval parasitism not predation by spiders was affected by tree species mixtures.

The abundance of natural enemies may be influenced by structural diversity independently of tree species diversity. In both the northwestern and the southeastern United States, structurally complex old-growth forest canopies support a much higher ratio of predatory insects to plant-eating insects than the more homogeneous canopies of plantations (Schowalter 1989). Various factors might explain that difference, including the fact that old-growth canopies support a larger diversity of epiphytes.

Earlier we saw that landscape patterns can strongly affect herbivore dynamics by presenting a homogenous or heterogeneous food source. Landscape patterns may also influence the abundance of predators, especially where forests exists as small islands within a sea of farms. Forests isolated within agricultural lands have relatively low ratios of parasitoids to their tent caterpillar prey (Roland and Taylor 1997; Hassell 2000) and of predatory beetles to their bark beetle prey (Ryall and Fahrig 2005). But one must be cautious about the concept of fragmentation and its implications for ecological processes. Depending on circumstances (especially the landscape context), fragmentation may inhibit insect infestations (e.g., loblolly pine existing as scattered stands within a sea of upland forests), or it may promote them by weakening the natural enemy complex. Tsharntke and Brandl (2004) review how fragmentation influences plant-insect interactions.

19.2.5.3 The Chemical Environment

Volatile chemicals given off by nonhost plants may confuse insects by masking olfactory cues that they use to locate hosts. According to Schowalter and Lohman (1999)[11]: "Herbivorous insects are highly sensitive to host spacing and to confusion by the presence of non-host plant species.

> What possible evolutionary advantage could a tree of one species gain by producing chemicals that repel the enemies of another tree species? In a community driven solely by competition among different tree species, that sort of "helping" behavior makes no evolutionary sense. Of course the tree producing the chemical could be doing it for quite another, selfish reason, with the repellent effect merely an incidental spin-off. However, "enemies of friends are enemies" comes to mind (chapter 11); recall from chapter 11 that a relatively high proportion of interactions among potential competitors may in fact be mutually beneficial.

[11] See Schowalter and Lohman (1999) for original references.

Different plant species produce different blends of volatile chemicals, which insect herbivores use as host cues. Volatiles from non-host species are non-attractive, or even repellent, to a given insect." Hardwoods have been found to release several volatile chemicals that repel conifer-feeding bark beetles (Jactel and Brockerhoff 2007).

19.2.6 Climate

Climate affects the growth of herbivore populations directly through its effect on fecundity (e.g., survival may be greater in warm than in cold winters, brood production may increase with longer warm seasons), or indirectly by altering the effectiveness of other regulatory mechanisms (e.g., transmission of fungal pathogens by insect vectors, tree defenses, populations or tracking behavior of natural enemies) (Ayres and Lombardero 2000). For example, levels of parasitism on herbivorous insects correlate closely with year-to-year variation in precipitation; areas with high variabililty in yearly precipitation have lower levels of parasitism than areas in which yearly precipitation is more uniform (Stireman et al. 2005).

Any factor that affects tree physiology can influence levels of chemical defenses or foliar nutrients and thus alter food quality for the herbivore (Ayres and Lombardero 2000). Earlier we saw that drought can result in increased nutritional quality of food and reduced tree defenses. However, the influence of precipitation on a given herbivore may vary from one area to another. Southern pine beetle epidemics in Georgia are associated with low summer rainfall; in Texas, Louisiana, and Mississippi; however, they are correlated with high winter rainfall, and in North Carolina with high spring rainfall (King 1972).

Climate can produce multiple, sometimes contradictory effects on herbivore population dynamics. For example, the extent and severity of spruce budworm outbreaks in eastern Canada are strongly influenced by two different temperature-related variables, but in opposite directions (Gray 2007). Outbreaks correlate positively with summer extreme temperatures and negatively with summer degree-days, which is a measure of total warmth received during a given summer. Gray (2007) speculates that total summer warmth may inhibit outbreaks through positive effects on the natural enemy complex. Populations of lepidopteran species tend to outbreak synchronously throughout the Northern Hemisphere, a phenomenon that Myers (1998) correlated with geographically synchronous cool springs, which were in turn correlated with troughs in the sunspot cycle. Selås et al. (2004), who also found a strong relationship between Lepidopteran numbers and the sunspot cycle, hypothesized that UV-B radiation was the link between the two. During periods of low sunspot activity, the atmospheric ozone layer is relatively thin, allowing more UV-B radiation to penetrate to the surface. Selas et al. suggest that trees divert energy from herbivore defenses to UV-B-protective pigments during these periods, thereby allowing moth populations to grow.

Climate may play an important role in synchronizing emergence of insects with bud flush. In interior Douglas-fir, the timing of budbreak is controlled by a combination of air and soil temperature, whereas the emergence of spruce budworm larvae is a function of air temperature only. In years with heavy snowpack, air temperatures warm at a faster rate than soil temperatures, so larvae emerge before buds are swelling and must either feed on older foliage (which is unpalatable) or search for another host tree. Both lead to high larval mortality (Beckwith and Burnell 1982). In years without heavy snowpack, larval emergence is more likely to be synchronized with the appearance of new, palatable foliage, and because soil moisture largely results from winter snowpack, late-summer drought likely increases the palatability of even the older foliage. These factors coming together could trigger epidemics (Perry and Pitman 1983).

Although climate plays a significant role, it is only one of a suite of factors that influence herbivore dynamics. The best we can say is that climate may alter the functioning of regulatory mechanisms in forests, but exactly how, and how much, varies from system to system. Despite uncertainties, however, there is widespread agreement that global climate change is likely to result in more pest outbreaks in forests. For example, warmer winters will probably increase the severity of pathogens in forest ecosystems because winter is a period of major mortality for pathogens (Harvell et al. 2002). Several plant diseases are more severe after mild winters, including Mediterranean oak decline, Dutch elm disease, and beech bark canker. With regard to insects, Logan et al. (2003) conclude that "to date, the majority of results assessing individual pest species' response to climate change indicate intensification in all aspects of outbreak behavior, and this certainly characterizes our work with the mountain pine beetle, gypsy moth, spruce beetle, and spruce budworm. Perhaps this is the result of a bias in reporting. More likely, the combination of attributes that first caused the insect species to be considered a pest also makes them sensitive to climate change."

In their assessment of potential climate change impacts on forest herbivores and pathogens, Ayres and Lombardero (2000) reached similar conclusions:

> Some scenarios are beneficial (e.g., decreased snow cover may increase winter mortality of some insect pests), but many are detrimental (e.g., warming tends to accelerate insect development rate and facilitate range expansions of pests and climate change tends to produce a mismatch between mature trees and their environment, which can increase vulnerability to herbivores and pathogens).

The implications of global change go beyond weather. Percy et al. (2002) found that higher levels of atmospheric carbon dioxide and ozone increased the numbers of aphids on quaking aspen (*Populus tremuloides*), as well as decreased the ratio of natural enemies to aphids (ozone had a particularly strong effect). Changes in foliar chemistry due to excessive nitrogen inputs (from nitrogenous pollutants) also alter plant-herbivore relationships, in general making foliage more palatable by increasing nitrogen concentration and decreasing the content of carbon-based defensive compounds (Throop and Lerdau 2004).

The future is now, especially in western North America. Warmer temperatures have allowed bark beetles to produce more broods and expand their ranges elevationally and latitudinally into areas that were formerly too cold, producing the largest outbreaks in recorded history (Carroll et al. 2003, Logan et al. 2003, Wilcox et al. 2006, Berg et al. 2006) A mountain pine beetle outbreak had covered 7 million ha in British Columbia as of 2005, and for the first time in recorded history substantial numbers have crossed the Continental Divide and attacked jack pine in Alberta. The stage is set for the mountain pine beetle to move eastward through the extensive jack pine stands of central and eastern Canada.

Since the early 1990's, large spruce bark beetle outbreaks have occurred throughout western North America. In Alaska, the largest spruce beetle epidemic on record covered 1.2 million hectares during the 1990's and spilled over into the Yukon Territory, which in the past had been too cold for significant bark beetle activity (Berg et al. 2006). Drought, which as we have seen lowers tree resistance to bark beetles, has also been a major contributor to outbreaks, especially in the southwestern United States, where a long-lasting and severe drought that is most likely related to climate change has resulted in bark beetle outbreaks and widespread mortality of ponderosa and pinyon pines (Breshears et al. 2005).

The unfolding drama in western North America is not restricted to bark beetles. Spruce budworm, which is also temperature-limited and drought-favored, has become a factor in interior Alaska only since the early 1990's, and by 2006 had defoliated thousands of hectares of white spruce in that state (USFS 2007). In British Columbia, an unprecedented epidemic of Dothistroma needle blight of lodgepole pine is associated with increased summer precipitation that may be related to global change (Woods et al. 2005) Woods et al. (2005) speculate that even a small change in climate could allow the pathogen to cross an environmental threshold that had previously restricted its spread. Such thresholds (or "tipping points"), in which small environmental changes have large ecological effects, may be common in nature (chapter 20).

19.3 COEVOLUTIONARY BALANCE IN FORESTS

We have seen that herbivore populations are regulated by a combination of five factors:

1. The nutritional quality of plant tissues
2. Secondary chemicals that either are toxic or reduce the digestibility of food
3. Predators and pathogens
4. The abundance of food at both the stand and the landscape scales
5. Climate

These factors should not be thought of as acting separately but rather as interacting to produce a coherent, interdependent whole. Chemical defenses are energy expensive, and toxins in particular have a limited evolutionary lifetime. Herbivores respond to plant chemical defenses by evolving detoxification mechanisms, plants evolve new defenses, and this leads the herbivore populations to coevolve again. Van Valen (1973) has called this notion of continual evolution and coevolutionary response the **Red Queen hypothesis,** after the Queen in *Alice and Wonderland* who had to "run and run just to stay in one place." One problem (at least in long-lived perennials such as trees) is that herbivore populations have much higher fecundity and therefore greater evolutionary potential than plants. However, other factors must be considered. Plants generate and maintain a diverse mosaic of defenses within both communities and individuals, thereby denying herbivores a single target toward which to evolve. Natural enemies work synergistically with plants to magnify control over herbivores. Mutualists that aid in plant defense (i.e., fungal endophytes, mycorrhizal fungi, and rhizosphere bacteria) have fast generation times and seem likely to confer on their plant hosts the ability to quickly counter new innovations on the part of herbivores (Carroll 1988). Because of the multifaceted nature of interactions in natural ecosystems, strong one-on-one coevolution is probably relatively rare and restricted to unapparent plants and their specialist herbivores (Stamp 2003, chapter 13).

Phytochemical coevolution theory deals with patterns arising from the ability of herbivores (especially insects) to evolve tolerance to plant defensive chemicals. The theory assumes that herbivores have two choices in dealing with plant chemical defenses (Cornell and Hawkins 2003): (1) they can become specialists by evolving tolerance to a limited number of chemicals and eating the plants that contain only those, or (2) they can become generalists but at the cost of lower feeding success on any particular plant. The theory generates predictions that have mostly been confirmed, indicating it is correct in its essentials (Cornell and Hawkins 2003).

In Atsatt and O'Dowd's (1976) view, regulating factors such as plant diversity and predation not only directly affect herbivore populations but also act to extend the lifetime of plant defensive chemicals by reducing the evolutionary pressure on them. Plant defensive chemicals, in turn, increase the efficacy of natural enemies by reducing growth rate and vigor of herbivores and by increasing the amount of time they must spend searching for appropriate food, and therefore the time that they are exposed to natural enemies.

The balance between individual plant defenses and ecological regulating factors is likely to vary from one system to another. Because chemical defenses are expensive and evolutionarily vulnerable, we might expect that they carry less of the regulatory load in systems where natural enemies or climate effectively regulate herbivores. On the other hand, systems characterized by low primary productivity may have few predators, and plant chemicals defenses correspondingly become more prevalent. In energy-poor systems, smaller populations are maintained at all trophic levels; however, because a relatively small proportion of energy is transferred from one trophic level to another, the predator populations should be reduced proportionally more than those of the lower trophic levels.

If primary production is low enough, many predators may drop out of the picture altogether. This seems to be the case in "blackwater" river areas of the tropics, where there is little evidence of predators and, predictably, plant chemical defenses are extraordinarily high (Janzen 1974). Less extreme but similar examples can be found in temperate forests. Douglas-fir in the harsh continental climates of interior North America has high levels of chemical defense against spruce budworm, whereas Douglas-fir in the mesic, productive habitats west of the crest of the Cascade Range apparently has evolved very little chemical defense (Perry and Pitman 1983). Forests in both areas contain spruce budworm, so we must conclude one of two things about Douglas-fir west of the Cascades: either there are chemical defenses not yet detected, or other factors have effectively regulated budworm populations so that herbivore pressures have not been significant enough to give an evolutionary advantage to the individual tree that spends energy for chemical defenses rather than growth.

Indigenous herbivores are integral parts of normally functioning ecosystems. They are elements of the system that do work and extract energy in payment. As we saw in chapter 13, plant chemical responses to herbivores can influence species composition and processes throughout communities and ecosystems,. The nutrient cycle is particularly effected by herbivory (Bardgett and Wardle 2003; Schowalter 2006). Herbivores have been found to increase, decrease, or have no effect on rates of decomposition and nutrient release from litter. According to Hunter et al. (2003), "In Southern Appalachian forests, outbreaks of insect herbivores have been shown repeatedly to increase the availability of nutrients in soil and the export of nitrate in forest streams." (Although heavy insect defoliation often does lead to increased nitrate leaching from the soil to streams, the majority of nitrogen transferred to the soil in insect frass is retained within the system [Lovett et al. 2002]). Insect herbivory has also been shown to alter litter quality and accelerate the nutrient cycle in semiarid pinyon pine woodland (Chapman et al. 2003) and lower montane tropical rain forest (Fonte and Schowalter 2005). In contrast, herbivory on riparian poplar (*Populus* sp.) by a leaf-galling aphid increased the concentration of secondary compounds in litter and therefore decreased rates of decomposition, although the effect varied with tree genotype (Schweitzer et al. 2005b). Bardgett and Wardle (2003) suggest that positive effects of herbivory on soil biota (e.g., decomposers) are most common in ecosystems with high soil fertility, and negative effects are most common in ecosystems with low soil fertility. While some data fit that pattern, others do not. As Fonte and Schowalter (2005) point out, "the role of phytophagous insects in ecosystem nutrient cycling remains poorly understood."

> During their years belowground, periodical cicadas siphon nutrition from the roots of oaks and other hardwoods into their bodies. Upon emergence they mate and die, returning their carcasses to the soil and its decomposers. The pulse of nutrients contained in cicada bodies is equivalent to as much as 500 kg/ha of high-quality nitrogen fertilizer (Ostfeld and Keesing 2004). Unlike fertilizer, however, cicada bodies are not an input from outside the system; rather, they represent a diversion, storage, and pulsed addition of nutrients already there.

When their numbers are sufficiently high, ungulates and other browsing mammals can change the vegetation type from preferred to nonpreferred plant species. Since browsers usually prefer plants that are high in nutrients and low in secondary chemicals, the spin-off result is a change from readily decomposable to slowly decomposable litter. Moose, for example, tend to convert stands from nutrient-rich hardwoods (their preferred food) to nutrient-poor conifers (Pastor et al. 1998).

Not surprisingly, the degree to which aboveground herbivory influences nutrient cycling depends on herbivore abundance, and insects at endemic levels appear to have a minor effect on the nutrient cycle. In three eucalyptus forests with defoliation less than 10 percent, only 4 percent of the nitrogen, phosphorus, and potassium cycling from foliage to the ground passed through insect bodies (Ohmart et al. 1983); in a Swedish Scots pine forest with low levels of defoliation, only 8 percent of the nitrogen cycled from foliage to soil was in insect frass or bodies (Larsson and Tenow 1980). Endemic (i.e., nonoutbreak) levels of caterpillars and aphids reduced the quality of foliar leachates from oaks, beech, and spruce, but soil solution chemistry was unaffected (Stadler et al. 2001).

Why do herbivore populations occasionally outbreak? The most accurate statement that can be made is that various factors may permit herbivore populations to climb past some threshold, beyond which natural controls are no longer effective. Any number of factors may trigger that, including an increase in the abundance of food plants, changes in climate that favor population growth, stresses that reduce plant chemical defenses, changes in habitat that reduce numbers of natural enemies, or decreased susceptibility to viruses or other infectious diseases. Most outbreaks of indigenous insects and pathogens probably result from some combination of the above factors. Reduced predation by birds, for example, weakens one constraint that by itself may not result in an insect outbreak but increases the probability of outbreak should another constraint be weakened (Holling 1988).

Forests that are declining in vigor through natural causes tend to become susceptible to insect herbivores, and the herbivore often is an agent of cleansing and renewal. However, forests also become susceptible when the ecological context within which they exist is significantly changed, which may include introduction of an exotic pest organism, an adaptive leap by native pest, a significant change in climate (such as abnormal drought or heat), simplification of the communities within which the trees exist, or an unnatural complexification (as occurred with fire exclusion in North American coniferous forests), The effect of the latter two depends on how widespread they are. Problems with native pathogens and defoliating insects have become commonplace in regions where humans have significantly altered forested landscapes (either purposely or inadvertently), including the western spruce budworm, Swiss needle cast[12] and several root rots in western North America, fusiform rust and southern pine beetle in the southeast, and eastern spruce budworm and ash yellows (a viral disease) in the northeastern United States and eastern Canada. As discussed, exotic plantations may not fit this pattern; however, that appears to be changing.

Although herbivores play many beneficial roles in ecosystems, the balance between their destructive potential and positive spin-offs of that potential can be tipped if factors maintaining that balance are altered. Humans appear to have inadvertently tipped this balance in many areas, setting the stage for infestations that are more destructive than the historic norm. How these changes play out in the long run remains to be seen.

[12]Despite the name, Swiss needle cast is native to coastal North America. The fungus caused so little damage in old-growth forests that it wasn't named until it appeared in plantations of Douglas-fir being grown in Switzerland. During the 1990s it reached epidemic proportions in Douglas-fir plantations in western Oregon.

19.4 SUMMARY

Herbivore species vary in the type and range of plant species they consume, the kind of plant tissues they eat, and the way that they eat them. A single herbivore species may also have considerable genetic variation, specializing on different plants in different areas. At endemic levels, herbivores consume only a small proportion of the food available to them, particularly in forest ecosystems; however, many herbivore species (especially insects) periodically outbreak to very high population levels. During these periods, a large proportion of available food may be eaten. While bark beetle epidemics usually result in tree death, mortality from defoliation is less common and may even benefit stands by relieving stresses due to overstocking.

Ecologists have long asked why the world is green. Balance between plants and herbivores is maintained by numerous factors, which can be grouped into properties of individual plants and properties of the ecosystem. The two most important individual-plant properties acting to damp herbivore populations are low concentrations of essential nutrients (particularly nitrogen) in plant tissues, and various secondary chemicals (e.g., tannins and alkaloids) that appear to be produced especially to discourage high levels of herbivory. Ecosystem-level properties include genetic diversity in the plants (within individuals, among individuals of the same species, and among different species), natural enemies of the herbivores, and climate (a property of the planetary ecosystem). Individual-plant and ecosystem-level factors operate in concert and are constantly tempered by evolution on the part of all the players and by climatic fluctuations. The result is a dynamic balance between various ecosystem components.

Though herbivores can be highly destructive, they play an important role in ecosystem function. In today's world of changing climate, altered forests and landscapes, and invasive species, the balance between positive and negative effects may be shifting. The best way that the forester can prevent large-scale outbreaks of indigenous herbivores is to maintain healthy forests that are structurally and genetically diverse.

20

Ecosystem Stability I

Introduction and Case Studies

MOVEMENT AND CHANGE CHARACTERIZE EVERY living system. Individuals grow and die, and throughout the lifetime of a higher organism its cells may be renewed many times. The number of individuals of a particular species may be high at one time within a given area and low at another. Ecosystems and landscapes are in constant flux; indeed, the differential adaptations of organisms to environmental variability (e.g., in climates, soils, and disturbances) accounts for much of the rich natural diversity that humans value so highly. The opposite of change is monotony and death.

Because disturbance and change are such integral parts of nature, the concept of stability can be quite elusive. Grimm and Wissel (1997) call it "one of the most nebulous terms in the whole of ecology." As Loreau et al. (2002) observe, "one of the difficulties of measuring stability in natural ecosystems is that [they] show a variety of complex dynamics."

Intuitively, we know that embedded within the apparent chaos of nature there must be some degree of constancy and predictability; otherwise life would not be possible. At the very minimum each organism must find food, protection, and (for those who reproduce sexually) a mate or face extinction. All higher trophic levels depend (either directly or indirectly) on a certain constancy in the plants, and the plants depend, in turn, on the rains, minerals, organisms that cycle nutrients, and temperature ranges necessary for physiological processes to occur. Moreover, the success of the whole system depends on herbivores and pathogens not eating all of the plants and therefore destroying the base upon which everything else rests.

As observed at the outset, no living system is completely constant in either composition or processes. Constancy at one level may (and usually does) mask considerable change at lower levels, the most obvious example being the mechanisms

In their review of the literature on ecological stability, Grimm and Wissel (1997) identified a total of 163 definitions from 70 different **stability concepts,** where they define a stability concept as "a concept that has something to do with stability, whatever stability is." Fortunately, however, Grimm and Wissel conclude that "out of all of the 163 definitions of the 70 terms there are only three fundamentally different properties: (1) 'staying essentially unchanged,' (2) 'returning to the reference state (or dynamic) after a temporary disturbance' and (3) 'persistence through time of an ecological system.' Note the difference between properties 1 and 3. 'Staying essentially unchanged' refers to a certain reference state or dynamics, which may be an equilibrium, or oscillations, or irregular but limited fluctuations. 'Persistence through time,' on the other hand, does not refer to any particular dynamic but only to the question whether a system persists as an identifiable entity."

Even these three are not distinct but, rather, have considerable overlap and interdependence. Concept (3) is a general form of concept (1). Concepts (1) and (3) both depend absolutely on concept (2), a property called **resilience** (section 20.2).

that repair damage and the very busy immune system that maintain the health of organisms. The stable ecosystem is one that is capable of constraining its fluctuations within certain bounds and, if not maintaining constancy in its species composition and productivity, at least maintaining constancy in certain potentials. Although at first glance the idea of simultaneous change and stability seems like a contradiction, there are many common examples. Consider a bicycle rider (Mollison 1990): the stability of the whole system (i.e., the body plus bicycle) is underpinned by constant feedbacks and adjustments among numerous muscles and nerves. As we have discussed, an ecosystem differs in many respects from an organism; however, the example of the bicycle rider embodies a general principle. Movement and change at one hierarchical level can lead to stability at a higher level. But what is the higher level of the ecosystem? How do we protect and preserve it? How do we gauge the health of a system? These are critically important questions for land managers, and never more so than in today's world. They are also exceedingly difficult questions whose answers lie in poorly understood interactions and interrelationships between organisms and their environment, among species within communities, and among the different hierarchical levels that characterize the organization of life on earth. We must draw on all of our knowledge about structure and process in nature in order to approach the question of stability.[1] This chapter first deals with some basic issues. In sections 20.1 and 20.2 we address the question of what it is we wish to stabilize and discuss some key concepts. In sections 20.3 through 20.6 we examine some instances in which ecosystems have been destabilized and degraded. Our discussion of stability continues in the following two chapters. In chapter 21, we review evidence for the stabilizing role of biodiversity, which, in chapter 22, leads to issues underlying the conservation of individual species. Management strategies for protecting species and whole systems are taken up in chapter 23.

20.1 STABILITY OF WHAT?

There are many different ideas about what should be protected and conserved in nature: endangered species, biodiversity as a whole,[2] ecological integrity, ecosystem processes, productive potential, soil fertility, and water quality, to name a few. These are all important to preserve; moreover, all are manifestations to one degree or another of healthy ecosystems. It is useful to distinguish three different sets to which the concept of stability can be applied. Keep in mind that in all of these (but particularly the first two) stability may include considerable change at the scale of local ecosystems and landscapes (as commonly occurs during disturbance and succession). Instability occurs when the system crosses some threshold from which recovery to a former state is either impossible (e.g., extinction), or, if possible, occurs only over relatively long time periods or with outside subsidies of energy and matter (e.g., loss of topsoil).

1. *Species stability,* or the maintenance of viable populations and metapopulations of individual species. (The concept of viable populations is discussed in chapter 22.)
2. *Structural stability,* or the stability of various aspects of ecosystem structure, such as the various interrelationships (e.g., food web organization, mutualisms), species diversity, potential biomass accumulation, or landscape patterns.
3. *Functional stability,* or the stability of processes that perform key functions, such as primary productivity, hydrology, and nutrient cycling.

[1]In trying to understand stability, we are really grappling with the issue of complexity, which physicist Daniel Stein describes as "almost a theological concept; many people talk about it, but nobody knows what 'it' really is" (Stein 1989). The ecologist Daniel Botkin puts it this way: "The biosphere has a great momentum, and we understand almost nothing about its functioning" (Botkin 1990). One thing we do know is that complexity is the way of the real world, and we must learn how to deal with it despite the many uncertainties.

[2]Recall that biodiversity has been defined as "the variety of life and its processes."

Note that these refer to different hierarchical levels; the first to populations, the second and third to structure and process, respectively, at the level of local ecosystems and higher. The stability of each is generally linked to the stability of the others (although this is not necessarily the case). The upshot is that the coin of stability has two sides and a middle: on one side are individual populations and metapopulations, which in final analysis are the basic units of survival; on the other side are the higher-order aspects of system structure (e.g., food webs, guilds, communities, landscapes, regions, and ultimately the planet as a whole); the middle that ties these two together comprises the numerous relationships and processes discussed throughout this book. Successful conservation requires that we attend to the coin as a whole, and this, in turn, means protecting links that may be far-flung in space and time. The multiplicity of both direct and indirect linkages that characterize nature raise the distinct possibility that to protect any one ecosystem, we must ultimately protect the integrity of the entire biosphere.

Different ecologists have applied different meanings to the term *resilience*. Holling (1973) defined it as "the amount of disturbance that a system can absorb without changing stability domains," which is equivalent to Loreau et al.'s (2002) *robustness*. Loreau et al. (2002) defined resilience as the *rate* at which a system returns to a prior state after a perturbation, while they applied the term **qualitative stability** to the ability to return with no consideration of rate (equivalent to resilience as we define it). However, introducing a rate term into the definition of ecological resilience is potentially confusing and we avoid it. Depending on which aspect of the ecosystem is under consideration, a fast return time may be either beneficial or detrimental to overall ecosystem health. For example, a rapid return to predisturbance nutrient retention is beneficial, while a fast return to closed canopy forest truncates the highly diverse early successional stages and reduces ecosystem richness.

20.2 RESISTANCE, RESILIENCE, ROBUSTNESS

Of the many different (and often overlapping) concepts related to this issue, three in particular capture the essence of ecosystem stability:

1. **Resistance.** The ability of a system to absorb small perturbations and prevent them from amplifying into large disturbances
2. **Resilience.** The ability of a system to return to its original state (e.g., level of productivity, species composition) after a perturbation
3. **Robustness.** The amount of perturbation that a system can tolerate before switching to another state (Loreau et al. 2002)

Resistance mechanisms may be thought of as those properties (both of the system and of individuals) that prevent organisms from succumbing to some stress. At the ecological level, resilience mechanisms come into play when organisms are weakened or killed. Consider a forest that experiences periodic drought. Trees resist the potential physiological stresses associated with low precipitation in various ways. Water is drawn from storage reservoirs such as small soil pores, decaying logs, and sapwood. Energy is shunted to root and mycorrhiza growth. Stomatal closure prevents excessive water loss (although at the cost of photosynthesis). If photosynthesis is reduced too much, trees cut back their production of defensive chemicals, and, depending on the efficacy of other controls over herbivore populations (e.g., natural enemies, climate; chapter 19), trees may become susceptible to insect outbreak.

If (as is currently happening in the southwestern United States) the drought is so severe that the capacity of the various resistance mechanisms is exceeded, trees will be weakened and perhaps killed. In western North America, for example, insect outbreaks that are triggered by drought may be followed at some interval by a catastrophic wildfire. At this point, the mechanisms of resilience come into play. Early successional trees and shrubs occupy the site, stabilizing soil nutrients and organisms and thereby preserving the productive potential of the site. In fact, resilience is best exemplified by the process of succession, in which species composition clearly changes but the system stays on a certain track that will eventually lead back to some reasonable approximation of where it was previously (so long as external driving variables, in particular climate, do not change too much).

20.3 POLLUTION

Widespread dieback of forests began occurring in the latter half of the twentieth century in various areas of the world, but especially in industrial countries. The German word for this phenomenon is *Waldsterben* ("forest death"). *Waldsterben* is most dramatic in central and eastern Europe and in eastern North America, where a minimum of 7 million ha had been affected as of 1987 (Hinrichsen 1987). This figure does not include the Soviet Union and Canada, both of which have extensive forest areas that are affected. The obvious symptom of *Waldsterben* (crown damage and eventual tree death) spread rapidly. In 1982, the Federal Republic of Germany estimated that 8 percent of its forests showed symptoms of decline. This had increased to 34 percent by 1983 and 50 percent by 1985 (Hinrichsen 1987).

No single factor seems to underlie the current dieback of forests throughout the world. In many areas of Europe, eastern North America, and Asia it is clearly related to acid rain (McLaughlin 1998; DeHayes et al. 1999), while in other cases the connection with pollution is not so clear. For example, recent analyses of tree rings show that growth of white oak *(Quercus alba)* began to decline throughout the eastern United States in the 1950s. The reasons are unknown, but acid rain does not appear to be a factor (Phipps and Whiton 1988). Red spruce began declining in the northeastern United States in 1800, possibly because of a warmer climate (Hamburg and Cogbill 1988); however, dieback of 25 to 50 percent of canopy trees during the 1970s and 1980s was almost certainly related to acidic deposition (Driscoll et al. 2001). Evidence suggests that extensive dieback of sugar maple in eastern North America is related to depletion of soil cations, which is related, in turn, to the leaching effect of acid rain (Driscoll et al. 2001). As we discussed earlier in the book, in some cases forest dieback is a natural cyclic process.

Even where pollution appears to be involved, demonstrating this with a high degree of scientific certainty can be difficult for two reasons. First, tree death is often caused by a complex of factors that are not directly related to pollution but are probably triggered by it (e.g., herbivorous insects and pathogens). Second, the numerous different pollutants that may be involved make it difficult or impossible in many cases to relate symptoms to any one. A formidable array of anthropogenic chemicals may occur in polluted air, including sulfur dioxide (largely from burning coal); nitrous oxides (from coal, automobile exhausts, and intensive agriculture); ozone; toxic metals such lead, cadmium, zinc, and copper that enter the atmosphere from smelters and power plants that burn fossil fuels; and a variety of organic compounds originating from pesticides, herbicides, and other sources. One study in Germany found traces of 400 organic compounds in 1 m^3 of air (P. Schutt, personal communication to D. Hinrichsen).

Severe pollution triggers a general unraveling of the ecosystem, leading ultimately to its collapse. Bormann (1985) describes the following stages:

- *Stage 0.* Anthropogenic pollutant levels insignificant. Pristine systems.
- *Stage I.* Anthropogenic pollution occurs at generally low levels. Ecosystems serve as a sink for some pollutants, but species and ecosystem functions are relatively unaffected.
- *Stage IIA.* Levels of pollutants are inimical to some aspect of the life cycle of sensitive species or individuals, which are therefore subtly and adversely affected. For example, sensitive plants may suffer reduced photosynthesis, a change in reproductive capacity, or a change in predisposition to insect or fungus attack.[3]
- *Stage IIB.* With increased pollution stress, populations of sensitive species decline, and their effectiveness as functional members of the ecosystem diminishes. Ultimately, these species may be lost from the system, but a more likely fate is that some individuals will remain as insignificant components.
- *Stage IIIA.* With still more pollution stress, due to higher concentrations or longer exposure at the same concentration, size becomes important to survival, and large plants, trees, and shrubs of all species die. The basic structure of the forest ecosystem changes (Woodwell 1970), and biotic regulation is affected. Both Gordon and Gorham (1963) and Woodwell (1970) describe the process as peeling off layers of forest structure: first the trees, followed by the tall shrubs, and finally under the severest of conditions, the short shrubs and herbs. The forest ecosystem becomes dominated by small scattered shrubs and herbs, including weedy species not previously present. Productivity drops as the ability of the system to repair itself by substituting tolerant for intolerant species is exceeded.[4] As layers of vegetation die, masses of highly flammable dead wood are left behind, increasing the probability that natural or human induced-fire will occur and conversion to open shrubland will be speeded. Toxic concentrations of some pollutants may limit many species (Jordan 1975; Hutchinson 1980). The capacity of the ecosystem to regulate energy flow and biogeochemical cycles is severely diminished. Runoff increases; the loss of nutrients previously held and recycled accelerates; erosion increases, and soil nutrients are exported to interconnected aquatic systems, which may be severely affected (Bormann et al. 1974; Likens 1984).
- *Stage IIIB.* Ecosystem collapse. The ecosystem at this point is so damaged by loss of species, ecosystem structure, nutrients, and soil that the capacity for self-repair is severely diminished. Even if the perturbing force is removed, the system may never return to predisturbance levels of structure and function or may need centuries or even millennia to do so. This type of degraded ecosystem occurs frequently around strong point sources of air pollution of preemission control days, such as Copper Hill, Tennessee, and Sudbury, Ontario.

One of the more thorough analyses of a declining forest was conducted by a multidisciplinary team of scientists from the University of Bayreuth, in the Federal Republic of Germany (Oren and Schulze 1989; Schulze 1989; Schulze et al. 1989a,b). They compared obviously declining with apparently healthy spruce stands in the Fichtelgebirge mountains of northeastern Bavaria, an area characterized by acid soils and high levels of pollution from SO_2, NO_x, and

[3]Mycorrhizal fungi are probably also weakened at this stage.

[4]Bormann refers here to pollution tolerance, not shade tolerance.

ozone. Following is a brief summary of the story pieced together by researchers at Fichtelgebirge.

20.3.1 The Process of Forest Decline at Fichtelgebirge

Acidity resulting from SO_2 and NO_x pollution affected soil chemistry at Fichtelgebirge in at least three ways: (1) leaching of basic cations was accelerated, particularly magnesium and calcium; (2) the ratio of aluminum to both calcium and magnesium increased in soil solution; and (3) acidity inhibited nitrifying bacteria, thereby increasing the ratio of NH_4^+ to NO_3^- in soil solution. Trees became magnesium and calcium deficient in part, at least, from lower concentrations on the exchange complex and in soil solution, a situation that was exacerbated because tree uptake of these elements was inhibited by the relatively high levels of aluminum and NH_4^+ in soil solution (chapter 18). The situation was further exacerbated because changes in soil chemistry caused trees to alter distribution of roots within the soil profile. Trees produced fewer root tips in mineral soil, probably because of the lower ratio of calcium to aluminum in soil solution (chapter 16). However, because calcium:aluminum ratios were less influenced in humus than in mineral soil, root-tip formation remained relatively stable in humus. This shift of rooting activity from mineral soil to humus layers probably reduced the ability of trees to gather nutrients and may have made them more susceptible to drought (because humus dries out more rapidly than mineral soil) (Schnieder et al. 1989). Aboveground, normally benign endophytic fungi (commonly present in tree foliage; chapter 11) responded to tree stress by becoming pathogenic.

As acid precipitation adds SO_4^{-2} and NO_3^- to soil solution, cations move from the exchange complex into solution to maintain electrical neutrality. As ionic concentration of the soil solution increases, so does the ratio of trivalent to divalent to monovalent ions. This means that cation concentrations in solution increase in the following order: Al^{3+}, Ca^{2+}, Mg^{2+}, K^+, Na^+, and H^+ (Reuss 1983; Johnson et al. 1989). But aluminum also has more affinity than calcium and magnesium for exchange sites, including those associated with roots and mycorrhizae. The net effect in acidifying soils is twofold: (1) aluminum blocks calcium and magnesium uptake by roots, and (2) deprived of both cation exchange sites and biological sinks, increasing amounts of calcium and magnesium are leached to streams. Experience in the northeastern United States indicates that forest vegetation is at risk when the molar calcium:aluminum ratio of soil water drops below 1, and base saturation of soils is less than 20 percent (Driscoll et al. 2001).

Yellow foliage, the most obvious symptom of nutrient deficiency exhibited by trees at Fichtelgebirge, was not due so much to a low supply of a given element, but to imbalances among the elements. Had the availability of all nutrients been affected evenly, trees would simply have reduced growth proportionally, but magnesium and calcium deficiencies were accompanied by an excess of available nitrogen (from the NO_x pollutants). Because new needle growth is determined by the concentration of nitrogen in older needles, trees had to translocate large amounts of magnesium from old needles to match the demand created by abundant foliar nitrogen. This created a nutrient imbalance in mature foliage and produced the yellowing that is characteristic of magnesium deficiency. Stem growth was more strongly affected than photosynthesis by the cation deficiencies; hence, trees in polluted stands became relatively inefficient in their stem growth per unit leaf area (Oren and Schulze 1989).

The patterns of decline at Fichtelgebirge do not necessarily characterize all forests that are affected by acid rain. Soils of that area are inherently more acid than those of at least some of the polluted forests in North America and, hence, less able to absorb and buffer the effects of acid precipitation. This difference is one of degree rather than kind, however. Continued inputs of acidity will eventually overwhelm the buffering capacity of most soils and lead to events similar to those that occurred in central Europe.

20.3.2 Similarities between Fichtelgebirge and Declining Red Spruce in North America

Patterns at Fichtelgebirge are quite similar in some respects to those in declining red spruce stands in the eastern United States, where 50 percent of soil calcium has been lost from some areas since the mid-twentieth century (Likens et al. 1996) and forest decline correlates with lower calcium:aluminum ratios in both fine roots and stems (Baes and McLaughlin 1986; Bondietti et al. 1989; Shortle and Smith 1988). However, in the United States there is also another effect that either did not occur or was not discovered at Fichtelgebirge. Research shows that acid deposition onto foliage of red spruce leaches soluble calcium associated with membranes,[5] disrupting membrane integrity and leading to various physiological problems, including lowered resistance to freezing and higher respiration rates (DeHayes et al. 1999). Several other North American tree species are also susceptible to acid-induced leaching of foliar calcium, including yellow birch, white spruce, red and sugar maples, and eastern white pine (DeHayes et al. 1999).

[5] Membrane-associated calcium is a very small proportion of the total foliar calcium pool but plays a unique and critical role.

Calcium differs from most other nutrients in certain respects that make calcium nutrition of trees particularly vulnerable to excessive acidity. As we saw in chapter 16, calcium is required for the growth of all meristems, including those of the fine roots and the cambium. Calcium differs from all other macronutrients in that it is quite immobile in plant tissues; therefore, the demands of new growth cannot be met by translocation from other tissues within the tree. One implication is that root growth can be sustained only where there is an adequate supply of calcium in the soil (or humus), and the ratio of calcium to other cations (especially aluminum, but also magnesium, potassium, and NH_4^+) is sufficiently high that calcium uptake is not inhibited. If either condition does not hold, root growth will be inhibited, reducing, in turn, the ability of trees to gather water and other nutrients, and probably affecting other factors such as mycorrhizal fungal populations, rhizosphere communities, and soil aggregation. Lower radial growth due to deficiency of calcium (or any other nutrient) results in less phloem to translocate photosynthates downward from the crown, and less sapwood area to transport water to crowns and to store carbohydrates (Shortle and Smith 1988; Waring 1989).

20.3.3 Effects of Pollution on Susceptibility to Pests and Pathogens

Stressed trees frequently reduce their production of defensive chemicals and may become susceptible to any number of herbivorous insects and pathogens (chapter 19) (fig. 20.1). Moreover, calcium plays an important role in defense against herbivorous insects and pathogens (McLaughlin and Wimmer 1999), and its leaching loss through acid deposition may increase tree susceptibility. McLaughlin and Wimmer (1999) discuss the experience in the eastern United States:

> Intensification of several types of diseases have . . . been noted at about the same time that growth declines have begun to appear in high-elevation red spruce forests. In the Great Smokey Mountains National Park in the southeastern USA, for example, a 1995 report listed seven distinct disease types that had accelerated notably in the park and in some cases within the region during the past two decades. . . . These included increases in foliar diseases, dogwood anthracnose and several fungal or insect related diseases of tree stems. Among the stem diseases were butternut canker and Dutch elm disease of American elms. Increased insect attacks from hemlock wooley adelgid, European mountain ash sawfly and southern pine beetle were also noted and, most significantly, extensive attack of Fraser fir with balsam wooley

Figure 20.1. These Fraser fir trees in Smokey Mountain National Park were killed by the balsam wooly adelgid, an introduced insect. The trees were first weakened by air pollution, however, which in all likelihood increased their susceptibility to the insect.

adelgid. In addition, decline of American beech increased in association with both a bark scale insect and Nectria canker.

In the San Bernadino National Forest of California, ponderosa pines stressed by pollution became more susceptible to bark beetles (Stark et al. 1968).

20.3.4 Effects of Pollution on the Belowground

Roots, and all the belowground organisms and processes that depend on them, may be especially sensitive to pollution (McLaughlin 1989). According to Schutt (1989), "Degradation of the fine-root feeder system of trees and decreased abundance of mycorrhizae are common symptoms of decline in all tree species in central Europe."

Figure 20.2 shows a revealing relationship between the production of mycorrhizal fruiting bodies and tree death in Hungarian oak stands. The two mycorrhizal fungi in question, *Cantherellus cibarius* and *Boletus edulis*, produce edible fruiting bodies that are gathered and sold in markets. Quantities of these fungi marketed in Budapest dropped by more than 90 percent between 1969 and 1979, most of this occurring in a relatively brief period during the early 1970s. Oak mortality commenced in the mid 1970s and accelerated sharply through the mid-1980s. The pattern shown in fig. 20.2 is not restricted to Hungary. Both edible and inedible mushrooms (mostly of mycorrhizal fungi) are in severe decline all over Europe (Cherfas 1991).

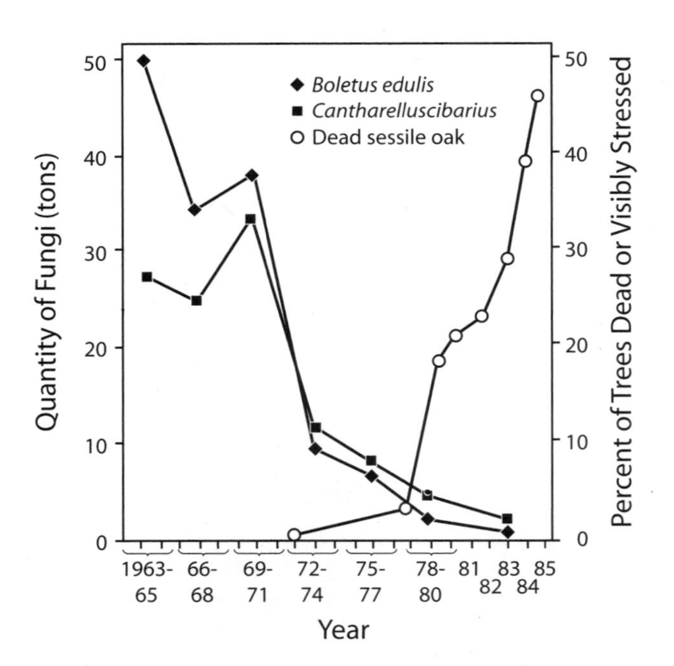

Figure 20.2. The quantities of mushrooms of two mycorrhizal fungi marketed in Budapest between 1963 and 1983 compared with the percentage mortality in Hungarian sessile oak stands *(Quercus petraea)*. The values for mushrooms are three-year averages. Tree mortality correlates closely with the proximity of stands to polluting industries. (Adapted from Jakucs 1988)

20.3.5 Chickens and Eggs

Although on its face the pattern shown in figure 20.2 suggests that stress in the mycorrhizal fungi led to stress in the trees, the truth could be quite the opposite. Subtle and undetected stresses in the tree may have stressed the fungi, eventually reflecting back to kill the tree. In the final analysis, it really doesn't matter which came first. Trees and the belowground ecosystem are tied so closely to one another through positive feedback (chapter 14) that what ever happens to one component will surely be mirrored in the other; they will reinforce themselves in health and sickness. This generalization extends beyond the tree-soil connection to encompass other parts of the ecosystem as well and leads to a general principle: stress in any part of an ecosystem is potentially reflected through positive feedback loops to other parts, magnified and reflected back.

> At this point, you may wish to ask the following question: if disturbances of one kind or another characterize all ecosystems, why don't these natural stresses magnify and spread through positive feedback loops to destabilize the system, as has happened in the case of pollution? The answer is that they will unless arrested in some way. In the healthy ecosystem this arresting is accomplished in part at least by biodiversity (Bormann 1985; Perry et al. 1989a), a point we return to in chapter 21.

20.3.6 Effects of Pollution on Streams and Lakes

Acid deposition can have profound negative effects on streams and lakes. According to Driscoll et al. (2001), "Decreases in pH and increased Al concentrations contribute to declines in species richness and in the abundance of zooplankton, macroinvertebrates, and fish." In the Adirondack Mountains (northeastern United States), 41 percent of lakes have exhibited chronic or episodic acidification, most of these because of acid deposition (Driscoll et al. 2001).

If acid deposition is reduced (e.g., through pollution controls), how long streams and lakes take to recover is largely unknown and depends on how far the system has been pushed in relation to its buffering capacity (Driscoll et al. 2001). Watersheds with base rich soils may recover quickly, while watersheds with naturally acidic soils (i.e., soils formed from base-poor rocks) may take many years to mobilize enough bases from rock weathering to neutralize decades of acid deposition.

20.4 DEGRADING FORESTS THROUGH MISMANAGEMENT

This very important issue can be quite elusive and difficult to deal with. To understand whether some management practice has degraded sites (or has the potential to degrade) requires two things: (1) a clear definition of what constitutes degradation; and (2) reliable measures of whether degradation has occurred. At least in theory, the first is relatively straightforward; the second can be quite difficult.

20.4.1 What Constitutes Degradation?

There are two general types of degradation: (1) deterioration of habitat (used here in the broad sense, including not only living space, but also environmental factors such as soil, air, and water quality); and (2) disruption of ecosystem processes.

Losses in habitat threaten the species that depend on that habitat. Logging forests on short rotations, for example, eliminates later successional stages and along with them unique habitats such as big trees, old dead wood, and, in many forest types, multiple canopy layers. In the intensively managed landscape, species that depend on old-growth forests are restricted to parks and preserves, where their numbers may be too low to prevent an ultimate slide toward extinction. Other types of management-related habitat degradation include elimination of native, noncrop plant species, and loss of topsoil.

Identifying reliable indicators of ecosystem health is one of the highest-priority tasks for ecologists during the coming years (Waring 1985, 1987; Costanza 1992; Riitters et al. 1992; Rapport et al. 1998). Rapport et al. (1985) conclude that ecosystems under stress manifest the following symptoms: "changes in nutrient cycling, productivity, the size of dominant species, species diversity, and a shift in species dominance to opportunistic shorter-lived forms." Change in nutrient cycling is only one of a larger set of stress symptoms that can be broadly classed as **loss of biotic regulation** (Bormann and Likens 1979a; chapter 18). Loss of biotic regulation manifests in two ways that were apparent in the examples of degradation discussed earlier: (1) plants lose their grip on the soil ecosystem, resulting in nutrient leaching to streams, loss of soil structure, and (in at least some cases) shifts in the dominance of soil communities from those that benefit to those that inhibit plants; and (2) pests and pathogens become more prevalent, and normally innocuous or even benign organisms become more destructive (as seen in the previous section, this is the case in pollution-stressed forests).

According to Rapport et al. (1998), ecosystem health can be gauged by measures of vigor, organization, and resilience, where *vigor* is measured as "activity, metabolism, primary productivity," *organization* as the number and diversity of interactions among system components, and *resilience* as the amount of stress a system can withstand without crossing a threshold and flipping to an alternate state (section 20.5.5). Angermeir and Karr (1994) define biological integrity as "a systems' wholeness, including presence of all appropriate elements and occurrence of all processes at appropriate rates." The gauges of health as defined by Rapport et al. and integrity as defined by Angermeir and Karr both express the same desired state, a system that is functioning well enough to keep it from collapsing or flipping to an alternate state in response to stress.

But what is "appropriate" or "well enough"? How do we assess the state of any given system? Clearly, a decline in productive capacity signals problems, because the primary producers provide the energy that runs the whole ecosystem. However, this may not be the best early-warning signal for at least two reasons. First, as we saw in the case of *Waldsterben*, obvious declines in the aboveground growth of trees may appear only after less obvious parts of the ecosystem (e.g., mycorrhizal fungi, soil nutrient balances) are significantly impacted. This is also the case for management-related stress: some early successional trees planted on sites where much of the nutrient capital has been removed may actually grow quite well for a few years, not showing signs of stress until they have utilized much of the remaining available nutrient capital. A better productivity-related gauge of forest health may be some measure of efficiency, such stem growth per unit of leaf area (Waring 1985, 1987). A second problem with using productivity to gauge ecosystem health is that it is a moving target, changing over successional sequences and from year to year, depending on climatic fluctuations.

Perhaps the clearest example of degradation in an ecosystem process is a loss of *productive capacity*, where that is defined as the average maximum gross primary productivity that is consistent with climate and resource supply and that can be sustained over long periods of time. Note that productive capacity is a potential and is not the same thing as *actual productivity*, which may vary widely over time depending on such things as season, successional stage, or fluctuations in precipitation. Note also that the relevant measure is gross, rather than net, primary productivity. This is because forests divert energy to structures and processes that don't appear as biomass increment (e.g. cell maintanenace the soil ecosystem) or if they do are difficult to measure using standard techniques (e.g. fine roots, mycorrhizal hyphae), but nevertheless contribute significantly to tree health. Therefore decreased growth (net primary productivity) does not necessarily imply that the ability of the forest to capture energy has also decreased. Loss of biotic regulation is another common form of process degradation that may manifest in various ways, such as inability to retain limiting nutrients, reduced storage and cycling of water, or a breakdown in ecosystem controls over herbivores. Other examples of degradation in processes include lowered capacity to adapt to environmental fluctuations, which happens when diversity (either within or among species) is excessively narrowed, and disruptions in the processes of migration and gene flow, which may be consequences of habitat fragmentation (chapter 22). In general, ecosystem processes are so closely intertwined that impacts on one will ramify through to affect others.

This section deals primarily with management impacts on productive capacity. Keep in mind, however, that productive capacity is closely linked to other processes occurring at both the ecosystem and the global scale, and processes are linked (through the organisms that drive them) to habitat quality at the scale of ecosystems and landscapes. We have seen many examples of this throughout the book. Recall, for example, the importance of wolves to ecosystem dynamics on Isle Royale (chapter 11), the role of cavity-nesting birds in controlling insect populations (chapter 19), and the dependence of primary productivity on the nutrients that are cycled and gathered by the belowground biota (chapter 18). We explore these interlinks in chapters 21 and 22.

20.4.2 Productive Capacity May Be Degraded through Effects on Either Ecosystem Resilience or Resistance

Reduced productive capacity is frequently manifested through lowered resilience following harvest: the regrowing forest does not return to the same levels of biomass and leaf area that it had prior to harvest, or does so very slowly. In extreme cases the ecosystem can be thrown completely off its normal successional trajectory and become dominated by weeds. As in polluted forests (section 20.3), reduced productive capacity may also be manifested through lowered resistance to pests and pathogens and perhaps to other disturbances, such as wind and fire.

There are clear examples of forests whose productive capacity has been degraded, particularly throughout the tropics and in high-elevation and high-latitude forests. As we discuss below, this is often related to management impacts on soils; however, it is not always straightforward to determine whether the long-term productive capacity of a site has been impaired. Relatively low productivity of early seral grass-forb-shrub-sapling stages does not imply that the system has been degraded relative to the later seral forest, as long as the system is on a trajectory that will return it to predisturbance gross primary productivity within a reasonable time. In Canada's boreal forest, for example, many areas logged over the past two decades have not returned to conifers and are occupied by alder or various ericaceous shrubs. Is this "degradation" in the sense that the productive capacity of the site has declined or the system has flipped to an alternate state (i.e., long-term dominance by hardwoods), or is it simply a normal successional sequence? This question cannot be answered without first understanding the natural temporal dynamic of the ecosystem of interest,[6] or, as discussed earlier, without identifying some measure or measures of overall ecosystem health (e.g., as body temperature measures the health of warm-blooded creatures).

Similarly, lowered resistance can be particularly difficult to detect without a good understanding of how the ecosystem works. Some aspects of resistance (e.g., production of defensive chemicals by trees) correlate reasonably well with tree growth; however, others (e.g., fire resistance and control of pests by predators) may not. Loblolly pines, for instance, grow better when herbaceous vegetation within plantations is controlled, but trees also become more susceptible to fusiform rust, and the rust eventually reduces tree growth (Zutter et al. 1987). Similarly, conifers in British Columbia grow better when associated hardwoods are removed but become more susceptible to *Armillaria* root disease (Baleshta et al. 2005). Hardwoods intermixed with conifers may also protect the conifers from fire (chapter 7). The upshot is that a plantation of healthy, rapidly growing young trees may in fact be unstable because it is vulnerable to pests, pathogens, fire, or severe droughts. This is not necessarily the case, but how do we tell? Our knowledge leaves much to be desired, particularly with regard to the stabilizing role of biodiversity.

20.4.3 Management Impacts on Soil Fertility

Degraded soils are a growing problem worldwide. As of 1990, an estimated 1.2 billion ha—an area as large as China and India combined—was classed as moderately, severely,

[6] A number of stand reconstructions in western Oregon indicate that forests recovering from a major disturbance went through long periods in which hardwoods dominated before conifers eventually emerged to form closed forests (e.g., Tappeiner et al. 1997; Sensinig 2002).

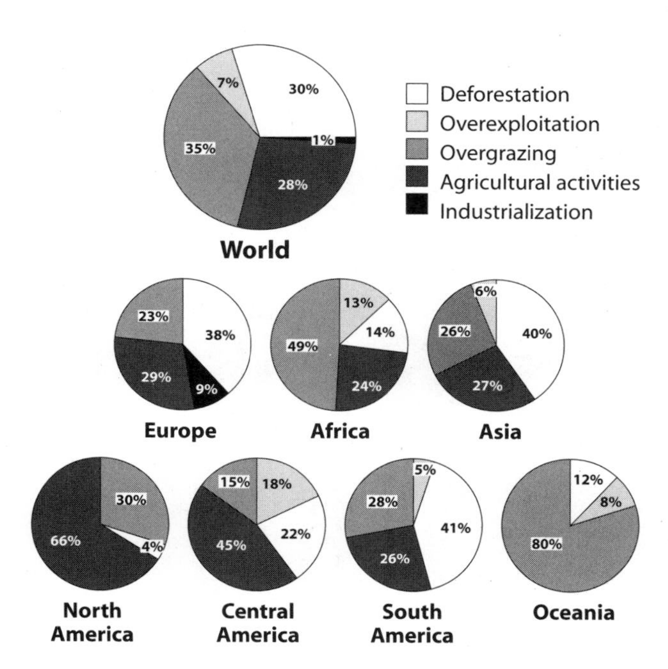

Figure 20.3. The causes of soil degradation. (Adapted from World Resources Institute 1992)

or extremely degraded (World Resources Institute 1992). This is slightly less than 11 percent of the earth's total vegetated area. The principal cause of degradation is water erosion, with wind erosion the second leading cause. Worldwide, 30 percent of degradation can be traced directly to deforestation (fig. 20.3). Many of the farms producing erosion in Asia, Africa, and Latin America are on what once was forest land (fig. 20.4).

Most of the world's soil degradation stems from deforestation (clearing for farms or ranches) and overgrazing, not from forest management. Nevertheless, improper management practices can, and in some cases have, reduced site productive capacity through affects on soil. This has been the subject of many scientific papers and symposia (Leaf 1979; Smith et al. 1986; Williams and Gresham 1987; Dyck and Mees 1989; Perry et al. 1989c; Gessel et al. 1990; Harvey and Neuenschwander 1991).

Soil degradation can usually be traced to one or more of the following: (1) excessive removal of nutrients and organic matter; (2) soil compaction; (3) loss of bioregulation, which means that the ability of plants to maintain critical soil processes and structures is weakened, reducing, in turn, the ability of soils to support vigorous plant growth. Erosion and nutrient leaching are two of the more common examples of weakened bioregulation, but complex changes in soil biology and hydrology are also involved.

20.4.3.1 Loss of Organic Matter

As we saw in chapter 17, forest harvest and associated postharvest treatments may either increase or decrease soil organic matter, depending on various factors including management practices, forest type, soil type, and whether nitrogen-fixing plants are part of the early successional community (Johnson and Curtis 2001). A widely replicated experiment in North America showed that even extreme levels of residue removal had no adverse effects on soil carbon and nitrogen during the first five treatment years (Fleming et al. 2006; chapter 17 in this volume); however, that study as well as others show that soils beneath broadleaved forests are more sensitive than those beneath conifers (Johnson and Curtis 2001).

Nevertheless, there are clear examples of management-related forest degradation that can be traced to excessive removal of nutrients and organic matter during harvest or site preparation, especially on soils that are already nutrient poor. Hot slash burns volatilize carbon, nitrogen, and sulfur. Whole-tree harvest removes large amounts of all nutrients. Piling or windrowing logging residues concentrates nutrients and organic matter on a small proportion of the site. Even raking litter from forest floor (when done every year) has been found to reduce growth of trees in otherwise undisturbed stands.[7]

Perhaps the best known example of lowered productivity due to management practices is the **second-rotation decline** of radiata pine grown in South Australia. Figure 20.5 shows change in site class from first to second rotation in the Penola State Forest (Powers 1989). Site class 1 is most productive (average 1,500 m^3/ha/yr) and site class 7 least productive (average 400 m^3/ha/yr). During the first rotation, the Penola forest was mostly site class 1 and 2, with a few areas of 3 and 4. In the second rotation, the forest was mostly site class 3 and 4, with some spots as low as 6 and 7. Quoting from Powers (1989), "Squire et al. (1985) traced growth declines to losses of forest-floor organic matter during site preparation. Slash burning following the first rotation exposes the soils surface to evaporation and promotes weed invasion. Together, these have a major impact in a region characterized by summer drought. Drought further reduces nutrient mineralization and transport to tree roots in soil solution. The end result is diminishing stand performance until crown closure. Maintaining surface organic matter not only prevents second-rotation decline, but may improve growth substantially over that noted in the first rotation (Squire et al. 1985)."

[7]In the past, German farmers regularly raked litter from pine stands to use as bedding for livestock. To test whether that might affect tree growth, German foresters set up a paired plot experiment in which litter was raked annually from one of the pair, while the other was left undisturbed. Results were assessed in terms of the German system of site class, in which site 1 is the most productive and site 5 the least. After several decades, site class of plots from which litter had been removed averaged more than one to more than two classes lower than plots without litter removal (Wiedemann 1935; Powers 1989). Similar effects of litter removal have been demonstrated in the United States (Cope 1925; Lutz and Chandler 1946).

Figure 20.4. A. These farms in Central America were established by clearing moist tropical rain forest and exposing soil to erosion. Silt from cleared forests in Panama is slowly filling in the Panama canal, which had to be closed in the 1980s to allow dredging. B. This photograph illustrates the importance of intact forest vegetation in controlling soil erosion. The stream on the right drains a watershed burned by wildfire a few days before the picture was taken; the light color is sediment. The stream on the left drains a watershed that was not burned. An intense rainstorm several days after the fire triggered the pulse of sediment to the stream. (Courtesy of Mike Amaranthus)

Machine piling provides some of the clearest examples of forest degradation resulting from management. Bulldozers frequently displace not only residues, but forest floor and even topsoil; therefore, an even greater proportion of site nutrients and organic matter ends up in windrows, where they are unavailable to most trees on a site (fig. 20.6). Soils are compacted and large aggregates (which as we saw in chapter 14 are critically important for numerous soil processes) may either be crushed or pushed up in windrows. Alegre and Cassel (1986) measured a 50 percent

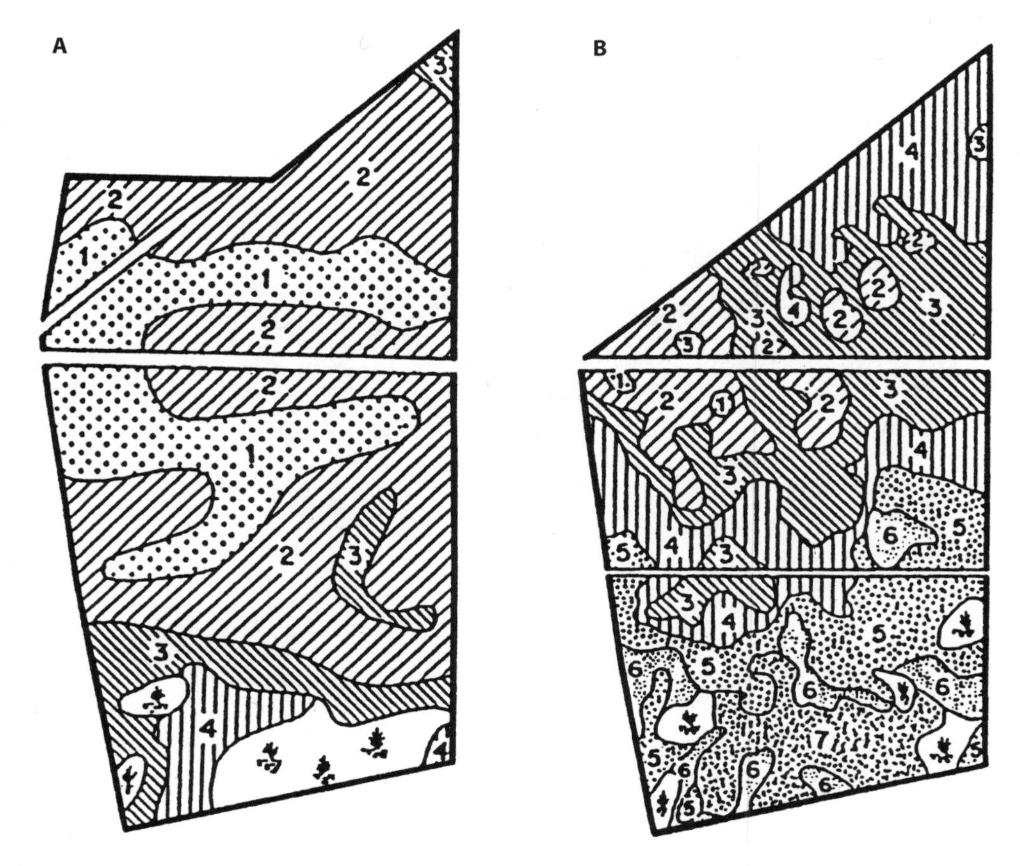

Figure 20.5. Mappings of site class for two successive rotations of Monterey pine on the Penola Forest Reserve, South Australia. A. The first rotation after clearing native vegetation. B. The second rotation. Site quality 1 is the most productive and 7 the least. Note that the area went from mostly site classes 1 and 2 in the first rotation to mostly site classes 3 and lower in the second rotation. Growth losses in this case were traced to losses of surface soil organic matter. (Powers 1989)

loss of large soil aggregates following bulldozer clearing of a Peruvian forest.

In chapter 17 we saw how windrowing had reduced productivity of both radiata pine stands in New Zealand and ponderosa pine stands in California (fig. 17.7). This is a common observation in many forest types around the world (e.g., Uhl et al. 1982; Cole and Schmidt 1986; Dyck and Beets 1987; Powers et al. 1988; Atzet et al. 1989; Fox et al. 1989; Harvey et al. 1989). Figure 20.7 shows a schematic of vegetation recovery in Amazonian caatinga forest that had been treated in different ways following logging. Extrapolating early rates of regrowth into the future, Uhl et al. (1982) estimated that areas where residues and top soil had been bulldozed would require 1,000 years to attain the biomass of the primary forest that had been logged.

In some cases, machine piling does not reduce early tree growth and may even benefit it (Tuttle et al. 1985; Haywood and Burton 1989), probably because the severe soil scarification reduces competition from weeds. Even in these instances, however, productivity is likely to decline as trees become larger and make greater demands on remaining site nutrient capital. A delayed effect of high levels of biomass removal has been seen in long-term experiments both in New Zealand (Smith et al. 2000) and Sweden (Egnell and Valinger 2003). In New Zealand, effects of residue removal on radiata pine growth did not become statistically significant until trees were nine years old, while in Sweden the effects of whole-tree harvest did not manifest until the fifteenth year. In both, effects appeared when trees were entering a phase of rapid growth and making high demands on soil resources. Results such as these emphasize the dangers of reaching conclusions too early in stand development. Patterns can change, and often do.

20.4.3.2 Compaction

Heavy machinery compacts soils, which reduces pore space and therefore water infiltration, aeration, and the ability of plants to root effectively (Greacen and Sands 1980; Froelich and McNabb 1984; Childs et al. 1989; Page-Dumroese et al. 2006). As little as one pass of heavy equipment over a piece of ground can compact soils, and effects can persist for decades.

Figure 20.6. After clear-cutting this Wyoming forest, logging residues and topsoil were scraped into piles by bulldozers. Trees planted on such areas sometimes grow well during the first few years, but the combination of lost topsoil and soil compaction (from the heavy equipment) in many cases eventually reduces productivity. It could take centuries to rebuild lost topsoil.

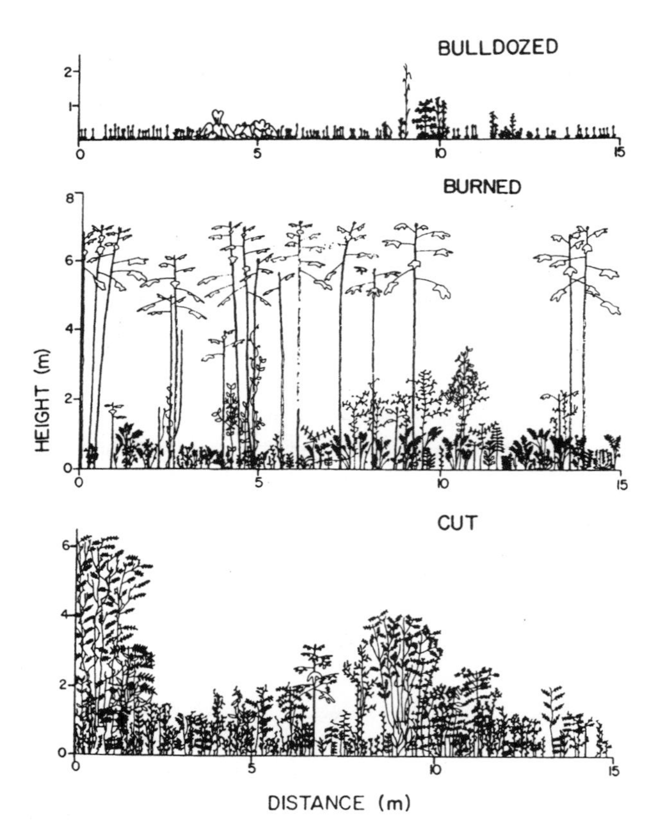

Figure 20.7. Profiles of vegetation in 1-m by 15-m strips three years after the forest occupying each of these sites in Amazonia was cut (bottom), cut and burned (middle), and cut and bulldozed clean (upper). (Uhl et al. 1982)

It has been commonly assumed that this reduces tree growth, however five-year results from a multireplicated experiment in North America showed little or no adverse effects on seedlings from compaction (Fleming et al. 2006). Neverthless, impacts are expected to manifest as trees age and enter a rapid growth phase. It seems prudent to avoid compaction or limit the compacted area. Where ground-based machinery must be used, mitigative techniques include using low pressure equipment, restricting machinery to a few skid trails that are designated before logging, or, where it is an option, by skidding and piling only on top of some protective layer, such as slash or snow.

20.4.3.3 Cumulative Effects: Harvesting Too Frequently

Harvest rotations that are more frequent than the natural disturbance cycle produce cumulative effects that build through time. Some of the clearest examples of degradation resulting from increased disturbance frequency are those related to shifting cultivation in the tropics. Shifting cultivation (sometimes called **swidden cultivation** or **slash-and-burn agriculture**) is an ancient technique in which small plots are cleared within the forest, crops are grown for a few years (usually no more than 3 or 4), then the plot is abandoned for a long, fallow period while the farmers move to clear another area. The system is driven by

Figure 20.8. Dry "miombo" forests in east Africa are cleared for farming by shifting cultivators. Soil fertility drops sharply after a few years and cannot be recovered, even with fertilizers. The original forest vegetation reestablishes very poorly on these sites if at all, resulting in miombo forest throughout east Africa being converted to semidesert scrub vegetation.

soil fertility. After a few years cropped soils lose their productive capacity, but this is regained after a sufficiently long fallow period (usually on the order of several decades).

Native people in both the moist and dry tropics have practiced shifting cultivation for centuries. Not only has it been a stable system, some researchers believe that the diversity of tropical forests reflects at least in part past cultivation patterns (chapter 10). At this point, however, much of the degradation that occurs in some areas of the tropics stems from shifting cultivation (fig. 20.8). What happened to cause a stable system of agroforestry to go awry? The answer lies in the frequency with which farmers return to a given spot. Increased populations and shrinking forest area (due to clearing for ranches, oil palm plantations, or other large-scale land uses) forced farmers to reduce the length of the fallow period. With shorter fallow, soils did not fully recover between periods of cropping. Eventually soils became so degraded that the forest could not recover vigorously (Nye and Greenland 1960; Arnason et al. 1982; Trenbath 1983; Adedeji 1984). There are also landscape-level cumulative effects, especially forest drying and increased fire susceptibility due to generally more open conditions.

Shifting agriculture may seem far removed from traditional forestry, but the ecological implications of frequent harvest are the same whether trees are cleared for farming or shipped to a pulp mill. In general, short rotation forestry exports large amounts of nutrients from sites and has a high probability of eventually reducing yields unless coupled with fertilization. Removing crowns (as in whole-tree harvest) greatly exacerbates this effect. Other cumulative stresses may build (e.g., soil compaction) that cannot be corrected by fertilization. It is more difficult, however, to generalize about what rotation length is too short. This will vary depending on forest type, management practices, soils, and the rates at which nutrients lost during harvest are replaced (by natural processes or fertilization). Nutrients are not the whole story, however. In some cases, such as the degraded miombo forests shown in figure 20.8, fertilizers alone do not restore site productivity. In such instances, which are common on highly weathered tropical soils, we must look beyond nutrients to soil organic matter, structure, and biology.

20.5 LOSS OF BIOREGULATION: BREAKING THE LINKS BETWEEN PLANTS AND SOILS

In earlier chapters we saw that plants divert a high proportion of the energy fixed in photosynthesis to roots, mycorrhizal fungi, and rhizospheres. In consequence, nutrient retention and cycling, soil physical structure, and the composition of the soil community are directly tied to the presence of plants, and in some cases to particular types of plants. This bioregulation of soil processes and properties by plants benefits plant growth, and plants and soils

Figure 20.9. The Siskiyou Mountains, Oregon. This clear-cut at a site called Cedar Camp (A) is one of a number on high-elevation granitic soils that were logged 20 to 25 years before the photograph was taken and unsuccessfully reforested despite having been planted an average of four to five times each. The inability to reforest clear-cuts has been a persistent problem in many high-elevation forests of the western United States. In sandy soils such as these from the forest adjacent to the Cedar Camp clear-cut (B) and from the clear-cut (C), soil structure is tied closely to the presence of roots and mycorrhizal hyphae.

become tied together by mutually reinforcing positive feedback (Perry et al. 1989a). When trees or ecologically equivalent plants are removed from forest ecosystems for a long enough period, the ability of soils to support those plants deteriorates. Adejuwon and Ekanade (1987) describe this phenomenon following conversion of Nigerian rain forest to agriculture: "when . . . forest is removed and replaced by field or tree crops, the balance between vegetation and soil breaks down, and this leads to instability and a considerable deterioration in soil quality."

Figure 20.9. *(Continued)*

20.5.1 A Case Study

Figures 20.9 and 20.10, which were taken in the Siskiyou Mountains of southwest Oregon and northern California, show what happened in one case when the energy supply to the soil ecosystem was eliminated by deforestation. That area (and many others like it) was clear-cut in the late 1960s and has not been successfully reforested despite many attempts. Naturally established forests growing adjacent to the site, however, are quite productive.

Comparison of soils between forest and clear-cut (named Cedar Camp) reveals that those beneath the forest are well structured, while those in the clear-cut are like beach sands (fig. 20.9B and C). Scanning electron microscopy shows an array of pore sizes in the forest soil: large pores to drain water and allow aeration, small pores to retain water (fig. 20.10; chapter 14). In contrast, the clear-cut soil has very little structural diversity. Although structural diversity in soils is generally associated with organic matter, in this case the two soils do not differ in total organic matter; the relevant difference is in *living* organic matter. More specifically, clear-cut soils have lost the fine roots and mycorrhizal hyphae that are important in binding soil microaggregates together into macroaggregates (chapter 14). The electron micrograph of figure 20.10C shows fungal hyphae in close association with soil minerals in the forest; the hyphae are probably mycorrhizal, but this is not certain.[8]

[8]Hyphae in the micrograph are associated with minerals, not organic matter, which means that their source of carbon is elsewhere. That "elsewhere" is more than likely a tree root, meaning, they are mycorrhizal.

Soil biology differs in several respects between the forest and the clear-cut at Cedar Camp. The ratio of bacteria to fungi is much higher in the clear-cut, as are the numbers of actinomycete colonies that produce allelopathic chemicals (Perry and Rose 1983; Friedman et al. 1990). In chapter 18 we discussed hydroxymate siderophores (HS), strong iron chelators that are produced by microbes and that are important in plant iron nutrition and defense against root pathogens. Concentrations of HS are much lower in clear-cut than in forest soils, probably reflecting lower numbers of the mycorrhizal fungi and rhizosphere bacteria that produce the chelator. Similar declines in HS were detected in 8 of 10 Oregon clear-cuts (Perry et al. 1984), suggesting this may be a general phenomenon.

Inability to reforest the Cedar Camp clear-cut was probably due to a combination of factors. The climate is severe, with long winters and short, droughty summers. In such an environment, establishing seedlings must gather resources and establish quickly or they do not survive the first growing season. The changes in the soil that occurred following clear-cutting made a difficult task even more difficult. The loss of soil structure in the Cedar Camp clear-cut slowed the rate at which water infiltrated into soils and reduced soil water storage capacity. Seedlings developed roots more slowly and formed fewer mycorrhizae than comparable seedlings planted in forest soil, reducing their ability to gather nutrients and water (Amaranthus and Perry 1987, 1989b). Reduced concentration of HS induced an iron deficiency (Perry et al. 1984), and the combination of fewer mycorrhizae and lower concentrations of HS may have reduced the ability of seedlings to fend off pathogens.

Figure 20.10. Scanning electron micrograph of soil from the Cedar Camp forest (A) and the clear-cut (B). Note the array of large and small pores in the forest soil and the absence of pores in the clear-cut soil. C. A 10-fold enlargement of the forest soil. The hyphae here are probably mycorrhizal, but this is not certain. It should not be construed that these types of changes happen in every clear-cut; it likely depends on the soil type and characteristics of the early successional community (which at Cedar Camp was dominated by an exotic annual grass). Few studies have been done, and it is unclear how far the results obtained at Cedar Camp can be generalized.

Many of these effects may have been either triggered or exacerbated by the spread of allelopathic actinomycetes (Friedman et al. 1990) and were initiated (or at least exacerbated) by herbiciding early successional shrubs that provided a soil environment supporting rapid root-tip formation by tree seedlings (Amaranthus et al. 1990b). The longer the clear-cut went without trees (or ecologically equivalent plants) the more the soil changed, and the more the soil changed, the more difficult it became for trees to establish.

Evidence suggests that loss of top-down regulation within the soil food web was also involved. Seedling growth and survival were significantly increased in the Cedar Camp clear-cut by adding small amounts of soil (about 150 ml) from established forests to planting holes at the time of planting (Amaranthus et al. 1990a; Colinas et al. 1994). Douglas-fir seedlings given soil from young Douglas-fir forests grew roots faster, formed more mycorrhizae, and survived and grew better than seedlings receiving no soil transfers (fig. 20.11). Soil transfers from an adjacent mature forest dominated by *Abies* sp. improved seedling growth, but not survival and mycorrhiza formation. Forest soils, especially those from stands of the same species as planted seedlings, contained something (or "somethings") that had been lost from the clear-cut, and that loss effectively destroyed system resiliency.

Figure 20.11. Douglas-fir seedlings after one growing season in the Cedar Camp clear-cut; seedlings were 1 year old when planted. The seedling on the left received 150 ml of forest soil at the time of planting (soil was added to the plant in hole); the seedlings on the right received none.

Further experiments at Cedar Camp showed that seedling growth in the clear-cut was nutrient limited; however, the benefits of forest soils were not directly related to the added nutrients, but rather to the addition of invertebrates that graze on microbes (Colinas et al. 1994). Think for a moment about what might be going on here. In chapter 18 we saw the importance of invertebrates in cycling nutrients. Now consider the fact that energy supply to the soil (in the form of fresh photosynthates) dropped once trees and early successional shrubs were removed. Which trophic levels are likely to be affected most severely as a consequence of reduced energy flow through the system (chapters 9 and 11)? Can you construct an ecologically plausible scenario out of these facts? (See DeAngelis [1992] for discussion and modeling of the relationships among nutrient cycling, food webs, and system stability.)

> Sharply reduced energy flow into the soil would be expected to affect higher trophic levels most (i.e., the numbers of invertebrates that graze on microbes would be reduced more than the microbes). Unpublished studies by Andy Moldenke of Oregon State University confirm that expectation; he finds that populations of microbial grazers are significantly lower in burned clear-cuts than in forests on the H. J. Andrews Experimental Forest. Studies in Colorado by Ingham et al. (1989) show that grasslands (which the Cedar Camp clear-cut had become) support fewer microarthopods than forests. Reduced populations of microbial grazers would lead, in turn, to lower rates of nutrient cycling and reduced tree growth (chapter 18).

The beneficial effect of forest soil transfers on seedling survival at Cedar Camp apparently stemmed from factors different than those that stimulated growth (Colinas et al. 1994). Survival on these sites relates more to rapid development of fine roots and mycorrhizae than to biomass increment per se (Amaranthus and Perry 1989b; Amaranthus et al. 1990a). In other words, gathering surface is more important than biomass for seedling establishment on these sites; the same would likely be true on any droughty or infertile site. Factor(s) in the forest soil transfers that stimulated early root development are still unidentified, but volatile chemicals emanating from forest soil organic matter may be involved. Both soil organic matter and microbes produce numerous volatile chemicals, some of which are known to stimulate root production and to inhibit plant pathogens (Graham and Linderman 1980;

Rodriguez-Kabana et al. 1987; Schisler and Linderman 1989).[9]

The unraveling of the soil ecosystem that occurred when plant-soil links were broken at Cedar Camp probably occurs to one degree or another in other degraded ecosystems, although there are not enough studies to say this for sure. It should be clear from the previous section that the phenomenon is complex and multifaceted. Even after more than a decade of study at Cedar Camp, many questions remain: exactly what changes occur, how quickly, and how do they feedback to affect seedling establishment? Generalizing beyond Cedar Camp requires that we ask still other questions. How does the rate at which soils deteriorate vary with soil type or other environmental factors? How do changes in the soil feedback to affect the ability of indigenous plants to grow? What is the connection between particular plant species and particular soil properties?

20.5.2 When Bioregulation Breaks Down: Generalities and Uncertainties

Weakened bioregulation of soils results in erosion, nutrient leaching below the root zone, shifts in the balances of different microbial types, and loss of soil structure.

20.5.2.1 Erosion

As we saw earlier, wind and water erosion are major causes of soil degradation worldwide. The implications of erosion extend much further than soil degradation: rivers and lakes fill with silt from eroding hillslopes, seriously damaging aquatic ecosystems and reducing the useful life of reservoirs. In 1962, the mouth of the Quileute River in Washington State was 15 m deep; by 1992 silt from upstream clear-cuts had filled the mouth until it was less than 3 m deep. Quileute Indians (whose traditional lands sit at the mouth) report that three species of fish have disappeared from the river. In addition, silt carried offshore has covered kelp beds until they are beginning to disappear. Populations of shorebirds declined, probably because their diet comprises fish that depend on kelp for food (Chris Morganroth, personal communication).

Steep slopes are particularly vulnerable to water erosion when forest cover is removed (Swanson et al. 1989). On the Siskiyou National Forest in Oregon, slopes steeper than 70 percent are 20 times more likely to fail than slopes between 50 and 70 percent, and 200 times more likely to fail than slopes less than 50 percent (Amaranthus et al. 1985). In the Pacific Northwest, topographic depressions on steep slopes are the most prone to mass failure, and clear-cutting these areas is not recommended. Logging roads significantly increase the probability of landslides on slopes that are otherwise stable (Amaranthus et al. 1985; Swanson et al. 1989).

Landslides transport large amounts of sediment to streams and significantly reduce productivity on areas where they occur. In one study in the Pacific Northwest, height growth of trees growing on landslide scars was 25 percent less than that of trees in nearby clear-cuts (Miles et al. 1984). In New Zealand, converting forest to pasture on steep slopes with soils developed from volcanic ash triggered small landslides that spread laterally with successive intense rains, the slides eventually covering 20 percent or more of the landscape (Crozier et al. 1980). As long as 75 years after sliding, productivity on old slide scars in New Zealand averaged 21 percent lower than on areas that had not slid (Trustram et al. 1983). On the other hand, fisheries biologists believe that, historically, periodic infusions of large wood from massive debris flows in the Oregon Coast Range created complexity that improved fish habitat in the long run (Reeves et al. 1995).

Although far less dramatic than landslides, surface erosion probably has a more widespread impact on ecosystem health, because "it is widespread and affects the nutrient-rich surface of the soil" (Swanson et al. 1989). Removing forest cover increases surface erosion for two reasons: (1) because of loss of tree cover and litter layers, rain impacts soils with much more force, and that extra energy is translated into moving soil; and (2) the mechanical holding power of roots is lost.

Surface erosion peaks soon after forest clearing and declines as plants reestablish, but it may take decades to return to predisturbance levels (Swanson et al. 1989). Figure 20.12 shows surface erosion from adjoining forested and logged watersheds on the H. J. Andrews Experimental Forest. The forest was logged in 1975, and by 1977 surface erosion rates (as measured by sediment inputs into streams) were about six hundred times higher in the logged than in the unlogged watershed (Swanson et al. 1989). Erosion dropped

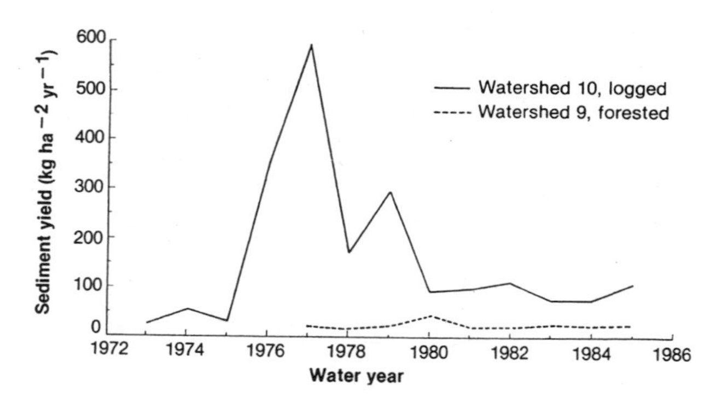

Figure 20.12. The rates of surface erosion from 1973 to 1985 for watersheds 9 (forested) and 10 (clear-cut June 1975) on the H. J. Andrews Experimental Forest, Oregon Cascades. (Swanson et al. 1989)

[9]The *type* of organic matter is probably more relevant than the *quantity*. Compared with clear-cut soils, which have had few fresh inputs of photosynthate in two decades, a higher proportion of organic matter in forest soils should be biologically active and therefore more likely to produce volatile chemicals.

sharply as vegetation regrew, but in 1985 was still three times higher in the clear-cut than in the forest. Surface erosion is exacerbated when logging slash is burned (Swanson et al. 1989), in some cases because of slower regrowth of early successional plants on burned than on unburned sites (this is not always true). Large amounts of soil may also be transported both during and immediately after the burn (Bennett 1982).

On soils with no unweathered rock left in the rooting zone, moderate amounts of erosion could actually improve productivity in the long run by lowering the soil surface, thereby bringing roots closer to fresh rock. On the other hand, soil carbon and nitrogen, which are stored predominantly in litter and surface soil organic matter, are lost in surface erosion, as are clays bound in organomineral complexes (Theng 1991). These losses reduce both fertility and the ability of soils to absorb and store water.

The degree to which a given amount of surface erosion reduces productivity depends on how deep the soil is: mountain soils are generally shallow and, hence, particularly vulnerable. Virtually all estimates of how surface erosion affects site productivity in forests are for crops established on former forest land, rather than for trees. As Theng (1991) stated in a review of soil science in the tropics, "For some shallow Alfisols in Nigeria, a 10 mm loss of soil has the effect of reducing corn yields by 50 percent. Indeed, with the possible exception of deep, naturally fertile Andisols derived from volcanic ash (Carson 1989), it is common for yields to approach zero when more than 75 mm of top soil has been eroded (Lal 1984)." Once enough topsoil has eroded, soil productivity may not be restored even through heavy application of fertilizers, reflecting a deterioration in structure due to lost organic matter (Lo 1989; Theng 1991).

In addition to nutrients and organic matter, mycorrhizal propagules may be lost when surface soils erode. Vesicular-arbuscular fungi appear to be particularly susceptible to erosion (Powell 1980; Habte et al. 1988; Habte 1989; Amaranthus and Trappe 1993). Amaranthus and Trappe (1993) collected soils that had eroded from slopes and accumulated behind check dams following a wildfire in southwest Oregon and used these to inoculate seedlings planted on slopes above the dams.[10] Seedlings of vesicular-arbuscular mycorrhizal cedar formed significantly more mycorrhizae when given soils from behind the dams, but Douglas-fir seedlings (an ectomycorrhizal tree) did not. In this case, with relatively mild erosion, inocula of vesicular-arbuscular mycorrhizal fungi appeared to be more vulnerable to loss in erosion than inocula of ectomycorrhizal fungi, perhaps because the latter were distributed more deeply in the soil profile. Douglas-fir seedlings, for example, form just as many mycorrhizae when grown in soils from a 10- to 30-cm depth as when grown in surface soils (Brainerd 1988). Propagules of ectomycorrhizal fungi can be lost if enough soil is eroded, however, as evidenced by the fact that survival of pines planted on highly eroded slopes in Mexico is improved by ectomycorrhizal inoculation (Valdes 1986).

[10]One technique used by European farmers to maintain soil fertility of steep lands that have been farmed for centuries is to periodically move soil that has accumulated on lower slopes back to upslope fields (Lowdermilk 1953).

20.5.2.2 Leaching

Significant nutrient leaching can occur in any ecosystem in which there is enough precipitation for water to move through the soil profile and sufficient anions to pair with base cations (chapter 17). Leaching is minimized by the biological dam—nutrient uptake and storage by plants and microorganisms—although nutrients do leach even from intact forests, something that has been exacerbated by acid rain but also occurs because of the acidity generated by normal biological activity and the inability of the biota to take up some complex organic molecules (Johnson et al. 1991, chapter 17). The resilience provided by rapid recovery of the plant community limits nutrient leaching following disturbance; however, this natural stabilizing mechanism is disrupted when early successional plants do not (or are not allowed to) regrow quickly (fig. 17.5).

Leaching can be a particularly serious problem after clearing forests in the moist tropics. The high rainfall rapidly leaches any soluble nutrient that is not tied up in biomass or held on the exchange complex. In the case of highly weathered Ultisols and Oxisols, the cation exchange capacity is low; therefore, soluble cations such as calcium, potassium, magnesium, and NH_4^+ are retained primarily by rapid cycling and fixation in tree biomass. When clear-cutting removes these biological sinks, large amounts of nutrients can be moved to streams unless rapid regrowth creates a new biological sink. Once nutrients are leached from these very old soils, there is no adequate source of replacement for those that enter the ecosystem primarily from rock weathering, because the rocks were long ago weathered of their nutrients. In this respect, a bare gravel surface left by a withdrawing glacier is a more hospitable environment for a colonizing plant, because the gravel at least contains elements that can (and will) be released by weathering. The old soil in an ecosystem where bioregulation has broken down does not. It is important to realize, however, that these comments cannot be generalized to all moist tropical forests. Some are growing on relatively young soils with abundant unweathered rock and good exchange capacity, and many have not been studied at all (Procter 1987).

20.5.2.3 Soil Structure

Except for compaction, little research has been done on interactions between forest health and soil structure; Kay (1990) reviews this topic with regard to agricultural soils. The stability of soil aggregates larger than about 0.25 mm is believed to be closely tied to roots and particularly to mycorrhizal hyphae (chapter 14), and the numbers of aggregates in that size class decline with reduced presence of roots and hyphae, resulting in increased bulk density and decreased soil porosity (Low 1955). Within 1 year of clearing forest and cultivating an Oxisol in southern Brazil, numbers of large aggregates had declined by more than one-half, soil organic matter had dropped by 25 percent, bulk density had increased nearly 50 percent, and soil porosity had dropped to only 7 percent of that in forest (Klamt et al. 1986). Cultivating is likely to have contributed significantly to soil deterioration in that case, so the results cannot be attributed solely to loss of bioregulation. Nevertheless, there is little doubt that loss of living roots and mycorrhizal hyphae lead to loss of soil structure.

Surface erosion leads to structural degradation because it removes organic matter and surface clays. In Nigeria, soils that have been converted from rain forest to either annual crops or tree crops (cocoa) have, respectively, 28 and 31 percent greater bulk density than rain forest soils, and 12 and 18 percent less porosity (Adejuwon and Ekanade 1987). The fact that soil structure deteriorated even when cocoa trees were growing on sites may be due to *clean weeding*, i.e., rows between trees were kept clear of other vegetation, in which case overall plant and litter cover would be lower than in natural forest. Wiersum (1984) measured 60 times more surface erosion beneath clean-weeded tree crops than beneath tree crops that had an understory cover crop or mulch.

20.5.2.4 Mycorrhizal Fungi and Other Soil Organisms

The effect of disturbance on mycorrhiza formation (and indeed on soil biology in general) is complex and poorly understood. Plants growing in disturbed soils often form a less diverse and sometimes quite different set of mycorrhizal types than plants growing in undisturbed forest (Pilz and Perry 1983; Schenck et al. 1987; Borchers and Perry 1990; Hagerman et al. 1999; Mah et al. 2001). Tree seedlings may form fewer mycorrhizae on sites that have experienced hot fires, lost top soil, or where host plants required by the fungi have been absent (Wright and Tarrant 1958; Harvey et al. 1980, 1983; Perry et al. 1982, 1987b; Parke 1984; Valdes 1986; Amaranthus and Perry 1987; Borchers and Perry 1990; Dahlberg 2002). Referring specifically to the tropics, Sieverding (1989) states, "There is little question that VAM [vesicular-arbuscular mycorrhizal] fungi have to be reintroduced to soils which are disturbed." On the other hand, in many instances seedlings growing in clear-cut soils form just as many mycorrhizae as those growing in forest soils, and sometimes they form more (Pilz and Perry 1983; Brainerd 1988; Fischer et al. 1994; Jha et al. 1992), a phenomenon that is almost certainly an adaptive response to lower soil fertility.[11] Mycorrhizae may be reduced on one tree species but not on another (Schoenberger and Perry 1982). In some instances, total mycorrhiza formation may eventually be as high in disturbed as in undisturbed soils, but the rate of formation is slower in the former (Amaranthus and Perry 1989b). Differences such as those discussed above can be partly explained by some combination of disturbance severity, the rapidity with which host plants recover, or the proximity to mature trees. Mine spoils, which certainly qualify as severely disturbed areas, provide some of the most striking examples of the benefits of mycorrhizal inoculation (Marx 1975; Jasper et al. 1989; Helm and Carling 1993; Rao and Tak 2001). In clear-cuts or other harvested areas, the ability of seedlings to access roots or mycorrhizal hyphae of mature trees (or other plants that share the same types of mycorrhizal fungi) strongly influences the diversity and types of mycorrhizae they form (Simard et al. 1997b; Hagerman et al. 1999; Dahlberg 2002; Outerbridge and Trofymow 2004). Even soil transfers from the rooting zone of the appropriate plant can profoundly alter mycorrhiza formation on seedlings (Amaranthus and Perry 1989a; Borchers and Perry 1990).

Physical, chemical, and/or biological factors may alter the ability of plants to form mycorrhizae, even when sufficient inoculum is present. In agricultural soils, for example, changes in soil structure produced by plowing sharply reduce the ability of plants to form vesicular-arbuscular mycorrhizae (Evans and Miller 1988, 1990; Jasper et al. 1990). Certain types of bacteria (called *helpers*) facilitate mycorrhiza formation (Bowen and Theodorou 1979; Garbaye and Bowen 1989); hence, disturbances that shift the composition of the soil microbial community could affect the ability of trees to form mycorrhizae.

Ectomycorrhizal fungi with aboveground fruiting bodies (e.g., mushrooms) produce copious amounts of airborne spores, but deposition of windborne spores depends on turbulence created by features such as established plants (Allen et al. 2005). Mycorrhizal fungi that produce belowground fruiting bodies[12] largely depend on animals for spore dispersal; hence, they are particularly dependent on the rapid recovery of host plants in order to maintain their presence on a site (e.g., Cuenca and Lovera 1990; Borchers and Perry 1990). Experience with the recovery of Mount Saint Helens shows that both large and small mammals are critically important in dispersing mycorrhizal propagules (Allen et al. 2005).

[11]Once again, we face the problem of nonlinearities and thresholds within ecosystems. Over certain ranges of soil degradation, trees form more mycorrhizae and hence buffer themselves against reduced fertility. At some point, however, soils degrade to the point that buffering is no longer possible, and the system is in danger of collapse.

[12]Recall from chapter 11 that all vesicular-arbuscular mycorrhizal fungi and an important subset of ectomycorrhizal fungi produce belowground fruiting bodies.

Figure 20.13. These Douglas-fir seedlings are part of an experiment in which seedlings were planted in three different types of vegetation, all occurring in a 0.5-ha area uniform in soils and topography (Amaranthus and Perry 1989a; Amaranthus et al. 1990a). The seedling on the right was grown in a recently cleared patch in a dense stand of the ectomycorrhizal shrub *Arctostaphylos ciscida* (white-leaved manzanita). The seedling in the middle was grown in an abandoned pasture occupied by annual grasses. The seedling on the left was grown beneath an open stand of ectomycorrhizal Oregon white oak. The differences in growth were accompanied by significant differences in the amounts and types of mycorrhiza and in associative nitrogen fixation within seedling rhizospheres. The effect apparently resulted from differences in soil biology rather than fertility; however, exactly what those differences were is not clear. Manzanita apparently forms ectomycorrhizae compatible with Douglas-fir, while Oregon white oak may form incompatible types (or the responses may have had nothing to do with mycorrhizae). How plant species influence one another through effects on soil organisms needs further research. It is interesting that members of the Kayapo tribe (Amazonia) distinguish plants that are "good neighbors" with one another and cultivate these together (Posey 1993). (Courtesy of Mike Amaranthus)

Survival or rapid reestablishment of host plants is particularly important in maintaining populations of obligate mycorrhizal fungi. Even where spores are abundant, they are less effective than mycorrhizal roots and hyphal fragments in inducing mycorrhiza formation (Ba et al. 1990). Establishing tree seedlings form the most mycorrhizae, and form them most quickly, when their roots are near those of an established plant with compatible mycorrhizal species (Amaranthus and Perry 1989a; Amaranthus et al. 1990a; Borchers and Perry 1990; Simard et al. 1997b; Hagerman et al. 1999; Dahlberg 2002; Outerbridge and Trofymow 2004) (fig. 20.13). This has important implications for the role of diverse plant species in ecosystem resilience, a point to which we return below.

How about other soil organisms? Clear-cutting has striking effects on composition of the soil community, especially when slash is broadcast burned or litter and surface soils otherwise disturbed. Various studies have found bacteria and actinomycetes to be more abundant, and fungi less abundant in clear-cut soils than in forests (Perry and Rose 1983; Friedman et al. 1989). In the Oregon Cascades, clear-cuts that have been broadcast burned and from which down logs have been removed have only one-tenth as many invertebrate individuals in soil, and one-half as many invertebrate species as old-growth forest soils (A. Moldenke, unpublished). These rather dramatic alterations result from a combination of destroying litter layers (where many of these organisms live), changes in both the amount and composition of organic matter (the food supply), and altered microclimates when canopies are removed (more extreme temperature and moisture fluctuations).

20.5.2.5 Soil Biology and Ecosystem Resilience: A Matter of Balance

Although it seems clear that losing too many mycorrhizal fungi can throw a system off track, generalizing about how other changes in soil biology affect system resilience is not so easy. Perhaps the best way to think about it is in terms of balance between two different food webs in the soil, one based in living plant material and the other in dead material (Richards 1987; chapter 9 in this volume). The **living web** (centered on rhizospheres) is driven by exudates and sloughing of cells from fine roots and mycorrhizae. It is very dynamic and includes organisms that feedback positively to plant growth (chapters 14 and 18). The **dead web** is based in litter and comprises microbes that are quite different from those in the rhizosphere (e.g., slower growing, more tolerant of environmental extremes, capable of long periods of dormancy). Invertebrates graze on microbes in both webs; however, since most microbes in forest soils occur in rhizospheres, grazers are likely to get most of their energy from the living web (except for the macroinvertebrates, such as earthworms, that graze heavily on litter).

Members of the dead web perform invaluable functions in ecosystems, but some produce antibiotics or are otherwise antagonistic to plants and/or plant mutualists (and also in some cases to plant pathogens). Antagonists include some species of actinomycetes and saprophytic fungi (e.g., *Penicillium* sp.). However, it does not follow that all species in these groups are antagonistic to plants; some benefit plants by inhibiting pathogens.

Healthy systems retain a balance between these two food webs that in theory can be tipped to favor the dead web in at least two ways: (1) excess accumulation of litter, which occurs naturally in some older forests (especially conifers), and is redressed by disturbances that accelerate mineralization; (2) extensive tree kill or stress, as occurs with clear-cutting, pollution, or some natural disturbances.

As long as succession proceeds unimpeded, recovering plants are likely to reestablish the living web quickly following a disturbance that kills the overstory. Legacy plants (e.g., surviving canopy trees, sprouters, or those that grow from buried seed) play a key role in maintaining the living web and providing the foci that catalyze reformation of the parent system. But what happens if recovery of the indigenous plant community is delayed for some reason? Although much research is needed on that question, the evidence available at this time indicates that populations of antagonistic microorganisms build, populations of beneficial organisms decline, nutrient cycling slows (because invertebrate numbers drop), and a vicious cycle is initiated in which poor recovery of indigenous plants perpetuates and exacerbates the conditions that inhibit recovery (Perry et al. 1989a). This is probably what happened at Cedar Camp, and perhaps on other degraded sites as well. In southwestern Montana, for example, root-tip production by tree seedlings is stimulated by the biota of forest soils, but inhibited by biota in the soils of old, unreforested clear-cuts (Perry et al. 1982).

> The live web and the dead web may also be thought of, respectively, as *fast* and *slow* energy channels. Rooney et al. (2006) show that such fast and slow channels are a typical feature of lower trophic levels in both soil and aquatic food webs and demonstrate that fast channels are characterized by strong trophic interactions and slow channels by weak interactions. Furthermore, as we saw in chapter 9, higher trophic levels feed from both channels, thereby acting as a functional link between the two. Based on their models, Rooney et al. (2006) go on to argue that the resulting strong asymmetries in energy flow and interaction strengths at lower trophic levels and the integrating effect of higher trophic levels are key determinants of ecological stability. For example, after a stand-destroying disturbance, the dead web stabilizes higher trophic levels (at least for a time) while recovering plants are reestablishing the live web. In cases like Cedar Camp, the stabilizing mechanism breaks down because the live web was not reestablished.

It should be clear by this point that nature seldom yields completely to tidy explanations such as those given here. (One of the primary reasons nature is so fascinating is that like art, it is very difficult to capture the essence.) Although details may vary, maintaining balance among differing functional groups is likely to be critically important in all ecosystems, as are biological legacies as mechanisms for maintaining threads of continuity and catalyzing system recovery (chapter 8).

20.5.3 Biodiversity Stabilizes the Soil Ecosystem

We have seen that when plant bioregulation over soil organisms and processes is weakened, soils deteriorate; this deterioration can, in turn, make it more difficult for the indigenous plant community to reestablish. In such instances (as happened at Cedar Camp), positive feedback drives the system rapidly toward degradation. However, disturbances that kill trees are common in nature. How is it that forests that are quite resilient against natural disturbances are sometimes destabilized by management? In general this can be traced to practices that disrupt resilience mechanisms, usually by reducing or eliminating structural and biological diversity (especially legacies). Consider, for example, the degraded Cedar Camp clear-cut discussed above. The healthy forest adjacent to the clear-cut was established following wildfire that (like logging) had killed all trees in the prior stand. It seems that nature knows something about reforestation that we do not. Why do forests of that area recover from wildfire but not clear-cutting? The answer probably lies in two factors. First, the standing dead snags left by wildfire ameliorated the climate for establishing seedlings, providing shade in the hot summer afternoons and a "greenhouse" cover that reduced heat loss during the cold nights. Second, hardwood trees and shrubs with the ability to sprout from roots—something that conifers do poorly or not at all—recovered quickly and reexerted biocontrol over soils: minimizing nutrient loss and stabilizing soil structure and those soil organisms that depend on plants for their continued survival (Amaranthus and Perry 1989b; Borchers and Perry 1990).

Cooperative interactions among perennial plant species that depend on the maintenance of certain soil properties are probably common in nature, particularly where variability in the severity or timing of disturbance produces uncertainty in which of several plant species will be positioned to quickly colonize disturbed sites (this kind of uncertainty characterizes many forest ecosystems). In essence the different plant species within a community are all riding in the same lifeboat (the soil), and it is in the interests of all to keep the boat in good working order.

20.6 LOSS OF BIOREGULATION: BREAKING THE TOP-DOWN LINKS

In chapters 11 and 19 we discussed the importance of top-down regulation and how its breakdown can lead to profound and often unhealthy changes in ecosystems. Without predators, herbivore populations can reach levels that

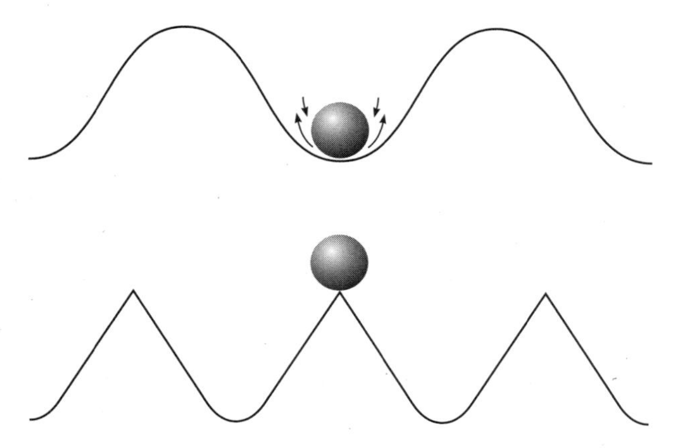

Figure 20.14. Two views of ecosystem stability. (Perry et al. 1990)

seriously damage plant communities. Without herbivores, plant communities can become overgrown and lose diversity. A balance is involved in healthy systems, a form of Goldilocks effect[13] in which too many or too few at any step in the trophic ladder can produce effects that ramify throughout the system.

20.7 BALLS, DANCERS, AND DANCES

In western society, we are taught the classical physicist's view of stability, which is exemplified by a ball resting at the bottom of an energy well (fig. 20.14): most ecology texts talk about ecosystem stability in the context of balls in wells of one kind or another. Translated to an ecosystem, the ball represents the **system state,** which is an abstract entity that encompasses the organisms, their interrelationships, the processes they drive, and all their modifications of and imprints on the abiotic components of the system (air, water, minerals). The energy well represents environmental givens such as macroclimate, topography, and rock type (i.e., physical factors that are relatively independent of the *local* biota). (We emphasize *local*, because macroclimate is influenced by the combined effects of living organisms acting at a global scale. Depending on rock type, it may have been strongly influenced by organisms in the geologic past.)

The fact that the ball rests at the bottom of the well implies that it is in equilibrium with its external environment. In fact, however, the stability of a complex system is less like that of a ball at the bottom of a well than it is like a ball balanced on the top of a peak.

There are three critical differences between the stability of a ball at the bottom of a well and that of a ball balanced on a peak. First, the probability that a ball in the bottom of a well will be destabilized by a given perturbation is proportional to the force that is applied, while destabilizing the ball on the peak has less to do with how hard it is pushed than with the way in which it is pushed. Stated somewhat differently, the effect of a given disturbance on system stability depends on its foreignness, or whether it has been common enough in the past that the species comprising the system have evolved adaptations to it. For example, according to Diamond (1986), "About one-half of all extinctions of island birds have been due to introduced mammals (especially rats and cats) on islands lacking native mammals and also lacking native land crabs (the invertebrate equivalent of rats)." The ground-nesting birds, safe in the absence of ground-dwelling predators, were especially vulnerable when foreign mammals appeared on the scene. In chapter 22 we see that a similar problem has arisen in eastern North America because of the fragmentation of large forest blocks into small islands.

> Swanson (F.J. Swanson, personal communication) distinguishes between *disturbance types* and *disturbance mechanisms*. Two quite different disturbance types may include some of the same mechanisms. For example, both fire and volcanic eruptions involve extreme heat. Species that have never (or rarely) experienced volcanoes throughout their evolutionary history may nevertheless be able to survive extreme heat because of adaptations to wildfire.

As another example of the destabilizing potential of foreign disturbances, a forest that is quite resilient following a highly destructive wildfire may be degraded by the seemingly much less disruptive treatment of eliminating the biological legacies. Although the fire is a very strong push, the ecosystem is adapted to deal with it; removing the legacies, however, disrupts these adaptations and throws the system off track (chapter 8). In some cases at least, relatively subtle shifts in disturbance regime, such as increasing the frequency of perturbations or altering the pattern of disturbance across the landscape (both of which commonly occur in forestry), can and have destabilized ecosystems.

The second difference between the stability of balls in wells and that of balls on peaks relates to **threshold changes.** While you get ample warning that the ball at the bottom of the well is being pushed away from its point of stability, the ball on the peak is subject to very rapid, threshold changes (i.e., it falls off the peak).[14] In recent years ecologists and earth systems scientists have increasingly recognized that threshold changes can be triggered when a

[13]The *Goldilocks effect* refers to the fact that physical conditions on earth are "just right" for the evolution of higher life-forms.

[14]Threshold changes are often modeled using **catastrophe theory.** DeAngelis (1992) describes a catastrophe in the mathematical sense as "a kind of structural instability in a system. A system is structurally instable when small changes in a driving variable or parameter can cause a sudden instability of the system." Models of catastrophic behavior in forests are given by Gatto and Rinaldi (1987) and DeAngelis (1992).

system's bounds of resilience are exceeded; indeed such changes have occurred in forests, arid grasslands, lakes, coral reefs, oceans, and the earth's climate system (earth system scientists call these **tipping points**[15]) (Perry 1995**;** Foster et al. 1997; Scheffer et al. 2001; Jennerjahn et al. 2004; Rietkerk et al. 2004; Tzedakis et al. 2004). A system may go over a cliff for one of two reasons: (1) the cliff erodes out from under it (i.e., resilience is eroded through factors such as loss of biodiversity), or (2) the cliff doesn't move, but a force the system is not prepared for pushes it over (i.e., *foreign* disturbances or new climatic regimes that exceed the bounds of resilience).

The third difference between balls in wells and balls on peaks has to do with **reversibility.** So long as it doesn't go too far, the effect of a perturbation on a ball in a well is easily reversed by removing the perturbing force. That is not the case for a ball on peak: once shoved off, its downward momentum takes over, and it is difficult or impossible to stop, even if the original perturbation is long gone. Consider, for example, Pimm's (1991) discussion of species extinctions on islands and within fragmented habitats: "Species extinctions create more species extinctions, the potential for species invasions, and yet more species extinctions. There seems to be no theoretical arguments or empirical data to suggest that as community composition collapses, further change becomes less likely. Indeed, just the opposite seems to be true; as community composition changes further changes seem inevitable."[16] The rapid and catastrophic deterioration in the health of many forests in interior North America during the 1980s is a good illustration of how a single perturbation can initiate a series of events that is difficult to stop. The perturbation in this case was the exclusion of fires beginning in the early 1900s. As we discussed in chapter 7, low-intensity ground fires were a normal, healthy part of the extensive ponderosa pine forests that once dominated the lower elevations of that region. Fire exclusion led to the spread of insects and diseases and allowed a buildup of fuels that has set the stage for catastrophic crown fires. These forests might eventually be returned to their original condition, but it won't be easy. Similarly, studies of forest response to past climate shifts show that "once an ecological threshold has been crossed, a return to the previous climatic conditions does not guarantee a similar reversal in vegetation" (Maslin 2004 discussing Tzadakis et al. 2004).

The behavioral differences discussed above can be traced to a single, more fundamental difference between balls in wells and balls on peaks. One ball exists in equilibrium with the local environment; the other is in disequilibrium. A complex system such as an organism or an ecosystem exists far from any kind of equilibrium with its external environment. This is easiest to see with an organism. Every organism takes in energy and converts it to order that is manifested in innumerable ways: the structure of molecules, the chemical gradients regulated by membranes, the architecture of cells and bodies, the specialized functions of various organs (e.g., flowers, phloem, eyes, livers, breasts), and so on. Neither are the systems that are created by organisms in any kind of equilibrium: the flow of energy through food webs creates order that appears as structures, relationships, fluxes of elements (i.e., all the patterns and processes that we have discussed in this book).

Ecosystems are always in disequilibrium with local environments, which is simply to say that organisms modify the physical environment to create conditions that would not exist in their absence. The earth's atmosphere is the clearest example of this at a global scale (chapter 3). One way in which disequilibrium is manifested on land is in the interaction between plants and soils. Consider nutrient cycling: tight cycling by the trees retains nutrients that would otherwise be leached to streams. Just how far from equilibrium the nutrient cycle of a given system is can be easily gauged by clearing the trees, preventing succession, and watching what happens to the nutrients. In earlier chapters we discussed just such an experiment on the Hubbard Brook Experimental Forest, in which there were very large nutrient losses to streams in the absence of vegetation. Recall, also from chapters 14 and 18 that fertility of moist tropical rain forests that grow on old, highly weathered soils depends on tight cycling by trees and mycorrhizal fungi. Equilibrium in these systems means that gravity moves all the nutrient capital to streams or deep soil layers; system health depends on remaining far from that equilibrium state. Maintaining stability under such conditions (i.e., staying on the peak) requires that key processes and functions be protected.

In truth, a ball on a peak is not an accurate metaphor for the stability of living systems, because it implies stasis and fragility. As we have seen, living systems are never static. The changes in species that accompany normal patterns of succession, for example, do not mean the ball has fallen off the peak, as long as certain processes and potential are maintained (e.g., nutrient retention and cycling, the potential to capture energy and reassemble the same or a similar community of organisms, the integrity of food webs). Nor are living systems fragile *as long as they are not pushed beyond the bounds of their adaptability.* The *dynamic balance* and *robustness yet threshold fragility* that characterize organisms and ecosystems are best illustrated by living systems themselves: the bicycle rider, dancers and the dances they create, or basketball players and the game in full tilt. Even these examples are not totally satisfactory, however, because they deal with systems that have clear boundaries. As we have seen, this is seldom the case for ecosystems. We may have to accept the fact that an accurate mental representation of

[15]One such tipping point is the melting of permafrost and release of the huge stores of methane (a strong greenhouse gas) that are stored beneath the ice. Permafrost is already melting in Siberia (Pearce 2005).

[16]We see in the next chapter that loss of a single species from a system does not necessarily cause the rest of the system to unravel: it depends on the species, the system, environmental context, and so on.

the dynamics of nature may not be possible. (As Pascal told us, "imagination tires before nature.")

20.8 SUMMARY

Stability does not mean "no change." Rather, the stability of complex systems is more like that of a dancer or a bicycle rider (i.e., dynamic movement that is constrained within certain bounds). The stable system has two levels of defense. First, it must be resistant, which means that small disturbances such as defoliating insects and periodic droughts are contained and prevented from killing large numbers of trees. Second, when, as periodically happens in many forests, resistance mechanisms are exceeded and large numbers of trees are killed, the system must be resilient; which means that nutrient cycling and other soil processes are stabilized and succession returns the system to a predisturbance community composition and level of productivity (within bounds; nothing ever returns to exactly what it was before). The species that compose a given ecosystem have evolved mechanisms to survive or recover from the disturbances that characterize that system (if they had not, they would not be there) and these adaptations make the ecosystem as a whole resistant and resilient.

The limits to stability (i.e., the point at which a system passes from stable to unstable) may be quite difficult to determine; however, certain general aspects can be identified:

- Destabilizing disturbances are "foreign" in the sense that they differ significantly from the historic disturbance regime. Industrial pollution clearly is a foreign disturbance, but other kinds of foreignness may be more subtle. Killing early successional plants, for example, eliminates an important resilience mechanism. A common disturbance in one system may be foreign to another: conifer forests are well adapted to fire, but tropical forests are not. A familiar type of disturbance may be perceived as foreign if it occurs with unprecedented severity of frequency (e.g., crown fires vs. surface fires).
- Bioregulation is compromised. This may involve such things as weakening the connection between native plants and the soil (leading to nutrient loss and altered soil biology and physical structure), disrupting top-down regulation (e.g., eliminating top predators), or altering landscape patterns in such a way that landscape-level controls are compromised (e.g., widespread monocultures that lead to pest problems).
- The initial foreign stress often triggers secondary stresses such as insect and pathogen outbreaks; in other words, the foreign stress imbalances the system in such a way that organisms normally innocuous (or even beneficial) behave destructively.
- Positive feedbacks among system components magnify the effects of the foreign disturbance and produce the potential for quite rapid degradation once some threshold of vulnerability is exceeded.

Identifying and protecting the mechanisms that confer resistance and resilience on ecosystems are critically important tasks for ecologists and land managers.

21

Ecosystem Stability II

The Role of Biodiversity

There exists little doubt that the Earth's biodiversity is declining. The Nature Conservancy, for example, has documented that one-third of the plant and animal species in the United States are now at risk of extinction. The problem is a monumental one, and forces us to consider in depth how we expect ecosystems, which ultimately are our life-support systems, to respond to reductions in diversity.

MCCANN 2000

Their humility as a people was evident by their praise towards life on earth, having recognized their success coming from the good will of all creatures.

FROM THE CULTURAL SERIES OF THE SHUSWAP NATION, BRITISH COLUMBIA

Managing forests inevitably means manipulating their structure, although the degree to which natural patterns might be altered by management ranges from slight to extreme. Evaluation of trade-offs boils down to maximizing wood production and economic returns (at least in the short run) through homogenization versus an array of sometimes intangible or deferred values that are provided by diverse forests and forested landscapes. Among the latter is the stabilizing role of biodiversity, which is the subject of this chapter.

21.1 MAY'S PARADOX

The role of biodiversity in stabilizing ecosystems has been much debated over the years (see McCann 2000). Up until the early 1970s, most ecologists believed that species diversity enhanced ecosystem stability and pointed to the moist tropical forests as an example of highly diverse ecosystems that were also quite stable. This belief was challenged by (among others) the theoretical ecologist, Robert May, whose models predicted essentially the opposite, that species-simple ecosystems were more stable than species-rich systems (May 1974). Following May's work the weight of ecological opinion shifted rather dramatically, and it became fashionable to argue that the species richness of the tropics resulted from, rather than caused, the stability of tropical forests.

Many ecologists realized that while the models of May and others had yielded important insights—and in particular had forced critical reexamination of an old idea (always a valuable service in science)—something was amiss. The prediction that simple systems were more stable than complex systems flew in the face of what we saw (or seemed to see) in nature; Orians (1975) called this disjunction between theory and apparent reality **May's paradox** (also see Margalef 1975). Consider, for example, that most simple of ecosystems, the modern agricultural field. When one accounts for the large amounts of pesticides and fertilizers required to maintain a productive cornfield, it is obviously far less stable than native tall grass prairie that it replaced. There is an important distinction here; the cornfield is not only a simple ecological system, it is more importantly a **simplified system** (i.e., one that contains far fewer species than the previous natural ecosystem on the same site). Moreover, the one plant species that is present (i.e., domesticated corn) is far more genetically homogeneous than a wild plant population.

21.2 INTENSIVE FOREST MANAGEMENT SIMPLIFIES NATURAL ECOSYSTEMS

Few managed forests are anywhere near as homogenized as a cornfield, but most are simpler than the natural unmanaged forest that they replace. The objective of intensive forest management is to maximize wood production. The cost of this is habitat complexity. McRae et al. (2001) compared patterns arising from wildfire with those created by forest management in Canada's boreal zone. They concluded:

> The scales of disturbance are different. . . . In particular, typical forestry does not result in the large numbers of small disturbances and the small number of extremely large disturbances created by wildfires. Moreover, the frequency of timber harvesting is generally different from typical fire return intervals. The latter varies widely, with stand replacing fires occurring in the range 20 to 500 years in Canada. In contrast, harvest frequencies are dictated primarily by the rotational age at merchantable size, which typically ranges from 40 to 100 years. Forest harvesting does not maintain the natural stand age distributions associated with wildfire in many regions, especially in the older age classes. [Wildfire] results in a complex mosaic of stand types and ages on the landscape. Timber harvesting does not generally emulate these ecological influences. . . . Wildfire leaves large numbers of snags and abundant coarse woody debris. . . . Successional pathways following logging and fire often differ. Harvesting tends to favor angiosperm trees and results in less dominance by conifers. . . . Understory species richness and cover do not always recover to the pre-harvest condition during the rotation period used in typical logging [and] animal species that depend on conifers or old-growth are affected negatively by forest harvesting in ways that may not occur after wildfire. Road networks . . . cause erosion, reduce the areas available for reforestation, fragment the landscape for some species and ecological functions, and allow easier access by humans, whereas there is no such equivalency in fire-disturbed forest.

McRae et al. (2001) point out that differences between harvesting and wildfire "vary a great deal among ecosystem types, harvesting practices, and scale of disturbance." Nevertheless, as a rule, harvesting simplifies both stands and landscapes.[1] Note that comparing species richness alone is unlikely to capture the scope of the differences between harvesting and natural disturbances. Nature abhors a vacuum, and within limits one species' loss will be another's gain. Therefore overall species richness may be less affected than particular sets of species (e.g., old-growth or early successional associates) and overall system complexity.[2] Moreover, shifts in species balances (especially among trophic levels) can significantly influence system behavior (chapters 9 and 11).

21.3 DOES BIODIVERSITY STABILIZE ECOSYSTEMS? YES, BUT . . .

We have encountered examples of the stabilizing role of diversity at various points throughout the book. The previous chapter closed with a discussion of how diverse adaptations of plants conferred resilience on ecosystems by stabilizing the soil following catastrophic disturbance. Resistance to insect and pathogen outbreaks is often enhanced by plant species diversity, and also by various protective mutualisms and by predators, the latter depending, in turn, on structural diversity in the plant community and on landscape diversity for habitat (chapters 10 and 19). The ability of trees to maintain physiological processes during drought is enhanced by factors such as diverse pore sizes in soil and old logs that act as water storage reservoirs (chapters 14 and 15). Through effects on the rate at which

[1] It is important to realize that none of this is a *necessary* consequence of forestry: managed forests and forested landscapes can be diverse, even to the point of reproducing habitats similar to old growth (chapter 23). Rather it is a consequence of focusing on a single forest value: wood.

[2] In theory, any reduction in complexity would lead to fewer niches and therefore lower species richness.

disturbances spread and the availability of colonizers to replace losses, both the resistance and resilience of any given local ecosystem depends on the landscape and regional diversity within which it is embedded.

How much can a forest or forested landscape be simplified before it loses resistance and resilience? Given the multiple aspects of diversity and the numerous possible feedbacks between structure and processes, that is not an easy question to answer, and our understanding leaves a great deal to be desired. Scientists have approached the problem with a combination of experiments, models, and observations (Pimm 1991; Naeem et al. 1999; McCann 2000; Loreau et al. 2002). For practical reasons, by far most experiments have been done in grasslands or laboratory microcosms, and many of these have dealt with the broader question of how diversity influences ecosystem functions (e.g., productivity, nutrient cycling). Through the end of the twentieth century, experiments on the relation between diversity and ecosystem functioning had produced the following generalizations (Naeem et al. 1999):

- Declining species richness can lead to declines in overall levels of ecosystem functioning, especially at lower levels of diversity.
- At least one species per functional group is essential to ecosystem functioning, and having more than one species per functional group is probably essential to ecosystem stability (section 21.4.3).
- The nature of an ecosystem's response to declining biodiversity depends on which species are lost.

In the 10-year grassland study reported by Tilman et al. (2006), stability of annual net primary productivity arose from two factors: (1) the so-called portfolio effect and (b) overyielding. The portfolio effect is a purely statistical process in which adding more species to a community lowers the average variance for the community as a whole (Doak et al. 1998). The descriptor *portfolio* refers by way of analogy to the investment strategy of holding a diverse portfolio of stocks in order to reduce financial fluctuations. The portfolio effect may or may not reduce variance at the community level, depending on the relationships between the mean and the variance for the individual species composing the community (Tilman et al. 1998). The overyielding effect refers to an ecological process in which diverse species utilize a greater proportion of available resources than any one species alone, which results in a positive relationship between diversity and productivity. When applied over time (which is necessary in order to ask questions about stability), overyielding implies that a diverse community more fully utilizes temporal fluctuations in resource availability than a simple community.

Questions concerning the stability of ecosystem functioning generally require long time periods to address in a meaningful way (e.g., to encompass a series of climatic fluctuations). Few long-term studies exist; however, 10-year results from an experimental grassland show that increasing plant diversity results in increased stability of community biomass (Tilman et al. 2006).

Suppose one finds an ecologically meaningful positive relationship between diversity and stability (i.e., not a statistical artifact). A central question arises: is it due to diversity per se or to the inclusion of certain species in the mix? The latter possibility is called the *sampling effect,* because the more species that are included in a mix the greater probability of including at least one with strong effects. In their meta-analysis of studies that have manipulated diversity, Cardinale et al. (2006) found that results across four different trophic groups (plants, herbivores, decomposers, and predators) were most consistent with the sampling effect. Cardinale et al. conclude that "average species loss does indeed affect the functioning of a wide variety of organisms and ecosystems, but the magnitude of these effects is ultimately determined by the identity of the species that are going extinct."

Models are generally based on the structure of food webs, reflecting that fact that the flow of energy through systems and the influences arising from the search for food are critical determinants of system dynamics. While greatly oversimplified, model food webs produce insights into factors affecting stability in the real world: the importance of alternative energy pathways, the sensitivity to loss of large predators, the importance of compartmentalization. For example, consider the hypothetical food web of figure 21.1. All herbivores feed on all plants; therefore losing one plant species has little or no ripple effect. All herbivores are controlled by a single top predator, however, the removal of

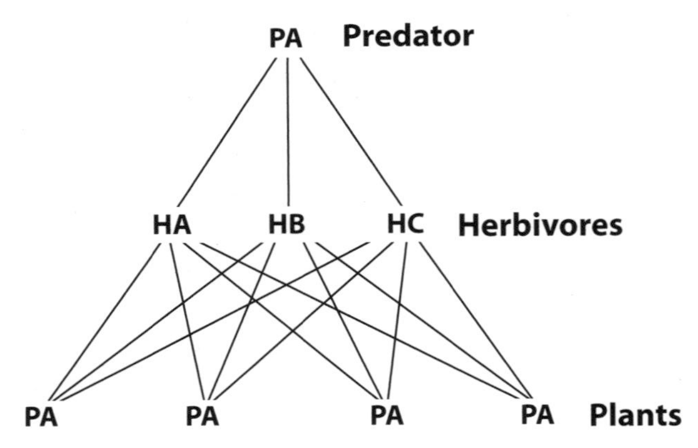

Figure 21.1. In a food web structure such as this, where redundancy is relatively high at the bottom and nonexistent at the top, the system is better buffered against the loss of a plant species than against loss of a top predator. This would not be true, however, if the lost plant species played some key role in the ecosystem or landscape.

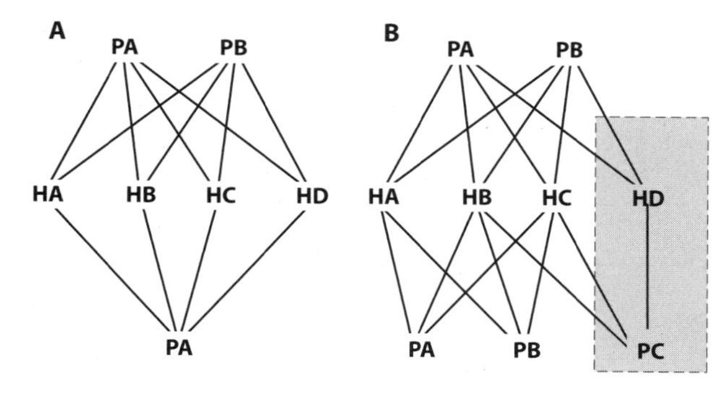

Figure 21.2. A. In this hypothetical food web, plant species PA is a keystone whose loss disrupts the whole system. B. This food web is an example of compartmentalization. The loss of plant species PC would lead to the loss of herbivore HD, which is completely dependent on PC. The negative effects on other herbivores in the system and on the predators, however, would be lessened, because they do not depend totally on PC or HD.

which has a very large ripple effect. Figure 21.2A shows a situation in which the plant is a keystone for the whole system (e.g., a plantation monoculture or a natural stand dominated by a single tree species). Figure 21.2B shows a hypothetical compartmentalized food web in which a plant species may be a keystone for some other species in the system, but not all.[3] (the degree of compartmentalization in real food webs is a matter of debate; however, in section 21.4.1 we see an example of compartmentalization in a real forest food web).

Food web models have grown increasingly sophisticated in the years since May's original work, incorporating more realistic structural aspects such as omnivory and weak interactions among species (i.e., lack of strong mutual dependence). With greater realism, the stabilizing role of biodiversity becomes more evident (McCann 2000). In all cases, the models validate what our own ecological intuition tells us: the critical question regarding ecosystem stability is whether system structure tends to absorb and dampened perturbations, or magnify them. Note that the importance of this question applies whether we are talking about ripple effects in food webs, the propagation of disturbances across landscapes (section 21.4.3), or cumulative effects through time (e.g., progressive nutrient loss).

Improvements that have been suggested for the structure of stability models of the food web and energy flow types include the following:

- *Linking cycles of nutrients and water to energy flow* (e.g., Jordan et al. 1972; O'Neill 1976; Pastor and Post 1988; Shukla et al. 1990; Wilson and Botkin 1990; DeAngelis 1992). The nutrient cycle provides a key feedback to food webs, because energy capture by green plants depends on it. The same is true on a somewhat different scale for the hydrologic cycle: evapotranspiration cycles water back to the air, whence it can fall again as precipitation (chapter 3).
- *Incorporating plant chemical defenses and animal parasites* (Freeland and Boulton 1992). Despite the importance of both of these factors in regulating herbivore populations, they seldom figure in traditional models.
- *Incorporating heterogeneity in space (i.e., the mosaic quality of ecosystems and landscapes) and time (e.g., seasonality in fruiting, or in the activities of mutualists).* Food-web models seldom incorporate the effects of spatial and temporal heterogeneity. Loreau et al. (2003) proposed that "biodiversity provides spatial insurance for ecosystem functioning by virtue of spatial exchanges among local ecosystems in heterogeneous landscapes." The efficiency of such exchanges (and therefore the efficacy of the insurance) strongly depends on landscape structure.

21.4 UNDERSTANDING STABILIZATION REQUIRES UNDERSTANDING STRUCTURE-FUNCTION INTERACTIONS

System stability arises from interactions between structure and function. Structure includes all those aspects of diversity discussed throughout the book: species richness and evenness, food web organization, unique habitats such as snags and beaver ponds, the varied niches that spin out of forest architecture and landscape complexity, the structural diversity of soils, and the myriad interconnections spanning local ecosystems through landscapes to the globe. Function refers to what organisms and other elements do in ecosystems, or what we have referred to previously as *influences*. Influences may include effects on either structure or processes (e.g., photosynthesis, the nutrient and hydrologic cycles, the dynamic aspects of food webs, succession, evolution, migration, and the movement of disturbances across landscapes). The inseparable dyad of structure and function comprises the system state: the ball-on-the-peak that we encountered in the last chapter.

Four aspects of structure-function that are both relevant to the stability of local ecosystems *and* under the control of land managers are keystones, complementarity, genetic diversity within populations, and landscape complexity.

21.4.1 Keystones

Webster's defines "keystone" as **"a part or force on which associated things depend for support."** In ecology, **keystones** are species, groups of species, habitats (e.g., large dead

[3]"Compartments exist in food webs if the interactions within the web are grouped into subsystems: that is, if species interact strongly only with species in their own sub-systems, and interact little, if at all, with species outside it" (Pimm and Lawton 1980).

wood) or abiotic factors (e.g., fire) that play a pivotal role in ecosystem (or landscape) processes and "upon which a large part of the community depends" (Noss 1991). Some landscape features, such as riparian zones or migration corridors, may also function as keystones. Loss of a keystone produces **cascade effects** (i.e., leads to the loss of other species or the disruption of processes). Mills et al. (1993) discuss the need to consider keystones in management and policy decisions.

There are undoubtedly keystones within ecosystems, but as we discussed in chapter 11, it is far from straightforward in many cases to say what they are. To determine the keystoneness of something we must understand how its loss affects system processes or structure, which in most cases requires considerably more knowledge about the structure of interactions and interdependencies than is available even for the most simple of systems. Clearly, trees compose a **keystone group** in forest ecosystems; they provide the energy and basic habitat structure for all organisms within the system. Other *potential* keystones can be identified for various ecosystems: moose, beavers, wolves, jaguars, fig trees, large dead wood, and so on. But verification requires an experiment in which the hypothesized keystone is removed, and these critical experiments have seldom been done in forest ecosystems. In some cases experiments are impossible or unethical (e.g., experimentally extirpating large predators). There are also "natural" experiments, such as the disappearance of large predators from much of their former range. Such anecdotal experiences are bound to be controversial, partly because they do not pass the scientific muster of being replicated experiments, and partly because it is frequently difficult or impossible to document all the factors that may have been involved. Unfortunately, however, many of the critical environmental questions we face today must be answered at spatial and temporal scales that preclude replicated experiments, and we must rely on what we see in nature (Holt 2006a). In the words of that great American sportsman and philosopher, Yogi Berra, "You can learn a lot just by looking." The keys are to observe carefully and to interpret

Ellison et al. (2005) define **foundation tree species** as those whose "architecture and functional ecology define forest structure and (whose) species specific traits control ecosystem dynamics." For example, beech is unique among northern hardwoods in its litter quality and seed production (Lovett et al. 2006). Similarly, hemlocks are "often the most important conifer in northern hardwood forests" (Lovett et al. 2006); they grow in pure stands in specific landscape positions, impose unique patterns on soil, and may play an important role in protecting stream integrity. Both species are threatened by exotic pests (beech by beech bark disease and hemlock by balsam wooly adelgid). Any tree species qualifies as a foundation when it is functionally unique (at some place or time) to the degree that its functions are irreplaceable by other trees in the system (see Ellison et al. [2005] and Lovett et al. [2006] for additional examples).

Chapin et al. (2000) point out that "there has been substantial debate over both the form of the relationship between species richness and ecosystem processes and the mechanisms underlying these relationships." Over 50 different hypotheses concerning the ecosystem consequences of biodiversity loss can be divided into three general classes (Naeem et al. 2002; see Schlapfer and Schmid [1999] for a more detailed classification):

1. *Species are primarily redundant.* So that (within limits) loss or addition of species does not influence ecosystem function. Two corollaries are that (a) species-simple systems are more vulnerable to species loss than species-rich systems (because they have less redundancy in key processes), and (b) higher trophic levels are more vulnerable than lower trophic levels (because they have fewer species and therefore less redundancy (Raffaelli 2004).
2. *Species are primarily singular.* Individual species make unique contributions to ecosystem function. Keystones that perform some unique function are one example; however, different species may perform the same function but under different conditions, in which case each contributes in a unique way (this is the concept of complementarity discussed later in the chapter). We refer to the former as **major keystones** and the latter as **conditional keystones.**
3. *Species functions are context dependent and therefore idiosyncratic or unpredictable.* The consequences of losing species depends on system-specific conditions and cannot be generalized. Effects may vary widely depending on the trophic level in which the loss occurs, the adaptability of remaining species, the particular community structure, or specific environmental conditions (Kondoh 2003; Raffaelli 2004; Zaveleta and Hulvey 2004). For example, experimental removals of both plant species and plant functional groups on 30 islands in northern Sweden showed that "although losses of functional groups and species often impaired key ecosystem processes, these effects were highly context dependent and strongly influenced by island size" (Wardle and Zackrisson 2005). (Island size influenced disturbance history and therefore successional stage.)

observations in an unbiased manner. New statistical techniques that can be used for large scale, unreplicated experiments are sorely needed (Miao and Carstenn 2006).

There are relative degrees of keystoneness: the loss of some species creates a ripple, while the loss of others creates a tidal wave. For example, prior to the arrival of chestnut blight (an introduced pathogen) the American chestnut was one of the most abundant trees in many forests of the eastern United States. However, its subsequent extirpation produced no vertebrate extinctions, and processes such as primary productivity, nutrient cycling, and hydrologic cycling apparently emerged intact (Pimm 1991). Squirrel populations apparently crashed initially, but as Motzkin and Foster (2004) observed, "It would be difficult to conclude that from the large numbers of gray, red, or flying squirrels in woods today." Perhaps seven species of moths (Lepidoptera) did go extinct following the loss of chestnut; however, that was only 12 percent of the Lepidoptera species that fed on chestnut (Pimm 1991). Loss of fig trees from tropical forests would create a much bigger wave, because numerous frugivorous species depend on them. The degree to which that wave spread throughout the system would depend at least in part on secondary effects arising from loss of the frugivores. In the final analysis, the most serious keystones (i.e., the ones whose loss creates a tidal wave) are those who play a singular role in key system processes, such as photosynthesis and food web dynamics, nutrient and water cycling, and top-down controls.

Better understanding the stabilizing role of keystones has considerable practical implications for sustainable forestry. Suppose, for example, that spotted owls disappear from forests of the Pacific Northwest, or old, decayed logs from the soil. How will the rest of the system respond? Plausible arguments can be made for various possibilities, but whether those arguments are correct is another question. Again, we run up against our ignorance of how ecosystems work (chapter 23 argues that intelligent land management includes being aware of ignorance and making decisions accordingly). With these caveats, let's look at some of the more obvious candidates for keystones.

21.4.1.1 Predators

Borrvall and Ebenman (2006) express the feelings of many ecologists: "The large vulnerability of top predators to human-induced disturbances on ecosystems is a matter of growing concern. . . . Their extinction can have far-reaching consequences for the structure and functioning of ecosystems." The likelihood that a predator is a major keystone increases as one moves up the trophic ladder for the simple reason that higher trophic levels contain fewer species than lower levels (Raffaelli 2004). For example, midlevel predators (e.g., lynx, coyotes) compose a redundant group in the boreal forest (i.e., removing one has little effect because the others compensate) (Krebs et al. 2001c). However, wolves play a unique role, and other predators cannot compensate for their loss. For example, a natural experiment in Alberta allowed researchers to compare nearby areas with and without wolves (Hebblewhite et al. 2005). Wolf exclusion induced a trophic cascade in which higher numbers of elk led to fewer aspen and willow, which, in turn, affected beavers and riparian songbirds (chapter 11). Similarly, extirpation of wolves and grizzly bears from the Greater Yellowstone Ecosystem allowed moose populations to grow, with consequent effects on willows and neotropical migrant birds that use willow communities (Berger et al. 2001). In chapter 11 we discussed the meltdown of ecosystems in Venezuela following the creation of predator-free islands by a hydroelectric project (Terborgh et al. 2001).

The birds, spiders, and insects that eat plant-eating insects are generally too abundant for any one species to be called a major keystone (although they certainly form a keystone functional group). However, they complement one another by feeding at different times and places (e.g., chapter 9) and, therefore, function as contextual keystones. In at least some instances a single predator may play a major keystone role in controlling phytophagous insects. To cite one example, damaging outbreaks of ants in Brazilian eucalyptus plantations are believed to result from local extirpation of anteaters, which were killed by humans for food (John Tappeiner, personal communication).

Large predators are particularly vulnerable in today's world for several reasons (Raffaelli 2004). Being at the top of the food chain, they are few in number to begin with; and they require more territory to make their living than species on lower trophic levels and, hence, are the first to feel the impact of habitat loss at a landscape scale. Moreover, top predators generally have no natural enemies, hence no innate flight reaction that might save them from humans. Finally, predators accumulate toxic materials that pass up the food chain.

21.4.1.2 Herbivores

Like predators, herbivores affect system structure and processes through what they eat. Also like predators, herbivores promote diversity by enforcing evenness on the system (i.e., by preventing one or a few plant species from dominating). A given herbivore plays a keystone role when it is so large or so abundant as to account for a major portion of the plant material consumed. Large mammals (e.g., elephants, moose, deer, wapiti [the North American elk]) are the most obvious examples (Naiman 1988), but small mammals can also play a keystone role when their numbers are large enough. For example, snowshoe hare is a keystone species in parts of the boreal forest (Krebs et al. 2001c). Invertebrates can also play a keystone role. Leafcutter ants, for example, are the principal consumers of vegetation in the neotropics (Wilson 1987). Termites are potential keystones in the tropics, not only because of their

Figure 21.3. Old, decaying logs are water reservoirs, sites of nitrogen fixation, and habitat for microbial species, invertebrates (including some important predators of defoliating insects), amphibians, small mammals, and even some large mammals such as pine marten. (Courtesy of Thom O'Dell)

consumption, but because of their unique impacts on soil structure (chapter 14).

21.4.1.3 Dead Wood

As we discussed in chapter 10, large dead wood plays a keystone role in forests. Snags are essential for cavity-nesting birds, and logs that fall into streams enhance fish diversity by creating pools and hiding places. Decayed logs on the forest floor and within the soil store water and provide habitat for microbes, invertebrates, and some vertebrates (fig. 21.3). Down logs and other coarse woody debris are also believed to be critical for the survival of martens, fishers, and sables (genus *Martes*), predators that are associated with old-growth forests throughout the boreal zone. Coarse woody debris is particularly important to these animals in the winter, providing "protection from predators, access to spaces beneath the snow where prey animals live, and protected sites where (animals) can minimize energetic costs while resting" (Buskirk 1992).

Harris and Maser (1984) estimate that removing snags would lead to the loss of about 10 species from Pacific Northwest forests, and removing both snags and downed logs would result in approximately 30 species being lost. Most cavity-nesting birds are insectivorous (Thomas et al. 1979); therefore, eliminating snags could eventually make forests more susceptible to insect outbreaks (Torgersen et al. 1990; chapter 19 in this volume). Bird boxes have been used to substitute for natural cavities, but Thomas et al. (1979) argue that bird boxes "are not generally a good alternative for woodpeckers, (which) seem to excavate cavities as a necessary part of their mating rituals." Moreover, they continue, because of the large number of species that use cavities, but require somewhat different types of cavity, "reliance on bird boxes requires the placement of many types and sizes of nest boxes at various heights and densities throughout the forest. The boxes must be constructed, installed, cleaned, and replaced periodically. The costs would be formidable and the results less successful than if snags were present. Imagine trying to provide nest boxes for all 62 cavity-using species in the Blue Mountains!"[4]

21.4.1.4 Builders

Builders, through activities that benefit themselves, inadvertently provide services to others. Noss (1991) discusses one of the more striking examples, the gopher tortoise, which is native to longleaf pine forests in the southeastern United States: "The gopher tortoise digs burrows up to 30 ft. long and 15 ft. deep, in which some 362 species of vertebrates and invertebrates have been found. Some of these species are . . . absolutely dependent upon the threatened tortoise." Beavers are keystones because of the unique habitat provided

[4]The Blue Mountains lie in northeastern Oregon and southeastern Washington.

Figure 21.4. The dead trees *(center and right)* are whitebark pines killed by white-pine blister rust, an exotic pest that has sharply reduced populations of white pines in the western United States. The seeds of whitebark pine are an important food for various animals, including grizzly bears; it is not known how extensively the loss of whitebark pine might affect the food web. This photograph is from Glacier National Park, Montana, but protection within a park does not guarantee that systems are free from foreign disturbances.

by their ponds. Various species of birds, especially woodpeckers, excavate tree holes that eventually might be used by any number of other species; these "primary cavity-nesters" form a keystone group. In tropical forests, leaf-cutter ants and termites build extensive homes that uniquely benefit other species and significantly affect ecosystem processes such as nutrient cycling. Termites use clays mined from deep soil layers as a kind of an adobe to provide strength to the aboveground portion of their nests (i.e., termiteria), which can be quite large. In old tropical soils whose clays and associated cations have leached below the rooting zone (chapter 14), termites play a keystone role by bringing these elements back to the surface. Tree seedlings have been observed to regenerate vigorously on abandoned termiteria (Salick et al. 1983), and in Borneo, colobine monkeys obtain nutrients by feeding on soil from termiteria (Davies and Baillie 1988). In Africa, humans benefit from termiteria in various ways, planting crops on them and using the clays to construct adobe homes.

21.4.1.5 Plants

Clearly trees are a keystone functional group in forest ecosystems. In monospecific forests, which occur naturally as well as because of human simplification, the health of the entire system turns on the health of a single tree species. A tree doesn't have to be in a monoculture to be a keystone, however; we have discussed at various points the importance of trees such as figs that fruit year-round in tropical forests. An example of a keystone food resource in montane forests of North America is whitebark pine (fig. 21.4), which is endangered because of an introduced pathogen (white pine blister rust). Moreover, with the warming climate it has become newly vulnerable to mountain pine beetle (previously its habitat was too high in elevation for the beetle). The large and nutritious whitebark pine seeds are important food for several vertebrates, including bears. Good crops of whitebark seeds are associated with good cub production and early weaning in grizzly bears *(Ursus horribilis);* while poor seed years result in increased bear mortality and conflicts with humans (Kendall and Arno 1989). The Russian brown bear *(Ursus arctos)* is similarly dependent on seed from Siberian stone pine. Kendall and Arno (1989) report that "during years of massive cone crop failure, large numbers of emaciated bears make long migrations in search of food, frequently entering villages and killing livestock and occasionally attacking people."

Plants may be keystones because they provide unique habitats or serve as centers of biological activity. For example, we saw in chapter 10 that epiphytes are keystone plants in some tropical forests: water collects in their cuplike leaves, the water attracts insects, the insects attract birds, and the birds attract monkeys that feed on bird eggs. Brazilian biologists working with the golden lion tamarin *(Leontopithicus rosalia)*—one of the world's most endangered primates—believe that epiphytes are an essential part of viable tamarin habitat.

Plants may also play a keystone role in processes other than photosynthesis. In forests that have only one major nitrogen-fixing plant species, that species is certainly a keystone. In an earlier chapter we discussed the importance of early successional nitrogen-fixing plants in the genus *Ceanothus* to the functioning of western North American forests. Long regarded a "pest" by many foresters, removing *Ceanothus* sp. eliminated the natural mechanisms by which nitrogen was replenished in soils following fire and potentially reduced long-term ecosystem productivity. In the Pacific Northwest, cedars and maples accumulate calcium, a nutrient that might otherwise be readily leached to streams.

21.4.2 Keystone Functional Groups

A keystone functional group is a collection of two or more species that together perform the same unique function (Walker 1992). One can easily think of trivial examples: green plants, for example, are a keystone functional group, as are decomposers, herbivores, and predators. Other examples are perhaps not so obvious. Consider groups based on mutualistic relationships: plant species that form only vesicular-arbuscular mycorrhiza would be one keystone group, while plants that form only ectomycorrhizae would be another. Nut-bearing trees, fruit-bearing trees, tree species with a flower structure that is accessible to only certain pollinators—each forms a separate keystone group regarding a particular function. An individual species may be part of one keystone group with regard to one function, and part of another keystone group with regard to another. For example, two tree species may host the same mycorrhizal fungi, but different pollinators.

The concept of keystone group also applies when we account for time. For example, during the three-month dry season in Peruvian tropical forests, 12 plant species flower or fruit. These 12 species (less than 1% of the total flora in those forests) compose a keystone group that supports the entire community of frugiverous mammals and birds (Terborgh 1986). If we look more closely, however, we see that these 12 species are not totally equivalent. Three species, for example, are palms whose nuts can be opened by only some of the mammals and birds; another three are the only plants that flower during that period and therefore provide the bulk of food for nectivorous birds. As illustrated by this example, the structure of species richness is equally important for ecosystem stability as overall richness, or perhaps even more so. Specifically, both the richness of functional groups and the richness of species within functional groups play significant role in ecosystem dynamics. The value of richness within groups can be traced to the phenomenon called **complementarity.**

21.4.3 The Structure of Functional Groups: Redundancy and Complementarity

> Historical studies suggest that all species have evolved in a dynamic environment and under conditions of change and that each has its own particular biology and response to these changes. This independence and individualistic characteristic of species confers great resilience to the forest overall.
>
> MOTZKIN AND FOSTER 2004

As we have seen, there are instances (e.g., the case of the American chestnut) in which a dominant tree species has been lost with relatively minor long-term effects on ecosystem processes or other species in the system. Species composition of forests has changed throughout history with changes in climate, yet we still find functioning ecosystems today (Davis 1981; Foster and Aber 2004). How is it that some species (even dominant ones) can be lost from systems with apparently no adverse effect on the rest of the system? The answer lies in the structure of functional groups.

Figure 21.5 illustrates (in a highly oversimplified way) two major features of structure within functional groups: **redundancy** and **complementarity.** The term *redundancy* is used rather loosely by ecologists but always to denote some form of repetition or "doing the same thing" (chapter 9). For example, all green plants photosynthesize; however, there are limits to that redundancy because many plants differ to some degree in the details of when and where they photosynthesize. When species within a functional group differ in the environmental details of performing the function, that is called *complementarity*. In figure 21.5, areas of overlap represent redundancy (for a given function), and areas without overlap represent complementarity. Species

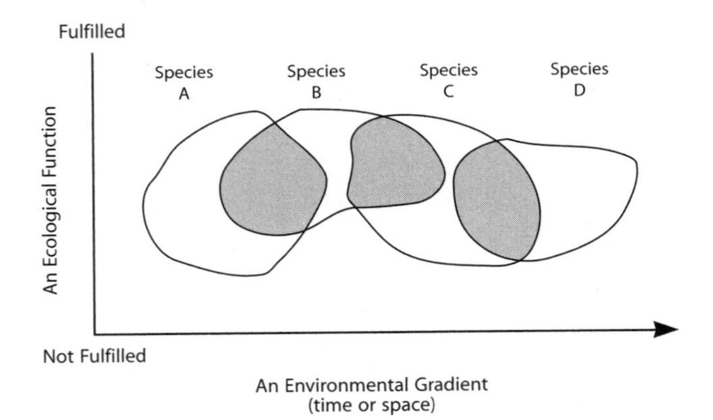

Figure 21.5. A schematic diagram illustrating how the interplay between redundancy and keystoness across an environmental gradient creates complementarity among species with regard to a given function. Assume some ecologic function, such as fruit production, and some environmental gradient, such as soil moisture content. The *y* axis denotes the degree to which that function is fulfilled (e.g., *not fulfilled* might be no fruit production, while *fulfilled* might be a bumper crop). Circles denote the range of the environmental gradient over which each species is fruiting and how much fruit it is producing. Over some ranges of the gradient, one or another species is a keystone, while over other ranges (the shaded areas) there is a redundancy because two species are fruiting. This of course is greatly oversimplified, but the basic point is valid: redundancy and keystoneness are frequently not invariant properties, but rather vary with environmental conditions.

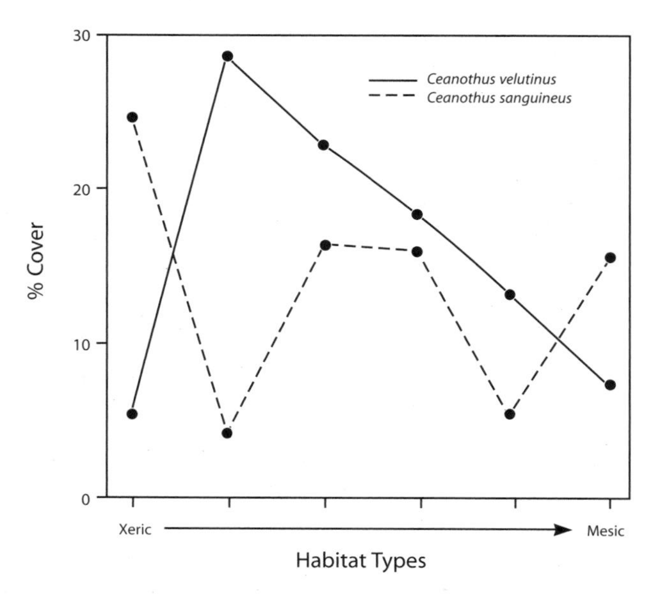

Figure 21.6. Variation in the cover of two species in the genus *Ceanothus* (early successional nitrogen-fixing shrubs) in clear-cuts on the H.J. Andrews Experimental Forest, Oregon Cascades. Habitat types (describe in chapter 5) were classed using late successional forbs and shrubs before clear-cutting, and *Ceanothus* cover was measured from 12 to 15 years after clear-cutting. Note that the more xeric sites had little overlap in the two species, while the more mesic sites had considerable overlap. (Adapted from Egeland 1982)

that function uniquely under some conditions but not others are **conditional keystones.**

Figure 21.6 shows a real-world example of different plant species that are both redundant and complementary in their occupation of the landscape. Two species of *Ceanothus* are common on the H. J. Andrews Experimental Forest. Both sprout from buried seed after fire and both fix nitrogen, but the two differ in the habitats they occupy. *Ceanothus sanguineus* is much more abundant than *C. velutinus* in the most xeric habitats, while the opposite is true in slightly moister habitats. The two are roughly equal in habitats that occupy the middle of the moisture gradient. Depending on where we look in that mountainous landscape, we find different levels of redundancy and complementarity for early successional nitrogen fixation. (Note that even where the species are redundant, as measured by their equal presence on a site, they may act in a complementary way in other aspects. For example, their favored positions on the landscape suggest the two differ in their moisture preferences. It follows that even where equally abundant they would complement one another as moisture varied seasonally or from year to year.)

A system without redundancy and complementarity in its key processes is unlikely to persist long in nature. Once again, the modern agricultural field serves as an example: without massive subsidies in the form of pesticides and fertilizers, these highly simplified ecosystems would not have a chance. There are numerous examples in natural ecosystems as well. We already talked about the loss of chestnut from deciduous forests of the eastern United States: those systems had redundancy and complementarity in photosynthesis, nutrient cycling, and all but a small portion of their food web structure.

Diffuse mutualisms (chapter 11) represent a common form of redundancy and complementarity within ecosystems. For example, the ability of a given fungus to form mycorrhizae with more than one tree species ensures continuity in hosts. There is considerable complementarity in controls over phytophagous insects as well, including an array of plant chemical and physical defenses, a diverse community of vertebrate and invertebrate predators, pathogens, parasites, and viruses. Similarly, Heemsbergen et al. (2004) showed that functional dissimilarity (i.e., complementarity) of macrodetritivores more strongly influenced leaf decomposition and soil respiration than did species number.

Redundancy and complementarity effectively spread the risk associated with uncertain environmental fluctuations and disturbance regimes and in so doing greatly enhance the resilience and resistance of the community as a whole. In fact, ecologists refer to the combined action of the two as the **insurance hypothesis** (Yachi and Loreau 1999). Consider again the Siskiyou Mountains of southwestern Oregon and northern California, where numerous tree and shrub species form ectomycorrhizae. These have a variety of adaptations to fire. Douglas-fir and ponderosa pine have thick bark that protects them from ground fires but not crown fires. Many of the hardwood trees (e.g., madrone, tanoak, chinquapin, various oaks) sprout from roots after their tops have been killed. Some shrubs have seed that remain dormant in the soil until triggered to germinate by heat. When conifers are destroyed by crown fires (or clear-cuts), hardwoods sprout from roots or spring up from buried seed and quickly stabilize the soil ecosystem (chapter 8). Maintaining a legacy of healthy soils is critically important for the reestablishment of Douglas-fir and ponderosa pine, whose seeds must come from living trees and, hence, may take considerable time to reoccupy large disturbances (seeds are rarely distributed further than a few hundred meters from a living tree). This is not altruism on the part of the hardwoods (in fact, a regime of more frequent crown fires or clear-cuts can tip a balance and lead to hardwoods dominating sites [Marty Main, personal communication]). More likely, it is a kind of risk-spreading that has evolved in the face of an unpredictable disturbance regime and the common dependence of all plants in the integrity of soil. The next fire that comes through could spare the conifers, and it would be they that keep the soils stable and healthy.

Although the disturbance regime is quite different, a similar pattern occurs in the gap-driven forests of the moist tropics, where the mix of tree species colonizing a gap varies widely depending on numerous factors (chapter 8). There, most plants form mycorrhizae with vesicular-arbuscular mycorrhizal fungi. As is the case in temperate forests, unpredictability favors the evolution of diffuse mutualisms. The fungus may have an entirely different tree as a partner tomorrow, next week, or next year, and the progeny of a given tree are very likely to establish on a piece of ground formerly occupied by a different tree species. It is like the old game in which the partners on a dance floor periodically change, the person left without a partner sitting down. The obligate mutualist left without a partner dies and does not pass its genes to the next generation. This is a powerful force that through evolutionary time selects for diffuse mutualisms.

Complementarity arises from adaptability. Since environments are highly variable in both time and space, functional constancy in ecosystems and landscapes arises from diverse adaptations. Consider diffuse mutualisms. Hundreds of plant species flower and fruit in tropical forests, but pollinators and frugivores are able to find food at different times during the year and from year to year because plant species differ with respect to the environmental cues that trigger them to produce flowers and fruits. Although hundreds (and sometimes thousands) of fungal species may be capable of forming mycorrhizae with a given tree species, not all of these do the same job for the tree. Forming mycorrhizae with fungi that differ in their responses to the environment enables a tree to maintain a certain constancy of physiological processes in the face of environmental fluctuations, and allows a more complete exploitation of microhabitat variability.

Other examples abound of species (and genotypes within species) that perform the same function, but differ in when or where. Sprouting hardwoods and conifers both maintain soil processes, but they also differ in their adaptations to disturbance and therefore function differently within the system. Or consider controls over herbivores. Under certain conditions predators and plant defenses will both be operating and backing one another up, but during drought, plant defenses drop and the burden shifts to predators.

21.4.4 Insurance Has Limits

The ecological insurance provided by redundancy and complementarity is not foolproof. One way to view this is in terms of probabilities rather than absolutes. For example, we have seen that populations of herbivorous insects are controlled by the combined effects of climate, plant chemical defenses, and natural enemies (chapter 19). Therefore, a drop in numbers of insect-eating birds, such as that now occurring in eastern North America, does not necessarily lead to insect outbreaks. However, it does make outbreaks more likely should other controlling factors be altered. Holling (1988) spoke to this in his analysis of whether the loss of birds could *by itself* lead to increased insect outbreaks in North America. He concluded that resilience provided by redundancies would prevent this from happening and went on to generalize: "This great resilience demonstrates a property common in ecological systems. First, the stability domains are large and the variables within them can fluctuate extensively. Second, the regulatory processes that are present are remarkably robust to external changes." Even so, Holling added the following important caveat: "This is not to say, however, that ecological systems are infinitely resilient nor that loss of robustness of regulation short of producing a qualitative flip in behavior has no costs. Loss of resilience from one cause can make the system more vulnerable to changes in other events that otherwise could have been absorbed. . . . Ironically, the great resilience of ecological systems mask slow erosion of their capacity to renew and in those circumstances leave managers ill-prepared for surprises."

Holling's point is well illustrated by the potential consequences of a warming climate. Insect populations have the potential to expand in temperate and boreal regions because a greater proportion of the year is suitable for their reproduction and growth. Whether outbreaks of herbivorous insects become more common will depend on the effectiveness of ecological controls: plant defenses, natural enemies, and the concentration of suitable food. Numbers of insect-eating birds are declining, however, weakening one important control agent. Suppose that warmer climates are accompanied by increasing droughtiness, reducing the ability of trees to produce defensive chemicals. The combined effects of warmer climates, declines in some predators, and reduced chemical defenses substantially increase the probability of insect outbreaks (a dynamic that is already happening; chapter 19).

21.4.5 The Stabilizing Role of Genetic Diversity within Populations

Genetic diversity is a fundamental component of biodiversity and is as critical to sustainability of our natural resources as are diversity of species and ecosystems.

BAGLEY ET AL. 2002

Populations of wild species usually maintain high levels of genetic diversity, a phenomenon that is particularly true of trees (chapter 9). A diversity of genotypes stabilizes populations in two ways: by preventing inbreeding depression, and by conferring adaptability.

21.4.5.1 Preventing Inbreeding Depression

Populations of normally outbreeding species (including all vertebrates and most higher plants) contain alleles that reduce fitness when they occur as homozygotes but not when in heterozygous combination with another allele. As genetic variability is reduced within a population, the probability of homozygotes that reduce fitness increases. The sum of alleles that reduce fitness when in homozygous combination in a given population is called **genetic load.** This phenomenon is believed to underlie what is commonly known as inbreeding depression, or the well-established tendency of inbred populations to lose vigor (Ledig 1986; Ralls et al. 1986; Hansson and Westerberg 2002).

Tree species that have been studied (mostly conifers) have a relatively large genetic load, leading Ledig (1986) to suggest that the high levels of heterozygosity present in populations of most tree species are maintained because of selection *against inbreeding* rather than selection *for the environmental buffering capacity of heterozygosity*.

21.4.5.2 Adaptability

Adaptability is a general term that denotes the ability of populations to track environmental changes in time and space, and to maintain parity in the evolutionary dance between eaters and the eaten. In both cases, genetic diversity must translate into phenotypic diversity, or what population geneticists refer to as *quantitative traits.* Variation in quantitative traits provides both raw material for evolutionary processes and a complex resource base that (in theory) slows the ability of consumers to evolve around defenses (e.g., trees and phytophagous insects; chapter 19). The relationship between genetic diversity and adaptability has been somewhat contentious; however, that has to do primarily with the genetic diversity we measure and what it tells us about adaptability. For example, diversity at the molecular level does not correlate well with quantitative traits and therefore tells little about adaptability (Reed and Frankham 2001). Few if any evolutionary ecologists would dispute that genetic variation in quantitative traits is the basis for adaptation (e.g., Bagley et al. 2002); in fact, that idea is one of the cornerstones of Darwin's theory. Variation in quantitative traits is likely to be especially important for population persistence in today's world of rapid climate change and other human influences that isolate populations and restrict their movements (e.g., habitat fragmentation), thereby hampering their ability to track environmental change by moving to another place (Lande and Shannon 1996).

In genetic parlance, **fitness** is a measure of how many progeny an individual leaves over its lifetime, which translates into how many copies of its genes it leaves behind. Fitness is an integrative measure that includes the effects both of genetic load and adaptability; although short-term, controlled studies are unlikely to give an accurate picture of the adaptability component of fitness. In their meta-analysis of studies on the relation between genetic variability and fitness, Reed and Frankham (2001) found that genetic variability explained 19 percent of the variance in fitness. They concluded that "the loss of adaptive genetic variation and inbreeding depression puts . . . populations at an increased risk of extinction. This increase can occur as a result of reduced reproductive fitness due to inbreeding depression, or from a failure to track the changing abiotic and biotic environment."

Modern techniques of molecular genetics have brought us to the threshold of a much deeper understanding of how the structure of genomes is reflected in quantitative traits, fitness, and even ecosystem processes. Unfortunately, because the techniques are time-consuming and expensive, progress is likely to be slow.

Finally, we may ask, in terms of adaptability, exactly what is the relevant genome? The two primary interfaces between plants and the environment, leaves and roots, are in many, if not most plants, characterized by symbioses with fungi (mycorrhizae within the roots and leaf endophytes within the leaves). The genetic diversity and relatively rapid evolutionary capacity of the fungal partners suggests that the plant gains a high degree of genetic adaptability through its symbionts (Carroll 1988).

21.4.5.3 How Forest Management Influences Genetic Diversity

Does forest management influence the genetic diversity of tree populations? The answer to that question depends on the type of forest management that is practiced. Silvicultural practices that promote natural regeneration should maintain genetic diversity, while artificial regeneration may or may not, depending on the source of planting stock. Genetic selection for "superior" genotypes (e.g., fast-growing trees) reduces genetic variability, but how much depends on the intensity of selection and whether provisions are made to retain heterozygosity within selected lines (McKeand et al. 2003). Cloning is a very efficient method to select for and reproduce a given trait, but it produces tree populations that are far less diverse genetically than those occurring under natural conditions. At present, most genetically-improved planting stock used in the southeast US is relatively diverse and no obvious problems due to a narrowed genetic base have been encountered so far (McKeand et al. 2003). However, the temptation to increase productivity by reducing genetic diversity is always present, leading McKeand et al. (2003) to assert that "a challenge for tree improvement research in the coming years will be to quantify acceptable levels of risk for plantations established across a range of landscapes".

Ledig (1988) argues that foresters should assure "adequate regeneration with seed from the in situ populations, including the entire diversity of woody and herbaceous species present. Preferably, natural regeneration would be used, but planting with seedlings from the native stand might be acceptable if done at high density, providing an opportunity for natural selection."

21.4.6 Landscape Complexity

The distribution of successional stages, vegetation types, and protected areas across the landscape influences

Figure 21.7. Tracking radio-collared mountain lions in the Eagle Cap Wilderness, northeastern Oregon. Many top predators were nearly extirpated from the western United States because of the widespread use of poison to control livestock predation. Mountain lion populations have begun to rebound since poisoning predators was outlawed in the early 1970s, thanks at least in part to protected areas like the Eagle Cap Wilderness, within which populations remained viable and which now serve as important source areas for mountain lions spreading out to recolonize former territory.

the stability of ecosystems and the viability of populations in three general ways: (1) by affecting the rate at which potentially destructive agents such as fire, wind, pests and diseases can spread, (2) by providing minimal habitat requirements and pathways of migration, and (3) by providing "source" areas, or sources of individuals to recolonize depleted habitats (fig. 21.7). This section deals with the first, the latter two being taken up in the following chapter on conserving species. Although these topics fall into different chapters, it is important to understand that the spread of destructive agents and the quality of habitats are interlinked phenomena that are not easily separable in nature. For instance, the spread of pests depends not only on the distribution of their food in space, but on the suitability of habitats within a given region for their natural enemies (Perry 1988; Torgensen et al. 1990). Moreover, as we shall discuss in the following chapter, the quality of habitat for a given population can be profoundly influenced by its accessibility to the pests and diseases of that particular species.

The effect of landscape pattern on the spread of disturbances is a complex affair. Although in general a heterogeneous landscape retards the spread of disturbance, while a homogeneous landscape enhances the spread (Forman 1987), whether a particular landscape is homogeneous or heterogeneous depends on the nature of the disturbance, the susceptibility of different community types and age classes to that disturbance, and the *grain* of the landscape, or the distribution of patch sizes (single tree gaps, 50-ha clear-cuts, etc.) (chapter 7).

Turner et al. (1989) distinguish two types of disturbances: (1) those that spread *within* the same cover type, such as a species-specific insect pest; and (2) those that cross ecosystem boundaries and spread *between* cover types, such as fire moving from pasture into forest. "Whether landscape heterogeneity enhances or retards the spread of disturbance may depend on which of these two modes of propagation is dominant. If the disturbance is likely to propagate within a community, high landscape heterogeneity is likely to retard the spread of the disturbance. If the disturbance is likely to move between communities, increased landscape heterogeneity should enhance the disturbance" (Turner et al. 1989).

21.4.6.1 What Constitutes a Heterogeneous Landscape?

It is important to understand that *cover type* as used here must be defined by what a particular pest or disturbance perceives as uniform, not what humans perceive as uniform. Consider two common disturbances in the boreal forests of Canada: spruce budworm and fire. Spruce budworm feeds on both balsam fir and white spruce, although it prefers the former. Hence, we might predict that the insect will spread more slowly through a landscape mixed with spruce and fir than through one consisting of pure fir, but more rapidly through a landscape that comprises both species than through one consisting of pure spruce.[5] On the other hand, jack pine, aspen, and birch are not attacked by budworm; hence, the presence of these cover types retards the spread of the insect (chapter 19). What about fire? Spruce, fir, and pine are all quite flammable and present a homogeneous landscape as far as fire is concerned; aspen and birch are less flammable, and stands of these species within the landscape matrix create heterogeneity that retards fire spread (Knight 1987).

So, a landscape composed of different forest types is not necessarily heterogeneous regarding the spread of disturbance. The converse is also true; a landscape dominated by a single species is not necessarily homogeneous to all disturbances, because some disturbances spread most readily through certain age classes or stand structures. For instance, mountain pine beetle attacks only lodgepole pines that are larger than a given diameter; for this insect, a landscape filled with lodgepole pine is *not* a uniform cover type if there is a patchy distribution of stands with different average tree sizes. (In chapter 19 we saw also that stands composed of single tree species may also present a genetically heterogeneous menu to tree-eaters.) The spread of fire also depends not just on species composition but on the homogeneity of age classes and stand structures across the landscape, although the nature of this relationship likely varies depending on forest type. Resistance of moist tropical rain forests and some old-growth conifer forests to the spread of crown fires results in part from the grain of their spatial heterogeneity, which generally consists of gaps ranging in size from a few tens to a few hundreds of meters. These gaps are sufficiently large to slow the spread of fire from crown to crown, but not so large that they significantly alter microclimate and lead to drying of the surrounding forest.[6] Openings on the scale of hectares (as is usually the case with clear-cuts or conversion of forest to pasture) create quite a different landscape grain and have quite different implications for fire.

21.4.6.2 Boundaries

Like landscape heterogeneity, what constitutes a **boundary** regarding the spread of disturbance must be evaluated specifically for each disturbance and each type of boundary. Wind is one of the clearest examples of a disturbance that spreads most readily across certain kinds of boundaries, especially the sharp edges created when large openings are created within previously intact forest (Franklin and Forman 1987); on the other hand, a boundary between two forest types may have little or no effect on wind damage. Fire moves readily from dry grass into a forest composed of flammable species, and such a forest is highly vulnerable when bordered by pasture (a particular problem in tropical forests). However, the boundary between flammable and inflammable vegetation types has quite different implications for fire spread.

21.4.6.3 Conclusions about Landscape Patterns and Stability

If the left hand don't get you then the right one will.

FROM AN OLD COAL MINING SONG

Forests are subject to a variety of disturbances—fire, wind, volcanoes, landslides, numerous species of phytophagous insects and pathogens, and a potpourri of viruses—most of which will respond somewhat differently to landscape pattern. For the forester, this presents considerable challenge in deciding how to best structure a landscape. The challenge is increased by the fact that the best approach for minimizing the spread of disturbances—relatively small-scale patchiness in different cover types (scale on the order of hectares; Turner et al. 1989)—may conflict with the habitat needs of mammals and birds (chapter 22).

Two things should be borne in mind when designing a forested landscape that reduces the probability of catastrophic spread of disturbances. First, each situation must be evaluated independently. What are the most threatening disturbances? Which plant communities and age classes are susceptible, and which are not? What is the natural grain of the landscape? Does it retard or promote the spread of common disturbances? Second, as a general principle we can assert that the more uniform a landscape, the greater the chance that *some* disturbance will spread catastrophically. Uniformity and risk increase as follows:

- Landscapes dominated by a single tree species
- Landscapes dominated by a single tree species with a relatively narrow age range
- Landscapes dominated by a single tree species that is genetically uniform or that does not have adequate tree-based defenses against pests and pathogens

Intensive forest management as practiced during much of the twentieth century (and nineteenth century in parts of Europe) produces either the second or third type of landscape. Single species are grown and harvested so as to maximize timber production. This is sometimes accompanied by vegetation control that reduces the cover of noncommercial plant species, and virtually always by harvest before a stand enters the later stages of succession (often the most biologically and structurally diverse) (chapters 9 and 10). Biological legacies are often removed as well. The result is greater landscape uniformity than occurred under natural

[5]Spruce stands intermixed with fir are probably more susceptible to budworm than a pure spruce landscape, because the buildup of insect populations on the more susceptible fir facilitates their spread into the spruce.

[6]Small-scale patchiness may have little effect on the spread of crown fires if the fire is sufficiently intense.

conditions, although some heterogeneity remains, and the basic question relates to the functional significance of reduced heterogeneity (section 21.4.7). The selection and propagation of one or a few genotypes (clones) eventually leads to the third pattern and creates an extremely vulnerable situation (chapter 19). Vulnerability is also created when trees that normally grow scattered within species-rich forests—such as most tropical trees—are planted in monocultures (chapter 19).

21.4.6.4 Fire

One of the biggest challenges of the future will be to design landscapes that retard (or at least do not promote) the spread of wildfire. Along with growth in populations of phytophagous insects, fire will become a major threat as greenhouse effects lead to warmer and drier regional climates. This has the greatest potential for disrupting forests in which trees have few adaptations to fire (e.g., those of the moist tropics); however, more frequent, severe fires could eventually degrade even fire-adapted forests. Landscape management to reduce fire hazard should include two goals:

1. Identification and inclusion of relatively noninflammable vegetation types. In conifer-dominated forests these often are the hardwoods, but no species should be assumed a priori to be noninflammable without verifying studies.
2. Creation, where possible, of a landscape grain that mitigates against excessive wind and heating within forest interiors; this retains a relatively moist microclimate and diminishes the chance that an ignition will turn into a conflagration.

The most common ignition source in the moist tropics is fire spreading from pastures or slash-and-burn agriculture into the forest. The best way to prevent that is to eschew those activities within the forest matrix; where that is not possible, constructing fire lines around pastures and fields prior to burning will perhaps help.

21.4.7. Ecosystem Viability Analysis

> Owing to interdependences among species in ecological communities, the loss of one species can trigger a cascade of secondary extinctions with potentially dramatic effects on the functioning and stability of the community.
>
> EBENMAN AND JONSSON 2005

Community viability analysis is a relatively new field that seeks to understand the structure of ecological linkages and how the loss of one species leads to the loss of others (i.e., secondary extinctions) (Ebenham and Jonsson 2005). The more general issue is how any alteration of ecosystem structure (e.g., loss of species or biological legacies, change in forest or landscape structure) ramifies to alter ecosystem structure and functioning (we refer to this as **ecosystem viability analysis**). The starting point is to ask questions about function, namely, what does the lost species, lost habitat, altered pattern, and so forth *do* in the ecosystem? For example, amphibians are precipitously declining throughout the world but especially in the neotropics. These organisms have a ubiquitous presence in stream and riparian food webs, many being primary consumers as juveniles and insectivores as adults. Whiles et al. (2006) conclude that "the handful of manipulative field studies to date show that primary production, nutrient cycling, leaf litter decomposition, and invertebrate populations change when tadpoles, frogs, and salamanders are removed or reduced in numbers . . . evidence to date suggests that amphibian declines will have large-scale and lasting ecosystem effects." Similarly, Sekercioğlu, et al. (2004) project that "by 2100, 6 to 14 percent of all bird species will be extinct and 7 to 25 percent will be functionally extinct. Important ecosystem processes, particularly decomposition, pollination, and seed dispersal, will likely decline as a

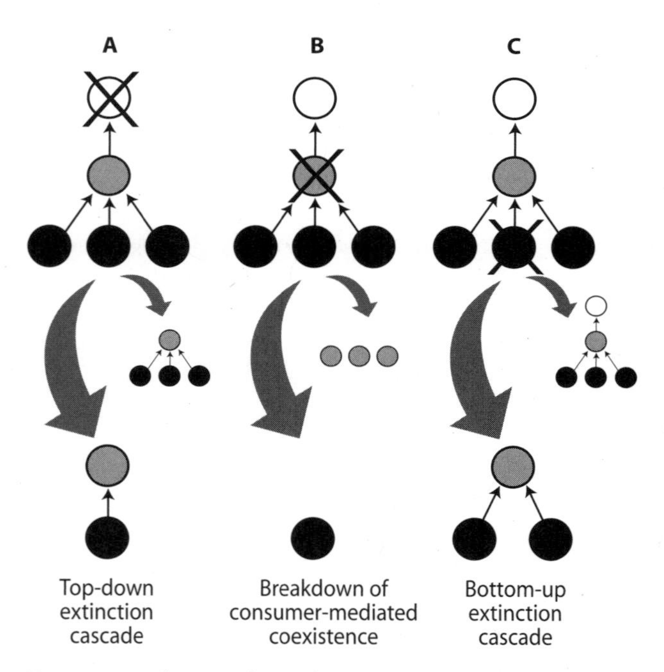

Figure 21.8. Mechanisms of secondary extinctions following the loss of a species (crossed-out circles). The large arrow points at the actual postextinction community that is predicted using a community viability analysis where changes in species densities are accounted for. The small arrow points at the postextinction community that would have been predicted using a community viability analysis where the dynamics of species is not taken into account. A. Loss of a top predator unleashing a top-down extinction cascade. B. Loss of a consumer (e.g., herbivore) leading to the breakdown of consumer-mediated coexistence among prey (e.g., plants). Dominating resource species out-compete subordinate ones in the absence of the consumer. C. Loss of a primary producer triggering a bottom-up extinction cascade. (Reprinted from Ebenham and Jonsson 2006, with permission from Elsevier)

result." Note the important concept of **functional extinction,** in other words, extinction of the services provided by a species precedes extinction of the species itself. Many other examples could be given and have been given at various points throughout the book. (In fact, the overall theme of this book can be boiled down to ecosystem viability analysis.)

Species function in various ways; however, the most ubiquitous has to do with eating and being eaten, and food webs are the logical starting point for analyzing whether a loss (or other structural alteration)[7] will create a ripple or a tidal wave (Berlow et al. 2004). Food web structure is emerging as a crucial aspect of stability. Rooney et al. (2006) argue that "strong asymmetries in energy flow and interaction strengths may be key determinants of ecological stability" (see also Holt 2006). Figure 21.8 shows a simplified scenario of how losses from different trophic levels might trigger secondary extinctions. In the figure, the small arrows denote predicted consequences if the only secondary extinctions are among species immediately adjacent to the primary extinction, while large arrows show predictions if further ramifications are taken into account. Loss of a top predator allows herbivore populations to grow and reduce plant diversity (fig. 21.8A) (e.g., moose in the absence of wolves affect willows). Loss of an herbivore disrupts herbivore-mediated coexistence and allows a single plant species to dominate (fig. 21.8B) (e.g., if we consider natural disturbances such as fire to be "consumers," removing them can lead to homogenized landscapes). Loss of a plant species could in theory cascade up the food chain and cause the loss of a top predator (fig. 21.8C).

The complexity of ecosystem viability analysis should not be underestimated, nor should the stakes. In the words of Woodruff (2001):

> Our inability to make clear predictions (beyond sweeping generalizations) about the future of life on earth has serious consequences for both biodiversity and the well being of humanity. We live at a geological instant when global rates of extinction are at an all time high for the last 65 million years and are increasing. Most extinctions go unrecognized; thus, estimates of overall rates have high errors. Currently, however, several million populations and 3,000–30,000 species go extinct annually of a global total of 10 million species. Probably at least 250,000 species went extinct in the last century, and 10–20 times that many are expected to disappear this century. Although we can identify the most threatened biomes and species in some groups, we cannot make acceptably rigorous predictions about the consequences of these extinctions for the future evolution of life or for the integrity of the biosphere's environmental services that we still take for granted.

[7]Species additions (e.g., invasive exotics) can have as much or more effect on ecosystem viability as species losses, as can factors such as an altered landscape structure that promote the spread of disturbance.

21.5 SUMMARY

The relationship between diversity and stability is a subject of much speculation and debate among ecologists. Modern agriculture, kept stable only by pesticides and fertilizers, provides clear evidence that oversimplification reduces stability. On the other hand, a dominant tree species has been extirpated from forests with only localized effects on the functioning of the remainder of the ecosystem. We can conclude that at least some (but not all) diversity is important to system stability. This leaves an immense middle ground of uncertainty about how much biological diversity can be lost from an ecosystem before its resilience and resistance are significantly reduced.

Understanding the relationship between diversity and stability requires that we understand how species and other structural elements and patterns function within ecosystems. Four aspects of system structure in particular must be taken into account: keystones, redundancy and complementarity, genetic diversity within populations, and landscape complexity. Keystones are species, groups of species, or structures that perform some unique function. Loss of a keystone has the potential to destabilize the rest of the systems. On the other hand, many ecosystem functions can be performed by more than one species. When species perform the same function but under different conditions this is called *complementarity;* when under the same conditions it is called *redundancy*. Because of complementarity and redundancy, some species can be lost without destabilizing the rest of the system. Unfortunately, ecologists have a poor understanding of the details of keystones, complementarity and redundancy within ecosystems, therefore it is seldom clear which species can be safely removed and which cannot.

Genetic diversity within wild populations is important to prevent inbreeding depression and to provide the adaptability that allows populations to better track variable environments, and landscape patterns influence the propagation of disturbances of all kinds. In general, disturbances propagate less readily through a heterogeneous than a homogeneous landscape, although the degree to which a given landscape is one or the other varies with the specific disturbance and forest type.

Ecosystem viability analysis seeks to understand the systemwide ramifications of factors such as species loss, invasive species, altered structure (e.g., of stands, landscapes, or food webs), and climate change. Changes in food web structure arising from changes in stand or landscape structure may significantly affect ecosystem function without outright species loss. Our current ability to make predictions concerning the consequences of the profound changes that earth is undergoing is extremely limited.

22

Ecosystem Stability III

Conserving Species

Earth is poised to become irreversibly poorer.

PIMM AND JENKINS 2005

The conservation of a significant proportion of forest biodiversity requires a comprehensive and multiscaled approach that includes both reserves and the matrix.

LINDENMAYER AND FRANKLIN 2002

With the advent of "intensive" plantation management (during the nineteenth century in central Europe and throughout much of the rest of the world since the end of the Second World War), species conservation became separated from forest management: to "conserve" was not to "utilize." Most everyone came to accept that the forests of the world would be divided into two quite different portions, the larger being intensively managed for products, the smaller consisting of scattered preserves where biodiversity would be maintained. However, several factors have acted to change that vision in recent years. Most notable was the realization that it wasn't working, that species diversity will not be protected in the long run through a system of nature reserves embedded within an intensively managed landscape. Coupled with that has been growing awareness of the critical global services provided by forests, and a sharply increased concern about their diversity and health (see Dombeck et al. [2003] for an excellent discussion of the degradation of America's public lands and the ongoing efforts to restore them). Nature reserves still play an important role in the emerging view, and in fact more are needed, but they must be coupled with forest-management techniques that reconcile commodity production with preservation of biodiversity (Sinclair et al. 2000; Scott et al. 2001; Lindenmayer and Franklin 2002).

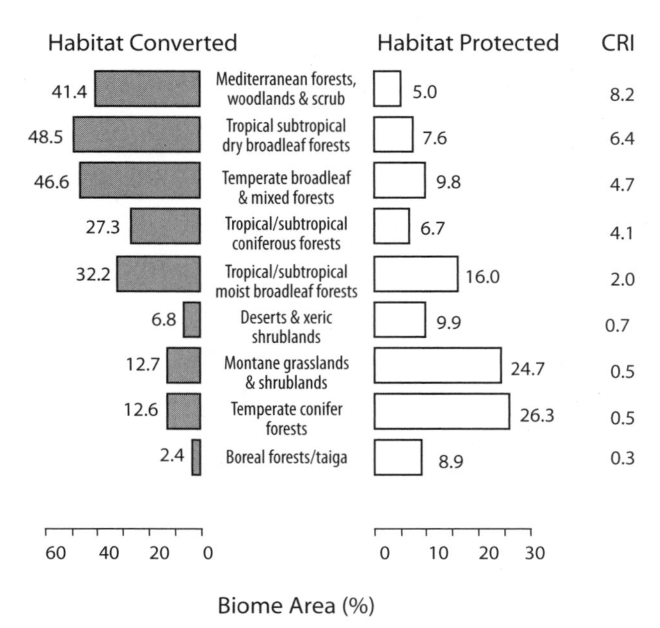

Figure 22.1. Habitat conversion and protection in 11 of the world's 13 terrestrial biomes. Biomes are ordered by their conservation risk index (CRI). CRI was calculated as the ratio of percent area converted to percent area protected as an index of relative risk of biomewide diversity loss. (Adapted from Hoekstra et al. 2005)

Some forest policy experts predict that by midcentury the world's demand for wood will be met by fast-growing plantations on a small proportion of the world's forest area. Sedjo (2001)[1] sums up these predictions as follows:

> Industrial wood production will be increasingly confined to high productivity sites where very intensive management is practiced. The total area involved in industrial plantation forests, although large in an aggregate sense, will be small on a global scale - perhaps consisting of a couple of hundred million ha, or perhaps about 5 per cent of the 3.4 billion has of land area currently in the global forest. Some natural forests will continue to be harvested for specialized types of industrial wood, but as of today, specialized wood types will constitute only a small fraction of total industrial wood requirements. This will leave approximately 90 percent of the world's forest for other purposes. In under a host of conditions these forests will be providing environmental and ecological services ranging from watershed protection to biodiversity reserves.

Even if the predictions of Sedjo and others prove accurate and many or all of the conflicts between forestry and conservation disappear, that will not become reality for several decades (and not at all for conflicts between agriculture and forest conservation). The intervening time will be a critical period for conservation, and protecting forest diversity will require that foresters add conservation principles to their tool kit.

The stakes are high. The 2006 IUCN Red List[2] (World Conservation Union/IUCN 2006a) contains 16,119 species threatened with extinction, a figure that is almost certainly an underestimate. Most are in tropical forests. Pimm et al. (2006) predict that if current rates of forest loss[3] continue, extinctions of bird species by the end of the twenty-first century will be 1,500 times greater than those before significant human impacts. Pimm et al. (2006) go on to argue that impacts on some other taxa are likely to be even greater (see also Pimm and Jenkins 2005). Entire biomes have been significantly affected by human uses. Figure 22.1 compares area protected to area converted to human use (especially agriculture) for each of the earth's major biomes. In biomes suitable for agriculture (including tropical forests, temperate broad-leaved forests, and Mediterranean forests and scrub), two to eight times more area has been converted than has been set aside in reserves. In contrast, more area has been set aside than converted in temperate conifer and boreal forests, although loss of forests to suburbanization is a serious problem in some areas (fig. 22.2). Moreover, large areas of temperate coniferous forests have been retained as forests but converted to plantations of the same or different species. As discussed later, managed forests can support significant levels of biodiversity; however, the conversion from old or naturally developing young forests to managed affects habitat for many species.

Achim Steiner, director general of IUCN, sums up the current situation: "biodiversity loss is increasing, not slowing down. The implications of this trend for the productivity and resilience of ecosystems and the lives and livelihoods of billions of people who depend on them are far-reaching. Reversing this trend is possible, as numerous conservation success stories have proven. To succeed on a global scale, we need new alliances across all sectors of society. Biodiversity cannot be saved by environmentalists alone—it must become the responsibility of everyone with the power and resources to act" (Word Conservation Union/IUCN 2006b).

The degree to which sensitive species can be maintained even within "gently" managed forests is still a matter of debate and certainly needs much more research. However, there is no intrinsic reason that a wide range of species cannot be maintained within forested landscapes that include areas managed for products. As we have seen, disturbances

[1]Also see The Great Restoration (http://greatrestoration.rockefeller.edu/) and Victor and Ausubel (2000).

[2]The Red List, produced by the World Conservation Union/IUCN, is considered the most authoritative survey of the status of species worldwide. It is available online as a searchable database (www.iucn.org/themes/ssc/redlist2006/redlist2006.htm).

[3]This refers primarily to forest clearing for agriculture and ranching in the tropics.

Figure 22.2. In the background is the riparian zone of the Rio Pedro River in southern Arizona. This riparian zone, a principal north-south route for neotropical migrant birds, is threatened by the growth of a nearby city. Increasing water use in this and other areas of the American Southwest is lowering water tables, which dries streams and kills the riparian trees.

are a healthy, diversifying force in nature as long as they are consistent with species adaptations. Forestry can never reproduce a natural disturbance for the simple reason that it takes wood away and the other does not. However, if we understand the needs of species and adopt a comprehensive, landscape-oriented approach, forestry can complement a reserve system by protecting critical habitats (Lindenmayer and Franklin 2002).

This chapter reviews some of the principles of conversation biology. How foresters can put these principles into practice is taken up in chapter 23. A single chapter can only scratch the surface of the vast literature dealing with this topic. We encourage readers to pursue citations given in the chapter, especially Lindenmayer and Franklin (2002), McComb (2008), and Raphael and Molina (2007).

22.1 CONSERVING SPECIES MEANS PROTECTING HABITAT

Local populations of any species become vulnerable to extinction when their numbers drop below some minimum level and they become inbred or lose too much genetic diversity, or when they become overwhelmed by predators, disease, invasive species, or other stresses. Evidence from extinctions during the recent past indicates that no one factor plays a decisive role, rather it is the synergistic interaction of multiple stressing factors (all related in one way or another to humans) that pushes species over the brink (Pimm 1996; LoGiudice 2006). For example, the Alleghany woodrat experienced, in succession, loss of habitat as forests were converted to agriculture, loss of a major food source when the American chestnut was wiped out by blight, fragmentation of forests in its northern range by human development (blocking dispersal pathways and making recolonization of empty forest patches less likely), loss of a another major food source when oaks were defoliated by gypsy moth, and increasing susceptibility to a nematode spread by raccoons (whose populations increased along with those of humans) (LoGiudice 2006).

Species with sufficient habitat may nevertheless be pushed to extinction; however, loss or degradation of significant amounts of habitat is usually the beginning of the end, making it all that more difficult to resist and recover from other stresses. Habitat must be considered at two different spatial scales: that required by the population, and that required by the metapopulation. In the most general terms, three conditions must be satisfied for any given species:

1. Local ecosystems and landscapes must contain sufficient habitat to support a viable breeding population. In general terms, viability entails a supply of basic ecological needs (e.g., resources, protection, and availability of mutualists) adequate to maintain some minimum

Three general modeling approaches are used to determine the habitat needs and population status of a given species or set of species (Norris 2004): statistical models, demographic models, and behavior-based models. Statistical models describe patterns of species occurrence in relation to any of several relevant environmental variables, including biotic and abiotic habitat (e.g., stand age and type, topoedaphic variables). Since the mid-1990s, statistical models have increasingly moved beyond the stand level and explored how landscape patterns at various scales influence species distributions (e.g., McGarigal and McComb 1995; Mitchell et al. 2006). Demographic models explicitly incorporate biology (as opposed to being purely correlative) by associating life-history variables (e.g., birth and death rates) with habitat. Demographic models specifically aimed at developing options for stabilizing and recovering endangered populations are often termed **population viability analyses** (or PVAs). Behavior-based models increase the biological content further yet by incorporating decision making (i.e., behavior) at the level of individuals, where behavioral decisions are assumed to be made so as to maximize individual fitness.

Each approach has strengths and weaknesses (Norris 2004). Statistical models are commonly used because large amounts of data can be gathered relatively quickly and cheaply. However, statistical models can only describe species response to existing habitat patterns and therefore lose reliability when considering the effects of new patterns (e.g., when a species is losing habitat). Demographic models are attractive because they incorporate biology. Although increasingly used in the management of endangered species, demographic models are also controversial because good data about demographics is not available for many species, and that raises concerns about the accuracy of the models. Furthermore, the demographic data that does exist is based on existing conditions; therefore, like statistical models, the reliability of demographic models in the face of new conditions is questionable. Being grounded in first-principles (i.e., evolutionary theory), behavior-based models are assumed to better predict species responses to environments beyond what currently exists, a feature that is especially important in identifying threshold responses to habitat loss. Behavior-based models are relatively new and have not yet been widely applied (Norris 2004).

Problems with limited information are especially acute for the many rare or little-known species, necessitating a fourth approach, which is called *population viability evaluation* (PVE) (Marcot and Molina 2007). Rather than relying on weak or nonexistent demographic data, a PVE uses expert panels to qualitatively rank threats to a species or set of species with similar needs. This was the approach taken by FEMAT, the plan to protect old-growth associates in the US Pacific Northwest. Given our poor state of knowledge about many species, PVE may be the most practical and useful approach in a majority of cases.

population size. Populations that drop below some minimum size (which varies depending on species and environment) slide irreversibly to local extinction. If enough local populations become extinct, the species slides towards extinction.

2. Regions must contain sufficient habitat to support a viable metapopulation that provides a large-enough gene pool to prevent genetic deterioration and a continuing source of immigrants both to colonize newly created habits (e.g., by disturbance and succession) and to replace extinctions of local populations. Large disturbances (e.g., fire, hurricanes) are a particular threat. Hurricane Hugo, for example, blew down a relatively large proportion of the habitat of the endangered red-cockaded woodpecker along the eastern seaboard of the United States.
3. Local habitats must be sufficiently interconnected at the regional level that individuals (or propagules) within the metapopulation can move between local populations. There must also be sufficient interconnection to provide pathways for migration (especially as global climate change unfolds). What constitutes sufficient interconnection depends on various factors, including in particular the mobility of the species in question. Species with limited dispersal ability are particularly threatened when regional habitats become fragmented.

Three kinds of extinction are recognized. **Local extinction** is the disappearance of a local population of metapopulation. **Global extinction** involves the loss of the entire species. **Ecological extinction** occurs when a population may persist but its numbers are too low for the species to function in its normal ecological role. Redford (1992), for example, suggests that populations of many large mammals in neotropical forests have declined to the point that they no longer effectively disperse tree seeds.

The size and diversity of a given area needed to meet minimum habitat requirements is the central issue for those who design nature reserves, and also for forest managers whose objectives include maintaining species diversity. For various reasons, however, translating general rules into specific guidelines is not easy. For species with large home ranges (e.g., wolves, jaguars, spotted owls), the area needed to support enough breeding pairs or packs to maintain a genetically viable population is likely to extend far beyond the bounds of even a large national park, and the habitat requirements of migratory species such as songbirds and elephants extends across international borders. Other problems arise when one deals with entire communities rather than single species. The varied (and sometimes conflicting) requirements on the part of individual species make the task of protecting entire communities a considerable challenge. Both the particular mix of habitats required and the area needed to contain them differ widely from one species to another, depending on their life histories, mobility, breeding characteristics, and probably many other factors. For example, consider animals that inhabit pine forests of the southern United States. Deer need early successional herb-shrub-sapling stages for food and closed forest for cover. The red-cockaded woodpecker requires pine forests with little hardwood understory (probably because a tall understory allows predators access to woodpecker nest cavities), but fox squirrels prefer hardwoods mixed with the pines (probably because of the mast, especially acorns, and rich insect fauna associated with hardwoods). Some birds nest in early successional stages and others in mature forests, some prefer pines while others prefer hardwoods. Small wetlands are important for herptofauna (amphibians and reptiles) (Russell et al. 2002). Accommodating the multiplicity of habitats required by diverse communities of organisms requires landscape and regional approaches.

How large should a nature preserve be? There has been much debate over the question of how to design nature reserves to protect the most species, and in particular whether the best strategy for a given region is a single large or several small reserves (known by the acronym SLOSS). Holsinger (2007) comments on the SLOSS controversy: "To a large extent, this whole debate seems to have missed the point. After all, we put reserves where we find species or communities that we want to save. We make them as large as we can, or as large as we need to protect the elements of our concern. We are not usually faced with the optimization choice poised in the debate. To the extent we have choices, the choices we face are more like those that Pressey et al. (1997) describe, i.e., how small an area can we get away with protecting and which are the most critical parcels?"

Schwartz (1999) discusses the issues surrounding reserve design.

The plot thickens; the welfare of one species may be (and probably usually is) linked to that of other species, which raises the specter of coextinctions. This is a particular risk with coevolved relationships. For example, Koh et al. (2004) estimate that extinctions of butterflies on the island of Singapore will rise exponentially with the extinction of their host plants. In Britain and the Netherlands, declines in plant species are linked to declines in their pollinators (Biesmeijer et al. 2006). In some cases one species may depend on another with quite different habitat requirements (Simberloff 1988). In the Amazon rain forest, for example, several species of frogs breed in pools that are created by peccaries (Zimmerman and Bierregaard 1986), thereby tying the welfare of narrow-ranging species directly to that of a broad-ranging species. There are many similar examples of close linkage between species that differ widely in life-form, life history, and habitat requirement, such as species that are tied to one another through mutualisms, or keystone species that perform some unique function within the community (Boogert et al. 2006; chapter 11 in this volume). Noss (1991) uses longleaf pine forests of the southeastern United States to exemplify the web of interactions that exists in ecosystems:

> [A] dominant and flammable groundcover plant, wiregrass, encourages frequent low intensity ground fires. These fires maintain an open, park like forest with one of the richest assemblages of herbaceous plants in the world (Clewell, 1989; Noss, 1988). Longleaf pine also plays a critical role in this process by converting lightning strikes into ground fires and by producing highly combustible duff (Platt et al., 1988). Animals such as the gopher tortoise, its many commensals, and the endangered red-cockaded woodpecker require the open habitat, maintained by fire. At least fourteen species of cavity-nesting birds and mammals, in turn, are dependent upon the red-cockaded woodpecker, which excavates cavities in living, old-growth pines. Thus an ecosystem may be characterized by several interdependent keystone species, all interacting with a keystone abiotic factor (in this case, fire).

22.2 WHAT KIND OF HABITAT? A MATTER OF BALANCE

Life fills every nook and cranny of the world. When one kind of habitat is destroyed, another is created, and the diversity of habitats produced by the continuing cycle of creation and destruction generates much of the world's biological diversity (chapter 10). What we must be concerned with in conservation ecology are maintaining a proper balance of habitats for indigenous species, and not creating the conditions that allow a disruptive species to invade and to exclude natives.

Table 22.1
Some characteristics of extinction-prone species

Top predators (vertebrates). Because of their position in the food chain, these animals have low population densities and large territories; hence, they are especially vulnerable to reductions in habitat. California condors, spotted owls, wolves, mountain lions, and jaguars are a few of the many examples. Some vertebrates that are lower on the food chain also require large territories, particularly food specialists such as frugivorous monkeys and birds.

Species with specialized habitat requirements. The situation depends on the habitat. The weedy species that frequently thrive around humans are hardly threatened in today's world, but many of those that depend on old-growth forests, large dead wood, riparian zones, wetlands, and other special habitats are.

Species that disperse poorly. These species are in danger when their populations become isolated and fragmented. Gene flow is reduced, and local extinctions (i.e., loss of local populations) may not be replaced by new colonizers. Poor dispersers will have particular problems with climate change. Note that dispersal of a given species depends not only on its intrinsic ability to cover space but also on the characteristics of the landscapes through which it must disperse. Even very mobile species may not move across landscapes where they are vulnerable to predators; the primary cause of mortality of Florida panthers, for example, is being run down by cars as they cross roads.

Migratory species. As Terborgh (1974) states, "Migratory species are exposed to double jeopardy because they are subject to the pressures of change at both ends of their routes, and may have to run a gauntlet of polluted waters and altered landscapes on the way." Migratory songbirds are one of the better examples. Neotropical migrants winter in Central American forests, which are being rapidly cut down, and those that summer in eastern North America have also lost habitat to farms, suburbs, and logging. Those that summer in western North America migrate through deserts and grasslands along riparian forests, at least some of which are threatened by overgrazing and dropping water tables (the latter resulting from heavy water use by sprawling suburbs) (fig. 22.2).

Species with a low intrinsic rate of population growth. These species are not capable of rebounding quickly once their numbers are reduced. This includes most (if not all) large animals at the top of food chains.

Species sought by humans for meat, trophies, or other commodities. These species are vulnerable to overhunting. Elephants and black rhinos are perhaps the best-known examples, but they are far from the only ones. Hunting has already sharply reduced the numbers of many large mammals in neotropical forests (Redford 1992).

Endemics. These are species that although locally abundant may have very restricted ranges. Island dwellers are a special kind of endemic that is particularly vulnerable to extinction.

Species with low genetic variability. These are particularly vulnerable to inbreeding. Moreover, they have low buffering power against environmental change.

Source: Adapted from Crow 1990 and Terborgh 1974.

22.2.1 Proper Balance

Species are not created equal regarding their ability to persist, and habitat conservation must begin with those that are most vulnerable to extinction. Table 22.1 lists some characteristics of vulnerable species, which can be summarized for forest dwellers as follows:

- Species requiring threatened habitats. That commonly includes species requiring large areas of forests that either are late successional or have a complex structure not found in plantations (and certainly not in farms and ranches). However, early successional habitats may also be at risk, especially those allowed to recover naturally following disturbance.[4] Bird species that use disturbance-mediated habitats are declining in North America (Brawn et al. 2001).
- Species subject to overhunting or overpredation.
- Species with restricted ranges (i.e., endemics) and that require special habitats not normally preserved in intensively managed forests.
- Species subject to cumulative or concentrated effects. Streams and rivers, for example, receive the concentrated impact of sedimentation from entire watersheds; this is one of several factors that has endangered salmonids in the Pacific Northwest.

[4]For example, sites are commonly salvaged following wildfire, and they may be artificially seeded with grasses and herbs (for erosion control) or replanted with trees at higher densities than occurred is a naturally recovering forest. In some cases naturally recovering vegetation may be suppressed.

> The best known example of a disturbance dependent forest bird is the Kirtlands warbler, which is closely tied to burned jack pine stands; but there are many others. In the northern Rocky Mountains, 15 bird species are more common in recently burned forests than in other habitats, and one (the black-backed woodpecker) appears to depend exclusively (or nearly so) on fire-killed snags (Hutto 1995). Fire-killed trees serve as the base of food chain extending from fungi to insects to birds. Extensive salvage following fire, which is a common practice, disrupts that chain. Jerry Franklin (personal communication) argues that naturally recovering early successional stages are the rarest of forest habitats.

22.3 FINE FILTERS, COARSE FILTERS, AND PLURALISM

How do we proceed in our efforts to preserve biological diversity? Three alterative approaches are possible:

How much area should be protected in order to conserve global diversity, and where should it be? The question of "how much is enough" has plagued conservation science ever since concerns for saving species first arose (Tear et al. 2005). One proposal that has received a great deal of attention from policy makers is setting aside 10 to 12 percent of the earth's land surface as reserves. Soule and Sanjayan (1998) comment on this target: "If successful, this campaign would double or triple the land area now designated as national parks or similar strict reserves. We are concerned, however, that these target percentages could become de facto ceilings of protection and imply that protecting 10% or so of the land is sufficient to prevent the predicted major extinction event."

In fact, species-area curves (chapter 9) predict that reducing habitat by 90 percent (i.e., the 10% protection target) would result in the loss of 50 percent of species dependent on the habitat in question (Soule and Sanjayan 1998; Svancara et al. 2005). In contrast to the policy-driven goal (i.e., 10 to 12%), recommendations based on scientific analyses range from 25 to 50 percent of a given area in reserves in order to meet conservation goals (Svancara et al. 2005). This varies according to numerous factors, particularly the species of concern and the degree to which the matrix contributes (Lindenmayer and Franklin 2002). It is probably safe to say that most such analyses assume the matrix contributes little or nothing to conserving habitat, which is mostly (but not always) true where the matrix is dominated by intensive human uses (e.g., agriculture, exotic tree plantations), but less true when the matrix is managed native forest (section 22.5).

As of 2004, the global network of protected areas covered over 12 percent of the earth's land surface (Chape et al. 2005). Among forested biomes, levels of protection varied from slightly more than 8 percent of biome area for temperate needle-leaved (i.e., conifers) and evergreen sclerophyllous forests to almost 19 percent of tropical humid forests (table 22.2) (biomes are defined using the system of Udvardy 1975). The degree of protection provided by this protected area network is limited for two reasons. First, many species are left unprotected (i.e., outside reserves) (Scott et al. 2001; Rodrigues et al. 2004), especially those with limited ranges (which account for the bulk of terrestrial diversity) (Pimm and Jenkins 2005). Second, the integrity of many parks is threatened, especially in the tropics where illegal logging and other forms of trespass are common occurrences (Gascon et al. 2000; Bruner et al. 2001; Curran et al. 2004).

Attention has focused on diversity **hot spots** as a way to maximize species protection at minimum cost. According to Myers et al. (2000), "as many as 44% of all species of vascular plants and 35% of all species in four vertebrate groups[5] are confined to 25 hotspots comprising only 1.4% of the land surface of the Earth."[6] In addition to plant diversity, Myers et al. defined hot spots by degree of threat (defined as having lost 70% or more of primary vegetation). Eleven of the 25 have already lost at least 90 percent of their primary vegetation, and three have lost 95 percent. Using the criteria of Myers et al. (2000), Conservation International later expanded the list to 34 hot spots (these are described at www.biodiversityhotspots.org/xp/Hotspots/). However, the situation is not as simple as it appears at first glance. Hot spots may be defined using different criteria, and there is an unfortunate lack of congruence among those defined on the basis of biological richness and those defined by either levels of endemism or numbers of rare or threatened species (Orme et al. 2005). This pattern seems to hold generally. Among the mammals of Mexico, for example, there is "very low correspondence among areas of high diversity, high endemicity, or high number of endangered species" (Ceballos et al. 1998). Similarly, hot spots of species richness do not correspond to hot spots of red-listed (i.e., threatened) species in Norway (Gjerde et al. 2004). On a positive note, however, a hot spot of endemism for one taxa of terrestrial vertebrates tends also to be a hot spot of endemism for other terrestrial vertebrates, and the same is true for richness hot spots (Lamoreux et al. 2006). Therefore, the same hot spots can protect centers of either richness or endemism for mammals, birds, reptiles, and amphibians. Moreover, although globally endemism and richness are rather poorly correlated, areas of high endemism also have greater richness than expected by chance (Lamoreux et al. 2006).

The science that deals with allocating scarce conservation resources (e.g., protected areas) falls under the umbrella of **conservation biogeography,** a new discipline broadly concerned with the application of biogeographical principles to conserving biodiversity (Possingham and Wilson 2005; Whittaker et al. 2005; Wilson et al. 2006). Brooks et al. (2006) review the various approaches to prioritizing areas for conservation across the globe.

1. A focus on individual species, called the **fine filter approach.**
2. A focus on whole ecosystems, called the **coarse filter approach.**
3. A combination of fine and coarse filters, which Noss (1991) calls the **pluralistic approach.**

Although most efforts have employed a fine filter, conservation biologists are increasingly troubled by this

[5]Mammals, birds, reptiles, and amphibians.

[6]The majority of hot spots are in the tropics, however temperate areas include New Zealand, the Caucasus, central Chile, and the California Floristic Province.

Table 22.2
Protected areas by forested biome

Udvardy biomes	Biome (km^2)	Extent of protected area (km^2)	% biome protected
Tropical humid forests	10,553,490	1,991,052	18.87
Subtropical/temperate rain forests/woodlands	3,961,627	539,155	13.61
Temperate needle-leaf forests/woodlands	17,032,915	1,424,311	8.36
Tropical dry forests/woodlands	17,316,029	2,302,192	13.30
Temperate broadleaf forests	11,278,456	1,159,314	10.28
Evergreen sclerophyllous forests	3,720,843	327,696	8.81

Source: Adapted from Chape et al. 2005.

approach. With the numerous species that are either currently or potentially threatened, it has become increasingly clear that trying to save all is tactically impossible when approached on a species-by-species basis (Franklin 1993). Moreover, the fine filter is unlikely to be fine enough: the species that get attention are those that are either easy to see and track (e.g., birds) or the "charismatic megavertebrates"—those animals that are particularly appealing or symbolic to humans (e.g., grizzly bears, panthers, eagles). Many of Edward Wilson's "little things that run the world" could easily slip through the mesh of the fine filter.

The basic premise of the coarse filter approach is that protecting species means protecting the structural and functional integrity of the system in which they are embedded. Rather than individual species, communities and landscapes are the focus of protection and (where necessary) restoration. Pioneered in the United States by the Nature Conservancy, this approach makes sense from a practical standpoint, but it is not without problems. Forest community types may be poor indicators of habitat suitability for at least some species (e.g. birds) (Cushman et al. 2008). Moreover, species can be lost from communities that are otherwise intact (or at least give the appearance of being intact). Earlier, for example, we saw that populations of large mammals had dropped to very low levels in "undisturbed" neotropical forests (Redford 1992). Mycorrhizal fungi began disappearing from European forests well before the trees began to die (chapter 20). Moreover, the question remains of how much of a given habitat or biome should be protected in order to meet conservation goals. That question can only be answered by monitoring those species that require the most area to maintain viable populations. The upshot is that while protecting whole communities (and indeed regional distributions of community types) is critically important if the goal is to maintain entire suites of regional biota, it is also necessary to protect and monitor individual species. This is the combination of fine and coarse filters that Noss (1991) calls the *pluralistic approach.* Noss (1990) further argues for a hierarchical approach to conservation and monitoring, with four levels in the hierarchy:

1. Regional landscape
2. Community or ecosystem
3. Population or species
4. Genetic

Table 22.3 lists the relevant aspects and monitoring tools for each of these levels. The proposal by Noss is one of the more thorough and potentially effective strategies, but implementing it will require major effort and expense. A few shortcuts have been suggested, most notably the use of **indicator species** (so-called because they are considered to indicate the health of the entire system of interest). An **umbrella species** is a type of indicator "whose conservation is expected to confer protection to a large number of co-occurring species" (Roberge and Angelstam 2004). Most suggested umbrellas are large mammals or birds, although small vertebrates and invertebrates have also been suggested. For example, Welsh and Droege (2001) argue that plethodontid salamanders[7] are useful indicators of biodiversity and integrity of North American forests. These lungless terrestrial salamanders are "tightly linked physiologically to microclimatic and successional processes that influence the distribution and abundance of numerous other hydrophilic but difficult-to-study forest-dwelling plants and animals" (Welsh and Droege 2001).

However, it is risky to assume that protecting, for example, spotted owls or jaguars or salamanders automatically protects associated species. In fact, the evidence suggests that "single-species umbrellas cannot ensure the protection of all co-occurring species because some species are inevitably limited by ecological factors that are not relevant to the umbrella" (Roberge and Angelstam 2004). Most conservation biologists now argue that a diversity of indicator species is required. For example, among the forest carnivores in the Rocky Mountains "habitat for grizzly bear has high overlap with that for wolverine, intermediate overlap with fisher, and low overlap with lynx" (Carroll et al. 2001), leading Carroll et al. (2001) to conclude that conservation

[7]Family Plethodontidae comprises lungless salamanders that respire through their skin and mouth lining.

Table 22.3
The indicator variables for inventorying and assessing terrestrial biodiversity at four levels of organization

Indicators	Composition	Structure	Function	Inventory and monitoring tools
Regional landscape	Identify distribution, richness, and proportions of patch (habitat) types and multipatch landscape types; collective patterns of species distributions (richness, endemism)	Heterogeneity: connectivity spatial linkage, patchiness porosity, contrast, grain size, fragmentation, configuration, juxtaposition, patch size frequency distribution perimeter–area ratio pattern of habitat layer distribution	Disturbance processes (aerial extent, frequency of return interval, rotation period, predictability, intensity, severity, seasonality), nutrient cycling rates, energy flow rates, patch persistence and turnover rates, rates of erosion and geomorphic and hydrologic processes, human land use trends	Aerial photographs (satellite and conventional aircraft) and other remote sensing data; geographic information system technology, tine series analysis, spatial statistics, mathematical indices (of pattern, heterogeneity, connectivity, layering, diversity, edge, morphology, autocorrelation, fractal dimension)
Community ecosystem	Identify relative abundance, frequency, richness, evenness, and diversity of species and guilds; proportions of endemic, exotic, threatened, and endangered species; dominance-diversity curves; life-form proportions; similarity coefficients; C_4-C_3 plant species ratios	Substrate and soil variables; slope and aspect; vegetation biomass and physiognomy; foliage density and layering; horizontal patchiness and gap proportions; abundance, density, and distribution of key physical features (e.g., cliffs, outcrops, sinks) and structural elements (snags, down logs); water and resources (e.g., mast) availability; snow cover	Biomass and resource productivity; herbivory, parasitism, and predation rates; colonization and local extinction rates; patch dynamics (fine-scale disturbance process); nutrient cycling rates; human intrusion rates and intensities	Aerial photographs and other remote-sensing data; ground-level photo stations; time series analysis; physical habitat measures and resource inventories; habitat suitability indices (multispecies), observations, censuses and inventories, captures, and other sampling methodologies; mathematical indices (e.g., of diversity, heterogeneity, layering dispersion, biotic integrity)
Population/ ecosystem	Absolute or relative abundance, frequency, importance or cover value, biomass, density	Dispersion (microdistribution), range (macrodistribution), population structure (gender ratio, age ratio), habitat variables (*see* community/ecosystem structure, earlier), within individual morphological variability adaption	Demographic process (fertility, recruitment rate, survivorship, mortality), metapopulation dynamics, population genetics (see below), population fluctuations, physiology, life history, phenology, growth rate (of individuals), acclimation	Censuses (observations, counts, captures, signs, radio tracking), remote sensing, habitat suitability index, species-habitat modeling, population viability analysis
Genetics	Allelic diversity, presence of particular rare alleles, deleterious recessives of karyotypic variants	Census and effective population size, heterozygosity, chromosomal or phenotypic polymorphism, generation overlap, heritability	Inbreeding depression, outbreeding rate, rate of genetic drift, gene flow, mutation rate, selection intensity	Electrophoresis, karyotypic analysis, DNA sequencing, offspring-parent regression, sib analysis, morphological analysis

Source: Noss 1990.

planning for carnivores in that region must consider the needs of several species.

A recent approach that builds on and extends the idea of umbrellas is that of **focal species** (Lambeck 1997), in which a suite of species is identified, each being an umbrella for a different limiting factor. For example, "the area requirements of species most limited by the availability of particular habitats will define the minimum suitable area for those habitat types; the requirements of the most dispersal-limited species will define the attributes of connecting vegetation; species reliant on critical resources will define essential compositional attributes; and species whose populations are limited by processes such as fire, predation, or weed invasion will define the levels at which these processes must be managed" (Lambeck 1997).

Identifying meaningful indicators requires understanding the relationship between the structure and function of natural systems at all hierarchical levels: community, region, and globe. What are the keystones, the most vulnerable points in system structure? What are the elements whose progressive loss can lead to unpleasant surprises? Unfortunately, our knowledge of these relationships is rudimentary at best. However, any approach that does not account for the entire suite of processes and forces acting on systems risks failure in the long run. This especially true in today's world of multiple stressors, many of which are unique in evolutionary history. For example, efforts to save old-growth-associated species in the Pacific Northwest have focused on a series of reserves, with the northern spotted owl as an umbrella defining area requirements of old-growth habitat (i.e., a pluralistic approach). However, these reserves are potentially threatened by future changes in climate and the fact that they are often embedded within in a landscape of densely stocked, younger forests that are vulnerable to crown fires (and that may well propagate fires into the old-growth reserves). That vulnerability is already playing out; the 2002 Biscuit fire in southwestern Oregon burned approximately 60,000 ha of late successional reserves.

22.4 VIABLE POPULATIONS

Deciding which species (or ecosystems) should have highest priority for protection is only the first step. Next, we must address the issue of how many. For example, how many mating pairs are needed for a healthy population? How much habitat does that number require?

A viable population is simply the number of individuals "that will insure (at some acceptable level of risk) that a population will exist in a viable state for a given interval of time" (Gilpin and Soule 1986). (Note that this may refer to either local populations or metapopulations, depending on the circumstances; for simplicity, we use *population* in the following discussion to cover both possibilities.) For example, saving enough old-growth Douglas-fir to support one hundred spotted owls will not save the owl, because it takes more than one hundred individuals to maintain a viable population (probably closer to several thousands). Populations that drop below some minimum size are sucked into what Gilpin and Soule (1986) call the **extinction vortex** (i.e., their fate is sealed unless extraordinary measures are taken).[8]

Populations that are losing individuals become at risk before this point, reaching a size in which they skirt the edge of the extinction vortex, liable to be pushed over by some random environmental event such as drought, a particularly hard winter, a devastating hurricane, or an outbreak of infectious disease. As first stressed by Caughley (1994), understanding and addressing the causes of decline is "pivotal to designing effective practical management" (Norris 2004). This has become known as the **declining population paradigm.**

How many individuals constitute a viable population? This is a very difficult question that depends on (among other things) (1) the population structure, social dynamic, and breeding characteristics of the species in question; (2) environmental fluctuations, particularly the possibility of catastrophic events that sharply reduce population size (Shaffer 1981); (3) environmental stresses, such as pollution, that reduce the vigor of individuals; and (4) various aspects of habitat quality at both the stand and landscape scales (section 22.5). Much of the information available to guide management decisions comes from models of population behavior, which play a useful role but cannot substitute for field biology. Moreover, the vast majority of these models deal with vertebrates and may or may not be relevant to the plants, insects, and microbes that compose the vast majority of the earth's biota (Raphael and Molina 2007). We return to the question of how many individuals constitute a viable population at the close of this section; first, we review some of the factors that place small populations at risk.

Population size affects the survival chances of individuals within the population (or metapopulation) in two general ways (Soule 1983; Lande 1988b). One relates to demography (i.e., the patterns of population growth and decline) and the other to population genetics and (in particular) the ability to retain genetic variability and to avoid inbreeding. Demography and population genetics are related. Small populations tend to lose genetic variability; however, the two factors also act independently.

22.4.1 Demography

Demographic factors come into play through the tendency of populations of most species to fluctuate depending on numerous factors of their environment.[9] The smaller a population, the greater the chances that one of its down cycles will either wipe it out completely or reduce it to a level where genetic or social deterioration triggers a slow but inexorable slide toward extinction.

> Populations of most species are moving on a big taco chip in space and time: high at one time, low at another; high in one place, low in another. The objective of conservation ecology is to keep a species from falling off the edge. We seldom know where that edge is, however, nor do we know where extinction vortices may lie on the surface of the chip itself. The more dramatic the topography of the taco chip (i.e., the greater the fluctuations in species numbers), the greater the chance that a species can be pushed off the edge when it is at a low point.

The population size at which such a slide begins is partly a question of population genetics, especially inbreeding and loss of genetic diversity, and will be discussed in the following section. However, nongenetic factors may also come into play. In some species, small populations can have rela-

[8] In the words of Soule (1987), "There are no hopeless cases, only people without hope and expensive cases."

[9] See Lande et al. (2003) for a thorough treatment of the ecology and conservation implications of population dynamics.

tively high genetic diversity yet still be at risk for purely demographic reasons, a phenomenon called the *Allee effect*.

Demographic risk is particularly high where factors such as infectious disease or large-scale natural events (e.g., hurricanes, wildfires) have the potential to sharply reduce population sizes. Small populations do not contain sufficient individuals to be buffered against such catastrophic losses. A particular threat is the possibility that two or more events that reduce population size may occur in quick succession, not allowing the population time to recover from one before it is further reduced by another. In today's world, the probability is growing that two or more forces that stress populations might occur simultaneously (e.g., acid rain plus climate change), which makes the buffering power of large, well-distributed populations even more vital than when under pristine conditions.

Purely demographic factors are particularly important in social animals, which depend on numbers for defense, foraging, and (in some cases) effective breeding. Such species can be thought as having a social threshold, below which the entire group becomes dysfunctional and unable to persist. In most cases, this social threshold is considerably higher than that determined solely by the demographics of population fluctuations or by genetics (Soule 1983).

Catastrophic events of one kind or another are possible (even inevitable) in most any environment, and local populations of many species may disappear from a given area periodically. A species is not threatened by these local demographic fluctuations so long as an intact, healthy metapopulation remains sufficiently interconnected to provide a source of immigrants that can replace local losses, but not so interconnected that the same stress threatens a large proportion of individuals within the species. When the metapopulation is in decline or uniformly threatened, however, or when remaining local populations are so fragmented and isolated that movements among them are inhibited, demographic losses within local ecosystems will not be replaced through natural mechanisms.

Depending on the size of the stage on which the shifting mosaic plays, the area required to buffer species against demographic calamities can be quite large. Shugart and Seagle (1985) simulated the relationship between temporal stability in the proportion of habitats within a given area and size of that area as measured by the number of habitat patches it encompassed. Habitat diversity fluctuated significantly within an area encompassing 200 patches, continued to fluctuate less severely in an area containing 2,000 patches, and remained constant in an area with 20,000 patches. In other words, for a forest type with an average patch size of 100 ha, their model predicts that somewhere between 200,000 and 2,000,000 ha would be required to maintain constancy in habitats. Of course nature is much more complicated than reflected by this (or any) model, but the general message is valid: large landscapes are required to buffer the effect of disturbance on habitat diversity.

Large areas are essential for supporting viable populations of top mammalian predators, and as we saw in chapter 11, effective top-down regulation is, in turn, crucial for the health and balance of the entire ecosystem (Terborgh et al. 1999). Scott et al. (1999) put it this way, "While we may have some chance of preserving field mice if we provide for the needs of grizzly bears, we have virtually no chance of preserving bears if we plan only for mice." A conference convened to address the science of conservation at the continental scale produced the following recommendations (Terborgh and Soule 1999): "More land must be protected. Land use must be prescribed on very large spatial scales. Management practices must be reformed. Top carnivores must be restored in many places where they have been extirpated. Alien species must be combated as a matter of public policy. Disturbed and degraded habitats must be made more natural. Free run must be given to physical processes, such as wildfires and floods, that rejuvenate plant communities and shape the landscape."

North and Central America would be tied together through "MegaLinkages" (fig. 22.3). Each MegaLinkage would consist of "core protected areas connected to one another by 'wildlife linkages,' mosaics of public and private lands that provide safe passageways for wildlife to travel freely from place to place. Private land owners within proposed conservation planning areas are not bound in any way by our recommendations but are encouraged to participate in voluntary actions to protect landscape linkages and native species" (Wildlands Project 2007).

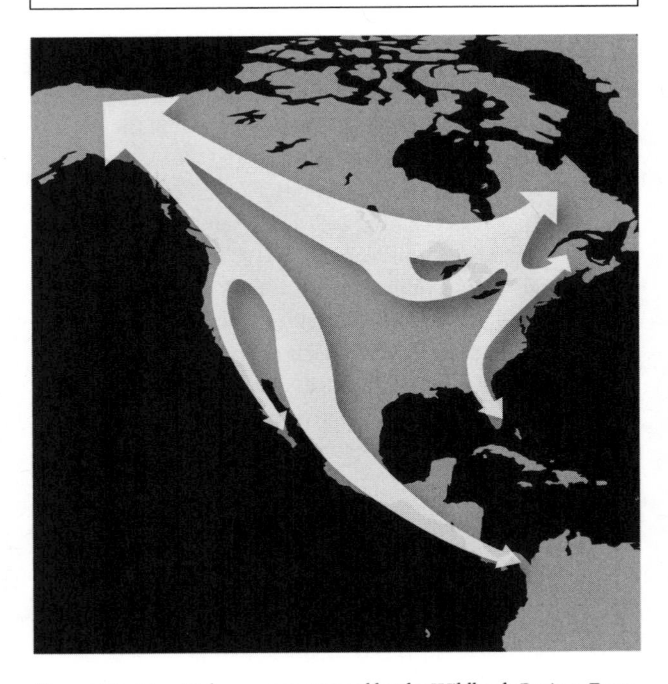

Figure 22.3. MegaLinkages as envisioned by the Wildlands Project. From west to east: Pacific, Spine of the Continent, Arctic-Boreal, and Atlantic. (Adapted from Wildlands Project 2007)

22.4.2 Population Genetics

Small, isolated populations are at risk genetically for two reasons. First, fewer individuals are exchanging fewer genes in small populations than in large populations, which leads to inbreeding that frequently results in loss of vigor and inability to resist stress (Lande 1988b; chapter 21 in this volume). Second, because a small population likely has fewer copies of a given allele than a large population, it tends to lose more alleles through strictly random processes (e.g., if the individual carrying the only copy of a given allele dies without successfully mating); hence, over time, small populations that are also isolated from other populations of the same species lose genetic variability. The genetic phenomenon underlying this loss of variability is called random drift, because gene frequencies within a population vary at least somewhat randomly from one generation to the next.

A key question regarding genetic drift (and one that bears directly on the issue of how habitats should be distributed across the landscape) is what constitutes isolation. The mathematics of population genetics predicts that genetic variability can be maintained by a very small amount of movement among subpopulations within a metapopulation. An old rule of thumb (based on many simplifying assumptions) holds that one migrant per generation into a subpopulation is sufficient to maintain genetic diversity and prevent inbreeding. This rule has been critically examined by Mills and Allendorf (1996), who argued that a minimum of 1 and maximum of 10 was a better rule, and by Wang (2004), who concluded that "in practice, it is difficult to provide a universally appropriate actual number of migrants per generation for the genetic management of fragmented populations."

22.4.3 Effective Population Size

In terms of population genetics, effective population size is equivalent to actual population size only under certain ideal breeding conditions that are rarely found in nature; such as equal numbers of males and females (all of which contribute equally to the gene pool) and no intergenerational fluctuation in population size (Franklin 1980). For most mammals and birds, effective population sizes are believed to be no more than 25 to 50 percent of actual population size (Noss R, personal communication). Social structure and breeding habits are important determinants of effective population size as it affects genetic integrity. For instance, animals that live in family groups may have only one mating pair per group. This is the case with many primates. Golden lion tamarins, for example, a rare species of monkey found in the remnant Atlantic rain forests of Brazil, live in groups of 8 to 12 individuals, but usually only one of several mated pairs within the group produces offspring. The effective population size for these animals is at least 7 to 10 times the actual population size. Whitmore (1975) estimates that 5,000 individuals would be required to maintain viable populations of hornbills and various species of monkeys native to Southeast Asia, which would require areas ranging from 167 to 10,000 ha, depending on species. On the other hand, tigers, which have quite a different social structure than primates or wolves, are believed to require a breeding pool of only about 400 individuals to maintain genetic integrity; however, this number of tigers needs roughly 40,000 ha of forest (Newman 1990).

In considering the genetic aspects of viable population sizes, several factors come into play with plants that are not encountered with higher animals. Being immobile, trees (and other plants) rely on some intermediary (either wind or animals) for outbreeding and seed dispersal. Wind pollination likely is relatively inefficient except in the immediate neighborhood of the contributing tree, while plants that are pollinated and whose seed are dispersed by animals depend totally on the health and welfare of the vector, including the area needed by the vector to maintain viable populations. In addition, trees (probably because they are long-lived) accumulate more genetic load (i.e., deleterious recessive alleles) than animals and, hence, are likely to require a relatively high level of outbreeding to prevent inbreeding depression (Ledig 1986; chapter 21 in this volume). Finally, management-related genetic selection for desirable growth traits has the potential to artificially narrow the gene pool of the entire species, both selected and wild trees, because selected individuals presumably contribute to the gene pool in proportion to their numbers (Ledig 1986).

22.4.4 Minimum Viable Populations

Theoretical considerations and empirical observations suggest that when one accounts for both genetic and demographic factors, the minimum population size required to maintain viable populations of animals and plants (i.e., **minimum viable population** [MVP]) is on the order of several thousands (Whitmore 1975; Soule 1987; Thomas 1990). For example, Reed et al. (2003) used population viability analyses to estimate MVPs for 102 vertebrate species, concluding that long-term persistence required habitat sufficient for 7,000 adults. Reed et al. (2003) also noted that because of population fluctuations, short-term studies significantly underestimate MVP.

However, no single number is universally applicable to all species, nor does the same number necessarily apply to any one species in all environmental situations (Gilpin and Soule 1986; Thomas 1990). Moreover, estimates of MVP are highly sensitive to assumptions about the degree to which a given species regulates its own populations. For example, Brook et al. (2006) used long-term data sets to model MVPs for 1,198 species (629 invertebrates, 529 vertebrates, 30 plants). They found values varying from a few hundred to tens of thousands depending on whether they assumed population growth was density dependent (i.e., some degree of endogenous regulation, resulting in lower MVP) or density independent (i.e., driven largely by environmental factors, resulting in higher MVP). In reality, populations are likely to be driven by some combination of density-dependent and density-independent factors, with the latter becoming stronger as external stresses become more pervasive (e.g., climate change, invasive species). Thomas (1990) cites several examples of the persistence (at least over several decades) of species reduced to far fewer than thousands of individuals (in some cases fewer than one hundred). "These examples," says Thomas (1990), "indicate that small populations should certainly not be abandoned as hopeless. It is often easier and less expensive, however, to ensure that a large population does not become small than to ensure that a small population does not become extinct."

The MVP concept expresses an ecological fact: populations lose viability and risk extinction when they fall below some threshold number of individuals; however, evidence is mixed regarding the utility of modeled MVPs as predictors of extinction risk. Using population viability analysis, O'Grady et al. (2004) identified population size and trend as the two most important factors influencing extinction risk in 45 bird and mammal species, thus supporting the usefulness of MVP. Among the species analyzed by Brook et al. (2006), the median MVP of species on the IUCN Red List was twice as high as that of species not on the Red List, thereby supporting the usefulness of MVP; however, there was also large variation around the medians, which resulted in low statistical correlation between the MVPs they calculated and extinction risk as determined by the IUCN. Regardless of the uncertainties, a management decision to reduce the numbers of a given species to or below some supposed minimum viable level carries considerable risk. If maintaining diversity is a management objective, as much habitat should be protected as possible.

The specifics of what constitutes suitable habitat will of course vary depending on species, but some generalities are possible:

- Suitable habitat for any one species encompasses all the other species and system processes on which the species of interest depends. In other words, to protect species it is necessary to protect the integrity of the systems within which they reside. It does not necessarily follow that to preserve one species we must protect all others that occupy the same ecosystem, but we must certainly protect some (and probably most) of them. As discussed in chapter 21, with few exceptions not enough is known about interconnections and interdependencies in nature to predict how loss of one species might ripple through the system to affect others. The safest course is to protect as much as we can.
- Considerably more area will be required to preserve many species than is likely to be set aside in even the largest reserves. If biological diversity is to be maintained, it must be done both in reserves and on managed lands (Lindenmayer and Franklin 2002; Ceballos et al. 2005). This is true not only for wide-ranging vertebrates but for some plants and invertebrates as well. According to Newman (1990), for instance, some tropical butterflies must range over areas of 100 km to collect the diversity of amino acids they require from different pollen sources. In the case of plant species that occur intermixed with other plant species at very low densities (as is the norm in many forest types), the several thousand individuals that may be required to maintain a viable population will be scattered over a vast area. In the lowland rain forests of Malaysia, for example, Whitmore (1975) found that the average density of mature individuals of 17 tree species varied from 33.0 to as low as 0.2 per 40 ha, depending on species. At these densities the areas required to preserve 5,000 mature individuals of a given species range from 6,000 to 1,000,000 ha! Although most common in the tropics, some higher-latitude trees also typically have few individuals per unit of area (e.g., western yew, a native of the Pacific Northwest that produces a compound used to treat some cancers).
- Regional distribution of habitat is a crucial factor. Attempting to maintain a species within a single large reserve is a risky strategy because a single event could wipe it out; however, small, isolated preserves are also highly vulnerable (section 22.5). The best course in many cases is to maintain populations that are separate from but also in contact within other populations within the metapopulation.

The plan for preserving the northern spotted owl is an example of the strategy of "separated but in contact" (Thomas et al. 1990; Murphy and Noon 1992). To give a brief and oversimplified version of this plan, single tracts large enough to maintain at least 20 pairs of owls are preserved; these are called **habitat conservation areas** (HCAs). Where no habitat this large remains, smaller HCAs are accepted. Individual HCAs that hold 20 or more pairs are separated by no more than 20 km (i.e., an average dispersal distance for young owls), while smaller HCAs are placed more closely to another HCA to better facilitate migration into them. Also to facilitate migration, at least 50 percent of the forest matrix outside HCAs is required to be in forest that provides acceptable foraging habitat for dispersing owls.

Changes in forest cover in the United States during the past 100 to 150 years have been influenced by two opposing dynamics. In some areas (especially the southeast and northeast), cover has increased due to forests reoccupying abandoned farmlands (Turner et al. 2003a; Fuller et al. 2004).[10] Overlain on this has been increasing fragmentation due to spreading urbanization since the midtwentieth century (Griffith et al. 2003; Radeloff et al. 2005). In their analysis of the southern Appalachians, Turner et al. (2003a) projected ongoing reforestation in less populated areas and declining forest cover in rapidly developing areas, a pattern likely to hold in other regions as well. As of the early twenty-first century, approximately 73 percent of all forests in the continental United States were in landscapes that were at least 60 percent forested (Riitters et al. 2003). However, 44 percent of forest locales were within 90 m of an edge (many of the edges stemming from widespread perforation with small openings [< 7 ha]). Riitters et al. (2003) conclude that "forests are connected over large regions, but . . . edge effects potentially influence ecological processes on most forest lands." The highest abundances of interior forest are in the southern Appalachians, New England, the Ozark Mountains, and the Pacific Northwest.

Road networks interrupt horizontal ecological flows, particularly hydrology (chapter 4) and the movement of some species (Forman and Alexander 1998). Roads may subdivide populations of some species, "with demographic and probably genetic consequences" (Forman and Alexander 1998). Roads also act as conduits for invasive species. For example, in northern hardwood forests of Wisconsin, exotic species occur primarily within 15 m of roads and only infrequently in the interior forest (Watkins et al. 2003). Roads also act as conduits for humans, a particular problem in intact tropical forests, where road building is quickly followed by settlement and clearing. Roads allow access for both legal and illegal hunting, which seems to have little impact on species with abundant habitat and high fecundity (deer are the classic example), but which creates problems for more vulnerable species. Studies from North America and Europe have shown that population sizes of bears, moose, wolves, and mountain lions are inversely proportional to road density (Brocke et al. 1989). For example, a 10-fold increase in road density in the Adirondack Mountains of New York is associated with a 10-fold decrease in the numbers of black bears (Brocke et al. 1989). The presence of wolf packs in the upper Midwest correlates closely with low road density, the best habitat being areas with road density less than 0.45 km/km^2 (Mladenoff et al. 1999).

Roads can significantly affect aquatic ecosystems. In their review, Trombulak and Frissell (2000) conclude that "not all species and ecosystems are equally affected by roads, but overall the presence of roads is highly correlated with changes in species composition, population sizes, and hydrologic and geomorphic processes that shape aquatic and riparian systems. More experimental research is needed to complement post-hoc correlative studies. Our review underscores the importance to conservation of avoiding construction of new roads in roadless or sparsely roaded areas and of removal or restoration of existing roads to benefit both terrestrial and aquatic biota."

22.5 LANDSCAPE PATTERNS: FRAGMENTATION, VARIEGATION, AND PERMEATION

> Current recommendations for biodiversity conservation focus on the need to conserve dynamic, multiscale ecological patterns and processes that sustain the full complement of biota and their supporting natural system.
>
> POIANI ET AL. 2000

> Thinking locally leads to small-minded decisions. . . . Managers need to consider a bigger picture.
>
> NOSS 2002

Landscape patterns profoundly influence biodiversity patterns (Gutzwiller 2002; Lindenmayer and Franklin 2002). Four issues have received particular attention: fragmentation, variegation, perforation, and permeation. In a **fragmented landscape,** large habitat areas are broken into smaller islands, something that has occurred widely over the past few decades in the tropics and parts of the temperate zone. Fragmented landscapes are basically binary, consisting of habitat islands within a sea of nonhabitat. A **variegated landscape** is characterized by a diversity of patch types representing gradients in habitat suitability rather than sharp boundaries. If fragmentation is the breaking of habitat, variegation might be considered the "bending" of habitat. Note that habitat for one species may be nonhabitat for another; hence, the same landscape may be either fragmented or variegated depending on the species in question. **Perforation** is the occurrence of relatively small openings within a forested matrix (a natural occurrence in many forest types and also the result of some human activities). **Permeation** refers to the incursion of roads, powerlines, or any linear "highway" allowing things such as pests, pathogens, exotic weeds, and humans and their pets easier access to the interior of large forest tracts. Permeation and anthropogenic perforation are usually the first steps in fragmentation.

[10]Forests regrowing on abandoned farms are likely to be depauperate in ancient forest herbs unless they are near surviving ancient forests to provide seed (Vellend 2003). Where they have survived, ancient forest remnants act as landscape-level legacies.

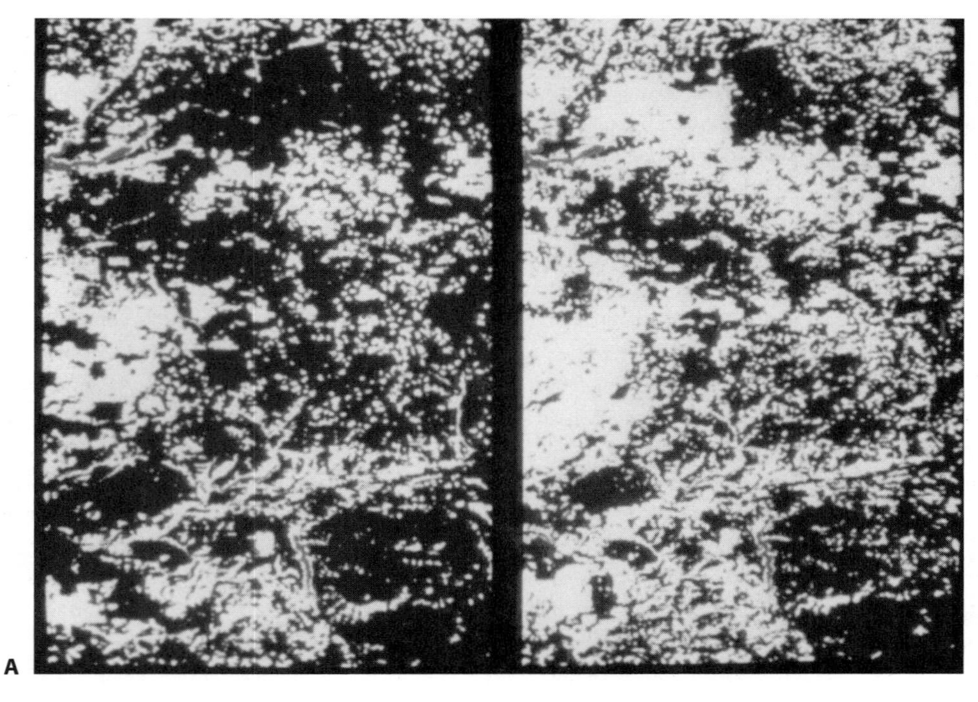

Figure 22.4. A. This remote image shows the change in the distribution of forest age classes over a 16-year period in a 3,000-km^2 area in the central Oregon Cascades. The image on the left was taken in 1972 and that on the right in 1988. Dark areas are closed-canopy forests, mostly old growth; light areas are open-canopied forests, mostly young plantations. Note the loss of total area of closed canopied forests and increasing fragmentation of what remains. B. A highly fragmented landscape in the Pacific Northwest. (A, adapted from Spies et al. 1992; B, courtesy of Warren Cohen)

In Europe, eastern North America, and the tropics, fragmentation results largely from the widespread conversion of forests to farms, ranches, or urban sprawl. Where forestry is the dominant use, the degree of fragmentation depends on the stand-level silvicultural approach (e.g., clear-cutting, retention harvests, partial cutting), the spatial pattern of harvest, and cutting rates. Interspersing small clear-cuts within a matrix of older forests create a landscape pattern that is much like taking bites from the middle of a pancake instead of the edges. At first, the bites exist as isolated islands surrounded by pancake. At some point, however, the situation is reversed, and it is the remaining pieces of pancake that are isolated islands (fig. 22.4). Furthermore, dispersed cutting requires that an extensive road system be

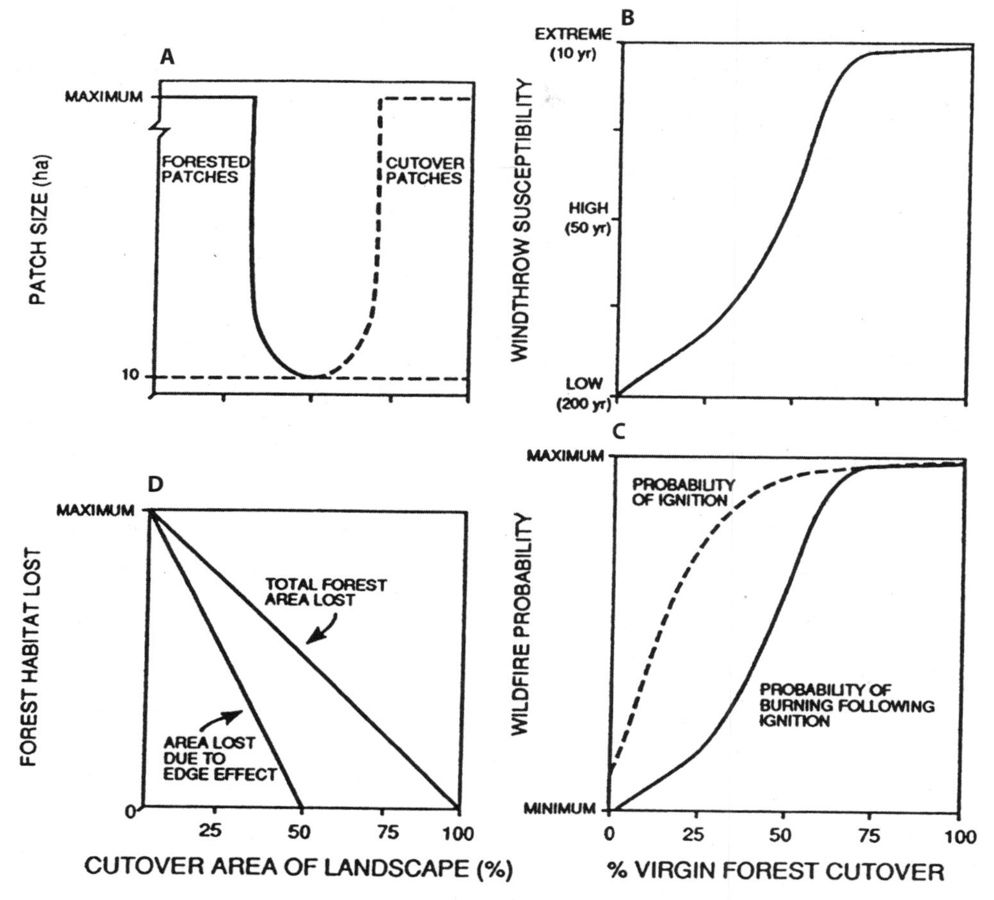

Figure 22.5. Dispersed patch clear-cutting leads to rapid fragmentation of forests (A) and increased susceptibility to catastrophic events such as windthrow (B) and wildfire (C). Fragmentation also reduces interior habitat for forest species: the "edge effect" curve (D). The edge effect is the effective loss of interior habitat, because proximity to an edge changes microclimate and increases vulnerability to predation and other outside disturbances not common to forest interiors (e.g., Chen et al. 1992, 1993). Note that in the model shown in D, no interior habitat remains when 50 percent of the landscape has been patch cut (i.e., as in Figure 22.3B). This pattern is highly dependent on landscape scale, which determines the size of forest fragments, and the degree to which edge effects extend into the forest (a rule of thumb for the latter is two to three tree heights). The curve depicting the probability of burning after ignition assumes that cutover areas and young plantations are more flammable than intact forest, an assumption likely to hold for many (if not most) types of forest but that depends on the vegetative community and amount of dead fuels in cutover areas. (Adapted from Franklin and Forman 1987)

established early on; hence, blocks of intact forest are both perforated and permeated. If the frequency of reentry exceeds the rate of regrowth on clear-cuts, the perforations inexorably lead to fragmentation (e.g., fig. 22.4). The transition from isolated cutover patches to isolated forest islands occurs very rapidly when more than 30 percent of a given landscape has been cut, with consequences for both habitat quality within remaining patches and their susceptibility to disturbance (Franklin and Forman 1987) (fig. 22.5). If cuts are dispersed sufficiently in time, a variegated landscape may result rather than fragmented one. In either case, however, the extensive road system alters a variety of ecological processes.

There is a huge literature on fragmentation and, more generally, the influence of landscape patterns on biodiversity (reviews by Debinski and Holt 2000; Lindenmayer and Franklin 2002; Noss 2002; Fahrig 2003; Kupfer et al. 2004). But in general, we can say that fragmentation influences biodiversity in four ways (fig. 22.6):

1. **Sample effects** refer to the chance that remnant fragments do not contain the full array of species present on the landscape prior to fragmentation (a particular risk where beta diversity is high).
2. **Area effects** refer to how the size of remnants effects their ability to provide suitable habitat for a given species.
3. **Isolation effects** refer to the proximity of remnants to one another and the harshness of the intervening matrix.
4. **Edge effects** have to do with the propagation of physical and biotic influences across edges, a function of the

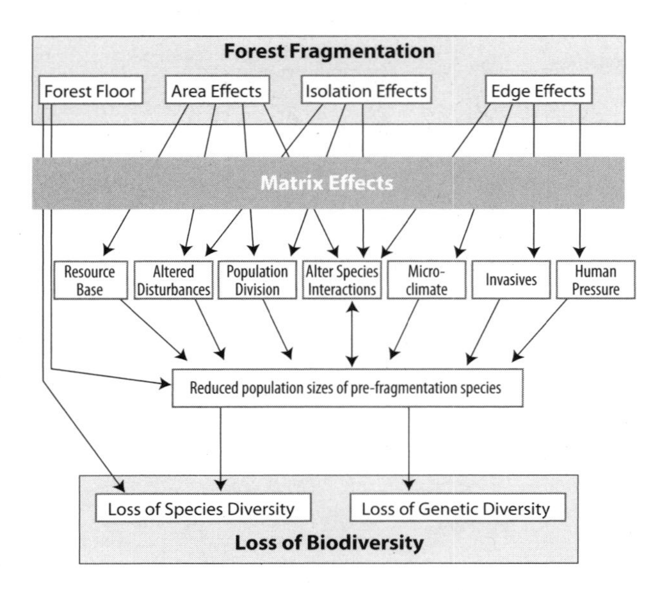

Figure 22.6. The effects of fragmentation. (Kupfer et al. 2004, as modified from Lindenmayer and Franklin 2002)

size and shape of remnant patches as well as the contrast between ecosystem characteristics on either side of the edge (Sisk and Haddad 2002). Recently, landscape ecologists have recognized the existence of edge effects at the landscape scale, which is to say that the aggregate of fragmentation, perforation, and permeation within a region have cumulative effects that alter landscape processes such as climate, fire, and trophic dynamics (Sih et al. 2000; Cochrane 2001; Cochrane and Laurance 2002).

The matrix acts as a filter modifying (to one degree or another) the strength of all effects (Lindenmayer and Franklin 2002).

Early thinking on fragmentation was largely shaped by two things: the theory of island biogeography (MacArthur and Wilson 1963) and Forman's (1995) "corridor-patch-matrix" model. In both, patches of habitat are seen as embedded within a hostile (nonhabitat) matrix, a view that is certainly true in some cases (e.g., where the matrix consists of farms or suburbs), but not all. Over time, a more complex view has emerged that recognizes a variety of landscape patterns are possible, many of which differ in their implications for biological diversity (McIntyre and Hobbs 1999; Lindenmayer and Franklin 2002). Kupfer et al. (2004) discuss this issue:

> Studies of mosaic landscapes containing a range of old-growth forest, successional habitats, and agriculture (e.g., shifting cultivation systems) . . . exemplify how landscapes can represent a range of conditions from deforestation to varying degrees of forest degradation in otherwise "intact" forest. Further, it is important to recognize that the dynamics of a forest fragmented by agriculture or urban development may be vastly different from those associated with the mosaic of mature and regenerating stands resulting from timber harvesting. The first situation represents cases where the modified habitat may be indefinitely affected and have little or no habitat value for forest species whereas the latter situation may involve only a temporary reduction in habitat for species that rely on mature forests.

Research on fragmentation falls into three broad categories: (1) studies of real landscapes in which the matrix is dominated by intensive human use (i.e., agriculture, suburbanization); (2) studies of real landscapes in which the matrix contains a significant amount of managed forests (these might be better characterized as variegation rather than fragmentation studies); and (3) experiments in which fragmentation is produced through designed manipulations (in forested landscapes, experimental fragments are usually embedded within clear-cuts).

22.5.1 Fragmentation experiments

At the turn of the twenty-first century there were 20 fragmentation experiments scattered around the globe, 6 of these in forests. Debinski and Holt (2000) summarize the experimental results (as of the mid-1990s)[11]: "Our comparisons showed a remarkable lack of consistency in results across studies, especially with regard to species abundance relative to fragment size. Arthropods showed the best fit with theoretical expectations of greater species richness on larger fragments. Highly mobile taxa such as birds and mammals, early-successional plant species, long-lived species, and generalist predators did not respond in the 'expected' manner. Reasons for these discrepancies included edge effects, competitive release in the habitat fragments, and the spatial scale of the experiments."

Island biogeography predicts total species richness declines as island size shrinks. However, fragmentation of terrestrial ecosystems does not produce an island surrounded by water but an island surrounded by other terrestrial habitats with their own set of characteristic species. Species composition may shift with fragmentation, especially at edges, while total species richness is unaffected or may even increase (e.g., Schmieglow et al. 1997). Henle et al. (2004) reviewed the literature on species sensitivity to fragmentation and identified five traits correlated with sensitivity: (1) population size and its converse, rarity; (2) population fluctuations and storage effects[12]; (3) traits associated with competitive ability and disturbance sensitivity in plants; (4) habitat specialization and matrix use; and (5) "relative biogeographic position." or location within a species' geographic range (species at the edge of their ranges are often considered to be more vulnerable to any stress). Others have

[11] McGarigal and Cushman (2002) compare different experimental approaches.

[12] Storage is basically a legacy effect and refers to the ability of an organism to "store" its presence on a site, especially persistence in some form through disturbance (e.g., clonal life-forms in plants).

Figure 22.7. Experimentally connected and isolated early successional longleaf pine patches within a matrix of pine plantations in South Carolina. (From Damschen et al. 2006. Reprinted with permission from AAAS)

found that area requirements, dispersal ability, and sensitivity to habitat degradation at the patch scale are important predictors of sensitivity to fragmentation (Vos et al. 2001; Lens et al. 2002). Lindenmayer and Franklin (2002) and Lens et al. (2002) stress that one must attend to both landscape patterns and quality of habitat in patches.

> There has been considerable debate about the efficacy of corridors (Beier and Noss 1998; Lindenmayer and Franklin 2002). Beier and Noss (1998) concluded that "the evidence from well designed studies suggests that corridors are valuable conservation tools." One of the most rigorous of the various corridor experiments showed that early successional longleaf pine patches within a matrix of densely stocked pine plantations in South Carolina had significantly greater plant richness when linked by corridors than when isolated (Damschen et al. 2006), in part at least because bluebirds dispersed seeds through the corridors (Levey et al. 2005) (fig. 22.7). However, the value of corridors does not necessarily extend to all species in all situations. For example, resident birds in Canada apparently benefit from corridors, while nonresidents do not (Hannon and Schmiegelow 2002). The northern spotted owl disperses randomly and benefits from matrix conditions that allow dispersal as opposed to corridors (Lindenmayer and Franklin 2002). In some cases "stepping stones" dispersed throughout the matrix facilitate species movement. In Australia, patches of native eucalyptus facilitate movement of native birds through exotic radiata pine plantations (Lindenmayer and Franklin 2002). In conifer-dominated forests of the Pacific Northwest, hardwoods left within harvested areas provide stepping stones for some old-growth-associated lichens (chapter 10).

Debinski and Holt went on to point out two significant lessons from the experiments: "One of the more consistently supported hypotheses was that movement and species richness are positively affected by corridors and connectivity. [Moreover] the three long-term studies revealed strong patterns that would have been missed in short-term investigations."[13]

22.5.2 The Biological Dynamics of Forest Fragments Project

The longest-running fragmentation study in forests is the Biological Dynamics of Forest Fragments Project (BDFFP) in central Amazonia where, in contrast to some other studies, species richness is significantly reduced even in large fragments (fig. 22.8). Laurance et al. (2002) discusses the 22-year results[14]:

> Many large mammals, primates, understory birds, and even certain beetle, ant, bee, termite, and butterfly species are highly sensitive to fragment area. A number of these species have disappeared from even the largest (100 ha) fragments in the study area. . . . In contrast, a few taxa have remained stable or even increased in species richness after fragment isolation. Frog richness increased because of an apparent resilience of most rain forest frogs to area and edge effects and an influx of non-rain-forest species from the surrounding matrix. Butterfly richness also rose after fragment isolation, largely from an invasion of generalist matrix species at the expense of forest-interior butterflies. Small-mammal richness has not declined in the BDFFP fragments,

[13]Short-term responses may include transitory effects arising from biotic adjustments to new conditions. For example, deep forest species may respond initially to habitat loss by packing into fragments. However, numbers will eventually shrink to what the reduced habitat can support.

[14]We have omitted original references from the quote. See Laurance et al. (2002) for sources.

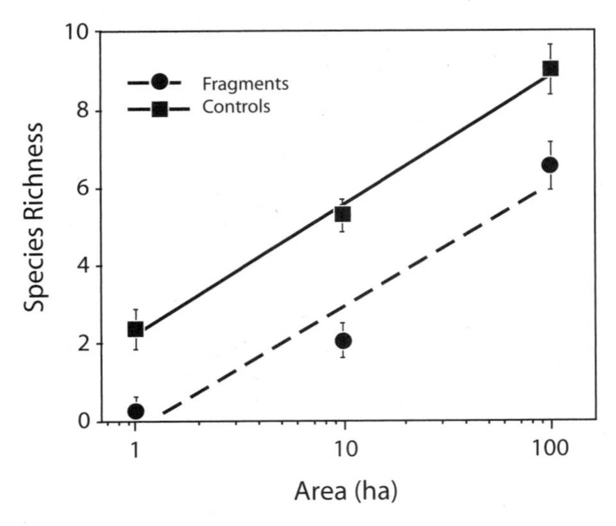

Figure 22.8. Species-area relationships for nine species of terrestrial insectivorous birds (mean + SE) in the Biological Dynamics of Forest Fragments Project study area. (Laurance et al. 2002, after Stratford and Stouffer 1999. Reprinted with permission of Blackwell Publishing)

because most species readily use edge and regrowth habitats. Collectively, BDFFP results reveal that the responses of different species and taxonomic groups to fragmentation are highly individualistic and suggest that species with small area needs which tolerate matrix and edge habitats are the least vulnerable.

Edge effects varied widely at BDFFP depending on the process in question, ranging from more than 300 m for increased wind speed and elevated tree mortality to only a few meters for reduced density of fungal fruiting bodies and invasion by disturbance-adapted plants (fig. 22.9).

As elsewhere, the nature of the matrix strongly influenced the ecology of BDFFP fragments. Quoting again from Laurance et al. (2002):

> Fragments surrounded by regrowth forest 5–10 m tall experienced less-intensive changes in microclimate and had lower edge-related tree mortality than did similar fragments adjoined by cattle pastures. Edge avoidance by mixed-species bird flocks was also reduced when

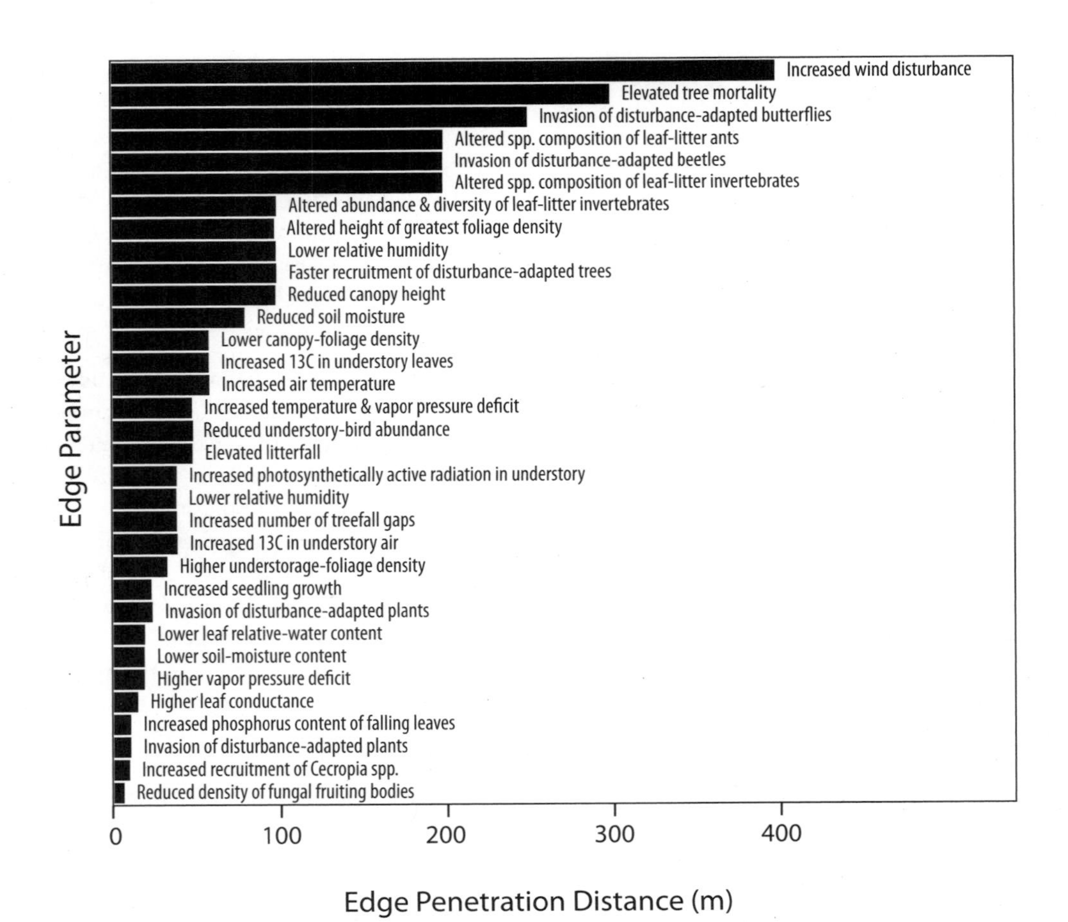

Figure 22.9. Penetration distances of different edge effects into the forest remnants of the Biological Dynamics of Forest Fragments Project. (Adapted from Laurance et al. 2002)

fragments were surrounded by regrowth rather than cattle pastures. Of even more significance is that the matrix influences fragment connectivity. Several species of primates, antbirds, obligate flocking birds, and euglossine bees that disappeared soon after fragment isolation recolonized fragments when regrowth regenerated in the surrounding landscape.

22.5.3 Empirical Studies

Nonexperimental studies of species responses to landscape patterns sacrifice precision and control for what may be a more accurate (but perhaps more difficult to interpret) picture of what plays out in "real" (i.e., nonexperimental) landscapes. The most consistent pattern emerging from such studies is that habitat loss has strong negative effects, but habitat fragmentation per se (i.e., the spatial arrangement of remaining habitat) has much weaker effects that may be either positive or negative depending on species and circumstances (Rochelle et al. 1999; Schmiegelow and Monkkonen 2002; Fahrig 2003).

As we have emphasized before, the nature of the matrix is crucial (Lindenmayer and Franklin 2002; Murphy and Lovett-Doust 2004). Songbird numbers are declining in forest fragments within human-dominated landscapes (farms, urban areas) in both North America and Europe (Wilcove 1985, 1989; Freemark and Merriam 1986; Small and Hunter 1988; Freemark and Collins 1992; Mortberg 2001). Declines appear to be related to the easier access to forest interiors that fragmentation and permeation provide to predators such as foxes, raccoons, corvids (i.e., crows, jays, ravens), domestic cats and dogs (e.g., Huhta et al. 2004), and nest parasites such as brown-headed cowbirds.[15] In a survey of breeding bird populations in deciduous forests of the eastern United States, Whitcomb et al. (1981) found that neotropical migrant birds accounted for 80 to 90 percent of breeding individuals in large forest tracts but less than one-half of breeding individuals in small tracts. They concluded that intact forests hundreds and perhaps thousands of hectares in size are required to maintain migrant bird communities.

The situation is quite different (at least for many mammals and birds) when the matrix comprises a heterogeneous and dynamic mixture of successional stages (and perhaps forest types) rather than a binary pattern of forest or nonforest. Several general patterns have emerged for birds (by far the most studied group) in diverse forested landscapes (Kremsater and Bunnell 1999; Rochelle et al. 1999; Schmiegelow and Monkkonen 2002; Fahrig 2003; Loehle et al. 2005a; Mitchell et al. 2006a; chapter 10 in this volume):

- Landscape patterns are significant but often account for less than 50 percent of total variability for a given species or guild.
- Total available habitat is more important than how the habitat is arranged.
- Many species and guilds see the landscape as variegated rather than fragmented (this is also true of plants [Murphy and Lovett-Doust 2004]).
- Most guilds, even those associated primarily with mature forests, respond positively to heterogeneity of stand ages and types.
- Species and guilds differ in the landscape scale to which they respond.
- Nest parasites and predators are less of a problem in a managed forest matrix than in an agricultural one.

Forestry produces a much more dynamic landscape than other intensive human land uses; farms are permanent (or semipermanent) clearings, harvested forests regrow, if not to old-growth at least to full forest cover. (Exotic plantations amidst native forests are a special situation). Moreover, leaving legacy structures in harvested stands (e.g., logs, snags, living trees, remnant patches of natural vegetation) can make the matrix much more hospitable for many species (Rochelle et al. 1999; Grove 2002; Lindenmayer and Franklin 2002;). To give a single example, habitat continuity in time is important for saproxylic insects, which depend on dead wood and large, old trees (Grove 2002). Saproxylic insects comprise one of the most species-rich groups in forests (saproxylic beetles alone include at least twice as many species as terrestrial vertebrates, and the beetles are only one of several saproxylic taxa) (Grove 2002). These insects function in cycling and serve as food for species such as the pileated woodpecker (which, in turn, is a keystone excavator of cavities used by many other species).

The fact that the managed matrix can be made more hospitable for many species does not imply that it can meet the needs of all old-forest specialists. For example, nearly one-third of all birds breeding in older boreal forests of Canada and Finland are old-forest specialists, and a substantial proportion of these are year-round residents with inherently low population densities (because of food limitations and harsh weather) (Schmiegelow and Monkkonen 2002). The combination of low densities and specialized needs are likely to make that group especially sensitive to habitat loss (Schmiegelow and Monkkonen 2002).

Birds probably don't tell a lot about less mobile organisms. Poor dispersers with specialized needs, such as flightless beetles, many amphibians, some lichens and fungi, and vernal herbs are of particular concern. The bird studies themselves have various limitations. Most are based on point counts (detections at a series of points), which may or may not correlate with fitness in a given habitat. Moreover,

[15] Brown-headed cowbirds lay their eggs in the nests of other birds. The owners of a parasitized nest obligingly feed any mouth that is opened to them, which more often than not is that of the aggressive cowbird chicks rather than their own brood. Airola (1986) found that parasitized nests contained 76% fewer host young than unparasitized nests. In Sweden, forest fragments within a matrix of agricultural lands support higher densities of hooded crows, a habitat generalist that preys on eggs and chicks of other bird species (Andren 1992).

A special type of fragmentation involves islands of native forest within exotic tree plantations (fig. 22.10). In a study of native forest islands within exotic pine plantations in Australia, 77 percent of birds found in the native eucalypt fragments were also present in the plantations (Tubelis et al. 2004); however, bird species richness in plantations dropped sharply as one moved from the native fragment edge into a plantation (fig. 22.11), which Tubelis et al. refer to as a **halo effect.** Older pine forests supported greater bird richness than younger pine forests, because older plantations have various features that might make them more attractive to native birds. Their height is similar to the native forest; they have a greater prevalence of native plant undergrowth than young plantations; their microclimate is more similar to the native forest; and the bark of older trees is deeper than that of younger and likely supports more insects for bark gleaners (Tubelis et al. 2004). Also, regardless of age, plantations adjacent to narrow native fragments had greater richness than plantations adjacent to wide native fragments, suggesting that narrow fragments are less suitable than broad fragments, and birds therefore make greater use of plantations. In the Maule region of central Chile, remaining native forests (*Nothofagus* sp.) occupy less than 10 percent of the landscape and exist as islands within a matrix of exotic pine plantations (Estades and Temple 1999). There is a halo effect for cavity-nesting birds (i.e., they are more common in pine plantations near native forests). However, the abundance of open-nesting species in pine plantations was more related to the presence of native vegetation in the pine understories (Estades and Temple 1999). A similar pattern occurs in Hawaii for some rare native insects, which occur on native plants in the understories of exotic plantations.

The studies cited above dealt with plantations established within native temperate forests. What about tropical ecosystems? Kanowski et al. (2005) used literature sources to assess the biodiversity consequences of various types of plantations in cleared rain forests of tropical and subtropical Australia. They concluded that "plantations of eucalypts and exotic pines . . . [have] little or no intrinsic value in rain-forest landscapes, provide poor quality habitat for rain-forest biota, and (particularly eucalypts) are characterized by a relatively open canopy which . . . favors the recruitment of weeds." In contrast, plantations consisting of rain-forest trees had moderately positive effects on biodiversity.

of necessity, studies deal with existing landscape patterns, which in many areas are so highly altered that sensitive species may have already disappeared or be too rare to detect in point counts. Such studies should be extrapolated with extreme care into more pristine regions. Additionally, Stratford and Robinson (2005) caution against extrapolating the results of temperate bird studies to the tropics.

Studies at the patch scale have found that amphibians tend to avoid recent clear-cuts and even selectively harvested stands, probably because of relatively dry microclimates (Martin and McComb 2003 and references cited therein), while many reptiles tend to favor more open conditions (Loehle et al. 2005b). However, as with most other broad taxonomic groups, herptofauna are highly individualistic in their habitat needs, leading various researchers to conclude that herptofaunal richness benefits from landscape patchiness (Martin and McComb 2003; Loehle et al. 2005b). Based on their studies in the Ozark Mountains, Renken et al. (2003) concluded that forest management was compatible with maintaining herptofaunal diversity, providing "relatively small amounts of the landscape are disturbed." In contrast. Loehle et al. (2005b) found that herptofaunal richness in the Ouachita Mountains of Arkansas was greater in managed than in unmanaged watersheds, a finding that probably reflects the high proportion of reptiles in their samples. Details such as the location of different patches may be crucial, especially for some species. For example, some amphibian species live in forests as adults and in streams as larvae. Without wide riparian zones or adequate corridors, these species must cross hostile territory to complete their life cycle and, as a result, they may account for the bulk of worldwide amphibian declines (Becker 2007). All in all, the results on herptofauna parallel those with birds: for conservation purposes, attention to species at risk is more appropriate than a focus on richness, although a proper balance of habitat types can benefit both.

One must be careful when drawing conclusions about the importance of total habitat loss as opposed to the spatial patterning of that loss, as responses are not necessarily linear. Theory suggests that as total habitat is reduced below 20 to 30 percent of the landscape, the effects of fragmentation per se (i.e., independently of habitat loss) should become more apparent. There are at least three reasons why that might be so:

Figure 22.10. In the background, native *Nothofagus* forest surrounded by exotic radiata pine plantations in Chile. Chile no longer permits native forest to be converted to exotic plantations.

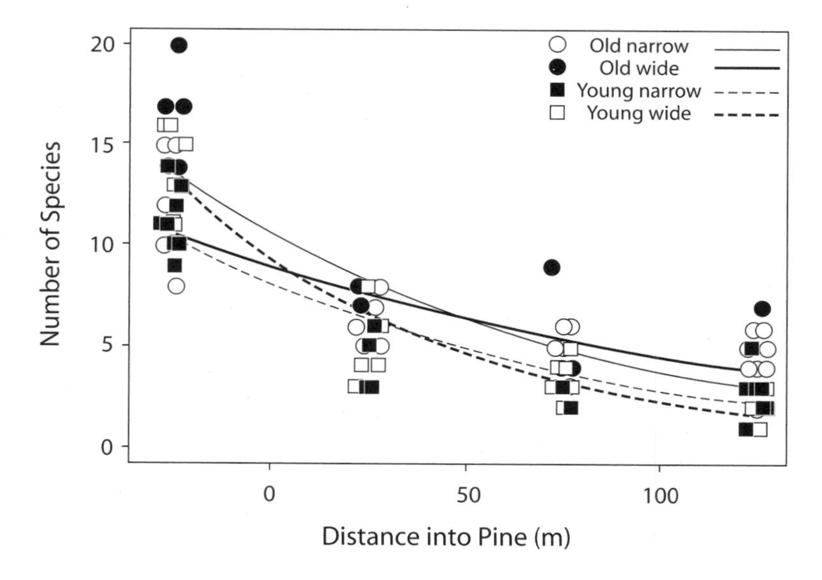

Figure 22.11. Mean bird species richness recorded per site in bands of distance from boundary lines for four boundary treatments at Tumut, New South Wales, Australia. The treatments were old/wide (old pine forest with wide eucalypt forest), old/narrow (old pine forest with narrow eucalypt forest), young/wide (young pine forest with wide eucalypt forest), and young/narrow (young pine forest with narrow eucalypt forest). From top to bottom on the right-hand side of the graph, the curves are old narrow, old wide, young narrow, and young wide. The decrease of richness with distance into pine is more rapid in young than in old pine, and for both ages more rapid when the adjacent eucalypt boundary is wide than when it is narrow. (Tubelis et al. 2004. Reprinted with permission of Blackwell Publishing)

1. Edge effects (i.e., micrometeorological and biotic changes propagating inward from edges) reduce the amount of interior habitat (e.g., a 10-ha remnant forest does not have 10 ha of interior habitat) (fig. 22.5D).
2. Cumulative landscape-level edge effects alter regional processes (e.g., climate and disturbance), thereby further altering habitat quality and susceptibility to disturbance.
3. Habitat patches become increasingly isolated from one another as the total habitat decreases. For dispersal-limited species, this increases the probability that at any given time the available habitat will not be fully occupied simply because it is too difficult to get there (e.g., Wiser et al. 1998).

Neutral landscape models have been most commonly used to explore the existence of thresholds with regard to the movement of organisms across landscapes (With 1997, 2002). Neutral landscape models are **percolation models** in which a two-dimensional surface is grided into cells and each cell assigned a value (e.g., occupied or unoccupied). The value taken by an individual cell depends, according to certain rules (which can vary), on the values of its neighbors. Models search for the number and arrangement of occupied cells that allow "percolation." in other words, connectance across the grid. In landscape research, these are called *neutral models* because they play out on a "neutral" two-dimensional surface, which means that real-world factors such as topography do not come into play. Thresholds are given as the proportion of a model landscape in suitable habitat; above the threshold, habitat is sufficiently connected for movement, below the threshold it is not. Depending on model assumptions, neutral landscape models predict movement thresholds ranging from 0.45 to 0.62 for organisms with limited dispersal, to 0.07 to 0.15 for organisms with good dispersal capabilities (With 2002). With (1997) gives several examples of the use of neutral landscape models in real-world management situations.

It is important to note that measures of total biodiversity may mask changes in relative abundance. Fragmentation alters the relative proportions of habitats and creates edges, therefore altering relative species abundances more than diversity per se. On the other hand, responses to edges are more complex than the old idea of "edge or interior species." Species respond to edges in a highly individualistic way that defies easy generalization (Sisk and Haddad 2002). Some deep-forest species may actually benefit from factors such as increased food production (e.g., insects, fruit) at edges (Schmiegelow and Monkkonen 2002).

Simberloff (1988) points out that in contrast to decreased habitat, the separation of habitat parcels may have both negative and positive effects on interior species. On the positive side, separation could (at least in theory) reduce the spread of damaging agents such as contagious diseases and introduced predators. On the negative side, separation could also lead to genetic isolation and therefore inbreeding and drift, although as we saw earlier, theory predicts that relatively little contact is required among populations to prevent inbreeding. Isolation may also increase genetic differentiation among populations, with uncertain effects on viability of the species as a whole. All of these arguments are based on models and theory that have little or no experimental verification, and they should be interpreted in that light.

Finally, we pass on some good advice from Schmiegelow and Monkkonen (2002): "System and species-specific considerations are important when assessing the potential outcome of habitat loss and fragmentation on regional biota. Indiscriminant application of conservation paradigms may lead to misguided research efforts and poor management guidelines."

22.6 SUMMARY

Populations (or metapopulations) that drop below some critical size become vulnerable to extinction because of both demographic and genetic factors, the latter including inbreeding depression and loss of genetic diversity. The size of populations (or metapopulations) necessary to maintain viability varies widely, depending on species and environment, but is on the order of several thousands of individuals for most species. Preserving adequate numbers of individuals for a given species requires that sufficient suitable habitat be maintained, including all components of the ecosystem on which a given species either directly or indirectly depends. Suitable habitat must be at the scale of landscapes and (for some species) regions, hemispheres, and the globe. Suitable habitats must be sufficiently interconnected across landscapes and regions so that individuals within one population can migrate to others within the metapopulation; interconnectance also provides pathways of migration in response to global warming. Landscapes may be fragmented, with forest habitats existing as islands within a sea of nonhabitat, or variegated, a more complex pattern of habitats of varying quality. In general, forest fragments and the species within them are more vulnerable to various biotic and abiotic disturbances than is the case with large forest tracts. However, there is considerable difference ecologically between forest fragments within a matrix of farms or suburbs and forested landscapes with diverse age and tree species.

For many species, more habitat will be required than is likely to be set aside, even in very large reserves. Preserving biological diversity will require maintaining suitable habitat on a significant proportion of managed forest lands. Ultimately, the welfare of all of earth's species (including humans) depends on the health of the planetary ecosystem.

23

The Future

> Waiting for the future is like waiting for a bus that never comes. Unless we start doing what we feel is important, we will get nowhere.
>
> TARTHANG TUILKU

> There is one outstandingly important fact regarding Spaceship Earth, and that is that no instruction book came with it.
>
> BUCKMINSTER FULLER

The issues facing the world in the twenty-first century are unique in human history: overpopulation, pollution, loss of biological diversity, climate change. It becomes increasingly clear that the earth and its residents compose a single ecosystem, a complex web of interaction and interdependence that is being seriously perturbed. Not surprisingly, the fate of the earth's forests has become a key issue. Forests harbor immense biological diversity and play a unique role within the global ecosystem as storehouses of carbon, recyclers of water, and protectors of watersheds. But more than this, humans have venerated forests as places of great spiritual power throughout history.

Despite the reverence in which they have been held (and the folk wisdom concerning their ecological role), forests have been among the first casualties as civilization spread over the past several thousand years. (chapter 1) Today, of course, not all logging results in long-term deforestation in the sense that forest is converted to nonforest, and there is a growing movement toward incorporating ecological values into managed forest landscapes (e.g., Lindenmayer and Franklin 2002). But the ancient problem of extraction without stewardship persists, especially in the tropics where it is exacerbated by forest clearing for subsistence farms, corporate farms (e.g., oil palm plantations), and ranching.

The earth has never seen human populations like we have today, and humans have never seen climate changes of the speed and magnitude that we are likely to experience in the coming years. However, one basic issue has not changed through the centuries: wood as a valuable commodity versus forests as repositories of numerous other tangible and intangible values. The vital importance of forests to the health of the biosphere is abundantly clear, but demand for wood continues to grow. The challenge before us is to find a sustainable balance.

This final chapter explores some of the more important issues that face both forests and forestry. First, we briefly discuss the threats posed to forests by climate change, and steps that can be taken to mitigate the effects of a changing environment and to ease the transition from one forest type to another. Then, we discuss strategies for maintaining biological diversity within managed forests and forested landscapes. These issues are of course interrelated, and both involve bringing the ecological relationships and dynamics that have been the subject of the book to bear on an applied problem: managing ecosystems rather than simply trees.

23.1 THE IMPLICATIONS OF GLOBAL WARMING

The earth is warming; there is no longer any doubt about that (IPCC 2007). Hansen et al. (2006) conclude that the globe "is as warm now as at the Holocene[1] maximum and within ~1°C of the maximum temperature of the past million years. We conclude that global warming of more than ~1°C, relative to 2000, will constitute 'dangerous' climate change as judged from likely effects on sea level and extermination of species."

Houghton (2005) sums up the current situation as follows: "Many of the likely characteristics of the resulting changes in climate (such as more frequent heat waves, increases in rainfall, increase in frequency and intensity of many extreme climate events) can be identified. Substantial uncertainties remain in knowledge of some of the feedbacks within the climate system (that affect the overall magnitude of change) and in much of the detail of likely regional change. Because of its negative impacts on human communities (including for instance substantial sea-level rise) and on ecosystems, global warming is the most important environmental problem the world faces."

Scholze et al. (2006) performed the first quantitative risk analysis of climate change effects. Among their findings are "high risk of forest loss . . . for Eurasia, eastern China, Canada, Central America, and Amazonia, with forest extensions into the Arctic and semiarid savannas; more frequent wildfire in Amazonia, the far north, and many semiarid regions; more runoff north of 50°N and in tropical Africa and northwestern South America; and less runoff in West Africa, Central America, southern Europe, and the eastern U.S."

As we have seen in previous chapters, nature is already responding. Broad changes are occurring in the earth's biota, including shifts in range, phenology, and other behaviors related to warming temperatures (Parmesan and Yohe 2003; Root et al. 2003; Jonzen et al. 2006). At least some of these shifts have a genetic basis, that is, species are evolving in response to changing climate (Balanyá et al. 2006; Bradshaw and Holzapfel 2006). More frequent, severe wildfires in western North America have been linked to global warming (Westerling et al. 2006), as has the large-scale dieback of overstory trees in the southwestern United States (Breshears et al. 2005), the increased destructiveness of tropical cyclones since the mid-1970s (Emanuel 2005; Webster et al. 2005), and the unprecedented levels of bark beetle infestations in North America. Anomalous drought in Amazonia has also been linked to warming oceans (Giles 2006), a phenomenon predicted by the Hadley Centre climate carbon cycle model and some other global change models. Cox et al. (2004) discuss results from the Hadley model:

> Amazonian temperature rises by more than 9 K over the 21st century, and rainfall drops by an alarming 64%. Together these changes lead to a 78% loss in vegetation carbon and a 72% loss in soil carbon. When the forest fraction begins to drop (from about 2040 onwards) C4 grasses initially expand to occupy some of the vacant lands. However, the relentless warming and drying make conditions unfavourable even for this plant functional type, and the Amazon ends as predominantly bare soil. As with the timing of dieback, the details of this simulation of the land-cover change should be treated with caution since they depend on known limitations and uncertainties in the . . . vegetation model. Despite these model deficiencies, it seems clear that the climate change in Amazonia (predicted by the Hadley model) would lead to rainforest loss (perhaps via increased fire frequency), and therefore drastic land-cover change.

There has been much recent discussion of **tipping points,** thresholds in which positive feedbacks within the internal system dynamics take over and "start to propel a change previously driven by external forces" (Walker 2006). One example is the ice-albedo feedback, in which surfaces exposed by melting ice (i.e., soil, water) have lower albedo than the ice and, therefore, absorb yet more heat to melt yet more ice. Another goes like this: more warmth and drought equals more forest fires, which pulse carbon into the atmosphere and, therefore, lead to yet more warmth (e.g., the previous discussion of Amazonian drying). Tipping points can lead to abrupt and significant climate

[1]The Holocene comprises the period since the end of the last glaciation, or approximately 10,000 years ago to the present.

changes. These have occurred in the past and may well occur again, severely challenging the ability of earth's biota to adapt.

Chapters 4 and 5 reviewed how vegetation zones might shift with warming, chapters 7 and 19 dealt with potential and current effects on disturbances, and chapter 15 discussed the possible effects on primary productivity. We are now in a position to identify the important questions regarding ecosystem stability should climates change as predicted. Will one community type be smoothly replaced by another? What are the chances that ecosystems will become stressed and destabilized if the predicted changes occur? How can resource managers mitigate the possible negative effects and facilitate a smooth transition from one type to another? We begin our consideration of these questions with what is perhaps the most critical component of ecosystem stability: biotic regulation as expressed in strong links between plants and soils and controls over disturbances of various kinds.

Whatever the future may hold, one indisputable fact is that the future is now, always has been, and always will be uncertain. This in itself has much to say about how we manage ecosystems, a point that we return to later.

23.1.1 Biotic Regulation of Soils during Species Transitions

The smoothness with which one community type changes to another will depend at least in part on what happens in the soil. We may think of this as the old residents of a given community passing the torch of maintaining a healthy soil to the newcomers. Will the torch be successfully passed? Will the old residents retain their biocontrol over soil structure and processes long enough for newcomers to establish themselves and take over the job? What happens if the torch is dropped—the current residents losing their grip on the soil before the newcomers arrive?

Most plant species are likely to efficiently bioregulate nutrients and soil structure, although species will differ in certain aspects of this, such as the particular nutrients that they concentrate, the rate at which they cycle nutrients, or the total nutrient sink that they provide (e.g., the biomass of grasslands and therefore their sink for nutrients is less than in forests). New species migrating onto a particular site may or may not require the same mycorrhizal fungi and rhizosphere bacteria as those supported by current residents. If the migrants require mutualists that are not present, the transition might be inhibited. On the other hand, if migrants find the mutualists they require, we might predict that the transition from one type to another will be facilitated so long as there is sufficient overlap between old residents and migrants (Perry et al. 1990). What are the chances that old residents and migrants will overlap sufficiently to stabilize soils? This is a difficult question that is unlikely to have any single answer. The degree of overlap will depend on two things: rates of migration (chapter 4), and the stability of resident communities as the environment changes. The latter will, in turn, depend on the disturbance regime.

23.1.2 Buffering Capacity

As we discussed at various points throughout the book, forests have a considerable ability to modify and ameliorate their environment, which suggests they can persist in disequilibrium with the climate for at least some period of time. The microclimate within an established forest is cooler and moister than that outside the forest. Trees invest considerable energy in soil properties that enhance water-holding capacity and facilitate resource gathering. Large trees store water in sapwood. In the process called "hydraulic lift," deep-rooted trees pump water from lower soil levels and release it to the upper soil.

> Evidence indicates that Amazonian rainforests were resilient to historic droughts (see the review by Malhi et al. 2008); however, the past is no longer a good guide to the future. Extensive clearing, especially in the eastern Amazon, has exacerbated climatically-related drought by reducing transpiration from the forest (hence local rainfall) and through drying induced by permeating the forest matrix with openings. Moreover and more significantly, clearing has introduced fire, a foreign disturbance to which rain forest trees are not adapted (chapter 7). Once burned, a forest burns more readily the next time. Mahli et al. (2008) suggest a tipping point exists in which repeated fires allow flammable grasses to penetrate forest interiors, making fire a permanent fixture and leading to widespread forest degradation (except in areas such as the northwestern Amazon that remain relatively pristine).

Species diversity and genetic diversity within populations will enhance buffering capacity if some species and genotypes are better adapted than others to new conditions (i.e., the "more adapted" helping to bridge between the old and the new). Tree species may have a great deal of within-population genetic buffering against changes in climate, especially those with wide distributions. That is the case with lodgepole pine (at least in a simple test of seedlings grown in two temperature regimes). In lodgepole populations scattered along a north-south gradient from northern Utah to central British Columbia, the greatest variability in response to temperature was among genotypes within populations rather than between the populations (Perry and Lotan 1978). Genotypes within populations also varied significantly in response to water stress (Perry et al. 1978).

As the prognosis for the Amazon indicates, maintaining biological diversity will not be helpful if other forces overwhelm its buffering capacity. However, diversity is an important tool to mediate an orderly transition of community types, and when coupled with protecting landscape integrity seems our best chance to avoid serious environmental degradation in the coming decades.

23.2 MAINTAINING BIOLOGICAL DIVERSITY IN MANAGED FORESTS

After ecstasy the laundry.

ANONYMOUS

There are basically two choices regarding how forests should be managed. The first is to homogenize forests and forested landscapes, thereby reducing natural diversity to concentrate on the economics of wood production. The second is to manage forested landscapes in such a way that natural diversity is preserved. (As we see later, these are not mutually exclusive when one takes a landscape view). The advantages of homogenizing are mostly economic: productivity is maximized (at least in the short run) by shortening rotations and growing only those species and genotypes that produce highest economic yields; management is relatively simple and therefore cheap; and products are homogenized, which makes for more efficient logging and milling.

As we discussed earlier, homogenizing forests has three primary disadvantages. First, it is not aesthetically appealing, a significant factor on public lands, whose owners (i.e., the "people") are making it increasingly clear that they do not want their lands managed as tree farms. (The public has not yet confronted their own demand for wood products.) Second, habitat is eliminated for some species and reduced for others, thereby making them vulnerable to local extinction (chapter 22). Third, because diversity serves as an important (but poorly understood) natural stabilizing mechanism within ecosystems, homogenization at the scale of landscapes risks making forests more vulnerable to various stresses and could actually decrease long-term yields (Perry 1988b; Franklin et al. 1989; Perry and Maghembe 1989; Torgerson et al. 1990; chapter 21 in this volume). Economically, the trade-off between homogenizing forests and reducing their stability is no different than that associated with investments in financial markets. High yields do not come free; they are paid for by higher risks. At a time of significant uncertainty (e.g., in where climate is headed, in how ecosystems work), the economic gains of homogenized forested landscapes may no longer be worth the risks.

Maintaining aesthetics, preserving species, and protecting natural stabilizing mechanisms are linked through the common thread of maintaining (or restoring) natural diversity within the managed landscape. Lindenmayer and Franklin (2002) dealt with this topic in detail (articles in Burton et al. [2003] specifically address sustainable management of boreal forests). The remainder of this chapter briefly reviews some general approaches to accomplishing this. We begin by discussing how practices that are aimed at maximizing wood production alter natural forests and landscapes, and then consider some of the key elements of the diverse managed forest and silvicultural techniques that might be applied to restore and to maintain diversity.

23.2.1 Natural Diversity and Intensive Forest Management

The key to maintaining diverse managed forests is to understand how the natural system works and to base management prescriptions on that (Lindenmayer and Franklin 2002; R.J. Mitchell et al. 2006). Because silviculture is largely the management of disturbance and succession, a logical place to begin is reviewing what we have already learned about natural patterns of disturbance and succession, then to evaluate how intensive management differs from those.

Most if not all natural forested landscapes fit Bormann and Likens's (1979a) description of the northern hardwood forest: shifting mosaics of irregular patches differing in composition and age, driven by the processes of disturbance, growth, and decay (chapter 5). The characteristic spatial and temporal scales of the mosaic will vary from one type to another, as well as over time, in any given type. The proportion of different successional stages and their spatial pattern (i.e., juxtaposition and interconnectedness) provide the habitat template on which regional diversity is built. Species are well adapted to disturbances that have been a normal part of their history, and far from being a negative factor, these "natural" events tend to both renew and diversify forested landscapes, so long as they do not become too severe or frequent.

By definition, disturbance and succession involve change; however, strong threads of continuity are embedded within the shifting mosaic of natural landscapes (chapter 8). Early successional communities contain numerous biological legacies of the previous stand (e.g., logs, snags, standing green trees, sprouting plants, mycorrhizal fungi) that provide structural complexity and play important roles in ecosystem resilience. Continuity in the proportion and interconnectedness of habitats within the shifting mosaic (i.e., at the scale of landscapes and regions) acts to maintain a regional biota with diverse habitat requirements (chapter 22).

In chapter 21 we discussed how intensive forest management departs from natural patterns. First and foremost, it focuses on one (or rarely, a few) species: the tree or trees that are the crop. Furthermore, this focus is restricted to only certain age classes of the crop. Techniques aimed at hastening full site occupation by commercial species shorten the time spent in early succession, moreover early successional communities following logging do not have the legacies that characterize early succession after natural disturbances. Absence of legacies disrupts continuity of habitats and may ultimately diminish system resilience. Harvesting when mean annual increment culminates completely eliminates the old-growth stage along with the unique and ecologically important habitats that it produces (e.g., big trees, large dead wood, multiple canopy layers, late successional tree species). In conifer forests, the closed canopy, midsuccessional stands produced by management, although important habitat for some species, generally support fewer species than either the early successional, shrub-forbs-sapling stage or the late successional, old-growth stage (Meslow 1978; Thomas et al. 1979; James and Warner 1982), although that is not always the case (Loehle et al. 2005a, 2005b). Added together, simplified stands form simplified landscapes, in which some components are completely missing while the relative proportion of those that remain may differ significantly from that of the unmanaged landscape (Haeussler and Kneeshaw 2003).

23.2.2 The Diverse Managed Forest: Some General Principles

Before we plow an unfamiliar patch
It is well to be informed about the winds
About the variations in the sky
The native traits and habits of the place
What each locale permits, and what denies.

VIRGIL, THE GEORGICS (CIRCA 30 B.C.)

If the objectives of forest management include maintaining (or restoring) biological diversity and sustaining productivity, it is necessary to understand the relationships among system structure, processes, and functions. That is of course much easier said than done. Structure comprises a numerous set that includes species composition, genetic diversity within species, food webs, age classes, soils, dead wood, and patterns at various scales of space and time. Process encompasses an even more numerous (and much less easily studied) set that includes nutrient cycles, demographics, two-way and higher-order interactions among species, and other factors too numerous to mention. Ultimately the dynamics of nature are defined by functions, or what all that multiplicity of structural elements and processes *do*. Moreover, different forest types differ in the way they manifest their complexity. In other words, complexity is in itself a complex affair.

One message that we hope comes through clearly in this book is that ecologists are a long way from a complete understanding of how nature works. Moreover, given the complexities that are involved, we will probably always be ignorant at some level. Nevertheless, enough is known (or at least suggested by the evidence) to guide decision making.

Two basic principles apply to any forest type:

1. *Spatial scale and context are important.* An individual stand should not be managed in isolation from the landscape in which it is embedded, nor should an individual watershed be managed in isolation from its region.
2. *Biological diversity is important.* A primary goal of ecosystem management at the stand level is to protect structural complexity, special habitats, keystones, redundancies, and biological legacies (chapter 8), which shifts the primary focus from what is taken during harvest to what is left (Franklin 1993; Lindenmayer and Franklin 2002).

23.2.2.1 Spatial Scale and Context: Landscapes and Regions[2]

As we discussed in chapters 10 and 22, landscape patterns determine the variety, integrity and interconnectedness of habitats within a region. Moreover, the amount and distribution of community types across the landscape strongly influence the rate at which virtually all types of forest disturbances propagate and the intensities that they attain (Turner 1987; Perry 1988b; Turner et al. 1989; chapter 7 in this volume). Insects, pathogens, fire, and wind (all forces that are relatively innocuous and even beneficial when at certain levels) can become highly destructive in a landscape that magnifies, rather than buffers and absorbs their destructive energy, a phenomenon that in all likelihood will be exacerbated by a warmer climate (Haeussler and Kneeshaw 2003). Finally, landscape patterns will determine how successfully many species migrate in response to climate change.

The primary objectives of landscape management correspond to the functional roles of pattern in the natural landscape, that is, to create (or maintain) a proper mix of habitats and to buffer and absorb the energy of destructive forces. These two objectives are not automatically consistent with one another; careful planning (with an eye to both) is necessary. (Bell and Apostol 2008 discuss principles and examples of forest landscape design).

23.2.2.2 Habitats

Before discussing how managers can restore and maintain habitat, it is useful to review some principles introduced in chapter 22 (McComb 2007 deals with this topic in detail). Two aspects of habitat are relevant at the scale of landscapes: (1) the relative amounts of different types, and (2) the degree of interconnectedness (or conversely, fragmentation

[2]Articles in Gutzwiller (2002) and Liu and Taylor (2002) deal with applying landscape ecology to management and conservation.

and isolation). The habitat template within a forested region comprises (1) different major forest types, (2) different successional stages within a given forest type, and (3) unique communities such as riparian zones. Species that typify different habitats (particularly seral stages) often differ in their life-history characteristics in ways that are significant for conservation strategies. It is not true (as one sometimes hears) that all early successional species are weeds, but many are opportunistic generalists whose populations grow rapidly and disperse widely. Also exotic weeds are most likely to successfully invade communities following disturbance. Compared with early successional communities, a higher proportion of late successional species have one or more characteristics that make them vulnerable to extinction (chapter 22). This is particularly true for those vertebrates (often at the top of the food chain) that require large amounts of territory or safe havens from hunters and poachers (Crow 1990).

Historic patterns of disturbance and recovery have increasingly been used as templates for management. Palik et al. (2002) and R.J. Mitchell et al. (2006) call this **natural-disturbance-based silviculture.** "A natural disturbance model for management assumes that species have evolved to local environmental conditions and a particular disturbance regime, such that there is a greater opportunity for sustaining biodiversity when the disparity between managed disturbances and natural disturbances is reduced" (R.J. Mitchell et al. 2006). R.J. Mitchell et al. (2006) give several examples from various regions of the United States.

At the landscape level, the concept of **natural range of variability** (also referred to as **historic range of variability**) has received considerable attention. It uses historic patterns of disturbance to guide harvesting patterns, as, for example, identifying areas best suited for longer rotations or small patch cuts and areas where shorter rotations or larger cuts would be appropriate (Landres et al. 1999). The natural range of variability concept has had the positive effect of making foresters think about history (reliably determining history is another matter); however, taken too literally the approach promotes the false belief that logging can fit within a "natural" range of variability. Logging of any kind—whether clear-cut or single-tree selection—is an unnatural event in the history of a forest, and the relevant scientific question is not whether a practice fits within natural variability, but how far management can depart from natural variability before compromising system integrity. Answering that question requires a functional approach that deals with the suite of processes underpinning integrity, which, in turn, requires a measurable definition of integrity, testable hypotheses concerning keystone functions and their links to system structure, and protocols for choosing among hypotheses when the systems of interest are highly dynamic in space and time and the information base is weak. That is a challenging agenda but progress is being made.

> The area occupied by old-growth forests has been sharply reduced throughout United States since EuroAmerican colonization, and because of the loss of biological legacies few younger forests are ecologically equivalent to those that result from natural disturbance. It follows that when a regional perspective is taken all forests except those in protected or inaccessible areas are outside of their historic range of variability (assuming history is not extended back to glactation) Moreover, as we discussed in chapter 22, where blocks of the original forest remain they are often isolated within sea of younger forests, with the characteristics of the matrix playing a profound role in the quality of processes and habitat within fragments.
>
> Therefore, if we want forests within their historic range of variability we must restore older forests on the landscape. Three caveats are necessary here. The first is obvious but bears repeating: "old growth" is not some homogeneous thing that is the same everywhere; an old taiga forest may be young by Pacific Northwest standards. Second, landscapes dominated by older forests generally are very heterogeneous regarding the age classes and species they comprise; in other words, they are mosaics. In our discussion of older forest habitat, it is really this mosaic quality (i.e., an old-growth matrix) that we refer to. The third is that a sole focus on preserving or restoring older forests can lead to deficiencies in important early successional habitat (especially naturally recovering early succession).

Applying habitat suitability models to real or modeled landscapes is a relatively new approach that shows considerable promise. For example, Spies et al. (2007) modeled habitat suitability for several focal species in the mixed public-private ownerships of the Oregon Coast Range. They found that moderately increasing the number and size of green trees retained on private lands increases habitat for red tree voles by 50 percent and for western blue birds by 400 percent. They also identified a likely future shortage of diverse early successional forests and associated species (including open forests with remnant trees and snags and shrubby open forests) and middle-age forests (50 to 150 years).[3]

In some areas, current land uses preclude restoring regional landscapes to even a close approximation of what they once were. In such cases the only possible approach is to ask functional questions: What landscape patterns provide the desired services and values?[4] For example, Loehle et al. (2006) used a

[3]Future shortages of these habitat types result from the sharply contrasting management on federal and private lands. The former manage for old-growth characteristics, while the latter manages on short rotations for fiber production.

[4]Palmer et al. (2004) refer to a function-based approach as "the science of ecosystem services."

When we came across this continent cutting the forests and plowing these prairies, we have never known what we were doing because we have never known what we were undoing.

WENDELL BERRY

There is a great deal of current activity aimed at restoring structure and function in forests and forested landscapes that have been altered from their historic condition, often with the loss of essential services. For example, on the J. W. Jones Ecological Research Center in the Coastal Plain of southwestern Georgia, an 88-ha slash pine plantation (established in 1939) is in the initial stages of conversion to a diverse longleaf pine forest. The approach is gradual, maintaining a cover of the overstory slash pine while fire and planting are used to promote longleaf pine regeneration (R.J. Mitchell et al. 2006). A combination of thinning (leaving the larger trees) and reintroduction of fire are being used to restore old-growth-type structures in many western U.S. forests (fig. 23.1A). Williams et al. (1997) describe several examples of watershed restoration projects in the United States.

In the tropics, a diverse coalition of international organizations, national governments and NGOs is promoting a landscape approach to forest restoration (FLR, or Forest Landscape Restoration): "Forest landscape restoration . . . focuses on restoring forest functionality: that is, the goods, services and ecological processes that forests can provide at the broader landscape level as opposed to solely promoting increased tree cover at a particular location" (Maginnis and Jackson 2005). Incorporating the needs and expertise of local human communities is central to FLR. Tropical forest landscape restoration projects include restoration of miombo woodland in Tanzania and dry forests in India (fig. 23.1B).

A similar community-based effort, but not necessarily restricted to restoration, is the International Model Forest Network (IMFN). IMFN defines a model forest as "both a geographic area and a specific partnership-based approach to sustainable forest management (SFM). Geographically, a model forest must encompass a land-base large enough to represent all of the forest's uses and values-it is a fully working landscape of forests and farms, protected areas, rivers and towns" (www.imfn.net/en/ev-22891-201-1-DO_TOPIC.html). Originated in Canada in the mid-1990s, as of 2007 46 model forests distributed among 22 countries were either established or under development.

Whether classified as restoration or not, there is abundant scope for community- based forest management strategies, especially (but not exclusively) in tropical countries. According to Johnson (2007), "policy shifts to recognize traditional and indigenous rights have resulted in a doubling of community-owned and administered forest lands in developing countries over the past two decades, to around 370 million hectares of natural forest (nearly one-quarter of all forests in these countries, three times the amount owned by individuals and firms). Current trends indicate that community tenure will double again by 2020 to more than 700 million hectares." As oof 2007, community forestry enterprises employed more than 110 million people worldwide (Johnson 2007).

Hilderbrand et al. (2005) offer some cautions concerning restoration projects.

> Failure to recognize limitations and tacit assumptions can lead to failures because of the over-application of over-simplified concepts to complex systems. . . . Any ecological restoration or management effort involves both explicit and implicit attempts to prescribe and predict the ecological future of a site. These efforts require extrapolating far beyond our predictive abilities, and we must be aware of our limitations as scientists, as well as our tendency as humans to rely on partial truths and assumptions. . . . Restorations should not be one-time events, but are likely to require periodic attention and adaptive management to increase the chances of responsive, adaptive, and successful projects.

combination of habitat-suitability models, information on stand type distributions, and a harvest scheduling model to explore options for conserving bird diversity and trade-offs with fiber yields on a corporate ownership in South Carolina. They found that setting aside pine and hardwood stands older than 40 years (about 24% of that particular land base, mostly hardwoods) benefited the richness of several focal bird species with a relatively small impact on overall timber yields. As with the landscape-level studies in the Oregon Coast Range (discussed above), these results are specific to a particular landscape and cannot be generalized to different ones. However, the general approach of applying habitat-suitability models to real landscapes provides a powerful tool for simulating how different management approaches effect biodiversity (a capability with particular relevance to private lands seeking forest certification[5]). Adding models that simulate effects

[5]According to the World Wildlife Fund, "Forest Certification is widely seen as the most important initiative of the last decade to promote better forest management" (www.panda.org/about_wwf/what_we_do/forests/our_solutions/responsible_forestry/certification/index.cfm). As of June 2006, nearly 76.5 million ha had been certified in 72 countries (see www.fsc.org/keepout/en/content_areas/92/1/files/ABU_REP_70_2006_06_21_FSC_certified_forests.pdf for a complete listing). Brown et al. (2001) discuss shortcomings in the assessment process at the landscape and regional scales and suggest improvements.

Figure 23.1. A. A thinned ponderosa pine forest on the Fremont-Winema National Forest, Oregon. Low-elevation forests in this droughty area are growing on soils that are typical of those formed under grasslands. B. This forest in the background was restored on degraded hill land by faculty and students of the Rishi Valley School, Andhra Pradesh, India. (Courtesy Murali Reddy www.mithya.com/learning/rishi.html)

on fiber yields allow private land managers to assess economic-ecologic trade-offs.

What scale of planning is appropriate if the objective is to maintain or restore biological diversity? Two factors are involved: (1) the area needed to maintain viable populations of umbrella species (chapter 22), and (2) the scale of natural disturbances. A single breeding unit of a typical vertebrate umbrella species likely needs thousands of hectares of suitable habitat. For example, a herd of white-lipped peccaries, which constitutes a single breeding unit, ranges over a minimum of 10,000 ha of Amazonian forest (Terborgh [1992] gives area requirements for some other Amazonian vertebrates), and one pair of spotted owls requires from 1,000 to perhaps 3,000 ha, depending on the abundance of prey. Protecting or restoring habitat for the several thousand breeding individuals needed to maintain minimum viable populations of these and other umbrellas requires regional-level planning. (In the case of migratory animals, hemispheric perspectives are necessary.)

As the scale of natural disturbances increases, the area needed to buffer late successional species against excessive habitat loss increases. Climate change introduces new considerations. Pathways of migration must be maintained. Because natural disturbance regimes are predicted to become more severe (chapter 7), the risk of losing already-limiting habitat is likely to increase in the future, thereby increasing the need for demographic buffering. A few plans have begun to address these scales, and in the future we will see more, increasingly integrating ecology and socioeconomics.[6]

23.2.2.3 Forest Protection

From the landscape perspective, the strategy for protecting forests is to create (or to maintain) landscapes that buffer and absorb disturbances rather than magnify them. Homogeneity will almost certainly magnify disturbances of one kind or another, and whether a particular heterogeneous pattern is absorbing or magnifying depends on the types of disturbances likely to move through it (Forman 1987; Perry 1988b; Turner et al. 1993). Once again, each case must be analyzed on its own merit. Roads facilitate the entry and spread of any number of potentially disruptive agents; in the Pacific Northwest, logging roads have facilitated the spread of at least five pest species (Strang et al. 1979; Daterman et al. 1986; Hansen et al. 1986).

Given the variety of disturbances possible in any forest type, and the need to balance forest protection with the requirements of many species for interconnectance within the habitat matrix, it is tempting to conclude that the best approach to protection and conservation is what we frequently see in natural landscapes: complex patterns of hierarchically nested patches, which is effectively a fractal structure. Whether this conjecture has validity, the scales at which heterogeneity occurs will be an important consideration. For instance, old-growth forests of the Pacific Northwest are (relatively) homogeneous on the scale of tens of kilometers but quite heterogeneous on the scale of tens to hundreds of meters. The large-scale homogeneity provides good habitat for interior forest species, while the low- and medium-scale heterogeneity diversifies the forest and helps to retard the spread of fire (Perry 1988b; Franklin et al. 1989). Densely stocked young stands are homogeneous at the smaller scales, which facilitates the spread of fire through crowns. A landscape filled with such stands not only provides poor habitat and is deadly boring, it is also a firetrap.

23.2.2.4 Stand Level: Biological Legacies

Although succession has a definite stochastic component; communities are not thrown about willy-nilly by natural disturbances; rather, they tend to maintain certain trajectories. This trajectory stability (Orians 1975) may be accomplished in various ways, but one recurrent, general theme is the importance of biological legacies—those components of biodiversity that are passed from old to new and provide continuity through time (chapter 8). The concept of biological legacy is very broad, but it can be summarized briefly as anything biological (or of biological origin) that persists through disturbances and as a consequence of its persistence helps maintain ecosystems and landscapes on a given trajectory. Legacies may be tangible objects (e.g., large dead wood), or they may be elements of pattern or system structure. They may function to maintain processes, habitats, or linkages; in the real world, these three elements are inseparable and work together to maintain properly functioning ecosystems. Following are some examples of legacies:

- Redundancies that stabilize critical processes, such as the nutrient cycle, or that maintain critical links, such as those between plants and plant mutualists such as mycorrhizal fungi and pollinators
- Keystones that provide some unique habitat or service, such as large dead wood and nitrogen-fixing plants
- Reservoirs, such as soil organic matter, large soil aggregates, and seed banks
- Stand-level structural patterns, such as multiple canopy layers.
- Landscape patterns.

Landscapes are the templates of survival for populations and metapopulations and, therefore, represent the intersection between spatial patterns and temporal continuity. No one piece of ground will always provide suitable habitat for spotted owls or red-cockaded woodpeckers, but so long as a legacy of sufficient suitable habitat occurs within a region, the species will persist and provide a continuing source of colonists. Maintenance of a diverse gene pool (provided by sufficient gene flow among subpopulations) is another legacy that depends on the distribution of habitats within regions.

[6]Lindenmayer and Franklin (2002) give several examples of multiscaled plans for biodiversity protection. In some areas, active participation by local people and other stakeholders has been solicited and even mandated. In 2000, the U.S. federal government passed legislation requiring multistakeholder resource advisory committees for any project (related to natural resource management) that gets federal funds. An analysis by the Sierra Institute found that these committees have succeeded in helping stakeholders find common ground (www.sierrainstitute.us/SecureRuralSchools.html). Janzen (1999) holds that "tropical wildlands and their biodiversity will survive in perpetuity only through their integration into human society." Integrated conservation and development projects in the tropics hold promise but in practice have proven difficult to implement (Alpert 1996; Brown 2003).

Now we review some specific pieces and patterns that should be protected within stands if management objectives include maintaining biological diversity and sustaining long-term productivity.

23.2.2.5 Large Senescent and Dead Trees

Snags and logs are habitat for many animals, including birds and insects that contribute to forest health by consuming defoliators and bark beetles (chapters 10 and 19). In the Blue Mountains of Oregon and Washington, for example, 39 bird and 23 mammal species use snags for nesting or shelter (Thomas et al. 1979). Harris and Maser (1984) estimate that a mature, unmanaged Douglas-fir forest in the Cascade Range supports approximately 90 terrestrial vertebrate species (excluding bats). Eliminating snags reduces the number to about 80; eliminating both snags and logs reduces it to about 60. Large senescent and dead trees (logs and snags) support a huge diversity of fungi and insects (Grove 2002). Old, decayed logs are rich sources of carbon for microbes and invertebrates and, in addition, are efficient water reservoirs, sites of nitrogen fixation, and (particularly during drought) may be centers of biological activity for organisms that cycle nutrients. Logs also create essential habitat for fish and invertebrates in streams and estuaries of the Pacific Northwest and (to a certain extent) even in the open ocean (Harmon et al. 1986; Maser et al. 1988a, 1988b).

23.2.2.6 Diverse Plant Species

Until recently, the majority of both applied and basic research on interactions among plant species has focused on competition. As a result, we know far less than we should about positive interactions, which are likely to be subtle, indirect, play out over long periods, or perhaps manifest themselves only at certain times during stand history (e.g., soil stabilization following disturbance, insect outbreaks). While plant species undoubtedly compete—everything needs resources to live—accumulating evidence shows that they may also benefit (either directly or indirectly) from one another's presence (chapter 11). One of the better-known examples is symbiotic nitrogen fixation, which is by far the major source of nitrogen for many forests (chapters 8 and 17). Other positive effects of noncommercial plant species on nutrient cycling are less well understood but likely are significant. Sprouting hardwoods have the capacity to rapidly recover following a stand-destroying disturbance, thereby stabilizing nutrients and probably beneficial soil biota (chapters 8 and 20). Some species (e.g., maples, cedars) cycle calcium, an element long associated with productive forest soils (chapters 14 and 15).

Retaining hardwoods within conifer forests can either improve or diminish important habitat, depending on forest type and exactly what habitat is desired. In chapter 10, we saw that numbers of birds in piedmont forests of the southeastern United States increase sharply as hardwoods establish beneath pines during later successional stages (Johnston and Odum 1956), and McComb et al. (1986a, 1986b) argue that retaining hardwoods within southern pine stands will be necessary to maintain populations of cavity-dependent species. On the other hand, the historic disturbance regime in longleaf pine forests[7] consisted of frequent ground fires that killed hardwoods and encouraged a highly diverse grass and forb layer, creating one of the most species-rich forest types in North America (and currently one of the most endangered) (Noss et al. 1995; Kirkman et al. 2001).

Mixtures of host and nonhost tree species lower the susceptibility of hosts to herbivores and pathogens (chapter19). The idea of plant-defense guilds (or "associational plant defenses") is not new (Vouté 1964; Atsatt and O'Dowd 1976; Hay 1986); organic gardeners routinely grow companion plants amidst other crops to provide protection against pests. Although some biologists reject the idea that one plant would somehow protect another, biologically plausible cause-effect mechanisms do exist (chapter 19). For instance, at least some species of flies and wasps that parasitize the larvae of defoliators require a source of nectar (i.e., flowering plants) in their diet (Vouté 1964). A diverse assemblage of volatile chemicals released by the plant community can confuse herbivores and make it more difficult for them to locate hosts. Simply reducing the concentration of hosts can have the same effect.

Another question (of considerable relevance to a hotter, drier future) is whether the relative inflammability of some hardwoods protects conifer stands from fire. Anecdotal evidence indicates they do (chapter 7). Research done 20 years ago showed that the chemical composition of the leaves of some plant species retards fire (Philpot 1970); this is a dangling scientific thread (one of many in ecology) that needs to be picked up and followed. It is interesting to note that following the Yacolt burn in southwestern Washington State early in this century, strips of alder were planted along with Douglas-fir as fire breaks. Did those old foresters know something that we have forgotten?

23.2.3 Silvicultural Approaches

To recapitulate, to restore and to maintain biological diversity, managers must restore a proper balance of habitats, where "balance" includes relative proportion, quality, and spatial arrangement; and maintain that balance by long-range planning and protecting the processes that keep systems healthy. How these objectives translate into specific management actions will vary depending on the ecology and history of given region. Nature provides a blueprint,

[7]The historic range of longleaf pine is generally south of the Piedmont, within and somewhat north of the Gulf Coastal Plain. Currently, longleaf occupies only a fraction of its former range.

but one that must be used carefully. For example, the pattern of legacies that natural disturbances leave provides a good guide for stand-level silviculture, but basing landscape-level harvest patterns on historic natural-disturbance patterns is inappropriate if done at the wrong scale. To illustrate why, consider the situation in the Pacific Northwest. That some past wildfires were quite large does not justify large cuts in watersheds not yet logged, because they are embedded in a regional landscape that has been highly altered by logging and bears no resemblance to historic patterns. Restoring an entire regional landscape toward historic patterns is another matter, because it mimics nature at the appropriate scale.

Various approaches may be used to create managed landscapes without destroying the basic fabric of the natural forest. As with many other things that relate to our future survival, modern humans have much to learn from preindustrial cultures in this regard. To indigenous peoples, the diversity of natural forests translated into a diversity of both tangible and intangible values. Forests provided wood, food, medicine, and spiritual enrichment. For example, Amerindians in Central and South America have traditionally practiced what Anderson (1990) refers to as "tolerant" forest management, in which "practices include favoring highly desirable tree species and eliminating or thinning less desirable competitors, while maintaining the essential forest structure and composition." (Also see Gomez-Pompa and Kaus [1990], and Browder [1992].)

Alcorn's (1990) seven strategies of indigenous agroforestry in the tropics is good advice for any forest type:

1. Take advantage of native trees and native tree communities;
2. rely on native successional processes;
3. use natural environmental variation;
4. incorporate numerous crop and native species;
5. [be] flexible;
6. spread risks by retaining diversity; and
7. maintain a reliable back-up to meet needs should other sources fail.

Gap-driven forest types, such as moist tropical, temperate deciduous, and some conifers, are uneven aged at a fairly fine scale (i.e., tens of meters) and include a significant proportion of tree species that tolerate at least moderate levels of shade. Harvesting trees in small groups (called **group selection**) or narrow strips maintains the natural structure of these forests and (in addition) protects soils from heat, erosion, and excessive nutrient leaching. Two silvicultural systems that are employed in moist tropical forests are the Palcazu system in Peru and the Celos system in Surinam. The Palcazu system uses the principles of gap dynamics to manage moist tropical forests for both wood and diversity (Tosi 1982). According to Hartshorn (1990): "The Palcazu model integrates local forest ownership and processing, harvesting of timber on long, narrow clear-cuts, animal traction for logging, nearly complete utilization of timber, and natural regeneration of native tree species. The forest management system involves clear-cut strips (30–40 m wide) rotated through a production unit on a 30–40 yr cycle." The location of each clear-cut strip in the Palcazu system is designed to maintain the matrix of high forest to serve as seed source and habitat. The Celos System involves selective harvest of commercial tree species using logging techniques carefully designed to avoid damage to residual trees. Noncommercial tree species are thinned to favor commercial species, of which there are 40 to 50 in those forests (de Graaf and Poels 1990).

Sist et al. (2003) offer seven guidelines for sustainable management of mixed dipterocarp forests in Southeast Asia:

1. Use reduced-impact logging practices.
2. Cut no more than 8 trees per ha on a cycle of 40–60 years.
3. Define minimum diameter cutting limits according to the structure, density, and diameter at reproduction of target species.
4. Avoid harvesting species with less than one adult tree per ha.
5. Minimize the size and connectivity of gaps (<600 m^2 wherever possible).
6. Refrain from treatments such as understory clearing.
7. Provide explicit protection for key forest species and the ecological processes they perform.

These guidelines are crucial if sustainable logging in tropical forests is to become a reality. Asner et al. (2006) used a time series of remote imagery (1999–2004) to quantify the effects of selection harvesting in Amazonia. They found that 21 to 24 percent of logging was sufficiently low impact to create gaps in less than 10 percent of the forest cover, while 61 to 68 percent of all logging operations created gaps in the 10 to 40 percent range, and an additional 8 to 17 percent resulted in greater than 40 percent opening. Ten percent of the forest in gaps is considered a threshold in Amazonia (and probably other moist tropical forests as well), beyond which the forest becomes more susceptible to drying and fire (Asner et al. 2006). The priming effect of roads was apparent in the Asner et al. (2006) study. Eighty percent of logging occurred within a 5-km belt around roads, and all occurred within 25 km of a road. Between 5 km and 25 km from a road there was a 35 percent probability that logged areas would be completely deforested by ranchers or farmers within four years of logging.

What about selective logging in other forest types? Regardless of where one is on the earth, sustainability depends on rates of harvest that do not exceed rates of regrowth. Lorimer (1989) discusses a silvicultural system that has been used successfully in the deciduous forests of the north-central United States: "Cutting is distributed over a wide range of size classes, and many medium and large trees are retained on the site. This procedure creates small, discrete gaps and the forest retains an old-growth appearance. The basal area removed in each harvest does not ordinarily exceed 30%. Removal of 20–30% of the basal area typically yields more than the 5.8 m usually considered to be the minimum needed for an economically viable logging operation." In the approach described by Lorimer, single-tree gaps are interspersed with multitree gaps at least 500 m in area. The larger gaps allow regeneration of moderately shade-intolerant tree species and create a diverse structural mosaic. Management by single-tree selection only would almost certainly lower the diversity of these forests.

Some old-growth conifer forests (e.g., ponderosa pine, Douglas-fir) may also be characterized by a patchy, multiaged structure, although the scale at which patchiness occurs as well as the size of opening needed to establish regeneration successfully differs from forests composed of more shade-tolerant species. Managing shade-intolerant species with single-tree selection is tricky and can result in conversion to shade-tolerant species unless canopy closure is also managed. However, a multiage structure is feasible using multitree gaps or periodic thinning that remove enough of the overstory trees to allow establishment of a younger age class. Gaps of 0.1 ha are sufficient to regenerate even shade-intolerant species such as ponderosa and longleaf pines (R.J. Mitchell et al. 2006). On the other hand, in forests that grow naturally with relatively open canopies (which would be the case on droughty or nutrient-limited sites), even shade-intolerant species may be successfully regenerated using single-tree selection: it is not how many trees you take out; it is how much light gets to the forest floor. Another important factor for natural regeneration is the suitability of the seedbed. In conifer forests that tend to develop heavy litter layers, successful regeneration within a multiaged management system will probably require controlled underburning or some other technique to expose mineral soil seedbeds.

In forest types traditionally managed on an even-aged basis (e.g., clear-cuts), leaving a legacy of large trees following regeneration harvests (i.e., **structural retention**) creates stand structures that mimic (at least to some degree) natural disturbance patterns, including the development of two or more canopy layers and large trees as a future source of snags and down logs (Lindenmayer and Franklin 2002; R.J. Mitchell et al. 2006) (fig. 23.2). According to Lindenmayer and Franklin (2002), "The types of structural features that should be considered for retention in logged forests include large living trees—particularly those exhibiting decadence and other age related features, such as large branches; large snags; overstory and understory plant species, which contribute to multiple vegetation layers and vertical heterogeneity; large logs; and intact areas of forest floor."

The numbers and spatial patterns of live trees that should be retained vary with objectives and forest type (see Lindenmayer and Franklin for a discussion). Prescriptions in the Pacific Northwest vary from less than 20 percent of the maximum basal area to 50 percent or more, while in temperate hardwood forests that are managed to retain old-growth structure, 70 percent or more of the basal area is retained at harvest (Lorimer 1989). In Sweden, the standard is 10 live trees per ha (Hazell and Gustafsson 1999) In Mountain ash forests of Australia, at least 5 to 10 cavity trees per hectare (in perpetuity) are required to maintain populations of the endangered marsupial, Leadbetter's possum, which is only one of hundreds of species in those forests that require cavities (Lindenmayer and Franklin 2002). Retention may be either dispersed, aggregated, or some combination of the two (fig. 23.2). Dispersing retained trees provides better protection for soils and establishing seedlings, while aggregating creates small islands with greatly reduced impacts on forest floor, soils, and the myriad organisms that require those habitats.

Density control is an important tool for diversifying many forest types (DeBell et al. 1997). Growing stands at wide spacing allows noncommercial plant species to coexist and produces large trees more quickly, and in dry forest types can reduce the risk of crown fire. When coupled with longer rotations, thinning can be used to manage the proportion of different seral-stage habitats within a landscape; for example, forests can be moved from an early successional to an old-growth structure without ever passing through a closed canopy stage (if that is desirable). On the other hand, entering stands too frequently risks tree damage and soil compaction, and in some cases opening stands too much allows aggressive clonal plant species to dominate the understory. Thinning too heavily on droughty sites can result in excessive drying. It is wise to proceed cautiously (including avoiding soil compaction and damage to residual trees). Thinning operations should be planned in a landscape context; for example, some species require closed forest, and the thinning strategy for a given area should account for their requirements.

Growing at least a portion of stands on long rotation will be essential to produce sufficient older forest habitat within the managed landscape and to avoid creating too much early successional habitat for animals such as deer that can become serious pests when too abundant (as has happened in Europe and the eastern and north-central United States). Should forests be cut shortly after attaining the desired structure, landscape-level goals will not be achieved. This point is illustrated by the following formula, which applies to any desired structure:

$$R = L / (1 - p)$$

where R is the rotation length, L is the length of time required to return to the desired stand structure following a regeneration cut, and p is the target for proportion of landscape in forest with the desired structure. For example, if the target is 60 percent of the landscape in forests with a

A

B

Figure 23.2. A mix of aggregated and dispersed retention, Vancouver Island, BC. (From Franklin et al. 2007)

complex, old-growth-like structure and 100 years is required to achieve that following a regeneration cut, rotations must be 250 years. If only 50 years is required to return to the desired structure, rotations of 125 years will allow the landscape target to be met. The time required to reach an old-growth-like structure depends on the silvicultural approach and, in particular, the degree of structural retention during harvest. Assuming that more structural retention allows a faster return to a complex stand structure, there is a nonlinear relationship between structural retention at the stand level and the rotation length required to achieve a given proportion of the landscape in complexly structured stands (fig. 23.3). (Note this could apply to any desired structural feature at the landscape scale, for example, the proportion of a landscape in a structure that fully regulates hydrology.)

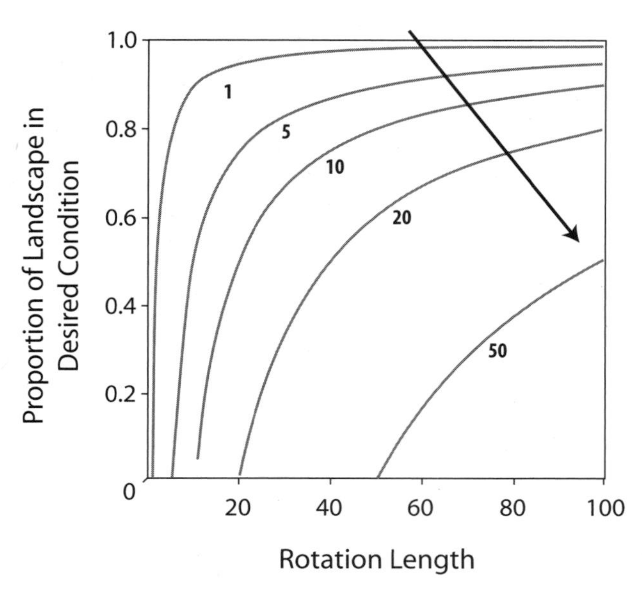

Figure 23.3. The effect of rotation length on the proportion of a landscape in some desired condition. Curves represent rates of recovery to the desired condition. The bold arrow indicates increasing recovery times. For example, assume the desired condition is hydrologic regulation equal to the preharvest forest. If it takes 20 years following harvest to recover and the rotation length is 50 years, in the fully regulated forest slightly more than 50 percent of the landscape will meet that condition. (Note that 20 years is illustrative only. Actual recovery times may be shorter or longer depending on conditions.)

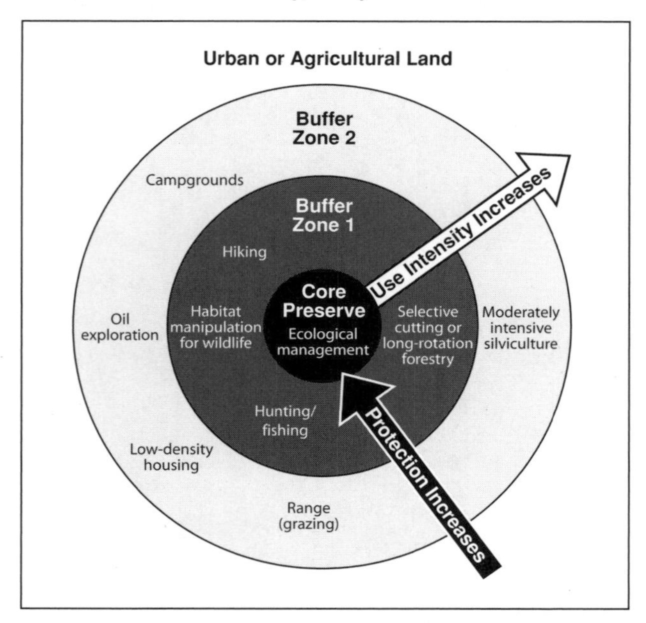

Figure 23.4. A multiple use model. A core preserve is surrounded by zones of increasingly intense use as one moves away from the preserve. (Noss 1987)

23.2.4 The Landscape of the Future

Assuming that the objective is to balance commodity production with maintenance of biological diversity and landscape stability, what should the forested landscape of the future look like? Harris (1984) proposed a basic scheme that he called the multiple use module (fig. 23.4) A multiple use module consists of four land classes: (1) a core reserve area, which is surrounded and buffered by (2) an area of "light-touch" management, such as selection forestry and long rotations (we might call these first-order buffers), which is surrounded and buffered by (3) an area of moderately intensive management (i.e., second-order buffers), which could include green-retention cuts and perhaps some selection cuts and long rotations. The fourth category consists of lands outside of these three that are highly altered by humans (e.g., intensively managed plantations, farms, suburbs, cities).

Any given region would have several to many interconnected multiple use modules, depending on the number of different unique habitats that require protection and the number of core reserves that are designated for each habitat. Regarding the first, a good example is the analysis by Diamond (1986) of the proper design of a reserve system in New Guinea:

> New Guinea's mountains rise to 16,500 feet. From lowland forest one ascends through hill forest, oak forest (*Castanopsis*), southern beech forest (*Nothofagus*), subalpine forest, and alpine grassland to glaciers on the highest peaks. An essential consideration in reserve design is that almost all species occupy only a fragment of this altitudinal gradient. For instance, 90 percent of New Guinea bird species have altitudinal ranges spanning less than 6000 feet, and many span less than 1000 feet. . . . Even species occupying broad bands require the whole span to complete their life cycle. . . . Hence the foremost habitat considerations in reserve design are that reserves must be selected to represent all altitudinal bands, and that the protected altitudinal bands in a given district must be joined as a continuous transect in order to permit migration.

Diamond points out that a mosaic of different forest types also exists within any given altitudinal band in New Guinea, thereby resulting in a vertical and horizontal mosaic of forest types that (as discussed in chapter 5) typifies forested landscapes throughout the world.

The regional landscape emerges when one combines three things: (1) core protected areas, including different forest types and different representatives of each type along with their buffers; (2) protected corridors; and (3) lands devoted to more or less intensive human use. The form such a multiple use landscape might take in any one region would depend on land use constraints such as human population density or the degree of private versus public land ownership. Where population density or land-ownership patterns restrict the amount of land that can be devoted to low-intensity use, the multiple use landscape might be as illustrated in figure 23.5A. Riparian zones are often designated as connecting corridors in such a scheme (Harris 1984) and are important special

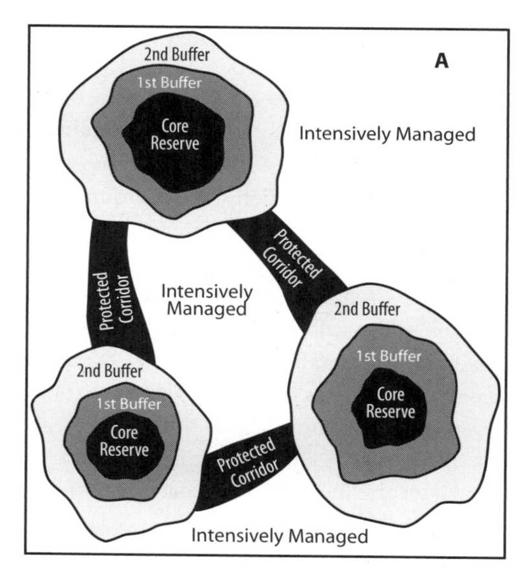

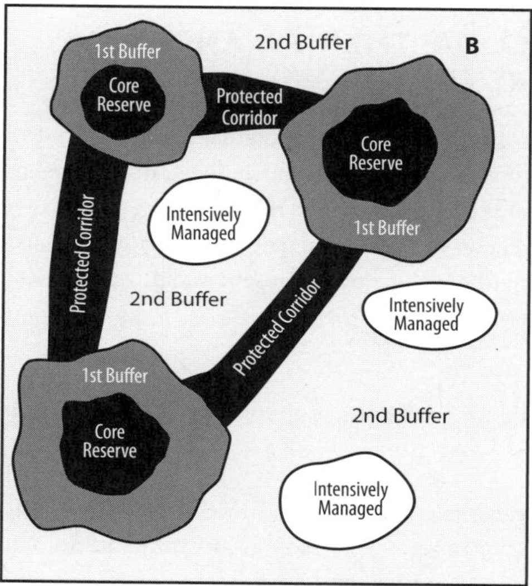

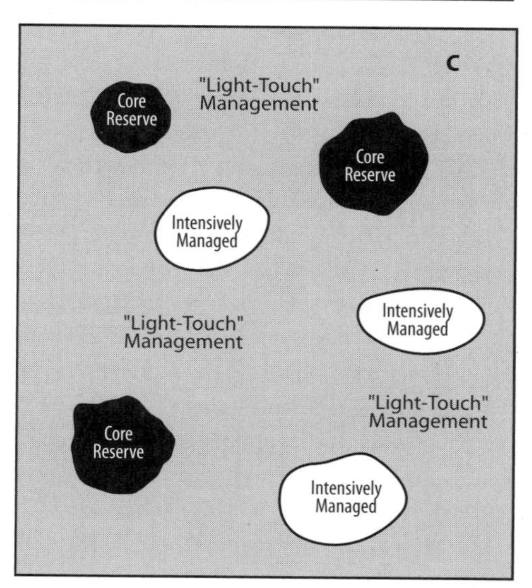

Figure 23.5. Various levels of multiple use landscapes. The protection of biological diversity increases from A through C, while commodity production decreases. The stability of commodity production could increase from A to C, however, depending on how closely it is tied to biodiversity.

Fischer et al. (2006) list 10 guiding principles for maintaining biodiversity, ecosystem function, and resilience in commodity production landscapes:

Pattern-Oriented Strategies

1. Maintain and create large, structurally complex patches of native vegetation.
2. Maintain structural complexity throughout the landscape.
3. Create buffers around sensitive areas.
4. Maintain or create corridors and stepping stones.
5. Maintain landscape heterogeneity and capture environmental gradients.

Process-Oriented Strategies

6. Maintain key species interactions and functional diversity.
7. Apply appropriate disturbance regimes.
8. Control aggressive, over-abundant, and invasive species.
9. Minimize threatening ecosystem processes.
10. Maintain species of particular concern.

In general, these and similar guidelines apply equally to salvage harvests in large disturbances, where environmental protections are no less important than in green-tree harvests (Lindenmayer et al. 2004). Beschta et al. (2004) and Karr et al. (2004) give guidelines specifically for salvage harvests. Also see articles in a special issue of *Conservation Biology*, volume 20, issue 4, 2006.

habitats deserving protection in their own right. The multiple use landscapes illustrated in figure 23.5B and C provide progressively better protection for biological diversity and (depending on the nature of land use in the matrix) greater overall landscape stability. The multiple use landscape of figure 23.5B has the same categories of land use as in figure 23.5A, but core reserves, their buffers, and areas of intense use all exist as islands within a sea of moderately intensive forestry. In other words, the second-order buffers, which might be characterized by structural retention along with some selection forestry and long-rotation areas, have been expanded at the expense of intensely managed areas. This approach is similar to that proposed for spotted-owl conservation on federal lands in the Pacific Northwest (chapter 22). The multiple use landscape of figure 23.5C goes a step further and creates landscapes where reserves and intensively managed areas exist as islands in a sea of "light-touch" management that is wholly characterized by density management, group selection (and perhaps some single-tree selection), and long rotations.

Each of the approaches in figure 23.5 involves trade-offs between wood production and a suite of alternative values, including habitat and watershed protection, aesthetics, narrow versus broad migration corridors, and perhaps long-term ecosystem stability. Compared with the landscape shown in figure 23.5A, reserves (and perhaps all categories of land use) in figure 23.5B and c are better buffered against catastrophic disturbance, and migration routes are spread broadly across the landscape rather than concentrated in narrow corridors. The approach in figure 23.5C expands the habitat for those late successional species that can tolerate some human activity and (because the dominant silvicultural approach includes long rotations) produces future late successional habitat. It is doubtful that long-term wood production can, at present, be reliably compared among these scenarios for any forest type, because not enough is known about the productivity of silvicultural systems other than plantations and the degree of risk that is associated with different silvicultural approaches (i.e., losses from fire, wind, biotic pests, or impacts on soils). From what we have seen in previous chapters, it is safe to say that the least-diverse landscape has the greatest risk. As we discussed in chapter 22, some predict that most of the globe's wood needs will eventually be met by fast-growing plantations on 10 percent or less of the landscape, which would make a global version of the figure 23.5C scenario quite plausible.

The scenarios of figure 23.5 represent three points along a continuum of possible landscape designs. Moreover, a mix of various multiple use landscapes might be appropriate for any given region. For example, the Coast Range of Oregon has a great deal of private lands mixed with federal lands. With the Northwest Forest Plan (NWFP) on federal lands, the Coast Range region falls somewhere between figure 23.5A and B. The Cascades Range (mostly federal land) is better suited (politically) for the more active species and landscape protection afforded by figure 23.5B and C. Regardless of the type of multiple use landscape that is employed, allocation of land to different uses should be based on biology and ecology as wel1 as on ownership. Gap analysis or a similar approach should be used to site core reserves and their buffers, and marginal lands (e.g., steep slopes, fragile soils) should be left untouched or (at most) lightly managed (Terborgh 1992).

All values produced by forests should be considered when evaluating economic trade-offs. As discussed early in the book, many forest values are intangible but nonetheless important to society, including watershed and soil protection, water cycling, carbon storage, recreation, and the psychological benefits to humans of diverse as opposed to monotonous environments. Furthermore, nontraditional forest products may be quite valuable in the marketplace.[8] For example, following their study of marketable goods produced by a tropical rain forest in Peru, Peters et al. (1989) concluded: "Tropical forests are worth considerably more than has been previously assumed, and . . . the actual market benefits of timber are very small relative to those of non-wood resources. Moreover, the total net revenues generated by the sustainable exploitation of 'minor' forest products are two to three times higher than those resulting from forest conversion (to strictly wood-producing plantations)." Peters and colleagues calculated that a moist tropical forest managed for yearly production of fruit and rubber, and periodic partial cutting for wood, would produce more than twice the value in marketable goods than if it were converted to plantation. (Browder [1992] provides a critical assessment of this analysis.) Note that the calculations by Peters et al. do not include the value of medicinal plants, which are common in many forests and may also be quite valuable in the marketplace (Balick and Mendelsohn 1991; Forlines et al. 1992; chapter 1 in this volume).[9]

23.3 CODA: THE NEW AND THE RENEWED

> The overall challenge of sustainability is to avoid crossing irreversible thresholds that damage the life systems of Earth while at the same time to create long-term economic, political, and moral arrangements that secure the well-being of present and future generations.
>
> ORR, 2002

> Every science begins as philosophy and ends as art.
>
> WILL DURANT

The basic principles discussed in this chapter are not new; rather, they renew a very old tradition: the forester who seeks to understand the natural rhythms of forested landscapes and to manage in harmony with those. As we pointed out at the beginning of this chapter, however, several things are new, and taken together, these argue forcefully for revaluating our management approach. The magnitude and rapidity of predicted climate change, and its potential to stress all ecosystems, is unprecedented. Also unprecedented is the degree to which virtually all aspects of society and nature have been homogenized, a fact that has forcefully invoked the common human need for diversity and beauty as well as material goods and has led to the strong environmental movements that are springing up throughout the world and that cut across all socioeconomic classes. Much knowledge is new; every day adds to our understanding of forest ecosystems: their complexity, and the vital role that they play within the global ecosystem. And, as is typically the case in the study

[8] See http://ncseonline.org/NCSSF/pageout.cfm?targetURL=http://www.ifcae.org/projects/ncssf/.

[9] An AP wire report on August 14, 2006, cited the problem of thieves stealing the bark from slippery elm trees in Kentucky. Slippery elm bark has been traditionally used in Appalachia for a variety of ailments and is now in demand on the mass herbal market. The article quotes John Garrison of the National Park Service as saying, "There's a huge market in botanicals going into herbal medicines. Virtually everything on public lands has a market."

of complex systems, the more we learn, the more we realize how much remains to be learned, recalling Aldo Leopold's first rule of intelligent tinkering—to keep all the pieces.

We are at a critical juncture in human history. The decisions that we make over the next few decades will determine the legacies of life, beauty, and security that are passed to the future. The unprecedented responsibility that has fallen to our generation demands solutions that rather than fragmenting and dividing, approach the earth and all of its inhabitants as an integrated and interdependent whole. The real world emerges from ecology, economics, sociology, philosophy, and art: real-world solutions must deal holistically with all these. Perry and Maghembe (1989) put it this way:

> Those of us charged with tending the Earth's life support systems can no longer focus solely on maximizing products or dollar values, but must manage for sustainability as well. To do this, our basic unit of management must be the ecosystem. When arguing for a view grounded in systems rather than individuals (crops), one must at some point specify the boundaries of the system. It seems increasingly clear that the system of relevance in the 20th century is the Earth, and that the old Cartesian view of humans standing outside and manipulating systems like puppeteers is no longer valid. We react to nature, nature reacts to us, and the ultimate outcome may not be predictable. . . . The technology of reductionism is not sufficient to cope with the "certain uncertainty" that paradoxically, has been magnified by our own power to manipulate. We must also invoke a technology of holism that draws on and learns from the robustness of nature. In doing so, we in a sense return to the ancient notion that the variety of ways in which humans interact with nature cannot be compartmentalized, but represents a whole that gives humankind vitality and balance.

Some have objected to the notion that ecosystems can be "managed," citing it as yet another example of human hubris—the "command and control" approach that has shown so little success with complex nature. Rather than command and control, however, ecosystem-based management should be viewed, and approached, like sailing. No sailor is foolish enough to believe the winds and currents can be controlled. Rather, a sailor learns the winds and currents and uses them to help reach a destination. In ecosystem-based management, one learns the forces that spring from and shape nature and uses those to help achieve sustainability.

This book has been about science, but we close with this anonymous piece of wisdom: "Science can be very useful to those who would be healers, but it cannot substitute for heart, intuition, and hands."

23.4 SUMMARY

Overpopulation, pollution, and climate change are major threats to forests and the biological diversity that they harbor. To maintain biological diversity (and along with it the health and integrity of entire ecosystems), silviculture must do two things: (1) protect species and habitats that have no market value, and (2) mimic (to the degree possible) natural disturbance and successional patterns at the scale of both stands and landscapes. Intensive forest management has a place, but when applied without benefit of broader goals, perspectives, and context, it does rather poorly in protecting ecosystem complexity and biological diversity. A more ecologically based management focuses on what it leaves behind rather than on what it takes. Biological legacies are protected and habitat imbalances redressed by restoring forested landscapes to fairly represent critical habitats. Rates of harvest do not exceed rates of recovery nor leave the forest vulnerable to the cumulative effects of additional disturbances. Simplified, intensively managed forests exist as islands within a more complex, naturally based matrix rather than vice versa, producing the shifting mosaic that characterized many natural forest landscapes. Silvicultural techniques for achieving this vary with forest type but include structural retention, partial harvest, density management, and long rotations. Ultimately, a sustainable future will be achieved only by considering the earth and all of its inhabitants as an integrated, interdependent whole.

BIBLIOGRAPHY

Aber, J.D., and Magill, A.H. 2004. Chronic nitrogen additions at the Harvard Forest (USA): the first 15 years of a nitrogen saturation experiment. *For. Ecol. Manage.* 196:1–5.

Aber, J.D., Currie, W., Castro, M., Martin, M., and Ollinger, S. 2004. Synthesis and extrapolation: models, remote sensing, and regional analyses. In *Forests in time.* Edited by D. R. Foster and J.D. Aber. Yale Univ. Press, New Haven, CT, chap. 17.

Aber, J. D., Melillo, J. M., Nadelhoffer, K. J., Pastor, J., and Boone, R.D. 1991. Factors controlling nitrogen cycling and nitrogen saturation in northern temperate forest ecosystems. *Ecol. Applic.* 1:303–315.

Aber, J., Neilson, R.P., McNulty, S., Lenihan, J.M., Bachelet, D., and Drapek, R.J. 2001. Forest processes and global environmental change: predicting the effects of individual and multiple stressors. *BioScience* 51:735–751.

Aber, J.D., Ollinger, S.V., and Driscoll, C.T. 1997. Modeling nitrogen saturation in forest ecosystems in response to land use and atmospheric deposition. *Ecol. Model.* 101:61–78.

Abrahamsen, G. 1983. Sulphur pollution: Ca, Mg and Al in soil and soil water and possible effects on forest trees. In *Effects of accumulation of air pollutants in forest ecosystems.* Edited by B. Ulrich and J. Pankrath. D. Reidel Publishing Company, Dordrecht.

Abrahamson, D.E. 1989. Global warming: the issue, impacts, responses. In *The challenge of global warming.* Edited by D.E. Abrahamson. Island Press, Washington, DC, 3–34.

Abrams, M.D. 2003. Where has all the white oak gone? *BioScience* 53:927–939.

Abuzinadah, R.A., and Read, D.J. 1986a. The role of proteins in the nitrogen nutrition of ectomycorrhizal plants. I. Utilization of peptides and proteins by ectomycorrhizal fungi. *New. Phytol.* 103:481–493.

Abuzinadah, R.A., and Read, D.J. 1986b. The role of proteins in the nitrogen nutrition of ectomycorrhizal plants. III. Protein utilization by *Betula*, *Picea* and *Pinus* in mycorrhizal association with *Hebeloma crustuliniforme. New Phytol.* 103:507–514.

Achard, F., Eva, H.D., Stibig, H.-J., Mayaux, P., Gallego, J., Richards, T., and Malingreau, J.-P. 2002. Determination of deforestation rates of the world's humid tropical forests. *Science* 297:999–1002.

Adams, J.M., and Woodward, F.I. 1989. Patterns in tree species richness as a test of the glacial extinction hypothesis. *Nature* 339:699–701.

Adams, M.B., Campbell, R.G., Alien, H.L., and Davey, C.B. 1987. Root and foliar nutrient concentrations in loblolly pine: effects of season, site, and fertilization. *For. Sci.* 33:984–996.

Adams, R.M., Rosenzweig, C., Peart, R.M., Ritchie, J.T., McCarl, B.A., Glyer, J.D., Curry, R.B., Jones, J.W., Boote, K.J., and Allen Jr., L.H. 1990. Global climate change and US agriculture. *Nature* 345:219–224.

Addington R.N., Donovan, L.A., Mitchell, R.J., Vose, J.M., Pecot, S.D., Jack, S.B., Hacke, U.G., Sperry, J.S., and Oren, R. 2006. Adjustments in hydraulic architecture of *Pinus palustris* maintain similar stomata conductance in xeric and mesic habitats. *Plant Cell Enviro.* 29:535–545.

Adedeji, F.O. 1984. Nutrient cycles and successional changes following shifting cultivation practice in moist semi-deciduous forests in Nigeria. *For. Ecol. Manage.* 9:87–99.

Adejuwon, J.O., and Ekanade, O. 1987. Edaphic component of the environmental degradation resulting from the replacement of tropical rain forest by tree crops in SW Nigeria. *Int. Tree Crops J.* 4:269–282.

Aerts, R. 1997. Climate, leaf litter chemistry and leaf litter decomposition in terrestrial ecosystems: a triangular relationship. *Oikos* 79:439–449.

Aerts, R. 2006. The freezer defrosting: global warming and litter decomposition rates in cold biomes. *J. Ecol.* 94:713–724.

Agee, J.K. 1993. *Fire ecology of Pacfic Northwest forests.* Island Press, Washington, DC.

Agee, J.K. 1997. The severe weather wildfire—too hot to handle? *Northwest Sci.* 71:153–156.

Agee, J.K., and Smith, L. 1984. Subalpine tree reestablishment after fire in the Olympic Mountains, Washington. *Ecology* 65:810–819.

Agee, J.K., Bahro, B., Finney, M.A., Omi, P.N., Sapsis, D.B., Skinner, C.N., van Wagtendonk, J.W., and Weatherspoon, C.P. 2000. The use of shaded fuel breaks in landscape fire management. *For. Ecol. Manage.* 127:55–66.

Agren, G.I. 2004. The C:N:P stoichiometry of autotrophs—theory and observations. *Ecol. Lett.* 7:185–191.

Aguilera, I.M., Griffiths, R.P., and Caldwell, B. 1993. Nitrgen in ectomycorrhizal mat and nonmat soils of different-age Douglas-fir forests. *Soil Biol. Biochem.* 25:1015–1019.

Aide, T.M. 1988. Herbivory as a selective agent on the timing of leaf production in a tropical understory community. *Nature* 336:574–575.

Airola, D.A. 1986. Brown-headed cowbird parasitism and habitat disturbance in the Sierra Nevada. *J. Wildl. Manage.* 50:571–575.

Akkermans, A. 1979. Symbiotic nitrogen fixers available for use in temperate forestry. In *Symbiotic nitrogen fixation in the management of temperate forests.* Edited by J.C. Gordon, C.T. Wheeler, and D.A. Perry. Forest Res. Lab., Oregon State Univ., Corvallis.

Alban, D.H. 1969. The influence of western hemlock and western red cedar on soil properties. *Soil Sci. Soc. Am. Proc.* 33:453–457.

Alban, D.H. 1982. Effects of nutrient accumulation by aspen, spruce, and pine on soil properties. *Soil Sci. Soc. Am. J.* 46:853–861.

Alban, D.H., Perala, D.A., and Schlaegel, B.E. 1978. Biomass and nutrient distribution in aspen, pine, and spruce stands on the same soil type in Minnesota. *Can. J. For. Res.* 8:290–299.

Albert, P.J., and Jerrett, P.A. 1981. Feeding preferences of spruce budworm (*Choristoneurafumiferana* Clem.) larvae to some host-plant chemicals. *J. Chem. Ecol.* 7:391–401.

Albrektson, A., Aronsson, A., and Tamm, C.O. 1977. The effects of forest fertilization on primary production and nutrient cycling in the forest ecosystem. *Silva Fennica* 11:233–239.

Alcorn, J.B. 1990. Indigenous agroforestry strategies meeting farmer's needs. In *Alternatives to deforestation.* Edited by A.B. Anderson. Columbia Univ. Press, New York, 141–151.

Alegre, J.C., and Cassel, D.K. 1986. Effect of land-clearing methods and postclearing management on aggregate stability and organic carbon content of a soil in the humid tropics. *Soil Sci.* 142:289–295.

Alexander, D.B. 2005. Bacteria and archaea. In *Principles and application of soil microbiology,* 2nd ed. Edited by D.M. Sylvia, J.L. Fuhrmann, P.G. Hartel, and D.A. Zuberer. Prentice Hall, Upper Saddle River, NJ, 101–139.

Alexander, I.J. 1985. The significance of ectomycorrhizas in the nitrogen cycle. In *Nitrogen as an ecological factor.* Edited by J.A. Lee, S. McNeill, and I.H. Rorison. Blackwell, Oxford, 69–93.

Alexander, J.D., Seavy, N.E., Ralph, C.J., and Hogoboom, B. 2006. Vegetation and topographical correlates of fire severity from two fires in the Klamath-Siskiyou region of Oregon and California. *Int. J. Wildl. Fire* 15:237–245.

Alexandre, D.Y. 1978. Le role disseminateur des elephant en Foret de Tai, Cote-D'Jvoire. *Terre Vie* 32:49–72.

Alfaro, R.I., Van Sickle, G.A., Thomson, A.J., and Wegurtz, E. 1982. Tree mortality and radial growth losses caused by the western spruce budworm in a Douglas-fir stand in British Columbia. *Can. J. For. Res.* 12:780–787.

Allen, C.D., Savage, M.S., Falk, D.A., Suckling, K.F., Swetnam, T.W., Schulke, T., Stacey, P.B., Morgan, P., Hoffman, M., and Klingle, J.T. 2002. Ecological restoration of southwestern ponderosa pine ecosystems: a broad perspective. *Ecol. Applic.* 12:1418–1433.

Allen, E.B., and Allen M.F. 1990. The mediation of competition by mycorrhizae in successional and patchy environments. In *Perspectives on plant competition.* Edited by J.B. Grace and D. Tilman. Academic Press, New York, 367–389.

Allen, E.B., Allen, M.F., Egerton-Warburton, L., Corkidi, L., and Gomez-Pompa, A. 2003. Impacts of early- and late-seral mycorrhizae during restoration in seasonal tropical forest, Mexico. *Ecol. Applic.* 13:1701–1717.

Allen, E.B., Allen, M.F., Helm, D.J., Trappe, J.M., Molina, R., and Rincon, E. 1995. Patterns and regulation of mycorrhizal plant and fungal diversity. *Plant Soil* 170:47–62.

Allen, J.C., and Barnes, D.F. 1985. The causes of deforestation in developing countries. *Ann. Assoc. Am. Geogr.* 75:163–184.

Allen, M.F. 1991. *The ecology of mycorrizae.* Cambridge Univ. Press, Cambridge.

Allen, M.F., Crisafulli, C.M., Morris, S.J., Egerton-Warburton, L.M., MacMahon, J.A., and Trappe, J.M. 2005. Mycorrhizae and Mount St. Helens: story of a symbiosis. In *Ecological responses to the 1980 eruption of Mount St. Helens.* Edited by V.H. Dale, F.J. Swanson, and C.M. Crisafulli. Island Press, Washington, DC, 221–232.

Allen, O.N., and Allen, E.K. 1981. *The Leguminosae.* Univ. of Wisconsin Press, Madison.

Allen, T.F.H., and Starr, T.B. 1982. *Hierarchy: perspectives for ecological complexity.* Univ. of Chicago Press, Chicago.

Allendorf, F.W., and Leary, R.F. 1986. Heterozygosity and fitness in natural populations of animals. In *Conservation biology: the science of scarcity and diversity.* Edited by M.E. Soule. Sinauer, Sunderland, MA, 57–76.

Alpert, P. 1996. Integrated conservation and development projects: examples from Africa. *BioScience* 46:845–855.

Alt, D.O., and Hyndman, D.W. 1978. *Roadside geology of Oregon.* Mountain Press, Missoula, MT.

Altman, S.A. 1989. The monkey and the fig. *Am. Sci.* 77(3):256–263.

Alvarez, L.W., Alvarez, W., Asaro, F., and Michel, H.V. 1980. Extraterrestrial cause for the Cretaceous-Tertiary extinction. *Science* 208:1095–1108.

Alverson, W.S., Waller, D.M., and Solheim, S.L. 1988. Forests to deer: edge effects in northern Wisconsin. *Conserv. Biol.* 2:348–358.

Amaranthus, M.P., and Perry, D.A. 1987. Effect of soil transfer on ectomycorrhiza formation and the survival and growth of conifer seedlings in disturbed forest sites. *Can. J. For. Res.* 17:944–950.

Amaranthus, M.P., and Perry, D.A. 1989a. Interaction effects of vegetation type and Pacific madrone soil inocula on survival, growth, and mycorrhiza formation of Douglas-fir. *Can. J. For. Res.* 19:550–556.

Amaranthus, M.P., and Perry, D.A. 1989b. Rapid root tip and mycorrhiza formation and increased survival of Douglas-fir seedlings after soil transfer. *New For.* 3:259–264.

Amaranthus, M.P., and Trappe, J.M. 1993. Effects of erosion on ecto- and VA-mycorrhizal inoculum potential of soil following forest fire in southwest Oregon. *Plant Soil* 150:41–49.

Amaranthus, M.P., Li, C.Y., and Perry, D.A. 1990a. Influence of vegetation type and madrone soil inoculum on associative nitrogen fixation in Douglas-fir rhizospheres. *Can. J. For. Res.* 20:368–371.

Amaranthus, M.P., Molina, R., and Perry, D.A. 1990b. Soil organisms, root growth, and forest regeneration. In *Forestry on the frontier.* Society of American Foresters, Washington, DC, 89–93.

Amaranthus, M.P., Parrish, D., and Perry, D.A. 1989. Decaying logs as moisture reservoirs following drought and wildfire. In *Stewardship of soil, water, and air resources.* Proc. Watershed 89, Juneau, Alaska. RlO-MB-77. Edited by E. Alexander. U.S. Department of Agriculture, Forest Service, Region 10 Juneau, Alaska.

Amaranthus, M.P., Perry, D.A., and Borchers, S.L. 1987. Reduction of native mycorrhizae reduce growth of Douglas-fir seedlings. In *Proc. 7th NACOM.* Edited by D.M. Sylvia and J.H. Graham. Institute of Food and Agricultural Sciences, Univ. of Florida, Gainesville, 80–81.

Amaranthus, M.P., Rice, R.M., Barr, N.R., and Ziemer, R.R. 1985. Logging and forest roads related to increased debris slides in southwestern Oregon. *J. For.* 83:229–233.

Amaranthus, M., Trappe, J.M., Bednar, L., and Arthur, D. 1994. Hypogeous fungal production in mature Douglas-fir forest fragments and surrounding plantations and its relation to coarse woody debris and animal mycophagy. *Can. J. For. Res.* 24:2157–2165.

Amaranthus, M.P., Trappe, J.M., and Perry, D.A. 1993. Soil moisutre, native revegetation, and *Pinus lambertiana* seedling growth, and mycorrhiza formation following wildfire and grass seeding. *Restor. Ecol.* (Sept.):188–195.

Amason, T., Lambert, J.D.H., Gale, J., Cal, J., and Vernon, H. 1982. Decline of soil fertility due to intensification of land use by shifting agriculturists in Belize, Central America. *Agroecosystems* 8:27–37.

Andersen, D.E., and MacMahon, J.A. 1981. Population dynamics and bioenergetics of a fossorial herbivore, *Thomomys talpoides* (Rodentia:Geomyidae) in a spruce-fir sere. *Ecol. Monogr.* 51:179–202.

Anderson, A.B. 1990. *Alternatives to deforestation.* Columbia Univ. Press, New York.

Anderson, D.W. 1988. The effect of parent material and soil development on nutrient cycling in temperate ecosystems. *Biogeochemistry* 5:71–97.

Anderson, I.E., Levine, J.S., Poth, M.A., and Riggan, P.J. 1988. Enhanced biogenic emissions of nitric oxide and nitrous oxide following surface biomass burning. *J. Geophys. Res.* 93:3893–3898.

Anderson, J.M. 1991.The effects of climate change on decomposition processes in grassland and coniferous forests. *Ecol. Applic.* 1:326–347.

Anderson, J.M., Ineson, P., and Huish, S.A. 1983. Nitrogen and cation mobilization by soil fauna feeding on leaf litter and soil organic matter from deciduous woodlands. *Soil Biol. Biochem.* 15:463–467.

Anderson, J.M., Leonard, M.A., Ineson, P., and Huish, S. 1985. Faunal biomass: a key component of a general model of nitrogen mineralization. *Soil Biol. Biochem.* 17:735–737.

Anderson, J.P.E., and Domsch, K.H. 1980. Quantities of plant nutrients in the microbial biomass of selected soils. *Soil Sci.* 130:211–216.

Anderson, L., Carlson, E.E., and Wakimoto, R.H. 1987. Forest fire frequency and western spruce budworm outbreaks in western Montana. *For. Ecol. Manage.* 22:251–260.

Anderson, R.M., and May, R.M. 1980. Infectious diseases and population cycles of forest insects. *Science* 210:658–661.

Andreae, M.O. 1986. The oceans as a source of biogenic gases. *Oceanus* 29(4):27–35.

Andreassian, V. 2004. Water and forests: from historical controversy to scientific debate. *J. Hydrol.* 291:1–27.

Andren, H. 1992. Corvid density and nest predation in relation to forest fragmentation: a landscape perspective. *Ecology* 73:794–804.

Andrewartha, H.G., and Birch, L.E. 1984. *The ecological web.* Univ. of Chicago Press, Chicago, 506.

Angermeir, P.L., and Karr, J.R. 1994. Biological integrity versus biological diversity as policy directives. *BioScience* 44:690–697.

Antonovics, J. 1992. Toward community genetics. In *Plant resistance to herbivores and pathogens: ecology, evolution, and genetics.* Edited by R.S. Frite and E.L. Simms. Univ. of Chicago Press, Chicago, 426–449.

April, R., and Keller, D. 1990. Mineralogy of the rhizosphere in forest soils of the eastern United States. *Biogeochemistry* 9:1–18.

Arevalo, J.R., deCoster, J.K., McAllister, S.D., and Palmer, M.W. 2000. Changes in two Minnesota forests during 14 years following catastrophic windthrow. *J. Veg. Sci.* 11:833–840.

Arms, K.P., and Feeny, R.E. 1974. Sodium: stimulus for puddling behavior by tiger swallowtail butterflies, *Papilio glaucus. Science* 185:372–374.

Armson, K.A. 1977. *Forest soils.* Univ. of Toronto Press, Toronto.

Arnold, A.E. 2002. Neotropical fungal endophytes: diversity and ecology. Ph.D. thesis. Univ. of Arizona, Tucson.

Arnold, A.E., Mejia, L.C., Kylio, D., Rojas, E.I., Maynard, Z., Robbins, N., and Herre, E.A. 2003. Fungal endophytes limit pathogen damage in a tropical tree. *PNAS* 100:15649–15654.

Arp, P.A., and Ouimet, R. 1986. Uptake of AL, CA, and P in black spruce seedlings: effect of organic versus inorganic AL in nutrient solutions. *Water Air Soil Pollut.* 31:367–375.

Asanao, Y., Ohte, N., and Uchida, T. 2004. Sources of weathering-derived solutes in two granitic catchments with contrasting forest growth. *Hydrol. Proc.* 18:651–666.

Ashby, W.R. 1956. *An introduction to cybernetics.* John Wiley & Sons, New York.

Ashok, K., Guan, Z., and Yamagata, T. 2001. Impact of the Indian Ocean dipole on the relationship between the Indian monsoon rainfall and ENSO. *Geophys. Res. Lett.* 28(23):4499–4502.

Ashton, I.W., Hyatt, L.A., Howe, K.M., Gurevitch, J., and Lerdau, M.T. 2005. Invasive species accelerate decomposition and litter nitrogen loss in a mixed deciduous forest. *Ecol. Applic.* 15:1263–1272.

Ashton, P.S. 1964. *Ecological studies in the mixed dipterocarp forests of Brunei State.* Clarendon Press, Oxford.

Ashton, P.S. 1988. Dipterocarp biology as a window to the understanding of tropical forest structure. *Annu. Rev. Ecol. Syst.* 19:347–370.

Askins, R.A., and Ewert, D.N. 1991. Impact of Hurricane Hugo on bird populations on St. John, U.S. Virgin Islands. *Biotropica* 23:481–487.

Asner, G.P., and Vitousek, P.M. 2005. Remote analysis of biological invasion and biogeochemical change. *PNAS* 102:4383–4386.

Asner, G.P., Broadbent, E.N., Oliveira, P.J.C., Keller, M., Knapp, D.E., and Silva, J.N.M. 2006. Condition and fate of logged forests in the Brazilian Amazon. *PNAS* 103:12947–12950.

Asner, G.P., Knapp, D.E., Broadbent, E.N., Oliveira, P.J.C., Keller, M., and Silva, J.N. 2005. Selective logging in the Brazilain Amazon. *Science* 310:480–482.

Assmann, E. 1970. *The principles of forest yield study.* Pergamon Press, Oxford.

Atsatt, P.R. 1988. Are vascular plants "inside-out" lichens? *Ecology* 69:17–23.

Atsatt, P.R. 1991. Fungi and the origin of land plants. In *Symbiosis as a source of evolutionary innovation. Speciation and morphogenesis.* Edited by L. Margulis and R. Fester. MIT Press, Cambridge, MA, 301–315.

Atsatt, P.R., and O'Dowd, D.J. 1976. Plant defense guilds. *Science* 193:24–29.

Atzet, T., Powers, R.F., McNabb, D.H., Amaranthus, M.P. and Gross, E.R. 1989. Maintaining long-term forest productivity in southwest Oregon and northern California. In *Maintaining the long-term productivity of Pacific Northwest forest ecosystems.* Edited by D.A. Perry, R. Meurisse, B. Thomas, et al. Timber Press, Portland, OR, 185–201.

Austin, M.P., Nicholls, A.O., and Margules, E.R. 1990. Measurement of the realized niche: environmental niches of five Eucalyptus species. *Ecol. Monogr.* 60; 161–177.

Axelsson, B. 1981. *Site differences in yield differences in biological production or in redistribution of carbon within trees.* Res. Rep. No. 9, Dept. Ecol. Environ., Res. Swed. Univ. Agric. Sci., Uppsala.

Axelrod, R., and Hamilton, W.D. 1981. The evolution of cooperation. *Science* 211:1390–1396.

Ayres, M.P., and Lombardero, M.J. 2000. Assessing the consequences of global change for forest disturbance from herbivores and pathogens. *Sci. Total Environ.* 262:263–286.

Ba, A.M., Garbaye, J., and Dexheimer, J. 1990. Influence of soil propagules on the sequence of ectomycorrhizal infection on *Afrelia africana* Sm. seedlings. In *Eighth North American conference on mycorrhizae.* Edited by M.E. Allen and S.E. Williams, Sept. 5–8, 1990. Jackson, Wyoming.

Bachelet, D. 2006. Modeled shifts in vegetation in the United States in Response to Climate Change. Presentation to National Commission on the Science of Sustainable Forestry's symposium on disturbance. Denver, CO. March 3–5, 2006.

Bachelet, D., and Neilson, R. 2000. Biome redistribution under climate change. In *The impact of climate change on America's forests.* Gen. Tech. Rep. RMRS-GTR-59. Edited by L.A. Joyce and R. Birdsey. U.S. Department of Agriculture, Forest Service. Fort Collins, CO.

Bachelet, D., Lenihan, J.M., Daly, C., Neilson, R.P., Ojima, D.S., and Parton, W.J. 2001. *MC1: a dynamic vegetation model for estimating the distribution of vegetation and associated carbon, nutrients, and water—technical documentation.* Version 1.0. Gen. Tech. Rep. PNW-GTR-508. U.S. Department of Agriculture, Forest Service, Pacific Northwest Res. Sta., Portland, OR.

Bachelet, D., Neilson, R.P., Hickler,T., Drapek, R.J., Lenihan, J.M. Sykes, M.T., Smith, B., Sitch, S., and Thonicke, K. 2003. Simulating past and future dynamics of natural ecosystems in the United States. *Global Biogeochem. Cycles* 17:14–1 to 14–21.

Bachelet, D., Neilson, R.P., Lenihan, J.M., and Drapek, R.J. 2004. Regional differences in the carbon source-sink potential of natural vegetation in the U.S. *Ecol. Manage.* 33(Supp 1).

Baes Jr., C.F. 1982. Effects of ocean chemistry and biology on atmospheric carbon dioxide. In *Carbon dioxide review: 1982.* Edited by W.C. Clark. Oxford Univ. Press, New York.

Baes III, C.F., and McLaughlin, S.B. 1986. *Multielemental analysis of tree rings: a survey of coniferous trees in the Great Smoky Mountains National Park.* Report ORNL-6155 ed. National Technical Information Service, Springfield, VA.

Bagley, M.J., Franson, S.E., Christ, S.A., Waits, E.R., and Toth, G.P. 2002. *Genetic diversity as an indicator of ecosystem condition and sustainability: utility for regional assessments of stream condition in the eastern United States.* U.S. Environmental Protection Agency, Cincinnati, OH.

Bahuguna, S.L. 1978. *Himalayan trauma: forests, faults, floods.* Ghandi Peace Foundation, New Delhi, 1978.

Bailey, J.D., and Tappeiner, J.C. 1998. Effects of thinning on structural development in 40– to 100–year old Douglas-fir stands. *For. Ecol. Manage.* 108:99–113.

Bailey, J.D., Mayrsohn, C., Doescher, P.S., St. Pierre, E., and Tappeiner, J.C. 1998. Understory vegetation in old and young Douglas-fir forests of western Oregon. *For. Ecol. Manage.* 112:289–302.

Bailey, R.G. 1995. *Ecosystem geography.* Springer, New York.

Baillie, I.C. 1989. Soil characteristics and classification in relation to the mineral nutrition of tropical wooded ecosystems. In *Mineral nutrients in tropicalforest and savanna ecosystems.* Edited by J. Proctor. Blackwell Scientific Publications, Oxford, 15–26.

Bak, P., Chen, K., and Tang, C. 1990. A forest-fire model and some thoughts on turbulence. *Phys. Lett. A* 147:297–300.

Bak, P., Tang, C., and Wiesenfeld, K. 1988. Self-organized criticality. *Phys. Rev. A* 38:364–374.

Baker, T.G., and Attiwill, P.M. 1984. Acetylene reduction in soil and litter from pine and eucalypt forests in south-eastern Australia. *Soil Biol. Biochem.* 16:241–245.

Baker, W.L., and Ehle, D. 2001. Uncertainty in surface fire history: the case of ponderosa pine forests in the western United States. *Can. J. For. Res.* 31:1205–1226.

Balanyá, J., Oller, J.M., Huey, R.B., Gilchrist, G.W., and Serra, L. 2006. Global genetic change tracks global climate warming in *Drosphila subobscura. Science* 313:1773–1775.

Balda, R.P. 1969. Foliage use by birds of the oak-juniper woodland ponderosa pine forest in southwestern Arizona. *Condor* 71:399–412.

Baldocchi, D., et al. 2001. FLUXNET: a new tool to study the temporal and spatial variablility of econsystem-scale carbon dioxide, water vapor, and energy flux densities. *Bull. Am. Meteorol. Soc.* 82:2415–2434.

Baldocchi, D.D. 2003 Assessing the eddy covariance technique for evaluating carbon dioxide exchange rates of ecosystems: past, present, and future. *Global Change Biol.* 9:479–492.

Baldwin, I.T., and Schultz, J.C. 1983. Rapid changes in tree leaf chemistry induced by damage: evidence for communication between plants. *Science* 221:277–278.

Baldwin, I.T., Halitschke, R., Paschold, A., von Dahl, C.V., and Preston, C.A. 2006. Volatile signaling in plan-plant interactions: "talking trees" in the genomics era. *Science* 311:812–815.

Baldwin, M., Kellogg, C.E., and Thorp, J. 1938. Soil classification. In *Soils and men: yearbook of agriculture.* U.S. Government Printing Office, Washington, DC, 979–1001.

Baleshta, K.E., Simard, S.W., Guy, R.D., and Chanway, C.P. 2005. Reducing paper birch density increases Douglas-fir growth rate and Armillaria root disease incidence in southern interior British Columbia. *For. Ecol. Manage.* 208:1–13.

Balick, M.J., and Mendelsohn, R. 1991. Asessing the economic value of traditional medicines from tropical rain forests. *Conserv. Biol.* 6:128–130.

Ballard, T.M. 1986. Overview of forest nutritional problems in the B.C. interior, and methods of diagnosis. In *Proceedings of the Interior Forest Fertilization Workshop, Stockmans Hotel, Kamloops.* Univ. of British Columbia, Vancouver, 10.

Ballard, T.M., and Carter, R.E. 1985. *Evaluating forest stand nutrient status.* Land Management Report No. 20. British Columbia Ministry of Forests, Victoria.

Balmford, A., Bruner, A., Cooper, P., Costanza, R., Farber, S., Green, R.E., Jenkins, M., Jefferiss, P., Madden, J., Munro, K., Myers, N., Naeem, S., Paavola, J., Rayment, M., Rosendo, S., Roughgarden, J., Trumper, K., and Turner, R.K. 2004. Economic reasons for conserving wild nature. *Science* 297:950–953.

Baltensweiler, W., Benz, G., Bovey, P., and Delucchi, V. 1977. Dynamics of larch bud moth populations, *Annu. Rev. Entomol.* 22:79–100.

Banin, A. 1986. Global budget of N_2O: the role of soils and their change. *Sci. Total Environ.* 55:27–38.

Barak, P. 2003. Essential elements for plant growth. Plant Nutrient Management course. www.soils.wisc.edu/courses/soils326/listofel.htm.

Barbosa, P., Krischik, V.A., and Jones, C.G., editors. 1991. *Microbial mediation of plant-herbivore interactions.* John Wiley & Sons, New York.

Barbosa, P., Segarra, A.E., Gross, P., Caldas, A., Ahlstrom, K., Carlson, R.W., Ferguson, D.C., Grissell, E.E., Hodges, R.W., Marsh, P.M., Poole, R.W., Schauff, M.E., Shaw, S.R., Whitfield, J.B., and Woodley, N.E. 2001. Differential parasitism of macrolepidopteran herbivores on two deciduous tree species. *Ecology* 82:698–704.

Bardgett, R.D., and Wardle, D.A. 2003. Herbivore-mediated linkages between aboveground and belowground communities. *Ecology* 84:2258–2268.

Barklund, P., and Unestam, T. 1988. Infection experiments with *Gremmeniella abietina* on seedlings of Norway spruce and Scots pine. *Eur. J. For. Pathol.* 18:409–420.

Barlow, C., and Volk, T. 1992. Gaia and evolutionary biology. *BioScience* 42:686–693.

Barro, S.C., and Conard, S.G. 1991. Fire effects on California chaparral systems: an overview. *Environ. Internat.* 17:135–149.

Bascompte, J., Jordano, P., Melian, C.J., and Olesen, J.M. 2003. The nested assembly of plant-animal mutualistic networks. *PNAS* 100:9383–9387.

Bascompte, J., Jordano, P., and Olesen, J.M. 2006. Asymmetric coevolutionary networks facilitate biodiversity maintenance. *Science* 312:431–433.

Batista, W.B., and Platt, W.J. 2003. Tree population responses to hurricane disturbance: syndromes in a south-eastern USA old-growth forest. *J. Ecology* 91(2):197–212.

Bawa, K.S. 1990. Plant-pollinator interactions in tropical rain forests. *Annu. Rev. Ecol. Syst.* 21:399–422.

Baxter, J.W., and Dighton, J. 2001. Ectomycorrhizal diversity alters growth and nutrient acquisition of grey birch (*Betula populifolia*) seedlings in host-symbiont culture conditions. *New Phytol.* 152:139–149.

Bazzaz, F. A. 1990. The response of natural ecosystems to the rising global CO_2 levels. *Annu. Rev. Ecol. Syst.* 21:167–196.

Bazzaz, F.A., Coleman, J.S., and Morse, S.R. 1990. Growth responses of seven major co-occurring tree species of the northeastern United States to elevated CO_2. *Can. J. For. Res.* 20:1479–1484.

Beadle, N.C.W. 1966. Soil phosphate and its role in molding segments of the Australian flora and vegetation, with special reference to xeromophy and sclerophylly. *Ecology* 47:992–1007.

Beadle, N.C.W., and White, G.J. 1968. The mineral content of the trunks of some Australian woody plants. *Proc. Ecol. Soc. Australia* 3:55–60.

Beardmore, J.A. 1983. Extinction, survival, and genetic varition. In *Genetics and conservation: a reference of managing wild animal and plant populations.* Edited by C.M. Schoenwald-Cox, S.M. Chambers, B. Macflryde, and W.L. Thomas. Benjamin/Cummings, London, 125–151.

Beaty, R.M., and Taylor, A.H. 2001. Spatial and temporal variation of fire regimes in a mixed conifer forest landscape, southern Cascades, California, USA. *J. Biogeogr.* 28(8): 955–966.

Becker, C.G., Fonseca, C.R., Haddad, C. F.B., Batista, R.F., and Prado, P.I. 2007. Habitat split and the global decline of amphibians. *Science* 318:1775–1777.

Becking, J.H. 1976. Nitrogen fixation in some natual ecosystems in Indonesia. In *Symbiotic nitrogen fixation in plants.* Edited by P.S. Nutman. Cambridge Univ. Press, Cambridge, 539–550.

Beckwith, R.C., and Burnell, D.G. 1982. Spring larval dispersal of the western spruce budworm (Lepidoptera: Tortricidae) in north central Washington. *Environ. Entomol.* 11:828–832.

Beedlow, P.A., Tingey, D.T., Phillips, D.L., Hogsett, W.E., and Olszyk, D.M. 2004. Rising atmospheric CO_2 and carbon sequestration in forests. *Front. Ecol. Environ.* 6:315–321.

Beerling, D.J., and Berner, R.A. 2005. Feedbacks and the coevolution of plants and atmospeheric CO_2. *PNAS* 102:1302–1305.

Beier, P., and Noss, R.F. 1998. Do habitat corridors provide connectivity? *Conserv. Biol.* 12:1241–1252.

Belanger, N., Cote, B., Fyles, J.W., Courchesne, F., and Hendershoot, W.H. 2004. Forest regrowth as the controlling factor of soil nutrient availability 75 years after fire in a deciduous forest of southern Quebec. *Plant Soil* 262:363–372.

Beletsky, L.D., and Orians, G.H. 1989. Familiar neighbors enhance breeding success in birds. *PNAS* 86:7933–7936.

Bell, G. 2001. Neutral macroecology. *Science* 293:2413–2418.

Bell, T., Ager, D., Song, J.-I., Newman, J.A., Thompson, I.P., Lilley, A.K., and van der Gast, C.J. 2005. Larger islands house more bacterial taxa. *Science* 308:1884–1886.

Bell, S., and Apostol, D.. 2007. *Designing sustainable forest landscapes*. Taylor & Francis, London, 368.

Bellingham, P.J., Tanner, E.V.J., and Healey, J.R. 1995. Damage and responsiveness of Jamaican montane tree species after disturbance by a hurricane. *Ecology* 76:2562–2580.

Belnap, J. 2005. Cyanobacteria and algae. In *Principles and application of soil microbiology,* 2nd ed. Edited by D.M. Sylvia, J.L. Fuhrmann, P.G. Hartel, and D.A. Zuberer. Prentice Hall, Upper Saddle River, NJ, 162–180.

Belovsky, G.E., and Slade, J.B. 2000. Insect herbivory accelerates nutrient cycling and increases plant production. *PNAS* 97:14414–14417.

Benkman, C.W., Holimon, W.C., and Smith, J.W. 2001. The influence of a competitor on the geogprahic mosaic of coevolution between crossbills and lodgepole pine. *Evolution* 55:282–294.

Benkman, C.W., Parchman, T.L., Favis, A., and Seipielski, A.M. 2003. Reciprocal selection causes and coevolutionary arms race between crossbills and lodgepole pine. *Am. Nat.* 162:182–194.

Bennett, K.A. 1982. Effects of slash burning on surface soil erosion in the Oregon Coast Range. Masters thesis. Oregon State Univ., Corvallis.

Bentley, B.L., and Carpenter, E.J. 1980. Effects of desiccation and rehydration on nitrogen fixation by echnical epiphylls in a tropical rainforest. *Microb. Ecol.* 6:109–113.

Bentley, B.L., and Carpenter, E.J. 1984. Direct transfer of newly-fixed nitrogen from free-living epiphyllous microorganisms to their host plant. *Oecologia (Berlin)* 63:52–56.

Berenbaum, M. 1980. Adaptive significance of midget pH in larvel *Lepidoptera*. *Am. Nat.* 115:138–146.

Berendse, F., Berg, B., and Bosatta, E. 1987. The effect of lignin and nitrogen on the decomposition of litter in nutrient-poor ecosystems: a theoretical approach. *Can. J. Bot.* 65: 1116–1120.

Berg, B. 1984. Decomposition of root litter and some factors regulating the process: long-term root litter decomposition in a Scots pine forest. *Soil Biol. Biochem.* 16:609–618.

Berg, B., and Meentemeyer, V. 2002. Litter quality in a north European transect versus carbon storage potential. *Plant Soil* 242:83–92.

Berg, B., and Staaf, H. 1980. Decomposition rate and chemical changes in Scots pine litter. II. Influence of chemical composition. In *Structure and function of northern coniferous forests—an ecosystem study.* Edited by T. Persson. *Ecol. Bull. (Stockholm)* 32: 373–390.

Berg, B., and Staaf, H. 1981. Leaching, accumulation and release of nitrogen in decomposing forest litter. In *Terrestrial nitrogen cycles.* Edited by F.E. Clark and T. Rosswall. *Ecol. Bull. (Stockholm)* 33:163–178.

Berg B., Johansson, M.B., and Meentemeyer, V. 2000. Litter decomposition in a transect of Norway spruce forests: substrate quality and climate control. *Can. J. For. Res.* 30:1136–1147.

Berg, E.E., Henry, J.D., Fastie, C.L, De Volder, A.D, and Matsuoka, S.M. 2006. Spruce beetle outbreaks on the Kenai Peninsula, Alaska, and Kluane National Park and Reserve, Yukon Territory: Relationship to summer temperatures and regional differences in disturbance regimes. *For. Ecol. Manage.* 227:219–232.

Berger, J., Stacey, P.B., Bellis, L., and Johnson, M.P. 2001. A mammalian predator-prey imbalance: grizzly bear and wolf extinction affect avian neotropical migrants. *Ecol. Applic.* 11:967–980.

Berger, J., Stacey, P.B., Bellis, L., and Johnson, M.P. 2001. A mammalian predator-prey imbalance: grizzly bear and wolf extinction affect neotropical migrants. *Ecol. Applic.* 11:947–960.

Bergeron, Y. 2000. Species and stand dynamics in the mixed woods of Quebec's southern boreal forest. *Ecology* 81:1500–1516.

Bergersen, F.J. 1970. The quantitative relationship between nitrogen fixation and the acetylene reduction assay. *Aust. J. Biol. Sci.* 23:1015–1025.

Bergh, O., Borsheim, K.Y., Bratbak, G., and Heldal, M. 1989. High abundance of viruses found in aquatic environments. *Nature* 340:467–468.

Berlow, E.L., Neutel, A.-M., Cohen, J.E., DeRuiter, P.C., Ebenman, B., Emmerson, M., Fox, J.W., Jansen, V.A.A., Jones, J.I., Kokkoris, G.D., Logofet, D.O., McKane, A.J., Montoya, J.M., and Petchey, O. 2004. Interaction strengths in food webs: issues and opportunities. *J. Anim. Ecol.* 73:585–598.

Bernal, J.D. 1951. *The physical basis of life.* Routledge and Kegan Paul, London.

Bernays, E., and Graham, M. 1988. On the evolution of host specificity in phytophagous arthropods. *Ecology* 69:886–892.

Berntson, G.M., and Aber, J.D. 2000. Fast nitrate immobilization in N saturated temperate forest soils. *Soil Biol. Biochem.* 32:151–156.

Berryman, A.A. 1988. Towards a unified theory of plant defense. In *Mechanisms of woody plant defenses against insects. Search for pattern.* Edited by W.J. Mattson, J. Levicux, and C. Bemard-Dagan. Springer-Verlag, New York, 39–55.

Beschta, R.L., Rhodes, J.J., Kauffman, J.B., Gresswell, R.E., Minshall, G.W., Karr, J.R., Perry, D.A., Hauer, F.R., and Frissell, C.A. 2004. Postfire management on forested public lands of the western US. *Conserv. Biol.* 18:957–967.

Bessie, W.C., and Johnson, E.A. 1995. The relative importance of fuels and weather on fire behavior in subalpine forests. *Ecology* 76:747–762.

Betts, R.A. 2000. Offset of the potential carbon sink from boreal forestation by decreases in surface albedo. *Nature* 408:187–190.

Bever, J.D., Schultz, P.A., Pringle, A., and Morton, J.B. 2001. Arbuscular mycorrhizal fungi: more diverse than meets the eye, and the ecological tale of why. *BioScience* 51:923–931.

Bhagwat, S.A. and Rutte, C. 2006. Sacred groves of India. www.edugreen.teri.res.in / explore / forestry / groves.htm.

Bicknell, S.H. 1982. Development of canopy stratification during early succession in northern hardwoods. *For. Ecol. Manage.* 4:41–51.

Bieleski, R.L. 1973. Phosphate pools, phosphate transport, and phosphate availability. *Annu. Rev. Plant Physiol.* 24:225–252.

Bienen, L. 2004. Deforestation and disease. *Front. Ecol. Environ.* 7:340.

Biernaskie, J.M., and Tyerman, J.G. 2005. The overextended phenotype. *Ecoscience* 12:3–4.

Bierregaard Jr., R.O., Lovejoy, T.E., Kapos, V., dos Santos, A.A., and Hutchings, R.W. 1992. The biological dynamics of tropical rainforest fragments. *BioScience* 42:859–866.

Biesmeijer, J.C., Roberts, S.P.M., Reemer, M., Ohlemuller, R., Edwards, M., Peeters, T., Schaffers, A.P., Potts, S.G., Kleukers, R., Thomas, C.D., Settele, J., and Kunin, W.E. 2006. Parallel declines in pollinators and insect-pollinated plants in Britain and the Netherlands. *Science* 313:351–354.

Bihan, B.L., Soulas, M.L., Camporota, P., Salerno, M.I., and Perrin, R. 1997. Evaluation of soil solar heating for control of damping-off fungi in two forest nurseries in France. *Biol. Fertil. Soils* 25:189–195.

Billings, S.A., and Ziegler, S.E. 2005. Linking microbial activity and soil organic matter transformations in forest soils under elevated CO_2. *Global Change Biol.* 11:203–212.

Binkley, D. 1982. Nitrogen fixation and net primary production in a young Sitka alder stand. *Can. J. Bot.* 60:281–284.

Binkley, D. 1983. Ecosystem production in Douglas-fir plantations: interaction of red alder and site fertility. *For. Ecol. Manage.* 5:215–227.

Binkley, D. 1986a. *Forest nutrition.* John Wiley & Sons, New York.

Binkley, D. 1986. Soil acidity in loblolly pine stands with interval burning. *Soil Sci. Soc. Am. J.* 50:1590–1594.

Binkley, D. 1992. Mixtures of nitrogen$_2$-fixing and non-nitrogen$_2$-fixing tree species. In *The ecology of mixed-species stands of trees.* Edited by M.G.R. Cannell, D.C. Malcolm, and P.A. Robertson. Blackwell, Oxford, 99–123.

Binkley, D., and Brown, T. 1993. Management impacts on water quality of forests and rangelands. Gen Tech. Rep. RM-239. U.S. Department of Agriculture, Forest Service, Rocky Mountain Forest and Range Exp. Sta., Fort Collins, CO.

Binkley, D., and Greene, S. 1983. Production in mixtures of conifers and red alder: the importance of site fertility. In *IUFRO symposium on forest site and continuous productivity.* Edited by R. Ballard and S.P. Gessel. Gen. Tech. Rep. PNW-163. U.S. Department of Agriculture, Forest Service, Portland, OR, 112–117.

Binkley, D., and Hart, S.C. 1989. The components of nitrogen availability in soils. *Adv. Soil Sci.* 10:57–112.

Binkley, D., and Menyailo, O., editors. 2005. Tree species effects on soils: implications for global change. In *Proceedings of the NATO Advanced Research Workshop on trees and soil interactions, implications to global climate change,* Aug. 2004, Kransnoyarsk, Russia. NATO Science Series, IV. Earth and Environmental Sciences Vol. 55. Springer, Dordrecht, The Netherlands.

Binkley, D., Ice, G.G., Kaye, J., and Williams, C.A. 2004. Nitrogen and phsopohorus concentrations in forest streams of the United States. *J. Am. Water Res. Assoc.* (October):1277–1291.

Binkley, D., Moore, M.M., Romme, W.H., and Brown, P.M. 2006. Was Aldo Leopold right about the Kaibab deer herd? *Ecosystems* 9:227–241.

Binkley, D., Senock, R., Bird, S., and Cole, T.G. 2003. Twenty years of stand development in pure and mixed stands of *Eucalyptus saligna* and nitrogen-fixing *Facaltaria moluccana*. *For. Ecol. Manage.* 182:93–102.

Binkley, D., Son, Y., and Valentine, D.W. 2000. Do forests receive occult inputs of nitrogen? *Ecosystems* 3:321–331.

Bird, J.A., and Torn, M.S. 2006. Fine roots vs. needles: a comparison of C-13 and N-15 dynamics in a ponderosa pine forest soil. *Biogeochemistry* 79:361–382.

Birkeland, P.W. 1999. Soils and geomorphology, 3rd ed. Oxford Univ. Press, New York.

Birkeland, P.W. 1984. *Soils and geomorphology.* Oxford Univ. Press, Oxford.

Birks, H.J.B., and Birks, H.H. 2004. The rise and fall of forests. *Science* 305:484–485.

Bisbee, K.E., Gower, S.T., Norman, J.M., and Nordheim, E.V. 2001. Environmental controls on ground cover species composition and productivity in a boreal black spruce forest. *Oecologia* 129:261–270.

Bisson, P.A., and Sedell, J.R. 1984. Salmonid populations in streams in clearcut vs. old-growth forests of western Washington. In *Fish and wildlfe relationships in old-growth forests.* Edited by W.R. Meehan, T.R. Merrell Jr., T.A. Hanley. American Institute of Fisheries Research Biologists, Juneau, AK, 121–129.

Bjornstad, O.N., Peltonen, M., Liebold, A.M., and Baltensweiler, W. 2002. Waves of larch budmoth outbreaks in the European Alps. *Science* 298:1020–1023.

Black, T.A. 1977. Micrometeorological studies of Douglas fir. In *Environmental effects on crop physiology.* Edited by J.J. Landsberg and C.V. Cutting. Academic Press, London, 57–73.

Blais, J.R. 1983. Trends in the frequency, extent, and severity of spruce budworm outbreaks in eastern Canada. *Can. J. For. Res.* 13:539–547.

Blake, D.R. 1989. Methane, CFCs, and other greenhouse gases. In *The challenge of global warming.* Edited by D.E. Abrahamson. Island Press, Washington, DC, 248–258.

Blaustein, A.R., Hoffman, P.D., Hokit, D.G., Kiesecker, J.M., Walls, S.C., and Hays, J.B. 1994. UV repair and resistance to solar UV-B in amphibian eggs: a link to population declines? *PNAS* 91:1791–1795.

Bledsoe, C.S., and Rygiewicz, P.T. 1986. Ectomycorrhizas effect ionic balance during ammonium uptake by Douglas-fir roots. *New Phytol.* 102:271–283.

Bledsoe, C.S., and Zasoski, R.J. 1983. Effects of ammonium and nitrate on growth and nitrogen uptake by mycorrhizal Douglas-fir seedlings. *Plant Soil* 71:445–454.

Blevins, D.G., and Lukaszewsk, K.M. 1998. Boron in plant structure and function. *Ann. Rev. Plant Physiol. Plant Molec. Biol.* 49:481–500.

Blum, J.D., Klaue, A., Nezat, C.A., Driscoll, C.T., Johnson, C.E., Siccama, T.G., Eager, C., Fahey, T.J., and Likens, G.E. 2002. Mycorrhizal weathering of apatite as an important calcium source in base-poor ecosystems. *Nature* 417:729–731.

Boege, K. 2004. Induced responses in three tropical dry forest plant species—direct and indirect effects on herbivory. *Oikos* 107:541–548.

Boettcher, S.E., and Kalisz, P.J. 1990. Single-tree influence on soil properties in the mountains of Eastern Kentucky. *Ecology* 71:1365–1372.

Bohlen, P.J., Groffman, P.M., Fahey, T.J., Fisk, M.C., Suárez, E., Pelletier, D.M., and Fahey, R.T. 2004a. Ecosystem consequences of exotic earthworm invasion of north temperate forests. *Ecosystems* 7:1–12.

Bohlen, P.J., Scheu, S., Hale, C.M., McLean, M.A., Migge, S., Groffman, P.M., and Parkinson, D. 2004b. Non-native invasive earthworms as agents of change in northern termeprate forests. *Front. Ecol. Environ.* 2:427–435.

Bohn, H.L., McNeal, B.L., and O'Connor, G.A. 1979. *Soil chemistry.* John Wiley & Sons, New York.

Bonan, G.B. 1999. Frost followed the plow: impacts of deforestation on the climate of the United States. *Ecol. Applic.* 9:1305–1315.

Bonan, G.B., and Shugart, H.H. 1989. Environmental factors and ecological processes in boreal forests. *Annu. Rev. Ecol. Syst.* 20:1–28.

Bondietti, E.A., Baes III, C.F., and McLaughlin, S.B. 1989. The potential of trees to record aluminum mobilization and changes in alkaline earth availability. In *Biological markers of air-pollution stress and damage in forests.* National Academy Press, Washington, DC, 281–292.

Bonet, J.A., Fischer, C.R., and Colinas, C. 2004. The relationship between forest age and aspect on the production of sporocarps of ectomycorrhizal fungi in Pinus sylvestris forests of the central Pyrenees. *For. Ecol. Manage.* 203:157–175.

Bonner, F.T. 1968. *Responses to soil moisture deficiency by seedlings of three hardwood species.* USDA For. Serv. Res. Note 50–70. U.S. Department of Agriculture, Forest Service, Southern Forest Exp. Sta. Asheville, NC.

Boogert, N.J., Paterson, D.M., and Laland, K.N. 2006. The implications of niche construction and ecosystem engineering for conservation biology. *BioScience* 56:570–578.

Boose, E.R., Serrano, M.I., and Foster, D.R. 2004. Landscape and regional impacts of hurricanes in Puerto Rico. *Ecol. Monogr.* 74:335–352.

Booth, M.G. 2004. Mycorrhizal networks mediate overstorey-understorey competition in a temperate forest. *Ecology Lett.* 7:538–54.

Borchers, J.G., and Perry, D.A. 1992. The influence of soil texture and aggregation on carbon and nitrogen dynamics in southwest Oregon forests and clearcuts. *Can. J. For. Res.* 22:298–305.

Borchers, S.L., and Perry, D.A. 1990. Growth and ectomycorrhiza formation of Douglas-fir seedlings grown in soils collected at different distances from pioneering hardwoods in southwest Oregon. *Can. J. For. Res.* 20:712–721.

Bork, J.L. 1985. Fire history in three vegetation types on the eastern side of the Oregon Cascades. PhD thesis. Oregon State Univ., Corvallis.

Bormann, B.T. 1989. Podzolization and windthrow: natural fluctuations in long-term productivity and implications for management. In *Maintaining the long-term productivity of Pacific Northwest forests.* Edited by D.A Perry, R. Meurise, B. Thomas, et al. Timber Press, Portland, OR.

Bormann, B.T., Bormann, F.H., Bowden, W.B., Piece, R.S., Hamburg, S.P., Wang, D., Snyder, M.C., Li, C.Y., and Ingersoll, R.C. 1993. Rapid N_2 fixation in pines, alder, and locust: evidence from the sandbox ecosystem study. *Ecology* 74:583–598.

Bormann, B.T., Keller, C.K., Wang, D., and Bormann, F.H. 2002. Lessons from the sandbox: is unexplained nitrogen real? *Ecosystems* 5:727–733.

Bormann, B.T., Wang, D., Bormann, F.H., Benoit, G., April, R., and Snyder, M.C. 1998. Rapid, plant-induced weathering in an aggrading experimental ecosystem. *Biogeochemistry* 43:129–155.

Bormann, F.H. 1985. Air pollution and forests: an ecosystem perspective. *BioScience* 35:434–441.

Bormann, F.H. 1991. The role of *Pinus* in the nitrogen economy of restored sites: a report on the cooperative sandbox study at Hubbard Brook, NH, USA. In *Abstract 1991 World Congress of Landscape Ecology.* International Association for Landscape Ecology, Ottawa.

Bormann, F.H., and Likens, G.E. 1979a. Catastrophic disturbance and the steady state in northern hardwood forests. *Am Sci.* 67:660–669.

Bormann, F.H., and Likens, G.E. 1979b. *Pattern and process in a forested ecosystem.* Springer-Verlag, New York.

Bormann, F.H., Likens, G.E., and Melillo, J.M. 1977. Nitrogen budget for an aggrading northern hardwood forest ecosystem. *Science* 196:981–983.

Bormann, F.H., Likens, G.E., Siccama, T.G., Pierce, R.S., and Eaton, J.S. 1974. The export of nutrients and recovery of stable conditions following deforestation at Hubbard Brook. *Ecol. Monogr.* 44:255–277.

Bormann, T., and DeBell, D.S. 1981. Nitrogen content and other soil properties related to age of red alder stands. *Soil Sci. Soc. Am. J.* 45:428–432.

Borneman, J., and Triplett, E.W. 1997. Molecular microbial diversity in soils from eastern Amazonia: evidence for unusual microorganisms and microbial population shifts associated with deforestation. *Appl. Environ. Microbiol.* 63:2647–2653.

Borrvall, C., and Ebenman, B. 2006. Early onset of secondary extinctions in ecological communities following the loss of top predators. *Ecol. Lett.* 9:435–442.

Bosatta, E., and Agren, G.I. 1991. Dynamics of carbon and nitrogen in the organic matter of the soil: a generic theory. *Am. Nat.* 138:227–245.

Bosatta, E., and Berendse, F. 1984. Energy or nutrient regulation of decomposition: implications for the mineralization-immobilization response to pertubations. *Soil Biol. Biochem.* 16:63–67.

Bosch, J.M., and Hewlett, J.D. 1982. A review of catchment experiments to determine the effect of vegetation changes on water yield and evapotranspiration. *J. Hydrol.* 55:3–23.

Bossema, I. 1979. Jays and oaks: an eco-ethological study of a symbiosis. *Behaviour* 70:1–117.

Bossier, P., Hofte, M., and Verstraete, W. 1988. Ecological significance of siderophores in soil. *Adv. Microb. Ecol.* 10:385–414.

Botkin, D.B., Estes, J.E., MacDonald, R.M., and Wilson, M.V. 1984. Studying the Earth's vegetation from space. *BioScience* 34:508–514.

Bottomley, P.J. 2005. Microbial ecology. In *Principles and application of soil microbiology,* 2nd ed. Edited by D.M. Sylvia, J.L. Fuhrmann, P.G. Hartel, and D.A. Zuberer.Prentice Hall, Upper Saddle River, NJ, 222–241.

Boucher, D.H., editor. 1985. *The biology of mutualism.* Oxford Univ. Press, New York.

Boucher, D.H. 1990. Growing back after hurricanes. *BioScience* 40:163–166.

Boucher, D.H., James, S., and Keeler, K.R. 1982. The ecology of mutualism. *Annu. Rev. Ecol. Syst.* 13:315–347.

Boudreault, C., Gauthier, S., and Bergeron, Y. 2000. Epiphytic lichens and bryophytes on *Populus tremuloides* along a chronosequence in the southwestern boreal forest of Quebec, Canada. *The Bryologist* 103:725–738.

Bourliere, F. 1983. Animal species diversity in tropical forests. In *Ecosystems of the world,* Vol. 14A, *Tropical rain forest ecosystems.* Edited by F.B. Golley. Elsevier, New York, 77–91.

Bousquet, P., Peylin, P., Cias, P., Le Quere, C., Friedlingstein, P., and Tans, P.P. 2000. Regional changes in carbon dioxide fluxes of land and oceans since 1980. *Science* 290:1342–1346.

Boutin, S., Wauters, L.A., McAdam, A.G., Humphries, M.M., Tosi, G., and Dhondt, A.E. 2006. Anticipatory reproduction and population growth in seed predators. *Science* 314:1928–1930.

Bowden, W.B. 1986. Gaseous nitrogen emissions from undisturbed terrestrial ecosystems: an assessment of their impacts on local and global nitrogen budgets. *Biogeochemistry* 2:249–279.

Bowden, W.P., and Bormann, F.R. 1986. Transport and loss of nitrous oxide in soil water after forest clear-cutting. *Science* 233:817–916.

Bowen, G.D., and Theodorou, C. 1979. Interactions between bacteria and ectomycorrhizal fungi. *Soil Biol. Biochem.* 11:119–126.

Bowen, H.J.M. 1979. *Environmental chemistry of the elements.* Academic Press, London,.

Bowers, M.A., and Sacchi, C.F. 1992. Fungal mediation of a plant-hervibore interaction in an early successional plant community. *Ecology* 72:1032–1037.

Bradford, K.J., and Hsiao, T.C. 1982. Physiological responses to moderate water stress. In *Encyclopedia of plant physiology,* Vol 12B. Edited by O.L. Lange, P.S. Nobel, C.B. Osmond, and R. Ziegler. Springer-Verlag, Berlin, 265–364.

Boyle, S.I., Hart, S.C., Kaye, J.P., and Waldrop, M.P. 2005. Restoration and canopy type influence soil microflora in a ponderosa pine forest. *Soil Sci. Soc. Am. J.* 69:1627–1638.

Bradshaw, W.E., and Holzapfel, C.M. 2006. Climate change: evolutionary response to rapid climate change. *Science* 312:1477–1478.

Brady, N.C., and Weil, R.R. 2002. The nature and properties of soils, 13th ed. Pearson Education, Upper Saddle River, NJ.

Brainerd, R. 1988. Mycorrhiza formation and diversity in undisturbed forest and clearcut and bumed areas in three forest types in Oregon. Masters thesis. Oregon State Univ., Corvallis.

Brasell, H.M., and Sinclair, D.F. 1983. Elements returned to forest floor in two rainforest and three N prep. plantation plots in tropical Australia. *J. Ecol.* 71:367–378.

Bratton, S.P. 1975. A comparison of the beta diversity functions of the overstory and herbaceous understory of a deciduous forest. *Bull. Torrey Bot. Club* 102:55–60.

Brawn, J.D., Robinson, S.K., and Thompson III, F.R. 2001. The role of disturbance in the ecology and conservation of birds. *Annu. Rev. Ecol. Syst.* 32:251–276.

Brenes-Arguedas, T., and Coley, P.D. 2005. Phenotypic variation and spatial structure of secondary chemistry in a natural population of a tropical tree species. *Oikos* 108:410–420.

Breshears, D.D., Cobb, N.S., Rich, P.M., Price, K.P., Allen, C.D., Balice, R.G., Romme, W.H., Kastens, J.H., Floyd, M.L., Belnap, J., Anderson, J.J., Myers, O.B., and Meyer, C.W. 2005. Regional vegetation die-off in response to global-change type drought. *PNAS* 102:15144–15148.

Breznak, J.A., Brill, W.J., Mertins, J.W., and Coppel, R.C. 1973. Nitrogen fixation in termites. *Nature* 244:577–580.

Briand, F., and Cohen, J.E. 1987. Environmental correlates of food chain length. *Science* 238:956–960.

Briffa, K.R. Bartholin, T.S., Eckstein, D., Jones, P.D., Karlén, W., Schweingruber, F.H., Zetterberg, P. 1990. A 1400 year tree-ring record of summer temperatures in Fennoscandia. *Nature* 346:434–439.

Briggs, J.C. 1991. A cretaceous-tertiary mass extinction? *BioScience* 41:619–624.

Brocke, R.H., O'Pezio, J.P., and Gustafson, K.A. 1989. A forest management scheme for mitigating impact of road networks on sensitive wildlife species. In *Is forest fragmentation a management issue in the northeast?* General Tech. Rep. NE-140. U.S. Forest Service, N.E. Forest Exp. Sta., 7–12. Newtown Square, PA.

Broecker, W.S. 1987. Unpleasant surprises in the greenhouse? *Nature* 328:123–126.

Broecker, W.S. 1989. Greenhouse surprises. In *The challenge of global warming.* Edited by D.E. Abrahamson. Island Press, Washington, DC, 196–212.

Broido, A., and Nelson, K. 1964. Ash content: its effect on combusion of corn plants. *Science* 146:652–653.

Brokaw, N., and Busing, R.T. 2000. Niche versus chance and tree diversity in forest gaps. *Tree* 15:183–188.

Brokaw, N.L., and Grear, J.S. 1991. Forest structure before and after Hurricane Hugo at three elevations in the Luquillo Mountains, Puerto Rico. *Biotropica* 23:386–392.

Brokaw, N.V.L. 1985. Treefalls, regrowth and community structure in tropical forests. In *The ecology of natural disturbance and patch dynamics.* Edited by S.T.A. Pickett, and P.S. White. Academic Press, New York.

Brokaw, N.V.L., and Schemer, S.M. 1989. Species composition in gaps and structure of a tropical forest. *Ecology* 70:538–541.

Brokaw, N.V.L., and Walker, L.R. 1991. Summary of the effects of Caribbean hurricanes on vegetation. *Biotropica* 23:442–447.

Brook, B.W., Sodhi, N.S., and Ng, P.K.L. 2003. Catastrophic extinctions follow deforestation in Singapore. *Nature* 424:420–426.

Brook, B.W., Traill, L.W., and Bradshaw, C.J.A. 2006. Minimum viable population sizes and global extinction risk are unrelated. *Ecol. Lett.* 9:375–382.

Brooks, P.C., Powison, D.S., and Jenkinson, D.S. 1985. The microbial biomass in soil. In *Ecological interactions in soil.* Edited by A.H. Fitter, U. Atkinson, D.J. Read, and M.B. Usher. Blackwell, Oxford, 123–126.

Brooks P.D., McKnight, D., and Elder, K. 2005. Carbon limitation of soil respiration under winter snowpacks: potential feedbacks between growing season and winter carbon fluxes. *Global Change Biol.* 11:231–238.

Brooks, T.M., Mittermeier, R.A., da Fonseca, G.A.B., Gerlach, J., Hoffmann, M., Lamoreux, J.F., Mittermeier, C.G., Pilgrim, J.D., and Rodrigues, A.S.L. 2006. Global biodiversity conservation priorities. *Science* 313:58–61.

Brouzes, R., Lasik, J., and Knowles, R. 1969. The effect of organic amendment, water content, and oxygen on the incorporation of N_2 by some agricultural and forest soils. *Can. J. Microbiol.* 15:899–905.

Browder, J.O. 1992. The limits of extractism. *BioScience* 42:174–182.

Brown Jr., W.L., and Wilson, E.O. 1956. Character displacement. *Syst. Zool.* 5:49–64.

Brown, A.H.D. 1979. Enzyme polymorphism in plant populations. *Theoret. Pop. Biol.* 15:1–42.

Brown, A.H.D., and Moran, G.F. 1981. Isozymes and the genetic resources of forest trees. In *Isozymes of North American forest trees and forest insects.* Edited by M.T. Conkle. General Tech. Rep. PSW-48. U.S. Department of Agriculture, Forest Service, Pacific S.W. Exp. Sta., Berkely, CA.

Brown, A.H.F. 1992. Functioning of mixed-species stands at Gisburn, N.W. England. In *The ecology of mixed species stands of trees.* Edited by M.G.R. Cannell, D.C. Malcolm, and PA. Robertson. Blackwell Scientific Publications, London, 125–150.

Brown, E.R., editor. 1985. *Management of wildlife and fish habitats in forests of western Oregon and Washington.* Pub. No. R6–F&WL-192 U.S. Department of Agriculture, Forest Service, Portland, OR.

Brown, J.H., Cruickshank, V.B., Gould, W.P., and Husband, T.P. 1988. Impact of gypsy moth defoliation in stands containing white pine. *North. J. Appl. For.* 5:108–111.

Brown, J.H., Gupta, V.K., Li, B.L., Milne, B.T., Respeto, C., and West, G.B. 2002. The fractal nature of nature: power laws, ecological complexity and biodiversity. *Phil. Trans. R. Soc. Lond. B* 357:619–626.

Brown, J.K. 1974. *Handbook for inventorying downed woody material.* Gen. Tech. Rep. INT-16. U.S. Department of Agriculture, Forest Service. Missoula, MT.

Brown, K. 2003. Integrating conservation and development: a case of institutional misfit. *Front. Ecol. Environ.* 1:479–487.

Brown, K.A., and Gurevitch, J. 2004. Long-term impacts of logging on forest diversity in Madagascar. *PNAS* 101:6045–6049.

Brown, N.R., Noss, R.F., Diamond, D.D., and Myers, M.N. 2001. Conservation biology and forest certification. *J. For.* (August):18–25.

Brown, R.T., Agee, J.K., and Franklin, J.F. 2004. Forest restoration and fire: principles in the context of place. *Conserv. Biol.* 18:903–912.

Brown, S., and Lugo, A.E. 1984. Biomass of tropical forests: a new estimate based on forest volumes. *Science* 223: 1290–1293.

Brown, S., Gillespie, A.J.R., and Lugo, A.E. 1989. Biomass estimation methods for tropical forests with applications to forest inventory data. *For. Sci.* 35:881–902.

Brown, S., and Lugo, A.E. 1992. Aboveground biomass estimates for tropical moist forests of the Brazilian Amazon. *Interciencia* 17:8–18.

Brubkaer, L.B. 1986. Responses of tree populations to climatic change. *Vegetatio* 67:119–130.

Brubaker, L.B. 1991. Climate change and the origin of old-growth Douglas-fir forests in the Puget Sound lowland. In *Wildlfe and vegetation of unmanaged Douglas-fir forests.* Edited by L.F. Riggiero, K.B. Aubry, A.B. Cary, and M.H. Huff. Gen. Tech. Rep. PNW-GT-285. U.S. Department of Agriculture, 17–24. Portland, OR.

Bruijnzeel, L.A. 1991. Nutrient input-output budgets of tropical forest ecosystems: a review. *J. Trop. Ecol.* 7:1–24.

Bruijnzeel, L.A. 2004. Hydrologic functions of tropical forests: not seeing the soil for the trees? *Agric. Ecosyst. Environ.* 104:185–228.

Bruner, A.G., Gullison, R.E., Rice, R.E., and da Fonseca, G.A.B. 2001. Effectiveness of parks in protecting tropical biodiversity. *Science* 291:125–128.

Brunig, E.F. 1970. Stand structure, physiognomy and environmental factors in some lowland forests in Sarawak. *Trop. Ecol.* 11:26–43.

Brunig, E.F. 1983. Vegetation structure and growth. In *Tropical rain forest ecosystems.* Edited by F.B. Golley. Elsevier, Amsterdam, 49–76.

Bruno, J.F., Stachowicz, J.J., and Bertness, M.D. 2003. Inclusion of facilitation into ecological theory. *Tree* 18:119–126.

Brush, G.S. 1982. An environmental analysis of forest patterns. *Am. Sci.* 70:18–25.

Bryan, R. 2002. From Rio to Joannesburg: ten years of progress in forest conservation? Proceedings of the Starker Lecture Series. Oregon State Univ. College of Forestry, Corvallis. www.cof.orst.edu/starkerlectures/transcripts/2002/2002.php.

Bryant, D., Nielsen, D., and Tangley, L. 1997. *The last frontier forests: ecosystems and economies on the edge.* World Resources Institute, Washington, DC.

Bryant, J.P., and Kuropat, P.L. 1980. Selection of winter forage by subartic browsing vertebrates: the role of plant chemistry. *Annu. Rev. Ecol. Syst.* 11:261–285.

Bryant, J.P., Chapin III, F.S., and Klein, D.R. 1983. Carbon/nutrient balance of boreal plants in relation to vertebrate herbivory. *Oikos* 40:357–368.

Bryant, J.P., Kuropat, P.J., Cooper, S.M., Frisby, K., and Owen-Smith, N. 1989. Resource availability hypothesis of plant anti-herbivore defence tested in a South African savanna ecosystem. *Nature* 340:227–229.

Bryant, J.P., Provenze, F.D., Pastor, J., Pastor, J., Reichardt, P.B., Clausen, T.P., and du Toit, J.T. 1991. Interactions between woody plants and browsing mammals mediated by secondary metabolites. *Annu. Rev. Ecol. Syst.* 22:431–446.

Buol, S.W., Hole, F.D., and McCracken, R.J. 1980. *Soil genesis and classification.* Iowa State Univ. Press, Ames.

Buol, S.W., Hole, F.D., and McCracken, R.J. 1989. *Soil genesis and classification*, 3rd ed. Iowa State Univ. Press, Ames.

Burdon, J.J., and Chilvers, G.A. 1974. Fungal and insect parasites contributing to niche differentation in mixed species stands of eucalypt saplings. *Aust. J. Bot.* 22:103–114.

Burger, J.A., and Pritchett, W.L. 1988. Site preparation effects on soil moisture and available nutrients in a pine plantation in the Florida flatwoods. *For. Sci.* 34:77–87.

Burgess, S., Burgess, S.O., Adams, M.A., Turner, N.C., and Ong, C.K. 1998. The redistribution of soil water by tree root systems. *Oecologia* 115:306–311.

Burke, M.J., Gusta, L.V., Quamme, H.A., Weiser, C.J., and Li, P.H. 1976. Freezing and injury in plants. *Annu. Rev. Plant Physiol.* 27:507–528.

Burton, P.J., Messier, C., Smith, D.W., and Adamowicz, W.L., editors. 2003. *Towards sustainable management of the boreal forest.* NRC Research Press, Ottawa, 261–306.

Buskirk, S.W. 1992. Conserving circumboreal forests for martens and fishers. *Conserv. Biol.* 6:318–320.

Butterfield, J., and Malvido, J.B. 1992. Effect of mixed-species tree planting on the distribution of soil invertebrates. In *The ecology of mixed-species stands of trees.* Edited by M.G.R. Cannell, D.C. Malcolm, and P.A. Robertson. Blackwell, Oxford, 255–265.

Busse, M.D., Beattie, S.E., Powers, R.F., Sanchez, F.G., and Tiarks, A.E. 2006. Microbial community responses in forest mineral soil to compaction, organic matter removal, and vegetation control. *Can. J. For. Res.* 36:577–588.

Cachier, H., and Ducret, J. 1991. Influence of biomass burning on equatorial African rains. *Nature* 352:228–230.

Cairney, J.W.G., and Ashford, A.E. 1989. Reducing activity at the root surface in *Eucalyptus pilularis–Pisolithus tinctorius* ectomycorrhizas. *Aust. J. Plant Physiol.* 16:99–105.

Cairns, J., Overbaugh, J., and Miller, S. 1988. The origin of mutants. *Nature* 335:142–145.

Calder, I.R., Rosier, P.T.W., Prasanna, K.T., and Parameswarappa, S. 1997. Eucalyptus ware use greater than rainfall input—a possible explanation from southern India. *Hydrol. Earth Syst. Sci.* 1:249–256.

Cale, W.G., Henebry, G.M., and Yeakley, J.A. 1989. Inferring process from pattern in natural communities. *BioScience* 39:600–605.

Calhoun, L.A., Findley, J.A., Miller, J.D., and Whitney, N.J. 1992. Metabolites toxic to spruce budworm from balsam fir needle endophytes. *Mycol. Res.* 96:281–286.

Callaway, R.M. 1997. Positive interactions in plant communities and the individualistic-continuum concept. *Oecologia* 112:143–149.

Callaway, R.M., and D'Antonio, C.M. 1991. Shrub facilitation of coast live oak establishment in central California. *Madrono* 38:158–169.

Callaway, R.M., Brooker, R.W., Choler, P., Kilkvidze, Z., Lortie, C.J., Michalet, R., Paolini, L., Pugnaire, F.L., Newingham, B., Aschehoug, E.T., Armas, C., Kikodze, D., and Cook, B.J. 2002. Positive interactions among alpine plants increase with stress. *Nature* 417:844–847.

Campbell, C.A. 1978. Soil organic carbon, nitrogen, and fertility. In *Soil organic matter.* Edited by M. Schnitzer and S.U. Kahn. Elsevier, Amsterdam, 172–271.

Campbell, J.L., Hornbeck, J.W., McDowell, W.H., Buso, D.C., Shanley, J.B., and Likens, G.E. 2000. Dissolved organic nitrogen budgets for upland, forested ecosystems in New England. *Biogeochemistry* 49:123–142.

Canham, C.D., and Loucks, O.L. 1984. Catastrophic windthrow in the presettlement forests of Wisconsin. *Ecology* 65:803–809.

Cannell, M.G.R. 1982. *World forest biomass and primary production data.* Academic Press, New York, 319.

Cannell, M.G.R. 1984. Woody biomass of forest stands. *For. Ecol. Manage.* 8:299–312.

Cannell, M.G.R. 1985. Dry matter partitioning in tree crops. In *Attributes of trees as crop plants.* Edited by M.G.R. Cannell and J.E. Jackson. Institute of Terrestrial Ecology, Huntingdon, England, 160–193.

Cannell, M.G.R., and Smith, R.J. 1986. Climatic warming, spring bud burst and frost damage on trees. *J. Appl. Ecol.* 23:177–191.

Cannell, M.G.R., Malcolm, D.C., and Robertson, P.A., editors. 1992. *The ecology of mixed-species stands of trees.* Special Publ. No. 11 of the British Ecological Society. Blackwell, Oxford.

Cannell, M.G.R., Sheppard, L.J., Ford, E.D., and Wilson, R.H.F. 1983. Clonal differences in dry matter distribution, wood specific gravity and foliage "efficiency" in *Picea sitchensis* and *Pinus contorta. Silvae Genet.* 32:195–202.

Carbone, C., and Gittleman, J.L. 2002. A common rule for the scaling of carnivore density. *Science* 295:2273–2276.

Cardille, J.A., and Foley, J.A. 2003. Agricultural land-use change in Brazilian Amazonia between 1980 and 1995: evidence from integrated satellite and census data. *Remote Sensing Environ.* 87:551–562.

Cardinale, B.J., Srivasta, D.S., Duffy, J.E., Wright, J.P., Downing, A.L., Sankaran, M., and Jouseau, C. 2006. Effects of biodiversity on the functioningh of trophic groups and ecosystems. *Nature* 443:989–992.

Carey, A.B. 1989. Wildlife associated with old-growth forests in the Pacific Northwest. *Natural Areas J.* 1989:9:151–162.

Carey, A.B. 2003. Biocomplexity and restoration of biodiversity in temperate coniferous forest: inducing spatial heterogeneity with variable-density thinning. *Forestry* 76:127–136.

Carey, A.B., and Curtis, R.O. 1996. Conservation of biodiversity: a useful paradigm for forest ecosystem management. *Wildlife Soc. Bull.* 24:610–620.

Carey, E.V., Sala, A., Keane, R., and Callaway, R.M. 2001. Are old forests underestimated as global carbon sinks? *Global Change Biol.* 7:339–344.

Carloni, K. 2005. The ecological legacy of Indian burning practices in southwestern Oregon. PhD Thesis. Oregon State Univ., Corvallis.

Carlyle, J.C., and Than, U.B.A. 1988. Abiotic controls of soil respiration beneath an eighteen-year-old *Pinus radiata* stand in south-eastern Australia. *J. Ecol.* 76:654–662.

Carroll, A.L., Taylor, S.W., Régnière, J., and Safranyik, L. 2003. Effects of climate change on range expansion by the mountain pine beetle in British Columbia. In *Mountain Pine Beetle Symposium: Challenges and Solutions.* October 30–31, 2003, Kelowna, British Columbia. Edited by T.L. Shore, J.E. Brooks, and J.E. Stone. Natural Resources Canada, Canadian Forest Service, Pacific Forestry Centre, Information Report BC-X-399, Victoria, BC, 298.

Carroll, C., Noss, R.F., and Paquet, P.C. 2001. Carnivores as focal species for conservation planning in the Rocky Mountain region. *Ecol. Applic.* 11:961–980.

Carroll, G. 1988. Fungal endophytes in stems and leaves: from latent pathogen to mutualistic symbiont. *Ecology* 69:2–9.

Carroll, G.C. 1991. Fungal associates of woody plants as insect antagonists in leaves and stems. In *Microbial mediation of plant-herbivore interactions.* Edited by P. Barbosa, V.A. Krischik, and G.G. Jones. John Wiley & Sons, New York, 253–271.

Carroll, G.C., and Carroll, F.E. 1978. Studies on the incidence of coniferous needle endophytes in the Pacific Northwest. *Can. J. Bot.* 56:3034–3043.

Carson, B. 1989. *Soil conservation strategies for upland areas in Indonesia.* Occasional Paper No. 9. East-West Environment and Policy Institute, Univ. of Hawaii, Honolulu.

Carter, R.E., Otchere-Boateng, J., and Klinka, K. 1983. Dieback of a 30–year-old Douglas-fir plantation in the Brittain River valley, British Columbia: symptoms and diagnosis. *For. Ecol. Manage.* 7:249–263.

Carter, T.C., Menzel, M.A., Chapman, B.R., and Miller, K.V. 2004. Paritioning of food resources by syntopic Eastern Red *(Lasiurus borealis),* Seminole *(L. semiolus),* and Evening *(Nycticeius humeralis)* bats. *Am. Midl. Nat.* 151:186–191.

Case, T.J., and Cody, M.L. 1987. Testing theories of island biogeography. *Am. Sci.* 75:402–412.

Caspersen, J.P., Pacala, S.W., Jenkins, J.C., Hurtt, G.C., Moorcroft, P.R., and Birdsey, R.A. 2000. Contributions of land-use history to carbon accumulation in U.S. forests. *Science* 290:1148–1155.

Castelletta, M., Sodhi, N.S., and Subaraj, R. 2000. Heavy extinction of forest avifauna in Singapore: lessons for biodiversity conservation in southeast Asia. *Conserv. Biol.* 14:1870–1880.

Cates, R., Redak, R., and Henderson, C.B. 1983. Natural product defensive chemistry of Douglas-fir, western spruce budworm success, and forest management practices. *Z. Angew. Entomol.* 96:173–182.

Cates, R.G., and Redak, R.A. 1988. Variation in the terpene chemistry of Douglas-fir and its relationship to western spruce budworm success. In *Chemical mediation of coevolution.* American Institute of Biological Sciences, Washington, DC, 317–344.

Cattelino, P.J., Noble, I.R., Slatyer, R.O., and Kessell, S.R. 1979. Predicting the multiple pathways of plant succession. *Environ. Manage.* 3(1):41–50.

Caughley, G. 1970. Eruption of ungulate populations, with emphasis on Himalayan thar in New Zealand. *Ecology* 51:53–72.

Caughley, G. 1994. Directions in conservation biology. *J. Anim. Ecol.* 63:215–244.

Cazares, E., and Smith, J.E. 1992. Occurrence of vesicular-arbuscular mycorrhiza on Douglas-fir and western hemlock seedlings. In *Mycorrizas in ecosystems.* Edited by D.J. Read, D.H. Lewis, A.H. Fitter, and I.J. Alexander. C.A.B. International, Wallingford, 374.

Cazares, E., Luoma, D.L., Amaranthus, M.P., Chambers, C.L., and Lehmkuhl, J.F. 1999. Interaction of fungal sporocarp production with small mammal abundance and diet in Douglas-fir stands of the southern Cascade Range. *NW Science* 73:64–76.

Ceballos, G., Erlich, P.R., Soberon, J., Salazar, I, and Fay, J.P. 2005. Global mammal conservation: what must we manage? *Science* 309:603–606.

Ceballos, G., Rodrigues, P., and Medellin, R.A. 1998. Assessing conservation priorities in megadiverse Mexico: mammalian diversity, endemicity, and endangerment. *Ecol. Applic.* 8:8–17.

Centre for Science and Environment. 1987. *The wrath of nature: the impact of environmental destruction on floods and droughts.* Centre for Science and the Environment, New Delhi.

Chadwick, O.A., Derry, L.A., Vitousek, P.M., Huebert, B.J., and Hedin, L.O. 1999. Changing sources of nutrients during four million years of ecosystem development. *Nature* 397:491–497.

Challinor, D. 1968. Alteration of surface soil characteristics by four tree species. *Ecology* 49:286–290.

Chan, B.C., Waiss Jr., A.C., and, Lukefahr. M. 1978. Condensed tannin, an antibiotic chemical from *Gossypium hirsutum*. *J. Insect Phys.* 24:113–118.

Chanasyk, D.S., Whitson, I.R., Mapfumo, E., Burke, J.M., and Prepas, E.E. 2003. The impacts of forest harvest and wildfire on soils and hydrology in temperate forests: a baseline to develop hypotheses for the Boreal Plain. *J. Environ. Eng. Sci./Rev. Gen. Sci. Environ.* 2(S1):S51–S62.

Chanway, C. 1992. Influence of soil biota on Douglas-fir *(Pseudotsuga menziesii)* seedling growth: the role of rhizosphere bacteria. *Can. J. Bot.* 70:1025–1031.

Chanway, C.P., and Holl, F.B. 1991. Biomass increase and associative nitrogen fixation of mycorrhizal *Pinus contorta* seedlings inoculaed with a plant growth promoting *Bacillus* strain. *Can. J. Bot.* 69:507–511.

Chanway, C.P., and Holl, F.B. 1992. Influence of soil biota on Douglas-fir (*Pseudotsuga menziesii* (Mirb. Franco)) seedling growth: the role of rhizosphere bacteria. *Can. J. Bot.* 70:1025–1031.

Chanway, C.P., Turkington, R., and Roll, F.B. 1991. Ecological implications of specificity between plants and rhizosphere micro-organisms. *Adv. Ecol. Res.* 21:121–169.

Chape, S., Harrison, J., Spalding, M., and Lysenko, I. 2005. Measuring the extent and effectiveness of protected areas as an indicator for meeting global biodiversity targets. *Phil. Trans. R. Soc. B* 2005. 360:443–455.

Chapin III, F.S. 1980. The mineral nutrition of wild plants. *Annu. Rev. Ecol. Syst.* 11:233–260.

Chapin III, F.S., Bloom, A.J., Field, C.B., and Waring, R.H. 1987. Plant responses to multiple environmental factors. *BioScience* 37:49–57.

Chapin III, F.S., Matson, P.A., and Mooney, H.A. 2002. *Principles of terrestrial ecosystem ecology.* Springer-Verlag, New York.

Chapin III, F.S., Vitousek, P.M., and Van Cleve, K. 1986. The nature of nutrient limitation in plant communities. *Am. Nat.* 127:48–58.

Chapin III, F.S., Zavaleta, E.S., Eviners, V.T., Naylor, R.L., Vitousek, P.M., Reynolds, H.L., Hooper, D.U., Lavoret, S., Sala, O.E., Hobbie, S.E., Mack, M.C., and Diaz, S. 2000. Consequences of changing biodiversity. *Nature* 405:234–242.

Chapman, S.K., Hart, S.C., Cobb, N.S., Whitham, T.G., and Koch, G.W. 2003. Insect herbivory increases litter quality and decomposition: an extension of the acceleration hypothesis. *Ecology* 84:2867–2876.

Chapman, S.K., Langley, J.A., Hart, S.C., and Koch, G.W. 2006. Plants actively control nitrogen cycling: uncorking the microbial bottleneck. *New Phytol.* 169:27–34.

Charison, R.J., Lovelock, J.E., Andreae, M.O., and Warren, S.C. 1967. Oceanic phytoplankton, atmospheric sulphur, cloud albedo and climate. *Nature* 326:655–661.

Chase, A. 1987. *Playing god in Yellowstone.* Atlantic Monthly Press, Boston.

Chase, J.M. 2003. Community assembly: when should history matter? *Oecologia* 136:489–498.

Chase, J.M., and Liebold, M.A. 2002. Spatial scale dictates the productivity-biodievsrity relationship. *Nature* 416:427–430.

Chase, J.M., and Leibold, M.A. 2003. Ecological *Niches: Linking Classical and Contemporary Approaches*. Univ. of Chicago Press, Chicago.

Chase, T.N., Pielke, R.A., Kittel, T.G.F., Nemani, R., and Running, S.W. 1996. Sensitivity of a general circulation model to global changes in leaf area. *J. Geophys. Res.* 101:7393–7408.

Chase, T.N., Pielke Sr., R.A., Kittel, T.G.F., Nemani, R.R., and Running, S.W. 2000. Simulated impacts of historical land cover changes on global climate in northern winter. *Clim. Dyn.* 16:93–105.

Chatarpaul, L., and Carlisle, A. 1983. Nitrogen fixation: a biotechnological opportunity for Canadian forestry. *For. Chron.* (October):249–259.

Chen, H., Wilkerson, C.G., Kuchar, J.A., Phinney, B.S., and Howe, G.A. 2005. Jasmonate-inducible plant enzymes degrade essential amino acids in the herbivore midgut. *PNAS* 102:19237–19242.

Chen, J., Franklin, J.F., and Spies, T.A. 1992. Vegetation responses to edge environments in old-growth Douglas-fir forests. *Ecol. Applic.* 2:387–396.

Chen, J., Franklin, J.F., and Spies, T.A. 1993. Contrasting microclimates among clearcut, edge, and interior of old-growth Douglas-fir forest. *Agric. For. Meteorol.* 63:219–237.

Cherfas, J. 1990. The fringe of the ocean—under siege from land. *Science* 1990:248:163–165.

Cherfas, J. 1991. Disappearing mushrooms: another mass extinction. *Science* 254:1458.

Chessin, M., and Zipf, A.E. 1990. Alarm systems in higher plants. *Bot. Rev.* 56:193–235.

Chesson, P. 2000. Mechanisms of maintenance of species diversity. *Annu. Rev. Ecol. Syst.* 31:343–366.

Childs, S.W., and Flint, A.L. 1990. Physical properties of forest soils containing rock fragments. In *Proceedings of the 7th North American Forest Soils Conference.* Edited by S. Gessel. Edmonton, Alberta, Canada, 95–121.

Childs, SW., Shade, S.P., Miles, D.W.R., Shepard, E., and Froehlich, H.A. 1989. Soil physical properties: importance to long-term forest productivity. In *Maintaining the long-term productivity of Pacific Northwest forest ecosystem.* Edited by D.A. Perry, R. Meurisse, B. Thomas, R. Miller, J. Boyle, J. Means, C.R. Perry, and R.F. Powers. Timber Press, Portland, OR, 53–66.

Chivian, E., editor. 2003. *Biodiversity: its importance to human health,* 2nd ed. Center for Health and the Global Environment, Harvard Medical School. Cambridge, MA.

Chornesky, E.A., Bartuska, A.M., Aplet, G.H., Britton, K.O., Cummings-Carlson, J., Davis, F.W., Eskow, J., Gordon, D.R., Gottschalk, K.W., Haack, R.A., Hansen, A.J., Mack, R.N., Rahel, F.J., Shannon, M.A., Wainger, L.A., and Wigley, T.B. 2005. Science priortites for reducing the threat of invasive species to sustainable forestry. *BioScience* 55:335–348.

Chow, W.S., Melis, A., and Anderson, J.M. 1990. Adjustments of photosystem stoichiometry in chloroplasts improve the quantum efficiency of photosynthesis. *PNAS* 87:7502–7506.

Christensen, N.L., Agee, J.K., Brussard, P.F., Hughes, J., Knight, D.H., Minshall, G.W., Peek, J. M., Pyne, S.J., Swanson, F.J., Thomas, J. W., Wells, S., Williams, S.E., and Wright, H.A. 1989. Interpreting the yellowstone fires of 1988. *BioScience* 39:678–685.

Ciais, P., Tans, P., Trolier, M., White, J.W.C., and Francey, R.J. 1995. A large Northern Hemisphere terrestrial hemisphere

CO_2 sink indicated by the $^{13}C/^{12}C$ ratio of atmospheric CO_2. *Science* 269:1098–1102.

Ciborowski, P. 1989. Sources, sinks, trends, and opportunities. In *The challenge of global warming.* Edited by D.E. Abrahamson. Island Press, Washington, DC, 213–230.

Clarholm, M. 1985. Interactions of bacteria, protozoa and plants leading to mineralization of soil nitrogen. *Soil Biol. Biochem.* 17:181–187.

Clark, C.L., Miller, J.D., and Whitney, N.J. 1989. Toxicity of conifer needle endophytes to spruce budworm. *Mycol. Res.* 93:508–512.

Clark, D.A., and Clark, D.B. 1984. Spacing dynamics of a tropical rain forest tree: evaluation of the Janzen-Connell model. *Am. Nat.* 124:769–788.

Clark, D.B., Palmer, M.W., and Clark, D.A. 1999. Edaphic factors and the landscape-scale distributions of tropical forest trees. *Ecology* 80:2662–2675.

Clark, J.S. 1988. Effect of climate change on fire regimes in northwestern Minnesota. *Nature* 334:233–235.

Clark, J.S. 1990a. Fire and climate change during the last 750 yr in northwestern Minnesota. *Ecol. Monogr.* 60:135–159.

Clark, J.S. 1990b. Twentieth-century climate change, fire suppression, and forest production and decomposition in northwestern Minnesota. *Can. J. For. Res.* 20:219–232.

Clark, J.S., and MacLachlan, J.S. 2003. Stability of forest biodiversity. *Nature* 423:635–637.

Clark, W.C., editor. 1982. *Carbon dioxide review 1982.* Clarendon Press, Oxford.

Clarkson, D.T., and Hanson, J.B. 1980. The mineral nutrition of wild plants. *Annu. Rev. Plant Physiol.* 31:239–298.

Claus, R., Hoppen, H.O., and Karg, H. 1981. The secret of truffles: a steroidal pherome? *Experientia* 37:1178–1179.

Clayton, J.L. 1979. Nutrient supply to soil by rock weathering. In *Impact of intensive harvesting on forest nutrient cycling.* Edited by A. Leaf. State Univ. of New York, Syracuse.

Clements, F.E. 1916. *Plant succession, an analysis of the development of vegetation.* Carnegie Institute, Washington, DC.

Cleveland C.C., Neff, J.C., Townsend, A.R., and Hood, E. 2004. Composition, dynamics, and fate of leached dissolved organic matter in terrestrial ecosystems: results from a decomposition experiment. *Ecosystems* 7:275–285.

Cleveland, C.C., Townsend, A.R., Schimel, D.S., Fisher, H., Howarth, R.W., Hedin, L.O., Perakis, S.S., Latty, E.F., von Fischer, J.C., Elseroad, A., and Wasson, M.F. 1999. Global patterns of terrestrial biological nitrogen (N_2) fixation in natural ecosystems. *Global Biogeochem. Cycles* 13:623–645.

Clewell, A.F. 1989. Natural history of wiregrass (*Aristida stricta* Michx., Gramineae), *Nat. Area J.* 9:223.

Cline, E.T., Ammirati, J.F., and Edmonds, R.L. 2005. Does proximity to mature trees influence ectomycorrhizal fungus communities of Douglas-fir seedlings? *New Phytol.* 166:993–1009.

Coblentz, B.E. 1977. Some range relationships of feral goats on Santa Catalina Island, California. *Range Manage.* 30:415–419.

Coblentz, B.E. 1978. Effects of feral goats *(Capra bircus)* on island ecosystems. *Biol. Conserv.* 13:279–286.

Coblentz, B.E. 1990. Exotic organisms: a dilemma for conservation biology. *Conserv. Biol.* 4:261–265.

Cochran, P.H. 1969. *Thermal properties and surface temperatures of seedbeds.* U.S. Department of Agriculture, Forest Service, Pacific N.W. Forest Range Exp. Sta., Portland, OR.

Cochrane, M.A. 2001. Synergistic interactions between habitat fragmentation and fire in evergreen tropical forests. *Conserv. Biol.* 15:1515–1521.

Cochrane, M.A., and Laurance, W.F. 2002. Fire as a large scale edge effect in Amazonian forests. *J. Trop. Ecol.* 8:311–325.

Cochrane, M.A., Alencar, A., Schulze, M.D., Souza Jr., C.M., Nepstad, D.C., Lefebvre, P., and Davidson, E.A. 1999. positive feedbacks in the fire dynamic of closed canopy tropical forests. *Science* 284:1832–1835.

Cohen, J.E., Briand, F., and Newman, C.M. 1986. A stochastic theory of community food webs. III. Predicted and observed lengths of food chains. *Proc. R. Soc. Lond. Biol.* 228:317–353.

Cole, C.V., and Heil, R.D. 1981. Phosphorus effects on terrestrial nitrogen cycling. In *Terrestrial nitrogen cycles.* Edited by F.E. Clark and T. Rosswall. *Ecol. Bull. (Stockholm)* 33:363–374.

Cole, D.M., and Schmidt, W.C. 1986. *Site treatments influence first 25 years development of mixed conifers in Montana.* Res. Paper INT-364. U.S. Department of Agriculture, Forest Service, Ogden, UT.

Cole, D.W. 1988. *Impact of whole-tree harvesting and residue removal on productivity and nutrient loss from selected soils of the Pacfic Northwest.* Univ. of Washington, Seattle.

Cole, D.W., and Rapp, M. 1981. Elemental cycling in forest ecosystems. In *Dynamic properties of forest ecosystems.* Edited by D.E. Reichle. Cambridge Univ. Press, Cambridge, 341–410.

Coleman, D.C. 1985. Through a ped darkly: an ecological assessment of root-soil-microbial-faunal interactions. In *Ecological interactions in soil.* Edited by A.H. Fiber, D. Atkinson, D.J. Read, and M.B. Usher. Blackwell, Oxford, 1–21.

Coleman, D.C., Anderson, R.V., Cole, C.V., McClellan, J.F., Woods, L.E., Trofymow, J.A., and Elliott, E.T. 1984. Roles of protozoa and nematodes in nutrient cycling. In *Soil Biology.* Edited by R.L. Todd. American Society for Agronomy, Madison, WI, 17–28.

Coleman, D.C., Crossley Jr., D.A., and Hendrix, P.F. 2004. *Fundamentals of soil ecology,* 2nd ed. Elsevier Academic Press, San Diego.

Coleman, D.C., Reid, C.P.P., and Cole, C.V. 1983. Biological strategies of nutrient cycling in soils. *Adv. Ecol. Res.* 13:1–55.

Coley, P.D. 1980. Effects of leaf age and plant life history patterns on herbivory. *Nature* 284:545–546.

Coley, P.D. 1987. Interspecific variation in plant anti-herbivore properties: the role of habitat quality and rate of disturbance. *New Phytol.* 106(Suppl):251–263.

Coley, P.D. 1988. Effects of plant growth rate and leaf lifetime on the amount and type of anti-herbivore defense. *Oecologia (Berlin)* 74:531–536.

Coley, P.D., Bryant, J.P., and Chapin III, F.S. 1985. Resource availability and plant antiherbivore defense. *Science* 230:895–899.

Colinas, C., Perry, D.A., Molina, R., and Amaranthus, M.P. 1994. Survival and growth of *Pseudotsuga menzzesii* seedlings inoculated with biocide-treated soils at planting in a degraded clearcut. *Can. J. For. Res.* 24:1741–1749.

Collins, J.P. 2003. What can we learn from community genetics? *Ecology* 84:574–577.

Comerford, N.B., and Fisher, R.F. 1984. Using foliar analysis to classify nitrogen-deficient sites. *Soil Sci. Am. J.* 48:910–913.

Compton, J.E., Watrud, L.S., Porteous, L.A., and DeGrood, S. 2004. Response of soil microbial biomass and community composition to chronic nitrogen additions at Harvard Forest. *For. Ecol. Manage.* 196:143–158.

Conard, S.C., Jaramillo, A.E., Cromack Jr., K., and Rose, S. 1985. *The role of the genus* Ceanothus *in western forest ecosystems.* Gen. Tech. Rpt. PNW-l82. U.S. Department of Agriculture, Forest Service. Portland, OR.

Condit, R., Ashton, P., Bunyavejchewin, S., Dattaraja, H.S., Davies, S., Esufali, S., Ewango, C., Foster, R., Gunatilleke, I.A.U.N., Gunatilleke, C.V.S., Hall, P., Harms, K.E., Hart, T., Hernandez, C., Hubbell, S., Itoh, A., Kiratiprayoon, S., LaFrankie, J., Loo de Lao, S., Makana, J.-R., Nur Supardi Noor, Md., Kassim, A.R., Russo, S., Sukumar, S., Samper, C., Suresh, H.S., Tan, S., Thomas, S., Valencia, R., Vallejo, M., Villa, G., and Zillio, T. 2006. The importance of demographic niches to tree diversity. *Science Express*, June 8.

Condit, R., Pitman, N., Leigh Jr., E.G., Chave, J., Terborgh, J., Foster R.B., Nunez, P., Aguilar, V.S., Valencia, R., Villa, G., Muller-Landau, H.C., Losos, E., and Hubbell, S.P. 2002. Beta-diversity in tropical forest trees. *Science* 295:666–668.

Conkle, M.T., technical coordinator. 1979. *Proceedings of the Symposium on Isozymes of North American Forest Trees and Forest Insects*, July 27, 1979, Berkeley, CA. U.S. Department of Agriculture, Forest Service, San Francisco, 64.

Connell, J.H. 1971. On the role of natural enemies in preventing competitive exclusion in some marine animals and in rain forest trees. In *Dynamics of populations. Proceedings of the Advanced Study Institute on Dynamics of Numbers in Populations*, 1970, Oosterbeek. Edited by P.J. den Boer and G.R. Gradwell. Center for Agricultural Publishing and Documentation, Wageningen, 298–310.

Connell, J.H. 1975. Some mechanisms producing structure in natural communities: a model and evidence from field experiments. In *Ecology and evolution of communities.* Edited by M.L. Cody and J.M. Diamond. Harvard Univ. Press, Cambridge, MA., 460–490.

Connell, J.H. 1978. Diversity in tropical rain forests and coral reefs. *Science* 199:1302–1309.

Connell, J.H. 1989. Some processes affecting the species composition in forest gaps. *Ecology* 70:560–562.

Connell, J.H., and Slatyer, R.O. 1977. Mechanisms of succession in natural communities and their role in community stability and organization. *Am. Nat.* 111:1119–1144.

Connell, J.H., Tracey, J.G., and Webb, J.G. 1984. Compensatory recruitment, growth, and mortality as factors maintaining rain forest tree diversity. *Ecol. Monogr.* 54:141–164.

Cool, J.C. 1980. *Stability and survival—the Himalayan challenge.* Ford Foundation, New York.

Coomes, D.A., Allen, R.B., Forsyth, D.M., and Lee, W.G. 2003. Factors preventing the recovery of New Zealand forests following control of invasive deer. *Conserv. Biol.* 17:1523–1739.

Cope, J.A. 1925. *Loblolly pine in Maryland.* Univ. of Maryland, College Park.

Cork, S.J., and Kenagy, G.J. 1989. Rates of gut passage and retention of hypogeous fungal spores in two forest-dwelling rodents. *J. Mamm.* 70:512–519.

Cornaby, B.W., Gist, C.S., and Crossley Jr., D.A. 1977. Resource partitioning in leaf-litter faunas from hardwood and hardwood-converted-to-pine forests. In *Mineral cycling in southeastern ecosystems.* ERDA Symposium Series (CONF-740513). Edited by F.G. Howell, J.B. Gentry, and M.H. Smith. U.S. Department of Energy. Washington, DC. 588–597.

Cornell, H.V., and Hawkins, B.A. 2003. Herbivore responses to plant secondary compounds: a test of phytochemical coevolution theory. *Am. Nat.* 161:507–522.

Cornish, P.M., and Vertessey, R.A. 2001. Forest age-induced changes in evapotranspiration and ware yield in a eucalypt forests. *J. Hydrol.* 242:43–63.

Costa, M.H., and Foley, J.A. 2000. Combined effects of deforestation and double atmospheric CO_2 concentrations on the climate of Amazonia. *J. Clim.* 12:18–35.

Costanza, R. 1992. Toward an operational definition of ecosystem health. In *Ecosystem health: new goals for environmental management.* Edited by R. Costanza, B.G. Norton, and B.D. Haskell. Island Press, 239–256. Washington, DC.

Costanza, R. 1996. Ecological economics: reintegrating the study of humans and nature. *Ecol. Applic.* 6:978–990.

Costanza, R. 2006. Nature: ecosystems without commodifying them. *Nature* 443:749.

Costanza, R., and Daly, H.E. 1992. Natural capital and sustainable development. *Conserv. Biol.* 6:37–46.

Costanza, R., d'Arge, R., de Groot, R., Farber, S., Grasso, M., Hannon, B., Limburg, K., Naeem, S., O'Neill, R.V., Paruelo, J., Raskin, R.G., Sutton, P., and van den Belt, M. 1997. The value of the world's ecosystem services and natural capital. *Nature* 387:253–260.

Cotton, P.A. 2003. Avian migration phenology and global climate change. *PNAS* 100:12219–12222.

Covington, W.W. 2000. Helping western forests heal: the prognosis is poor for United States forest ecosystems. *Nature* 408: 135–136.

Cowen, R. 1990. Medicine on the wild side. *Sci. News* 136:280–282.

Cowling, E., Galloway, J., Furiness, C., Barber, M., Bresser, T., Cassman, K., Erisman, J.W., Haeuber, R., Howarth, R.I., Melillo, J., Moomaw, W., Mosier, A., Sanders, K., Seitzinger, S., Smeulders, S., Socolow, R., Walters, D., West, F., and Zhu, Z. 2001. Summary statement. In *Proceedings of the 2nd International Nitrogen Conference.* Ecological Society of America, Washington, DC.

Cox, P.M., Betts, R.A., Collins, M., Harris, P.P., Huntingford, C., and Jones, C.D. 2004. Amazonian forest dieback under climate-carbon cycle projections for the 21st century. *Theor. Appl. Climatol.* 78:137–156.

Cox, P.M., Betts, R.A., Jones, C.D., Spall, S.A., and Totterdell, I.J. 2000. Acceleration of global warming due to carbon-cycle feedbacks in a coupled climate model. *Nature* 408:184–187.

Coyle, D.R., Nebeker, T.E., Hart, E.R., and Mattson, W.J. 2005. Biology and management of insect pests in North American intensively managed hardwood forest systems. *Ann. Rev. Entomol.* 50:1–29.

Crawford, D.L., and Crawford, R.L. 1980. Microbial degradation of lignin. *Enzyme Microb. Technol.* 2:11–22.

Crawford, H.S., and Jennings, D.T. 1989. Predation by birds on spruce budworm *Choristoneura fumiferana*: functional, numerical, and total responses. *Zoology* 70:152–163.

Crews, T.E., Kitayama, K., Fownes, J.H., Riley, R.H., Herbert, D.A., Mueller-Dombois, D., and Vitousek, P.M. 1995. Changes in soil phosphorus fractions and ecosystem dynamics across a long chronosequence in Hawaii. *Ecology* 76:1407–1424.

Crisafulli, C.M., Swanson, F.J., and Dale, V.H. 2005. Overview of ecological responses to the eruption of Mount St. Helens: 1980–2005. In *Ecological responses to the 1980 eruption of Mount St. Helens.* Edited by V.H. Dale, F.J. Swanson, and C.M. Crissafulli. Springer, New York, 287–299.

Crocker, R.L., and Major, J. 1955. Soil development in relation to vegetation and surface age at Glacier Bay, Alaska. *J. Ecol.* 43:427–448.

Cromack Jr., K., Fichter, B.L., Moldenke, A.M., Entry, J.A., and Ingham, E.R. 1988. Interactions between soil animals and ectomycorrhizal fungal mats. *Agric. Ecosyst. Environ.* 24:161–168.

Cromack Jr., K., Sollins, P., Graustein, W.C., Speidel, K., Todd, A.W., Spycher, G., Li, C.Y., and Todd, R.L. 1979. Calcium oxalate accumulation and soil weathering in mats in the hypogeous fungus Hysterangium crassum. *Soil Biol. Biochem.* 11:463–468.

Cromack Jr., K., Sollins, P., Todd, R.L., Crossley Jr., D.A., Fender, W.M., Fogel, R., and Todd, A.W. 1977. Soil microorganism-arthropod interactions: fungi as major calcium and sodium sources. In *The role of arthropods in forest ecosystems.* Edited by W.J. Mattson. Springer-Verlag, New York, 78–84.

Crome, F.H.J., and Richards, G.C. 1988. Bats and gaps: microchiropteran community structure in a Queensland rain forest. *Ecology* 69:1960–1969.

Crossley Jr., D.A. 1977. Oribatid mites and nutrient cycling. In *Biology of oribatid mites.* Edited by D.L. Dindal. SUNY College of Environmental Science and Forestry, Syracuse, NY, 71–85.

Crow, J.F., and Kimura, M. 1970. *An introduction to population genetics.* Harper & Row, New York,.

Crow, T.R. 1990. Old growth and biological diversity: a basis for sustainable forestry. In *Old growth forests. What are they? How do they work?* Faculty of Forestry, Univ. of Toronto, Canadian Scholars Press, Toronto, 49–62.

Crozier, M.J., Eyles, R.J., Marx, S.L., McConchie, J.A., and Owen, R.C. 1980. Distribution of landslips in the Wairarapa hill country. *N. Z. J. Geol. Geophys.* 23:579–586.

Crutchfleld, J.P., Farmer, J.D., Packard, N.H., and Shaw, R.S. 1986. Chaos. *Sci. Am.* 254:46–57.

Cuenca, G., and Lovera, M. 1990. Recolonization by VA-mycorrhizal propagules of severely disturbed areas of the Canaima National Park in Venezuala. In *Eighth North American conference on mycorrhizae.* Edited by M. Allen and S.E. Williams. Univ. of Wyoming, Laramie, 62.

Cuevas, E., and Medina, E. 1986. Nutrient dynamics within amazonian forest ecosystems. 1. Nutrient flux in fine litter fall and efficacy of nutrient utilization. *Oceologia (Berlin)* 68:466–472.

Cuevas, E., and Medina, E. 1988. Nutrient dynamics with amazonian forests. II. Fine root growth, nutrient availability and leaf litter decomposition. *Oecologia* 76:222–235.

Curran, L.M., Trigg, S.N., McDonald, A.K., Astiani, D., Hardiono, Y.M., Siregar, P., Caniago, I., and Kasischke, E. 2004. Lowland forest loss in protected areas of Indonesian Borneo. *Science* 303:1000–1004.

Currie, C.R., Poulsen, M., Mendenhall, J., Boomsma, J.J., and Billen, J. 2006. Coevolved crypts and exocrine glands support mutualistic bacteria in fungus-growing ants. *Science* 311:81–83.

Currie, D.J. 1991. Energy and large-scale patterns of animal- and plant-species richness. *Am. Nat.* 137:27–49.

Currie, D.J., and Paquin, V. 1987. Large-scale biogeographical patterns of species richness in trees. *Nature* 329:326–327.

Currie, D.J., Mittelbach, G.G., Cornell, H.V., Field, R., Guegan, J.-F., Hawkins, B.A., Kaufman, D.M., Kerr, J.T., Oberdorff, T., O'Brien, E., and Turner, J.R.G. 2004. Predictions and tests of climate-based hypotheses of broad-scale variation in taxonomic richness. *Ecol. Lett.* 7:1121–1134.

Currie W.S., and Aber, J.D. 1997. Modeling leaching as a decomposition process in humid Montane forests. *Ecology* 78:1844–1860.

Cushman, S.A., McKelvey, K.S., Flather, C.H., and McGarigal, K. 2008. Do forest community types provide a sufficient basis to evaluate biological diversity? *Front. Ecol. Environ.* 6:13–17.

Cwynar, L.C. 1987. Fire and the forest history of the North Cascade Range. *Ecology* 68:791–802.

Daehler, C.C. 2003. Performance comparisons of co-occurring native and alien invasive plants: implications for conservation and restoration. *Annu. Rev. Ecol. Syst.* 34:183–211.

Daft, M.J., Clelland, D.M., and Gardner, I.C. 1985. Symbiosis with endomycorrhizas and nitrogen-fixing organisms. *Proc. R. Soc. Edinburgh* 85B:283–298.

Dahlberg, A. 2002. Effects of fire on ectomycorrhizal fungi in Fennoscandian boreal forests. *Silva Fennica* 36(1):69–80.

Dailey, G.C. 1999. Developing a scientific basis for managing earth's life support systems. *Conserv. Ecol.* 3(2):14. www.consecol.org/vol3/iss2/art14.

Daily, G.C., and Ehrlich, P.R. 1992. Population, sustainability, and Earth's carrying capacity. *BioScience* 42:761–771.

Daily, G.C., Alexander, S., Erhlich, P.R., Goulder, L., Lubchenco, J., Matson, P.A., Mooney, H.A., Postel, S., Schneider, S.H., Tilman, D., and Woodwell, G.M. 1997. Ecosystem services: benefits supplied to human societies by natural ecosystems. In *Issues in Ecology*, 2. Ecological Society of America, Washington, DC, 16.

Daily, G.C., Soderqvist, T., Aniyar, S., Arrow, K., Dasgupta, P., Ehrlich, P.R., Folke, C., Jannson, A., Jannson, B.-O., Kautsky, N., Levin, S., Lubchenco, J., Maler, K.-G., Simpson, D., Starrett, D., Tilman, D., and Walker, B. 2000. The value of nature and the nature of value. *Science* 289:395–396.

Dale, V.H., Campbell, D.R., Adams, W.M., Crissafulli, C.M., Dains, V.I., Frenzen, P.M., and Holland, R.F. 2005a. Plant succession on the Mount St. Helens debris-avalanche deposit. In *Ecological Responses to the 1980 Eruption of Mount St. Helens.* Edited by V.H. Dale, F.J. Swanson, and C.M. Crisafulli. Springer, New York, 59–73.

Dale, V.H., Crisafulli, C.M., and Swanson, F.J. 2005b. 25 years of ecological change at Mount St. Helens. *Science* 308: 961–962.

Dale, V.H., Joyce, L.A., McNulty, S., Neilson, R.P., Ayres, M.P., Flannigan, M.D., Hanson, P.J., Irland, L.C., Lugo, A.E., Peterson, C.J., Simberloff, D., Swanson, F.J., Stocks, B.J., and Wotton, B.M. 2001. Climate change and forest disturbances. *BioScience* 51:723–734.

Dale, V.H., Swanson, F.J., and Crisafulli, C.M., editors. 2005c. *Ecological responses to the 1980 eruption of Mount St. Helens.* Springer, New York.

Damschen, E.I., Haddad, N.M., Orrock, J.L., Tewksbury, J.J., and Levey, D.J. 2006. Corridors increase plant species richness at large scales. *Science* 313:1284–1286.

Daneilsen, F., Sorensen, M.K., Olwig, M.F., Selvam, V., Parish, F., Burgess, N.D., Hiraishi, T., Karunagaran, V.M., Rasmussen, M.S., Hansen, L.B., Quarto, A., and Suryadiptura, N. 2005. The Asian tsunami: a protective role for coastal vegetation. *Science* 310:643.

Danielson, R.M., and Visser, S. 1989. Host response to inoculation and behaviour of introduced and indigenous ectomycorrhizal fungi of jack pine grown on oil-sands tailings. *Can. J. For. Res.* 19:1412–1421.

D'Antonio, C.M., Jackson, N.E., Horvitz, C.C., and Hedberg, R. 2004. Invasive plants in wildland ecosystems: merging the study of invasion processes with management needs. *Front. Ecol. Environ.* 2:513–521.

D'Antonio, C.M., and Vitousek, P.M. 1992. Biological invasions by exotic grasses, the grass/fire cycle, and global change. *Annu. Rev. Ecol. Syst.* 23:63–87.

D'Arrigo, R., Jacoby, G.C., and Fung, I.Y. 1987. Boreal forests and atmosphere-biosphere exchange of carbon dioxide. *Nature* 329:321–323.

Darwin, C. 1871. *The descent of man and selection in relation to sex,* 2nd ed. Murray, London.

Daterman, G.E., Miller, J.C., and Hanson, P.E. 1986. Potential for gypsy moth problems in southwest Oregon. In *Forest pest management in southwest Oregon.* Edited by O.T. Helgerson. Oregon State Univ., Corvallis, 37–40.

Davidson, E.A., and Janssens, I.A. 2006. Temperature sensitivity of soil carbon decomposition and feedbacks to climate change. *Nature* 440:165–173.

Davidson, E.A., and Swank, W.T. 1986. Environmental parameters regulating gaseous nitrogen losses from two forested ecosystems via nitrification and denitrification. *Appl. Environ. Microbiol.* 52:1287–1292.

Davidson, E.A., Hart, S.A., and Firestone, M.K. 1992. Internal cycling of nitrate in soils of a mature coniferous forest. *Ecology* 73:1148–1156.

Davidson, E.A., Hart, S.C., Shanks, C.A., and Firestone, M.K. 1991. Measuring gross nitrogen mineralization, immobilization, and nitrification by ^{15}N isotope dilution in intact soil cores. *J. Soil Sci.* 42:335–349.

Davidson E.A., Savage, K., Bolstad, P., Clark, D.A., Curtis, P.S., Ellsworth, D.S., Hanson, P.J., Law, B.E., Luo, Y., Pregitzer, K.S., Randolph, J.C., and Zak, D. 2002. Belowground carbon allocation in forests estimated from litterfall and IRGA-based soil respiration measurements. *Agric. For. Meteorol.* 113:39–51.

Davies, A.G., and Baille, I.C. 1988. Soil-eating by red leaf monkeys *(Presbytis rubicunda)* in Sabah, Northern Borneo. *Biotropica* 20:252–258.

Davis, M.B. 1981. Quaternary history and the stability of forest communities. In *Forest succession.* Edited by D.C. West, H.H. Shugart, and D.B. Botkin. Springer-Verlag, New York, 132–153.

Davis, M.B. 1989. Lags in vegetation response to greenhouse warming. *Clim. Changes* 15:75–82.

Davis, M.B., and Shaw, R.G. 2001. Range shifts and adaptive responses to Quaternary climate change. *Science* 292:673–679.

Davis, M.B., Woods, K.D., Webb, S.L., and Futyma, R.P. 1986. Dispersal versus climate: expansion of *Fagus* and *Tsuga* into the Upper Great Lakes region. *Vegetatio* 67:93–103.

Dawkins, R. 1982. *The extended phenotype.* Oxford Univ. Press, Oxford.

Dawson, J.O. 1983. Dinitrogen fixation in forest ecosystems. *Can. J. Microbiol.* 29:979–989.

Dawson, T.E. 1993. Hydraulic lift and water use by plants: implications for water balance, performance and plant-plant interactions. *Oecologia* 95:565–574.

Day, T., and Young, K.A. 2004. Competitive and facilitative evolutionary diversification. *BioScience* 54:101–109.

de Graf, N.R., and Pods, R.L.H. 1990. The Celos management system: a polycyclic method for sustained timber production in South American Rainforest. In *Alternative to deforestation: steps toward sustainable use of the Amazon basin.* Edited by A.B. Anderson. Columbia Univ. Press, New York, 116–127.

De Las Heras, J., Guerra, J., and Herranz, J.M. 1990. Bryophyte colonization of soils damaged by fire in south-east Spain: a preliminary report on dynamics. *J. Bryol.* 16:275–288.

de Olivera Carvalheiro, K,. and Nepstad, D.C. 1996. Deep soil heterogeneity and fine root distribution in forests and pastures of eastern Amazonia. *Plant Soil* 182:279–285.

DeAngelis, D.L. 1992. *Dynamics of nutrient cycling and foodwebs.* Chapman and Hall, London.

DeAngelis, D.L., Post, W.M., and Travis, C.C. 1986. *Positive feedback in natural systems.* Springer-Verlag, Berlin.

DeBach, P. 1974. *Biological control by natural enemies.* Cambridge Univ. Press, Cambridge.

DeBano, L.F. 1990. The effect of fire on soil properties. In *Proceedings: management and productivity of western montane forest soils.* General Technical Report INT-280. Edited by A.E. Harvey and L.F. Neuenschwander. U.S. Department of Agriculture, Forest Service, Intermountain. Res. Sta. Ogden, UT, 15:1–156.

DeBell, D.S., Curtis, R.O., Harrington, C.A., and Tappeiner, J.C. 1997. Shaping stand development through silvicultural practices. In *Creating a forestry for 21st century.* Edited by K.A. Kohm and J.F. Franklin. Island Press, Washington, DC, 141–149.

Debinski, D.M., and Holt, R.D. 2000. A survey and overview of habitat fragmentation experiments. *Conserv. Biol.* 14:342–355.

DeHayes, D.H., Schaberg, P.G., Hawley, G.J., and Strimbeck, G.R. 1999. Acid rain impacts on calcium nutrition and forest health. *BioScience* 49:789–800.

Del Vecchio, T.A., Gehring, C.A., Cobb, N.S., and Whitham, T.G. 1993. Negative effects of scale insect herbivory on the ectomycorrhizae of juvenile pinyon pine. *Ecology* 74:2297–2302.

Delcourt, H.R., Delcourt, P.A., and Webb, T. 1983. Dynamic plant ecology: the spectrum of vegetational change in space and time. *Quat. Sci. Rev.* 1:153–175.

DeLuca, T.H., Zackrisson, O., Nilsson, M.-C., and Sellstedt, A. 2002. Quantifying nitrogen-fixation in feather moss carpets of boreal forests. *Nature* 419:917–920.

Denison, W.C. 1979. *Lobaria oregana,* a nitrogen-fixing lichen in old-growth Douglas fir forests. In *Symbiotic fixation in the management of temperate forests.* Edited by J.C. Gordon, C.T. Wheeler, and D.A. Perry. Oregon State Univ., Corvallis, 266–275.

Dennis, R. Mayer, J. , Applegate, G., Chokkalingam, U., Pierce Colfer, C.J., Kurniawan, I., Lachowski, H., Maus, P., Permana, R.P., Ruchiat, Y. , Stolle, F., Suyanto, and Tomich, T.P. 2005. Fire, people, and pixels: linking social science and remote sensing to understanding underlying causes and impacts of fires in Indonesia. *Human Ecology* 33:465–504.

Denslow, J.S. 1987. Tropical rainforest gaps and tree species diversity. *Annu. Rev. Ecol. Syst,* 18:431–451.

Denslow, J.S., and Gomez-Diaz, A.E. 1990. Seed rain to tree-fall gaps in a Neotropical rain forest. *Can. J. For. Res.* 20:642–648.

Denslow, J.S., Vitousek, P.M., and Schultz, J.C. 1987. Bioassays of nutrient limitation in a tropical rain forest soil. *Oecologia (Berlin)* 74:370–376.

DeSpain, D. 1973. Vegetation of the Big Horn Mountains, Wyoming. *Ecol. Monogr.* 43:329–355.

Dessert, C., Dupre, B., Gaillardet, J., Francois, L.M., and Allegre, C.J. 2003. Basalt weathering laws and the impact of basalt weathering on the global carbon cycle. *Chem. Geol.* 202:257–273.

Devoto, M., Medan, D., and Montaldo, N.H. 2005. Patterns of interaction between plants and pollinators along an environmental gradient. *Oikos* 109:461–472.

Diamond, J. 1986. The design of a nature reserve system for Indonesian New Guinea. In *Conservation biology: the science of scarcity and diversity.* Edited by M.E. Soule. Sinauer Associates, Sunderland, MA, 485–503.

Diamond, J. 2005a. *Collapse: how societies choose to fail or succeed.* Viking Press, New York.

Diamond, J. 2005b. *The ends of the world as we know them. New York Times,* January 1.

Diamond, J., and Case, T., editors. 1986. *Community ecology.* Harper and Row, New York.

Dicke, M., and van Loon, J.J.A. 2000. Multitrophic effects of herbivore-induced plant volatiles in an evolutionary context. *Entomol. Exper. Applic.* 97:237–249.

Dickie, I.A., and Reich, P.B. 2005. Ectomycorrhizal fungal communities at forest edges. *J. Ecol.* 93:244–255.

Dickie, I.A., Xu, B., and Koide, R.T. 2002. Vertical niche differentiation of ectomycorrhizal hyphae in soil as shown by T-RFLP analysis. *New Phytol.* 156:527–535.

Dighton, J. 1991. Acquisition of nutrients from organic resources by mycorrhizal autotrophic plants. *Experientia* 47:362–369.

Dighton, J., and Harrison, A.F. 1983. Phosphorus nutrition of lodgepole pine and Sitka spruce as indicated by a root bioassay. *Forestry* 56:33–43.

Dighton, J., Thomas, E.D., and Latter, P.M. 1987. Interactions between tree roots, mycorrhizas, a saprotrophic fungus and the decomposition of organic substrates in a microcosm. *Biol. Fertil. Soils* 4:145–150.

Doak, D.F., Bigger, D., Harding, E.K., Marvier, M.A., O'Malley, R.E., and Thomson, D. 1998. The statistical inevitability of stability-diversity relationships in community ecology. *Am. Nat.* 151:264-276.

Dollinger, P.M., Ehrlich, P.R., Fitch, W.L., and Breedlove, D.E. 1973. Alkaloid and predation patterns in Colorado lupine populations. *Oecologia* 13:191–204.

Dombeck, M.P., Wood, C.A., and Williams, J.E. 2003. *From conquest to conservation: our public lands legacy.* Island Press, Washington, DC.

Donnelly, P.K., Caldwell, B.A., and Crawford, D.L. 1990. Lignin degradation by mycorrhizal fungi. In *Eigth North American conference on mycorrhizae.* Edited by M.F. Allen and S.F. Williams. Univ. of Wyoming, Laramie.

Donovan, T.M., and Flather, C.H. 2002. Relationships among North American songbird trends, habitat fragmentation, and landscape occupancy. *Ecol. Applic.* 12:364–374.

Douglas, R.B., Thomas, V.T., and Cullings, K.W. 2005. Belowground ectomycorrhizal community structure of mature lodepole pine and mixed conifer stands in Yellowstone National Park. *For. Ecol. Manage.* 208:303–317.

Drake, E.T., editor. 1968. *Evolution and environment. A symposium presented on the occasion of the 100th anniversary of the foundation of Peabody Museum of Natural History at Yale University.* Yale Univ. Press, New Haven, CT.

Drake, J.A. 1990. The mechanics of community assembly and succession. *J. Theor. Biol.* 147:213–233.

Drake, J.B., Dubayah, R.O., Knox, R.G., Clark, D.B., and Blair, J.B. 2002. Sensitivity of large-footprint Lidar to canopy structure and biomass in a neotropical rainforest. *Remote Sensing Environ.* 81:378–392.

Drew, T.J., and Flewelling, J.W. 1977. Some recent Japanese theories of yield-density relationships and their application to Monterey pine plantations. *For. Sci.* 23:517–534.

Drew, T.J., and Flewelling, J.W. 1979. Stand density management an alternative approach and its application to Douglas-fir plantations. *For. Sci.* 25:518–530.

Driscoll, C.T., Lawrence, G.B., Bulger, A.J., Butler, T.J., Cronan, C.S., Eager, C., Lambert, K.F., Likens, G.E., Stoddard, J.L., and Weathers, K.C. 2001. Acidic deposition in the northeastern United States: sources and inputs, ecosystem effects, and management strategies. *BioScience* 51:180–198.

Duddridge, J.A., Finlay, R.D., Read, D.J., and Soderstrom, B. 1988. The structure and function of the vegetative mycelium of ectomycorrhizal plants. III. Ultrastructural and autoradiographic analysis of inter-plant carbon distribution through intact mycelial systems. *New Phytol.* 108:183–188.

Duddridge, J.A., Malibari, A.S., and Read, D.J. 1980. Structure and function of mycorrhizal rhizomorphs with special reference to their role in water transport. *Nature* 287:834–836.

Dunn, A.L., Barford, C.C., Wofsy, F.C., Goulden, M.C., and Daube, D.C. 2007. A long-term record of carbon exchange in a boreal black spruce forest: means, response to interannual variability, and decadal trends. *Global Change Biol.* 13:577–590.

Dunne, J.A. 2005. The network structure of food webs. In *Ecological networks: linking structure to dynamics in food webs.* Edited by M. Pascual and J.A. Dunne. Oxford Univ. Press, Oxford.

Dunne, J.A., Williams, R.J., and Martinez, N.D. 2002. Food-web structure and network theory: the role of connectance and size. *PNAS* 99:12917–12922.

Duvigneaud, P., and Denaeyer-DeSmet, S. 1970. Biological cycling of minerals in temperate deciduous forests. In *Analysis of temperate forest ecosystems.* Edited by D.E. Reichle. Springer- Verlag, New York, 199–225.

Dwyer, G., Dushoff, J., and Yee, S.H. 2004. The combined effects of pathogens and predators on insect outbreaks. *Nature* 430:341–345.

Dyck, W.J., and Mees, C.A., editors. 1989. *Research strategies for long-term site productivity.* Bulletin No. 152. Forest Research Institute, Rotorua, New Zealand.

Dyck, W.J., and Skinner, M.F. 1990. Potential productivity decline in New Zealand radiata pine forests. In *Sustained productivity of forest soils.* Edited by S. P. Gessel et al. Faculty of Forestry Publication, Univ. of British Columbia, Vancouver, 318–332.

Dyck, W.J., Mees, C.A., and Comerford, N.B. 1989. Medium term effects of mechanical site preparation on radiata pine productivity in New Zealand—a retrospective approach. In *Research strategies for long-term site productivity.* Edited by W.J. Dyck and C.A. Mees. Bull. No. 152, Forest Research Institute, Rotorua, New Zealand.

Dyck, W.J., Mees, C.A., and Hodgkiss, P.D. 1987. Nitrogen availability and comparison to uptake in two New Zealand *Pinus radiata* forests. *N. Z. J. For. Sci.* 17:338–352.

Dyck, W.M., and Beets, P.M. 1987. Managing for long-term site productivity. *N. Z. For.* (Nov):23–26.

Eamus, D., and Jarvis, P.G. 1989. The direct effects of increase in the global atmospheric CO_2 concentration on natural and commercial temperate trees and forests. *Adv. Ecol. Res.* 19:1–55.

Ebenman, B. and Jonsson, T. 2005. Using community viability analysis to identify fragile systems and keystone species. *Trends Ecol. Evol.* 20:568–575.

Eckburg, P.B., Bik, E.M., Bernstein, C.N., Purdom, E., Dethlefsen, L., Sargent, M., Gill, S.R., Nelson, K.E., and Relman, D.A. 2005. Diversity of the human intestinal microbial flora. *Science* 308:1635–1638.

Edmonds R.I., and Hsiang, T. 1987. Forest floor and soil influence on response of Douglas-fir to urea. *Soil Sci. Soc. Am. J.* 51:1332–1337.

Edmunds Jr., G.F., and Alstad, D.N. 1978. Coevolution in insect herbivores and conifers. *Science* 199:941–945.

Edwards, C.A., Reichie, D.E., and Crossley Jr., D.A. 1970. The role of soil invertebrates in turnover of organic matter and nutrients. In *Temperate forest ecosystems.* Edited by D.E. Reichle. Springer-Verlag, New York, 147–172.

Edwards, P.J., and Wratten, S.D. 1983. Wound induced defenses in plants and their consequences for patterns of insect grazing. *Oecologia* 59:88–93.

Egeland, D.M. 1982. Vegetation-environmental relationships on two clearcuts on the western slopes of the Oregon Cascades. Masters thesis. Oregon State Univ., Corvallis.

Egerton-Warburton, L., Graham, R.C., and Hendrix, P.F. 2005. Soil ecosystem indicators of post-fire recovery in the California Chaparral. Final report to National Commission on the Science of Sustainable Forestry. http://ncseonline.org/NCSSF/cms.cfm?id=425#research.

Egnell, G., and Valinger, E. 2003. Survival, growth, and growth allocation of planted Scots pine trees after different levels of biomass removal in clear felling. *For. Ecol. Manage.* 177:65–74.

Ehrenfeld, J.G. 1990. Dynamics and processes of barrier island vegetation. *Rev. Aquatic Sci.* 2:437–480.

Ehrlich, P.R. 1980. The strategy of conservation, 1980–2000. In *Conservation biology: an evolutionary-ecological perspective.* Edited by M.E. Soule and B.A. Wilcox. Sinauer Associates, Sunderland, MA, 329–344.

Ehrlich, P.R. 1986. *The machinery of nature.* Simon and Schuster, New York.

Ehrlich, P.R., and Ehrlich, A.H. 1981. *Extinction: the causes of consequences of the disappearance of species.* Random House, New York.

Ehrlich, P.R., and Raven, P. 1964. Butterflies and plants: a study in coevolution. *Evolution.* 18:586–608.

Eldredge, N. 1986. Information, economics and evolution. *Annu. Rev. Ecol. Syst.* 17:351–370.

Eliasson, P.E., McMurtrie, R.E., Pepper, D.A., Strömgren, M., Linder, S., and Ågren, G.I. 2005. The response of heterotrophic CO_2 flux to soil warming. *Global Change Biol.* 11:167–181.

Elliott, E.T., Anderson, R.V., Coleman, D.C., and Cole, C.V. 1980. Habitable pore space and microbial trophic interactions. *Oikos* 35:327–335.

Ellison, A.M., Bank, M.S., Clinton, B.D., Colburn, E.A., Elliot, K., Ford, C.R., Foster, D.R., Kloeppel, B.D., Knoepp, J.D., Lovett, G.M., Mohan, J., Orwig, D.A., Rodenhouse, N.L., Sobczak, W.V., Stinson, K.A., Stone, J.K., Swan, C.M., Thompson, J., Van Holle, B., and Webster, J.R. 2005. Loss of foundation species: consequences for the structure and dynamics of forested ecosystems. *Front. Ecol. Environ.* 3:479–486.

Ellwood, D.C., Hedger, J.N., Latham, M.J., Lynch, J.M., and Slater, J.M., editors. 1980. *Contemporary microbial ecology.* Academic Press, London, 283–304.

Ellwood, M.D.F., and Foster, W.A. 2004. Doubling the estimate of invertebrate biomass in a rainforest canopy. *Nature* 429:549–551.

Elmerich, C., Newton, W.E., editors. 2006. *Associative and endophytic nitrogen-fixing bacteria and cyanobacterial associations.* Springer, New York.

Elser, J.J., Fagan, W.F., Denno, R.F., Dobberfuhl, D.R., Folarin, A., Huberty, A., Interlandi, S., Kilham, S.S., McCauley, E., Schulz, K.L., Siemann, E.H., and Sterner, R.W. 2000. Nutritional constraints in terrestrial and freshwater foodwebs. *Nature* 408:578–580.

Elton, C. 1927. *Animal ecology.* Sidgwick and Jackson, London.

Elton, C.S. 1958. *The ecology of invasions by plants and animals.* Methuen, London.

Emanuel, K. 2005. Increasing destructiveness of tropical cyclones over the past 30 years. *Nature* 436:686–688.

Emanuel, W.R., Shugart, H.H., and Stevenson, M.P. 1985. Climatic change and the broad-scale distribution of terrestrial ecosystem complexes. *Climat. Change* 7:29–43.

Emerson, B.C., and Kolm, N. 2005. Species diversity can drive speciation. *Nature* 434:1015–1017.

Emlen, S.T., and Wrege, PH. 1988. The role of kinship in helping decisions among white-fronted bee-eaters. *Behav. Ecol. Sociobiol.* 23:305–315.

Endler, A., and McLellan, T. 1988. The process of evolution: toward a newer synthesis. *Annu. Rev. Ecol. Syst.* 19:395–422.

Enquist, B.J., and Niklas, K.J. 2001. Invariant scaling relations across tree-dominated communities. *Nature* 410:655–741.

Enquist, B.J., Brown, J.H., and West, G.B. 1998. Allometric scaling of plant energetics and population density. *Nature* 395:163–165.

Enright, J.T. 1976. Climate and population regulation. *Oecologia* 24:295–310.

Entry, J.A., Rose, C.L., and Cromack Jr., K. 1991. Litter decomposition and nutrient release in ectomycorrhizal mat soils of a Douglas fir ecosystem. *Soil Biol. Biochem.* 23:285–290.

Epstein, E. 1972. *Mineral nutrition of plants: principles and perspectives.* Wiley, New York.

Ericsson, A., Hellkvist, J., Hillerdal-Hagstromer, K., Larsson, S., Mattson-Djos, E., and Tenow, O. 1980. Consumpton and pine growth—hypotheses on effects on growth processes by needle-eating insects. In *Structure and function of northern coniferous forests—an ecosystem study.* Edited by T. Persson. *Ecol. Bull. (Stockholm)* 32:537–545.

Erwin, T.L. 1988. The tropical forest canopy. In *Biodiversity.* Edited by E.O. Wilson and F.M. Peter. National Academy Press, Washington, DC, 123–129.

Erwin, T.L. 1991. An evolutionary basis for conservation strategies. *Science* 253:750–752.

Espinosa-Bancalari, M.A., and Perry, D.A. 1987. Distribution and increment of biomass in adjacent young Douglas-fir stands with different early growth rates. *Can. J. For. Res.* 17:722–730.

Estades, C.F., and Temple, S.A. 1999. Deciduous-forest bird communities in a fragmented landscape dominated by exotic pine plantations. *Ecol. Applic.* 9:573–585.

Evans, D.G., and Miller, M.R. 1988. Vesicular-arbuscular mycorrhizas and the soil-disturbance- induced reduction of nutrient absorption in maize. I. Causal relations. *New Phytol.* 110:67–74.

Evans, D.G., and Miller, M.H. 1990. The role of the external mycelial network in the effect of soil disturbance upon vesicular-arbuscular mycorrrhizal colonization of maize. *New Phytol.* 114:65–71.

Evans, J. 1982. *Plantation forestry in the tropics.* Clarendon Press, Oxford.

Evans, J. 2003. The southern pine beetle and forestry on the Cumberland Plateau in Tennessee: uniformity and vulnerability. *Distant Thunder* 16:10–11.

Evans, K.L., Greenwood, J.J.D., and Gaston, K.J. 2005b. The role of extinction and colonization in generating species-energy relationships. *J. Anim. Ecol.* 74:498–507.

Evans, K.L., Warren, P.H., and Gaston, K.J. 2005a. Species-energy relationships at the macroecological scale: a review of the mechanisms. *Biol. Rev.* 80:1–25.

Eviner, V.T., and Chapin III, F.S. 2003. Functional matrix: a conceptual framework for predicting multiple plant effects on ecosystems. *Annu. Rev. Ecol. Syst.* 34:455–485.

Ewel, J., Berish, C., Brown, B., Price, N., and Raich, J. 1981. Slash and burn impacts on a Costa Rican wet forest site. *Ecology.* 62:816–829.

Ewel, J.J., Mazzarino, M.J., and Berish, C.W. 1991. Tropical soil fetility changes under monocultures and successional communities of different structure. *Ecol. Applic.* 1:289–302.

Faegri, K., and van der Pijl, L. 1971. *The principles of pollination ecology.* Pergamon Press, Oxford.

Fagan, W.F., Lewis, M.A., Neubert, M.G., and van den Driessche, P. 2002. Invasion theory and biological control. *Ecol. Lett.* 5:148.

Fagerstrom, T., Larsson, S., and Tenow, O. 1987. On optimal defence in plants. *Funct. Ecol.* 1:73–81.

Fahey, T.J. 1983. Nutrient dynamics of aboveground detritus in lodgepole pine *(Pinus contorta* spp.*latfolia)* ecosystems, southeastern Wyoming. *Ecol. Monogr.* 53:51–72.

Fahey, T.J. 1992. Mycorrhizae and forest ecosystems. *Mycorrhiza* 1:83–89.

Fahey, T.J., and Knight, D.H. 1986. Lodgepole pine ecosystems. *BioScience* 36:610–617.

Fahrig, L. 2003. Effects of habitat fragmentation on biodiversity. *Annu. Rev. Ecol. Syst.* 34:487–515.

Falkowski, P., Scholes, R.J., Boyle, E., Canadell, J., Canfield, D., Elser, J., Gruber, N., Hibbard, K., Hogberg, P., Linder, S., Mackenzie, F.T., Moore III, B., Pedersen, T., Rosenthal, Y., Sieyzinger, S., Smetacek, V., and Steffen, W. 2004. The global carbon cycle: a test of our knowledge of Earth as a system. *Science* 290:291–296.

Fanning, K.A. 1989. Influence of atmospheric pollution on nutrient limitation in the ocean. *Nature* 339:460–463.

FAO. 2001. *State of the world's forests, 2001.* Food and Agriculture Organization of the United Nations, Rome.

Farmer, E.E., and Ryan, C.A. 1990. Interplant communication: airborne methyl jasmonate induces synthesis of proteinase inhibitors in plant leaves. *PNAS* 87:7713–7716.

Fearnside, P.M. 1989. Extractive reserves in Brazilian Amazonia. *BioScience* 39:387–393.

Fearnside, P.M. 1992. Forest biomass in Brazilian Amazonia: comments on the estimate by Brown and Lugo. *Interciencia* 17:19–27.

Federer, C.A., Hornbeck, J.W., Tritton, L.M., Martin, C.W., Pierce, R.S., and Smith, C.T. 1988. Long-term depletion of calcium and other nutrients in eastem U.S. forests. In *Proc. Int. Symposium on Acidic Deposition and Forest Decline,* Oct 20, 1988. State Univ. of New York, Syracuse, 26.

Feeny, P. 1970. Seasonal changes in oak leaf tannins and nutrients as a cause of spring feeding by winter moth caterpillars. *Ecology* 51:565–581.

Feeny, P. 1976. Plant apparency and chemical defense. In *Biochemical interactions between plants and insects.* Edited by J.W. Wallace and R.L. Mansell. Plenum, New York, 1–40.

Feeny, P.P. 1975. Biochemical evolution between plants and their insect herbivores. In *Coevolution of animals and plants.* Edited by L.E. Gilkert and P.R. Raven. Univ. of Texas Press, Austin, 1–19.

Felker, P., and Bandurski, R.S. 1979. Uses and potential uses of leguminous trees for minimal energy input agriculture. *Econ. Bot.* 33:172–184.

Feller, M.C. 1982. *The ecological effects of slash burning with particular reference to British Columbia, a literature review.* Rep. No. 13. British Columbia Ministry of Forest and Land Management, Victoria.

Feller, M.C., and Kimmins, J.P. 1984. Effects of clearcutting and slash burning on stream water chemistry and watershed nutrient budgets in southwestem British Columbia. *Water Resourc.* 29:29–40.

Felton, G.W. 2005. Indigestion is a plant's best defense. *PNAS* 102:18771–18772.

FEMAT. 1993. *Forest ecosystem management: an ecological, economic, and social assessment.* U.S. Department of Agriculture, Forest Service, Washington, DC.

Fiadero, M.E. 1983. Physical-chemical processes in the open ocean, in *The major biogeochemical cycles and their interactions.* Edited by B. Bolin and R.B. Cook. John Wiley & Sons, Chichester, 461–476.

Field, C.B., and Raupach, M.R., editors. 2004. *The global carbon cycle SCOPE 62.* Island Press, Washington, DC.

Fierer, N., and Jackson, R.B. 2006. The diversity and biogeography of soil bacteria communities. *PNAS* 103:626–631.

Fife, D.N., and Nambiar, E.K.S. 1982. Accumulation and retranslocation of mineral nutrients in developing needles in relation to seasonal growth of young radiata pine trees. *Ann. Bot.* 50:817–829.

Fine, P.V.A., Mesones, I., and Coley, P.D. 2004. Herbivored promote habitat specialization by trees in Amazonian forests. *Science* 305:663–665.

Finegan, B. 1984. Forest succession. *Nature* 312:109–114.

Finlay, R.D., and Read, D.J. 1986. The structure and function of the vegetative mycehum of ectomycorrhizal plants. I. Translocation of ^{14}C-labeled carbon between plants interconnected by a common mycelium. *New Phytol.* 103:143–156.

Finzi A.C., DeLucia, E.H., Hamilton, J.G., Richter, D.D., and Schlesinger, W.H. 2002. The nitrogen budget of a pine forest under free air CO_2 enrichment. *Oecologia*, 132:567–578.

Fischer, A.G. 1960. Latitudinal variations in organic diversity. *Evolution* 14:64–81.

Fischer, C., Janos, D., Perry, D.A., Linderman, R.G., and Sollins, P. 1994. Mycorrhiza inoculum potentials in tropical secondary successions. *Biotropica* 26:369–377.

Fischer, J., Lindenmayer, D.B., and Manning, A.D. 2006. Biodioversity, ecosystem function, and resilience: ten guiding principles for commodity production landscapes. *Front. Ecol. Environ.* 4:80–86.

Fisher, D.G., Hart, S.C., Whitham, T.G., Martinsen, G.D., and Keim, P. 2004. Ecosystem implications of genetic variation in water-use of a dominat riparian tree. *Oecologia* 139:288–297.

Fisher, R.F. 1972. Spodosol development and nutrient distribution under Hydnaceae fungal mats. *Soil Sci. Soc. Am. Proc.* 36:492–495.

Fisher, R.F., and Binkley, D. 2000. Ecology and management of forest soils, 3rd ed. John Wiley and Sons, New York.

Fisher, R.F., Woods, R.A., and Glavicic, M.R. 1978. Allelopathic effects of goldenrod and aster on young sugar maple. *Can. J. For. Res.* 8:1–9.

Fisk, M.C., Zak, D.R., and Crow, T.R. 2002. Nitrogen storage and cycling in old- and second-growth northern hardwood forests. *Ecology* 81:73–87.

Flannigan, M.D., Bergeron, Y., Engelmark, O., and Wotton, B.M. 1998. Future wildfire in circumboreal forests in relation to global warming. *J. Veg. Sci.* 9:469–476.

Fleming, R.L., Powers, R.F., Foster, N.W., Kranabetter, J.M., Scott, D.A., Ponder Jr., F., Berch, S., Chapman, W.K.,

Kabzems, R.D., Ludovici, K.H., Morris, D.M., Page-Dumroese, D.S., Sanborn, P.T., Sanchez, F.G., Stone, D.M., and Tiarks, A.E. 2006. Effects of organic matter removal, soil compaction, and vegetation control on 5–year seedling performance: a regional comparison of long-term soil productivity sites. *Can. J. For. Res.* 36:529–550.

Fleming, T.H., Breitwisch, R., and Whitesides, G.H. 1987. Pattems of tropical vertebrate frugivore diversity. *Annu. Rev. Ecol. Syst.* 18:111–136.

Flinn, K.M., and Vellend, M. 2005. Recovery of forest plant communities in post-agricultural landscapes. *Front. Ecol. Environ.* 3:243–250.

Flohn, H. 1969. *Climate and weather.* McGraw-Hill, New York.

Florence, L.Z., and Cook, F.D. 1984. Asymbiotic N_2-flxing bacteria associated with three boreal conifers. *Can. J. For. Res.* 14:595–597.

Florence, R.G. 1965. Decline of old-growth redwood forests in relation to some soil microbiological processes. *Ecology:* 52–64.

Fogel, R. 1980. Mycorrhizae and nutrient cycling in natural forest ecosystems. *New Phytol.* 86:199–212.

Fogel, R., and Cromack Jr., K. 1977. Effect of habitat and substrate quality on Douglas-fir litter decomposition in western Oregon. *Can. J. Bot.* 55:1632–1640.

Fogel, R., and Hunt, C. 1983. Contribution of mycorrhizae and soil fungi to nutrient cycling in a Douglas-fir ecosystem. *Can. J. For. Res.* 13:219–232.

Fonte, S.J., and Schowalter, T.D. 2005. The influence of a neotropical herbivore (*Lamponius portoricensis*) on nutrient cycling and soil processes. *Oecologia* 146:423–431.

Forcier, L.K. 1975. Reproductive strategies and the co-occurrence of climax tree species. *Science* 189:808–810.

Ford, E.D. 1985. Branching, crown structure and the control of timber production. In *Attributes of trees as crop plants.* Edited by M.G.R. Cannel and J.E. Jackson. Institute of Terrestrial Ecology, Huntingdon, England, 228–252.

Ford, E.D., and Newbould, P.J. 1971. The leaf canopy of a coppiced deciduous woodland. I. Development and structure. *J. Ecol.* 59:843–862.

Forlines, D.R., Tavenner, T., Malan, J.C.S., and Karchesy, J.J. 1992. Plants of the Olympic coastal forests: ancient knowledge of materials and medicines and future heritage. In *Plant polyphenols.* Edited by R.W. Remmingway and P.E. Laks. Plenum Press, New York, 767–782.

Forman, R.T.T. 1975. Canopy lichens with blue-green algae: a nitrogen source in a Colombian rain forest. *Ecology* 56:1176–1184.

Forman, R.T.T. 1987. The ethics of isolation, the spread of disturbance, and landscape ecology. In *Landscape heterogeneity and disturbance.* Edited by M.G. Turner. Springer-Verlag, New York, 213–229.

Forman, R.T. 1995. *Land mosaics: the ecology of landscapes and regions.* Cambridge Univ. Press, New York.

Forman, R.T., and Alexander, L.E. 1998. Roads and their major ecological effects. *Annu. Rev. Ecol. Syst.* 29:207–231.

Forrester D.I., Bauhus, J., Cowie, A.L., and Vanclay, J.K. 2006. Mixed-species plantations of *Eucalyptus* with nitrogen-fixing trees: a review. *For. Ecol. Manage.* 233:211–230.

Foster, D., Motzkin, G., O'Keefe, J., Boose, E., Orwig, D., Fuller, J., and Hall, B. 2004. The environmental and human history of New England. In *Forests in time.* Edited by D.R. Foster and J.B. Aber. Yale Uuiv. Press, New Haven, CT.

Foster, D., Swanson, F., Aber, J., Burke, I., Brokaw, N., Tilman, D., and Knapp, A. 2003. The importance of land-use legacies to ecology and conservation. *BioScience* 53:77–88.

Foster, D.R., Aber, J.D., Melillo, J.M., Bowden, R.D., and Bazzaz, F.A. 1997. Forest reposnse to disturbance and anthropogenic stress. *BioScience* 47:437–445.

Foster, D.R., Schoonmaker, P.K., and Pickett, S.T.A. 1990. Insights from paleoecology to community ecology. *Trends Ecol. Evol.* 5:119–122.

Foster, N.W. 1974. Annual macroelement transfer from *Pinus banksiana* Lamb. forest to soil. *Can. J. For. Res.* 4:470–476.

Foster, N.W., and Gessel, S.P. 1992. The natural addition of nitrogen, potassium and calcium to a *Pinus banksiana* Lamb. forest floor. *Can. J. For. Res.* 2:448–455.

Foster, R.C. 1986. The ultrastructure of the rhizoplane and rhizosphere. *Annu. Rev. Phytopathol.* 24:211–234.

Fowler, S.V., and Lawton, J.R. 1985. Rapidly induced defenses and talking trees: the devil's advocate position. *Am. Nat.* 126:181–195.

Fox, L.R. 1988. Diffuse coevolution within complex communities. *Ecology* 69:906–907.

Fox. L.R., and Macauley, B.J. 1977. Insect grazing on eucalyptus in response to variation in leaf tannins and nitrogen. *Oecologia* 29:145–162.

Fox, L.R., and Morrow, P.A. 1981. Specialization: species property on local phenomenon. *Science* 211:887–893.

Fox, T.R. 2000. Sustained productivity in intensively managed forest plantations. *For. Ecol. Manage.* 138:187–202.

Fox, T.R., Morris, L.A., and Maimone, R.A. 1989. Windrowing reduces growth in a loblolly pine plantation in the North Carolina Piedmont. In *Proceedings of the Fifth Biennal Southern Silvicultural Research Conference.* Edited by J.H. Miller. U.S. Department of Agriculture, Forest Service, New Orleans, 133–139.

Francis, A.P., and Currie, D.J. 2003. A globally consistent richness-climate relationship for angiosperms. *a.* 161: 523–536.

Francis, R., and Read, D.J. 1984. Direct transfer of carbon between plants connected by vesicular arbuscular mycorrhizal mycelium. *Nature* 307:53–56.

Francis, R., Finlay, R.D., and Read, D.J. 1986. Vesicular-arbuscular mycorrhiza in natural vegetation systems. IV. Transfer of nutrients in inter- and intra-specific combinations of host plants. *New Phytol.* 102:103–111.

Frank, D.A., and McNaughton, S.J. 1991. Stability increases with diversity in plant communities: empirical evidence from the 1988 Yellowstone drought. *Oikos* 62:360–362.

Frankie, G.W. 1975. Tropical forest phenology and pollinator plant coevolution. In *Coevolution of animals and plants.* Edited by L.E. Gilbert and P.R. Raven. Univ. of Texas Press, Austin, 192–209.

Franklin, I.R. 1980. Evolutionary change in small populations. In *Conservation biology: an evolutionary-ecological perspective.* Edited by M.E. Soule and B.A. Wilcox. Sinauer Associates, Sunderland, MA, 135–149.

Franklin, J.F. 1989. Toward a new forestry. *Am. For.,* (Nov/Dec): 1–8.

Franklin, J.F. 1993. Preserving biodiversity: species, ecosystems or landscapes. *Ecol. Applic.* 3:202-205.

Franklin, J.F., and Forman, R.T.T. 1987. Creating landscape patterns by forest cutting: ecological consequences and principles. *Landscape Ecol.* 1:5–18.

Franklin, J.F., and Hemstrom, M.A. 1981. Aspects of succession in the coniferous forests of the Pacific Northwest. In *Forest succession*. Edited by D.C. West, H.H. Shugart, and D.B. Botkin. Springer-Verlag, New York, 212–229.

Franklin, J.F., and MacMahon, J.A. 2000. Messages from a mountain. *Science* 288:1183–1185.

Franklin, J.F., and Maser, C. 1988. *Looking ahead: some options for public lands*. Gen. Tech. Report PNW-GTR-229. U.S. Department of Agriculture, Forest Service, Portland, OR,113–122.

Franklin J.F., and Van Pelt, R. 2004. Spatial aspects of structural complexity in old-growth forests. *J. For.* 102:22–27.

Franklin, J.F., Cromack Jr., K., Denison, W., McKee, A., Maser, C., Sedell, J., Swanson, F., and Juday, G. 1981. *Ecological characteristics of old-growth Douglas-fir forests*. Gen. Tech. Rpt. PNW-118. U.S. Department of Agriculture, Forest Service, Portland, OR.

Franklin, J.F., Lindenmayer, D.B., MacMahon, J.A., McKee, A., Magnusson, J., Perry, D.A., Waide, R., and Foster, D.R. 2000. Threads of continuity: ecosystem disturbances, biological legacies, and ecosystem recovery. *Conserv. Biol. Pract.* 1:8–16.

Franklin, J.F., MacMahon, J.A., Swanson, F.J., and Sedell, J.R. 1985. Ecosystem responses to the eruption of Mount St. Helens. *Natl. Geogr. Res.* (Spring):198–216.

Franklin, J.F., Mitchell, R.J., and Palik, B. 2007. *Natural disturbance and stand development principles for ecological forestry*. Gen. Tech. Report NRS-19. U.S. Department of Agriculture, Forest Service, Newtown Square, PA.

Franklin, J.F., Perry, D.A., Schowalter, T.D., Harmon, M.E., McKee, A., and Spies, T.A. 1989. Importance of ecological diversity in maintaining long-term site productivity. In *Maintaining the long-term productivity of Pacific Northwest forest ecosystems*. Edited by D.A. Perry, R. Meurisse, B. Thomas, et al. Timber Press, Portland, OR, 82–97.

Franklin, J.F., Spies, T., Perry, D.A., Harmon, M., and McKee, A. 1987. Modifying Douglas-fir management regimes for nontimber objectives. In *Stand management for the future*. Edited by C.D. Oliver, D.P. Hanley, and J.A. Johnson. College of Forest Resources, Univ. of Washington, Seattle, 373–379.

Franklin, J.F., Spies, T.A., Van Pelt, R., Carey, A.B., Thornburgh, D.A., Berg, D.R., Lindenmayer, D.B., Harmon, M.E., Keeton, W.S., Shaw, D.C., Bible, K., and Chen, J. 2002. Disturbances and structural development of natural forest ecosystems with silvicultural implications, using Douglas-fir forests as an example. *For. Ecol. Manage.* 155:399–423.

Franklin, J.F., Swanson, F.J., Harmon, M.E., Perry, D.A.; Spies, T.A.; Dale, V.H., McKee, A. Ferrell, W.K., and Means, J.E. 1992. Effects of global climatic change on forests in northwestern North America. In *Global warming and biological diversity*. Edited by R.L. Peters and T.E. Lovejoy. Yale Univ. Press, New Haven, CT, 244–257.

Frazer, J.G. 1981. *The golden bough*. Avenel Books, New York.

Freeland, W.J., and Boulton, W.J. 1992. Coevolution of food webs: parasites, predators and plant secondary compounds. *Biotropica* 24:309–327.

Freemark, K. 1989. Landscape ecology of forest birds in the northeast. In *Is forest fragmentation a management issue in the northeast?* Gen. Tech. Rep. NE-140. U.S. Department of Agriculture, Forest Service, Durham, NH, 7–12.

Freemark, K., and Collins, B. 1992. Landscape ecology of birds breeding in temperate forest fragments. In *Ecology and conservation of neotropical migrant landbirds*. Edited by J. Hagan and D.M. Johnston. Smithsonian Institution Press, Washington, DC.

Freemark, K.E., and Merriam, H.G. 1986. Importance of area and habitat heterogeneity to bird assemblages in temperate forest fragments. *Biol. Conserv.* 36:115–141.

Fretwell, S.D. 1987. Food chain dynamics: the central theory of ecology? *Oikos* 50:291–301.

Frey, S.D., Knorr, M., Parrent, J.L., and Simpson, R.T. 2004. Chronic nitrogen enrichment affects the structure and function of the soil microbial community in temperate hardwood and pine forests. *For. Ecol. Manage.* 196:159–171.

Fried, J.S., Boyle, J.R., Tappeiner II, J.C., and Cromack Jr., K. 1990. Effects of bigleaf maple on soils in Douglas-fir forests. *Can. J. For. Res.* 20:259–266.

Friedman, J., Hutchins, A., Li, C.Y., and Perry, D.A. 1989. Actinomycetes inducing phytotoxic or fungistatic activity in a Douglas-fir forest and in an adjacent area of repeated regeneration failure in southwestern Oregon. *Biologia Plantarum (Praha)* 31:487–495.

Friend, A.L., Ohmann, S.M., and Hinckley, T.M. 1987. Fine root and hyphal growth in Douglas-fir stands: response to nitrogen stress. Abstract. *Annu. Mtg. Soil Sci. Soc. America* 1987:256.

Froelich, H.A., and McNabb, D.H. 1984. Minimizing soil compaction in Pacific Northwest forests. In *Forest soils and treatment impacts*. Edited by E.L. Stone. Univ. of Tennessee, Knoxville, 159–192.

Fryer, G.I., and Johnson, E.A. 1988. Reconstructing the fire behavior and effects in a subalpine forest. *J. Appl. Ecol.* 25:1063–1072.

Fujimori, T. 1971. *Primary productivity of a young* Tsuga heterophylla *stand and some speculations about biomass offorest communities on the Oregon coast*. Res. Pap. PNW-123. U.S. Department of Agriculture, Forest Service, Portland, OR.

Fujita, T.T. 1987. *U.S. tornadoes. Part 1. 70–year statistics*. Satellite and Mesometeorology Research Project (SMRP), Res. Paper Number 218. Univ. of Chicago, Chicago.

Fukami, T., and Morin, P.J. 2003. Productivity-biodiversity relationships depend on the history of community assembly. *Nature* 424:423–425.

Fukuoka, K. 1985. *The economics of coexistence between man and nature*. Rissho Univ. Institute for Economic Research, Tokyo.

Fuller, B. 1981. *Critical path*. St. Martin's Press, New York.

Fuller, J., Foster, D., Motzkin, G., McLachlan, J., and Barry, S. 2004. Broadscale forest response to land use and climate change. In *Forests in time*. Edited by D.R. Foster and J.D. Aber. Yale Univ. Press, New Haven, 101–141.

Furnier, G.R., Knowles, P., Clyde, MA., and Dancik, B.P. 1987. Effects of avian seed dispersal on the genetic structure of whitebark pine populations. *Evolution* 41:607–612.

Futuyama, D.J. 1976. Food plant specialization and environmental predictability in Lepidoptera. *Am. Nat.* 110:285–292.

Futuyma, D.J., and Moreno, G. 1988. The evolution of ecological specialization. *Annu. Rev. Ecol. Syst.* 19:207–233.

Futuyma, D.J., and Slatkin, M. 1983. *Coevoluation*. Sinauer Associates, Sunderland, MA.

Fyfe, W.S., Kronberg, B.L., Leonardos, O.H., and Olorunfemi, N. 1983. Global tectonics and agriculture: a geochemical perspective. *Agric. Ecosyst. Environ.* 9:383–399.

Gadgil, R.L., and Gadgil, P.D. 1971. Mycorrhiza and litter decomposition. *Nature* 233:133.

Gadgil, R.L., and Gadgil, P.D. 1974. Suppression of litter decomposition by mycorrhizal roots of Pinus radiata. *N. Z. J. For. Sci.* 5:33–41.

Galloway, J.N. 1998. The global nitrogen cycle: changes and consequences. *Environ. Pollut.* 102(S1):15–24.

Galloway, J.N., Dentener, F.J., Capone, D.G., Boyer, E.W., Howarth, R.W., Seitzinger, S.P., Asner, G.P., Cleveland, C.C., Green, P.A., Holland, E.A., Karl, D.M., Michaels, A.F., Porter, J.H., Townsend, A.R., and Vorosmarty, C.J. 2004. Nitrogen cycles: past, present, and future. *Biogeochemistry* 70:153–226.

Gamier, B.J., and Ohmura, A. 1968. A method of calculating the direct short wave radiation income of slopes. *J. Appl. Meteorol.* 7:796–800.

Gange, A.C., Brown, V.K., and Alpin, D.M. 2003. Multitrophic links between arbuscular mycorrhizal fungi and insect parasitoids. *Ecol. Lett.* 6:1051–1055.

Ganley, R.J., Brunsfield, S.J., and Newcombe, G. 2004. A community of unknown, endophytic fungi in western white pine. *PNAS* 101:10107–10112.

Garbaye, J., and Bowen, G.D. 1987. Effect of different microflora on the success of ectomycorrhizal inoculation of *Pinus radiata, Can. J. For. Res.* 17:941–943.

Garbaye, J., and Bowen, G.D. 1989. Stimulation of ectomycorrhizal infection of *Pinus radiata* by some microorganisms associated with the mantle of ectomycorrhizas. *New Phytol.* 112: 383–388.

Gardner, M. 1970. Mathematical games. *Sci. Am.* 123:120–124.

Garfinkle, H.L., and Brubaker, L.B. 1980. Modern climate-tree-growth relationships and climatic reconstruction in subarctic Alaska. *Nature* 286:872–874.

Garstang, M., Greco, S., and Swap, R. 1991. Quoted in News paragraph. *BioScience* 41:439.

Gartlan, J.S., Newbery, D.M., Thomas, D.W., and Waterman, P.G. 1986. The influence of topography and soil phosphorus on the vegetation of Korup Forest Reserve, Cameroun. *Vegetatio* 65:131–148.

Gartner, T.B., and Cardon, Z.G. 2004. Decomposition dynamics in mixed-species leaf litter. *Oikos* 104:230–246.

Garwood, N.C., Janos, D.P., and Brokaw, N. 1979. Earthquake-caused landslides: a major disturbance to tropical forests. *Science* 205:997–999.

Gascon, C., Williamson, C.B., and da Fonseca, G.A.B. 2000. Receding forest edges and vanishing reserves. *Science* 288:1356–1358.

Gaston, K. 2000. Global patterns in biodiversity. *Nature* 405:220–227.

Gates, J.E., and Evans, D.R. 1998. Cowbirds breeding in the central Appalachians: spatial and temporal patterns and habitat selection. *Ecol. Applic.* 8:27–40.

Gatto, M., and Rinaldi, S. 1987. Some models of catastrophic behavior in exploited forests. *Vegetatio* 69:213–222.

Gautier-Hion, A. 1989. Are figs always keystone resources for tropical frugivorous vertebrates? A test in Gabon. *Ecology* 70:1826–1823.

Gedalof, Z., Peterson, D.L., and Mantua, N.J. 2005. Atmospheric, climatic, and ecological controls on extreme wildfire years in the northwestern united states. *Ecol. Applic.* 15:154–174.

Gehring, C.A., and Connell, J.H. 2006. Arbuscular mycorrhizal fungi in the tree seedlings of two Australian rain forests: occurrence, colonization, and relationships with plant performance. *Mycorrhiza* 16:89–98.

Gehring, C.A., and Whitham, T.G. 1991. Herbivore driven mycorrhizal mutualism in insect susceptible pinyon pine. *Nature* 353:556–557.

Gehring, C.A., and Whitham, T.G. 1992. Reduced mycorrhizae on *Juniperus monosperma* with mistletoe: the influence of environmental stress and tree gender on a plant parasite and a plant fungal mutuahsm. *Oecologia* 89:298–303.

Gehring, C.A., and Whitham, T.G. 2002. Mycorrhizae-herbivore interactions: population and community consequences. In: *Mycorrhizal Ecology.* Ecological Studies, Vol. 157. Edited by M.G.A. van der Heijden and I.R. Sanders. Springer-Verlag, Berlin, 295–320.

Geiszler, D.R., Gara, R.I., Driver, C.H., Gallucci, V.F., and Martin, R.E. 1980. Fire, fungi, and beetle influences on a lodgepole pine ecosystem of south-central Oregon. *Oecologia* 46:239–243.

Gent Jr., J.A., and Morris, L.A. 1986. Soil compaction from harvesting and site preparation in the Upper Gulf Coastal Plain. *Soil Sci. Soc. Am. J.* 50:443–446.

Gent Jr., J.A., Ballard, R., Hassan, A.E., and Cassel, D.K. 1984. Impact of harvesting and site preparation on physical properties of Piedmont forest soils. *Soil Sci. Soc. Am. J.* 48:173–177.

Gentry, A.H. 1982. Patterns of neotropical plant species diversity. *Evol. Biol.* 15:1–84.

Gentry, A.H., and Lopez-Parodi, J. 1980. Deforestation and increased flooding of the upper Amazon. *Science* 210: 1354–1356.

Gessel, S.P., Lacate, D.S., Weetman, G.F., and Powers, R.F., editors. 1990. *Sustained productivity of forest soils. Proceedings of the seventh North American Forest Soils Conference, 1988.* Univ. of British Columbia, Vancouver.

Gholz, H.L., Wedin, D.A., Smitherman, S.M., Harmon, M.E., and Parton, W. 2000. Long-term dynamics of pine and hardwood litter in contrasting environments: toward a global model of decomposition. *Global Change Biol.* 6:751–765.

Gibson, I.A.S., and Jones, T. 1977. Monoculture as the origin of major forest pests and diseases. In *Origins of pest, parasite, disease, and weed problems.* Edited by J.M. Cherrett and G.R. Sujor. Blackwell, Oxford, 139–161.

Gibson, I.A.S., Burley, J., and Spreight, M.R. 1982. The adoption of heritable resistance to pests and pathogens in forest crops. In *Resistance to diseases and pests in forest trees.* Edited by H.M. Heybrock, B.R. Stephan, and K. von Weissenberg. Pudoc, Wageningen, 9.

Gilbert, B., and Lechowicz, M.J. 2004. Neutrality, niches, and dispersal in a temperate forest understory. *PNAS* 101:7651–7656.

Gilbert, F.S. 1980. The equilibrium theory of island biogeography: fact or fiction? *J. Biogeogr.* 7:209–235.

Gilbert, L.E. 1975. Ecological consequences of a coevolved mutualism between butterflies and plants. In *Coevolution of animals and plants.* Edited by L.E. Gilbert and P.H. Raven. Univ. of Texas Press, Austin, 210–240.

Gilbert, L.E. 1980. Food web organization and the conservation of neotropical diversity. *Conservation biology: an evolutionary-ecological perspective.* Edited by M.E. Soule and B.A. Wilcox. Sinauer Associates, Sunderland, MA, 11–33.

Giles, J. 2006. The outlook for Amazonia is dry. *Nature* 442: 726–727.

Gill, R.A., and Jackson, R.B. 2000. Global patterns of root turnover for terrestrial ecosystems. *New Phytol.* 147:13–31.

Gillis, A.M. 1991. Can organisms direct their evolution? *BioScience* 41:202–205.

Gilpin, M.E. 1979. Spiral chaos in a predator-prey model. *Am. Nat.* I 13:306–308.

Gilpin, M.E., and Soule, M.E. 1986. Minimum viable populations: processes of species extinction. In *Conservation biology: the science of scarcity and diversity.* Edited by M.E. Soule. Sinauer Associates, Sunderland, MA, 19–34.

Gjerde, I., Saetersdal, M., Rolstad, J., Blom, H.H., and Storaunet, K.O. 2004. Fine-scale diversity and rarity hotspots in northern forests. *Conserv. Biol.* 18:1032–1042.

Gleason, H.A. 1926. The individualistic concept of the plant association. *Bull. Torrey Bot. Club* 53:7–26.

Gleik, P. 1987. *Chaos*. Penguin, New York.

Glitzenstein, J.S., Harcombe, P.A., and Strong, D.R. 1986. Disturbance, succession, and maintenance of species diversity in an east Texas forest. *Ecol. Monogr.* 56:243–258.

Global Invasive Species Database. 2005. *Passiflora tarminiana*. www.issg.org/database/species/ecology.asp?si=336fr=1sts=sss.

Goebel, P.C., Palik, B.J., and Pregitizer, K.S. 2003. Plant diversity contributions of riparian areas in watersheds of the northern Lake States, USA. *Ecol. Applic.* 13:1595–1609.

Goldberg, D.E. 1982. The distribution of evergreen and deciduous trees relative to soil type: an example from the Sierra Madre, Mexico, and a general model. *Ecology* 63:942–951.

Goldfarb, B., Nelson, E.E., and Hansen, E.M. 1989. *Trichoderma* spp.: growth rates and antagonism to *Phellinus weirii* in vitro. *Mycologia* 81:375–381.

Goller, R., Wilcke, W., Fleishbein, K., Valarezo, C., and Zech, W. 2006. Dissolved nitrogen, phosphorus, and sulfur forms in the ecosystem fluxes of a montane forest in Ecuador. *Biogeochemistry* 77:57–89.

Golley, F.B. 1972. Energy flux in ecosystems. In *Ecosystem structure and function*. Edited by J.A. Weins. Oregon State Univ. Press, Corvallis, 69–90.

Golley, F.B. 1983a. Decomposition. In *Tropical rain forest ecosystems*. Edited by F.B. Golley. Elsevier, Amsterdam, 157–166.

Golley, F.B. 1983b. Nutrient cycling and nutrient conservation. In *Tropical rain forest ecosystems*. Edited by F.B. Golley. Elsevier, Amsterdam, 137–156.

Gomez-Aparicio, L., Zamora, R., Gomes, J.M., Hodar, J.A., Castro, J., and Baraza, E. 2004. Applying plant facilitation to forest restoration: a meta-analysis of the use of shrubs as nurse plants. *Ecol. Applic.* 14:1128–1138.

Gomez-Pompa, A., and Kaus, A. 1990. Traditional management of tropical forests in Mexico. In *Alternatives to deforestation*. Edited by A.B. Anderson. Columbia Univ. Press, New York, 43–64.

Gomez-Pompa, A., and Vazquez-Yanes, C. 1981. Successional studies of a rain forest in Mexico. In *Forest succession, concepts and application*. Edited by D.C. West, R.H. Shugart, and D.B. Botkin. Springer-Verlag, New York, 246–266.

Gonor, J.J., Sedell, J.R., and Benner, P.A. 1988. *What we know about large trees in estuaries, in the sea, and on coastal beaches*. Gen. Tech. Rep. PNW-GTR-229. U.S. Department of Agriculture, Forest Service, Portland, OR, 83–112.

Gonzalez. V.G., Gonzalez, O.C., Millan, H.J., and Escobar, R.R. 1983. *Estudio de fertilizacion en plantaciones de* Pinus radiata. *Primeros resultados*. FO: DP/CHI/76/003, Documento de Trabajo No. 51, p. 159. Corporacion Nacional Forestal, Organizacion De Las Naciones Unidas, Para La Agricultura y La Alimentacion. Investigación y Desarrollo Forestal. Documento de Trabajo (Chile). No. 51. Santiago, Chile.

Goodale, C.L., Aber, J.D., and McDowell, W.H. 2000. The long-term effects of disturbance on organic and inorganic nitrogen export in the White Mountains, New Hampshire. *Ecosystems* 3:433–50.

Goodale, C.L., Lajtha, K., Nadelhoffer, K.J., Boyer, E.W., and Jaworski, N.A. 2002. Forest nitrogen sinks in large eastern U.S. watersheds: estimates from forest inventory and an ecosystem model. *Biogeochemistry* 57/58:239–266.

Goodman, D.M., and J.A. Trofymo. 1998. Comparison of communities of ectomycorrhizal fungi in old-growth and mature stands of Douglas-fir at two sites on southern Vancouver Island. *Can. J. For. Res.* 28:574–581.

Goodnight, C.J., and Stevens, L. 1997. Experimental studies of group selection: what do they tell us about group selection in nature? *Am. Nat.* 150:S59–S79.

Gordon, A.G., and Gorham, E. 1963. Ecological aspects of air pollution from an iron-sintering plant at Wawa, Ontario. *Can. J. Bot.* 41:1063–1978.

Gordon, J.C., and Wheeler, C.T. 1983. *Biological nitrogen fixation in forest ecosystems*. M. Nighoff/W. Junk, The Hague.

Gordon, J.C., Wheeler, C.T., and Perry, D.A., editors. 1979. *Symbiotic nitrogen fixation in the management of temperate forests*. Oregon State Univ., Corvallis, 501.

Gordon, L.J., Steffen, W., Jonsson, B.F., Folke, C., Falkenmark, M., and Johannessen, A. 2005. Human modification of of global water vapor flows from the land surface. *PNAS* 102:7612–7617.

Gordon, W.S., and Jackson, R.B. 2000. Nutrient concentrations in fine roots. *Ecology* 81:275–280.

Goreau, T.J., and de Mello, W.Z. 1988. Tropical deforestation: some effects on atmospheric chemistry. *Ambio* 17:275–281.

Gorham, E., Vitousek, P.M., and Reiners, W.A. 1979. The regulation of chemical budgets over the course of terrestrial ecosystem succession. *Annu. Rev. Ecol. Syst.* 10:53–84.

Gosz, J.R. 1992. Gradient analysis of ecological change in time and space: implications for forest management. *Ecol. Applic.* 2:248–261.

Gosz, J.R., and White, C.S. 1986. Seasonal and annual variation in nitrogen mineralization and nitrification along an elevational gradient in New Mexico. *Biogeochemistry* 2:281–297.

Gosz, J.R., Holmes, R.T., Likens, G.E., and Bormann, F.H. 1978. The flow of energy in a forest ecosystem. *Sci. Am.* 238:93–101.

Gosz, J.R., Likens, G.E., and Bormann, F.H. 1972. Nutrient content of litter fall on the Hubbard Brook Experimental Forest, New Hampshire. *Ecology* 54:769–784.

Gould, F. 1983. Genetics of plant-herbivore systems: interactions between applied and basic study. In *Variable plants and herbivores in natural and managed systems*. Edited by R.F. Denno and R. McClure. Academic Press, New York, 599–653.

Gould, F., and Hodgson, E. 1980. Mixed function oxidase and glutathione transferase activity in last instar *Heliothes verescens* larvae. *Pestic. Biochem. Physiol.* 13:34–40.

Gould, S.J. 1984. Caring groups and selfish genes. In *Conceptual issues in evolutionary biology*. Edited by E. Sober. MIT Press, Cambridge, MA, 119–124.

Gould, S.J., and Eldredge, N. 1977. Punctuated equilibria: the tempo and mode of evolution reconsidered. *Paleobiology* 3:115–151.

Goyer, R.A., and Finger, C.K. 1980. Relative abundance and seasonal distribution of the major hymenopterous parasites of the southern pine beetle, *Dendroctuns frontalis* Zimmerman, and loblolly pine. *Environ. Entomol.* 9:97–100.

Graham, J.H., and Linderman, R.G. 1980. Ethylene production by ectomycorrhizal fungi, *Fusarium oxysporum* f. sp. *pini*, and by aseptically synthesized ectomycorrhizae and *Fusarium*-infected Douglas-fir roots. *Can. J. Microbiol.* 26:1340–1347.

Graham, R.L., Turner, M.G., and Dale, V.H. 1990. How increasing CO_2 and climate change affect forests. *BioScience* 40:575–587.

Granhall, U., and Lindberg, T. 1978. Nitrogen fixation in some coniferous forest ecosystems. *Ecol. Bull. (Stockholm)* 26:178–192.

Grant, P.R., and Grant, B. R. 2006. Evolution of character displacement in Darwin's finches. *Science* 313:224–226.

Graumlich, L.J., Brubaker, L.B., and Grier, C.C. 1989. Long-term trends in forest net primary productivity: Cascade Mountains, Washington. *Ecology* 70:405–410.

Graustein, W.C., Cromack Jr., K., and Sollins, P. 1977. Calcium oxalate: occurrence in soils and effect on nutrient and geochemical cycles. *Science* 198:1252–1254.

Graveland, J., van der Wal, R., van Balen, J.H., and van Noordwiijk, A.J. 1994. Poor reproduction in forest passerines from decline of snail abundance on acidified soils. *Nature* 368:446–448.

Graves, G.R., and Rahbek, C. 2005. Source pool geometry and the assembly of continental avifaunas. *PNAS* 102:7871–7876.

Gray, A.N., and Spies, T.A. 1997. Microsite controls on tree seedling establishment in conifer forest canopy gaps. *Ecology* 78:2458–2473.

Gray, D. 2007. The relationship between climate and outbreak characteristics of the spruce budworm in eastern Canada. *Climatic Change.* DOI 10.007/s10584-007-9317-5

Gray, W.M. 1990. Strong association between west African rainfall and U.S. landfall of intense hurricanes. *Science* 249:1251–1255.

Greacen, E.L., and Sands, R. 1980. Compaction of forest soils. A review. *Aust. J. Soil Res.* 18:163–189.

Green, G.M., and Sussman, R.W. 1990. Deforestation history of the eastern rain forests of Madagascar from satellite images. *Science* 248:212–215.

Green, J.L., Holmes, A.J., Westoby, M., Oliver, I., Briscoe, D., Dangerfield, M., Gillings, M., and Beattie, A. 2004. Spatial scaling of microbial eukaryote diversity. *Nature* 432:747–750.

Green, R.E., Cornell, S.J., Scharlemann, J.P.W., and Balmford, A. 2005. Farming and the fate of wild nature. *Science* 307:550–555.

Greenburg, R., and Gradwohl, J. 1986. Constant density and stable territoriality in some tropical insectivorous birds. *Oecologia* 69:618–625.

Greene, D.F., and Johnson, E.A. 2000. Tree recruitment from burn edges. *Can. J. For. Res.* 30:1264–1274.

Gregorich, E.G., and Janzen, H.H. 1996. Storage of soil carbon in the light fraction and macroorganic matter. In *Structure and organic matter storage in agricultural soils.* Edited by M.R. Carter and B.A. Stewart. Lewis Publishers, Boca Raton, FL, 167–190.

Grene, M. 1987. Hierarchies in biology. *Am. Sci.* 75:504–510.

Gresham, C.A., Williams, T.M., and Lipscomb, D.J. 1991. Hurricane Hugo wind damage to southeastern U.S. Coastal forest tree species. *Biotropica* 23:420–426.

Grier, C.C. 1975. Wildfire effects on nutrient distribution and leaching in coniferous ecosystems. *Can. J. For. Res.* 5:599–607.

Griffith, J.A., Stehman, S.V., and Loveland, T.R. 2003. Landscape trends in Mid-Atlantic and southeastern United States ecoregions. *Environ. Manage.* 32:572–588.

Griffiths, R.P., and Swanson, A.K. 2001. Forest soil characteristics in a chronosequence of harvested Douglas-fir forests. *Can. J. For. Res.* 31:1871–1879.

Griffiths, R.P., Baham, J.E., and Caldwell. B.A. 1994. Soil solution chemistry of ectomycorrhizal mats in forest soil. *Soil Biol. Biochem.* 26:331–337.

Griffiths, R.P., Bradshaw, G.A., Marks, B., and Lienkaemper, G.W. 1996. Spatial distribution of ectomycorrhizal mats in coniferous forests of the Pacific Northwest, USA. *Plant Soil* 80:147–158.

Griffiths, R.P., Caldwell, B.A., Cromack Jr., K., and Morita, R.Y. 1990. Douglas-fir forest soils colonized by ectomycorrhizal mats. I. Seasonal variation in nitrogen chemistry and nitrogen cycle transformation rates. *Can. J. For. Res.* 20:211–218.

Grigal, D.F. 2000. Effects of extensive forest management on soil productivity. *For. Ecol. Manage.* 138:167–185.

Grime, J.P., Mackey, J.M.L., Hillier, S.H., and Read, D.J. 1987. Floristic diversity in a model system using experimental microcosms. *Nature* 328:420–422.

Grimm, V., and Wissel, C. 1997. Babel, or the ecological stability discussions: an inventory and analysis of terminology and a guide for avoiding confusion. *Oecologia* 109:323–334.

Grimm, V., Revilla, E., Berger, U., Jeltsch, F., Mooij, W.M., Railsback, S.F., Thulke, H.-H., Weiner, J., Wiegand, T., and DeAngelis, D.L. 2005. Pattern-oriented modeling of agent-based complex systems: lessons from ecology. *Science* 310:987–991.

Grinnell, J. 1914. An account of the mammals and birds of the Lower Colorado Valley with especial reference to the distributional problems presented. *U. Calif. Pub. Zool.* 12:51–294.

Grissino-Mayer, H.D., and Swetnam, T.W. 2000. Century-scale climate forcing of fire regimes in the American southwest. *The Holocene* 10:213–220.

Groffman, P.M., Driscoll, C.T., Fahey, T.J., Hardy, J.P., Fitzhugh, R.D., and Tierney, G.L. 2001. Colder soils in a warmer world: a snow manipulation study in a northern hardwood forest ecosystem. *Biogeochemistry* 56:135–150.

Groffman, P.M., Driscoll, C.T., Likens, G.E., Fahey, T.J., Holmes, R.T., Eagar, C., and Aber, J.D. 2004. Nor gloom of night: a new conectual model for the Hubbard Brook Ecosystem Study. *BioScience* 54:139–148.

Groner, E., and Novoplansky, A. 2003. Reconsidering diversity-productivity relationships: directness of productivity estimates matters. *Ecol. Lett.* 6:695–699.

Grove, S.J. 2002. Saproxylic insect ecology and the sustainable management of forests. *Annu. Rev. Ecol. Syst.* 33:1–23.

Grubb, P.J. 1977. The maintenance of species-richness in plant communities: the importance of the regeneration niche. *Biol. Rev.* 52:107–145.

Grubb, P.J. 1986. Problems posed by sparse and patchily distributed species in species-rich plant communities. In *Community ecology.* Edited by J. Diamond and T.J. Case. Harper & Row, New York, 207–225.

Grubb, P.J. 1989. The role of mineral nutrients in the tropics: a plant ecologists view. In *Mineral nutrients in tropical forest and savanna ecosystems.* Edited by J. Proctor. Blackwell Scientific Publications, Oxford, 417–440.

Gruner, D.S. 2004. Attenuation of top-down and bottom-up forces in a complex terrestrial community. *Ecology* 85:3010–3022.

Gunderson, L.H. 2000. Ecological resilience—in theory and practice. *Annu. Rev. Ecol. Syst.* 31:425–439.

Gunnarsson, T., and Tunlid, A. 1986. Recycling of fecal pellets in isopods: microorganisms and nitrogen compounds as potential food for *Oniscus asellus* L. *Soil Biol. Biochem.* 18:595–600.

Gupta, G.P. 1975. Sediment production: status report on data collection and utilization. *Soil Conserv. Digest* 3:10–21.

Gusewell, S. 2004. N:P ratios in terrestrial plants: vasriation and functional significance. *New Phytol.* 164:243–266.

Guthrie, R.L., and Witty, J.E. 1982. New designations for soil horizons and layers and the new Soil Survey Manual. *Soil Sci. Soc. Am. J.* 46:443–444.

Gutschick, V.P. 1981. Evolved strategies in nitrogen acquisition by plants. *Am. Nat.* 118:607–637.

Gutzwiller, K.J., editor. 2002. *Applying landscape ecology in biological conservation.* Springer, New York.

Habte, M. 1989. Impact of simulated erosion on the abundance and activity of indigenous vesicular-arbuscular mycorrhizal endorphytes in an oxisol. *Biol. Fertil. Soils* 7:164–467.

Habte, M., Fox, R.L., Aziz, T., and El-Swaify, S.A. 1988. Interaction of vesicular-arbuscular mycorrhizal fungi with erosion in an oxisol. *Appl. Environ. Microbiol.* 54:945–950.

Hacke, U.G., Sperry, J.S., Ewers, B.E., Schäfer, K.V.R., Oren, R., and Ellsworth, D.S. 2000. Influence of soil porosity on water use in *Pinus taeda*. *Oecologia* 124:495–505.

Hacke, U.G., Sperry, J.S., Pockman, W.P., Davis, S.D., and McCulloh, K.A. 2001. Trends in wood density and structure are linked to prevention of xylem embolism by negative pressure. *Oecologia* 126:457–461.

Haeussler, S., and Kneeshaw, D. 2003. Comparing forest management to natural processes. In *Towards sustainable management of the boreal forest.* Edited by P.J. Burton, C. Messier, D.W. Smith, and W.L. Adamowicz. NRC Research Press, Ottawa, 368.

Hagan III, J.M., and Johnston, D.W. 1992. *Ecology and conservation and neotropical migrant landbirds.* Smithsonian Institution Press, Washington, DC.

Hagerman S.M., Jones, M.D., Bradfield, G.E., Gillespie, M., and Durall, D.M. 1999. Effects of clear-cut logging on the diversity and persistence of ectomycorrhizae at a subalpine forest. *Can. J. For. Res.* 29:124–134.

Haigh, M.J. 1981. Road construction and rural stress in the Indian Himalaya. *Nordia* 16:175–180.

Haines, S.B., and DeBell, D.S. 1979. Use of nitrogen-fixing plants to improve and maintain productivity of forest soils. In *Proc. impact of intensive harvesting on forest nutrient cycling.* Edited by A.L. Leaf. School of Forestry, Syracuse, NY, 279–303.

Haines, S.G., Haines, L.W., and White, G. 1979. Nitrogen-fixing plants in southeastern United States forestry. In *Symbiotic nitrogen fixation in the management of temperate forests.* Edited by J.C. Gordon, C.T. Wheeler, and D.A. Perry. Oregon State Univ., Corvallis, 429–443.

Hairston Jr., N.G. 1995. Commentary. *Ecology* 76:1371.

Hairston, N.G., Smith, F.E., and Slobodkin, L.B. 1960. Community structure, population control, and competition. *Am. Nat.* 94:421–425.

Hall, B.G. 1990. Spontaneous point mutations that occur more often when advantageous than when neutral. *Genetics* 126:5–16.

Hall, F.G., Botkin, D.B., Strebel, D.E., Woods, K.D., and Goetz, S.J. 1991. Large-scale patterns of forest succession as determined by remote sensing. *Ecology* 72:628–640.

Hallinger, D.Y. 1987. Gas exchange and dry matter allocation responses to elevation of atmospheric CO_2 concentration in seedlings of three tree species. *Tree Physiol.* 3:193–202.

Halpern, C.B. 1988. Early successional pathways and the resistance and resilience of forest communities. *Ecology* 69:1703–1715.

Ham, F.P. 1987. Interactions of insects, trees, and air pollutants. *Tree Physiol.* 3:93–102.

Hamburg, S.P., and Cogbill, C.V. 1988. Historical decline of red spruce populations and climatic warming. *Nature* 331:428–431.

Hammond, P.C., and Miller, J.C. 1998. Comparison of the biodiversity of Lepidoptera within three forests ecosystems. *Ann. Entomol. Soc. Am.* 91:323–328.

Hamrick, J.L., and Loveless, M.D. 1986. Isozyme variation in tropical trees: procedures and preliminary results. *Biotropica* 18:201–207.

Hamrick, J.L., Linhart, Y.B., and Mitton, J.B. 1979. Relationships between life history characteristics and electrophoretically-detectable genetic variation in plants. *Annu. Rev. Ecol. Syst.* 10:173–200.

Hanawalt, R.B. 1969a. Environmental factors influencing the sorption of atmospheric ammonia by soils. *Soil Sci. Soc. Am. Proc.* 33:231–238.

Hanawalt, R.B. 1969b. Soil properties affecting the sorption of atmospheric ammonia. *Soil Sci. Soc. Am. Proc.* 33:725–729.

Hankioja, E., and Niemela, P. 1979. Birch leaves as a resource for herbivores: seasonal occurrence of increased resistance in foliage after mechanical damage of adjacent leaves. *Oecologia* 39:151–159.

Hanley, T.A., and Taber, R.D. 1980. Selective plant species inhibition by elk and deer in three conifer communities in western Washington. *For. Sci.* 26:97–107.

Hannon, S.J., and Schmiegelow, F.K.A. 2002. Corridors may not improve the conservation value of small reserves for most boreal birds. *Ecol. Applic.* 12:1457–1468.

Hansen, A., Urban D., and Marks, B. 1992. Avian community dynamics: the interplay of landscape trajectories and species life histories. In *Landscape boundaries: consequences for biodiversity and ecological flows.* Edited by A. Hansen and F.J. Castri. Springer-Verlag, New York, 170–195.

Hansen, A.J., Garman, S.L., and Marks, B. 1993. An approach for managing vertebrate diversity across multiple-use landscapes. *Ecol. Appl.* 3:481–496.

Hansen, A.J., Neilson, R.P., Dale, V.H., Flather, C.H., Iverson, L.R., Currie, D.J., Shafer, S., Cook, R., and Bartlein, P.J. 2001. Global change in forests: responses of species, communities, and biomes. *BioScience* 51:765–779.

Hansen, A.J., Spies, T.A., Swanson, F.J., and Ohmann, J.L. 1991. Conserving biological diversity in managed forests. *BioScience* 41:382–392.

Hansen, E.H., and Munns, D.N. 1988. Effects of $CaSO_4$ and NaC1 on growth and nitrogen fixation of *Leucaena leucocephala. Plant Soil* 107:95–99.

Hansen, E.M., Goheen, D.J., Hessburg, P.F., Witcosky, J.J., and Schowalter, T.D. 1986. Biology and management of black-stain root disease in Douglas-fir. In *Forest pest management in southwest Oregon.* Edited by O.T. Helgerson. Oregon State Univ., Corvallis, 13–19.

Hansen, J., Fung, I., Lacis, A., Rind, D., Lebedeff, S., Ruedy, R., Russell, G., and Stone, P. 1988. Global climate changes as forecast by Goddard Institute for Space Studies three-dimensional model. *J. Geophys. Res.* 93:9341–9364.

Hansen, J., Nazarenko, L., Ruedy, R., Sato, M., Willis, J., Del Genio, A., Koch, D., Lacis, A., Lo, K., Menon, S., Novakov, T., Perlwitz, J., Russell, G., Schmidt, G.A., and Tausnev, N. 2005. Earth's energy imbalance: confirmation and implications. *Science* 308:1431–1435.

Hansen, J., Sato, M., Ruedy, R., Lo, K., Lea, D.W., and Medina-Elizade, M. 2006. Global temperature change. *PNAS* 103:14288–14293.

Hansen, M.C., and DeFries, R.S. 2004. Detecting long-term global forest change using continuous fields of tree-cover maps from 8-km advanced very high resolution radiometer (AVHRR) data for the years 1982–99. *Ecosystems* 7:695–716.

Hansson, B., and Westerberg, L. 2002. On the correlation between heterozygosity and fitness in natural populations. *Molecular Ecology* 11:2467–2474.

Hardwick, R.C. 1987. The nitrogen content of plants and the self-thinning rule of plant ecology: a test of the core-skin hypothesis. *Ann. Bot.* 60:439–446.

Hardy, J.T., Groffman, P.M., Fitzhugh, R.D., Henry, K.S., Welman, A.T., Demers, J.D., Fahey, T.J., Driscoll, C.T., Tierney, G.L., and Nolan, S. 2001. Snow depth manipulation and its influence on the soil frost and water dynamics in a northern hardwood forest. *Biogeochemistry* 56:151–174.

Hardy, K. 1981. *Aspects of microbiology series*, No. 4, *Bacterial plasmids*. American Society for Microbiology, Washington, DC.

Harmon, M.E., Baker, G.A., Spycher, G., and Greene, S. 1990a. Leaf-litter decomposition in *Picea Tsuga* forests of Olympic National Park, Washington, USA. *For. Ecol. Manage.* 31:55–66.

Harmon, M.E., Ferrell, W.K., and Franklin, S.F. 1990b. Effects on carbon storage of conversion of old-growth forests to young forests. *Science* 247:699–702.

Harmon, M.E., Franklin, J.F., Swanson, F.J., Sollins, P., Gregory, S.V., Lattin, J.D., Anderson, N.H., Cline, S.P., Aumen, N.G., Sedell, J.R., Lienkaemper, G.W., Cromack Jr., K., and Cummins, K.W. 1986. Ecology of coarse woody debris in temperate ecosystems. *Adv. Ecol. Res.* 15:133–302.

Harmon, M.E., Nadelhoffer, K.J., and Blair, J.M. 1999. Measuring decomposition, nutrient turnover, and stores in plant litter. In *Standard soil methods for long-term ecological research*. Edited by G.P. Robertson, D.C. Coleman, C.S. Bledsoe, and P, Sollins. Oxford Univ. Press, New York, 202–240.

Harper, J.L. 1977. *Population biology of plants*. Academic Press, London.

Harper, K.A., MacDonald, S.E., Burton, P.J., Chen, J., Brosofke, K.D., Saunders, S.C., Euskirchen, E.S., Roberts, D., Jaiteh, M.S., and Esseen, P.-A. 2005. Edge influence on forest structure and composition in fragmented landscapes. *Conserv. Biol.* 19:768–782.

Harr, R.D. 1976. *Forest practices and streamflow in western Oregon*. Gen. Tech. Rep. PNW-49. U.S. Department of Agriculture, Forest Service, Portland, OR.

Harr, R.D., Harper, W.C., Krygier, J.T., and Hsieh, F.S. 1975. Changes in storm hydrographs after road building and clear-cutting in the Oregon Coast Range. *Water Resourc. Res.* 11:436–444.

Harrington, J.B. 1987. Climatic change: a review of causes. *Can. J. For. Res.* 17:1313–1339.

Harrington, R.A., Fownes, J.H., and Vitousek, P.M. 2001. Production and resource use efficiencies in N- and P-limited tropical forests: a comparison of responses to long-term fertilization. *Ecosystems* 4:646–657.

Harris, L.D. 1984. *The fragmented forest*. Univ. of Chicago Press, Chicago.

Harris, L.D., and Maser, C. 1984. Animal community characteristics. In *The fragmented forest*. Edited by L.D. Harris. Univ. of Chicago Press, Chicago, 44–68.

Harris, W.F., Kinerson, R.S., and Edwards, N.T. 1978. Comparisons of belowground biomass of natural deciduous forest and loblolly pine plantations. *Pedobiologia* 17:369–381.

Harris, W.F., Santantonio, D., and McGinty, D. 1979. The dynamic belowground ecosystem. In *Forests: fresh perspectives from ecosystem analysis*. Edited by R.H. Waring. Oregon State Univ. Press, Corvallis, 119–130.

Harrison, J.L. 1962. The distribution of feeding habits among animals in a tropical rain forest. *J. Anim. Ecol.* 31:53–63.

Harsfield, D. 1977. Relationship between feeding of *Philaenus spumarius* (L.) and the amino acid concentration in the xylem sap. *Ecol. Entomol.* 2:259–266.

Hart, S.C. 1999. Nitrogen transformations in fallen tree boles and mineral soil of an old-growth forest. *Ecology* 80:1385–1394.

Hart, S.C. 2006. Potential impacts of climate change on nitrogen transformations and greenhouse gas fluxes in forests: a soil transfer study. *Global Change Biol.* 12:1032–1046.

Hart, S.C., and Firestone, M.K. 1991. Forest floor-mineral soil interactions in the internal nitrogen cycle of an old-growth forest. *Biogeochemistry* 12:103–127.

Hart, S.C., and Perry, D.A. 1999. Transferring soils from high- to low-elevation forests increases nitrogen cycling rates: climate change implications. *Global Change Biol.* 5:23–32.

Hart, S.C., Classen, A.T., and Wright, R.J. 2005a. Long-term interval burning alters fine root and mycorrhizal dynamics in a ponderosa pine forest. *J. Appl. Ecol.* 42:752–761.

Hart, S.C., DeLuca, T.H., Newman, G.S., MacKenzie, M.D., and Boyle, S.I. 2005b. Post-fire vegetative dynamics as drivers of microbial community structure and function in forest soils. *For. Ecol. Manage.* 220:166–184.

Hart, S.C., Nason, G.E., Myrold, D.D., and Perry, D.A. 1994. Dynamics of gross nitrogen transformations in an old-growth forest soil during a long-term laboratory incubation: the carbon connection. *Ecology* 75:880–891.

Hart, S.C., Stark, J.M., Davidson, E.A., and Firestone, M.K. 1992. Nitrogen mineralization, immobilization, and nitrification. In *Methods of soil analysis: biochemical and microbiological properties*. Edited by R.W. Weaver, C. Angle, and P. Bottomley. Third edition. *Soil Sci*. Soc. Am., Madison, WI.

Hart, T.B., Hart, J.A., and Murphy, P.G. 1989. Monodominant and species-rich forests of the humid tropics: causes for their co-occurrence. *Am. Nat.* 133:613–633.

Hartenstein, R. 1962. Soil Oribatei. I. Feeding specificity among forest soil Oribatei (Acarina). *Annu. Entomol. Soc. Am.* 55:202–206.

Hartshorn, G.S. 1989. Application of gap theory to tropical forest management: natural regeneration on strip clear-cuts in the Peruvian Amazon. *Ecology* 70:567–569.

Hartshorn, G.S. 1990. Natural forest management by the Yanesha Forestry Cooperative in Peruvian Amazon. In *Alternatives to deforestation: steps toward sustainable use of the Amazon Basin*. Edited by A.B. Anderson. Columbia Univ. Press, New York, 128–138.

Harvell, C.D. 1990. The ecology and evolution of inducible defenses. *Q. Rev. Biol.* 65:323–340.

Harvell, C.D., Mitchell, C.E., Ward, J.R., Altizer, S., Dobson, A.P., Ostfeld, R.S., and Samuel, M.D. 2002. Climate warming and disease risks for terrestrial and marine biota. *Science* 296:2158–2162.

Harvey, A.E., and Neuenschwander, L., editors. 1991. *Proceedings: management and productivity of western montane forest soils*. Gen. Tech. Rep. INT-280. U.S. Department of Agriculture, Forest Service, Ogden, UT.

Harvey, A.E., Jurgensen, M.F., and Larsen, M.J. 1983. Effect of soil organic matter on regeneration in northern Rocky

Mountain forests. In *IUFRO symposium on forest site and continuous productivity.* Gen. Tech. Rep. PNW-163. U.S. Department of Agriculture, Forest Service, Portland, OR, 239–242.

Harvey, A.E., Jurgensen, M.E., Larsen, M.J., and Graham, R.T. 1987. Relationship among soil microsite, ectomycorrhizae, and natural conifer regeneration of old-growth forests in westem Montana. *Can. J. For. Res.* 17:58–62.

Harvey, A.E., Larsen, M.J., and Jurgensen, M.F. 1979. Comparative distribution of ectomycorrhizae in soils of three western Montana forest habitat types. *For. Sci.* 25:450–458.

Harvey, A.E., Larsen, M.J., and Jurgensen, M.F. 1980. Clearcut harvesting and ectomycorrhizae: survival of activity on residual roots and influence on a bordering stand in western Montana. *Can. J. For. Res.* 10:300–303.

Harvey, A.E., Meurisse, R.T., Geist, J.M., Jurgensen, M.F., McDonald, G.I., Graham, R.T., and Stark, N. 1989. Managing productivity processes in the inland northwest—mixed conifers and pines. In *Maintaining the longterm productivity of Pacfic Northwest forest ecosystems.* Edited by D.A. Perry, R. Meurisse, B. Thomas, et al. Timber Press, Portland, OR, 164–184.

Hassell, M.P. 2000. *The spatial and temporal dynamics of host-parasitoid interactions.* Oxford Univ. Press, New York.

Hassell, M.P., Comins, H.N., and May, R.M. 1994. Species coexistence and self-organizing spatial dynamics. *Nature* 370:290–292.

Hasson, E. 1967. Untersuchungen uber die Bedeutund der Krautund Strauchschicht Nahrungsquelle fur Imagines Entomophagen Hymenopteren. *Z. Ang. Entomol.* 60:238–265.

Hastings, A., Horn, C.L., Ellner, S., Turchin, P., and Godfray, H.C.J. 1993. Chaos in ecology: is Mother Nature a strong attractor? *Annu. Rev. Ecol. Syst.* 24:1–33.

Hatton, J.C., and Smart, N.O.E. 1984. The effect of long-term exclusion of large herbivores on soil nutrient status in Murchison Falls National Park, Uganda. *Afr. J. Ecol.* 22:23–30.

Hattenschwiler, S., and Gasser, P. 2005. Soil animals alter plant litter diversity effects on decomposition. *PNAS* 102:1519–24.

Hattenschwiler, S., and Vitousek, P.M. 2000. The role of polyphenols in terrestrial ecosystem nutrient cycling. *Trends Ecol. Evol.* 15:238–43.

Hattenschwiler S., Tiunov, A.V., and Scheu, S. 2005. Biodiversity and litter decomposition interrestrial ecosystems. *Annu. Rev. Ecol. Evol. Syst.* 36:191–218.

Hattori, T., and Hattori, R. 1976. The physical environment in soil microbiology: an attempt to extend principles of microbiology to soil microorganisms. *CRC Crit. Rev. Microbiol.* (May):423– 461.

Haukioja, E. 1990. Induction of defenses in trees. *Annu. Rev. Entomol.* 36:25–42.

Haukioja, E., and Niemela, P. 1979. Birch leaves as a resource for herbivores: seasonal occurrence and of increased resistance in foliage after mechanical damage of adjacent leaves. *Oecologia* 39:151–159.

Hawai'i Center for Volcanology. 2004. General information. www.soest.hawaii.edu/GG/HCV/mloa-eruptions.html.

Hawkins, B.A., and Porter, E.E. 2003 Water-energy balance and the geographic pattern of species richness of western Palearctic butterflies. *Ecol. Entomol.* 28(6):678–686.

Hawkins, B.A., and Lawton, J.H. 1987. Species richness for parasitoids of British phytophagous insects. *Nature* 326:788–790.

Hawkins, B.A., Field, R., Cornell, H.V., Currie, D.J., Guegan, J.-F., Kaufman, D.M., Kerr, J.T., Mittelbach, G.G., Oberdorff, T., O'Brien, E.M., Porter, E.E., and Turner, J.R.G. 2003. Energy, water, and broadscale geographic patterns of species richness. *Ecology* 84:3105–3117.

Hawkins, B.A., Porter, E.E., and Diniz-Filho, J.A.F. 2003. Productivity and history as predictors of the latitudinal diversity gradient of terrestrial birds. *Ecology* 84:1608–1623.

Haxeltine, A., and Prentice, I.C. 1996. BIOME3: an equilibrium terrestrial biosphere model based on ecophysiological constraints, resource availability and competition among plant functional types. *Global Biogeochem. Cycles* 10:693–709.

Hay, M.E. 1986. Associational plant defenses and the maintenace of species diversity: turning competitors into accomplices. *Am. Nat.* 128:617–641.

Hayes, J.P., and Hagar, J.C. 2002. Ecology and management of wildlife and their habitats in the Oregon Coast Range. In. *Forest and stream management in the Oregon Coast Range.* Edited by S.D. Hobbs, J.P. Hayes, R.L. Johnson, G.H. Reeves, T.A. Spies, and J.C. Tappeiner. Oregon State Univ. Press, Corvallis, 99–134.

Haynes, R.J. 1986. Origin, distribution, and cycling of nitrogen in terrestrial ecosystems. In *Mineral nitrogen in the plant-soil system.* Edited by R.J. Haynes. Academic Press, Orlando, FL, 1–51.

Haywood, J.D., and Burton, J.D. 1989. Loblolly pine plantation development is influenced by site preparation and soils in the West Gulf Coastal Plain. *So. J. Appl. For.* 13:17–21.

Hazell, P., and Gustafsson, L. 1999. Retention of trees at final harvest—evaluation of a conservation technique using epiphytic bryophyte and lichen transplants. *Biol. Conserv.* 90:133–142.

He, X.T., Stevenson, F.J., Mulvaney, R.L., and Kelley, K.R. 1988. Incorporation of newly immobilized ^{15}N into stable organic forms in soil. *Soil Biol. Biochem.* 20:75–81.

Heal, O.W., and Dighton, J. 1985. Resource quality and tropic structure in the soil system. In *Ecological interactions in soil.* Edited by A.H. Fitter. Blackwell, Oxford, 339–354.

Heath, B., Sollins, P., Perry, D.A., and Cromack Jr., K. 1988. Asymbiotic nitrogen fixation in litter from Pacific Northwest forests. *Can J. For. Res.* 18:68–74.

Hebblewhite, M., White, C.A., Nietvelt, C.G., McKenzie, J.A., Hurd, T.E., Fryxell, J.M., Bayley, S.E., and Paquet, P.C. 2005. Human activity mediates a trophic cascade caused by wolves. *Ecology* 86:2135–2144.

Hedin, L.O., Armesto, J.J., and Johnson, A.H. 1995. Patterns of nutrient loss from unpolluted, old-growth temperate forests—evaluation of biogeochemical theory. *Ecology* 76:493–509.

Hedrick, P.W. 1986. Genetic polymorphism in heterogeneous environments: a decade later. *Annu. Rev. Ecol. Syst.* 17:535–566.

Heemsbergen, D.A., Berg, M.P., Loreau, M, van Hal, J.R., Faber, J.H., and Verhoef, H.A. 2004. Biodiversity effects on soil processes explained by interspecfic functional dissimilarity. *Science* 306:1019–1020.

Heinemann, J.A., and Sprague Jr., J.A. 1989. Bacterial conjugative plasmids mobilize DNA transfer between bacteria and yeast. *Nature* 340:205–209.

Heinselman, M.L. 1973. Fire in the virgin forests of the Boundary Waters Canoe Area, Minnesota. *Quat. Res.* 3:329–382.

Heinselman, M.L. 1981. Fire and succession in the conifer forests of northern North America. In *Forest succession, concepts, and application*. Edited by D.C. West, H.H. Shugart, and D.B. Botkin. Springer-Verlag, New York, 375–405.

Helfield, J.M., and Naiman, R.J. 2006. Keystone interactions: salmon and bear in riparian forests of Alaska. *Ecosystems* 9:167–180.

Helm, D.J., and Carling, D.E. 1993a. Use of soil transfer for reforestation on abandoned mine lands in Alaska. I. Effects of soil transfer and phosphorus on growth and mycorrhizal formation by *Populus balsamifera*. *Mycorrhiza* 3:97–106.

Helm, D.J., and Carling, D.E. 1993b. Use of soil transfer for reforestation on abandoned mine lands in Alaska. II. Effects of soil transfers from different successional stages on growth and mycorrhizal formation by *Populus balsamifera* and *Alnus crispa*. *Mycorrhiza* 3:107–114.

Hemstrom, M.A. 1979. A recent disturbance history of forest ecosystems at Mount Rainier National Park. Ph.D. thesis. Oregon State Univ., Corvallis.

Henderson, G.S., Swank, W.T., Waide, J.B., and Grier, C.C. 1978. Nutrient budgets of Appalachian and Cascade region watersheds: a comparison. *For. Sci.* 24:385–397.

Hendrickson, O.Q., and Burgess, D. 1989. Nitrogen-fixing plants in a cut-over lodgepole pine stand of southern British Columbia. *Can. J. For. Res.* 19:936–939.

Hendrix, P.F., Crossley Jr., D.A., Blair, J.M., and Coleman, D.C. 1990. Soil biota as components of sustainable agroecosystems. In *Sustainable agricultural systems*. Edited by C.A. Edwards, R. Lal, P. Madden, R.H. Miller, and G. House, Soil and Water Conservation Society, Ankeny, IA, 639–654

Henle, K., Davies, K.F., Kleyer, M., Margules, C., and Settele, J. 2004. Predictors of species sensitivity to fragmentation. *Biodivers. Conserv.* 13:207–251.

Herbert, B.E., and Bertsch, P.M. 1995. Characterization of dissolved and colloidal organic matter in soil solution: a review. In *Carbon forms and functions in forest soils*. Edited by W.W. McFee and J.M. Kelly. Soil Science Society of America, Madison, WI, 63–88.

Herms, D.A., and Mattson, W.J. 1992. The dilemma of plants: to grow or defend. *Q. Rev. Biol.* 67:283–335.

Herrera, R. 1979. Nutrient distribution and cycling in an Amazon caatinga forest on spodosols in southern Venezuela. Ph.D. thesis, Univ. of Reading, Reading, England.

Hessburg, P.F., Agee, J.K., and Franklin, J.F. 2005. Dry forests and wildland fires of the inland Northwest USA: contrasting the landscape ecology of the pre-settlement and modern eras. *For. Ecol. Manage.* 211:117–139.

Hesselman, H. 1926. Studier over barrskogens humustache, dess egenskaper och beroende av skogsvarden. *Staens Skogsforsoksant Meddel* 22:169–552.

Heyerdahl, E.K., Brubaker, L.B., and Agee, J.K. 2001. Spatial controls of historical fire regimes: a multiscale example from the interior west, USA. *Ecology* 82:660–678.

Hibbs, D.E. 1983. Forty years of forest succession in central New England. *Ecology* 64:1394–1401.

Higashi, S., and Yaumachi, K. 1979. (No title available.) *Jpn. J. Ecol.* 29:257.

Highwater, J. 1981. *The primal mind*. Meridian, New York.

Hilderbrand, R.H., Watts, A.C., and Randle, A.M. 2005. The myths of restoration ecology. *Ecol. Soc.* 10(1):19. www.ecologyandsociety.org/vol10/iss1/art19/.

Hillebrand, H., and Blenckner, T. 2002. Regional and local impact on species diversity—from pattern to process. *Oecologia* 132:479–491.

Hinkley, T.M. 1978. Leaf conductance and photosynthesis in four species of the oak-hickory forest type. *For. Sci.* 24:73–84.

Hinrichsen, D. 1987. The forest decline enigma. *BioScience* 37:542–265.

Hinsinger, P. 2001. Bioavailability of soil inorganic P in the rhizosphere as affected by root-induced chemical changes: a review. *Plant Soil* 237:173–195.

Hintikka, V., and Naykki, O. 1967. Notes on the effects of the fungus *Hydnellumfrrrungineum* (Fr.) Karst. on forest soil and vegetation. *Commun. Inst. For. Fenn.* 62:1–23.

Hobbie, S.E., and Vitousek, P.M. 2000. Nutrient regulation of decomposition in Hawaiian forests: do the same nutrients limit production and decomposition? *Ecology* 81:1867–1877.

Hobson, K.A., Bayne, E.M., and van Wilgenburg, S.L. 2002. Large-Scale conversion of forest to agriculture in the boreal plains of Saskatchewan. *Conserv. Biol.* 16:1530–154.

Hoekstra, J.M., Boucher, T.M., Ricketts, T.H., and Roberts, C. 2005. Confronting a biome crisis: global disparities of habitat loss and protection. *Ecol. Lett.* 8:23–29.

Hoffland, E., Kuyper, T.W., Wallender, H., Plassard, C., Gorbushina, A.A., Haselwander, K., Holmstrom, S., Landeweert, R., Lundstrom, U.S., Rosling, A., Sen, R., Smits, M.M., van Hees, P.A.W., and van Breeman, N. 2004. The role of fungi in weathering. *Front. Ecol. Environ.* 2:258–264.

Hogberg, P. 1986. Nitrogen-fixation and nutrient relations in savanna woodland trees (Tanzania). *J. Appl. Ecol.* 23:675–688.

Hogberg, P. 1989. Root symbioses of trees in savannas. In *Mineral nutrients in tropicalforest and savanna ecosystems*. Edited by J. Proctor. Blackwell Scientific Publications, Oxford, 121–136.

Hole, F.D. 1981. Effects of animals on soil. *Geoderma* 25:75–112.

Holland, E. A., Braswell, B.H., Selzman, J., and Lamarque, J.-F. 2005. Nitrogen deposition onto the United States and western Europe: synthesis of observations and models. *Ecol. Applic.* 15:38–57.

Holling, C.S. 1959. The components of predation as revealed by a study of small-marnmal predation of the European pine sawfly. *Can. Entomol.* 91:293–320.

Holling, C.S. 1973. Resilience and stability of ecological systems. *Annu. Rev. Ecol. Syst.* 4:1–23.

Holling, C.S. 1988. Temperate forest insect outbreaks, tropical deforestation and migratory birds. *Mem. Entomol. Soc. Can.* 146:21–32.

Hollinger, D.Y. 1986. Herbivory and the cycling of nitrogen and phosphorus in isolated California oak trees. *Oecologia (Berlin)* 70:291–297.

Hollinger, D.Y. 1987. Gas exchange and dry matter allocation responses to elevation of atmospheric CO_2 concentration in seedlings of three tree species. *Tree Physiol.* 3:193–202.

Holloway, J.T. 1954. Forest and climate of the South Island of New Zealand. *R. Soc. N. Z. Trans.* 82:329–410.

Holm, L.R., Plucknett, D.L., Pancho, J.V., and Herberger, J.P. 1977. *The world's worst weeds*. Univ. of Hawaii Press, Honolulu.

Holmgren, M., Scheffer, M., Ezcurra, E., Gutierrez, J.R., and Mohren, G.M.J. 2001. El Nino effects on the dynamics of terrestrial ecosystems. *Tree* 16:89–94.

Holmgren, M., Scheffer, M., and Huston, M.A. 1997. The interplay of facilitation and competition in plant communities. *Ecology* 78:1966–1975.

Holsinger, K. 2007. Theory and design of nature reserves: the SLOSS debate. http://darwin.eeb.uconn.edu/eeb310/lecture-notes/reserves/node8.html.

Holt, R.D. 2006. Asymmetry and stability. *Nature* 442:252–253.

Holt, R.D. 2006a. Making a virtue out of a necessity: hurricanes and the resilience of community organization. *PNAS* 103:2005–2006.

Holt, R.D., and Hoopes, M.F. 2005. Food web dynamics in a metacommunity context. In *Metacommunities: spatial dynamics and ecological communities.* Edited by M. Holyoak, M.A. Leibold, and R.D. Holt. Univ. of Chicago Press, Chicago, 68–93.

Holyoak, M., Leibold, M.A., Holt, R.D., and Hoopes, M.F. 2005. Metacommunities: A framework for large-scale community ecology. In *Metacommunities: spatial dynamics and ecological communities.* Edited by M. Holyoak, M.A. Leibold, and R.D. Holt. Univ. of Chicago Press, Chicago, 1–32.

Hoover, M.D. 1949. Hydrologic characteristics of South Carolina piedmont forest soils. *Soil Sci. Soc. Am. Proc.* 14:353–358.

Hoover, M.D. 1949. Hydrologic characteristics of South Carolina piedmont forest soils. *Soil Sci. Soc. Am. Proc.* 14:353–358.

Hopwood, D. 1991. *Principles and practices of new forestry: a guide for British Columbians.* Land Manage. Rep. No. 71. Bntish Columbia Ministry of Forests, Vancouver.

Horbeck, J.W., and Swank, W.T. 1992. Watershed ecosystem analysis as a basis for multiple-use management of eastern forests. *Ecol. Applic.* 2:238–247.

Horner, J.D., Gosz, J.R., and Cates, R.G. 1988. The role of carbon-based plant secondary metabolites in decomposition in terrestrial ecosystems. *Am. Nat.* 132:869–883.

Horsfield, D. 1977. Relationship between feeding of *Philaenus spumarius* (L.) and the amino acid concentration in the xylem sap. *Ecol. Entomol.* 2:259–266.

Horsley, S.B. 1977. Allelopathic inhibition of black cherry by fern, grass, goldenrod, and aster. *Can. J. For. Res.* 7:205–216.

Horton, J., and S. Hart. 1998. Hydraulic lift: a potentially important ecosystem process. *Tree* 13:232–236.

Horton, T.R., Bruns, T.D., and Parker, V.T. 1999. Ectomycorrhizal fungi associated with Arctostaphylous contribute to Pseudotsuga menziesii establishment. Can. J. Bot. 77:93–102.

Houghton, J. 2005. Gobal warming. *Rep. Prog. Phys.* 68:1343–1403.

Houghton, R.A. 1989. Emissions of greenhouse gases. Deforestation rates in tropical forests and their climatic implications. In *A Friends of the Earth report.* Norman Myers, London, 53–62.

Houghton, R.A., and Hackler, J.L. 2004. Carbon flux to the atmosphere from land-use changes. http://cdiac.esd.ornl.gov/trends/landuse/houghton/houghton.html.

Houghton, R.A., and Skole, D.L. 1990. Changes in the global carbon cycle between 1700 and 1895. In *The Earth transformed by human action.* Edited by B.L. Turner. Cambridge Univ. Press, Cambridge.

Houghton, R.A., and Woodwell, G.M. 1989a. Global climatic change. *Sci. Am.* 260:2–10.

Houghton, R.A., and Woodwell, G.M. 1989b. Global climatic change. *Sci. Am.* 260:36–44.

Houlahan, J.E., and Findley, C.S. 2004. Effect of invasive plant species on temperate wetland plant diversity. *Conserv. Biol.* 18:1132–1138.

Howe, H.F., and Smallwood, J. 1982. Ecology of seed dispersal. *Annu. Rev. Ecol. Syst.* 13:201–228.

Howlett, R. 1990. A chaotic synthesis. *Nature* 346:104–105.

Hoyle, M.C., and Mader, D.L. 1964. Relationships of foliar nutrients to growth of red pine in western Massachusetts. *For. Sci.* 10:337–347.

Hubbell, D.H., and Kidder, G. 2003. Biological nitrogen fixation. Fact Sheet SL-16. Soil and Water Science Department, Florida Cooperative Extension Service, Institute of Food and Agricultural Sciences, University of Florida, Gainesville. http://edis.ifas.ufl.edu/SS180.

Hubbell, S.P. 1979. Tree dispersion, abundance, and diversity in a tropical dry forest. *Science* 203:1299–1309.

Hubble, S.P. 2001. *The unified neutral theory of biodiversity and biogeography.* Princeton Univ. Press, Princeton, NJ.

Hughen, K.A., Eglinton, T.I., Xu, L., and Makou, M. 2004. Abrupt tropical vegetation response to rapid climate changes. *Science* 304:1955–1959.

Hughes, J.B., Hellmann, J.J., Ricketts, T.H., and Bohannan, B.J.M. 2001. Counting the uncountable: statistical approaches to estimating microbial diversity. *Appl. Environ. Microbiol.* 67:4399–4406.

Hughes, R.F., and Denslow, J.S. 2005. Invasion by a N_2-fixing tree alters function and structure in wet lowland forests of Hawaii. *Ecol. Appl.* 15:1615–1628.

Huhta, E., Ajo, T., Jantti, A., Suorsa, P., Kuitunen, M., Nikula, A., and Hakkarainen, H. 2004. Forest fragmentation increases nest predation in the Eurasian trecreeper. *Conserv. Biol.* 18:148–155.

Hung, L.L., and Trappe, J.M. 1983. Growth variation between and within species of ectomycorrhizal fungi in response to pH in vitro. *Mycologia* 72:234–241.

Hunt, G.A., and Fogel, R. 1983. Fungal hyphal dynamics in a western Oregon Douglas-fir stand. *Soil Biol. Biochem.* 15:641–649.

Hunter, A.F., and Aarssen, L.W. 1988. Plants helping plants. *BioScience* 38:34–40.

Hunter, M.D. 2001. Insect population dynamics meets ecosystem ecology: effects of herbivory on soil nutrient dynamics. *Agic. For. Entomol.* 3:77–84.

Hunter, M.D., and Price, P.W. 1992. Playing chutes and ladders: heterogeneity and the relative roles of bottom-up and top-down forces in natural communities. *Ecology* 73:724–732.

Hunter M.D., Adl, S., Pringle, C.M., and Coleman, C. 2003. Relative effects of macro invertebrates and habitat on the chemistry of litter during decomposition. *Pedobiologia* 47:101–115.

Hunter, M.D., Linnen, C.R., and Reynolds, B.C. 2003. Effects of endemic densities of canopy herbivores on nutrient dynamics along a gradient in elevation in the southern Appalachians. *Pedobiologia* 47:231–244.

Hunter Jr., M.L., Jacobson Jr., G.L., and Webb III, T. 1988. Paleoecology and the coarse-filter approach to maintaining biological diversity. *Conserv. Biol.* 2:375–385.

Huntly, N. 1991. Herbivores and the dynamics of cummunities and ecosystems. *Annu. Rev. Ecol. Syst.* 22:477–503.

Hurtt, G.C., and S.W. Pacala. 1995. The consequences of recruitment limitation: Reconciling chance, history, and competitive differences between plants. *J. Theor. Biol.* 176:1–12.

Huss-Danell, K. 1986. Growth and production of leaf litter nitrogen by *Alnus incana* in response to liming and fertilization on degenerated forest soil. *Can. J. For. Res.* 16:847–853.

Huston, M. 1979. A general hypothesis of species diversity. *Am. Nat.* 113:81–101.

Huston, M. 1980. Soil nutrients and tree species richness in Costa Rican forests. *J. Biogeogr.* 7:147–157.

Huston, M., and Smith, T. 1987. Plant succession: life history and competition. *Am. Nat.* 130:168–198.

Huston, M., DeAngelis, D., and Post, W. 1988. New computer models unify ecological theory. *BioScience* 38:682–691.

Huston, M., Gomezdelcampo, E., and Nesteruk, R.S. 2004. *Linking topography, hydrology, and biodiversity to understand terrestrial impacts on aquatic systems.* Final report to National Commission on Sustainable Forestry. www.ncseonline.org/ewebeditpro/items/O62F3298.pdf.

Hutchins, H.E., and Lanner, R.M. 1982. The central role of Clark's nutcracker in the dispersal and establishment of whitebark pine. *Oecologia* 55:192–201.

Hutchinson, G.E. 1959. Homage to Santa Rosalia or why are there so many kinds of animals. *Am. Nat.* 93:145–159.

Hutchinson, G.E. 1957. Concluding remarks. *Cold Spring Harbor Symp. Quant. Biol.* 22:41 5–427.

Hutchinson, T.C. 1980. Impact of heavy metals on terrestrial and aquatic ecosystems. In *Proceedings of effects of air pollutants on mediterranean and temperate forest ecosystems: an international symposium.* Gen. Tech. Rpt. PSW-43. Edited by P. Miller. U.S. Forest Service Pacific SW Res. Sta., Berkeley, CA, 158–164.

Hutto, R.L. 1995. Composiiton of bird communities following stand-replacement fires in Northern Rocky Mountains (USA) conifer forests. *Conserv. Biol.* 9:1041–1058.

Igor Volkov, I., Banavar, J.R., Hubbell, S.P., and Maritan, A. 2003 Neutral theory and relative species abundance in ecology. *Nature* 424:1035–1037.

Ikeda, T., Matsumura, F., and Benjamin, D.M. 1977. Mechanism of feeding discrimination between mature and juvenile foliage by two species of pine sawflies. *J. Chem. Ecol.* 3:677–694.

Ineson, P., and Anderson, J.M. 1985. Aerobically isolated bacteria associated with the gut and faeces of the litter feeding macroarthropods *Oniscus asellus* and *Glomeris marginata*. *Soil Biol. Biochem.* 17:843–849.

Ingestad, T. 1979. Mineral nutrient requirements of *Pinus silvestris* and *Picea abies* seedlings. *Physiol. Plant* 45:373–380.

Ingestad, T. 1982. Relative addition rate and external concentration: driving variables used in plant nutrition research. In *Plant, cell and environment.* Blackwell, London, 443–453.

Ingestad, T., and Lund, A.B. 1979. Nitrogen stress in birch seedlings. I. Growth technique and growth. *Physiol. Plant* 45:137–148.

Ingestad, T., Aronsson, A., and Agren, G.I. 1981. Nutrient flux density model of mineral nutrition in conifer ecosystems. *Studia Forestalia Suecica* 160:61–71.

Ingham, E.R., and Molina, R. 1991. Interactions among mycorrhizal fungi, rhizosphere organisms, and plants. In *Microbial mediation of plant-herbivore interactions.* Edited by P. Barbosa, V.A. Krischik, and C.G. Jones. Wiley, New York, 169–197.

Ingham, E.R., Cambardella, C., and Coleman, D.C. 1986. Manipulation of bacteria, fungi and protozoa by biocides in lodgepole pine forest soil microcosms: effects on organism interactions and nitrogen mineralization. *Can. J. Soil Sci,* 66:261–272.

Ingham, E.R., Coleman, D.C., and Moore, J.C. 1989. An analysis of food-web structure and function in a shortgrass praire, a mountain meadow, and a lodgepole pine forest. *Biol. Fertil. Soils* 8:29–37.

Ingham, R.E., Trofymow, J.S., Ingham, E.R., and Coleman, D.C. 1985. Interactions of bacteria, fungi, and their nematode grazers: effects on nutrient cycling and plant growth. *Ecol. Monogr.* 55:119–140.

Innes, J.L., and Er, K.B.H. 2002. Questionable utility of the frontier forest concept. *BioScience* 52:1095–1109.

IPCC. 2001. Climate change 2001: synthesis report. Intergovernmental Panel on Climate Change, Geneva.

Isack, H.A., and Reyer, H.U. 1989. Honey guides and honey gatherers: interspecific communication in a symbiotic relationship. *Science* 243:1343–1346.

Ishii, H., Tanabe. S.-I., and Hiura, T. 2004. Exploring the relationships among canopy structure, stand productivity, and biodiversity of temperate forest ecosystems. *For. Sci.* 50:342–355.

ISI Web of Science Database. http://portal.isiknowledge.com

Isichei, A.O. 1980. Nitrogen fixation by blue-green algal soil crusts in Nigerian savanna. In *Nitrogen cycling in West African ecosystems.* Edited by T. Rosswall. Royal Swedish Academy of Sciences, Stockholm, 191–198.

IUCN. 2005. *Forest landscape restoration: broadening the vision of West African forests.* The World Conservation Union. Gland, Switzerland.

Iverson, L.R., and Prasad, A.M. 2002. Potential redistribution of tree species habitat under five climate change scenarios in the eastern US. *For. Ecol. Manage.* 155:205–222.

Jablonski, D. 1991. Extinctions: a paleontological perspective. *Science* 253:754–757.

Jackson, R.B., Mooney, H.A., and Schulze, E.-D. 1997. A global budget for fine root biomass, surface area, and nutrient contents. *PNAS* 94:7362–7366.

Jactel, H., and Brockerhoff, E.G. 2007. Tree diversity reduces herbivory by forest insects. *Ecol. Lett.* 10:835–848.

Jaenike, J. 1990. Host specialization in phytophagous insects. *Annu. Rev. Ecol. Syst.* 21:243–273.

Jakucs, P. 1988. Ecological approach to forest decay in Hungary. *Ambio* 17:267–274.

James, E.K. Sprent, J.I. Dilworth, M.J., and Newton, W.E., editors. 2007. *Nitrogen-fixing leguminous symbioses. Nitrogen fixation: origins, applications, and research progress*, Vol. 7. Springer New York/Berlin.

James, F.C., and Warner, N.O. 1982. Relationships between temperate forest bird communities and vegetation structure. *Ecology* 63:19–171.

Jandl, R., and Sollins, P. 1997. Water extractable soil carbon in relation to the belowground carbon cycle. *Biol. Fertil. Soils* 25:196–201.

Janos, D.P. 1980a. Mycorrhizae influence tropical succession. *Biotropica* 12(suppl):56–64.

Janos, D.P. 1980b. Vesicular-arbuscular mycorrhizae affect lowland tropical rain forest plant growth. *Ecology* 61:151–162.

Janos, D.P. 1983. Tropical mycorrhizas, nutrient cycles and plant growth. In *Tropical rain forest: ecology and management.* Edited by S.L. Sutton, T.C. Whitmore, and A.C. Chadwick. Blackwell, Oxford, 327–345.

Janos, D.P. 1985. Mycorrhizal fungi: agents or symtpoms of tropical community composition? In *Proc. 6th North American conference on mycorrhizae.* Edited by R. Molina. Forest Res. Lab., Oregon State Univ., Corvallis, 98–103.

Janos, D.P. 1987. VA mycorrhizas in humid tropical systems. In *Ecophysiology of VA mycorrhizal plants.* Edited by G.R. Safir. CRC Press, Boca Raton, FL, 107–134.

Janos, D.P. 1988. Mycorrhiza applications in tropical forestry: are temperate-zone approaches appropriate? In *Trees and mycorrhizae.* Edited by F.S.P. Ng. Forest Research Inst., Kuala Lumpur, Malaysia, 133–188.

Janzen, D. 1999. Gardenification of tropical conserved wildlands: multitasking, multicropping, and multiusers. *PNAS* 96:5987–5994.

Janzen, D.H. 1966. Coevolution of mutualism between ants and acacias in Central America. *Evolution* 20:249–275.

Janzen, D.H. 1970. Herbivores and the number of tree species in tropical forests. *Am. Nat.* 104:501–527.

Janzen, D.H. 1974. Tropical blackwater rivers, animals, and mast fruiting by the dipterocarpaceae. *Biotropica* 6:69–103.

Janzen, D.H. 1983. Food webs: who eats what, why, how and with what effects in tropical forests? In *Tropical rain forest ecosystems. Structure and function.* Edited by R.B. Golley. Edgeview, Amsterdam, 167–182.

Janzen, D.H. 1985. The natural history of mutualisms. In *The biology of mutualism.* Edited by D.H. Boucher. Oxford Univ. Press, New York, 40–49.

Janzen, D.H. 1988. On the broadening of insect-plant research. *Ecology* 69:905.

Jarvis, P.G. 1985. Transpiration and assimilation of tree and agricultural crops: the "omega factor." In *Attributes of trees as crop plants.* Edited by M.G.R. Cannell and J.E. Jackson. Institute of Terrestrial Ecology, Huntingdon, England, 460–480.

Jarvis, P.G. 1989. Atmospheric carbon dioxide and forests. *Phil. Trans. R. Soc. Lond.* 324:369–392.

Jarvis, P.G., and Leverenz, J.W. 1983. Productivity of temperate, deciduous and evergreen forests. *J. Plant Physiol.* 12D:234–274.

Jasper, D.A., Abbott, L.K., and Robson, AD. 1989. The loss of VA mycorrhizal infectivity during bauxite mining may limit the growth of *Acacia pulchella* R. *Aust. J. Bot.* 37:33–42.

Jasper, D.A., Abbott, L.K., and Robson, A.D. 1990. Soil disturbance and VA mycorrhizal fungi in native perennial ecosystems. In *The Eighth North American conference on mycorrhizae.* Edited by M.F. Allen and S.E. Williams. Univ. of Wyoming Agic. Exp. Sta., Laramie, 156.

Jasper, D.A., Abbott, L.K., and Robson, A.D. 1991. The effect of soil disturbance on vesicular arbuscular mycorrhizal fungi in soils from different vegetation types. *New Phytol.* 118:471–476.

Jeffries, P., Gianinazzi, S., Perroto, S., Turnau, K., and Barea, J.-M. 2003. The contribution of arbuscular mycorrhizal fungi in sustainable maintenance of plant health and soil fertility. *Biol. Fertil. Soils* 37:1–16.

Jenkinson, D.S., Harkness, D.D., Vance, E.D., Adams, D.E., and Harrison, A.F. 1992. Calculating net primary production and annual input of organic matter to soil from the amount and radiocarbon content of soil organic matter. *Soil Biol. Biochem.* 24:295–308.

Jennerjahn, T.C., Ittekkot, V., Arz, H.W., Behling, H., Patzold, J., and Wefer, G. 2004. Asynchronous terrestrial and marine signals of climate change during Heinrich events. *Science* 306:2236–2239.

Jenny, H. 1941. *Factors of soil formation.* McGraw-Hill, New York.

Jenny, H. 1980. *The soil resource. Origin and behavior.* Springer-Verlag, New York.

Jenny, H. 1985. *Meeting the expectations of the land: essays in sustainable agriculture and stewardship.*

Edited by W. Jackson, W. Berry, and B. Colman. Farrar Straus & Giroux, New York.

Jepson, P., Jarvie, J.K., MacKinnon, K., and Monk, K.A. 2001. The end of Indonesia's lowland forests? *Science* 292:859–861.

Jeschke, J.M., and Strayer, D.L. 2005. Invasion success of vertebrates in Europe and North America. *PNAS* 102:7198–7202.

Jha, D.K., Sharma, G.D., and Mishra, R.R. 1992. Ecology of soil microflora and mycorrhizal symbionts in degraded forests at two altitudes. *Biol. Fertil. Soils* 12:272–278.

Johnson, C.N. 1996. Interactions between mammals and ectomycorrhizal fungi. TREE 11:503–507.

Johnson, D.W. 1983. The effects of harvesting intensity on nutrient depletion in forests. In *IUFRO symposium on forest site and continuous productivity.* Tech. Rep. PNW-163. Edited by R. Ballard and S.P. Gessel. U.S. Department of Agriculture, Forest Service, Portland, OR, 157–166.

Johnson, D.W. 1984. Sulfur cycling in forests. *Biogeochemistry* 1:29–43.

Johnson, D.W., and Cole, D.W. 1980. Anion mobility: relevance to nutrient transport from forest ecosystems. *Environ. Int.* 3:79–90.

Johnson, D.W., and Curtis, P.S. 2001. Effects of forest management on soil C and N storage: meta analysis. *For. Ecol. Manage.* 140:227–238.

Johnson, D.W., and Todd, D.E. 1987. Nutrient export by leaching and whole-tree harvesting in a loblolly pine and mixed oak forest. *Plant Soil* 102:99–109.

Johnson, D.W., Cresser, M.S., Nilsson, S.I., Turner, J., Ulrich, B., Binkley, D., and Cole, D.W. 1991. Soil changes in forest ecosystems: evidence for and probable causes. *Proc. R. Soc. Edinburgh* 97B:81–116.

Johnson, D.W., Van Miegroet, H., and Swank, W.T. 1989. Markers of air pollution in forests: nutrient cycling. In *Biologic markers of air-pollution stress and damage in forests.* National Academy Press, Washington, DC, 133–142.

Johnson, D.W., West, D.C., Todd, D.E., and Mann, L.K. 1982. Effects of sawlog vs. whole-tree harvesting on the nitrogen, phosphorus, potassium, and calcium budgets of an upland mixed oak forest. *Soil Sci. Soc. Am. J.* 46:1304–1309.

Johnson, E.A., and Larsen, C.P.S. 1991. Climatically induced change in fire frequency in the southern Canadian Rockies. *Ecology* 72:194–201.

Johnson, E.A., Morin, H., Miyanishi, K., Gagnon, R., and Greene, D.F. 2003. A process approach to understanding disturbance and forest dynamics for sustainable forestry. In *Towards sustainable management of the boreal forest.* NRC Research Press, Ottawa, 261–306.

Johnson, K.N., Franklin, J.F., Thomas, J.W., and Gordon, J. 1991. Alternatives for management of late successional forests of the Pacific Northwest. Unpublished Report to the Agriculture Committee and the Merchant Marine and Fisheries Committee of the U.S. House of Representatives.

Johnson, R. 1998. The forest cycle and low river flows: a review of UK and international studies. *For. Ecol. Manage.* 109:1–7.

Johnson, V.S., Litvaitis, J.A., Lee, T.D., and Frey, S.D. 2006. The role of spatial and temporal scale in colonization and spread of invasive shrubs in early successional habitats. *For. Ecol. Manage.* 228, 124–134.

Johnson, W.C., and Adkinson, S. 1985. Dispersal of beech nuts blue jays in fragmented landscapes. *Am. Midl. Nat.* 113:319–324.

Johnson, W.C., and Adkisson, C.S. 1986. Air-lifting the oaks. *Nat. Hist.* 95:40–47.

Johnson, S. 2007. The road forks for tropical forests. *Tropical Forest Update* 17:1–2.

Johnston, C.A., Groffman, P., Breshears, D.D., Cardon, Z.G., Currie, W., Emanuel, W., Gaudinski, J., Jackson, R.B., Lajtha, K., Nadelhoffer, K., Nelson, D., Post, W.M., Retallack, G., and Wielopolski, L. 2004. Carbon cycling in soil. *Front. Ecol. Environ.* 2:522–528.

Johnston, D.W., and Odum, E.P. 1956. Breeding bird populations in relation to plant succession in the piedmont of Georgia. *Ecology* 37:50–62.

Jones, J.A. 1989. Environmental influences on soil chemistry in central semiarid Tanzania. *Soil Sci. Soc. Am. J.* 53:1748–1758.

Jones, J.A. 1990. Termites, soil fertility and carbon cycling in dry tropical Africa: a hypothesis. J. *Trop. Ecol.* 6:291–305.

Jones, J.A., and Grant, G.G.E. 1996. Peak flow responses to clearcutting and roads in small and large basins, western Cascades, Oregon. *Water Resourc. Res.* 32:959–974.

Jones, J.A., and Post, D.A. 2004. Seasonal and successional streamflow response to forest cutting and regrowth in the northwest and eastern United States. *Water Resourc. Res.* 40, W05203, doi:10.1029/2003WR002952:2004.

Jones, K. 1982. Nitrogen fixation in the canopy of temperate forest trees: a re-examination. *Ann. Bot.* 50:329–334.

Jones, K., and Bangs, D. 1985. Nitrogen fixation by free-living heterotrophic bacteria in an oak forest: the effect of liming. *Soil Biol. Biochem.* 17:705–709.

Jones, M.D., Durall, D.M., and Cairney, J.W.G. 2003. Ectomycorrhizal fungal communities in young forest stands regenerating after clearcut logging. *New Phytol.* 157:399–422.

Jonsson, L., Dahlberg, A., Nilsson, M.-C., Zackrisson, O., and Karen, O. 1999. Ectomycorrhizal fungal communities in late-successional Swedish boreal forests, and their composition following wildfire. *Molec. Ecol.* 8:205–215.

Jonsson L., Nilsson, M.-C., Wardle, D.A., Zackrisson, O. 2001. Context dependent effects of ectomycorrhizal species richness on tree seedling productivity. *Oikos* 93:353–364.

Jonzen, N., Linden, A., Ergon, T., Knudsen, E., Vik, J.O., Rubolini, D., Piacentini, D., Brinch, C., Spina, F., Karlsson, L., Stervander, M., Andersson, A., Waldenstrom, J., Lehikoinen, A., Edvardsen, E., Solvang, R., and Stenseth, N.C. 2006. Rapid advance of spring arrival dates in long-distance migratory birds. *Science* 312:1959–1961.

Jordan, C.F. 1982. The nutrient balance of an Amazonian rain forest. *Ecology* 63:647–654.

Jordan, C.F. 1983. Productivity of tropical rain forest ecosystems and the implications for their use as future wood and energy sources. In *Tropical rain forest ecosystems.* Edited by F.B. Galley. Elsevier, New York, 117–136.

Jordan, C.F. 1985. *Nutrient cycling in tropical forest ecosystems.* John Wiley & Sons, New York,.

Jordan, C.F., and Herrera, R. 1981. Tropical rain forests: are nutrients really critical? *Am. Nat.* 117:167–180.

Jordan, C.F., Golley, F., Hall, J.D., and Hall, J. 1979. Nutrient scavenging of rainfall by the canopy of an Amazonian rain forest. *Biotropica* 12:61–66.

Jordan, C.F., Kline, J.R., and Sasscer, D.S. 1972. Relative stability of mineral cycles in forest ecosystems. *Am. Nat.* 106:237–253.

Jordan, M.J. 1975. Effects of zinc smelter emissions and fire on a chestnut-oak woodland. *Ecology* 56:78–91.

Jordano, P. 1987. Patterns of mutualistic interactions in pollination and seed dispersal: connectance, dependence asymmetries, and coevolution. *Am. Nat.* 129:657–677.

Jordano, P., Bascompte, J., and Olesen, J.M. 2003. Invariant properties in coevolutionary networks of plant-animal interactions. *Ecology Lett.* 6:69–81.

Joshi, S.C., editor. 1984. *Rural development in the Himalaya: problems and prospects.* Gyanodara Prakashan Publishers, Naintal, India.

Joslin, J.D., Gaudinski, J.B., Torn, M.S., Riley, W.J., and Hanson, P.J. 2006. Fine-root turnover patterns and their relationship to root diameter and soil depth in a ^{14}C-labeled hardwood forest. *New Phytol.* 172:523–535.

Jouzel, J., Lorius, C., Petit, J.R., Genthon, C., Barkov, N.I., Kotlyakov, V.M., and Petrov, V.M. 1987. Vostok ice core: a continuous isotope temperature record over the last climatic cycle (160,000 years). *Nature* 329:403–408.

Joyce, L., Aber, J., McNulty, S., Dale, V., Hansen, A., Irland, L., Neilson, R., and Skog, K. 2001. Potential consequences of climate variability and change for the forests of the United States. In *Climate change impacts in the United States.* Report for the U.S. Global Change Research Program. Edited by National Assessment Synthesis Team Cambridge Univ. Press. 489–523.

Jukes, M.R., Ferris, R., and Peace, A.J. 2002. The influence of stand structure and composition on diversity of canopy Coleoptera in coniferous plantations in Britain. *For. Ecol. Manage.* 163:27–41.

Jumpponen, A., Mattson, K., Trappe, J.M., and Ohtonen, R. 1999. Effects of established willows on primary succession on Lyman Glacier forefront, North Cascade Range, Washington, USA: evidence for simultaneous canopy inhibition and soil facilitation. *Arctic Alpine Res.* 30:31–39

Jumpponen, A., Trappe, J.M., and Cazares, E. 1999. Ectomycorrhizal fungi in Lyman Lake Basin: a comparison between primary and secondary successional sites. *Mycologia* 91:575–582.

Jumpponen, A., Trappe, J.M., and Cazares, E. 2002. Occurrence of ectomycorrhizal fungi on the forefront of retreating Lyman Glacier (Washington, USA) in relation to time since deglaciation. *Mycorrhiza* 12:43–49.

Jurgensen, M.E., Larsen, M.J., Graham, R.T., and Harvey, A.E. 1987. Nitrogen fixation in woody residue of northern Rocky Mountain conifer forests. *Can. J. For. Res.* 17:1283–1288.

Kaereiva, P. 1983. Influence of vegetation texture on herbivore populations: resource concentration and herbivore movement. In *Variable plants and herbivores in natural and managed systems.* Edited by R.F. Denno and M.S. McClure, 259–290. Academic, New York.

Kaimowitz, D. 2005a. Clearing for commodities. Posted to The Center for International Forestry Research (CIFOR), POLEX Listserve. April 24, 2005. polex@cgiar.org.

Kaimowitz, D. 2005b. Playing with fire. Posted to The Center for International Forestry Research (CIFOR), POLEX Listserve. Sept. 13, 2005. polex@cgiar.org.

Kaiser, J. 2002. Satellites spy more forest than expected. *Science* 297:919.

Kallio, S., and Kallio, P. 1975. Nitrogen fixation in lichens at Kevo, North Finland. In *Fennoscandian tundra ecosystems, part I. Plants and microorganisms.* Edited by F.E. Wielgolaski. Springer, New York, 292–304.

Kanowski, J., Catterall, C.P., and Wardell-Johnson, G.W. 2005. Consequences of broadscale timber plantations for biodiversity in cleared rainforest landscapes of tropical and subtropical Australia. *For. Ecol. Manage.* 208:359–372.

Kaplan, R. 1997. End of the earth. Vintage Books, New York.

Karban, R. 1989. Fine-scale adaptation of herbivorous thrips to individual host plants. *Nature* 340:60–61.

Karban, R., and Baldwin, I.T. 1997. Induced responses to herbivory. Univ. of Chicago Press, Chicago.

Karban, R., and Myers, J.R. 1989. Induced plant responses to herbivory. *Annu. Rev. Ecol. Syst.* 20:331–348.

Kareiva, P.M., and Bertness, M.D. 1997. Re-examing the role of positive interactions in communities. *Ecology* 78:1945.

Karki, L., and Tigerstedt, P.M.A. 1985. Definition and exploitation of forest tree ideotypes in Finland. In *Attributes of trees as crop plants.* Edited by M.G.R. Cannel and J.E. Jackson.

Institute of Terrestrial Ecology, Huntingdon, England, 102–109.

Karr, J.R., Rhodes, J.J., Minshall, G.W., Hauer, F.R., Beschta, R.L., Frissell, C.A., and Perry, D.A. 2004. The effects of postfire salvage logging on aquatic ecosystems in the American West. *BioScience* 54:1029–1033.

Katznelson, H., Lochhead, A.G., and Timonin, M.I. 1948. Soil microorganisms and the rhizosphere. *Bot. Rev.* 14:543–587.

Kauffman, J.B. 1990. Ecological relationships of vegetation and fire in Pacific Northwest forests. In *Natural and prescribed fire in Pacific Northwest forests.* Edited by J. Walstad, et al. Oregon State Univ. Press, Corvallis, OR, 39–52.

Kauffman, J.B. 1991. Survival by sprouting following fire in tropical forests of the eastern Amazon. *Biotropica* 22:219–224.

Kauffman, J.B., and Uhl, C. 1990. Interactions and consequences of deforestation and fire in the rainforests of the Amazon Basin. In *Fire in the tropical and subtropical biota.* Edited by J.G. Goldhammer. Springer-Verlag, Berlin, 117–134.

Kauffman, J.B., Till, K.M., and Shea, R.W. 1992. *Biogeochemistry* of deforestation and biomass burning. *ACS Symp. Ser.* 483:428–456.

Kauffman, J.B., Uhl, C., and Cummings, D.L. 1988. Fire in the Venezuelan Amazon. 1: fuel biomass and fire chemistry in the evergreen rainforest of Venezuela. *Oikos* 53:167–175.

Kaushal, S.S., and Lewis, Jr., W.M. 2005. Fate and transport of organic nitrogen in minimally disturbed Montane streams of Colorado, USA. *Biogeochemistry* 74:303–321.

Kay, B.D. 1990. Rates of change of soil structure under different cropping systems. *Adv. Soil Sci.* 12:1–52.

Kayastha, S.L., and Juyal, B.N. 1978. Forests, environment and development in the Himalaya: *Man and Nature* 1:17–28.

Keeling, C.D. 1986. *Atmospheric CO_2* concentrations—Mauna Loa Observatory, Hawaii 1958–1986. Oak Ridge National Laboratory, Oak Ridge, TN.

Keeney, DR. 1980. Prediction of soil nitrogen availability in forest ecosystems: a literature review. *For. Sci.* 26:159–171.

Keller, M., Goreau, T.J., Wofsy, S.C., Kaplan, W.A., and McElroy, M.B. 1983. Production of nitrous oxide and consumption of methane by forest soils. *Geophys. Res. Lett.* 10:1156–1159.

Kellman, M., and Carty, A. 1986. Magnitude of nutrient influxes from atmospheric sources to a central American *Pinus caribaea* woodland. *J. Appl. Ecol.* 23:211–226.

Kellman, M., and Miyanishi, K. 1982. Forest seedling establishment in neotropical savannas: observations and experiments in the Mountain Pine Ridge savanna, Belize. *J. Biogeogr.* 9:193–206.

Kellogg, W.W. 1983. Feedback mechanisms in the climate system affecting future levels of carbon dioxide. *J. Geophys. Res.* 88:1263–1269.

Kellogg, W.W., and Schware, R. 1981. *Climate change and society.* Westview Press, Boulder, CO.

Kellomaki, S., Hanninen, H., and Kolstrom, T. 1988. Model computations on the impacts of the climatic change on the productivity and silvicultural management of the forest ecosystem. *Silva Fennica* 22:293–305.

Kelly, C.K., and Bowler, M.C. 2002. Coexistence and relative abundance in forest trees. *Nature* 417:437–440.

Kendall, K.C., and Arno, S.F. 1989. Whitebark pine—an important but endangered wildlife resource. Paper presented at the symposium on Whitebark Pine Ecosystems: Ecology and Management of a High Mountain Resource, Montana State Univ., Bozeman.

Kendrick, B. 1991. Fungal symbioses and evolutionary innovations. In *Symbiosis as a source of evolutionary innovation.* Edited by L. Margulis and R. Fester. MIT Press, Cambridge, MA, 249–261.

Kennedy, A.C. 2005. Rhizosphere. In *Principles and application of soil microbiology,* 2nd ed. Edited by D.M. Sylvia, J.L. Fuhrmann, P.G. Hartel, and D.A. Zuberer. Prentice Hall, Upper Saddle River, NJ, 242–262.

Kennedy, M., Droser, M., Mayer, L.M., Pevear, D., and Mrofka, D. 2006. Late Precambrian oxygentation; inception of the clay mineral factory. *Science* 311:1446–1449.

Kennedy, M.J., Hedin, L.O., and Derry, L.A. 2002. Decoupling of unpolluted temperate forests from rock nutrient sources revealed by natural $^{87}Sr/^{86}Sr$ tracer addition. *PNAS* 99:9639–9644.

Kepler, C.B., and Scott, J.M. 1985. Conservation of island ecosystems. In *Conservation of island birds.* Tech. Publ. No. 3. Edited by P.O. Moors. International Council of Bird Preservation, Cambridge, UK, 255–271.

Kerns, B.K. and Ohmann, J.L. 2004. Evaluation and prediction of shrub cover in coastal Oregon forests (USA). *Ecol. Indic.* 4:83–98.

Kerr, C., Nixon, C.J., and Matthews, R.W. 1992. Siviculture and yield of mixed-species stands: the UK experience. In *The ecology of mixed-species stands of trees.* Edited by M.G.R. Cannell, D.C. Malcom, and P.A. Robertson. Blackwell Scientific Publications, London, 35–52.

Kerr, J.T., Southwood, T.R.E., and Cihlar, J. 2001. Remotely sensed habitat diversity predicts butterfly species richness and community similarity in Canada. *PNAS* 98:11365–11370.

Kerr, R.A. 1990. Hurricane-drought link bodes ill for U.S. coast. *Science* 247:162.

Kessler, A., and Baldwin, I.T. 2002. Plant responses to insect herbivory: the emerging molecular analysis. *Annu. Rev. Plant Biol.* 53:299–328.

Kessler, A., Halischke, R., and Baldwin, I.T. 2004. Silencing the jasmonate cascades: induced plant defenses and insect populations. *Sciencexpress,* July 1, 2004, 1–4.

Kessler, W.B., Salwasser, H., Cartwright Jr., C.W., and Caplan, J.A. 1992. New perspectives for sustainable natural resources management. *Ecol. Applic.* 2:221–225.

Keyes, M.R., and Grier, C.C. 1981. Above- and below-ground net production in 40-year-old Douglas-fir stands on low and high productivity sites. *Can. J. For. Res.* 11:599–605.

Keystone Center. 1991. *Biological diversity on federal lands. Report of a Keystone policy dialogue.* The Keystone Center, Keystone, CO.

Khanna, P.R., and Ulrich, B. 1984. Soil characteristics influencing nutrient supply in forests. In *Nutrition of plantation forests.* Edited by G.D. Bowen and E.K.S. Nambiar. Academic Press, London, 79–117.

Kienast, F., and Luxmoore, R.J. 1988. Tree-ring analysis and conifer growth responses to increased atmospheric CO_2 levels. *Oecologia* 76:487–495.

Killsgaard, C.W., Greene, S.E., and Stafford, S.G. 1987. Nutrient concentrations in litterfall from some western conifers with special reference to calcium. *Plant Soil* 102:223–227.

Kilbertus, G. 1980. Etude des microhabitats contenus dans les agregats du sol: leur relation avec la biomass bacterienne et la taille des procaryotes presents. *Rev. Ecol. Biol. Soil* 17:543–558.

Killham, K., Firestone, M.K., and McColl, J.G. 1983. Acid rain and soil microbial activity: effects and their mechanisms. *J. Environ. Qual.* 12:133–137.

Kimble, J.M., Heath, L.S., Birdsey, R.A., and Lal, R., editors. 2003. *The potential of U.S. forest soils to sequester carbon and mitigate the greenhouse effect.* CRC Press, Boca Raton, FL,.

Kimmerer, R.W., and Lake, F.K. 2001. The role of indigenous burning in land mamanement. *J. For.* 99:36–41.

Kimmins, J.P. 1977. Evaluation of the consequences for future tree productivity of the loss of nutrients in whole-tree harvesting. *For. Ecol. Manage.* 1:169–183.

Kimmins, J.P. 2004. *Forest ecology,* 3rd ed. Prentice Hall, Upper Saddle River, NJ.

Kimmins, J.P., and Krumlik, G.J. 1976. On the question of nutrient losses accompanying whole-tree harvesting. In *Oslo biomass studies.* Univ. of Maine, Orono, 41–53.

King, E.W. 1972. Rainfall and epidemics of the southern pine beetle. *Environ. Entomol.* 1:279–285.

King, G.A., Winjum, J.K., Dixon, R.K., and Arnaut, L.Y., editors. 1990. *Response and feedbacks of forest systems to global climate change.* Research Report EPA/600/3–90/080. U.S. Environmental Protection Agency, Washington, DC.

Kira, T. 1975. Primary production of forests. In *Photosynthesis and productivity in different environments.* Edited by J.P. Cooper. Cambridge Univ. Press, Cambridge, 5–40.

Kira, T., and Kumura, A. 1983. Dry matter production and efficiency in various types of plant canopies. In *Plant research and agroforestry.* Edited by PA. Huxley. International Council for Research in Agroforestry, Nairobi, 347–364.

Kira, T., Shinozaki, K., Hozumi, K. 1969. Stucture of forest canopies as related to their primary productivity. *Plant Cell Physiol.* 10:129–142.

Kirkman, L.K., Mitchell, R.J., Helton, R.C., and Drew, M.B. 2001. Productivity and species richness across an environmental gradient in a fire-dependent ecosystem. *Am. J. Bot.* 88:2119–2128.

Kitchell, J.F., O'Neill, R.V., Webb, D., Gallepp, G.W., Bartell, S.M., Koonce, J.F., and Ausmus, B.S. 1979. Consumer regulation of nutrient cycling. *BioScience.* 29:28–34.

Kitzberger, T., Swetnam, T.W., and Veblen, T.T. 2001. Inter-hemispheric synchrony of forest fires and the El Niño–Southern Oscillation. *Global Ecol. Biogeogr.* 10:315–326.

Klarreich, E. 2004. Generous players. *Sci. News* 166:58–60.

Klamt, E., Mielniczuk, J., and Schneider, P. 1986. Degradation of properties of red Brazilian subtropical soils by management. In *Proceedings of the international symposium on red soils.* Edited by Institute of Soil Science, Academia Sinica. Elsevier, Amsterdam, 523–542.

Kleber, M., Sollins, P., and Sutton, R. 2007. A conceptual model of organo-mineral interactions in soils: self-assembly of organic molecular fragments into zonal structures on mineral surfaces. *Biogeochemistry* 85:9–24.

Klein, B.C. 1989. Effects of forest fragmentation on dung and carrion beetle communitiel in central Amazonia. *Ecology* 70:1715–1725.

Klein, D.A., Salzwedel, J.L., and Dazzo, R.B. 1990. Microbial colonization of plant roots. In *Biotechnology of plant microbe interactions.* Edited by J.P. Nakas and C. Hagedorn. McGraw-Hill, 189–225.

Klein, R.M., and Perkins, T.D. 1988. Primary and secondary causes and consequences of contemporary forest decline. *Bot. Rev.* 54:2–43.

Klemmedson, J.O. 1987. Influence of oak in pine forests of central Arizona on selected nutrients of forest floor and soil. *Soil Sci. Soc. Am. J.* 51:1623–1628.

Klinge, H. 1985. Foliar nutrient levels of native tree species from Central Amazonia. *Amazoniana* 9:281–295.

Klinge, H., Furch, K., Harms, E., and Revilla, J. 1983. Foliar nutrient levels of native tree species from Central Amazonia. *Amazonia.* 8:19–45.

Klironomos, J.N. 2002. Feedback with soil biota contributes to plant rarity and invasiveness in communities. *Nature* 417:67–70.

Klironomos, J.N., McCune, J., Hart, M., and Neville, J. 2000. The influence of arbuscular mycorrhizae on the relationship between plant diversity and productivity. *Ecol. Lett.* 3:137–141.

Knight, D.H. 1987. Parasites, lightning, and the vegetation mosaic in wilderness landscapes. In *Landscape heterogeneity and disturbance.* Edited by MG. Turner. Springer-Verlag, New York, 60–83.

Knight, D.H., Fahey, T.J., and Running, S.W. 1985. Water and nutrient outflow from contrasting lodgepole pine forests in Wyoming. *Ecol. Monogr.* 55:29–48.

Knight, T.M., McCoy, M.W., Chase, J.M., McCoy, K.A., and Holt, R.D. 2005. Trophic cascades across ecosystems. *Nature* 437:880–883.

Knops, J.M.H., and Tilman, D. 2000. Dynamics of soil nitrogen and carbon for 61 years after agricultural abandonment. *Ecology* 81:88–98.

Knutti, R., Stocker, T.F., and Plattner, G.K. 2002. Constraints on radiative forcing and future climate change from observations and climate model ensembles. *Nature* 416:719–723.

Kochy, M., and Wilson, S.D. 2001. Nitrogen deposition and forest expansion in the northern Great Plains. *J. Ecol.* 89:807–819.

Koehn, R.K., and Hilbish, T.J. 1987. The adaptive importance of genetic variation. *Am. Sci.* 75:134–141.

Koenig, W.D., and Liebhold, A.M. 2003. Regional impacts of periodical cicadas on oak radial increment. *Can. J. For. Res.* 33:1084–1089.

Koestler, A. 1969. Beyond atomism and holism—the concept of the holon. In *Beyond reductionism.* Edited by A. Koestler and J.R. Smythies. Hutchinson, London, 192–232.

Koh, L.P., Sodhi, N.S., and Brook, B.W. 2004. Co-extinctions of tropical butterflies and their host plants. *Biotropica* 36:272–274.

Kohm, K.A., and Franklin, J.F., editors. 1997. Creating a forestry for the 21st century. Island Press, Washington, DC.

Koide, R.T., and Wu, T. 2003. Ectomycorrhizas and retarded decomposition in a *Pinus resinosa* plantation. *New Phytol.* 158:401–407.

Kolb, W., and Martin, P. 1988. Influence of nitrogen on the number of N_2-fixing and total bacteria in the rhizosphere. *Soil Biol. Biochem.* 20:221–225.

Kondoh, M. 2003. Foraging adaptation and the relationship between food-web complexity and stability. *Science* 299:1388–1391.

Koricheva, J., Vehviläinen, H., Riihimäki, J., Ruohomäki, K., Kaitaniemi, P., and Ranta, H. 2006. Diversification of tree stands as a means to manage pests and diseases in boreal forests: myth or reality? *Can. J. For. Res.* 36:324–336.

Kramer, P.J. 1981. Carbon dioxide concentration photosynthesis, and dry matter production. *BioScience* 31:29–33.

Kramer, P.J. 1983. *Water relations of plants.* Academic Press, New York.

Kramer, P.J., and Kozlowski, T.T. 1979. *Physiology of woody plants.* Academic Press, New York.

Krebs, C.J. 1972. *Ecology: the experimental analysis of distribution and abundance.* Harper & Row, New York.

Krebs, C.J. 1988. *The message of ecology.* Harper & Row, New York.

Krebs, C.J., and Boonstra, R. 2001.The Kluane Region. Pp 9–24 In *Ecosystem dynamics of the boreal forest,* edited by C.J. Krebs, S. Boutin, and R. Boonstra. Oxford Univ. Press, Oxford.

Krebs, C.J., Boonstra, R., Boutin, S., and Sinclair, A.R.E. 2001c. Conclusions and future directions. Chap. 20 in *Ecosystem dynamics of the boreal forest.* Edited by C.J. Krebs, S. Boutin, and R. Boonstra. Oxford Univ. Press, Oxford, 490–501.

Krebs, C.J., Boutin, S., and Boonstra, R., editors. 2001b. *Ecosystem dynamics of the boreal forest.* Oxford Univ. Press, Oxford.

Krebs, C.J., Dale, M.R.T., Nams, V.O., Sinclair, A.R.E., and O'Donoghue, M. 2001a. Shrubs. Chap 6 in *Ecosystem dynamics of the boreal forest.* Edited by C.J. Krebs, S. Boutin, and R. Boonstra. Oxford Univ. Press, Oxford, 92–115.

Kreft, H., and Jetz, W. 2007. Global patterns and determinants of vascular plant diversity. *PNAS* 104:5925–5930.

Kremen, C. 1992. Assessing the indicator properties of species assemblages for natural area monitoring. *Ecol. Applic.* 2:203–217.

Kremsater, L., and Bunnell, F.L. 1999. Edge effects: theory, evidence and implications to management of western North American forests. In *Forest fragmentation: wildlife and management implications.* Edited by J.A. Rochelle, L.A. Lehmann, and J. Wisniewski. Brill, Leiden, The Netherlands, 117–153.

Kroll, J.C., and Fleet, R.R. 1979. Impact of woodpecker predation on overwintering within-tree populations of southern pine beetle (*Dendroctunus frontalis*). In *The role of insectivorous birds in forest ecosystems.* Edited by J.G. Dickson. Academic Press, New York, 269–281.

Krumbein, W.E., editor. 1978. *Environmental biogeochemistry and geomicrobiology.* Ann Arbor *Science*, Ann Arbor, MI.

Krupa, S.V. 2003. Effects of atmospheric ammonia (NH_3) on terrestrial vegetation: a review. *Environ. Pollut.* 124:179–221.

Kucharik, C.J., Foley, J.A., Delire, C., Fisher, V.A., Coe, M.T., Lenters, J.D., Young-Molling, C., Ramankutting, N., Norman, J.M., and Gower, S.T. 2000. Testing the performance of a dynamic global ecosystem model: water balance, carbon balance, and vegetation structure. *Global Biogeocehm. Cycles* 14:795–825.

Kuerer, G., and Atwood, C.E. 1973. Diprinoid sawflies: polymorphism and speciation. *Science* 179:1090–1099.

Kulman, H.M. 1971. Effects of insect defoliation on growth and mortality of trees. *Annu. Rev. Entomol.* 16:289–324.

Kumar, K.K., Rajagopalan, B., Hoerling, M., Bates, G., and Cane, M. 2006. Unraveling the mystery of Indian monsoon failure during El Niño. *Science* 314:115–119.

Kung, E.W. 1972. Rainfall and epidemics of the southern pine beetle. *Environ. Entomol.* 1:279–285.

Kunz, W.A., and Kimmins, J.P. 1987. Analysis of some sources of error in methods used to determine fine root production in forest ecosystems: a simulation approach. *Can. J. For. Res.* 17:909–912.

Kupfer, J., Malanson, G.P, and Franklin, S.B. 2004. Identifying the biodiversity research needs related to forest fragmentation. A report prepared for the National Commission on Science for Sustainable Forestry. www.ncseonline.org/ewebeditpro/items/O62F3754.pdf.

Kurtz, W.A., and Kimmins, J.P. 1987. Analysis of some sources of error in methods used to determine fine root production in forest ecostystems: a simulation approach. *Can. J. For. Res.* 17:909–912.

Kytoviita, M.-M., Vestberg, M., and Tuomi, J. 2003. A test of mutual aid in common mycorrhizal networks: established vegetation negates benefit in seedlings. *Ecology* 84:898–906.

Lafferty, K.D., Dobson, A.P., and Kuris, A.M. 2006. Parasites dominate food web links. *PNAS* 103:11211–11216.

Lafond, A., and Laflamme, Y. 1968. Relative concentrations of iron and manganese: a factor affecting jack pine regeneration and jack pine-black spruce succession. In *Tree growth and forest soils.* Edited by C.T. Youngberg. Oregon State Univ. Press, Corvallis, OR, 305–312.

Laiho, R., and Prescott, C.E. 2004. Decay and nutrient dynamics of coarse woody debris in northern coniferous forests: a synthesis. *Can. J. For. Res.* 34:763–777.

Lajtha, K., Seely, B., and Valiela, I. 1995. Retention and leaching losses of atmospherically derived nitrogen in the aggrading coastal watershed of Waquoit Bay, MA. *Biogeochemistry* 28:33–54.

Lal, R. 1984. Soil erosion from tropical arable lands and its control. *Adv. Agron.* 37:183–248.

LaMarche Jr., V.C., Graybil, D.A., Fritts, H.C., and Rose, M.R. 1984. Increasing atmospheric carbon dioxide: tree ring evidence for growth enhancement in natural vegetation. *Science* 225:1019–1021.

Lamb, D. 1980. Soil nitrogen mineralisation in a secondary rainforest succession. *Oecologia (Berlin)* 47:257–263.

Lambers, J.H.R., Clark, J.S., and Beckage, B. 2002. Density-dependent mortality and the latitudinal gradient in species diversity. *Nature* 417:732–735.

Lambeck, R.J. 1997. Focal species: a multi-species umbrella for nature conservation. *Conserv. Biol.* 11:849–856.

Lamoreux, J.F., Morrison, J.C., Ricketts, T.H., Olson, D.M., Dinerstein, E., McKnight, M.W., and Shugart, H.H. 2006. Global tests of biodievsrity concordance and the importance of endemism. *Nature* 440:212–214.

Lande, R. 1988a. Demographic models of the northern spotted owl (*Strix occidentalis caurina*). *Oecologia (Berlin)* 75:601–607.

Lande, R. 1988b. Genetics and demography in biological conservation. *Science* 241:1455–1460.

Lande, R., Engen, S., and Saether, B.-E. 2003. *Stochastic Population Dynamics.in Ecology and Conservation.* Oxford Univ. Press.

Lande, R. and Shannon, S. 1996. The role of genetic variation in adaptation and population persistence in a changing environment. *Evolution* 50:434–437.

Landres, P.B., and MacMahon, J.A. 1983. Community organization of arboreal birds in some oak woodlands of western North America. *Ecol. Monogr.* 53:183–208.

Landres, P.B., Morgan, P., and Swanson, F.J. 1999. Overview of the use of natural variability concepts in managing ecological systems. *Ecol. Applic.* 9:1179–1188.

Landres, P.B., Verner, J., and Thomas, J.W. 1988. Ecological uses of vertebrate indicator species: a critique. *Conserv. Biol.* 2:316–329.

Lanfranco, L., Delpero, M., and Bonfante, P. 1999. Intrasporal variability of ribosomal sequences in the endomycorrhizal fungus *Gigaspora margarita. Molec. Ecol.* 8:37–45.

Lang, E., and Jagnow, G. 1986. Fungi of a forest soil nitrifying at low pH values. *FEMS Microbiol. Ecol.* 38:257–265.

Langford, A.O., and Fehsenfeld, F.C. 1992. Natural vegetation as a source or sink for atmospheric ammonia: a case study. *Science* 255:581–583.

Langley J.A., Chapman, S.K., and Hungate, B.A. 2006. Ectomycorrhizal colonization slows root decomposition: the postmortem fungal legacy. *Ecol. Lett.* 9:955–959.

Lanner, R.M. 1985. Some attributes of nut-bearing trees of temperate forest origin. In *Attributes of tress as crop plants.* Edited by M.G.R. Cannel and J.E. Jackson. Institute of Terrestrial Ecology, Huntingdon, England, 426–437.

Larcher, W. 2003. *Physiological plant ecology,* 4th edition, Springer-Verlag, New York.

Larsen, J.A. 1983. *The boreal ecosystem.* Academic Press, New York.

Larsson, S., and Tenow, O. 1980. Needle eating insects and grazing dynamics in a mature Scots pine forest in central Sweden. In *Structure and function of northern conferous forests—an ecosystem study.* Edited by T. Persson. *Ecol. Bull. (Stockholm)* 32:269–306.

Lashof, D.A. 1989. The dynamic greenhouse: feedback processes that may influence future concentrations of atmospheric trace gases and climatic change. *Climat. Change* 14:213–242.

Lashof, D.A., and Ahuja, D.R. 1990. Relative contributions of greenhouse gas emissions to global warming. *Nature* 344:529–531.

Laszlo, P. 1987. Chemical reactions on clays. *Science* 235:234–236.

Latif, M., and Barnett, T.P. 1994. Causes of decadal climate variability over the North Pacific and North America. *Science* 266:634–637.

Laurance, W.F., and Fearnside, P.M. 1999. Amazon burning. *Tree* 14:457.

Laurance, W.F., Cochrane, M.A., Bergen, S., Fearnside, P.M., Delamonica, P., Barber, C., D'Angelo, S., and Fernandes, T. 2001. The future of the Brazilian Amazon. *Science* 291:438–439.

Laurance, W.F., Lovejoy, T.E., Vasconcelos, H.L., Bruna, E.M., Didham, R.K., Stouffer, P.C., Gascon, C., Bierregaard, R.O., Laurance, S.G., and Sampaio, E. 2002. Ecosystem decay of Amazonian forest fragments: a 22-year investigation. *Conserv. Biol.* 16:605–618.

Lavelle, P. 1987. Biological processes and productivity of soils in the humid tropics. In *Geophysiology of Amazonia.* Edited by R.E. Dickinson. Wiley, New York, 175–223.

LaVelle, P., Lattaud, C., Trigo, D. and Barois, I. 1995. Mutualism and biodiversity in soils. In *The significance and regulation of soil biodiversity.* Edited by H.P. Collins, G.P. Robertson, and M.J. Klug. Kluwer Academic. 23–33.

Lavender, D.P. 1984. Greenhouse effect. *J. For.* 82:245.

Lavrinenko, D.D. 1972. *Interaction of wood species in different types of forests.* English translation by Indian National Scientific Documentation Center. U.S. Department of Commerce, Springfield, VA.

Law, R., and Blackford, J.C. 1992. Self-assembling food webs: a global viewpoint of coexistence of species in Lotka-Volterra communities. *Ecology* 73:567–578.

Lawrence, R., Potts, B.M., and Whitham, T.G. 2003. Relative importance of plant ontogeny, host genetic variation, and leaf age for a common herbivore. *Ecology* 84:1171–1178.

Lawton, J. 2001. Earth system science. *Science* 292:1965.

Lawton, J.H. 1989. Food webs. In *Ecological concepts.* Edited by J.M. Cherrett. Blackwell, Oxford, 43–78.

Lawton, J.H. 1999. Are there general laws in ecology? *Oikos* 84:177–192.

Lawton, J.H., and Brown, V.K. 1992. Redundancy in ecosystems. In *Biodiversity and ecosystem function.* Edited by E.-D. Shulze and H.A. Mooney. Springer-Verlag, New York, 255–270.

Lawton, J.H., and Strong Jr., D.R. 1981 Community patterns and competition in folivorous insects. *Am. Nat.* 118:317–338.

Lawton, R.O. 1990. Canopy gaps and light penetration into a wind-exposed tropical lower montane rain forest. *Can. J. For. Res.* 20:659–667.

Lawton, R.O., Nair, U.S., Pielke Sr., R.A., and Welch, R.M. 2001. Climatic impact of tropical lowland deforestation on nearby montane cloud forests. *Science* 294:584–587.

Le Corff, J., Marquis, R.J., and Whitfield, J.B. 2000. Temporal and spatial variation in a parasitoid community associated with the herbivores that feed on Missouri *Quercus. Environ. Entomol.* 29:181–194.

Le Quere, C., and Metzi, N. 2004. Natural processes regulating the ocean uptake of CO_2. In *The global carbon cycle, SCOPE 62.* Edited by C.B. Field and M.R. Raupach. Island Press, Washington, DC, 243–255.

Leaf, A., editor. 1979. *Impact of harvesting on forest nutrient cycling.* State Univ. of New York, Syracuse.

Leake, J., Johnson, D., Donnely, D., Muckle, G., Boddy, L., and Read, D. 2004. Networks of power and influence: the role of mycorrhizal mycelium in controlling plant communities and agroecosystem function. *Can. J. Bot.* 82:1016–1045.

Lean, J., and Warrilow, D.A. 1989. Simulation of the regional climatic impact of Amazon deforestation. *Nature* 342:411–413.

Ledig, F.T. 1986. Heterozygosity, heterosis, and fitness in outbreeding plants. In *Conservation biology: the science of scarcity and diversity.* Edited by M.E. Soule. Sinauer Associates, Sunderland, MA, 77–104.

Ledig, F.T. 1988. The conservation of diversity in forest trees. *BioScience* 38:471–479.

Ledig, F.T., Guries, R.P., and Bonefeld, B.A. 1983. The relation of growth to heterozygosity in pitch pine. *Evolution* 37:1227–1238.

Lee, J.-E., Oliveira, R.S., Dawson, T.E., and Fung, I. 2005. Root functioning modifies seasonal climate. *PNAS* 102:17576–17581.

Lee, K.G., and Wood, T.G. 1971. *Termites and soils.* Academic Press, New York.

Lefsky, M.A., Cohen, W.B., Acker, S.A., Parker, G.G., Spies, T.A., and Harding, D.J. 1999. Lidar remote sensing of the canopy structure and biophysical properties of Douglas-fir western hemlock forests. *Remote Sensing Environ.* 76:339–361.

Lefsky, M.A., Cohen, W.B., Parker, G.G., and Harding, D.J. 2002. Lidar remote sensing for ecosystem studies. *BioScience* 52:19–30.

Lehmann, R.G., Cheng, H.H., and Harsh, J.B. 1987. Oxidation of phenolic acids by soil iron and manganese oxides. *Soil Sci. Soc. Am. J.* 51:352–356.

Leibold, M.A. 1989. Resource edibility and the effects of predators and productivity on the outcome of trophic interactions. *Am. Nat.* 134:922–949.

Leibold, M.A., Holyoak, M., Mouquet, N., Amarasekare, P., Chase, J.M., Hoopes, M.F., Holt, R.D., Shurin, J.B., Law, R., Tilman, D., Loreau, M., and Gonzalez, A. 2004. The metacommunity concept: a framework for multi-scale community ecology. *Ecology Lett.* 7:601–613.

Leigh Jr., E.G., and Ziegler, C. 2002. *A magic web: the tropical forest of Barro Colorado Island.* Oxford Univ. Press.

Leighton, M., and Leighton, D.R. 1983. Vertebrae responses to fruiting seasonality within a Bornean rain forest. In *Tropical rain forest: ecology and management.* Edited by S.L. Sutton, T.C. Whitmore, and A.C. Chadwick. Blackwell, Oxford, 181–196.

Lens, L., Van Dongen, S., Norris, K., Githru, M., and Matthysen, E. 2002. Avian persistence in fragmented rainforest. *Science* 290:1236–1238.

Lensing, J.R., and Wise, D.M. 2006. Predicted climate change alters the indirect effect of predators on an ecosystem process. *PNAS* 103:15502–15505.

Leopold, A. 1943. Deer irruptions. *Wisconsin Conserv. Bull.* August. Reprinted in *Wisc. Conserv. Dept. Publ.* 321:3–11.

Leopold, A. 1966. *A Sand County almanac.* Oxford Univ. Press, Oxford.

Leopold, A.C., and Ardrey, R. 1972. Toxic substances in plants and the food habits of early man. *Science* 176:512–513.

Lepers, E., Lambin, E.F., Janetos, A.C., DeFries, R., Achard, F., Ramankutty, N., and Scholes, R.J. 2005. A sysnthesis of information on rapid land-cover change for the period 1981–2000. *BioScience* 55:115–124.

Lertzman, K.P. 1989. *Gap-phase community dynamics in a subalpine old growth forest.* Ph.D. dissertation. Univ. of British Columbia, Vancouver.

Leslie, A. 2005. What will we want from the forest? *Trop. For. Update* 15:14–16.

Leverenz, J.W., and Lev, D.J. 1987. Effects of carbon dioxide-induced climate changes on the natural ranges of six major commercial tree species in the western United States. In *The greenhouse effect, climate change, and U.S. forests.* Edited by W.E. Shands and J.S. Hoffman. The Conservation Foundation, Washington, DC, 123–155.

Levey, D.J., Bolker, B.M., Tewksbury, J.J., Sargent, S., and Haddad, N.M. 2005. Effects of landscape corridors on seed dispersal by birds. *Science* 309:146–148.

Levin, D.A. 1976. Alkaloid-bearing plants: an ecogeographic perspective. *Am. Nat.* 110:261–284.

Levin, D.A., and York Jr., B.M. 1978. The toxicity of plant alkaloids: an ecogeographic perspective. *Biochem. Syst. Ecol.* 6:61–76.

Levins, R., and Lewontin, R. 1985. *The dialectical biologist.* Harvard Univ. Press, Cambridge, MA.

Lewin, R. 1986. In ecology, change brings stability. *Science* 234:1071–1073.

Lewinsohn, T.M., Prado, P.I., Jordano, P., Bascompte, J., and Olesen, J.M. 2006. Structure in plant-animals interaction assemblages. *Oikos* 113:174–186.

Lewis, D.H. 1985. Symbiosis and mutualisms: crisp concepts and soggy semantics. *The biology of mutualism.* Edited by D.H. Boucher. Oxford Univ. Press, New York, 29–39.

Lewis, D.H. 1991. Mutualistic symbioses in the origin and evolution of land plants. In *Symbiosis as a source of evolutionary innovation. Speciation and morphogenesis.* Edited by L. Margulis and R. Fester. The MIT Press, Cambridge, MA, 288–300.

Lewis, G.P., and Likens, G.E. 2000. Low stream nitrate concentrations associated with oak forests on the Allegheny High Plateau of Pennsylvania. *Water Resourc. Res.* 36:3091–3094.

Lewis Jr., W.M. 1986. Nitrogen and phosphorus runoff losses from a nutrient-poor tropical moist forest. *Ecology* 67:1275–1282.

Lewis Jr., W.M., Melack, J.M., McDowell, W.H., McClain, M., and Richey, J.E. 1999. Nitrogen yields from undisturbed watersheds in the Americas. *Biogeochemistry* 46:149–162.

Lewontin, R. 2000. *The triple helix.* Harvard Univ Press, Cambridge, MA.

Li, C.Y., and Castellano, M.A. 1987. *Azospirillum* isolated from within sporocarps of the mycorrhizal fungi *Hebeloma crustuliniforme, Laccaria laccata* and *Rhizopogon vinicolor. Trans. Br. Mycol. Soc.* 88:563–566.

Li, C.Y., and Hung, L.L. 1987. Nitrogen-fixing (acetylene-reducing) bacteria associated with ectomycorrhizae of Douglas-fir. *Plant Soil* 98:425–428.

Li, C.Y., Maser, C., and Fay, H. 1986. Initial survey of acetylene reduction and selected microorganisms in the feces of 19 species of mammals. *Great Basin Nat.* 46:646–650.

Li, P., and Adams, W.T. 1989. Range-wide patterns of allozyme variation in Douglas-fir (*Pseudotsuga menziesii*). *Can. J. For. Res.* 19:149–161.

Li, Q., Allen, H.L., and Wollum II, A.G. 2004. Microbial biomass and bacterial functional diversity in forest soils: effects of organic matter removal, compaction, and vegetation control. *Soil. Biol. Biochem.* 36:571–579.

Lichstein, J. W., Simons, T.R., and Franzreb, K.E. 2002. Landscape effects on breeding songbird abundance in managed forests. *Ecol. Applic.* 12:836–857.

Lieth, H., and Whittaker, R.H., editors. 1975. *Ecological studies, 14, primary productivity of the biosphere.* Springer-Verlag, New York.

Lieutier, F., and Berryman, A.A. 1988. Elicitation of defensive reactions in conifers. In *Mechanisms of woody plant defenses against insects. Search for pattern.* Edited by W.J. Matson, J. Levieux, and C. Bernard-Dagan. Springer-Verlag, New York, 314–319.

Likens, G.E. 1983. A priority for ecological research, *Bull. Ecol. Soc. Am.* 64:234–243.

Likens, G.E. 1984. Beyond the shoreline: a watershed-ecosystem approach. *Verh. Int. Verein Limnol.* 22:1–22.

Likens, G.E., Bormann, F.H., Pierce, R.S., Eaton, J.S., and Johnston, N.M. 1977. *Biogeochemistry of a forested ecosystem.* Springer-Verlag, New York.

Likens, G.E., Bormann, F.H., Pierce, R.S., and Reiners, W.A. 1978. Recovery of a deforested ecosystem. *Science* 199:492–496.

Likens, G.E., Driscoll, C.T., and Buso, D.C. 1996. Long-term effects of acid rain: response and recovery of a forest ecosystem. *Science* 272:244–246.

Lill, J.T., Marquis, R.J., and Ricklefs, R.E. 2002. Host plants influence parasitism of forest caterpillars. *Nature* 417:170–173.

Lilleskov E.A., Bruns, T.D., Horton, T.R., Taylor, D.L., and Grogan, P. 2004. Detection of forest stand-level spatial structure in ectomycorrhizal fungal communities. *FEMS Microbiol. Ecol.* 49:319–332.

Lim, K., Treitz, P., Wulder, M., St.-Onge, B., and Flood, M. 2003. LiDAR remote sensing of forest structure. *Progr. Phys. Geogr.* 27:88–106.

Lindberg, S.E., Lovett, G.M., Richter, D.D., and Johnson, D.W. 1986. Atmospheric deposition and canopy interactions of major ions in a forest. *Science* 231:141–145.

Lindbladh, M., Niklasson, M., and Nilsson, S.G. 2003. Long-time record of fire and open canopy in a high biodiversity forest in southeast Sweden. *Biol. Conserv.* 114:231–243.

Lindenmayer, D.B. 1999. Future directions for biodiversity conservation in managed forests: Indicator species, impact studies, and monitoring programs. *For. Ecol. Manage.* 115:277–287.

Lindenmayer, D., and Beaton, E. 2001. *Life in the tall eucalypt forest.* New Holland, London.

Lindenmayer, D.B., and Franklin, J.F. 2002. *Conserving forest biodiversity.* Island Press, Washington, DC.

Lindenmayer, D.B., Cunningham, R.B., Donnelly, C.F., and Franklin, J.F. 2000. Structural features of old-growth Australian montane ash forests. *For. Ecol. Manage.* 134:189–204.

Lindenmayer, D.B., Cunningham, R.B., Donnelly, C.F., Tanton, M.T., and Nix, H.A. 1993. The abundance and development of cavities in Eucalyptus trees: a case study in the montane forests of Victoria, southeastern Australia. *For. Ecol. Manage.* 60:77–104.

Lindenmayer, D.B., Foster, D.R., Franklin, J.F., Hunter, M.L., Noss, R.F., Schmiegelow, F.A., and Perry, D. 2004. Ecology: enhanced: salvage harvesting policies after natural disturbance. *Science* 303:1303.

Linder, S., Benson, M.L., Myers, B.J., and Raison, R.J. 1987. Canopy dynamics and growth of *Pinus radiata*. I. Effects of irrigation and fertilization during a drought. *Can. J. For. Res.* 17:1157–1165.

Linderman, R.G. 1989. Organic amendments and soil-borne diseases. *Can. J. Plant Pathol.* 11:180–183.

Lindh, B.C., and Muir, P.S. 2004. Understory vegetation in young Douglas-fir forests: does thinning help restore old-growrh composition? *For. Ecol. Manage.* 192:285–296.

Little, S.M., and Ohmann, I.L. 1988. Estimating nitrogen loss from forest floor during prescribed fires in Douglas-fir/ western hemlock clearcuts. *For. Sci.* 34:152–164.

Litton, C.M., Raich, J.W., and Ryan, M.G. 2007. Carbon allocation in forest ecosystems. *Global Change Biol.* 13:2089–2109.

Liu, J., and Taylor, W.W., editor. 2002. *Integrating landscape ecology into natural resource management*. Cambridge Univ. Press, Cambridge.

Lo, K.F.A. 1989. Erosion-productivity interrelationships for tropical soils. In *Soil Fertil. Taiwan*. 19–30.

LoBuglio, K.F., and Wilcox, H.E. 1988. Growth and survival of ectomycorrhizal and ectendomycorrhizal seedlings of *Pinus resinosa* on iron tailings. *Can. J. Bot.* 66:55–60.

Lodge, D.J., Fisher, P.J., and Sutton, B.C. 1996. Endophytic fungi of *Manilkara bidentata* leaves in Puerto Rico. *Mycologia* 88:733–738.

Loehle, C. 2004. Applying landscape principles to fire hazard reduction. *For. Ecol. Manage.* 198:261–267.

Loehle, C., Van Deusen, P.B, Wigley, T.B., Mitchell, M.S., Rutzmoser, S.H., Aggett, J., Beebe, J.A. and Smith, M.L. 2006. A method for landscape analysis of forestry guidelines using bird habitat models and the Habplan harvest scheduler. *For. Ecol. Manage.* 232:55–67.

Loehle, C., Bently Wigley, T., Rutzmoser, S., Gerwin, J.A., Keyser, P.D., Lancia, R.A., Reynolds, C.J., Thill, R.E., Weih, R., White Jr., D., and Bohall Wood, P. 2005a. Managed forest landscape structure and avian species richness in the southeastern US. *For. Ecol. Manage.* 214:279–293.

Loehle, C., Wigley, T.B., Shipman, P.A., Fox, S.F., Rutzmoser, S., Thill, R.E., and Melchiors, M.A. 2005b. Herpetofaunal species richness responses to forest landscape structure in Arkansas. *For. Ecol. Manage.* 209:293–308.

Logan, J.A., Regniere, J., and Powell, J.A. 2003. Assessing the impacts of global warming on forest pest dynamics. *Front. Ecol. Environ.* 1:130–137.

LoGiudice, K. 2006. Toward a synthetic view of extinction: a history lesson from a North American rodent. *BioScience* 56:687–693.

Lohm, U., Larsson, K., and Nommik, H. 1984. Acidification and liming of confierous forest soil: long-term effects on turnover rates of carbon and nitrogen during an incubation experiment. *Soil Biol. Biochem.* 16:343–346.

Lõhmus, P. 2003. Composition and substrata of forest lichens in Estonia: a meta-analysis. *Folia Cryptog. Estonica Fasc.* 40:19–38.

Lonsdale, W.M. 1990. The self-thinning rule: dead or alive? *Ecology* 71:1373–1388.

Loomis, W.E. 1932. Growth-differentiation balance vs. carbohydrate-nitrogen ratio. *Proc. Am. Soc. Hort. Sci.* 29:240–245.

Lord, C., editor. 1968. *Biochemists handbook*. Blackwell, London.

Loreau, M., Downing, A., Emmerson, M., Gonzalez, A., Hughes, J., Incausti, P., Joshi, J., Norberg, J., and Sala, O. 2002. A new look at the relationship between diversity and stability. In *Bidiversity and ecosystem functioning: synthesis and perspectives*. Edited by Loreau, M., Naeem, S., and Inchausti, P. Oxford Univ. Press, 79–91.

Loreau, M., Mouquet, N., and Holt, R.D. 2003. Meta-ecosystems: a theoretical framework for a spatial ecosystem ecology. *Ecol. Lett.* 6:673–679.

Loreau, M., Mouquet, N., and Holt, R.D. 2005. From metacommunities to metaecosystems. In *Metacommunities: spatial dynamics and ecological communities*. Edited by M. Holyoak, M.A. Leibold, and R.D. Holt. Univ. of Chicago Press, Chicago, 418–438.

Lorimer, C.G. 1989. Relative effects of small and large disturbances on temperate hardwood forest structure. *Ecology* 70:565–567.

Lorimer, C.G. 2001. Historical and ecological roles of disturbance in eastern North American forests: 9,000 years of change. *Wildl. Soc. Bull.* 29:425–439.

Lorimer, C.G., and White, A.S. 2003. Scale and frequency of natural disturbances in the northeastern US: implications for early successional forest habitats and regional age distributions. *For. Ecol. Manage.* 185:41–64.

Lotan, J.E., and Perry, D.A. 1983. *Ecology and regeneration of lodgepole pine*. Agriculture Handbook No. 606. U.S. Department of Agriculture, Forest Service, Washington, DC.

Lousier, J.D. 1984. Population dynamics and production studies of species of Euglyphidae (testacea, rhizopoda, protoza) in an aspen woodland soil. *Pedobiologia* 26:309–330.

Lousier, J.D. 1985. Population dynamics and production studies of species of centropyxidae (testacea, rhizopida) in an aspen woodland soil. *Arch. Protistenk.* 130:165–178.

Lousier, J.D., and Parkinson, D. 1976. Litter decomposition in a cool temperate deciduous forest. *Can. J. Bot.* 54:419–436.

Lousier, J.D., and Parkinson, D. 1984. Annual population dynamics and production ecology of testacea (protoza, rhizopoda) in an aspen woodland soil. *Soil Biol. Biochem.* 16:103–114.

Loveless, M.D., and Hamrick, J.L. 1984. Ecological determinants of genetic structure in plant populations. *Annu. Rev. Ecol. Syst.* 15:65–95.

Lovelock, J. 1988. *The ages of Gaia*. W.W. Norton, New York.

Lovelock, J. 2004. Reflections on Gaia. In *Scientists debate Gaia*. Edited by S.H. Schnieder, J.R. Miller, E. Crist, and P.J. Boston. MIT Press, Cambridge, MA, 1–5.

Lovelock, J.E. 1979. *Gaia: a new look at life on Earth*. Oxford Univ. Press, Oxford.

Lovett, G.M. 1994. Atmospheric deposition of nutrients and pollutants in North America: an ecological perspective. *Ecol. Applic.* 4:629–650.

Lovett, G.M., and Mitchell, M.J. 2004. Sugar maple and nitrogen cycling in the forests of eastern North America. *Front. Ecol. Environ.* 2:81–88.

Lovett, G.M., Canham, C.D., Arthur, M.A., Weathers, K.C., and Fitzhugh, R.D. 2006. Forest ecosystem reponses to exotic pests and pathogens in eastern North America. *BioScience* 56:395–405.

Lovett, G.M., Christenson, L.M., Groffman, P.M., Jones, C.G., Hart, J.E., and Mitchell, M.J. 2002. Insect defoliation and nitrogen cycling in forests. *BioScience* 52:335–341.

Lovett, G.M., Reiners, W.A., and Olson, R.K. 1982. Cloud droplet deposition in subalpine balsam fir forests: hydrological and chemical inputs. *Science* 218:1303–1304.

Lovett, G.M., Weathers, K.C., and Arthur, M.A. 2002. Control of nitrogen loss from forested watersheds by soil carbon:nitrogen ratio and tree species composition. *Ecosystems* 5:712–718.

Lovett, G.M., Weathers, K.C., and Sobczak, W.V. 2000. Nitrogen saturation and retention in forested watersheds of the Catskill Mountains, New York. *Ecol. Applic.* 10:73–84.

Low, A.J. 1955. Improvements in the structural state of soils under leys. *Soil Sci.* 6:179–199.

Lowdermilk, W.C. 1953. *Conquest of the land through 7000 years.* USDA Soil Cons. Serv. Ag. Info. Bull. No. 99. U.S. Department of Agriculture, Washington, DC.

Lowman, M.D., and Heatwole, H. 1992. Spatial and temporal variability in defoliators of Australian eucalypts. *Ecology* 73:129–142.

Lugo, A.E. 2000. Effects and outcomes of Caribbean hurricanes in a climate change scenario. *Sci. Total Environ.* 262:243–251.

Lugo, A.E. 2004. The outcome of alien tree invasions in Puerto Rico. *Front. Ecol. Environ.* 2:265–273.

Lugo, A.E., and Scatena, F.N. 1996. Background and catastrophic tree mortality in tropical moist, wet, and rain forests. *Biotropica* 28:585–599.

Lumsden, R.D., Garcia, E.R., Lewis, J.A., and Frias, T.G.A. 1987. Suppression of damping-off caused by *Pythium* spp. in soil from the indigenous Mexican chinampa agricultural system. *Soil Biol. Biochem.* 19:501–508.

Lundgren, B. 1978. *Soil conditions and nutrient cycling under natural and plantation forests in Tanzanian highlands.* Reports in Forest Ecology and Forest Soils 31. Swedish Univ. of Agricultural Science, Uppsala.

Lutz, H.J., and Chandler Jr., R.F. 1946. *Forest soils.* John Wiley & Sons, New York.

Lyle Jr., E.S. 1969. Mineral deficiency symptoms in loblolly pine seedlings. *Agron. J.* 61:395–398.

Lyle Jr., E.S., and Adams, F. 1971. Effect of available soil calcium on taproot elongation of loblolly pine *(Pinus taeda)* seedlings. *Soil Sci. Soc. Am. Proc.* 35:800–805.

Lynch, J.F. 1991. Effects of Hurricane Gilbert on birds in a dry tropical forest in the Yucatan Peninsula. *Biotropica* 23:488–496.

Lynch, J.M. 1983. *Soil biotechnology: microbial factors in crop production.* Blackwell Scientific Publications, Oxford.

Lynch, J.M., and Bragg, E. 1985. Microorganisms and soil aggregate stability. *Adv. Soil Sci.* 2:133–171.

Macadam, A.M. 1987. Effects of broadcast slash burning on fuels and soil chemical properties in the sub-boreal spruce zone of central British Columbia. *Can. J. For. Res.* 17:1577–1584.

MacArthur, R.H. 1958. Population ecology of some warblers of northeastern coniferous forests. *Ecology* 39:599–619.

MacArthur, R.H. 1964. Environmental factors affecting bird species diversity. *Am. Nat.* 98:387–397.

MacArthur, R.H. 1972. *Geographical ecology.* Harper & Row, New York. p. 269.

MacArthur, R.H., and MacArthur, J.W. 1961. On bird species diversity. *Ecology* 41:594–598.

MacArthur, R.H., and Wilson, E.O. 1963. An equilibrium theory of insular zoogeography. *Evolution* 17:373–387.

MacArthur, R.H., and Wilson, E.O. 1967. *The theory of island biogeography.* Princeton Univ. Press, Princeton, NJ.

MacFaden, S.W., and Capen, D.E. 2002. Avian habitat relationships ate multiple scales in a New England forest. *For. Sci.* 48:243–253.

Mack, M.C., D'Antonio, C.M., and Ley, R.E. 2001. Alteration of ecosystem nitrogen dynamics by exotic plants: a case study of C4 grasses in Hawaii. *Ecol. Applic.* 11:1323–1335.

Mackasey, P.A. 1992. Sustainable resource development plan framework for the Neskonlith Indian Band, British Columbia. Masters thesis. Oregon State Univ., Corvallis.

MacMahon, J.A. 1987. Disturbed lands and ecological theory: an essay about a mutualistic association. In *Restoration ecology.* Edited by W.R. Jordan, M.E. Gilpin, and J.D. Aber. Cambridge Univ. Press, Cambridge, 221–237.

MacPhee, D. 1993. Directed evolution reconsidered. *Am. Sci.* 81:554–561.

Madgwick, H.A.I., Jackson, D.S., and Knight, P.J. 1977. Aboveground dry matter, energy, and nutrient contents of trees in an age series of *Pinus radiata* plantations. *N. Z. J. For. Sci.* 7:445–468.

Magill, A.H., Aber, J.D., Currie, W.S., Nadelhoffer, K.J., Martin, M.E., McDowell, W.H., Melillo, J.M., and Steudler, P. 2004. Ecosystem response to 15 years of chronic nitrogen additions at the Harvard Forest LTER, Massachusetts, USA. *For. Ecol. Manage.* 196:7–28.

Maginnis, S., and Jackson, W. 2005. Restoring forest landscapes. www.sur.iucn.org/bosques/documentos/publicaciones/restoring_forest_landscapes.pdf.

Mah, K., Tackaberry, L.E., Egger, K.N., and Massicotte, H.B. 2001. The impacts of broadcast burning after clear-cutt ing on the diversity of ectomycorrhizal fungi associated with hybrid spruce seedlings in central British Columbia. *Can. J. For. Res.* 31:224–235.

Maharjan, R., Seeto, S., Notley-McRobb, L., and Ferenci, T. 2006. Clonal adaptive radiation in a constant environment. *Science* 313:514–517.

Maherali, H., and DeLucia, E.H. 2001. Influence of climate-driven shifts in biomass allocation on water transport and storage in ponderosa pine. *Oecologia* 129:481–491.

Malajczuk, N. 1979. The microflora of unsuberized roots of *Eucalyptus calophylla* R. Br. and *Eucalyptus marginata* Donn ex Sm. seedlings grown in soils suppressive and conducive to *Phytophthora cinnamomi* Rands. II. Mycorrhizal roots and associated microflora. *Aust. J. Bot.* 27:255–272.

Malajczuk, N., Trappe, J.M., and Molina, R. 1987. Interrelationships among some ectomycorrhizal trees, hypogeous fungi and small mammals: western Australian and northwest American parallels. *Aust. J. Ecol.* 12:53–55.

Malamud, B.D., Millington, J.D.A., and Perry, G.L.W. 2005. Characterizing wildfire regimes in the United States. *PNAS* 102:4694–4699.

Malamud, B.D., Morein, G., and Turcotte, D.L. 1998. Forest fires: an example of self-organized critical behavior. *Science* 281:1840–1841.

Malavasi, U.C., and Perry, D.A. 1993. Genetic variation in competitive ability of some shade-tolerant and shade-intolerant Pacific coast (USA) conifers. *For. Ecol. Manage.* 56:69–81.

Malcolm, D.C., Hooker, J.E., and Wheeler, C.T. 1985. *Frankia* symbiosis as a source of nitrogen in forestry: a case study of symbiotic nitrogen-fixation in a mixed *Alnus-Picea* plantation in Scotland. *Proc. R. Soc. Edinburgh* 85B:263–282.

Malcolm, J.R., Markham, A., Neilson, R.P., and Garaci, M. 2002. Estimating migration rates under scenarios of global climate change. *J. Biogeogr.* 29:835–849.

Malhi, Y., Roberts, J. T., Betts, R. A., Killeen, T. J., Li, W., and Nobre, C. A. 2008. Climate change, deforestation, and the fate of the Amazon. *Science* 319:169–172.

Malizewska, W., and Moreau, R. 1960. A study of the fungal microflora in the rhizosphere of fir (*Abies alba* Mill.). In *The ecology of soil fungi.* Edited by D. Parkinson and J.S. Waid. Liverpool Univ. Press, Liverpool, 209–218.

Malloch, D.W., Pirozynyski, K.A., and Raven, P.H. 1980. Ecological and evolutionary significance of mycorrhizal symbioses in vascular plants. *PNAS* 77:2112–2118.

Malo, B.A., and Purvis, E.R. 1964. Soil absorption of atmospheric ammonia. *Soil Sci.* 97:242–247.

Mandlebrot, B.B. 1983. *The fractal geometry of nature.* W.H. Freeman and Co., New York.

Manjunath, A., and Habte, M. 1988. Development of vesicular-arbuscular mycorrhizal infection and the uptake of immobile nutrients in *Leucaena leucocephala. Plant Soil* 106:97–103.

Mann, C. 1991. Extinction: are ecologists crying wolf? *Science* 253:736–738.

Marcot, B., and Molina, R. 2007. Special considerations for the science, conservation, and management of rare or little-known species. In *Conservation of rare or little-known species.* Edited by M.G. Raphael, and R. Molina. Island Press, Washington, DC. 93–124.

Margalef, R. 1975. Diversity, stability and maturity in natural ecosystems. In *Unjfying concepts in ecology.* Edited by W.F. van Dobben and R.H. Lowe-McConnell. Dr. W. Junk B.V. Publishers, The Hague, 151–160.

Margulis, L. 1981. *Symbiosis in cell evolution: life and its environment on the early earth.* Freeman.

Margulis, L. 1990. Kingdom animalia: the zoological malaise from a microbial perspective. *Am. Zool.* 30:861–875.

Margulis, L. 1991. Symbiogenesis and symbionticism. In *Symbiosis as a source of evolutionary innovation. Speciation and morphogenesis.* Edited by L. Margulis and R. Fester. MIT Press, Cambridge, MA, 1–14.

Margulis, L. 2004. Gaia by any other name. In *Scientists debate Gaia.* Edited by S.H. Schnieder, J.R. Miller, E. Crist, and P.J. Boston. MIT Press, Cambridge, MA, 7–12.

Margulis, L., and Lovelock, J.E. 1974. Biological modulation of the earth's atmosphere. *Icarus* 21:471–489.

Margulis, L., and Lovelock, J. 1989. Gaia and geognosy. In *Global ecology: toward a science of the biosphere.* Edited by M.B. Rambler, L. Margulis, and R. Fester. Academic Press, New York.

Margulis, L., and Sagan, D. 1986. *Microcosmos.* Summit Books, New York.

Markewitz, D., Richter, D.D., Allen, H.L., and Urrego, J.B. 1998. Three decades of observed soil acidification in the Calhoun Experimental Forest: has acid rain made a difference? *Soil Sci. Soc. Am. J.* 62:1428–1439.

Marks, P.L. 1974. The role of pin cherry (*Prunus pensylvanica* L.) in the maintenance of stability in northern hardwood ecosystems. *Ecol. Monogr.* 44:73–88.

Marland, G., Boden, T.A., and Andres, R.J. 2003. Global, regional, and national CO_2 emissions. In *Trends: a compendium of data on global change.* Carbon Dioxide Information Analysis Center, Oak Ridge National Laboratory, U.S. Department of Energy, Oak Ridge, TN.

Marquest, P.A., Fernandez, M., Navarrete, S.A., and Valdovinos, C. 2004. Diversity emerging: toward a deconstruction of biodiversity patterns. In Frontiers of Biogeography. Edited by M.V. Lomlino and L.R. Heaney. Sinauer, Sunderland MA, chap. 10.

Marquis, R.J. 2004. Herbivores rule. *Science* 305:619–621.

Marquis, R.J., and Whelan, C.J. 1994. Insectivorous birds increase growth of white oak through consumption of leaf-chewing insects. *Ecology* 75:2007–2014.

Marra, P.P., Griffing, S., Caffrey, C., Kilpatrick, A.M., McLean, R., Brand, C., Saito, E., Dupuis, A.P., Kramer, L., and Novak, R. 2004. West Nile virus and wildlife. *BioScience* 54:393–402.

Marshall, K.C. 1976. *Interfaces in microbial ecology.* Harvard Univ. Press, Cambridge, MA.

Martin, H.G., and Goldenfeld, N. 2006. On the origin and robustness of power-law species-area relationships in ecology. *PNAS* 103:10310–10315.

Martin, J.H., Gordon, R.M., Fitzwater, S., and Broenkow, W.W. 1989. phytoplankton/iron studies in the Gulf of Alaska. *Deep-Sea Res.* 36:649–680.

Martin, K.J., and McComb, B.C. 2003. Amphibian habitat associations at patch and landscape scales in the central Oregon Coast Range. *J. Wildl. Manage.* 67:672–683.

Martin, M.M., Rockholm, D.C., and Martin, J.S. 1985. Effects of surfactants, pH, and certain cations on precipitation of proteins by tannins. *J. Chem. Ecol.* 11:485–494.

Martinez-Ramos, M., Alvarez-Buylla, E., and Sarukhan, J. 1989. Tree demography and gap dynamics in a tropical rain forest. *Ecology* 70:555–558.

Martinez-Ramos, M., Alvarez-Buylla, E., Sarukhan, J., and Pinero, D. 1988. Treefall age determination and gap dynamics in a tropical forest. *J. Ecol.* 76:700–716.

Marx, D.R. 1972. Ectomycorrhizae as biological deterrents to pathogenic root infections. *Annu. Rev. Phytopathol.* 10:429–453.

Marx, D.R. 1975. Mycorrhiza and establishment of trees on strip-mined land. *Ohio J. Sci.* 75:288–297.

Maschinski, J., and Whitham, T.G. 1989. The continuum of plant responses to herbivory: the influence of plant association, nutrient availability, and timing. *Am. Nat.* 134:1–19.

Maser, C. 1990. *The redesigned forest.* Stoddart Publishing, Toronto.

Maser, C., Anderson, R.G., Cromack Jr., K., Williams, J.T., and Martin, R.E. 1979. Dead and down woody material. In *Wildlife habitats in managed forests: the Blue Mountains of Oregon and Washington.* Agriculture Handbook No. 553. Edited by J.W. Thomas. U.S. Department of Agriculture, Forest Service, Washington, DC, 78–95.

Maser, C., Cline, S.P., Cromack Jr., K., Trappe, J.M., and Hansen, E. 1988b. What we know about large trees that fall to the forest floor. In *From the forest to the sea: a story of fallen trees.* Gen. Tech. Rep. PNW-GTR-229. Edited by C. Maser, R.F. Tarrant, J.M. Trappe, and J.F. Franklin. U.S. Department of Agriculture, Forest Service, Portland, OR, 25–46.

Maser, C., Tarrant, R.F., Trappe, J.M., and Franklin, J.F., editors. 1988a. *From the forest to the sea: a story of fallen trees.* Gen. Tech. Rep. PNW-GTR-229. U.S. Department of Agriculture, Forest Service, Portland, OR.

Maser, C., Trappe, J.M., and Nussbaum, R.A. 1978. Fungal-small mammal interrelationships with emphasis on Oregon coniferous forests. *Ecology* 59:799–809.

Maslin, M. 2004. Ecological versus climatic thresholds. *Science* 306:2197–2198.

Mason, R.R., and Luck, R.F. 1978a. Population growth and regulation. In *The Douglas-fir tussock moth: a synthesis.* USDA Forest Service Tech. Bull. 1585. Edited by M.H. Brookes, R.W. Stark, and R.W. Campbell. U.S. Department of Agriculture, Washington, DC, 41–47.

Mason, R.R., and Luck, R.R. 1978b. Population ecology. In *The Douglas-fir tussock moth: a synthesis*USDA Forest Service Tech. Bull. 1585 . Edited by M.H. Brookes, R.W. Stark, and R.W. Campbell. U.S. Department of Agriculture, Washington, DC, 39–46.

Mason, R.R., Torgersen, T.R., Wickman, B.E., and Paul, H.G. 1983. Natural regulation of a Douglas-fir tussock moth (Lepidoptera: Lymantriidae) population in the Sierra Nevada. *Envrion. Entomol.* 12:587–594.

Mathur, H.M. 1975. Landslides in the Himalayan ranges and their control. *Indian Farming* 25:17–19.

Matson, P.A., Vitousek, P.M., Ewel, J.J., Mazzarino, M.J., and Robertson, G.P. 1987. Nitrogen transformations following tropical forest felling and burning on a volcanic soil. *Ecology* 68:491–502.

Mattson Jr., W.J. 1980. Herbivory in relation to plant nitrogen content. *Annu. Rev. Ecol. Syst.* 11:119–161.

Mattson, W.J., and Addy, N.D. 1975. Phytophagous insects as regulators of forest primary production. *Science* 190:515–522.

Mattson, W.J., and Haack, R.A. 1987. The role of drought in outbreaks of plant-eating insects. *BioScience* 37:110–118.

Mattson, W.J., Lawrence, R.K., Haack, R.A., Herms, D.A., and Charles, P.-J. 1988. Defensive strategies of woody plants against different insect-feeding guilds in relation to plant ecological strategies and intimacy of association with insects. In *Mechanisms of woody plant defenses against insects. Search for pattern.* Edited by W.J. Mattson, J. Levieux, and C. Bernard-Dagano, Springer-Verlag, New York, 1–38.

Matzner, E., Murach, D., and Fortmann, H. 1986. Soil acidity and its relationship to root growth in declining forest stands in Germany. *Water Air Soil Pollut.* 31:273–282.

Maxwell, A.L. 2004. Fire regimes in north-eastern Cambodian monsoonal forests, with a 9300–year sediment charcoal record. *J. Biogeogr.* 31:225–239.

May, R.M. 1974. *Stability and complexitiy in model ecosystems.* Princeton Univ. Press, Princeton, NJ.

May, R.M. 1975. Stability in ecosystems: some comments. In *Unifying concepts in ecology.* Edited by W.H. van Dobben and R.H. Lowe-McConnell. Dr. W. Junk B.V. Publishers, The Hague, 161–168.

May, R.M. 1988. How many species are there on earth? *Science* 241:1441–1449.

May, R.M. 1989. Honeyguides and humans. *Ecology* 338:707–708.

Mayle, F.E., Burbridge, R., and Killeen, T.J. 2000. Millennial-scale dynamics of southern Amazonian rainforests. *Science* 290:2291–2294.

McArthur, J.V., Kovacic, D.A., and Smith, M.H. 1988. Genetic diversity in natural populations of a soil bacterium across a landscape gradient. *PNAS* 85:9621–9624.

McBrayer, J.F., Ferris, J.M., Metz, L.J., Gist, C.S., Cornaby, B.W., Kitazawa, Y., Kitazawa,T., Wernz, J.G., Krantz, G.W., and Jensen, H. 1977. Decomposer invertebrate populations in U.S. forest biomes. *Pedobiologia* 17:S89–96.

McCann, K.S. 2000. The diversity stability debate. *Nature* 405:228–233.

McCarthy, G.W., and Bremner, J.M. 1986. Effects of phenolic compounds on nitrification in soil. *Soil Sci. Soc. Am. J.* 50:920–923.

McCarthy HR, Oren R, Finzi AC, Johnsen KH (2006) Canopy leaf area constrains [CO_2]-induced enhancement of productivity and partitioning among aboveground carbon pools. *Proc. Natl. Acad. Sci. U.S.A.* 103:19356–19361.

McCarthy, J. 2001. Gap dynamics of forest trees: a review with particular attention to boreal forests. *Environ. Rev.* 9:1–59.

McClaugherty, C., and Berg, B. 1987. Cellulose, lignin and nitrogen concentrations as rate regulating factors in late stages of forest litter decomposition. *Pedobiologia* 30:101–112.

McClaugherty, C.A., Aber, J.D., and Melillo, J.M. 1982. The role of fine roots in the organic matter and nitrogen budgets of two forested ecosystems. *Ecology* 63:1481–1490.

McClaugherty, C.A., Aber, J.D., and Melillo, J.M. 1984. Decomposition dynamics of fine roots in forested ecosystems. *Oikos* 42:378–386.

McClaugherty, C.A., Pastor, J., Aber, J., and Melillo, J.M. 1985. Forest litter decomposition in relation to soil nitrogen dynamics and litter quality. *Ecology* 66:266–275.

McClure, M.S. 1977. Resurgence of the scale, *Fiorimia externa* (Homoptera: Diaspididae) on hemlock following insecticide application. *Environ. Entomol.* 6:480–484.

McClure, M.S. 1986. Role of predators in regulation of endemic populations of *Matsucoccus matsumurae* (Homoptera: Margarodidae) in Japan. *Environ. Entomol.* 15:976–983.

McColl, J.G. 1973. Soil moisture influence on growth, transpiration, and nutrient uptake of pine seedlings. *For. Sci.* 19:281–288.

McComb, B.C. 2007. *Wildlife habitat management: concepts and applications in forestry.* Taylor and Francis Publishers, Boca Raton, FL.

McComb, W.C., Bonney, S.A., Sheffield, R.M., and Cost, N.D. 1986a. Den tree characteristics and abundance in Florida and South Carolina. *J. Wildl. Manage.* 50:584–591.

McComb, W.C., Bonney, S.A., Sheffield, R.M., and Cost, N.D. 1986b. Snag resources in Florida—are they sufficient for average populations of primary cavity-nesters? *Wildl. Soc. Bull.* 14:40–48.

McComb, W.C., and Muller, R.N. 1983. Snag densities in old-growth and second-growth Appalachian forests. *J. Wildl. Manage.* 47:376–382.

McComb, W.C., McGarigal, K., Fraser, J.D., and Davis, W.H. 1989. Planning for basin-level cumulative effects in the Appalachian coal field. *Trans. N. Am. Wildl. and Natur. Resour. Conf.* 54:102–112.

McCullough, D.G., Werner, R.A., and Neumann, D. 1998. Fire and insects in northern and boreal forests ecosystems of North America. *Annu. Rev. Entomol.* 43:107–127.

McCune, B. 1993. Gradients in epiphytic biomass in three *Pseudotsuga-Tsuga* forests of different ages in western Oregon and Washington. *Bryologist* 96:405–411.

McCune, B., Amsberry, K.A., Camacho, F.J., Clery, S., Cole, C., Emerson, C., Felder, G., French, P., Greene, D., Harris, R., Hutten, M., Larson, B., Lesko, M., Majors, S., Markwell, T., Parker, G.G., Pendergrass, K., Peterson, E.B., Peterson, E.T., Platt, J., Proctor, J., Rambo, T., Rosso, A., Shaw, D., Turner, R., and Widmer, M. 1997. Vertical profile of epiphytes in a Pacific Northwest old-growth forest. *Northwest Sci.* 71145–152.

McCune, B., Rosentreter, R., Ponzetti, J.M., and Shaw, D. 2000. Epiphyte habitats in an old conifer forest in western Washington, U.S.A. *Bryologist* 103:417–427.

McFee, W.W., and Stone, E.L. 1965. Quantity, distribution and variability of organic matter and nutrients in a forest podzol in New York. *Soil Sci.* Soc. Am. Proc. 29:32–436.

McGarigal, K., and Cushman, S.A. 2002. Comparative evaluation of experimental approaches to the study of habitat fragmentation effects. *Ecol. Applic.* 12:335–345.

McGarigal, K., and McComb, W.C. 1995. relationships between landscape structure and breeding birds in the Oregon Coast Range. *Ecol. Monogr.* 65:235–260.

McGee, G., Leopold, D.J., and Nyland, R.D. 1999. Structural characteristics of old-growth, maturing, and partially cut northern hardwood forests. *Ecol. Applic.* 9:1316–1329.

McGill, B.J. 2003. A test of the unified neutral theory of biodiversity. *Nature* 422:881–885.

McGill, B.J., Hadley, E.A., and Maurer, B.A. 2005. Community inertia of Quaternary small mammal assemblages in North America. *PNAS* 102:16701–16706.

McGill, W.B., and Cole, C.V. 1981. Comparative aspects of cycling of organic C, N, S and P through soil organic matter. *Geoderma* 26:267–286.

McGrath, D.A., Smith, C.K., Gholz, H.L., and de Assis Oliveira, F. 2001. Effects of land-use change on soil nutrient dynamics in Amazonia. *Ecosystems* 4:625–645.

McGroddy, M.E., Daufresne, T., and Hedin, L.O. 2004. Scaling of C:N:P stoichiometry in forests worldwide: implications of terrestrial redfield-type ratios. *Ecology* 85:2390–2401.

McIntosh, R.P. 1981. Succession and ecological theory. In *Forest succession, concepts and application.* Edited by D.C. West, R.H. Shugart, and D.B. Botkin. Springer-Verlag, New York, 10–23.

McIntosh, R.P. 1987. Pluralism in ecology. *Annu. Rev. Ecol. Syst.* 18:321–341.

McIntyre, S. and Hobbs, R. 1999. A framework for conceptualizing human effects on landscapes and its relevance to management and research models. *Conserv. Biol.* 13:1282–1292.

McIntyre, P.J., and Whitham, T.G. 2003. Plant genotype affects long-term herbivore population dynamics and extinction: conservation implications. *Ecology* 84:311–322.

McKey, D. 1975. The ecology of coevolved seed dispersal systems. In *Coevolution of animals and plants.* Edited by L.E. Gilbert and P.R. Raven. Univ. of Texas Press, Austin, 159–191.

McKey, D., Waterman, P.G., Mbi, C.N., Gartlan, J.S., and Struhsaker, T.T. 1978. Phenolic content of vegetation in two African rain forests: ecological implications. *Science* 202:61–64.

McKeand, S., Mullin, T., Byram, T., and White, T. 2003. Deployment of genetically improved loblolly and slash Pines in the South. *J. Forest.* April/May:32–35.

McLaren, B.E., and Peterson, R.O. 1994. Wolves, moose, and tree rings on Isle Royale. *Science* 266:1555–1558.

McLauchlan, K.K., Craine, J.M., Oswald, W.W., Leavitt, P.R., and Likens, G.E. 2007. Changes in nitrogen cycling during the past century in a northern harwood forest. *PNAS* 104:7466–7470.

McLaughlin, D. 1998. A decade of forest tree monitoring in Canada: evidence of air pollution effects. *Environ. Rev.* 6:151–171.

McLaughlin, S.B. 1989. Carbon allocation processes as indicators of pollutant impacts on forest trees. In *Biological markers of airpollution stress and damage in forests.* National Academy Press, Washington, DC, 293–302.

McLaughlin, S.B., and Wimmer, R. 1999. Calcium physiology and terrestrial ecosystem processes. *New Phytol.* 142:373–417.

McNabb, D.H., and Cromack Jr., K. 1990. Effects of prescribed fire on nutrients and soil productivity. In *Natural and prescribed fire in Pacific Northwest forests.* Edited by J.D. Walstad, S.R. Radosevich, and D.V. Sandberg. Oregon State Univ., Corvallis, OR.

McNabb, D.H., D. Startsev, and H. Nguyen. 2001. Soil wetness and traffic level effects on bulk density and air-filled porosity of compacted boreal forest soils. *Soil Sci. Soc. Am. J.* 65:1238–1247.

McNaughton, S.J., Oesterheld, M., Frank, D.A., and Williams, K.J. 1989. Ecosystem-level patterns of primary productivity and herbivory in terrestrial habitats. *Nature* 341:142–144.

McNeely, J.A., Miller, K.R., Reid, W.V., Mittermeier, R.A., and Werner, T.B. 1990. *Conserving the world's biological diversity.* The World Bank, Washington, DC.

McRae, D.J., Duchesne, L.C., Freedman, B., Lynham, T.J., and Woodley, S. 2001. Comparisons between wildfire and forest harvesting and their implications in forest management. *Environ. Rev.* 9:223–260.

Medina, E. 1983. Adaptations of tropical trees to moisture stress. In *Tropical rain forest ecosystems.* Edited by F.B. Galley. Elsevier, Amsterdam, 225–238.

Medina, E. 1984. Nutrient balance and physiological processes at the leaf level. In *Physiological ecology of plants of the wet tropics.* Edited by E. Medina, H.A. Mooney, and C. Vazquez-Yanes. Dr. W. Junk, The Hague, 139–154.

Medina, E., Mooney, H.A., and Vazquez-Yanes, C., editors. 1984. *Physiological ecology of plants of the wet tropics.* Dr. W. Junk, The Hague.

Meentemeyer, V., and Berg, B. 1986. Regional variation in rate of mass loss of *Pinus sylvestris* needle litter in Swedish pine forests as influenced by climate and litter quality. *Scand. J. For. Res.* 1:167–180.

Meier, C.E., Grier, C.C., and Cole, D.W. 1985. Below- and aboveground N and P use by *Abies amabilis* stands. *Ecology* 66:1928–1942.

Melián, C.J., and Bascompte, J. 2004. Food web cohesion. *Ecology* 85:352–358.

Melillo, J.M., and Aber, J.D. 1984. Nutrient immobilization in decaying litter: an example of carbon-nutrient interactions. In *Trends in ecologic research for the 1980s.* Edited by J. H. Cooley and F.B. Golley. Plenum Pub. Corp.

Melillo, J.M., and Gosz, J.R. 1983. Interactions of biogeochemical cycles in forest ecosystems. In *The major biogeochemical cycles and their interations.* Edited by B. Bolin and R.B. Cook. John Wiley & Sons, New York, 177–222.

Melillo, J.M., Aber, J.D., and Muratore, J.F. 1982. Nitrogen and lignin control of hardwood leaf litter decomposition dynamics. *Ecology* 63:621–626.

Melillo, J.M., Aber, J.D., Steudler, P.A., and Schimel, J.P. 1983. Denitrification potentials in a successional sequence of northern hardwood forest stands. *Ecol. Bull. (Stockholm)* 35:217–228.

Melillo, J.M., Steudler, P.A., Aber, J.D., Newkirk, K., Lux, H., Bowles, F.P., Catricala, C., Magill, A., Ahrens, T., and Morrisseau, S. 2002. Soil warming and carbon-cycle feedbacks to the climate system. *Science* 298:2173–2176.

Melosh, H.J., Schneider, N.M., Zahnie, K.J., and Latham, D. 1990. Ignition of global wildfires at the Cretaceous/Tertiary boundary. *Nature* 343:251–254.

Mencuccini, M. 2003. The ecological significance of long-distance water transport: short-term regulation, long-term acclimation and the hydraulic costs of stature across plant life forms. *Plant Cell Environ.* 26:163–182.

Mengel, K. 1985. Dynamics and availability of major nutrients in soils. *Adv. Soil Sci.* 2:65–131.

Mengel, K., and Kirby, E.A. 2001. *Principles of plant nutrition.* 5th ed. Kluwer Academic, The Netherlands.

Meslow, E.C. 1978. The relationship of birds to habitat structure-plant communities and successional stages. In *Proceedings of a workshop on nongame bird management in coniferous forests of the western United States.* Gen. Tech. Rep. PNW-64. Edited by R.M. de Graff. U.S. Department of Agriculture, Forest Service, Portland, OR, 12–18.

Meyer, J.R., and Linderman, R.G. 1986. Response of subterranean clover to dual inoculation with vesicular-arbuscular mycorrhizal fungi and a plant-growth promoting bacterium, *Pseudomonas putida. Soil. Biol. Biochem.* 18:185–190.

Meyn, A., White, P.S., Bukh, C., and Jentsch, A. 2007. Environmental drivers of large, infrequent wildfires: the emerging conceptual model. *Progr. Phys. Geogr.* 31:287–312.

Miao, S., and Carstenn, S. 2006. A new direction for large-scale experimental design and analysis. *Front. Ecol. Environ.* 4:227.

Mikammal, E.I. 1971. Some aspects of radiant energy in a red pine forest. *Arch. Met. Giophys. Bioklimatol. Ser. B* 19:29–52.

Mikola, J., Bardgett, R.D., and Hedlund, K. 2002. Biodiversity, ecosystem functioning, and soil decomposer food webs. In *Biodiversity and ecosystem functioning.* Edited by M. Loreau, S. Naemm, and P. Inchausti. Oxford Univ. Press, 169–180.

Mikola, P. 1970. Mycorrhizal inoculation in aforestation. *Int. Rev. For. Res.* 3:123–196.

Mikutta, R., Kleber, M., Torn, M.S., and Jahn, R. 2006. Stabilization of soil organic matter: association with minerals or chemical recalcitrance? *Biogeochemistry* 77:25–56.

Miles, D., Swanson, F.J., and Youngberg, C.T. 1984. Effects on landslide erosion on subsequent Douglas-fir growth and stock levels in the Western Cascades. *Soil Sci. Soc. Am. J.* 48:667–671.

Millbank, J.W. 1985. Lichens and plant nutrition. *Proc. R. Soc. Edinburgh* 85B:253–261.

Millennium Ecosystem Assessment. 2005a. *Ecosystems and human well-being: synthesis.* Island Press, Washington, DC.

Millennium Ecosystem Assessment. 2005b. *Ecosystems and human health: summary.* Island Press, Washington, DC.

Miller, H.G., and Miller, J.D. 1976. Effect of nitrogen supply on net primary production in Corsican pine. *J. Appl. Ecol.* 13:249–256.

Miller, H.G., Miller, J.D., and Cooper, J.M. 1981. Optumum foliar nitrogen concentration in pine and its change with stand age. *Can. J. For. Res.* 11:563–572.

Miller, J.D. 1986. Toxic metabolities of epiphytic and endophytic fungi of conifer needles. In *Microbiology of the phyllosphere.* Edited by N.J. Fokkema and J. van den Heuvel. Cambridge Univ. Press, Cambridge, 221–231.

Miller, R.E., and Murray, M.D. 1978. The effects of red alder on growth of Douglas-fir. In *Utilization and management of alder.* Gen. Tech. Rpt. PNW-70. Edited by D.G. Briggs, D.S. DeBell, and W.A. Atkinson. U.S. Department of Agriculture, Forest Service, Portland, OR, 203–306.

Miller, R.M. 1987. Mycorrhizae and succession. In *Restoration ecology: a synthetic approach to ecological research.* Edited by W.R. Jordan III, M.E. Gilpin, and J.D. Aber. Cambridge Univ. Press, Cambridge, 205–219.

Miller, W.F., Dougherty, P.M., and Switzer, G.L. 1987. Effect of rising carbon dioxide and potential climate change on loblolly pine distribution, growth, survival, and productivity. In *The greenhouse effect, climate change, and U.S. forests.* Edited by W.E. Shands and J.S. Hoffman. The Conservation Foundation, Washington, DC, 157–187.

Millot, G. 1979. The calcerous epigenesis phenomenon and its role in the weathering process. *Bull. Assoc. Fr. Etud. Sol. Sci. Sol.* 2/3:259–261.

Mills, L.S., and Allendorf, F.W. 1996. The one-migrant-per-generation rule in conservation and management. *Conserv. Biol.* 10:1509–1518.

Mills, L.S., Soule, M.E., and Doak, D.F. 1993. The keystone-species concept in ecology and conservation. *BioScience* 43:219–224.

Minderman, G. 1956. Addition, decomposition, and accumulation of organic matter in forests. *J. Ecol.* 56:355–362.

Minnich, R.A., Barbour, M.G., Burk, J.H., and Sosa-Ramírez, J. 2000. Californian mixed-conifer forests under unmanaged fire regimes in the Sierra San Pedro Mártir, Baja California, Mexico. *J. Biogeogr.* 27:105–129.

Mishra, B., and Srivastava, L.L. 1986. Degradation of humic acid of a forest soil by some fungal isolates. *Plant Soil* 96:413–416.

Mitchell, M.S., Rutzmoser, S.H., Wigley , T.B., Loehle, C., Gerwin, J.A., Keyser, P.D., Lancia, R.A., Perry, R.W., Reynolds, C.J., Thill, R.E., Weih, R., White, D., and Wood, P.B. 2006a. Relationships between avian richness and landscape structure at multiple scales using multiple landscapes. *For. Ecol. Manage.* 221:155–169.

Mitchell, R.G., Waring, R.H., and Pitman, G.B. 1983. Thinning lodgepole pine increases tree vigor and resistance to mountain pine beetle. *For. Sci.* 29:204–211.

Mitchell, R.J., Franklin, J.F., Palik, B.J., Kirkman, L.K., Smith, L.L., Engstrom, R.T., and Hunter Jr., M.L. 2006b. *Natural disturbance-based silviculture for restoration and maintenance of biological diversity.* Final report to The National Commission on the Science of Sustainable Forestry. www.ncseonline.org/ewebeditpro/items/O62F3299.pdf.

Mitchell, R.J., Garrett, H.E. Cox, G.S., Atalay, A., and Dixon, R.K. 1987. Boron fertilization, ectomycorrhizal colonization, and growth of *Pinus echinata* seedlings. *Can. J. For. Res.* 17:1153–1156.

Mittelbach, G.G., Steiner, C.F., Scheiner, S.M., Gross, K.L., Reynolds, H.L., Waide, R.B., Willg, M.R., Dodson, S.I., and Gough, L. 2001. What is the observed relationship between species richness and productivity? *Ecology* 82:2381–2396.

Mitter, C., Futuyama, D.H., Schneider, J.C., and Hare, J.D. 1979. Genetic variation and host plant relations in a parthenogenetic moth. *Evolution* 33:777–789.

Mitton, J.B., and Grant, M.C. 1984. Associations among protein heterozygosity, growth rate, and developmental homeostasis. *Annu. Rev. Ecol. Syst.* 15:479–499.

Miyasaka, S.C., and Habte, M. 2001. Plant mechanisms and mycorrhizal symbioses to increase phosphorus uptake efficiency. *Commun. Soil Sci. Plant Anal.* 32:1101–1147.

Mladenoff, D.J., Sickley, T.A., and Wydeven, A.P. 1999. Predicting gray wolf landscape recolonization: logistic regression models vs. new field data. *Ecol. Applic.* 9:37–44.

Mohren, G.M.J., Van Den Burg, J., and Burger, F.W. 1986. Phosphorus deficiency induced by nitrogen input in Douglas-fir in the Netherlands. *Plant Soil* 95:191–200.

Moldenke, A. 1990. One hundred twenty thousand little legs. *Wings* 1990(Summer):11–14.

Molina, R., Massicotte, H., and Trappe, J.M. 1992. Specificity phenomena in mycorrhizal symbioses: community-ecological consequences and practical implications. In *Mycorrhizal functioning. An integrative plant-fungal process.* Edited by M.F. Allen. Chapman & Hall, New York, 357–423.

Molino, J.-F., and Sabatier, D. 2001. Tree diversity in tropical rainforests: a validation of the intermediate disturbance hypothesis. *Science* 294:1702–1704.

Mollison, B. 1990. *Permaculture. A practical guide for a sustainable future.* Island Press, Washington, DC.

Monastersky, R. 1989. Cloudy concerns: will clouds prevent or promote a drastic global warming? *Sci. News* 136:106–107.

Monastersky, R. 1992. Closing in on the killer. *Sci. News* 141:56–58.

Monastersky, R. 1993. The deforestation debate. *Sci. News* 144:26–27.

Monsi, M., and Saeki, T. 1953. Uber den Lichtfaktor in den Pflanzengesellschaften und seine Bedeutung fur die Stoffproduktion. *Jpn. J. Bot.* 14:22–52.

Monsi, M., Uchijima, Z., and Oikawa, T. 1973. Structure of foliage canopies and photosynthesis. *Annu. Rev. Ecol. Syst.* 4:301–327.

Monteith, J.L. 1975. *Principles of environmental physics.* Whitsable, Kent.

Montoya, J.M., and Solé, R.V. 2003. Topogical properties of food webs: from real data to community assembly models. *Oikos* 102:614–622.

Montoya, J.M., Pimm, S.L., and Solé, R. 2006. Ecological networks and their fragility. *Nature* 442:259–264.

Mooney, H., Cropper, A., and Reid, W. 2005. Confronting the human dilemma. *Nature* 434:561–562.

Mooney, H.A., Drake, B.G., Luxmoore, R.J., Oechel, W.C., and Pitelka, L.F. 1991. Predicting ecosystem responses to elevated CO_2 concentrations. *BioScience* 41:96–104.

Mooney, K.A., and Tillberg, C.V. 2005. Temporal and spatial variation to ant omnivory in pine forests. *Ecology* 86:1225–1235.

Moore, B.D., and Foley, W.J. 2005. Tree use by koalas in a chemically complex landscape. *Nature* 435:488–490.

Moore, G.W., Bond, B.J., Jones, J.A., Phillips, N., and Meinzer, F.C. 2004. Structural and compositional controls on transpiration in 40– and 450–year-old riparian forests in western Oregon. *Tree Physiol.* 24:481–491.

Moore, J.C., and Hunt, H.W. 1988. Resource compartmentation and the stability of real ecosystems. *Nature* 333:261–263.

Moore, J.C., de Ruiter, P.C., Hunt, H.W., Coleman, D.C., and Freckman, D.W. 1996. Microcosms and soil ecology: critical linkages between fields studies and modeling food webs. *Ecology* 77:694–705.

Moore, P. 1986. What makes rainforests so special? *New Sci.* 111:38–40.

Moore, T.R., Trofymow, J.A., Taylor, B., Prescott, C., Camire, C., Duschene, L., Fyles, J., Kozak, L., Kranabetter, M., Morrison, I., Siltanen, M., Smith, S., Tistus, B., Visser, S., Wein, R., and Zoltai, S. 1999. Litter decomposition rates in Canadian forests. *Global Change Biol.* 5:75–82.

Moorhead, D.L., and Sinsabaugh, R.L. 2006. A theoretical model of litter decay and microbial interaction. *Ecol. Monogr.*76:151–174.

Morgan, J.L., Campbell, J.M., and Malcom, D.C. 1992. Nitrogen relations of mixed-species stands on oligotrophic soils. In *The ecology of mixed-species stands of trees.* Edited by M.G.R. Cannell, D.C. Malconi, and P.A. Robertson. Blackwell Scientific Publications, London, 65–86.

Mori, S.A., and Prance, G.T. 1987. Species diversity, phenology, plant-animal interactions, and their correlation with climate, as illustrated by the Brazil nut family (Lecythidaceae). In *The geophysiology of Amazonia.* Edited by R.E. Dickinson. John Wiley & Sons, New York, 69–89.

Moritz, M.A., Morais, M.E., Summerell, L.A., Carlson, J.M., and Doyle, J. 2005. Wildfires, complexity, and highly optimized tolerance. *PNAS* 102:17912–17917.

Morowitz, H.J. 1968. *Energy flow in biology.* Academic Press, New York.

Morris, D.W. 2005. On ethe roles of time, space and habitat in aboreal small mammal assemblage: predictably stochastic assembly. *Oikos* 109:223–238.

Morris, L.A., and Miller, R.E. 1994. Evidence for long-term productivity changes as provided by field trials. In *Impacts of forest harvesting on long-term site productivity.* Edited by W.J. Dyck, D.W. Cole, and N.B Comerford. Chapman and Hall, London, 41–80.

Morrison, P.H. 1986. The history and role of fire in forest ecosystems of the central western Cascades of Oregon determined by forest stand analysis. Masters thesis. Univ. of Washington, Seattle.

Morrison, P.H., and Swanson, F.J. 1990. *Fire history and pattern in a Cascade Range landscape.* Gen. Tech. Rpt. PNW-GTR-254. U.S. Department of Agriculture, Forest Service, Portland, OR.

Morrow, P.A. 1977. The significance of phytophagous insects in the eucalyptus forests of Australia. In *The role of arthropods in forest ecosystems.* Edited by W.J. Mattson. Springer-Verlag, 19–29.

Mortberg, U.M. 2001. Resident bird species in urban forest remnants; landscape and habitat perspectives. *Landscape Ecol.* 16:193–203.

Morton, J.B. 2005. Fungi. In *Principles and application of soil microbiology,* 2nd ed. Edited by D.M. Sylvia, J.L. Fuhrmann, P.G. Hartel, and D.A. Zuberer. Prentice Hall, Upper Saddle River, NJ, 141–161.

Mosse, B. 1977. The role of mycorrhiza in legume nutrition on marginal soils. In *Exploiting the legume-rhizobium symbiosis in tropical agriculture.* Edited by J.M. Vincent, A.S., Whitney, and J. Bose. Univ. of Hawaii, Honolulu, 275–292.

Mosugelo, D.K., Moe, S.R., Ringrose, S., and Nellemann, C. 2002. Vegetation changes during a 36–year period in northern Chobe National Park, Botswana. *Afr. J. Ecol.* 40:232–240.

Motzkin, G., and D. Foster. 2004. Insights for ecology and conservation. In *Forests in time.* Edited by D.R. Foster and J.B. Aber. Yale Uuiv. Press, New Haven, CT. 367–379.

Motzkin, G., Foster, D., Allen, A., Donohue, K., and Wilson, P. 2004. Forest landscape patterns, structure, and composition. In *Forests in time.* Edited by D.R. Foster and J.D. Aber. Yale Univ. Press, New Haven, CT, 171–201.

Mountainspring, S., and Scott, J.M. 1985. Interspecific competition among Hawaiian forest birds. *Ecol. Monogr.* 55:219–239.

Moyle, P.B., Li, H.W., and Barton, B.A. 1986. The Frankenstein effect: impact of introduced fishes on native fishes in North America. In *Fish culture in fisheries management.* Edited by R.H. Stroud. American Fisheries Society, Bethesda, MD.

Mueller-Dombois, D. 1972. Crown distortion and elephant distribution in the woody vegetations of Rubruna National Park, Ceylon. *Ecology* 53:208–226.

Mueller-Dombois, D. 1986. Perspectives for an etiology of stand-level dieback. *Annu. Rev. Ecol. Syst.* 17:221–243.

Muir, W. M. 1995. Group selection for adaptation to multihen cages: selection program and direct responses. *Poultry Science* 75:447–458.

Murdoch, W.W., and Briggs, C.J. 1996. Theory for biological control: recent developments *Ecology* 77:2001–2013.

Murphy, D.D., and Noon, B.R. 1992. Integrating scientific methods with habitat conservation planning: reserve design for northern spotted owls. *Ecol. Applic.* 2:3–17.

Murphy, H.T., and Lovett-Doust, J. 2004. Context and connectivity in plant metapopulations and landscape mosaics: does the matrix matter? *Oikos* 105:3–14.

Mutch, R.W. 1970. Wildland fires and ecosystems—a hypothesis. *Ecology* 51:1046–1051.

Myers, J. 1993. Population outbreaks in forest Lepidoptera. *Am. Sci.* 81:240–251.

Myers, J. 1998. Synchrony in outbreaks of forest Lepidoptera: a possible example of the Moran effect. *Ecology* 79:1111–1117.

Myers, J.H. 1988. Can a general hypothesis explain population cycles of forest Leidoptera? *Adv. Ecol. Res.* 18:179–284.

Myers, N. 1984. Genetic resources in jeopardy. *Ambio* 13:171–174.

Myers, N. 1986. Environmental repercussions of deforestation in the Himalayas. *J. World For. Res. Manage.* 2:63–72.

Myers, N. 1988. Tropical deforestation and climate change. *Environ. Conserv.* 15:293–298.

Myers, N., Mittermeier, R.A., Mittermeier, C.G., da Fonseca, G.A.B., and Kent, J. 2000. Biodiversity hotspots for conservation priorities. *Nature* 403:853–858.

Myrold, D.D. 1987. Relationship between microbial biomass nitrogen and a nitrogen availability index. *Soil Sci. Soc. Am. J.* 51:1047–1049.

Nadelhoffer, K.J., and Raich, J.W. 1992. Fine root production estimates and belowground carbon allocation in forest ecosystems. *Ecology* 73:1139–1147.

Nadelhoffer, K.J., Aber, J.D., and Mellilo, J.M. 1984. Seasonal patterns of ammornum and nitrate uptake in nine temperate forest ecosystems. *Plant Soil* 80:321–335.

Nadelhoffer, K.J., Aber, J.D., and Melillo, J.M. 1985. Fine roots, net primary production, and soil nitrogen availability: a new hypothesis. *Ecol.* 66:1377–1390.

Nadelhoffer, K.J., Colman, B.P., Currie, W.S., Magill, A., and Aber, J.D. 2004. Decadal scale fates of ^{15}N tracers added to oak and pine stands under ambient and elevated N inputs at the Harvard Forest (USA). *For. Ecol. Manage.* 196:89–107.

Nadkarni, N. 1994. Diversity of species and interactions in the upper tree canopy of forest ecosystems. *Am. Zool.* 34:70–78.

Naeem, S., Chapin III, F.S., Costanza, R., Erlich, P.R., Golley, F.B., Hooper, D.U., Lawton, J.H., O'Neill, R.V., Mooney, H.A., Sala, O.E., Symstad, A.J., and Tilman, D. 1999. Biodiversity and ecosystem functioning: maintaining natural life support processes. *Issues Ecol.* 4:1–6.

Naeem, S., Loreau, M., and Inchausti, P. 2002. Biodiversity and ecosystem functioning: the emergence of a synthetic ecologtical framework. In *Biodiversity and ecosystem functioning: maintaining natural life support processes.* Edited by M. Loreau, S. Naeem, and P. Inchausti. Oxford Univ. Press, Oxford. 3–11.

Naiman, R.J. 1988. Animal influences on ecosystem dynamics. *BioScience* 38:750–752.

Naiman, R.J. 1991. *New perspectives for watershed management.* Springer-Verlag, New York.

Nambiar, E.K.S. 1987. Do nutrients retranslocate from fine roots? *Can. J. For. Res.* 17:913–918.

Nambiar, E.K.S., and Fife, D.N. 1987. Growth and nutrient retranslocation in needles of radiata in relation to nitrogen supply. *Ann. Bot.* 60:147–156.

Nardon, P., and Grenier, A.M. 1991. Serial endosymbiosis theory and weevil evolution: the role of symbiosis. In *Symbiosis as a source of evolutionary innovation. Speciation and morphogenesis.* Edited by L. Margulis and R. Fester. MIT Press, Cambridge, MA, 153–169.

NAS. 2002. *Abrupt climate change: inevitable surprises.* National Academy Press, Washington, DC.

NASA. 1988. *Earth system science: a closer view.* National Aeronautics and Space Administration, Washington, DC.

Natural Resources Canada. 2007a. The Model Forests Program. http://cfs.nrcan.gc.ca/index/cmfp.

Natural Resources Canada. 2007b. The Mountain Pine Beetle in British Columbia. http://mpb.cfs.nrcan.gc.ca/biology/introduction_e.html.

NCSSF. 2003. Second symposium. Fire, forest health, and biodiversity. National Commission for the Science of Sustainable Forestry. www.ncseonline.org/ewebeditpro/items/O62F3312.pdf.

Neal, J.L. Bollen, W.B., and Zak, B. 1964. Rhizosphere microflora associated with mycorrhizae of Douglas-fir. *Can. J. Microbiol.* 10:259–265.

Needham, A.E. 1965. *The uniqueness of biological materials.* Pergamon, Oxford.

Needham, T., Kershaw, J.A., MacLean, D.A., and Su, Q. 1999. Effects of mixed stand management to reduce impacts of spruce budworm defoliation on balsam fir stand-level growth and yield. *North. J. Appl. For.* 16:19–24.

Neff, J.C., and Asner, G.P. 2001. Dissolved organic carbon in terrestrial ecosystems: synthesis and a model. *Ecosystems* 4:29–48.

Neff, J.C., Chapin III, F.S., and Vitousek, P.M. 2003. Breaks in the cycle: dissolved organic nitrogen in terrestrial ecosystems. *Front. Ecol. Environ.* 1:205–211.

Neff, J.C., Holland, E.A., Dentener, F.J., McDowell, W.H., and Russell, K.M. 2002. The origin, composition and rates of organic nitrogen deposition: a missing piece of the nitrogen cycle? *Biogeochemistry* 57/58:99–136.

Neftel, A., Moore, E., Oeschger, H., and Stauffer, B. 1985. Evidence from polar ice cores for the increase in atmospheric CO_2 in the past two centuries. *Nature* 315:45–47.

Neilson, R.P. 1995. A model for predicting continental scale vegetation distribution and water balance. *Ecol. Applic.* 5:362–385.

Neilson, R.P., and Drapek, R.J. 1998. Potentially complex biosphere responses to transient global warming. *Global Change Biol.* 4:505–521.

Neilson, R.P., King, G.A., and DeVelice, R.L. 1989. *Sensitivity of ecological landscapes and regions to global climatic change.* Environmental Research Laboratory, Corvallis, OR.

Neilson, R.P., King, G.A., DeVelice, R.L., and Lenihan, J.M. 1992. Regional and local vegetation patterns: the responses of vegetation diversity to subcontinental air masses. In *Landscape boundaries: consequences for biotic diversity and ecological flows. Ecological studies 92.* Edited by A.J. Hansen and F. di Castri. Springer-Verlag, New York, 129–149.

Neilson, R.P., King, G.A., and Koerper, G. 1991. Toward a rule-based biome model. In *Proceedings of the Seventh Annual Pacific Climate (PACLIM) Workshop.* Interagency Ecological Studies Program Tech. Rep. 26. Edited by J.L. Betancourt and V.L. Tharp. California Department of Water Resources, Sacramento, 37–57.

Neilson, R.P., Prentice, I.C., Smith, B., Kittel, T.G.F., and Viner, D. 1998. Simulated changes in vegetation distribution under global warming. In *The regional impacts of climate change: an assessment of vulnerability.* Edited by R.T. Watson, M.C. Zinyowera, R.H. Moss, and D.J. Dokken. Cambridge Univ. Press, Cambridge, 439–456.

Neitlich, P.N., and McCune, B. 1997. Hotspots of epiphytic lichen diversity in two young managed forests. *Conserv. Biol.* 11:172–182.

Nepstad, D., Uhl, C., and Serrao, E.A. 1990. Surmounting barriers to forest regeneration in abandoned, highly degraded pastures: a case study from Paragorninas, Para, Brazil. In *Alternatives to deforestation.* Edited by A.B. Anderson. Columbia Univ. Press, New York, 215–229.

Neufeld, J.D., and Mohn, W.W. 2005. Unexpectedly high bacterial diversity in Arctic tundra relative to boreal forest soils, revealed by serial analysis of ribosomal sequence tags. *Appl. Environ. Microbiol.* 71:5710–5718.

Neuhauser, C., Andow, D.A., Heimpel, G.E., May, G., Shaw, R.G., and Wagenius, S. 2003. Community genetics: expanding the synthesis of ecology and genetics. *Ecology* 84:545–558.

Neutel, A.-M., Heesterbeek, J.A.P., and de Ruiter, P.C. 2002. Stability in real food webs: weak links in long loops. *Science* 296:1120–1123.

Nevo, E. 1978. Genetic variation in natural populations: patterns and theory. *Theoret. Pop. Biol.* 13:121–177.

Nevo, E., Beileiles, A., and Ben-Shiomo, R. 1984. The evolutionary significance of genetic diversity: ecological, demographic, and environmental correlates. In *Evolutionary dynamics of genetic diversity. Lecture notes in biomathematics 53.* Edited by G.S. Mani. Springer-Verlag, New York, 13–21.

Newbery, D.M., Alexander, I.J., Thomas, D.W., and Gartlan, J.S. 1988. Ectomycorrhizal rain- forest legumes and soil phosphorus in Korup National Park, Cameroon. *New Phytol.* 109:433– 450.

Newman, A. 1990. *Tropical rain forest.* Facts on File, New York.

Newman, E.I. 1988. Mycorrhizal links between plants: their functioning and ecological significance. *Adv. Ecol. Res.* 18:243–270.

Newman, M. 2000. The power of design. *Nature* 405:412–413.

Newman, R.H., and Tate, K.R. 1984. Use of alkaline soil extracts for ^{13}C n.m.r. characterization of humic substances. *J. Soil Sci.* 34:47–54.

Newmark, W.D. 1987. A land-bridge island perspective on mammalian extinctions in western North American parks. *Nature* 325:430–432.

Newton, M., and Cole, E.C. 1987. A sustained yield scheme for old-growth Douglas-fir. *W. J. Appl. For.* 2:22–25.

Ng, D. 1988. A novel level of interactions in plant-insect systems. *Nature* 334:611–613.

Niesten, E.T., Rice, R.E., Ratay, S.M., and Paratore, K. 2005. *Commodities and conservation: the need for greater habitat protection in the tropics.* Conservation International, Washington, DC.

NIFC. 2007. Summary statistics, 1997–2006. National Interagency Fire Center. Boise, Idaho. www.nifc.gov/fire_info/nfn.htm.

Nilsson, I. 1978. The influence of *Dasychira pudibunda* (Lepidoptera) on plant nutrient transports and tree growth in a beech *Fagus sylvatica* forest in southern Sweden. *Oikos* 30:133–148.

Nilsson, L.O., Giesler. R., Baath, E., and Wallander, H. 2005. Growth and biomass of mycorrhizal mycelia in coniferous forests along natural nutrient gradients. *New Phytol.* 165:613–622.

Nilsson, M.-C. 1992. The mechanisms of biological interference by Empetrum hermaphroditum on tree seedling establsihement in boreal forest ecosystems. Ph.D. dissertation. Swedish Univ. of Agricultural Sciences, Umea.

Nilsson, M-C., and D.A. Wardle. 2005. Understory vegetation as a forest ecosystem driver: evidence from the northern Swedish boreal forest. *Front. Ecol. Environ.* 3:421–428.

Nilsson, M.-C., and Zackrisson, O. 1992. Inhibition of Scots pine seedling establishment by Empetrum hermaphroditum. *J. Chem. Ecol.* 18:1857–1870.

Nilsson, S.I., Miller, H.G., and Miller, J.D. 1982. Forest growth as a possible cause of soil and water acidification: an examination of the concepts. *Oikos* 39:40–49.

Nishida, R. 2002. Sequestration of defensive substances from plants by Leidoptera. *Annu. Rev. Entomol.* 47:57–92.

NOAA. 2005. Teleconnections: introduction. National Oceanic and Atmospheric Administration. www.cpc.ncep.noaa.gov/data/teledoc/teleintro.shtml.

Noms, D.M. 1992. Quoted in *Sci. News* 141:94.

Norby, R.J., and Luo, Y. 2004. Evaluating ecosystem responses to rising atmospheric CO_2 and global warming in a multi-factor world. *New Phytol.* 162:281–293.

Norby, R.J., DeLucia, E.H., Gielen, B., Calfapietra, C., Giardina, C.P., King, J.S., Ledford, J., McCarthy, H.R., Moore, D.J.P., Ceulemans, R., De Angelis, P., Finzi, A.C., Karnosky, D.F., Kubiske, M.E., Lukac, M., Pregitzer, K.S., Scarascia-Mugnozza, G.E., Schlesinger, W.H., and Oren, R. 2005. Forest response to elevated CO_2 is conserved across a broad range of productivity. *PNAS* 102:18052–18056.

Norby, R.J., Ledford, J., Reilly, C.D., Miller, N.E., and O'Neill, E.G. 2004. Fine-root production dominates response of a deciduous forest to atmospheric CO_2 enrichment. *PNAS* 101:9689–9693.

Norby, R.J., O'Neill, E.G., Hood, W.G., and Luxmoore, R.J. 1987. Carbon allocation, root exudation and mycorrhizal colonization of *Pinus echinata* seedlings grown under CO_2 enrichment. *Tree Physiol.* 3:203–210.

Norby, R.J., Pastor, J., and Melillo, J.M. 1986. Carbon-nitrogen interactions in CO_2-enriched white oak: physiological and long-term perspectives. *Tree Physiol.* 2:233–241.

Norman, J.M., Jarvis, P.G. 1974. Photosynthesis in Sitka spruce (*Picea sitchensis* [Bong.] Carr.) III. Measurements of canopy structure and interception of radiation. *J. Appl. Ecol.* 11:375–398.

Norman, S.P., and Taylor, A.H. 2003. Tropical and north Pacific teleconnections influence fire regimes in pine-dominated forests of northeastern California, USA. *J. Biogeogr.* 30:1081–1092.

Norris, K. 2004. Managing threatened species: the ecological toolbox, evolutionary theory and declining-population paradigm. *J. Appl. Ecol.* 41:413–426.

Norris, S. 2003. Neutral theory: a new, unified model for ecology. *BioScience* 53:124–129.

Norse, E. 1990. *Ancient forests of the Pacific Northwest.* Island Press, Washington, DC.

Norton, J.M., Smith, J.L., and Firestone, M.K. 1990. Carbon flow in the rhizosphere of ponderosa pine seedlings. *Soil Biol. Biochem.* 22:449–456.

Noss, R. 2002. Context matters. *Conserv. Pract.* 3:10–19.

Noss, R.F. 1983. A regional landscape approach to maintain diversity. *BioScience* 33:700–706.

Noss, R.F. 1987. Protecting natural areas in fragmented landscapes. *Nat. Areas J.* 7:2–13.

Noss, R.F. 1988. The longleaf pine landscape of the Southeast: almost gone and almost forgotten. *Endang. Species Update* 5:1–8.

Noss, R.F. 1990. Indicators for monitoring biodiversity: a hierarchial approach. *Conserv. Biol.* 4:355–364.

Noss, R.F. 1991. From endangered species to a biodiversity. In *Balancing on the brink of extinction: the Endangered Species Act and lessons for the future.* Edited by K. Kohm. Island Press, Washington, DC, 227–245.

Novotny, V., and Basset, Y. 2005. Host specificity of insect herbivores in tropical forests. *Proc. R. Soc. B* 272:1083–1090.

Novotny, V., Basset, Y., Miller, S.E., Weiblen, G.D., Bremer, B., Cizek, L., and Drozd, P. 2002. Low host specificity of herbivorous insects ina tropical forest. *Nature* 416:841–844.

Novotny, V., Drozd, P., Miller, S.E., Kulfan, M., Janda, M., Basset, Y., and Weiblen, G.D. 2006. Why are there so many species of herbivorous insects in tropical rainforests? *Science* 313:1115–1118.

Nowak, M.A. 2006. Five rules for the evoloution of cooperation. *Science* 314:1560–1563.

Nye, I.H., and Greenland, D.J. 1960. *The soil under shifting cultivation.* Tech. Commun. 51. Commonwealth Ag. Bureaux, Farnham Royal, Bucks, UK.

Nye, P.H., and Tinker, P.B. 1977. *Solute movement in the soil-root system.* Blackwell, Oxford.

Nykänen, H., and Koricheva, J. 2004. Damage-induced changes in woody plants and their effects on insect herbivore performance: a meta-analysis. *OIKOS* 104:247–268.

Oades, J.M. 1984. Soil organic matter and structural stability: mechanism and implications for management. *Plant Soil* 76:319–337.

Oades. J.M. 1988. The retention of organic matter in soils. *Biogeochemistry* 5:35–70.

Odion, D.C., Frost, E.J., Strittholt, J.R., Jiang, H., Dellasalla, D.A., and Morowitz, M.A. 2004. Patterns of fire severity and forest conditions in the western Klamath Mountains, California. *Conserv. Biol.* 18:927–936.

Odling-Smee, F.J., Laland, K.N., and Feldman, M.W. 2003. *Niche construction: the neglected process in evolution.* Princeton Univ. Press, Princeton, NJ.

Odum, E.P. 1969. The strategy of ecosystem development. *Science* 164:262–270.

Odum, E.P. 1971. *Fundamentals of ecology.* W.B. Saunders, Philadelphia.

Odum, E.P. 1985. Trends expected in stressed ecosystems. *BioScience* 35:419–422.

Oekland, B., Bakke, A., Haagvar, S., and Kvamme, T. 2005. What factors influence the diversity of saproxylic beetles? A multi-scaled study of spruce forests in southern Norway. *Biodivers. Conserv.* 5:75–101.

Office of Technology Assessment.1989. An analysis of the Montreal protocol on substances that deplete the ozone layer. In *The challenge of global warming.* Edited by Timothy Wirth and D.E. Abrahamson. Island Press, Washington, DC, 291–304.

O'Grady, J.J., Reed, D.H., Brook, B.W., and Frankham, R. 2004. What are the best correlates of predicted extinction risk? *Biol. Conserv.* 118:513–520.

O'Hara, G.W., Boonkerd, N., and Dilworth, M.J. 1988. Mineral constraints to nitrogen fixation. *Plant Soil* 108:93–110.

Ohmart, C.P., Stewart, L.G., and Thomas J.R. 1983. Leaf consumption by insects in three Eucalyptus forest types in southeastern Australia and their role in short-term nutrient cycling. *Oecologia* 59:322–330.

Ohtonen, R., Fritze, H., Pennanen, T., Jumpponen, A., and Trappe, J. 1999. Ecosystem properties and microbial community changes in primary succession on a glacier forefront. *Oecologia* 119:239–246.

Okasanen, L., Fretwell, D., Arruda, J., and Niemela, P. 1981. Exploitation ecosystems in gradients of primary productivity. *Am. Nat.* 118:240–261.

Oki, T., and Kanae, S. 2006. Global hydrological cycles and world water resources. *Science* 313:1068–1072.

Oksanen, L. 1988. Ecosystem organization: mutualism and cybernetics or plain Darwinian struggle for existence? *Am. Nat.* 131:424–444.

Olesen, J.M., and Jornado, P. 2002. Geographic patterns in plant-pollinator mutualistic networks. *Ecology* 83:2416–2424.

Oliva, P., Viers, J., and Dupre, B. 2003. Chemical weathering in granitics environments. *Chem. Geol.* 202:225–256.

Oliver, C. and Larson, B. 1990. *Forest stand dynamics.* Wiley.

Oliver, C.D. 1981. Forest development in North America following major disturbances. *For. Ecol. Manage.* 3:153–168.

Oliver, K.M., Moran, N.A., and Hunter, M.S. 2005. Variation in resistance to parasitism in aphids is due to symbionts not host genotype. *PNAS* 102:12795–12800.

Olson, D.M., Dinerstein, E., Wikramanyake, E.D., Burgess, N.D., Powell, G.V.N., Underwood, E.C., D'Amico, J.A., Itoua, I., Strand, H.E., Morrison, J.C., Loucks, C.J., Allnut, T.F., Ricketts, T.H., Kura, Y., Lamoreux, J.F., Wettengel, W.W., Hedao, P., and Kassem, K.R. 2001. Terrestrial ecoregions of the world: A new map of life on Earth. *BioScience* 51:933–938.

Olson, J.S. 1966. Energy storage and the balance of producers and decomposers in ecological systems. *Ecology* 44:322–331.

Olson, R.K., and Reiners, W.A. 1983. Nitrification in subalpine balsam fir soils: tests for inhibitory factors. *Soil Biol. Biochem.* 15:413–418.

O'Neill, E.G., Luxmoore, R.J., and Norby, R.J. 1987a. Elevated atmospheric CO_2 effects on seedling growth, nutrient uptake, and rhizosphere bacterial populations of *Liriodendron tulipifera* L. *Plant Soil* 104:3–11.

O'Neill, E.G., Luxmoore, R.J., and Norby, R.J. 1987b. Increases in mycorrhizal colonization and seedling growth in *Pinus echinata* and *Quercus alba* in an enriched CO_2 atmosphere. *Can. J. For. Res.* 17:878–883.

O'Neill, R.V. 1976. Ecosystem persistence and heterotrophic regulation. *Ecology* 57:1244–1253.

O'Neill, R.V., and DeAngelis, D.L. 1981. Comparative productivity and biomass relations of forest ecosystems. In *Dynamic properties of forest ecosystems.* Edited by D.E. Reichle. Cambridge Univ. Press, Cambridge, MA, 411–449.

O'Neill, R.V., DeAngelis, D.L., Waide, J.B., and Allen, T.F.H. 1986. *A hierarchial concept of ecosystems.* Princeton Univ. Press, Princeton, NJ.

O'Neill, R.V., Ross-Todd, B.M., and O'Neill, F.G. 1980. Synthesis of terrestrial microcosm studies. In *Microcosms as potential screening tools for transport of toxic substances:final report.* ORNL/TM-7028. Edited by V.F. Harris. Oak Ridge National Laboratory, Oak Ridge, TN, 239–257.

Öpik, M., Moora, M., Liira, J., and Zobel, M. 2006. Composition of root-colonizing arbuscular mycorrhizal fungal communities in different ecosystems around the globe. *J. Ecol.* 94:778–790.

Oren, R. 1996. Nutritional disharmony—soil and weather effects on source-sink interactions. In *Plant response to air pollution.* Edited by M. Iqbal, and M. Yunus. John Wiley & Sons, Chichester, England, 75–98.

Oren, R., and Schulze, E.-D. 1989. Nutritional disharmony and forest decline: a conceptual model. In *Forest decline and air pollution: a study of spruce* (Picea abies (L.) Karst.) *on acid soils.*

Ecological Studies Series, Vol. 77. Edited by E.-D. Schulze, O.L. Lange, and R. Oren. Springer-Verlag, Berlin, 425–443.

Oren, R., Ewers, B.E., Todd, P., Phillips, N., and Katul, G. 1998. Water balance delineates the soil layer in which moisture affects canopy conductance. *Ecol. Applic.* 8:990–1002.

Oren, R., Schulze, E.D., Werk, K.S., and Meyer, J. 1988. Performance of two *Picea abies* (L.) Karst. stands at different stages of decline. VII. Nutrient relations and growth. *Oecologia* 77:163–173.

Oren, R., Sperry, J.S., Katul, G.G., Pataki, D.E., Ewers, B.E., Phillips, N., and Schäfer, K.V.R. 1999. Survey and synthesis of intra- and interspecific variation in stomatal sensitivity to vapour pressure deficit. *Plant Cell Environ.* 22:1515–1526.

Orians, G.H. 1975. Diversity, stability and maturity in natural ecosystems. In *Unifying concepts in ecology.* Edited by W.H. von Dobben and R.H. Lowe-McConnell. Dr. W. Junk By. Publishers, The Hague, 139–150.

Orme, C.D.L., Davies, R.G., Burgess, M., Eigenbrod, F., Pickup, N., Olson, V.A., Webster, A.J., Ding, T.-S., Rasmussen, P.C., Ridgeley, R.S., Stattersfield, A.J., Bennett, P.M., Blackburn, T.M., Gaston, K.J., and Owens, I.P.F. 2005. Global hotspots of species richness are not congruent with endemism or threat. *Nature* 436:1016–1019.

Orr, D. 2002. Four challenges of sustainability. *Conserv. Biol.* 16:1457–1460.

Ostfeld, R.S., and Holt, R.D. 2004. Are predators good for your health? Evaluating evidence for top-down regulation of zoonotic disease reservoirs. *Front. Ecol. Environ.* 2:13–20.

Ostfeld, R.S., and Keesing, F. 2004. Oh the locusts sand, then they dropped dead. *Science* 306:1488–1489.

Ostling, A. 2005. Neutral theory tested by birds. *Nature* 436:635–636.

Outerbridge, R.A., and Trofymow, J.A. 2004. Diversity of ectomycorrhizae on experimentally planted Douglas-fir seedlings in variable retention forestry sites on southern Vancouver Island. *Can. J. Bot.* 82:1671–1681.

Overpeck, J.T., Bartlein, P.J., and Webb III, T. 1991. Potential magnitude of future vegetation change in eastern North America: comparisons with the past. *Science.* 254:692–695.

Overpeck, J.T., Rind, D., and Goldberg, R. 1990, Climate-induced changes in forest disturbance and vegetation. *Nature* 343:51–53.

Ovington, J.D. 1956. The composition of tree leaves. *Forestry* 29:22–28.

Ovington, J.D. 1959. The circulation of minerals in plantations of *Pinus sylvestris* L. *Ann. Bot.* 23:229–239.

Ovington, J.D. 1983. *Temperate broad-leaved evergreen forests.* Ecosystems of the World, Vol. 10. Elsevier, Amsterdam.

Owen, D.F. 1978. Why do aphids synthesize melezitose? *Oikos* 31:264–267.

Owen, D.F. 1980. How plants may benefit from the animals that eat them. *Oikos* 35:230–235.

Pacala, S.W., and Crawley, M.J. 1992. Herbivores and plant diversity. *Am. Nat.* 140:2433–260.

Pace, M.L., Cole, J.J., Carpenter, S.R., and Kitchell, J.F. 1999. Trophic cascades revealed in diverse ecosystems. *Trends Ecol. Evol.* 14:483–488.

Pace, N.R. 1996. New perspective on the natural microbial world: molecular microbial ecology. *Am. Soc. Microbiol. News* 62:463–470.

Packer, A., and Clay, K. 2000. Soil pathogens and spatial patterns of seedling mortality in a temperate tree. *Nature* 404:278–281.

Padilla, F.M., and Pugnaire, F.I. 2006. The role of nurse plants in the restoration of degraded environments. *Front. Ecol. Environ.* 4:196–202.

Page, S.E., Siegert, F., Riely, J.O., Boehm, H.-D.V., Jaya, A., and Limin, S. 2000. The amount of carbon released from peat and forest fires in Indonesia in 1997. *Nature* 420:61–65.

Page-Dumroese, D.S., Jurgensen, M.F., Tiarks, A.E., Ponder Jr., F., Sanchez, F.G., Fleming, R.L., Kranabetter, J.M., Powers, R.F., Stone, D.M., Elioff, J.D., and Scott, D.A. 2006. Soil physical property changes at the North American Long-Term Soil Productivity study sites: 1 and 5 years after compaction. *Can. J. For. Res.* 36:551–564.

Paine, R.T. 1974. Intertidal community structure. Experimental studies on the relationship between a dominant competitor and its principal predator. *Oecologia* 15:93–120.

Paine, R.T., Tegner, M.J., and Johnson, E.A. 1998. Compounded perturbations yield ecological surprises. *Ecosystems* 1:535–545.

Paine, T.D., Raffa, K.F., and Harrington, T.C. 1997. Interactions among scolytid bark beetles, their associated fungi, and live host conifers. *Annu. Rev. Entomol.* 42:179–206.

Paine, T.D., Stephen, F.M., and Gates, R.G. 1988. Phenology of an induced response in loblolly pine following inoculation of fungi associated with the southern pine beetle. *Can. J. For. Res.* 18:1556–1562.

Palik, B.J., Mitchell, R.J., and Hiers, J.K. 2002. Modeling silviculture after natural disturbance to maintain biological diversity: balancing complexity and implementation. *For. Ecol. Manage.* 155:347–356.

Palm, CA., Houghton, R.A., Melillo, J.M., and Skole, D.L. 1986. Atmospheric carbon dioxide from deforestation in southeast Asia. *Biotropica.* 18:177–188.

Palmer, M., Bernhardt, E., Chornesky, E., Collins, S., Dobson, A., Duke, C., Gold, B., Jacobsen, R., Kingsland, S., Kranz, R., Mappin, M., Martinez, M.L., Micheli, F., Morse, J., Pace, M., Pascual, M., Palumbi, S., Riechman, O.J., Simons, A., Townsend, A., and Turner, M. 2004. Ecology for a crowded planet. *Science* 304:1251–1252.

Palmer, T.N., and Brankovic, C. 1989. The 1988 US drought linked to anomalous sea surface temperature. *Nature* 338:54–56.

Palmiotto, P.A., Davies, S.J., Vogt, K.A., Ashton, M.S., Vogt, D.J., and Ashton, P.S. 2004. Soil-related habitat specialization in dipterocarp rain forest tree species in Borneo. *J. Ecol.* 92:609–625.

Palmroth, S., Maier, C.A., McCarthy, H.R., Oishi, A.C., Kim, H.-S., Johnsen, K., Katul, G.G., and Oren, R. 2005. Contrasting responses to drought of forest floor CO_2 efflux in a loblolly pine plantation and a nearby oak-hickory forest. *Global Change Biol.* 11:1–14.

Palmroth, S., Oren, R., McCarthy, H.R., Johnsen, K.H., Finzi, A.C., Butnor, J.R., Ryan, M,G., and Schlesinger, W.H. 2006. Aboveground sink strength in forests controls the allocation of carbon belowground and its $[CO_2]$-induced enhancement. *Proc. Natl. Acad. Sci. U.S.A.* 103:19362–19367.

Palov, M.P., and Zhdano, A.P. 1972. On migrations and aggressive behavior of brown bears in the Siberian Taiga. In *Proceedings of conference on bear ecology, morphology, protection and utilization.* Soviet Union Academy of Sciences, Moscow, 64–66.

Pankow, W., Boller, T., and Wiemken, A. 1991. The significance of mycorrhizas for protective ecosystems. *Experimentia* 47:391–394.

Parchman, T.L., and Benkman, C.W. 2002. Diversifying coevolution between crossbills and black spruce on Newfoundland. *Evolution* 56:1663–1672.

Parke, J.L. 1984. Inoculum potential of ectomycorrhizal fungi in forest soil from southwest Oregon and northern California. *For. Sci.* 30:300–304.

Parke, J.L., Linderman, R.G., and Black, C.H. 1983b. The role of ectomycorrhizas in drought tolerance of Douglas-fir seedlings. *New Phytol.* 95:83–95.

Parke, J.L., Linderman, R.G., and Trappe, J.M. 1983a. Effects of forest litter on mycorrhiza development and growth of Douglas-fir and western red cedar seedlings. *Can. J. For. Res.* 13:666–671.

Parker, G.G., and Russ, M.E. 2004. The canopy surface and stand development: assessing forest canopy structure and complexity with near-surface altimetry. *For. Ecol. Manage.* 189:307–315.

Parker, G.G., Harmon, M.E., Lefsky, M.A., Chen, J., Van Pelt, R., Weiss, S.B., Thomas, S.C., Winner, W.E., Shaw, D.C., and Franklin, J.F. 2004. Three-dimensional structure of an old-growth Pseudotsuga-tsuga canopy and its implications for radiation balance, microclimate, and gas exchange. *Ecosystems* 7:440–453.

Parker, J.D., Burkepile, D.E., and Hay, M.E. 2006. Opposing effects of native and exotic herbivores on plant invasions. *Science* 311:1459–1461.

Parker, V.T., and Yoder-Williams, M.P. 1989. Reduction of survival and growth of young *Pinus jeffreyi* by an herbaceous perennial, *Wyethia mollis*. *Am. Midl. Nat.* 121:105–111.

Parkinson, D., Visser, S., and Whittaker, J.B. 1979. Effects of collembolan grazing on fungal colonization of leaf litter. *Soil Biol. Biochem.* 11:529–535.

Parmesan, C., and Yohe, G. 2003. A globally coherent fingerprint of climate change impacts across natural systems. *Nature* 421:37–42.

Parsons, J.W., and Tinsley, J. 1975. Nitrogenous substances. In *Soil components*, Vol. 1. Edited by J.E. Gieseking. Springer, New York, 263–331.

Pastor, J., and Post, W.M. 1988. Response of northern forests to CO_2-induced climate change. *Nature* 334:55–58.

Pastor, J., Aber, J.D., McClaugherty, C.A., and Melillo, J.M. 1984. Aboveground production and N and P cycling along a nitrogen mineralization gradient on Blackhawk Island, Wisconsin. *Ecology* 65:256–268.

Pastor, J., Dewey, B., Moen, R., Mladenoff, D.J., White, M., and Cohen, Y. 1998. Spatial patterns in the moose-forest-soil ecosystem on Isle Royale, Michigan, USA. *Ecol. App.* 8:411–424.

Pastor, J., Gardner, R.H., Dale, V.H., and Post, W.M. 1987. Successional changes in nitrogen availability as a potential factor contributing to spruce declines in boreal North America. *Can. J. For. Res.* 17:1394–1400.

Pastor, J., Naiman, R.J., Dewey, B., and McInnes, P. 1988. Moose, microbes and the boreal forest. *BioScience* 38:770–777.

Patric, J.H., Evans, J.O., and Helvey, J.D. 1984. Summary of sediment yield data from forested land in the United States. *J. For.* 82:101–104.

Patridge, L., and Harvey, P.H. 1988. The ecological context of life history evolution. *Science* 241:1449–1455.

Patriquin, D.G., Dobereiner, J., and Jain, D.K. 1983. Sites and processes of association between diazotrophs and grasses. *Can. J. Microbiol.* 29:900–915.

Patten, B.C., and Odum, E.P. 1981. The cybernetic nature of ecosystems. *Am. Nat.* 118:886–895.

Paul, E. 1984. Dynamics of organic matter in soils. *Plant Soil* 76:275–285.

Paul, E.A., and Clark, F.E. 1989. *Soil microbiology and biochemistry.* Academic Press, San Diego.

Paul, E.A., and Clark, F.E. 1996. *Soil microbiology and biochemistry*, 2nd ed. Academic Press, San Diego.

Paul, E.A., and Juma, N.G. 1986. Mineralization and immobilization of soil nitrogen by microorganisms. *Ecol. Bull. (Stockholm)* 33:179–195.

Paustian, K., Parton, W.J., and Persson, J. 1992. Modeling soil organic matter in organic-amended and nitrogen-fertilized long-term plots. *Soil Sci. Soc. Am. J.* 56:476–488.

Pawlowski, K., and Newton, W.E., editors. 2006. *Nitrogen-fixing actinorhizal symbioses*. Nitrogen fixation: origins, applications, and research progress, Vol. 6. Springer.

Payette, S., Fortin, M.-J., and Gamache, I. 2001. The subarctic forest-tundra: the structure of a biome in a changing climate. *BioScience* 51:709–718.

Pearce, F. 2005. Climate warning as Siberia melts. *New Sci.* 187:12–12.

Peck, J., and McCune, B. 1998. Commercial moss harvest in northwest Oregon: biomass and accumulation of epiphytes. *Biol. Conserv.* 86:299–305.

Peck, J.E., Grabner, J., Ladd, D., and Larsen, D.R. 2004. Microhabitat affinities of Missouri Ozarks lichens. *Bryologist* 107:47–61.

Peet, R.K., and Christensen, N.L. 1980. Succession: a population process. *Vegetatio* 43:131–140.

Pendall, E., Bridgham, S., Hanson, P.J., Hungate, B., Kicklighter, D.W., Johnson, D.W., Law, B.E., Luo, Y., Megonigal, J.P., Olsrud, M., Ryan, M.G., and Wan, S. 2004. Below-ground process responses to elevated CO_2 and temperature: a discussion of observations, measurement methods, and models. *New Phytol.* 162:311–322.

Pennisi, E. 1992. Hairy harvest. *Sci. News* 141:366–367.

Pennisi, K. 2005. Sky-high experiments. *Science* 309:1314–1315.

Percy, K.E., Awmack, C.S., Lindroth, R.L., Kubiske, M.E., Kopper, B.J., Isebrands, J.G., Pregitzer, K.S., Hendry, G.R., Dickson, R.E., Zak, D.R., Oksanen, E., Sober, J., Harrington, R., and Karnosky, D.F. 2002. Altered performance of forest pests under atmospheres enriched by CO_2 and O_3. *Nature* 420:403–407.

Perkins, T.E., and Matlack, G.R. 2002. Human-generated pattern in commercial forests of southern Missippi and consequences for the spread of pests and pathogens. *For. Ecol. Manage.* 157:143–154.

Perlin, J.A. 1989. *Forest journey. The role of wood in the development of civilization.* W.W. Norton, New York.

Perry, D., Rose, S.L., Pilz, D., and Schoenberger, M. 1984. Reduction of natural FE^{3+} chelators in disturbed forest soils. *Soil Sci. Soc. Am. J.* 48:379–382.

Perry, D.A. 1984. A model of physiological and allometric factors in the self-thinning curve. *J. Theor. Biol.* 106:383–401.

Perry, D.A. 1985. The competition process in forest stands. In *Attributes of trees as crop plants.* Edited by M.G.R. Cannell and J.E. Jackson. Institute of Terrestrial Ecology, Huntingdon, England, 481–506.

Perry, D.A. 1988a. An overview of sustainable forestry. *J. Pestic. Reform* 8:8–12.

Perry, D.A. 1988b. Landscape patterns and forest pests. *Northwest Environ. J.* 4:213–228.

Perry, D.A. 1995. Self-organizing systems across scales. *Tree* 10:241–244.

Perry, D.A. 1998. The scientific basis of forestry. *Annu. Rev. Ecol. Syst.* 29:435–466.

Perry, D.A., and Borchers, J.G. 1990. Climate change and ecosystem response. *Northwest Environ. J.* 6:293–313.

Perry, D.A., and Lotan, J.E. 1978. Variation in lodgepole pine (*Pinus contorta* var. *Latifolia)*: greenhouse response of wind pollinated families from five populations to day length and temperature-soil. *Can. Jour. For. Res.* 8:81–89.

Perry, D.A., Lotan, J.E., Hinz, P., and Hamilton, M. 1978. Variation in lodgepole pine: family response to stress induced by polyethylene glycol 6000. *Forest Sci.* 24:523–526.

Perry, D.A., and Lotan, J.E. 1979. A model of fire selection for serotiny in lodgepole pine. *Evolution* 33:958–968.

Perry, D.A., and Maghembe, J. 1989. Ecosystem concepts and current trends in forest management time for reappraisal. *For. Ecol. Manage.* 26:123–140.

Perry, D.A., and Pitman, G.B. 1983. Genetic and environmental influences in host resistance to herbivory: Douglas-fir and the western spruce budworm. *Z. Ange. Entomol.* 96:217–228.

Perry, D.A., and Rose, S.L. 1983. Soil biology and forest productivity: opportunities and constraints. In *IUFRO symposium onfrrest site and continuous productivity.* Serv. Gen. Tech. Rep. PNW-163. Edited by R. Ballard and S.P. Gessel. U.S. Department of Agriculture, Forest Service, Portland OR, 229–238.

Perry, D.A., Amaranthus, M.P., Botchers, J.G., Borchers, S.L., and Brainerd, R.E. 1989a. Bootstrapping in ecosystems. *BioScience* 39:230–237.

Perry, D.A., Bell, T., and Amaranthus, M.A. 1992. Mycorrhizal fungi in mixed-species forests and other tales of positive feedback, redundancy and stability. In *The ecology of mixed-species stands on trees.* Special Publ. No. 11 of the British Ecological Society. Edited by M.G.R. Cannel, D.C. Malcolm, and P.A. Robertson. Blackwell, London, 151–179.

Perry, D.A., Borchers, J.G., Borchers, S.L., and Amaranthus, M.P. 1990. Species migrations and ecosystem stability during climate change: the belowground connection. *Conserv. Biol.* 4:266– 274.

Perry, D.A., Borchers, J.G., Turner, D.P., Gregory, S.V., Perry, C.R., Dixon, R.K., Hart, S.C., Kauffman, B., Neilson, R.P., and Sollins, P. 1991. Biological feedbacks to climate change: terrestrial ecosystems as sinks and sources of carbon and nitrogen. *Northwest Environ. J.* 7:203–232.

Perry, D.A., Choquette, C., and Schroeder, P. 1987a. Nitrogen dynamics in conifer-dominated forests with and without hardwoods. *Can. J. For. Res.* 17:1434–1441.

Perry, D.A., Jing, H., Youngblood, A., and Oetter, D.R. 2004. Forest structure and fire susceptibility in volcanic landscapes of the eastern High Cascades, Oregon. *Cons. Biol.* 18:913–926.

Perry, D.A., Margolis, H., Choquette, C., Molina, R., and Trappe, J.M. 1989b. Ectomycorrhizal mediation of competition between coniferous tree species. *New Phytol.* 112:501–511.

Perry, D.A., Meurisse, R., Thomas, B., Miller, R., and Boyle, J., editors. 1989c. *Maintaining long term productivity of Pacific Northwest forest ecosystems.* Timber Press, Portland, OR.

Perry, D.A., Meyer, M.M., Egeland, D., Rose, S.L., and Pilz, D. 1982. Seedling growth and mycorrhizal formation in clearcut and adjacent, undisturbed soils in Montana: a greenhouse bioassay. *For. Ecol. Manage.* 4:261–273.

Perry, D.A., Molina, R., and Amaranthus, M.P. 1987b. Mycorrhizae, mycorrhizospheres, and reforestation: current knowledge and research needs. *Can. J. For. Res.* 17:929–940.

Petelle, M. 1980. Aphids and melezitose: a test of Owen's 1978 hypothesis. *Oikos* 35:127–128.

Petelle, M. 1984. Aphid honeydew sugars and soil nitrogen. *Soil Biol. Biochem.* 3:203–206.

Peters, C.M., Gentry, A.H., and Mendelsohn, R.O. 1989. Valuation of an Amazonian rainforest. *Nature* 339:655–656.

Peters, R.L., and Lovejoy, T.E., editors. 1992. *Global warming and biological diversity.* Yale Univ. Press, New Haven, CT.

Petersen, H. 1982. The total soil fauna and its composition. *Oikos* 39:330–339.

Petersen, H., and Luxton M. 1982. A comparation analysis of soil fauna populations and their role in decomposition processes. *Oikos* 39:288–388.

Peterson, C.J. 2000. Catastrophic wind damage to North American forests and the potential impact of climate change. *Sci. Total Environ.* 262:287–311.

Peterson, D.L., Arbaugh, M.J., Robinson, L.J., and Derderian, B.R. 1990. Growth trends of whitebark pine and logepole pine in a subalpine Sierra Nevada Forest, California, U S A. *Arctic Alpine Res.* 3:233–243.

Peterson, R.L. 1955. *North american moose.* Univ. of Toronto Press, Toronto.

Peterson, R.O. 1993. Implications of long-term research on wolves and moose in Isle Royale National Park. In *Abstracts, 78th Annual ESA Meeting.* Ecological Society of America, 392.

Petraitis, P.S., Latham, R.E., and Niesenbaum, R.A. 1989. The maintenance of species diversity by disturbance. *Q. Rev. Biol.* 64:393–418.

Petranka, J.W., and McPherson, J.K. 1979. The role of *Rhus copallina* in the dynamics of the forest-prairie ectone in north-central Oklahoma. *Ecology* 60:956–965.

Pfister, R.D., Kovaichik, B., Arno, S., and Presby, R. 1977. *Forest habitat types of Montana.* USDA Forest Service Tech. Rep. INT-34. U.S. Department of Agriculture, Forest Service, Intermountain Res. Sta., Ogden, UT.

Pfister, R.D. 1976. Land capability assessment by habitat types. In *America's renewable resource potential, 1975: the turning point.* Society of American Foresters, Washington, DC, 312–325.

Philander, X. 1989. El Nino and La Nina. *Am. Sci.* 77:451–459.

Phillips, F.J. 1909. A study of pinyon pine. *Bot. Gazette* 48:216–223.

Philpot, C.W. 1970. Influence of mineral content on the pyrolysis of plant materials. *For. Sci.* 16:461–471.

Phipps, R.L., and Whiton, J.C. 1988. Decline in long-term growth trends of white oak. *Can. J. For. Res.* 18:24–32.

Pickard, W.F. 1983. Three interpretations of the self-thinning rule. *Ann. Bot.* 51:749–757.

Pickett, S.T.A. 1976. Succession: An evolutionary interpretation. *Am. Nat.* 110:107–119.

Pickett, S.T.A., and McDonnell, M.J. 1989. Changing perspectives in community dynamics: a theory of successional forces. *Trends Ecol. Evol.* 4:241–245.

Pickett, S.T.A., and White, P.S., editors. 1986. *The ecology of natural disturbance and patch dynamics.* Academic Press, Orlando, FL.

Pickett, S.T.A., Collins, S.L., and Armesto, J.J. 1987a. A hierarchical consideration of causes and mechanisms of succession. *Vegetatio* 69:109–114.

Pickett, S.T.A., Collins, S.L., and Armesto, J.J. 1987b. Models, mechanisms and pathways of succession. *Bot. Rev.* 53:335–371.

Pickett, S.T.A., Parker, V.T., and Fiedler, P.L. 1992. The new paradigm in ecology: implications for conservation biology above the species level. In *Consrvation biology.* Edited by P.L. Fiedler and S.K. Jain. Chapman and Hall, New York, 65–88.

Pielke, R.A. 2002. The influence of land use change and landscape dynamics on the climate system: relevance for climate change policy beyond the radiative effect of greenhouse gases. *Phil. Trans. R. Soc. A* 360:1705–1719.

Piene, H. 1989a. Spruce budworm defoliation and growth loss in young balsam fir: defoliation in spaced and unspaced stands and individual tree survival. *Can. J. For. Res.* 19:1211–1217.

Piene, H. 1989b. Spruce budworm defoliation and growth loss in young balsam fir: recovery of growth in spaced stands. *Can. J. For. Res.* 19:1616–1624.

Pierce, J.L., Meyer, G.A., and Juli, J.T. 2004. Fire-induced erosion and millennial-scale climate change in northern ponderosa pine forests. *Nature* 432:87–90.

Pilz, D.P., and Perry, D.A. 1983. Impact of clearcutting and slash burning on ectomycorrhizal associations of Douglas-fir seedlings. *Can. J. For. Res.* 14:94–100.

Pimentel, D., L. Lach, R. Zuniga, and D. Morrison. 2000. Environmental and economic costs associated with non-indigenous species in the United States. *BioScience* 50:53–65.

Pimm, S., Raven, P., Peterson, A., Sekercioglu, C.H., and Erlich, P.R. 2006. Human impacts on the rates of recent, present, and future bird extinctions. *PNAS* 103:10941–10946.

Pimm, S.L. 1982. *Food webs.* Chapman and Hall, London.

Pimm, S.L. 1984. The complexity and stability of ecosystems. *Nature* 307:321–326.

Pimm, S.L. 1986. Community stability and structure. In *Conservation biology: the science of scarcity and diversity.* Edited by M.E. Soule. Sinauer Associates, Sunderland, MA, 309–329.

Pimm, S.L. 1991. *The balance of nature? Ecological issues in the conservation of species and communities.* Univ. of Chicago Press, Chicago.

Pimm, S.L. 1996. Lessons from a kill. *Biodivers. Conserv.* 5:1059–1067.

Pimm, S.L., and Jenkins, C. 2005. Sustaining the variety of life. *Sci. Am.* 293:66–73.

Pimm, S.L., and Lawton, J.H. 1980. Are food webs divided into compartments? *J. Anim. Ecol.* 49:879–898.

Pimm, S.L., Jones, H.L., and Diamond J. 1988. On the risk of extinction. *Am. Nat.* 132:757–785.

Pirozynski, K.A. 1981. Interactions between fungi and plants through the ages. *Can. J. Bot.* 59:1824–1827.

Pirozynski, K.A., and Malloch, D.W. 1975. The origin of land plants: a matter of mycotrophism. *Biosystems* 6:153–164.

Pitman, N.C.A., Terborgh, N., Silman, M.R., and Nunez, P.V. 1999. Tree species distributions in an upper Amazonian forest. *Ecology* 80:2651–2661.

Platt, W.G., Evans, G.W., and Rathbun, S.L. 1988. The population dynamics of a long-lived conifer *(Pinus palustris)*. *Am. Nat.* 131:491–525.

Poage, N.J., and Tappeiner, J.C. 2002. Long-term patterns of diameter and basal area growth of old-growth Douglas-fir trees in western Oregon. *Can. J. For. Res.* 32:1232–1243.

Pohlman, A.A., and McColl, J.G. 1988. Soluble organics from forest litter and their role in metal dissolution. *Soil Sci. Soc. Am. J.* 52:265–271.

Poiani, K.A., Richter, B.D., Anderson, M.G., and Richter, H.E. 2000. Biodiversity conservation at multiple scales: functional sites, landscapes, and networks. *BioScience* 50:133–146.

Polis, G.A., Sears, A.L.W., Huxel, G.R., Strong, D.R., and Maron, J. 2000. When is a trophic cascade a trophic cascade? *Trends Ecol. Evol.* 15:473–475.

Pool, R. 1989. Ecologists flirt with chaos. *Science* 243:310–313.

Popma, J., Bongers, F., Martinez-Ramos, M., and Veneklaas, E. 1988. Pioneer species distribution in treefall gaps in neotropical rain forest: a gap definition and its consequences. *J. Trop. Ecol.* 4:77–88.

Porder, S., Asner, G.P., and Vitousek, P.M. 2005. Ground-based and remotely sensed nutrient availability across a tropical landscape. *PNAS* 102:10909–10912.

Posey, D.A. 1993. Indigenous knowledge in the conservation and use of world forests. In *World forests for the future.* Edited by K. Ramakrishna and G.M. Woodwell. Yale Univ. Press, New Haven, CT, 59–77.

Possingham, H.P., and Wilson, K.A. 2005. Turning up the heat on hotspots. *Nature* 436:919–920.

Post, D.A., and Jones, J.A. 2001. Hydrologic regimes of forested, mountainous, headwater basins in New Hampshire, North Carolina, Oregon, and Puerto Rico. *Adv. Water Resourc.* 24:1195–1210.

Post, D.M. 2002. The long and short of food-chain length. *Tree* 17:269–277.

Post, W.M., Emanuel, W.R., Zinke, P.J., and Stangenberger, A.G. 1982. Soil carbon pools and world life zones. *Nature* 298:156–159.

Post, W.F., Pastor, J., Zinke, P.J., Stanenberger, A.G. 1985. Global patterns of nitrogen storage. *Nature* 317:613–616.

Post, W.M., Peng, T.H., Emanuel, W.R., King, S.W., Dale, V.H., and DeAngelis, D.L. 1990. The global carbon cycle. *Am. Sci.* 78:310–326.

Post, W.M., Travis, C.C., and DeAngelis, D.L. 1985. Mutualism, limited competition, and positive feedback. Chapter 12 in *The biology of mutalism.* Edited by D.H. Boucher. Oxford Univ. Press, New York, 305–325.

Postel, S., and Heise, L. 1988. Reforesting the earth. In *State of the world 1988.* W.W. Norton, New York.

Postel, S.L., and Thompson Jr., B.H. 2005. Watershed protection: capturing the ebenefits of nature's water supply services. *Nat. Resourc. Forum* 29:98–108.

Postgate, J.R. 1982. *The fundamentals of nitrogen fixation.* Cambridge Univ. Press, Cambridge, England.

Potter, C.S. 1999. Terrestrial biomass and the effects of deforestation on the global carbon cycle. *BioScience* 49:769–778.

Potter, G.L., Ellsaesser, H.W., MacCracken, M.C., and Luther, F.M. 1975. Possible climatic impact of tropical deforestation. *Nature* 258:697–698.

Potts, M.D., Davies, S.J., Bossert, W.H., Tan, S., and Nur Supardi, M.N. 2004. Habitat heterogeneity and niche structure of trees in two tropical rainforests. *Oecologia* 139:446–453.

Powell, C.L. 1980. Mycorrhizal infectivity of eroded soils. *Soil Biol. Biochem.* 12:247–250.

Powell, J.R., and Taylor, C.E. 1979. Genetic variation in ecologically diverse enviornments. *Am. Sci.* 67:590–596.

Power, M.E. 1992. Top-down and bottom-up forces in food webs: do plants have primacy? *Ecology* 73:733–746.

Power, M.E. 2001. Field biology, food web models, and management: challenges of context and scale. *Oikos* 94:118–129.

Powers, J.S., Treseder, K.K., and Lerdau, M.T. 2005. Fine roots, arbuscular mycorrhizal hyphae and soil nutrients in four neotropical rain forests: patterns across large geographical distances. *New Phytol.* 165:913–921.

Powers, R.F. 1980. Mineralizable soil nitrogen as an index of nitrogen availability to forest trees. *Soil Sci. Soc. Am. J.* 44:1314–1320.

Powers, R.F. 1983. Estimating soil nitrogen availability through soil and foliar analysis. In *Forest soils and treatment impacts.* Edited by E.L. Stone. Univ. of Tennessee, Knoxville, 353–379.

Powers, R.F. 1984. Estimating soil nitrogen availability through soil and foliar analysis. In *Proc. 6th North American Forest Soils*

Conference. Edited by E. Stone. Univ. of Tennessee, Knoxville, 353–379.

Powers, R.F. 1989. Maintaining long-term forest productivity in the Pacific Northwest: defining the issues. In *Maintaining the long-term productivity of Pacgic Northwest forest ecosystems.* Edited by D.A. Perry, R. Meurisse, B. Thomas, et al. Timber Press, Portland, OR, 3–16.

Powers, R.F. 1999. On the sustainable productivity of planted forests. *New For.* 17:263–306.

Powers, R.F. 2006. Long-term soil productivity: genesis of the concept and principles behind the program. *Can. J. For. Res.* 36:519–528.

Powers, R.F., Alban, D.R., Miller, R.E., Tiarks, A.E., Wells, C.G., Avers, P.E., Cline, R.G., Fitzgerald, R.D., and Loftus Jr, N.S. 1990. Sustaining site productivity in North American forests: problems and prospects. In *Sustained productivity of forest soils.* Edited by S.P. Gessel, D.S. Locate, G.E. Westman, and R.F. Powers. Univ. of British Columbia, Vancouver, 49–79.

Powers, R.F., Webster, S.R., and Cochran, P.R. 1988. Estimating the response of ponderosa pine forests to fertilization. In *Future forests of the mountain west: a stand culture symposium.* Gen. Tech. Rep. INT-243. Edited by W.C. Schmidt. U.S. Department of Agriculture, Forest Service, Ogden, UT, 219–225.

Prado, P.I., and Lewinsohn, T.M. 2004. Compartments in insect-plant associations and their consequences for community structure. *J. Anim. Ecol.* 73:1168–1178.

Prahl, F.G. 1990. Prospective use of molecular paleontology to test for iron limitation on marine primary productivity. Paper presented at the Organic Geochemistry Meeting Honolulu, 1990.

Pregitzer, K.S., and Barnes, B.V. 1982. The use of ground flora to indicate edaphic factors in upland ecosystems of the McCormick Experimental Forest, Upper Michigan. *Can. J. For. Res.* 12:661–672.

Pregitzer, K.S., Barnes, B.V., and Lemme, G.D. 1983, Relationship of topography to soils and vegetation in an upper Michigan ecosystem. *Soil Sci. Soc. Am. J. 47*:117–123.

Prendergast, J.R., Quinn, R.M., Lawton, J.H., Eversham, B.C., and Gibbons, D.W. 1993. Rare species, the coincidence of diversity hotspots and conservation strategies. *Nature* 365:335–337.

Prentice, K.C., and Fung, J.Y. 1990. The sensitivity of terrestrial carbon storage to climate change. *Nature* 346:48–50.

Prepas, E.E., Pinel-Alloul, B., R.Steedman, J., Planas, D., and Charette, T. 2003. Impacts of forest disturbance on boreal surface waters in Canada. In *Towards sustainable management of the boreal forest.* Edited by P.J. Burton, C. Messier, D.W. Smith, and W.L. Adamowicz. NRC Research Press, Ottawa, 369–393.

Prescott-Allen, R., and Prescott-Allen, C. 1978. *Sourcebook for a world conservation strategy: threatened vertebrates.* International Union for Conservation of Nature and Natural Resources, Gland, Switzerland.

Pressey, R.L., Possingham, H.P., and Day, J.R. 1997. Effectiveness of alternative heuristic algorithms for identifying minimum requirements for conservation reserves. *Biol. Conserv.* 80:207–219:1997.

Price, J.P. 2004. Floristic biogeography of the Hawaiian Islands: influences of area, environment and paleogeography. *J. Biogeogr.* 31:487–500.

Price, J.P., and Wagner, W.L. 2004. Speciation in Hawaiian angiosperm lineages: cause, consequence, and mode. *Evolution* 58:2185–2200.

Price, K., and Hochachka, G. 2001. Epiphytic lichen abundance: effects of stand age and composition in coastal British Columbia. *Ecol. Applic.* 11:904–913.

Price, P.W. 1980. *Evolutionary biology of parasites.* Princeton Univ. Press, Princeton, NJ.

Price, P.W. 1988. An overview of organismal interactions in ecosystems and evolutionary and ecological time. *Agric. Ecosyst. Environ.* 24:369–377.

Price, P.W. 1991. The web of life: development over 3.8 billion years of trophic relationships. In *Symbiosis as a source of evolutionary innovation.* Edited by L. Margulis and R. Fester. MIT Press, Cambridge, MA, 262–272.

Price, P.W. 2002. Resource-driven terrestrial food webs. *Ecol. Res.* 17:241–247.

Price, P.W., Westoby, M., Rice, B., Atsatt, P.R., Fritz, R.S., Thompson, J.N., and Mobley, K. 1986. Parasite mediation in ecological interactions. *Annu. Rev. Ecol. Syst.* 17:487–505.

Prigogine, I., and Stengers, I. 1984. *Order out of chaos.* Bantam, Toronto.

Primack, R.B., and Kang, H. 1989. Measuring fitnes and natural selection in wild plant populations. *Annu. Rev. Ecol. Syst.* 20:367–396.

Pritchett, W.L. 1979. *Properties and management of forest soils.* John Wiley & Sons, New York.

Pritchett, W.L., and Fisher, R.F. 1987. *Properties and management of forest soils,* 2nd ed. John Wiley & Sons, New York.

Probst, J.R., and Crow, T.R. 1991. Integrating biological diversity and resource rnanagement. *J. For.* (February):12–17.

Proctor, J. 1987. Nutrient cycling in primary and old secondary rainforests. *Appl. Geogr.* 7:135–152.

Prospero, J.M., Glaccum, R.A., and Nees, R.T. 1981. Atmospheric transport of soil dust from Africa to South America. *Nature* 289:570–572.

Puga, C. 1985. Influence of vesicular-arbuscular mycorrhizae on competition between corn and weeds in Panama. M.S. thesis, Univ. of Miami, Coral Gables, FL.

Pyne, S. 1995. *World fire: the culture of fire on earth.* Henry Holt and Co., New York.

Qian, H., Ricklefs, R.E., and White, P.S. 2005. Beta diversity of angiosperms in temperate floras of eastern Asia and eastern North America. *Ecol. Lett.* 8:15–22.

Quinn, J.F., and Hastings, A. 1987. Extinction in subdivided habitats. *Conserv. Biol.* 1:198–208.

Quinn, J.F., and Hastings, A. 1988. Extinction in subdivided habitats: reply to Gilpin. *Conserv. Biol.* 2:293–296.

Rachie, K.O. 1983. Intecropping tree legumes with annual crops. In *Plant research and agroforestry.* Edited by PA. Huxley. Int. Council for Res. in Agroforestry, Nairobi, 103–114.

Rackham, O. 1992. Mixtures, mosaics and clones: the distribution of trees within European woods and forests. In *The ecology of mixed-species stands of trees.* Edited by M.G.R. Cannel, D.C. Malcolm, and P.A. Robertson. Blackwell, Oxford, 1–20.

Radeloff, V.C., Hammer, R.B., and Stewart, S.I. 2005. Rural and suburban sprawl in the U.S. Midwest from 1940 to 2000 and its relation to forest fragmentation. *Conserv. Biol.* 19:793–805.

Rafes, P.M. 1971. Pests and the damage which they cause to forests. In *Productivity of forest ecosystems.* UNESCO, Rome, 357–367.

Raffaelli, D. 2004. How extinction patterns affect ecosystems. *Science* 306:1141–1142.

Raich, J.W., and Nadelhoffer, K.J. 1989. Belowground carbon allocation in forest ecosystems: global trends. *Ecology* 70:1346–1354.

Raison, R.J., Khanna, P.J., and Woods, P.V. 1985. Mechanisms of element transfer to the atmosphere during vegetation fires. *Can. J. For. Res.* 15:132–140.

Rajaniemi, T.K., Allison, V.J., and Goldberg, D.E. 2003. Root competition can cause a decline in diversity with increased productivity. *J. Ecol.* 91:3:407–416.

Ralls, K., Harvey, P.H., and Lyles, A.M. 1986. Inbreeding in natural populations of birds and mammals. In *Conservation biology: the science of scarcity and diversity.* Edited by M.E. Soule. Sinauer Associates, Sunderland, MA, 35–56.

Raloff, J. 1992. Garden-variety tonic for stress. *Sci. News* 141:94–95.

Ralph, C.J., Sauer, J.R., and Droege, S., tech. editors. 1995. *Monitoring bird populations by point counts.* U.S. Forest Service Gen. Tech. Rep. PSW-GTR-149. Pacific Southwest Res. Sta., Albany, CA.

Ramanathan, V. 1988. The greenhouse theory of climate change: a test by an inadvertent global experiment. *Science* 240:293–299.

Ramanathan, V. 1989. Observed increases in greenhouse gases and predicted climatic changes. In *The challenge of global warming.* Edited by D.E. Abrahamson. Island Press, Washington, DC, 239–247.

Rambelli, A. 1973. The rhizopshere of mycorrhizae. In *Ectomycorrhizae: their ecology and physiology.* Edited by A.C. Marks and T.T. Kozlowski. Academic Press, London, 229–349.

Randerson, J.T., Liu, H., Flanner, M.G., Chambers, S.D., Jin, Y., Hess, P.G., Pfister, G., Mack, M.C., Treseder, K.K., Welp, L.R., Chapin, F.S., Harden, J.W., Goulden, M.L., Lyons, E., Neff, J.C., Schuur, E.A.G., and Zender, C.S. 2006. The impact of boreal forest fire on climate warming. *Science* 314:1130–1132.

Rao, A.V., and Tak, R. 2001. Influence of mycorrhizal fungi on the growth of different tree species and their nutrient uptake in gypsum mine spoil in India. *Appl. Sopil Ecol.* 17:279–284.

Raphael, M.G., and Molina, R. *Conservation of rare or little-known species*. Island Press, Washington, DC. 375.

Rapport, D.J. 1989. What constitutes ecosystem health? *Pers. Biol. Med.* 33:120–132.

Rapport, D.J., Costanza, R., and McMichael, A.J. 1998. Assessing ecosystem health. *Tree* 13:397–402.

Rapport, D.J., Regier, H.A., and Hutchinson, T.C. 1985. Ecosystem behavior under stress. *Am. Nat.* 125:617–640.

Rashid, G.H., and Shaefer, R. 1986. Denitrification studies in a temperate forest catena soil. I. Formation of ammonium during nitrate reduction. *Plant Soil* 93:367–372.

Rasmussen, D.I. 1941. Biotic communities of Kaibab Plateau, Arizona. *Ecol. Monogr.* 3:229–275.

Rastetter, E.B., Perakis, S.S., Shaver, G.R., and Agren, G.I. 2005. Terrestrial C sequestration at elevated CO_2 and temperature: the role of dissolved organic N loss. *Ecol. Applic.* 15:71–86.

Rauner, J.L. 1976. Deciduous forests. In *Vegetation and the atmosphere Vol II. Case studies*. Edited by J.L. Monteith. Academic Press, London, New York, 241–264.

Raupp, M.J., and Denno, R.E. 1983. Leafage as a predictor of herbivore distribution and abundance. In *Variable plants and herbivores in natural and managed systems.* Edited by R.E. Denno and M.S. McClure. Academic, New York, 91–124.

Rausher, M.D. 1988. Is coevolution dead? *Ecology* 69:898–901.

Raval, A., and Ramanathan, V. 1989. Observational determination of the greenhouse effect. *Nature* 342:758–761.

Read, D.J. 1987. In support of Frank's organic nitrogen theory. *Angew. Botanik.* 61:25–37.

Read, D.J. 1988. Development and function of mycorrhizal hyphae in soil. In *Mycorrhizae in the next decade.* Edited by D.M. Sylvia, L.L. Hung, and J.H. Graham. Univ. of Florida Press, Gainesville, FL, 178–180.

Read, D.J. 1991a. The mycorrhizal mycelium. In *Biotecnologia forestal: micorrhizas bosques, erosion y agricultura.* Edited by E. Barreno and M. Honorubia. UIMP, Valencia.

Read, D.J. 1991b. Mycorrhizas in ecosystems. *Experientia* 47:376–391.

Read, D.J., and Boyd, R. 1986. Water relations of mycorrhizal fungi and their host plants. In *Fungi and plants.* Edited by P.G. Ayres and L. Boddy. Cambridge Univ. Press, Cambridge, 287–303.

Read, D.J., Francis, R., and Finlay, R.D. 1985. Mycorrhizal mycelia and nutrient cycling in plant communities. In *Ecological interactions in soil.* Edited by A.H. Fitter, D. Atkinson, D.J. Read, and M.B. Usher. Blackwell Oxford, 193–218.

Reagan, D.P., and R.B. Waide, editors. 1996. *The food web of a tropical rain forest.* Univ. of Chicago Press, Chicago.

Real, L.A., and Levin, S.A. 1991. Theoretical advances: the role of theory in the rise of modern ecology. In *Foundations of ecology: classic papers with commentaries.* Edited by L.A. Real and J.H. Brown Univ. of Chicago Press, 177–191.

Recher, H.F. 1969. Bird species diversity and habitat diversity in Australia and North America. *Am. Nat.* 103:75–80.

Reddy, A.S.N. 2001. Calcium: silver bullet in signaling. *Plant Sci.* 160:381–404.

Redfield, A.C. 1958. The biological control of chemical factors in the environment. *Am. Sci.* 46:205–221.

Redford, K. 1992. The empty forest. *BioScience* 42:412–422.

Reed, D.H., and Frankham, R. 2001. How closely correlated are molecular and quantitative measures of genetic variation? A meta-analysis. *Evolution* 55:1095–1103.

Reed, D.H., O'Grady, J.J., Brook, B.W., Ballou, J.D., and Frankham, R. 2003. Estimates of minimum viable population sizes for vertebrates and factors influencing those estimates. *Biol. Conserv.* 113:2334.

Reeves, G.H., Benda, L.E., Burnett, K.M., Bisson, P.A., 5, and Sedell, R. 1995. A disturbance-based ecosystem approach to maintaining and restoring freshwater habitats of evolutionarily significant units of anadromous salmonids in the Pacific Northwest. In *Evolutionand the aquatic ecosystem: defining unique units in population conservation.* Edited by J.L. Nielsen. American Fisheries Society Symposium 17, Bethesda, MD. 334–349.

Reeves, G.H., Bottom, D.L., and Brookes, M.H., editors. 1992. *Ethical questions for resource managers.* Gen. Tech. Rpt. PNW-GTR-288. U.S. Department of Agriculture, Forest Service, Portland, OR.

Reeves, G.H., Burnett, K.M., and McGarry, E.V. 2003. Sources of large wood in the main stem of a fourth-order watershed in coastal Oregon. *Can. J. For. Res.* 33:1363–1370.

Regal, P.J. 1982. Pollination by wind and animals: ecology of geographic patterns. *Annu. Rev. Ecol. Syst.* 13:497–524.

Rehfuess, K.E. 1987. Perceptions on forest diseases in central Europe. *Forestry* 60:1–11.

Rehfuess, K.E. 1987a. Underplanting of pines with legumes in Germany. In *Symbiotic nitrogen fixation in the management of temperate forests.* Edited by J.C. Gordon, C.T. Wheeler, and D.A. Perry. Oregon State Univ., Corvallis, 374–387.

Reich, P.B., Oleksyn, J., Modrzynski, J., Mrozinski, P., Hobbie, S.E., Eisenstat, D.M., Chorover, J., Chadwick, O.A., Hale, C.M., and Tjoelker, M.J. 2005. Linking litter calcium, earthworms, and soil properties: a common garden test with 14 tree species. *Ecol. Lett.* 8:811–818.

Reichle, D.E. 1981. *Dynamic properties of forest ecosystems.* Cambridge Univ. Press, Cambridge.

Reid, C.P.P. 1984. Mycorrhizae: a root-soil interface in plant nutrition. In *Microbial-plant interactions.* Soil Science Society of America, Madison, WI, 29–49.

Reid, C.P.P., and Mexal, J.G. 1977. Water stress effects on root exudation by lodgepole pine. *Soil Biol. Biochem.* 9:417–422.

Reid, W.V. 2001. Capturing the value of ecosystem services to protect biodiversity. In *Managing human dominated ecosystems monographs in systematic botany*, Vol 84. Edited by G. Chichilnisky, G.C. Daily, P. Ehrlich, G. Heal, J.S. Miller. Missouri Botanical Garden, St. Louis,197–225.

Reid, W.V., and Miller, K.R. 1989. *Keeping options alive. The scientific basis for conserving biodiversity.* World Resources Institute, Washington, DC.

Reidel, C.R., editor. 1982. Air pollution stress and energy policy. In *New England prospects: critical choices in a time of change.* Univ. Press of New England, Hanover, NH, 85–140.

Reiners, W.A. 1986. Complementary models for ecosystems. *Am. Nat.* 127:59–73.

Reiners, W.A., and Olson, R.K. 1984. Effects of canopy components on throughfall chemistry: an experimental analysis. *Oecologia* 63:320–330.

Reiter, M.L., and Beschta, R.L. 1995. Effects of forest practices on water. In *Cumulative effects of forest practices in Oregon.* Edited by R.L. Beschta, J.R. Boyle, C.C. Chambers, W.P. Gibson, S.V. Gregory, et al. Report for Oregon Dept. of Forestry, Salem, chap. 7.

Remezov, N.P., and Pogrebnyak, P.S. 1969. *Forest soil science.* Translated by A. Gourevitch and edited by Y. Haver. Israel Program for Scientific Translation Press, Jerusalem, Israel.

Remmert, H., editor. 1991. *The mosaic cycle concept of ecosystems.* Springer-Verlag, Berlin.

Renken, R.B., Gram, W.K., Fantz, D.K., Richter, S.C., Miller, T.J., Ricke, K.B., Russell, B., and Wang, X.-Y. 2004. Effects of forest management on amphibians and reptiles in Missouri Ozark forests. *Conserv. Biol.* 18:174–188.

Reuss, J.O. 1983. Implications of the calcium-aluminum system for the effect of acid precipitation on soils. *J. Environ. Qual.* 12:591–595.

Reuss, L., Schutz, K., Haubert, D., Haggblom, M.M., Kandeler, E., and Scheu, S. 2005. Application of lipid analysis to understand trophic interactions in soil. *Ecology* 86:2075–2082.

Revelle, R.R., and Waggoner, P.E. 1989. Effects of climatic change on water supplies in the western United States. In *The challenge of global warming.* Edited by D.E. Abrahamson. Island Press, Washington, DC, 151–160.

Rhoades, C., and Binkley, D. 1996. Factors influencing decline in soil pH in Hawaiian *Eucalyptus* and *Albizia* plantations. *For. Ecol. Manage.* 80:47–56.

Rhoades, D.F. 1979. Evolution of plant chemical defence against herbivores. In *Herbivores: their interaction with secondary plant metabolites.* Edited by G.A. Rosenthal and D.R. Janzen. Academic Press, New York, 3–54.

Rhoades, D.F. 1983. Herbivore population dynamics and plant chemistry. In *variable plants and herbivores in natural and managed systems.* Edited by R.F. Denno and M.S. McClure. Academic, New York, 155–220.

Rhoades, D.F., and Cates, R.G. 1976. Toward a general theory of plant antiherbivore chemistry. In *Biochemical interactions between plants and insects. Recent advances in phytochemistry,* Vol. 10. Edited by J.W. Wallace and R.L. Mansell. Plenum Publishing, New York, 168–213.

Rhoades, D.F., and Hedin, P., editors. 1982. *Plant resistance to insects.* American Chemical Society, Washington, DC, 55–68.

Rice, E. 1984. *Allelopathy.* Academic Press, Orlando, FL.

Rice, E.L., and Pancholy, S.K. 1972. Inhibition of nitrification by climax ecosystems. *Am. J. Bot.* 59:1033–1040.

Richards, B.N. 1964. Fixation of atmospheric N in coniferous forests. *Aust. For.* 28:68–74.

Richards, B.N. 1973. Nitrogen fixation in the rhizosphere of conifers. *Soil Biol. Biochem.* 5:149–152.

Richards, B.N. 1987. *The microbiology of terrestrial ecosystems.* John Wiley & Sons, New York.

Richards, B.N.J., and Charley, N.L. 1983. Mineral cycling processes and system stability in the eucalypt forest. *For. Ecol. Manage.* 7:31–47.

Richter Jr., D.D., and Markewitz, D. 2001. *Understanding soil change.* Cambridge Univ. Press., Cambridge.

Richter, D.L., Zuellig, T.R., Bagley, S.T., and Bruhn, J.N. 1989. Effects of red pine (*Pinus resinosa* Ait.) mycorrhizoplane-associated actinomycetes on *in vitro* growth of ectomycorrhizal fungi. *Plant Soil* 115:109–116.

Riihimaki, J., Kataniemi, P., Koricheva, J., and Vehviloainen, H. 2005. Testing the enemies hypothesis in forest stands: the important role of tree species composition. *Oecologia* 142:90–97.

Riitters, K.R., Law, B.E., Kucera, R.C., Gallant, A.L., DeVelice, R.L., and Palmer, C.J. 1992. A selection of forest condition indicators for monitoring. *Environ. Monit. Assess.* 20:21–33.

Ricklefs, R.E. 1987. Community diversity: relative roles of local and reginal processes. *Science* 235:167–171.

Ricklefs, R.E. 2000. Rarity and diversity in Amazonian forest trees. *Tree* 15:83–84.

Ricklefs, R.E. 2004. A comprehensive framework for global patterns in biodiversity. *Ecol. Lett.* 7:1–15.

Ricotta, C., Avena, G., and Marchetti, M. 1999. The flaming sandpile: self-organized criticality and wildfires. *Ecol. Model.* 119:73–77.

Riegel, G.M., Svecar, T.J., and Busse, M.D. 2002. Does the presence of *Wyethia* mollis affect growth of *Pinus jeffreyi* seedlings? *West. North Am. Nat.* 62:141–150.

Rietkerk, M., Dekker, S.C., de Ruiter, P.C., and van de Koppel, J. 2004. Self-organized patchiness and catatrophic shifts in ecosystems. *Science* 305:1926–1929.

Riitters, K.H., Wickham, J.D., O'Neill, R.V., Jones, K.B., Smith, E.R., Coulston, J.W., Wade, T.G., and Smith, J.H. 2002. Fragmentation of continental United States forests. *Ecosystems* 5:815–822.

Rillig, M.C. 2004. Arbuscular mycorrhizae and terrestrial ecosystem processes. *Ecol. Lett.* 7:740–754.

Rillig, M.C., Wright, S.F., and Eviner, V.T. 2002. The role of arbuscular mycorrhizal fungi and glomalin in soil aggregation: comparing effects of five plant species. *Plant Soil* 238:325–333.

Rillig, M.C., Wright, S.F., Nichols, K.A., Schmidt, W.F., and Torn, M.S. 2001. Large contribution of arbuscular mycorrhizal fungi to soil carbon pools in tropical forest soils. *Plant Soil* 233:167–177.

Ripper, W.E. 1956. Effect of pesticides on balance of arthropod populations. *Annu. Rev. Entomol.* 1:403–438.

Ripple, W.J., and Beshcta, R.L. 2005. Linking wolves and plants: Aldo Leopold on trophic cascades. *BioScience* 55:613–621.

Rissing, S.W., Pollock, G.B., Higgins, M.R., Hagen, R.H., and Smith, D.R. 1989. Foraging specialization without relatedness or dominance among co-founding ant queens. *Nature* 338:420–422.

Rizzo, D.M., and Garbelotto, M. 2003. Sudden oak death: endangering California and Oregon forest ecosystems. *Front Ecol. Environ.* 1:197–204.

Roberge, J.-M., and P. Angelstam. 2004. Usefulness of the umbrella species concept as a conservation tool. *Conserv. Biol.* 18:76–85.

Roberts, A., and Tregonning, K. 1980. The robustness of natural systems. *Nature* 288:265–266.

Robertson, G.P. 1984. Nitrification and nitrogen mineralization in a lowland rainforest succession in Costa Rica, Central America. *Oecologia (Berlin)* 61:99–104.

Robertson, G.P., and Tiedje, J.M. 1984. Denitrification and nitrous oxid e production in successional and old-growth Michigan forests. *Soil Sci. Soc. Am. J.* 48:383–389.

Robichaux, R.H., Rundel, P.W., Stemmermann, L., Canfield, J.E., Morse, S.R., and Friedman, W.E. 1984. Tissue water deficits and plant growth in wet tropical enviroments. In *Physiological ecology of the wet tropics.* Edited by E. Medina, H.A. Mooney, and C. Vazquez-Yanes. Dr. W. Junk, The Hague, 99–112.

Roche, L. 1979. Forestry and the conservation of plants and animals in the tropics. *For. Ecol. Manage.* 2:103–122.

Rochelle, J.A., Lehmann, L.A., and Wisniewski, J., editors. 1999. *Forest fragmentation. Wildlife and management implications.* Brill, Leiden.

Rodewald, A.D., and Yahner, R.H. 2001. Influence of landscape composition on avian community structure and associated mechanisms. *Ecology* 82:3493–3504.

Rodin, L.E., and Bazilevich, N.I. 1967. *Production and mineral cycling in terrestrial ecosystems.* Oliver and Boyd, Edinburgh.

Rodrigues, A.S.L., Andelman, S.J., Bakarr, M.I., Boitani, L., Brooks, T.M., Cowling, R.M., Fishpool, L.D.C., da Fonseca, G.A.B., Gaston, K.J., Hoffman, M., Long, J.S., Marquet, P.A., Pilgrim, J.D., Pressey, R.L., Schipper, J., Sechrest, W., Stuart, S.N., Underhill, L.G., Waller, R.W., Watts, M.E.J., and Yan, X. 2004. Effectiveness of the global protected area network in representing species diversity. *Nature* 428:640–642.

Rodriguez-Iturbe, I., and Rinaldo, A. 1997. *Fractal river networks.* Cambridge Univ. Press, Cambridge.

Rodriguez-Kabana, R., Morgan-Jones, G., and Chet, I. 1987. Biological control of nematodes: soil amendments and microbial antagonists. *Plant Soil* 100:237–247.

Rogers, R.H., Bingham, G.E., Cure, J.D., Smith, J.M., and Surano, K.A. 1983. Responses of selected plant species to elevated carbon dioxide in the field. *J. Environ. Qual.* 12:569–574.

Rohde, K. 1992. Latitudinal gradients in species diversity: the search for a primary cause. *Oikos* 65:514–527.

Rojas, N.S., Perry, D.A., Li, C.Y., and Ganio, L.M. 2002. Interactions among soil biology, nutrition, and performance of actinorhizal plant species in the H.J. Andrews Experimental Forest of Oregon. *Appl. Soil Ecol.* 19:13–26.

Rojas, N.S., Perry, D.A., Li, C.Y., and Friedman, J. 1992. Influence of actinomycetes on *Frankia* infection, nitrogenase activity and seedling growth of red alder. *Soil Biol. Biochem.* 24:1043–1049.

Roland, J., and Taylor. P.D. 1997. Insect parasitoid species respond to forest structure at different spatial scales. *Nature* 386:710–713.

Rolstad, J., Gjerde, I., Storaunet, K.O., and Rolstad, E. 2001. Epiphytic lichens in norwegian coastal spruce forest: historic logging and present forest structure. *Ecol. Applic.* 11:412–436.

Romme, W.H. 1982. Fire and landscape diversity of subalpine forests in Yellowstone National Park. *Ecol. Monogr.* 52:199–221.

Romme, W.H., Everham, E.H., Frelich, L.E., Moritz, M.A., and Sparks, R.E. 1998. Are large, infrequent disturbances qualitatively different from small, frequent disturbances. *Ecosystems* 1:524–534.

Rook, D.A., Grace, J.C., Beets, P.N., Whitehead, D., Santantonio, D., and Madgwick, H.A.I. 1985. Forest canopy design: biological models and management implications. In *Attributes of trees as crop plants.* Edited by M.G.R. Cannel and J.E. Jackson. Terrestrial Ecology, Huntingdon, England, 507–524.

Rooney, N., McCann, K., Gellner, G., and Moore, J.C. 2006. Structural asymmetry and the stability of diverse food webs. *Nature* 442:265–269.

Root, R.B. 1967. The niche exploitation pattern of the blue-gray gnatcatcher. *Ecol. Monogr.* 37:317–350.

Root, R.B. 1973. Organization of a plant-arthropod association in simple and diverse habitats: the fauna of collards (*Brassica oleracea*). *Ecol. Monogr.* 45:95–124.

Root, T. 1988. Environmental factors associated with avian distributional boundaries. *J. Biogeography* 15:489–505.

Root, T.L., Price, J.T., Hall, K.R., Schneider, S.H., Rosenzweig, C., and Pounds, J.A. 2003. Fingerprints of global warming on wild animals and plants. *Nature* 421:57–60.

Rose, S.L. 1988. Above and belowground community development in a marine sand dune ecosystem. *Plant Soil* 109:215–226.

Rose, S.L., Li, C.-Y., and Hutchins, A.S. 1980. A streptomycete antagonist to *Phellinus weirii, Fomes annosus,* and *Phytophtora cinnamoni. Can. J. Microbiol.* 26:583–587.

Rose, S.L., Perry, D.A., Pilz, D., and Schoenberger, M.M. 1983. Allelopathic effects of litter on the growth and colonization of mycorrhizal fungi. *J. Chem. Ecol.* 9:1153–1162.

Rosenberg, D.K., Fraser, J.D., and Stauffer, D.F. 1988. Use and characteristics of snags in young and old forest stands in southwest Virginia. *For. Sci.* 34:224–228.

Rosenfeld, A.R., and Botkin, D.B. 1990. Trees can sequester carbon, or die, decay, and amplify the threat of global warming. Unpublished manuscript.

Roskoski, J.P. 1980. Nitrogen fixation in hardwood forests of the northeastern United States. *Plant Soil* 54:33–44.

Roslin, T., Gripenberg, S., Salminen, J.-P., Karonen, M., O'Hara, R.B., Pihlaja, K., and Pulkkinen, P. 2006. Seeing the trees for the leaves—oaks as mosaics for a host-specific moth. *Oikos* 113:106–120.

Rosswall, T. 1983. The nitrogen cycle. In *The major biogeochemical cycles and their interactions.* Edited by B. Bolin and R.B. Cook. John Wiley & Sons, Chichester, 46–50.

Rothermel, R.C. 1991. Predicting behavior and size of crown fires in the northern Rocky Mountains. Res. Paper INT-438, U.S. Department of Agriculture, Forest Service, Intermountain Res. Sta., Ogden, UT.

Roughgarden, J. 1979. *Theory of population genetics and evolutionary ecology: an introduction.* MacMillan, New York.

Rowe, J.S., and Scotter, G.W. 1973. Fire in the boreal forest. *Quat. Res.* 3:444–464.

Roy, K., Jablonski, D., and Valentine, J.W. 2004. Beyond species richness: biogeographic patterns and biodiversity dynamics using other metrics of diversity. In *Frontiers of biogeogrpahy*. Edited by M.V. Lomolino and L.A. Heaney. Sinauer, Sunderland, MA, 151–170.

Runkle, J.R. 1982. Patterns of disturbance in some old-growth mesic forests of the eastern United States. *Ecology* 62:1533–1546.

Running, S.W. 1976. Environmental control of leaf water conductance in conifers. *Can. J. For. Res.* 6:104–112.

Runyon, J., Waring, R.H., Goward, S.N., and Welles, J.M. 1994. Environmental limits on net primary production and light-use efficiency across the Oregon transect. *Ecol. Applic.* 4:1654–1659.

Russell, K.R., Guynn Jr., D.C., and Hanlin, H.G. 2002. Importance of small isolated wetlands for herptofaunal diversity in managed, young growth forests in the Coastal Plain of South Carolina. *For. Ecol. Manage.* 163:43–59.

Russo, S.E., Davies, S.J., King, D.A., and Tan, S. 2005. Soil-related performance variation and distributions of tree species in a Bornean rain forest. *J. Ecol.* 93:879–889.

Rustad, L.E., Campbell, J.L., Marion, G.M., Norby, R.J., Mitchell, M.J., Hartley, A.E., Cornelissen, J.H.C., and Gurevitch, J. 2001. A meta-analysis of the response of soil respiration, net nitrogen mineralization, and aboveground plant growth to experimental ecosystem warming. *Oecologia* 126:543–562.

Ruusila, V., Morin, J.-P., van Ooik, T., Saloniemi, I., Ossipov, V., and Haukioja, E. 2005. A short-lived herbivore on a long-lived host: tree resistance to herbivory depends on leaf age. *Oikos* 108:99–104.

Ryall, K. L., and Fahrig, L. 2005. Habitat loss decreases predator-prey ratios in a pine bark beetle system. *Oikos* 110:265–270.

Ryan, C.A. 1983. Insect-induced chemical signals regulating natural plant protection responses. In *Variable plants and herbivores in natural and managed systems.* Edited by R.S. Denno and M. McClure. Academic Press, New York, 43–60.

Ryan, M.G. 1991. A simple method for estimating gross carbon budgets for vegetation in forest ecosystems. *Tree Physiol.* 9:255–266.

Ryan, P.J., Gessel, S.P., and Zasoski, R.J. 1986. Acid tolerance of Pacific Northwest conifers in solution culture. II. Effect of varying aluminum concentration at constant pH. *Plant Soil* 96:259–272.

Ryan, R.B. 1987. Classical biological control. *J. For.* 85:29–31.

Ryan, R.B., Tunnock, S., and Ebel, F.W. 1987. The larch casebearer in North America. *J. For.* 85:33–39.

Ryerson, D.E., Swetnam, T.W., and Lynch, A.M. 2003. A tree ring reconstruction of western spruce budworm outbreaks in the San Juan Mountains, Colorado, USA. *Can. J. For. Res.* 33:1010–1028.

Rygiewicz, P.T., and Bledsoe, C.S. 1984. Mycorrhizal effects on potassium fluxes by northwest coniferous seedlings. *Plant Physiol.* 76:918–923.

Sabine, C.L., Feely, R.A., Gruber, N., Key, R.M., Lee, K., Bullister, J.L., Wanninkhof, R., Wong, C.S., Wallace, D.W.R., Tilbrook, B., Millero, F.J., Peng, T.-H., Kozyr, A., Ono, T., Rios, A.F. 2004b. The oceanic sink for anthropogenic CO_2. *Science* 305:367–371.

Sabine, C.L., Heimann, M., Artaxo, P., Bakker, D.C.E., Chen, C.-T.A., Field, C.B., Gruber, N., Le Quere, C., Prinn, R.G., Richey, J.E., Lankao, P.R., Sathaye, J.A., and Valentini, R. 2004a. Current status and past trends on the global carbon cycle. In *The global carbon cycle SCOPE 62.* Edited by C.B. Field and M.R. Raupach. Island Press, Washington, DC, 17–44

Sader, S.A., and Joyce, A.T. 1988. Deforestation rates and trends in Costa Rica, 1940 to 1983. *Biotropica* 20:11–19.

Safir, G., editor. 1987. Phylogenetic and ecological aspects of mycotrophy in the magnoliophyta (angiosperms) from an evolutionary standpoint. In *Ecophysiology of vesicular-arbuscular mycorrhizal plants.* CRC Press, Boca Raton, FL, 5–25.

Safranyik, L., Shrimpton, D.M., and Whitney, H.S. 1975. An interpretation of the interaction between lodgepole pine, the mountain pine beetle and its associated blue stain fungi in westem Canada. In *Management of lodgepole pine ecosystems.* Edited by D.M. Baumgartner. Washington State Univ. Coop. Ext. Service, Pullman, 406–428.

Sagan, C., Toon, O.B., and Pollack, J.B. 1979. Anthropogenic albedo changes and the Earth's climate. *Science* 206:1363–1368.

Sahrawat, K.L., and Keeney, D.R. 1986. Nitrous oxide emission from soil. *Adv. Soil Sci.* 4:103–148.

Sakai, A.K., Allendorf, F.W., Holt, J.S., Lodge, D.M., Molofsky, J., With, K.A., Baughman, S., Cabin, R.J., Cohen, J.E., Ellstrand, N.C., McCauley, D.E., O'Neil, P., Parker, I.M., Thompson, J.N., and Weller, S.G. 2001. The population biology of invasive species. *Annu. Rev. Ecol. Syst.* 32:305–332.

Salati, E. 1987. The forest and the hydrological cycle. In *The geophysiology of Amazonia.* Edited by R.E. Dickinson. John Wiley & Sons, New York.

Saldarriaga, J.G., and West, D.C. 1986. Holocene fires in the northern Amazon Basin. *Quat. Res.* 26:358–366.

Salick, J., Herrera, R., and Jordan, C.F. 1983. Termitaria: nutrient patchiness in nutrient-deficient rainforests. *Biotropica* 15:1–7.

Salo, J., Kalliola, R. Hakkinen, I., Mäkinen, Y., Niemelä, P., Puhakka, M., and Coley, P.D. 1986. River dynamics and the diversity of Amazon lowland forest. *Nature* 322:254–258.

Salthe, S. 1990. The evolution of the biosphere: towards a new mythology. *World Futures* 30:53–67.

Salwasser, H. 1991. In search of an ecosystem approach to endangered species conservation. In *Balancing on the brink of extinction: the Endangered Species Act and lessons for the future.* Edited by K. Kohm. Island Press, Washington, DC, 247.

Salwasser, H., and Tappeiner II, J.C. 1981. An ecosystem approach to integrated timber and wildlife habitat management. In *Transactions of the 46th North American Wildlife and Natural Resources Conference.* Wildlife Management Institute, Washington, DC, 473–487.

Sanchez, F.G., Tiarks, A.E., Kranabetter, J.M., Page-Dumroese, D.S., Powers, R.F., Sanborn, P.T., and Chapman, W.K. 2006. Effects of organic matter removal and soil compaction on fifth-year mineral soil carbon and nitrogen contents for sites across the United States and Canada. *Can. J. For. Res.* 36:565–576.

Sanchez, P.A., Palm, C.A., Davey, C.B., Szott, L.T., and Russell, C.E. 1985. Trees as soil improvers in the humid tropics? In *The Ecology of mixed-species stands of trees.* Edited by M.G.R. Cannell, D.C. Malcolm, and P.A. Robertson. Blackwell Scientific Publications, London, 327–358.

Sanders, I.R., Koch, A., and Kuhn, G. 2003. Arbuscular mycorrhizal fungi: genetics of multigenomic, clonal networks and its ecological consequences. *Biol. J. Linnean Soc.* 79:59–60.

Sanford Jr., R.L., Saldarriaga, J., Clark, K.E., Uhl, C., and Herreera, R. 1985. Amazon rain-forest fires. *Science* 227:53–55.

Santantonio, D., and Grace, J.C. 1987. Estimating fine-root production and turnover from biomass and decomposition data: a compartment-flow model. *Can. J. For. Res.* 17:900–908.

Santharam, V., and S. Rangaswami. 2002. Habitat renewal and recolonisation of birds in the Rishi Valley campus. Presented at the National Seminar on Conservation of Eastern Ghats, March 24–26, 2002, Tirupati, Andhra Pradesh.

Sapp, J. 1991. Living together: symbiosis and cytoplasmic inheritance. In *Symbiosis as a source of evolutionary innovation.* Edited by L. Margulis and R. Fester. MIT Press, Cambridge, MA, 15–25.

Satchell, J.E., and Lowe, D.G. 1967. Selection of leaf litter by *Lumbricus terrestris.* In *Progress in soil biology.* Edited by O. Graff and J.E. Satchell North Holland, Amsterdam, 102–119.

Sayer, J., Ishwaran, N., Thorsell, J., and Sigaty, T. 2000. Tropical forest biodiversity and the World Heritage Convention. *Ambio* 29:302–309.

Scatena, F.N., Moya, S., Estrada, C., and Chinea, J.D. 1996. The first five years in the reorganization of aboveground biomass and nutrient use following hurricane Hugo in the Bisley Experimental Watersheds, Luquillo Experimental Forest, Puerto Rico. *Biotropica* 28:424–440.

Schaberg, P.G., DeHayes, D.H., and Hawley, G.J. 2001. Anthropogenic calcium depletion: a unique threat to forest ecosystem health? *Ecosyst. Health* 7:214–228.

Schäfer, K.V.R., Oren, R., Ellsworth, D.S., Lai, C.-T., Herrick, J.D., Finzi, A.C., Richter, D.D., and Katul, G.G. 2003. Exposure to an enriched CO_2 atmosphere alters carbon assimilation and allocation in a pine forest ecosystem. *Global Change Biol.* 9:1378–1400.

Schaffer, W.M., and Kot, M. 1985. Do strange attractors govern ecological systems? *BioScience* 35:342–350.

Scheffer, M., Carpenter, S., Foley, J.A., Folkes, C., and Walker, B. 2001. Catastrophic shifts in ecosystems. *Nature* 413:591–596.

Schellnhuber, J., and Sahagian, D. 2002. The twenty-three GIAM questions. *Global Change News Lett.* 49:20–21.

Schemske, D.W., and Horvitz, C.C. 1984. Variation among floral visitors in pollination ability: a precondition for mutualism specialization. *Science* 225:519–521.

Schenck, N.C., Siqueira, J.O., and Oliveria, E. 1987. Changes in the incidence of VA mycorrhizal fungi with changes in ecosystems. In *Interrelations between microorganisms and plants in soil.* Edited by V. Vancura and F. Kunc. Elsevier.

Schiermeier, Q. 2005. That sinking feeling. *Nature* 435:732–733.

Schimel, D., and Baker, D. 2000. The wildfire factor. *Nature* 420:29–30.

Schimel, J.P., and Bennett, J. 2004. Nitrogen mineralization: challenges of a changing paradigm. *Ecology* 85:591–602.

Schimel, J.P., and Firestone, M.K. 1989. Nitrogen incorporation and flow through a confierous forest soil profile. *Soil Sci. Am. J.* 53:779–784.

Schimel, J.P., Cates, R.G., and Ruess, R. 1998. The role of balsam poplar secondary chemicals in controlling soil nutrient dynamics through succession in the Alaskan taiga. *Biogeochemistry* 42:221–234.

Schimel, J.P., Firestone, M.K., and Killham, K.S. 1984. Identification of heterotrophic nitrification in a Sierran forest soil. *Appl. Environ. Microbiol.* 48:802–806.

Schimper, A.F.W. 1903. *Plant geography upon a physiological basis.* Clarendon, Oxford.

Schisler, D.A., and Linderman, R.G. 1984. Evidence for the involvement of the soil microbiota in the exclusion of *Fusarium* from coniferous forest soils. *Can. J. Microbiol.* 30:142–150.

Schisler, D.A., and Linderman, R.G. 1989. Influence of humic-rich organic amendments to coniferous nursery soils on Douglas-fir growth, damping-off and associated soil microorganisms. *Soil Biol. Biochem.* 21:403–408.

Schlapfer, F., and Schmid, B. 1999. Ecosystem effects of biodiversity: a classification of hypotheses and exploration of empirical results. *Ecol. Applic.* 9:893–912.

Schlesinger, W.H. 1990. Vegetation an unlikely answer. *Nature* 348:679.

Schlesinger, W.H. 1997. *Biogeocemistry. An analysis of global change.* Academic Press, San Diego.

Schlesinger, W.H., Reynolds, J.F., Cunningham, G.L., Huenneke, L.F., Jarrell, W.M., Virginia, R.A., and Whitford, W.G. 1990. Biological feedbacks in global desertification. *Science* 247:1043–1048.

Schmid, H.P. 1994. Source areas for scalars and scalar fluxes. *Boundary-Layer Meteorol.* 67:293–318.

Schmid, P.E., Tokeshi, M., and Schmid-Araya, J.M. 2000. Relation between population density and body size in stream communities. *Science* 289:1557–1560.

Schmiegelow, F.K.A., and Monkkonen, M. 2002. Habitat loss and fragmentation in dynamic landscapes: avian perspectives from the boreal forest. *Ecol. Applic.* 12:375–389.

Schmiegelow, F.K.A., Machtans, C.S., and Hannon, S.J. 1997. Are boreal birds resilient to fragmentation? An experimental study of short-term community responses. *Ecology* 78:1914–1932.

Schmitz, O.J., Hamback, P.A., and Beckerman, A.P. 2000. Trophic cascades in terrestrial systems: a review of the effects of carnivore removals on plants. *Am. Nat.* 155:141–153.

Schneider, B.U., Meyer, J., Schulze, E.-D., and Zech, W. 1989. Root and mycorrhizal development in healthy and declining Norway spruce stands. In *Forest decline and air pollution.* Edited by E.-D. Schulze, O.L., Lange, and R. Oren. Springer-Verlag, New York.

Schneider, S.H., and Boston, P.J., editors. 1991. *Scientists on Gaia.* MIT Press, Cambridge, MA.

Schneider, S.H., and Londer, R. 1984. *The coevolution of climate and life,* Sierra Club Books, San Francisco.

Schnieder, S.H., Miller, J.R., Crist, E., and Boston, P.J., editors. 2004. *Scientists debate Gaia.* MIT Press, Cambridge, MA.

Schoenau, J.J., and Bettany, J.R. 1987. Organic matter leaching as a component of carbon, nitrogen, phosphorus, and sulfur cycles in a forest, grassland, and gleyed soil. *Soil Sci. Soc. Am. J.* 51:646–651.

Schoenberger, M.M., and Perry, D.A. 1982. The effect of soil disturbance on growth and mycorrhizal formation of Douglas-fir and western hemlock seedlings: a greenhouse bioassay. *Can. J. For. Res.* 12:343–353.

Schoener, T.W. 1989. Food webs from the small to the large. *Ecology* 70:1559–1589.

Schoener, T.W. 1989. The ecological niche. In *Ecological concepts.* Edited by J.M. Cherrett. Blackwell, Oxford, 79–113.

Schoenly, K., Beaver, R.A., and Heumier, T.A. 1991. On the trophic relations of insects: a food-web approach. *Am. Nat.* 173:597–638.

Schoennagel, T., Veblen, T.T., and Romme, W.H. 2004. The interaction of fire, fuels, and climate across Rocky Mountain forests. *BioScience* 54:661–676.

Scholander, P.F., Hammel, H.T., Bradstreet, E.D., and Hemmingsten, E.A. 1965. Sap pressure in vascular plants. *Science* 148:339–346.

Scholze, M., Knorr, W., Arnell, N.W., and Prentice, I.C. 2006. A climate-change risk analysis for world ecosystems. *PNAS* 103:13116–13120.

Schonewald-Cox, C.M., Chambers, S.M., MacBryde, B., and Thomas, L., editors. 1983. *Genetics and conservation.* Benjamin/Cummings, London.

Schowalter, T.D. 1981. Insect herbivore relationship to the state of the host plant: biotic regulation of ecosystem nutrient cycling through ecological succession. *Oikos* 37:126–130.

Schowalter, T.D. 1989. Canopy arthropod community structure and herbivory in old-growth and regenerating forests in western Oregon. *Can. J. For. Res.* 19:318–322.

Schowalter, T.D. 2006. Insect Ecology: An Ecosystem Approach, 2nd edition. Academic Press, San Diego.

Schowalter, T.D., and Ganio, L.M. 1998. Vertical and seasonal variation in canopy arthropod communities in an old-growth conifer forest in in southwestern Washington. *Bull. Entomol. Res.* 88:633–640.

Schowalter, T.D., and Lowman, M.D. 1999. Forest herbivory: insects. In *Ecosystems of disturbed ground.* Edited by L.R. Walker. Elsevier, Amsterdam, chap. 9.

Schowalter, T.D., and Sabin, T.E. 1991. Litter microarthropod responses to canopy herbivory, season and decomposition in litterbags in a regenerating conifer ecosystem in western Oregon. *Biol. Fertil. Soils.* 11:93–96.

Schowalter, T.D., and Turchin, P. 1993. Southern pine beetle infestation development: interaction between pine and hardwood basal areas. *For. Sci.* 39:201–210.

Schowalter, T.D., Coulson, R.N., and Crossley Jr., D.A. 1981. Role of southem pine beetle and fire in maintenance of structure and function of the southeastern coniferous forest. *Environ Entomol.* 10:821–825.

Schowalter, T.D., Sabin, T.E., Stafford, S.G., and Sexton, J.M. 1991. Phytophage effects on primary production, nutrient turnover, and litter decomposition of young Douglas-fir in western Oregon. *For. Ecol. Manage.* 42:229–243.

Schroth, M.N., and Hancock, J.G. 1982. Disease-suppressive soil and root-colonizing bacteria. *Science* 216:1376–1381.

Schroth, M.N., and Weinhold, A.R. 1986. Root-colonizing bacteria and plant health. *Hortscience* 21:1295–1298.

Schultz, J.C. 1983. Habitat selection and foraging tactics of catepillars in heterogeneous trees. In *Variable plants and herbivores in natural and managed systems.* Edited by R.F. Denno and M.S. McClure. Academic Press, New York, 61–90.

Schultz, J.C. 1988. Plant responses induced by herbivores. *Trends Ecol. Evol.* 3(2):45–49.

Schulze, E.D. 1989. Air pollution and forest decline in a spruce (*Picea abies*) forest. *Science* 244:776–783.

Schulze, E.D., Lange, O.L., and Oren, R., editors. 1989. *Forest decline and air pollution.* Ecological studies, Vol. 77. Springer-Verlag, Berlin.

Schulze, E.D., Oren, R., and Lange, O.L. 1989. Processes leading to forest decline: a synthesis. In *Forest decline and air pollution. A study of spruce* (Picea abies) *on acid soils.* Ecological Studies, Vol. 77. Edited by E.-D. Schulze, O.L. Lange, and R. Oren. Springer-Verlag, Berlin, 459–468.

Schulze, E.D., Robichaux, R.H., Grace, J., Rundel, P.W., and Ehleringer, J.R. 1987. Plant water blance. *BioScience* 37:30–37.

Schutt, P. 1989. Symptoms as bioindicators of decline in European forests. In *Biological markers of air-pollution stress and damage in forests.* National Academy Press, Washington, DC, 119–124.

Schwartz, M.W. 1999. Choosing the appropriate scale of reserves for conservation. *Annu. Rev. Ecol. Syst.* 30:83–108.

Schwartzman, D.W., and Volk, T. 1989. Biotic enhancement of weathering and the habitability of Earth. *Nature* 340:457–460.

Schweitzer, J.A., Bailey, J.K., Hart, S.C., and Whitham, T.G. 2005a. Nonadditive effects of mixing cottonwood genotype on litter decomposition and nutrient dynamics. *Ecology* 86:2834–2840.

Schweitzer, J.A., Bailey, J.K., Hart, S.C., Wimp, G.M., Chapman, S.K., and Whitham, T.G. 2005b. The interaction of plant genotype and herbivory decelerate leaf litter decomposition and alter nutrient dynamics. *Oikos* 110:133–145.

Schweitzer, J.A., Bailey, J.K., Rehill, B.J., Martinsen, G.D., Hart, S.C., Lindroth, R.L., Keim, P., and Whitham, T.G. 2004. Genetically based trait in a dominant tree affects ecosystem processes. *Ecol. Lett.* 7:127–134.

Schwemmler, W. 1991. Symbiogenesis in insects as a model for cell differentiation, morphogenesis, and speciation. In *Symbiosis as a source of evolutionary innovation.* Edited by L. Margulis and R. Fester. MIT Press, Cambridge, MA, 178–204.

Scott, A.I. 1974. Biosynthesis of natural products. *Science* 184:760–764.

Scott, D.A., Norris, A.Mc., and Burger, J.A. 2005. Rapid indices of potential nitrogen mineralization for intensively managed hardwood plantations. *Comm. Soil Sci. Plant Anal.* 36:1421–1434.

Scott, J.M., Csuti, B., Jacobi, J.D., and Estes, J.E. 1987. Species richness. *BioScience* 37:782–788.

Scott, J.M., Davis, F.W., McGhie, R.G., Wright, R.G., Groves, C., and Estes, J. 2001. Nature reserves: do they capture the full range of America's biological diversity? *Ecol. Applic.* 11:999–1007.

Scott, J.M., Norse, E.A., Arita, H., Dobson, A., Estes, J.A., Foster, M., Gilbert, B., Jensen, D.B., Knight, R.L., Mattson, D., and Soule, M.E. 1999. The issue of scale in selecting and designing biological reserves. In *Continental conservation.* Edited by M.E. Soule and J. Terborgh. Island Press, Washington, DC, 19–37.

Scott, N.A., and Binkley, D. 1997. Foliage litter quality and annual net N mineralization: comparison across North American forest sites. *Oecologia* 111:151–159.

Seagle, S.W., and Sturtevant, B.R. 2005. Forest productivity predicts invertebrate biomass and ovenbird (*Seiurus aurocapillus*) reproduction in Appalachian landscapes. *Ecology* 86:1531–1539.

Seastedt, T.R. 1984. The role of microarthropods in decomposition and mineralization processes. *Annu. Rev. Entomol.* 29:25–46.

Seastedt, T.R., and Crossley Jr., D.A. 1981. Sodium dynamics in forested ecosystems and the animal starvation hypothesis. *Am. Nat.* 117:1029–1034.

Seastedt, T.R., Crossley Jr., D.A., and Hargrove, W.W. 1983. The effects of low-level consumption by canopy arthropods on the growth and nutrient dynamics of black locust and red maple trees in the southern Appalachians. *Ecology* 64:1040–1048.

Sedell, J., Sharpe, M., Dravnieks Apple, D., Copenhagen, M., and Furniss, M. 2000. *Water and the Forest Service.* FS-660. U.S. Department of Agriculture, Forest Service, Washington, DC.

Sedell, J.R., Bison, P.A., Swanson, F.J., and Gregory, S.V. 1988. What we know about large trees that fall into streams and rivers. Gen. Tech. Rep. PNW-GTR-229 ed. U.S. Department of Agriculture, Forest Service, Portland, Oregon, 47–82.

Sedjo, R.J. 2001. S.J. Hall Lecture delivered September 28, 2001. College of Natural Resources, Univ. of California, Berkeley. www.cnr.berkeley.edu/forestry/old_files/lectures/hall/2001 sedjo/2001sedjo.html.

Sekercioğlu, C., Daily, G.C., and Ehrlich, P.R. 2004. Ecosystem consequences of bird declines. *PNAS* 101:18042–18047.

Selås, V., Hogstad, O., Kobro, S., and Rafoss, T. 2004. Can sunspot activity and ultraviolet-B radiation explain cyclic outbreaks of forest moth pest species? *Proc. Royal Soc. B.* 271:1897–1901.

Selmants, P.C. 2007. Carbon, nitrogen, and phosphorus dynamics across a three million year substrate age gradient in northern Arizona, USA. Doctoral dissertation, Northern Arizona Univ., Flagstaff.

Sensinig, T. 2002. Development, fire history, and current and past growth of old-growth and young-growth forest stands in the Cascade, Siskiyou, and Mid-Coast Mountains of southwestern Oregon. Ph.D. thesis. Oregon State Univ., Corvallis.

Setala, H. 2002. Sensitivity of ecosystem functioning to changes in trophic structure, functional group composition and species diversity in belowground food webs. *Ecol. Res.* 17:207–215.

Setala, H., and Huhta, V. 1991, Soil fauna increase *Betula pendula* growth: laboratory experiments with coniferous forest floor. *Ecology* 72:665–671.

Setälä, H., and McLean, M.A. 2004. Decomposition rate of organic substrates in relation to the species diversity of soil saprophytic fungi. *Oecologia*, 139:98–107.

Setala, H., Haimi, J., and Huhta, V. 1988. A microcosm study on the respiration and weight loss in birch litter and raw humus as influenced by soil fauna. *Biol. Fertil. Soils.* 5:282–287.

Seymour, R.S., and Hunter Jr., M.L. 1992. *New forestry in eastern spruce-fir forests: principles and applications to Maine.* Miscellaneous Publ. 716. Univ. of Maine, Orono.

Seymour, R.S., White, A.S., and deMaynadier, P.G. 2002. Natural disturbance regimes in northeastern North America—evaluating Silvicultural systems using natural scales and frequencies. *For. Ecol. Manage.* 155:357–367.

Shaffer, M.L. 1981. Minimum population sizes for species conservation. *BioScience* 31:131–134.

Shands, W.E., and Hoffman, J.S., editors. 1987. *The greenhouse effect, climate change, and U.S. forests.* The Conservation Foundation, Washington, DC.

Sharitz, R.R., Boring, L.R., Van Lear, D.H., and Pinder III, J.E. 1992. Integrating ecological concepts with natural resource management of southern forests. *Ecol. Applic.* 2:226–237.

Sharma, E. 1988. Alitudinal variation in nitrogenase activity of the Himalayan alder naturally regenerating on landslide-affected sites. *New Phytol.* 108:411–416.

Sharma, G.D. 1983. Influence of jhumming on the structure and function of microorganisms in a forested ecosystem. *Hill Geogr.* 2:1–11.

Shaw, C., and Pawluk, S. 1986. The development of soil structure by *Octolasion tyrtaeum, Apporrectodea turgida* and *Lumbricus terrestris* in parent materials belonging to different textural classes. *Pedobiologia* 29:327–339.

Shaw, G.C., Little, H.A., and Durzan, D.J. 1978. Effect of fertilization of balsam fir trees on spruce budworm nutrition and development. *Can. J. For. Res.* 8:364–374.

Shearer, G., and Kohl, D.R. 1986. N_2-fixation in field settings: estimations based on natural ^{15}N abundance. *Aust. J. Plant Physiol.* 13:699–756.

Shelford, V.E. 1963. *The ecology of North America.* Univ. of Illinois Press, Urbana.

Shemakhanova, N.M. 1967. *Mycotrophy of woody plants.* Translated by S. Nenchonok. Edited by E. Rabinovitz. U.S. Department of Commerce, Springfield, VA.

Sheppard, B. 1990. The politics of new forestry. Paper presented to the 1990 meeting of the Coos Chapter, Society of American Foresters, Coos Bay, OR.

Shestak, C.J., and Busse, M.D. 2005. Compaction alters physical but not biological indices of soil health. *Soil Sci. Soc. Am. J.* 69:236–246.

Shinozaki, K., and Kira, T. 1977. Canopy structure and light utilization. In *Primary productivity of Japanese forests.* Edited by T. Shidei and T. Kira. Univ. of Tokyo, Tokyo, 75–86.

Shortle, W.C., and Smith, K.T. 1988. Aluminum-induced calcium deficiency syndrome in declining red spruce. *Science* 240:1017–1018.

Shugart, H.H., and Seagle, S.W. 1985. Modeling forest landscapes and the role of disturbance in ecosystems and communities. In *The ecology of natural disturbance and patch dynamics.* Edited by S.T.A. Pickett and P.S. White. Academic Press, Orlando, FL, 353–368.

Shugart, H.H., Reichle, D.E., Edwards, N.T., and Kercher, J.R. 1976. A model of calcium-cycling in an east Tennessee *Liriodendron* forest: model structure, parameters and frequency response analysis. *Ecology* 57:99–109.

Shukla, J., and Mintz, Y. 1982. Influence of land-surface evapotranspiration on the Earth's climate. *Science* 215:1498–1500.

Shukla, J., Nobre, C., and Sellers, P. 1990. Amazon deforestation and climate change. *Science* 247:1322–1325.

Shultz, J.C. 1983. Habitat selection and foraging tactics of caterpillars in heterogeneous trees. In *Variable plants and herbivores in natural and managed systems.* Edited by R.F. Denno and M.S. McClure. Academic Press, London, 61–90.

Shumway, D.L., Abrams, M.D., and Ruffner, C.M. 2001. A 400–year history of fire and oak recruitment in an old-growth oak forest in western Maryland. *Can. J. For. Res.* 31:1437–1443.

Shumway, J.S., and Atkinson, W.A. 1977. *Measuring and predicting growth response in unthinned stands of Douglas-fir by paired tree analysis and soil testing.* Note No. 15. Department of Natural Resources, Washington State, Olympia.

Shuster, S.M., Lonsdorf, E.V., Wimp., G.M., Bailey, J.K., and Whitham, T.G. 2006. Community heritability measures the evolutionary consequences of indirect genetic effects on community structure. *Evolution* 60:991–1003.

Sickman, J.O., Melack, J.M., and Stoddard. J.L. 2002. Regional analysis of inorganic nitrogen yield and retention in high-elevation ecosystems of the Sierra Nevada and Rocky Mountains. *Biogeochemistry* 57/58:341–374.

Siegert, F., Ruecker, G., Hinrichs, A., and Hoffmann, A.A. 2001. Increased damage from fires in logged forests during droughts caused by El Niños. *Nature* 414:437–440.

Siepielski, A.M., and Benkman, C.W. 2004. Interactions among moths, crossbills, squirrels, and lodgepole pine in a geographic selection mosaic. *Evolution* 58:95–101.

Sieverding, E. 1989. Ecology of VAM fungi in tropical agrosystems. *Agric. Ecosyst. Environ.* 29:369–390.

Sih, A., Jonsson, B.G., and Luikart, G. 2000. Do edge effects occur over large spatial scales? *Tree* 15:134–135.

Sillett, S.C., McCune, B., Peck, J.E., Rambo, T.R., and Ruchty, A. 2000. Dispersal limitations of epiphytic lichens result in species dependent on old-growth forests. *Ecol. Applic.* 10:789–799.

Silva, I.R., Novais, R.F., Jahm, G.M., Barros, N.F., Gebrim, F.O., Nunes, F.N., Neves, J.C., and Leite, F.P. 2004. Responses of eucalypt species to aluminum: the possible involvement of low molecular weight organic acids in the Al tolerance mechanism. *Tree Physiol.* 11:1267–77.

Silver, W., Neff, J., McGroddy, M., Veldkamp, E., Keller, M., and Cosme, R. 2000. Effects of soil texture on belowground carbon and nutrient storage in a lowland Amazonian forest ecosystem. *Ecosystems* 3:193–209.

Silver, W.L., and Miya, R.K. 2001. Global patterns in root decomposition: comparisons of climate and litter quality effects. *Oecologia* 129:407–419.

Silver, W.L., Thompson, A.W., McGroddy, M.E., Varner, R.K., Dias, J.D., Silva, H., Crill, P.M., and Keller, M. 2005. Fine root dynamics and trace gas fluxes in two lowland tropical forest soils. *Global Change Biol.* 11:290–306.

Silvester, W.B. 1989. Molybdenum limitation of asymbiotic nitrogen fixation in forests of Pacific Northwest America. *Soil. Biol. Biochem.* 21:283–289.

Silvester, W.B., and Bennett, K.J. 1973. Acetylene reduction by roots and associated soil of New Zealand conifers. *Soil Biol. Biochem.* 5:171–179.

Silvester, W.B., and Smith, D.R. 1969. Nitrogen fixation by *Gunnera-Nostoc* symbiosis. *Nature* 224:1231.

Simard, S.W., Hagerman, S.M., Sachs, D.L., Heineman, J.L., and Mather, W.J. 2005. Conifer growth, *Armillaria ostoyae* root disease, and plant diversity responses to broadleaf competition reduction in mixed forests of southern interior British Columbia. *Can. J. For. Res.* 35, 843–859.

Simard, S.W., Jones, M.D., and Durall, D.M. 2003. Carbon and nutrient fluxes within and between mycorrhizal plants. *Ecol. Studies* 137:33–74.

Simard, S.W., Perry, D.A., Jones, M.D., Myrold, D.D., and Durall, D.M. 1997a. Net transfer of carbon between trees species with shared ectomycorrhizal fungi. *Nature* 388:579–582.

Simard, S.W., Perry, D.A., Smith, J.E., and Molina, R. 1997b. Effects of soil trenching on ectomycorrhizas of *Pseudotsuga menzesii* seedlings grown in mature forests of *Betula papyrifera* and *Pseudotsuga menzesii. New Phytol.* 136:327–340.

Simberloff, D. 1988. The contribution of population and community biology to conservation science. *Annu. Rev. Ecol. Syst.* 19:473–511.

Simberloff, D., and Dayan, T. 1991. The guild concept and the structure of ecological communities. *Annu. Rev. Ecol. Syst.* 22:115–143.

Simberloff, D., and Stiling, P. 1996. Risks of species introduced for biological control. *Biol. Conserv.* 78:185–192.

Simon, H.A. 1962. The architecture of complexity. *Proc. Am. Philos. Soc.* 106:467–482.

Sinclair, A.R.E., and Krebs, C.J. 2001. Trophic interactions, community organization, and the Kluane ecosystem. Chapter 3 in *Ecosystem dynamics of the boreal forest*. Edited by C.J. Krebs, S. Boutin, and R. Boonstra. Oxford Univ. Press, Oxford, 25–48.

Sinclair, A.R.E., Ludwig, D., and Clark, C.W. 2000. Conservation in the real world. *Science* 289:1875.

Singer, M.J., and Munns, D.N. 2006. *Soils: an introduction*, 6th ed. Prentice Hall, Upper Saddle River, NJ.

Singh, J.S., Raghubanshi, A.S., Singh, R.S., and Srivastava, S.C. 1989. Microbial biomass acts as a source of plant nutrients in dry tropical forest and savanna. *Nature* 338:499–500.

Sirois, L., and Payette, S. 1991. Reduced postfire tree regeneration along a boreal forest-forest tundra transect in northern Quebec. *Ecology* 72:619–627.

Sisk, T.D., and Haddad, N.M. 2002. Incorporating the effects of habitat edges into landscape models: effective area model for cross-boundary management. In *Integrating landscape ecology into natural resource management*. Edited by J. Liu and W.W. Taylor. Cambridge Univ. Press, Cambridge.

Sist, P., Fimbel, R., Sheil, D., Nasi, R., and Chevallier, M.-H. 2003. Towards sustainable management of mixed dipterocarp forests of South-east Asia: moving beyond minimum diameter cutting limits. *Environ. Conserv.* 30:364–374.

Sitch, S., Brovkin, V., von Bloh, W., van Vuuren, D., Eickout, B., and Ganopolski, A. 2005. Impacts of future land cover changes on atmosphereic CO_2 and climate. *Global Biogeochem. Cycles* 19:GB2013.

Skinner, M.E., and Bowen, G.D. 1974. The uptake and translocation of phosphate by mycelial strands of pine mycorrhizas. *Soil Biol. Biochem.* 6:53–56.

Skole, D., and Tucker, C. 1993. Tropical deforestation and habitat fragmentation in the Amazon: satellite data from 1978 to 1988. *Science* 260:1905–1909.

Skujins, J., Tann, C.C., and Borjesson, I. 1987. Dinitrogen fixation in a montane forest sere determined by $^{15}N_2$ assimilation and in situ acetylene-reduction methods. *Soil Biol. Biochem.* 19:465–471.

Skutch, A.F. 1987. *Helpers at bird's nests.* Univ. of Iowa Press, Iowa City.

Slankis, V. 1974. Soil factors influencing formation of mycorrhizae. *Annu. Rev. Phytopathol.* 12:437– 457.

Small, M.F., and Hunter, M.L. 1988. Forest fragmentation and avian nest predation in forested landscapes. *Oecologia* 76:62–64.

Smethurst, P.J., Turvey, N.D., and Attiwill, P.M. 1986. Effect of *Lupinus* spp. on soil nutrient availability and the growth of *Pinus radiata* D. Don seedlings on a sandy podzol in Victoria, Australia. *Plant Soil* 95:183–190.

Smith, C.T., Lowe, A.T., Skinner, M.F., Beets, P.N., Schoenholtz, S.H., and Fang, S. 2000. Response of radiata pine forests to residue management and fertilization across a fertility gradient in New Zealand. *For. Ecol. Manage.* 138:203–223.

Smith Jr., C.T., McCormack Jr., M.L., Hombeck, J.W., and Martin, C.W. 1986. Nutrient and biomass removals from a red spruce-balsam fir whole-tree harvest, *Can. J. For. Res.* 16:381–388.

Smith, D.R., editor. 1975. *Proceedings of the symposium on management of forest and range habitats for nongame birds, May 6–9, 1975, Tucson, Arizona.* Gen. Tech. Rep. WO-1. U.S. Department of Agriculture, Forest Service, Washington, DC.

Smith, J.E., Molina, R., Huso, M.M.P., and Larsen, M.J. 2000. Occurrence of *Piloderma fallax* in young, rotation-age, and old-growth stands of Douglas-fir (*Pseudotsuga menziesii*) in the Cascade Range of Oregon, U.S.A. *Can. J. Bot.* 78:995–1001.

Smith, J.E., Molina, R., Huso, M.M.P., Luoma, D.L., McKay, D., Castellano, M.A., Lebel, T., and Valachovic, Y. 2002. Species richness, abundance, and composition of hypogeous and epigeous ectomycorrhizal fungal sporocarps in young, rotation-age, and old-growth stands of Douglas-fir (*Pseudotsuga menziesii*) in the Cascade Range of Oregon, U.S.A. *Can. J. Bot.* 80:186–204.

Smith, J.L., and Paul, E.A. 1990. The significance of soil microbial biomass estimations. In *Soil biochemistry,* Vol. 6. Edited by J.M. Bollag and G. Stotzky. Marcel Decker, New York, 357–396.

Smith, K.W. 1992. Bird populations: effects of tree species mixtures. In *The ecology of mixed-species stands of trees.* Edited by M.G.R. Cannel, D.C. Malcolm, and P.A. Robertson. Blackwell, Oxford, 312.

Smith, M.A., Woodley, N.E., Janzen, D.H., Hallwachs, W., and Hebert, P.D.N. 2006. DNA barcodes reveal cryptic host-specificity within the presumed polyphagous members of a genus of parasitoid flies (Diptera: Tachinidae). *PNAS* 103:3657–3662.

Smith, R.B., Waring, R.H., and Perry, D.A. 1981. Interpreting foliar analyses from Douglas-fir as weight per unit of leaf area. *Can. J. For. Res.* 11:593–598.

Smith, S.E., and D.J. Read. 1997. *Mycorrhizal symbiosis*. Academic, New York.

Smith, T., and Huston, M. 1989. A theory of the spatial and temporal dynamics of plant communities. *Vegetatio* 83:49–69.

Smith, T.M., Leemans, R., and Shugart, H.H. 1992. Sensitivity of terrestnal carbon storage to CO_2 induced climate change: comparison of four scenarios based on general circulation models. *Climat. Change* 21:367–384.

Smithwick, E.A.H., Turner, M.G., Mack, M.C., and Chapin III, F.S. 2005. Postfire soil N cycling in northern conifer forests affected by severe, stand-replacing wildfires. *Ecosystems* 8:163–181.

Snow, D.W. 1971. Evolutionary aspects of fruit-eating by birds. *Ibis* 113:194–202.

Soil Survey Division Staff. 1993. *Soil survey manual.* U.S. Department of Agriculture Handbook 18. Soil Conservation Service. Soil Survey Staff. 1975. Soil taxonomy: a basic system of soil classification for making and interpreting soil surveys. USDA Handbook 436. U.S. Govt. Printing Office, Washington, DC.

Soil Survey Staff. 1999. Soil taxonomy, 2nd ed. USDA NRCS Agriculture Handbook No. 436. U.S. Gov. Printing Office, Washington, DC.

Solé, R.V., and Manrubia, S.C. 1995. Are rainforests self-organized in a critical state? *J. Theor. Biol.* 173:31–40.

Solé, R.V., Bartumeus, F., and Gamarra, J.G.P. 2005. Gap percolation in rainforests. *Oikos* 110:177–185.

Solé, R.V., Manrubia, S.C., Benton, M., Kauffman, S., and Bak, P. 1999. Criticality and scaling in evolutionary ecology. *Tree* 14:156–160.

Sollins, P. 1998. Factors influencing species composition in tropical lowland rainforest: does soil matter? *Ecology* 79:23–30.

Sollins, P., and Radulovich, R. 1988. Effects of soil physical structure on solute transport in a weathered tropical soil. *Soil Sci. Soc. Am. J.* 52:1168–1173.

Sollins, P., Cline, S.P., Verhoeven, T., Sachs, D., and Spycher, G. 1987. Patterns of log decay in old-growth Douglas-fir forests. *Can. J. For. Res.* 17:1585–1595.

Sollins, P., Cromack Jr., K., McCorison, F.M., Waxing, R.H., and Harr, R.D. 1981. Changes in nitrogen cycling at an old-growth Douglas-fir site after disturbance. *J. Environ. Qual.* 10:37–42.

Sollins, P., Glassman, C., Paul, E.A., Swanston, C., Lajtha, K., Heil, J.W., and Elliott, E.T. 1999. Soil carbon and nitrogen pools and fractions. In *Standard soil methods for long-term ecological research.* Edited by G.P. Robertson, D.C. Coleman, C.S. Bledsoe, and P. Sollins. Oxford Univ. Press, New York, 89–105.

Sollins, P., Grier, C.C., and McCorison, F.M. 1980. The internal element cycles of an old-growth Douglas-fir ecosystem in western Oregon. *Ecol. Monogr.* 50:261–285.

Sollins, P., Robertson, G.P., and Uehara, G. 1988. Nutrient mobility in variable- and permanent-charge soils. *Biogeochemistry* 6:181–199.

Sollins, P., Spycher, G., and Glassman, C.A. 1984. Net nitrogen mineralization from light- and heavy-fraction forest soil organic matter. *Soil Biol. Biochem.* 16:31–37.

Sollins, P., Swanston, C., and Kramer, M. 2007. Stabilization and destabilization of soil organic matter—a new focus. *Biogeochemistry* 85:1–7.

Solomon, A.M., and West, D.C. 1987. Simulating forest ecosystem responses to expected climate change in eastern North America: applications to decision making in the forest industry. In *The greenhouse effect, climate change, and U.S. Forests.* Edited by W.E. Shands and J.S. Hoffman. The Conservation Foundation, Washington, DC, 189–217.

Sonea, S., and Panisset, M. 1983. *A new bacteriology.* Jones & Bartlett.

Song, W., Weicheng, F., Binghong, W., and Jianjun. Z. 2001. Self-organized criticality of forest fire in China. *Ecol. Model.* 145:61–68.

Soule, M.E. 1980. Thresholds for survival: maintaining fitness and evolutionary potential. In *Consewation biology: an evolutionary ecological perspective.* Edited by M.E. Soule and B.A. Wilcox. Sinauer Associates, Sunderland, MA, 151–169.

Soule, M.E. 1983. What do we really know about extinctions? In *Genetics and conservation.* Edited by C.M. Schonewald-Cox. Benjamin/Cummings, London, 111–124.

Soule, M.E., editor. 1986. *Conservation biology: the science of scarcity and diversity.* Sinauer Associates, Sunderland, MA.

Soule, M.E. 1987. *Viable populations for conservation.* Cambridge Univ. Press, Cambridge.

Soule, M.E., and Kohm, K.A., editors. 1989. *Research priorities for conservation biology.* Island Press, Washington, DC.

Soule, M.E., and Sanjayan, M.A. 1998. Conservation targets: do they help? *Science* 279:2060–2061.

Soule, M.E., and Simberloff, D. 1986. What do genetics and ecology tell us about the design of nature reserves? *Biol. Conserv.* 35:19–40.

Soule, M.E., and Terborgh, J., editors. 1999. *Continental conservation.* Island Press, Washington, DC.

Soule, M.E., and Wilcox, B.A., editors. 1980. *Conservation biology: an evolutionary ecological perspective.* Sinauer Associates, Sunderland, MA.

Southwood, T.R.E. 1960. The abundance of Hawaiian trees and the number of their associated insect species. *Proc. Hawaii Entomol. Soc.* 17:299–303.

Southwood, T.R.E. 1961. The number of species of insects associated with various trees. *J. Anim. Ecol.* 30:1–8.

Southwood, T.R.E. 1972. The insect/plant relationship—an evolutionary perspective. In *Insect/plant relationships.* Edited by H.F. van Eulen. Blackwell, Oxford, 3–30.

Southwood, T.R.E. 1988. Tactics, strategies and templets. *Oikos* 52:3–18.

Speer, J.H., Swetnam, T.W., Wickman, B.E., and Youngblood, A. 2001. Changes in Pandora moth outbreak dynamics during the past 622 years. *Ecology* 82:679–697.

Speir, T.W., Cowling, S.C., and Sparling, G.P. 1986. Effects of microwave radiation on the microbial biomass, phosphatase activity and levels of extractable N and P in a low fertility soil under pasture. *Soil Biol. Biochem.* 18:377–382.

Spencer, C.N., McClelland, B.R., and Stanford, J.A. 1991. Shrimp stocking, salmon collapse, and eagle displacement. *BioScience* 41:14–21.

Sperry, J.S., Hacke, U.G., Oren, R., and Comstock, J.P. 2002. Water deficits and hydraulic limits to leaf water supply. *Plant Cell Environ.* 25:251–263.

Spiedel, D.R., and Agnew, A.F. 1982. *The natural geochemistry of our environment.* Westview Press, Boulder, CO.

Spiers, G.A., Gagnon, D., Nason, G.E., Packee, E.C., and Lousier, J.D. 1986. Effects and importance of indigenous earthworms on decomposition and nutrient cycling in coastal forest ecosystems. *Can. J. For. Res.* 16:983–989.

Spies, T.A. 1998. Forest structure: a key to the ecosystem. *Northwest Sci.* 72(Special Issue 2):34–39.

Spies, T.A. 2004. Ecological concepts and diversity of old-growth forests. *J. For.* 102:14–20.

Spies, TA., and Cline, S.P. 1988. Coarse woody debris in forests and plantations of coastal Oregon. Gen. Tech. Rep. PNW-GTR-229 ed. U.S. Department of Agriculture, Forest Service, Portland, OR, 5–24.

Spies, T.A., Franklin, J.F., and Klopsch, M. 1990. Canopy gaps in Douglas-fir forests of the Cascade Mountains. *Can. J. For. Res.* 20:649–658.

Spies, T.A., McComb, B.C., Kennedy, R.S.H., McGrath, M.T., Olsen, K., and Pabst, R.J. 2007. Potential effects of forest policies on terrestrial biodiversity in a multi-ownership province. *Ecol. Applic.* 17:48–65.

Spies, T.A., Reeves, G.H., Burnett, K.M., McComb, W.C., Johnson, K.N., Grant, G., Ohmann, J.L., Garman, S.L., and Bettinger, P. 2002. Assessing the ecological consequences of forest policies in a multi-ownership province in Oregon. In *Integrating landscape ecology into natural resource management.* Edited by J. Liu and W.W. Taylor. P 179–202. Cambridge Univ. Press, Cambridge.

Spies, T.A., Ripple, W.J., and Bradshaw, G.A. 1994. Dynamics and pattern of a managed coniferous forest landscape in Oregon. *Ecol. Appl.* 4:555–568.

Sprent, J.I., and Silvester, W.B. 1973. Nitrogen fixation by *Lupinus arboreus* grown in the open and under different aged stands of *Pinus radiata. New Phytol.* 72:991–1003.

Squire, R.O.P., Farrell, W., Flinn, D.W., and Aeberli, B.C. 1985. Productivity of first and second rotation stands of radiate pine on sandy soils. II. Height and volume growth at five years. *Aust. For.* 48:127–137.

SSSA. 1997. *Soil Science Society of America: glossary of soil science terms.* 1996. Soil Science Society of America, Madison, WI.

St. John, T.V., and Coleman, D.C. 1982. The role of mycorrhizae in plant ecology. *Can. J. Bot.* 61:1005–1014.

Staaf, H., and Berg, B. 1977. Mobilization of plant nutrients in a Scots pine forest moor in central Sweden. *Silva Fenn.* 11:210–217.

Staddon, P.L., Ramsey, C.B., Ostle, N., Ineson, P., and Fitter, A.H. 2003. Rapid turnover of hyphae of mycorrhizal fungi determined by AMS microanalysis of ^{14}C. *Science* 300:1138–1140.

Stadler, B., Solinger, S., and Michalzik, B. 2001. Insect herbivores and the nutrient flow from the canopy to the soil in coniferous and deciduous forests. *Oecologia* 126:104–113.

Stago, P. 1989. Apache resource initiative strengthens self-determination. In *National native forestry symposium—ethic to reality.* Vancouver, B.C., November 22–24, 1989, 13.

Stamp, N. 2003. Out of the quagmire of plant defense hypotheses. *Quart. Rev. Biol.* 78:23–55.

Stark, N., and Jordan, C.F. 1978. Nutrient retention by the root mat of an Amazonian rain forest. *Ecology* 59:434–437.

Stark, N., Spitzner, C., and Essig, D. 1985. Xylem sap analysis for determining nutritional status of trees: *Pseudotsuga menziesii. Can. J. For. Res.* 15:429–437.

Stark, R.W., Miller, P.P., Cobb Jr., F.W., Wood, D.L., and Parmeter Jr., J.R. 1968. Incidence of bark beetle infestation on injured trees. *Italgardia* 39:121.

Starkey, R.L. 1929. Some influences of the development of higher plants upon the microorganisms in the soil. I. Historical and introductory. *Soil Sci.* 27:319–334.

Starkey, R.L. 1958. Interrelations between microorganisms and plant roots in the rhizosphere. *Bacter. Rev.* 22:154–172.

Stauffer, B., Fischer, G., Neftel, A., and Oeschger, H. 1985. Increase of atmospheric methane recorded in antarctic ice core. *Science* 229:1386–1388.

Stednick, J.D. 2000. Timber management. In *Drinking water from forests and grasslands: a synthesis of the scientific literature.* Gen. Tech. Rep. SRS-39. Edited by G.E. Dissmeyer. U.S. Department of Agriculture, Forest Service, South Res. Sta., Asheville, NC.

Stehli, F.G. 1968. Taxonomic diversity gradients in pole location: the recent model. In *Evolution and Environment.* Edited by E.T. Drake. Yale Univ. Press, New Haven, CT,163–227.

Stein, D. 1988. *Lectures in the sciences of complexity.* Addison-Wesley, Redwood City, CA.

Steinberg, P.D., and Rillig, M.C. 2003. Differential decomposition of arbuscular mycorrhizal fungal hyphae and glomalin. *Soil Biol. Biochem.* 35:191–194.

Stenberg P. 1998. Implications of shoot structure on the rate of photosynthesis at different level in a coniferous canopy using a model incorporating grouping and penumbra. *Funct. Ecol.* 12:82–91.

Stenstrom, E., Ek, M., and Unestam, T. 1990. Variation in field response of *Pinus sylvestris* to nursery inoculation with four different ectomycorrhizal fungi. *Can. J. For. Res.* 20:1796–1803.

Stephens, B., Gurney, K.R., Tans, P.P., Sweeney, C., Peters, W., Bruhwiler, L., Ciais, P., Ramonet, M., Bousquet, P., Nakazawa, T., Aoki, S., Machida, T., Inoue, G., Vinnichenko, N., Lloyd, J., Jordan, A., Heimann, M., Shibistova, O., Langenfelds, R., Steele, L.P., Francey, R.J., and Denning, A.S. 2007. Weak northern and strong tropical land carbon uptake from vertical profiles of atmospheric CO_2. *Science* 316:1732–1738.

Stephens, G.R. 1981. *Defoliation and mortality in Connecticut forests.* Bull. 796. Connecticut Ag. Exp. Sta., New Haven, CT.

Stephens, S.L., and Moghaddas, J.J. 2005. Silvicultural and reserve impacts on potential fire behavior and forest conservation: twenty-five years of experience from Sierra Nevada mixed conifer forests. *Biol. Conserv.* 125:369–379.

Stern, K., and Roche, L. 1974. *Genetics of forest ecosystems.* Springer-Verlag, Berlin.

Stevenson, F.J. 1986. *Cycles of soil.* John Wiley & Sons, New York.

Stevenson, F.J. 1994. *Humus chemistry*, 2nd ed. John Wiley & Sons, New York.

Stewart, B.W., Capo, R.C., and Chadwick, O.A. 2001. Effects of rainfall on weathering rate, base cation provenance, and Sr isotope composition of Hawaiian soils. *Geochem. Cosmochem. Acta* 65:1087–1099.

Stewart, J.W.B., and Tiessen, H. 1987. Dynamics of soil organic phosphorus. *Biogeochemistry* 4:41–60.

Stockwell, C.A., Hendry, A.P., and Kinnison, M.T. 2003. Contemporary evolution meets conservation biology. *Tree* 18:94–101.

Stireman, J.O., III., Dyer, L.A., Janzen, D.H., et al. 2005. Climatic unpredictability and parasitism of caterpillars: implications of global warming. *Proc. Natl. Acad. Sci. U.S.A.* 102:17384–17387.

Stoddard, M.A., and Hayes, J.P. 2005. The influence of forest management on headwater stream amphibians at multiple spatial scales. *Ecol. Applic.* 15:811–823.

Stoecker, D.K., Michaels, A.E., and Davis, L.H. 1987. Large proportion of marine planktonic ciliates found to contain functional chloroplasts. *Nature* 326:790–791.

Stone, J. K., Bacon, C.W., and White Jr., J.F. 2000. An overview of endophytic microbes: endophytism defined. In *Microbial endophytes.* Edited by C.W. Bacon and J.F. White Jr. Dekker, New York, 3–29.

Stone, L. 2004. A three-player solution. *Nature* 430:299–300.

Stone, L., and Roberts, A. 1991. Conditions for a species to gain advantage from the presence of competitors. *Ecology* 72:1964–1972.

Stott, P.A., and Kettleborough, J.A. 2002. Origins and estimates of uncertainty in twenty-first century temperature rise. *Nature* 416:723–726.

Stott, P.A., Kettleborough, J.A., and Allen, M.R. 2006. Uncertainty in continental-scale temperature predictions. *Geophys. Res. Lett.* 33:L02708, doi:10.1029/2005GL024423.

Strang, R.M., Lindsay, K.M., and Price, R.S. 1979. Knapweeds: British Columbia's undesirable aliens. *Rangelands* 1:141–143.

Stratford, J.A., and Robinson, W.D. 2005. Gulliver travels to the fragmented tropics: geographic variation in mechanisms of avian extinction. *Front. Ecol. Environ.* 3:91–98.

Strauss, S.H. 1986. Heterosis at allozyme loci under inbreeding and crossbreeding in *Pinus attenuata*. *Genetics* 113:115–134.

Strauss, S.H. 1987. Heterozygosity and developmental stability under inbreeding and crossbreeding in *Pinus attenuata*. *Evolution* 41:331–339.

Strauss, S.Y., Webb, C.O., and Salamin, N. 2006. Exotic taxa less related to native species are more invasive. *PNAS* 15:5841–5845.

Street-Perrott, F.A., and Perrott, R.A. 1990. Abrupt climate fluctuations in the tropics: the influence of Atlantic Ocean circulation. *Nature* 343:607–612.

Stretch, R.C., and Viles, H.A. 2002. The nature and rate of weathering by lichens on lava flows in Lazarote. *Geomorphology* 47:87–94.

Strickland, T.C., and Fitzgerald, J.W. 1984. Formation and mineralization of organic sulfur in forest soils. *Biogeochemistry* 1:79–95.

Strickler, G.S., and Edgerton, P.J. 1976. Emergent seedlings from coniferous litter and soil in eastern Oregon. *Ecology* 57:801–807.

Strömgren, M., and Linder, S. 2002. Effects of nutrition and soil warming on stemwood production in a boreal Norway spruce stand. *Global Change Biol.* 8:1194–1204.

Strong, D.R. 1992. Are trophic cascades all wet? Differentiation and donor-control in speciose ecosystems. *Ecology* 73:747–754.

Strong, D.R., Simberloff, D., Abele, L.G., and Thistle, A.B. 1984. *Ecological communities: conceptual issues and the evidence.* Princeton Univ. Press, Princeton, NJ.

Strong Jr., D.R. 1979. Biogeographic dynamics of insect-host plant communities. *Annu. Rev. Entomol.* 24:89–119.

Stroo, H.F., Klein, T.M., and Alexander, M. 1986. Heterotrophic nitrification in an acid forest soil by an acid-tolerant fungus. *Appl. Environ. Microbiol.* 52:1107–1111.

Strullu, D.G., Grellier, B., Garrec, J.P., McCready, C.C., and Harley, J.L. 1986. Effects of monovalent and divalent cations on phosphate absorption by beech mycorrhizas. *New Phytol.* 103:403–416.

Strzelczyk, E., Pokojska, A., Rozycki, H., Perry, D.A., and Li, C.Y. 1993. Growth of mycorrhizal fungi in dixenic cultures with bacteria in media of different composition. *Acta Microbiol. Polonica* 42:41–50.

Su, J.C., and Woods, S.A. 2001. Importance of sampling along a vertical gradient to compare the insect fauna in managed forests. *Environ. Entomol.* 30:400–408.

Su, Q., MacLean, D.A., and Needham, T.D. 1996. The influence of hardwood content on balsam fir defoliation by spruce budworm. *Can. J. For. Res.* 26:1620–1628.

Sucoff, E. 1979. Estimate of nitrogen fixation on leaf surfaces of forest species in Minnesota and Oregon. *Can. J. For. Res.* 9:474–477.

Sukhinin, A.I., French, N.H.F., Kasischke, E.S., Hewson, J.H., Soja, A.J., Csiszar, I.A., Hyer, E.J., Loboda, T., Conrad, S.G., Romasko, V.I., Pavlichenko, E.A., Miskiv, S.I., and Slinkina, O.A. 2004. AVHRR-based mapping of fires in Russia: new products for fire management and carbon-cycle studies. *Remote Sensing Environ.* 93:546–564.

Sunda, W.G., and Kieber, D.J. 1994. Oxidation of humic substances by manganese oxides yields low-molecular-weight organic substrates. *Nature* 367:62–64.

Suter II, G.W. 1987. Species interactions. In *Methods for assessing the effects of mixtures of chemicals.* Edited by V.B. Vouk, G.C. Butler, A.C. Upton, et al. SCOPE Report. John Wiley and Sons, New York. 745–758.

Sutton, R.T., and Hodson, D.L.R. 2005. Atlantic ocean forcing of North American and European summer climate. *Science* 309:115–118.

Svancara, L.K., Brannon, R., Scott, J.M., Groves, C.R., Noss, R.F., and Pressey, R.L. 2005. Plocy-driven versus evidence-based conservation: a review of political targets and biological needs. *BioScience* 55:989–995.

Swank, W.T., and Henderson, G.S. 1976. Atmospheric input of some cations and anions to forest ecosystems in North Carolina and Tennessee. *Water Resourc. Res.* 12:541–546.

Swank, W.T., and Van Lear, D.R. 1992. Ecosystem perspectives of mutliple-use management. *Ecol. Applic.* 2:219–220.

Swank, W.T., and Vose, J.M. 1997. Long-term nitrogen dynamics of coweeta forested watersheds in the southeastern United States of America. *Global Biogeochem. Cycles* 11:657–671.

Swank, W.T., Vose, J.M., and Elliot, K.J. 2001. Long-term hydrologic and water quality responses following commercial clearcutting of mixed hardwoods on a southern Appalachian catchment. *For. Ecol. Manage.* 143:163–178.

Swank, W.T., Waide, J.B., Crossley Jr., D.A., and Todd, R.L. 1981. Insect defoliation enhances nitrate export from forest ecosystems. *Oecologia (Berlin)* 51:297–299.

Swanson, F.J., and Franklin, J.F. 1992. New forestry principles from ecosystem analysis of Pacific Northwest forests. *Ecol. Applic.* 2:262–274.

Swanson, F.J., Clayton, J.L., Megahan, W.F., and Bush, G. 1989. Erosional processes and long-term site productivity. In *Maintaining the long-term productivity of Pacific Northwest forest ecosystems.* Edited by D.A. Perry, R. Meurisse, B. Thomas, et al. Timber Press, Portland, OR, 67–81.

Swenson, J.J., and Waring, R.H. 2006. Modeled photosynthesis predicts woody plant richness at three geographic scales across the north-western United States. *Global Ecol. Biogeogr.* DOI: 10.1111/j.1466–822x.2006.00242.x, 1–16.

Swenson, R. 1991. End-directed physics and evolutionary ordering: obviating the problem of the population of one. In *The cybernetics of complex systems: self-organization, evolution, and social change.* Edited by F. Geyer. Intersystems Publications, Salinas, CA.

Swenson, W., Wilson, D.S., and Elias, R. 2000. Artificial ecosystem selection. *PNAS* 97:9110–9114.

Swetnam, T.W. 1990. *Fire history and climate in the southwestern United States.* Gen. Tech. Rep. RM-191. U.S. Department of Agriculture, Forest Service, Fort Collins, CO, 6–17.

Swetnam, T.W., and Betancourt, J.L. 1989. Enso variability and forest fires in the southwestern United States. Paper presented at the Sixth Annual Pacific Climate (PACLIM) Workshop, Asilomar, CA, March 5–8, 1989.

Swetnam, T.W., and Betancourt, J.L. 1990. Fire-southern oscillation relations in the southwestern United States. *Science* 249:1017–1020.

Swetnam, T.W., and Lynch, A.M. 1989. A tree-ring reconstruction of western spruce budworm history in the southern Rocky Mountains. *For. Sci.* 35:962–986.

Swift, M.J., Heal, O.W., and Anderson, J.M. 1979. *Decomposition in terrestrial ecosystems.* Univ. of California Press, Berkeley.

Switzer, G.L., and Nelson, L.E. 1972. Nutrient accumulation and cycling in loblolly pine (*Pinus taeda* L.) plantation ecosystems: the first twenty years. *Soil Sci. Soc. Am. Proc.* 36:143–147.

Syers, J.K., and Springett, J.A. 1984. Earthworms and soil fertility. *Plant Soil* 76:93–104.

Sylvia, D.M., Fuhrmann, J.L., Hartel, P.G., and Zuberer, D.A., editors. 2005. *Principles and application of soil microbiology*, 2nd ed. Prentice Hall, Upper Saddle River, NJ.

Tamm, C.O. 1921. On the influence of rock substrata on forest soils. *Medd. Stat. Skogsforsoks.* 18:105–164.

Tamm, C.O. 1964. Determination of nutnent requirements of forest stands. *Int. Rev. For. Res.* 115–170.

Tamm, C.O. 1979. Nutrient cycling and productivity of forest ecosystems. In *Impact of intensive harvesting on forest nutrient cycling.* Edited by A.L. Leaf. State Univ. of New York, Syracuse, 2–21.

Tan, C.S., and Black, T.A. 1976. Factors affecting the canopy resistance of a Douglas-fir forest. *Boundary-Layer Meterol.* 10:475–488.

Tans, P.P., Fung, I.Y., and Takahashi, T. 1990. Observational constraints on the global atmospheric CO_2 budget. *Science* 247:1431–1438.

Tappeiner, J., Zasada, J., Ryan, P., and Newton, M. 1991. Salmonberry clonal and population structure: the basis for persistent cover. *Ecology* 72:609–618.

Tappeiner, J.C., and Alm, A.A. 1975. Undergrowth vegetation effects on the nutrient content of litterfall and soils in red pine and birch stands in northern Minnesota. *Ecology* 56:1193–1200.

Tappeiner, J.C., Huffman, D., Marshall, D., Spies, T.A., and Bailey, J.D. 1997. Density, ages, and growth rates in old-growth and young-growth forests in coastal Oregon. *Can. J. For. Res.* 27:638–648.

Tappeiner, J.C., Zasada, J.C., Huffman, D.W., and Ganio, L.M. 2001. Salmonberry and salal annual aerial stem production: the maintenance of shrub cover in forest stands. *Can. J. For. Res.* 31:1629–1638.

Tarrant, R.E., and Trappe, J.M. 1971. The role of *Alnus* in improving the forest environment. In *Biological nitrogen fixation in natural and agricultural habitats.* Edited by T.A. Lie, and E.G. Mulder. *Plant Soil* Special Volume, 335–348.

Tate, K.R. 1984. The biological transformation of P in soil. *Plant Soil* 76:245–256.

Tate, K.R., and Salcedo, I. 1988. Phosphorus control of soil organic matter accumulation and cycling. *Biogeochemistry* 5:99–107.

Taylor, A.H., and Beaty, R.M. 2005. Climatic influences on fire regimes in the northern Sierra Nevada mountains, Lake Tahoe Basin, Nevada, USA. *J. Biogeogr.* 32:425–438.

Taylor, A.H., and Skinner, C.N. 1998. Fire history and landscape dynamics in a late-successional reserve, Klamath Mountains, California, USA. *For. Ecol. Manage.* 111:285–301.

Taylor, A.H., and Skinner, C.N. 2003. Spatial patterns and controls on historical fire regimes and forest structure in the Klamath Mountains. *Ecol. Applic.* 13:704–719.

Taylor, B.R., Parkinson, D., and Parsons, W.F.J. 1989. Nitrogen and lignin content as predictors of litter decay rates: a microcosm test. *Ecology* 70:97–104.

Taylor, C.M., and Hastings, A. 2005. Allee effects in biological invasions. *Ecology Lett.* 8:8:895–908.

Tear, T.H., Kareiva, P., Angermeier, P.L., Comer, P., Czech, B., Kautz, R., Landon, L., Mehlman, D., Murphy, K., Ruckelshaus, M., Scott, J.M., and Wilhere, G. 2005. How much is enough? The recurrent problem of setting measurable objectives in conservation. *BioScience* 55:835–849.

Tenow, O., and Larsson, S. 1987. Consumption by needle-eating insects on Scots pine in relation to season and stand age. *Holarctic Ecol.* 10:249–260.

Tepfer, D. 1984. Transformation of several species of higher plants by *agrobacterium* rhizogenes: sexual transmission of the transformed genotype and phenotype. *Cell* 37:959–967.

ter Steege, H., Pitman, N.C.A., Phillips, O.L., Chave, J., Sabatier, D., Duque, A., Molono, J.-F., Prevost, M.-F., Spichiger, R., Castellanos, H., von Hildebrand, P., and Vasquez, R. 2006. Continental-scale patterns of canopy tree composition and function across Amazonia. *Nature* 443:444–448.

Terborgh, J. 1974. Preservation of natural diversity: the problem of extinction prone species. *BioScience* 24:71 5–722.

Terborgh, J. 1986. Keystone plant resources in the tropical forest. In *Conservation biology: the science of scarcity and diversity.* Edited by M.E. Soule. Sinauer Associates, Sunderland, MA, 330–344.

Terborgh, J. 1988. The big things that run the world—a sequel to E.Q. Wilson. *Conserv. Biol.* 2:402–403.

Terborgh, J. 1992. Maintenance of diversity in tropical forests. *Biotropica* 24:283–292.

Terborgh, J., and Soule, M.E. 1999. Why we need megareserves: large scale reserve networks and how to design them. In *Continental conservation.* Edited by M.E. Soule and J. Terborgh. Island Press, Washington, DC, 199–209.

Terborgh, J., and Winter, B. 1983. A method for siting parks and reserves with special reference to Colombia and Ecuador. *Biol. Conserv.* 27:45–58.

Terborgh, J., Estes, J.A., Paquet, P., Ralls, K., Boyd-Heger, D., Miller, B.J., and Noss, R.F. 1999. The role of top carnivores in regulating terrestrial ecosystems. In *Continental conservation*. Edited by M.E. Soule and J. Terborgh. Island Press, Washington, DC, 39–64.

Terborgh, J., Feeley, K., Silman, M., Nuñez, P., and Balukjian, B. 2006. Vegetation dynamics of predator-free land-bridge islands. *J. Ecol.* 94:253–263.

Terborgh, J., Lopez, L., Nunez V., P., Rao, M., Shafabuddin, G., Orihuela, G., Riveros, M., Ascanio, R., Adler, G.H., Lambert, T.D., and Balbas, L. 2001. Ecological meltdown in predator-free forest fragments. *Science* 294:1923–1926.

Tevis Jr., L. 1956. Pocket gophers and seedlings of red fir. *Ecol.* 37:379–381.

Thatcher, R.C., Searcy, J.L., Coster, J.E., and Hertel, G.D., editors. 1980. *The southern pine beetle.* USDA Forest Service Tech. Bull. 1631. U.S. Department of Agriculture, Forest Service, Washington, DC.

Theng, K.G. 1991. Soil science in the tropics—the next 75 years. *Soil Sci.* 151:76–90.

Theodorou, C., and Bowen, G.D. 1971. Effects of non-host plants on growth of mycorrhizal fungi of radiata pine. *Aust. For.* 35:17–22.

Thielges, B.A. 1963. Altered polyphenol metabolism in the foliage of *Pinus sylvestris* associated with European pine sawfly attack. *Can. J. Bot.* 46:724–725.

Thomas, C.D. 1990. What do real population dynamics tell us about minimum viable population sizes? *Conserv. Biol.* 4:324–327.

Thomas, J.A., Telfer, M.G., Roy, D.B., Preston, C.D., Greenwood, J.J.D., Asher, J., Fox, R., Clarke, R.T., and Lawton, J.H. 2004. Comparative losses of British butterflies, birds, and plants and the global extinction crisis. *Science* 303:1879–1881.

Thomas, J.W., editor. 1979. *Wildlife habitats in managed forests of the Blue Mountains of Oregon and Washington.* Agriculture Handbook No. 553. U.S. Department of Agriculture, Forest Service, Washington, DC.

Thomas, J.W., Anderson, R.G., Maser, C., and Bull, E.L. 1979. Snags. In *Wildlife habitats in managed forests of the Blue Mountains of Oregon and Washington.* Edited by J.W. Thomas. Agriculture Handbook No. 553. U.S. Department of Agriculture, Forest Service, Washington, DC, 60–77.

Thomas, J.W., Forsman, E.D., Lint, J.B., Meslow, E.C., Noon, B.R., and Verner, J. 1990. *A conservation strategy for the northern spotted owl.* U.S. Fish and Wildlife Service, Portland, OR.

Thompson, G.W., and Medve, R.J. 1984. Effects of aluminum and manganese on the growth of ectomycorrhizal fungi. *Appl. Environ. Microbiol.* 48:556–560.

Thompson, J.N. 1982. *Interaction and coevolution.* John Wiley & Sons, New York.

Thompson, J.N. 1988. Coevolution and alternative hypotheses on insect/plant interactions. *Ecology* 69:893–895.

Thompson, J.N. 2005. *The geographic mosaic of coevolution.* Univ. of Chicago Press, Chicago.

Thompson, J.N. 2006. Mutualistic webs of species. *Science* 312:372–373.

Thompson, J.R., Spies, T.A., and Ganio, L.M. 2007. Re-burn severity in managed and unmanaged vegetation in a large wildfire. *PNAS* 104:10743–10748.

Thompson, R.M., and Townsend, C.R. 2005. Food-web topology varies with spatial scale in a patchy environment. *Ecology* 86:1916–1925.

Throop, H.L., and Lerdau, M.T. 2004. Effects of nitrogen deposition on insect herbivory: implications for community and ecosystem processes ecosystems. *Ecosystems* 7:109–133.

Tiivel, T. 1991. Cell symbiosis, adaptation, and evolution: insect-bacteria examples. In *Symbiosis as a source of evolutionary innovation. Speciation and morphogenesis.* Edited by L. Margulis and R. Fester. MIT Press, Cambridge, MA, 170–177.

Tilman, D. 1982. *Resource competition and community structure.* Princeton Univ. Press, Princeton, N.J.

Tilman, D. 1985. The resource-ratio hypothesis of plant succession. *Am. Nat.* 125:827–852.

Tilman, D. 1986. Evolution and differentiation in terrestrial plant communities: the importance of the soil resource: light gradient. In *Community ecology.* Edited by J. Diamond and T.J. Case. Harper and Row, New York, 359–380.

Tilman, D. 2004. Niche tradeoffs, neutrality, and community structure: a stochastic theory of resource competition, invasion, and community assembly. *PNAS* 101:18854–10861.

Tilman, D., and Downing, J.A. 1994. Biodiversity and stability in grasslands. *Nature* 367:363–365.

Tilman, D., Lehman, C.L., and Bristow, C.E. 1998. Diversity-stability relationships: statistical inevitability or ecological consequence? *Am. Nat.* 151:277–288.

Tilman, D., Reich, P.B., and Knops, J.M.H. 2006. Biodiversity and ecosystem stability in a decade long grassland experiment. *Nature* 441:629–632.

Tilman, G.D. 1984. Plant dominance along an experimental nutrient gradient. *Ecology* 65:1445–1453.

Timmer, V.R., and Stone, E.L. 1978. Comparative foliar analysis of young balsam fir fertilized with nitrogen, phosphorus, potassium, and lime. *Soil Sci. Soc. Am. J.* 42:125–130.

Tinley, K.L. 1982. The influence of soil moisture balance on ecosystem patterns in southern Africa. In *Ecological studies 42. Ecology of tropical savannas.* Edited by B.J. Huntley and B.H. Walker. Springer-Verlag, New York, 175–192.

Tisdall, J.M., and Oades, J.M. 1982. Organic matter and water stable aggregates in soils. *Soil Sci.* 33:141–164.

Tiunov, A.V., and Scheu, S. 2005. Facilitative interactions rather than resource partitioning drive diversity-functioning relationships in laboratory fungal communities. *Ecol. Lett.* 8:618–625.

Tjepkema, J. 1979. Oxygen relations in leguminous and actinorhizal nodules. In *Symbiotic nitrogen fixation in the management of temperate forests.* Edited by J.C. Gordon, C.T. Wheeler, and D.A. Perry. Oregon State Univ., Corvallis, 175–186.

Tkeda, T., Matsumura, F., and Benjamin, D.M. 1977. Mechanism of feeding discrimination between matured and juvenile foliage by two species of pine sawflys. *J. Chem. Ecol.* 3:677–694.

Todd, R.I., Meyer, R.D., and Waide, J.B. 1978. Nitrogen fixation in a deciduous forest in the southeastern United States. *Ecol. Bull. (Stockholm)* 26:172–177.

Torgersen, T.R., and Dahlsten, D. L. 1978. Natural enemies. In *The Douglas-fir tussock mosth: a synthesis.* USDA Forest Service Tech. Bull. 1585. Edited by T.R. Torgensen, D.L. Dahlsten, M. H. Brooks et al. U.S. Department of Agriculture, Forest Service, Washington, DC. 47–53.

Torgersen, T.R., and Dahlsten, D.L. 1978. Natural mortality. In *The Douglas-fir tussock moth: a synthesis.* USDA Forest Service Tech. Bull. 1585. Edited by Edited by M.H. Brookes, R.W. Stark, and R.W. Campbell. U.S. Department of Agriculture, Forest Service, Washington, DC, 47–53.

Toljander, Y.K., Lindahl, B.D., Holmer, L., and Hogberg, N.O.S. 2006. Environmental fluctuations facilitate species co-existence and increase decomposition in communities of wood decay fungi. *Oecologia* 148:625–631.

Tolley, L.C., and Strain, B.R. 1984. Effects of CO_2 enrichment on growth of *Liquidambar styraciflua* and *Pinus taeda* seedlings under different irradiance levels. *Can. J. For. Res.* 14:343–350.

Torchin, M.E., and Mitchell, C.E. 2004. Parasites, pathogens, and invasions by plants and animals. *Front. Ecol. Environ.* 2:183–190.

Torgersen, T.R., Mason, R.R., and Campbell, R.W. 1990. Predation by birds and ants on two forest insect pests in the Pacific Northwest. *Stud. Avian Biol.* 13:14–19.

Torrey, J.G. 1976. Initiation and development of root nodules of *Casuarina* (Casuarinaceae). *Am. J. Bot.* 63:335–344.

Torrey, J.G. 1978. Nitrogen fixation by actinomycete-nodulated angiosperms. *BioScience* 28:586–592.

Torsvik, V., Goksøyr, J., Daae, F.L., Sørheim, R., Michalsen, J., and Salte, K. 1994. Use of DNA analysis to determine the diversity of microbial communities. In *Beyond the biomass.* Edited by K. Ritz, J. Dighton, and K.E. Giller. John Wiley and Sons, Chichester, 39–48.

Tosi, J.A. 1982. *Sustained yield management of natural forests.* Report to U.S. AID / Peru. U.S. Agency for International. Development, Washington, DC.

Toutain, F. 1987. Les litieres siege de systemes interactifs et moteurs des ces interactions. *Rev. Ecol. Biol. Sol.* 24:231–242.

Trabaud, L. 2003. In Forum, "Towards a sounder fire ecology." *Front. Ecol. Environ.* 5:274–275.

Tranquilini, W. 1979. *Physiological ecology of the alpine timberline.* Springer-Verlag, New York.

Trapnell, C.G. 1959. Ecological results of woodland burning experiments in northern Rhodesia. *J. Ecol.* 47:129–168.

Trappe, J.M. 1962. Fungus associates of ectotrophic mycorrhizae. *Bot. Rev.* 28:538–606.

Trappe, J.M. 1987. Phylogenetic and ecological aspects of mycotrophy in the angiosperms from an evolutionary standpoint. In *Ecophysiology of VA mycorrhizal plants.* Edited by G.R. Safir. CRC Press, Boca Raton, FL, 5–25.

Trappe, J.M., and Maser, C. 1977. Ectomycorrhizal fungi: interactions of mushrooms and truffles with beasts and trees. In *Mushrooms and man. An interdisciplinary approach to mycology.* Edited by T. Walters. Linn Benton Community College, Albany, OR, 165–179.

Traulsen, A., and Nowak, M.A. 2006. Evolution of cooperation by multilevel selection. *PNAS* 103:10952–10955.

Trejo, L. 1976. Diseminacion de semillas por ayes en "Los Tuxtias" Veracruz. In *Regeneracion de Selvas.* Edited by A. Gomez-Pompa, C. Vazquez-Yanes, S. del Amo, and A. Butanda. CECSA, Mexico, D.F., 447–470.

Trenbath, B.R. 1983. Decline of soil fertility and the collapse of shifting cultivation systems under intensification. In *Tropical rain forest: the Leeds symposium.* British Ecological Soc., London, 279–292.

Treseder, K.K., and Cross, A. 2006. Global distributions of arbuscular mycorrhizal fungi. *Ecosystems* 9:305–316.

Triquet, A.M., McPeek, G.A., and McComb, W.C. 1990. Songbird diversity in clearcuts with and without a riparian buffer strip. *J. Soil Water Conserv.* 45:500–503.

Trivers, R.L. 1971. The evolution of reciprocal altruism. *Quart. Rev. Biol.* 46:35–57.

Trojanowski, J., Haider, K., and Huttermann, A. 1984. Decomposition of ^{14}C-labelled lignin, holocellulose and lignocellulose by mycorrhizal fungi. *Arch. Microbiol.* 139:202–206.

Trombulak, S.C., and Frissell, C.A. 2000. Review of ecological effects of roads on terrestrial and aquatic communities. *Conserv. Biol.* 14:18–30.

Truman, R.A., Humphreys, F.R., and Ryan, P.J. 1986. Effect of varying solution ratios of Al to Ca and Mg on the uptake of phosphorus by *Pinus radiata. Plant Soil* 96:109–123.

Trumbore, S. 2006. Carbon respired by terrestrial ecosystems – recent progress and challenges. *Global Change Biol.* 12:141–153.

Trustrum, N.A., Lambert, M.G., and Thomas, V.J. 1983. The impact of soil slip erosion on hill country pasture production in New Zealand. In *Second International Conference on Soil Erosion and Conservation.* Edited by C.L. O'Loughlin and A.J. Pearce. Univ. of Hawaii, Honolulu.

Tsharntke, T., and Brandl, R. 2004. Plant-insect interactions in fragmented landscapes. *Annu. Rev. Entomol.* 49:405–430.

Tubelis, D.P., Lindenmayer, D.B., and Cowling, A. 2004. Novel patch-matrix interactions: patch width influences matrix use by birds. *Oikos* 107:634–644.

Tuomi, J., Niemelä, P., Chapin III, F.S., Bryant, J.P., and Siren, S. 1988. Defensive responses of trees in relation to their carbon / nutrient balance. In Mechanisms of woody plant defenses against insects: search for pattern. Edited by W.J. Mattson et al. Springer, New York, 57–72.

Tuomasto, H., Ruokolainen, K., and Yli-Halla, M. 2003. Dispersal, environment, and floristic variation of western Amazonian forests. *Science* 299:241–244.

Tupas, G.L., and Sajise, P.E. 1977. The role of savanna trees in plant succession. I. Ecological conditions associated with tree clumps. *Philip. J. Biol.* 6:229–244.

Turcotte, D.L., and Rundle, J.B. 2002. Self-organized complexity in the physical, biological, and social sciences. *PNAS* 99:2463–2465.

Turcotte, D.L., Malamud, B.D., Guzzeti, F., and Reichenbach, P. 2002. Self-organization, the cascade model, and natural hazards. *PNAS* 99:2530–2537.

Turlings, T.C.J., Tumlinson, J.R., and Lewis, W.J. 1990. Exploitation of herbivore-induced plant odors by host-seeking parasitic wasps. *Science* 250:1251–1253.

Turner, D.P., and Franz, E.R. 1985. The influence of western hemlock and western red cedar on microbial numbers, nitrogen mineralization, and nitrification. *Plant Soil* 88:259–267.

Turner, J., and Gessel, S.P. 1990. Forest productivity in the southern hemisphere with particular emphasis on managed forests. In Sustained productivity of forest soils. Edited by S.P. Gessel et al. Univ. of British Columbia, Faculty of Forestry, Vancouver, 23–39.

Turner, J., and Lambert, M.J. 1983. Nutrient cycling within a 27–year-old *Eucalyptus grandis* plantation in New South Wales. *For. Ecol. Manage.* 6:156–168.

Turner, J., and Lambert, M.J. 1986a. Fate of applied nutrients in a long-term superphosphate trial in *Pinus radiata. Plant Soil* 93:373–382.

Turner, J., and Lambert, M.J. 1986b. Nutrition and nutritional relationships of *Pinus radiata. Annu. Rev. Ecol. Syst.* 17:325–350.

Turner, J., Lambert, M.J., and Gessel, S.P. 1977. Use of foliage sulphate concentrations to predict response to urea application by Douglas-fir. *Can. J. For. Res.* 7:476–480.

Turner, J.R.G., and Hawkins, B.A. 2004. The global diversity gradient. In *Frontiers of biogeography*. Edited by M.V. Lomolino and L.A. Heaney. Sinauer, Sunderland, MA, 171–189.

Turner, M.G. 1987. *Landscape heterogeneity and disturbance.* Springer-Verlag, New York.

Turner, M.G., Baker, W.L., Peterson, C.J., and Peet, R.K. 1998. Factors influencing succession: lessons from large, infrequent natural disturbances. *Ecosystems* 1:511–523.

Turner, M.G., Gardner, R.H., Dale, V.R., and O'Neill, R.V. 1989. Predicting the spread of disturbance across heterogeneous landscapes. *Oikos* 55:121–129.

Turner, M.G., Gardner, R.H., and O'Neill, R.V. 2003a. *Landscape ecology in theory and practice.* Springer-Verlag, New York.

Turner, M.G., Pearson, S.M., Bolstad, P., and Wear, D.N. 2003b. Effects of land-cover change on spatial pattern of forest communities in the Southern Appalachian Mountains (USA). *Landscape Ecol.* 18:449–464.

Turner, M.G., Romme, W.R., Gardner, R.H., O'Neill, R.V., and Kratz, T.K. 1993. A revised concept of landscape equilibrium: disturbance and stability on scaled landscapes. *Landscape Ecol.* 8:213–227.

Turner, M.G., Romme, W.H., and Tinker, D.B. 2003c. Surprises and lessons from the 1988 Yellowstone fires. *Front. Ecol. Environ.* 1:351–358.

Turner, M.G., Tinker, D.B., Romme, W.H., Kashian, D.M., and Litton, C.M. 2004. Landscape patterns of sapling density, leaf area, and aboveground net primary production in postfire lodgepole pine forests, Yellowstone National Park (USA). *Ecosystems* 7:751–775.

Tuttle, C.L., Golden, M.S., and Meldahl, R.S. 1985. Surface soil removal and herbicide treatment: effects on soil properties and loblolly pine early growth. *Soil Sci. Soc. Am. J.* 49:1558–1562.

Tzedakis, P.C., Lawson, I.T., Frogley, M.R., Hewitt, G.M., and Preece, R.C. 2004. Buffered tree population changes in a Quaternary refugium: evolutionary implications. *Science* 297:2044–2046.

Udvardy, M. 1975. A classification of the biogeographical provinces of the world. Prepared as a contribution to UNESCO's Man and the Biosphere Programme Project No. 18. IUCN, Morges, Switzerland.

Uhl, C., and Jordan, C.F. 1984. Succession and nutrient dynamics following forest cutting and burning in Amazonia. *Ecology* 65:1476–1490.

Uhl, C., and Kauffman, J.B. 1990. Deforestation, fire susceptibility, and potential tree responses to fire in the eastern Amazon. *Ecology* 71:437–449.

Uhl, C., Jordan, C., Clark, K., Clark, H., and Herrera, R. 1982. Ecosystem recovery in Amazon caatinga forest after cutting, cutting and burning, and bulldozer clearing treatments. *Oikos* 38:313–320.

Uhl, C., Kauffman, J.B., and Cummings, D.L. 1988. Fire in the Venezuelan Amazon. 2: environmental conditions necessary for forest fires in the evergreen rainforest of Venezuela. *Oikos* 53:176–184.

Ulrich, B. 1969. Investigations on cycling of bioelements in forests of central Europe. In *Productivity of forest ecosystems. Proceedings of the Brussels symposium.* UNESCO, Rome, 502–507.

Ulrich, B. 1983. Soil acidity and its relation to acid deposition. In *Effects of accumulation of air pollutants in forest ecosystems.* Edited by B. Ulrich and J. Pankrath. D. Reidel Publishing Co., Dordrecht, 127–146.

Ulrich, B., Benecke, P., Harris, W.F., Khanna, P.K., and Mayer, R. 1981. Soil processes. In *Dynamic properties of forest ecosystems.* Edited by D.E. Reichle. Cambridge Univ. Press, Cambridge, England, 265–340.

Ungs, J.M. 1981. Distribution of light within the crown of an open grown Douglas-fir. PhD Thesis. Oregon State Univ., Corvallis.

Urabe, J. 1993. N and P cycling coupled by grazers' activities: food quality and nutrient release by zooplankton. *Ecology* 74:2337–2350.

Uribe, E.G., and Luttge, U. 1984. Solute transport and the life functions of plants. *Am. Sci.* 72:567–573.

U.S. Commitee for an International Geosphere-Biosphere; Program Commission on Physical *Sciences*, Mathematics and Resources National Research Council. 1986. *Global change in the geosphere-biosphere.* National Academy Press, Washington, DC.

U.S. Forest Service. 1988. *Silver Fire Recovery Project. Draft environmental impact statement. Siskiyou National Forest. Grants Pass, Oregon.* U.S. Department of Agriculture, Forest Service, Pacific Northwest Region, Portland, OR.

U.S. Forest Service. 2007. Forest health conditions in Alaska-2006. U.S. Department of Agriculture, Forest Service, Alaska Region, Juneau, AK. www.dnr.state.ak.us/forestry/pdfs/07for_h_conditions_2006_web.pdf.

Ustinov, S.K. 1972. Cannibalism and attacks on humans by brown bears in eastern Siberia. In *Proceedings of confrrence on brown bear ecology, morphology, protection and utilization.* Soviet Union Academy of Sciences, Moscow, 85–87.

Valdes, M. 1986. Survival and growth of pines with specific ectomycorrhizae after 3 years on a highly eroded site. *Can. J. Bot.* 64:885–888.

Vamosi , J.C., Knight, T.M., Steets, J.A., Mazer, S.J., Burd, M., and Ashman, T.-L. 2006. Pollination decays in biodiversity hotspots. *PNAS* 103:956–961.

van Breeman, N., Boyer, E.W., Goodale, C.L., Jaworski, N.A., Paustian, K., Seitzinger, S.P., Lajtha, K., Mayer, B., van Dam, D., Howarth, R.W., Nadelhoffer, K.J., Eve, M., and Billen, G. 2002. Where did all the nitrogen go? Fate of nitrogen inputs to large watersheds in the northeastern U.S.A. *Biogeochemistry* 57/58 267–293.

Van Cleve, K., and Harrison, A.F. 1985. Bioassay of forest floor phosphorus supply for plant growth. *Can. J. For. Res.* 15:156–162.

Van Cleve, K., and Viereck, LA. 1986. Forest succession in relation to nutrient cycling in the boreal forest of Alaska. In *Forest succession, concepts and application.* Edited by D.C. West, R.H. Shugart, and D.B. Botkin. Springer-Verlag, New York, 185–211.

Van Cleve, K., Barney, R., and Schlenter, R. 1981. Evidence of temperature control of production and nutrient cycling in two interior Alaska black spruce ecosystems. *Can. J. For. Res.* 11:258–273.

Van Cleve, K., Chapin III, F.S., Dyrness, C.T., and Viereck, L.A. 1991. Element cycling in taiga forests: state-factor control. *BioScience* 41:78–88.

Van Cleve, K., Oliver, L., Schlentner, R., Viereck, L.A., and Dyrness, C.T. 1983. Productivity and nutrient cycling in Taiga forest ecosystems. *Can. J. For. Res.* 13:747–766.

Van Den Driessche, R. 1974. Prediction of mineral nutrient status of trees by foliar analysis. *Bot. Rev.* 40:347–394.

Van Den Driessche, R., and Webber, J.E. 1977. Seasonal variations in a Douglas-fir stand in total and soluble nitrogen in inner bark and root and in total and mineralizable nitrogen in soil. *Can. J. For. Res.* 7:641–647.

Van der Heijden, M.G.A., Klironomos, J.N., Ursic, M., Moutoglis, P., Streitwolf-Engel, R., Boller, T., Wiemken, A., and Sanders. I.R. 1998. Mycorrhizal fungal diversity determines plant biodiversity, ecosystem variability and productivity. *Nature* 396:69–72.

Van der Molen, M.K. 2002. Meteorlogical impacts of land use changes in the maritime tropics. Ph.D. thesis. Vrije Univ., Amsterdam.

van Gemerden, B. S., Han, O., Parren, M.P.E., and Bongers, F. 2003. The pristine rain forest? Remnants of historical human impacts on current tree species composition and diversity. *J. Biogeogr.* 30:1381–1397.

Van Horne, B., Hanley, T.A., Cates, R.G., McKendrick, J.D., and Horner, J.D. 1988. Influence of seral stage and season on leaf chemistry of southeastern Alaska deer forage. *Can. J. For. Res.* 18:94–103.

Van Ommeren, R.J., and Whitham, T.G. 2002. Changes in interactions between juniper and mistletoe mediated by shared avian frugivores: parasitism to potential mutualism. *Oecologia* 130:281–288.

Van Valen, L. 1973. A new evolutionary law. *Evol. Theory* 1:1–30.

Van Veen, F.J., Morris, R.J., and Godfray, H.C.J. 2006. Apparent competition, quantitative food webs, and the structure of phytophagous insect communities. *Annu. Rev. Entomol.* 51:187–208.

Van Veen, F.J.F., and Murrell, D.J. 2005. A simple explanation for universal scaling relations in food webs. *Ecology* 86:3258–3263.

van Veen, J.A., Ladd, J.N., Martin, J.K., and Amato, M. 1987. Turnover of carbon, nitrogen and phosphorus through the microbial biomass in soils incubated with ^{14}C-, ^{15}N- and ^{32}P-labelled bacterial cells. *Soil Biol. Biochem.* 19:559–565.

Van Voris, P., O'Neill, R.V., Emanuel, W.R., and Shugart Jr., H.R. 1980. Functional complexity and ecosystem stability. *Ecology* 61:1352–1360.

Van Wagner, C.E. 1977. Conditions for the start and spread of crown fires. *Can. J. For. Res.* 7:23–35.

Van Wagner, C.E. 1984. Fire behavior in northern conifer forests and shrublands. In *The role of fire in northern circumpolar ecosystems.* Edited by R.W. Wein and D.A. MacLean. John Wiley & Sons, New York, 65–80.

Vander Wall, S.B., and Balda, R.P. 1977. Coadaptations of the Clark's nutcracker and the pinon pine for efficient seed harvest and dispersal. *Ecol. Monogr.* 47:89–111.

Vandermeer, J. 1990. Indirect and diffuse interactions: complicated cycles in a population embedded in a large community. *J. Theor. Biol.* 142:429–442.

Vandermeer, J., Boucher, D., Perfecto, I., and Granzow de la Cerda, I. 1996. A theory of disturbance and species diversity: evidence from Nicaragua after Hurricane Joan. *Biotropica* 28:600–613.

Vandermeer, J., Granzow de la Cerda, I., Boucher, D., Perfecto, I., and Ruiz, J. 2000. Hurricane disturbance and tropical tree species diversity. *Science* 290:788–791.

Vandermeer, J., Hazlett, B., and Rathcke, B. 1985. Indirect facilitation and mutualism. *The biology of mutualism.* Edited by D.H. Boucher. Oxford Univ. Press, New York, 326–343.

Vasechko, G.I. 1983. An ecological approach to forest protection. *For. Ecol. Manage.* 5:133–168.

Veblen, T.T. 1985. Stand dynamics in Chilean *Nothofagus* forests. In *The ecology of natural disturbance and patch dynamics.* Edited by S.T.A. Pickett and P.S. White. Academic Press, Orlando, FL, 35–52.

Veblen, T.T., and Ashton, D.R. 1978. Catastrophic influences on the vegetation of the Valdivian Andes, Chile. *Vegetatio* 36:149–167.

Veblen, T.T., Kitzberger, T., and Donnegan, J. 2000. Climatic and human influences on fire regimes in ponderosa pine forests in the Colorado Front Range. *Ecol. Applic.* 10:1178–1195.

Velazquez-Martinez, A. 1990. *Interacting effects of stand density, site frctors, and nutrients on productivity and productive efficiency of Douglas-fir plantations in the Oregon Cascades.* Ph.D. thesis. Department of Forest *Science*, College of Forestry, Oregon State Univ., Corvallis.

Velasquez-Martinez, A., and Perry, D.A. 1997. Factors influencing the availability of nitrogen in thinned and unthinned Douglas-fir stands in the central Oregon Cascades. *For. Ecol. Manage.* 93:195–203.

Velazquez-Martinez, A., Perry, D.A., and Bell, T.E. 1992. Response of aboveground biomass increment, growth efficiency, and foliar nutrients to thinning, fertilization, and pruning in young Douglas-fir plantations in the central Oregon Cascades. *Can. J. For. Res.* 22:1278–1289.

Vellend, M. 2003. Habitat loss inhibits recovery of plant diversity as forests regrow. *Ecology* 84:1158–1164.

Verhoef, R.A., and De Goede, R.G.M. 1985. Effects of collembolan grazing on nitrogen dynamics in a coniferous forest. In *Ecological interactions in soil.* Edited by A.H. Fitter. Blackwell, Oxford, 367–376.

Vertessy, R.A., Watson, F.G.R., and O'Sullivan, S.K. 2001. Factors determining relations between stand age and catchment water balance in mountain ash forests. *For. Ecol. Manage.* 143:13–26.

Victor, D.G., and Ausebel, J.H. 2000. Restoring the forests. *Foreign Affairs* 79:127–144.

Viereck, L.A. 1973. Wildfire in the taiga of Alaska. *Quat. Res.* 3:465–495.

Viereck, L.A., Dyrness, C.T., and Van Cleve, K. 1983. Vegetation, soils, and forest productivity in selected forest types in interior Alaska. *Can. J. For. Res.* 13:703–720.

Virginia, R.A. 1986. Soil development under legume tree canopies. *For. Ecol. Manage.* 16:69–79.

Visser, S., and Parkinson, D. 1975. Litter respiration and fungal growth under low temperature conditions, In *Biodegradation et humification.* Edited by G. Kilbertus, O. Resinger, A. Mourey, and J.A. Cancela da Fonseca. Pierron Editeur, Sarreguemines, France, 88–97.

Vitousek, P.M. 1984a. Anion fluxes in three Indiana forests. *Oecologia (Berlin)* 61:105–108.

Vitousek, P.M. 1984b. Litterfall, nutrient cycling, and nutrient limitation in tropical forests. *Ecology* 65:285–298.

Vitousek, P.M. 2004. *Nutrient cycling and limitation.* Princeton Univ. Press, Princeton, NJ. 223 p.

Vitousek, P.M., and Andariese, S.W. 1986. Microbial transformations of labelled nitrogen in a clear-cut pine plantation. *Oecologia (Berlin)* 68:601–605.

Vitousek, P.M., and Hobbie, S.E. 2000. The control of heterotrophic nitrogen fixation in decomposing litter. *Ecology* 81:2366–2376.

Vitousek, P.M., and Matson, P.A. 1985. Disturbance, nitrogen availability, and nitrogen losses in an intensively managed loblolly pine plantation. *Ecology* 66:1360–1376.

Vitousek, P.M., and Matson, P.A. 1988. Nitrogen transformations in a range of tropical forest soils. *Soil Biol. Biochem.* 20:361–367.

Vitousek, P.M., and Melillo, J.M. 1979. Nitrate losses from disturbed forests: patterns and mechanisms. *For. Sci.* 25:605–615.

Vitousek, P.M., and Reiners, W.A. 1975. Ecosystem succession and nutrient retention: a hypothesis. *BioScience* 25:376–381.

Vitousek, P.M., and Sanford Jr., R.L. 1986. Nutrient cycling in moist tropical forest. *Annu. Rev. Ecol. Syst.* 17:137–168.

Vitousek, P.M., Aber, J.D., Howarth, R.W., Likens, G.E., Matson, P.A., Schindler, D.W., Schelsinger, W.H., and Tilman, D.G. 1996b. Human alteration of the global nitrogen cycle: sources and consequences. *Ecol. Applic.* 7:737–750.

Vitousek, P.M., Cassman, K., Cleveland, C., Crews, T., Field, C.B., Grimm, N.B., Howarth, R.W., Marino, R., Martinelli, L., Rastetter, E.B., and Sprent, J.I. 2002. Towards and ecological understanding of biological nitrogen fixation. *Biogeochemistry* 57/58:1–45.

Vitousek, P.M., D'Antonio, C.M., Loope, L.L., and Westbrooks, R. 1996a. Biological invasions as global environmental change. *Am. Sci.* 84:468–478.

Vitousek, P.M., Gosz, J.R., Grier, C.C., Melillo, J.M., and Reiners, W.A. 1982. A comparative analysis of potential nitrification and nitrate mobility in forest ecosystems. *Ecol. Monogr.* 52:155–177.

Vitousek, P.M., Hedin, L.O., Matson, P.A., Fownes, J.H., and Neff, J.C. 1998. Within-system element cycles, input-output budgets, and nutrient limitation. In *Successes, limitations, and frontiers in ecosystem science.* Edited by M.L. Pace, P.M. Groffman. Springer-Verlag, New York, 432–451.

Vitousek, P.M., Matson, P.A., and Turner, D.R. 1988. Elevational and age gradients in Hawaiian montane rainforest: foliar and soil nutrients. *Oecologia.* 77:565–570.

Vogt, K., Moore, E., Gower, S., Vogt, D., Sprugel, D., and Grier, C. 1989. Productivity of upper slope forests in the Pacific Northwest. In *Maintaining the long-term productivity of Pacific Northwest forest ecosystems.* Edited by D.A. Perry, R. Meurisse, B. Thomas, et al. Timber Press, Portland, OR, 137–163.

Vogt, K.A., Dahlgren, R.A., Ugolini, F., Zabowski, D., Moore, E.E., and Zasosk i, R. 1987. Aluminum, Fe, Ca, Mg, K, Mn, Cu, Zn, and P in above- and below-ground biomass. I. Concentrations in subalpine *Abies amabilis* and *Tsuga mertensiana. Biogeochemistry* 4:277–294.

Vogt, K.A., Edmonds, R.L., Grier, C.C., and Piper, S.R. 1980. Seasonal changes in mycorrhizal and fibrous-textured root biomass in 23– and 180–year-old Pacific silver fir stands in western Washington. *Can. J. For. Res.* 10:523–529.

Vogt, K.A., Grier, C.C., Meier, C.E., and Edmonds, R.L. 1982. Mycorrhizal role in net primary production and nutrient cycling in *Abies amabilis* ecosystems in western Washington. *Ecology* 63:370–380.

Vogt, K.A., Grier, C.C., and Vogt, D.J. 1986. Production, turnover, and nutrient dynamics of above- and belowground detritus of world forests. *Adv. Ecol. Res.* 15:303–377.

Vogt, K.A., Vogt, D.J., Moore, E.E., Littke W., Grier, C.C., and Leney, L. 1985. Estimating Douglas-fir fine root biomass and production from living bark and starch. *Can. J. For. Res.* 15:177–179.

Vogt, K.A., Vogt, D.J., Moore, E.E., Fatuga, B.A., Redlin, M.R., and Edmonds, R.L. 1987. Conifer and angiosperm fine-root biomass in relation to stand age and site productivity in Douglas-fir forests. *J. Ecol.* 75:857–870.

Vorontozov, G.I. 1963. *Biological basis for forest proection.* Visshaya Shkola Press, Moscow. Vorontozov, G.I. 1967. Criteria for chemical insect control in deciduous forest stands. *Vopr. Zaschiti Lesa* 15:19–29.

Vos, C.C., Verboom, J., Opdam, P.F.M., and Ter Braak, C.J.F. 2001. Toward ecologically scaled landscape indices. *Am. Nat.* 183:24–41.

Vouté, A.D. 1964. Harmonious control of forest insects. *Int. Rev. For. Res.* 1:325–383.

Wada, K. 1985. The distinctive properties of andosols. *Adv. Soil Sci.* 2:173–229.

Wade, M. J. 2003. Community genetics and species interactions. *Ecology* 84:583–585.

Waggoner, P.E., Turner, N.C. 1971. Transpiration and its control by stomata on a pine forest. *Bull. Conn. Agric. Exp. Stn.* 726:1–87.

Wagner, M.R. 1988. Induced defenses in ponderosa pine against defoliating insects. In *Mechanisms of woody plant defenses against insects. Search for pattern.* Edited by W.J. Mattson, J. Levieux, and C. Bernard-Dagan. Springer-Verlag, New York, 142–155.

Waide, R.B. 1991a. The effect of Hurricane Hugo on bird populations in the Luquillo Experimental Forest, Puerto Rico. *Biotropica* 23:475–480.

Waide, R.B. 1991b. Summary of the response of animal populations to hurricanes in the Caribbean. *Biotropica* 23:508–512.

Waide, R.B., Willig, M.R., Steiner, C.F., Mittelbach, G., Gough, L., Dodson, S.I., Juday, G.P., and Parmenter, R. 1999. The relationship between productivity and species richness. *Annu. Rev. Ecol. Syst.* 30:257–300.

Waldbauer, G.P. 1968. The consumption and utilization of food by insects. *Adv. Insect Physiol* 5:229–288.

Walker, B.H. 1992. Biodiversity and ecological redundancy. *Conserv. Biol.* 6:18–23.

Walker, G. 2006. The tipping point of the iceberg. *Nature* 441:802–805.

Walker, L.R. 1991. Tree damage and recovery from Hurricane Hugo in Luquillo Experimental Forest, Puerto Rico. *Biotropica* 23:379–385.

Walla, T.R., Engen, S., DeVries, P.J., and Lande, R. 2004. Modeling vertical beta diversity in tropical butterfly communities. *Oikos* 107:610–618.

Wallander, H., Nilsson, L.O., Hagerberg, D., and Bååth, E. 2001. Estimation of the biomass and seasonal growth of external mycelium of ectomycorrhizal fungi in the field. *New Phytol.* 151:753–760.

Wallner, W.E., and Walton, G.S. 1979. Host defoliation: a possible determinant of gypsy moth population quality. *Ann. Entomol. Soc. Am.* 72:62–67.

Walter, H. 1985. *Vegetation of the Earth,* 3rd ed. Springer-Verlag, Berlin.

Walters, C.J. 1986. *Adaptive management of renewable resources.* McGraw-Hill, New York.

Walters, J.R. 1991. Application of ecological principles to the management of endangered species: the case of the red-cockaded woodpecker. *Annu. Rev. Ecol. Syst.* 22:505–523.

Walther, G.-R., Post, E., Convey, P., Menzel, A., Parmesan, C., Beebee, T.J.C., Fromentin, J.-M., Hoegh-Guldberg, O., and Bairlein, F. 2002. Ecological responses to climate change. *Nature* 416:389–395.

Walworth, J.L., and Summer, M.E. 1987. The diagnosis and recommendation integrated system (DRIS). *Adv. Soil Sci.* 6:149–188.

Wan, S., Hui, D., and Luo, Y. 2001. Fire effects on nitrogen pools and dynamics in terrestrial ecosystems: a meta-analysis. *Ecol. Applic.* 11:1349–1365.

Wang, G., and Schimel, D. 2003. Climate change, climate modes, and climate impacts. *Annu. Rev. Environ. Resourc.* 28:1–28.

Wang, J. 2004. Application of the one-migrant-per-generation rule to conservation and management. *Conserv. Biol.* 18:332–343.

Wang, X., Auler, A.S., Edwards, R.L., Cheng, H., Cristalli, P.S., Smart, P.L., Richards, D.A., and Shen, C.-C. 2004. Wet periods in northeastern Brazil over the past 210 kyr linked to distant climate anomalies. *Nature* 432:740–743.

Wang, Y., Cheng, H., Edwards, R.L., He, Y., Kong, X., An, Z., Wu, J., Kelly, M.J., Dykoski, C.A., and Li, X. 2005. The Holocene Asian monsoon: links to solar changes and North Atlantic climate. *Science* 308(5723):854–857.

Wardle, D.A. 2002. Communities and ecosystems. Princeton Univ. Press, Princeton, NJ.

Wardle, D.A., and Bardgett, R.D. 2004. Human-induced changes in large herbivorous mammal density: the consequences for decomposers. *Front. Ecol. Environ.* 3:145–153.

Wardle, D.A., and Zackrisson, O. 2005. Effects of species and functional groups loss on island ecosystem properties. *Nature* 435:806–810.

Wardle, D.A., Bardgett, R.D., Klironomos, J.N., Setala, H., van der Putten, W.H., and Wall, D.H. 2004. Ecological linkages between aboveground and belowground biota. *Science* 304:1629–1633.

Wardle, D.A., Nilsson, M.-C., Zackrisson, O., and Gallet, C. 2003. Determinants of litter mixing effects in a Swedish boreal forest. *Soil Biol. Biochem.* 35:827–35.

Wardle, D.A., Walker, L.R., and Bardgett, R.D. 2004b. Ecosystem properties and forest decline in long term chronosequences. *Science* 305:509–513.

Waring, R.H. 1985. Imbalanced forest ecosystems: assessments and consequences. *For. Ecol. Manage.* 12:93–112.

Waring, R.H. 1987. Characteristics of trees predisposed to die. *BioScience* 37:569–574.

Waring, R.H. 1989. Resource allocation in trees and ecosystems. In *Biological markers of air-pollution stress and damage in forests.* National Academy Press, Washington, DC, 127–132.

Waring, R.H., and Franklin, J.F. 1979. The evergreen forests of the Pacific Northwest. *Science* 204:1380–1386.

Waring, R.H., and Running, S.W. 1976. Water uptake, storage and transpiration by conifers: a physiological model. *Ecol. Stud. Anal. Synth.* 19:189–202.

Waring, R.H., and Running, S.W. 1978. Sapwood water storage: its contribution to transpiration and effect upon water conductance through the stems of old-growth Douglas-fir. *Plant Cell Environ.* 1:131-140.

Waring, R.H., and Schlesinger, W.H. 1985. *Forest ecosystems: concepts and management.* Academic Press, Orlando, FL.

Waring, R.H., Coops, N.C., Fan, W., and Nightingale, J.M. 2006. MODIS enhanced vegetation index predicts tree species richness across forested ecoregions in the contiguous U.S.A. *Remote Sensing Environ.* 103:218–226.

Waring, R.H., Coops, N.C., Ohmann, J.L., and Sarr, D.A. 2002. Interpreting woody plant richness from seasonal ratios of photosymthesis. *Ecology* 83:2964–2970.

Waterman, P.C., Ross, J.A.M., Bennett, E.L., and Davies, A.G. 1988. A comparison of the floristics and leaf chemistry of the tree flora in two malaysian rain forests and the influence of leaf chemistry on populations of colobine monkeys in the Old World. *Biol. J. Linn. Soc.* 34:1–32.

Watkins, R.Z., Chen, J., Pickens, J., and Brosofske, K.D. 2003. Effects of forest roads on understory plants in a managed hardwood landscape. *Conserv. Biol.* 17:411–419.

Watmough, S.A., and Dillon, P.J. 2003. Mycorrhizal weathering in base-poor forests. *Nature* 423:823–824.

Watt, A.D. 1992. Insect pest population dynamics: effects of tree species diversity. In *The ecology of mixed-species stands of trees.* Edited by M.G.R. Cannel, D.C. Malcolm, and P.A. Robertson. Blackwell, Oxford, 267–275.

Watts, J.A. 1982.The carbon dioxide question: a data sampler. In *Carbon dioxide review: 1982.* Edited by W.C. Clark. Oxford Univ. Press, New York, 429–469.

Weast, R.C. 1982. *Handbook of chemistry and physics.* 60th ed. CRC Press, Boca Raton, FL.

Weathers, K.C., Simkin, S.M., Lovett, G.M., and Lindberg, S.E. 2006. Empirical modeling of atmospheric deposition in mountainous landscapes. *Ecol. Applic.* 16:1590–1607.

Weatherspoon, C.P., and Skinner, C.N. 1995. An assessment of factors associated with damage to tree crowns from the 1987 wildfires in northern California. *For. Sci.* 41:430–451.

Webb III, T. 1992. Past changes in vegetation and climate: lessons for the future. In *Global warming and biological diversity.* Edited by R.L. Peters and T.E. Lovejoy. Yale Univ. Press, New Haven, CT, 59–75.

Webb, U. 1968. Environmental relationships of the structural types of Australian rain forest vegetation. *Ecology* 49:296–311.

Webb, W.L., Lauenroth, W.K., Szarek, S.R., and Kinerson, R.S. 1983. Primary production and abiotic controls in forests, grasslands, and desert ecosystems in the United States. *Ecology* 64:134–151.

Weber, J. 1981. A natural control of Dutch elm disease. *Nature (London)* 292:449–451.

Webster, P.J., Holland, G.J., Curry, J.A., and Chang, H.-R. 2005. Changes in tropical cyclone number, duration, and intensity in a warming environment. *Science* 309:1844–1846.

Weetman, G.F., and Algar, D. 1974. Jack pine nitrogen fertilization and nutrition studies: three year results. *Can. J. For. Res.* 4:381–398.

Weetman, G.F., and Webber, B. 1972. The influence of wood harvesting on the nutrient status of two spruce stands. *Can. J. For. Res.* 2:351–369.

Wei, X., and Kimmins, J.P. 1998. Asymbiotic nitrogen fixation in harvested and wildfire-killed lodgepole pine forests in the central interior of British Columbia. *For. Ecol. Manage.* 109:343–53.

Weier, K.L., and Gilliam, J.W. 1986. Effect of acidity on denitrification and nitrous oxide evolution from Atlantic coastal plain soils. *Soil Sci. Soc. Am. J.* 50:1202–1205.

Weir, J.T., and Schluter, D. 2007. The latitudinal gradient in recent speciation and extinction rates of birds and mammals. *Science* 315:1574–1576.

Weisberg, P. J. 2004. Importance of non-stand-replacing fire for development of forest structure in the Pacific Northwest, USA. *For. Sci.* 50:245–258.

Welch, T.G., and Klemmedson, J.O. 1975. Influence of the biotic factor and parent material on distribution of nitrogen and carbon in ponderosa pine ecosystems. In *Forest soils and forest land management.* Proc. 4th North Am. Forest Soils Conf., Laval Univ., Québec, Aug. 1973. Edited by B. Bernier and C.H. Winget. Les Presses de L'Université Laval, Québec, Canada, 159–178.

Weller, D.E. 1987. Self-thinning exponent correlated with allometric measures of plant geometry. *Ecology* 68:813–821.

Wells, J.M., and Boddy, L. 1990. Wood decay, and phosphorus and fungal biomass allocation, in mycelial cord systems. *New Phytol.* 116:285–295.

Wells, J.M., Hughes, C., and Boddy, L. 1990. The fate of soil-derived phosphorus in mycelial cord systems of *Phanerochaete velutina* and *Phallus impudicus*. *New Phytol.* 114:595–606.

Welsh, H.H., and Droege, S. 2001. A case for using plethodontid salamanders for monitoring biodiversity and ecosystem integrity of North American forests. *Conserv. Biol.* 15:558–573.

Went, F.W., and Stark, N. 1968. The biological and mechanical role of soil fungi. *PNAS* 60:497–504.

Werner, P. 1984. Changes in soil properties during tropical wet forest succession in Costa Rica. *Biotropica* 16:43–50.

Werner, R.A., Holsten, E.H., Matsuoka, S.M., and Burnside, R.E. 2006. Spruce beetles and forest ecosystems in south-central Alaska: a review of 30 years of research. *For. Ecol. Manage.* 227:195–206.

Werth, D., and Avissar, R. 2002. The local and global effects of Amazonian deforestation. *J. Geophys. Res.* 107:55–1 to 55–8.

Wessman, CA., Aber, J.D., Peterson, D.L., and Melillo, J.M. 1988. Remote sensing of canopy chemistry and nitrogen cycling in temperate forest ecosystems. *Nature* 335:154–156.

West, D.C., Shugart, H.H., and Botkin, D.B., editors. 1981. *Forest succession*. Springer-Verlag, New York.

Westall, J., and Stumm, W. 1980. The hydrosphere. In *The handbook of environmental chemistry. Vol. 1, Part A.* Edited by O. Hutzinger. Springer-Verlag, Berlin, 17–49.

Westerling, A.L., and Swetnam, T.W. 2003. Interannual to decadal drought and wildfire in the western United States. *EOS* 84:545–560.

Westerling, A.L., Hidalgo, H.G., Cayan, D.R., and Swetnam, T.W. 2006. Warming and earlier Spring increases western U.S. forest wildfire activity. *Scienceexpress*, July 6. www.scienceexpress.org.

Westoby, M. 1984. The self-thinning rule. *Adv. Ecol. Res.* 14:167–225.

Wheelwright, N.T. 1988. Fruit-eating birds and bird-dispersed plants in the tropics and temperate zone. *Tree* 3:270:274.

Whigham, D.F., Olmsted, I., Cano, E.G., and Harmon, M.E. 1991. The impact of Hurricane Gilbert on trees, litterfall, and woody debris in a dry tropical forest in the northeastern Yucatan Peninsula, *Biotropica* 23:434–441.

Whiles, M.R., Lips, K.R., Pringle, C.M., Kilham, S.S., Bixby, R.J., Brenes, R., Connelly, S., Colon-Gaud, J.C., Hunte-Brown, M., Huryn, A.D., Montgomery, C., and Peterson, S. 2006. The effects of amphibian population declines on the structure and function of Neotropical stream ecosystems. *Front. Ecol. Environ.* 4:28–34.

Whipps, J.M., and Lynch, J.M. 1986. The influence of the rhizosphere on crop productivity. *Adv. Microb. Ecol.* 9:187–244.

Whitcomb, R.F., Robbins, C.S., Lynch, J.F., Whitcomb, B.L., Klimkiewicz, M.K., and Bystrak, D. 1981. Effects of forest fragmentation on avifauna of the eastern deciduous forest. In *Forest island dynamics in man-dominated landscapes.* Edited by R.L. Burgess and D.M. Sharpe. Springer-Verlag, Berlin, 125–206

White, A.F., and Brantley, S.L. 2003. The effect of time on the weathering of silicate minerals: why do weathering rates differ in the laboratory and field? *Chem. Geol.* 202:479–506.

White, C.A., Feller, M.C., and Bayley, S. 2003. Predation risk and the functional response of elk-aspen herbivory. *For. Ecol. Manage.* 181:77–97.

White, C.S. 1986a. Effects of prescribed fire on rates of decomposition and nitrogen mineralization in a ponderosa pine ecosystem. *Biol. Fertil. Soils.* 2:87–95.

White, C.S. 1986b. Volatile and water-soluable inhibitors of nitrogen mineralization and nitrification in a ponderosa pine ecosystem. *Biol. Fertil. Soils* 2:97–104.

White, C.S., and Gosz, J.R. 1987. Factors controlling nitrogen mineralization and nitrification in forest ecosystems in New Mexico. *Biol. Fertil. Soils.* 5:195–202.

White, C.S., Moore, D.J., Horner, J.D., and Gosz, J.R. 1988. Nitrogen mineralization-immobilization response to field N or C perturbations: an evaluation of a theoretical model. *Soil Biol. Biochem.* 20:101–105.

White, E.M., Thompson, W.W., and Gartner, F.R. 1973. Heat effects on nutrient release from soils under ponderosa pine. *J. Range Manage.* 26:22–24.

White, J. 1981. The allometric interpretation of the self-thinning rule. *J. Theor. Biol.* 89:475–500.

White, J., and Harper, J.L. 1970. Correlated changes in plant size and number in plant populations. *J. Ecol.* 58:467–485.

White, T.C.R. 1984. The abundance of invertebrate herbivores in relation to the availability of nitrogen in stressed food plants. *Oecologia* 63:90–105.

Whitham, T.C. 1983. Host manipulation of parasites: within-plant variation as a defense against rapidly evolving pests. In *Variable plants and herbivores in natural and managed systems.* Edited by R.F. Denno and M.S. McClure. Academic Press, London, 15–41.

Whitham, T.G., and Slobodchikoff, C.N. 1981. Evolution by individuals, plant-herbivore interactions, and mosaics of genetic variability: the adaptive significance of somatic mutations in plants. *Oecologia* 49:287–292.

Whitham, T.G., Bailey, K., Schweitzer, J.A., Shuster, S.M., Bangert, R.K., LeRoy, C.J., Lonsdorf, E.V., Allan, G.J., DiFazio, S.P., Potts, B.M., Fischer, D.G., Gehring, C.A., Lindroth, R.L., Marks, J.C., Hart, S.C., Wimp, G.M., and Wooley, S.C. 2006. A framework for community and ecosystem genetics: from genes to ecosystems. *Nat. Rev.* 7:510–523.

Whitham, T.G., Lonsdorf, E.L., Schweitzer, J.A., Bailey, J.K., Fischer, D.G., Shuster, S.M., Lindroth, R.L., Hart, S.C., Allan, G.J., Gehring, C.A., Keim, P., Potts, B.M., Marks, J., Rehill, B.J., DiFazio, S.P., LeRoy, C.J., Wimp, G.M., and Woolbright, S. 2005. "All effects of a gene on the world": extended phenotypes, feebacks, and multilevel selection. *Ecoscience* 12:5–7.

Whitham, T.G., Young, W.P., Martinsen, G.D., Gehring, C.A., Schweitzer, J.A., Shuster, S.M., Wimp, G.M., Fischer, D.G., Bailey, J.K., Lindroth, R.L., Woolbright, S., and Kuske, C.R. 2003. Community genetics: a consequence of the extended phenotype. *Ecology* 84:559–573.

Whiting, G.J., Gandy, E.L., and Yoch, D.C. 1986. Tight coupling of root-associated nitrogen fixation and plant photosynthesis in the salt marsh grass *Spartina alterniflora* and carbon dioxide enhancement of nitrogenase activity. *Appl. Environ. Microbiol.* 52:108–113.

Whitlock, C., Shafer, S.L., and Marlon, J. 2003. The role of climate and vegetation change in shaping past and future fire regimes in the northwestern US and the implications for ecosystem management. *For. Ecol. Manage.* 178:5–21.

Whitman, A.A., and Hagan, J.M. 2007. An index to identify late-successional forest in temperate and boreal zones. *For. Ecol. Manage.* 246:144–154.

Whitmore, T.C. 1984. *Tropical rain forests of the Far East.* Clarendon Press, Oxford.

Whitmore, T.C. 1989. Canopy gaps and the two major groups of forest trees. *Ecology* 70:536–538.

Whittaker, R.H. 1956. Vegetation of the Great Smoky Mountains. *Ecol. Monogr.* 26:1–80.

Whittaker, R.H. 1960. Vegetation of the Siskiyou Mountains, Oregon and California. *Ecol. Monogr.* 30:279–338.

Whittaker, R.H. 1965. Dominance and diversity in land plant communities. *Science* 147:250–260.

Whittaker, R.H. 1967. Gradient analysis of vegetation. *Biol. Rev.* 42:207–264.

Whittaker, R.H. 1969. Evolution of diversity in plant communities. In *Diversity and stability in ecological systems.* Brookhaven Symposium. Biol. 22. Brookhaven National Laboratory. Brookhaven, New York

Whittaker, R.H. 1972. Evolution and measurement of species diversity. *Taxon* 21:213–251.

Whittaker, R.H. 1975. *Communities and ecosystems.* Princeton Univ. Press, Princeton, NJ.

Whittaker, R.H., and Niering, W.A. 1968. Vegetation of the Santa Catalina Mountains, Arizona. IV. Limestone and acid soils. *J. Ecol.* 56:523–544.

Whittaker, R.H., and Niering, WA. 1965. Vegetation of the Santa Catalina Mountains, Arizona: a gradient analysis of the south slope. *Ecology.* 46:429–452.

Whittaker, R.H., Levin, S.A., and Root, RB. 1973. Niche, habitat, and ecotope. *Am. Nat.* 107:321–338.

Whittaker, R.J., Araújo, M.B., Jepson, P., Ladle, R.J., Watson, J.E.M., Willis, K.J. 2005. Conservation biogeography: assessment and prospect. *Divers. Distribut.* 11:3–23.

Wickman, B. 1963. *Mortality and growth reduction of white fir following defoliation by the Douglas-fir tussock moth.* U.S.F.S. Res. Paper PSW-7. U.S. Department of Agriculture, Forest Service, Berkeley, CA.

Wickman, B.E., Mason, R.R., and Swetnam, T.W. 1994. Searching for long-term patterns of forest insect outbreaks. In S. R. Leather, K. F. A. Walters, N. J.Mills, and A. D. Watt, editors. Individuals, populations,and patterns in ecology. Intercept, Andover, UK. 251–261.

Wickman, B.E., Mason, R.R., and Thompson, C.C. 1973. *Major outbreaks of the Douglas-fir tussock moth in Oregon and California.* USDA For. Serv. Gen. Tech. Rep. PNW-5. U.S. Department of Agriculture, Forest Service, Pacific Northwest Exp. Sta., Portland, OR.

Wiedemann, E. 1935. Die Ergebrisse 40–jahriger Vorr at spflege in den preussischen forst lichen Versuchsflachen. *Forstarchiv* 11:65–108.

Wiegand, T., Gunatilleke, C.V.S., Gunatilleke, I.A.U.N., and Huth, A. 2007. How individual species structure diversity in tropical forests. *PNAS* 104:19029–19033.

Wiens, J.A. 1984. On understanding a non-equilibrium world: myth and reality in community patterns and processes. In *Ecological communities: conceptual issues and evidence.* Edited by D.R. Strong, D. Simberloff, L.G. Abele, and A.B. Thistle. Princeton Univ. Press, Princeton, NJ, 439–457.

Wiens, J.J., and Donoghue, M.J. 2004. Historical biogeography, ecology, and species richness. *Tree* 19:639–645.

Wiersum, R.F. 1984. Surface erosion under various tropical agroforestry systems. In *Symposium on effects of forest land use on erosion and slope stability.* Edited by C.L. O'Loughlin and A.J. Pearce. East-West Center, Honolulu, 231–239.

Wilcove, D. 1990. Empty skies, *Nat. Conserv. Mag.* 40:4–13.

Wilcove, D.S. 1985. Nest predation in forest tracts and the decline of migratory songbirds. *Ecology* 66:1211–1214.

Wilcove, D.S. 1989. Forest fragmentation as a wildlife management issue in the eastern United States. In *Is forest fragmentation a management issue in the northeast?* Gen. Tech. Rep. NE-140. U.S. Department of Agriculture, Forest Service, Durham, NH, 1–5.

Wilcove, D.S. 1989. Protecting biodiversity in multiple-use lands: lessons from the U.S. Forest Service. *Trends Ecol. Evol.* 4:385–388.

Wilcove, D.S., McLellan, C.R., and Dobson, A.P. 1986. Habitat fragmentation in the temperate zone. In *Conservation biology: the science of scarcity and diversity.* Edited by M.E. Soule. Sinauer Associates, Sunderland, MA, 237–256.

Wilcove, D.S., Rothstein, D., Dubow, D., Phillips, J., and Losos, A. 1998. Quantifying threats to imperiled species in the United States. *BioScience* 408:607–615.

Wilcox, B.A., and Murphy, D.D. 1985. Conservation strategy: the effects of fragmentation on extinction. *Am. Nat.* 125:879–887.

Wilde, S.A. 1946. *Forest soils and forest growth.* Chronica Botanica Co, Waltham, MA.

Wilde, S.A. 1958. *Forest soils.* Ronald Press, New York.

Wildlands Project. 2007. What we do. www.twp.org/cms/page1090.cfm.

Wilk, J. 2000. Local perceptions about forests and water in two tropical catchments. *GeoJournal* 50:339–347.

Wilkinson, D.M. 1997. Plant colonization: are wind dispersed seeds really dispersed by birds at larger spatial and temporal scales? *J. Biogeogr.* 24:61–65.

Wilkinson, G.S. 1990. Food sharing in vampire bats. *Sci. Am.* 1990(February):76–82.

Will, T. 1991. Birds on a severely hurricane-damaged Atlantic Coast rain forest in Nicaragua. *Biotropica* 23:497–507.

Williams, B.L. 1972. Nitrogen mineralization and organic matter decomposition in Scots pine humus. *Forestry* 45:177–188.

Williams, J.E., Wood, C.A., and Dombeck, M.P., editors. 1997. *Watershed restoration: principles and practices.* American Fisheries Society, Bethesda, MD.

Williams, R.J. , Berlow, E.L., Dunne, J.A., Barabasi, A.-L., and Martinez, N.D. 2002. Two degrees of separation in complex food webs. *PNAS* 99:12913–12916.

Williams, S.T., Davies, F.L., Mayfield, C.I., and Khan, M.R. 1971. Studies on the ecology of actinomycetes in soil. II. The pH requirements of streptomycetes from two acid soils. *Soil Biol. Biochem.* 3:187–195.

Williams, T.M., and Gresham, C.A., editors. 1987. *Predicting consequences of intensive forest harvesting on long-term productivity by site classi fication.* lEA/BE Project A3 (CPC-10), Report No. 6, Rotorua, New Zealand. International Energy Agency, Paris.

Williamson, M. 1996. *Biological invasions.* Chapman & Hall, London.

Willig, M.R., Kaufman, D.M., and Stevens, R.D. 2003. Latitudinal gradients of biodiversity. *Annu. Rev. Ecol. Syst.* 34:273–309.

Willis, K.J., Gillson, L., and Brncic, T.M. 2004. How "virgin" is virgin rainforest? *Science* 304:402–403.

Wills, C., Condit, R., Foster, R.B., and Hubbell, S.P. 1997. Strong density- and diversity-related effects help to maintain tree species diversity in a neotropical forest. *PNAS* 94:1252–1257.

Wills, C., Harms, K.E., Condit, R., King, D., Thompson, J., He, F., Muller-Landau, H.C., Ashton, P., Losos, E., Comita, L., Hubbell, S., LaFrankie, J., Bunyavejchewin, S., Dattaraja, H.S., Davies, S., Esufali, S., Foster, R., Gunatilleke, N., Gunatilleke, S., Hall, P., Itoh, A., John, R., Kiratiprayoon, S., Loo de Lao, S., Massa, M., Nath, C., Nur Supardi Noor, Md.,

Rahman Kassim, A., Sukumar, R., Satyanarayana Suresh, H., Sun, I.-F., Tan, S., Yamakura, T., and Zimmerman, J. 2006. Nonrandom processes maintain diversity in tropical forests. *Science* 311:527–531.

Wilmers, C.C., E. Post, R. O. Peterson, and J. A. Vucetich. 2006. Predator disease out-break modulates top-down, bottom-up and climatic effects on herbivore population dynamics. *Ecol. Lett.* 9:383–389.

Wilson, A.T. 1978. Pioneer agriculture explosion and CO_2 levels in the atmosphere. *Nature* 273:40–41.

Wilson, D.S. 1980. *The natural selection of populations and communities.* Benjamin/Cummings Publishing Company, Menlo Park, CA.

Wilson, D.S. 1986. Adaptive indirect effects. In *Community ecology.* Edited by J. Diamond and T.J. Case. Harper & Row, New Yok, 437–444.

Wilson, D.S. 1997. Biological communities as functionally organized units. *Ecology* 78:2018–2024.

Wilson, D.S., and Swenson, W. 2003. Community genetics and community selection. *Ecology* 84:586–588.

Wilson, E.O. 1985. The sociogenesis of insect colonies. *Science* 228:1489–1495.

Wilson, E.O. 1987. The little things that run the world. *Conserv. Biol.* 1:344–346.

Wilson, E.O. 1988. The current state of biological diversity. In *Biodiversity.* Edited by E.O. Wilson. National Academy Press, Washington, DC, 3–18.

Wilson, E.O., and Peters, F.M., editors. 1988. *Biodiversity.* National Academy Press, Washington, DC.

Wilson, K.A., McBride, M.F., Bode, M., and Possingham, H.P. 2006. Prioritizing global conservation efforts. *Nature* 440:337–340.

Wilson, M.V. 1982. *Microhabitat influences on species distributions and community dynamics in the conifer woodland of the Siskiyou Mountains, Oregon.* Ph.D. thesis. Cornell Univ., Ithaca, NY.

Wilson, M.V., and Botkin, D.B. 1990. Models of simple microcosms: emergent properties and the effect of complexity on stability. *Am. Nat.* 135:414–434.

Windsor, G.M., Rand, A.S., and Rand, W.M. 1986. *Variation in rainfall on Barrow Colorado Island, Balboa, Panama.* Smithsonian Tropical Research Institute. Barro Colorado Island, Panama.

Wingfield, M.J. 2003. Increasing threat of diseases to exotic plantation forests in the Southern Hemisphere: lessons from Cryphonectria canker. *Austr. Plant Pathol.* 32:133–139.

Wise, M.D., and Abrahamson, W.G. 2005. Beyond the compensatory continuum: environmental resource levels and plant tolerance of herbivory. *Oikos* 109:417–428.

Wiser, S.K., Peet, R.K., and White, P.S. 1998. Prediction of rare plant occurrence: a Southern Appalachian example. *Ecol. Applic.* 8:909–920.

With, K.A. 1997. The application of neutral landscape models in conservation biology. *Conserv. Biol.* 11:1069–1080.

With, K.A. 2002. Using percolation theory to assess landscape connectivity and effects of habitat fragmentation. In *Applying landscape ecology in biological conservation.* Edited by K.J. Gutzwiller. Springer, 105–130.

Witty, J.H., Graham, R.C., Hubbert. K.R., Doolittle, J.A., and Wald, J.A. 2003. Contributions of water supply from the weathered bedrock zone to forest soil quality. *Geoderma* 114:389–400.

Wolbach, W.S., Gilmour, I., Anders, E., Orth, C.J., and Brooks, R.R. 1988. Global fire at the Cretaceous-Tertiary boundary. *Nature* 334:665–669.

Wolf, E.C. 1985. Challenges and priorities in conserving biological diversity. *Interciencia* 5:236–242.

Wolfe, B.E., and Klironomos, J.N. 2005. Breaking new ground: soil communities and exotic plant invasion. *BioScience* 55:477–487.

Wolff, J.O., and Dueser, R.D. 1986. Noncompetitive coexistence between *Peromyscus* species and *Clethrionomys gapperi. Can. Field Nat.* 100:186–191.

Wollum II, A.G. 1999. Habitat and organisms. In *Principles and application of soil microbiology,* 2nd ed. Edited by D.M. Sylvia, J.L. Fuhrmann, P.G. Hartel, and D.A. Zuberer. Prentice Hall, Upper Saddle River, NJ, 3–20.

Wood, N. 1974. *Many winters: prose and poetry of the Pueblos.* Doubleday, Garden City, NY.

Wood, T.G., and Sands, W.A. 1978. The role of termites in ecosystems. In *Production ecology of ants and termites.* Edited by M.V. Brian. Cambridge Univ. Press, Cambridge, MA, 245–292.

Woodman, J.N. 1990. Climate change and the distribution of tree species. Presented at the Convention of the Society of American Foresters, Washington, DC, 1990.

Woodruff, D.S. 2001. Declines of biomes and biotas and the future of evolution. *PNAS* 98:5471–5476.

Woods, A., Coates, K.D., and Hamann, A. 2005. Is an unprecedented Dothistroma needle blight epidemic related to climate change? *BioScience* 55:761–769.

Woods, K.D. 2004. Intermediate disturbance in a late-successional hemlock-northern hardwood forest. *J. Ecol.* 92:464–476.

Woods, K.D., and Whittaker, R.H. 1981. Canopy-understory interaction and the internal dynamics of mature hardwood and hemlock-hardwood forests. In *Forest succession, concepts and application.* Edited by D.C. West, H.H. Shugart, and D.B. Botkin. Springer-Verlag, New York, 305–323.

Woodward, F.I. 1987. Stomatal numbers are sensitive to increases in CO_2 from pre-industrial levels. *Nature* 327:617–618.

Woodwell, G.M. 1970. Effects of pollution on the structure and physiology of ecosystems. *Science* 168:429–433.

Woodwell, G.M., Hobbie, J.E., Houghton, R.A., Melillo, J.M., Moore, B., Peterson, B.J., and Shaver, G.R. 1983. Global deforestation: contribution to atmospheric carbon dioxide. *Science* 222:1081–1086.

Woodwell, G.M., Whittaker, R.H., Reiners, WA., Likens, G.E., Delwiche, C.C., and Botkin, D.B. 1978. The biota and the world carbon budget. *Science* 199:141–146.

Wooster, M.J., and Strub, N. 2002. Study of the 1997 Borneo fires: quantitative analysis using global area coverage (GAC) satellite data. *Global Biogeochem. Cycles* 16:9–1 to 9–12.

Wooten, J.T. 2005. Field parameterization and experimental test of the neutral theory of biodiversity. *Nature* 433:309–312.

World Conservation Union/IUCN. 2006a. 2006 IUCN Red List. www.iucn.org/themes/ssc/redlist2006/redlist2006.htm.

World Conservation Union/IUCN. 2006b. Release of the 2006 IUCN Red List of Threatened Species reveals ongoing decline of the status of plants and animals. News release. www.iucn.org/en/news/archive/2006/05/02_pr_red_list_en.htm.

World Resources Institute. 2004. Earth trends environmental information. www.earthtrends.wri.org/.

World Resources Institute. 1992. *World resources 1992–1993.* Oxford Univ. Press, Oxford.

World Resources Institute/The World Conservation Union/United Nations Environment Program. 1992. *Global biodiversity strategy.* World Resources Institute, Washington, DC.

Worrall, J.J., and Harrington, J.C. 1988. Etiology of canopy gaps in spruce-fir forests at Crawford Notch, New Hampshire. *Can. J. For. Res.* 18:1463–1469.

Wright, A.H., and Bailey, A.W. 1982. *Fire ecology.* John Wiley & Sons, New York.

Wright, E., and Tarrant, R.F. 1958. *Occurrence of mycorrhizae after logging and slash burning in the Douglas-fir forest type.* USDA For. Serv. Res. Note PNW-160. U.S. Department of Agriculture, Forest Service, Pacific Northwest Exp. Sta., Portland, OR.

Wright, S. 1969. *Evolution and the genetics of population, the theory of gene frequencies,* Vol. 2. Univ. of Chicago Press, Chicago.

Wright, S., Keeling, J., and Gillman, L. 2006. The road from Santa Rosalia: a faster tempo of evolution in tropical climates. *PNAS* 103:7718–7722.

Wright, S.J. 2002. Plant diversity in tropical forests: a review of mechanisms of species coexistence. *Oecologia* 130:1–14.

Wurz, T.L., Wiita, A.L., Weber, N.S., and Pilz, D. 2005. Harvesting morels after wildfire in Alaska. PNW-RN-546. U.S. Department of Agriculture, Forest Service, Pacific Northwest Res. Sta., Portland, OR.

Wynne-Edwards, V.C. 1962. *Animal dispersion in relation to social behavior.* Oliver & Boyd, London.

Yachi, S., and Loreau, M. 1999. Biodiversity and ecosystem productivity in a fluctuating environment: the insurance hypothesis. *PNAS* 96:1463–1468.

Yamanaka, T., Li, C.-Y., Bormann, B.T., and Okabe, H. 2003. Tripartite associations in an alder: effects of *Frankia* and *Alpova diplophloeus* on the growth, nitrogen fixation and mineral acquisition of *Alnus tenufolia. Plant Soil* 254:179–186.

Young, J.L. 1964. Ammonia and ammonium reactions with some Pacific Northwest soils. *Soil Sci. Soc. Am. Proc.* 28:339–345.

Youngberg, C.T., Wollum, A.G., and Scott, W. 1979. *Ceanothus* in Douglas-fir clear-cuts: nitrogen accretion and impact on regeneration. In *Symbiotic nitrogen fixation in the management of temperate forests.* Oregon State Univ., Corvallis, OR, 224–233.

Zak, D.R., Groffman, P.M., Pregitzer, K.S., Christensen, S., and Tiedje, J.M. 1990. The vernal dam: plant-microbe competition for nitrogen in northern hardwood forests. *Ecology* 71:651–656.

Zak, D.R., Holmes, W.E., Finzi, A.C., Norby, R.J., and Schlesinger, W.H. 2003. Soil nitrogen cycling under elevated CO_2: A synthesis of forest face éxperiments. *Ecol. Applic.* 13:1508–1514.

Zambryski, P., Tempe, J., and Schell, J. 1989. Transfer and function of T-DNA genes from agrobacterium Ti and Ri plasmids in plants. *Cell* 56:193–201.

Zangaro, W., Bononi, V.L.R., and Trufen, S.B. 2001. Mycorrhizal dependency, inoculum potential, and habitat preference of native woody species in south Brazil. *J. Trop. Ecol.* 16:603–622.

Zarin, D.J., Davidson, E.A., Brondizio, E., Vieira, I.C.G., Sa, T., Feldpausch, T., Schurrr, E.A.G., Mesquita, R., Moran, E., Delamonica P. , Ducey, M.J., Hurtt, G.C., Salimon, C., and Denich, M. 2005. Legacy of fire slows carbon accumulation in Amazonian forest regrowth. *Front. Ecol. Environ.* 3:365–369.

Zavaleta, E.S., and Hulvey, K.B. 2004. Realistic species losses disproportionately reduce grassland resistance to biological invaders. *Science* 306:1175–1177.

Zavitkovski, J., and Newton, M. 1968. Ecological importance of snowbrush. *(Ceanothus velutinus)* in the Oregon Cascades. *Ecology* 49:1134–1145.

Zedler, P.H., Gautier, C.R., and McMaster, G.S. 1983. Vegetation change in response to extreme events: the effect of a short interval between fires in California chapparal and coastal scrub. *Ecology* 64:809–818.

Zelles, L., Scheunert, I., and Kreutzer, K. 1987. Effect of artificial irrigation, acid precipitation and liming on the microbial activity in soil of a spruce forest. *Biol. Fertil. Soils.* 4:137–143.

Zhao, M., Pitman, A.J., and Chase, T. 2001a. The impact of land cover change on the atmospheric circulation. *Clim. Dyn.* 17:467–477.

Zhao, Z.-W., Zia, Y.-M., Qin, X.-Z., Li, X.-W., Cheng, L.-Z., Sha, T., and Wang, G.-H. 2001b. Arbuscular mycorrhizal status of plants and the spore density of arbuscular mycorrhizal fungi in the tropical rain forest of Xishuangbanna, southwest China. *Mycorrhiza* 11:159–162.

Zhou, G., Liu, S., Li, Z., Zhang, D., Tang, X., Zhou, C., Yan, J., and Mo, J. 2006. Old-growth forests can accumulate carbon in soils. *Science* 314:1417.

Zhu, H., Higginbotham, K.O., Dancik, B.P., and Navratil, S. 1988. Intraspecific genetic variability of isozymes in the ectomycorrhizal fungus Suillus tomentosus. *Can. J. Bot.* 66:588–594.

Zhu, Y.G., and Miller, R.M. 2003. Carbon cycling by arbuscular mycorrhizal fungi in soil-plant systems. *Trends Plant Sci.* 8:407–409.

Ziemer RR, Lewis J, Rice RM, Lisle TE. 1991a. Modelling the cumulative watershed effects of forest management strategies. *J. Environ. Qual.* 20:36–42.

Ziemer RR, Lewis J, Lisle TE, Rice RM. 1991b. Long-term sedimentation effects of different patterns of timber harvesting. *Int. Assoc. Hydrol. Sci.* 203:143–50.

Zimmer, C. 2003. Rapid evolution can foil even the best-laid plans. *Science* 300:895–896.

Zimmerman, B.L., and Bierregaard, R.O. 1986. Relevance of the equilibrium theory of island biogeography and species-area relations to conservation with a case from Amazonia. *J. Biogeography* 13:133–143.

Zlotin, R.I. 1985. Role of soil fauna in the formation of soil properties. *Intecol. Bull.* 12:39–42.

Zobel, D.B., McKee, A., Hawk, G.M., and Dryness, C.T. 1976. Relationships of environment to composition, structure, and diversity of forest communities of the central western Cascades of Oregon. *Ecol. Monogr.* 46:135–156.

Zou, X., Valentine, D.W., Sanford Jr., R.L., and Binkley, D. 1992. Resin-core and buried-bag estimates of nitrogen transformations in Costa Rican lowland rainforests. *Plant Soil* 139:275–283

Zutter, B.R., Gjerstad, D.H., and Glover, G.R. 1987. Fusiform rust incidence and severity in loblolly pine plantations following herbaceous weed control. *For. Sci.* 33:790–800.

INDEX

Page numbers followed by italic *f* indicate illustrations; those followed by italic *t* indicate tables.